succeed in your neuroscience course

sylvius 4

by S. Mark Williams, Leonard E. White, and

An Interactive Atlas and Visual Glossary of Human Neuroanatomy

Each new copy of *Neuroscience*, Fourth Edition includes an access code for a free download of *Sylvius* 4. This new version of *Sylvius* provides a unique digital learning environment for exploring and understanding the structure of the human central nervous system. *Sylvius* features fully annotated surface views of the human brain, as well as interactive tools for dissecting the central nervous system and viewing fully annotated cross-sections of preserved specimens and living subjects imaged by magnetic resonance. This newly expanded and reconfigured *Sylvius* is more than a conventional atlas; it incorporates a comprehensive, visually-rich, searchable database of more than 500 neuroanatomical terms that are concisely defined and visualized in photographs, magnetic resonance images, and illustrations from *Neuroscience*, Fourth Edition.

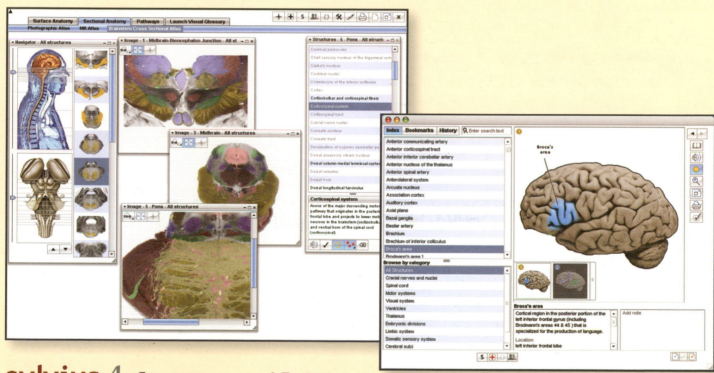

sylvius 4 Components and Features

- **Surface Anatomy Atlases:** Photographic, Magnetic Resonance Image, Brainstem Model

- **Sectional Anatomy Atlases:** Photographic, Magnetic Resonance Image (newly expanded), Brainstem and Spinal Cord

- **Pathways:** Allow you to follow the flow of information in several important long-tract pathways of the central nervous system

*(Features and **download instructions** continued on next page)*

- **Visual Glossary:** Searchable glossary that provides visual representations, concise anatomical and functional definitions, and audio pronunciation of neuroanatomical structures
- **Over 500 neuroanatomical structures and terms** are identified and described
- **Categories of anatomical structures and terms** (e.g., cranial nerves, spinal cord tracts, lobes, cortical areas, etc.) that can be browsed easily
- **A notes feature** that allows you to type and save notes for any selected structure
- **A self-quiz mode** that allows assessment of structure identification and functional information

sylvius 4 *Free Download Instructions*

Sylvius 4 is provided as a downloadable program that you save on your local computer. You only need to download the program once—it will then run from your computer and does not require an Internet connection after the initial installation.

To download sylvius 4, follow these instructions:
1. Scratch below to reveal your unique download code.
2. Go to http://www.sinauer.com/sylvius4
3. Follow the instructions to download and register the program. (Requires a valid email address.)

Scratch below to reveal your download code:

Important Note: If this code has been revealed, it may have already been used, and may no longer be valid. Each code can be used by only one person. If this code is no longer valid, you can purchase a new code online at: http://www.sinauer.com/sylvius4.

Download codes contain only the following characters: A–Z, a–z, 0–9, as well as a dash (-) between each group of characters, with the following exceptions: there are no capital O's or lowercase L's. Also, zeros have a slash through them (0).

NEUROSCIENCE

FOURTH EDITION

Neuroscience

FOURTH EDITION

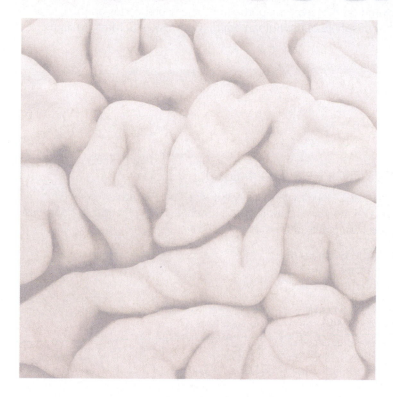

EDITED BY

Dale Purves

George J. Augustine

David Fitzpatrick

William C. Hall

Anthony-Samuel LaMantia

James O. McNamara

Leonard E. White

Sinauer Associates, Inc. • *Publishers*
Sunderland, Massachusetts U.S.A.

THE COVER

Pictured on the cover is a portion of the cerebral cortex on the lateral side of the left hemisphere of the human brain. (Courtesy of S. Mark Williams.)

NEUROSCIENCE, Fourth Edition

Copyright © 2008 by Sinauer Associates, Inc. All rights reserved.
This book may not be reproduced in whole or in part without permission.
Address inquiries and orders to

Sinauer Associates, Inc.
23 Plumtree Road
Sunderland, MA 01375 U.S.A.

www.sinauer.com
FAX: 413-549-1118
orders@sinauer.com
publish@sinauer.com

Library of Congress Cataloging-in-Publication Data
Neuroscience / edited by Dale Purves ... [et al.]. — 4th ed.
    p. ; cm.
 Includes bibliographical references and index.
 ISBN 978-0-87893-697-7 (hardcover : alk. paper)
1. Neurosciences. I. Purves, Dale.
 [DNLM: 1. Nervous System Physiology. 2. Neurochemistry. WL 102
N50588
2008]

QP355.2.N487 2008
612.8—dc22   2007024950

Printed in U.S.A.

5 4 3

The insight and enthusiasm of Larry Katz informed and enlivened the first two editions of this book, and his accomplishments have influenced an entire generation of neuroscientists.

The authors and editors dedicate this Fourth Edition of Neuroscience *to his memory.*

Contributors

George J. Augustine, Ph.D.

David Fitzpatrick, Ph.D.

William C. Hall, Ph.D.

Anthony-Samuel LaMantia, Ph.D.

James O. McNamara, M.D.

Richard D. Mooney, Ph.D.

Michael L. Platt, Ph.D.

Dale Purves, M.D.

Sidney A. Simon, Ph.D.

Leonard E. White, Ph.D.

S. Mark Williams, Ph.D.

UNIT EDITORS

UNIT I: George J. Augustine

UNIT II: David Fitzpatrick

UNIT III: William C. Hall and Leonard E. White

UNIT IV: Anthony-Samuel LaMantia

UNIT V: Dale Purves

Contents in Brief

Contents

Unit I

Neural Signaling

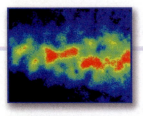

Unit II

Sensation and Sensory Processing

CHAPTER 12
Central Visual Pathways 289

CHAPTER 13
The Auditory System 313

CHAPTER 14
The Vestibular System 343

Unit III
Movement and Its Central Control

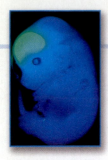

Unit IV

The Changing Brain

Unit V

Complex Brain Functions

APPENDIX
Survey of Human Neuroanatomy 815

ATLAS
The Human Central Nervous System 843

Glossary G–1

Illustration Credits IC–1

Index I–1

Preface

Whether judged in molecular, cellular, systemic, behavioral, or cognitive terms, the human nervous system is a stupendous piece of biological machinery. Given its accomplishments—all the artifacts of human culture, for instance—there is good reason for wanting to understand how the brain and the rest of the nervous system works. The debilitating and costly effects of neurological and psychiatric disease add a further sense of urgency to this quest. The aim of this book is to highlight the intellectual challenges and excitement—as well as the uncertainties—of what many see as the last great frontier of biological science. The information presented should serve as a starting point for undergraduates, medical students, graduate students in the neurosciences, and others who want to understand how the human nervous system operates. Like any other great challenge, neuroscience should be, and is, full of debate, dissension, and considerable fun. All these ingredients have gone into the construction of the fourth edition of this book; we hope they will be conveyed in equal measure to readers at all levels.

Acknowledgments

We are grateful to numerous colleagues who provided helpful contributions, criticisms, and suggestions to this and previous editions. We particularly wish to thank Ralph Adolphs, David Amaral, Eva Anton, Gary Banker, Bob Barlow, Marlene Behrmann, Ursula Bellugi, Dan Blazer, Bob Burke, Roberto Cabeza, Jim Cavanaugh, John Chapin, Milt Charlton, Michael Davis, Rob Deaner, Bob Desimone, Allison Doupe, Sasha du Lac, Jen Eilers, Anne Fausto-Sterling, Howard Fields, Elizabeth Finch, Nancy Forger, Jannon Fuchs, Michela Gallagher, Dana Garcia, Steve George, the late Patricia Goldman-Rakic, Jennifer Groh, Mike Haglund, Zach Hall, Kristen Harris, Bill Henson, John Heuser, Jonathan Horton, Ron Hoy, Alan Humphrey, Jon Kaas, Kai Kaila, Jagmeet Kanwal, Herb Killackey, Len Kitzes, Andrew Krystal, Arthur Lander, Story Landis, Simon LeVay, Darrell Lewis, Jeff Lichtman, Alan Light, Steve Lisberger, Arthur Loewy, Ron Mangun, Eve Marder, Robert McCarley, Greg McCarthy, Jim McIlwain, Chris Muly, Vic Nadler, Ron Oppenheim, Larysa Pevny, Franck Polleux, Scott Pomeroy, Rodney Radtke, Louis Reichardt, Sidarta Ribiero, Marnie Riddle, Jamie Roitman, Steve Roper, John Rubenstein, Ben Rubin, David Rubin, Josh Sanes, Cliff Saper, Lynn Selemon, Carla Shatz, Bill Snider, Larry Squire, John Staddon, Peter Strick, Warren Strittmatter, Joe Takahashi, Richard Weinberg, Jonathan Weiner, Christina Williams, and Joel Winston. It is understood, of course, that any errors are in no way attributable to our critics and advisors. We also wish to thank our colleagues Nell Cant, Dona M. Chikaraishi, Michael D. Ehlers, Gillian Einstein, Erich Jarvis, Lawrence C. Katz, Julie Kauer, Donald Lo, Miguel A. L. Nicolelis, Peter H. Reinhart, J. H. Pate Skene, James Voyvodic, and Fulton Wong for their contributions to earlier editions.

We also thank the students at Duke University as well as many other students and colleagues who provided suggestions for improvement of the last edition. Finally, we owe special thanks to Andy Sinauer, Graig Donini, Carol Wigg, Christopher Small, Jefferson Johnson, Janice Holabird, and the rest of the staff at Sinauer Associates for their outstanding work and high standards.

Supplements to Accompany
Neuroscience, Fourth Edition

For the Student

Sylvius 4:
An Interactive Atlas and Visual Glossary of Human Neuroanatomy

(Free download; access code provided with every new copy of the text)
S. Mark Williams, Leonard E. White, and Andrew C. Mace

The new *Sylvius* provides a unique digital learning environment for exploring and understanding the structure of the human central nervous system. *Sylvius* features fully annotated surface views of the human brain, as well as interactive tools for dissecting the central nervous system and viewing fully annotated cross-sections of preserved specimens and living subjects imaged by magnetic resonance. (See the inside front cover for additional information and download instructions.)

Companion Website (www.sinauer.com/neuroscience4e)

New for the Fourth Edition, the *Neuroscience* companion website features review and study tools to help students master the material presented in the neuroscience course. Access to the site is free of charge and requires no access code. (Instructor registration required for quizzes.) The site includes chapter summaries, an extensive collection of detailed animations, online quizzes, and flashcard activities for every chapter. (See the inside front cover for additional information.)

For the Instructor

Instructor's Resource Library

Expanded for the Fourth Edition, the *Neuroscience* Instructor's Resource Library includes a variety of resources to help in developing your course and delivering your lectures. The Library includes:
- Textbook Figures & Tables: All the figures and tables from the textbook, provided in JPEG format (both high- and low-resolution).
- PowerPoint® Presentations: All the figures and tables from each chapter, making it easy to add figures to your own presentations.
- *Sylvius* Image Library: A range of images from the companion program *Sylvius* 4, in PowerPoint format.
- Animations: An extensive new set of detailed animations.
- Quiz Questions: All of the quiz questions from the companion website.
- Review Questions: Short-answer style review questions for each chapter.

Online Quizzing

New for the Fourth Edition, instructors have access to a bank of online quizzes that they can choose to assign or let their students use for self-review purposes. Instructors can use the quizzes as-is, or they can create their own quizzes using any combination of publisher-provided questions and their own questions. The online grade book stores quiz results, which can be downloaded for use in grade book programs. (*Note*: Instructor registration is required in order for students to have access to the quizzes.)

NEUROSCIENCE

FOURTH EDITION

Chapter 1

Studying the Nervous System

Overview

Neuroscience encompasses a broad range of questions about how the nervous systems of humans and other animals develop, how they are organized, and how they function to generate behavior. These questions can be explored using the tools of genetics, molecular and cell biology, systems anatomy and physiology, behavioral biology, and psychology. The major challenge facing students of neuroscience is to integrate the diverse knowledge derived from these various levels of analysis into a more or less coherent understanding of brain structure and function (one has to qualify "coherent understanding" because so many questions remain unanswered). Many of the issues that have been explored successfully concern how the principal cells of all animal nervous systems—neurons and glia—perform their basic functions in anatomical, electrophysiological, cellular, and molecular terms. The varieties of neurons and supporting glial cells that have been identified are assembled into ensembles called neural circuits, and these circuits are the primary components of neural systems that process specific types of information. These systems in turn serve one of three general functions. Sensory systems represent information about the state of the organism and its environment; motor systems organize and generate actions; and associational systems link the sensory and motor components of the nervous system, providing the basis for "higher order" brain functions such as perception, attention, cognition, emotions, language, rational thinking, and the complex neural processes at the core of understanding human beings, their behavior, their history, and perhaps their future.

Genetics, Genomics, and the Brain

The complete sequence of the human genome is perhaps the most logical starting point for studying the brain and the rest of the nervous system; after all, this inherited information is also the starting point of each of us as individuals. The relative ease of obtaining, analyzing, and correlating gene sequences with neurobiological observations in humans and other animals has facilitated a wealth of new insights into the biology of the nervous system. In parallel with studies of normal nervous systems, the genetic analysis of human pedigrees with various brain diseases implies that it may soon be possible to understand and treat disorders long considered beyond the reach of science and medicine.

A **gene** is a sequence of the DNA nucleotides adenine (A), thymine (T), cytosine (C), and guanine (G). Within each gene, segments of the sequence called **exons** are transcribed into messenger RNA and subsequently into a chain of amino acids that specifies a given protein. Between the exons are sequence segments called **introns**. Although intron sequences are removed from the final gene transcript, they frequently influence how the exons are expressed and thus

Figure 1.1   Estimates of the number of genes in the human genome, as well as in the genomes of the mouse, the fruit fly *Drosophila melanogaster*, and the nematode worm *Caenorhabditis elegans*. Note that the number of genes does not correlate with organismal complexity; the simpler nematode has more genes than the fruit fly, and current analysis indicates that mice and humans actually have about the same number of genes. Much genetic activity is dependent on transcription factors that regulate when and to what degree a given gene is expressed.

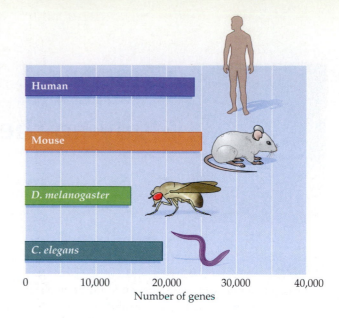

the nature of the resulting protein. In addition, the set of exons that defines the gene's mRNA transcript and resulting protein is flanked by upstream (5′) and downstream (3′) regulatory sequences (which can be promotional or inhibitory) that control the timing, location, and level of a gene's transcription.

A majority of the approximately 25,000 genes in the human genome are expressed in the developing and adult brain. The same is true in mice, fruit flies, and worms—the species commonly used in modern genetics, and increasingly used in neuroscience (Figure 1.1). Very few genes are *uniquely* expressed in neurons, however, indicating that nerve cells share most of the basic structural and functional properties of other cells. Accordingly, a great deal of "brain-specific" genetic information must reside in the introns and regulatory sequences that control the timing, quantity, variability, and cellular specificity of gene expression.

One of the most promising dividends of sequencing the human genome has been the realization that one or a few genes, when altered (mutated), can explain at least some aspects of neurological and psychiatric diseases. Before gene sequencing became routine, there often was little sense of how or why the normal biology of the nervous system was compromised in brain pathologies. The identification of genes correlated with disorders such as Huntington's disease, Parkinson's disease, Alzheimer's disease, clinical depression, and schizophrenia has provided a promising start to understanding these pathological processes in a much deeper way, and eventually devising rational therapies.

But genomic information alone cannot explain how the brain works in normal individuals, or how disease processes disrupt normal brain functions. To achieve these goals one must also understand the cell biology, anatomy, and physiology of the brain in both health and disease.

"Model" Organisms in Neuroscience

Many of the goals of modern neuroscience focus on understanding the organization and function of the human nervous system, as well as the pathological bases of neurological and psychiatric diseases. These issues, however, are often difficult to address by studying the human brain; therefore, neuroscientists have relied on the nervous systems of other animals to guide their studies. Over the last two centuries or so, critical information about the detailed anatomy, biochemistry, physiology, and cell biology of neural systems has been gleaned by studying the brains of a wide variety of species. Often the choice of species studied reflects assumptions about enhanced functional capacity in that species; for example, from the 1950s through the 1970s, cats were the subjects of pioneering studies on visual function. Cats were chosen for this research because they are highly "visual" animals, and therefore could be expected to have well-developed brain regions devoted to vision—regions similar to those found in primates, including humans. Much of what is currently known about human vision is based on studies carried out in cats. Studies on invertebrates such as the squid and the sea snail *Aplysia californica* yielded similarly critical insights into basic cell biology of neurons, synaptic transmission, and synaptic plasticity (the basis of learning and memory). In each case, the animal chosen offered advantages that made it possible to answer some of the key questions about neuroscience that we address throughout this text.

Today biochemical, cellular, anatomical, physiological, and behavioral studies continue to be conducted on a wide range of animals. However, the complete sequencing of the genomes of a small number of invertebrate, vertebrate, and mammalian species has led to the informal adoption of four "model" organisms by many neuroscientists. These are the nematode worm *Caenorhabditis elegans*; the fruit fly *Drosophila melanogaster*; the zebrafish *Danio rerio*; and the mouse *Mus musculus*. Despite certain limitations in each of these species, the relative ease of their genetic analysis and manipulation, as well as the availability of their complete genome sequences, facilitates research on a wide range of neuroscientific questions at the molecular, cellular, anatomical, and physiological levels.

Of course, other species are studied as well. Avians and amphibians, in the form of chickens and frogs, continue to be particularly useful for studying early neural development, and mammals such as the rat are used extensively for neuropharmacological and behavioral studies of adult brain function. Finally, nonhuman primates (the rhesus monkey in particular) provide opportunities to study complex functions that closely approximate those carried out in the human brain.

The Cellular Components of the Nervous System

Early in the nineteenth century, the cell was recognized as the fundamental unit of all living organisms. It was not until well into the twentieth century, however, that neuroscientists agreed that nervous tissue, like all other organs, is made up of these fundamental units. The major reason was that the first generation of "modern" neurobiologists in the nineteenth century had difficulty resolving the unitary nature of nerve cells with the microscopes and cell staining techniques then available. The extraordinarily complex shapes and extensive branches of individual nerve cells—all of which are packed together and thus difficult to resolve from one another—further obscured their resemblance to the geometrically simpler cells of other tissues (Figure 1.2). Some biologists of that era even concluded that each nerve cell was connected to its neighbors by protoplasmic links, forming a continuous nerve cell network, or *reticulum* (Latin for "net"). The Italian pathologist Camillo Golgi articulated and championed this "reticular theory" of nerve cell communication. Golgi made many important contributions to medical science, including identifying the cellular organelle eventually called the Golgi apparatus; the critically important cell staining technique that bears his name (see Figure 1.2); and an understanding of the pathophysiology of malaria. His reticular theory of the nervous system, however, eventually fell from favor and was replaced by what came to be known as the "neuron doctrine." The major proponents of the neuron doctrine were the Spanish neuroanatomist Santiago Ramón y Cajal and the British physiologist Charles Sherrington.

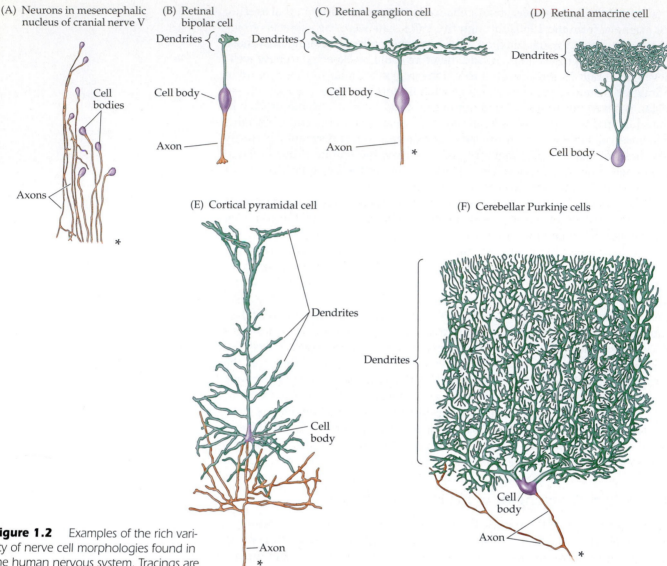

(A) Neurons in mesencephalic nucleus of cranial nerve V

Cell bodies

Axons

*

(B) Retinal bipolar cell

Dendrites {

Cell body

Axon

(C) Retinal ganglion cell

Dendrites {

Cell body

Axon

*

(D) Retinal amacrine cell

Dendrites {

Cell body

(E) Cortical pyramidal cell

Dendrites

Cell body

Axon

*

(F) Cerebellar Purkinje cells

Dendrites {

Cell body

Axon

*

Figure 1.2 Examples of the rich variety of nerve cell morphologies found in the human nervous system. Tracings are from actual nerve cells stained by impregnation with silver salts (the so-called Golgi technique, the method used in the classical studies of Golgi and Cajal). Asterisks indicate that the axon runs on much farther than shown. Note that some cells, like the retinal bipolar cell, have a very short axon, and that others, like the retinal amacrine cell, have no axon at all. The drawings are not all at the same scale.

The spirited debate occasioned by the contrasting views represented by Golgi and Cajal in the early twentieth century set the course of modern neuroscience. Based on light microscopic examination of nervous tissue stained with silver salts according to Golgi's pioneering cell-staining method, Cajal argued persuasively that nerve cells are discrete entities, and that they communicate with one another by means of specialized contacts, which Sherrington called *synapses*. Despite the ultimate triumph of Cajal's view over that of Golgi, both were awarded the Nobel Prize in Physiology or Medicine for their essential contribution to understanding the basic organization of the brain.

The subsequent work of Sherrington and others demonstrating the transfer of electrical signals at synaptic junctions between nerve cells provided strong support of the neuron doctrine, although occasional challenges to the autonomy of individual neurons remained. It was not until the advent of electron microscopy in the 1950s that any lingering doubts about the discreteness of neurons were resolved. The high-magnification, high-resolution pictures obtained with the electron microscope (see Figure 1.3) clearly established that nerve cells are functionally independent units; such micrographs also identified the specialized cellular junctions that Sherrington had named synapses. Perhaps as belated consolation to Golgi, however, these electron microscopic studies also demonstrated specialized—although relatively rare—intercellular continuities called gap junctions (similar to those found between cells in epithelia like the lung and intestine) between some neurons. These junctions do indeed allow for cytoplasmic continuity and direct transfer of electrical and chemical signals between cells in the nervous system.

The histological studies of Cajal, Golgi, and a host of successors led to the consensus that the cells of the nervous system can be divided into two broad categories: **nerve cells**, or **neurons**, and supporting cells called **neuroglia** (or simply **glia**). Nerve cells are specialized for electrical signaling over long distances. Understanding this process represents one of the more dramatic success stories in modern biology and is the subject of Unit I. Glial cells, in contrast to nerve cells, are not capable of significant electrical signaling; nevertheless, glia have several essential functions in the developing and adult brain. They are also essential contributors to repair of the damaged nervous system—in some cases promoting regrowth of damaged neurons, and in others preventing such regeneration (see Unit IV).

Neurons and glia share the complement of organelles found in all cells, including the endoplasmic reticulum and Golgi apparatus, mitochondria, and a variety of vesicular structures. In neurons, however, these organelles are often more prominent in distinct regions of the cell. Mitochondria, for example, tend to be concentrated at synapses, while protein-synthetic organelles such as the endoplasmic reticulum are largely excluded from axons and dendrites. In addition to the distribution of organelles and subcellular components, neurons and glia are in some measure different from other cells in the specialized fibrillar or tubular proteins that constitute the cytoskeleton (see Figure 1.4). Although many of these proteins—isoforms of actin, tubulin, myosin, and several others—are found in other cells, their distinctive organization in neurons is critical for the stability and function of neuronal processes and synaptic junctions. The various filaments, tubules, subcellular motors, and scaffolding proteins of the neuronal cytoskeleton orchestrate many functions, including the growth of axons and dendrites; the trafficking and appropriate positioning of membrane components, organelles, and vesicles; and the active processes of exocytosis and endocytosis underlying synaptic communication. Understanding the ways in which these molecular components are used to insure the proper development and function of neurons and glia remains a primary focus of modern neurobiology.

Neurons

Neurons are clearly distinguished by specialization for intercellular communication. This attribute is apparent in their overall morphology, in the specific organization of their membrane components for electrical signaling, and in the structural and functional intricacies of the synaptic contacts between neurons (Figure 1.3). The most obvious morphological sign of neuronal specialization for communication via electrical signaling is the extensive branching of neurons. The most salient aspect of this branching for typical nerve cells is the elaborate arborization of **dendrites** that arise from the neuronal cell body in the form of *dendritic branches* (or *dendritic processes*; see Figure 1.3E). Dendrites are the primary target for synaptic input from other neurons and are distinguished by their high content of ribosomes, as well as by specific cytoskeletal proteins.

The spectrum of neuronal geometries ranges from a small minority of cells that lack dendrites altogether to neurons with dendritic branches that rival the

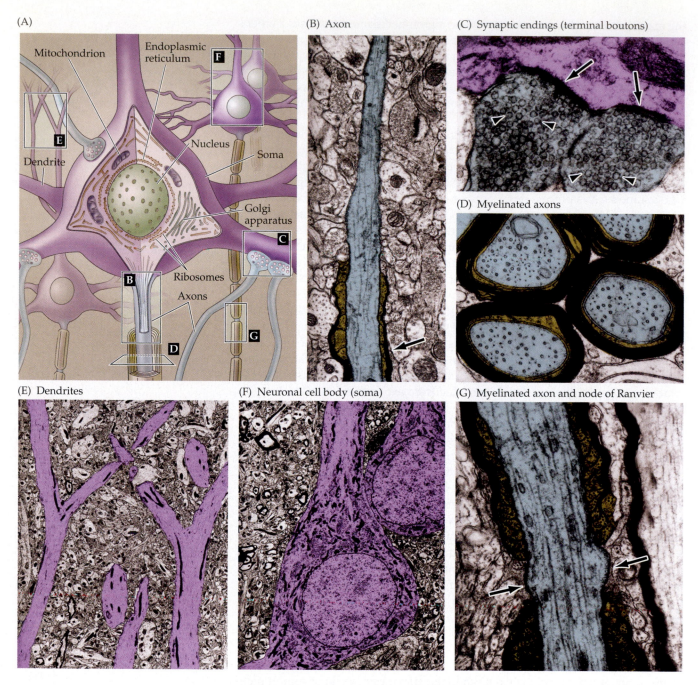

(A)

Mitochondrion

Endoplasmic reticulum

F

E

Dendrite

Nucleus

Soma

Golgi apparatus

C

B

Ribosomes

Axons

D

G

(B) Axon

(C) Synaptic endings (terminal boutons)

(D) Myelinated axons

(E) Dendrites

(F) Neuronal cell body (soma)

(G) Myelinated axon and node of Ranvier

Figure 1.3   The major light and electron microscopical features of neurons. (A) Diagram of nerve cells and their component parts. (B) Axon initial segment (blue) entering a myelin sheath (green). (C) Terminal boutons (blue) loaded with synaptic vesicles (arrowheads) forming synapses (arrows) with a dendrite (purple). (D) Transverse section of axons (blue) ensheathed by the processes of oligodendrocytes (gold). (E) Apical dendrites (purple) of cortical pyramidal cells. (F) Nerve cell bodies (purple) occupied by large round nuclei. (G) Portion of a myelinated axon (blue) illustrating the intervals between adjacent segments of myelin (gold) referred to as nodes of Ranvier (arrows). (Micrographs from Peters et al., 1991.)

complexity of a mature tree (see Figure 1.2). The number of inputs that a particular neuron receives depends on the complexity of its dendritic arbor: nerve cells that lack dendrites are innervated by just one or a few other nerve cells, whereas neurons with increasingly elaborate dendritic branches are innervated by a commensurately larger number of other neurons. The *number of inputs* to a single neuron reflects the degree of **convergence**, while the *number of targets* innervated by any one neuron represents its **divergence**.

The synaptic contacts made on dendrites (and, less frequently, on neuronal cell bodies) comprise a special elaboration of the secretory apparatus found in most polarized epithelial cells. Typically, the **presynaptic terminal** is immediately adjacent to a **postsynaptic specialization** of the target cell. For the majority of synapses, there is no physical continuity between these pre- and postsynaptic elements. Instead, pre- and postsynaptic components communicate via the secretion of molecules from the presynaptic terminal that bind to receptors in the postsynaptic specialization. These molecules must traverse an interval of extracellular space between pre- and postsynaptic elements called the **synaptic cleft**. The synaptic cleft, however, is not simply an empty space to be traversed; rather, it is the site of extracellular proteins that influence the diffusion, binding, and degradation of molecules secreted by the presynaptic terminal. The number of synaptic inputs received by each nerve cell in the human nervous system varies from 1 to about 100,000. This range reflects a fundamental purpose of nerve cells, namely to integrate information from other neurons. The number of synaptic contacts from different presynaptic neurons onto any particular cell is therefore an especially important determinant of neuronal function.

The information conveyed by synapses on the neuronal dendrites is integrated and "read out" at the origin of the **axon**, the portion of the nerve cell specialized for relaying electrical signals (see Figure 1.3B). The axon is a unique extension from the neuronal cell body that may travel a few hundred micrometers (μm, also called microns) or much farther, depending on the type of neuron and the size of the species. Moreover, the axon also has a distinct cytoskeleton whose elements are central for its functional integrity (Figure 1.4). Many nerve cells in the human brain have axons no more than a few millimeters long, and a few have no axons at all.

Relatively short axons are a feature of **local circuit neurons**, or **interneurons**, throughout the brain. The axons of **projection neurons**, however, extend to distant targets. For example, the axons that run from the human spinal cord to the foot are about a meter long. The event that carries signals over such distances is a self-regenerating wave of electrical activity called an **action potential** that propagates from its point of initiation at the cell body (the **axon hillock**) to the terminus of the axon, at which point synaptic contacts are made. The target cells of neurons—sites where axons terminate and synapses are made—include other nerve cells in the brain, spinal cord, and autonomic ganglia, as well as the cells of muscles and glands throughout the body.

The chemical and electrical process by which the information encoded by action potentials is passed on at synaptic contacts to the next cell in a pathway is called **synaptic transmission**. Presynaptic terminals (also called *synaptic endings, axon terminals,* or *terminal boutons;* see Figure 1.3C) and their postsynaptic specializations are typically **chemical synapses**, the most abundant type of synapse in the nervous system. Another type, the **electrical synapse** (facilitated by the gap junctions mentioned above), is far more rare (see Chapter 5).

The secretory organelles in the presynaptic terminal of chemical synapses are **synaptic vesicles**, spherical structures filled with **neurotransmitter molecules**. The positioning of synaptic vesicles at the presynaptic membrane and their fusion to initiate neurotransmitter release is regulated by a number of proteins either

Figure 1.4 Distinctive arrangement of cytoskeletal elements in neurons. (A) The cell body, axons, and dendrites are distinguished by the distribution of tubulin (green throughout cell) versus other cytoskeletal elements—in this case, the microtubule-binding protein Tau (red), which is found only in axons. (B) The localization of actin (red) to the growing tips of axonal and dendritic processes is shown here in a cultured neuron taken from the hippocampus. (C) In contrast, in a cultured epithelial cell, actin (red) is distributed in fibrils that occupy most of the cell body. (D) In astroglial cells in culture, actin (red) is also seen in fibrillar bundles. (E) Tubulin (green) is seen throughout the cell body and dendrites of neurons. (F) Although tubulin is a major component of dendrites, extending into spines, the head of the spine is enriched in actin (red). (G) The tubulin component of the cytoskeleton in non-neuronal cells is arrayed in filamentous networks. (H–K) Synapses have a distinct arrangement of cytoskeletal elements, receptors, and scaffold proteins. (H) Two axons (green; tubulin) from motor neurons are seen issuing two branches each to four muscle fibers. The red shows the clustering of postsynaptic receptors (in this case for the neurotransmitter acetylcholine). (I) A higher power view of a single motor neuron synapse shows the relationship between the axon (green) and the postsynaptic receptors (red). (J) Proteins in the extracellular space between the axon and its target muscle are labeled green. (K) Scaffolding proteins (green) localize receptors (red) and link them to other cytoskeletal elements. The scaffolding protein shown here is dystrophin, whose structure and function are compromised in the many forms of muscular dystrophy. (A courtesy of Y. N. Jan; B courtesy of E. Dent and F. Gertler; C courtesy of D. Arneman and C. Otey; D courtesy of A. Gonzales and R. Cheney; E from Sheng, 2003; F from Matus, 2000; G courtesy of T. Salmon et al.; H–K courtesy of R. Sealock.)

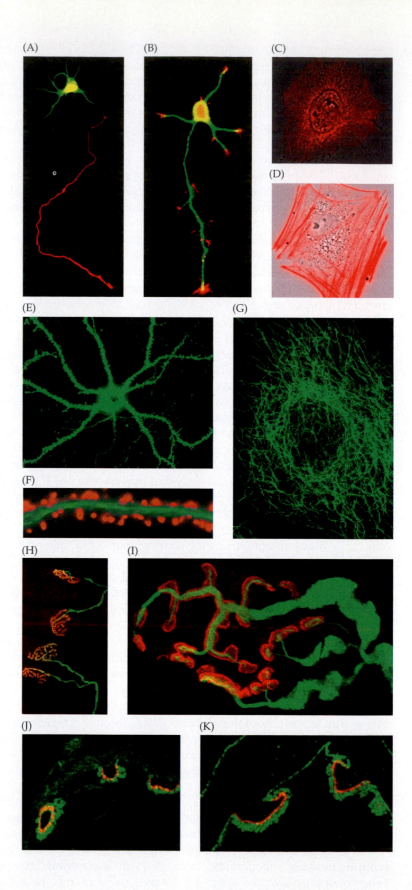

within or associated with the vesicle. The neurotransmitters released from synaptic vesicles modify the electrical properties of the target cell by binding to **neurotransmitter receptors** localized primarily at the postsynaptic specialization.

The intricate and concerted activity of neurotransmitters, receptors, related cytoskeletal elements, and signal transduction molecules are the basis for nerve cells communicating with one another, and with effector cells in muscles and glands.

Neuroglial Cells

Neuroglial cells—also referred to as glial cells or simply glia—are quite different from nerve cells. Glia are more numerous than neurons in the brain, outnumbering them by a ratio of perhaps 3 to 1. Despite their superior numbers, glia do not participate directly in synaptic interactions and electrical signaling, although their supportive functions help define synaptic contacts and maintain the signaling abilities of neurons. Glial cells also have complex processes extending from their cell bodies, but these processes are generally less prominent than neuronal branches and do not serve the same purposes as axons and dendrites.

The term *glia* (Greek, "glue") reflects the nineteenth-century presumption that these cells "held the nervous system together" in some way. The word has survived, despite the lack of any evidence that glial cells actually bind nerve cells together. Glial roles that *are* well established include maintaining the ionic milieu of nerve cells; modulating the rate of nerve signal propagation; modulating synaptic action by controlling the uptake of neurotransmitters at or near the synaptic cleft; providing a scaffold for some aspects of neural development; and aiding (or in some instances impeding) recovery from neural injury.

There are three types of glial cells in the mature central nervous system: astrocytes, oligodendrocytes, and microglial cells (Figure 1.5). **Astrocytes**, which are restricted to the central nervous system (that is, the brain and spinal cord),

Figure 1.5 *Varieties of neuroglial cells. Tracings of an astrocyte (A), an oligodendrocyte (B), and a microglial cell (C) visualized using the Golgi method. The images are at approximately the same scale. (D) Astrocytes in tissue culture, labeled (red) with an antibody against an astrocyte-specific protein. (E) Oligodendroglial cells (green) in tissue culture labeled with an antibody against an oligodendroglial-specific protein. (F) Peripheral axons are ensheathed by myelin (labeled red) except at nodes of Ranvier (see Figure 1.3G). The green label indicates ion channels concentrated in the node; the blue label indicates a molecularly distinct region called the paranode. (G) Microglial cells from the spinal cord, labeled with a cell type-specific antibody. Inset: Higher-magnification image of a single microglial cell labeled with a macrophage-selective marker. (A–C after Jones and Cowan, 1983; D, E courtesy of A.-S. LaMantia; F courtesy of M. Bhat; G courtesy of A. Light; inset courtesy of G. Matsushima.)*

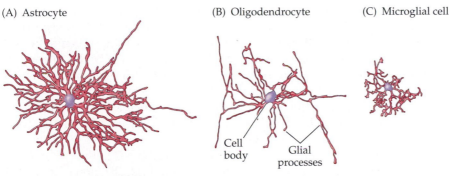

(A) Astrocyte   (B) Oligodendrocyte   (C) Microglial cell

Cell body   Glial processes

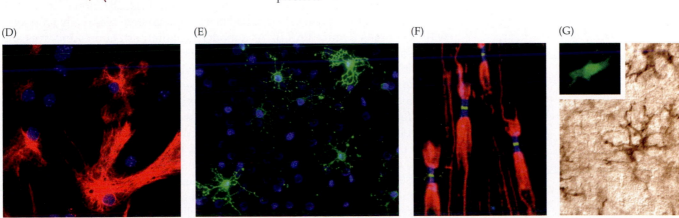

(D)   (E)   (F)   (G)

have elaborate local processes that give these cells a starlike ("astral") appearance. A major function of astrocytes is to maintain, in a variety of ways, an appropriate chemical environment for neuronal signaling. In addition, recent observations suggest that a subset of astrocytes in the adult brain may retain the characteristics of neural stem cells—that is, the capacity to enter mitosis and generate all of the cell classes found in the nervous tissue (see Unit IV).

Oligodendrocytes, which are also restricted to the central nervous system, lay down a laminated, lipid-rich wrapping called **myelin** around some, but not all, axons (see Figures 1.3D,G). Myelin has important effects on the speed of the transmission of electrical signals (see Chapter 3). In the peripheral nervous system, the cells that elaborate myelin are called **Schwann cells**.

Finally, **microglial cells** are derived primarily from hematopoietic precursor cells (although some may be derived directly from neural precursor cells). They share many properties with macrophages found in other tissues, and are primarily scavenger cells that remove cellular debris from sites of injury or normal cell turnover. In addition, microglia, like their macrophage counterparts, secrete signaling molecules—particularly a wide range of cytokines that are also produced by cells of the immune system—that can modulate local inflammation and influence cell survival or death. Indeed, some neurobiologists prefer to categorize microglia as a type of macrophage. Following brain damage, the number of microglia at the site of injury increases dramatically. Some of these cells proliferate from microglia resident in the brain, while others come from macrophages that migrate to the injured area and enter the brain via local disruptions in the cerebral vasculature.

Cellular Diversity in the Nervous System

Although the cellular constituents of the human nervous system are in many ways similar to those of other organs, they are unusual in their extraordinary numbers: the human brain is estimated to contain 100 *billion* neurons and several times as many supporting cells. More importantly, the nervous system has a greater range of distinct cell types—whether categorized by morphology, molecular identity, or physiological activity—than any other organ system (a fact that presumably explains why, as mentioned at the start of this chapter, so many different genes are expressed in the nervous system). The cellular diversity of any nervous system undoubtedly underlies the capacity of the system to form increasingly complicated networks and to mediate increasingly sophisticated behaviors. For much of the twentieth century, neuroscientists relied on the set of techniques developed by Cajal, Golgi, and other pioneers of histology and pathology to describe and categorize the diverse cell types in the nervous system. The staining method named for Golgi permitted visualization of individual nerve cells and their processes that had been impregnated, seemingly randomly, with silver salts (Figure 1.6A,B). As a modern counterpart, fluorescent dyes and other soluble molecules injected into single neurons, often after physiological recording to identify the function of the cell, provide an alternative approach to visualizing single nerve cells and their processes (Figure 1.6C,D).

As a complement to these approaches (which provide a random sample of only a few neurons and glia), other stains are used to demonstrate the distribution of all cell bodies—but not their processes or connections—in neural tissue. The widely used Nissl method is one such method; this technique stains the nucleolus and other structures (e.g., ribosomes) where DNA or RNA is found (Figure 1.6E). Such stains demonstrate that the size, density and distribution of the total population of nerve cells is not uniform from region to region within the brain. In some regions, such as the cerebral cortex, cells are arranged in layers (Figure

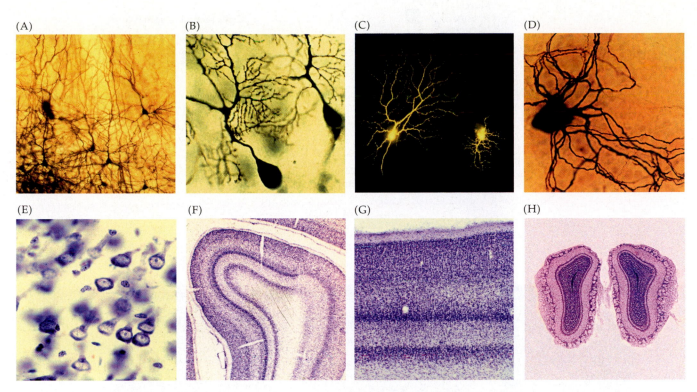

Figure 1.6 Visualizing nerve cells and their connections. (A) Cortical neurons stained using the Golgi method (impregnation with silver salts). (B) Golgi-stained Purkinje cells in the cerebellum. Purkinje cells have a single, highly branched apical dendrite. (C) Intracellular injection of fluorescent dye labels two retinal neurons that vary dramatically in the size and extent of their dendritic arborizations. (D) Intracellular injection of an enzyme labels a neuron in a ganglion of the autonomic (involuntary) nervous system. (E) The dye cresyl violet stains RNA in all cells in a tissue, labeling the nucleolus (but not the nucleus) as well as the ribosome-rich endoplasmic reticulum. Dendrites and axons are not labeled, explaining the "blank" spaces between neurons. (F) Nissl-stained section of the cerebral cortex, showing cell bodies arranged into layers of differing cell densities. (G) Higher magnification of one area of cerebral cortex shows that differences in cell density define boundaries between layers of this visual cortex. (H) Nissl stain of the olfactory bulbs reveals a distinctive distribution of cell bodies, particularly those cells arranged in rings on the outer surface of the bulb. These structures, including the cell-sparse tissue contained within each ring, are called glomeruli. (C courtesy of C. J. Shatz; all others courtesy of A.-S. LaMantia and D. Purves.)

1.6F,G), each layer recognized by distinctive differences in cell density. Structures like the olfactory bulb display even more complicated arrangements of cell bodies (Figure 1.6H). Additional approaches, detailed later in the chapter, have provided further definition of differences between nerve cells from region to region. These include the identification of how subsets of neurons are connected to one another, and how molecular differences further distinguish classes of nerve cells in a variety of brain regions (see Figure 1.11).

Neural Circuits

Neurons never function in isolation; they are organized into ensembles called **neural circuits** that process specific kinds of information and provide the foundation of sensation, perception, and behavior. The synaptic connections that define neural circuits are typically made in a dense tangle of dendrites, axon terminals, and glial cell processes that together constitute what is called **neuropil** (from the Greek word *pilos*, "felt"; see Figure 1.3). The neuropil constitutes the regions between nerve cell bodies where most synaptic connectivity occurs.

Although the arrangement of neural circuits varies greatly according to the function being served, some features are characteristic of all such ensembles. Preeminent is the direction of information flow in any particular circuit, which is obviously essential to understanding its purpose. Nerve cells that carry information from the periphery *toward* the brain or spinal cord (or deeper centrally within the spinal cord and brain) are called **afferent neurons**; nerve cells that carry information *away* from the brain or spinal cord (or away from the circuit in question) are called **efferent neurons**. Interneurons (local circuit neurons; see above) only participate in the local aspects of a circuit, based on the short distances over

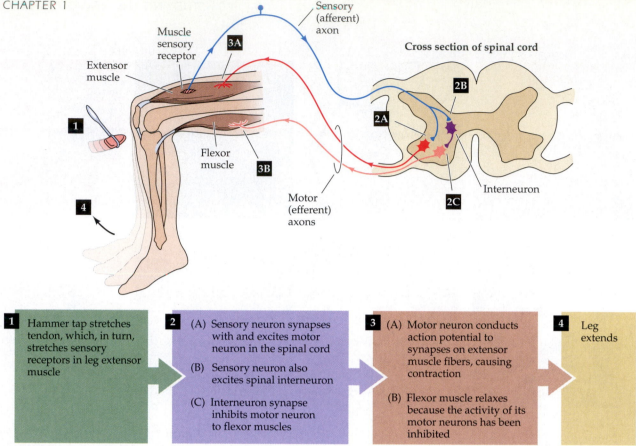

Figure 1.7 A simple reflex circuit, the knee-jerk response (more formally, the myotatic reflex), illustrates several points about the functional organization of neural circuits. Stimulation of peripheral sensors (a muscle stretch receptor in this case) initiates receptor potentials that trigger action potentials that travel centrally along the afferent axons of the sensory neurons. This information stimulates spinal motor neurons by means of synaptic contacts. The action potentials triggered by the synaptic potential in motor neurons travel peripherally in efferent axons, giving rise to muscle contraction and a behavioral response. One of the purposes of this particular reflex is to help maintain an upright posture in the face of unexpected changes.

which their axons extend. These three functional classes—afferent neurons, efferent neurons, and interneurons—are the basic constituents of all neural circuits.

A simple example of a neural circuit is one that mediates the **myotatic spinal reflex**, commonly known as the "knee-jerk" reflex (Figure 1.7). The afferent neurons that control the reflex are sensory neurons whose cell bodies lie the peripheral **dorsal root ganglia** and whose peripheral axons terminate in sensory endings in skeletal muscles. (The ganglia that serve this same of function for much of the head and neck are called **cranial nerve ganglia**; see the Appendix.) The central axons of these afferent sensory neurons enter the spinal cord, where they terminate on a variety of central neurons concerned with the regulation of muscle tone—most obviously on the **motor neurons** that determine the activity of the related muscles. These motor neurons constitute the efferent neurons. One group of motor neurons in the ventral horn of the spinal cord projects to the flexor muscles in the limb, the other to extensor muscles. Spinal cord interneurons are the third element of this circuit. The interneurons receive synaptic contacts from sensory afferent neurons and make synapses on the efferent motor neurons that project to the flexor muscles; thus they are capable of modulating the input–output linkage. The excitatory synaptic connections between the sensory afferents and the extensor efferent motor neurons cause the extensor muscles to contract; at the same time, interneurons activated by the afferents are inhibitory, and their activation diminishes electrical activity in flexor efferent motor neurons and causes the flexor muscles to become less active. The result is a complementary activation and inactivation of the synergistic and antagonistic muscles that control the position of the leg.

A more detailed picture of the events underlying the myotatic or any other neural circuit can be obtained by electrophysiological recording. There are two

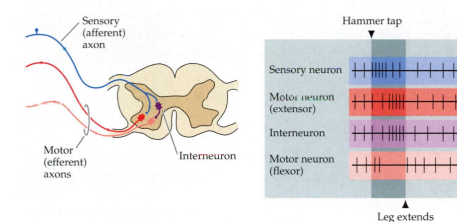

Figure 1.8 Relative frequency of action potentials (indicated by individual vertical lines) in different components of the myotatic reflex as the reflex pathway is activated. Notice the modulatory effect of the interneuron.

approaches to measuring the electrical activity of a nerve cell: **extracellular recording**, where an electrode is placed *near* the nerve cell of interest to detect its activity; and **intracellular recording**, where the electrode is placed *inside* the cell. Extracellular recordings primarily detect **action potentials**, the all-or-nothing changes in the potential (voltage) across nerve cell membranes that convey information from one point to another in the nervous system; action potentials are described in detail in Chapter 2. Extracellular recording is particularly useful for detecting temporal patterns of action potential activity and relating those patterns to stimulation by other inputs, or to specific behavioral events. Intracellular recording can detect the smaller, graded potential changes that trigger action potentials, and thus allow a more detailed analysis of communication between neurons within a circuit. These graded triggering potentials can arise at either sensory receptors or synapses and are called **receptor potentials** or **synaptic potentials**, respectively.

For the myotatic circuit, electrical activity can be measured both extracellularly and intracellularly, thus defining the functional relationships among the neurons in the circuit. With electrodes placed near, but still outside of individual cells, the pattern of action potential activity can be extracellularly recorded for each element of the circuit (i.e., afferents, efferents, and interneurons) before, during, and after a stimulus (Figure 1.8). By comparing the onset, duration, and frequency of action potential activity in each cell, a functional picture of the circuit emerges. As a result of the stimulus, the sensory neuron is triggered to fire at higher frequency (i.e., more action potentials per unit time). This increase triggers a higher frequency of action potentials in both the extensor motor neurons and the interneurons. Concurrently, the inhibitory synapses made by the interneurons onto the flexor motor neurons cause the frequency of action potentials in these cells to decline. Using intracellular recording, it is possible to observe directly the changes in membrane potential underlying the synaptic connections of each element of the myotatic reflex circuit (Figure 1.9).

The Organization of the Human Nervous System

When considered together, circuits that process similar types of information make up **neural systems** that serve broader behavioral purposes. The most general functional distinction divides such collections into **sensory systems** that acquire and process information from the environment (e.g., the visual system or the auditory system, both described in Unit II), and **motor systems** that respond to such information by generating movements and other behavior (Unit III). There are, however, large numbers of cells and circuits that lie

Figure 1.9   *Intracellularly recorded responses underlying the myotatic reflex. (A) Action potential measured in a sensory neuron. (B) Postsynaptic triggering potential recorded in an extensor motor neuron. (C) Postsynaptic triggering potential in an interneuron. (D) Postsynaptic inhibitory potential in a flexor motor neuron. Such intracellular recordings are the basis for understanding the cellular mechanisms of action potential generation, and the sensory receptor and synaptic potentials that trigger these conducted signals.*

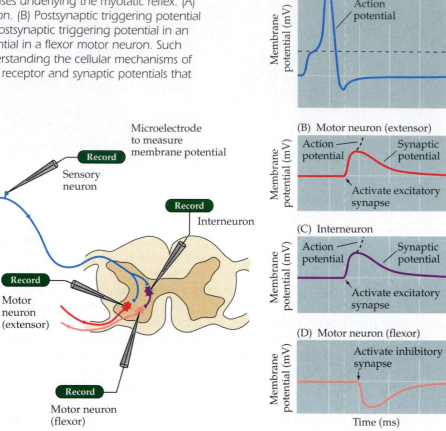

between these relatively well-defined input and output systems. These are collectively referred to as **associational systems**, and they mediate the most complex and least well-characterized brain functions (Unit V).

In addition to these broad functional distinctions, neuroscientists and neurologists have conventionally divided the vertebrate nervous system anatomically into central and peripheral components (Figure 1.10). The **central nervous system**, typically referred to as the **CNS**, comprises the **brain** (cerebral hemispheres, diencephalon, cerebellum, and brainstem) and the **spinal cord** (see Appendix A for more information about the gross anatomical features of the CNS). The **peripheral nervous system** (**PNS**) includes the sensory neurons that link sensory receptors on the body surface or deeper within it with relevant processing circuits in the central nervous system. The motor portion of the peripheral nervous system in turn consists of two components. The motor axons that connect the brain and spinal cord to skeletal muscles make up the **somatic motor division** of the peripheral nervous system, whereas the cells and axons that innervate smooth muscles, cardiac muscle, and glands make up the **visceral** or **autonomic motor division**.

Those nerve cell bodies that reside in the peripheral nervous system are located in **ganglia**, which are simply local accumulations of nerve cell bodies and supporting cells. Peripheral axons are gathered into bundles called **nerves**, many of which are enveloped by the glial cells of the peripheral nervous system; as mentioned earlier, these peripheral glia are called **Schwann cells**.

Nerve cells in the central nervous system are arranged in two different ways. **Nuclei** are local accumulations of neurons that have roughly similar connec-

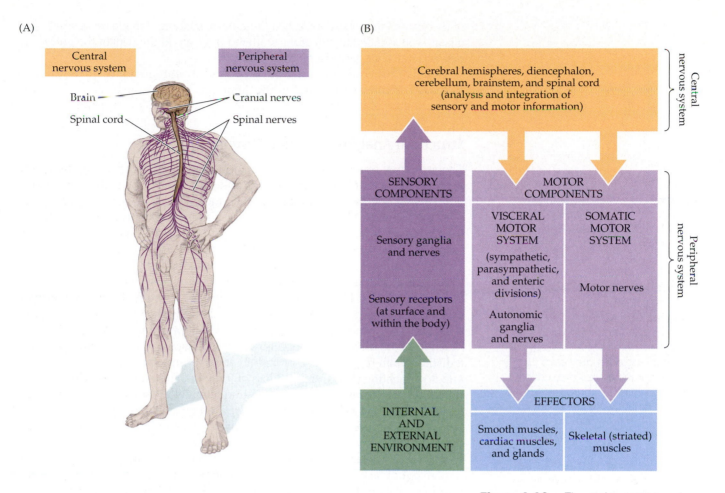

(A)

Central nervous system

Peripheral nervous system

Brain

Spinal cord

Cranial nerves

Spinal nerves

(B)

Central nervous system

Cerebral hemispheres, diencephalon, cerebellum, brainstem, and spinal cord (analysis and integration of sensory and motor information)

Peripheral nervous system

SENSORY COMPONENTS

Sensory ganglia and nerves

Sensory receptors (at surface and within the body)

MOTOR COMPONENTS

VISCERAL MOTOR SYSTEM

(sympathetic, parasympathetic, and enteric divisions)

Autonomic ganglia and nerves

SOMATIC MOTOR SYSTEM

Motor nerves

INTERNAL AND EXTERNAL ENVIRONMENT

EFFECTORS

Smooth muscles, cardiac muscles, and glands

Skeletal (striated) muscles

Figure 1.10 The major components of the nervous system and their functional relationships. (A) The CNS (brain and spinal cord) and PNS (spinal and cranial nerves). (B) Diagram of the major components of the central and peripheral nervous systems and their functional relationships. Stimuli from the environment convey information to processing circuits within the brain and spinal cord, which in turn interpret their significance and send signals to peripheral effectors that move the body and adjust the workings of its internal organs.

tions and functions; such collections are found throughout the cerebrum, brainstem, and spinal cord. In contrast, **cortex** (plural, *cortices*) describes sheet-like arrays of nerve cells (consult Appendix A for additional information and illustrations). The cortices of the cerebral hemispheres and of the cerebellum provide the clearest example of this organizational principle.

Axons in the central nervous system are gathered into **tracts** that are more or less analogous to nerves in the periphery. Tracts that cross the midline of the brain are referred to as **commissures**. Two gross histological terms distinguish regions rich in neuronal cell bodies versus regions rich in axons. **Gray matter** refers to any accumulation of cell bodies and neuropil in the brain and spinal cord (e.g., nuclei or cortices), whereas **white matter** (named for its relatively light appearance, the result of the lipid content of myelin) refers to axon tracts and commissures.

The organization of the visceral motor division of the peripheral nervous system (the nerve cells that control the functions of the visceral organs including the heart, the lungs, the gastrointestinal tract, and the genitalia) is a bit more complicated (see Chapter 21). Visceral motor neurons in the brainstem and spinal cord—the so-called preganglionic neurons—form synapses with peripheral motor neurons that lie in the **autonomic ganglia**. The peripheral motor neurons in autonomic ganglia innervate smooth muscle, glands, and cardiac muscle, thus controlling most involuntary (visceral) behavior. In the **sympathetic division** of the autonomic motor system, the ganglia lie along or in front of the vertebral column and send their axons to a variety of peripheral targets. In the **parasympathetic division**, the ganglia are found within or adjacent to the organs they innervate. Another component of the visceral motor system, called the **enteric system**, is made up of small ganglia as well as individual neurons scattered throughout the wall of the gut. These neurons influence gastric motility and secretion.

Details about the physical structures and overall anatomy of the human nervous system can be found in the Appendix to this book, as well as in the abridged atlas of the central nervous system that follows the Appendix.

Structural Analysis of Neural Systems

The structural organization of the brain and peripheral nervous system—the anatomical details of its ganglia, nuclei, and cortices, and the pattern of connections defined by its nerves and tracts—is fundamental for understanding nervous system function. By observing differences in tissue appearance, particularly the distribution of gray and white matter, the regional anatomy of a human brain can be discerned. These anatomical differences were of great use for the earliest pathologists of the nervous system, who made inferences of functional localization (that is, which subregion of the nervous system subserves which behavioral ability) by correlating gross damage to brain structures observed postmortem with functional deficits recorded during the lifetime of the individual. This use of structure to infer function was adopted for experimentation, and much of neuroscience rests upon observations made by purposefully damaging a distinct brain region, nerve, or tract in an experimental animal and observing and documenting the subsequent loss of function. These **lesion studies** provide the foundation for much of our current understanding of neuroanatomy. Parallel with this effort, neuroanatomists correlated gross differences in brain structure with differences in cell density and packing discerned in histological material stained to show cell bodies (see Figure 1.6).

The current detailed view of connectional neuroanatomy emerged only after the advent of techniques to trace neural connections from their source to their termination (**anterograde tracing**) or vice versa (**retrograde tracing**). These

Figure 1.11 Cellular and molecular approaches for studying connectivity and molecular identity of nerve cells. (A–C) Tracing connections and pathways in the brain. (A) Radioactive amino acids can be taken up by one population of nerve cells (in this case, injection of a radioactively labeled amino acid into one eye) and be transported to the axon terminals of those cells in the target region in the brain. (B) Fluorescent molecules injected into nerve tissue are taken up by the axon terminals at the site of the injection. The molecules are then transported, labeling the cell bodies and dendrites of the nerve cells that project to the injection site. (C) Tracers that label axons can reveal complex pathways in the nervous system. In this case, a dorsal root ganglion has been injected, showing the variety of axon pathways from the ganglion into the spinal cord. (D–G) Molecular differences among nerve cells. (D) A single glomerulus in the olfactory bulb (see Figure 1.6H) has been labeled with an antibody against the inhibitory neurotransmitter GABA. The label shows up as a red stain, revealing GABA to be localized in subsets of neurons around the glomerulus as well as at synaptic endings in the neuropil of the glomerulus. (E) The cerebellum has been labeled with an antibody that recognizes subsets of dendrites (green). (F) Here the cerebellum has been labeled with a probe (blue) for a specific gene that is expressed only by Purkinje cells. (A courtesy of P. Rakic; B courtesy of B. Schofield; C courtesy of W. D. Snider and J. Lichtman; D–F courtesy of A.-S. LaMantia, D. Meechan and T. Maynard.)

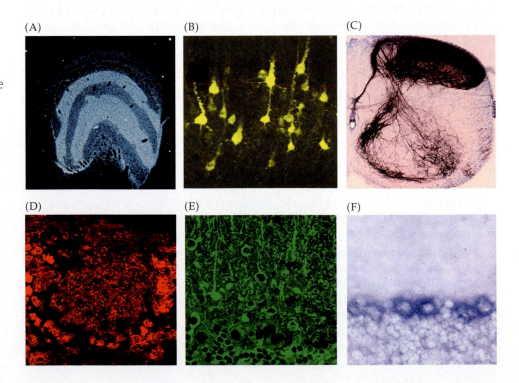

(A)   (B)   (C)

(D)   (E)   (F)

approaches permit detailed assessment of connections between various regions of the nervous system, thus facilitating the "mapping" of connections between neurons in one structure (for example, the eye) and their targets in the brain. Initially these techniques relied on physically injecting visualizable molecules into the brain to be taken up by local cell bodies and transported to axon terminals or taken up by local axons and terminals and transported back to the parent cell body (Figure 1.11A,B). Additional tracers can demonstrate an entire network of axonal projections from nerve cells exposed to the tracer (Figure 1.11C). These approaches permit assessment of the extent of connections from a single population of nerve cells to their targets throughout the nervous system.

The analysis of connectivity in neural systems has been augmented by molecularly based histochemical techniques that demonstrate biochemical and genetic distinctions of nerve cells and their processes. Whereas general cell staining methods primarily show differences in cell size and distribution, antibody stains can recognize specific proteins that are found in different regions of a nerve cell, or in different classes of nerve cells. These approaches have clarified the distribution of synapses, dendrites, and other molecular distinctions among nerve cells in a variety of brain regions (Figure 1.11D,E). In addition, antibodies for various proteins, as well as probes for specific mRNA transcripts (which detect gene expression in the relevant cells), can be used to make molecular distinctions between apparently equivalent nerve cells (Figure 1.11F). Even more recently, molecular genetic and neuroanatomical methods have been combined to visualize the expression of fluorescent or other tracer molecules under the control of regulatory sequences of neural genes. This approach illuminates individual cells in fixed or living tissue in remarkable detail, allowing nerve cells and their processes to be identified by both their transcriptional state (i.e., which genes are being transcribed in that cell) as well as their structure and connections. Molecular and genetic engineering approaches allow researchers to trace connections between molecularly defined populations of neurons and their targets (Figure 1.12). The use of several approaches—pathway tracing, analysis of molecular identity of nerve cells, and genetic approaches to identifying cell identities and connections—is now routine in the ongoing study of how neural tissue is organized into functional circuits and systems.

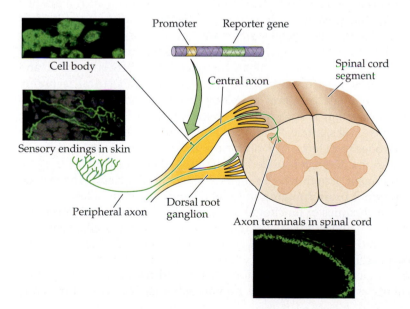

Figure 1.12 Genetic engineering is used to show pathways within the nervous system. A "reporter gene" that codes for some visualizable substance (e.g., green fluorescent protein, GFP) is inserted into the genome under the control of a cell type-specific promoter (a DNA sequence that turns the gene "on" in specific tissue and cell types). The reporter is expressed only in those cell types, revealing the cell bodies, axons, and dendrites of all cells in the nervous system that express the gene. Here the reporter is under the control of a promoter DNA sequence that is activated only in a subset of dorsal root ganglion neurons. Photographs show that the reporter labels neuronal cell bodies; the axons that project to the skin as free nerve endings; and the axon that projects to the dorsal root of the spinal cord to relay this sensory information from the skin to the brain. (Photographs from Zylka et al., 2005.)

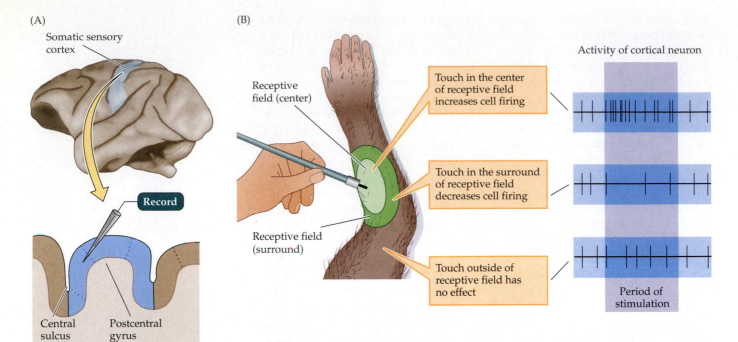

(A)

Somatic sensory cortex

Record

Central sulcus   Postcentral gyrus

(B)

Receptive field (center)

Receptive field (surround)

Activity of cortical neuron

Touch in the center of receptive field increases cell firing

Touch in the surround of receptive field decreases cell firing

Touch outside of receptive field has no effect

Period of stimulation

Figure 1.13 Single-unit electrophysiological recording from cortical pyramidal neuron, showing the firing pattern in response to a specific peripheral stimulus. (A) Typical experiment set-up, in which a recording electrode is inserted into the brain. (B) Defining neuronal receptive fields.

Functional Analysis of Neural Systems

A wide range of physiological methods is now available to evaluate the electrical (and metabolic) activity of the neuronal circuits that make up a neural system. Two approaches, however, have been particularly useful in defining how neural systems represent information. The most widely used method is **single-cell**, or **single-unit**, **electrophysiological recording** with microelectrodes. This method often records from several nearby cells in addition to the one selected, providing further useful information. The use of microelectrodes to record action potential activity provides a cell-by-cell analysis of the organization of topographic maps, and can give specific insight into the type of stimulus to which the neuron is "tuned" (i.e., the stimulus that elicits a maximal change in action potential activity from the baseline state). Single-unit analysis is often used to define a neuron's **receptive field**—the region in sensory space (e.g., the body surface, or a specialized structure such as the retina) within which a specific stimulus elicits the greatest action potential response (Figure 1.13). This approach to understanding neural systems was introduced by Stephen Kuffler and Vernon Mountcastle in the early 1950s and has since been used by several generations of neuroscientists to evaluate the relationship between stimuli and neuronal responses in both sensory and motor systems. Electrical recording techniques at the single-cell level have now been extended and refined to include single and simultaneous multiple cell analysis in animals performing complex cognitive tasks, intracellular recordings in intact animals, and the use of patch electrodes to detect and monitor the activity of the individual membrane molecules that ultimately underlie neural signaling (see Unit I).

The second major area in which remarkable technical advances have been made is **functional brain imaging** in human subjects (and to a lesser extent animals). The techniques of functional brain imaging have revolutionized our understanding of neural systems over the last two decades, as well as our ability to diagnose and describe functional abnormalities (Box 1A). Unlike electrical methods of recording neural activity, which expose the brain and insert electrodes into it, functional imaging is noninvasive and thus applicable to both clinical patients and normal, healthy human subjects. Moreover, functional

BOX 1A   Brain Imaging Techniques

Computerized Tomography (CT)

In the 1970s, **computerized tomography**, or **CT**, opened a new era in noninvasive imaging by introducing the use of computer processing technology to help probe the living brain. Prior to CT, the only brain imaging technique available was standard X-ray film, which has poor soft tissue contrast and involves relatively high radiation exposure.

The CT approach uses a narrow X-ray beam and a row of very sensitive detectors placed on opposite sides of the head to probe just a small portion of tissue at a time with limited radiation exposure (Figure A). In order to make an image, the X-ray tube and detectors rotate around the head to collect radiodensity information from every orientation around a narrow slice. Computer processing techniques then calculate the radiodensity of each point within the slice plane, producing a tomographic image (*tomo* means "cut" or "slice"). If the patient is slowly moved through the scanner while the X-ray tube rotates in this way, a three-dimensional radiodensity matrix can be created, allowing images to be computed for any plane through the brain. CT scans can readily distinguish gray matter and white matter, differentiate the ventricles quite well, and show many other brain structures with a spatial resolution of several millimeters.

Magnetic Resonance Imaging (MRI)

Brain imaging took a huge step forward in the 1980s with the development of **magnetic resonance imaging** (**MRI**). MRI is based on the fact that the nuclei of some atoms act as spinning magnets, and that if placed in a strong magnetic field these atoms will line up with the field and spin at a frequency that is dependent on the field strength. If a brief radiofrequency pulse tuned to atoms' spinning frequency is applied, they are knocked out of alignment with

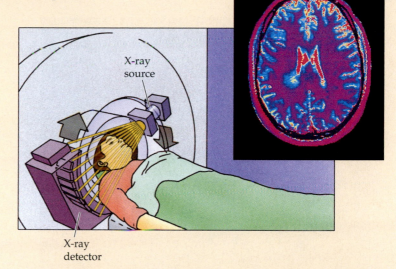

(A) In computerized tomography, the X-ray source and detectors are moved around the patient's head. The inset shows a horizontal CT section of a normal adult brain.

the field and subsequently emit energy in an oscillatory fashion as they gradually realign themselves with the field. The strength of the emitted signal depends on how many atomic nuclei are affected by this process.

In MRI, the magnetic field is distorted slightly by imposing magnetic gradients along three different spatial axes so that only nuclei at certain locations are tuned to the detector's frequency at any given time. Almost all MRI scanners use detectors tuned to the radio frequencies of spinning hydrogen nuclei in water molecules, creating images based on the distribution of water in different tissues. Careful manipulation of magnetic field gradients and radiofrequency pulses make it possible to construct extraordinarily detailed spatial images of the brain at any location and orientation, with submillimeter resolution (Figure B).

The strong magnetic field and radiofrequency pulses used in MRI scanning are harmless, making this technique completely noninvasive (although metal objects in or near a scanner are a safety concern). MRI is also extremely versatile because, by changing the scanning parameters, images based on a

wide variety of different contrast mechanisms can be generated. For example, conventional MR images take advantage of the fact that hydrogen in different types of tissue (e.g., gray matter, white matter, cerebrospinal fluid) have slightly different realignment rates, meaning that soft tissue contrast can be manipulated simply by adjusting when the realigning hydrogen signal is measured. Different parameter settings can also be used to generate images in which gray and white matter are invisible but the brain vasculature stands out in sharp detail. Safety and versatility have made MRI the technique of choice for imaging brain structure in most applications.

Functional Brain Imaging

Imaging functional variations in the living brain has become possible with the recent development of techniques for detecting small, localized changes in metabolism or cerebral blood flow. To conserve energy, the brain regulates its blood flow such that active neurons with relatively high metabolic demands receive more blood than relatively inactive neu-

(Continued on next page)

BOX 1A   (Continued)

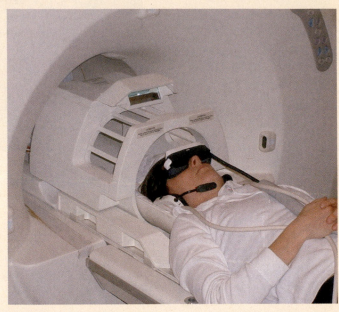

(B) In MRI scanning, the head is placed in the center of a large magnet. A radiofrequency antenna coil is placed around the head for exciting and recording the magnetic resonance signal. For fMRI, stimuli can be presented using virtual reality video goggles and stereo headphones while inside the scanner.

rons. Detecting and mapping these local changes in cerebral blood flow forms the basis for three widely used functional brain imaging techniques: **positron emission tomography** (**PET**), **single-photon emission computerized tomography** (**SPECT**), and **functional magnetic resonance imaging** (**fMRI**).

In PET scanning, unstable positron-emitting isotopes are incorporated into different reagents (including water, precursor molecules of specific neurotransmitters, or glucose) and injected into the bloodstream. Labeled oxygen and glucose quickly accumulate in more metabolically active areas, and labeled trans-

mitter probes are taken up selectively by appropriate regions. As the unstable isotope decays, it results in the emission of two positrons moving in opposite directions. Gamma ray detectors placed around the head register a "hit" only when two detectors 180° apart react simultaneously. Images of tissue isotope density can then be generated (much the way CT images are calculated) showing the location of active regions with a spatial resolution of about 4 mm. Depending on the probe injected, PET imaging can be used to visualize activity-dependent changes in blood flow, tissue metabolism, or biochemical activity. SPECT imaging is similar to PET in that it involves injection or inhalation of radio-labeled compound (for example, ^{133}Xe or ^{123}I-labeled iodoamphetamine), which produce photons that are detected by a gamma camera moving rapidly around the head.

Functional MRI, a variant of MRI, currently offers the best approach for visualizing brain function based on local metabolism. fMRI is predicated on the fact that hemoglobin in blood slightly distorts the magnetic resonance properties of hydrogen nuclei in its vicinity, and the amount of magnetic distortion changes depending on whether the hemoglobin has oxygen bound to it.

imaging allows the simultaneous evaluation of multiple brain structures (which is possible but difficult with electrical recording methods).

Over the last 20 years, ever more powerful noninvasive methods have allowed neuroscientists to evaluate the representation of an enormous number of complex human behaviors, and at the same time have provided diagnostic tools whose now-routine use often overshadows the truly extraordinary nature of the information they provide. Many patients today can take for granted accurate and precise diagnoses and treatments that 20 years ago would have been well-educated guesses on the part of physicians. It remains interesting, however, that many of the observations resulting from the new technologies have in fact confirmed inferences about functional localization and the organization of neural systems that were originally based on the study of neurological patients who exhibited altered behavior after stroke or other forms of brain injury.

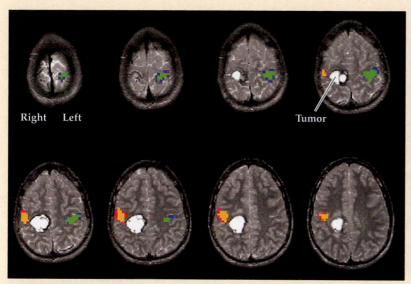
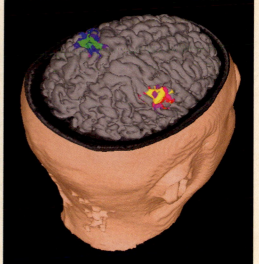

(C) MRI images of an adult patient with a brain tumor, with fMRI activity during a hand motion task superimposed (left hand activity is shown in yellow, right hand activity in green). At right is a three-dimensional surface reconstructed view of the same data.

When a brain area is activated by a specific task it begins to use more oxygen and within seconds the brain microvasculature responds by increasing the flow of oxygen-rich blood to the active area. These changes in the concentration of oxygen and blood flow lead to localized blood oxygenation level-dependent (BOLD) changes in the magnetic resonance signal. Such fluctuations are detected using statistical image processing techniques to produce maps of task-dependent brain function (Figure C).

Because fMRI uses signals intrinsic to the brain without any radioactivity, repeated observations can be made on the same individual—a major advantage over imaging methods such as PET. The spatial resolution (2–3 mm) and temporal resolution (a few seconds) of fMRI are also superior to other functional imaging techniques. MRI has thus emerged as the technology of choice for probing both the structure and function of the living human brain.

References

HUETTEL, S. A., A. W. SONG AND G. MCCARTHY (2004) *Functional Magnetic Resonance Imaging.* Sunderland, MA: Sinauer Associates.

OLDENDORF, W. AND W. OLDENDORF JR. (1988) *Basics of Magnetic Resonance Imaging.* Boston: Kluwer Academic Publishers.

RAICHLE, M. E. (1994) Images of the mind: Studies with modern imaging techniques. Ann. Rev. Psychol. 45: 333–356.

SCHILD, H. (1990) *MRI Made Easy (...Well, Almost).* Berlin: H. Heineman.

Analyzing Complex Behavior

Many of the most widely heralded advances in modern neuroscience have involved reducing the complexity of the brain to more readily analyzed components—i.e., genes, molecules, and cells. Nevertheless, the brain functions as a whole, and the study of more complex (and, some might argue, more interesting) brain functions such as perception, language, emotion, memory, and consciousness remain a central challenge for contemporary neuroscientists. In recognition of this challenge, over the last 25 years or so a field known as **cognitive neuroscience** has emerged that is specifically devoted to understanding these issues (see Unit V). This field's evolution has also rejuvenated the study of neuroethology (which is devoted to observing complex behaviors of animals in their native environments—for example, social communication in birds and non-human primates), and has encouraged the development of tasks to better evaluate the genesis of complex behaviors in human subjects. When used in combination with functional imaging, well-designed behavioral tasks can facilitate identification of brain networks devoted to specific complex functions, including language skills, mathematical and musical ability, emotional responses, aesthetic judgments, and abstract thinking. Carefully constructed behavioral tasks can also be used to study the pathology of complex brain diseases that compromise cognition, such Alzheimer's disease, schizophrenia, and depression.

In short, new or revitalized efforts to study higher brain functions with increasingly powerful techniques offer ways of beginning to understand even the most complex aspects of human behavior.

Summary

The brain can be studied by methods that range from genetics and molecular biology to behavioral testing of normal human subjects. In addition to an ever-increasing store of knowledge about the anatomical organization of the nervous system, many of the most notable successes of modern neuroscience have come from understanding nerve cells as the structural and functional unit of the nervous system. Studies of the cellular architecture and molecular components of neurons and glia have revealed much about their individual functions at a remarkably detailed level, providing a basis for understanding how nerve cells are organized into circuits, and circuits into systems that process specific types of information pertinent to perception and action. Among the goals that remain are understanding how basic molecular genetic phenomena are linked to cellular, circuit, and system functions; understanding how these processes go awry in neurological and psychiatric diseases; and understanding the especially complex functions of the brain that make us human.

Additional Reading

BRODAL, P. (1992) *The Central Nervous System: Structure and Function.* New York: Oxford University Press.

GIBSON, G. AND S. MUSE (2001) *A Primer of Genome Science.* Sunderland, MA: Sinauer Associates.

NATURE VOL. 409, NO. 6822 (2001) Issue of February 16. Special issue on the human genome.

PETERS, A., S. L. PALAY AND H. DE F. WEBSTER (1991) *The Fine Structure of the Nervous System: Neurons and Their Supporting Cells,* 3rd Ed. New York: Oxford University Press.

POSNER, M. I. AND M. E. RAICHLE (1997) *Images of Mind,* 2nd Ed. New York: W. H. Freeman & Co.

RAMÓN Y CAJAL, S. (1984) *The Neuron and the Glial Cell.* (Transl. by J. de la Torre and W. C. Gibson.) Springfield, IL: Charles C. Thomas.

RAMÓN Y CAJAL, S. (1990) *New Ideas on the Structure of the Nervous System in Man and Vertebrates.* (Transl. by N. Swanson and L. W. Swanson.) Cambridge, MA: MIT Press.

SCIENCE VOL. 291, NO. 5507 (2001) Issue of February 16. Special issue on the human genome.

SHEPHERD, G. M. (1991) *Foundations of the Neuron Doctrine.* History of Neuroscience Series, No. 6. Oxford: Oxford University Press.

NEURAL SIGNALING

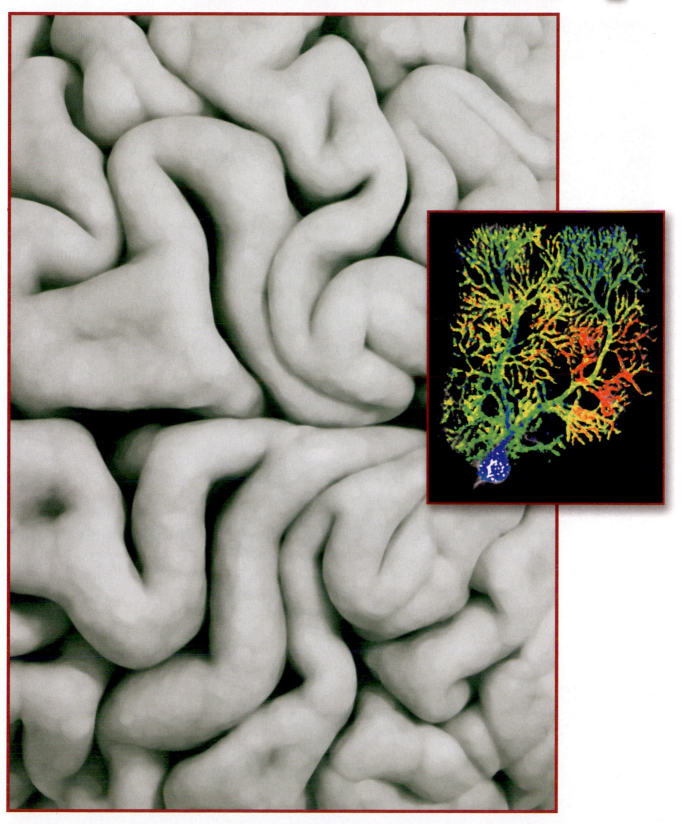

UNIT I NEURAL SIGNALING

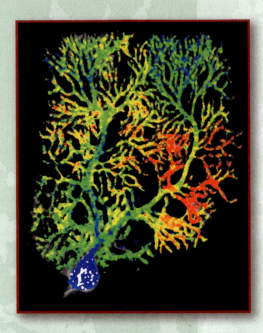

Calcium signaling in a cerebellar Purkinje cell. A fluorescent indicator dye revealed that activation of the climbing fiber synapse increases the concentration of calcium ions inside the Purkinje cell. This rise in Ca^{2+} concentration serves as a second messenger signal that produces a form of synaptic plasticity called long-term synaptic depression. (Courtesy of Keiko Tanaka and George J. Augustine.)

The brain is remarkably adept at acquiring, coordinating, and disseminating information about the body and its environment. Such information must be processed within milliseconds, yet it also can be stored away as memories that endure for years. Neurons within the central and peripheral nervous systems perform these functions by generating sophisticated electrical and chemical signals. This unit describes these signals and how they are produced. It explains how one type of electrical signal, the action potential, allows information to travel along the length of a nerve cell. It also explains how other types of signals—both electrical and chemical—are generated at synaptic connections between nerve cells. Synapses permit information transfer by interconnecting neurons to form the circuitry on which neural processing depends. Finally, this unit describes the intricate biochemical signaling events that take place within neurons. Understanding these fundamental forms of neuronal signaling provides a foundation for appreciating the higher-level functions considered in the rest of the book.

The cellular and molecular mechanisms that give neurons their unique signaling abilities are also targets for disease processes that compromise the function of the nervous system. A working knowledge of the cellular and molecular biology of neurons is therefore fundamental to understanding a variety of brain pathologies, and for developing novel approaches to diagnosing and treating these all too prevalent problems.

Chapter 2

Electrical Signals of Nerve Cells

Overview

Nerve cells generate a variety of electrical signals that transmit information. Although neurons are not intrinsically good conductors of electricity, they have evolved elaborate mechanisms for generating these signals based on the flow of ions across their plasma membranes. Ordinarily, neurons generate a negative potential called the resting membrane potential, which can be measured by recording the voltage inside nerve cells relative to the outside. The action potential transiently abolishes the negative resting potential and makes the transmembrane potential positive. Action potentials are propagated along the length of axons and are the fundamental signals that carry information from one place to another in the nervous system. Still other types of electrical signals are produced by the activation of synaptic contacts between neurons or by the actions of external forms of energy on sensory neurons. All of these electrical signals arise from ion fluxes brought about by nerve cell membranes being selectively permeable to different ions, and by the non-uniform distribution of these ions across the membrane.

Electrical Potentials across Nerve Cell Membranes

Neurons employ several different types of electrical signals to encode and transfer information. The best way to observe these signals is to use an intracellular microelectrode to measure the electrical potential across the neuronal plasma membrane. A typical microelectrode is a piece of glass tubing pulled to a very fine point (with an opening of less than 1 μm diameter), which is filled with a good electrical conductor, such as a concentrated salt solution. This conductive core can then be connected to a voltmeter, such as an oscilloscope, to record the transmembrane voltage of the nerve cell.

The first type of electrical phenomenon can be observed as soon as a microelectrode is inserted through the membrane of the neuron. Upon entering the cell, the microelectrode reports a negative potential, indicating that neurons have a means of generating a constant voltage across their membranes when at rest. This voltage, called the **resting membrane potential**, depends on the type of neuron being examined, but it is always a fraction of a volt (typically −40 to −90 mV).

The electrical signals produced by neurons are caused by responses to stimuli, which then change the resting membrane potential. **Receptor potentials** are due to the activation of sensory neurons by external stimuli such as light, sound, or heat. For example, touching the skin activates Pacinian corpuscles, receptor neurons that sense mechanical disturbances of the skin. These neurons respond to touch with a receptor potential that changes the resting potential for a fraction of a second (Figure 2.1A). Such transient changes in the membrane potential of these receptor neurons are the first step in generating the sensation of vibrations (or

Figure 2.1 Types of neuronal electrical signals. In all cases, microelectrodes are used to measure changes in the resting membrane potential during the indicated signals. (A) A brief touch causes a receptor potential in a Pacinian corpuscle in the skin. (B) Activation of a synaptic contact onto a hippocampal pyramidal neuron elicits a synaptic potential. (C) Stimulation of a spinal reflex produces an action potential in a spinal motor neuron.

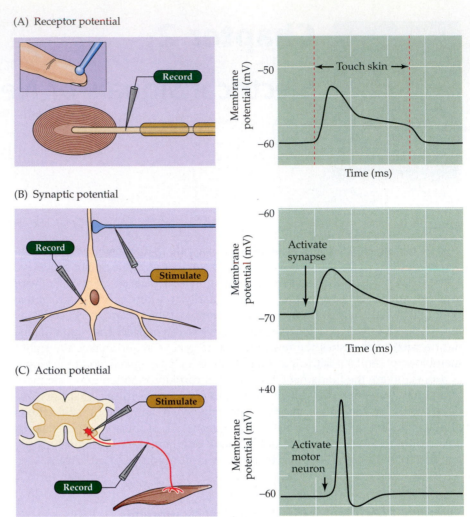

(A) Receptor potential

(B) Synaptic potential

(C) Action potential

"tickles") of the skin in the somatic sensory system (see Chapter 9). Similar sorts of receptor potentials are observed in all other sensory neurons during transduction of sensory signals, as described in Unit II.

Another type of electrical signal is associated with communication between neurons at synaptic contacts. Activation of these synapses generates **synaptic potentials**, which allow transmission of information from one neuron to another. An example of such a signal is shown in Figure 2.1B. In this case, activation of a synaptic terminal innervating a hippocampal pyramidal neuron causes a very brief change in the resting membrane potential in the pyramidal neuron. Synaptic potentials serve as the means of exchanging information in complex neural circuits in both the central and peripheral nervous systems (see Chapter 5).

The use of electrical signals—as in sending electricity over wires to provide power or information—presents a series of problems in electrical engineering. A fundamental problem for neurons is that their axons, which can be quite long (remember that a spinal motor neuron can extend for a meter or more), are not good electrical conductors. Although neurons and wires are both capable of passively conducting electricity, the electrical properties of neurons compare poorly to an ordinary wire. To compensate for this deficiency, neurons have evolved a "booster system" that allows them to conduct electrical signals over great distances despite their intrinsically poor electrical characteristics. The electrical signals produced by this booster system are called **action potentials** (which are also referred to as "spikes" or "impulses"). An example of an action potential recorded from the axon of a spinal motor neuron is shown in Figure 2.1C.

One way to elicit an action potential is to pass an electrical current across the membrane of the neuron. In normal circumstances, this current would be generated by receptor potentials or by synaptic potentials. In the laboratory, however, electrical current suitable for initiating an action potential can be readily produced by inserting a second microelectrode into the same neuron and then connecting the electrode to a battery (Figure 2.2A). If the current delivered in this way makes the membrane potential more negative (**hyperpolarization**), nothing very dramatic happens. The membrane potential simply changes in proportion to the magnitude of the injected current (Figure 2.2B, top panel). Such hyperpolarizing responses do not require any unique property of neurons and are therefore called passive electrical responses. A much more interesting phenomenon is seen if current of the opposite polarity is delivered, so that the membrane potential of the nerve cell becomes more positive than the resting potential (**depolarization**). In this case, at a certain level of membrane potential, called the **threshold potential**, an action potential occurs (see the right side of Figure 2.2B).

The action potential, which is an active response generated by the neuron, is a brief (about 1 ms) change from negative to positive in the transmembrane potential. Importantly, the amplitude of the action potential is independent of the magnitude of the current used to evoke it; that is, larger currents do not elicit larger action potentials. The action potentials of a given neuron are therefore said to be *all-or-none*, because either they occur fully or they do not occur at all. If the amplitude or duration of the stimulus current is increased sufficiently, multiple action potentials occur, as can be seen in the responses to the three different current intensities shown on the right side in Figure 2.2B. It follows, therefore, that the intensity of a stimulus is encoded in the frequency of action potentials rather than

Figure 2.2 Recording passive and active electrical signals in a nerve cell. (A) Two microelectrodes are inserted into a neuron; one of these measures membrane potential while the other injects current into the neuron. (B) Inserting the voltage-measuring microelectrode into the neuron reveals a negative potential— the resting membrane potential. Injecting current through the current-passing microelectrode alters the neuronal membrane potential. Hyperpolarizing current pulses produce only passive changes in the membrane potential. While small depolarizing currents also elicit only passive responses, depolarizations that cause the membrane potential to meet or exceed threshold additionally evoke action potentials. Action potentials are active responses in the sense that they are generated by changes in the permeability of the neuronal membrane.

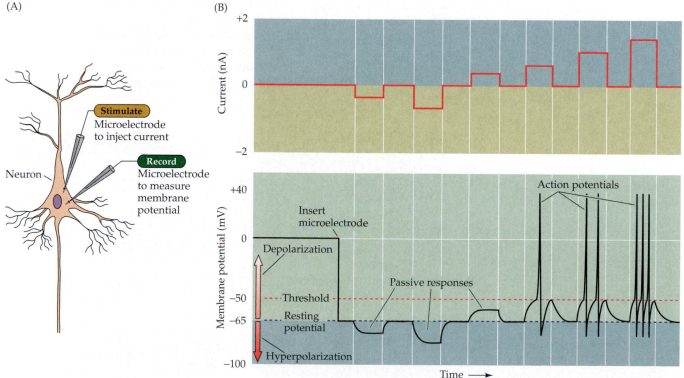

(A)

Stimulate
Microelectrode to inject current

Neuron

Record
Microelectrode to measure membrane potential

(B)

Current (nA)

Membrane potential (mV)

Insert microelectrode

Depolarization

Action potentials

Passive responses

Threshold

Resting potential

Hyperpolarization

Time →

in their amplitude. This arrangement differs dramatically from receptor potentials, whose amplitudes are graded in proportion to the magnitude of the sensory stimulus; or from synaptic potentials, whose amplitudes vary according to the number of synapses activated and the previous amount of synaptic activity.

Because electrical signals are the basis of information transfer in the nervous system, it is essential to understand how these signals arise. Remarkably, all of the electrical signals described above are produced by similar mechanisms that rely upon the movement of ions across the neuronal membrane. The remainder of this chapter addresses the question of how nerve cells use ions to generate electrical potentials. Chapter 3 explores more specifically the means by which action potentials are produced and how these signals solve the problem of long-distance electrical conduction within nerve cells. Chapter 4 examines the properties of the membrane molecules—channels and transporters—responsible for electrical signaling. Finally, Chapters 5 through 8 consider how electrical signals are transmitted from one nerve cell to another at synaptic contacts.

How Ionic Movements Produce Electrical Signals

Electrical potentials are generated across the membranes of neurons—and, indeed, all cells—because (1) there are *differences in the concentrations* of specific ions across nerve cell membranes, and (2) the membranes are *selectively permeable* to some of these ions. These two facts depend in turn on two different kinds of proteins in the cell membrane (Figure 2.3). The ion concentration gradients are established by proteins known as **active transporters**, which, as their name suggests, actively move ions into or out of cells against their concentration gradients. The selective permeability of membranes is due largely to **ion channels**, proteins that allow only certain kinds of ions to cross the membrane in the direction of their concentration gradients. Thus, channels and transporters basically work against each other, and in so doing, they generate all electrical signals produced by neurons, including the resting membrane potential, action potentials, synaptic potentials, and receptor potentials. The structure and function of these channels and transporters are described in Chapter 4.

To appreciate the role of ion gradients and selective permeability in generating a membrane potential, consider a simple system in which an artificial membrane separates two compartments containing solutions of ions. In such a system, it is possible to determine the composition of the two solutions and, thereby, control

Figure 2.3 Ion transporters and ion channels are responsible for ionic movement across neuronal membranes. Transporters create concentration differences by actively transporting ions against their chemical gradients. Channels take advantage of these concentration gradients, allowing selected ions to move, via diffusion, down their chemical gradients and thereby generate electrical signals.

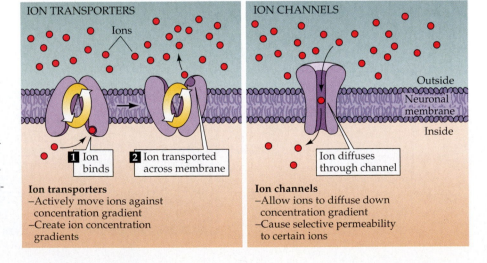

(A)

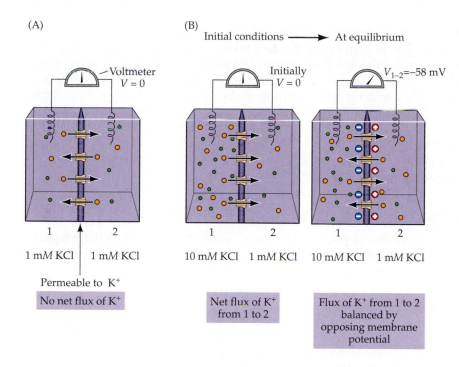

(B)   Initial conditions ⟶ At equilibrium

(C)

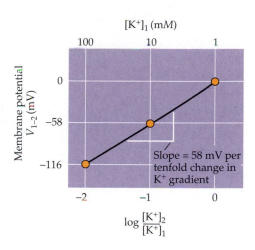

Figure 2.4 Electrochemical equilibrium. (A) A membrane permeable only to K^+ (yellow spheres) separates compartments 1 and 2, which contain the indicated concentrations of KCl (Cl^- is shown as green spheres.) (B) Increasing the KCl concentration in compartment 1 to 10 mM initially causes a small movement of K^+ into compartment 2 (initial conditions) until the electromotive force acting on K^+ balances the concentration gradient, and the net movement of K^+ becomes zero (at equilibrium). (C) The relationship between the transmembrane concentration gradient ($[K^+]_2/[K^+]_1$) and the membrane potential. As predicted by the Nernst equation, this relationship is linear when plotted on semilogarithmic coordinates, with a slope of 58 mV per tenfold difference in the concentration gradient.

the ion gradients across the membrane. For example, take the case of a membrane that is permeable only to potassium ions (K^+). If the concentration of K^+ on each side of this membrane is equal, then no electrical potential will be measured across it (Figure 2.4A). However, if the concentration of K^+ is not the same on the two sides, then an electrical potential will be generated. For instance, if the concentration of K^+ on one side of the membrane (compartment 1) is 10 times higher than the K^+ concentration on the other side (compartment 2), then the electrical potential of compartment 1 will be negative relative to compartment 2 (Figure 2.4B). This difference in electrical potential is generated because potassium ions flow down their concentration gradient and take their electrical charge (one positive charge per ion) with them as they go. Because neuronal membranes contain pumps that accumulate K^+ in the cell cytoplasm, and because potassium-permeable channels in the plasma membrane allow a transmembrane flow of K^+, an analogous situation exists in living nerve cells. A continual resting efflux of K^+ is therefore responsible for the resting membrane potential.

In the hypothetical case just described, equilibrium will quickly be reached. As K^+ moves from compartment 1 to compartment 2 (the initial conditions on the left of Figure 2.4B), a potential is generated that tends to impede further flow of K^+. This impediment results from the fact that the potential gradient across the membrane tends to repel the K^+ that would otherwise move across the membrane. Thus, as compartment 2 becomes positive relative to compartment 1, the increasing positivity makes compartment 2 less attractive to the positively charged K^+. The net movement (or flux) of K^+ will stop at the point (at equilibrium on the right of Figure 2.4B) where the potential change across the membrane (the relative positivity of compartment 2) exactly offsets the concentration gradient (the tenfold excess of K^+ in compartment 1). At this **electrochemical equilibrium**, there is an exact balance between two opposing forces: (1) the concentration gradient that causes K^+ to move from compartment 1 to compartment 2, taking along positive charge, and (2) an opposing electrical gradient that increasingly tends to stop K^+ from moving across the membrane (see Figure 2.4B). The number of ions that needs to flow to generate this electrical potential is very small (approximately 10^{-12} moles of K^+ per cm^2 of membrane, or 10^{12} potassium ions).

This last fact is significant in two ways. First, it means that the concentrations of permeant ions on each side of the membrane remain essentially constant, even after the flow of ions has generated the potential. Second, the tiny fluxes of ions required to establish the membrane potential do not disrupt chemical electroneutrality because each ion has an oppositely charged counter-ion (e.g., the chloride ions in the example shown in Figure 2.4) to maintain the neutrality of the solutions on each side of the membrane. The concentration of K^+ remains equal to the concentration of chloride ions (Cl^-) in the solutions in compartments 1 and 2, meaning that the separation of charge that creates the potential difference is restricted to the immediate vicinity of the membrane.

The Forces That Create Membrane Potentials

The electrical potential generated across the membrane at electrochemical equilibrium, the **equilibrium potential**, can be predicted by a simple formula called the **Nernst equation**. This relationship is generally expressed as

$$E_X = \frac{RT}{zF} \ln \frac{[X]_2}{[X]_1}$$

where E_X is the equilibrium potential for any ion X, R is the gas constant, T is the absolute temperature (that is, the temperature on the Kelvin scale), z is the valence (electrical charge) of the permeant ion, and F is the Faraday constant (the amount of electrical charge contained in one mole of a univalent ion). The brackets indicate the concentrations of ion X on each side of the membrane, and ln indicates the natural logarithm of the concentration gradient. Because it is easier to perform calculations using base 10 logarithms and to perform experiments at room temperature, this relationship is usually simplified to

$$E_X = \frac{58}{z} \log \frac{[X]_2}{[X]_1}$$

where log indicates the base 10 logarithm of the concentration ratio. Thus, for the example in Figure 2.4B, the potential across the membrane at electrochemical equilibrium is

$$E_K = \frac{58}{z} \log \frac{[K]_2}{[K]_1} = 58 \log \frac{1}{10} = -58 \, mV$$

The equilibrium potential is conventionally defined in terms of the potential difference between the reference compartment (side 2 in Figure 2.4) and the other side. This approach is also applied to biological systems. In this case, the outside of the cell is the conventional reference point, defined as zero potential. Thus, when the concentration of K^+ is higher inside than outside, an inside-negative potential is measured across the K^+-permeable neuronal membrane.

For a simple hypothetical system with only one permeant ion species, the Nernst equation allows the electrical potential across the membrane at equilibrium to be predicted exactly. For example, if the concentration of K^+ on side 1 is increased to 100 mM, the membrane potential will be –116 mV. More generally, if the membrane potential is plotted against the logarithm of the K^+ concentration gradient ($[K]_2/[K]_1$), the Nernst equation predicts a linear relationship with a slope of 58 mV (actually 58/z) per tenfold change in the K^+ gradient (Figure 2.4C). (The slope becomes 61 mV at mammalian body temperatures.)

To reinforce and extend the concept of electrochemical equilibrium, consider some additional experiments on the influence of ion species and ion permeability that could be performed on the simple model system in Figure 2.4. What would happen to the electrical potential across the membrane (the potential of side 1 relative to side 2) if K^+ on side 2 were replaced with 10 mM sodium ion(s) (Na^+) and the K^+ in compartment 1 were replaced by 1 mM Na^+? No potential would be generated, because no Na^+ could flow across the membrane (which was defined as being permeable only to K^+). However, if, under these ionic conditions (10 times more Na^+ in compartment 2) the K^+-permeable membrane were to be magically replaced by a membrane permeable only to Na^+, a potential of +58 mV would be measured at equilibrium. If 10 mM calcium ion(s) (Ca^{2+}) were present in compartment 2 and 1 mM Ca^{2+} in compartment 1, and a Ca^{2+}-selective membrane separated the two sides, what would happen to the membrane potential? A potential of +29 mV would develop, because the valence of calcium is +2. Finally, what would happen to the membrane potential if 10 mM Cl^- were present in compartment 1 and 1 mM Cl^- were present in compartment 2, with the two sides separated by a Cl^--permeable membrane? Because the valence of this anion is –1, the potential would again be +58 mV.

The balance of chemical and electrical forces at equilibrium means that the electrical potential can determine ion fluxes across the membrane, just as the ionic gradient can determine the membrane potential. To examine the influence of membrane potential on ionic flux, imagine connecting a battery across the two sides of the membrane to control the electrical potential across the membrane without changing the distribution of ions on the two sides (Figure 2.5). As long as the battery is off, things will be just as shown in Figure 2.4, with the flow of K^+ from compartment 1 to compartment 2 causing a negative membrane potential (Figure 2.5A, left). However, if the battery is used to make compartment 1 initially more negative relative to compartment 2, there will be less K^+ flux because the negative potential will tend to keep K^+ in compartment 1. How negative will side 1 need to be before there is no net flux of K^+? The answer is –58 mV, the voltage needed to counter the tenfold difference in K^+ concentrations on the two sides of the membrane (Figure 2.5A, center). If compartment 1 is initially made more negative than –58 mV, then K^+ will flow from compartment 2 into compartment 1, because the positive ions will be attracted to the more negative potential of compartment 1 (Figure 2.5A, right). This example demonstrates that both the direction and magni-

Figure 2.5 Membrane potential influences ion fluxes. (A) Connecting a battery across the K^+-permeable membrane allows direct control of membrane potential. When the battery is turned off (*left*), K^+ (yellow) simply flows according to its concentration gradient. Setting the initial membrane potential (V_{1-2}) at the equilibrium potential for K^+ (*center*) yields no net flux of K^+, while making the membrane potential more negative than the K^+ equilibrium potential (*right*) causes K^+ to flow against its concentration gradient. (B) The relationship between membrane potential and the direction of K^+ flux.

(A)

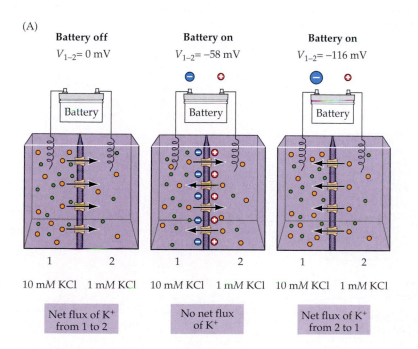

(B)

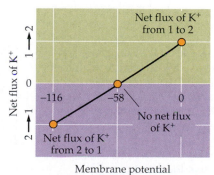

tude of ion flux depend on the membrane potential. Thus, in some circumstances, the electrical potential can overcome an ionic concentration gradient.

The ability to alter ion flux experimentally by changing either the potential imposed on the membrane (Figure 2.5B) or the transmembrane concentration gradient for an ion (see Figure 2.4C) provides convenient tools for studying ion fluxes across the plasma membranes of neurons, as will be evident in many of the experiments described in the following chapters.

Electrochemical Equilibrium in an Environment with More Than One Permeant Ion

Now consider a somewhat more complex situation in which Na^+ and K^+ are unequally distributed across the membrane, as in Figure 2.6A. What would happen if 10 mM K^+ and 1 mM Na^+ were present in compartment 1, and 1 mM K^+ and 10 mM Na^+ were present in compartment 2? If the membrane were permeable only to K^+, the membrane potential would be –58 mV; if the membrane were permeable only to Na^+, the potential would be +58 mV. But what would the potential be if the membrane were permeable to both K^+ and Na^+? In this case, the potential would depend on the relative permeability of the membrane to K^+ and Na^+. If it were more permeable to K^+, the potential would approach –58 mV, and if it were more permeable to Na^+, the potential would be closer to +58 mV. Because there is no permeability term in the Nernst equation, which only considers the simple case of a single permeant ion species, a more elaborate equation is needed that takes into account both the concentration gradients of the permeant ions and the relative permeability of the membrane to each permeant species.

Such an equation was developed by David Goldman in 1943. For the case most relevant to neurons, in which K^+, Na^+, and Cl^- are the primary permeant ions, the **Goldman equation** is written

$$V = 58 \log \frac{P_K[K]_2 + P_{Na}[Na]_2 + P_{Cl}[Cl]_1}{P_K[K]_1 + P_{Na}[Na]_1 + P_{Cl}[Cl]_2}$$

where V is the voltage across the membrane (again, compartment 1 relative to the reference compartment 2), and P indicates the permeability of the membrane to each ion of interest. The Goldman equation is thus an extended version of the Nernst equation that takes into account the relative permeabilities of each of the

Figure 2.6 Resting and action potentials arise from membrane permeabilities to different ions. (A) Hypothetical situation in which a membrane variably permeable to Na^+ (red) and K^+ (yellow) separates two compartments that contain both ions. For simplicity, Cl^- ions are not shown in the diagram. (B) Schematic representation of the membrane ionic permeabilities associated with resting and action potentials. At rest, neuronal membranes are more permeable to K^+ (yellow) than to Na^+ (red); accordingly, the resting membrane potential is negative and approaches the equilibrium potential for K^+, E_K. During an action potential, the membrane becomes very permeable to Na^+ (red); thus the membrane potential becomes positive and approaches the equilibrium potential for Na^+, E_{Na}. The rise in Na^+ permeability is transient, however, so that the membrane again becomes primarily permeable to K^+ (yellow), causing the potential to return to its negative resting value.

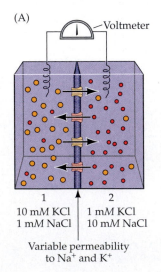

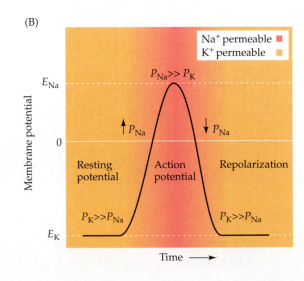

TABLE 2.1 **Extracellular and Intracellular Ion Concentrations**

Ion	Concentration (mM)	
	Intracellular	Extracellular
Squid neuron		
Potassium (K$^+$)	400	20
Sodium (Na$^+$)	50	440
Chloride (Cl$^-$)	40–150	560
Calcium (Ca^{2+})	0.0001	10
Mammalian neuron		
Potassium (K$^+$)	140	5
Sodium (Na$^+$)	5–15	145
Chloride (Cl$^-$)	4–30	110
Calcium (Ca^{2+})	0.0001	1–2

ions involved. The relationship between the two equations becomes obvious in the situation where the membrane is permeable only to one ion, say, K$^+$; in this case, the Goldman expression collapses back to the simpler Nernst equation. In this context, it is important to note that the valence factor (z) in the Nernst equation has been eliminated; this is why the concentrations of negatively charged chloride ions (Cl$^-$) have been inverted relative to the concentrations of the positively charged ions (K$^+$ and Na$^+$) [remember that –log (A/B) = log (B/A)].

If the membrane in Figure 2.6A is permeable to K$^+$ and Na$^+$ only, the terms involving Cl$^-$ drop out because P_{Cl} is 0. In this case, solution of the Goldman equation yields a potential of –58 mV when only K$^+$ is permeant, +58 mV when only Na$^+$ is permeant, and some intermediate value if both ions are permeant. For example, if K$^+$ and Na$^+$ were equally permeant, then the potential would be 0 mV.

With respect to neural signaling, it is particularly pertinent to ask what would happen if the membrane started out being permeable to K$^+$, and then temporarily switched to become most permeable to Na$^+$. In this circumstance, the membrane potential would start out at a negative level, become positive while the Na$^+$ permeability remained high, and then fall back to a negative level as the Na$^+$ permeability decreased again. As it turns out, this last case essentially describes what goes on in a neuron during the generation of an action potential. In the resting state, P_K of the neuronal plasma membrane is much higher than P_{Na}, since, as a result of the action of ion transporters, there is always more K$^+$ inside the cell than outside (Table 2.1), the resting potential is negative (Figure 2.6B). As the membrane potential is depolarized (by synaptic action, for example), P_{Na} increases. The transient increase in Na$^+$ permeability causes the membrane potential to become even more positive (red region in Figure 2.6B), because Na$^+$ rushes in (there is much more Na$^+$ outside a neuron than inside, again as a result of ion pumps). Because of this positive feedback loop, an action potential occurs. The rise in Na$^+$ permeability during the action potential is transient, however; as the membrane permeability to K$^+$ is restored, the membrane potential quickly returns to its resting level.

Armed with an appreciation of these simple electrochemical principles, it will be much easier to understand the following, more detailed account of how neurons generate resting and action potentials.

The Ionic Basis of the Resting Membrane Potential

The action of ion transporters creates substantial transmembrane gradients for most ions. Table 2.1 summarizes the ion concentrations measured directly in an exceptionally large nerve cell found in the nervous system of the squid (Box 2A). Such measurements directly indicate that there is much more K⁺ inside the neuron than outside, and much more Na⁺ outside than inside. Similar concentration gradients occur in the neurons of most animals, including humans. However, because the ionic strength of mammalian blood is lower than that of sea-dwelling animals such as squid, in mammals the concentrations of each ion are several times lower. These transporter-dependent concentration gradients are, indirectly, the source of the resting neuronal membrane potential and the action potential.

Once the ion concentration gradients across various neuronal membranes are known, the Nernst equation can be used to calculate the equilibrium potential for K⁺ and other major ions. Since the resting membrane potential of the squid neuron is approximately –65 mV, K⁺ is the ion that is closest to being in electrochemical equilibrium when the cell is at rest. This fact implies that the resting membrane is more permeable to K⁺ than to the other ions listed in Table 2.1, and that this permeability is the source of resting potentials.

It is possible to test this guess, as Alan Hodgkin and Bernard Katz did in 1949, by asking what happens to the resting membrane potential if the concentration of K⁺ outside the neuron is altered. If the resting membrane were permeable only to K⁺, then the Goldman equation (or even the simpler Nernst equation) predicts that the membrane potential will vary in proportion to the logarithm of the K⁺ concentration gradient across the membrane. Assuming that the internal K⁺ concentration is unchanged during the experiment, a plot of membrane potential against the logarithm of the external K⁺ concentration should yield a straight line with a slope of 58 mV per tenfold change in external K⁺ concentration at room temperature (see Figure 2.4C).

When Hodgkin and Katz carried out this experiment on a living squid neuron, they found that the resting membrane potential did indeed change when the external K⁺ concentration was modified, becoming less negative as external K⁺ concentration was raised (Figure 2.7A). When the external K⁺ concentration was raised high enough to equal the concentration of K⁺ inside the neuron, thus mak-

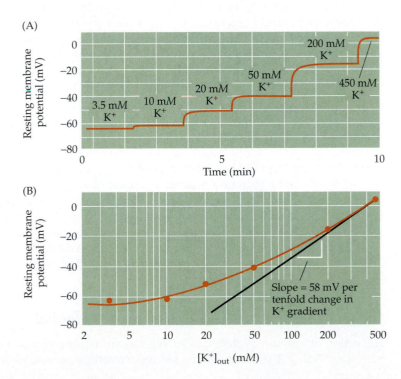

Figure 2.7 Experimental evidence that the resting membrane potential of a squid giant axon is determined by the K⁺ concentration gradient across the membrane. (A) Increasing the external K⁺ concentration makes the resting membrane potential more positive. (B) Relationship between resting membrane potential and external K⁺ concentration, plotted on a semi-logarithmic scale. The straight line represents a slope of 58 mV per tenfold change in concentration, as given by the Nernst equation. (After Hodgkin and Katz, 1949.)

BOX 2A   The Remarkable Giant Nerve Cells of Squid

Many of the initial insights into how ion concentration gradients and changes in membrane permeability produce electrical signals came from experiments performed on the extraordinarily large nerve cells of the squid. The axons of these nerve cells can be up to 1 mm in diameter—100 to 1000 times larger than mammalian axons. Thus, squid axons are large enough to allow experiments that would be impossible on most other nerve cells. For example, it is not difficult to insert simple wire electrodes inside these giant axons and make reliable electrical measurements. The relative ease of this approach yielded the first intracellular recordings of action potentials from nerve cells and, as discussed in the next chapter, the first experimental measurements of the ion currents that produce action potentials. It also is practical to extrude the cytoplasm from giant axons and measure its ionic composition (see Table 2.1). In addition, some giant nerve cells form synaptic contacts with other giant nerve cells, producing very large synapses that have been extraordinarily valuable in understanding the fundamental mechanisms of synaptic transmission (see Chapter 5).

Giant neurons evidently evolved in squid because they enhanced survival. These neurons participate in a simple neural circuit that activates the contraction of the mantle muscle, producing a jet propulsion effect that allows the squid to move away from predators at a remarkably fast speed. As discussed in Chapter 3, larger axonal diameter allows faster conduction of action potentials.

Thus, presumably these huge nerve cells help squid escape more successfully from their numerous enemies.

Today—nearly 70 years after their discovery by John Z. Young at University College London—the giant nerve cells of squid remain useful experimental systems for probing basic neuronal functions.

References

LLINÁS, R. (1999) *The Squid Synapse: A Model for Chemical Transmission*. Oxford: Oxford University Press.

YOUNG, J. Z. (1939) Fused neurons and synaptic contacts in the giant nerve fibres of cephalopods. *Phil. Trans. R. Soc. Lond.* 229(B): 465–503.

(A) Diagram of a squid, showing the location of its giant nerve cells. Different colors indicate the neuronal components of the escape circuitry. The first- and second-level neurons originate in the brain, while the third-level neurons are in the stellate ganglion and innervate muscle cells of the mantle. (B) Giant synapses within the stellate ganglion. The second-level neuron forms a series of fingerlike processes, each of which makes an extraordinarily large synapse with a single third-level neuron. (C) Structure of a giant axon of a third-level neuron lying within its nerve. The enormous difference in the diameters of a squid giant axon and a mammalian axon are shown below.

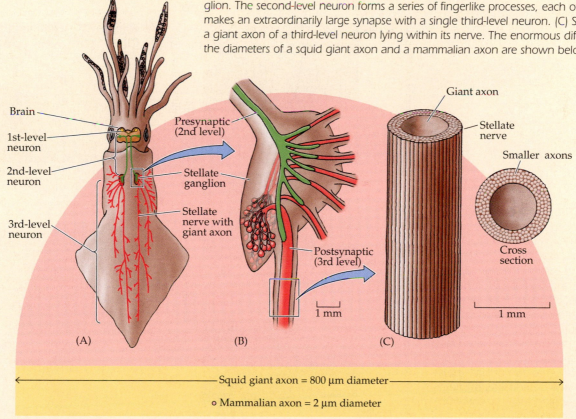

Brain
1st-level neuron
2nd-level neuron
3rd-level neuron
(A)

Presynaptic (2nd level)
Stellate ganglion
Stellate nerve with giant axon
Postsynaptic (3rd level)
1 mm
(B)

Giant axon
Stellate nerve
Smaller axons
Cross section
1 mm
(C)

Squid giant axon = 800 µm diameter
Mammalian axon = 2 µm diameter

ing the K^+ equilibrium potential 0 mV, the resting membrane potential was also approximately 0 mV. In short, the resting membrane potential varied as predicted with the logarithm of the K^+ concentration, with a slope that approached 58 mV per tenfold change in K^+ concentration (Figure 2.7B). The value obtained was not exactly 58 mV because other ions, such as Cl^- and Na^+, are also slightly permeable, and thus influence the resting potential to a small degree. The contribution of these other ions is particularly evident at low external K^+ levels, again as predicted by the Goldman equation. In general, however, manipulation of the external concentrations of these other ions has only a small effect, emphasizing that K^+ permeability is indeed the primary source of the resting membrane potential.

In summary, Hodgkin and Katz showed that the inside-negative resting potential arises because (1) the membrane of the resting neuron is more permeable to K^+ than to any of the other ions present, and (2) there is more K^+ inside the neuron than outside. The selective permeability to K^+ is caused by K^+-permeable membrane channels that are open in resting neurons, and the large K^+ concentration gradient is, as noted, produced by membrane transporters that selectively accumulate K^+ within neurons. Many subsequent studies have confirmed the general validity of these principles.

The Ionic Basis of Action Potentials

What causes the membrane potential of a neuron to depolarize during an action potential? Although a general answer to this question has been given in Figure 2.6B (increased permeability to Na^+), it is well worth examining some of

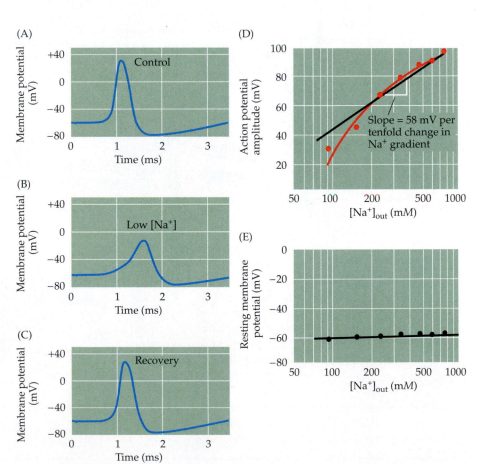

Figure 2.8 The role of Na^+ in the generation of an action potential in a squid giant axon. (A) An action potential evoked with the normal ion concentrations inside and outside the cell. (B) Both the amplitude and rate of rise of the action potential diminish when external Na^+ concentration is reduced to one-third of normal, but (C) recover when the Na^+ is replaced. (D) While the amplitude of the action potential is quite sensitive to the external concentration of Na^+, the resting membrane potential (E) is little affected by changing the concentration of this ion. (After Hodgkin and Katz, 1949.)

the experimental support for this concept. Given the data presented in Table 2.1, one can use the Nernst equation to calculate that the equilibrium potential for Na^+ (E_{Na}) in neurons, and indeed in most cells, is positive. Thus, if the membrane were to become highly permeable to Na^+, the membrane potential would approach E_{Na}. Based on these considerations, Hodgkin and Katz hypothesized that the action potential arises because the neuronal membrane becomes temporarily permeable to Na^+.

Taking advantage of the same style of ion substitution experiment they used to assess the resting potential, Hodgkin and Katz tested the role of Na^+ in generating the action potential by asking what happens to the action potential when Na^+ is removed from the external medium. They found that lowering the external Na^+ concentration, to lower E_{Na}, reduces both the rate of rise of the action potential and its peak amplitude (Figure 2.8A–C). Indeed, when they examined this Na^+ dependence quantitatively, they found a more-or-less linear relationship between the amplitude of the action potential and the logarithm of the external Na^+ concentration (Figure 2.8D). The slope of this relationship approached a value of 58 mV per tenfold change in Na^+ concentration, as expected for a membrane selectively permeable to Na^+. In contrast, lowering Na^+ concentration had very little effect on the resting membrane potential (Figure 2.8E). Thus, while the resting neuronal membrane is only slightly permeable to Na^+, the membrane becomes extraordinarily permeable to Na^+ during the **rising phase** and **overshoot phase** of the action potential. (Box 2B provides an explanation of action potential nomenclature.) This temporary increase in Na^+ permeability results from the opening of Na^+-selective channels that are essentially closed in the resting state. Membrane pumps maintain a large electrochemical gradient for Na^+, which is in much higher concentration outside the neuron than inside. When the Na^+ channels open, Na^+ flow into the neuron, causing the membrane potential to depolarize and approach E_{Na}.

The time that the membrane potential lingers near E_{Na} (about +58 mV) during the overshoot phase of an action potential is brief because the increased membrane permeability to Na^+ itself is short-lived. The membrane potential rapidly repolarizes to resting levels and is actually followed by a transient **undershoot**. As will be described in Chapter 3, these latter events in the action potential are due to an inactivation of the Na^+ permeability and an increase in the K^+ permeability of the membrane. During an undershoot, the membrane potential is transiently hyperpolarized because K^+ permeability becomes even greater than it is at rest. The action potential ends when this phase of enhanced K^+ permeability subsides, and the membrane potential thus returns to its normal resting level.

The ion substitution experiments carried out by Hodgkin and Katz provided convincing evidence that the resting membrane potential results from a high resting membrane permeability to K^+, and that depolarization during an action potential results from a transient rise in membrane Na^+ permeability. Although these experiments identified the ions that flow during an action potential, they did not establish *how* the neuronal membrane is able to change its ionic permeability to generate the action potential, or what mechanisms trigger this critical change. The next chapter addresses these issues, documenting the surprising conclusion that the neuronal membrane potential itself affects membrane permeability.

Summary

Nerve cells generate electrical signals to convey information over substantial distances and to transmit it to other cells by means of synaptic connections. These signals ultimately depend on changes in the resting electrical potential across the neuronal membrane. A resting potential occurs because nerve cell membranes are

BOX 2B Action Potential Form and Nomenclature

The action potential of the squid giant axon has a characteristic shape, or waveform, with a number of different phases (Figure A). During the rising phase, the membrane potential rapidly depolarizes. In fact, action potentials cause the membrane potential to depolarize so much that the membrane potential transiently becomes positive with respect to the external medium, producing an overshoot. The overshoot of the action potential gives way to a falling phase in which the membrane potential rapidly repolarizes. Repolarization takes the membrane potential to levels even more negative than the resting membrane potential for a short time; this brief period of hyperpolarization is called the undershoot.

Although the waveform of the squid action potential is typical, the details of the action potential form vary widely from neuron to neuron in different animals. In myelinated axons of vertebrate motor neurons (Figure B), the action potential is virtually indistinguishable from that of the squid axon. However, the action potential recorded in the cell body of this same motor neuron (Figure

C) looks rather different. Thus, the action potential waveform can vary even within the same neuron. More complex action potentials are seen in other central neurons. For example, action potentials recorded from the cell bodies of neurons in the mammalian inferior olive (a region of the brain stem involved in motor control) last tens of milliseconds (Figure D). These action potentials exhibit a pronounced plateau during their falling phase, and their undershoot lasts even longer than that of the motor neuron. One of the most dramatic types of action potentials occurs in the cell bodies of cerebellar Purkinje neurons (Figure E). These potentials have several complex phases that result from the summation of multiple, discrete action potentials.

The variety of action potential waveforms could mean that each type of neuron has a different mechanism of action potential production. Fortunately, however, these diverse waveforms all result from relatively minor variations in the scheme used by the squid giant axon. For example, plateaus in the repolarization phase result from the presence of

ion channels that are permeable to Ca^{2+}, and long-lasting undershoots result from the presence of additional types of membrane K^+ channels. The complex action potential of the Purkinje cell results from these extra features plus the fact that different types of action potentials are generated in various parts of the Purkinje neuron—cell body, dendrites, and axons—and are summed together in recordings from the cell body. Thus, the lessons learned from the squid axon are applicable to, and indeed, essential for, understanding action potential generation in all neurons.

References

BARRETT, E. F. AND J. N. BARRETT (1976) Separation of two voltage-sensitive potassium currents, and demonstration of a tetrodotoxin-resistant calcium current in frog motoneurones. *J. Physiol. (Lond.)* 255: 737–774.

DODGE, F. A. AND B. FRANKENHAEUSER (1958) Membrane currents in isolated frog nerve fibre under voltage clamp conditions. *J. Physiol. (Lond.)* 143: 76–90.

HODGKIN, A. L. AND A. F. HUXLEY (1939) Action potentials recorded from inside a nerve fibre. *Nature* 144: 710–711.

LLINÁS, R. AND M. SUGIMORI (1980) Electrophysiological properties of *in vitro* Purkinje cell dendrites in mammalian cerebellar slices. *J. Physiol. (Lond.)* 305: 197–213.

LLINÁS, R. AND Y. YAROM (1981) Electrophysiology of mammalian inferior olivary neurones *in vitro*. Different types of voltage-dependent ionic conductances. *J. Physiol. (Lond.)* 315: 549–567.

(A) The phases of an action potential of the squid giant axon. (B) Action potential recorded from a myelinated axon of a frog motor neuron. (C) Action potential recorded from the cell body of a frog motor neuron. The action potential is smaller and the undershoot prolonged in comparison to the action potential recorded from the axon of this same neuron (B). (D) Action potential recorded from the cell body of a neuron from the inferior olive of a guinea pig. This action potential has a pronounced plateau during its falling phase. (E) Action potential recorded from the cell body of a Purkinje neuron in the cerebellum of a guinea pig. (A after Hodgkin and Huxley, 1939; B after Dodge and Frankenhaeuser, 1958; C after Barrett and Barrett, 1976; D after Llinás and Yarom, 1981; E after Llinás and Sugimori, 1980.)

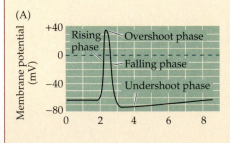
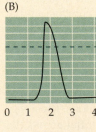
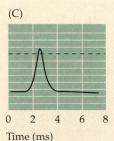
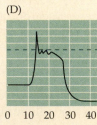
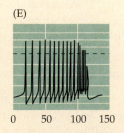

Time (ms)

permeable to one or more ion species subject to an electrochemical gradient. More specifically, a negative membrane potential at rest results from a net efflux of K^+ across neuronal membranes that are predominantly permeable to K^+. In contrast, an action potential occurs when a transient rise in Na^+ permeability allows a net flow of Na^+ in the opposite direction across the membrane that is now predominantly permeable to Na^+. The brief rise in membrane Na^+ permeability is followed by a secondary, transient rise in membrane K^+ permeability that repolarizes the neuronal membrane and produces a brief undershoot of the action potential. As a result of these processes, the membrane is depolarized in an all-or-none fashion during an action potential. When these active permeability changes subside, the membrane potential returns to its resting level because of the high resting membrane permeability to K^+.

Additional Reading

Reviews

HODGKIN, A. L. (1951) The ionic basis of electrical activity in nerve and muscle. *Biol. Rev.* 26: 339–409.

HODGKIN, A. L. (1958) The Croonian Lecture: Ionic movements and electrical activity in giant nerve fibres. *Proc. R. Soc. Lond. (B)* 148: 1–37.

Important Original Papers

BAKER, P. F., A. L. HODGKIN AND T. I. SHAW (1962) Replacement of the axoplasm of giant nerve fibres with artificial solutions. *J. Physiol. (London)* 164: 330–354.

COLE, K. S. AND H. J. CURTIS (1939) Electric impedance of the squid giant axon during activity. *J. Gen. Physiol.* 22: 649–670.

GOLDMAN, D. E. (1943) Potential, impedance, and rectification in membranes. *J. Gen. Physiol.* 27: 37–60.

HODGKIN, A. L. AND P. HOROWICZ (1959) The influence of potassium and chloride ions on the membrane potential of single muscle fibres. *J. Physiol. (London)* 148: 127–160.

HODGKIN, A. L. AND B. KATZ (1949) The effect of sodium ions on the electrical activity of the giant axon of the squid. *J. Physiol. (London)* 108: 37–77.

HODGKIN, A. L. AND R. D. KEYNES (1953) The mobility and diffusion coefficient of potassium in giant axons from *Sepia. J. Physiol. (London)* 119: 513–528.

KEYNES, R. D. (1951) The ionic movements during nervous activity. *J. Physiol. (London)* 114: 119–150.

Books

HODGKIN, A. L. (1967) *The Conduction of the Nervous Impulse*. Springfield, IL: Charles C. Thomas.

HODGKIN, A. L. (1992) *Chance and Design*. Cambridge: Cambridge University Press.

JUNGE, D. (1992) *Nerve and Muscle Excitation*, 3rd Ed. Sunderland, MA: Sinauer Associates.

KATZ, B. (1966) *Nerve, Muscle, and Synapse*. New York: McGraw-Hill.

MOORE, J. W. AND A. E. STUART (2007) *Neurons in Action: Tutorials and Simulations Using NEURON, Version 2*. Sunderland, MA: Sinauer Assoicates.

Chapter 3

Voltage-Dependent Membrane Permeability

Overview

The action potential is the primary electrical signal generated by nerve cells and arises from changes in the permeability of the nerve cell's axonal membranes to specific ions. Present understanding of these changes in ionic permeability is based on evidence obtained by the voltage clamp technique, which permits detailed characterization of permeability changes as a function of membrane potential and time. For most types of axons, these changes consist of a rapid and transient rise in sodium ion (Na^+) permeability, followed by a slower but more prolonged rise in permeability to potassium ions (K^+). Both permeabilities are voltage-dependent, increasing as the membrane potential depolarizes. The kinetics and voltage dependence of Na^+ and K^+ permeabilities provide a complete explanation of action potential generation. Depolarizing the membrane potential to the threshold level causes a rapid, self-sustaining increase in Na^+ permeability that produces the rising phase of the action potential; however, the Na^+ permeability increase is short-lived and is followed by a slower increase in K^+ permeability that restores the membrane potential to its usual negative resting level. A mathematical model that describes the behavior of these ionic permeabilities predicts virtually all of the observed properties of action potentials. Importantly, this same ionic mechanism permits action potentials to be propagated along the length of neuronal axons, explaining how electrical signals are conveyed throughout the nervous system.

Ionic Currents Across Nerve Cell Membranes

The previous chapter introduced the idea that nerve cells generate electrical signals by virtue of a membrane that is differentially permeable to various ion species. In particular, a transient increase in the permeability of the neuronal membrane to Na^+ initiates the action potential. This chapter considers exactly how this increase in Na^+ permeability occurs. A key to understanding this phenomenon is the observation that action potentials are initiated *only* when the neuronal membrane potential becomes more positive than a threshold level. This observation suggests that the mechanism responsible for the increase in Na^+ permeability is sensitive to the membrane potential. Therefore, if one could understand how a change in membrane potential activates Na^+ permeability, it should be possible to explain how action potentials are generated.

The fact that the Na^+ permeability that generates the membrane potential change is itself sensitive to the membrane potential presents both conceptual and practical obstacles to studying the mechanism of the action potential. A practical problem is the difficulty of systematically varying the membrane potential to study the permeability change: such changes in membrane potential will produce an action potential, thus causing further uncontrolled changes in the membrane potential. Historically, then, it was not possible to understand action potentials until a new technology, the **voltage clamp technique**, was developed that allowed experimenters to control membrane potential *and* simultaneously measure the underlying permeability changes (Box 3A). These features of the voltage clamp technique provide the information needed to define the ionic permeability of the membrane at any level of membrane potential.

In the late 1940s, Alan Hodgkin and Andrew Huxley working at the University of Cambridge used the voltage clamp technique to work out the perme-

---

BOX 3A The Voltage Clamp Technique

Breakthroughs in scientific research often rely on the development of new technologies. In the case of the action potential, detailed understanding came only after of the invention of the voltage clamp technique by Kenneth Cole in the 1940s. This device is called a voltage clamp because it controls, or clamps, membrane potential, or voltage, at any level desired by the experimenter. As described in the figure below, voltage clamping measures the membrane potential with a microelectrode (or other type of electrode) placed inside the cell (1) and electronically compares this voltage to the voltage to be maintained, called the *command voltage* (2). The clamp circuitry then passes a current back into the cell though another intra-

cellular electrode (3). This electronic feedback circuit holds the membrane potential at the desired level, even in the face of permeability changes that would normally alter the membrane potential (such as those generated during the action potential). Most importantly, the device permits the simultaneous measurement of the current needed to keep the cell at a given voltage (4). This current is exactly equal to the amount of current flowing across the neuronal membrane, allowing direct measurement of these membrane currents. Therefore, the voltage clamp technique can indicate how membrane potential influences ionic current flow across the membrane. This information gave Hodgkin and Huxley the key insights

that led to their model for action potential generation.

Today the voltage clamp method remains widely used to study ionic currents in neurons and other cells. The most popular contemporary version of this approach is the patch clamp technique, a variation of voltage clamping that can be applied to virtually any cell and has a resolution high enough to measure the minute electrical currents flowing through single ion channels (see Box 4A).

Reference

COLE, K. S. (1968) *Membranes, Ions and Impulses: A Chapter of Classical Biophysics.* Berkeley, CA: University of California Press.

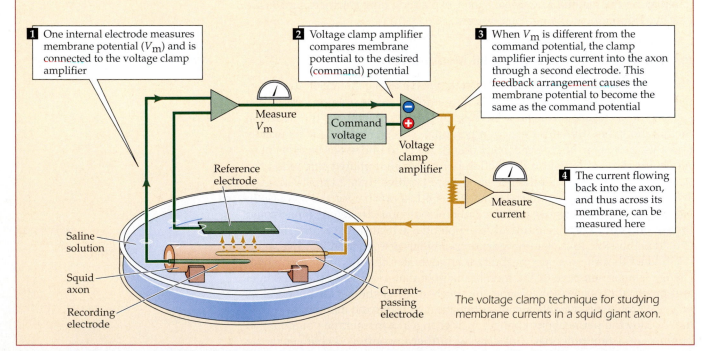

1 One internal electrode measures membrane potential (V_m) and is connected to the voltage clamp amplifier

2 Voltage clamp amplifier compares membrane potential to the desired (command) potential

3 When V_m is different from the command potential, the clamp amplifier injects current into the axon through a second electrode. This feedback arrangement causes the membrane potential to become the same as the command potential

4 The current flowing back into the axon, and thus across its membrane, can be measured here

Measure V_m

Command voltage

Voltage clamp amplifier

Reference electrode

Measure current

Saline solution

Squid axon

Recording electrode

Current-passing electrode

The voltage clamp technique for studying membrane currents in a squid giant axon.

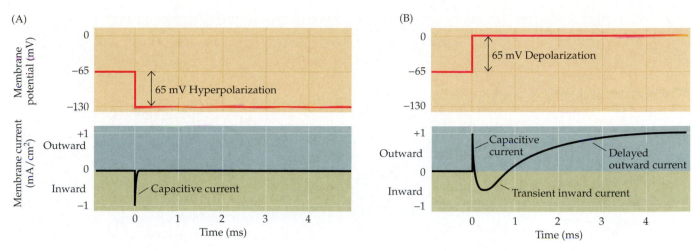

Figure 3.1   Current flow across a squid axon membrane during a voltage clamp experiment. (A) A 65 mV hyperpolarization of the membrane potential produces only a very brief capacitive current. (B) A 65 mV depolarization of the membrane potential also produces a brief capacitive current, which is followed by a longer lasting but transient phase of inward current and a delayed but sustained outward current. (After Hodgkin et al., 1952a.)

ability changes underlying the action potential. They again chose to use the giant neuron of the squid because its large size (up to 1 mm in diameter; see Box 2A) allowed insertion of the electrodes necessary for voltage clamping. They were the first investigators to test the hypothesis that potential-sensitive Na^+ and K^+ permeability changes are both necessary and sufficient to produce action potentials.

Hodgkin and Huxley's first goal was to determine whether neuronal membranes do, in fact, have voltage-dependent permeabilities. To address this issue, they asked whether ionic currents flow across the membrane when its potential is changed. The result of one such experiment is shown in Figure 3.1. Figure 3.1A illustrates the currents produced by a squid axon when its membrane potential, V_m, is hyperpolarized from the resting level of –65 mV to –130 mV. The initial response of the axon results from the redistribution of charge across the axonal membrane. This *capacitive current* is nearly instantaneous, ending within a fraction of a millisecond. Aside from this brief event, however, very little current flows when the membrane is hyperpolarized. But when the membrane potential is depolarized from –65 mV to 0 mV, the response is quite different (Figure 3.1B). Following the capacitive current, the axon produces a rapidly rising inward ionic current (inward refers to a positive charge entering the cell—that is, cations in or anions out), which gives way to a more slowly rising, delayed outward current. The fact that membrane depolarization elicits these ionic currents establishes that the membrane permeability of axons is indeed voltage-dependent.

Two Types of Voltage-Dependent Ionic Current

While the results shown in Figure 3.1 demonstrate that the ionic permeability of neuronal membranes is voltage-sensitive, these experiments do not identify how many types of permeability exist, or which ions are involved. As discussed in Chapter 2 (see Figure 2.5), varying the potential across a membrane makes it possible to deduce the equilibrium potential for the fluxes of different ions through the membrane, and thus to identify the ions that are flowing. Because the voltage clamp method allows the membrane potential to be changed while ionic currents are being measured, it was a straightforward matter for Hodgkin and Huxley to determine ionic permeability by examining how the properties of the early inward and late outward currents changed as the membrane potential was varied (Figure 3.2).

As already noted, no appreciable ionic currents flow at membrane potentials more negative than the resting potential. At more positive potentials, however,

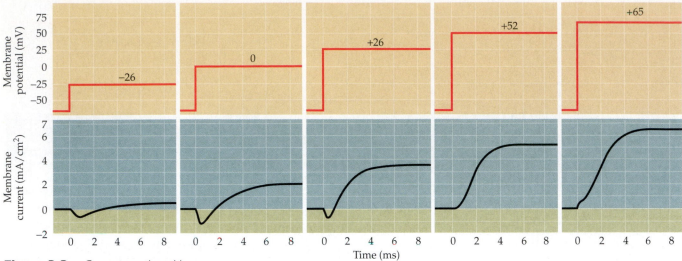

Figure 3.2 Current produced by membrane depolarizations to several different potentials. The early current first increases, then decreases in magnitude as the depolarization increases; note that this current is actually reversed in polarity at potentials more positive than about +55 mV. In contrast, the late current increases monotonically with increasing depolarization. (After Hodgkin et al., 1952a.)

the currents not only flow but they also change in magnitude. The early current has a U-shaped dependence on membrane potential, increasing over a range of depolarizations up to approximately 0 mV but decreasing as the potential is depolarized further. In contrast, the late current increases monotonically with increasingly positive membrane potentials. These different responses to membrane potential can be seen more clearly when the magnitudes of the two current components are plotted as a function of membrane potential, as in Figure 3.3.

The voltage sensitivity of the early inward current gives an important clue about the nature of the ions carrying the current—namely, that no current flows when the membrane potential is clamped at +52 mV. For the squid neurons studied by Hodgkin and Huxley, the external Na^+ concentration is 440 mM, and the internal Na^+ concentration is 50 mM. For this concentration gradient, the Nernst equation predicts that the equilibrium potential for Na^+ should be +55 mV. Recall further from Chapter 2 that at the equilibrium potential for Na^+ (E_{Na}) there is no net flux of Na^+ across the membrane, even if the membrane is highly permeable to Na^+. Thus, the experimental observation that no current flows at the membrane potential where Na^+ cannot flow is a strong indication that the early inward current is carried by entry of Na^+ into the axon. The fact that the early current reverses its direction at potentials more positive than E_{Na} provides another indication that this current is carried by Na^+: increasing membrane permeability to Na^+ at such potentials should cause current to flow outward as Na^+ leaves the axon, due to the reversed electrochemical gradient.

An even more demanding way to test whether Na^+ carries the early inward current is to examine the behavior of this current after removing external Na^+. Removing the Na^+ outside the axon makes E_{Na} more negative, which should reverse the electrochemical gradient for Na^+ and cause the current to become outward. When Hodgkin and Huxley performed this experiment, they observed just such a reversal of the early current (Figure 3.4). Removing external Na^+

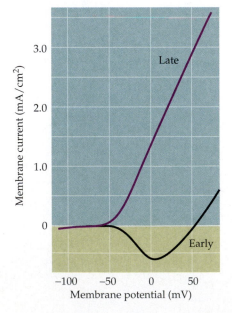

Figure 3.3 Relationship between current amplitude and membrane potential, taken from experiments such as the one shown in Figure 3.2. While the late outward current increases steeply with increasing depolarization, the early inward current first increases in magnitude, but then decreases and reverses to outward current at about +55 mV (the sodium equilibrium potential). (After Hodgkin et al., 1952a.)

Figure 3.4   Dependence of the early inward current on sodium. In the presence of normal external concentrations of Na⁺, depolarization of a squid axon to 0 mV produces an inward initial current. However, removal of external Na⁺ causes the initial inward current to become outward, an effect that is reversed by restoration of external Na⁺. (After Hodgkin and Huxley, 1952a.)

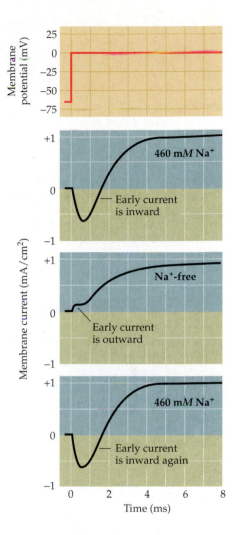

caused the early inward current to reverse its polarity and become an outward current at the same membrane potential that gave rise to an inward current when external Na⁺ was present. This result demonstrates convincingly that the early inward current measured when Na⁺ is present in the external medium must be due to Na⁺ entering the neuron.

Notice in Figure 3.4 that the experimental removal of external Na⁺ has little effect on the outward current that flows after the neuron has been kept at a depolarized membrane voltage for several milliseconds. This further result shows that the late outward current must be due to the flow of an ion other than Na⁺. Several lines of evidence presented by Hodgkin, Huxley, and others showed that this late outward current is caused by potassium ions exiting the neuron. Perhaps the most compelling demonstration of K⁺ involvement is that the amount of K⁺ efflux from the neuron (measured by loading the neuron with radioactive K⁺) is closely correlated with the magnitude of the late outward current.

Taken together, these experiments using the voltage clamp show that changing the membrane potential to a level more positive than the resting potential produces two effects: an *early influx of Na⁺* into the neuron, followed by a *delayed efflux of K⁺*. The early influx of Na⁺ produces a transient inward current, whereas the delayed efflux of K⁺ produces a sustained outward current. The differences in the time course and ionic selectivity of the two fluxes suggest that two different

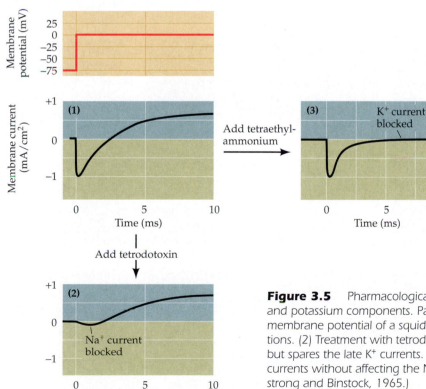

Figure 3.5   Pharmacological separation of Na⁺ and K⁺ currents into sodium and potassium components. Panel (1) shows the current that flows when the membrane potential of a squid axon is depolarized to 0 mV in control conditions. (2) Treatment with tetrodotoxin causes the early Na⁺ currents to disappear but spares the late K⁺ currents. (3) Addition of tetraethylammonium blocks the K⁺ currents without affecting the Na⁺ currents. (After Moore et al., 1967 and Armstrong and Binstock, 1965.)

ionic permeability mechanisms are activated by changes in membrane potential. Confirmation that there are indeed two distinct mechanisms has come from pharmacological studies of drugs that specifically affect these two currents (Figure 3.5). **Tetrodotoxin** (**TTX**), an alkaloid neurotoxin found in certain puffer fish, tropical frogs, and salamanders, blocks the Na^+ current without affecting the K^+ current. Conversely, **tetraethylammonium ions** block K^+ currents without affecting Na^+ currents. The differential sensitivity of Na^+ and K^+ currents to these drugs provides strong additional evidence that Na^+ and K^+ flow through independent permeability pathways. It is now known that these pathways are membrane proteins known as ion channels that are selectively permeable to either Na^+ or K^+. In fact, tetrodotoxin, tetraethylammonium, and other drugs that interact with specific types of ion channels have been extraordinarily useful tools in characterizing these channel molecules, as discussed in Chapter 4.

Two Voltage-Dependent Membrane Conductances

The next goal Hodgkin and Huxley set for themselves was to describe Na^+ and K^+ permeability changes mathematically. To do this, they assumed that the ionic currents are due to a change in **membrane conductance**, defined as the reciprocal of the membrane resistance. Membrane conductance is thus closely related, although not identical, to membrane permeability. When evaluating ionic movements from an electrical standpoint, it is convenient to describe them in terms of ionic conductances rather than ionic permeabilities. For present pur-

Figure 3.6 *Membrane conductance changes underlying the action potential are time- and voltage-dependent. Depolarizations to various membrane potentials (A) elicit different membrane currents (B). Below are shown the Na^+ (C) and K^+ (D) conductances calculated from these currents. Both peak Na^+ conductance and steady-state K^+ conductance increase as the membrane potential becomes more positive. In addition, the activation of both conductances, as well as the rate of inactivation of the Na^+ conductance, occur more rapidly with larger depolarizations. (After Hodgkin and Huxley, 1952b.)*

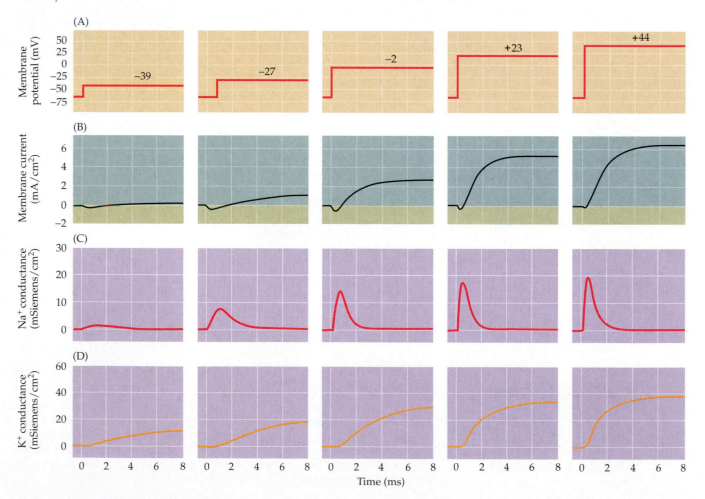

poses, permeability and conductance can be considered synonymous. If membrane conductance (g) obeys Ohm's Law (which states that voltage is equal to the product of current and resistance), then the ionic current that flows during an increase in membrane conductance is given by

$$I_{ion} = g_{ion}\,(V_m - E_{ion})$$

where I_{ion} is the ionic current, V_m is the membrane potential, and E_{ion} is the equilibrium potential for the ion flowing through the conductance, g_{ion}. The difference between V_m and E_{ion} is the electrochemical driving force acting on the ion.

Hodgkin and Huxley used this simple relationship to calculate the dependence of Na^+ and K^+ conductances on time and membrane potential. They knew V_m, which was set by their voltage clamp device (Figure 3.6A). They could determine E_{Na} and E_K from the ionic concentrations on the two sides of the axonal membrane (see Table 2.1). The currents carried by Na^+ and K^+—I_{Na} and I_K—could be determined separately from recordings of the membrane currents resulting from depolarization (Figure 3.6B) by measuring the difference between currents recorded in the presence and absence of external Na^+ (as shown in Figure 3.4). From these measurements, Hodgkin and Huxley were able to calculate g_{Na} and g_K (Figure 3.6C,D), from which they drew two fundamental conclusions. Their first conclusion was that the Na^+ and K^+ conductances change over time. For example, both conductances require some time to **activate**, or turn on. In particular, the K^+ conductance has a pronounced delay, requiring several milliseconds to reach its maximum (Figure 3.6D), while the Na^+ conductance reaches its maximum more rapidly (Figure 3.6C). The more rapid activation of the Na^+ conductance allows the resulting inward Na^+ current to precede the delayed outward K^+ current (see Figure 3.6B).

Although the Na^+ conductance rises rapidly, it quickly declines, even when the membrane potential is kept at a depolarized level. This fact shows that depolarization not only causes the Na^+ conductance to activate, but also causes it to decrease over time, or **inactivate**. The K^+ conductance of the squid axon does not inactivate in this way; thus, while the Na^+ and K^+ conductances share the property of time-dependent activation, only the Na^+ conductance inactivates. (Inactivating K^+ conductances have since been discovered in other types of nerve cells; see Chapter 4.) The time courses of the Na^+ and K^+ conductances are voltage-dependent, with the speed of both activation and inactivation increasing at more depolarized potentials. This finding accounts for more rapid time courses of membrane currents measured at more depolarized potentials.

The second conclusion derived from Hodgkin and Huxley's calculations is that both the Na^+ and K^+ conductances are voltage-dependent—that is, both increase progressively as the neuron is depolarized. Figure 3.7 illustrates this by plotting

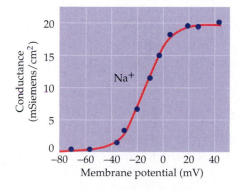

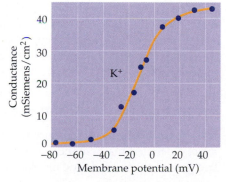

Figure 3.7   Depolarization increases Na^+ and K^+ conductances of the squid giant axon. The peak magnitude of Na^+ conductance and steady-state value of K^+ conductance both increase steeply as the membrane potential is depolarized. (After Hodgkin and Huxley, 1952b.)

(A)

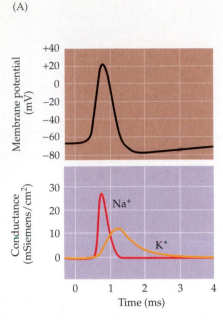

(B)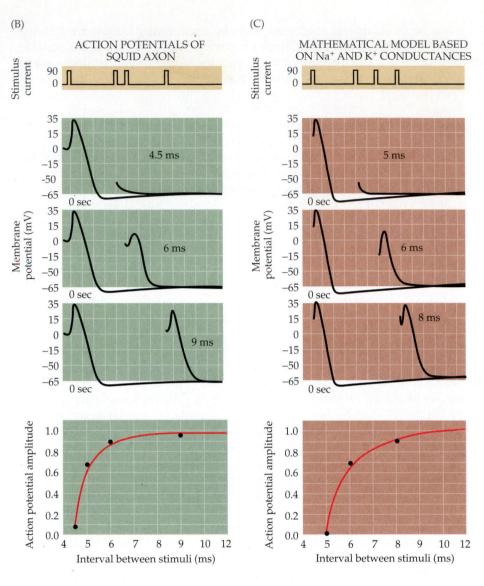

Figure 3.8 *Mathematical reconstruction of the action potential. (A) Reconstruction of an action potential (black curve) together with the underlying changes in Na⁺ (red curve) and K⁺ (yellow curve) conductance. The size and time course of the action potential were calculated using only the properties of g_{Na} and g_K measured in voltage clamp experiments. (B) The refractory period can be observed by stimulating an axon with two current pulses separated that are by variable intervals. While the first stimulus reliably evokes an action potential, during the refractory period the second stimulus generates only a small action potential (or no response at all). (C) The mathematical model accurately simulates responses of the axon during the refractory period. (After Hodgkin and Huxley, 1952d.)*

the relationship between peak value of the conductances (from Figure 3.6C and D) against the membrane potential. Note the similar voltage dependence for each conductance; both conductances are quite small at negative potentials, maximal at very positive potentials, and exquisitely dependent on membrane voltage at intermediate potentials. The observation that Na⁺ and K⁺ conductances are sensitive to changes in membrane potential shows that the mechanism underlying these conductances somehow "senses" the voltage across the membrane.

All told, the voltage clamp experiments carried out by Hodgkin and Huxley showed that the ionic currents that flow when the neuronal membrane is depolarized are due to three different voltage-sensitive processes: (1) activation of Na⁺ conductance; (2) activation of K⁺ conductance; and (3) inactivation of Na⁺ conductance.

Reconstructing the Action Potential

From their experimental measurements, Hodgkin and Huxley were able to construct a detailed mathematical model of the Na⁺ and K⁺ conductance changes. The goal of these modeling efforts was to determine whether the Na⁺ and K⁺

conductances alone are sufficient to produce an action potential. Using this information, they could in fact generate the form and time course of the action potential with remarkable accuracy (Figure 3.8A). The Hodgkin-Huxley model could simulate many other features of action potential behavior in the squid axon. For example, it was well known that, following an action potential, the axon experiences a brief **refractory period** during which it is resistant to further excitation (Figure 3.8B). The model was capable of closely mimicking such behavior (Figure 3.8C).

The Hodgkin-Huxley model also provided many insights into how action potentials are generated. Figure 3.8A shows a reconstructed action potential, together with the time courses of the underlying Na^+ and K^+ conductances. The coincidence of the initial increase in Na^+ conductance with the rapid rising phase of the action potential demonstrates that a selective increase in Na^+ conductance is responsible for action potential initiation. The increase in Na^+ conductance causes Na^+ to enter the neuron, thus depolarizing the membrane potential, which approaches E_{Na}. The rate of depolarization subsequently falls, both because the electrochemical driving force on Na^+ decreases and because the Na^+ conductance inactivates. At the same time, depolarization slowly activates the voltage-dependent K^+ conductance, causing K^+ to leave the cell and repolarizing the membrane potential toward E_K. Because the K^+ conductance becomes temporarily higher than it is in the resting condition, the membrane potential actually becomes briefly more negative than the normal resting potential (the undershoot). The hyperpolarization of the membrane potential causes the voltage-dependent K^+ conductance (and any Na^+ conductance not inactivated) to turn off, allowing the membrane potential to return to its resting level. The relatively slow time course of turning off the K^+ conductance, as well as the persistence of Na^+ conductance inactivation, is responsible for the refractory period (see also Figure 3.12 below).

This mechanism of action potential generation represents a positive feedback loop: Activating the voltage-dependent Na^+ conductance increases Na^+ entry into the neuron, which makes the membrane potential depolarize, which leads to the activation of still more Na^+ conductance, more Na^+ entry, and still further depolarization (Figure 3.9). Positive feedback continues unabated until Na^+ conductance inactivation and K^+ conductance activation restore the membrane potential to the resting level. Because this positive feedback loop, once initiated, is sustained by the intrinsic properties of the neuron—namely, the voltage dependence of the ionic conductances—the action potential is self-supporting, or **regenerative**. This regenerative quality explains why action potentials exhibit all-or-none behavior (see Figure 2.2), and why they have a threshold (see Box 3B). The delayed activation of the K^+ conductance represents a negative feedback loop that eventually restores the membrane to its resting state.

Hodgkin and Huxley's reconstruction of the action potential and all its features shows that the properties of the voltage-sensitive Na^+ and K^+ conductances, together with the electrochemical driving forces created by ion transporters, are sufficient to explain action potentials. Their use of both empirical and theoretical methods brought an unprecedented level of rigor to a long-standing problem, setting a standard of proof that is achieved only rarely in biological research.

Long-Distance Signaling by Means of Action Potentials

The voltage-dependent mechanisms of action potential generation also explain the long-distance transmission of these electrical signals. Recall from Chapter 2 that neurons are relatively poor conductors of electricity, at least compared to a

Figure 3.9 Feedback cycles responsible for membrane potential changes during an action potential. Membrane depolarization rapidly activates a positive feedback cycle fueled by the voltage-dependent activation of Na^+ conductance. This phenomenon is followed by the slower activation of a negative feedback loop as depolarization activates a K^+ conductance, which helps to repolarize the membrane potential and terminate the action potential.

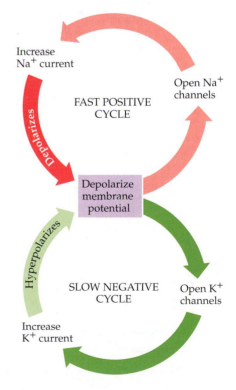

BOX 3B   Threshold

An important—and potentially puzzling—property of the action potential is its initiation at a particular membrane potential, called **threshold**. Indeed, action potentials never occur without a depolarizing stimulus that brings the membrane to this level. The depolarizing "trigger" can be one of several events: a synaptic input, a receptor potential generated by specialized receptor organs, the endogenous pacemaker activity of cells that generate action potentials spontaneously, or the local current that mediates the spread of the action potential down the axon.

Why the action potential "takes off" at a particular level of depolarization can be understood by comparing the underlying events to a chemical explosion (Figure A). Exogenous heat (analogous to the initial depolarization of the membrane potential) stimulates an exothermic chemical reaction, which produces more heat, which further enhances the reaction (Figure B). As a result of this positive feedback loop, the rate of the reaction builds up exponentially—the definition of an explosion.

In any such process, however, there is a threshold, that is, a point up to which heat can be supplied without

resulting in an explosion. The threshold for the chemical explosion diagrammed here is the point at which the amount of heat supplied exogenously is just equal to the amount of heat that can be dissipated by the circumstances of the reaction (such as escape of heat from the beaker).

The threshold of action potential initiation is, in principle, similar (Figure C). There is a range of "subthreshold" depolarization, within which the rate of increased sodium entry is less than the rate of potassium exit (remember that the membrane at rest is highly permeable to K+, which therefore flows out as the membrane is depolarized). The point at which Na+ inflow just equals K+ outflow represents an unstable equilibrium analogous to the ignition point of an explosive mixture.

The behavior of the membrane at threshold reflects this instability: the membrane potential may linger at the threshold level for a variable period before either returning to the resting level or flaring up into a full-blown action potential. In theory at least, if there is a net internal gain of a single

Na+ ion, an action potential occurs; conversely, the net loss of a single K+ ion leads to repolarization. A more precise definition of threshold, therefore, is that value of membrane potential, in depolarizing from the resting potential, at which the current carried by Na+ entering the neuron is exactly equal to the K+ current that is flowing out. Once the triggering event depolarizes the membrane beyond this point, the positive feedback loop of Na+ entry on membrane potential closes and the action potential "fires."

Because the Na+ and K+ conductances change dynamically over time, the threshold potential for producing an action potential also varies as a consequence of the previous activity of the neuron. For example, following an action potential, the membrane becomes temporarily refractory to further excitation because the threshold for firing an action potential transiently rises. There is, therefore, no specific value of membrane potential that defines the threshold for a given nerve cell in all circumstances.

A positive feedback loop underlying the action potential explains the phenomenon of threshold.

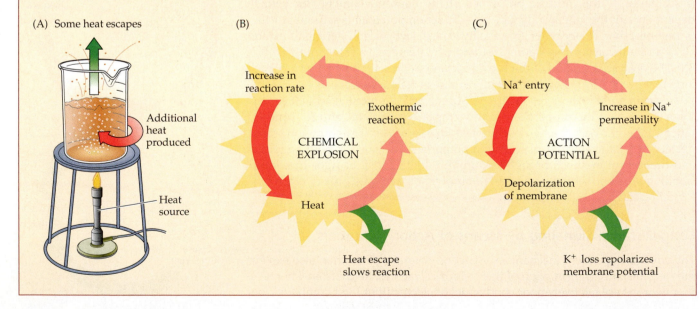

(A) Some heat escapes

Additional heat produced

Heat source

(B) Increase in reaction rate

Exothermic reaction

CHEMICAL EXPLOSION

Heat

Heat escape slows reaction

(C) Na+ entry

Increase in Na+ permeability

ACTION POTENTIAL

Depolarization of membrane

K+ loss repolarizes membrane potential

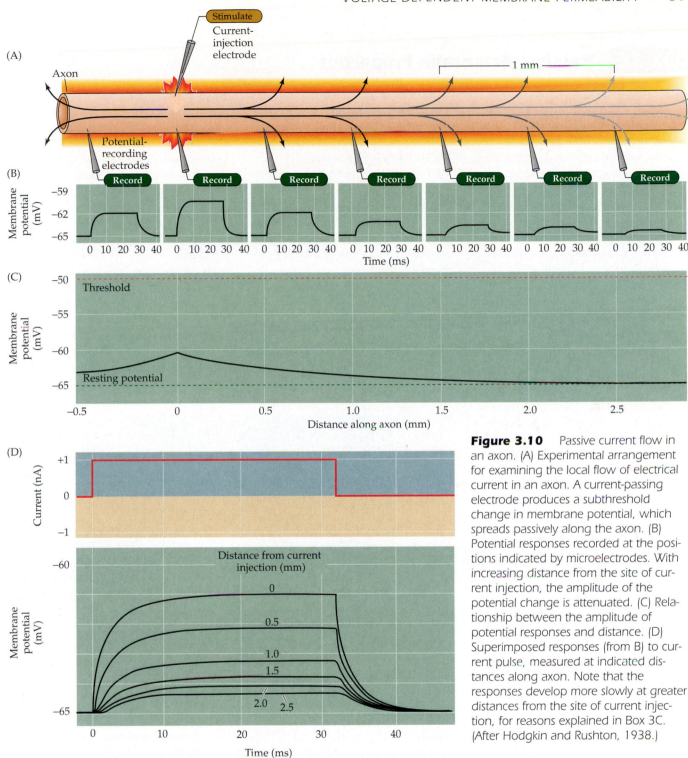

Figure 3.10 Passive current flow in an axon. (A) Experimental arrangement for examining the local flow of electrical current in an axon. A current-passing electrode produces a subthreshold change in membrane potential, which spreads passively along the axon. (B) Potential responses recorded at the positions indicated by microelectrodes. With increasing distance from the site of current injection, the amplitude of the potential change is attenuated. (C) Relationship between the amplitude of potential responses and distance. (D) Superimposed responses (from B) to current pulse, measured at indicated distances along axon. Note that the responses develop more slowly at greater distances from the site of current injection, for reasons explained in Box 3C. (After Hodgkin and Rushton, 1938.)

wire. Current conduction by wires, and by neurons in the absence of action potentials, is called **passive current flow** (Figure 3.10). The passive electrical properties of a nerve cell axon can be determined by measuring the voltage change resulting from a current pulse passed across the axonal membrane (Figure 3.10A). If this current pulse is not large enough to generate action potentials, the magnitude of the resulting potential change decays exponentially with increasing distance from the site of current injection (Figure 3.10B). Typically, the

BOX 3C   Passive Membrane Properties

The passive flow of electrical current plays a central role in action potential propagation, synaptic transmission, and all other forms of electrical signaling in nerve cells. Therefore, it is worthwhile understanding in quantitative terms how passive current flow varies with distance along a neuron. For the case of a cylindrical axon, such as the one depicted in Figure 3.10, subthreshold current injected into one part of the axon spreads passively along the axon until the current is dissipated by leakage out across the axon membrane. The decrement in the current flow with distance (Figure A) is described by a simple exponential function:

$$V_x = V_0\, e^{-x/\lambda}$$

where V_x is the voltage response at any distance x along the axon, V_0 is the voltage change at the point where current is injected into the axon, e is the base of natural logarithms (approximately 2.7), and λ is the length constant of the axon. As evident in this relationship, the length constant is the distance where the initial voltage response (V_0) decays to 1/e (or

37%) of its value. The length constant is thus a way to characterize how far passive current flow spreads before it leaks out of the axon, with leakier axons having shorter length constants.

The length constant depends upon the physical properties of the axon, in particular the relative resistances of the plasma membrane (r_m), the intracellular axoplasm (r_i), and the extracellular medium (r_0). The relationship between these parameters is

$$\lambda = \sqrt{\frac{r_m}{r_o + r_i}}$$

Hence, to improve the passive flow of current along an axon, the resistance of the plasma membrane should be as high as possible and the resistances of the

axoplasm and extracellular medium should be low.

Another important consequence of the passive properties of neurons is that currents flowing across a membrane do not immediately change the membrane potential. For example, when a rectangular current pulse is injected into the axon shown in the experiment illustrated in Figure 3.10A, the membrane potential depolarizes slowly over a few milliseconds and then repolarizes over a similar time course when the current pulse ends (see Figure 3.10D). These delays in changing the membrane potential are due to the fact that the plasma membrane behaves as a capacitor, storing the initial charge that flows at the beginning and end of the current pulse. For the

(A) Spatial decay of membrane potential along a cylindrical axon. A current pulse injected at one point in the axon (0 mm) produces voltage responses (V_x) that decay exponentially with distance. The distance where the voltage response is 1/e of its initial value (V_0) is the length constant, λ.

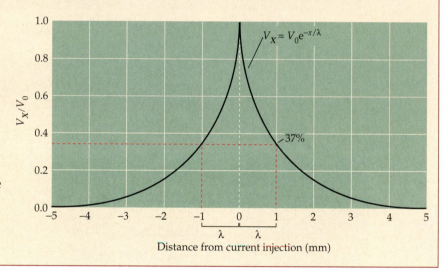

potential falls to a small fraction of its initial value at a distance of no more than a couple of millimeters away from the site of injection (Figure 3.10C). The progressive decrease in the amplitude of the induced potential change occurs because the injected current leaks out across the axonal membrane; accordingly, less current is available to change the membrane potential farther along the axon (Box 3C). Thus, the leakiness of the axonal membrane prevents effective passive transmission of electrical signals in all but the shortest axons (those 1 mm or less in length). Likewise, the leakiness of the membrane slows the time course of the responses measured at increasing distances from the site where current was injected (Figure 3.10D).

If the experiment shown in Figure 3.10 is repeated with a depolarizing current pulse large enough to produce an action potential, the result is dramatically different (Figure 3.11A). In this case, an action potential occurs without decre-

case of a cell whose membrane potential is spatially uniform, the change in the membrane potential at any time, V_t after beginning the current pulse (Figure B) can also be described by an exponential relationship:

$$V_t = V_\infty(1 - e^{-t/\tau})$$

where V_∞ is the steady-state value of the membrane potential change, t is the time after the current pulse begins, and τ is the membrane time constant. The time constant is thus defined as the time when the voltage response (V_t) rises to

$1 - (1/e)$ (or 63%) of V_∞. After the current pulse ends, the membrane potential change also declines exponentially according to the relationship

$$V_t = V_\infty e^{-t/\tau}$$

During this decay, the membrane potential returns to $1/e$ of V_∞ at a time equal to t. For cells with more complex geometries than the axon in Figure 3.10, the time courses of the changes in membrane potential are not simple exponentials, but nonetheless depend on the membrane time constant. Thus, the time

constant characterizes how rapidly current flow changes the membrane potential. The membrane time constant also depends on the physical properties of the nerve cell, specifically on the resistance (r_m) and capacitance (c_m) of the plasma membrane such that:

$$\tau = r_m c_m$$

The values of r_m and c_m depend, in part, on the size of the neuron, with larger cells having lower resistances and larger capacitances. In general, small nerve cells tend to have long time constants and large cells brief time constants.

References

HODGKIN, A. L. AND W. A. H. RUSHTON (1938) The electrical constants of a crustacean nerve fibre. *Proc. R. Soc. Lond.* 133: 444–479.

JOHNSTON, D. AND S. M.-S. WU (1995) *Foundations of Cellular Neurophysiology*. Cambridge, MA: MIT Press.

RALL, W. (1977) Core conductor theory and cable properties of neurons. In *Handbook of Physiology*, Section 1: *The Nervous System*, Vol. 1: *Cellular Biology of Neurons*. E. R. Kandel (ed.). Bethesda, MD: American Physiological Society, pp. 39–98.

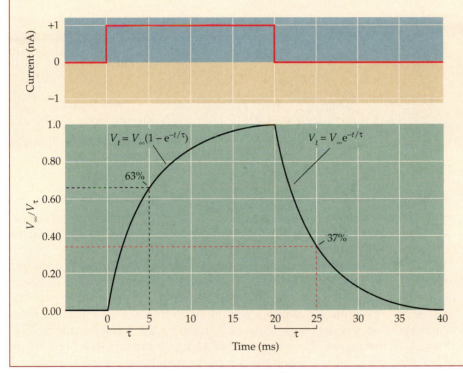

(B) Time course of potential changes produced in a spatially uniform cell by a current pulse. The rise and fall of the membrane potential (V_τ) can be described as exponential functions, with the time constant τ defining the time required for the response to rise to $1 - (1/e)$ of the steady-state value (V_∞), or to decline to $1/e$ of V_∞.

ment along the entire length of the axon, which in humans may be a distance of a meter or more (Figure 3.11B). Thus, action potentials somehow circumvent the inherent leakiness of neurons.

How do action potentials traverse great distances along such a poor passive conductor? A partial answer is provided by the observation that the amplitude of the action potentials recorded at different distances is constant. This all-or-none behavior indicates that more than simple passive flow of current must be involved in action potential propagation. A second clue comes from examination of the time of occurrence of the action potentials recorded at different distances from the site of stimulation: action potentials occur later and later at greater distances along the axon (see Figure 3.11B). Thus, the action potential has a measurable rate of transmission, called the **conduction velocity**. The delay in the arrival of the action potential at successively more distant points

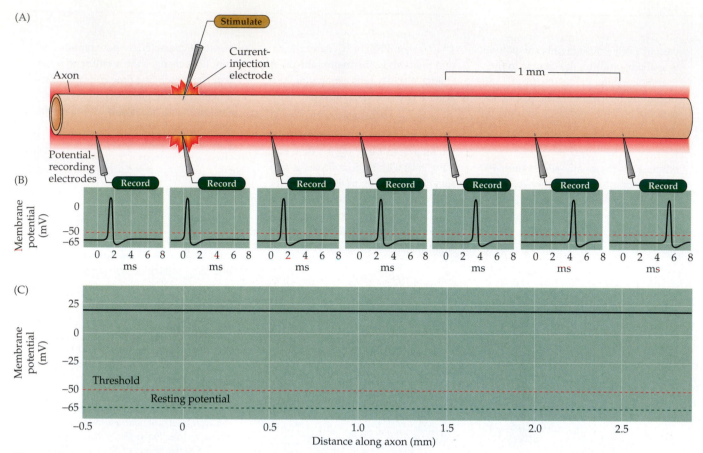

Figure 3.11 Propagation of an action potential. (A) In this experimental arrangement, an electrode evokes an action potential by injecting a suprathreshold current. (B) Potential responses recorded at the positions indicated by microelectrodes. The amplitude of the action potential is constant along the length of the axon, although the time of appearance of the action potential is delayed with increasing distance. (C) The constant amplitude of an action potential (solid black line) measured at different distances.

along the axon differs from the case shown in Figure 3.10, in which the electrical changes produced by passive current flow occur at more or less the same time at successive points.

The mechanism of action potential propagation is easy to grasp once one understands how action potentials are generated and how current passively flows along an axon (Figure 3.12). A depolarizing stimulus—a synaptic potential or a receptor potential in an intact neuron, or an injected current pulse in an experiment—locally depolarizes the axon, opening the voltage-sensitive Na$^+$ channels in that region. The opening of Na$^+$ channels causes inward movement of Na$^+$, and the resultant depolarization of the membrane potential generates an action potential at that site. Some of the local current generated by the action potential then flows passively down the axon, in the same way that subthreshold currents spread along the axon (see Figure 3.10). Note that this passive current flow does not require the movement of Na$^+$ along the axon, but instead occurs by a shuttling of charge, somewhat similar to what happens when wires passively conduct electricity by transmission of electron charge. This passive current flow depolarizes the membrane potential in the adjacent region of the axon, thus opening the Na$^+$ channels in the neighboring membrane. The local depolarization triggers an action potential in this region, which then spreads again in a continuing cycle until the end of the axon is reached. Thus, action potential propagation requires the coordinated action of two forms of current flow—the passive flow of current as well as active currents flowing through voltage-dependent ion channels. The regenerative properties of Na$^+$ channel opening allow action potentials to propagate in an all-or-none fashion by acting

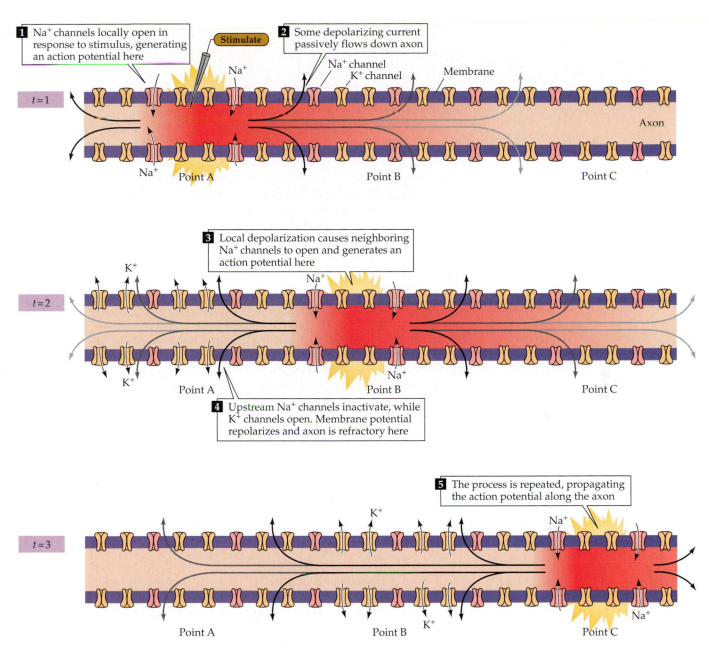

1 Na⁺ channels locally open in response to stimulus, generating an action potential here

Stimulate

2 Some depolarizing current passively flows down axon

Na⁺ channel
K⁺ channel
Membrane

Na⁺

$t = 1$

Axon

Na⁺
Point A   Point B   Point C

3 Local depolarization causes neighboring Na⁺ channels to open and generates an action potential here

K⁺

Na⁺

$t = 2$

K⁺
Point A   Na⁺
Point B   Point C

4 Upstream Na⁺ channels inactivate, while K⁺ channels open. Membrane potential repolarizes and axon is refractory here

5 The process is repeated, propagating the action potential along the axon

K⁺   Na⁺

$t = 3$

Point A   K⁺
Point B   Na⁺
Point C

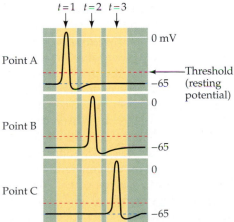

$t = 1$ $t = 2$ $t = 3$

Point A

0 mV

Threshold (resting potential)

−65

Point B

0

−65

Point C

0

−65

Figure 3.12   Action potential conduction requires both active and passive current flow. Depolarization opens Na⁺ channels locally and produces an action potential at point A of the axon (time $t = 1$). The resulting inward current flows passively along the axon, depolarizing the adjacent region (point B) of the axon. At a later time ($t = 2$), the depolarization of the adjacent membrane has opened Na⁺ channels at point B, resulting in the initiation of the action potential at this site and additional inward current that again spreads passively to an adjacent point (point C) farther along the axon. At a still later time ($t = 3$), the action potential has propagated even farther. This cycle continues along the full length of the axon. Note that as the action potential spreads, the membrane potential repolarizes due to K⁺ channel opening and Na⁺ channel inactivation, leaving a "wake" of refractoriness behind the action potential that prevents its backward propagation (panel 4). The panel to the left of this figure legend shows the time course of membrane potential changes at the points indicated.

as a booster at each point along the axon, thus ensuring the long-distance transmission of electrical signals.

The Refractory Period Ensures Polarized Propagation of Action Potentials

Recall that axons become refractory following an action potential: that is, the generation of an action potential briefly makes it harder for the axon to produce subsequent action potentials (see Figure 3.8B). Refractoriness limits the number of action potentials that a neuron can produce per unit time, with different types of neurons having different maximum rates of action potential firing (a result of their different types and densities of ion channels). As described in a previous section, the refractory period arises because the depolarization that produces Na^+ channel opening also results in the delayed activation of K^+ channels and Na^+ channel inactivation, which temporarily makes it more difficult for the axon to produce another action potential. The action potential sweeps along the length of an axon, and in its wake Na^+ channels are inactivated and K^+ channels activated for a brief time. Refractoriness of the membrane region where an action potential has been generated prevents subsequent re-excitation of this membrane (see Figure 3.12)—an important feature that prevents action potentials from propagating backward (i.e., back toward their point of initiation). Thus, refractory behavior ensures polarized propagation of action potentials from their usual point of initiation near the neuronal cell body, toward the synaptic terminals at the distal end of the axon.

Myelination Results in Increased Conduction Velocity

The rate of action potential conduction limits the flow of information within the nervous system. It is not surprising, then, that various mechanisms have evolved that optimize the propagation of action potentials. Because action potential conduction requires passive and active flow of current (see Figure 3.12), the rate of action potential propagation is determined by both of these phenomena. One way of improving passive current flow is to increase the diameter of an axon, which effectively decreases the internal resistance to passive current flow (see Box 3C). The consequent increase in action potential conduction velocity presumably explains why giant axons evolved in invertebrates such as squid, and why rapidly conducting axons in all animals tend to be larger than slowly conducting ones.

Another strategy to improve the passive flow of electrical current is to insulate the axonal membrane, reducing the ability of current to leak out of the axon and thus increasing the distance along the axon that a given local current can flow passively. Among vertebrates, this strategy is evident in the **myelination** of axons, a process by which oligodendrocytes in the central nervous system (and Schwann cells in the peripheral nervous system) wrap the axon in **myelin**, which consists of multiple layers of closely opposed glial membranes (Figure 3.13A). By acting as an electrical insulator, myelin greatly speeds up action potential conduction (Figure 3.14). For example, whereas unmyelinated axon conduction velocities range from about 0.5 to 10 m/s, myelinated axons can conduct at velocities of up to 150 m/s. The major reason underlying this marked increase in speed is that the time-consuming process of action potential generation occurs only at specific points along the axon, called **nodes of Ranvier**, where there is a gap in the myelin wrapping. If the entire surface of an axon were insulated, there would be no place for current to flow out of the axon and action potentials could not be generated. Thus, the voltage-gated Na^+ channels required for action

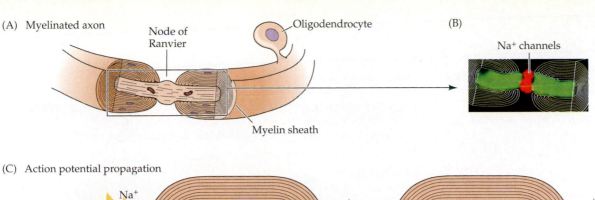

(A) Myelinated axon

Node of Ranvier

Oligodendrocyte

(B)

Na⁺ channels

Myelin sheath

(C) Action potential propagation

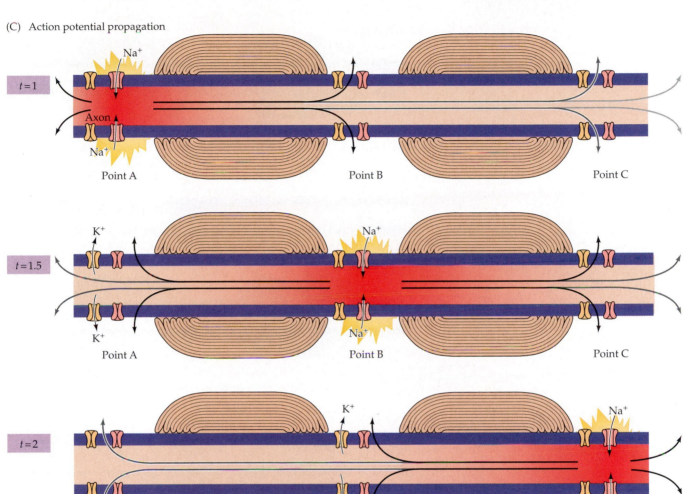

t = 1

Na⁺

Axon

Na⁺

Point A   Point B   Point C

t = 1.5

K⁺

Na⁺

K⁺

Na⁺

Point A   Point B   Point C

t = 2

K⁺

Na⁺

K⁺

Na⁺

Point A   Point B   Point C

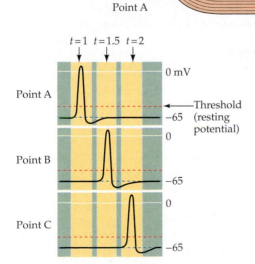

t = 1 t = 1.5 t = 2

Point A

0 mV

Threshold (resting potential)

−65

Point B

0

−65

Point C

0

−65

Figure 3.13 Saltatory action potential conduction along a myelinated axon. (A) Diagram of a myelinated axon. (B) Localization of voltage-gated Na channels (red) at a node of Ranvier in a myelinated axon of the optic nerve. Green color indicates a protein called Caspr, which is located adjacent to the node of Ranvier. (C) Local current in response to action potential initiation at a particular site flows locally, as described in Figure 3.12. However, the presence of myelin prevents the local current from leaking across the internodal membrane; it therefore flows farther along the axon than it would in the absence of myelin. Moreover, voltage-gated Na⁺ channels are present only at the nodes of Ranvier (K⁺ channels are present at the nodes of some neurons but not others). This arrangement means that the generation of active, voltage-gated Na⁺ currents need only occur at these unmyelinated regions. The result is a greatly enhanced velocity of action potential conduction. The panel to the left of this figure legend shows the time course of membrane potential changes at the points indicated. (B from Chen et al., 2004.)

Figure 3.14 Comparison of speed of action potential conduction in unmyelinated (upper) and myelinated (lower) axons. For clarity, passive current flow is shown only for the direction of action potential propagation.

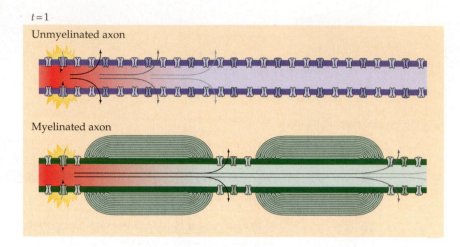

t = 1

Unmyelinated axon

Myelinated axon

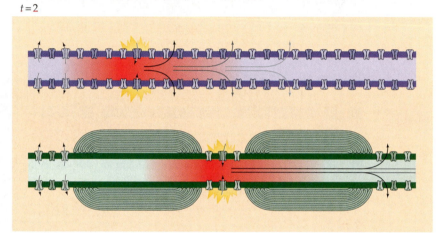

t = 2

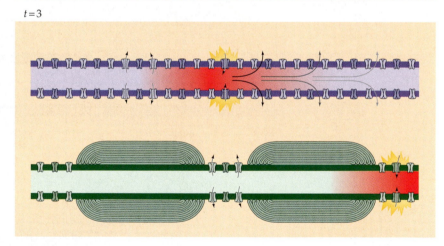

t = 3

potentials are found only at these nodes of Ranvier (Figure 3.13B). An action potential generated at one node of Ranvier elicits current that flows passively within the myelinated segment until the next node is reached. This local current flow then generates an action potential in the neighboring segment, and the cycle is repeated along the length of the axon. Because current flows across the neuronal membrane only at the nodes (Figure 3.13C), this type of propagation is called **saltatory**, meaning that the action potential jumps from node to node. Not surprisingly, loss of myelin, as occurs in diseases such as multiple sclerosis, causes a variety of serious neurological problems (Box 3D).

BOX 3D Multiple Sclerosis

Multiple sclerosis (MS) is a disease of the central nervous system characterized by a variety of clinical problems arising from multiple regions of demyelination and inflammation along axonal pathways. The disorder commonly begins between ages 20 and 40 and is characterized by the abrupt onset of neurological deficits that typically persist for days or weeks and then subside. The clinical course ranges from patients with no persistent neurological loss, some of whom experience only occasional later recurrences, to others who progressively deteriorate as a result of extensive and relentless central nervous system involvement.

The signs and symptoms of MS are determined by the location of the affected regions. Particularly common are monocular blindness (due to lesions of the optic nerve), motor weakness or paralysis (due to lesions of the corticospinal tracts), abnormal somatic sensations (due to lesions of somatic sensory pathways, often in the posterior columns), double vision (due to lesions of medial longitudinal fasciculus), and dizziness (due to lesions of vestibular pathways). Abnormalities are often apparent in the cerebrospinal fluid, which usually contains an abnormal number of cells associated with inflammation and an increased content of antibodies (a sign of an altered immune response).

The diagnosis of MS generally relies on the presence of a neurological problem that remits and then returns at an unrelated site. Confirmation can sometimes be obtained from magnetic resonance imaging (MRI) or from functional evidence of lesions in a particular pathway by abnormal evoked potentials. The histological hallmark of MS (seen at postmortem exam) is multiple lesions at different sites showing loss of myelin associated with infiltration of inflammatory cells and, in some instances, loss of axons themselves.

The concept of MS as a demyelinating disease is deeply embedded in the clinical literature, although precisely how the demyelination translates into functional deficits is poorly understood. The loss of myelin sheath surrounding many axons compromises action potential conduction, and the abnormal patterns of nerve conduction that result presumably produce most of the clinical deficits in the disease. However, MS may have effects that extend beyond loss of the myelin sheath. It is clear that some axons are actually destroyed, probably as a result of the inflammation targeting the overlying myelin and/or loss of trophic support of the axon by oligodendrocytes. Thus, axon loss also contributes to the functional deficits in MS, especially in the chronic, progressive forms of the disease.

The ultimate cause of MS remains unclear. The immune system undoubtedly contributes to the damage, and immunomodulatory therapies provide substantial benefits to some patients. Precisely how the immune system is activated to cause the injury is not known. The most popular hypothesis is that MS is an autoimmune disease (i.e., a disease in which the immune system attacks the body's proper constituents). The fact that immunization of experimental animals with any one of several molecular constituents of the myelin sheath can induce a demyelinating disease (called experimental allergic encephalomyelitis) shows that an autoimmune attack on the myelin membrane is sufficient to produce a clinical picture similar to that of MS. A possible explanation of the human disease is that a genetically susceptible individual becomes transiently infected (by a minor viral illness, for example) with a microorganism that expresses a molecule structurally similar to a component of myelin. An immune response to this antigen is mounted to attack the invader, but the failure of the immune system to discriminate between the foreign protein and self

results in destruction of otherwise normal myelin; this exact scenario occurs in mice infected with Theiler's virus.

An alternative hypothesis is that MS is caused by a persistent infection by a virus or other microorganism. In this scenario, the immune system's ongoing efforts to get rid of the pathogen cause the damage to myelin. Tropical spastic paraparesis (TSP) provides a precedent for this idea. TSP is a disease characterized by the gradual progression of weakness of the legs and impaired control of bladder function associated with increased deep tendon reflexes and a positive Babinski sign (see Chapter 17). This clinical picture is similar to that of rapidly advancing MS, and TSP is known to be caused by persistent infection with a retrovirus (human T lymphotropic virus-1). This precedent notwithstanding, proving the persistent viral infection hypothesis for MS requires unambiguous demonstration of the presence of a virus. Despite periodic reports of a virus associated with MS, convincing evidence has not been forthcoming, and multiple sclerosis remains a daunting clinical challenge.

References

ADAMS, R. D. AND M. VICTOR (2005) *Principles of Neurology*, 8th Ed. New York: McGraw-Hill, pp. 771–796.

FROHMAN, E. M., M. K. RACKE AND C. S. RAINE (2006) Multiple sclerosis: The plaque and its pathogenesis. *N. Engl. J. Med.* 354: 942–955.

HAUSER, S. L. AND J. R. OKSENBERG (2006) The neurobiology of multiple sclerosis: Genes, inflammation, and neurodegeneration. *Neuron* 52: 61–76.

MILLER, D. H. AND 9 OTHERS (2003) A controlled trial of natalizumab for relapsing multiple sclerosis. *N. Engl. J. Med.* 348: 15–23.

WAXMAN, S. G. (2006) Ions, energy, and axonal injury: Towards a molecular neurology of multiple sclerosis. *Trends Mol. Med.* 12: 192–195.

ZANVIL, S. S. AND L. STEINMAN (2003) Diverse targets for intervention during inflammatory and neurodegenerative phases of multiple sclerosis. *Neuron* 38: 685–688.

Summary

The action potential and all its complex properties can be explained by time- and voltage-dependent changes in the Na^+ and K^+ permeabilities of neuronal membranes. This conclusion derives primarily from evidence obtained by a device called the voltage clamp. The voltage clamp technique is an electronic feedback method that allows control of neuronal membrane potential and, simultaneously, direct measurement of the voltage-dependent fluxes of Na^+ and K^+ that produce the action potential. Voltage clamp experiments show that a transient rise in Na^+ conductance activates rapidly and then inactivates during a sustained depolarization of the membrane potential. Such experiments also demonstrate a rise in K^+ conductance that activates in a delayed fashion and, in contrast to the Na^+ conductance, does not inactivate. Mathematical modeling of the properties of these conductances indicates that they, and they alone, are responsible for the production of all-or-none action potentials in the squid axon. Action potentials propagate along the nerve cell axons initiated by the voltage gradient between the active and inactive regions of the axon by virtue of the local current flow. In this way, action potentials compensate for the relatively poor passive electrical properties of nerve cells and enable neural signaling over long distances. These classic electrophysiological findings provide a solid basis for considering the functional and ultimately molecular variations on neural signaling described in the next chapter.

Additional Reading

Reviews

ARMSTRONG, C. M. AND B. HILLE (1998) Voltage-gated ion channels and electrical excitability. *Neuron* 20: 371–380.

NEHER, E. (1992) Ion channels for communication between and within cells. *Science* 256: 498–502.

SALZER, J. L. (2003) Polarized domains of myelinated axons. *Neuron* 40: 297–318.

Important Original Papers

ARMSTRONG, C. M. AND L. BINSTOCK (1965) Anomalous rectification in the squid giant axon injected with tetraethylammonium chloride. *J. Gen. Physiol.* 48: 859–872.

CHEN, C. AND 17 OTHERS (2004) Mice lacking sodium channel β1 subunits display defects in neuronal excitability, sodium channel expression, and nodal architecture. *J. Neurosci.* 24: 4030–4042.

HODGKIN, A. L. AND A. F. HUXLEY (1952a) Currents carried by sodium and potassium ions through the membrane of the giant axon of *Loligo*. *J. Physiol.* 116: 449–472.

HODGKIN, A. L. AND A. F. HUXLEY (1952b) The components of membrane conductance in the giant axon of *Loligo*. *J. Physiol.* 116: 473–496.

HODGKIN, A. L. AND A. F. HUXLEY (1952c) The dual effect of membrane potential on sodium conductance in the giant axon of *Loligo*. *J. Physiol.* 116: 497–506.

HODGKIN, A. L. AND A. F. HUXLEY (1952d) A quantitative description of membrane current and its application to conduction and excitation in nerve. *J. Physiol.* 116: 507–544.

HODGKIN, A. L. AND W. A. H. RUSHTON (1938) The electrical constants of a crustacean nerve fibre. *Proc. R. Soc. Lond.* 133: 444–479.

HODGKIN, A. L., A. F. HUXLEY AND B. KATZ (1952) Measurements of current–voltage relations in the membrane of the giant axon of *Loligo*. *J. Physiol.* 116: 424–448.

HUXLEY, A. F. AND R. STAMPELI (1949) Evidence for saltatory conduction in peripheral myelinated nerve fibres. *J. Physiol.* 108: 315–339.

MOORE, J. W., M. P. BLAUSTEIN, N. C. ANDERSON AND T. NARAHASHI (1967) Basis of tetrodotoxin's selectivity in blockage of squid axons. *J. Gen. Physiol.* 50: 1401–1411.

TASAKI, I. AND T. TAKEUCHI (1941) Der am Ranvierschen Knoten entstenhende Aktionßtrom und seine Bedeutung für die Erregungsleitung. *Pflügers Arch.* 244: 696–711.

Books

AIDLEY, D. J. AND P. R. STANFIELD (1996) *Ion Channels: Molecules in Action*. Cambridge: Cambridge University Press.

HILLE, B. (2001) *Ion Channels of Excitable Membranes*, 3rd Ed. Sunderland, MA: Sinauer Associates.

JOHNSTON, D. AND S. M.-S. WU (1995) *Foundations of Cellular Neurophysiology*. Cambridge, MA: MIT Press.

JUNGE, D. (1992) *Nerve and Muscle Excitation*, 3rd Ed. Sunderland, MA: Sinauer Associates.

MATTHEWS, G. G. (2003) *Cellular Physiology of Nerve and Muscle*, 4th Ed. Malden, MA: Blackwell Publishing.

MOORE, J. W. AND A. E. STUART (2007) *Neurons in Action: Tutorials and Simulations Using NEURON, Version 2*. Sunderland, MA: Sinauer Assoicates.

Chapter 4

Channels and Transporters

Overview

The generation of electrical signals in neurons requires that plasma membranes establish concentration gradients for specific ions and that the membranes undergo rapid and selective changes in their permeability to these ions. The membrane proteins that create and maintain ion gradients are called active transporters, whereas other proteins called ion channels give rise to selective ion permeability changes. As their name implies, ion channels are transmembrane proteins that contain a specialized structure, called a pore, that permits particular ions to cross the neuronal membrane. Some of these channels also contain other structures that are able to sense the electrical potential across the membrane. Such voltage-gated channels open or close in response to the magnitude of the membrane potential, allowing the membrane permeability to be regulated by changes in this potential. Other types of ion channels are gated by extracellular chemical signals such as neurotransmitters, and some by intracellular signals such as second messengers. Still others respond to mechanical stimuli, temperature changes, or a combination of stimuli. Many types of ion channels have now been characterized at both the gene and protein level, resulting in the identification of a large number of ion channel subtypes that are expressed differentially in neuronal and non-neuronal cells. The specific expression pattern of ion channels in each cell type can generate a wide spectrum of electrical characteristics. In contrast to ion channels, active transporters are membrane proteins that produce and maintain ion concentration gradients. The most important of these is the Na^+ pump, which hydrolyzes ATP to regulate the intracellular concentrations of both Na^+ and K^+. Other active transporters produce concentration gradients for the full range of physiologically important ions, including Cl^-, Ca^{2+}, and H^+. From the perspective of electrical signaling, active transporters and ion channels are complementary: transporters create the concentration gradients that help drive ion fluxes through open ion channels, thus generating electrical signals.

Ion Channels Underlying Action Potentials

Although Hodgkin and Huxley had no knowledge of the physical nature of the conductance mechanisms underlying action potentials, they nonetheless proposed that nerve cell membranes have channels that allow ions to pass selectively from one side of the membrane to the other (see Chapter 3). Based on the ionic conductances and currents measured in voltage-clamp experiments, the postulated channels had to have several properties. First, because the ionic currents are quite large, the channels had to be capable of allowing ions to move across the membrane at high rates. Second, because the ionic currents depend on the electrochemical gradient across the membrane, the channels had to make use of these gradients. Third, because Na^+ and K^+ flow across the mem-

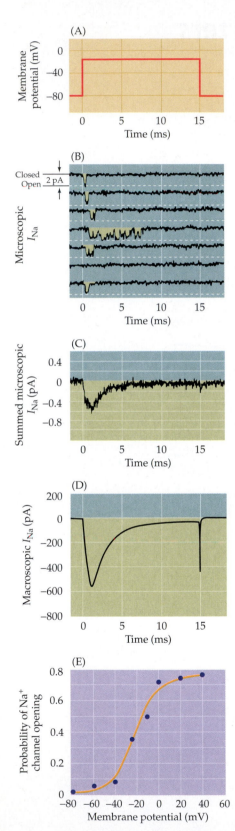

brane independently of each other, different channel types had to be capable of discriminating between Na⁺ and K⁺, allowing only one of these ions to flow across the membrane under the relevant conditions. Finally, given that the conductances are voltage-dependent, the channels had to be able to sense the voltage drop across the membrane, opening only when the voltage reached appropriate levels. While this concept of channels was highly speculative in the 1950s, later experimental work established beyond any doubt that transmembrane proteins called voltage-sensitive ion channels indeed exist and are responsible for all of the ionic conductance phenomena described in Chapter 3.

The first direct evidence for the presence of voltage-sensitive, ion-selective channels in nerve cell membranes came from measurements of the ionic currents flowing through individual ion channels. The voltage-clamp apparatus used by Hodgkin and Huxley could only resolve the *aggregate* current resulting from the flow of ions through many thousands of channels. A technique capable of measuring the currents flowing through single channels was devised in 1976 by Erwin Neher and Bert Sakmann at the Max Planck Institute in Goettingen. This remarkable approach, called patch clamping (Box 4A), revolutionized the study of membrane currents. In particular, the patch clamp method provided the means to test directly Hodgkin and Huxley's proposals about the characteristics of ion channels.

Currents flowing through Na⁺ channels are best examined in experimental circumstances that prevent the flow of current through other types of channels that are present in the membrane (e.g., K⁺ channels). Under such conditions, depolarizing a patch of membrane from a squid giant axon causes tiny inward currents to flow, but only occasionally (Figure 4.1). The size of these currents is minuscule—approximately 1–2 pA (i.e., 10^{-12} ampere), which is orders of magnitude smaller than the Na⁺ currents measured by voltage clamping the entire axon. The currents flowing through single channels are called **microscopic currents** to distinguish them from the **macroscopic currents** flowing through a large number of channels distributed over a much more extensive region of surface membrane. Although microscopic currents are certainly small, a current of 1 pA nonetheless reflects the flow of thousands of ions per millisecond. Thus, as predicted, a single channel can let many ions pass through the membrane in a very short time.

Several observations further proved that the microscopic currents in Figure 4.1B are due to the opening of single, voltage-activated Na⁺ channels. First, the currents are carried by Na⁺; thus, they are directed inward when the membrane potential is more negative than E_{Na}, reverse their polarity at E_{Na}, are outward at more positive potentials, and are reduced in size when the Na⁺ concentration of the external medium is decreased. This behavior exactly parallels that of the macroscopic Na⁺ currents described in Chapter 3. Second, the channels have a

Figure 4.1 Patch clamp measurements of ionic currents flowing through single Na⁺ channels in a squid giant axon. In these experiments, Cs⁺ was applied to the axon to block voltage-gated K⁺ channels. Depolarizing voltage pulses (A) applied to a patch of membrane containing a single Na⁺ channel result in brief currents (B, downward deflections) in the seven successive recordings of membrane current (I_{Na}). (C) The sum of many such current records shows that most channels open in the initial 1–2 ms following depolarization of the membrane, after which the probability of channel openings diminishes because of channel inactivation. (D) A macroscopic current measured from another axon shows the close correlation between the time courses of microscopic and macroscopic Na⁺ currents. (E) The probability of an Na⁺ channel opening depends on the membrane potential, increasing as the membrane is depolarized. (B,C after Bezanilla and Correa, 1995; D after Vandenburg and Bezanilla, 1991; E after Correa and Bezanilla, 1994.)

BOX 4A   The Patch Clamp Method

A wealth of new information about ion channels resulted from the invention of the patch clamp method in the 1970s. This technique is based on a very simple idea. A glass pipette with a very small opening is used to make tight contact with a tiny area, or patch, of neuronal membrane. After the application of a small amount of suction to the back of the pipette, the seal between the pipette and membrane becomes so tight that no ions can flow between the pipette and the membrane. Thus, all the ions that flow when a single ion channel opens must flow into the pipette. The resulting electrical current, though small, can be measured with an ultrasensitive electronic amplifier connected to the pipette. This arrangement is the *cell-attached patch clamp recording method*. As with the conventional voltage clamp method, the patch clamp method allows experimental control of the membrane potential to characterize the voltage dependence of membrane currents.

Minor technical modifications allow other recording configurations. For example, if the membrane patch within the pipette is disrupted by briefly applying strong suction, the interior of the pipette becomes continuous with the cell cytoplasm. This arrangement allows measurements of electrical potentials and currents from the entire cell and is therefore called *whole-cell recording*. The whole-cell configuration also allows diffusional exchange between the pipette and the cytoplasm, producing a convenient way to inject substances into the interior of a "patched" cell.

Two other variants of the patch clamp method originate

from the finding that once a tight seal has formed between the membrane and the glass pipette, small pieces of membrane can be pulled away from the cell without disrupting the seal; this yields a preparation that is free of the complications imposed by the rest of the cell. Simply retracting a pipette that is in the cell-attached configuration causes a small vesicle of membrane to remain attached to the pipette. By exposing the tip of the pipette to air, the vesicle opens to yield a small patch of membrane with its (former) intracellular surface exposed. This arrangement, called the *inside-out patch recording configuration*, allows the measurement of single-channel currents with the added benefit of making it possible to change the medium to which the

intracellular surface of the membrane is exposed. Thus, the inside-out configuration is particularly valuable when studying the influence of intracellular molecules on ion channel function.

Alternatively, if the pipette is retracted while it is in the whole-cell configuration, a membrane patch is produced that has its extracellular surface exposed. This arrangement, called the *outside-out recording configuration*, is optimal for studying how channel activity is influenced by extracellular chemical signals, such as neurotransmitters (see Chapter 5). This range of possible configurations makes the patch clamp method an unusually versatile technique for studies of ion channel function.

Four configurations in patch clamp measurements of ionic currents.

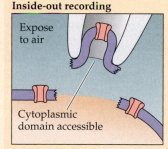
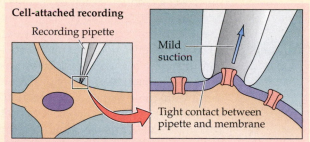
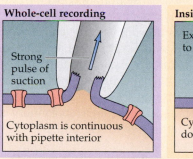
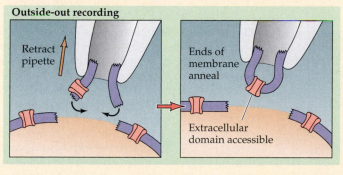

Cell-attached recording — Recording pipette — Mild suction — Tight contact between pipette and membrane

Whole-cell recording — Strong pulse of suction — Cytoplasm is continuous with pipette interior

Inside-out recording — Expose to air — Cytoplasmic domain accessible

Outside-out recording — Retract pipette — Ends of membrane anneal — Extracellular domain accessible

References

HAMILL, O. P., A. MARTY, E. NEHER, B. SAKMANN AND F. J. SIGWORTH (1981) Improved patch-clamp techniques for high-resolution current recording from cells and cell-free membrane patches. *Pflügers Arch.* 391: 85–100.

LEVIS, R. A. AND J. L. RAE (1998) Low-noise patch-clamp techniques. *Meth. Enzym.* 293: 218–266.

SAKMANN, B. AND E. NEHER (1995) *Single-Channel Recording,* 2nd Ed. New York: Plenum Press.

time course of opening, closing, and inactivating that matches the kinetics of macroscopic Na$^+$ currents. This correspondence is difficult to appreciate in the measurement of microscopic currents flowing through a single open channel, because individual channels open and close in a stochastic (random) manner, as can be seen by examining the individual traces in Figure 4.1B. However, repeated depolarization of the membrane potential causes each Na$^+$ channel to open and close many times. When the current responses to a large number of such stimuli are averaged together, the collective response has a time course that looks much like the macroscopic Na$^+$ current (Figure 4.1C). In particular, the channels open mostly at the beginning of a prolonged depolarization, showing that they subsequently inactivate in the way seen for the macroscopic Na$^+$ current (compare Figures 4.1C and D). Third, both the opening and closing of the channels are voltage-dependent; thus, the channels are closed at –80 mV but open when the membrane potential is depolarized. In fact, the probability that any given channel will be open varies with membrane potential (Figure 4.1E), again as predicted from the macroscopic Na$^+$ conductance (see Figure 3.7). Finally, tetrodotoxin, which blocks the macroscopic Na$^+$ current (see Box 4C), also blocks microscopic Na$^+$ currents. Taken together, these results show that the macroscopic Na$^+$ current measured by Hodgkin and Huxley does indeed arise from the aggregate effect of many thousands of microscopic Na$^+$ currents, each representing the opening of a single voltage-sensitive Na$^+$ channel.

Patch clamp experiments have also revealed the properties of the channels responsible for the macroscopic K$^+$ currents associated with action potentials. When the membrane potential is depolarized (Figure 4.2A), microscopic outward currents (Figure 4.2B) can be observed under conditions that block Na$^+$ channels. The microscopic outward currents exhibit all the features expected for currents flowing through action potential-related K$^+$ channels. Thus, the microscopic currents (Figure 4.2C), like their macroscopic counterparts (Figure 4.2D), fail to inactivate during brief depolarizations. Moreover, these single-channel currents are sensitive to ionic manipulations and drugs that affect the macroscopic K$^+$ currents and, like the macroscopic K$^+$ currents, are voltage-dependent (Figure 4.2E). This and other evidence shows that macroscopic K$^+$ currents associated with action potentials arise from the opening of many voltage-sensitive K$^+$ channels.

In summary, patch clamping has allowed direct observation of microscopic ionic currents flowing through single ion channels, confirming that voltage-sensitive Na$^+$ and K$^+$ channels are responsible for the macroscopic conductances and currents that underlie the action potential. Measurements of the behavior of single ion channels has also provided some insight into the molecular attributes of these channels. For example, single-channel studies show that the membrane of the squid axon contains at least two types of channels—one selectively permeable to Na$^+$ and a second selectively permeable to K$^+$. Thus,

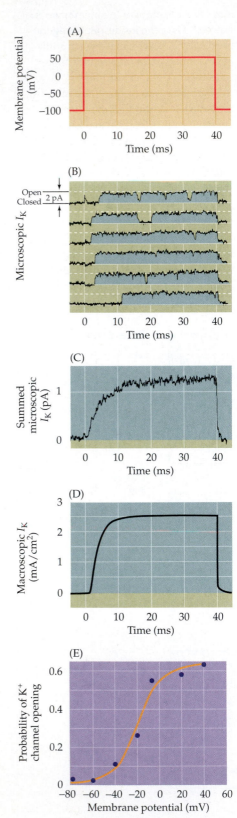

Figure 4.2 Patch clamp measurements of ionic currents flowing through single K$^+$ channels in a squid giant axon. In these experiments, tetrodotoxin was applied to the axon to block voltage-gated Na$^+$ channels. Depolarizing voltage pulses (A) applied to a patch of membrane containing a single K$^+$ channel results in brief currents (B, upward deflections) whenever the channel opens. (C) The sum of such current records shows that most channels open with a delay, but remain open for the duration of the depolarization. (D) A macroscopic current measured from another axon shows the correlation between the time courses of microscopic and macroscopic K$^+$ currents. (E) The probability of a K$^+$ channel opening depends on the membrane potential, increasing as the membrane is depolarized. (B and C after Augustine and Bezanilla, in Hille 2001; D after Augustine and Bezanilla, 1990; E after Perozo et al., 1991.)

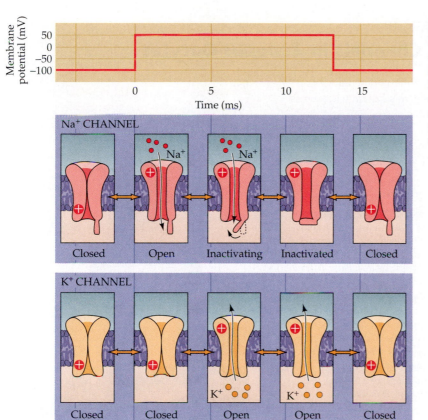

Figure 4.3   Functional states of voltage-gated Na⁺ and K⁺ channels. The gates of both channels are closed when the membrane potential is hyperpolarized. When the potential is depolarized, voltage sensors (indicated by +) allow the channel gates to open—first the Na⁺ channels and then the K⁺ channels. Na⁺ channels also inactivate during prolonged depolarization, whereas many types of K⁺ channels do not.

these channels have **ion selectivity**, meaning that they are able to discriminate between Na⁺ and K⁺. Because their opening is influenced by membrane potential, both channel types are **voltage-gated**. For each channel, depolarization increases the probability of channel opening, whereas hyperpolarization closes them (see Figures 4.1E and 4.2E). Thus, both channel types must have a **voltage sensor** that detects the potential across the membrane (Figure 4.3). However, these channels differ in important respects. In addition to their different ion selectivities, depolarization also inactivates the Na⁺ channel but not the K⁺ channel, causing Na⁺ channels to pass into a non-conducting state. The Na⁺ channel must therefore have an additional molecular mechanism responsible for **inactivation**. And, as expected from the macroscopic behavior of the Na⁺ and K⁺ currents described in Chapter 3, the kinetic properties of the gating of the two channels differs. This information about the physiology of single channels set the stage for subsequent studies of the molecular diversity of ion channels in various cell types, and of their detailed functional characteristics.

The Diversity of Ion Channels

Molecular genetic studies, in conjunction with the patch clamp method and other techniques, have led to many additional advances in understanding ion channels. Genes encoding Na⁺ and K⁺ channels, as well as many other channel types, have been identified and cloned. A surprising fact that has emerged from these molecular studies is the diversity of genes that code for ion channels. Well over 100 ion channel genes have now been discovered, a number that could not have been anticipated from early studies of ion channel function. To understand the functional significance of this multitude of ion channel genes, the channels

BOX 4B Expression of Ion Channels in *Xenopus* Oocytes

Bridging the gap between the sequence of an ion channel gene and understanding channel function is a challenge. To meet this challenge, it is essential to have an experimental system in which the gene product can be expressed efficiently, and in which the function of the resulting channel can be studied with methods such as the patch clamp technique. Ideally, the vehicle for expression should be readily available, have few endogenous channels, and be large enough to permit mRNA and DNA to be microinjected with ease. Oocytes (immature eggs) from the clawed African frog, *Xenopus laevis* (Figure A), fulfill all these demands. These huge cells (approximately 1 mm in diameter; Figure B) are easily harvested from the female *Xenopus*. Work performed in the 1970s by John Gurdon, a developmental biologist, showed that injection of exogenous mRNA into frog oocytes causes them to synthesize foreign protein in prodigious quantities. In the early 1980s, Ricardo Miledi, Eric Barnard, and other neurobiologists demonstrated that *Xenopus* oocytes could express exogenous ion channels, and that physiological methods could be used to study the ionic currents generated by the newly-synthesized channels (Figure C).

As a result of these pioneering studies, heterologous expression experiments have now become a standard way of studying ion channels. The approach has been especially valuable in deciphering the relationship between channel structure and function. In such experiments, defined mutations (often affecting a single nucleotide) are made in the part of the channel gene that encodes a structure of interest; the resulting channel proteins are then expressed in oocytes to assess the functional consequences of the mutation.

The ability to combine molecular and physiological methods in a single cell system has made *Xenopus* oocytes a powerful experimental tool. Indeed, this system has been as valuable to contemporary studies of voltage-gated ion channels as the squid axon was to such studies in the 1950s and 1960s.

References

GUNDERSEN, C. B., R. MILEDI AND I. PARKER (1984) Slowly inactivating potassium channels induced in *Xenopus* oocytes by messenger ribonucleic acid from *Torpedo* brain. *J. Physiol. (Lond.)* 353: 231–248.

GURDON, J. B., C. D. LANE, H. R. WOODLAND AND G. MARBAIX (1971) Use of frog eggs and oocytes for the study of messenger RNA and its translation in living cells. *Nature* 233: 177–182.

STÜHMER, W. (1998) Electrophysiological recordings from *Xenopus* oocytes. *Meth. Enzym.* 293: 280–300.

SUMIKAWA, K., M. HOUGHTON, J. S. EMTAGE, B. M. RICHARDS AND E. A. BARNARD (1981) Active multi-subunit ACh receptor assembled by translation of heterologous mRNA in *Xenopus* oocytes. *Nature* 292: 862–864.

(A)

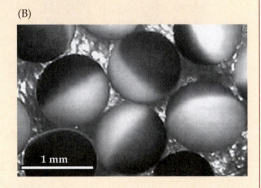

(B)

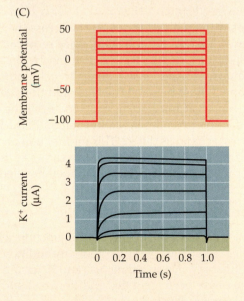

1 mm

(C)

(A) The clawed African frog, *Xenopus laevis*. (B) Several oocytes from *Xenopus* highlighting the dark coloration of the original pole and the lighter coloration of the vegetal pole. (Courtesy of P. Reinhart.) (C) Results of a voltage clamp experiment showing K+ currents produced following injection of K+ channel mRNA into an oocyte. (After Gundersen et al., 1984.)

can be selectively expressed in well-defined experimental systems, such as in cultured cells or frog oocytes (Box 4B), and then studied with patch clamping and other physiological techniques. Channel genes can also be deleted from genetically tractable organisms, such as mice or fruit flies, to determine the role that these channels play in the intact organism. Such studies have found many voltage-gated channels that respond to membrane potential in much the same way as the Na^+ and K^+ channels that underlie the action potential. Other channels, however, are gated by chemical signals that bind to extracellular or intracellular domains on these proteins and are insensitive to membrane voltage. Still others are sensitive to mechanical displacement, or to changes in temperature.

Further magnifying the diversity of ion channels are a number of mechanisms that can produce functionally different types of ion channels from a single gene. Ion channel genes contain a large number of coding regions that can be spliced together in different ways, so a single gene can give rise to channel proteins that can have dramatically different functional properties. RNAs encoding ion channels also can be edited, modifying their base composition after transcription from the gene. For example, editing the RNA of some receptors for the neurotransmitter glutamate (see Chapter 6) changes a single amino acid within the receptor, a change that gives rise to channels that differ in their selectivity for cations and in their conductance. Channel proteins can also undergo posttranslational modifications, such as phosphorylation by protein kinases (see Chapter 7), which can further change their functional characteristics. Thus, although the basic electrical signals of the nervous system are relatively stereotyped, the proteins responsible for generating these signals are remarkably diverse, conferring specialized signaling properties to many of the neuronal cell types that populate the nervous system. These channels also are involved in a broad range of neurological diseases.

Voltage-Gated Ion Channels

Voltage-gated ion channels that are selectively permeable to each of the major physiological ions—Na^+, K^+, Ca^{2+}, and Cl^-—have now been discovered (Figure 4.4A–D). Indeed, many different genes have been discovered for each type of voltage-gated ion channel. An example is the identification of ten human Na^+ channel genes. This finding was unexpected because Na^+ channels from many different cell types have similar functional properties, consistent with their origin from a single gene. It is now clear, however, that all of these Na^+ channel genes (called *SCN genes*) produce proteins that differ in their structure, function, and

Figure 4.4 *Types of voltage-gated ion channels. Examples of voltage-gated channels include those selectively permeable to Na^+ (A), Ca^{2+} (B), K^+ (C), and Cl^- (D). Ligand-gated ion channels include those activated by the extracellular presence of neurotransmitters, such as glutamate (E). Other ligand-gated channels are activated by intracellular second messengers, such as Ca^{2+} (F) or the cyclic nucleotides, cAMP and cGMP (G).*

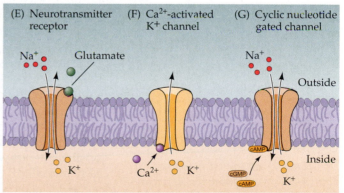

VOLTAGE-GATED CHANNELS

(A) Na^+ channel   (B) Ca^{2+} channel   (C) K^+ channel   (D) Cl^- channel

LIGAND-GATED CHANNELS

(E) Neurotransmitter receptor   (F) Ca^{2+}-activated K^+ channel   (G) Cyclic nucleotide gated channel

distribution in specific tissues. For instance, in addition to the rapidly inactivating Na^+ channels discovered by Hodgkin and Huxley in squid axon, a voltage-sensitive Na^+ channel that does not inactivate has been identified in mammalian neurons. As might be expected, this channel gives rise to action potentials of long duration and is a target of local anesthetics such as benzocaine and lidocaine.

Other electrical responses in neurons entail the activation of voltage-gated Ca^{2+} channels (Figure 4.4B). In some neurons, voltage-gated Ca^{2+} channels give rise to action potentials in much the same way as voltage-sensitive Na^+ channels. In other neurons, Ca^{2+} channels control the shape of action potentials generated primarily by Na^+ conductance changes. More generally, by affecting intracellular Ca^{2+} concentration, the activity of Ca^{2+} channels regulates an enormous range of biochemical signaling processes within cells (see Chapter 7). Perhaps the most important brain process regulated by voltage-sensitive Ca^{2+} channels is the release of neurotransmitters at synapses (see Chapter 5). Given these crucial functions, it is perhaps not surprising that 16 different Ca^{2+} channel genes (called *CACNA genes*) have been identified. Like Na^+ channels, Ca^{2+} channels differ in their activation and inactivation properties, allowing subtle variations in both electrical and chemical signaling processes mediated by Ca^{2+}. As a result, drugs that block voltage-gated Ca^{2+} channels are especially valuable in treating a variety of conditions ranging from heart disease to anxiety disorders.

By far the largest and most diverse class of voltage-gated ion channels are the K^+ channels (Figure 4.4C). Nearly a hundred K^+ channel genes are known, and these fall into several distinct groups that differ substantially in their activation, gating, and inactivation properties. Some take minutes to inactivate, as in the case of squid axon K^+ channels studied by Hodgkin and Huxley (Figure 4.5A); others inactivate within milliseconds, as is typical of most voltage-gated Na^+ channels (Figure 4.5B). These properties influence the duration and rate of action potential firing, with important consequences for axonal conduction and synaptic transmission. Perhaps the most important function of K^+ channels is the role they play in generating the resting membrane potential (see Chapter 2). At least two families of K^+ channels that are open at substantially negative membrane voltage levels contribute to setting the resting membrane potential (Figure 4.5D).

Finally, several types of voltage-gated Cl^- channel have been identified (Figure 4.4D). These channels are present in every type of neuron, where they control excitability, contribute to the resting membrane potential, and help regulate cell volume.

Ligand-Gated Ion Channels

Many types of ion channels respond to chemical signals (ligands) rather than to changes in the membrane potential. The most important **ligand-gated ion channels** in the nervous system are those activated by neurotransmitters (Figure 4.4E). These channels are essential for synaptic transmission and other forms of cell-cell signaling phenomena discussed in Chapters 5–7. Whereas the voltage-gated ion channels underlying the action potential typically allow only one type of ion to permeate, channels activated by extracellular ligands are usually less selective, often allowing two or more types of ions to pass through the channel pore.

Other ligand-gated channels are sensitive to chemical signals arising within the cytoplasm of neurons (see Chapter 7), and can be selective for specific ions such as K^+ or Cl^-, or permeable to all physiological cations. Such channels are distinguished by ligand-binding domains on their *intracellular* surfaces; these domains interact with second messengers such as Ca^{2+}, the cyclic nucleotides cAMP and cGMP, or protons. Examples of channels that respond to intracellular cues include Ca^{2+}-activated K^+ channels (Figure 4.4.F), the cyclic nucleotide-

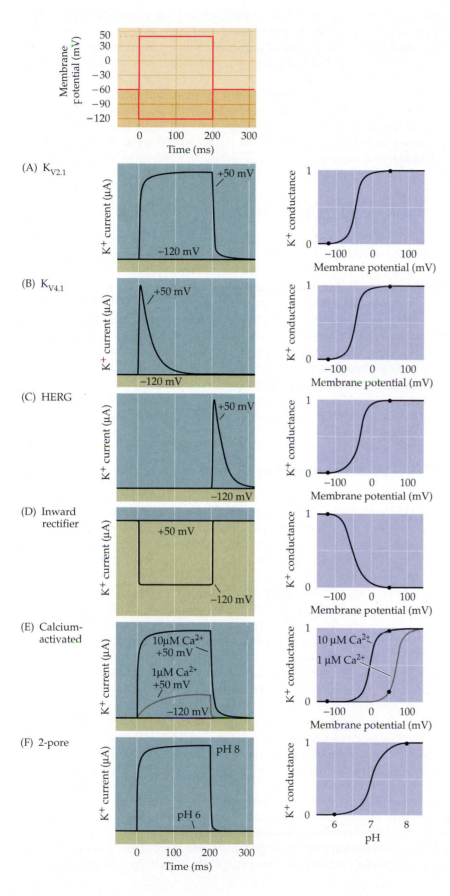

Figure 4.5 Diverse properties of K$^+$ channels. Different types of K$^+$ channels were expressed in *Xenopus* oocytes (see Box 4B), and the voltage clamp method was used to change the membrane potential (top) and measure the resulting currents flowing through each type of channel. These K$^+$ channels vary markedly in their gating properties, as evident in their currents (left) and conductances (right). (A) K$_{V2.1}$ channels show little inactivation and are closely related to the delayed rectifier K$^+$ channels involved in action potential repolarization. (B) K$_{V4.1}$ channels inactivate during a depolarization. (C) HERG channels inactivate so rapidly that current flows only when inactivation is rapidly removed at the end of a depolarization. (D) Inward rectifying K$^+$ channels allow more K$^+$ current to flow at hyperpolarized potentials than at depolarized potentials. (E) Ca^{2+}-activated K$^+$ channels open in response to intracellular Ca^{2+} ions and, in some cases, membrane depolarization. (F) K$^+$ channels with two pores usually respond to chemical signals, such as pH, rather than changes in membrane potential.

gated cation channel (Figure 4.4G), or acid-sensing ion channels (ASICs). The main function of these channels is to convert intracellular chemical signals into electrical information. This process is particularly important in sensory transduction, where channels gated by cyclic nucleotides convert odors and light, for example, into electrical signals.

Although many ligand-gated ion channels are located in the cell surface membrane, others are found in the membranes of intracellular organelles such as mitochondria or the endoplasmic reticulum. Some of these latter channels are selectively permeable to Ca^{2+} and regulate the release of Ca^{2+} from the lumen of the endoplasmic reticulum into the cytoplasm, where this second messenger can then trigger a spectrum of cellular responses, as described in Chapter 7.

Stretch- and Heat-Activated Channels

Still other ion channels respond to heat or membrane deformation. Heat-activated ion channels, such as some members of the transient receptor potential (TRP) gene family, contribute to the sensations of pain and temperature and help mediate inflammation (see Chapter 10). These channels are often specialized to detect specific temperature ranges, and some are even activated by cold. Other ion channels respond to mechanical distortion of the plasma membrane and are the basis of stretch receptors and neuromuscular stretch reflexes (see Chapters 9, 16, and 17). A specialized form of these channels apparently enables hearing by allowing auditory hair cells to respond to sound waves (see Chapter 13).

In summary, this tremendous variety of ion channels allows neurons to generate electrical signals in response to changes in membrane potential, synaptic input, intracellular second messengers, light, odors, heat, sound, touch, and many other stimuli.

The Molecular Structure of Ion Channels

Understanding the physical structure of ion channels is obviously key to sorting out how they actually work. Until recently, most information about channel structure was derived indirectly, from studies of the amino acid composition and physiological properties of these proteins. For example, a great deal has been learned about the functions of particular amino acids within the proteins by using **mutagenesis** to explore the expression of such channels in *Xenopus* oocytes (see Box 4B). Such studies have revealed a general transmembrane architecture common to all the major ion channel families. These molecules are all integral membrane proteins that span the plasma membrane repeatedly. Na^+ and Ca^{2+} channel proteins consist of repeating motifs of 6 membrane-spanning regions that are repeated 4 times, for a total of 24 transmembrane regions (Figure 4.6A,B). Na^+ or Ca^{2+} channels can be produced by just one of these proteins, although other accessory proteins, called β subunits, can regulate the function of these channels.

K^+ channel proteins typically span the membrane 6 times (Figure 4.6C), though there are some K^+ channels (including one bacterial channel and several mammalian channels) that span the membrane only twice (Figure 4.6D), and others that span the membrane 4 times (Figure 4.6F) or 7 times (Figure 4.6E). Each of these proteins serves as a K^+ channel subunit, with 4 of these subunits typically aggregating to form a single functional ion channel.

Imaginative mutagenesis experiments have provided information about how these proteins function. Two membrane-spanning domains of all ion channels appear to form a central **pore** through which ions can diffuse, and one of these domains contains a protein loop that confers ion selectivity, allowing only cer-

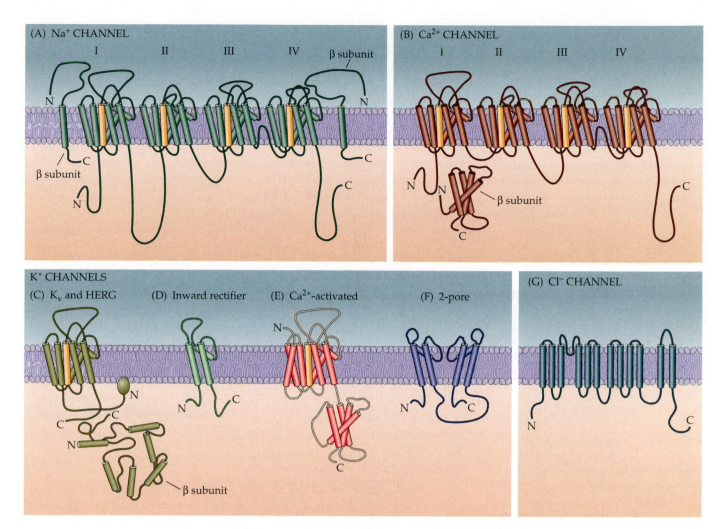

(A) Na$^+$ CHANNEL

I   II   III   IV   β subunit

N   N

C   C

β subunit

N   C

N

(B) Ca^{2+} CHANNEL

I   II   III   IV

N   N

β subunit

C   C

K$^+$ CHANNELS

(C) K$_v$ and HERG

N   C   N

C   β subunit

(D) Inward rectifier

N   C

(E) Ca^{2+}-activated

N

C

(F) 2-pore

N   C

(G) Cl$^-$ CHANNEL

N   C

Figure 4.6 *Topology of the principal subunits of voltage-gated Na$^+$, Ca^{2+}, K$^+$, and Cl$^-$ channels. Repeating motifs of Na$^+$ (A) and Ca^{2+} (B) channels are labeled I, II, III, and IV; (C–F) K$^+$ channels are more diverse. In all cases, four subunits combine to form a functional channel. (G) Chloride channels are structurally distinct from all other voltage-gated channels.*

tain ions to diffuse through the channel pore (Figure 4.7). As might be expected, the amino acid composition of the pore loop differs according to which ion the channel conducts. These distinct structural features of channel proteins also provide unique binding sites for drugs and for various neurotoxins known to block specific subclasses of ion channels (Box 4C). Furthermore, many voltage-gated ion channels display a distinct type of transmembrane helix containing a number of positively charged amino acids along one face of the helix (the yellow structures in Figures 4.6 and 4.7). This structure is evidently the voltage sensor that detects changes in the electrical potential across the membrane. Membrane depolarization allows the channel pore to open by influencing the charged amino acids and causing the helix to change its position, though the nature of this position change is not yet clear. Other mutagenesis experiments have demonstrated that one end of certain K$^+$ channels plays a key role in channel inactivation. This intracellular structure (labeled "N" in Figure 4.6C) can plug the channel pore during prolonged depolarization.

More recently, direct information about the structural underpinnings of ion channel function has come from **X-ray crystallography** studies of K$^+$ channels. The first information about the structure of the K$^+$ channel pore came from studies of a bacterial K$^+$ channel. This channel was chosen for analysis because the large quantity of channel protein needed for crystallography could be obtained by growing large numbers of bacteria expressing this molecule. The

BOX 4C   Toxins That Poison Ion Channels

Given the importance of Na^+ and K^+ channels for neuronal excitation, it is not surprising that a number of organisms have evolved channel-specific toxins as mechanisms for self-defense or for capturing prey. A rich collection of natural toxins selectively target the ion channels of neurons and other cells. These toxins are valuable not only for survival, but for studying the function of cellular ion channels. The best-known channel toxin is *tetrodotoxin*, which is produced by certain puffer fish and other animals. Tetrodotoxin produces a potent and specific obstruction of the Na^+ channels responsible for action potential generation, thereby paralyzing the animals unfortunate enough to ingest it. *Saxitoxin*, a chemical homologue of tetrodotoxin produced by dinoflagellates, has a similar action on Na^+ channels. The potentially lethal effects of eating shellfish that have ingested these "red tide" dinoflagellates are due to the potent neuronal actions of saxitoxin.

Scorpions paralyze their prey by injecting a potent mix of peptide toxins that also affect ion channels. Among these are the α-*toxins*, which slow the inactivation of Na^+ channels (Figure A1); exposure of neurons to these toxins prolongs the action potential (Figure A2), thereby scrambling informa-

tion flow within the nervous system of the soon-to-be-devoured victim. Other peptides in scorpion venom, called β-*toxins*, shift the voltage dependence of Na^+ channel activation (Figure B). These toxins cause Na^+ channels to open at potentials much more negative than normal, disrupting action potential generation. Some alkaloid toxins combine these actions, both removing inactivation *and* shifting activation of Na^+ channels. One such toxin is *batrachotoxin*, produced by a species of frog; some tribes of South American Indians use this poison on their arrow tips. A number of plants produce similar toxins, including *aconitine*, from buttercups; *veratridine*, from lilies; and a number of insecticidal toxins produced by plants such as chrysanthemums and rhododendrons.

Potassium channels have also been targeted by toxin-producing organisms. Peptide toxins affecting K^+ channels include *dendrotoxin*, from wasps; *apamin*,

from bees; and *charybdotoxin*, yet another toxin produced by scorpions. All of these toxins block K^+ channels as their primary action; no toxin is known to affect the activation or inactivation of these channels, although such agents may simply be awaiting discovery.

References

CAHALAN, M. (1975) Modification of sodium channel gating in frog myelinated nerve fibers by *Centruroides sculpturatus* scorpion venom. *J. Physiol. (Lond.)* 244: 511–534.

CATTERALL, W. A., S. CESTELE, V. YAROV-YAROVOY, F. H. YU, K. KONOKI AND T. SCHEUER (2007) Voltage-gated ion channels and gating modifier toxins. *Toxicon* 49: 124–141.

NARAHASHI, T. (2000) Neuroreceptors and ion channels as the basis for drug action: Present and future. *J. Pharmacol. Exptl. Therapeutics* 294: 1–26.

SCHMIDT, O. AND H. SCHMIDT (1972) Influence of calcium ions on the ionic currents of nodes of Ranvier treated with scorpion venom. *Pflügers Arch.* 333: 51–61.

(A) Effects of toxin treatment on frog axons. (1) α-Toxin from the scorpion *Leiurus quinquestriatus* prolongs Na^+ currents recorded with the voltage clamp method. (2) As a result of the increased Na^+ current, α-toxin greatly prolongs the duration of the axonal action potential. Note the change in timescale after treating with toxin. (B) Treatment of a frog axon with β-toxin from another scorpion, *Centruroides sculpturatus*, shifts the activation of Na^+ channels, so that Na^+ conductance begins to increase at potentials much more negative than usual. (A after Schmidt and Schmidt, 1972; B after Cahalan, 1975.)

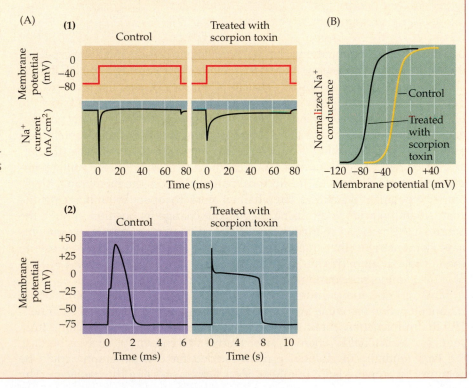

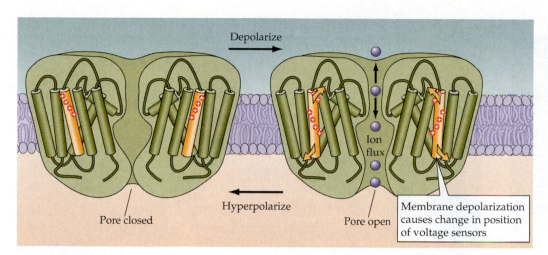

Figure 4.7 A charged voltage sensor permits voltage-dependent gating of ion channels. The process of voltage activation involves the movement of a positively-charged voltage sensor within the membrane. This movement causes a change in the conformation of the channel pore, enabling the channel to conduct ions.

results of such studies showed that the channel is formed by subunits that each cross the plasma membrane twice; between these two membrane-spanning structures is a loop that inserts into the plasma membrane (Figure 4.8A). Four of these subunits are assembled together to form a channel (Figure 4.8B). In the center of the assembled channel is a narrow opening through the protein that allows K^+ to flow across the membrane. This opening is the channel pore and is formed by the protein loop, as well as by the membrane-spanning domains. The structure of the pore is well suited for conducting K^+ ions (Figure 4.8C). The narrowest part is near the outside mouth of the channel and is so constricted that only a nonhydrated K^+ ion can fit through the bottleneck. Larger cations, such as Cs^+, cannot traverse this region of the pore, and smaller cations such as Na^+ cannot enter the pore because the "walls" of the pore are too far apart to stabilize a dehydrated Na^+ ion. This part of the channel complex is responsible for the selective permeability to K^+ and is therefore called the **selectivity filter**. The sequence of amino acids making up part of this selectivity filter is often referred to as the K^+ channel "signature sequence," and it differs from the sequences found in channels that are permeable to other cations.

Deeper within the channel is a water-filled cavity that connects to the interior of the cell. This cavity evidently collects K^+ from the cytoplasm and, utilizing negative charges from the protein, allows K^+ ions to become dehydrated so they can enter the selectivity filter. These "naked" ions are then able to move through four K^+ binding sites within the selectivity filter to eventually reach the extracellular space (recall that the normal concentration gradient drives K^+ out of cells). The presence of multiple (up to four) K^+ ions within the selectivity filter causes electrostatic repulsion between the ions that helps to speed their transit through the selectivity filter, thereby permitting rapid ion flux through the channel.

Recent crystallographic studies have also determined the structure of a mammalian voltage-gated K^+ channel. These studies have yielded many insights into how voltage-dependent gating of ion channels occurs. As was the case for the bacterial K^+ channels described above, four subunits assemble to form the voltage-gated K^+ channel (Figure 4.9A). While the pore region of this channel is very similar to that of the bacterial K^+ channels (Figure 4.9B), the voltage-gated channel has additional structures on its cytoplasmic side, such as a β subunit and a T1 domain that links the β subunit to the channel. Holes between the T1 domain and the part of the channel embedded in the membrane allow K^+ to enter the channel and permit insertion of parts of the channel involved in inac-

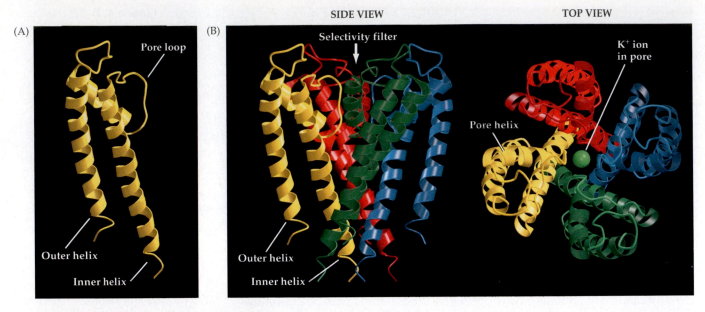

(A) Pore loop / Outer helix / Inner helix

SIDE VIEW   TOP VIEW

(B) Selectivity filter / Outer helix / Inner helix / Pore helix / K⁺ ion in pore

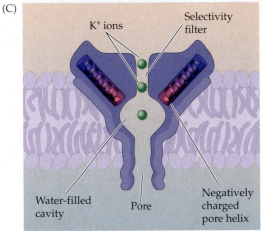

(C) K⁺ ions / Selectivity filter / Water-filled cavity / Pore / Negatively charged pore helix

Figure 4.8   *Structure of a simple bacterial K⁺ channel determined by crystallography. (A) Structure of one subunit of the channel, which consists of two membrane-spanning domains and a pore loop that inserts into the membrane. (B) Three-dimensional arrangement of four subunits (each in a different color) to form a K⁺ channel. The top view illustrates a K⁺ ion (green) within the channel pore. (C) The permeation pathway of the K⁺ channel consists of a large aqueous cavity connected to a narrow selectivity filter. Helical domains of the channel point negative charges (red) toward this cavity, allowing K⁺ ions (green) to become dehydrated and then move through the selectivity filter. (From Doyle et al., 1998.)*

tivation (Figure 4.9C). Most importantly, this channel has voltage sensors that operate as separate domains embedded in the plasma membrane (Figure 4.9B). These sensors have positively charged amino acids that allow them to respond to changes in membrane potential. These sensors apparently move in response to membrane depolarization, thereby exerting force on a helical structure, termed the S4–S5 linker, that either pulls the channel pore open or pushes it closed (Figure 4.9D). Thus, voltage-dependent gating of ion channels can now be understood in structural terms. The precise movements of the voltage sensor that occur during membrane depolarization are not yet clear. Based on the paddlelike structure of the sensor, it has been proposed that depolarization causes the sensor to flip from one side of the membrane to the other (Figure 4.9E).

In summary, then, ion channels are integral membrane proteins with characteristic features that allow them to assemble into multimolecular aggregates. Collectively, these structures allow channels to conduct ions, to sense the transmembrane potential, to inactivate, and to bind to various neurotoxins. A combination of physiological, molecular biological, and crystallographic studies has begun to provide a detailed physical picture of K⁺ channels. This work has provided considerable insight into how ions are conducted from one side of the plasma membrane to the other, how a channel can be selectively permeable to a single type of ion, how channel proteins are able to sense changes in membrane voltage, and how the channel pore is gated. It is likely that the other types of ion channels will be similar in their functional architecture. This sort of work also has illuminated how mutations in ion channel genes can lead to a variety of neurological disorders (Box 4D).

Active Transporters Create and Maintain Ion Gradients

Up to this point, the discussion of the molecular basis of electrical signaling has taken for granted the remarkable fact that nerve cells maintain ion concentration gradients across their surface membranes. None

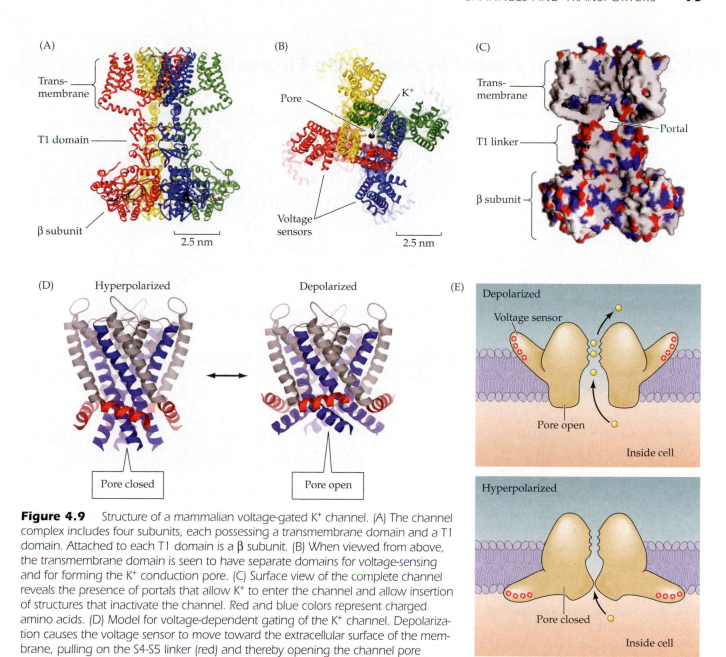

Figure 4.9 Structure of a mammalian voltage-gated K+ channel. (A) The channel complex includes four subunits, each possessing a transmembrane domain and a T1 domain. Attached to each T1 domain is a β subunit. (B) When viewed from above, the transmembrane domain is seen to have separate domains for voltage-sensing and for forming the K+ conduction pore. (C) Surface view of the complete channel reveals the presence of portals that allow K+ to enter the channel and allow insertion of structures that inactivate the channel. Red and blue colors represent charged amino acids. (D) Model for voltage-dependent gating of the K+ channel. Depolarization causes the voltage sensor to move toward the extracellular surface of the membrane, pulling on the S4-S5 linker (red) and thereby opening the channel pore (blue). Conversely, hyperpolarization causes this linker to push down and shut the channel pore. (E) Paddle model for movement of the voltage sensor. Depolarization causes the paddle-like voltage sensor domain to move toward the extracellular surface of the membrane, while hyperpolarization causes it to move toward the intracellular surface of the membrane. (A–C from Long et al., 2005a; D from Long et al., 2005b; E after Lee, 2006.)

of the ions of physiological importance (Na+, K+, Cl−, and Ca2+) are in electrochemical equilibrium. Because channels produce electrical effects by allowing one or more of these ions to diffuse down their electrochemical gradients, there would be a gradual dissipation of these concentration gradients unless nerve cells could restore ions displaced during the current flow that occurs as a result of both neural signaling and the continual ionic leakage that occurs at rest. The

BOX 4D   Diseases Caused by Altered Ion Channels

Several genetic diseases, collectively called *channelopathies*, result from small but critical alterations in ion channel genes (Figure A). The best characterized of these diseases are those that affect skeletal muscle cells. In these disorders, alterations in ion channel proteins produce either myotonia (muscle stiffness due to excessive electrical excitability) or paralysis (due to insufficient muscle excitability). Other disorders arise from ion channel defects in heart, kidney, and the inner ear.

Channelopathies associated with ion channels localized in brain are much more difficult to study. Nonetheless, voltage-gated Ca^{2+} channels have recently been implicated in a range of neurological diseases. These include episodic ataxia, spinocerebellar degeneration, night blindness, and migraine headaches. *Familial hemiplegic migraine* (FHM) is characterized by migraine attacks that typically last one to three days. During such episodes, patients experience severe headaches and vomiting. Several mutations in a human Ca^{2+} channel have been identified in families with FHM, each having different clinical symptoms. For example, a mutation in the pore-forming region of the channel produces hemiplegic migraine with progressive cerebellar ataxia, whereas other mutations cause only the usual FHM symptoms. How these altered Ca^{2+} channel properties lead to migraine attacks is not known.

Episodic ataxia type 2 (EA2) is a neurological disorder in which affected individuals suffer recurrent attacks of abnormal limb movements and severe ataxia. These problems are sometimes accompanied by vertigo, nausea, and headache. Usually, attacks are precipitated by emotional stress, exercise, or alcohol and last for a few hours. The mutations in EA2 cause Ca^{2+} channels to be truncated at various sites, which may cause the clinical manifestations of the disease by preventing the normal assem-

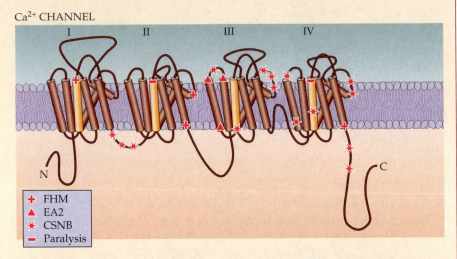

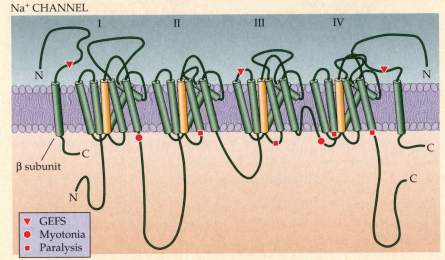

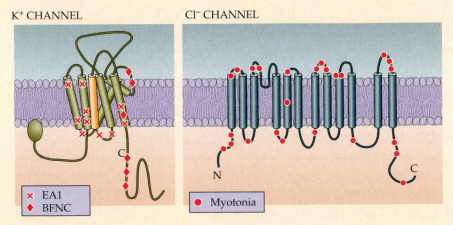

(A) Genetic mutations in Ca^{2+} channels, Na^+ channels, K^+ channels, and Cl^- channels that result in diseases. Red symbols indicate the sites and specific pathologies of these mutations. (After Lehmann-Horn and Jurkat-Kott, 1999.)

bly of Ca^{2+} channels in the membrane.

X-linked *congenital stationary night blindness* (CSNB) is a recessive retinal disorder that causes night blindness, decreased visual acuity, myopia, nystagmus, and strabismus. Complete CSNB causes retinal rod photoreceptors to be nonfunctional. Incomplete CSNB causes subnormal (but measurable) functioning of both rod and cone photoreceptors. Like EA2, the incomplete type of CSNB is caused by mutations producing truncated Ca^{2+} channels. Abnormal retinal function may arise from decreased Ca^{2+} currents and neurotransmitter release from photoreceptors (see Chapter 11).

A defect in brain Na^+ channels causes *generalized epilepsy with febrile seizures* (GEFS) that begins in infancy and usually continues through early puberty. This defect has been mapped to two mutations: one on chromosome 2 that encodes an α subunit for a voltage-gated Na^+ channel, and the other on chromosome 19 that encodes a Na^+ channel β subunit. These mutations cause a slowing of Na^+ channel inactivation (Figure B), which may explain the neuronal hyperexcitability underlying GEFS.

Another type of seizure, *benign familial neonatal convulsion* (BFNC), is due to K^+ channel mutations. This disease is characterized by frequent brief seizures commencing within the first week of life

and disappearing spontaneously within a few months. The mutation has been mapped to at least two voltage-gated K^+ channel genes. A reduction in K^+ current flow through the mutated channels probably accounts for the hyperexcitability associated with this defect. A related disease, episodic ataxia type 1 (EA1), has been linked to a defect in another type of voltage-gated K^+ channel. EA1 is characterized by brief episodes of ataxia. Mutant channels inhibit the function of other, non-mutant K^+ channels and may produce clinical symptoms by impairing action potential repolarization. Mutations in the K^+ channels of cardiac mus-

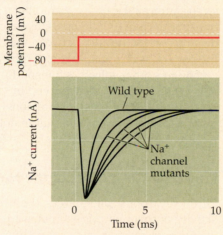

(B) Mutations in Na^+ channels slow the rate of inactivation of Na^+ currents. (After Barchi, 1995.)

cle are responsible for the irregular heartbeat of patients with long Q-T syndrome. Numerous genetic disorders affect the voltage gated channels of skeletal muscle and are responsible for a host of muscle diseases that either cause muscle weakness (*paralysis*) or muscle contraction (*myotonia*).

References

ASHCROFT, F. M. (2000) *Ion Channels and Disease.* Boston: Academic Press.

BARCHI, R. L. (1995) Molecular pathology of the skeletal muscle sodium channel. *Annu. Rev. Physiol.* 57: 355–385.

BERKOVIC, S. F. AND I. E. SCHEFFER (1997) Epilepsies with single gene inheritance. *Brain Develop.* 19: 13–28.

COOPER, E. C. AND L. Y. JAN (1999) Ion channel genes and human neurological disease: Recent progress, prospects, and challenges. *Proc. Natl. Acad. Sci. USA* 96: 4759–4766.

DAVIES, N. P. AND M. G. HANNA (1999) Neurological channelopathies: Diagnosis and therapy in the new millennium. *Ann. Med.* 31: 406–420.

JEN, J. (1999) Calcium channelopathies in the central nervous system. *Curr. Opin. Neurobiol.* 9: 274–280.

KHOSRAVANI, H. AND G. W. ZAMPONI (2006) Voltage-gated calcium channels and idiopathic generalized epilepsies. *Physiol. Rev.* 86: 941–966.

LEHMANN-HORN, F. AND K. JURKAT-ROTT (1999) Voltage-gated ion channels and hereditary disease. *Physiol. Rev.* 79: 1317–1372.

work of generating and maintaining ionic concentration gradients for particular ions is carried out by a group of plasma membrane proteins known as **active transporters**.

Active transporters carry out this task by forming complexes with the ions that they are translocating. The process of ion binding and unbinding for transport typically requires several milliseconds. As a result, ion translocation by active transporters is much slower than ion movement through channels (recall that ion channels can conduct thousands of ions across a membrane each *millisecond*). In short, active transporters gradually store energy in the form of ion concentration gradients, whereas the opening of ion channels rapidly dissipates this stored energy during relatively brief electrical signaling events.

Several types of active transporter have now been identified. Although the specific jobs of these transporters differ, all must translocate ions against their electrochemical gradients. Moving ions uphill requires the consumption of energy, and neuronal transporters fall into two classes based on their energy sources. Some transporters acquire energy directly from the hydrolysis of ATP and are called **ATPase pumps** (Figure 4.10, left panel). The most prominent example of an ATPase pump is the **Na^+ pump** (or, more properly, the Na^+/K^+ ATPase pump), which is responsible for maintaining transmembrane concentration gradients for both Na^+ and K^+ (Figure 4.10A). Another is the Ca^{2+} pump, which provides one of the main mechanisms for removing Ca^{2+} from cells (Figure 4.10B).

The second class of active transporter does not use ATP directly, but depends instead on the electrochemical gradients of other ions as an energy source. This type of transporter carries one or more ions *up* its electrochemical gradient while simultaneously taking another ion (most often Na^+) *down* its gradient. Because at least two species of ions are involved in such transactions, these transporters are usually called **ion exchangers** (Figure 4.10, right). An example of such a transporter is the Na^+/Ca^{2+} exchanger, which shares with the Ca^{2+} pump the important job of keeping intracellular Ca^{2+} concentrations low (Figure 4.10C). Two other exchangers in this category regulate intracellular Cl^- concentration by translocating Cl^- along with extracellular Na^+ and/or K^+. These exchangers are the $Na^+/K^+/Cl^-$ co-transporter, which transports Cl^- along with Na^+ and K^+ into cells (Figure 4.10D), and the Na^+/Cl^- co-transporter, which swaps intracellular Cl^- for extracellular K^+ (Figure 4.10E). Because these two co-transporters move Cl^- in opposite directions, the net intracellular–concentration depends on the balance between their activities. Other ion exchangers, such as the Na^+/H^+ exchanger (Figure 4.10F), also regulate intracellular pH. Other ion exchangers are involved in transporting neurotransmitters into synaptic terminals (Figure 4.10G), as described in Chapter 6. Although the electrochemical gradient of Na^+ (or other counter ions) is the proximate source of energy for ion exchangers, these gradients ultimately depend on the hydrolysis of ATP by ATPase pumps, such as the Na^+/K^+ ATPase pump.

Functional Properties of the Na^+/K^+ Pump

Of these various transporters, the best understood is the Na^+/K^+ pump. The activity of this pump is estimated to account for 20–40 percent of the brain's energy consumption, indicating its crucial importance. The Na^+ pump was discovered in neurons in the 1950s, when Richard Keynes at Cambridge University used radioactive Na^+ to demonstrate the energy-dependent efflux of Na^+ from squid giant axons. Keynes and his collaborators found that this efflux ceased when the supply of ATP in the axon was interrupted by treatment with metabolic poisons (Figure 4.11A, point 4). (Other conditions that lower intracellular

Figure 4.10 Examples of ion transporters found in cell membranes. (A,B) Some transporters are powered by the hydrolysis of ATP (ATPase pumps), whereas others (C–G) use the electrochemical gradients of co-transported ions as a source of energy (ion exchangers).

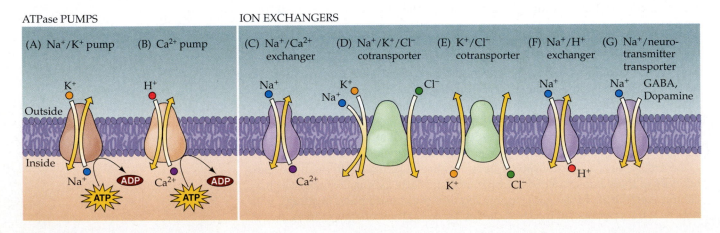

Figure 4.11   Ion movements due to the Na⁺/K⁺ pump. (A) Measurement of radioactive Na⁺ efflux from a squid giant axon. This efflux depends on external K⁺ and intracellular ATP. (B) A model for the movement of ions by the Na⁺/K⁺ pump. Uphill movements of Na⁺ and K⁺ are driven by ATP, which phosphorylates the pump. These fluxes are asymmetrical, with three Na⁺ carried out for every two K⁺ brought in. (A after Hodgkin and Keynes, 1955; B after Lingrel et al., 1994.)

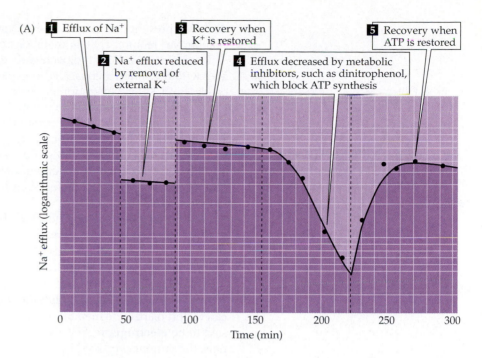

(A)

1. Efflux of Na⁺

2. Na⁺ efflux reduced by removal of external K⁺

3. Recovery when K⁺ is restored

4. Efflux decreased by metabolic inhibitors, such as dinitrophenol, which block ATP synthesis

5. Recovery when ATP is restored

Na⁺ efflux (logarithmic scale)

Time (min)

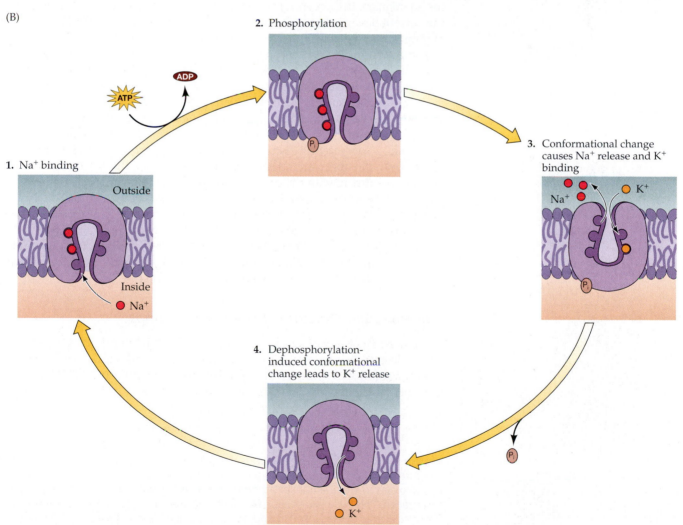

(B)

1. Na⁺ binding

Outside

Inside

● Na⁺

2. Phosphorylation

ATP

ADP

Pᵢ

3. Conformational change causes Na⁺ release and K⁺ binding

Na⁺

K⁺

Pᵢ

4. Dephosphorylation-induced conformational change leads to K⁺ release

K⁺

Pᵢ

ATP also prevent Na+ efflux.) These experiments showed that removing intracellular Na+ requires cellular metabolism.

Further studies with radioactive K+ demonstrated that Na+ efflux is associated with the simultaneous, ATP-dependent influx of K+. These opposing fluxes of Na+ and K+ are operationally inseparable; removal of external K+ greatly reduces Na+ efflux (Figure 4.11, point 2), and vice versa. These energy-dependent movements of Na+ and K+ implicated an ATP-hydrolyzing Na+/K+ pump in the generation of the transmembrane gradients of both Na+ and K+. The exact mechanism responsible for these fluxes of Na+ and K+ is still not entirely clear, but the pump is thought to alternately shuttle these ions across the membranes in a cycle fueled by the transfer of a phosphate group from ATP to the pump protein (Figure 4.11B).

Additional quantitative studies of the movements of Na+ and K+ indicate that the two ions are not pumped at identical rates. The K+ influx is only about two-thirds the Na+ efflux. Thus, the pump apparently transports 2 K+ into the cell for every 3 Na+ that are removed (see Figure 4.11B). This stoichiometry causes a net loss of one positively charged ion from inside the cell during each round of pumping, meaning that the pump generates an electrical current that can hyperpolarize the membrane potential. For this reason, the Na+/K+ pump is said to be **electrogenic**. Because pumps act much more slowly than ion channels, the current produced by the Na+/K+ pump is quite small. For example, in the squid axon, the net current generated by the pump is less than 1 percent of the current flowing through voltage-gated Na+ channels and affects the resting membrane potential by only a millivolt or less.

Although the electrical current generated by the activity of the Na+/K+ pump is small, under special circumstances the pump can significantly influence the membrane potential. For instance, prolonged stimulation of small unmyelinated axons produces a substantial hyperpolarization (Figure 4.12). During the period of stimulation, Na+ enters through voltage-gated channels and accumulates within the axons. As the pump removes this extra Na+, the resulting current generates a long-lasting hyperpolarization. Support for this interpretation comes from the observation that conditions that block the Na+/K+ pump—for example, treatment with ouabain, a plant glycoside that specifically inhibits the pump—prevent the hyperpolarization. The electrical contribution of the Na+/K+ pump is particularly significant in these small-diameter axons because their large surface-to-volume ratio causes intracellular Na+ concentration to rise to higher levels than it would in other cells. Nonetheless, it is important to emphasize that, in most circumstances, the Na+/K+ pump plays no part in generating the action potential and has very little *direct* effect on the resting potential.

The Molecular Structure of the ATPase Pumps

The preceding observations imply that the Na+ and K+ pump must exhibit several molecular properties: (1) it must bind both Na+ and K+; (2) it must possess sites that bind ATP and receive a phosphate group from this ATP; and (3) it must bind ouabain, the toxin that blocks this pump (Figure 4.13A).

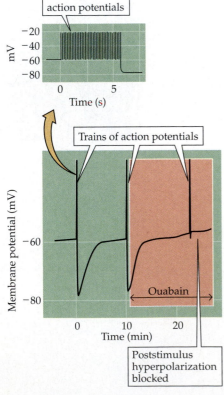

Figure 4.12 The electrogenic transport of ions by the Na+/K+ pump can influence membrane potential. Measurements of the membrane potential of a small unmyelinated axon show that a train of action potentials is followed by a long-lasting hyperpolarization. This hyperpolarization is blocked by ouabain, indicating that it results from the activity of the Na+/K+ pump. (After Rang and Ritchie, 1968.)

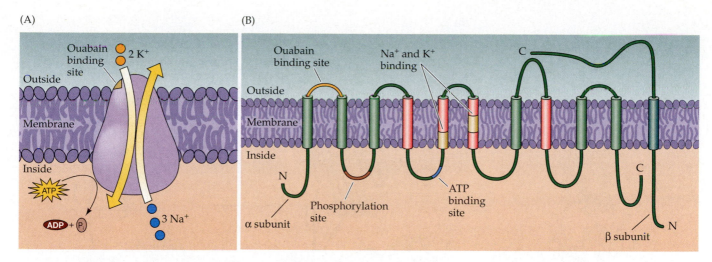

(A)

Ouabain binding site

2 K⁺

Outside

Membrane

Inside

ATP

ADP + Pᵢ

3 Na⁺

(B)

Ouabain binding site

Na⁺ and K⁺ binding

C

Outside

Membrane

Inside

N

α subunit

Phosphorylation site

ATP binding site

C

N

β subunit

Figure 4.13   *Molecular organization of the Na⁺/K⁺ pump. (A) General features of the pump. (B) The molecule spans the membrane 10 times. Amino acid residues thought to be important for binding of ATP, K⁺, and ouabain are highlighted. (After Lingrel et al., 1994.)*

A variety of studies have now identified regions of the Na⁺/K⁺ pump that account for these properties. This pump is a large, integral membrane protein made up of at least two subunits, called α and β. The primary sequence shows that the α subunit spans the membrane ten times, with most of the molecule found on the cytoplasmic side, whereas the β subunit spans the membrane only once and is predominantly extracellular (Figure 4.13B). Although the detailed structure of the Na⁺/K⁺ pump is not yet known, analysis of its amino acid sequence has identified structures involved in some of the pump's functions. One intracellular domain of the protein is required for ATP binding and hydrolysis, and the amino acid phosphorylated by ATP has been identified. Another extracellular domain may represent the binding site for ouabain. However, the sites involved in the most critical function of the pump—the movement of Na⁺ and K⁺—have not yet been defined. Nonetheless, altering certain membrane-spanning domains (red in Figure 4.13B) impairs ion translocation; moreover, kinetic studies indicate that both ions bind to the pump at the same site. Because these ions move across the membrane, it is likely that this site traverses the plasma membrane; it is also likely that the site has a negative charge, since both Na⁺ and K⁺ are positively charged. The observation that removing negatively charged residues in a membrane-spanning domain of the protein (pale yellow in Figure 4.13B) greatly reduces Na⁺ and K⁺ binding provides at least a hint about the ion-translocating domain of the transporter molecule.

The relationship between transporter structure and function has been elucidated in much greater detail for the case of a Ca²⁺ pump that is closely related to the Na⁺/K⁺ pump. This pump uses ATP hydrolysis to power the translocation of Ca²⁺ from cytoplasm across the membrane of the sarcoplasmic reticulum, an intracellular Ca²⁺ storage organelle in muscle that is analogous to the endoplasmic reticulum used for storage and release of Ca²⁺ within neurons and glia (see Chapter 7). The structure of this ATPase pump has been determined by the same X-ray crystallography techniques that elucidated the atomic structure of K⁺ channels and many other proteins. Such studies reveal that the Ca²⁺ pump, like the Na⁺/K⁺ pump, is a very large protein than spans the membrane ten times and consists of several domains (Figure 4.14A). One of these domains binds ATP and is called the nucleotide-binding domain, while other domains are involved in pump phosphorylation or in ion translocation. As is also the case for the Na⁺/K⁺ pump, the Ca²⁺ pump undergoes phosphorylation that fuels a cycle of conformational changes.

(A)

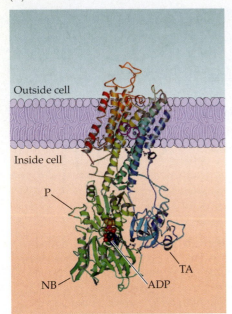

(B)

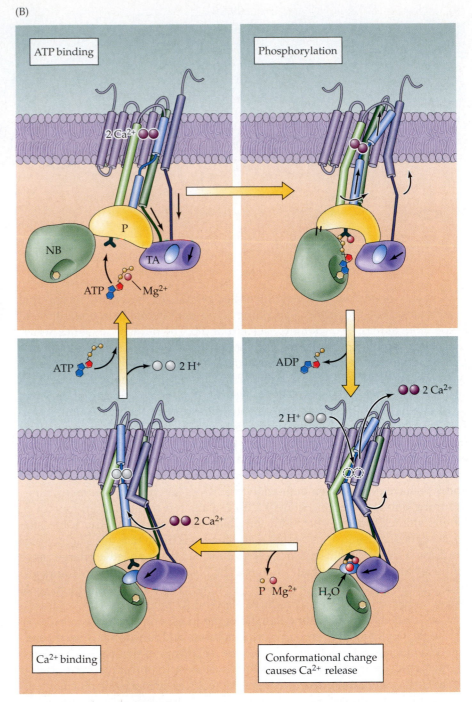

Figure 4.14 Molecular structure of the Ca^{2+} pump. (A) Structure of the Ca^{2+} pump. Domains responsible for nucleotide binding (NB), phosphorylation (P), and ion translocation activity (TA) are indicated. Panel (A) and the first panel in (B) show the structure of the pump when bound to ADP; in this state, two Ca^{2+} are sequestered within the membrane-spanning regions of the pump. (B) Hypothetical sequence of structural changes associated with translocation of Ca^{2+} by the Ca^{2+} pump. Analogous to the sequence of events involved in the function of the Na^+/K^+ pump (see Figure 4.11B), the Ca^{2+} pump undergoes a cycle of phosphorylation and dephosphorylation that causes conformational changes (black arrows) that drive Ca^{2+} across the membrane. (After Toyoshima et al., 2004.)

By examining the structure of the Ca^{2+} pump at different stages of this cycle, the mechanism of Ca^{2+} translocation has been clarified. Ca^{2+} first binds to the cytoplasmic side of the pump. The ions are then driven through the membrane as a result of phosphorylation-induced conformational changes in the membrane-spanning domains that ultimately result in Ca^{2+} being released on the other side of the membrane (Figure 4.14B). Unlike ion channels, in which ion translocation occurs by diffusion-based movement through an aqueous pore, translocation of Ca^{2+} by the pump occurs by sequestering the ion deep within the protein, away from the aqueous medium. This explains how the pump is able to move Ca^{2+} uphill, against the steep electrochemical gradient for Ca^{2+} present across the membrane.

Summary

Ion transporters and channels have complementary functions. The primary purpose of transporters is to generate transmembrane concentration gradients, which are then exploited by ion channels to generate electrical signals. Ion channels are responsible for the voltage-dependent conductances of nerve cell membranes. The channels underlying the action potential are integral membrane proteins that open or close ion-selective pores in response to the membrane potential, allowing specific ions to diffuse across the membrane. The flow of ions through single open channels can be detected as tiny electrical currents, and the synchronous opening of many such channels generates the macroscopic currents that produce action potentials. Molecular studies show that such voltage-gated channels have highly conserved structures that are responsible for features such as ion permeation and voltage sensing, as well as the features that specify ion selectivity and toxin sensitivity. Other types of channels are sensitive to chemical signals, such as neurotransmitters or second messengers, or to heat or membrane deformation. A large number of ion channel genes create channels with a correspondingly wide range of functional characteristics, thus allowing different types of neurons to have a remarkable spectrum of electrical properties. Ion transporter proteins are quite different in both structure and function. The energy needed for ion movement against a concentration gradient (e.g., in maintaining the resting potential) is provided either by the hydrolysis of ATP or by the electrochemical gradient of co-transported ions. The Na^+/K^+ pump produces and maintains the transmembrane gradients of Na^+ and K^+, while other transporters are responsible for the electrochemical gradients for other physiologically important ions, such as Cl^-, Ca^{2+}, and H^+. Together, transporters and channels provide a reasonably comprehensive molecular explanation for the ability of neurons to generate electrical signals.

Additional Reading

Reviews

ARMSTRONG, C. M. AND B. HILLE (1998) Voltage-gated ion channels and electrical excitability. *Neuron* 20: 371–380.

BEZANILLA, F. AND A. M. CORREA (1995) Single-channel properties and gating of Na^+ and K^+ channels in the squid giant axon. In *Cephalopod Neurobiology*, N. J. Abbott, R. Williamson and L. Maddock (eds.). New York: Oxford University Press, pp. 131–151.

CATTERALL, W. A. (2000) From ionic currents to molecular mechanisms: The structure and function of voltage-gated sodium channels. *Neuron* 26: 13–25.

GOUAUX, E. AND R MACKINNON (2005) Principles of selective ion transport in channels and pumps. *Science* 310: 1461-1465.

JAN, L.Y. AND Y. N. JAN (1997) Voltage-gated and inwardly rectifying potassium channels. *J. Physiol.* 505: 267–282.

JENTSCH, T. J., M. POET, J. C. FUHRMANN AND A. A. ZDEBIK (2005) Physiological functions of CLC Cl^- channels gleaned from human genetic disease and mouse models. *Annu. Rev. Physiol.* 67: 779–807.

KAPLAN, J. H. (2002) Biochemistry of Na,K-ATPase. *Annu. Rev. Biochem.* 71: 511–535.

KRISHTAL, O. (2003). The ASICs: Signaling molecules? Modulators? *Trends Neurosci,* 26: 477–483.

LINGREL, J. B., J. VAN HUYSSE, W. O'BRIEN, E. JEWELL-MOTZ, R. ASKEW AND P. SCHULTHEIS (1994) Structure-function studies of the Na, K-ATPase. *Kidney Internat.* 45: S32–S39.

MACKINNON, R. (2003) Potassium channels. *FEBS Lett.* 555: 62–65.

MOLLER, J.V., P. NISSEN, T. L. SORENSEN AND M. LE MAIRE (2005) Transport mechanism of the sarcoplasmic reticulum Ca^{2+}-ATPase pump. *Curr. Opin. Struct. Biol.* 15: 387–393.

NEHER, E. (1992) Nobel lecture: Ion channels for communication between and within cells. *Neuron* 8: 605–612.

PATAPOUTIAN, A., A. M. PEIER, G. M. STORY AND V. VISWANATH (2003). ThermoTRP channels and beyond: Mechanisms of temperature sensation. *Nature Rev. Neurosci.* 4: 529–539.

SEEBURG, P. H. (2002). A-to-I editing: New and old sites, functions and speculations. *Neuron* 35: 17–20.

Important Original Papers

BOULTER, J. AND 6 OTHERS (1990) Molecular cloning and functional expression of glutamate receptor subunit genes. *Science* 249: 1033–1037.

CATERINA, M. J., M. A. SCHUMACHER, M. TOMINAGA, T. A. ROSEN, J. D. LEVINE AND D. JULIUS (1997) The capsaicin receptor: A heat-activated ion channel in the pain pathway. *Nature* 389: 816–824.

CHA, A., G. E. SNYDER, P. R. SELVIN AND F. BEZANILLA (1999) Atomic scale movement of the voltage-sensing region in a potassium channel measured via spectroscopy. *Nature* 402: 809–813.

CHANDA, B., O. K. ASAMOAH, R. BLUNCK, B. ROUX AND F. BEZANILLA (2005) Gating charge displacement in voltage-gated ion channels involves limited transmembrane movement. *Nature* 436: 852–856.

CUELLO, L. G., D. M. CORTES AND E. PEROZO (2004) Molecular architecture of the KvAP voltage-dependent K$^+$ channel in a lipid bilayer. *Science* 306: 491–495.

DOYLE, D. A. AND 7 OTHERS (1998) The structure of the potassium channel: Molecular basis of K$^+$ conduction and selectivity. *Science* 280: 69–77.

FAHLKE, C., H. T. YU, C. L. BECK, T. H. RHODES AND A. L. GEORGE JR. (1997) Pore-forming segments in voltage-gated chloride channels. *Nature* 390: 529–532.

HO, K. AND 6 OTHERS (1993) Cloning and expression of an inwardly rectifying ATP-regulated potassium channel. *Nature* 362: 31–38.

HODGKIN, A. L. AND R. D. KEYNES (1955) Active transport of cations in giant axons from *Sepia* and *Loligo. J. Physiol.* 128: 28–60.

HOSHI, T., W. N. ZAGOTTA AND R. W. ALDRICH (1990) Biophysical and molecular mechanisms of *Shaker* potassium channel inactivation. *Science* 250: 533–538.

JIANG, Y. AND 6 OTHERS (2003) X-ray structure of a voltage-dependent K$^+$ channel. *Nature* 423: 33–41.

LEE, A. G. (2006) Ion channels: A paddle in oil. *Nature* 444: 697.

LLANO, I., C. K. WEBB AND F. BEZANILLA (1988) Potassium conductance of squid giant axon. Single-channel studies. *J. Gen. Physiol.* 92: 179–196.

LONG, S. B., E. B. CAMPBELL AND R. MACKIN-NON (2005a) Crystal structure of a mammalian voltage-dependent Shaker family K$^+$ channel. *Science* 309: 897–903.

LONG, S. B., E. B. CAMPBELL AND R. MACKIN-NON (2005b) Voltage sensor of Kv1.2: Structural basis of electromechanical coupling. *Science* 309: 903–908.

MIKAMI, A. AND 7 OTHERS (1989) Primary structure and functional expression of the cardiac dihydropyridine-sensitive calcium channel. *Nature* 340: 230–233.

NODA, M. AND 6 OTHERS (1986) Expression of functional sodium channels from cloned cDNA. *Nature* 322: 826–828.

NOWYCKY, M. C., A. P. FOX AND R. W. TSIEN (1985) Three types of neuronal calcium channel with different calcium agonist sensitivity. *Nature* 316: 440–443.

PAPAZIAN, D. M., T. L. SCHWARZ, B. L. TEMPEL, Y. N. JAN AND L. Y. JAN (1987) Cloning of genomic and complementary DNA from *Shaker*, a putative potassium channel gene from *Drosophila. Science* 237: 749–753.

RANG, H. P. AND J. M. RITCHIE (1968) On the electrogenic sodium pump in mammalian non-myelinated nerve fibres and its activation by various external cations. *J. Physiol.* 196: 183–221.

SIGWORTH, F. J. AND E. NEHER (1980) Single Na$^+$ channel currents observed in cultured rat muscle cells. *Nature* 287: 447–449.

THOMAS, R. C. (1969) Membrane current and intracellular sodium changes in a snail neurone during extrusion of injected sodium. *J. Physiol.* 201: 495–514.

TOYOSHIMA, C., H. NOMURA AND T. TSUDA (2004) Luminal gating mechanism revealed in calcium pump crystal structures with phosphate analogues. *Nature* 432: 361–368.

VANDERBERG, C. A. AND F. BEZANILLA (1991) A sodium channel model based on single channel, macroscopic ionic, and gating currents in the squid giant axon. *Biophys. J.* 60: 1511–1533.

WALDMANN, R., G. CHAMPIGNY, F. BASSILANA, C. HEURTEAUX AND M. LAZDUNSKI (1997) A proton-gated cation channel involved in acid-sensing. *Nature* 386: 173–177.

WEI, A. M., A. COVARRUBIAS, A. BUTLER, K. BAKER, M. PAK AND L. SALKOFF (1990) K$^+$ current diversity is produced by an extended gene family conserved in *Drosophila* and mouse. *Science* 248: 599–603.

YANG, N., A. L. GEORGE JR. AND R. HORN (1996) Molecular basis of charge movement in voltage-gated sodium channels. *Neuron* 16: 113–22.

Books

AIDLEY, D. J. AND P. R. STANFIELD (1996) *Ion Channels: Molecules in Action.* Cambridge: Cambridge University Press.

ASHCROFT, F. M. (2000) *Ion Channels and Disease.* Boston: Academic Press.

HILLE, B. (2001) *Ion Channels of Excitable Membranes*, 3rd Ed. Sunderland, MA: Sinauer Associates.

JUNGE, D. (1992) *Nerve and Muscle Excitation*, 3rd Ed. Sunderland, MA: Sinauer Associates.

Chapter 5

Synaptic Transmission

Overview

The human brain contains at least 100 billion neurons, each with the ability to influence many other cells. Clearly, sophisticated and highly efficient mechanisms are needed to enable communication among this astronomical number of elements. Such communication is made possible by synapses, the functional contacts between neurons. Two categories of synapses—electrical and chemical—can be distinguished on the basis of their mechanism of transmission. At electrical synapses, current flows through gap junctions, which are specialized membrane channels that connect two cells. In contrast, chemical synapses enable cell-to-cell communication via the secretion of neurotransmitters; these chemical agents released by the presynaptic neurons produce secondary current flow in postsynaptic neurons by activating specific receptor molecules. The total number of neurotransmitters is not known, but is well over 100. Virtually all neurotransmitters undergo a similar cycle of use that includes: synthesis and packaging into synaptic vesicles; release from the presynaptic cell; binding to postsynaptic receptors; and, finally, rapid removal and/or degradation. The secretion of neurotransmitters is triggered by the influx of Ca^{2+} through voltage-gated channels, which gives rise to a transient increase in Ca^{2+} concentration within the presynaptic terminal. The rise in Ca^{2+} concentration causes synaptic vesicles to fuse with the presynaptic plasma membrane and release their contents into the space between the pre- and postsynaptic cells. Although it is not yet understood exactly how Ca^{2+} triggers exocytosis, specific proteins on the surface of the synaptic vesicle and elsewhere in the presynaptic terminal mediate this process. Neurotransmitters evoke postsynaptic electrical responses by binding to members of a diverse group of neurotransmitter receptors. There are two major classes of receptors: those in which the receptor molecule is also an ion channel, and those in which the receptor and the ion channel are separate molecules. These distinct receptors give rise to electrical signals by transmitter-induced opening or closing of the ion channels. Whether the postsynaptic actions of a particular neurotransmitter are excitatory or inhibitory is determined by the ionic permeability of the ion channel affected by the transmitter, and by the concentration of permeant ions inside and outside the cell.

Electrical Synapses

The many different kinds of synapses within the human brain can be divided into two general classes: electrical synapses and chemical synapses. Although they are a distinct minority, electrical synapses are found in all nervous systems, permitting direct, passive flow of electrical current from one neuron to another.

The structure of an electrical synapse is shown schematically in Figure 5.1A. The "upstream" neuron, which is the source of current, is called the **presynap-**

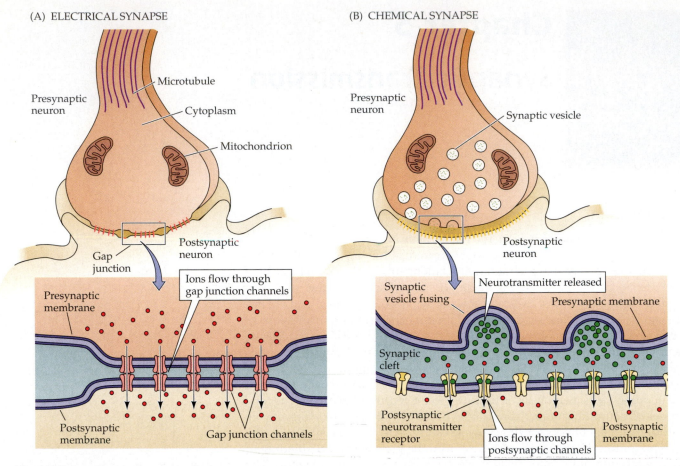

(A) ELECTRICAL SYNAPSE

Presynaptic neuron
Microtubule
Cytoplasm
Mitochondrion
Gap junction
Postsynaptic neuron

Presynaptic membrane
Ions flow through gap junction channels
Postsynaptic membrane
Gap junction channels

(B) CHEMICAL SYNAPSE

Presynaptic neuron
Synaptic vesicle
Postsynaptic neuron

Synaptic vesicle fusing
Neurotransmitter released
Presynaptic membrane
Synaptic cleft
Postsynaptic neurotransmitter receptor
Ions flow through postsynaptic channels
Postsynaptic membrane

Figure 5.1 Electrical and chemical synapses differ fundamentally in their transmission mechanisms. (A) At electrical synapses, gap junctions between pre- and postsynaptic membranes permit current to flow passively through intercellular channels. This current flow changes the postsynaptic membrane potential, initiating (or in rare instances inhibiting) the generation of postsynaptic action potentials. (B) At chemical synapses, there is no intercellular continuity, and thus no direct flow of current from pre- to postsynaptic cell. Synaptic current flows across the postsynaptic membrane only in response to the secretion of neurotransmitters, which open or close postsynaptic ion channels after binding to receptor molecules.

tic element, and the "downstream" neuron into which this current flows is termed **postsynaptic**. The membranes of the two communicating neurons come extremely close at the synapse and are actually linked together by an intercellular specialization called a **gap junction**. Gap junctions contain precisely aligned, paired channels in the membrane of the pre- and postsynaptic neurons, such that each channel pair forms a pore (Figure 5.2A). The pore of a gap junction channel is much larger than the pores of the voltage-gated ion channels described in the previous chapter. As a result, a variety of substances can simply diffuse between the cytoplasm of the pre- and postsynaptic neurons. In addition to ions, substances that diffuse through gap junction pores include molecules with molecular weights as great as several hundred daltons. This permits ATP and other important intracellular metabolites, such as second messengers (see Chapter 7), to be transferred between neurons.

Electrical synapses thus work by allowing ionic current to flow passively through the gap junction pores from one neuron to another. The usual source of this current is the potential difference generated locally by the action potential (see Chapter 3). Communication via gap junctions has a number of interesting consequences. One is that transmission can be bidirectional; that is, current can flow in either direction across the gap junction, depending on which member of the coupled pair is invaded by an action potential (although some types of gap junctions have special features that render their transmission unidirectional). Another important feature of the electrical synapse is that transmission is extraordinarily fast: because passive current flow across the gap junction is vir-

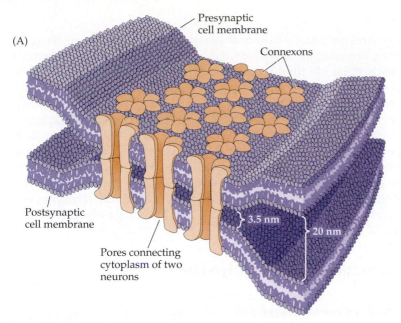

(A)

Presynaptic
cell membrane

Connexons

Postsynaptic
cell membrane

Pores connecting
cytoplasm of two
neurons

3.5 nm

20 nm

Figure 5.2 Structure and function of gap junctions at electrical synapses. (A) Gap junctions consist of hexameric complexes formed by the coming together of subunits called connexons, which are present in both the pre- and postsynaptic membranes. The pores of the channels connect to one another, creating electrical continuity between the two cells. (B) Rapid transmission of signals at an electrical synapse in the crayfish. An action potential in the presynaptic neuron causes the postsynaptic neuron to be depolarized within a fraction of a millisecond. (C) Electrical synapses allow synchronization of electrical activity in hippocampal interneurons. In a pair of interneurons connected by electrical synapses, generation of an action potential in one neuron often results in the synchronized firing of an action potential in another neuron (red asterisks). (B after Furshpan and Potter, 1959; C after Beierlein et al., 2000.)

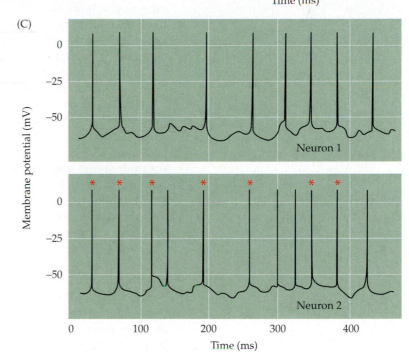

tually instantaneous, communication can occur without the delay that is characteristic of chemical synapses.

These features are apparent in the operation of the first electrical synapse to be discovered. At this synapse, found in the crayfish nervous system, a postsynaptic electrical signal is observed within a fraction of a millisecond after the generation of a presynaptic action potential (Figure 5.2B). In fact, at least part of this brief synaptic delay is caused by propagation of the action potential into the presynaptic terminal, so there may be essentially no delay at all in the transmission of electrical signals across the synapse. Such synapses interconnect many of the neurons within the circuit that allows the crayfish to escape from its predators, thus minimizing the time between the presence of a threatening stimulus and a potentially life-saving motor response.

A more general purpose of electrical synapses is to synchronize electrical activity among populations of neurons. For example, the brainstem neurons that generate rhythmic electrical activity underlying breathing are synchronized by electrical synapses, as are populations of interneurons in the cerebral cortex, thalamus, cerebellum, and other brain regions (Figure 5.2C). Electrical transmission between certain hormone-secreting neurons within the mammalian hypothalamus ensures that all cells fire action potentials at about the same time, thus facilitating a burst of hormone secretion into the circulation. The fact that gap junction pores are large enough to allow molecules such as ATP and second messengers to diffuse intercellularly also permits electrical synapses to coordinate the intracellular signaling and metabolism of coupled cells. This property may be particularly important for glial cells, which form large intracellular signaling networks via their gap junctions.

Signal Transmission at Chemical Synapses

The general structure of a chemical synapse is shown schematically in Figure 5.1B. The space between the pre- and postsynaptic neurons is substantially greater at chemical synapses than at electrical synapses and is called the **synaptic cleft**. However, the key feature of all chemical synapses is the presence of small, membrane-bounded organelles called **synaptic vesicles** within the presynaptic terminal. These spherical organelles are filled with one or more **neurotransmitters**, the chemical signals secreted from the presynaptic neuron, and it is these chemical agents acting as messengers between the communicating neurons that gives this type of synapse its name.

Transmission at chemical synapses is based on the elaborate sequence of events depicted in Figure 5.3. The process is initiated when an action potential invades the terminal of the presynaptic neuron. The change in membrane potential caused by the arrival of the action potential leads to the opening of voltage-gated calcium channels in the presynaptic membrane. Because of the steep concentration gradient of Ca^{2+} across the presynaptic membrane (the external Ca^{2+} concentration is approximately 10^{-3} M, whereas the internal Ca^{2+} concentration is approximately 10^{-7} M), the opening of these channels causes a rapid influx of Ca^{2+} into the presynaptic terminal, with the result that the Ca^{2+} concentration of the cytoplasm in the terminal transiently rises to a much higher value. Elevation of the presynaptic Ca^{2+} concentration, in turn, allows synaptic vesicles to fuse with the plasma membrane of the presynaptic neuron. The Ca^{2+}-dependent fusion of synaptic vesicles with the terminal membrane causes their contents, most importantly neurotransmitters, to be released into the synaptic cleft.

Following exocytosis, transmitters diffuse across the synaptic cleft and bind to specific receptors on the membrane of the postsynaptic neuron. The binding of neurotransmitter to the receptors causes channels in the postsynaptic membrane to open (or sometimes to close), thus changing the ability of ions to flow across the postsynaptic membrane. The resulting neurotransmitter-induced current flow alters the conductance and (usually) the membrane potential of the postsynaptic neuron, increasing or decreasing the probability that the neuron will fire an action potential. In this way, information is transmitted from one neuron to another.

Properties of Neurotransmitters

The notion that electrical information can be transferred from one neuron to the next by means of chemical signaling was the subject of intense debate through the first half of the twentieth century. A key experiment that supported this idea

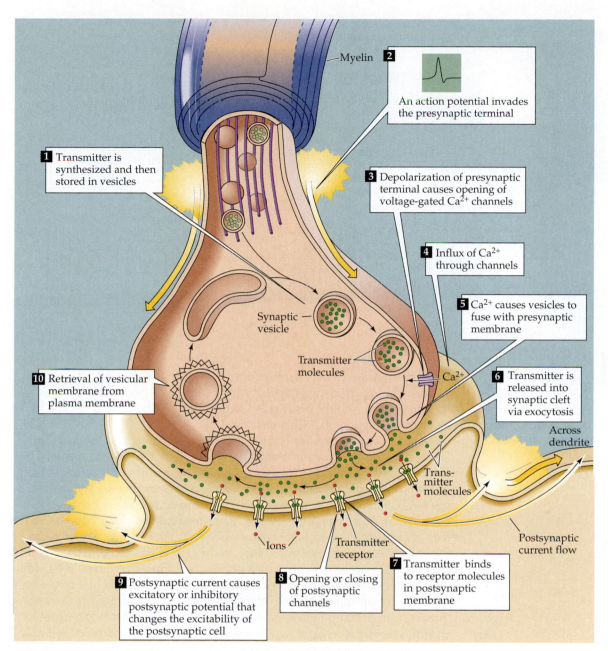

Myelin

2 An action potential invades the presynaptic terminal

1 Transmitter is synthesized and then stored in vesicles

3 Depolarization of presynaptic terminal causes opening of voltage-gated Ca^{2+} channels

4 Influx of Ca^{2+} through channels

5 Ca^{2+} causes vesicles to fuse with presynaptic membrane

Synaptic vesicle

Transmitter molecules

Ca^{2+}

6 Transmitter is released into synaptic cleft via exocytosis

10 Retrieval of vesicular membrane from plasma membrane

Across dendrite

Transmitter molecules

Postsynaptic current flow

Ions

Transmitter receptor

7 Transmitter binds to receptor molecules in postsynaptic membrane

9 Postsynaptic current causes excitatory or inhibitory postsynaptic potential that changes the excitability of the postsynaptic cell

8 Opening or closing of postsynaptic channels

Figure 5.3   Sequence of events involved in transmission at a typical chemical synapse.

was performed in 1926 by German physiologist Otto Loewi. Acting on an idea that allegedly came to him in the middle of the night, Loewi proved that electrical stimulation of the vagus nerve slows the heartbeat by releasing a chemical signal. He isolated and perfused the hearts of two frogs, monitoring the rates at which they were beating (Figure 5.4). His experiment collected the perfusate flowing through the stimulated heart and transferred this solution to the second heart. When the vagus nerve to the first heart was stimulated, the beat of this heart slowed. Remarkably, even though the vagus nerve of the second heart had not been stimulated, its beat also slowed when exposed to the perfusate from the first heart. This result showed that the vagus nerve regulates the heart rate by releasing a chemical that accumulates in the perfusate. Originally referred to as "vagus substance," the agent was later shown to be **acetylcholine (ACh)**.

(A)

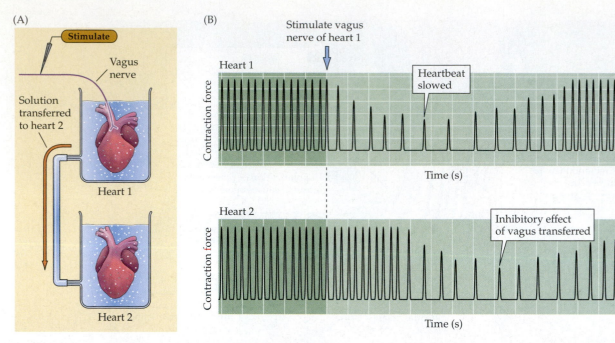

(B)

Figure 5.4 Loewi's experiment demonstrating chemical neurotransmission. (A) Diagram of experimental setup. (B) Where the vagus nerve of an isolated frog's heart was stimulated, the heart rate decreased (upper panel). If the perfusion fluid from this heart was transferred to a second heart, the second heart's rate decreased as well (lower panel).

ACh is now known to be a neurotransmitter that acts not only in the heart but at a variety of postsynaptic targets in the central and peripheral nervous systems, preeminently at the neuromuscular junction of striated muscles and in the visceral motor system (see Chapters 6 and 21).

Over the years, a number of formal criteria have emerged that definitively identify a substance as a neurotransmitter (Box 5A). These criteria have led to the identification of more than 100 different neurotransmitters, which can be classified into two broad categories: small-molecule neurotransmitters and neuropeptides (see Chapter 6). Having more than one transmitter diversifies the physiological repertoire of synapses. Multiple neurotransmitters can produce different types of responses on individual postsynaptic cells. For example, a neuron can be excited by one type of neurotransmitter and inhibited by another type of neurotransmitter. The speed of postsynaptic responses produced by different transmitters also differs, allowing control of electrical signaling over different time scales. In general, small-molecule neurotransmitters mediate rapid synaptic actions, whereas neuropeptides tend to modulate slower, ongoing synaptic functions.

Until relatively recently, it was believed that a given neuron produced only a single type of neurotransmitter. It is now clear, however, that many types of neurons synthesize and release two or more different neurotransmitters. When more than one transmitter is present within a nerve terminal, the molecules are called **co-transmitters**. Because different types of transmitters can be packaged in different populations of synaptic vesicles, co-transmitters need not be released simultaneously. When peptide and small-molecule neurotransmitters act as co-transmitters at the same synapse, they are differentially released according to the pattern of synaptic activity: low-frequency activity often releases only small neurotransmitters, whereas high-frequency activity is required to release neuropeptides from the same presynaptic terminals. As a result, the chemical signaling properties of such synapses change according to the rate of activity.

Effective synaptic transmission requires close control of the concentration of neurotransmitters within the synaptic cleft. Neurons have therefore developed a sophisticated ability to regulate the synthesis, packaging, release, and degradation (or removal) of neurotransmitters to achieve the desired levels of transmit-

BOX 5A   Criteria That Define a Neurotransmitter

Three primary criteria have been used to confirm that a molecule acts as a neurotransmitter at a given chemical synapse.

1. *The substance must be present within the presynaptic neuron.* Clearly, a chemical cannot be secreted from a presynaptic neuron unless it is present there. Because elaborate biochemical pathways are required to produce neurotransmitters, showing that the enzymes and precursors required to synthesize the substance are present in presynaptic neurons provides additional evidence that the substance is used as a transmitter. Note, however, that since the transmitters glutamate, glycine, and aspartate are also needed for protein synthesis and other metabolic reactions in all neurons, their presence is *not* sufficient evidence to establish them as neurotransmitters.

2. *The substance must be released in response to presynaptic depolarization, and the release must be Ca^{2+}-dependent.* Another essential criterion for identifying a neurotransmitter is to demonstrate that it is released from the presynaptic neuron in response to presynaptic electrical activity, and that this release requires Ca^{2+} influx into the presynaptic terminal. Meeting this criterion is technically challenging, not only because it may be difficult to selectively stimulate the presynaptic neurons, but also because enzymes and transporters efficiently remove the secreted neurotransmitters.

3. *Specific receptors for the substance must be present on the postsynaptic cell.* A neurotransmitter cannot act on its target unless specific receptors for the transmitter are present in the postsynaptic membrane. One way to demonstrate the presence of receptors is to show that application of an exogenous transmitter mimics the postsynaptic effect of presynaptic stimulation. A more rigorous demonstration is to show that ago-

nists and antagonists that alter the normal postsynaptic response have the same effect when the substance in question is applied exogenously. High-resolution histological methods can also be used to show that specific receptors are present in the postsynaptic membrane.

Fulfilling these criteria establishes unambiguously that a substance is used as a transmitter at a given synapse. Practical difficulties, however, have prevented these standards from being applied at many types of synapses. It is for this reason that so many substances must be referred to as "putative" neurotransmitters.

Demonstrating the identity of a neurotransmitter at a synapse requires showing (1) its presence, (2) its release, and (3) the postsynaptic presence of specific receptors.

(1)

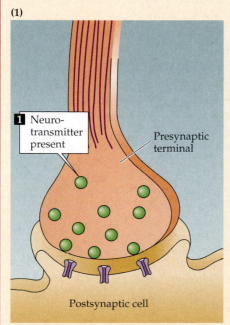

1 Neurotransmitter present

Presynaptic terminal

Postsynaptic cell

(2)

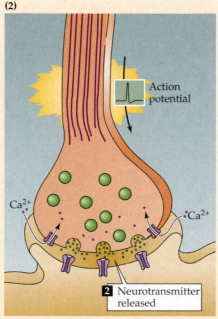

Action potential

Ca^{2+}

Ca^{2+}

2 Neurotransmitter released

(3)

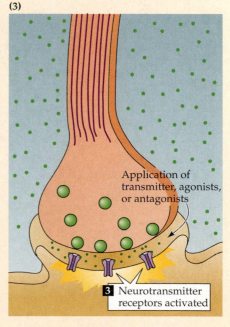

Application of transmitter, agonists, or antagonists

3 Neurotransmitter receptors activated

ter molecules. The synthesis of small-molecule neurotransmitters occurs locally within presynaptic terminals (Figure 5.5A). The enzymes needed to synthesize these transmitters are produced in the neuronal cell body and transported to the nerve terminal cytoplasm at 0.5–5 millimeters a day by a mechanism called **slow axonal transport**. The precursor molecules required to make new molecules of neurotransmitter are usually taken into the nerve terminal by transporters found in the plasma membrane of the terminal. The enzymes synthesize neurotransmitters in the cytoplasm of the presynaptic terminal and the transmitters are then loaded into synaptic vesicles via transporters in the vesicular membrane (see Chapter 4). For some small-molecule neurotransmitters, the final steps of synthesis occur inside the synaptic vesicles. Most small-molecule neurotransmitters are packaged in vesicles 40 to 60 nm in diameter, the centers of which appear clear in electron micrographs; accordingly, these vesicles are referred to as **small clear-core vesicles** (Figure 5.5B). Neuropeptides are synthesized in the cell body of a neuron, meaning that the peptide is produced a long distance away from its site of secretion (Figure 5.5C). To solve this problem, peptide-filled vesicles are transported along an axon and down to the synaptic terminal via **fast axonal transport**. This process carries vesicles at rates up to 400 mm/day along cytoskeletal elements called microtubules (in contrast to the slow axonal transport of the enzymes that synthesize small-molecule transmitters). Microtubules are long, cylindrical filaments, 25 nm in diameter, present throughout neurons and other cells. Peptide-containing vesicles are moved along these microtubule "tracks" by ATP-requiring "motor" proteins such as kinesin. Neuropeptides are packaged into synaptic vesicles that range from 90 to 250 nm in diameter. These vesicles are electron-dense in electron micrographs—hence they are referred to as **large dense-core vesicles** (Figure 5.5D).

After a neurotransmitter has been secreted into the synaptic cleft, it must be removed to enable the postsynaptic cell to engage in another cycle of synaptic transmission. The removal of neurotransmitters involves diffusion away from the postsynaptic receptors, in combination with reuptake into nerve terminals or surrounding glial cells, degradation by specific enzymes, or a combination of these mechanisms. Specific transporter proteins remove most small-molecule neurotransmitters (or their metabolites) from the synaptic cleft, ultimately delivering them back to the presynaptic terminal for reuse.

Figure 5.5 *Metabolism of small-molecule and peptide transmitters. (A) Small-molecule neurotransmitters are synthesized at nerve terminals. The enzymes necessary for neurotransmitter synthesis are made in the cell body of the presynaptic cell (1) and are transported down the axon by slow axonal transport (2). Precursors are taken up into the terminals by specific transporters, and neurotransmitter synthesis and packaging take place within the nerve endings (3). After vesicle fusion and release (4), the neurotransmitter may be enzymatically degraded. The reuptake of the neurotransmitter (or its metabolites) starts another cycle of synthesis, packaging, release, and removal (5). (B) Small clear-core vesicles at a synapse between a presynaptic terminal and a dendritic spine in the central nervous system. Such vesicles typically contain small-molecule neurotransmitters. (C) Peptide neurotransmitters, as well as the enzymes that modify their precursors, are synthesized in the cell body (1). Enzymes and propeptides are packaged into vesicles in the Golgi apparatus. During fast axonal transport of these vesicles to the nerve terminals (2), the enzymes modify the propeptides to produce one or more neurotransmitter peptides (3). After vesicle fusion and exocytosis, the peptides diffuse away and are degraded by proteolytic enzymes (4). (D) Large dense-core vesicles in a central presynaptic terminal synapsing onto a dendrite. Such vesicles typically contain neuropeptides or, in some cases, biogenic amines. (B and D from Peters, Palay, and Webster, 1991.)*

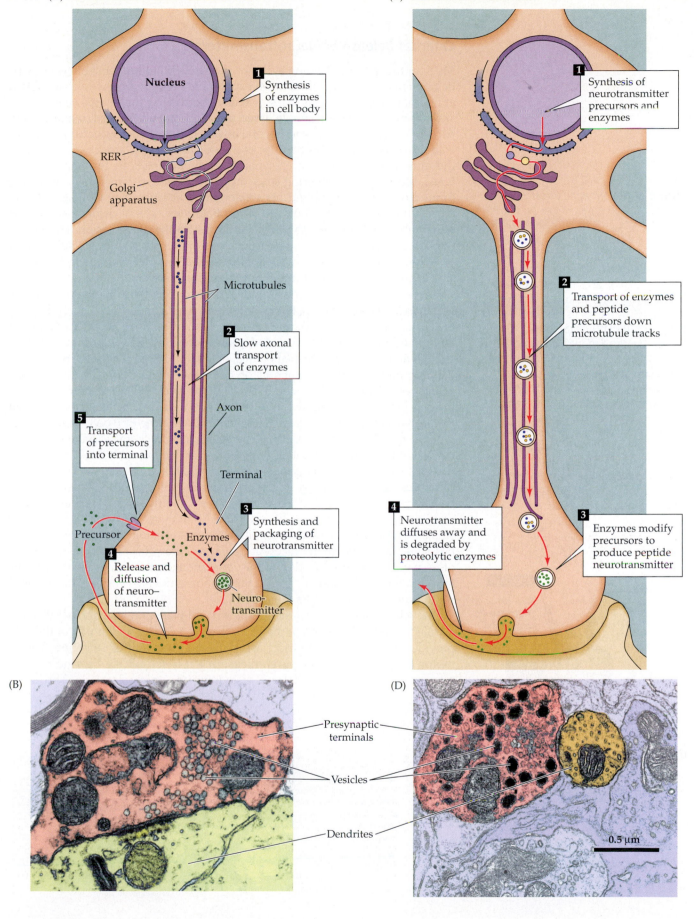

(A) SMALL-MOLECULE TRANSMITTER

Nucleus

1 Synthesis of enzymes in cell body

RER

Golgi apparatus

Microtubules

2 Slow axonal transport of enzymes

Axon

5 Transport of precursors into terminal

Terminal

Precursor

Enzymes

3 Synthesis and packaging of neurotransmitter

4 Release and diffusion of neuro–transmitter

Neuro-transmitter

(C) PEPTIDE TRANSMITTER

Nucleus

1 Synthesis of neurotransmitter precursors and enzymes

2 Transport of enzymes and peptide precursors down microtubule tracks

4 Neurotransmitter diffuses away and is degraded by proteolytic enzymes

3 Enzymes modify precursors to produce peptide neurotransmitter

(B)

Presynaptic terminals

Vesicles

Dendrites

(D)

Vesicles

0.5 μm

Quantal Release of Neurotransmitters

Much of the evidence leading to the present understanding of chemical synaptic transmission was obtained from experiments examining the release of ACh at neuromuscular junctions. These synapses between spinal motor neurons and skeletal muscle cells are simple, large, and peripherally located, making them particularly amenable to experimental analysis. Such synapses occur at specializations called **end plates** because of the saucer-like appearance of the site on the muscle fiber where the presynaptic axon elaborates its terminals (Figure 5.6A). Most of the pioneering work on neuromuscular transmission was performed by Bernard Katz and his collaborators at University College London during the 1950s and 1960s, and Katz has been widely recognized for his remarkable contributions to understanding synaptic transmission. Though he worked primarily on the frog neuromuscular junction, numerous subsequent experiments have confirmed the applicability of his observations to transmission at chemical synapses throughout the nervous system.

When an intracellular microelectrode is used to record the membrane potential of a muscle cell, an action potential in the presynaptic motor neuron can be seen to elicit a transient depolarization of the postsynaptic muscle fiber. This change in membrane potential, called an **end plate potential** (**EPP**), is normally large enough to bring the membrane potential of the muscle cell well above the threshold for producing a postsynaptic action potential (Figure 5.6B). The postsynaptic action potential triggered by the EPP causes the muscle fiber to contract. Unlike the case for electrical synapses, there is a pronounced delay between the time that the presynaptic motor neuron is stimulated and when the EPP occurs in the postsynaptic muscle cell. Such a delay is characteristic of all chemical synapses.

One of Katz's seminal findings, in studies carried out with Paul Fatt in 1951, was that spontaneous changes in muscle cell membrane potential occur even in the absence of stimulation of the presynaptic motor neuron. These changes have the same shape as EPPs but are much smaller (typically less than 1 mV in amplitude, compared to an EPP of perhaps 40 or 50 mV). Both EPPs and these small, spontaneous events are sensitive to pharmacological agents that block postsynaptic acetylcholine receptors, such as curare (see Box 6B). These and other parallels between EPPs and the spontaneously occurring depolarizations led Katz and his colleagues to call these spontaneous events **miniature end plate potentials**, or **MEPPs** (Figure 5.6C).

The relationship between the full-blown end plate potential and MEPPs was clarified by careful analysis of the EPPs (Figure 5.7). The magnitude of the EPP provides a convenient electrical assay of neurotransmitter secretion from a motor neuron terminal; however, measuring it is complicated by the need to prevent muscle contraction from dislodging the microelectrode. The usual

Figure 5.6 *Synaptic transmission at the neuromuscular junction. (A) Experimental arrangement, typically using the muscle of a frog or rat. The axon of the motor neuron innervating the muscle fiber is stimulated with an extracellular electrode, while an intracellular microelectrode is inserted into the postsynaptic muscle cell to record its electrical responses. (B) End plate potentials evoked by stimulation of a motor neuron are normally above threshold and therefore produce an action potential in the postsynaptic muscle cell. (C) Spontaneous miniature EPPs (MEPPs) occur in the absence of presynaptic stimulation. (D) When the neuromuscular junction is bathed in a solution that has a low concentration of Ca^{2+}, stimulating the motor neuron evokes EPPs whose amplitudes are reduced to about the size of MEPPs. (After Fatt and Katz, 1952.)*

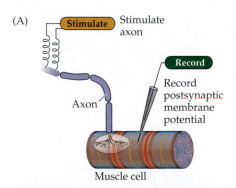

(A)

Stimulate
Stimulate
axon

Record

Record postsynaptic membrane potential

Axon

Muscle cell

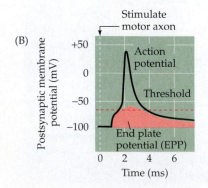

(B)

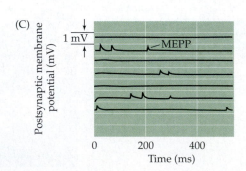

(C)

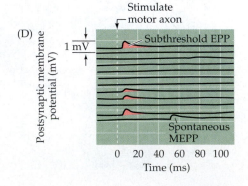

(D)

means of eliminating muscle contractions is either to lower Ca^{2+} concentration in the extracellular medium or to partially block the postsynaptic ACh receptors with curare. As expected from the scheme illustrated in Figure 5.3, lowering the Ca^{2+} concentration reduces neurotransmitter secretion, thus reducing the magnitude of the EPP below the threshold for postsynaptic action potential production and allowing it to be measured more precisely. Under such conditions, stimulation of the motor neuron produces very small EPPs that fluctuate in amplitude from trial to trial (Figure 5.6D). These fluctuations give considerable insight into the mechanisms responsible for neurotransmitter release. In particular, the variable evoked response in low Ca^{2+} is now known to result from the release of unit amounts of ACh by the presynaptic nerve terminal. Indeed, the amplitude of the smallest evoked response is strikingly similar to the size of single MEPPs (compare Figures 5.6C and D). Further supporting this similarity, increments in the EPP response (Figure 5.7A) occur in units about the size of single MEPPs (Figure 5.7B). These "quantal" fluctuations in the amplitude of EPPs indicated to Katz and colleagues that EPPs are made up of individual units, each equivalent to a MEPP.

The idea that EPPs represent the simultaneous release of many MEPP-like units can be tested statistically. A method of statistical analysis based on the independent occurrence of unitary events (called Poisson statistics) predicts what the distribution of EPP amplitudes would look like during a large number of trials of motor neuron stimulation, under the assumption that EPPs are built up from unitary events that are like MEPPs (see Figure 5.7B). The distribution of EPP amplitudes determined experimentally was found to be just that expected if transmitter release from the motor neuron is indeed quantal (the red curve in Figure 5.7A). Such analyses confirmed the idea that release of acetylcholine does indeed occur in discrete packets, each equivalent to a MEPP. In short, a presynaptic action potential causes a postsynaptic EPP because it synchronizes the release of many transmitter quanta.

Release of Transmitters from Synaptic Vesicles

The discovery of the quantal release of packets of neurotransmitter immediately raised the question of how such quanta are formed and discharged into the synaptic cleft. At about the time Katz and his colleagues were using physiological methods to discover quantal release of neurotransmitter, electron microscopy revealed, for the first time, the presence of synaptic vesicles in presynaptic terminals. Putting these two discoveries together, Katz and others proposed that synaptic vesicles loaded with transmitter are the source of the quanta. Subsequent biochemical studies confirmed that synaptic vesicles are the

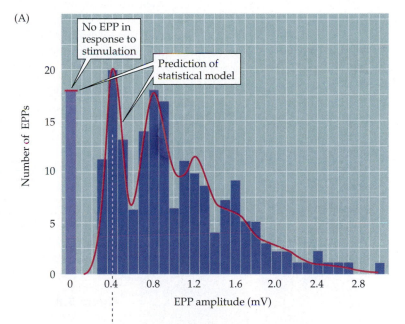

(A)

No EPP in response to stimulation

Prediction of statistical model

Number of EPPs

EPP amplitude (mV)

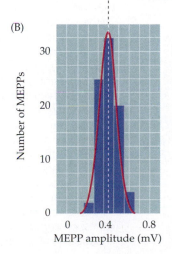

(B)

Number of MEPPs

MEPP amplitude (mV)

Figure 5.7 Quantized distribution of EPP amplitudes evoked in a low Ca^{2+} solution. Peaks of EPP amplitudes (A) tend to occur in integer multiples of the mean amplitude of MEPPs, whose amplitude distribution is shown in (B). The leftmost bar in the EPP amplitude distribution shows trials in which presynaptic stimulation failed to elicit an EPP in the muscle cell. The red curve indicates the prediction of a statistical model based on the assumption that the EPPs result from the independent release of multiple MEPP-like quanta. The observed match, including the predicted number of failures, supports this interpretation. (After Boyd and Martin, 1955.)

repositories of transmitters. These studies have shown that ACh is highly concentrated in the synaptic vesicles of motor neurons, where it is present at a concentration of about 100 mM. Given the diameter of a small, clear-core synaptic vesicle (~50 nm), approximately 10,000 molecules of neurotransmitter are contained in a single vesicle. This number corresponds quite nicely to the amount of ACh that must be applied to a neuromuscular junction to mimic an MEPP, providing further support for the idea that quanta arise from discharge of the contents of single synaptic vesicles.

To prove that quanta are caused by the fusion of individual synaptic vesicles with the plasma membrane, it is necessary to show that each fused vesicle causes a single quantal event to be recorded postsynaptically. This challenge was met in the late 1970s, when John Heuser, Tom Reese, and their colleagues correlated measurements of vesicle fusion with the quantal content of EPPs at the neuromuscular junction. In their experiments, the number of vesicles that fused with the presynaptic plasma membrane was measured by electron microscopy in terminals that had been treated with variable concentrations of a drug (4-aminopyridine, or 4-AP) that enhances the number of vesicle fusion events produced by single action potentials (Figure 5.8A). Parallel electrical measurements were made of the quantal content of the EPPs elicited in this way. A comparison of the number of synaptic vesicle fusions observed with the electron microscope and the number of quanta released at the synapse showed a good correlation between these two measures (Figure 5.8B). These results

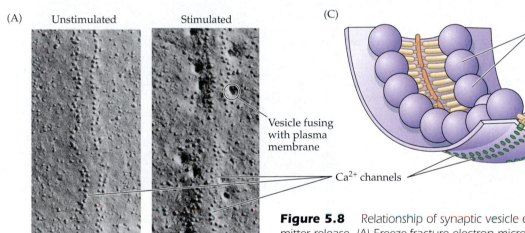

(A)

Unstimulated   Stimulated

Vesicle fusing with plasma membrane

Ca^{2+} channels

(C)

Synaptic vesicles

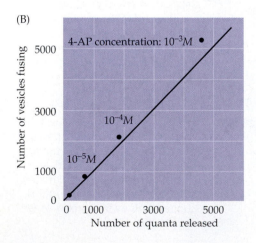

(B)

Number of vesicles fusing

4-AP concentration: $10^{-3}M$

$10^{-4}M$

$10^{-5}M$

Number of quanta released

Figure 5.8   Relationship of synaptic vesicle exocytosis and quantal transmitter release. (A) Freeze-fracture electron microscopy was used to visualize the fusion of synaptic vesicles in presynaptic terminals of frog motor neurons. This view looks down on the release sites from outside the presynaptic terminal. *Left:* Plasma membrane of an unstimulated presynaptic terminal. *Right:* Plasma membrane of a terminal stimulated by an action potential. Stimulation causes the appearance of dimple-like structures that represent the fusion of synaptic vesicles with the presynaptic membrane. (B) Comparison of the number of observed vesicle fusions to the number of quanta released by a presynaptic action potential. Transmitter release was varied by using a drug (4-AP) that affects the duration of the presynaptic action potential, thus changing the amount of calcium that enters during the action potential. The diagonal respresents the 1:1 relationship expected if each vesicle that opened released a single quantum of transmitter. (C) Fine structure of vesicle fusion sites of frog presynaptic terminals. Synaptic vesicles are arranged in rows and are connected to each other and to the plasma membrane by a variety of proteinaceous structures (yellow). Green structures in the presynaptic membrane, corresponding to the rows of particles seen in (A), are thought to be Ca^{2+} channels. (A and B from Heuser et al., 1979; C after Harlow et al., 2001)

remain one of the strongest lines of support for the idea that a quantum of transmitter release is due to a synaptic vesicle fusing with the presynaptic membrane. Subsequent evidence, based on other means of measuring vesicle fusion, has left no doubt about the validity of this general interpretation of chemical synaptic transmission. Very recent work has identified structures within the presynaptic terminal that connect vesicles to the plasma membrane and may be involved in membrane fusion (Figure 5.8C).

Local Recycling of Synaptic Vesicles

The fusion of synaptic vesicles causes new membrane to be added to the plasma membrane of the presynaptic terminal, but the addition is not permanent. Although a bout of exocytosis can dramatically increase the surface area of presynaptic terminals, this extra membrane is removed within a few minutes. Heuser and Reese performed another important set of experiments showing that the fused vesicle membrane is actually retrieved and taken back into the cytoplasm of the nerve terminal (a process called endocytosis). The experiments, again carried out at the frog neuromuscular junction, were based on filling the synaptic cleft with horseradish peroxidase (HRP), an enzyme that can be made to produce a dense reaction product that is visible in an electron microscope. Under appropriate experimental conditions, endocytosis could then be visualized by the uptake of HRP into the nerve terminal (Figure 5.9). To activate endocytosis, the presynaptic terminal was stimulated with a train of

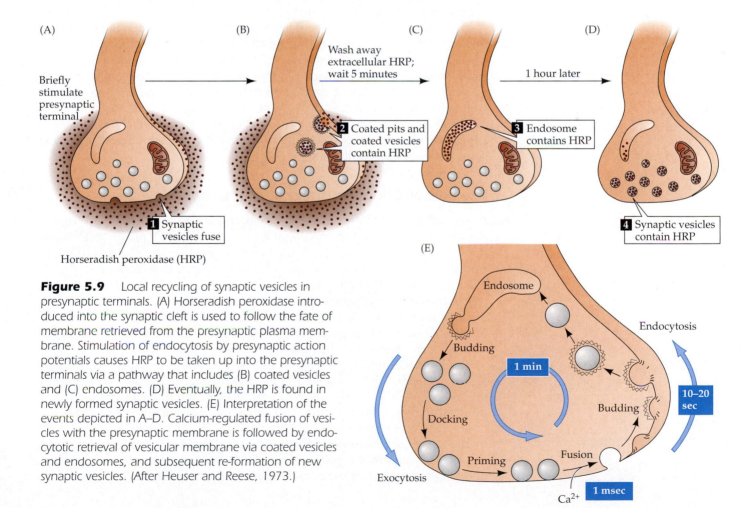

Figure 5.9   Local recycling of synaptic vesicles in presynaptic terminals. (A) Horseradish peroxidase introduced into the synaptic cleft is used to follow the fate of membrane retrieved from the presynaptic plasma membrane. Stimulation of endocytosis by presynaptic action potentials causes HRP to be taken up into the presynaptic terminals via a pathway that includes (B) coated vesicles and (C) endosomes. (D) Eventually, the HRP is found in newly formed synaptic vesicles. (E) Interpretation of the events depicted in A–D. Calcium-regulated fusion of vesicles with the presynaptic membrane is followed by endocytotic retrieval of vesicular membrane via coated vesicles and endosomes, and subsequent re-formation of new synaptic vesicles. (After Heuser and Reese, 1973.)

action potentials, and the subsequent fate of the HRP was followed by electron microscopy. Immediately following stimulation, the HRP was found within special endocytotic organelles called coated vesicles (Figure 5.9A,B). A few minutes later, however, the coated vesicles had disappeared and the HRP was found in a different organelle, the endosome (Figure 5.9C). Finally, within an hour after stimulating the terminal, the HRP reaction product appeared inside synaptic vesicles (Figure 5.9D).

These observations indicate that synaptic vesicle membrane is recycled within the presynaptic terminal via the sequence summarized in Figure 5.9E. In this process, called the **synaptic vesicle cycle**, the retrieved vesicular membrane passes through a number of intracellular compartments—such as coated vesicles and endosomes—and is eventually used to make new synaptic vesicles. After synaptic vesicles are re-formed, they are stored in a reserve pool within the cytoplasm until they need to participate again in neurotransmitter release. These vesicles are mobilized from the reserve pool, docked at the presynaptic plasma membrane, and primed to participate in exocytosis once again. More recent experiments, employing a fluorescent label rather than HRP, have determined the time course of synaptic vesicle recycling. These studies indicate that the entire vesicle cycle requires approximately 1 minute, with membrane budding during endocytosis requiring 10–20 seconds of this time. As can be seen from the 1-millisecond delay in transmission following excitation of the presynaptic terminal (see Figure 5.6B), membrane fusion during exocytosis is much more rapid than budding during endocytosis. Thus, all of the recycling steps interspersed between membrane budding and subsequent re-fusion of a vesicle are completed in less than a minute.

The precursors to synaptic vesicles *originally* are produced in the endoplasmic reticulum and Golgi apparatus in the neuronal cell body. Because of the long distance between the cell body and the presynaptic terminal in most neurons, transport of vesicles from the soma would not permit rapid replenishment of synaptic vesicles during continuous neural activity. Thus, local recycling is well suited to the peculiar anatomy of neurons, giving nerve terminals the means to provide a continual supply of synaptic vesicles. As might be expected, defects in synaptic vesicle recycling can cause severe neurological disorders, some of which are described in Box 5B.

The Role of Calcium in Transmitter Secretion

As was apparent in the experiments of Katz and others described in the preceding sections, lowering the concentration of Ca^{2+} outside a presynaptic motor nerve terminal reduces the size of the EPP (compare Figure 5.6B and D). Moreover, measurement of the number of transmitter quanta released under such conditions shows that the reason the EPP gets smaller is that lowering Ca^{2+} concentration decreases the number of vesicles that fuse with the plasma membrane of the terminal. An important insight into *how* Ca^{2+} regulates the fusion of synaptic vesicles was the discovery that presynaptic terminals have voltage-sensitive Ca^{2+} channels in their plasma membranes (see Chapter 4).

The first indication of presynaptic Ca^{2+} channels was provided by Katz and Ricardo Miledi. They observed that presynaptic terminals treated with tetrodotoxin (which blocks Na^+ channels; see Chapter 3) could still produce a peculiarly prolonged type of action potential. The explanation for this surprising finding was that current was still flowing through Ca^{2+} channels, substituting for the current ordinarily carried by the blocked Na^+ channels. Subsequent voltage clamp experiments, performed by Rodolfo Llinás and others at a giant presynaptic terminal of the squid (Figure 5.10A), confirmed the presence of

BOX 5B   Diseases That Affect the Presynaptic Terminal

Various steps in the exocytosis and endocytosis of synaptic vesicles are targets of a number of rare but debilitating neurological diseases. Many of these are myasthenic syndromes, in which abnormal transmission at neuromuscular synapses leads to weakness and fatigability of skeletal muscles. One of the best understood examples of such disorders is the *Lambert-Eaton myasthenic syndrome* (LEMS), an occasional complication in patients with certain kinds of cancers. Biopsies of muscle tissue removed from LEMS patients allow intracellular recordings identical to those shown in Figure 5.6. Such recordings have shown that when a motor neuron is stimulated, the number of quanta contained in individual EPPs is greatly reduced, although the amplitude of spontaneous MEPPs is normal. Thus, LEMS impairs evoked neurotransmitter release, but does not affect the size of individual quanta.

Several lines of evidence indicate that this reduction in neurotransmitter release is due to a loss of voltage-gated Ca^{2+} channels in the presynaptic terminal of motor neurons. This defect in neuromuscular transmission can be overcome by increasing the extracellular concentration of Ca^{2+}, and anatomical studies indicate a lower density of Ca^{2+} channel proteins in the presynaptic plasma membrane. The loss of presynaptic Ca^{2+} channels in LEMS apparently arises from a defect in the immune system. The blood of LEMS patients has a very high concentration of antibodies that bind to Ca^{2+} channels, and it seems likely that these antibodies are the primary cause of LEMS; removal of Ca^{2+} channel antibodies from the blood of LEMS patients by plasma exchange reduces muscle weakness. Similarly, immunosuppressant drugs can alleviate LEMS symptoms. Perhaps most telling, injecting antibodies from the blood of LEMS patients into experimental animals elicits muscle weakness and abnormal neuromuscular transmission.

Why the immune system generates antibodies against Ca^{2+} channels is not clear. Most LEMS patients have small-cell carcinoma, a form of lung cancer that may somehow initiate the immune response to Ca^{2+} channels. Whatever the origin, the binding of antibodies to Ca^{2+} channels causes a reduction in Ca^{2+} channel currents. It is this antibody-induced defect in presynaptic Ca^{2+} entry that accounts for the muscle weakness associated with LEMS.

Genetic disorders termed *congenital myasthenic syndromes* also cause muscle weakness by affecting neuromuscular transmission. Some of these syndromes affect the acetylcholinesterase that degrades acetylcholine in the synaptic cleft, whereas others arise from autoimmune attack of acetylcholine receptors (see Box 6B). However, a number of these congenital syndromes arise from defects in acetylcholine release due to altered synaptic vesicle traffic within the motor neuron terminal. Neuromuscular synapses in some patients have EPPs with reduced quantal content, a deficit that is especially prominent when the synapse is activated repeatedly. Electron microscopy shows that presynaptic motor nerve terminals have a greatly reduced number of synaptic vesicles. The defect in neurotransmitter release evidently results from an inadequate number of synaptic vesicles available for release during sustained presynaptic activity. The origins of this shortage of synaptic vesicles is not clear, but could result either from an impairment in endocytosis in the nerve terminal (see figure) or from a reduced supply of vesicles from the motor neuron cell body.

Patients suffering from *familial infantile myasthenia* appear to have neuromuscular weakness that arises from reductions in the size of individual quanta rather than the number of quanta released. Motor nerve terminals

from these patients have synaptic vesicles that are normal in number, but smaller in diameter. This finding suggests a genetic lesion that somehow alters formation of new synaptic vesicles following endocytosis, thereby leading to less acetylcholine in each vesicle.

Another disorder of synaptic transmitter release results from poisoning by anaerobic *Clostridium* bacteria. Microorganisms belonging to this genus produce some of the most potent toxins known, including several botulinum toxins and tetanus toxin. Both botulism and tetanus are potentially deadly disorders.

Botulism can occur by consuming food containing *Clostridium* bacteria or by infection of wounds with the spores of these ubiquitous organisms. In either case, the presence of the toxin can cause

(Continued on next page)

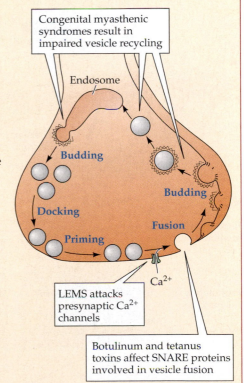

Congenital myasthenic syndromes result in impaired vesicle recycling

Endosome

Budding

Budding

Docking

Fusion

Priming

Ca^{2+}

LEMS attacks presynaptic Ca^{2+} channels

Botulinum and tetanus toxins affect SNARE proteins involved in vesicle fusion

The presynaptic targets of several neurological disorders.

BOX 5B (Continued)

paralysis of peripheral neuromuscular synapses due to abolition of neurotransmitter release. This interference with neuromuscular transmission causes skeletal muscle weakness, in extreme cases producing respiratory failure due to paralysis of the diaphragm and other muscles required for breathing. Botulinum toxins also block synapses innervating the smooth muscles of several organs, giving rise to visceral motor dysfunction.

Tetanus typically results from the contamination of puncture wounds by *Clostridium* bacteria that produce tetanus toxin. In contrast to botulism, tetanus poisoning blocks the release of inhibitory transmitters from interneurons in the spinal cord. This effect causes a loss of synaptic inhibition on spinal

motor neurons, producing hyperexcitation of skeletal muscle and tetanic contractions in affected muscles (hence the name of the disease).

Although their clinical consequences are dramatically different, clostridial toxins have a common mechanism of action (see figure). Tetanus toxin and botulinum toxins work by cleaving the SNARE proteins involved in fusion of synaptic vesicles with the presynaptic plasma membrane (see Box 5C). This proteolytic action presumably accounts for the block of transmitter release at the afflicted synapses. Their differing actions on synaptic transmission at excitatory motor versus inhibitory synapses apparently results from the fact that these toxins are taken up by different types of neurons.

Whereas botulinum toxins are taken up by motor neurons, tetanus toxin is preferentially targeted to interneurons. The basis for this differential uptake of toxins is not known, but presumably arises from the presence of different types of toxin receptors on the two types of neurons.

References

ENGEL, A. G. (1991) Review of evidence for loss of motor nerve terminal calcium channels in Lambert-Eaton myasthenic syndrome. *Ann. N.Y. Acad. Sci.* 635: 246–258.

ENGEL, A. G. (1994) Congenital myasthenic syndromes. *Neurol. Clin.* 12: 401–437.

LANG, B. AND A. VINCENT (2003) Autoantibodies to ion channels at the neuromuscular junction. *Autoimmun. Rev.* 2: 94–100.

MASELLI, R. A. (1998) Pathogenesis of human botulism. *Ann. N.Y. Acad. Sci.* 841: 122–139.

voltage-gated Ca^{2+} channels in the presynaptic terminal (Figure 5.10B). Such experiments showed that the amount of neurotransmitter released is very sensitive to the exact amount of Ca^{2+} that enters. Further, blockade of these Ca^{2+} channels with drugs also inhibits transmitter release (Figure 5.10B, right panels). These observations all confirm that the voltage-gated Ca^{2+} channels are directly involved in neurotransmission. Thus, presynaptic action potentials open voltage-gated Ca^{2+} channels, with a resulting influx of Ca^{2+}.

That Ca^{2+} entry into presynaptic terminals causes a rise in the concentration of Ca^{2+} within the terminal has been documented by microscopic imaging of terminals filled with Ca^{2+}-sensitive fluorescent dyes (Figure

(A)

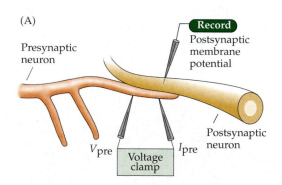

(B)

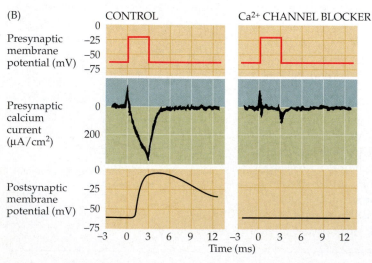

Figure 5.10 The entry of Ca^{2+} through the specific voltage-dependent calcium channels in the presynaptic terminals causes transmitter release. (A) Experimental setup using an extraordinarily large synapse in the squid. The voltage clamp method detects currents flowing across the presynaptic membrane when the membrane potential is depolarized. (B) Pharmacological agents that block currents flowing through Na^+ and K^+ channels reveal a remaining inward current flowing through Ca^{2+} channels. This influx of calcium triggers transmitter secretion, as indicated by a change in the postsynaptic membrane potential. Treatment of the same presynaptic terminal with cadmium, a calcium channel blocker, eliminates both the presynaptic calcium current and the postsynaptic response. (After Augustine and Eckert, 1984.)

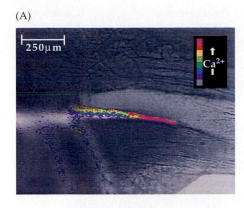

(A)

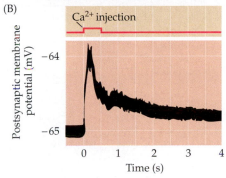

(B)

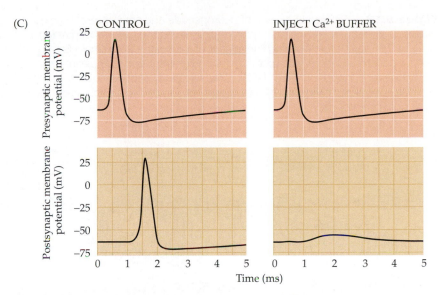

(C)

Figure 5.11 Evidence that a rise in presynaptic Ca^{2+} concentration triggers transmitter release from presynaptic terminals. (A) Fluorescence microscopy measurements of presynaptic Ca^{2+} concentration at the squid giant synapse (see Figure 5.8A). A train of presynaptic action potentials causes a rise in Ca^{2+} concentration, as revealed by the dye fura-2, which fluoresces more strongly as Ca^{2+} concentration increases. (B) Microinjection of Ca^{2+} into a squid giant presynaptic terminal triggers transmitter release, measured as a depolarization of the postsynaptic membrane potential. (C) Microinjection of BAPTA, a Ca^{2+} chelator, into a squid giant presynaptic terminal prevents transmitter release. (A from Smith et al., 1993; B after Miledi, 1971; C after Adler et al., 1991.)

5.11A). The consequences of the rise in presynaptic Ca^{2+} concentration for neurotransmitter release has been directly shown in two ways. First, microinjection of Ca^{2+} into presynaptic terminals triggers transmitter release in the absence of presynaptic action potentials (Figure 5.11B). Second, presynaptic microinjection of calcium chelators (chemicals that bind Ca^{2+} and keep its concentration buffered at low levels) prevents presynaptic action potentials from causing transmitter secretion (Figure 5.11C). These results prove beyond any doubt that a rise in presynaptic Ca^{2+} concentration is both necessary and sufficient for neurotransmitter release. Thus, as is the case for many other forms of neuronal signaling (see Chapter 7), Ca^{2+} serves as a second messenger during transmitter release.

While Ca^{2+} is a universal trigger for transmitter release, not all transmitters are released with the same speed. For example, while secretion of ACh from motor neurons requires only a fraction of a millisecond (see Figure 5.6), release of neuropeptides require high-frequency bursts of action potentials for many seconds. These differences in the rate of release probably arise from differences in the spatial arrangement of vesicles relative to presynaptic Ca^{2+} channels. This perhaps is most evident in cases where small molecules and peptides serve as co-transmitters (Figure 5.12). Whereas the small, clear-core vesicles containing small-molecule transmitters are typically docked at the plasma membrane in advance of Ca^{2+} entry, large dense core vesicles containing peptide transmitters are farther away from the plasma membrane (see Figure 5.5D). At low firing frequencies, the concentration of Ca^{2+} may increase only locally at the presynaptic plasma membrane, in the vicinity of open Ca^{2+} channels, limiting release to small-molecule transmitters from the docked small, clear-core vesicles. Prolonged high-frequency stimulation increases the Ca^{2+}

Figure 5.12 Differential release of neuropeptide and small-molecule co-transmitters. Low-frequency stimulation preferentially raises the Ca^{2+} concentration close to the membrane, favoring the release of transmitter from small clear-core vesicles docked at presynaptic specializations. High-frequency stimulation leads to a more general increase in Ca^{2+}, causing the release of peptide neurotransmitters from large dense-core vesicles, as well as small-molecule neurotransmitters from small clear-core vesicles.

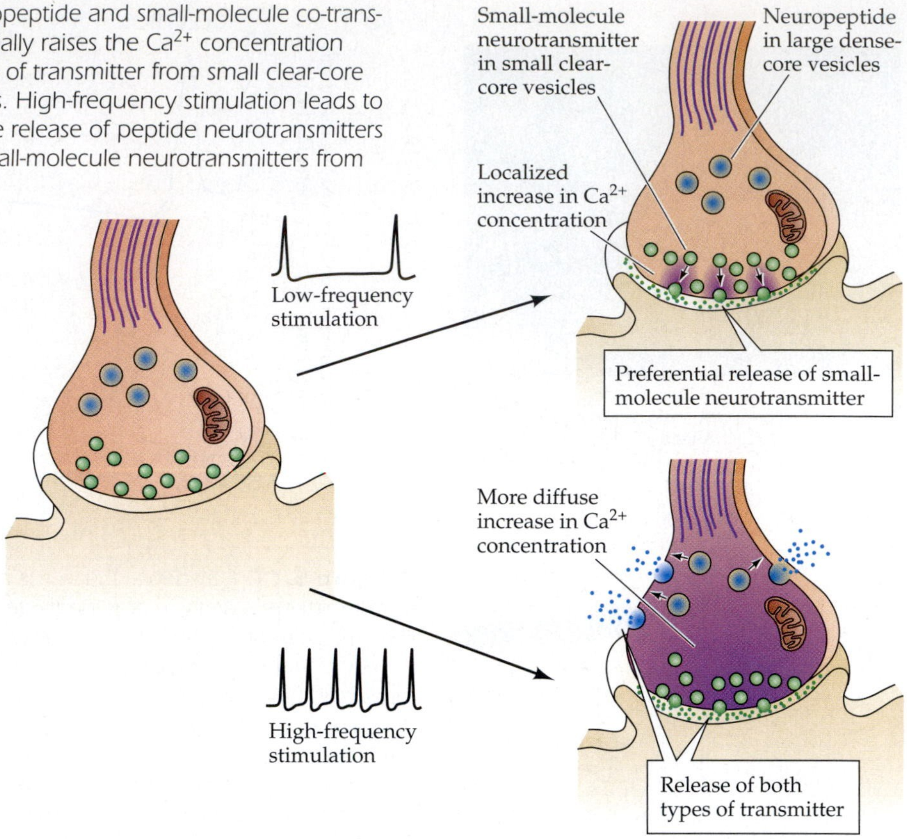

concentration throughout the presynaptic terminal, thereby inducing the slower release of neuropeptides.

Molecular Mechanisms of Transmitter Secretion

Precisely how an increase in presynaptic Ca^{2+} concentration goes on to trigger vesicle fusion and neurotransmitter release is not understood. However, many important clues have come from molecular studies that have identified and characterized the proteins found on synaptic vesicles (Figure 5.13A) and their binding partners on the presynaptic plasma membrane and cytoplasm. Most, if not all, of these proteins act at one or more steps in the synaptic vesicle cycle. Although a complete molecular picture of neurotransmitter release is still lacking, the roles of several proteins involved in vesicle fusion have been deduced (Figure 5.13B).

Several lines of evidence indicate that the protein **synapsin**, which reversibly binds to synaptic vesicles, may cross-link vesicles to actin filaments in the cytoskeleton to keep the vesicles tethered within the reserve pool. Mobilization of these reserve pool vesicles is the result of phosphorylation of synapsin by protein kinases, most notably the **Ca^{2+}/calmodulin-dependent protein kinase, type II** (CaMKII; see Chapter 7). Phosphorylation by CaMKII allows synapsin to dissociate from the vesicles. Once vesicles are free from their reserve pool tethers, they make their way to the plasma membrane, to which they are then attached by poorly understood docking reactions. A series of priming reactions then prepares the vesicular and plasma membranes for fusion.

A large number of proteins are involved in priming, including some that are also involved in other types of membrane fusion events common to all cells (Figure 5.13B). For example, two proteins originally found to be important for

(A)

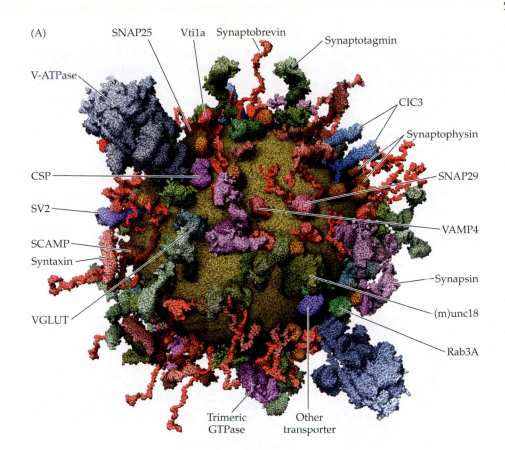

SNAP25 Vti1a Synaptobrevin Synaptotagmin

V-ATPase

CSP

SV2

SCAMP

Syntaxin

VGLUT

CIC3

Synaptophysin

SNAP29

VAMP4

Synapsin

(m)unc18

Rab3A

Trimeric GTPase   Other transporter

Figure 5.13 Presynaptic proteins and their roles in synaptic vesicle cycling. (A) Model of the molecular organization of a synaptic vesicle. The cytoplasmic surface of the vesicle membrane is densely covered by proteins, only 70 percent of which are shown here. (B) The vesicle trafficking cycle shown in Figure 5.9E is now known to be mediated by a number of presynaptic proteins, including some of those shown in (A), with different proteins participating in different reactions. (A from Takamori et al., 2006.)

(B)

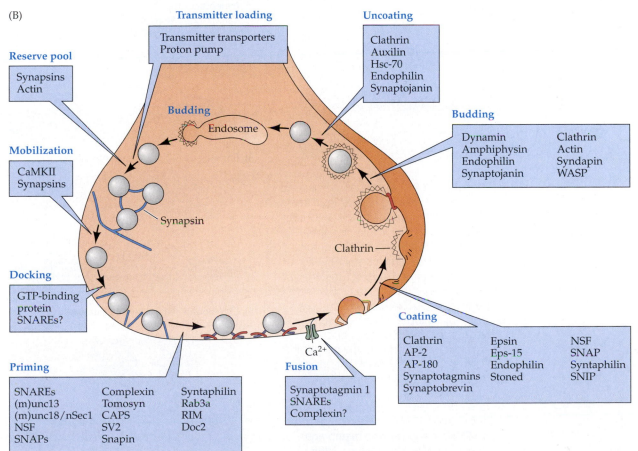

Transmitter loading

Transmitter transporters
Proton pump

Uncoating

Clathrin
Auxilin
Hsc-70
Endophilin
Synaptojanin

Reserve pool

Synapsins
Actin

Budding

Endosome

Budding

Dynamin	Clathrin
Amphiphysin	Actin
Endophilin	Syndapin
Synaptojanin	WASP

Mobilization

CaMKII
Synapsins

Synapsin

Clathrin

Docking

GTP-binding protein
SNAREs?

Coating

Ca^{2+}

Clathrin	Epsin	NSF
AP-2	Eps-15	SNAP
AP-180	Endophilin	Syntaphilin
Synaptotagmins	Stoned	SNIP
Synaptobrevin		

Priming

SNAREs	Complexin	Syntaphilin
(m)unc13	Tomosyn	Rab3a
(m)unc18/nSec1	CAPS	RIM
NSF	SV2	Doc2
SNAPs	Snapin	

Fusion

Synaptotagmin 1
SNAREs
Complexin?

the fusion of vesicles with membranes of the Golgi apparatus, the ATPase **NSF** (*NEM-sensitive fusion protein*) and **SNAPs** (*soluble NSF-attachment proteins*), are also involved in priming synaptic vesicles for fusion. These two proteins work by regulating the assembly of other proteins that are called **SNAREs** (*SNAp REceptors*). Many of the other proteins involved in priming—including (m)unc-13, nSec-1, complexin, snapin, syntaphilin, and tomosyn—also interact with the SNAREs.

One of the main purposes of priming seems to be to organize SNARE proteins into the correct conformation for membrane fusion. One of the SNARE proteins, **synaptobrevin**, is in the membrane of synaptic vesicles, while two other SNARE proteins called **syntaxin** and **SNAP-25** are found primarily on the plasma membrane. These SNARE proteins can form a macromolecular complex that spans the two membranes, thus bringing them into close apposition (Figure

(A)

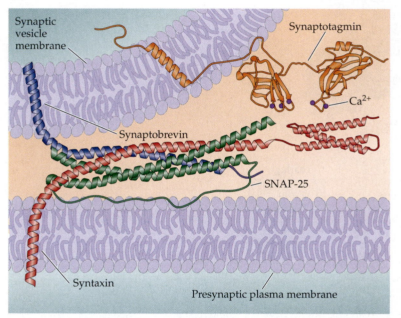

(B) (1) Vesicle docks

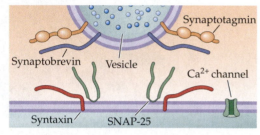

(2) SNARE complexes form to pull membranes together

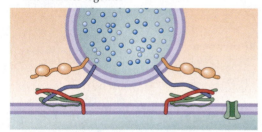

(3) Entering Ca^{2+} binds to synaptotagmin

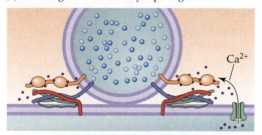

(4) Ca^{2+}-bound synaptotagmin catalyzes membrane fusion

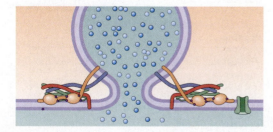

Figure 5.14 Molecular mechanisms of exocytosis during neurotransmitter release. (A) Structure of the SNARE complex. The vesicular SNARE, synaptobrevin (blue), forms a helical complex with the plasma membrane SNAREs syntaxin (red) and SNAP-25 (green). Also shown is the structure of synaptotagmin, a vesicular Ca^{2+}-binding protein. (B) A model for Ca^{2+}-triggered vesicle fusion. SNARE proteins on the synaptic vesicle and plasma membranes form a complex (as in A) that brings together the two membranes. Ca^{2+} then binds to synaptotagmin, causing the cytoplasmic region of this protein to insert into the plasma membrane, bind to SNAREs and catalyze membrane fusion. (A after Sutton et al., 1998.)

5.14A). Such an arrangement is well suited to promote the fusion of the two membranes, and several lines of evidence suggest that this is what actually occurs. One important observation is that toxins that cleave the SNARE proteins block neurotransmitter release (Box 5C). In addition, putting SNARE proteins into artificial lipid membranes and allowing these proteins to form complexes with each other causes the membranes to fuse.

Because the SNARE proteins do not bind Ca^{2+}, still other molecules must be responsible for Ca^{2+} regulation of neurotransmitter release. Several presynaptic proteins, including calmodulin, CAPS, and (m)unc-13, are capable of binding Ca^{2+}. However, it appears that Ca^{2+} regulation of neurotransmitter release is conferred by **synaptotagmin**, a protein found in the membrane of synaptic vesicles (Figure 5.14A). Synaptotagmin binds Ca^{2+} at concentrations similar to those required to trigger vesicle fusion within the presynaptic terminal; this property allows synaptotagmin to act as a Ca^{2+} sensor, signaling the elevation of Ca^{2+} within the terminal and thus triggering vesicle fusion. In support of this idea, disruption of synaptotagmin in the presynaptic terminals of mice, fruit flies, squid, and other experimental animals impairs Ca^{2+}-dependent neurotransmitter release. In fact, deletion of even one of the 19 synaptotagmin genes of mice is a lethal mutation, causing the mice to die soon after birth. How Ca^{2+} binding to synaptotagmin leads to exocytosis is not yet clear. It is known that Ca^{2+} changes the chemical properties of synaptotagmin, allowing it to insert into membranes and to bind to other proteins, most notably the SNAREs. A plausible model is that SNARE proteins bring the two membranes close together, and that Ca^{2+}-induced changes in synaptotagmin then produce the final fusion of these membranes (Figure 5.14B).

Still other proteins appear to be involved at the endocytosis steps of the synaptic vesicle cycle (Figure 5.15). The most important protein involved in endocytotic budding of vesicles from the plasma membrane is **clathrin**. Clathrin

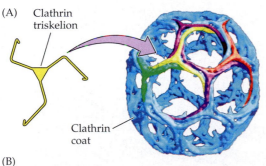

(A) Clathrin triskelion

Clathrin coat

(B)

Figure 5.15 Molecular mechanisms of endocytosis following neurotransmitter release. (A) Individual clathrin triskelia (left) assemble together to form membrane coats (right) involved in membrane budding during endocytosis. (B) A model for membrane budding during endocytosis. Following addition of synaptic vesicle membrane during exocytosis, clathrin triskelia attach to the vesicular membrane. Their attachment is aided by adaptor proteins (such as AP-2 and AP180). Polymerization of clathrin causes the membrane to curve, allowing dynamin to pinch off the coated vesicle. Subsequent uncoating of the vesicle by the ATPase Hsc-70 and a co-factor, auxilin, yields a synaptic vesicle. (A after Marsh and McMahon, 2001.)

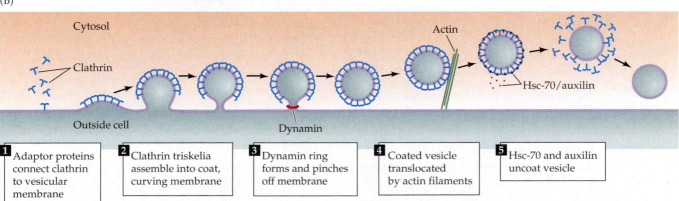

Cytosol

Actin

Clathrin

Hsc-70/auxilin

Outside cell

Dynamin

| **1** Adaptor proteins connect clathrin to vesicular membrane | **2** Clathrin triskelia assemble into coat, curving membrane | **3** Dynamin ring forms and pinches off membrane | **4** Coated vesicle translocated by actin filaments | **5** Hsc-70 and auxilin uncoat vesicle |

BOX 5C Toxins That Affect Transmitter Release

Several important insights into the molecular basis of neurotransmitter secretion have come from analyzing the actions of a series of biological toxins produced by a fascinating variety of organisms. One family of such agents is the clostridial toxins responsible for botulism and tetanus (see Box 5B). These substances exert their toxic actions by inhibiting exocytotic neurotransmitter release from presynaptic terminals. Clever and patient biochemical work has shown that these toxins are highly specific proteases that cleave presynaptic SNARE proteins (see figure). Tetanus toxin and botulinum toxin (types B, D, F, and G) specifically cleave the vesicle SNARE protein synaptobrevin. Other botulinum toxins are proteases that cleave syntaxin (type C) and SNAP-25 (types A and E), SNARE proteins found on the presynaptic plasma membrane. Destruction of these presynaptic proteins is the basis for the inhibitory actions of the toxins on neurotransmitter release. These observations, as well as other evidence described in the text, provide proof that the three synaptic SNARE proteins are important for fusion of synaptic vesicles with the presynaptic plasma membrane.

Another toxin that targets neurotransmitter release is α-latrotoxin, a protein found in the venom of the female black widow spider. Application of this molecule to neuromuscular synapses causes a massive discharge of synaptic vesicles, even when Ca^{2+} is absent from the extracellular medium. While it is not yet clear how this toxin triggers Ca^{2+}-independent exocytosis, α-latrotoxin binds to two different types of presynaptic proteins that may mediate its actions. One group of binding partners for α-latrotoxin is the neurexins, a group of integral membrane proteins found in presynaptic terminals. Several lines of evidence implicate binding to neurexins in at least some of the actions of α-latrotoxin. Because the

neurexins bind to synaptotagmin, a vesicular Ca^{2+}-binding protein that is known to be important in exocytosis, this interaction may allow α-latrotoxin to bypass the usual Ca^{2+} requirement for triggering vesicle fusion. Another type of presynaptic protein that can bind to α-latrotoxin is called CL1 (based on its previous names, Ca^{2+}-independent receptor for latrotoxin and *latrophilin-1*). CL1 is a relative of the G-protein-coupled receptors that mediate the actions of neurotransmitters and other extracellular chemical signals (see Chapter 7). Thus, the binding of α-latrotoxin to CL1 is thought to activate an intracellular signal transduction cascade that may be involved in the Ca^{2+}-independent actions of α-latrotoxin. While more work is needed to establish the roles of neurexins and CL1 in the actions of α-latrotoxin definitively, effects on these two proteins probably account for the potent presynaptic actions of this toxin.

Still other toxins produced by snakes, snails, spiders, and other predatory animals are known to affect transmitter release, but their sites of action have yet to be identified. Based on the precedents described here, it is likely that these biological poisons will continue to provide valuable tools for elucidating the molecular basis of neurotransmitter release, just as they will continue to enable the predators to feast on their prey.

References

GRUMELLI, C., C. VERDERIO, D. POZZI, O. ROSSETTO, C. MONTECUCCO AND M. MATTEOLI (2005) Internalization and mechanism of action of clostridial toxins in neurons. *Neurotoxicology* 26: 761–767.

HUMEAU, Y., F. DOUSSAU, N. J. GRANT AND B. POULAIN (2000) How botulinum and tetanus neurotoxins block neurotransmitter release. *Biochimie* 82: 427–446.

SUDHOF, T. C. (2001) α-Latrotoxin and its receptors: Neurexins and CIRL/latrophilins. *Annu. Rev. Neurosci.* 24: 933–962.

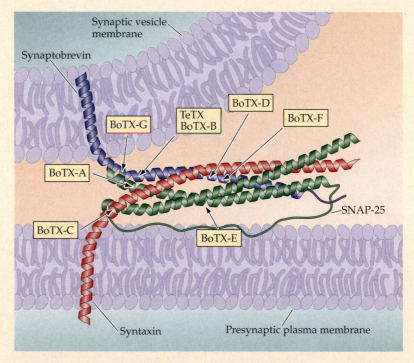

Cleavage of SNARE proteins by clostridial toxins. Indicated are the sites of proteolysis by tetanus toxin (TeTX) and various types of botulinum toxin (BoTX). (After Sutton et al., 1998.)

has a unique structure, called a *triskelion* because of its three-legged appearance (Figure 5.15A). During endocytosis, clathrin triskelia attach to the vesicular membrane that is to be retrieved (Figure 5.15B). A number of adaptor proteins, such as AP-2 and AP180, connect clathrin to the proteins and lipids of this membrane. These adaptor proteins, as well as other proteins such as amphiphysin, epsin, and Eps-15, help to assemble individual triskelia into structures that resemble geodesic domes (see Figure 5.15A). The domelike structures form coated pits that initiate membrane budding, increasing the curvature of the budding membrane until it forms a coated, vesicle-like structure. Another protein, called **dynamin**, causes the final pinching-off of membrane that completes the production of coated vesicles. The clathrin coats then are removed by the ATPase **Hsc-70**, with another protein, **auxilin**, serving as a co-factor. Other proteins, such as **synaptojanin**, are also important for vesicle uncoating. Uncoated vesicles can then continue their journey through the recycling process, eventually becoming refilled with neurotransmitter due to the actions of neurotransmitter transporters in the vesicle membrane. These transporters exchange protons within the vesicle for neurotransmitter; the acidic interior of the vesicle is produced by a proton pump that also is located in the vesicle membrane.

In summary, a complex cascade of proteins, acting in a defined temporal and spatial order, allows neurons to secrete transmitters. Although our understanding of the detailed molecular mechanisms responsible for transmitter secretion is not completely clear, rapid progress is being made toward this goal.

Neurotransmitter Receptors

The generation of postsynaptic electrical signals is also understood in considerable depth. Such studies began in 1907, when the British physiologist John N. Langley introduced the concept of **receptor molecules** to explain the specific and potent actions of certain chemicals on muscle and nerve cells. Much subsequent work has shown that receptor molecules do indeed account for the ability of neurotransmitters, hormones, and drugs to alter the functional properties of neurons. While it has been clear since Langley's day that receptors are important for synaptic transmission, their identity and detailed mechanism of action remained a mystery until quite recently. It is now known that neurotransmitter receptors are proteins embedded in the plasma membrane of postsynaptic cells. Domains of receptor molecules that extend into the synaptic cleft bind neurotransmitters that are released by the presynaptic neuron. The binding of neurotransmitters, either directly or indirectly, causes ion channels in the postsynaptic membrane to open or close. Typically, the resulting ion fluxes change the membrane potential of the postsynaptic cell, thus mediating the transfer of information across the synapse.

Postsynaptic Membrane Permeability Changes during Synaptic Transmission

Just as studies of the neuromuscular synapse paved the way for understanding neurotransmitter release mechanisms, this peripheral synapse has been equally valuable for understanding the mechanisms that allow neurotransmitter receptors to generate postsynaptic signals. The binding of ACh to postsynaptic receptors opens ion channels in the muscle fiber membrane. This effect can be demonstrated directly by using the patch clamp method (see Box 4A) to measure the minute postsynaptic currents that flow when two molecules of individual ACh bind to receptors, as Erwin Neher and Bert Sakmann first did in 1976. Exposure of the extracellular surface of a patch of postsynaptic membrane to

ACh causes single-channel currents to flow for a few milliseconds (Figure 5.16A). This shows that ACh binding to its receptors opens ligand-gated ion channels, much in the way that changes in membrane potential open voltage-gated ion channels (see Chapter 4).

The electrical actions of ACh are greatly multiplied when an action potential in a presynaptic motor neuron causes the release of millions of molecules of ACh into the synaptic cleft. In this more physiological case, the transmitter molecules bind to many thousands of ACh receptors packed in a dense array on the postsynaptic membrane, transiently opening a very large number of postsynaptic ion channels. Although individual ACh receptors only open briefly, (Figure 5.16B1), the opening of a large number of channels is synchronized by the brief duration during which ACh is secreted from presynaptic terminals (Figure 5.16B2,3). The macroscopic current resulting from the summed opening of many ion channels is called the **end plate current**, or **EPC**. Because the current flowing during the EPC is normally inward, it causes the postsynaptic membrane potential to depolarize. This depolarizing change in potential is the EPP (Figure 5.16C), which typically triggers a postsynaptic action potential by opening voltage-gated Na^+ and K^+ channels (see Figure 5.6B).

(A) Patch clamp measurement of single ACh receptor current

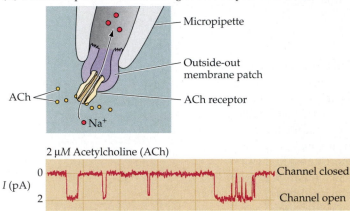

(B) Currents produced by:

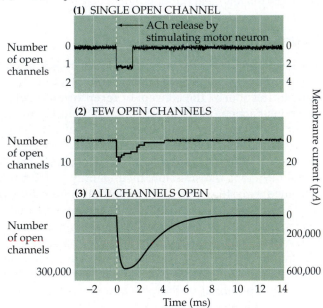

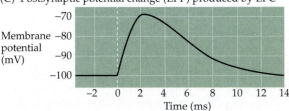

Figure 5.16 *Activation of ACh receptors at neuromuscular synapses. (A) Outside-out patch clamp measurement of single ACh receptor currents from a patch of membrane removed from the postsynaptic muscle cell. When ACh is applied to the extracellular surface of the membrane clamped at negative voltages, the repeated brief opening of a single channel can be seen as downward deflections corresponding to inward current (i.e., positive ions flowing into the cell). (B) Synchronized opening of many ACh-activated channels at a synapse being voltage-clamped at negative voltages. (1) If a single channel is examined during the release of ACh from the presynaptic terminal, the channel opens transiently. (2) If a number of channels are examined together, ACh release opens the channels almost synchronously. (3) The opening of a very large number of postsynaptic channels produces a macroscopic EPC. (C) In a normal muscle cell (i.e., not being voltage-clamped), the inward EPC depolarizes the postsynaptic muscle cell, giving rise to an EPP. Typically, this depolarization generates an action potential (not shown).*

The identity of the ions that flow during the EPC can be determined via the same approaches used to identify the roles of Na^+ and K^+ fluxes in the currents underlying action potentials (see Chapter 3). Key to such an analysis is identifying the membrane potential at which no current flows during transmitter action. When the potential of the postsynaptic muscle cell is controlled by the voltage clamp method (Figure 5.17A), the magnitude of the membrane potential clearly affects the amplitude and polarity of EPCs (Figure 5.17B). Thus, when the postsynaptic membrane potential is made more negative than the resting potential, the amplitude of the EPC becomes larger, whereas this current is reduced when the membrane potential is made more positive. At approximately 0 mV, no EPC is

Figure 5.17 The influence of the postsynaptic membrane potential on end plate currents. (A) A postsynaptic muscle fiber is voltage clamped using two electrodes while the presynaptic neuron is electrically stimulated to cause the release of ACh from presynaptic terminals. This experimental arrangement allows the recording of macroscopic EPCs produced by ACh. (B) Amplitude and time course of EPCs generated by stimulating the presynaptic motor neuron while the postsynaptic cell is voltage clamped at four different membrane potentials. (C) The relationship between the peak amplitude of EPCs and postsynaptic membrane potential is nearly linear, with a reversal potential (the voltage at which the direction of the current changes from inward to outward) close to 0 mV. Also indicated on this graph are the equilibrium potentials of Na^+, K^+, and Cl^-. (D) Lowering the external Na^+ concentration causes EPCs to reverse at more negative potentials because E_{Na} is less positive. (E) Raising the external K^+ concentration makes the reversal potential more positive because E_K is less negative. (After Takeuchi and Takeuchi, 1960.)

(A) Scheme for voltage clamping postsynaptic muscle fiber

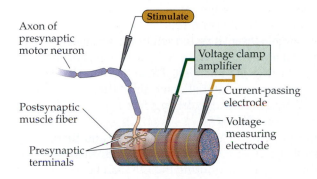

(B) Effect of membrane voltage on postsynaptic end plate currents (EPCs)

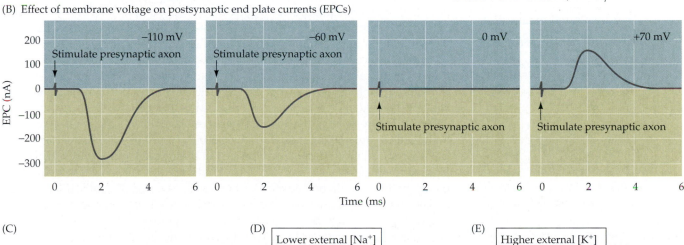

(C)

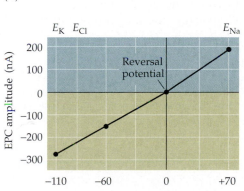

(D)

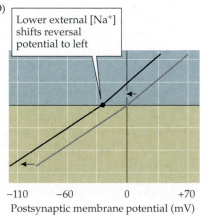

(E)

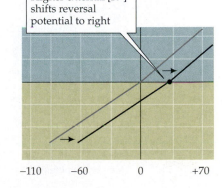

Postsynaptic membrane potential (mV)

(A)

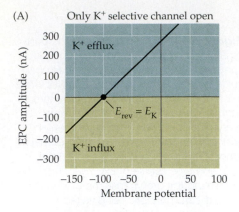

Only K⁺ selective channel open

(B)

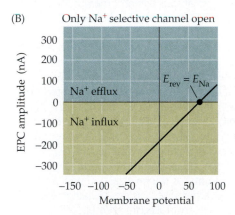

Only Na⁺ selective channel open

(C)

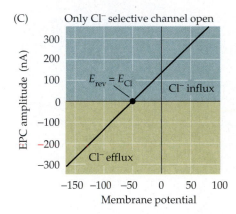

Only Cl⁻ selective channel open

(D)

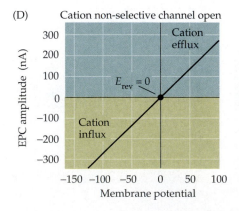

Cation non-selective channel open

detected, and at even more positive potentials, the current reverses its polarity, becoming outward rather than inward (Figure 5.17C). The potential where the EPC reverses, about 0 mV in the case of the neuromuscular junction, is called the **reversal potential**.

As was the case for currents flowing through voltage-gated ion channels (see Chapter 3), the magnitude of the EPC at any membrane potential is given by the product of the ionic conductance activated by ACh (g_{ACh}) and the electrochemical driving force on the ions flowing through ligand-gated channels. Thus, the value of the EPC is given by the relationship

$$EPC = g_{ACh}(V_m - E_{rev})$$

where E_{rev} is the reversal potential for the EPC. This relationship predicts that the EPC will be an inward current at potentials more negative than E_{rev} because the electrochemical driving force, $V_m - E_{rev}$, is a negative number. Further, the EPC will become smaller at potentials approaching E_{rev} because the driving force is reduced. At potentials more positive than E_{rev}, the EPC is outward because the driving force is reversed in direction (that is, positive). Because the channels opened by ACh are largely insensitive to membrane voltage, g_{ACh} will depend only on the number of channels opened by ACh, which depends in turn on the concentration of ACh in the synaptic cleft. Thus, the magnitude and polarity of the postsynaptic membrane potential determines the direction and amplitude of the EPC solely by altering the driving force on ions flowing through the receptor channels opened by ACh.

When V_m is at the reversal potential, $V_m - E_{rev}$ is equal to 0 and there is no net driving force on the ions that can permeate the receptor-activated channel. As a result, the identity of the ions that flow during the EPC can be deduced by observing how the reversal potential of the EPC compares to the equilibrium potential for various ion species (Figure 5.18). For example, if ACh were to open an ion channel permeable only to K⁺, then the reversal potential of the EPC would be at the equilibrium potential for K⁺, which for a muscle cell is close to −100 mV (Figure 5.18A). If the ACh-activated channels were permeable only to Na⁺, then the reversal potential of the current would be approximately +70 mV, the Na⁺ equilibrium potential of muscle cells (Figure 5.18B); if these channels were permeable only to Cl⁻, then the reversal potential would be approximately −50 mV (Figure 5.18C). By this reasoning, ACh-activated channels cannot be permeable to only one of these ions, because the reversal potential of the EPC is not near the equilibrium potential for any of them (see Figure 5.17C). However, if these channels were permeable to both Na⁺ and K⁺, then the reversal potential of the EPC would be between +70 mV and −100 mV (Figure 5.18D).

The fact that EPCs reverse at approximately 0 mV is therefore consistent with the idea that ACh-activated ion channels are almost equally permeable to both Na⁺ and K⁺. This was tested in 1960 by Akira and Noriko Takeuchi, who performed experiments showing that, as predicted, the magnitude and reversal

Figure 5.18 The effect of ion channel selectivity on the reversal potential. Voltage clamping a postsynaptic cell while activating presynaptic neurotransmitter release reveals the identity of the ions permeating the postsynaptic receptors being activated. (A) The activation of postsynaptic channels permeable only to K⁺ results in currents reversing at E_K, near −100 mV. (B) The activation of postsynaptic Na⁺ channels results in currents reversing at E_{Na}, near +70 mV. (C) Cl⁻-selective currents reverse at E_{Cl}, near −50 mV. (D) Ligand-gated channels that are about equally permeable to both K⁺ and Na⁺ show a reversal potential near 0 mV.

potential of the EPC is changed by altering the concentration gradients of Na^+ and K^+. Experimentally lowering the external Na^+ concentration, which makes E_{Na} more negative, produces a negative shift in E_{rev} (see Figure 5.17D), whereas elevating external K^+ concentration, which makes E_K more positive, causes E_{rev} to shift to a more positive potential (Figure 5.17E). Such experiments confirmed that the ACh-activated ion channels are in fact permeable to both Na^+ and K^+.

Even though the channels opened by the binding of ACh to its receptors are permeable to both Na^+ and K^+, at the resting membrane potential the EPC is generated primarily by Na^+ influx. If the membrane potential is kept at E_K, the EPC arises entirely from an influx of Na^+ because at this potential there is no driving force on K^+ (Figure 5.19A). At the usual muscle fiber resting membrane potential of –90 mV, there is a small driving force on K^+, but a much greater one on Na^+. Thus, during the EPC, much more Na^+ flows into the muscle cell than K^+ flows out (Figure 5.19B); it is the net influx of positively charged Na^+ that constitutes the inward current measured as the EPC. At the reversal potential of about 0 mV, Na^+ influx and K^+ efflux are exactly balanced, so no current flows during the opening of channels by ACh binding (Figure 5.19C). At potentials more positive than E_{rev} the balance reverses; for example, at E_{Na} there is no influx of Na^+ and a large efflux of K^+ because of the large driving force on K^+ (Figure 5.19D). Even more positive potentials cause efflux of both Na^+ and K^+ and produce an even larger outward EPC.

Were it possible to measure the EPP at the same time as the EPC (the voltage clamp technique prevents this by keeping membrane potential constant), the EPP would be seen to vary in parallel with the amplitude and polarity of the EPC (Figure 5.19E,F). At the usual postsynaptic resting membrane potential of –90 mV, the large inward EPC causes the postsynaptic membrane potential to become more depolarized, as seen in Figure 5.19F. However, at 0 mV, the EPP reverses its polarity, and at more positive potentials, the EPP is hyperpolarizing. Thus, the polarity and magnitude of the EPC depend on the electrochemical driving force, which in turn determines the polarity and magnitude of the EPP. EPPs will depolarize when the membrane potential is more negative than E_{rev}, and hyperpolarize when the membrane potential is more positive than E_{rev}. The general rule, then, is that *the action of a transmitter drives the postsynaptic membrane potential toward E_{rev} for the particular ion channels being activated.*

Although this discussion has focused on the neuromuscular junction, similar mechanisms generate postsynaptic responses at all chemical synapses. The general principle is that transmitter binding to postsynaptic receptors produces a postsynaptic conductance change as ion channels are opened (or sometimes closed). The postsynaptic conductance is increased if—as at the neuromuscular junction—channels are opened, and decreased if channels are closed. This conductance change typically generates an electrical current, the **postsynaptic current (PSC)**, which in turn changes the postsynaptic membrane potential to produce a **postsynaptic potential (PSP)**. As in the specific case of the EPP at the neuromuscular junction, PSPs are depolarizing if their reversal potential is more positive than the postsynaptic membrane potential and hyperpolarizing if their reversal potential is more negative.

The conductance changes and the PSPs that typically accompany them are the ultimate outcome of most chemical synaptic transmission, concluding a sequence of electrical and chemical events that begins with the invasion of an action potential into the terminals of a presynaptic neuron. In many ways, the events that produce PSPs at synapses are similar to those that generate action potentials in axons: conductance changes produced by ion channels lead to ionic current flow that changes the membrane potential (see Figure 5.19).

Figure 5.19   Na⁺ and K⁺ movements during EPCs and EPPs. (A–D) Each of the postsynaptic potentials (V_{post}) indicated at the left results in different relative fluxes of net Na⁺ and K⁺ (ion fluxes). These ion fluxes determine the amplitude and polarity of the EPCs, which in turn determine the EPPs. Note that at about 0 mV the Na⁺ flux is exactly balanced by an opposite K⁺ flux, resulting in no net current flow, and hence no change in the membrane potential. (E) EPCs are inward currents at potentials more negative than E_{rev} and outward currents at potentials more positive than E_{rev}. (F) EPPs depolarize the postsynaptic cell at potentials more negative than E_{rev}. At potentials more positive than E_{rev}, EPPs hyperpolarize the cell.

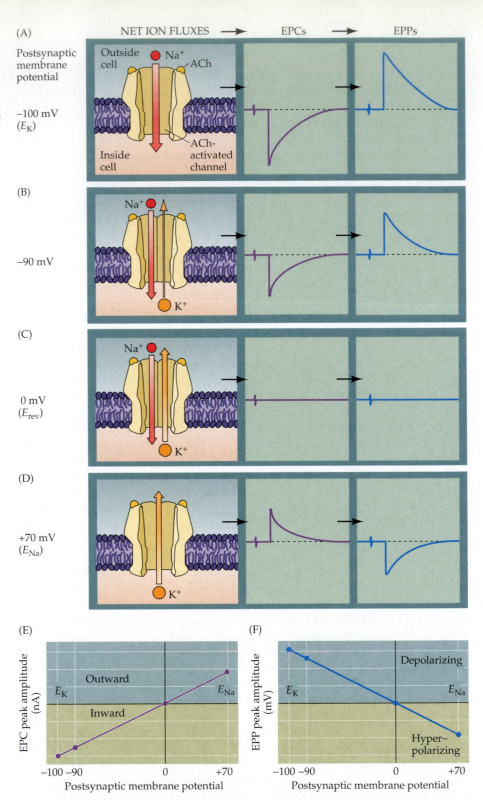

Excitatory and Inhibitory Postsynaptic Potentials

PSPs ultimately alter the probability that an action potential will be produced in the postsynaptic cell. At the neuromuscular junction, synaptic action increases the probability that an action potential will occur in the postsynaptic muscle cell; indeed, the large amplitude of the EPP ensures that an action potential is always triggered. At many other synapses, PSPs similarly increase the probability of firing a postsynaptic action potential. However, still other synapses *decrease* the probability that the postsynaptic cell will generate an action potential. PSPs are called **excitatory** (**EPSPs**) if they increase the likelihood of a postsynaptic action potential occurring, and **inhibitory** (**IPSPs**) if they decrease this likelihood. Given that most neurons receive inputs from both excitatory and inhibitory synapses, it is important to understand the mechanisms that determine whether a particular synapse excites or inhibits its postsynaptic partner.

The principles of excitation just described for the neuromuscular junction are pertinent to all excitatory synapses. The principles of postsynaptic inhibition are much the same as for excitation, and are also quite general. In both cases, neurotransmitters binding to receptors open or close ion channels in the postsynaptic cell. Whether there is an EPSP or an IPSP depends on the type of channel that is coupled to the receptor, and on the concentration of permeant ions inside and outside the cell. In fact, the only distinction between postsynaptic excitation and inhibition is the reversal potential of the PSP in relation to the threshold voltage for generating action potentials in the postsynaptic cell.

Consider, for example, a neuronal synapse that uses glutamate as the transmitter. Many such synapses have receptors that, like the ACh receptors at neuromuscular synapses, open ion channels that are nonselectively permeable to cations (see Chapter 6). When these glutamate receptors are activated, both Na^+ and K^+ flow across the postsynaptic membrane, yielding an E_{rev} of approximately 0 mV for the resulting postsynaptic current. If the resting potential of the postsynaptic neuron is –60 mV, the resulting EPSP will depolarize by bringing the postsynaptic membrane potential toward 0 mV. For the hypothetical neuron shown in Figure 5.20A, the action potential threshold voltage is –40 mV. Thus, a glutamate-induced EPSP will increase the probability that this neuron produces an action potential, defining the synapse as excitatory.

As an example of inhibitory postsynaptic action, consider a neuronal synapse that uses GABA as its transmitter. At such synapses, the GABA receptors typically open channels that are selectively permeable to Cl^- and the action of GABA causes Cl^- to flow across the postsynaptic membrane. Consider a case where E_{Cl} is –70 mV, as is typical for many neurons, so that the postsynaptic resting potential of –60 mV is less negative than E_{Cl}. The resulting positive electrochemical driving force ($V_m - E_{rev}$) will cause negatively charged Cl^- to flow into the cell and produce a hyperpolarizing IPSP (Figure 5.20B). This hyperpolarizing IPSP will take the postsynaptic membrane away from the action potential threshold of –40 mV, clearly inhibiting the postsynaptic cell.

Figure 5.20 *Reversal potentials and threshold potentials determine postsynaptic excitation and inhibition. (A) If the reversal potential for a PSP (0 mV) is more positive than the action potential threshold (–40 mV), the effect of a transmitter is excitatory, and it generates EPSPs. (B) If the reversal potential for a PSP is more negative than the action potential threshold, the transmitter is inhibitory and generate IPSPs. (C) IPSPs can nonetheless depolarize the postsynaptic cell if their reversal potential is between the resting potential and the action potential threshold. (D) The general rule of postsynaptic action is: If the reversal potential is more positive than threshold, excitation results; inhibition occurs if the reversal potential is more negative than threshold.*

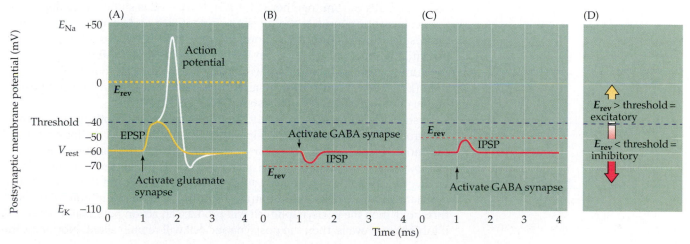

Surprisingly, inhibitory synapses need not produce hyperpolarizing IPSPs. For instance, if E_{Cl} were –50 mV instead of –70 mV, then the negative electrochemical driving force would cause Cl^- to flow out of the cell and produce a depolarizing IPSP (Figure 5.20C). However, the synapse would still be inhibitory: Given that the reversal potential of the IPSP still is more negative than the action potential threshold (–40 mV), the depolarizing IPSP would inhibit because the postsynaptic membrane potential would be kept more negative than the threshold for action potential initiation. Another way to think about this peculiar situation is that if another excitatory input onto this neuron brought the cell's membrane potential to –41 mV, just below threshold for firing an action potential, the IPSP would then hyperpolarize the membrane potential toward –50 mV, bringing the potential away from the action potential threshold. Thus, while EPSPs depolarize the postsynaptic cell, IPSPs can hyperpolarize or depolarize; indeed, an inhibitory conductance change may produce no potential change at all and still exert an inhibitory effect by making it more difficult for an EPSP to evoke an action potential in the postsynaptic cell.

Although the particulars of postsynaptic action can be complex, a simple rule distinguishes postsynaptic excitation from inhibition: An EPSP has a reversal potential more positive than the action potential threshold, whereas an IPSP has a reversal potential more negative than threshold (Figure 5.20D). Intuitively, this rule can be understood by realizing that an EPSP will tend to depolarize the membrane potential so that it exceeds threshold, whereas an IPSP will always act to keep the membrane potential more negative than the threshold potential.

Summation of Synaptic Potentials

The PSPs produced at most synapses in the brain are much smaller than those at the neuromuscular junction; indeed, EPSPs produced by individual excitatory synapses may be only a fraction of a millivolt and are usually well below the threshold for generating postsynaptic action potentials. How, then, can such synapses transmit information if their PSPs are subthreshold? The answer is that neurons in the central nervous system are typically innervated by thousands of synapses, and the PSPs produced by each active synapse can *sum together*—in space and in time—to determine the behavior of the postsynaptic neuron.

Consider the highly simplified case of a neuron that is innervated by two excitatory synapses, each generating a subthreshold EPSP, and by an inhibitory synapse that produces an IPSP (Figure 5.21A). While activation of either one of the excitatory synapses alone (E1 or E2 in Figure 5.21B) produces a subthreshold EPSP, activation of *both* excitatory synapses at about the same time causes the two EPSPs to sum together (E1 + E2). If E1 + E2 is sufficient to depolarize the postsynaptic neuron to the threshold potential, a postsynaptic action potential results. **Summation** thus allows subthreshold EPSPs to influence action potential production. Likewise, an IPSP generated by an inhibitory synapse (I) can sum (algebraically speaking) with a subthreshold EPSP to reduce its amplitude (E1 + I), or it can sum with suprathreshold EPSPs to prevent the postsynaptic neuron from reaching threshold (E1 + I + E2).

In short, the summation of EPSPs and IPSPs by a postsynaptic neuron permits a neuron to integrate the electrical information provided by all the inhibitory and excitatory synapses acting on it at any moment. Whether the sum of active synaptic inputs results in the production of an action potential depends on the balance between excitation and inhibition. If the sum of all EPSPs and IPSPs results in a depolarization of sufficient amplitude to raise the membrane potential above threshold, then the postsynaptic cell will produce an action potential. Conversely, if inhibition prevails, then the postsynaptic cell will remain silent. Normally, the

(A)

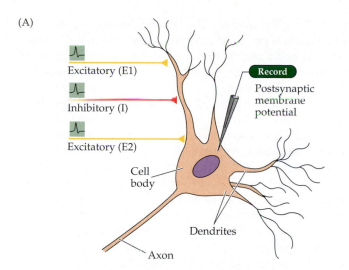

Excitatory (E1)

Inhibitory (I)

Excitatory (E2)

Record

Postsynaptic membrane potential

Cell body

Dendrites

Axon

Figure 5.21 Summation of postsynaptic potentials. (A) A microelectrode records the postsynaptic potentials produced by the activity of two excitatory synapses (E1 and E2) and an inhibitory synapse (I). (B) Electrical responses to synaptic activation. Stimulating either excitatory synapse (E1 or E2) produces a subthreshold EPSP, whereas stimulating both synapses at the same time (E1 + E2) produces a suprathreshold EPSP that evokes a postsynaptic action potential (shown in blue). Activation of the inhibitory synapse alone (I) results in a hyperpolarizing IPSP. Summing this IPSP (dashed red line) with the EPSP (dashed yellow line) produced by one excitatory synapse (E1 + I) reduces the amplitude of the EPSP (orange line), while summing it with the suprathreshold EPSP produced by activating synapses E1 and E2 keeps the postsynaptic neuron below threshold, so that no action potential is evoked.

(B)

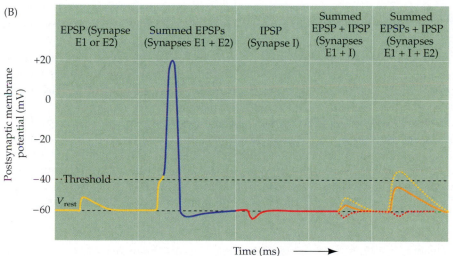

EPSP (Synapse E1 or E2) | Summed EPSPs (Synapses E1 + E2) | IPSP (Synapse I) | Summed EPSP + IPSP (Synapses E1 + I) | Summed EPSPs + IPSP (Synapses E1 + I + E2)

Postsynaptic membrane potential (mV)

Time (ms)

balance between EPSPs and IPSPs changes continually over time, depending on the number of excitatory and inhibitory synapses active at a given moment and the magnitude of the current at each active synapse. Summation is therefore a neurotransmitter-induced tug-of-war between all excitatory and inhibitory postsynaptic currents; the outcome of the contest determines whether or not a postsynaptic neuron fires an action potential and, thereby, becomes an active element in the neural circuits to which it belongs (Figure 5.22).

Two Families of Postsynaptic Receptors

The opening or closing of postsynaptic ion channels is accomplished in different ways by two broad families of receptor proteins. The receptors in one family—called **ionotropic receptors**—are linked directly to ion channels (the

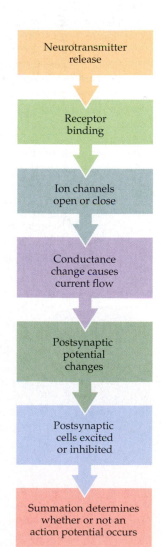

Neurotransmitter release

Receptor binding

Ion channels open or close

Conductance change causes current flow

Postsynaptic potential changes

Postsynaptic cells excited or inhibited

Summation determines whether or not an action potential occurs

Figure 5.22 Events from neurotransmitter release to postsynaptic excitation or inhibition. Neurotransmitter release at all presynaptic terminals on a cell results in receptor binding, which causes the opening or closing of specific ion channels. The resulting conductance change causes current to flow, which may change the membrane potential. The postsynaptic cell sums (or integrates) all of the EPSPs and IPSPs, resulting in moment-to-moment control of action potential generation.

Greek word *tropos* means "to turn" and signifies movement in response to a stimulus). These receptors contain two functional domains: an extracellular site that binds neurotransmitters, and a membrane-spanning domain that forms an ion channel (Figure 5.23A). Thus ionotropic receptors combine transmitter-binding and channel functions into a single molecular entity; reflecting this concatenation, ionotropic receptors are also known as **ligand-gated ion channels**. Such receptors are multimers made up of at least four or five individual protein subunits, each of which contributes to the pore of the ion channel.

The second family of neurotransmitter receptors are the **metabotropic receptors**, so called because the eventual movement of ions through a channel depends on one or more metabolic steps. These receptors do not have ion channels as part of their structure; instead, they affect channels by the activation of intermediate molecules called **G-proteins** (Figure 5.23B). For this reason, metabotropic receptors are also called **G-protein-coupled receptors**. Metabotropic receptors are monomeric proteins with an extracellular domain that contains a neurotransmitter binding site and an intracellular domain that binds to G-proteins. Neurotransmitter binding to metabotropic receptors activates G-proteins, which then dissociate from the receptor and interact directly with ion channels or bind to other effector proteins, such as enzymes, that make intracellular messengers that open or close ion channels. Thus, G-proteins can be thought of as transducers that couple neurotransmitter binding to the regulation of postsynaptic ion channels. The postsynaptic signaling events initiated by metabotropic receptors are taken up in detail in Chapter 7.

These two families of postsynaptic receptors give rise to PSPs with very different time courses, producing postsynaptic actions that range from less than a millisecond to minutes, hours, or even days. Ionotropic receptors generally mediate rapid postsynaptic effects. Examples are the EPP produced at neuromuscular synapses by ACh (see Figure 5.16), EPSPs produced at certain glutamatergic synapses (see Figure 5.19A), and IPSPs produced at certain GABAergic synapses

Figure 5.23 A neurotransmitter can affect the activity of a postsynaptic cell via two different types of receptor proteins: ionotropic receptors (also called ligand-gated ion channels); and metabotropic or G-protein-coupled receptors. (A) Ligand-gated ion channels combine receptor and channel functions in a single protein complex. (B) Metabotropic receptors usually activate G-proteins, which modulate ion channels directly or indirectly through intracellular effector enzymes and second messengers.

(A) LIGAND-GATED ION CHANNELS

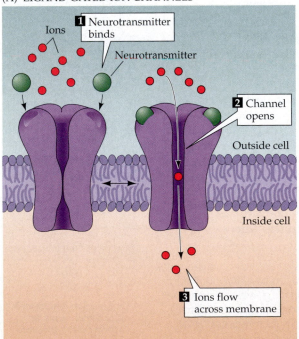

(B) G-PROTEIN-COUPLED RECEPTORS

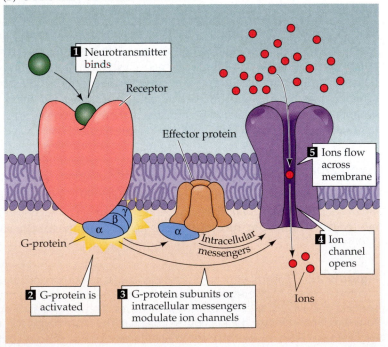

(see Figure 5.20B). In all three cases, the PSPs arise within a millisecond or two of an action potential invading the presynaptic terminal and last for only a few tens of milliseconds or less. In contrast, the activation of metabotropic receptors typically produces much slower responses, ranging from hundreds of milliseconds to minutes or even longer. The comparative slowness of metabotropic receptor actions reflects the fact that multiple proteins need to bind to each other sequentially in order to produce the final physiological response. Importantly, a given transmitter may activate both ionotropic and metabotropic receptors to produce both fast and slow PSPs at the same synapse.

Perhaps the most important principle to keep in mind is that the response elicited at a given synapse depends upon the neurotransmitter released and the postsynaptic complement of receptors and associated channels. The molecular mechanisms that allow neurotransmitters and their receptors to generate synaptic responses are considered in the next chapter.

Summary

Synapses communicate the information carried by action potentials from one neuron to the next in neural circuits. The cellular mechanisms that underlie synaptic transmission are closely related to the mechanisms that generate other types of neuronal electrical signals, namely ion flow through membrane channels. In the case of electrical synapses, these channels are gap junctions; direct but passive flow of current through the gap junctions is the basis for transmission. In the case of chemical synapses, channels with smaller and more selective pores are activated by the binding of neurotransmitters to postsynaptic receptors after release from the presynaptic terminal. The large number of neurotransmitters in the nervous system can be divided into two broad classes: small-molecule transmitters and neuropeptides. Neurotransmitters are synthesized from defined precursors by regulated enzymatic pathways, packaged into one of several types of synaptic vesicle, and released into the synaptic cleft in a Ca^{2+}-dependent manner. Many synapses release more than one type of neurotransmitter, and multiple transmitters can even be packaged within the same synaptic vesicle. Transmitter agents are released presynaptically in units or quanta, reflecting their storage within synaptic vesicles. Vesicles discharge their contents into the synaptic cleft when the presynaptic depolarization generated by the invasion of an action potential opens voltage-gated calcium channels, allowing Ca^{2+} to enter the presynaptic terminal. How calcium triggers neurotransmitter release is not yet established, but synaptotagmin, SNAREs, and a number of other proteins found within the presynaptic terminal are clearly involved. Postsynaptic receptors are a diverse group of proteins that transduce binding of neurotransmitters into electrical signals by opening or closing postsynaptic ion channels. The postsynaptic currents produced by the synchronous opening or closing of ion channels changes the conductance of the postsynaptic cell, thus increasing or decreasing its excitability. Conductance changes that increase the probability of firing an action potential are excitatory, whereas those that decrease the probability of generating an action potential are inhibitory. Because postsynaptic neurons are usually innervated by many different inputs, the integrated effect of the conductance changes underlying all EPSPs and IPSPs produced in a postsynaptic cell at any moment determines whether or not the cell fires an action potential. Two broadly different families of neurotransmitter receptors have evolved to carry out the postsynaptic signaling actions of neurotransmitters. The postsynaptic effects of neurotransmitters are terminated by the degradation of the transmitter in the synaptic cleft, by transport of the transmitter back into cells, or by diffusion out of the synaptic cleft.

Additional Reading

Reviews

AUGUSTINE, G. J. AND H. KASAI (2007) Bernard Katz, quantal transmitter release, and the foundations of presynaptic physiology. *J. Physiol. (Lond.)* 578: 623–625.

BRODSKY, F. M., C.Y. CHEN, C. KNUEHL, M. C. TOWLER AND D. E. WAKEHAM (2001) Biological basket weaving: Formation and function of clathrin-coated vesicles. *Annu. Rev. Cell. Dev. Biol.* 17: 517–568.

BRUNGER, A. T. (2005) Structure and function of SNARE and SNARE-interacting proteins. *Q. Rev. Biophys.* 38: 1–47.

CARLSSON, A. (1987) Perspectives on the discovery of central monoaminergic neurotransmission. *Annu. Rev. Neurosci.* 10: 19–40.

CONNORS, B. W. AND M. A. LONG (2004) Electrical synapses in the mammalian brain. *Annu. Rev. Neurosci.* 27: 393-418.

DE CAMILLI, P. (2004–2005) Molecular mechanisms in membrane traffic at the neuronal synapse: Role of protein-lipid interactions. Harvey Lecture 100: 1–28.

EMSON, P. C. (1979) Peptides as neurotransmitter candidates in the CNS. *Prog. Neurobiol.* 13: 61–116.

GALARRETA, M. AND S. HESTRIN (2001) Electrical synapses between GABA-releasing interneurons. *Nature Rev. Neurosci.* 2: 425–433.

JACKSON, M. B. AND E. R. CHAPMAN (2006) Fusion pores and fusion machines in Ca^{2+}-triggered exocytosis. *Annu. Rev. Biophys. Biomol. Struct.* 35: 135–160.

MARSH, M. AND H. T. MCMAHON (1999) The structural era of endocytosis. *Science* 285: 215–220.

PIERCE, K. L., R. T. PREMONT AND R. J. LEFKOWITZ (2002) Seven-membrane receptors. *Nature Rev. Mol. Cell Biol.* 3: 639–650.

RIZZOLI, S. O. AND W. J. BETZ (2005) Synaptic vesicle pools. *Nature Rev. Neurosci.* 6: 57–69.

ROTHMAN, J. E. (2002) The machinery and principles of vesicle transport in the cell. *Nature Med.* 8: 1059–1062.

SÜDHOF, T. (2004) The synaptic vesicle cycle. *Annu. Rev. Neurosci.* 27: 509–547.

Important Original Papers

ADLER, E., M. ADLER, G. J. AUGUSTINE, M. P. CHARLTON AND S. N. DUFFY (1991) Alien intracellular calcium chelators attenuate neurotransmitter release at the squid giant synapse. *J. Neurosci.* 11: 1496–1507.

AUGUSTINE, G. J. AND R. ECKERT (1984) Divalent cations differentially support transmitter release at the squid giant synapse. *J. Physiol. (Lond.)* 346: 257–271.

BEIERLEIN, M., J. R. GIBSON AND B. W. CONNORS (2000) A network of electrically coupled interneurons drives synchronized inhibition in neocortex. *Nature Neurosci.* 3: 904–910.

BOYD, I. A. AND A. R. MARTIN (1955) The end-plate potential in mammalian muscle. *J. Physiol. (Lond.)* 132: 74–91.

CURTIS, D. R., J. W. PHILLIS AND J. C. WATKINS (1959) Chemical excitation of spinal neurons. *Nature* 183: 611–612.

DALE, H. H., W. FELDBERG AND M.VOGT (1936) Release of acetylcholine at voluntary motor nerve endings. *J. Physiol.* 86: 353–380.

DEL CASTILLO, J. AND B. KATZ (1954) Quantal components of the end plate potential. *J. Physiol. (Lond.)* 124: 560–573.

FATT, P. AND B. KATZ (1951) An analysis of the end plate potential recorded with an intracellular electrode. *J. Physiol. (Lond.)* 115: 320–370.

FATT, P. AND B. KATZ (1952) Spontaneous subthreshold activity at motor nerve endings. *J. Physiol. (Lond.)* 117: 109–128.

FURSHPAN, E. J. AND D. D. POTTER (1959) Transmission at the giant motor synapses of the crayfish. *J. Physiol. (Lond.)* 145: 289–325.

GEPPERT, M. AND 6 OTHERS (1994) Synaptotagmin I: A major Ca^{2+} sensor for transmitter release at a central synapse. *Cell* 79: 717–727.

HARRIS, B. A., J. D. ROBISHAW, S. M. MUMBY AND A. G. GILMAN (1985) Molecular cloning of complementary DNA for the alpha subunit of the G protein that stimulates adenylate cyclase. *Science* 229: 1274–1277.

HEUSER, J. E., T. S. REESE, M. J. DENNIS,Y., JAN, L. JAN AND L. EVANS (1979) Synaptic vesicle exocytosis captured by quick freezing and correlated with quantal transmitter release. *J. Cell Biol.* 81: 275–300.

HEUSER, J. E. AND T. S. REESE (1973) Evidence for recycling of synaptic vesicle membrane during transmitter release at the frog neuromuscular junction. *J. Cell Biol.* 57: 315–344.

HÖKFELT, T., O. JOHANSSON, A. LJUNGDAHL, J. M. LUNDBERG AND M. SCHULTZBERG (1980) Peptidergic neurons. *Nature* 284: 515–521.

JONAS, P., J. BISCHOFBERGER AND J. SANDKUHLER (1998) Co-release of two fast neurotransmitters at a central synapse. *Science* 281: 419–424.

LOEWI, O. (1921) Über humorale übertragbarkeit der herznervenwirkung. *Pflügers Arch.* 189: 239–242.

MILEDI, R. (1973) Transmitter release induced by injection of calcium ions into nerve terminals. *Proc. R. Soc. Lond. B* 183: 421–425.

NEHER, E. AND B. SAKMANN (1976) Single-channel currents recorded from membrane of denervated frog muscle fibres. *Nature* 260: 799–802.

REKLING, J. C., X. M. SHAO AND J. L. FELDMAN (2000) Electrical coupling and excitatory synaptic transmission between rhythmogenic respiratory neurons in the pre-Botzinger complex. *J. Neurosci.* 20, RC113: 1–5.

SMITH, S. J., J. BUCHANAN, L. R. OSSES, M. P. CHARLTON AND G. J. AUGUSTINE (1993) The spatial distribution of calcium signals in squid presynaptic terminals. *J. Physiol. (Lond.)* 472: 573–593.

SUTTON, R. B., D. FASSHAUER, R. JAHN AND A. T. BRÜNGER (1998) Crystal structure of a SNARE complex involved in synaptic exocytosis at 2.4 Å resolution. *Nature* 395: 347–353.

TAKAMORI, S. AND 21 OTHERS (2006) Molecular anatomy of a trafficking organelle. *Cell* 127: 831–846.

TAKEUCHI, A. AND N. TAKEUCHI (1960) One the permeability of end-plate membrane during the action of transmitter. *J. Physiol. (Lond.)* 154: 52–67.

WICKMAN, K. AND 7 OTHERS (1994) Recombinant $G_{\beta\gamma}$ activates the muscarinic-gated atrial potassium channel I_{KACh}. *Nature* 368: 255–257.

Books

COOPER, J. R., F. E. BLOOM AND R. H. ROTH (2002) *The Biochemical Basis of Neuropharmacology.* New York: Oxford University Press.

HALL, Z. (1992) *An Introduction to Molecular Neurobiology.* Sunderland, MA: Sinauer Associates.

KATZ, B. (1966) *Nerve, Muscle, and Synapse.* New York: McGraw-Hill.

KATZ, B. (1969) *The Release of Neural Transmitter Substances.* Liverpool: Liverpool University Press.

LLINÁS, R. R. (1999) *The Squid Giant Synapse: A Model for Chemical Synaptic Transmission.* Oxford: Oxford University Press.

NICHOLLS, D. G. (1994) *Proteins, Transmitters, and Synapses.* Oxford: Blackwell.

PETERS, A., S. L. PALAY AND H. DEF. WEBSTER (1991) *The Fine Structure of the Nervous System: Neurons and their Supporting Cells,* 3rd ed. Oxford: Oxford University Press.

Chapter 6

Neurotransmitters and Their Receptors

Overview

For the most part, neurons in the human brain communicate with one another by releasing chemical messengers called neurotransmitters. A large number of neurotransmitters are now known and more remain to be discovered. The main excitatory neurotransmitter in the brain is the amino acid glutamate, while the main inhibitory neurotransmitter is γ-aminobutyric acid, or GABA. These and all other neurotransmitters evoke their postsynaptic electrical responses by binding to and activating members of an even more diverse group of proteins called neurotransmitter receptors. Most neurotransmitters are capable of activating several different receptors, further increasing the possible modes of synaptic signaling. After activating their postsynaptic receptors, neurotransmitters are removed from the synaptic cleft by neurotransmitter transporters or by degradative enzymes. Abnormalities in the function of neurotransmitter systems contribute to a wide range of neurological and psychiatric disorders. As a result, many neuropharmacological therapies are based on drugs that affect neurotransmitters, their receptors, and/or the proteins responsible for removal of neurotransmitters from the synaptic cleft.

Categories of Neurotransmitters

More than 100 different agents are known to serve as neurotransmitters. This large number of transmitters allows for tremendous diversity in chemical signaling between neurons. It is useful to separate this panoply of transmitters into two broad categories based simply on size (Figure 6.1). **Neuropeptides** are relatively large transmitter molecules composed of 3 to 36 amino acids. Individual amino acids, such as glutamate and GABA, as well as the transmitters acetylcholine, serotonin, and histamine, are much smaller than neuropeptides and have therefore come to be called **small-molecule neurotransmitters**. Within the small-molecule neurotransmitters, the **biogenic amines** (dopamine, norepinephrine, epinephrine, serotonin, and histamine) are often discussed separately because of their similar chemical properties and postsynaptic actions. The particulars of synthesis, packaging, release, and removal differ for each neurotransmitter (Table 6.1). This chapter will describe some of the main features of these transmitters and their postsynaptic receptors.

Acetylcholine

As mentioned in the previous chapter, acetylcholine (ACh) was the first substance identified as a neurotransmitter. In addition to the action of ACh as the neurotransmitter at skeletal neuromuscular junctions (see Chapter 5), as well as the neuromuscular synapse between the vagus nerve and cardiac muscle fibers, ACh serves as a transmitter at synapses in the ganglia of the visceral motor sys-

SMALL-MOLECULE NEUROTRANSMITTERS

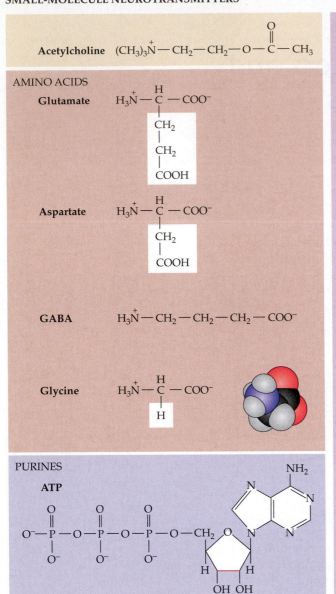

Acetylcholine $(CH_3)_3\overset{+}{N}-CH_2-CH_2-O-\overset{\displaystyle O}{\overset{\|}{C}}-CH_3$

AMINO ACIDS

Glutamate $H_3\overset{+}{N}-\overset{\displaystyle H}{\underset{\displaystyle CH_2}{\overset{|}{\underset{|}{C}}}}-COO^-$ $\underset{CH_2}{\underset{|}{\overset{|}{}}}$ $COOH$

Aspartate $H_3\overset{+}{N}-\overset{\displaystyle H}{\underset{\displaystyle CH_2}{\overset{|}{\underset{|}{C}}}}-COO^-$ $COOH$

GABA $H_3\overset{+}{N}-CH_2-CH_2-CH_2-COO^-$

Glycine $H_3\overset{+}{N}-\overset{\displaystyle H}{\underset{\displaystyle H}{\overset{|}{\underset{|}{C}}}}-COO^-$

PURINES

ATP

$O^--\overset{\displaystyle O}{\overset{\|}{\underset{\displaystyle O^-}{\underset{|}{P}}}}-O-\overset{\displaystyle O}{\overset{\|}{\underset{\displaystyle O^-}{\underset{|}{P}}}}-O-\overset{\displaystyle O}{\overset{\|}{\underset{\displaystyle O^-}{\underset{|}{P}}}}-O-CH_2$

BIOGENIC AMINES

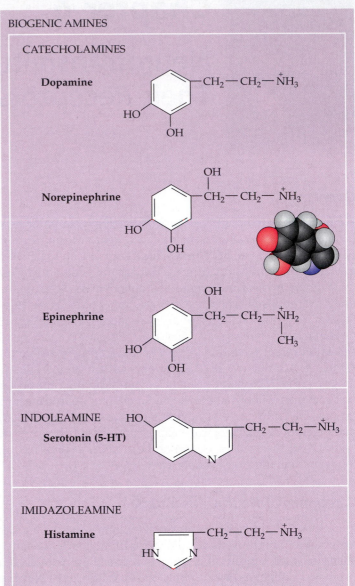

CATECHOLAMINES

Dopamine

Norepinephrine

Epinephrine

INDOLEAMINE
Serotonin (5-HT)

IMIDAZOLEAMINE
Histamine

PEPTIDE NEUROTRANSMITTERS (more than 100 peptides, usually 3–30 amino acids long)

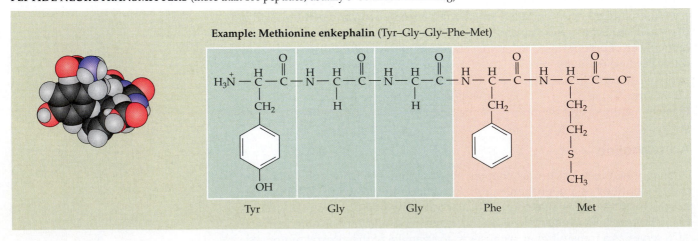

Example: Methionine enkephalin (Tyr–Gly–Gly–Phe–Met)

Tyr   Gly   Gly   Phe   Met

TABLE 6.1 **Functional Features of the Major Neurotransmitters**

Neurotransmitter	Postsynaptic effect[a]	Precursor(s)	Rate-limiting step in synthesis	Removal mechanism	Type of vesicle
ACh	Excitatory	Choline + acetyl CoA	CAT	AChEase	Small, clear
Glutamate	Excitatory	Glutamine	Glutaminase	Transporters	Small, clear
GABA	Inhibitory	Glutamate	GAD	Transporters	Small, clear
Glycine	Inhibitory	Serine	Phosphoserine	Transporters	Small, clear
Catecholamines (epinephrine, norepinephrine, dopamine)	Excitatory	Tyrosine	Tyrosine hydroxylase	Transporters, MAO, COMT	Small dense-core, or large irregular dense-core
Serotonin (5-HT)	Excitatory	Tryptophan	Tryptophan hydroxylase	Transporters, MAO	Large, dense-core
Histamine	Excitatory	Histidine	Histidine decarboxylase	Transporters	Large, dense-core
ATP	Excitatory	ADP	Mitochondrial oxidative phosphorylation; glycolysis	Hydrolysis to AMP and adenosine	Small, clear
Neuropeptides	Excitatory and inhibitory	Amino acids (protein synthesis)	Synthesis and transport	Proteases	Large, dense-core
Endocannabinoids	Inhibits inhibition	Membrane lipids	Enzymatic modification of lipids	Hydrolysis by FAAH	None
Nitric oxide	Excitatory and inhibitory	Arginine	Nitric oxide synthase	Spontaneous oxidation	None

[a]The most common postsynaptic effect is indicated; the same transmitter can elicit postsynaptic excitation or inhibition depending on the nature of the ion channels affected by transmitter binding (see Chapter 5).

tem, and at a variety of sites within the central nervous system. Whereas a great deal is known about the function of cholinergic transmission at neuromuscular junctions and ganglionic synapses, the actions of ACh in the central nervous system are not as well understood.

Acetylcholine is synthesized in nerve terminals from the precursors acetyl coenzyme A (acetyl CoA, which is synthesized from glucose) and choline, in a reaction catalyzed by choline acetyltransferase (CAT; Figure 6.2). Choline is present in plasma at a high concentration (about 10 mM) and is taken up into cholinergic neurons by a high-affinity Na$^+$/choline transporter. After synthesis in the cytoplasm of the neuron, a vesicular ACh transporter loads approximately 10,000 molecules of ACh into each cholinergic vesicle.

In contrast to most other small-molecule neurotransmitters, the postsynaptic actions of ACh at many cholinergic synapses (the neuromuscular junction in

◀ **Figure 6.1** Examples of small-molecule and peptide neurotransmitters. Small-molecule transmitters can be subdivided into acetylcholine, the amino acids, purines, and biogenic amines. The catecholamines, so named because they all share the catechol moiety (i.e., a hydroxylated benzene ring), make up a distinctive subgroup within the biogenic amines. Serotonin and histamine contain an indole ring and an imidazole ring, respectively. Size differences between the small-molecule neurotransmitters and the peptide neurotransmitters are indicated by the space-filling models for glycine, norepinephrine, and methionine enkephalin. (Carbon atoms are black, nitrogen atoms blue, and oxygen atoms red.)

Figure 6.2 Acetylcholine metabolism in cholinergic nerve terminals. The synthesis of acetylcholine from choline and acetyl CoA requires choline acetyltransferase. Acetyl CoA is derived from pyruvate generated by glycolysis, while choline is transported into the terminals via a Na⁺-dependent transporter. Acetylcholine is loaded into synaptic vesicles via a vesicular transporter. After release, acetylcholine is rapidly metabolized by acetylcholinesterase, and choline is transported back into the terminal.

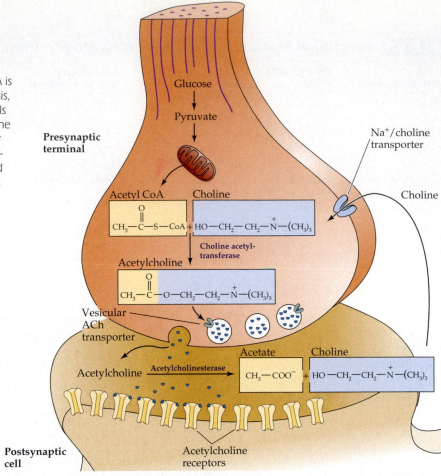

particular) is not terminated by reuptake but by a powerful hydrolytic enzyme, acetylcholinesterase (AChE). This enzyme is concentrated in the synaptic cleft, ensuring a rapid decrease in ACh concentration after its release from the presynaptic terminal. AChE has a very high catalytic activity (about 5000 molecules of ACh per AChE molecule per second) and hydrolyzes ACh into acetate and choline. The choline produced by ACh hydrolysis is transported back into nerve terminals and used to resynthesize ACh.

Among the many interesting drugs that interact with cholinergic enzymes are the organophosphates, which include some potent chemical warfare agents. One such compound is the nerve gas "Sarin," made notorious after a group of terrorists released the gas in Tokyo's underground rail system. Organophosphates can be lethal because they inhibit AChE, allowing ACh to accumulate at cholinergic synapses. This buildup of ACh depolarizes the postsynaptic cell and renders it refractory to subsequent ACh release, causing neuromuscular paralysis and other effects. The high sensitivity of insects to these AChE inhibitors has made organophosphates popular insecticides.

Many of the postsynaptic actions of ACh are mediated by the nicotinic ACh receptor (nAChR), so named because the CNS stimulant, nicotine, also binds to these receptors. Nicotine consumption produces some degree of euphoria, relaxation, and eventually addiction, effects believed to be mediated in this case by nAChRs. Nicotinic receptors are the best-studied type of ionotropic neurotransmitter receptor. As described in Chapter 5, nAChRs are nonselective cation channels that generate excitatory postsynaptic responses. A number of biological toxins specifically bind to and block nicotinic receptors (Box 6A). The avail-

BOX 6A Neurotoxins that Act on Postsynaptic Receptors

Poisonous plants and venomous animals are widespread in nature. The toxins they produce have been used for a variety of purposes, including hunting, healing, mind-altering, and, more recently, research. Many of these toxins have potent actions on the nervous system, often interfering with synaptic transmission by targeting neurotransmitter receptors. The poisons found in some organisms contain a single type of toxin, whereas others contain a mixture of tens or even hundreds of toxins.

Given the central role of ACh receptors in mediating muscle contraction at neuromuscular junctions in numerous species, it is not surprising that a large number of natural toxins interfere with transmission at this synapse. In fact, the classification of nicotinic and muscarinic ACh receptors is based on the sensitivity of these receptors to the toxic plant alkaloids nicotine and muscarine, which activate nicotinic and muscarinic ACh receptors, respectively. Nicotine is derived from the dried leaves of the tobacco plant *Nicotinia tabacum*, and

muscarine is from the poisonous red mushroom *Amanita muscaria*. Both toxins are stimulants that produce nausea, vomiting, mental confusion, and convulsions. Muscarine poisoning can also lead to circulatory collapse, coma, and death.

The poison α-bungarotoxin, one of many peptides that together make up the venom of the banded krait, *Bungarus multicinctus* (Figure A), blocks transmission at neuromuscular junctions and is used by the snake to paralyze its prey. This 74-amino-acid toxin blocks neuromuscular transmission by irreversibly binding to nicotinic ACh receptors, thus preventing ACh from opening postsynaptic ion channels. Paralysis ensues because skeletal muscles can no longer be activated by motor neurons. As a result of its specificity and its high affinity for nicotinic ACh receptors, α-bungarotoxin has contributed greatly to understanding the ACh receptor molecule. Other snake toxins that block nicotinic ACh receptors are cobra α-neurotoxin and the sea snake peptide erabutoxin. The same strategy used by

these snakes to paralyze prey was adopted by South American Indians who used curare, a mixture of plant toxins from *Chondodendron tomentosum*, as an arrowhead poison to immobilize their quarry. The active agent in curare, δ-tubocurarine, blocks nicotinic ACh receptors.

Another interesting class of animal toxins that block postsynaptic receptors are the peptides produced by fish-hunting marine cone snails (Figure B). These colorful snails kill small fish by "shooting" venomous darts into them. The venom contains hundreds of peptides, known as the conotoxins, many of which target proteins that are important in synaptic transmission. There are conotoxin peptides that block ACh and glutamate receptors, as well as voltage-gated Ca^{2+} and Na^+ channels. The array of physiological responses produced by these peptides all serve to immobilize any prey unfortunate enough to encounter the cone snail. Many other organisms, including other mollusks,

(Continued on next page)

(A)

(B)

(C)

(A) The banded krait *Bungarus multicinctus.* (B) A marine cone snail (*Conus* sp.) uses venomous darts to kill a small fish. (C) Betel nuts, *Areca catechu*, growing in Malaysia. (A, Robert Zappalorti/Photo Researchers, Inc.; B, Zoya Maslak and Baldomera Olivera, University of Utah; C, Fletcher Baylis/Photo Researchers, Inc.)

BOX 6A   (Continued)

corals, worms and frogs, also utilize toxins containing specific blockers of ACh receptors.

Other natural toxins have mind- or behavior-altering effects and in some cases have been used for thousands of years by shamans and, more recently, physicians. Two examples are plant alkaloid toxins that block muscarinic ACh receptors: atropine from deadly nightshade (belladonna), and scopolamine from henbane. Because these plants grow wild in many parts of the world, exposure is not unusual, and poisoning by either toxin can also be fatal.

Another postsynaptic neurotoxin that, like nicotine, is used as a social drug is found in the seeds from the betel nut, *Areca catechu* (Figure C). Betel nut chewing, although unknown in the United States, is practiced by up to 25 percent of the population in India, Bangladesh, Ceylon, Malaysia, and the Philippines. Chewing these nuts produces a euphoria caused by arecoline, an alkaloid agonist of nicotinic ACh receptors. Like nicotine, arecoline is an addictive central nervous system stimulant.

Many other neurotoxins alter transmission at noncholinergic synapses. For example, amino acids found in certain mushrooms, algae, and seeds are potent glutamate receptor agonists. The excitotoxic amino acids kainate, from the red

alga *Digenea simplex*, and quisqualate, from the seed of *Quisqualis indica*, are used to distinguish two families of non-NMDA glutamate receptors (see text). Other neurotoxic amino acid activators of glutamate receptors include ibotenic acid and acromelic acid, both found in mushrooms, and domoate, which occurs in algae, seaweed, and mussels. Another large group of peptide neurotoxins blocks glutamate receptors. These include the α-agatoxins from the funnel web spider, NSTX-3 from the orb weaver spider, jorotoxin from the Joro spider, and β-philanthotoxin from wasp venom, as well as many cone snail toxins.

All the toxins discussed so far target excitatory synapses. The inhibitory GABA and glycine receptors, however, have not been overlooked by the exigencies of survival. Strychnine, an alkaloid extracted from the seeds of *Strychnos nux-vomica*, is the only drug known to have specific actions on transmission at glycinergic synapses. Because the toxin blocks glycine receptors, strychnine poisoning causes overactivity in the spinal cord and brainstem, leading to seizures. Strychnine has long been used commercially as a poison for rodents, although alternatives such as the anticoagulant coumadin are now more popular because they are safer for humans. Neurotoxins that block GABA$_A$ receptors

include plant alkaloids such as bicuculline from Dutchman's breeches and picrotoxin from *Anamerta cocculus*. Dieldrin, a commercial insecticide, also blocks these receptors. These agents are, like strychnine, powerful central nervous system stimulants. Muscimol, a mushroom toxin that is a powerful depressant as well as a hallucinogen, activates GABA$_A$ receptors. A synthetic analogue of GABA, baclofen, is a GABA$_B$ agonist that reduces EPSPs in some brainstem neurons and is used clinically to reduce the frequency and severity of muscle spasms.

Chemical warfare between species has thus given rise to a staggering array of molecules that target synapses throughout the nervous system. Although these toxins are designed to defeat normal synaptic transmission, they have also provided a set of powerful tools to understand postsynaptic mechanisms.

References

ADAMS, M. E. AND B. M. OLIVERA (1994) Neurotoxins: Overview of an emerging research technology. *Trends Neurosci.* 17: 151–155.

HUCHO, F. AND Y. OVCHINNIKOV (1990) *Toxins as Tools in Neurochemistry.* Berlin: Walter de Gruyer.

LEWIS, R. L. AND L. GUTMANN (2004) Snake venoms and the neuromuscular junction. *Seminars Neurol.* 24: 175–179.

ability of these highly specific ligands—particularly a component of snake venom called α-bungarotoxin—has provided a valuable way to isolate and purify nAChRs. This pioneering work paved the way to cloning and sequencing the genes encoding the various subunits of the nAChR.

Based on these molecular studies, the nAChR is now known to be a large protein complex consisting of five subunits arranged around a central membrane-spanning pore (Figure 6.3). In the case of skeletal muscle AChRs, the receptor pentamer contains two α subunits, each of which binds one molecule of ACh. Because both ACh binding sites must be occupied for the channel to open, only relatively high concentrations of this neurotransmitter lead to channel activation. These subunits also bind other ligands, such as nicotine and α-bungarotoxin. At the neuromuscular junction, the two α subunits are combined with up to four other types of subunit—β, γ, δ, ε—in the ratio 2α:β:ε:δ. Neuronal

(A)

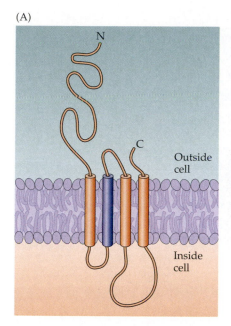

(B)

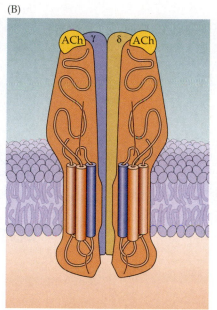

(C)

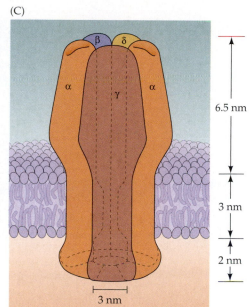

Figure 6.3 The structure of the nACh receptor/channel. (A) Each receptor subunit crosses the membrane four times. The membrane-spanning domain that lines the pore is shown in blue. (B) Five such subunits come together to form a complex structure containing 20 transmembrane domains that surround a central pore. (C) The openings at either end of the channel are very large—approximately 3 nm in diameter; even the narrowest region of the pore is approximately 0.6 nm in diameter. By comparison, the diameter of Na$^+$ or K$^+$ is less than 0.3 nm. (D) An electron micrograph of the nACh receptor, showing the actual position and size of the protein with respect to the membrane. (D from Toyoshima and Unwin, 1990.)

(D)

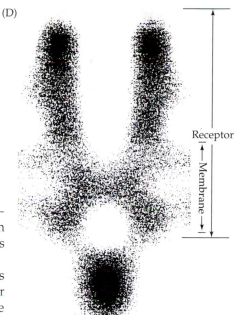

nAChRs typically differ from those of muscle in that they lack sensitivity to α-bungarotoxin, and comprise only two receptor subunit types (α and β), which are present in a ratio of 3α:2β. In all cases, however, five individual subunits assemble to form a functional, cation-selective nACh receptor.

Each subunit of the nAChR molecule contains four transmembrane domains that make up the ion channel portion of the receptor, and a long extracellular region that makes up the ACh-binding domain (Figure 6.3A). Unraveling the molecular structure of this region of the nACh receptor has provided insight into the mechanisms that allow ligand-gated ion channels to respond rapidly to neurotransmitters: The intimate association of the ACh binding sites with the pore of the channel presumably accounts for the rapid response to ACh (Figure 6.3B–D). Indeed, this general arrangement is characteristic of *all* of the ligand-gated ion channels at fast-acting synapses, as summarized in Figure 6.4. Thus, the nicotinic receptor has served as a paradigm for studies of other ligand-gated ion channels, at the same time leading to a much deeper appreciation of several neuromuscular diseases (Box 6B).

A second class of ACh receptors is activated by muscarine, a poisonous alkaloid found in some mushrooms (see Box 6A), and thus they are referred to as muscarinic ACh receptors (mAChRs). mAChRs are metabotropic and mediate most of the effects of ACh in brain. Several subtypes of mAChR are known (Figure 6.5). Muscarinic ACh receptors are highly expressed in the striatum and

(A)

(B)

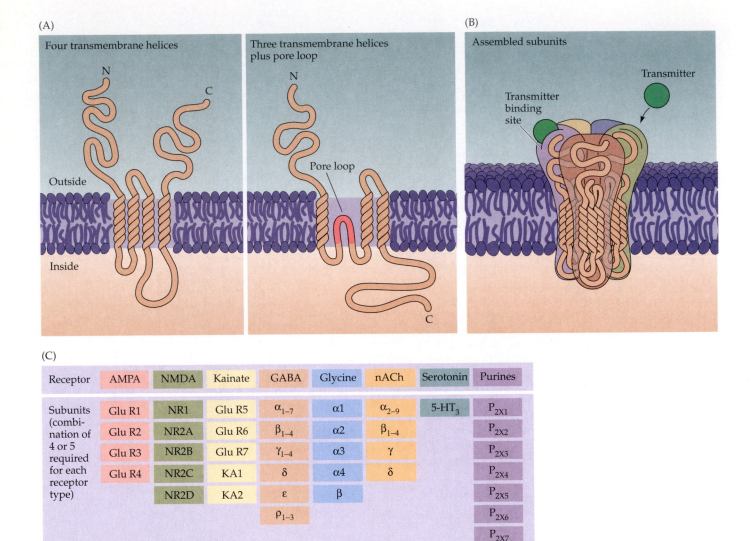

(C)

Receptor	AMPA	NMDA	Kainate	GABA	Glycine	nACh	Serotonin	Purines
Subunits (combination of 4 or 5 required for each receptor type)	Glu R1	NR1	Glu R5	α_{1-7}	$\alpha1$	α_{2-9}	5-HT$_3$	P$_{2X1}$
	Glu R2	NR2A	Glu R6	β_{1-4}	$\alpha2$	β_{1-4}		P$_{2X2}$
	Glu R3	NR2B	Glu R7	γ_{1-4}	$\alpha3$	γ		P$_{2X3}$
	Glu R4	NR2C	KA1	δ	$\alpha4$	δ		P$_{2X4}$
		NR2D	KA2	ϵ	β			P$_{2X5}$
				ρ_{1-3}				P$_{2X6}$
								P$_{2X7}$

Figure 6.4 The general architecture of ligand-gated receptors. (A) One of the subunits of a complete receptor. The long N-terminal region forms the ligand-binding site, while the remainder of the protein spans the membrane either four times (left) or three times (right). (B) Assembly of either four or five subunits into a complete receptor. (C) A diversity of subunits come together to form functional ionotropic neurotransmitter receptors.

various other forebrain regions, where they exert an inhibitory influence on dopamine-mediated motor effects. These receptors are also found in the ganglia of the peripheral nervous system. Finally, they mediate peripheral cholinergic responses of autonomic effector organs—such as heart, smooth muscle, and exocrine glands—and are responsible for the inhibition of heart rate by the vagus nerve. Numerous drugs act as mACh receptor agonists or antagonists, but most of these do not discriminate between different types of muscarinic receptors and often produce side effects. Nevertheless, mACh blockers that are therapeutically useful include atropine (used to dilate the pupil), scopolamine (effective in preventing motion sickness), and ipratropium (useful in the treatment of asthma).

Glutamate

Glutamate is the most important transmitter in normal brain function. Nearly all excitatory neurons in the central nervous system are glutamatergic, and it is estimated that over half of all brain synapses release this agent. Glutamate plays an especially important role in clinical neurology because elevated concentra-

BOX 6B Myasthenia Gravis

Myasthenia gravis is a disease that interferes with transmission between motor neurons and skeletal muscle fibers (see Box 5B) and afflicts approximately 1 of every 200,000 people. Originally described by the British physician Thomas Willis in 1685, the hallmark of the disorder is muscle weakness, particularly during sustained activity (Figure A). Although the course is variable, myasthenia commonly affects muscles controlling the eyelids (resulting in drooping of the eyelids, or ptosis) and eye movements (resulting in double vision, or diplopia). Muscles controlling facial expression, chewing, swallowing, and speaking are other common targets.

An important indication of the cause of myasthenia gravis came from the clinical observation that the muscle weakness improves following treatment with inhibitors of acetylcholinesterase, the enzyme that normally degrades acetylcholine at the neuromuscular junction. Studies of muscle from myasthenic patients showed that both end plate potentials (EPPs) and miniature end plate potentials (MEPPs) are much smaller than normal (Figure B; also see Chapter 5). Because both the frequency of MEPPs and the quantal content of EPPs are normal, it seemed likely that myasthenia gravis affects the postsynaptic muscle cells. Indeed, electron microscopy shows that the structure of neuromuscular junctions is altered, obvious changes being a widening of the synaptic cleft and an apparent reduction in the number of acetylcholine receptors in the postsynaptic membrane.

A chance observation in the early 1970s led to the discovery of the underlying cause of these changes. Jim Patrick and Jon Lindstrom, then at the Salk Institute, were attempting to raise antibodies to nicotinic ACh receptors by immunizing rabbits with the receptors. Unexpectedly, the immunized rabbits developed muscle weakness that improved after treatment with acetylcholinesterase inhibitors. Subsequent work showed that the blood of myasthenic patients contains antibodies directed against the ACh receptor, and that these antibodies are present at neuromuscular synapses. Removal of antibodies by plasma exchange improves the weakness. Finally, injecting the serum of myasthenic patients into mice produces myasthenic effects (because the serum carries circulating antibodies).

These findings indicate that myasthenia gravis is an autoimmune disease that targets nicotinic ACh receptors. The immune response reduces the number of functional receptors at the neuromuscular junction and can eventually destroy them altogether, diminishing the efficiency of synaptic transmission; muscle weakness occurs because motor neurons are less capable of exciting the postsynaptic muscle cells. This causal sequence also explains why cholinesterase inhibitors alleviate the signs and symptoms of myasthenia—the inhibitors increase the concentration of acetylcholine in the synaptic cleft, allowing more effective activation of those postsynaptic receptors not yet destroyed by the immune system.

It is still not clear what triggers the immune system to produce an autoimmune response to acetylcholine receptors. However, these insights allowed many patients to be treated with combinations of immunosuppression and cholinesterase inhibitors.

References

ELMQVIST, D., W. W. HOFMANN, J. KUGELBERG AND D. M. J. QUASTEL (1964) An electrophysiological investigation of neuromuscular transmission in myasthenia gravis. *J. Physiol. (Lond.)* 174: 417–434.

PATRICK, J. AND J. LINDSTROM (1973) Autoimmune response to acetylcholine receptor. *Science* 180: 871–872.

VINCENT, A. (2002) Unravelling the pathogenesis of myasthenia gravis. *Nature Rev. Immunol.* 2: 797–804.

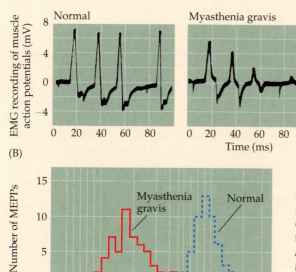

(A)

(B)

(A) Myasthenia gravis reduces the efficiency of neuromuscular transmission. Electromyographs show muscle responses elicited by stimulating motor nerves. In normal individuals, each stimulus in a train evokes the same contractile response. In contrast, transmission rapidly fatigues in myasthenic patients, although it can be partially restored by administration of acetylcholinesterase inhibitors such as neostigmine. (B) Distribution of MEPP amplitudes (note logarithmic scale) in muscle fibers from myasthenic patients (solid line) and controls (dashed line). The smaller size of MEPPs in myasthenics is due to a diminished number of postsynaptic receptors. (A after Harvey et al., 1941; B after Elmqvist et al., 1964.)

Figure 6.5 Structure and function of metabotropic receptors. (A) The transmembrane architecture of metabotropic receptors. These monomeric proteins contain seven transmembrane domains. Portions of domains II, III, VI, and VII make up the neurotransmitter-binding region. G-proteins bind to both the loop between domains III and IV and to portions of the C-terminal region. (B) Varieties of metabotropic neurotransmitter receptors.

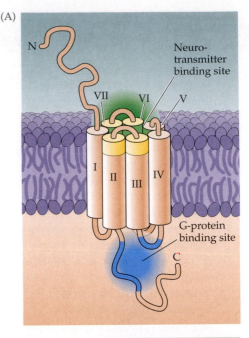

(B)

Receptor class	Glutamate	GABA_B	Dopamine	NE, Epi	Histamine	Serotonin	Purines	Muscarinic
Receptor subtype	Class I	GABA_B R1	D1_A	α1	H1	5-HT 1	A type	M1
	mGlu R1	GABA_B R2	D1_B	α2	H2	5-HT 2	A1	M2
	mGlu R5		D2	β1	H3	5-HT 4	A2a	M3
	Class II		D3	β2		5-HT 5	A2b	M4
	mGlu R2		D4	β3		5-HT 6	A3	M5
	mGlu R3					5-HT 7	P type	
	Class III						P2x	
	mGlu R4						P2y	
	mGlu R6						P2z	
	mGlu R7						P2t	
	mGlu R8						P2u	

tions of extracellular glutamate, released as a result of neural injury, are toxic to neurons (Box 6C).

Glutamate is a nonessential amino acid that does not cross the blood-brain barrier and therefore must be synthesized in neurons from local precursors. The most prevalent precursor for glutamate synthesis is glutamine, which is released by glial cells. Once released, glutamine is taken up into presynaptic terminals and metabolized to glutamate by the mitochondrial enzyme glutaminase (Figure 6.6). Glutamate can also be synthesized by transamination of 2-oxoglutarate, an intermediate of the tricarboxylic acid cycle. Hence, some of the glucose metabolized by neurons can also be used for glutamate synthesis.

The glutamate synthesized in the presynaptic cytoplasm is packaged into synaptic vesicles by transporters, termed VGLUT. At least three different VGLUT genes have been identified. Once released, glutamate is removed from the synaptic cleft by the excitatory amino acid transporters (EAATs). Five different types of high-affinity glutamate transporters exist, some of which are present in glial cells and others in presynaptic terminals. Glutamate taken up by glial cells is converted into glutamine by the enzyme glutamine synthetase; glutamine is then transported out of the glial cells and into nerve terminals. In this way,

BOX 6C   Excitotoxicity Following Acute Brain Injury

Excitotoxicity refers to the ability of glutamate and related compounds to destroy neurons by excessive activation of glutamate receptors. Normally, the concentration of glutamate released into the synaptic cleft rises to high levels (approximately 1 mM), but it remains at this concentration for only a few milliseconds. If abnormally high levels of glutamate accumulate in the cleft, the excessive activation of neuronal glutamate receptors can literally excite neurons to death.

The phenomenon of excitotoxicity was discovered in 1957 when D. R. Lucas and J. P. Newhouse serendipitously found that feeding sodium glutamate to infant mice destroys neurons in the retina. Roughly a decade later, John Olney at Washington University extended this discovery by showing that regions of glutamate-induced neuronal loss can occur throughout the brain. The damage was evidently restricted to the postsynaptic cells—the dendrites of the target neurons were grossly swollen—while the presynaptic terminals were spared. Olney also examined the relative potency of glutamate analogs and found that their neurotoxic actions paralleled their ability to activate postsynaptic glutamate receptors. Furthermore, glutamate receptor antagonists were effective in blocking the neurotoxic effects of glutamate. In light of this evidence, Olney postulated that glutamate destroys neurons by a mechanism similar to transmission at excitatory glutamatergic synapses, and coined the term *excitotoxic*

to refer to this pathological effect.

Evidence that excitotoxicity is an important cause of neuronal damage after brain injury has come primarily from studying the consequences of reduced blood flow. The most common cause of reduced blood flow to the brain (ischemia) is the occlusion of a cerebral blood vessel (i.e., a stroke; see the Appendix). The idea that excessive synaptic activity contributes to ischemic injury emerged from the observation that concentrations of glutamate and aspartate in the extracellular space around neurons increase during ischemia. Moreover, microinjection of glutamate receptor antagonists in experimental animals protects neurons from ischemia-induced damage. Together, these findings imply that extracellular accumulation of glutamate during ischemia activates glutamate receptors excessively, and that this somehow triggers a chain of events that leads to neuronal death. The reduced supply of oxygen and glucose during ischemia presumably elevates extracellular glutamate levels by slowing the energy-dependent removal of glutamate at synapses.

Excitotoxic mechanisms have now been shown to be involved in other acute forms of neuronal insult, including hypoglycemia, trauma, and repeated intense seizures (called *status epilepticus*). Understanding excitotoxicity therefore has important implications for treating a variety of neurological disorders. For instance, blockade of glutamate receptors could, in principle, protect neurons

from injury due to stroke, trauma, or other causes. Unfortunately, clinical trials of glutamate receptor antagonists have not led to much improvement in the outcome of stroke. The ineffectiveness of this quite logical treatment is probably due to several factors, one of which is that substantial excitotoxic injury occurs quite soon after ischemia, prior to the typical initiation of treatment. It is also likely that excitotoxicity is only one of several mechanisms by which ischemia damages neurons; other candidates include damage secondary to inflammation. Pharmacological interventions that target all these mechanisms nonetheless hold considerable promise for minimizing brain injury after stroke and other causes.

References

LUCAS, D. R. AND J. P. NEWHOUSE (1957) The toxic effects of sodium L-glutamate on the inner layers of the retina. *Arch. Opthalmol.* 58: 193–201.

OLNEY, J. W. (1969) Brain lesions, obesity and other disturbances in mice treated with monosodium glutamate. *Science* 164: 719–721.

OLNEY, J. W. (1971) Glutamate-induced neuronal necrosis in the infant mouse hypothalamus: An electron microscopic study. *J. Neuropathol. Exp. Neurol.* 30: 75–90.

ROTHMAN, S. M. (1983) Synaptic activity mediates death of hypoxic neurons. *Science* 220: 536–537.

SYNTICHAKI, P. AND N. TAVERNARAKIS (2003) The biochemistry of neuronal necrosis: Rogue biology? *Nature Neurosci. Rev.* 4: 672–684.

---

synaptic terminals cooperate with glial cells to maintain an adequate supply of the neurotransmitter. This overall sequence of events is referred to as the **glutamate-glutamine cycle** (see Figure 6.6).

Several types of glutamate receptors have been identified. Three of these are ionotropic receptors called, respectively, **NMDA receptors**, **AMPA receptors**, and **kainate receptors** (see Figure 6.4C). These glutamate receptors are named after the agonists that activate them: NMDA (*N*-methyl-D-aspartate), AMPA (α-amino-3-hydroxyl-5-methyl-4-isoxazole-propionate), and kainic acid. All of the ionotropic glutamate receptors are nonselective

Figure 6.6 Glutamate synthesis and cycling between neurons and glia. The action of glutamate released into the synaptic cleft is terminated by uptake into neurons and surrounding glial cells via specific transporters. Within the nerve terminal, the glutamine released by glial cells and taken up by neurons is converted back to glutamate. Glutamate is transported into cells via excitatory amino acid transporters (EAATs) and loaded into synaptic vesicles via vesicular glutamate transporters (VGLUT).

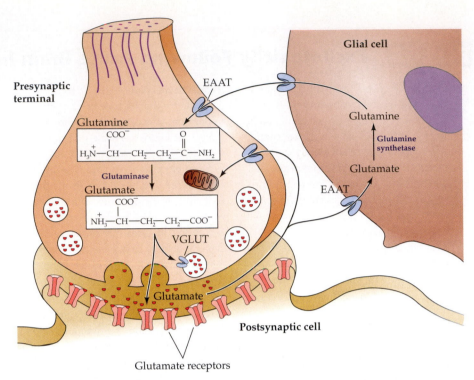

cation channels similar to the nAChR, allowing the passage of Na^+ and K^+, and in some cases small amounts of Ca^{2+}. Hence AMPA, kainate, and NMDA receptor activation always produces excitatory postsynaptic responses. Like other ionotropic receptors, AMPA/kainate and NMDA receptors are also formed from the association of several protein subunits that can combine in many ways to produce a large number of receptor isoforms.

NMDA receptors have especially interesting properties (Figure 6.7A). Perhaps most significant is the fact that NMDA receptor ion channels allow the entry of Ca^{2+} in addition to monovalent cations such as Na^+ and K^+. As a result, EPSPs produced by NMDA receptors can increase the concentration of Ca^{2+} within the postsynaptic neuron; the Ca^{2+} concentration change can then act as a second messenger to activate intracellular signaling cascades (see Chapter 7). Another key property is that they bind extracellular Mg^{2+}. At hyperpolarized membrane potentials, this ion blocks the pore of the NMDA receptor channel. Depolarization, however, pushes Mg^{2+} out of the pore, allowing other cations to flow. This property provides the basis for a voltage-dependence to current flow through the receptor (dashed line in Figure 6.7B) and means that NMDA receptors pass cations (most notably Ca^{2+}) only during depolarization of the postsynaptic cell, due to either activation of a large number of excitatory inputs and/or by repetitive firing of action potentials in the presynaptic cell. These properties are widely thought to be the basis for some forms of synaptic plasticity, as described in Chapter 8. Another unusual property of NMDA receptors is that opening the channel of this receptor requires the presence of a co-agonist, the amino acid glycine (Figure 6.7A,B). There are at least five forms of NMDA receptor subunits (NMDA-R1, and NMDA-R2A through NMDA-R2D); different synapses have distinct combinations of these subunits, producing a variety of NMDA receptor-mediated postsynaptic responses.

Whereas some glutamatergic synapses have only AMPA or NMDA receptors, most possess both AMPA and NMDA receptors. An antagonist of NMDA receptors, APV (2-amino-5-phosphono-valerate), is often used to differentiate between

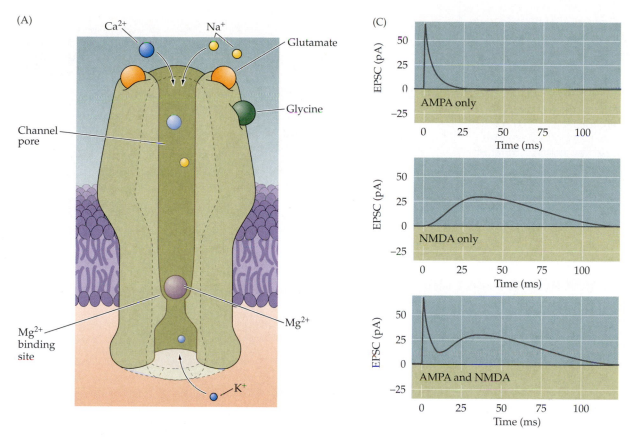

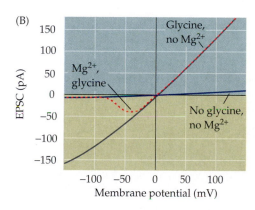

Figure 6.7 NMDA and AMPA/kainate receptors. (A) NMDA receptors contain binding sites for glutamate and the co-activator glycine, as well as an Mg^{2+}-binding site in the pore of the channel. At hyperpolarized potentials, the electrical driving force on Mg^{2+} drives this ion into the pore of the receptor and blocks it. (B) Current flow across NMDA receptors at a range of postsynaptic voltages, showing the requirement for glycine, and Mg^{2+} block at hyperpolarized potentials (dotted line). (C) The differential effects of glutamate receptor antagonists indicate that activation of AMPA or kainate receptors produces very fast EPSCs (top panel) and activation of NMDA receptors causes slower EPSCs (middle panel), so that EPSCs recorded in the absence of antagonists have two kinetic components due to the contribution of both types of response (bottom panel). Currents are recorded at a potential of +50 mV, so that the block by Mg^{2+} is removed. As a result, EPSCs are outward in polarity, even though they are inward at physiological resting potentials.

the two receptor types. The use of this drug has also revealed differences between the electrical signals produced by NMDA and those produced by AMPA receptors, such as the fact that the synaptic currents produced by NMDA receptors are slower and longer-lasting than the those produced by AMPA/kainate receptors (see Figure 6.7C). The physiological roles of kainate receptors are less well-defined; in some cases, these receptors are found on presynaptic terminals and serve as a feedback mechanism to regulate glutamate release.

In addition to these ionotropic glutamate receptors, there are three types of metabotropic glutamate receptor (mGluRs) (Figure 6.5). These receptors, which modulate postsynaptic ion channels indirectly, differ in their coupling to intracellular signal transduction pathways (see Chapter 7) and in their sensitivity to pharmacological agents. Unlike the excitatory ionotropic glutamate receptors, mGluRs cause slower postsynaptic responses that can either increase or

Figure 6.8 Synthesis, release, and reuptake of the inhibitory neurotransmitters GABA and glycine. (A) GABA is synthesized from glutamate by the enzyme glutamic acid decarboxylase, which requires pyridoxal phosphate. (B) Glycine can be synthesized by a number of metabolic pathways; in the brain, the major precursor is serine. High-affinity transporters terminate the actions of these transmitters and return GABA or glycine to the synaptic terminals for reuse, with both transmitters being loaded into synaptic vesicles via the vesicular inhibitory amino acid transporter (VIATT).

(A)

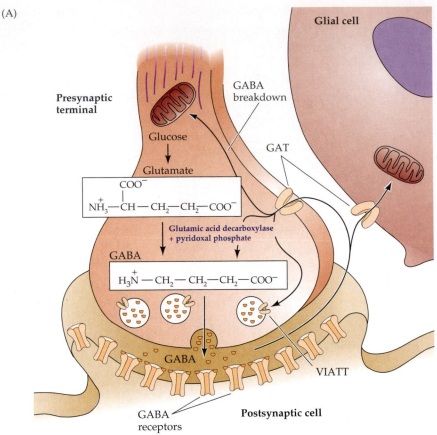

(B)

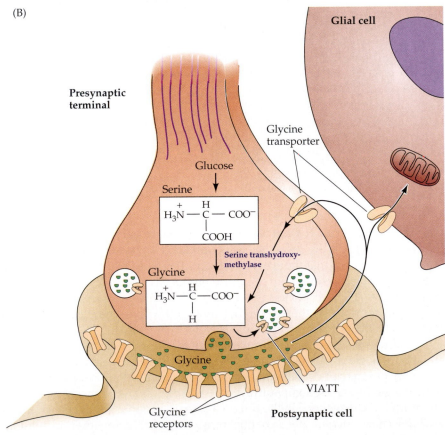

decrease the excitability of postsynaptic cells. As a result the physiological roles of mGluRs are quite varied.

GABA and Glycine

Most inhibitory synapses in the brain and spinal cord use either γ-aminobutyric acid (GABA) or glycine as neurotransmitters. Like glutamate, GABA was identified in brain tissue during the 1950s. The details of its synthesis and degradation were worked out shortly thereafter by the work of Ernst Florey and Eugene Roberts. During this era, David Curtis and Jeffrey Watkins first showed that GABA can inhibit action potential firing in mammalian neurons. Subsequent studies by Edward Kravitz and colleagues established that GABA serves as an inhibitory transmitter at lobster neuromuscular synapses. It is now known that as many as a third of the synapses in the brain use GABA as their inhibitory neurotransmitter. GABA is most commonly found in local circuit interneurons, although cerebellar Purkinje cells (see Chapter 19) provide an example of a GABAergic projection neuron.

The predominant precursor for GABA synthesis is glucose, which is metabolized to glutamate by the tricarboxylic acid cycle enzymes (pyruvate and glutamine can also act as precursors). The enzyme glutamic acid decarboxylase (GAD), which is found almost exclusively in GABAergic neurons, catalyzes the conversion of glutamate to GABA (Figure 6.8A). GAD requires a cofactor, pyridoxal phosphate, for activity. Because pyridoxal phosphate is derived from vitamin B_6, a deficiency in this vitamin can lead to diminished GABA synthesis. The significance of this became clear after a disastrous series of infant deaths was linked to the omission of vitamin B_6 from infant formula. Lack of B_6 resulted in a large reduction in the GABA content of the brain, and the subsequent loss of synaptic inhibition caused seizures that in some cases were fatal. Once GABA is synthesized, it is transported into synaptic vesicles via a vesicular inhibitory amino acid transporter (VIATT).

The mechanism of GABA removal is similar to that for glutamate. Both neurons and glia contain high-affinity transporters for GABA, termed GATs; several forms of GAT have been identified. Most GABA is eventually converted to succinate, which is metabolized further in the tricarboxylic acid cycle that mediates cellular ATP synthesis. The enzymes required for this degradation, GABA transaminase and succinic semialdehyde dehydrogenase, are mitochondrial enzymes. Inhibition of GABA breakdown causes a rise in tissue GABA content and an increase in the activity of inhibitory neurons. There are also other pathways for degradation of GABA. The most noteworthy of these results in the production of γ-hydroxybutyrate, a GABA derivitive that has been abused as a "date rape" drug. Oral administration of γ-hydroxybutyrate can cause euphoria, memory deficits, and unconsciousness. Presumably these effects arise from actions on GABAergic synapses in the CNS.

Inhibitory synapses employing GABA as their transmitter can exhibit three types of postsynaptic receptors, called $GABA_A$, $GABA_B$, and $GABA_C$. $GABA_A$ and $GABA_C$ are ionotropic receptors, while $GABA_B$ receptors are metabotropic. The ionotropic GABA receptors are usually inhibitory because their associated channels are permeable to Cl^- (Figure 6.9). The reversal potential for Cl^- usually is more negative than the threshold for neuronal firing (see Figure 5.20) due to the action of the K^+/Cl^- cotransporter (see Figure 4.10), which keeps intracellular Cl^- concentration low. The resulting influx of negatively charged Cl^- through ionotropic GABA receptors inhibits postsynaptic cells. In cases where Cl^- concentration within the postsynaptic cell is high (for example in developing brains), $GABA_A$ receptors can excite their postsynaptic targets (Box 6D).

(A)

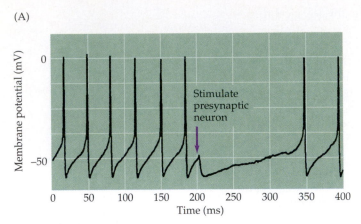

(B)

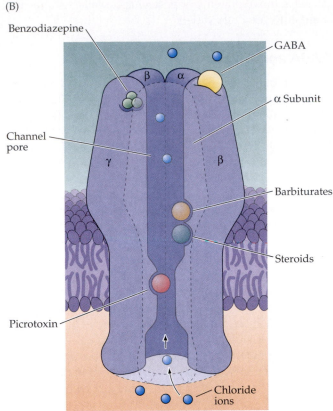

Figure 6.9 Ionotropic GABA receptors. (A) Stimulation of a presynaptic GABAergic interneuron, at the time indicated by the arrow, causes a transient inhibition of action potential firing in its postsynaptic target. This inhibitory response is caused by activation of postsynaptic GABA$_A$ receptors. (B) Ionotropic GABA receptors contain two binding sites for GABA (only one of which is visible in this illustration) and numerous sites at which drugs bind to and modulate these receptors. (A after Chavas and Marty, 2003.)

Like other ionotropic receptors, GABA receptors are pentamers assembled from a combination of five types of subunits (α, β, γ, δ, and ρ; see Figure 6.4C). As a result of this subunit diversity, as well as variable stoichiometry of subunits, the functions of GABA$_A$ receptors differ widely among neuronal types. Drugs that act as agonists or modulators of postsynaptic GABA receptors, such as benzodiazepines and barbiturates, are used clinically to treat epilepsy and are effective sedatives and anesthetics. Binding sites for GABA, barbiturates, steroids, and picrotoxin are all located within the pore domain of the channel (Figure 6.9B). Another site, called the benzodiazepine binding site, lies outside the pore and modulates channel activity. Benzodiazepines, such as diazepam (Valium®) and chlordiazepoxide (Librium®), are tranquilizing (anxiety-reducing) drugs that enhance GABAergic transmission by binding to the α and δ subunits of GABA$_A$ receptors. Barbiturates, such as phenobarbital and pentobarbital, are hypnotics that bind to the α and β subunits of some GABA receptors and are used therapeutically for anesthesia and to control epilepsy. Another drug that can alter the activity of GABA-mediated inhibitory circuits is alcohol; at least some aspects of drunken behavior are caused by alcohol-mediated alterations in ionotropic GABA receptors.

Metabotropic GABA receptors (GABA$_B$) are also widely distributed in the brain. Like the ionotropic GABA$_A$ receptors, GABA$_B$ receptors are inhibitory. Rather than activating Cl⁻ selective channels, however, GABA$_B$-mediated inhibition is due to the activation of K⁺ channels. A second mechanism for GABA$_B$-mediated inhibition is by blocking Ca²⁺ channels, which tends to hyperpolarize postsynaptic cells. Unlike some other metabotropic receptors, GABA$_B$ receptors assemble as heterodimers of GABA$_B$ R1 and R2 subunits.

BOX 6D Excitatory Actions of GABA in the Developing Brain

In the mature brain, GABA normally functions as an inhibitory neurotransmitter. In the developing brain, however, GABA excites its target cells. This remarkable reversal of action arises from developmental changes in intracellular Cl^- homeostasis.

In young cortical neurons, intracellular Cl^- concentration is controlled mainly by the $Na^+/K^+/Cl^-$ cotransporter. This transporter pumps Cl^- into the neurons and yields a high $[Cl^-]$ inside the cell (Figure A, left). As the neurons continue to develop, they begin to express a K^+/Cl^- cotransporter that pumps Cl^- out of the neurons, thus lowering $[Cl^-]_i$ (Figure A, right). Such shifts in Cl^- homeostasis can cause $[Cl^-]_i$ to drop several-fold over the first 1 to 2 weeks of postnatal development (Figure B).

Because ionotropic GABA receptors are Cl^--permeable channels, ion flux through these receptors varies according to the electrochemical driving force on Cl^-. In the young neurons, where $[Cl^-]_i$ is high, E_{Cl} is more positive than the resting potential. As a result, GABA depolarizes these neurons. E_{Cl} often is more positive

than the threshold for firing action potentials, so that GABA can excite these neurons to fire action potentials (Figure C, left). As described in the text, the lower $[Cl^-]_i$ of mature neurons causes E_{Cl} to be more negative than the action potential threshold (and often more negative than the resting potential), yielding inhibitory responses to GABA (Figure C, right).

Why does GABA undergo such a switch in its postsynaptic actions? While the logic of this phenomenon is not yet completely clear, it appears that depolarizing GABA responses produce electrical activity that controls neuronal proliferation, migration, growth, and maturation, as well as determining synaptic connectivity. Once these developmental processes are completed, full functioning of the resulting neural circuitry requires *inhibitory* transmission—which can also be provided by GABA. Further work will be needed to appreciate to full significance of the excitatory actions of GABA, as well as to understand the mechanisms underlying the expression of the K^+/Cl^- cotransporter that ends the brief career of GABA as an excitatory neurotransmitter.

References

CHERUBINI, E., J. L. GAIARSA AND Y. BEN-ARI (1991) GABA: An excitatory transmitter in early postnatal life. *Trends Neurosci.* 14: 515–519.

OBATA, K, M. OIDE AND H. TANAKA (1978) Excitatory and inhibitory actions of GABA and glycine on embryonic chick spinal neurons in culture. *Brain Res.* 144: 179–184.

PAYNE, J. A., C. RIVERA, J. VOIPIO AND K. KAILA (2003) Cation-chloride co-transporters in neuronal communication, development and trauma. *Trends Neurosci.* 26: 199–206.

RIVERA, C. AND 8 OTHERS (1999) The K^+/Cl^- co-transporter KCC2 renders GABA hyperpolarizing during neuronal maturation. *Nature* 397: 251–255.

(A) Developmental switch in expression of Cl^- transporters lowers $[Cl^-]_i$, thereby reversing direction of Cl^- flux through GABA receptors. (B) Imaging $[Cl^-]_i$ between postnatal (P) days 5 and 20 (left) demonstrates a progressive reduction in $[Cl^-]_i$ (right). (C) Developmental changes in $[Cl^-]_i$ cause GABA responses to shift from depolarizing in young (6-day-old) neurons (left) to hyperpolarizing in older (10-day-old) neurons (right) cultured from the chick spinal cord. (B courtesy of T. Kuner and G. Augustine; C after Obata et al., 1978.)

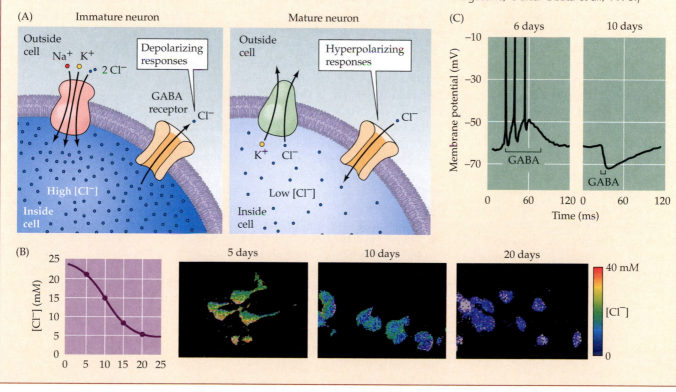

BOX 6E Biogenic Amine Neurotransmitters and Psychiatric Disorders

The regulation of the biogenic amine neurotransmitters is altered in a variety of psychiatric disorders. Indeed, most psychotropic drugs (defined as drugs that alter behavior, mood, or perception) selectively affect one or more steps in the synthesis, packaging, or degradation of biogenic amines. Sorting out how these drugs work has been extremely useful in beginning to understand the molecular mechanisms underlying some of these diseases.

Based on their effects on humans, psychotherapeutic drugs can be divided into several broad categories: antipsychotics, antianxiety drugs, antidepressants, and stimulants. The first antipsychotic drug used to ameliorate disorders such as schizophrenia was reserpine. Reserpine was developed in the 1950s and initially used as an antihypertensive agent; it blocks the uptake of norepinephrine into synaptic vesicles and therefore depletes the transmitter at aminergic terminals, diminishing the ability of the sympathetic division of the visceral motor system to cause vasoconstriction (see Chapter 21). A major side effect in hypertensive patients treated with reserpine—behavioral depression—suggested the possibility of using it as an antipsychotic agent in patients suffering from agitation and pathological anxiety. (Its ability to cause depression in mentally healthy individuals also suggested that aminergic transmitters are involved in mood disorders; see Box 29E.)

Although reserpine is no longer used as an antipsychotic agent, its initial success stimulated the development of antipsychotic drugs such as chlorpromazine, haloperidol, and benperidol, which over the last several decades have radically changed the approach to treating psychotic disorders. Prior to the discovery of these drugs, psychotic patients were typically hospitalized for long periods, sometimes indefinitely, and in the 1940s were subjected to desperate measures such as frontal lobotomy (see Box 26B). Modern antipsychotic drugs now allow most patients to be treated on an outpatient basis after a brief hospital stay. Importantly, the clinical effectiveness of these drugs is correlated with their ability to block brain dopamine receptors, implying that activation of dopamine receptors contributes to some types of psychotic illness. A great deal of effort continues to be expended on developing more effective antipsychotic drugs with fewer side effects, and on discovering the mechanism and site of action of these medications.

The second category of psychotherapeutic drugs is the antianxiety agents. Anxiety disorders are estimated to afflict 10–35 percent of the population, making them the most common psychiatric problem. The two major forms of pathological anxiety—panic attacks and generalized anxiety disorder—both respond to drugs that affect aminergic transmission. The agents used to treat panic disorders include inhibitors of the enzyme monoamine oxidase (MAO inhibitors) required for the catabolism of the amine neurotransmitters, and blockers of serotonin receptors. The most effective drugs in treating generalized anxiety disorder have been benzodiazepines, such as chlordiazepoxide (Librium®) and diazepam (Valium®). In contrast to most other psychotherapeutic drugs, these agents increase the efficacy of transmission at $GABA_A$ synapses rather than acting at aminergic synapses.

Antidepressants and stimulants also affect aminergic transmission. A large number of drugs are used clinically to treat depressive disorders. The three major classes of antidepressants—MAO inhibitors, tricyclic antidepressants, and serotonin uptake blockers such as fluoxetine (Prozac®) and trazodone—all influence various aspects of aminergic transmission. MAO inhibitors such as phenelzine block the breakdown of amines, whereas the tricyclic antidepressants such as desipramine block the reuptake of norepinephrine and other amines. The extraordinarily popular antidepressant fluoxetine (Prozac®) selectively blocks the reuptake of serotonin without affecting the reuptake of catecholamines. Stimulants such as amphetamine are also used to treat some depressive disorders. Amphetamine stimulates the release of norepinephrine from nerve terminals; the transient "high" resulting from taking amphetamine may reflect the emotional opposite of the depression that sometimes follows reserpine-induced norepinephrine depletion.

Despite the relatively small number of aminergic neurons in the brain, this litany of pharmacological actions emphasizes that these neurons are critically important in the maintenance of mental health.

References

FRANKLE, W. G., J. LERMA AND M. LARUELLE (2003) The synaptic hypothesis of schizophrenia. *Neuron* 39: 205–216.

FREEDMAN, R. (2003) Schizophrenia. *N. Engl. J. Med.* 349: 1738–1749.

LEWIS, D. A. AND P. LEVITT (2002) Schizophrenia as a disorder of neurodevelopment. *Annu. Rev. Neurosci.* 25: 409–432.

NESTLER, E. J., M. BARROT, R. J. DILEONE, A. J. EISCH, S. J. GOLD AND L. M. MONTEGGIA (2002) Neurobiology of depression. *Neuron* 34: 13–25.

ROSS, C. A., R. L. MARGOLIS, S. A. READING, M. PLETNIKOV AND J. T. COYLE (2006) Neurobiology of schizophrenia. *Neuron* 52: 139–153.

The distribution of the neutral amino acid glycine in the central nervous system is more localized than that of GABA. About half of the inhibitory synapses in the spinal cord use glycine; most other inhibitory synapses use GABA. Glycine is synthesized from serine by the mitochondrial isoform of serine hydroxymethyltransferase (see Figure 6.8B), and is transported into synaptic vesicles via the same vesicular inhibitory amino acid transporter that loads GABA into vesicles. Once released from the presynaptic cell, glycine is rapidly removed from the synaptic cleft by the plasma membrane glycine transporters. Mutations in the genes coding for some of these enzymes result in hyperglycinemia, a devastating neonatal disease characterized by lethargy, seizures, and mental retardation.

The receptors for glycine are also ligand-gated Cl⁻ channels, their general structure mirroring that of the $GABA_A$ receptors. Glycine receptors are pentamers consisting of mixtures of the 4 gene products encoding glycine-binding α subunits, along with the accessory β subunit. Glycine receptors are potently blocked by strychnine, which may account for the toxic properties of this plant alkaloid (see Box 6A).

The Biogenic Amines

Biogenic amine transmitters regulate many brain functions and are also active in the peripheral nervous system. Because biogenic amines are implicated in such a wide range of behaviors (ranging from central homeostatic functions to cognitive phenomena such as attention), it is not surprising that defects in biogenic amines function are implicated in most psychiatric disorders. The pharmacology of amine synapses is critically important in psychotherapy, with drugs affecting the synthesis, receptor binding, or catabolism of these neurotransmitters being among the most important agents in the armamentarium of modern pharmacology (Box 6E). Many drugs of abuse also act on biogenic amine pathways.

There are five well-established biogenic amine neurotransmitters: the three **catecholamines—dopamine, norepinephrine (noradrenaline)**, and **epinephrine (adrenaline)**—along with **histamine** and **serotonin** (see Figure 6.1). All the catecholamines (so named because they share the catechol moiety) are derived from a common precursor, the amino acid tyrosine (Figure 6.10). The first step in catecholamine synthesis is catalyzed by tyrosine hydroxylase in a reaction requiring oxygen as a co-substrate and tetrahydrobiopterin as a cofactor to synthesize dihydroxyphenylalanine (DOPA). Histamine and serotonin are synthesized via other routes, as described below.

• *Dopamine* is present in several brain regions (Figure 6.11A), although the major dopamine-containing area of the brain is the corpus striatum, which receives major input from the substantia nigra and plays an essential role in the coordination of body movements. In Parkinson's disease, for instance, the dopaminergic neurons of the substantia nigra degenerate, leading to a characteristic motor dysfunction (see Box 18A). Dopamine is also believed to be involved in motivation, reward, and reinforcement, and many drugs of abuse work by affecting dopaminergic synapses in the CNS (Box 6F). In addition to these roles in the CNS, dopamine also plays a poorly understood role in some sympathetic ganglia.

Dopamine is produced by the action of DOPA decarboxylase on DOPA (see Figure 6.10). Following its synthesis in the cytoplasm of presynaptic terminals,

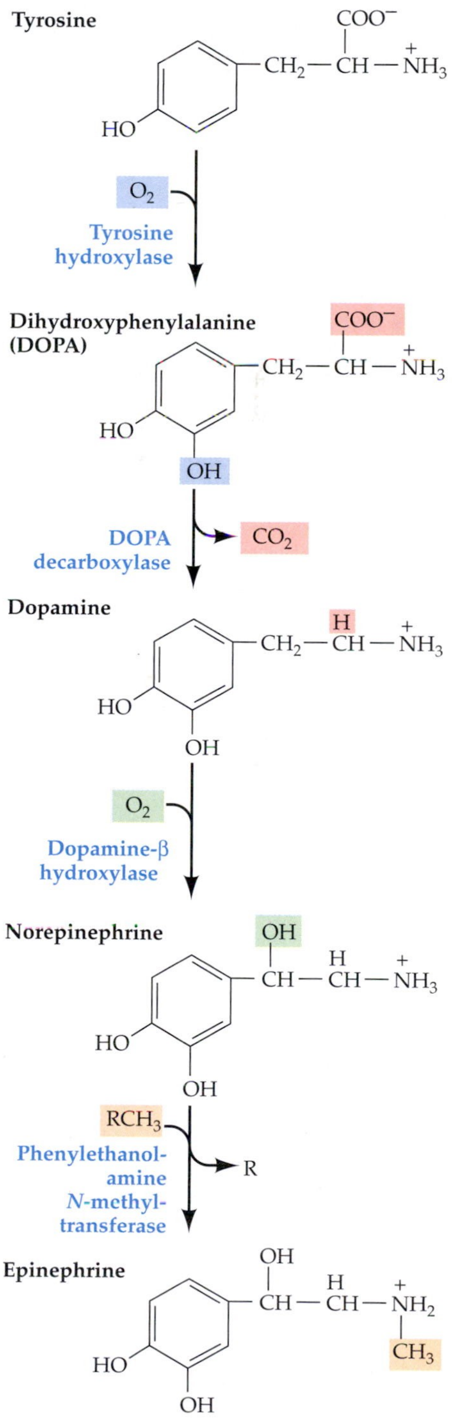

Figure 6.10 *The biosynthetic pathway for the catecholamine neurotransmitters. The amino acid tyrosine is the precursor for all three catecholamines. The first step in this reaction pathway, catalyzed by tyrosine hydroxylase, is rate-limiting.*

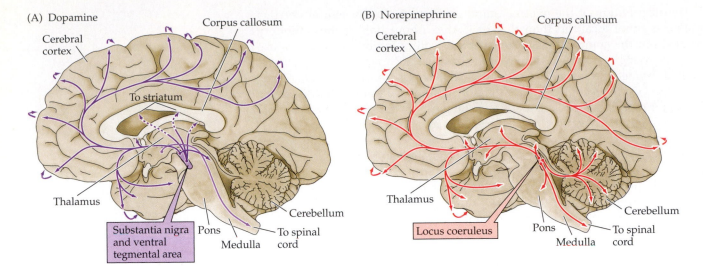

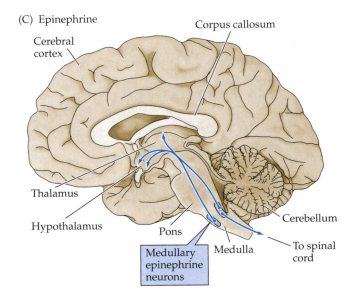

Figure 6.11 The distribution in the human brain of neurons and their projections (arrows) containing catecholamine neurotransmitters. Curved arrows along the perimeter of the cortex indicate the innervation of lateral cortical regions not shown in this midsagittal plane of section.

dopamine is loaded into synaptic vesicles via a vesicular monoamine transporter (VMAT). Dopamine action in the synaptic cleft is terminated by reuptake of dopamine into nerve terminals or surrounding glial cells by a Na^+-dependent dopamine transporter, termed DAT. Cocaine apparently produces its psychotropic effects by binding to and inhibiting DAT, yielding a net increase in dopamine concentration in the synaptic cleft. Amphetamine, another addictive drug, also inhibits DAT as well as the transporter for norepinepherine (see below). The two major enzymes involved in the catabolism of dopamine are monoamine oxidase (MAO) and catechol O-methyltransferase (COMT). Both neurons and glia contain mitochondrial MAO and cytoplasmic COMT. Inhibitors of these enzymes, such as phenelzine and tranylcypromine, are used clinically as antidepressants (see Box 6E).

Once released, dopamine acts exclusively by activating G-protein-coupled receptors. Most dopamine receptor subtypes (see Figure 6.5B) act by either activating or inhibiting adenylyl cyclase (see Chapter 7). Activation of these receptors generally contribute to complex behaviors; for example, administration of dopamine receptor agonists elicits hyperactivity and repetitive, stereotyped

BOX 6F Addiction

Drug addiction is a chronic, relapsing disease with obvious medical, social, and political consequences. Addiction (also called substance dependence) is a persistent disorder of brain function in which drug use escapes control, becoming compulsive despite serious negative consequences for the afflicted individual. Addiction can be defined in terms of both *physical* dependence and *psychological* dependence, in which an individual continues the drug-taking behavior despite obviously maladaptive consequences.

The range of substances that can generate this sort of dependence is wide; the primary agents of abuse at present are opioids, cocaine, amphetamines, marijuana, alcohol, and nicotine. Importantly, the phenomenon of addiction is not limited to human behavior, but is demonstrable in laboratory animals. Most of these same agents are self-administered if primates, rodents, or other species are provided with the opportunity to do so.

In addition to a compulsion to take the agent of abuse, a major feature of addiction for many drugs is a constellation of negative physiological and emotional features, loosely referred to as "withdrawal syndrome," that occur when the drug is not taken. The signs and symptoms of withdrawal are different for each agent of abuse, but in general are characterized by effects opposite those of the positive experience induced by the drug itself. Consider, as an example, cocaine, a drug that was estimated to be in regular use by 5 to 6 million Americans during the decade of the 1990s, with about 600,000 regular users either addicted or at high risk for addiction. The positive effects of the drug, whether it is smoked or inhaled as a powder (in the form of the alkaloidal free base), is a "high" that is nearly immediate but generally lasts only a few minutes, typically leading to a desire for additional drug in as little as 10 minutes

to half an hour. The "high" is described as a feeling of well-being, self confidence, and satisfaction. Conversely, when the drug is not available, frequent users experience depression, sleepiness, fatigue, drug-craving, and a general sense of malaise.

Another aspect of addiction to cocaine or other agents is tolerance, defined as a reduction in the response to the drug upon repeated administration. Tolerance occurs as a consequence of the persistent use of multiple drugs but is particularly significant in drug addiction, since it progressively increases the dose needed to experience the desired effects.

Although it is fair to say that the neurobiology of addiction is incompletely understood, for cocaine and many other agents of abuse the addictive effects involve activation of dopamine receptors in critical brain regions involved in motivation and emotional reinforcement (see Chapter 29). The most important of these areas is the midbrain dopamine system, especially its projections from the ventral tegmentum to the nucleus acumbens. Agents such as cocaine appear to act by raising dopamine levels in these areas, making this transmitter more available to receptors by interfering with re-uptake of synaptically released dopamine by the dopamine transporter. The reinforcement and motivation of drug-taking behaviors is thought to be related to the projections to the nucleus acumbens.

The most common opioid drug of abuse is heroin. Heroin is a derivative of the opium poppy and is not legally available for clinical purposes in the United States. The number of heroin addicts in the United States is estimated to be between 750,000 and a million individuals. The positive feelings produced by heroin, generally described as the "rush," are often compared to the feeling of sexual orgasm and begin in less than a minute after intravenous injec-

tion. There is then a feeling of general well-being (referred to as "on the nod") that lasts about an hour. The symptoms of withdrawal can be intense; these are restlessness, irritability, nausea, muscle pain, depression, sleeplessness, and a sense of anxiety and malaise. The reinforcing aspects of the drug entail the same dopaminergic circuitry in the ventral tegmental area and nucleus acumbens as does cocaine, although additional areas are certainly involved, particularly the sites of opioid receptors described in Chapter 10.

The treatment of any form of addiction is difficult and must be tailored to the circumstances of the individual. In addition to treating acute problems of withdrawal and "detoxification," patterns of behavior must be changed that may take months or years. Addiction is thus a chronic disease state that requires continual monitoring during the lifetime of susceptible individuals.

References

AMERICAN PSYCHIATRIC ASSOCIATION (1994) *Diagnostic and Statistical Manual of Mental Disorders*, 4th Edition (DSM IV). Washington, D.C.

HYMAN, S. E. AND R. C. MALENKA (2001) Addiction and the brain: The neurobiology of compulsion and its persistence. *Nature Rev. Neurosci.* 2: 695–703.

KALIVAS, P. W., N. VOLKOW AND J. SEAMANS (2005) Unmanageable motivation in addiction: A pathology in prefrontal-accumbens glutamate transmission. *Neuron* 45: 647–650.

LAAKSO, A., A. R. MOHN, R. R. GAINETDINOV AND M. G. CARON (2002) Experimental genetic approaches to addiction. *Neuron* 36: 213–228.

MADRAS, B. K., C. M.COLVIS, J. D. POLLOCK, J. L. RUTTER, D. SHURTLEFF AND M. VON ZASTROW (2006) *Cell Biology of Addiction*. Cold Spring Harbor: Cold Spring Harbor Laboratory Press.

O'BRIEN, C. P. (2006) Goodman and Gilman's *The Pharmaceutical Basis of Therapeutics*, 11th Edition. New York: McGraw-Hill, Chapter 23, pp. 607–627.

behavior in laboratory animals. Activation of another type of dopamine receptor in the medulla inhibits vomiting. Thus, antagonists of these receptors are used as emetics to induce vomiting after poisoning or a drug overdose. Dopamine receptor antagonists can also elicit catalepsy, a state in which it is difficult to initiate voluntary motor movement, suggesting a basis for this aspect of some psychoses.

• *Norepinephrine* (also called noradrenaline) is used as a neurotransmitter in the locus coeruleus, a brainstem nucleus that projects diffusely to a variety of forebrain targets (Figure 6.11B) and influences sleep and wakefulness, attention, and feeding behavior. Perhaps the most prominent noradrenergic neurons are sympathetic ganglion cells, which employ norepinephrine as the major peripheral transmitter in this division of the visceral motor system (see Chapter 21).

Norepinephrine synthesis requires dopamine β-hydroxylase, which catalyzes the production of norepinephrine from dopamine (see Figure 6.10). Norepinephrine is then loaded into synaptic vesicles via the same VMAT involved in vesicular dopamine transport. Norepinepherine is cleared from the synaptic cleft by the norepinepherine transporter (NET), which also is capable of taking up dopamine. As mentioned, NET serves as a molecular target of amphetamine, which acts as a stimulant by producing a net increase in the concentration of released norepinepherine and dopamine. A mutation in the NET gene is a cause of orthostatic intolerance, a disorder that produces lightheadedness while standing up. Like dopamine, norepinepherine is degraded by MAO and COMT.

Norepinepherine, as well as epinephrine, acts on α- and β-adrenergic receptors (see Figure 6.5B). Both types of receptor are G-protein-coupled; in fact, the β-adrenergic receptor was the first identified metabotropic neurotransmitter receptor. Two subclasses of α-adrenergic receptors are now known. Activation of α_1 receptors usually results in a slow depolarization linked to the inhibition of K^+ channels, while activation of α_2 receptors produces a slow hyperpolarization due to the activation of a different type of K^+ channel. There are three subtypes of β-adrenergic receptor, two of which are expressed in many types of neurons. Agonists and antagonists of adrenergic receptors, such as the β blocker propanolol (Inderol®), are used clinically for a variety of conditions ranging from cardiac arrhythmias to migraine headaches. However, most of the actions of these drugs are on smooth muscle receptors, particularly in the cardiovascular and respiratory systems (see Chapter 21).

• *Epinephrine* (also called adrenaline) is found in the brain at lower levels than the other catecholamines and also is present in fewer brain neurons than other catecholamines. Epinephrine-containing neurons in the central nervous system are primarily in the lateral tegmental system and in the medulla and project to the hypothalamus and thalamus (Figure 6.11C). The function of these epinepherine-secreting neurons is not known.

The enzyme that synthesizes epinephrine, phenylethanolamine-*N*-methyltransferase (see Figure 6.10), is present only in epinephrine-secreting neurons. Otherwise, the metabolism of epinepherine is very similar to that of norepinepherine. Epinepherine is loaded into vesicles via the VMAT. No plasma membrane transporter specific for epinepherine has been identified, though the NET is capable of transporting epinepherine. As already noted, epinepherine acts on both α- and β-adrenergic receptors.

• *Histamine* is found in neurons in the hypothalamus that send sparse but widespread projections to almost all regions of the brain and spinal cord (Figure 6.12A). The central histamine projections mediate arousal and attention, similar to central ACh and norepinephrine projections. Histamine also controls the reactivity of the vestibular system. Allergic reactions or tissue damage cause release of histamine from mast cells in the bloodstream. The close proximity of

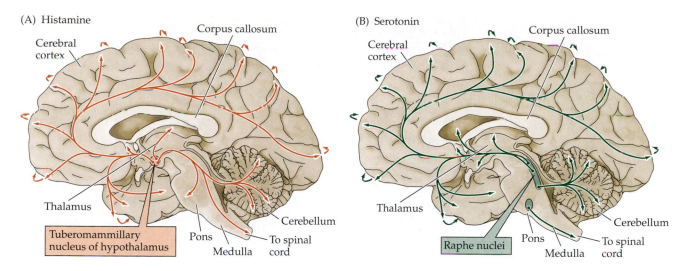

(A) Histamine

Corpus callosum

Cerebral cortex

Thalamus

Tuberomammillary nucleus of hypothalamus

Pons

Medulla

To spinal cord

Cerebellum

(B) Serotonin

Corpus callosum

Cerebral cortex

Thalamus

Raphe nuclei

Pons

Medulla

To spinal cord

Cerebellum

Figure 6.12 The distribution in the human brain of neurons and their projections (arrows) containing histamine (A) or serotonin (B). Curved arrows along the perimeter of the cortex indicate the innervation of lateral cortical regions not shown in this midsagittal plane of section.

mast cells to blood vessels, together with the potent actions of histamine on blood vessels, also raises the possibility that histamine may influence brain blood flow.

Histamine is produced from the amino acid histidine by a histidine decarboxylase (Figure 6.13A) and is transported into vesicles via the same VMAT as the catecholamines. No plasma membrane histamine transporter has been identified yet. Histamine is degraded by the combined actions of histamine methyltransferase and MAO.

There are three known types of histamine receptors, all of which are G-protein-coupled receptors (see Figure 6.5B). Because of the importance of histamine receptors in the mediation of allergic responses, many histamine receptor antagonists have been developed as antihistamine agents. Antihistamines that cross the blood-brain barrier, such as diphenhydramine (Benadryl®), act as sedatives by interfering with the roles of histamine in CNS arousal. Antagonists of the H_1 receptor also are used to prevent motion sickness, perhaps because of the role of histamine in controling vestibular function. H_2 receptors control the secretion of gastric acid in the digestive system, allowing H_2 receptor antagonists to be used in the treatment of a variety of upper gastrointestinal disorders (e.g., peptic ulcers).

• *Serotonin*, or 5-hydroxytryptamine (5-HT), was initially thought to increase vascular tone by virtue of its presence in serum (hence the name serotonin). Serotonin is found primarily in groups of neurons in the raphe region of the pons and upper brainstem, which have widespread projections to the forebrain (see Figure 6.12B) and regulate sleep and wakefulness (see Chapter 28). Serotonin occupies a place of prominence in neuropharmacology because a large number of antipsychotic drugs that are valuable in the treatment of depression and anxiety act on serotonergic pathways (see Box 6E).

Serotonin is synthesized from the amino acid tryptophan, which is an essential dietary requirement. Tryptophan is taken up into neurons by a plasma membrane transporter and hydroxylated in a reaction catalyzed by the enzyme tryptophan-5-hydroxylase (Figure 6.13B), the rate-limiting step for 5-HT synthesis. Loading of 5-HT into synaptic vesicles is done by the VMAT that is also responsible for loading of other monoamines into synaptic vesicles. The synaptic effects of serotonin are terminated by transport back into nerve terminals via a specific serotonin transporter (SERT). Many antidepressant drugs are selective serotonin reuptake inhibitors (SSRIs) that inhibit transport of 5-HT by SERT.

(A)
Histidine

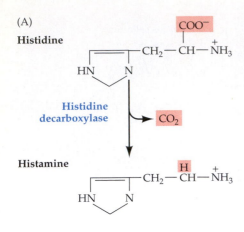

Histamine

(B)
Tryptophan

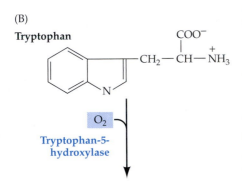

5-Hydroxytryptophan

Serotonin (5-hydroxytryptamine)

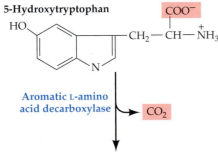

Figure 6.13 *Synthesis of histamine and serotonin. (A) Histamine is synthesized from the amino acid histidine. (B) Serotonin is derived from the amino acid tryptophan by a two-step process that requires the enzymes tryptophan-5-hydroxylase and a decarboxylase.*

Perhaps the best known example of an SSRI is Prozac®. The primary catabolic pathway for 5-HT is mediated by MAO.

A large number of 5-HT receptors have been identified. Most 5-HT receptors are metabotropic (see Figure 6.5B). These have been implicated in behaviors, including the emotions, circadian rhythms, motor behaviors, and state of mental arousal. Impairments in the function of these receptors have been implicated in numerous psychiatric disorders, such as depression, anxiety disorders, and schizophrenia (see Chapter 29), and drugs acting on serotonin receptors are effective treatments for a number of these conditions. Activation of 5-HT receptors also mediates satiety and decreased food consumption, which is why serotonergic drugs are sometimes useful in treating eating disorders.

Only one group of serotonin receptors, called the 5-HT$_3$ receptors, are ligand-gated ion channels (see Figure 6.4C). These are non-selective cation channels and therefore mediate excitatory postsynaptic responses. Their general structure, with functional channels formed by assembly of multiple subunits, is similar to the other ionotropic receptors described in the chapter. Two types of 5-HT$_3$ subunit are known, and form functional channels by assembling as a heteromultimer. 5-HT receptors are targets for a wide variety of therapeutic drugs including ondansetron (Zofran®) and granisetron (Kytril®), which are used to prevent postoperative nausea and chemotherapy-induced emesis.

ATP and Other Purines

Interestingly, all synaptic vesicles contain ATP, which is co-released with one or more "classical" neurotransmitters. This observation raises the possibility that ATP acts as a co-transmitter. It has been known since the 1920s that the extracellular application of ATP (or its breakdown products AMP and adenosine) can elicit electrical responses in neurons. The idea that some purines (so named because all these compounds contain a purine ring; see Figure 6.1) are also neurotransmitters has now received considerable experimental support. ATP acts as an excitatory neurotransmitter in motor neurons of the spinal cord, as well as sensory and autonomic ganglia. Postsynaptic actions of ATP have also been demonstrated in the central nervous system, specifically for dorsal horn neurons and in a subset of hippocampal neurons. Adenosine, however, cannot be considered a classical neurotransmitter because it is not stored in synaptic vesicles or released in a Ca^{2+}-dependent manner. Rather, it is generated from ATP by the action of extracellular enzymes. A number of enzymes, such as apyrase and ecto-5' nucleotidase, as well as nucleoside transporters are involved in the rapid catabolism and removal of purines from extracellular locations. Despite the relative novelty of this evidence, it suggests that excitatory transmission via purinergic synapses is widespread in the mammalian brain.

In accord with this evidence, receptors for both ATP and adenosine are widely distributed in the nervous system, as well as many other tissues. Three classes of these purinergic receptors are now known. One of these classes consists of ligand-gated ion channels (see Figure 6.4C); the other two are G-protein-coupled metabotropic receptors (see Figure 6.5B). Like many ionotropic transmitter receptors, the ligand-gated channels are nonselective cation channels that mediate excitatory postsynaptic responses. The genes encoding these channels, however, are unique in that they appear to have only two transmembrane domains. Ionotropic purinergic receptors are widely distributed in central and peripheral neurons. In sensory nerves they evidently play a role in mechanosensation and pain; their function in most other cells, however, is not known.

The two types of metabotropic receptors activated by purines differ in their sensitivity to agonists: One type is preferentially stimulated by adenosine, whereas the other is preferentially activated by ATP. Both receptor types are found throughout the brain, as well as in peripheral tissues such as the heart, adipose tissue, and the kidney. Xanthines such as caffeine and theophylline block adenosine receptors, and this activity is thought to be responsible for the stimulant effects of these agents.

Peptide Neurotransmitters

Many peptides known to be hormones also act as neurotransmitters. Some peptide transmitters have been implicated in modulating emotions (see Chapter 29). Others, such as substance P and the opioid peptides, are involved in the perception of pain (see Chapter 10). Still other peptides, such as melanocyte-stimulating hormone, adrenocorticotropin, and β-endorphin, regulate complex responses to stress.

The mechanisms responsible for the synthesis and packaging of peptide transmitters are fundamentally different from those used for the small-molecule neurotransmitters and are much like the synthesis of proteins that are secreted from non-neuronal cells (pancreatic enzymes, for instance). Peptide-secreting neurons generally synthesize polypeptides in their cell bodies that are much larger than the final, "mature" peptide. Processing these polypeptides in their cell bodies, which are called **pre-propeptides** (or pre-proproteins), takes place by a sequence of reactions in several intracellular organelles. Pre-propeptides are synthesized in the rough endoplasmic reticulum, where the signal sequence of amino acids—that is, the sequence indicating that the peptide is to be secreted—is removed. The remaining polypeptide, called a **propeptide** (or proprotein), then traverses the Golgi apparatus and is packaged into vesicles in the *trans*-Golgi network. The final stages of peptide neurotransmitter processing occur after packaging into vesicles and involve proteolytic cleavage, modification of the ends of the peptide, glycosylation, phosphorylation, and disulfide bond formation.

Propeptide precursors are typically larger than their active peptide products and can give rise to more than one species of neuropeptide (Figure 6.14). This means that multiple neuroactive peptides can be released from a single vesicle. In addition, neuropeptides often are co-released with small-molecule neurotransmitters. Thus, peptidergic synapses often elicit complex postsynaptic responses. Peptides are catabolized into inactive amino acid fragments by enzymes called peptidases, usually located on the extracellular surface of the plasma membrane.

The biological activity of the peptide neurotransmitters depends on their amino acid sequence (Figure 6.15). Based on their amino acid sequences, neuropeptide transmitters have been loosely grouped into five categories: the brain/gut peptides, opioid peptides, pituitary peptides, hypothalamic releasing hormones, and a catch-all category containing other peptides that are not easily classified.

Substance P is an example of the first of these categories (Figure 6.15A). The study of neuropeptides actually began more than 60 years ago with the accidental discovery of substance P, a powerful hypotensive agent. (The peculiar name derives from the fact that this molecule was an unidentified component of *powder* extracts from brain and intestine.) This 11-amino-acid peptide is present in

Figure 6.14 Proteolytic processing of the pre-propeptides pre-proopiomelanocortin (A) and pre-proenkephalin A (B). For each pre-propeptide, the signal sequence is indicated at the left; the locations of active peptide products are indicated by different colors. The maturation of the pre-propeptides involves cleaving the signal sequence and other proteolytic processing. Such processing can result in a number of different neuroactive peptides such as ACTH, γ-lipotropin, and β-endorphin (A), or multiple copies of the same peptide, such as met-enkephalin (B).

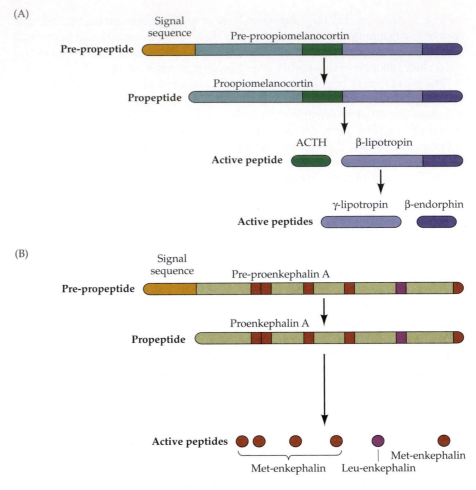

high concentrations in the human hippocampus, neocortex, and also in the gastrointestinal tract; hence its classification as a brain/gut peptide. It is also released from C fibers, the small-diameter afferents in peripheral nerves that convey information about pain and temperature (as well as postganglionic autonomic signals). Substance P is a sensory neurotransmitter in the spinal cord, where its release can be inhibited by opioid peptides released from spinal cord interneurons, resulting in the suppression of pain (see Chapter 10). The diversity of neuropeptides is highlighted by the finding that the gene coding for substance P also encodes a number of other neuroactive peptides, including neurokinin A, neuropeptide K, and neuropeptide Y.

An especially important category of peptide neurotransmitters is the family of opioids (Figure 6.15B). These peptides are so named because they bind to the same postsynaptic receptors activated by opium. The opium poppy has been cultivated for at least 5000 years, and its derivatives have been used as an analgesic since at least the Renaissance. The active ingredients in opium are a variety of plant alkaloids, predominantly morphine. Morphine, named for Morpheus, the Greek god of dreams, remains one of the most effective analgesics in use today, despite its addictive potential (see Box 6A). Synthetic opiates such as meperidine and methadone are also used as analgesics, and fentanyl, a drug with 80 times the analgesic potency of morphine, is widely used in clinical anesthesiology.

The opioid peptides were discovered in the 1970s during a search for *endorphins*, *endo*genous compounds that mimicked the actions of *morphine*. It was hoped that such compounds would be analgesics, and that understanding them

(A) Brain–gut peptides

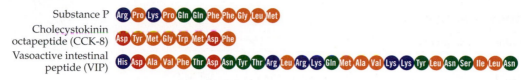

(B) Opioid peptides

Amino acid properties
- Hydrophobic (orange)
- Polar, uncharged (green)
- Acidic (red)
- Basic (blue)

(C) Pituitary peptides

(D) Hypothalamic–releasing peptides

(E) Miscellaneous peptides

Figure 6.15 *Neuropeptides vary in length, but usually contain between 3 and 36 amino acids. The sequence of amino acids determines the biological activity of each peptide.*

would shed light on drug addiction. The endogenous ligands of the opioid receptors have now been identified as a family of more than 20 opioid peptides that fall into three classes: the endorphins, the enkephalins, and the dynorphins (Table 6.2). Each of these classes are liberated from an inactive pre-propeptide (pre-proopiomelanocortin, pre-proenkephalin A, and pre-prodynorphin), derived from distinct genes (see Figure 6.14). Opioid precursor processing is carried out by tissue-specific processing enzymes that are packaged into vesicles, along with the precursor peptide, in the Golgi apparatus.

Opioid peptides are widely distributed throughout the brain and are often co-localized with other small-molecule neurotransmitters, such as GABA and 5-HT. In general, these peptides tend to be depressants. When injected intracerebrally in experimental animals, they act as analgesics; on the basis of this and other evidence, opioids are likely to be involved in the mechanisms underlying acupuncture-induced analgesia. Opioids are also involved in complex behaviors such as sexual attraction and aggressive/submissive behaviors. They have also been implicated in psychiatric disorders such as schizophrenia and autism, although the evidence for this is debated. Unfortunately, the repeated administration of opioids leads to tolerance and addiction.

Virtually all neuropeptides initiate their effects by activating G-protein-coupled receptors. The study of these metabotropic peptide receptors in the brain has been difficult because few specific agonists and antagonists are known. Peptides activate their receptors at low (n*M* to µ*M*) concentrations compared to the

TABLE 6.2	Endogenous Opioid Peptides
Name	Amino acid sequence[a]
Endorphins	
α-Endorphin	*Tyr-Gly-Gly-Phe*-Met-Thr-Ser-Glu-Lys-Ser-Gln-Thr-Pro-Leu-Val-Thr
α-Neoendorphin	*Tyr-Gly-Gly-Phe*-Leu-Arg-Lys-Tyr-Pro-Lys
β-Endorphin	*Tyr-Gly-Gly-Phe*-Met-Thr-Ser-Glu-Lys-Ser-Gln-Thr-Pro-Leu-Val-Thr-Leu-Phe-Lys-Asn-Ala-Ile-Val-Lys-Asn-Ala-His-Lys-Gly-Gln
γ-Endorphin	*Tyr-Gly-Gly-Phe*-Met-Thr-Ser-Glu-Lys-Ser-Gln-Thr-Pro-Leu-Val-Thr-Leu
Enkephalins	
Leu-enkephalin	*Tyr-Gly-Gly-Phe*-Leu
Met-enkephalin	*Tyr-Gly-Gly-Phe*-Met
Dynorphins	
Dynorphin A	*Tyr-Gly-Gly-Phe*-Leu-Arg-Arg-Ile-Arg-Pro-Lys-Leu-Lys-Trp-Asp-Asn-Gln
Dynorphin B	*Tyr-Gly-Gly-Phe*-Leu-Arg-Arg-Gln-Phe-Lys-Val-Val-Thr

[a] Note the initial homology, indicated by italics.

concentrations required to activate receptors for small-molecule neurotransmitters. These properties allow the postsynaptic targets of peptides to be quite far removed from presynaptic terminals and to modulate the electrical properties of neurons that are simply in the vicinity of the site of peptide release. Neuropeptide receptor activation is especially important in regulating the postganglionic output from sympathetic ganglia and the activity of the gut (see Chapter 21). Peptide receptors, particularly the neuropeptide Y receptor, are also implicated in the initiation and maintenance of feeding behavior leading to satiety or obesity.

Other behaviors ascribed to peptide receptor activation include anxiety and panic attacks, and antagonists of cholecystokinin receptors are clinically useful in the treatment of these afflictions. Other useful drugs have been developed by targeting the opiate receptors. Three well-defined opioid receptor subtypes (μ, δ, and κ) play a role in reward mechanisms as well as addiction. The μ-opiate receptor has been specifically identified as the primary site for drug reward mediated by opiate drugs

Unconventional Neurotransmitters

In addition to the conventional neurotransmitters already described, some unusual molecules are also used for signaling between neurons and their targets. These chemical signals can be considered as neurotransmitters because of their roles in interneuronal signaling and because their release from neurons is regulated by Ca^{2+}. However, they are unconventional, in comparison to other neurotransmitters, because they are not stored in synaptic vesicles and are not released from presynaptic terminals via exocytotic mechanisms. In fact, these unconventional neurotransmitters need not be released from presynaptic terminals at all and are often associated with "retrograde" signaling from postsynaptic cells back to presynaptic terminals.

• *Endocannabinoids* are a family of related endogenous signals that interact with cannabinoid receptors. These receptors are the molecular targets of Δ^9-tetrahydrocannabinol, the psychoactive component of the marijuana plant, *Cannabis* (Box 6G). While some members of this emerging group of chemical signals remain to be determined, anandamide and 2-arachidonylglycerol (2-AG) have been established as endocannabinoids. These signals are unsaturated

BOX 6G Marijuana and the Brain

Medicinal use of the marijuana plant, *Cannabis sativa* (Figure A), dates back thousands of years. Ancient civilizations—including both Greek and Roman societies in Europe, as well as Indian and Chinese cultures in Asia—appreciated that this plant was capable of producing relaxation, euphoria, and a number of other psychopharmacological actions. In more recent times, medicinal use of marijuana has largely subsided (although it remains useful in relieving the symptoms of terminal cancer patients); the recreational use of marijuana (Figure B) has become so popular that some societies have decriminalized its use.

Understanding the brain mechanisms underlying the actions of marijuana was advanced by the discovery that a cannabinoid, Δ^9-tetrahydrocannabinol (THC; Figure C), is the active component of marijuana. This finding led to the development of synthetic derivatives, such as WIN 55,212-2 and rimonabant (see Figure 6.16), that have served as valuable tools for probing the brain actions of THC. Of particular interest is that receptors for these cannabinoids exist in the brain and exhibit marked regional variations in distribution, being especially enriched in the brain areas—such as substantia nigra and caudate putamen—that have been implicated in drug abuse (Figure D). The presence of these brain receptors for cannabinoids led in turn to a search for endogenous cannabinoid compounds in the brain, culminating in the discovery of endocannabinoids such as 2-AG and anandamide (see Figure 6.16). This path of discovery closely parallels the identification of endogenous opioid peptides, which resulted from the search for endogenous morphine-like compounds in the brain (see text and Table 6.2).

Given that THC interacts with brain endocannabinoid receptors, particularly the CB1 receptor, it is likely that such actions are responsible for the behavioral consequences of marijuana use. Indeed, many of the well-documented effects of marijuana are consistent with the distribution and actions of brain CB1 receptors. For example, marijuana effects on perception could be due to CB1 receptors in the neocortex, effects on psychomotor control due to endocannabinoid receptors in the basal ganglia and cerebellum, effects on short-term memory due to cannabinoid receptors in the hippocampus, and the well-known effects of marijuana on stimulating appetite due to hypothalamic actions. While formal links between these behavioral consequences and underlying brain mechanisms are still being forged, studies of the drug's actions have shed substantial light on basic synaptic mechanisms, which promise to further elucidate the mode of action of one of the world's most popular drugs.

References

ADAMS, A. R. (1941) Marihuana. *Harvey Lect.* 37: 168–168.

FREUND, T. F., I. KATONA AND D. PIOMELLI (2003) Role of endogenous cannabinoids in synaptic signaling. *Physiol. Rev.* 83: 1017–1066.

GERDEMAN, G. L., J. G. PARTRIDGE, C. R. LUPICA AND D. M. LOVINGER (2003) It could be habit forming: Drugs of abuse and striatal synaptic plasticity. *Trends Neurosci.* 26: 184–192.

HOWLETT, A.C. (2005) Cannabinoid receptor signaling. *Hndbk. Exp. Pharmacol.* 168: 53–79.

IVERSEN, L. (2003) Cannabis and the brain. *Brain* 126: 1252–1270.

MECHOULAM, R. (1970) Marihuana chemistry. *Science* 168: 1159–1166.

(A)

Cannabis sativa

(B)

(C)

Δ^9-Tetrahydrocannabinol (THC)

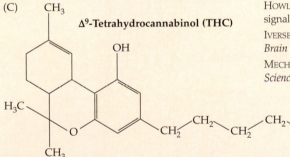

(D)

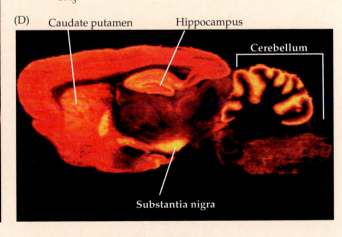

(A) Leaf of *Cannabis sativa*, the marijuana plant. (B) Smoking ground-up *Cannabis* leaves is a popular method of achieving the euphoric effects of marijuana. (C) Structure of THC (Δ^9-tetrahydrocannabinol), the active ingredient of marijuana. (D) Distribution of brain CB1 receptors, visualized by examining the binding of CP-55,940, a CB1 receptor ligand. (B photo © Henry Diltz/Corbis; C after Iversen, 2003; D courtesy of M. Herkenham, NIMH.)

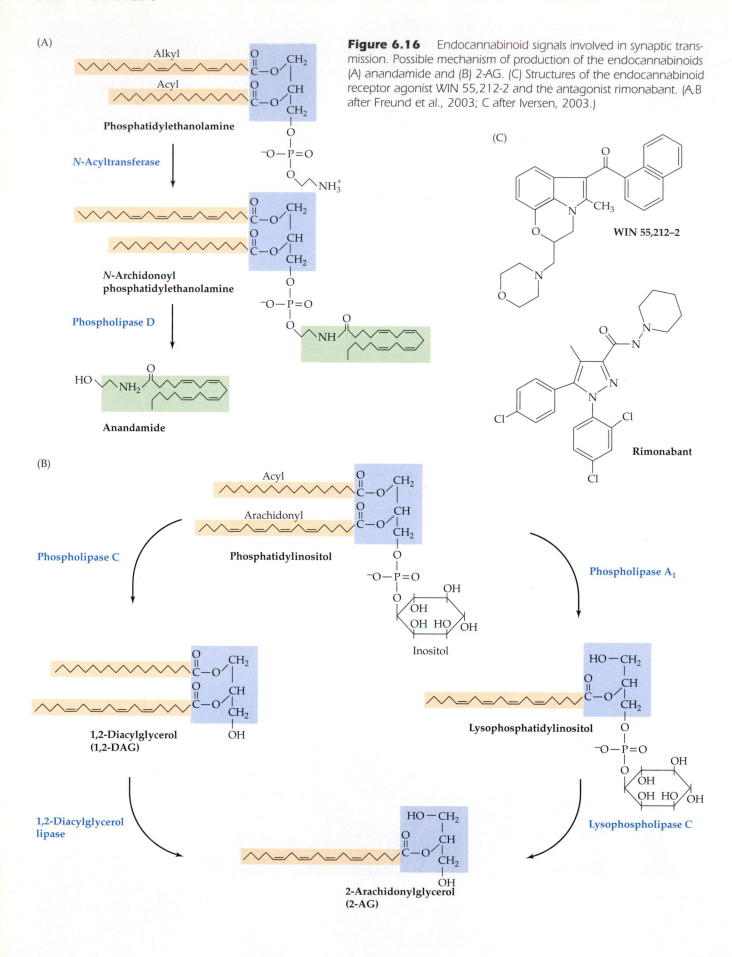

(A)

Alkyl

Acyl

Phosphatidylethanolamine

N-Acyltransferase

N-Archidonoyl
phosphatidylethanolamine

Phospholipase D

HO ⌒ NH₂

Anandamide

(B)

Acyl

Arachidonyl

Phospholipase C

Phosphatidylinositol

Inositol

Phospholipase A₁

1,2-Diacylglycerol
(1,2-DAG)

Lysophosphatidylinositol

1,2-Diacylglycerol
lipase

Lysophospholipase C

2-Arachidonylglycerol
(2-AG)

(C)

WIN 55,212–2

Rimonabant

Figure 6.16 Endocannabinoid signals involved in synaptic transmission. Possible mechanism of production of the endocannabinoids (A) anandamide and (B) 2-AG. (C) Structures of the endocannabinoid receptor agonist WIN 55,212-2 and the antagonist rimonabant. (A,B after Freund et al., 2003; C after Iversen, 2003.)

fatty acid with polar head groups and are produced by enzymatic degradation of membrane lipids (Figure 6.16A,B). Production of endocannabinoids is stimulated by a second messenger signal within postsynaptic neurons, typically a rise in postsynaptic Ca^{2+} concentration. Although the mechanism of endocannabinoid release is not entirely clear, it is likely that these hydrophobic signals diffuse through the postsynaptic membrane to reach cannabinoid receptors on other nearby cells. Endocannabinoid action is terminated by carrier-mediated transport of these signals back into the postsynaptic neuron. There they are hydrolyzed by the enzyme fatty acid hydrolase (FAAH).

At least two types of cannabinoid receptor have been identified, with most actions of endocannabinoids in the CNS mediated by the type termed CB1. CB1 is a G-protein-coupled receptor that is related to the metabotropic receptors for ACh, glutamate, and the other conventional neurotransmitters. Several compounds that are structurally related to endocannabinoids and that bind to the CB1 receptor have been synthesized (Figure 6.16C). These compounds act as agonists or antagonists of the CB1 receptor and serve as both tools for elucidating the physiological functions of endocannabinoids and as targets for developing therapeutically useful drugs.

Endocannabinoids participate in several forms of synaptic regulation. The best-documented action of these agents is to inhibit communication between postsynaptic target cells and their presynaptic inputs. In both the hippocampus and the cerebellum, among other regions, endocannabinoids serve as retrograde signals to regulate GABA release at certain inhibitory terminals. At such synapses, depolarization of the postsynaptic neuron causes a transient reduction in inhibitory postsynaptic responses (Figure 6.17). Depolarization reduces synaptic transmission by elevating the concentration of Ca^{2+} within the postsynaptic neuron. This rise in Ca^{2+} triggers synthesis and release of endocannabinoids from the postsynaptic cells. The endocannabinoids then make their way to the presynaptic terminals and bind to CB1 receptors on these terminals. Activation of the CB1 receptors inhibits the amount of GABA released in response to presynaptic action potentials, thereby reducing inhibitory transmission. These mechanisms responsible for the reduction in GABA release are not entirely clear, but probably involve effects on voltage-gated Ca^{2+} channels and/or K^+ channels in the presynaptic neurons.

• *Nitric oxide* (NO) is an unusual but especially interesting chemical signal. NO is a gas that is produced by the action of nitric oxide synthase, an enzyme that converts the amino acid arginine into a metabolite (citrulline) and simultaneously generates NO (Figure 6.18). Neuronal NO synthase is regulated by Ca^{2+} binding to the Ca^{2+} sensor protein calmodulin (see Chapter 7). Once pro-

Figure 6.17 Endocannabinoid-mediated retrograde control of GABA release. (A) Experimental arrangement. Stimulation of a presynaptic interneuron causes release of GABA onto a postsynaptic pyramidal neuron. (B) IPSCs elicited by the inhibitory synapse (control) are reduced in amplitude following a brief depolarization of the postsynaptic neuron. This reduction in the IPSC is due to less GABA being released from the presynaptic interneuron. (C) The reduction in IPSC amplitude produced by postsynaptic depolarization lasts a few seconds and is mediated by endocannabinoids, because it is prevented by the endocannabinoid receptor antagonist rimonabant. (B,C after Ohno-Shosaku et al., 2001.)

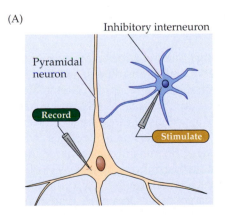

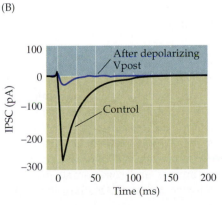

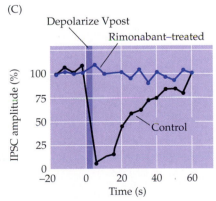

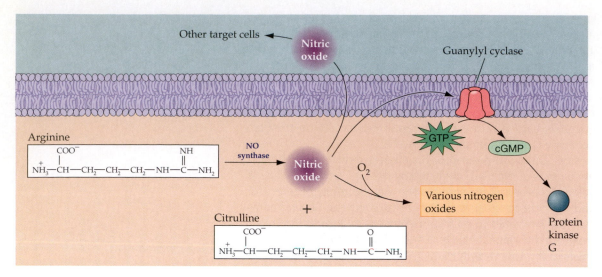

Figure 6.18   *Synthesis, release, and terminations of NO.*

duced, NO can permeate the plasma membrane, meaning that NO generated inside one cell can travel through the extracellular medium and act within nearby cells. Thus, this gaseous signal has a range of influence that extends well beyond the cell of origin, diffusing a few tens of micrometers from its site of production before it is degraded. This property makes NO a potentially useful agent for coordinating the activities of multiple cells in a very localized region and may mediate certain forms of synaptic plasticity that spread within small networks of neurons.

All of the known actions of NO are mediated within its cellular targets; for this reason, NO often is considered a second messenger rather than a neurotransmitter. Some of these actions of NO are due to the activation of the enzyme guanylyl cyclase, which then produces the second messenger cGMP within target cells (see Chapter 7). Other actions of NO are the result of covalent modification of target proteins via nitrosylation, the addition of a nitryl group to selected amino acids within the proteins. NO decays spontaneously by reacting with oxygen to produce inactive nitrogen oxides. As a result, NO signals last for only a short time, on the order of seconds or less. NO signaling evidently regulates a variety of synapses that also employ conventional neurotransmitters; so far, glutamatergic synapses are the best-studied target of NO in the central nervous system. NO may also be involved in some neurological diseases. For example, it has been proposed that an imbalance between nitric oxide and superoxide generation underlies some neurodegenerative diseases.

Summary

The complex synaptic computations occurring at neural circuits throughout the brain arise from the actions of a large number of neurotransmitters, which act on an even larger number of postsynaptic neurotransmitter receptors. Glutamate is the major excitatory neurotransmitter in the brain, whereas GABA and glycine are the major inhibitory neurotransmitters. The actions of these small-molecule neurotransmitters are typically faster than those of the neuropeptides. Thus, most small-molecule transmitters mediate synaptic transmission when a rapid response is essential, whereas the neuropeptide transmitters, as well as the biogenic amines and some small-molecule neurotransmitters, tend to modulate ongoing activity in the brain or in peripheral target tissues in a more gradual and ongoing way. Two broadly different families of neurotransmitter receptors have evolved to carry out the postsynaptic signaling actions of neurotransmitters. Ionotropic or ligand-gated ion channels combine the neurotransmitter receptor and ion channel in one molecular entity, and therefore give rise to rapid postsynaptic electrical responses. Metabotropic receptors regulate the activity of postsynaptic ion channels indirectly, usually via G-proteins, and induce slower and longer-lasting electrical responses. Metabotropic receptors are especially important in regulating behavior, and drugs targeting these receptors have been clinically valuable in treating a wide range of behavioral disorders. The postsynaptic response at a given synapse is determined by the combination of receptor subtypes, G-protein subtypes, and ion channels that are expressed in the postsynaptic cell. Because each of these features can vary both within and among neurons, a tremendous diversity of transmitter-mediated effects is possible. Drugs that influence transmitter actions have enormous importance in the treatment of neurological and psychiatric disorders, as well as in a broad spectrum of other medical problems.

Additional Reading

Reviews

BARNES, N. M. AND T. SHARP (1999) A review of central 5-HT receptors and their function. *Neuropharm.* 38: 1083–1152.

BOURIN, M., G. B. BAKER AND J. BRADWEJN (1998) Neurobiology of panic disorder. *J. Psychosomatic Res.* 44: 163–180.

BURNSTOCK, G. (2006) Purinergic signalling: An overview. *Novartis Found. Symp.* 276: 26–48.

CARLSSON, A. (1987) Perspectives on the discovery of central monoaminergic neurotransmission. *Annu. Rev. Neurosci.* 10: 19–40.

CIVELLI, O. (1998) Functional genomics: The search for novel neurotransmitters and neuropeptides. *FEBS Letters* 430: 55–58.

FREUND, T. F., I. KATONA AND D. PIOMELLI (2003) Role of endogenous cannabinoids in synaptic signaling. *Physiol Rev.* 83: 1017–1066.

HÖKFELT, T. D. AND 10 OTHERS (1987) Coexistence of peptides with classical neurotransmitters. *Experientia* Suppl. 56: 154–179.

HOWLETT, A. C. (2005) Cannabinoid receptor signaling. *Handbook Exp. Pharmacol.* 168: 53–79.

HYLAND, K. (1999) Neurochemistry and defects of biogenic amine neurotransmitter metabolism. *J. Inher. Metab. Dis.* 22: 353–363.

IVERSEN, L. (2003) Cannabis and the brain. *Brain* 126: 1252–1270.

KOOB, G. F., P. P. SANNA AND F. E. BLOOM (1998) Neuroscience of addiction. *Neuron* 21: 467–476.

LAUBE, B., G. MAKSAY, R. SCHEMM AND H. BETZ (2002) Modulation of glycine receptor function: A novel approach for therapeutic intervention at inhibitory synapses? *Trends Pharmacol. Sci.* 23: 519–527.

LERMA, J. (2006) Kainate receptor physiology. *Curr. Opin. Pharmacol.* 6: 89–97.

MASSON, J., C. SAGN, M. HAMON AND S. E. MESTIKAWY (1999) Neurotransmitter transporters in the central nervous system. *Pharmacol. Rev.* 51: 439–464.

NAKANISHI, S. (1992) Molecular diversity of glutamate receptors and implication for brain function. *Science* 258: 597–603.

PERRY, E., M. WALKER, J. GRACE AND R. PERRY (1999) Acetylcholine in mind: A neurotransmitter correlate of consciousness? *Trends Neurosci.* 22: 273–280.

PIERCE, K. L., R. T. PREMONT AND R. J. LEFKOWITZ (2002) Seven-transmembrane receptors. *Nature Rev. Mol. Cell Biol.* 3: 639–650.

SCHWARTZ, J. C., J. M. ARRANG, M. GARBARG, H. POLLARD AND M. RUAT (1991) Histaminergic transmission in the mammalian brain. *Physiol. Rev.* 71: 1–51.

SCHWARTZ, M. W., S. C. WOODS, D. PORTE JR., R. J. SEELEY AND D. G. BASKIN (2000) Central nervous system control of food intake. *Nature* 404: 661–671.

STAMLER, J. S., E. J. TOONE, S. A. LIPTON AND N. J. SUCHER (1997) (S)NO Signals: Translocation, regulation, and a consensus motif. *Neuron* 18: 691–696.

TUCEK, S., J. RICNY AND V. DOLEZAL (1990) Advances in the biology of cholinergic neurons. *Adv. Neurol.* 51: 109–115.

WEBB, T. E. AND E. A. BARNARD (1999) Molecular biology of P2Y receptors expressed in the nervous system. *Prog. Brain Res.* 120: 23–31.

WILSON, R. I. AND R. A. NICOLL (2002) Endocannabinoid signaling in the brain. *Science* 296: 678–682.

Important Original Papers

BRENOWITZ, S. D. AND W. G. REGEHR (2003) Calcium dependence of retrograde inhibition by endocannabinoids at synapses onto Purkinji cells. *J. Neurosci.* 23: 6373–6384.

CHAVAS, J. AND A. MARTY (2003) Coexistence of excitatory and inhibitory GABA synapses in the cerebellar interneuron network. *J. Neurosci.* 23: 2019–2031.

CHEN, Z. P., A. LEVY AND S. L. LIGHTMAN (1995) Nucleotides as extracellular signalling molecules. *J. Neuroendocrinol.* 7: 83–96.

CURTIS, D. R., J. W. PHILLIS AND J. C. WATKINS (1959) Chemical excitation of spinal neurons. *Nature* 183: 611–612.

DALE, H. H., W. FELDBERG AND M. VOGT (1936) Release of acetylcholine at voluntary motor nerve endings. *J. Physiol.* 86: 353–380.

DAVIES, P. A. AND 6 OTHERS (1999) The 5-HT3B subunit is a major determinant of serotonin-receptor function. *Nature* 397: 359–363.

GOMEZA, J. AND 6 OTHERS (2003) Inactivation of the glycine transporter 1 gene discloses vital role of glial glycine uptake in glycinergic inhibition. *Neuron* 40: 785–796.

GU, J. G. AND A. B. MACDERMOTT (1997) Activation of ATP P2X receptors elicits glutamate release from sensory neuron synapses. *Nature* 389: 749–753.

HÖKFELT, T., O. JOHANSSON, A. LJUNGDAHL, J. M. LUNDBERG AND M. SCHULTZBERG (1980) Peptidergic neurons. *Nature* 284: 515–521.

HOLLMANN, M., C. MARON AND S. HEINEMANN (1994) N-glycosylation site tagging suggests a three transmembrane domain topology for the glutamate receptor GluR1. *Neuron* 13: 1331–1343.

HUGHES, J., T. W. SMITH, H. W. KOSTERLITZ, L. A. FOTHERGILL, B. A. MORGAN AND H. R. MORRIS (1975) Identification of two related pentapeptides from the brain with potent opiate agonist activity. *Nature* 258: 577–580.

KAUPMANN, K. AND 10 OTHERS (1997) Expression cloning of GABAβ receptors uncovers similarity to metabotropic glutamate receptors. *Nature* 386: 239–246.

LEDEBT, C. AND 9 OTHERS (1997) Aggressiveness, hypoalgesia and high blood pressure in mice lacking the adenosine A2a receptor. *Nature* 388: 674–678.

NAVEILHAN, P. AND 10 OTHERS (1999) Normal feeding behavior, body weight and leptin response require the neuropeptide YY2 receptor. *Nature Med.* 5: 1188–1193.

OHNO-SHOSAKU, T., T. MAEJIMA, AND M. KANO (2001) Endogenous cannabinoids mediate retrograde signals from depolarized postsynaptic neurons to presynaptic terminals. *Neuron* 29: 729–738.

ROSENMUND, C., Y. STERN-BACH AND C. F. STEVENS (1998) The tetrameric structure of a glutamate receptor channel. *Science* 280: 1596–1599.

THOMAS, S. A. AND R. D. PALMITER (1995) Targeted disruption of the tyrosine hydroxylase gene reveals that catecholamines are required for mouse fetal development. *Nature* 374: 640–643.

UNWIN, N. (1995) Acetylcholine receptor channels imaged in the open state. *Nature* 373: 37–43.

WANG, Y. M. AND 8 OTHERS (1997) Knockout of the vesicular monoamine transporter 2 gene results in neonatal death and supersensitivity to cocaine and amphetamine. *Neuron* 19: 1285–1296.

Books

BALAZS, R., R. J. BRIDGES AND C. W. COTMAN (2006) *Excitatory Amino Acid Transmission in Health and Disease*. New York: Oxford University Press.

COOPER, J. R., F. E. BLOOM AND R. H. ROTH (2003) *The Biochemical Basis of Neuropharmacology*. New York: Oxford University Press.

FELDMAN, R. S., J. S. MEYER AND L. F. QUENZER (1997) *Principles of Neuropharmacology*, 2nd Edition. Sunderland, MA: Sinauer Associates.

HALL, Z. (1992) *An Introduction to Molecular Neurobiology*. Sunderland, MA: Sinauer Associates.

HILLE, B. (2002) *Ion Channels of Excitable Membranes*, 3rd Edition. Sunderland, MA: Sinauer Associates.

MYCEK, M. J., R. A. HARVEY AND P. C. CHAMPE (2000) *Pharmacology*, 2nd Edition. Philadelphia, New York: Lippincott/ Williams and Wilkins Publishers.

NICHOLLS, D. G. (1994) *Proteins, Transmitters, and Synapses*. Boston: Blackwell Scientific.

SIEGEL, G. J., B. W. AGRANOFF, R. W. ALBERS, S. K. FISHER AND M. D. UHLER (1999) *Basic Neurochemistry*. Philadelphia: Lippincott-Raven.

Chapter 7

Molecular Signaling within Neurons

Overview

As is apparent in the preceding chapters, electrical and chemical signaling mechanisms allow one nerve cell to receive and transmit information to another. This chapter focuses on the related events within neurons and other cells that are triggered by the interaction of a chemical signal with its receptor. This intracellular processing typically begins when extracellular chemical signals, such as neurotransmitters, hormones, and trophic factors, bind to specific receptors located either on the surface or within the cytoplasm or nucleus of the target cells. Such binding activates the receptors and in so doing stimulates cascades of intracellular reactions involving GTP-binding proteins, second messenger molecules, protein kinases, ion channels, and many other effector proteins whose modulation temporarily changes the physiological state of the target cell. These same intracellular signal transduction pathways can also cause longer-lasting changes by altering the transcription of genes, thus affecting the protein composition of the target cells on a more permanent basis. The large number of components involved in intracellular signaling pathways allows precise temporal and spatial control over the function of individual neurons, thereby allowing the coordination of electrical and chemical activity in the related populations of neurons that comprise neural circuits and systems.

Strategies of Molecular Signaling

Chemical communication coordinates the behavior of individual nerve and glial cells in physiological processes that range from neural differentiation to learning and memory. Indeed, molecular signaling ultimately mediates and fine-tunes all brain functions. To carry out such communication, a series of extraordinarily diverse and complex chemical signaling pathways has evolved. The preceding chapters have described in some detail the electrical signaling mechanisms that allow neurons to generate action potentials for conduction of information. These chapters also described synaptic transmission, a special form of chemical signaling that transfers information from one neuron to another. But chemical signaling is not limited to synapses (Figure 7.1A). Other well-characterized forms of chemical communication include **paracrine** signaling, which acts over a longer range than synaptic transmission and involves the secretion of chemical signals onto a group of nearby target cells, and **endocrine** signaling, which refers to the secretion of hormones into the bloodstream where they can affect targets throughout the body.

Chemical signaling of any sort requires three components: a molecular *signal* that transmits information from one cell to another; a *receptor* molecule that transduces the information provided by the signal; and a *target* molecule that mediates the cellular response (Figure 7.1B). The part of this process that takes place within the confines of the target cell is called **intracellular signal trans-**

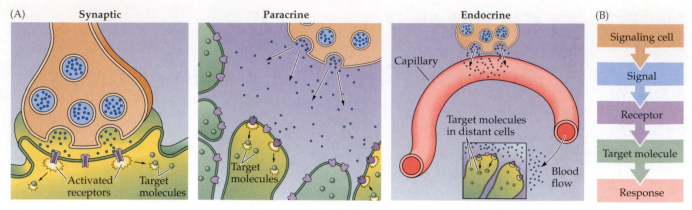

Figure 7.1   *Chemical signaling mechanisms. (A) Forms of chemical communication include synaptic transmission, paracrine signaling, and endocrine signaling. (B) The essential components of chemical signaling are: cells that initiate the process by releasing signaling molecules; specific receptors on target cells; second messenger target molecules; and subsequent cellular responses.*

duction. A good example of transduction in the context of *intercellular* communication is the sequence of events triggered by chemical synaptic transmission: neurotransmitters serve as the signal, neurotransmitter receptors serve as the transducing receptor, and the target molecule is an ion channel that is altered to cause the electrical response of the postsynaptic cell. In many cases, however, synaptic transmission activates additional *intracellular* pathways that have a variety of functional consequences. For example, the binding of the neurotransmitter norepinephrine to its receptor activates GTP-binding proteins, which produces second messengers within the postsynaptic target, activates enzyme cascades, and eventually changes the chemical properties of numerous target molecules within the affected cell.

A general advantage of chemical signaling in both intercellular and intracellular contexts is **signal amplification**. Amplification occurs because individual signaling reactions can generate a much larger number of molecular products than the number of molecules that initiate the reaction. In the case of norepinephrine signaling, for example, a single norepinephrine molecule binding to its receptor can generate many thousands of second messenger molecules (such as cyclic AMP), yielding activation of tens of thousands of molecules of the target protein (Figure 7.2). Similar amplification occurs in all signal transduction pathways. Because the transduction processes often are mediated by a sequential set of enzymatic reactions, each with its own amplification factor, a small number of signal molecules ultimately can activate a very large number of target molecules. Such amplification guarantees that a physiological response is evoked in the face of other, potentially countervailing, influences.

Another rationale for these complex signal transduction schemes is to permit precise control of cell behavior over a wide range of times. Some molecular interactions allow information to be transferred rapidly, while others are slower and longer lasting. For example, the signaling cascades associated with synaptic transmission at neuromuscular junctions allow a person to respond to rapidly changing cues, such as the trajectory of a pitched ball, while the slower responses triggered by adrenal medullary hormones (epinephrine and norepinephrine) secreted during a challenging game produce slower (and longer lasting) effects on muscle metabolism (see Chapter 21) and emotional state (see Chapter 29). To encode information that varies so widely over time, the concentration of the relevant signaling molecules must be carefully controlled. On one hand, the concentration of every molecule within the signaling cascade must return to subthreshold values before the arrival of another stimulus. On the other hand, keeping the intermediates in a signaling pathway activated is critical for a sustained response. Having multiple levels of molecular interactions facilitates the intricate timing of these events.

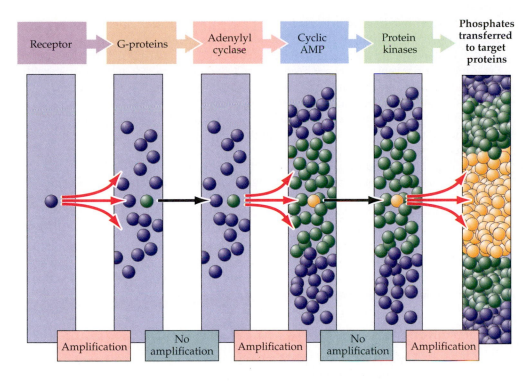

Receptor	G-proteins	Adenylyl cyclase	Cyclic AMP	Protein kinases	Phosphates transferred to target proteins

Amplification	No amplification	Amplification	No amplification	Amplification

Figure 7.2 Amplification in signal transduction pathways. The activation of a single receptor by a signaling molecule, such as the neurotransmitter norepinephrine, can lead to the activation of numerous G-proteins inside cells. These activated proteins can bind to other signaling molecules, such as the enzyme adenylyl cyclase. Each activated enzyme molecule generates a large number of cAMP molecules. cAMP binds to and activates another family of enzymes—the protein kinases—that can phosphorylate many target proteins. While not every step in this signaling pathway involves amplification, overall the cascade results in a tremendous increase in the potency of the initial signal.

The Activation of Signaling Pathways

The molecular cascades of signal transduction pathways are always initiated by a chemical signaling molecule. Such signaling molecules can be grouped into three classes: **cell-impermeant**, **cell-permeant**, and **cell-associated signaling molecules** (Figure 7.3). The first two classes are secreted molecules and thus can act on target cells removed from the site of signal synthesis or release. Cell-impermeant signaling molecules typically bind to receptors associated with cell membranes. Hundreds of secreted molecules have now been identified, including the neurotransmitters discussed in Chapter 6, as well as proteins such as neurotrophic factors (see Chapter 23), and peptide hormones such as glucagon, insulin, and various reproductive hormones. These signaling molecules are typi-

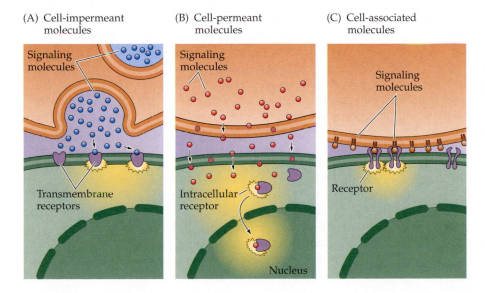

(A) Cell-impermeant molecules

Signaling molecules

Transmembrane receptors

(B) Cell-permeant molecules

Signaling molecules

Intracellular receptor

Nucleus

(C) Cell-associated molecules

Signaling molecules

Receptor

Figure 7.3 Three classes of cell signaling molecules. (A) Cell-impermeant molecules, such as neurotransmitters, cannot readily traverse the plasma membrane of the target cell and must bind to the extracellular portion of transmembrane receptor proteins. (B) Cell-permeant molecules are able to cross the plasma membrane and bind to receptors in the cytoplasm or nucleus of target cells. (C) Cell-associated molecules are presented on the extracellular surface of the plasma membrane. These signals activate receptors on target cells only if they are directly adjacent to the signaling cell.

cally short-lived, either because they are rapidly metabolized or because they are internalized by endocytosis once bound to their receptors.

Cell-permeant signaling molecules can cross the plasma membrane to act directly on receptors that are inside the cell. Examples include numerous steroid (glucocorticoids, estradiol, and testosterone) and thyroid (thyroxin) hormones, and retinoids. These signaling molecules are relatively insoluble in aqueous solutions and are often transported in blood and other extracellular fluids by binding to specific carrier proteins. In this form, they may persist in the bloodstream for hours or even days.

The third group of chemical signaling molecules, cell-associated signaling molecules, are arrayed on the extracellular surface of the plasma membrane. As a result, these molecules act only on other cells that are physically in contact with the cell that carries such signals. Examples include proteins such as the integrins and neural cell adhesion molecules (NCAMs) that influence axonal growth (see Chapter 23). Membrane-bound signaling molecules are more difficult to study, but are clearly important in neuronal development and other circumstances where physical contact between cells provides information about cellular identities.

Receptor Types

Regardless of the nature of the initiating signal, cellular responses are determined by the presence of receptors that specifically bind the signaling molecules. Binding of signal molecules causes a conformational change in the receptor, which then triggers the subsequent signaling cascade within the affected cell. Given that chemical signals can act either at the plasma membrane or within the cytoplasm (or nucleus) of the target cell, it is not surprising that receptors are actually found on both sides of the plasma membrane. The receptors for impermeant signal molecules are membrane-spanning proteins. The extracellular domain of such receptors includes the binding site for the signal, while the intracellular domain activates intracellular signaling cascades after the signal binds. A large number of these receptors have been identified and are grouped into families defined by the mechanism used to transduce signal binding into a cellular response (Figure 7.4).

Channel-linked receptors (also called ligand-gated ion channels) have the receptor and transducing functions as part of the same protein molecule. Interaction of the chemical signal with the binding site of the receptor causes the opening or closing of an ion channel pore in another part of the same molecule. The resulting ion flux changes the membrane potential of the target cell and, in some cases, can also lead to entry of Ca^{2+} ions that serve as a second messenger signal within the cell. Good examples of such receptors are the ionotropic neurotransmitter receptors described in Chapters 5 and 6.

Enzyme-linked receptors also have an extracellular binding site for chemical signals. The intracellular domain of such receptors is an enzyme whose catalytic activity is regulated by the binding of an extracellular signal. The great majority of these receptors are **protein kinases** that phosphorylate intracellular target proteins, often tyrosine residues on these proteins, thereby changing the physiological function of the target cells. Noteworthy members of this group of receptors are the Trk family of neurotrophin receptors (see Chapter 23) and other receptors for growth factors.

G-protein-coupled receptors regulate intracellular reactions by an indirect mechanism involving an intermediate transducing molecule, called the **GTP-binding proteins** (or **G-proteins**). Because these receptors all share the structural feature of crossing the plasma membrane seven times, they are also

(A) Channel-linked receptors

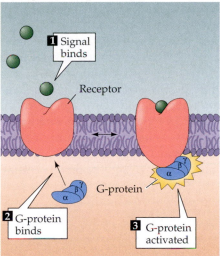

(B) Enzyme-linked receptors

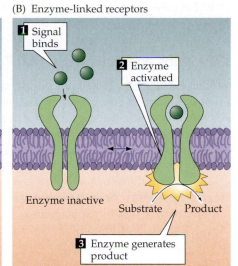

(C) G-protein-coupled receptors

(D) Intracellular receptors

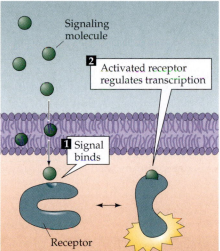

Figure 7.4 *Categories of cellular receptors. Membrane-impermeant signaling molecules can bind to and activate either channel-linked receptors (A), enzyme-linked receptors (B), or G-protein-coupled receptors (C). Membrane permeant signaling molecules activate intracellular receptors (D).*

referred to as 7-transmembrane receptors (or metabotropic receptors; see Chapter 5). Hundreds of different G-protein-linked receptors have been identified. Well-known examples include the β-adrenergic receptor, the muscarinic type of acetylcholine receptor, metabotropic glutamate receptors, receptors for odorants in the olfactory system, and many types of receptors for peptide hormones. Rhodopsin, a light-sensitive, 7-transmembrane protein in retinal photoreceptors, is another form of G-protein-linked receptor (see Figure 11.9).

Intracellular receptors are activated by cell-permeant or lipophilic signaling molecules (Figure 7.4D). Many of these receptors lead to the activation of signaling cascades that produce new mRNA and protein within the target cell. Often such receptors comprise a receptor protein bound to an inhibitory protein complex. When the signaling molecule binds to the receptor, the inhibitory complex dissociates to expose a DNA-binding domain on the receptor. This activated form of the receptor can then move into the nucleus and directly interact with nuclear DNA, resulting in altered transcription. Some intracellular receptors are located primarily in the cytoplasm, while others are in the nucleus.

In either case, once these receptors are activated they can affect gene expression by altering DNA transcription.

G-Proteins and Their Molecular Targets

Both G-protein-linked receptors and enzyme-linked receptors can activate biochemical reaction cascades that ultimately modify the function of target proteins. For both these receptor types, the coupling between receptor activation and their subsequent effects are the GTP-binding proteins. There are two general classes of GTP-binding protein (Figure 7.5). **Heterotrimeric G-proteins** are composed of three distinct subunits (α, β, and γ). There are many different α, β, and γ subunits, allowing a bewildering number of G-protein permutations. Regardless of the specific composition of the heterotrimeric G-protein, its α subunit binds to guanine nucleotides, either GTP or GDP. Binding of GDP then allows the α subunit to bind to the β and γ subunits to form an inactive trimer. Binding of an extracellular signal to a G-protein-coupled receptor in turn allows the G-protein to bind to the receptor and causes GDP to be replaced with GTP (Figure 7.5A). When GTP is bound to the G-protein, the α subunit dissociates from the βγ complex and activates the G-protein. Following activation, both the GTP-bound α subunit and the free βγ complex can bind to downstream effector molecules that mediate a variety of responses in the target cell.

The second class of GTP-binding proteins are **monomeric G-proteins** (also called **small G-proteins**). These monomeric GTPases also relay signals from activated cell surface receptors to intracellular targets such as the cytoskeleton and the vesicle trafficking apparatus of the cell. The first small G-protein was discovered in a virus that causes *rat* sarcoma tumors and was therefore called **ras**. Ras is a molecule that helps regulate cell differentiation and proliferation by relaying signals from receptor kinases to the nucleus; the viral form of ras is defective, which accounts for the ability of the virus to cause the uncontrolled cell proliferation that leads to tumors. Since then, a large number of small

Figure 7.5 *Types of GTP-binding protein. (A) Heterotrimeric G-proteins are composed of three distinct subunits (α, β, and γ). Receptor activation causes the binding of the G-protein and the α subunit to exchange GDP for GTP, leading to a dissociation of the α and βγ subunits. The biological actions of these G-proteins are terminated by hydrolysis of GTP, which is enhanced by GTPase-activating (GAP) proteins. (B) Monomeric G-proteins use similar mechanisms to relay signals from activated cell surface receptors to intracellular targets. Binding of GTP stimulates the biological actions of these G-proteins, and their activity is terminated by hydrolysis of GTP, which is also regulated by GAP proteins.*

(A) Heterotrimeric G-proteins

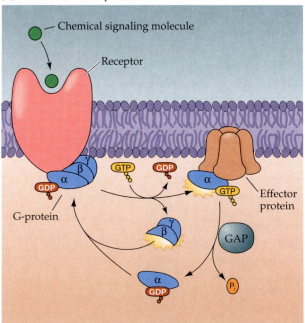

(B) Monomeric G-proteins

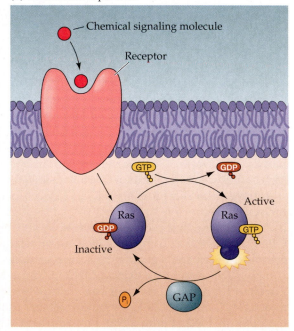

GTPases have been identified and can be sorted into five different subfamilies with different functions. For instance, some are involved in vesicle trafficking in the presynaptic terminal or elsewhere in the neuron, while others play a central role in protein and RNA trafficking in and out of the nucleus.

Signaling by both heterotrimeric and monomeric G-proteins is terminated by hydrolysis of GTP to GDP. The rate of GTP hydrolysis is an important property of G-proteins; this rate can be regulated by other proteins, termed GTPase-activating proteins (GAPs). By replacing GTP with GDP, GAPs return G-proteins to their inactive form. GAPs were first recognized as regulators of small G-proteins, but recently similar proteins have been found to regulate the α subunits of heterotrimeric G-proteins. Hence, monomeric and trimeric G-proteins function as molecular timers that are active in their GTP-bound state, and become inactive when they have hydrolyzed the bound GTP to GDP (Figure 7.5B).

Activated G-proteins alter the function of many downstream effectors. Most of these effectors are enzymes that produce intracellular second messengers. Effector enzymes include adenylyl cyclase, guanylyl cyclase, phospholipase C, and others (Figure 7.6). The second messengers produced by these enzymes trigger the complex biochemical signaling cascades discussed in the next section. Because each of these cascades is activated by specific G-protein subunits, the pathways activated by a particular receptor are determined by the specific identity of the G-protein subunits associated with it.

As well as activating effector molecules, G-proteins can also directly bind to and activate ion channels. For example, some neurons, as well as heart muscle cells, have G-protein-coupled receptors that bind acetylcholine. Because these receptors are also activated by the agonist muscarine, they are usually called muscarinic receptors (see Chapters 6 and 21). Activation of muscarinic receptors

Figure 7.6 Effector pathways associated with G-protein-coupled receptors. In all three examples shown here, binding of a neurotransmitter to such a receptor leads to activation of a G-protein and subsequent recruitment of second messenger pathways. G_s, G_q, and G_i refer to three different types of heterotrimeric G-protein.

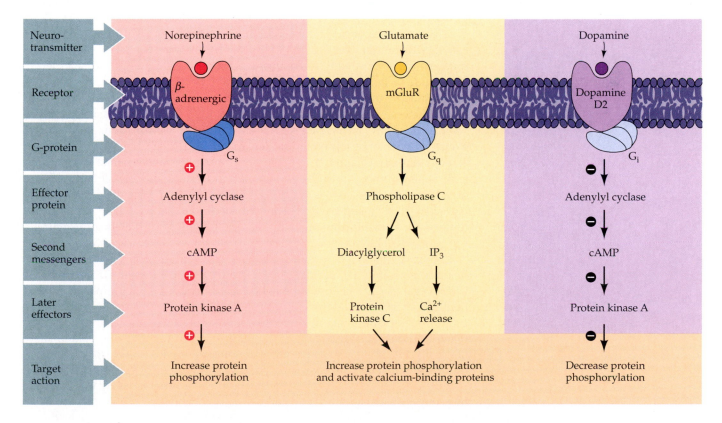

can open K$^+$ channels, thereby inhibiting the rate at which the neuron fires action potentials, or slowing the heartbeat of muscle cells. These inhibitory responses are believed to be the result of βγ subunits of G-proteins binding to the K$^+$ channels. The activation of α subunits can also lead to the rapid closing of voltage-gated Ca^{2+} and Na$^+$ channels. Because these channels participate in action potential generation, closing them makes it more difficult for target cells to fire (see Chapters 3 and 4).

In summary, the binding of chemical signals to their receptors activates cascades of signal transduction events in the cytosol of target cells. Within such cascades, G-proteins serve a pivotal function as the molecular transducing elements that couple membrane receptors to their molecular effectors within the cell. The diversity of G-proteins and their downstream targets leads to many types of physiological responses. By directly regulating the gating of ion channels, G-proteins can influence the membrane potential of target cells.

Second Messengers

Neurons use many different second messengers as intracellular signals. These messengers differ in the mechanism by which they are produced and removed, as well as their downstream targets and effects (Figure 7.7A). This section summarizes the attributes of some of the principal second messengers.

• *Calcium.* The calcium ion (Ca^{2+}) is perhaps the most common intracellular messenger in neurons. Indeed, few neuronal functions are immune to the influence—direct or indirect—of Ca^{2+}. In all cases, information is transmitted by a transient rise in the cytoplasmic calcium concentration, which allows Ca^{2+} to bind to a large number of Ca^{2+}-binding proteins that serve as molecular targets. One of the most thoroughly studied targets of Ca^{2+} is **calmodulin**, a Ca^{2+}-binding protein abundant in the cytosol of all cells. Binding of Ca^{2+} to calmodulin activates this protein, which then initiates its effects by binding to still other downstream targets, such as protein kinases.

Ordinarily the concentration of Ca^{2+} ions in the cytosol is extremely low, typically 50–100 nanomolar (10^{-9} M). The concentration of Ca^{2+} ions outside neurons—in the bloodstream or cerebrospinal fluid, for instance—is several orders of magnitude higher, typically several millimolar ($10^{-3}M$). This steep Ca^{2+} gradient is maintained by a number of mechanisms (Figure 7.7B). Most important in this maintenance are two proteins that translocate Ca^{2+} from the cytosol to the extracellular medium: an ATPase called the **calcium pump**, and an **Na$^+$/Ca^{2+} exchanger**, which is a protein that replaces intracellular Ca^{2+} with extracellular sodium ions (see Chapter 4). In addition to these plasma membrane mechanisms, Ca^{2+} is also pumped into the endoplasmic reticulum and mitochondria. These organelles can thus serve as storage depots of Ca^{2+} ions that are later released to participate in signaling events. Finally, nerve cells contain other Ca^{2+}-binding proteins—such as **calbindin**—that serve as Ca^{2+} buffers. Such buffers reversibly bind Ca^{2+} and thus blunt the magnitude and kinetics of Ca^{2+} signals within neurons.

The Ca^{2+} ions that act as intracellular signals enter cytosol by means of one or more types of Ca^{2+}-permeable ion channels (see Chapter 4). These can be voltage-gated Ca^{2+} channels or ligand-gated channels in the plasma membrane, both of which allow Ca^{2+} to flow down the Ca^{2+} gradient and into the cell from the extracellular medium. In addition, other channels allow Ca^{2+} to be released from the interior of the endoplasmic reticulum into the cytosol. These intracellular Ca^{2+}-releasing channels are opened or closed in response to various intracellular signals. One such channel is the **inositol trisphosphate (IP$_3$) receptor**. As the name implies, these channels are regulated by IP$_3$, a second messenger described in more detail below. A second type of intracellular Ca^{2+}-

(A)

Second messenger	Sources	Intracellular targets	Removal mechanisms
Ca²⁺	Plasma membrane: Voltage-gated Ca²⁺ channels, Various ligand-gated channels Endoplasmic reticulum: IP₃ receptors, Ryanodine receptors	Calmodulin, Protein kinases, Protein phosphatases, Ion channels, Synaptotagmin, Many other Ca²⁺-binding proteins	Plasma membrane: Na⁺/Ca²⁺ exchanger, Ca²⁺ pump Endoplasmic reticulum: Ca²⁺ pump Mitochondria
Cyclic AMP	Adenylyl cyclase acts on ATP	Protein kinase A, Cyclic nucleotide-gated channels	cAMP phosphodiesterase
Cyclic GMP	Guanylyl cyclase acts on GTP	Protein kinase G, Cyclic nucleotide-gated channels	cGMP phosphodiesterase
IP₃	Phospholipase C acts on PIP₂	IP₃ receptors on endoplasmic reticulum	Phosphatases
Diacylglycerol	Phospholipase C acts on PIP₂	Protein kinase C	Various enzymes

Figure 7.7 Neuronal second messengers. (A) Mechanisms responsible for producing and removing second messengers, as well as the downstream targets of these messengers. (B) Proteins involved in delivering calcium to the cytoplasm and in removing calcium from the cytoplasm. (C) Mechanisms of production and degradation of cyclic nucleotides. (D) Pathways involved in production and removal of diacylglycerol (DAG) and IP₃.

(B)

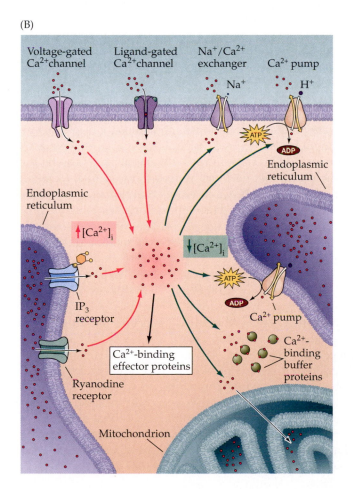

(C)

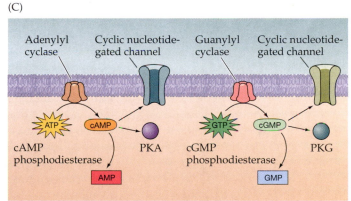

(D)

releasing channel is the **ryanodine receptor**, named after a drug that binds to and partially opens these receptors. Among the biological signals that activate ryanodine receptors are cytoplasmic Ca^{2+} and, at least in muscle cells, depolarization of the plasma membrane.

These various mechanisms for elevating and removing Ca^{2+} ions allow precise control of both the timing and location of Ca^{2+} signaling within neurons, which in turn permit Ca^{2+} to control many different signaling events. For example, voltage-gated Ca^{2+} channels allow Ca^{2+} concentrations to rise very rapidly and locally within presynaptic terminals to trigger neurotransmitter release, as already described in Chapter 5. Slower and more widespread rises in Ca^{2+} concentration regulate a wide variety of other responses, including gene expression in the cell nucleus.

• *Cyclic nucleotides.* Another important group of second messengers are the cyclic nucleotides, specifically cyclic adenosine monophosphate (cAMP) and cyclic guanosine monophosphate (cGMP) (Figure 7.7C). Cyclic AMP is a derivative of the common cellular energy storage molecule, ATP. Cyclic AMP is produced when G-proteins activate adenylyl cyclase in the plasma membrane. This enzyme converts ATP into cAMP by removing two phosphate groups from the ATP. Cyclic GMP is similarly produced from GTP by the action of guanylyl cyclase. Once the intracellular concentration of cAMP or cGMP is elevated, these nucleotides can bind to two different classes of targets. The most common targets of cyclic nucleotide action are protein kinases, either the cAMP-dependent protein kinase (PKA) or the cGMP-dependent protein kinase (PKG). These enzymes mediate many physiological responses by phosphorylating target proteins, as described in the following section. In addition, cAMP and cGMP can bind to certain ligand-gated ion channels, thereby influencing neuronal signaling. These cyclic nucleotide-gated channels are particularly important in phototransduction and other sensory transduction processes, such as olfaction. Cyclic nucleotide signals are degraded by phosphodiesterases, enzymes that cleave phosphodiester bonds and convert cAMP into AMP or cGMP into GMP.

• *Diacylglycerol and IP$_3$.* Remarkably, membrane lipids can also be converted into intracellular second messengers (Figure 7.7D). The two most important messengers of this type are produced from phosphatidylinositol bisphosphate (PIP$_2$). This lipid component is cleaved by phospholipase C, an enzyme activated by certain G-proteins and by calcium ions. Phospholipase C splits the PIP$_2$ into two smaller molecules that each act as second messengers. One of these messengers is diacylglycerol (DAG), a molecule that remains within the membrane and activates protein kinase C, which phosphorylates substrate proteins in both the plasma membrane and elsewhere. The other messenger is inositol trisphosphate (IP$_3$), a molecule that leaves the cell membrane and diffuses within the cytosol. IP$_3$ binds to IP$_3$ receptors, channels that release Ca^{2+} from the endoplasmic reticulum. Thus, the action of IP$_3$ is to produce yet another second messenger (perhaps a third messenger, in this case!) that triggers a whole spectrum of reactions in the cytosol. The actions of DAG and IP$_3$ are terminated by enzymes that convert these two molecules into inert forms that can be recycled to produce new molecules of PIP$_2$.

The intracellular concentration of these second messengers changes dynamically over time, allowing very precise control over their downstream targets. These signals can also be localized to small compartments within single cells or can spread over great distances, even spreading between cells via gap junctions (Chapter 5). Understanding of the complex temporal and spatial dynamics of these second messenger signals has been greatly aided by the development of imaging techniques that visualize second messengers and other molecular signals within cells (Box 7A).

BOX 7A Dynamic Imaging of Intracellular Signaling

Dramatic breakthroughs in our understanding of the brain often rely on development of new experimental techniques. This certainly has been true for our understanding of intracellular signaling in neurons, which has benefited enormously from the invention of imaging techniques that allow direct visualization of signaling processes within living cells. The first advance—and arguably the most significant—came from development, by Roger Tsien and his colleagues, of the fluorescent dye fura-2 (Figure A). Calcium ions bind to

fura-2 and cause the dye's fluorescence properties to change. When fura-2 is introduced inside cells and the cells are then imaged with a fluorescence microscope, this dye serves as a reporter of intracellular Ca^{2+} concentration. Fura-2 imaging has allowed us to detect the spatial and temporal dynamics of the Ca^{2+} signals that trigger innumerable processes within neurons and glial cells; for example, fura-2 was used to obtain the image of Ca^{2+} signaling during neurotransmitter release shown in Figure 5.11A.

Subsequent refinement of the chemical structure of fura-2 has yielded many other fluorescent Ca^{2+} indicator dyes with different fluorescence properties and different sensitivities to Ca^{2+}. One of these dyes is Calcium Green, which

was used to image the dynamic changes in Ca^{2+} concentration produced within the dendrites of cerebellar Purkinje cells by the intracellular messenger IP_3 (Figure B). Further developments have led to indicators for visualizing the spatial and temporal dynamics of other second messenger signals, such as cAMP.

Another tremendous advance in the dynamic imaging of signaling processes came from the discovery of a green fluorescent protein first isolated from the jellyfish *Aequorea victoria*. As its name indicates, the green fluorescent protein, or GFP, is brightly fluorescent (Figure C). Molecular cloning of the *GFP* gene allows imaging techniques to visualize expression of gene products labeled with GFP fluorescence. The first such use of GFP was in experiments with the worm *Caenorhabditis elegans*, in which Martin Chalfie and his colleagues rendered neurons fluorescent by inducing GFP expression in these cells. Many subsequent experiments have used

(Continued on next page)

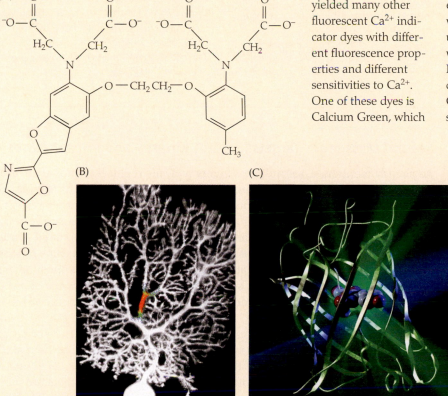

(A) Chemical structure of the Ca^{2+} indicator dye fura-2. (B) Imaging of changes in intracellular Ca^{2+} concentration produced in a cerebellar Purkinje neuron by the actions of the second messenger IP_3. (C) Molecular model of green fluorescent protein. GFP is shaped like a can, with the fluorescent moiety contained inside the can. (D) Expression of GFP in a pyramidal neuron in the mouse cerebral cortex. (A after Grynkiewicz et al., 1985; B from Finch and Augustine, 1998; C © Armand Tepper, Leiden University; D from Feng et al., 2000.)

BOX 7A *(Continued)*

expression of GFP in the mammalian brain to image the structure of individual neurons (Figure D).

Molecular genetic strategies make it possible to attach GFP to almost any protein, thereby allowing fluorescence microscopy to image the spatial distribution of labeled proteins. In this way, it has been possible to visualize dynamic changes in the location of neuronal proteins during signaling events.

As was the case for fura-2, subsequent refinement of GFP has led to numerous improvements. One such improvement is the production of proteins that fluoresce in colors other than green, thus permitting the simultaneous imaging of multiple types of proteins and/or multiple types of neurons. Further refinements have led to techniques that harness the power of GFP to monitor the enzymatic activity of protein kinases and other signaling proteins.

Just in the way that development of the Golgi staining technique opened our eyes to the cellular composition of the brain (Chapter 1), study of intracellular signaling in the brain has been revolutionized by fura-2, GFP, and other dynamic imaging tools. There is no end in sight for the potential of such imaging methods to illuminate new and important aspects of brain signaling.

References

BACSKAI, B. J. AND 6 OTHERS (1993) Spatially resolved dynamics of cAMP and protein kinase A subunits in *Aplysia* sensory neurons. *Science* 260: 222–226.

CHALFIE, M., Y. TU, G. EUSKIRCHEN, W. W. WARD AND D. C. PRASHER (1994) Green fluorescent protein as a marker for gene expression. *Science* 263: 802–805.

CONNOR, J. A. (1986) Digital imaging of free calcium changes and of spatial gradients in growing processes in single mammalian central nervous system cells. *Proc. Natl. Acad. Sci. USA* 83: 6179–6183.

FENG, G. AND 8 OTHERS (2000) Imaging neuronal subsets in transgenic mice expressing multiple spectral variants of GFP. *Neuron* 28: 41–51.

FINCH, E. A. AND G. J. AUGUSTINE (1998) Local calcium signaling by IP_3 in Purkinje cell dendrites. *Nature* 396: 753–756.

GIEPMANS, B. N., S. R. ADAMS, M. H. ELLISMAN AND R. Y. TSIEN (2006) The fluorescent toolbox for assessing protein location and function. *Science* 312: 217–224.

GRYNKIEWICZ, G., M. POENIE AND R. Y. TSIEN (1985) A new generation of Ca^{2+} indicators with greatly improved fluorescence properties. *J. Biol. Chem.* 260: 3440–3450.

MEYER, T. AND M. N. TERUEL (2003) Fluorescence imaging of signaling networks. *Trends Cell Biol.* 13: 101–106.

MIYAWAKI, A. (2005) Innovations in the imaging of brain functions using fluorescent proteins. *Neuron* 48: 189–199.

TSIEN, R. Y. (1998) The green fluorescent protein. *Annu. Rev. Biochem.* 67: 509–544.

Second Messenger Targets: Protein Kinases and Phosphatases

As already mentioned, second messengers typically regulate neuronal functions by modulating the phosphorylation state of intracellular proteins (Figure 7.8). Phosphorylation (the addition of phosphate groups) rapidly and reversibly changes protein function. Proteins are phosphorylated by a wide variety of **protein kinases**; phosphate groups are removed by other enzymes called **protein phosphatases**. The degree of phosphorylation of a target protein thus reflects a balance between the competing actions of protein kinases and phosphatases, thus integrating a host of cellular signaling pathways. The substrates of protein kinases and phosphatases include enzymes, neurotransmitter receptors, ion channels, and structural proteins.

Protein kinases and phosphatases typically act either on the serine and threonine residues (Ser/Thr kinases or phosphatases) or the tyrosine residues (Tyr kinases or phosphatases) of their substrates. Some of these enzymes act specifically on only one or a handful of protein targets, while others are multifunctional and have a broad range of substrate proteins. The activity of protein kinases and phosphatases can be regulated either by second messengers, such as cAMP or Ca^{2+}, or by extracellular chemical signals, such as growth factors. Typically, second messengers activate Ser/Thr kinases, whereas extracellular signals activate Tyr kinases. Although thousands of protein kinases are expressed in the brain, a relatively small number function as regulators of neuronal signaling.

• *cAMP-dependent protein kinase (PKA).* The primary effector of cAMP is the cAMP-dependent protein kinase (PKA). PKA is a tetrameric complex of two

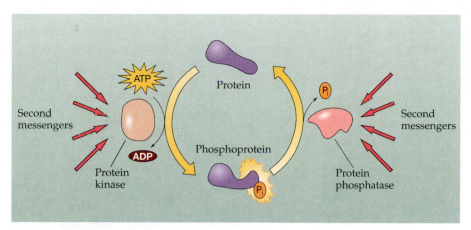

Figure 7.8 Regulation of cellular proteins by phosphorylation. Protein kinases transfer phosphate groups (P_i) from ATP to serine, threonine, or tyrosine residues on substrate proteins. This phosphorylation reversibly alters the structure and function of cellular proteins. Removal of the phosphate groups is catalyzed by protein phosphatases. Both kinases and phosphatases are regulated by a variety of intracellular second messengers.

catalytic subunits and two inhibitory (regulatory) subunits. cAMP activates PKA by binding to the regulatory subunits and causing them to release active catalytic subunits. Such displacement of inhibitory domains is a general mechanism for activation of several protein kinases by second messengers (Figure 7.9A). The catalytic subunit of PKA phosphorylates serine and threonine residues of many different target proteins. Although this subunit is similar to the catalytic domains of other protein kinases, distinct amino acids allow the PKA to bind to specific target proteins, thus allowing only those targets to be phosphorylated in response to intracellular cAMP signals.

• *Ca²⁺/calmodulin-dependent protein kinase type II (CaMKII).* Ca^{2+} ions binding to calmodulin can regulate protein phosphorylation/dephosphorylation. In neurons, the most abundant Ca²⁺/calmodulin-dependent protein kinase is CaMKII, a multifunctional protein kinase that phosphorylates serine and threonine residues on target proteins. CaMKII is composed of approximately 14 subunits, which in the brain are the α and β types. Each subunit contains a catalytic domain and a regulatory domain, as well as other domains that allow the enzyme to oligomerize and target to the proper region within the cell. Ca²⁺/calmodulin activates CaMKII by displacing the inhibitory domain from the catalytic site (Figure 7.9B). CaMKII phosphorylates a large number of substrates, including ion channels and other proteins involved in intracellular signal transduction.

• *Protein kinase C (PKC).* Another important group of Ser/Thr protein kinases is protein kinase C (PKC). PKCs are diverse monomeric kinases activated by the second messengers DAG and Ca^{2+}. DAG causes PKC to move from the cytosol to the plasma membrane, where it also binds Ca^{2+} and phosphatidylserine, a membrane phospholipid (Figure 7.9C). These events relieve autoinhibition and cause PKC to phosphorylate various protein substrates. PKC also diffuses to sites other than the plasma membrane—such as the cytoskeleton, perinuclear sites, and the nucleus—where it phosphorylates still other substrate proteins. Prolonged activation of PKC can be accomplished with phorbol esters, tumor-promoting compounds that activate PKC by mimicking DAG.

• *Protein tyrosine kinases.* Two classes of protein kinases transfer phosphate groups to tyrosine residues on substrate proteins. Receptor tyrosine kinases are transmembrane proteins with an extracellular domain that binds to protein ligands (growth factors, neurotrophic factors, or cytokines) and an intracellular catalytic domain that phosphorylates the relevant substrate proteins. Nonreceptor tyrosine kinases are cytoplasmic or membrane-associated enzymes that are indirectly activated by extracellular signals. Tyrosine phosphorylation is less common than Ser/Thr phosphorylation, and it often serves to recruit signal-

Figure 7.9 *Mechanism of protein kinase activation.* Protein kinases contain several specialized domains with specific functions. Each of the kinases has homologous catalytic domains responsible for transferring phosphate groups to substrate proteins. These catalytic domains are kept inactive by the presence of an autoinhibitory domain that occupies the catalytic site. Binding of second messengers, such as cAMP, DAG, and Ca^{2+}, to the appropriate regulatory domain of the kinase removes the autoinhibitory domain and allows the catalytic domain to be activated. For some kinases, such as PKC and CaMKII, the autoinhibitory and catalytic domains are part of the same molecule. For other kinases, such as PKA, the autoinhibitory domain is a separate subunit.

(A) PKA

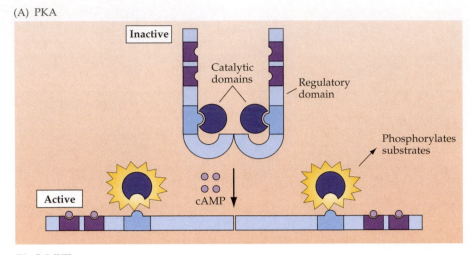

(B) CaMKII

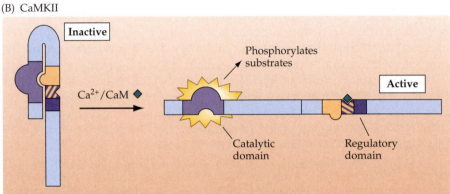

(C) PKC

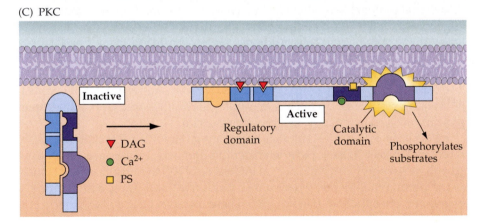

ing molecules to the phosphorylated protein. Tyrosine kinases are particularly important for cell growth and differentiation (see Chapters 22 and 23).

• *Mitogen-activated protein kinase (MAPK).* In addition to protein kinases that are directly activated by second messengers, some of these kinases are activated by phosphorylation by another protein kinase. Important examples of such protein kinases are the mitogen-activated protein kinases (MAPKs), also called extracellular signal-regulated kinases (ERKs). MAPKs were first identified as participants in the control of cell growth and are now known to have many other signaling functions. MAPKs are normally inactive in neurons but become activated when they are phosphorylated by other kinases. In fact, MAPKs are part of a kinase cascade in which one protein kinase phosphorylates and activates the next protein kinase in the cascade. The extracellular signals that trigger these kinase cascades are often extracellular growth factors that bind to receptor tyrosine kinases that, in turn, activate monomeric G-proteins such as ras. Once activated, MAPKs can phosphorylate transcription factors, proteins that regulate gene expression (see below). Among the wide variety of other MAPK substrates are various enzymes, including other protein kinases, and cytoskeletal proteins.

The best-characterized protein phosphatases are the Ser/Thr phosphatases PP1, PP2A, and PP2B (also called calcineurin). In general, protein phosphatases display less substrate specificity than protein kinases. Their limited specificity may arise from the fact that the catalytic subunits of the three major protein phosphatases are highly homologous, though each still associates with specific targeting or regulatory subunits. PP1 dephosphorylates a wide array of substrate proteins and is probably the most prevalent Ser/Thr protein phosphatase in mammalian cells. PP1 activity is regulated by several inhibitory proteins expressed in neurons. PP2A is a multisubunit enzyme with a broad range of substrates that overlap with PP1. PP2B, or calcineurin, is present at high levels in neurons. A distinctive feature of this phosphatase is its activation by Ca^{2+}/calmodulin. PP2B is composed of a catalytic and a regulatory subunit. Ca^{2+}/calmodulin activates PP2B primarily by binding to the catalytic subunit and displacing the inhibitory regulatory domain. PP2B generally does not have the same molecular targets as CaMKII, even though both enzymes are activated by Ca^{2+}/calmodulin.

In summary, activation of membrane receptors can elicit complex cascades of enzyme activation, resulting in second messenger production and protein phosphorylation or dephosphorylation. These cytoplasmic signals produce a variety of rapid physiological responses by transiently regulating enzyme activity, ion channels, cytoskeletal proteins, and many other cellular processes. At excitatory synapses, these signaling components are often contained within dendritic spines, which appear to serve as specialized signaling compartments within neurons (Box 7B). In addition, such signals can propagate to the nucleus to cause long-lasting changes in gene expression.

Nuclear Signaling

Second messengers elicit prolonged changes in neuronal function by promoting the synthesis of new RNA and protein. The resulting accumulation of new proteins requires at least 30–60 minutes, a time frame that is orders of magnitude slower than the responses mediated by ion fluxes or phosphorylation. Likewise, the reversal of such events requires hours to days. In some cases, genetic "switches" can be thrown to permanently alter a neuron, as in neuronal differentiation (see Chapter 22).

The amount of protein present in cells is determined primarily by the rate of transcription of DNA into RNA (Figure 7.10). The first step in RNA synthesis is the decondensation of the structure of chromatin to provide binding sites for the RNA polymerase complex and for **transcriptional activator proteins**, also called **transcription factors**. Transcriptional activator proteins attach to binding sites that are present on the DNA molecule near the start of the target gene sequence; they also bind to other proteins that promote unwrapping of DNA. The net result of these actions is to allow RNA polymerase, an enzyme complex, to assemble on the **promoter** region of the DNA and begin transcription. In addition to clearing the promoter for RNA polymerase, activator proteins can stimulate transcription by interacting with the RNA polymerase complex or by interacting with other activator proteins that influence the polymerase.

BOX 7B Dendritic Spines

Many synapses in the brain involve small protrusions from dendritic branches known as spines (Figure A). Spines are distinguished by the presence of globular tips called spine heads; when spines are present, the synapses innervating dendrites are made from these heads. Spine heads are connected to the main shafts of dendrites by narrow links called spine necks (Figure B). Just beneath the site of contact between the terminals and the spine heads are intracellular structures called postsynaptic densities (Figure C). The number, size, and shape of dendritic spines are quite variable and can, at least in some cases, change dynamically over time (see Figure 8.12).

Since the earliest description of these structures by Santiago Ramón y Cajal in the late 1800s, dendritic spines have fascinated generations of neuroscientists, inspiring many speculations about their function. One of the earliest conjectures was that the narrow spine neck electrically isolates synapses from the rest of the neuron. Given that the size of spine necks can change, such a mechanism could cause the physiological effect of individual synapses to vary over time, thereby providing a cellular mechanism for forms of synaptic plasticity such as LTP and LTD. However, subsequent measurements of the properties of spine necks indicate that these structures would be relatively ineffective in attenuating the flow of electrical current between spine heads and dendrites.

Another theory—currently the most popular functional concept—postulates that spines create biochemical compartments. This idea is based on the supposition that the spine neck could prevent diffusion of biochemical signals from the spine head to the rest of the dendrite. Several observations are consistent with this notion. First, measurements show that the spine neck does indeed serve as a barrier to diffusion, slowing

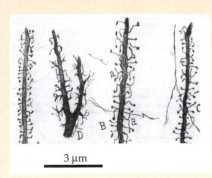
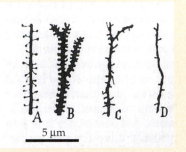

(A)

(A) Cajal's classic drawings of dendritic spines. *Left,* Dendrites of cortical pyramidal neurons. *Right,* higher-magnification images of several different types of dendritic spines.

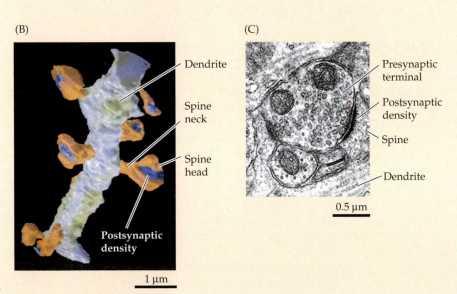

(B) High-resolution electron microscopic reconstruction of a small region of the dendrite of a hippocampal pyramidal neuron. (C) Electron micrograph of a cross section through an excitatory synapse. (B from Harris, 1994; C from Kennedy, 2000.)

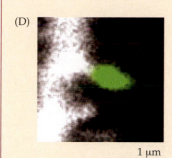

(D)

1 µm

(D) Localized Ca²⁺ signal (green) produced in the spine of a hippocampal pyramidal neuron following activation of a glutamatergic synapse. (E) Postsynaptic densities include dozens of signal transduction molecules, including glutamate receptors (NMDA-R; mGluR), tyrosine kinase receptors (RTK), and intracellular signal transduction molecules, most notably the protein kinase CaMKII. (D from Sabatini et al., 2002; E after Sheng and Kim, 2002.)

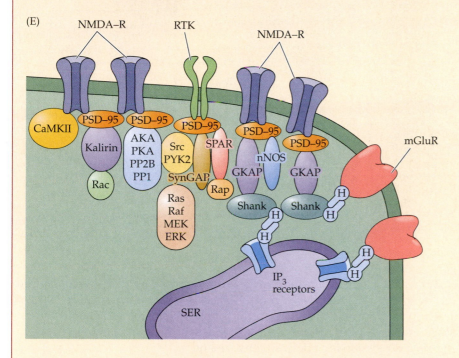

(E)

teins involved in intracellular signal transduction (Figure E). According to this view, the spine head is the destination for these signaling molecules during the assembly of synapses, as well as the target of the second messengers that are produced by the local activation of glutamate receptors. Recent work indicates that spines also can trap molecules that are diffusing along the dendrite, which could be a means of concentrating these molecules within spines.

Although the function of dendritic spines remains enigmatic, Cajal undoubtedly would be pleased at the enormous amount of attention that these tiny synaptic structures continue to command, and the real progress that has been made in understanding what they are capable of doing.

References

GOLDBERG, J. H., G. TAMAS, D. ARONOV AND R. YUSTE (2003) Calcium microdomains in aspiny dendrites. *Neuron* 40: 807–821.

HARRIS, K. M. (1994) Serial electron microscopy as a complement to confocal microscopy for the study of synapses and dendritic spines. In *Three-Dimensional Confocal Microscopy*. New York: Academic Press.

KENNEDY, M. B. (2000) Signal-processing machines at the postsynaptic density. *Science* 290: 750–754.

MIYATA, M. AND 9 OTHERS (2000) Local calcium release in dendritic spines required for long-term synaptic depression. *Neuron* 28: 233–244.

NIMCHINSKY, E. A., B. L. SABATINI AND K. SVOBODA (2002) Structure and function of dendritic spines. *Annu. Rev. Physiol.* 64: 313–353.

SABATINI, B. L., T. G. OERTNER AND K. SVOBODA (2002) The life cycle of Ca²⁺ ions in dendritic spines. *Neuron* 33: 439–452.

SANTAMARIA, F., S. WILS, E. DE SCHUTTER AND G. J. AUGUSTINE (2006) Anomalous diffusion in Purkinje cell dendrites caused by spines. *Neuron* 52: 635–648.

SHENG, M. AND M. J. KIM (2002) Postsynaptic signaling and plasticity mechanisms. *Science* 298: 776–780.

YUSTE, R. AND D. W. TANK (1996) Dendritic integration in mammalian neurons, a century after Cajal. *Neuron* 16: 701–716.

the rate of molecular movement by a factor of 100 or more. Second, spines are found only at excitatory synapses, where it is known that synaptic transmission generates many diffusible signals, most notably the second messenger Ca²⁺. Finally, fluorescence imaging shows that synaptic Ca²⁺ signals can indeed be restricted to dendritic spines (Figure D).

Nevertheless, there are counterarguments to the hypothesis that spines provide relatively isolated biochemical compartments. For example, it is known that other second messengers, such as IP₃, can diffuse out of the spine head and into the dendritic shaft. Presumably this difference in diffusion is due to the fact that IP₃ signals last longer than

Ca²⁺ signals, allowing IP₃ sufficient time to overcome the diffusion barrier of the spine neck. Another relevant point is that postsynaptic Ca²⁺ signals are highly localized, even at excitatory synapses that do not have spines. Thus, in at least some instances, spines are neither necessary nor sufficient for localization of synaptic second messenger signaling.

A final and less controversial idea is that the purpose of spines is to serve as reservoirs where signaling proteins, such as the downstream molecular targets of Ca²⁺ and IP₃, can be concentrated. Consistent with this possibility, glutamate receptors are highly concentrated on spine heads, and the postsynaptic density comprises dozens of pro-

Figure 7.10 Steps involved in transcription of DNA into RNA. Condensed chromatin is decondensed into a beads-on-a-DNA-string array in which an upstream activator site (UAS) is free of proteins and is bound by a sequence-specific transcriptional activator protein (transcription factor). The transcriptional activator protein then binds co-activator complexes that enable the RNA polymerase with its associated factors to bind at the start site of transcription and initiate RNA synthesis.

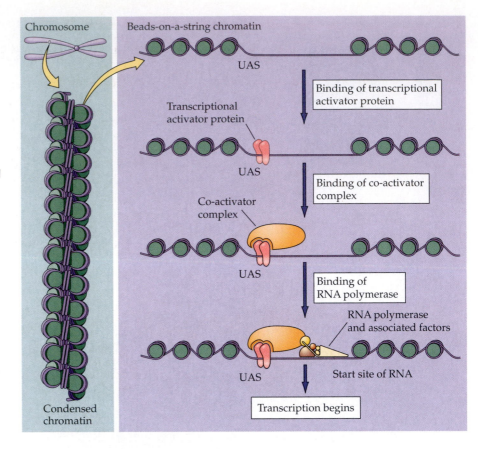

Intracellular signal transduction cascades regulate gene expression by converting transcriptional activator proteins from an inactive state to an active state in which they are able to bind to DNA. This conversion comes about in several ways. The key activator proteins and the mechanisms that allow them to regulate gene expression are briefly summarized here.

• *CREB*. The *c*AMP *r*esponse *e*lement *b*inding protein, usually abbreviated **CREB**, is a ubiquitous transcriptional activator (Figure 7.11). CREB is normally bound to its binding site on DNA (called the cAMP response element, or CRE), either as a homodimer or bound to another, closely related transcription factor. In unstimulated cells, CREB is not phosphorylated and has little or no transcriptional activity. However, phosphorylation of CREB greatly potentiates transcription. Several signaling pathways are capable of causing CREB to be phosphorylated. Both PKA and the ras pathway, for example, can phosphorylate CREB. CREB can also be phosphorylated in response to increased intracellular calcium, in which case the CRE site is also called the CaRE (calcium response element) site. The calcium-dependent phosphorylation of CREB is primarily caused by Ca^{2+}/calmodulin kinase IV (a relative of CaMKII) and by MAP kinase, which leads to prolonged CREB phosphorylation. CREB phosphorylation must be maintained long enough for transcription to ensue, even though neuronal electrical activity only transiently raises intracellular calcium concentration. Such signaling cascades can potentiate CREB-mediated transcription by inhibiting a protein phosphatase that dephosphorylates CREB. CREB is thus an example of the convergence of multiple signaling pathways onto a single transcriptional activator.

Many genes whose transcription is regulated by CREB have been identified. CREB-sensitive genes include the immediate early gene, *c-fos* (see below), the neurotrophin BDNF (see Chapter 23), the enzyme tyrosine hydroxylase (which

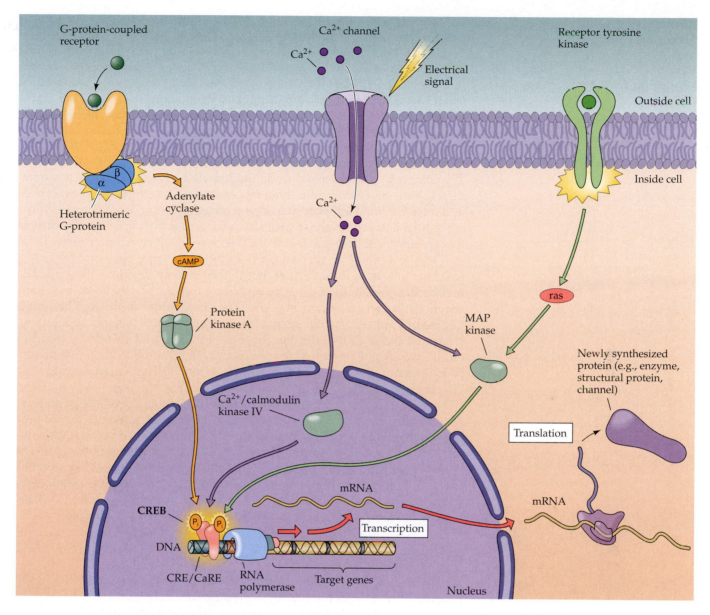

Figure 7.11 *Transcriptional regulation by CREB. Multiple signaling pathways converge by activating kinases that phosphorylate CREB. These include PKA, Ca^{2+}/calmodulin kinase IV, and MAP kinase. Phosphorylation of CREB allows it to bind co-activators (not shown in the figure), which then stimulate RNA polymerase to begin synthesis of RNA. RNA is then processed and exported to the cytoplasm, where it serves as mRNA for translation into protein.*

is important for synthesis of catecholamine neurotransmitters; see Chapter 6), and many neuropeptides (including somatostatin, enkephalin, and corticotropin releasing hormone). CREB also is thought to mediate long-lasting changes in brain function. For example, CREB has been implicated in spatial learning, behavioral sensitization, long-term memory of odorant-conditioned behavior, and long-term synaptic plasticity (see Chapter 8).

• *Nuclear receptors.* Nuclear receptors for membrane-permeant ligands also are transcriptional activators. The receptor for glucocorticoid hormones illustrates one mode of action of such receptors. In the absence of glucocorticoid hormones, the receptors are located in the cytoplasm. Binding of glucocorticoids causes the receptor to unfold and move to the nucleus, where it binds a specific recognition site on the DNA. This DNA binding activates the relevant RNA polymerase complex to initiate transcription and subsequent gene expression. Thus, a critical regulatory event for steroid receptors is their translocation to the nucleus to allow DNA binding.

The receptors for thyroid hormone (TH) and other non-steroid nuclear receptors illustrate a second mode of regulation. In the absence of TH, the receptor is bound to DNA and serves as a potent repressor of transcription. Upon binding TH, the receptor undergoes a conformational change that ultimately opens the promoter for polymerase binding. Hence, TH binding switches the receptor from being a repressor to being an activator of transcription.

• *c-fos.* A different strategy of gene regulation is apparent in the function of the transcriptional activator protein **c-fos**. In resting cells, c-fos is present at a very low concentration. Stimulation of the target cell causes c-fos to be synthesized, and the amount of this protein rises dramatically over 30–60 minutes. Therefore, *c-fos* is considered to be an **immediate early gene** because its synthesis is directly triggered by the stimulus. Once synthesized, c-fos protein can act as a transcriptional activator to induce synthesis of second-order genes. These are termed **delayed response genes** because their activity is delayed by the fact that an immediate early gene—*c-fos* in this case—must be activated first.

Multiple signals converge on *c-fos*, activating different transcription factors that bind to at least three distinct sites in the promoter region of the gene. The regulatory region of the *c-fos* gene contains a binding site that mediates transcriptional induction by cytokines and ciliary neurotropic factor. Another site is targeted by growth factors such as neurotrophins through ras and protein kinase C, and a CRE/CaRE that can bind to CREB and thereby respond to cAMP or calcium entry resulting from electrical activity. In addition to synergistic interactions among these *c-fos* sites, transcriptional signals can be integrated by converging on the same activator, such as CREB.

Nuclear signaling events typically result in the generation of a large and relatively stable complex composed of a functional transcriptional activator protein, additional proteins that bind to the activator protein, and the RNA polymerase and associated proteins bound at the start site of transcription. Most of the relevant signaling events act to "seed" this complex by generating an active transcriptional activator protein by phosphorylation, by inducing a conformational change in the activator upon ligand binding, by fostering nuclear localization, by removing an inhibitor, or simply by making more activator protein.

Examples of Neuronal Signal Transduction

Understanding the general properties of signal transduction processes at the plasma membrane, in the cytosol, and within the nucleus make it possible to consider how these processes work in concert to mediate specific functions in the brain. Three important signal transduction pathways illustrate some of the roles of intracellular signal transduction processes in the nervous system.

• *NGF/TrkA.* The first of these is signaling by the **nerve growth factor** (**NGF**). This protein is a member of the neurotrophin growth factor family and is required for the differentiation, survival, and synaptic connectivity of sympathetic and sensory neurons (see Chapter 23). NGF works by binding to a high-affinity tyrosine kinase receptor, TrkA, found on the plasma membrane of these target cells (Figure 7.12). NGF binding causes TrkA receptors to dimerize, and the intrinsic tyrosine kinase activity of each receptor then phosphorylates its partner receptor. Phosphorylated TrkA receptors trigger the ras cascade, resulting in the activation of multiple protein kinases. Some of these kinases translocate to the nucleus to activate transcriptional activators, such as CREB. This ras-based component of the NGF pathway is primarily responsible for inducing and maintaining differentiation of NGF-sensitive neurons. Phosphorylation of TrkA also causes this receptor to stimulate the activity of phospholipase C, which increases production of IP_3 and DAG. IP_3 induces release of Ca^{2+} from the

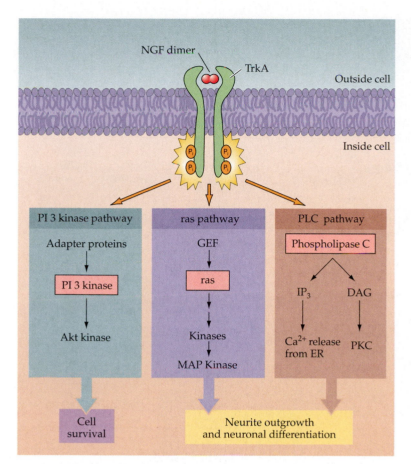

Figure 7.12   *Mechanism of action of NGF. NGF binds to a high-affinity tyrosine kinase receptor, TrkA, on the plasma membrane to induce phosphorylation of TrkA at two different tyrosine residues. These phosphorylated tyrosines serve to tether various adapter proteins or phospholipase C, which, in turn, activate three major signaling pathways: the PI 3 kinase pathway leading to activation of Akt kinase, the ras pathway leading to MAP kinases, and the PLC pathway leading to release of intracellular Ca^{2+} and activation of PKC. The ras and PLC pathways primarily stimulate processes responsible for neuronal differentiation, while the PI 3 kinase pathway is primarily involved in cell survival.*

endoplasmic reticulum, and diacylglycerol activates PKC. These two second messengers appear to target many of the same downstream effectors as ras. Finally, activation of TrkA receptors also causes activation of other protein kinases (such as Akt kinase) that inhibit cell death. This pathway, therefore, primarily mediates the NGF-dependent survival of sympathetic and sensory neurons described in Chapter 23.

• *Long-term depression (LTD)*. The interplay between several intracellular signals can be observed at the excitatory synapses that innervate Purkinje cells in the cerebellum. These synapses are central to information flow through the cerebellar cortex, which in turn helps coordinate motor movements (see Chapter 19). One of the synapses is between the parallel fibers (PFs) and their Purkinje cell targets. LTD is a form of synaptic plasticity that causes the PF synapses to become less effective (see Chapter 8). When PFs are active, they release the neurotransmitter glutamate onto the dendrites of Purkinje cells. This activates AMPA-type receptors, which are ligand-gated ion channels (see Chapter 6), and causes a small EPSP that briefly depolarizes the Purkinje cell. In addition to this electrical signal, PF synaptic transmission also generates two second messengers within the Purkinje cell (Figure 7.13). The glutamate released by PFs activates metabotropic glutamate receptors, which stimulates phospholipase C to produce IP_3 and DAG. When the PF synapses alone are active, these intracellular signals are insufficient to open IP_3 receptors or to stimulate PKC.

LTD is induced when PF synapses are activated at the same time as the glutamatergic climbing fiber synapses that also innervate Purkinje cells. The climbing fiber synapses produce large EPSPs that strongly depolarize the membrane

Figure 7.13 *Signaling at cerebellar parallel fiber synapses. Glutamate released by parallel fibers activates both AMPA-type and metabotropic receptors. The latter produces IP$_3$ and DAG within the Purkinje cell. When paired with a rise in Ca^{2+} associated with activity of climbing fiber synapses, the IP$_3$ causes Ca^{2+} to be released from the endoplasmic reticulum, while Ca^{2+} and DAG together activate protein kinase C. These signals together change the properties of AMPA receptors to produce LTD.*

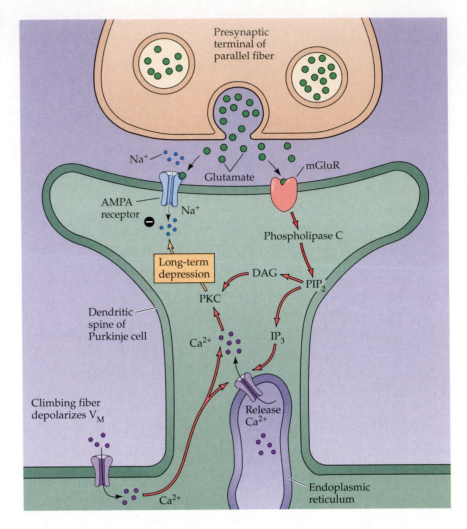

potential of the Purkinje cell. This depolarization allows Ca^{2+} to enter the Purkinje cell via voltage-gated Ca^{2+} channels. When both synapses are simultaneously activated, the rise in intracellular Ca^{2+} concentration caused by the climbing fiber synapse enhances the sensitivity of IP$_3$ receptors to the IP$_3$ produced by PF synapses and allows the IP$_3$ receptors within the Purkinje cell to open. This releases Ca^{2+} from the endoplasmic reticulum and further elevates Ca^{2+} concentration locally near the PF synapses. This larger rise in Ca^{2+}, in conjunction with the DAG produced by the PF synapses, activates PKC. PKC in turn phosphorylates a number of substrate proteins. Ultimately, these signaling processes change AMPA-type receptors at the PF synapse, so that these receptors produce smaller electrical signals in response to the glutamate released from the PFs. This weakening of the PF synapse is the final cause of LTD.

In short, transmission at Purkinje cell synapses produces brief electrical signals and chemical signals that last much longer. The temporal interplay between these signals allows LTD to occur only when both PF and climbing fiber synapses are active. The actions of IP$_3$, DAG and Ca^{2+} also are restricted to small parts of the Purkinje cell dendrite, which is a more limited spatial range than the EPSPs, which spread throughout the entire dendrite and cell body of the Purkinje cell. Thus, in contrast to the electrical signals, the second messenger signals can impart precise information about the location of active synapses and allow LTD to occur only in the vicinity of active PFs.

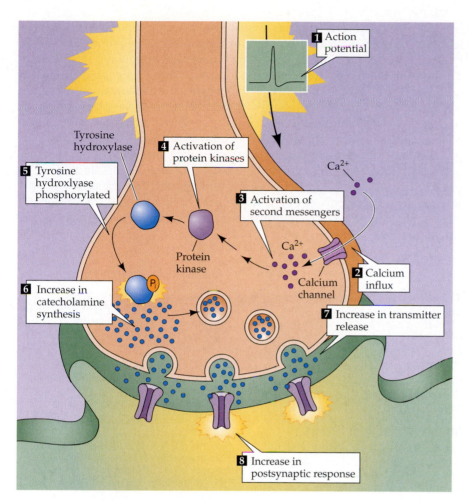

Tyrosine hydroxylase

5 Tyrosine hydroxlyase phosphorylated

4 Activation of protein kinases

1 Action potential

Ca^{2+}

3 Activation of second messengers

Protein kinase

Ca^{2+}

2 Calcium influx

Calcium channel

6 Increase in catecholamine synthesis

P

7 Increase in transmitter release

8 Increase in postsynaptic response

Figure 7.14 Regulation of tyrosine hydroxylase by protein phosphorylation. This enzyme governs the synthesis of the catecholamine neurotransmitters and is stimulated by a number of intracellular signals. In the example shown here, neuronal electrical activity (1) causes influx of Ca^{2+} (2). The resultant rise in intracellular Ca^{2+} concentration (3) activates protein kinases (4), which phosphorylate tyrosine hydroxylase (5) to stimulate catecholamine synthesis (6). This, in turn, increases release of catecholamines (7) and enhances the postsynaptic response produced by the synapse (8).

• *Phosphorylation of tyrosine hydroxylase.* A third example of intracellular signaling in the nervous system is the regulation of the enzyme tyrosine hydroxylase. Tyrosine hydroxylase governs the synthesis of the catecholamine neurotransmitters: dopamine, norepinephrine, and epinephrine (see Chapter 6). A number of signals, including electrical activity, other neurotransmitters, and NGF, increase the rate of catecholamine synthesis by increasing the catalytic activity of tyrosine hydroxylase (Figure 7.14). The rapid increase of tyrosine hydroxylase activity is largely due to phosphorylation of this enzyme.

Tyrosine hydroxylase is a substrate for several protein kinases, including PKA, CaMKII, MAP kinase, and PKC. Phosphorylation causes conformational changes that increase the catalytic activity of tyrosine hydroxylase. Stimuli that elevate cAMP, Ca^{2+}, or DAG can all increase tyrosine hydroxylase activity and thus increase the rate of catecholamine biosynthesis. This regulation by several different signals allows for close control of tyrosine hydroxylase activity and illustrates how several different pathways can converge to influence a key enzyme involved in synaptic transmission.

Summary

Diverse signal transduction pathways exist within all neurons. Activation of these pathways typically is initiated by chemical signals such as neurotransmitters and hormones, which bind to receptors that include ligand-gated ion channels, G-protein-coupled receptors and tyrosine kinase receptors. Many of these receptors activate either heterotrimeric or monomeric G-proteins that regulate intracellular enzyme cascades and/or ion channels. A common outcome of the activation of these receptors is the production of second messengers, such as cAMP, Ca^{2+}, and IP_3, that bind to effector enzymes. Particularly important effectors are protein kinases and phosphatases that regulate the phosphorylation state of their substrates, and thus their function. These substrates can be metabolic enzymes or other signal transduction molecules, such as ion channels, protein kinases, or transcription factors that regulate gene expression. Examples of transcription factors include CREB, steroid hormone receptors, and c-fos. This plethora of molecular components allows intracellular signal transduction pathways to generate responses over a wide range of times and distances, greatly augmenting and refining the information-processing ability of neuronal circuits and ultimately systems.

Additional Reading

Reviews

AUGUSTINE, G. J., F. SANTAMARIA AND K. TANAKA (2003) Local calcium signaling in neurons. *Neuron* 40: 331–346.

DEISSEROTH, K., P. G. MERMELSTEIN, H. XIA AND R. W. TSIEN (2003) Signaling from synapse to nucleus: The logic behind the mechanisms. *Curr. Opin. Neurobiol.* 13: 354–365.

EXTON, J. H. (1998) Small GTPases. *J. Biol. Chem.* 273: 19923.

FISCHER, E. H. (1999) Cell signaling by protein tyrosine phosphorylation. *Adv. Enzyme Regul. Review* 39: 359–369.

GILMAN, A. G. (1995) G proteins and regulation of adenylyl cyclase. *Biosci. Rep.* 15: 65–97.

GRAVES, J. D. AND E. G. KREBS (1999) Protein phosphorylation and signal transduction. *Pharmacol. Ther.* 82: 111–121.

GREENGARD, P. (2001) The neurobiology of slow synaptic transmission. *Science* 294: 1024–1030.

ITO, M. (2002) The molecular organization of cerebellar long-term depression. *Nature Rev. Neurosci.* 3: 896–902.

KENNEDY, M. B., H. C. BEALE, H. J. CARLISLE AND L.R. WASHBURN (2005) Integration of biochemical signalling in spines. *Nature Rev. Neurosci.* 6: 423–434.

KUMER, S. AND K. VRANA (1996) Intricate regulation of tyrosine hydroxylase activity and gene expression. *J. Neurochem.* 67: 443–462.

NISHIZUKA, Y. (1992) Intracellular signaling by hydrolysis of phospholipids and activation of protein kinase C. *Science* 258: 607–614.

REICHARDT, L. F. (2006) Neurotrophin-regulated signalling pathways. *Philos. Trans. Roy. Soc. London B* 361: 1545–1564.

RODBELL, M. (1995) Signal transduction: Evolution of an idea. *Bioscience Reports* 15: 117–133.

SHENG, M. AND M. J. KIM (2002) Postsynaptic signaling and plasticity mechanisms. *Science* 298: 776–780.

WEST, A. E. AND 8 OTHERS (2001) Calcium regulation of neuronal gene expression. *Proc. Natl. Acad. Sci. USA* 98: 11024–11031.

Important Original Papers

BURGESS, G. M., P. P. GODFREY, J. S. MCKINNEY, M. J. BERRIDGE, R. F. IRVINE AND J.W. PUTNEY, JR. (1984) The second messenger linking receptor activation to internal Ca release in liver. *Nature* 309: 63–66.

DE KONINCK, P. AND H. SCHULMAN (1998) Sensitivity of CaM kinase II to the frequency of Ca^{2+} oscillations. *Science* 279: 227–230.

DE ZEEUW, C. I., C. HANSEL, F. BIAN, S. K. E. KOEKKOEK, A. M. VAN ALPHEN, D. J. LINDEN AND J. OBERDICK (1998). Expression of a protein kinase C inhibitor in Purkinje cells blocks cerebellar long-term depression and adaptation of the vestibulo-ocular reflex. *Neuron* 20: 495–508

FINCH, E. A. AND G. J. AUGUSTINE (1998) Local calcium signaling by IP_3 in Purkinje cell dendrites. *Nature* 396: 753–756.

LINDGREN, N. AND 8 OTHERS (2000) Regulation of tyrosine hydroxylase activity and phosphorylation at ser(19) and ser(40) via activation of glutamate NMDA receptors in rat striatum. *J. Neurochem.* 74: 2470–2477.

MILLER, S. G. AND M. B. KENNEDY (1986) Regulation of brain type II Ca^{2+}/calmodulin-dependent protein kinase by autophosphorylation: A Ca^{2+}-triggered molecular switch. *Cell* 44: 861–870.

NORTHUP, J. K., P. C. STERNWEIS, M. D. SMIGEL, L. S. SCHLEIFER, E. M. ROSS AND A. G. GILMAN (1980) Purification of the regulatory component of adenylate cyclase. *Proc. Natl. Acad. Sci. USA* 77: 6516–6520.

ROSENBERG, O. S., S. DEINDL, R. J. SUNG, A. C. NAIRN AND J. KURIYAN (2005) Structure of the autoinhibited kinase domain of CaMKII and SAXS analysis of the holoenzyme. *Cell* 123: 849-860.

SAITOH, T. AND J. H. SCHWARTZ (1985) Phosphorylation-dependent subcellular translocation of a Ca^{2+}/calmodulin-dependent protein kinase produces an autonomous enzyme in *Aplysia* neurons. *J. Cell Biol.* 100: 835–842.

SHEN, K., M. N. TERUEL, J. H. CONNOR, S. SHENOLIKAR AND T. MEYER (2000) Molecular memory by reversible translocation of calcium/calmodulin-dependent protein kinase II. *Nature Neurosci.* 3: 881–886.

SHIFMAN, J. M., M. H. CHOI, S. MIHALAS, S. L. MAYO AND M. B. KENNEDY (2006) Ca^{2+}/calmodulin-dependent protein kinase II is activated by calmodulin with two bound calciums. *Proc. Natl. Acad. Sci. USA* 103: 13968–13973.

SU, Y. AND 7 OTHERS (1995) Regulatory subunit of protein kinase A: Structure of deletion mutant with cAMP binding domains. *Science* 269: 807–813.

TAO, X., S. FINKBEINER, D. B. ARNOLD, A. J. SHAYWITZ AND M. E. GREENBERG (1998) Ca^{2+} influx regulates BDNF transcription by a CREB family transcription factor-dependent mechanism. *Neuron* 20: 709–726.

TESMER, J. J., R. K. SUNAHARA, A. G. GILMAN AND S. R. SPRANG (1997) Crystal structure of the catalytic domains of adenylyl cyclase in a complex with $G_{s\alpha}$-$GTP_{\gamma}S$. *Science* 278: 1907–1916.

Books

ALBERTS, B., A. JOHNSON, J. LEWIS, M. RAFF, K. ROBERTS AND P. WALTER (2002) *Molecular Biology of the Cell*, 4th Ed. New York: Garland Science.

CARAFOLI, E. AND C. KLEE (1999) *Calcium as a Cellular Regulator*. New York: Oxford University Press.

Chapter 8

Synaptic Plasticity

Overview

Synaptic connections between neurons provide the basic "wiring" of the brain's circuitry. However, in contrast to the wiring of an electronic device such as a computer, synaptic connectivity between neurons is a dynamic entity that is constantly changing in response to neural activity and other influences. Such changes in synaptic transmission arise from a number of forms of plasticity that vary in time scale from milliseconds to years. Most short-term forms of synaptic plasticity affect the amount of neurotransmitter released from presynaptic terminals in response to a presynaptic action potential. Several forms of short-term synaptic plasticity—including facilitation, augmentation, and potentiation—enhance neurotransmitter release and are caused by persistent actions of calcium ions within the presynaptic terminal. Another form of short-term plasticity is synaptic depression, which decreases the amount of neurotransmitter released and appears to result from an activity-dependent depletion of synaptic vesicles that are ready to undergo exocytosis. Long-term forms of synaptic plasticity alter synaptic transmission over time scales of 30 minutes or longer. Examples of such long-lasting plasticity include long-term potentiation and long-term depression. These long-lasting forms of synaptic plasticity arise from molecular mechanisms that vary over time: the initial changes in synaptic transmission arise from posttranslational modifications of existing proteins, most notably changes in the trafficking of glutamate receptors, while later phases of synaptic modification result from changes in gene expression. These changes in gene expression produce enduring changes in synaptic transmission, including growth of synapses, that can yield essentially permanent modifications of brain function.

Short-Term Synaptic Plasticity

Chemical synapses are capable of undergoing plastic changes that either strengthen or weaken synaptic transmission. Synaptic plasticity mechanisms occur on time scales ranging from milliseconds to days, weeks, or longer. The short-term forms of plasticity—those lasting for a few minutes or less—are readily observed during repeated activation of any chemical synapse. Repetitive activity produces several forms of short-term synaptic plasticity that differ in their time courses and their underlying mechanisms.

Synaptic facilitation is a rapid increase in synaptic strength that occurs when two or more action potentials invade the presynaptic terminal within a few milliseconds of each other (Figure 8.1A). By varying the time interval between presynaptic action potentials, it can be seen that facilitation lasts for tens of milliseconds (Figure 8.1B). Many lines of evidence indicate that facilitation is the result of prolonged elevation of presynaptic calcium levels following

(A)

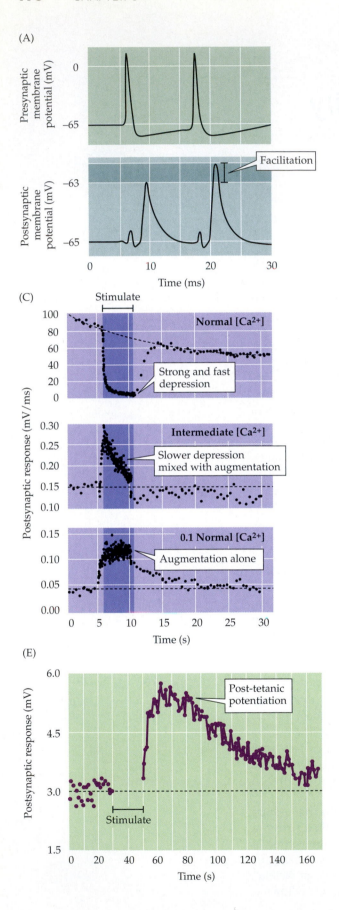

(B)

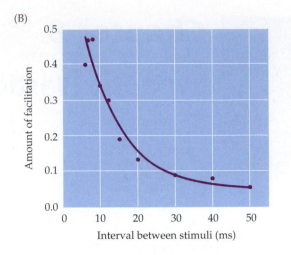

(C)

(D)

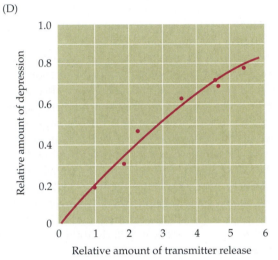

(E)

Figure 8.1 Forms of short-term synaptic plasticity. (A) Facilitation at the squid giant synapse. A pair of presynaptic action potentials elicits two EPSPs. Because of facilitation, the second EPSP is larger than the first. (B) By varying the time interval between pairs of presynaptic action potentials, facilitation can be seen to decay over a time course of tens of milliseconds. (C) In normal physiological conditions, a high-frequency tetanus (bar) causes pronounced depression of EPSPs at the squid giant synapse (top). Lowering the external Ca^{2+} concentration to an intermediate level reduces transmitter release and causes a mixture of depression and augmentation (middle). Further reduction of external Ca^{2+} eliminates depression and leaves only augmentation (bottom). (D) Synaptic depression at the frog neuromuscular synapse increases in proportion to the amount of transmitter released from the presynaptic terminal. (E) Application of a high-frequency tetanus (bar) to presynaptic axons innervating a spinal motor neuron causes a post-tetanic potentiation that persists for a couple of minutes after the tetanus ends. (A,B after Charlton and Bittner, 1978; C after Swandulla et al., 1991; D after Betz, 1970; E after Lev-Tov et al., 1983.)

synaptic activity. Although the entry of Ca^{2+} into the presynaptic terminal occurs within a millisecond or two after an action potential invades (see Figure 5.10B), the mechanisms that return Ca^{2+} to resting levels are much slower. Thus, when action potentials arrive close together in time, Ca^{2+} builds up within the terminal and allows more neurotransmitter to be released by a subsequent presynaptic action potential. The target of this residual Ca^{2+} signal is not yet clear; one possibility is partial occupancy of the Ca^{2+} binding sites of synaptotagmin, the Ca^{2+} sensor protein that triggers neurotransmitter release.

Opposing facilitation is **synaptic depression**, which causes neurotransmitter release to decline during sustained synaptic activity. An important clue to the cause of synaptic depression comes from observations that depression depends on the amount of neurotransmitter that has been released. For example, lowering the external Ca^{2+} concentration (and thus reducing the number of quanta released by each presynaptic action potential) causes the rate of depression to be slowed (Figure 8.1C). Likewise, the amount of depression is proportional to the total amount of transmitter released from the presynaptic terminal (Figure 8.1D). These results have led to the idea that depression is caused by progressive depletion of a pool of synaptic vesicles that are available for release. When rates of release are high, these vesicles deplete rapidly and cause a lot of depression. Depletion slows as the rate of release is reduced, yielding less depression. According to this *vesicle depletion hypothesis*, depression causes the strength of transmission to decline until synaptic vesicle supply is replenished by mobilization of vesicles from a reserve pool. Consistent with this explanation are observations that more depression is observed after the size of the reserve pool is reduced by impairing synapsin, a protein that maintains vesicles in this pool (see Chapter 5).

Still other forms of synaptic plasticity, such as synaptic **potentiation** and **augmentation**, also are elicited by repeated synaptic activity and serve to increase the amount of transmitter released from presynaptic terminals. Both augmentation and potentiation enhance the ability of incoming Ca^{2+} to trigger fusion of synaptic vesicles with the plasma membrane, but they work over different time scales. While augmentation rises and falls over a few seconds (Figure 8.1C, lower panel), potentiation acts over a time scale of tens of seconds to minutes (Figure 8.1E). As a result of its slower time course, potentiation can greatly outlast the tetanus and is often called **post-tetanic potentiation** (**PTP**). While both augmentation and potentiation are thought to arise from prolonged elevation of presynaptic calcium levels during synaptic activity, the molecular mechanisms responsible for these forms of plasticity are poorly understood. It has been proposed that augmentation results from Ca^{2+} enhancing the actions of the presynaptic protein munc-13, while potentiation may arise from Ca^{2+} activating presynaptic protein kinases that go on to phosphorylate substrates like synapsin, which regulate transmitter release.

During repetitive synaptic activity, these forms of short-term plasticity can interact with each other, causing synaptic transmission to change in complex ways. For example, at the peripheral neuromuscular synapse, repeated activity first causes facilitation and then augmentation to enhance synaptic transmission (Figure 8.2). Depletion of synaptic vesicles then causes depression to dominate and weaken the synapse. Presynaptic action potentials that occur within a minute or two after the end of the tetanus release more neurotransmitter because of the persistence of post-tetanic potentiation. Although their relative contributions vary from synapse to synapse, these forms of short-term synaptic plasticity cause transmission at all chemical synapses to change dynamically as a consequence of the recent history of synaptic activity.

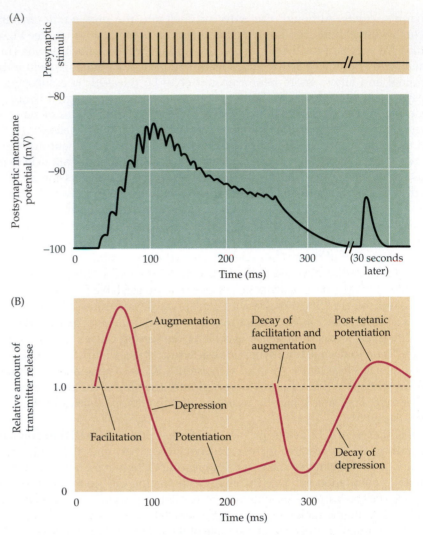

Figure 8.2 *Short-term plasticity at the neuromuscular synapse. (A) A train of electrical stimuli (top) applied to the presynaptic motor nerve produces changes in EPP amplitude (below). (B) Dynamic changes in transmitter release caused by the interplay of several forms of short-term plasticity. Facilitation and augmentation of the EPP occurs at the beginning of the stimulus train and are followed by a pronounced depression of the EPP. Potentiation begins late in the stimulus train and persists for many seconds after the end of the stimulus, leading to post-tetanic potentiation. (A after Katz, 1966; B after Malenka and Siegelbaum, 2001.)*

Long-Term Synaptic Plasticity Underlies Behavioral Modification in *Aplysia*

Facilitation, depression, augmentation, and potentiation modify synaptic transmission over time scales of a couple of minutes or less. While these mechanisms probably are responsible for many short-lived changes in brain circuitry, they cannot provide the basis for changes in brain function that persist for weeks, months, or years. Many synapses exhibit long-lasting forms of synaptic plasticity that are plausible substrates for more permanent changes in brain function. Because of their duration, these forms of synaptic plasticity may be cellular correlates of learning and memory. Thus, a great deal of effort has gone into understanding how they are generated.

An obvious obstacle to exploring synaptic plasticity in the brains of humans and other mammals is the enormous number of neurons and the complexity of synaptic connections. One way to circumvent this dilemma is to examine plasticity in far simpler nervous systems. The assumption in this strategy is that plasticity is so fundamental that its essential cellular and molecular underpinnings are likely to be conserved in the nervous systems of very different organisms. This approach has been successful in identifying several forms of long-term synaptic plasticity, and in demonstrating that such forms of synaptic plasticity underlie simple forms of learning.

Eric Kandel and his colleagues at Columbia University addressed such questions using the marine mollusk *Aplysia californica* (Figure 8.3A). This sea slug has only a few tens of thousands of neurons, many of which are quite large (up to 1 mm in diameter) and are found in stereotyped locations within the ganglia that make up the animal's nervous system (Figure 8.3B). These attributes make it practical to monitor the electrical activity of specific, identifiable nerve cells and to define the synaptic circuits involved in mediating the limited behavioral repertoire of *Aplysia*.

Aplysia exhibit several elementary forms of behavioral plasticity. One form is **habituation**, a process that causes the animal to become less responsive to repeated occurrences of a stimulus. Habituation is found in many other species, including humans. For example, when dressing we initially experience tactile sensations as the clothes stimulate our skin, but habituation quickly causes these sensations to fade. Similarly, a light touch to the siphon of an *Aplysia* results in withdrawal of the animal's gill, but habituation from repeated siphon stimulation causes the gill withdrawal to weaken (Figure 8.3C).

The gill withdrawal response of *Aplysia* also exhibits a form of plasticity called **sensitization**. Sensitization allows an animal to generalize an aversive response elicited by a noxious stimulus to a variety of other, non-noxious stimuli. In *Aplysia* that have habituated to siphon touching, sensitization of gill withdrawal is elicited by pairing a strong electrical stimulus to the animal's tail with another light touch of the siphon. This pairing causes the siphon stimulus to again elicit a strong withdrawal of the gill (Figure 8.3C, right) because the noxious stimulus to the tail sensitizes the gill withdrawal reflex to light touch. After a single stimulus to the tail, the gill withdrawal reflex remains enhanced for at least an hour (Figure 8.3D). With repeated pairing of tail and siphon stimuli, this behavior can be altered for days or weeks (Figure 8.3E), demonstrating a simple form of long-term memory.

The small number of neurons in the *Aplysia* nervous system makes it possible to define the synaptic circuits involved in gill withdrawal and to monitor the activity of individual neurons and synapses in these circuits. Although hundreds of neurons are ultimately involved in producing this simple behavior, the activities of only a few different types of neurons can account for gill withdrawal and its plasticity during habituation and sensitization. These critical neurons include mechanosensory neurons that innervate the siphon, motor neurons that innervate muscles in the gill, and interneurons that receive inputs from a variety of sensory neurons (Figure 8.4A). Touching the siphon activates the mechanosensory neurons, which form excitatory synapses that release glutamate onto both the interneurons and the motor neurons; thus, touching the siphon increases the probability that both these postsynaptic targets will produce action potentials. The interneurons form excitatory synapses on motor neurons, further increasing the likelihood of the motor neurons firing action potentials in response to mechanical stimulation of the siphon. When the motor neurons are activated by the summed synaptic excitation of the sensory neurons and interneurons, they release acetylcholine that excites the muscle cells of the gill, producing gill withdrawal.

Figure 8.3 *Short-term sensitization of the* Aplysia *gill withdrawal reflex. (A) Diagram of the animal (commonly known as a sea slug). (B) The abdominal ganglion of* Aplysia. *The cell bodies of many of the neurons involved in gill withdrawal can be recognized by their size, shape, and position within this ganglion. (C) Changes in the gill withdrawal behavior due to habituation and sensitization. The first time that the siphon is touched, the gill contracts vigorously. Repeated touches elicit smaller gill contractions due to habituation. Subsequently pairing a siphon touch with an electrical shock to the tail restores a large and rapid gill contraction, due to short-term sensitization. (D) A short-term sensitization of the gill withdrawal response is observed following the pairing of a single tail shock with a siphon touch. (E) Repeated applications of tail shocks causes prolonged sensitization of the gill withdrawal response. (After Squire and Kandel, 1999.)*

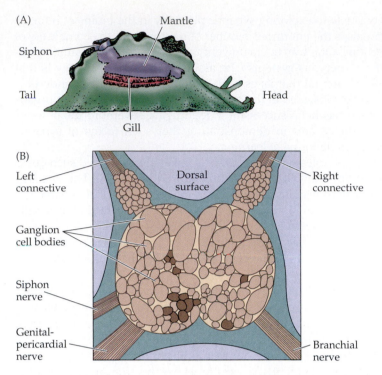

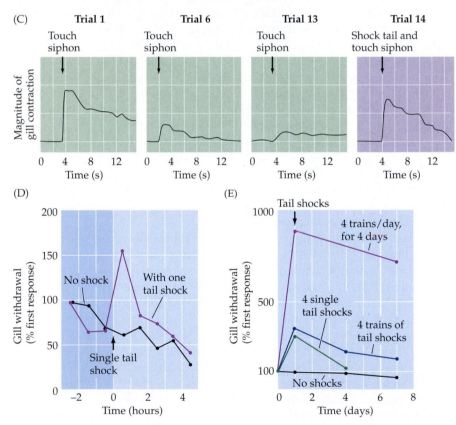

(A)

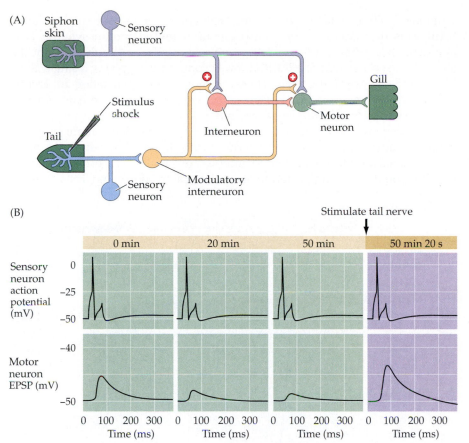

Figure 8.4 Synaptic mechanisms underlying short-term sensitization. (A) Neural circuitry involved in sensitization. Normally, touching the siphon skin activates sensory neurons that excite interneurons and gill motor neurons, yielding a contraction of the gill muscle. A shock to the animal's tail stimulates modulatory interneurons that alter synaptic transmission between the siphon sensory neurons and gill motor neurons, resulting in sensitization. (B) Changes in synaptic efficacy at the sensory-motor synapse during short-term sensitization. Prior to sensitization, activating the siphon sensory neurons causes an EPSP to occur in the gill motor neurons. Activation of the serotonergic modulatory interneurons enhances release of transmitter from the sensory neurons onto the motor neurons, increasing the EPSP in the motor neurons and causing the motor neurons to more strongly excite the gill muscle. (C) Time course of the serotonin-induced facilitation of transmission at the sensory motor synapse. (After Squire and Kandel, 1999.)

(C)

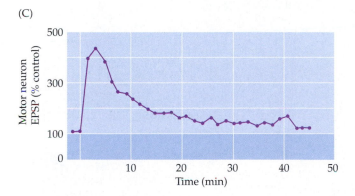

Both habituation and sensitization appear to arise from plastic changes in synaptic transmission in this circuit. During habituation, transmission at the glutamatergic synapse between the sensory and motor neurons is depressed (Figure 8.4B, left). This synaptic depression is thought to be responsible for the decreasing ability of siphon stimuli to evoke gill contractions during habituation. Much like the short-term form of synaptic depression described in the preceding section, this depression is presynaptic and is due to a reduction in the number of synaptic vesicles available for release. In contrast, sensitization modifies the function of this circuit by recruiting additional neurons. The tail shock that evokes sensitization activates sensory neurons that innervate the tail. These sensory neurons in turn excite modulatory interneurons that release serotonin on to the presynaptic terminals of the sensory neurons of the siphon (see Figure 8.4A). Serotonin enhances transmitter release from the siphon sensory neuron terminals, leading to increased synaptic excitation of the motor neurons (Figure 8.4B). This modulation of the sensory neuron-motor neuron synapse lasts approximately an hour (Figure 8.4C), which is similar to the duration of the short-term sensitization of gill withdrawal produced by applying a single stimulus to the tail (see Figure 8.3D). Thus, the short-term sensitization apparently is due to recruitment of additional synaptic elements that modulate synaptic transmission in the gill withdrawal circuit.

The mechanism thought to be responsible for the enhancement of glutamatergic transmission during short-term sensitization is shown in Figure 8.5A. Serotonin released by the facilitatory interneurons binds to G-protein-coupled receptors on the presynaptic terminals of the siphon sensory neurons (step 1),

(A)

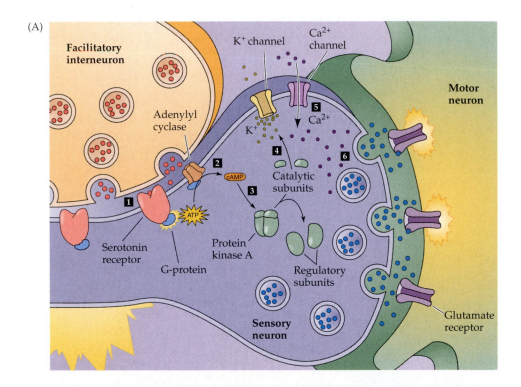

(B)

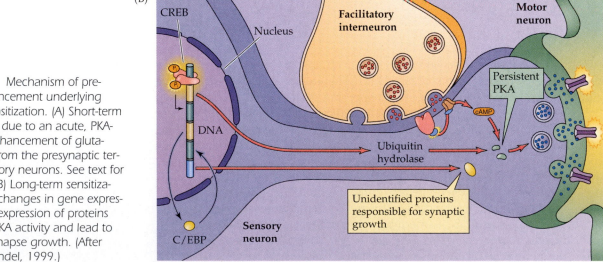

Figure 8.5 Mechanism of presynaptic enhancement underlying behavioral sensitization. (A) Short-term sensitization is due to an acute, PKA-dependent enhancement of glutamate release from the presynaptic terminals of sensory neurons. See text for explanation. (B) Long-term sensitization is due to changes in gene expression, causing expression of proteins that change PKA activity and lead to changes in synapse growth. (After Squire and Kandel, 1999.)

which stimulates production of the second messenger, cAMP (step 2). cAMP binds to the regulatory subunits of protein kinase A (PKA; step 3), liberating catalytic subunits of PKA that are then able to phosphorylate several proteins, probably including K^+ channels (step 4). The net effect of the action of PKA is to reduce the probability that the K^+ channels open during a presynaptic action potential. This effect prolongs the presynaptic action potential, thereby opening more presynaptic Ca^{2+} channels (step 5). Finally, the enhanced influx of Ca^{2+} into the presynaptic terminals increases the amount of transmitter released onto motor neurons during a sensory neuron action potential (step 6). In summary, short-term sensitization of gill withdrawal is mediated by a signal transduction cascade that involves neurotransmitters, second messengers, one or more protein kinases, and ion channels. This cascade ultimately enhances synaptic transmission between the sensory and motor neurons within the gill withdrawal circuit.

The same serotonin-induced enhancement of glutamate release that mediates short-term sensitization is also thought to underlie long-term sensitization. However, during long-term sensitization this circuitry is affected for up to several weeks. The prolonged duration of this form of plasticity is evidently due to changes in gene expression and thus protein synthesis (Figure 8.5B). With repeated training (i.e., additional tail shocks), the serotonin-activated PKA involved in short-term sensitization now also phosphorylates—and thereby activates—the transcriptional activator CREB. As described in Chapter 7, CREB binding to the cAMP responsive elements (CREs) in regulatory regions of nuclear DNA increases the rate of transcription of downstream genes. Although the changes in genes and gene products that follow CRE activation have been difficult to sort out, several consequences of gene activation have been identified. First, CREB stimulates the synthesis of an enzyme, ubiquitin hydroxylase, that stimulates degradation of the regulatory subunit of PKA. This causes a persistent increase in the amount of free catalytic subunit, meaning that some PKA is persistently active and no longer requires serotonin to be activated. CREB also stimulates another transcriptional activator protein, called C/EBP. C/EBP stimulates transcription of other, unknown genes that cause addition of synaptic terminals, yielding a long-term increase in the number of synapses between the sensory and the motor neurons. Such structural increases are not seen following short-term sensitization and may represent the ultimate cause of the long-lasting change in overall strength of the relevant circuit connections that produce a long-lasting enhancement in the gill withdrawal response.

Another protein involved in the long-term synaptic facilitation is a cytoplasmic polyadenylation element binding protein, somewhat confusingly called CPEB. CPEB activates mRNAs and may be important for local control of protein synthesis. Most intriguingly, CPEB has self-sustaining properties similar to those of prion proteins (see Box 19A), which could allow CPEB to remain active in perpetuity and thereby mediate permanent changes in synaptic transmission.

Studies of *Aplysia*, and related work on other invertebrates such as the fruit fly (Box 8A), have led to at least two generalizations about synaptic plasticity. First, synaptic plasticity clearly can lead to changes in circuit function and, ultimately, to behavioral plasticity. This conclusion has triggered intense interest in understanding synaptic plasticity mechanisms. Second, these plastic changes in synaptic function can be either short-term effects that rely on posttranslational modification of existing synaptic proteins, or long-term changes that require changes in gene expression, new protein synthesis, and growth of new synapses (or the elimination of existing ones). Thus, it appears that short-term and long-term changes in synaptic function have different mechanistic underpinnings. As

BOX 8A Genetics of Learning and Memory in the Fruit Fly

As part of a renaissance in the genetic analysis of simple organisms in the mid-1970s, several investigators recognized that the genetic basis of learning and memory might be effectively studied in the fruit fly, *Drosophila melanogaster*. Although learning and memory is certainly one of the more difficult problems *Drosophila* geneticists have tackled, their efforts have been surprisingly successful. A number of genetic mutations have been discovered that affect learning and memory, and the identification of these genes has provided a valuable framework for studying the cellular mechanisms of these processes.

The initial problem in this work was to develop behavioral tests that could identify abnormal learning and/or memory defects in large populations of flies. This challenge was met by Seymour

Benzer and his colleagues Chip Quinn and Bill Harris at the California Institute of Technology, who developed the olfactory and visual learning tests that have become the basis for most subsequent analyses of learning and memory in the fruit fly. Behavioral paradigms pairing odors or light with an aversive stimulus allowed Benzer and his colleagues to assess associative learning in flies. The design of an ingenious testing apparatus controlled for non-learning-related sensory cues that had previously complicated such behavioral testing. Moreover, the apparatus allowed large numbers of flies to be screened relatively easily, expediting the analysis of mutagenized populations (see figure).

These studies led to the identification of an ever-increasing number of single gene mutations that disrupt learning

and/or memory in flies. The behavioral and molecular studies of the mutants (given whimsical but descriptive names like *dunce*, *rutabaga*, and *amnesiac*) suggested that a central pathway for learning and memory in the fly is signal transduction mediated by the cyclic nucleotide cAMP. Thus, the gene products of the *dunce*, *rutabaga*, and *amnesiac* loci are, respectively, a phosphodiesterase (which degrades cAMP), an adenylyl cyclase (which converts ATP to cAMP), and a peptide transmitter that stimulates adenylyl cyclase. This conclusion about the importance of cAMP has been confirmed by the finding that genetic manipulation of the CREB transcription factor also interferes with learning and memory in normal flies.

These observations in *Drosophila* accord with conclusions reached in studies of *Aplysia* and mammals (see text) and have emphasized the importance of cAMP-mediated learning and memory in a wide range of additional species.

References

DAVIS, R. L. (2004) Olfactory learning. *Neuron* 44: 31–48.

QUINN, W. G., W. A. HARRIS AND S. BENZER (1974) Conditioned behavior in *Drosophila melanogaster*. *Proc. Natl. Acad. Sci. USA* 71: 708–712.

TULLY, T. (1996) Discovery of genes involved with learning and memory: An experimental synthesis of Hirshian and Benzerian perspectives. *Proc. Natl. Acad. Sci. USA* 93: 13460–13467.

WADDELL, S. AND W. G. QUINN (2001) Flies, genes, and learning. *Annu. Rev. Neurosci.* 24: 1283–1309.

WEINER, J. (1999) *Time, Love, Memory: A Great Biologist and His Quest for the Origins of Behavior.* New York: Knopf.

(A) (B)

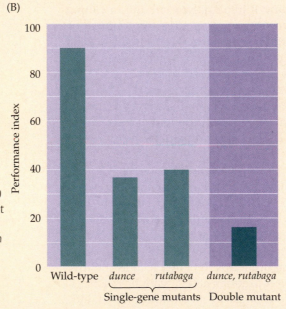

(A) The fruit fly *Drosophila melanogaster*. (B) Performance of normal and mutant flies on an olfactory learning task. The performance of both *dunce* and *rutabaga* mutants on this task is diminished by at least 50 percent. Flies that are mutant at both the *dunce* and *rutabaga* locus show a larger decrease in performance, suggesting that the two genes disrupt different but related aspects of learning. (B after Tully, 1996.)

will be seen in the following sections, these generalizations apply to synaptic plasticity in the mammalian brain and, in fact, have helped guide our understanding of these forms of synaptic plasticity.

Long-Term Potentiation at a Hippocampal Synapse

Long-term synaptic plasticity has also been identified within the mammalian brain. Here, some patterns of synaptic activity produce a long-lasting increase in synaptic strength known as **long-term potentiation** (**LTP**), whereas other patterns of activity produce a long-lasting decrease in synaptic strength, known as **long-term depression** (**LTD**). LTP and LTD are broad terms that describe only the direction of change in synaptic efficacy; in fact, different cellular and molecular mechanisms can be involved in producing LTP or LTD at different synapses throughout the brain. In general, LTP and LTD are produced by different histories of activity, and are mediated by different complements of intracellular signal transduction pathways in the nerve cells involved.

Long-term synaptic plasticity has been most thoroughly studied at excitatory synapses in the mammalian hippocampus. The hippocampus (Figure 8.6, top) is a brain area that is especially important in the formation and/or retrieval of some forms of memory (see Chapter 31). In humans, functional imaging shows that the human hippocampus is activated during certain kinds of memory tasks, and that damage to the hippocampus results in an inability to form certain types of new memories. In rodents, hippocampal neurons fire action potentials only when an animal is in certain locations. Such "place cells" appear to encode spatial memories, an interpretation supported by the fact that hippocampal damage prevents rats from developing proficiency in spatial learning tasks (see Figure 31.8). Although many other brain areas are involved in the complex process of memory formation, storage, and retrieval, these observations have led many investigators to study long-term synaptic plasticity of hippocampal synapses.

Work on LTP began in the late 1960s, when Terje Lomo and Timothy Bliss, working in the laboratory of Per Andersen in Oslo, Norway, discovered that a

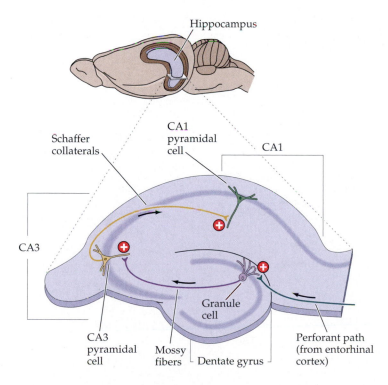

Figure 8.6 Diagram of a section through the rodent hippocampus showing the major regions, excitatory pathways, and synaptic connections. Long-term potentiation has been observed at each of the three synaptic connections shown here.

Figure 8.7 Long-term potentiation of Schaffer collateral-CA1 synapses. (A) Arrangement for recording synaptic transmission; two stimulating electrodes (1 and 2) each activate separate populations of Schaffer collaterals, thus providing test and control synaptic pathways. (B) Left: Synaptic responses recorded in a CA1 neuron in response to single stimuli of synaptic pathway 1, minutes before and one hour after a high-frequency train of stimuli. The high-frequency stimulus train increases the size of the EPSP evoked by a single stimulus. Right: Responses produced by stimulating synaptic pathway 2, which did not receive high-frequency stimulation, is unchanged. (C) The time course of changes in the amplitude of EPSPs evoked by stimulation of pathways 1 and 2. High-frequency stimulation of pathway 1 causes a prolonged enhancement of the EPSPs in this pathway (purple). This potentiation of synaptic transmission in pathway 1 persists for several hours, while the amplitude of EPSPs produced by pathway 2 (orange) remains constant. (D) Recordings of EPSPs from the living hippocampus reveal that high-frequency stimulation can produce LTP that lasts for more than a year. (A–C after Malinow et al., 1989; D after Abraham et al., 2002.)

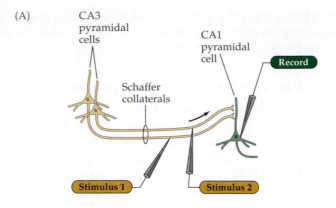

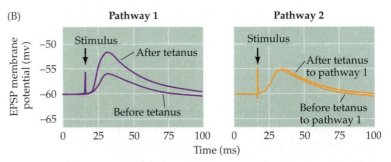

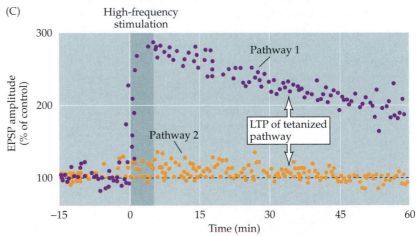

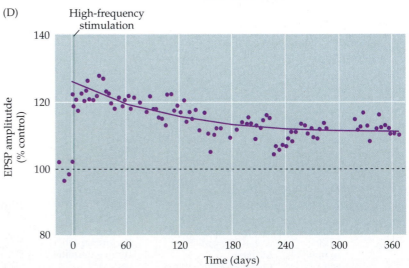

few seconds of high-frequency electrical stimulation can enhance synaptic transmission in the hippocampus of intact rabbits for days or even weeks. More recently, however, progress in understanding the mechanism of LTP has relied heavily on in vitro studies of slices of living hippocampus. The arrangement of neurons allows the hippocampus to be sectioned such that most of the relevant circuitry is left intact. In such preparations, the cell bodies of the pyramidal neurons lie in a single densely packed layer that is readily apparent (Figure 8.6, bottom). This layer is divided into several distinct regions, the major ones being CA1 and CA3. "CA" refers to *cornu Ammon*, Latin for Ammon's horn—the ram's horn that resembles the shape of the hippocampus. The dendrites of pyramidal cells in the CA1 region form a thick band (the stratum radiatum), where they receive synapses from Schaffer collaterals, the axons of pyramidal cells in the CA3 region. Much of the work on LTP has focused on the synaptic connections between the Schaffer collaterals and CA1 pyramidal cells. Electrical stimulation of Schaffer collaterals generates EPSPs in the postsynaptic CA1 cells (Figure 8.7A,B). If the Schaffer collaterals are stimulated only two or three times per minute, the size of the evoked EPSP in the CA1 neurons remains constant. However, a brief, high-frequency train of stimuli to the same axons causes LTP, which is evident as a long-lasting increase in EPSP amplitude (Figure 8.7B,C). While the maximum duration of LTP is not known, LTP can last for more than a year in some cases (Figure 8.7D). The long duration of LTP shows that this form of synaptic plasticity is capable of serving as a mechanism for long-lasting storage of information. LTP occurs at each of the three excitatory synapses of the hippocampus shown in Figure 8.6, as well as at synapses in a variety of brain regions, including the cortex, amygdala, and cerebellum.

LTP of the Schaffer collateral synapse exhibits several properties that make it an attractive neural mechanism for information storage. First, LTP is *state-dependent*: The state of the membrane potential of the postsynaptic cell determines whether or not LTP occurs (Figure 8.8). If a single stimulus to the Schaf-

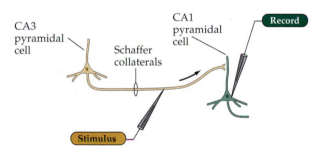

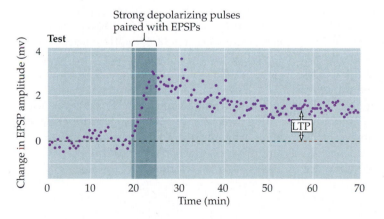

Figure 8.8 Pairing presynaptic and postsynaptic activity causes LTP. Single stimuli applied to a Schaffer collateral synaptic input evokes EPSPs in the postsynaptic CA1 neuron. These stimuli alone do not elicit any change in synaptic strength. However, when the CA1 neuron's membrane potential is briefly depolarized (by applying current pulses through the recording electrode) in conjunction with the Schaffer collateral stimuli, there is a persistent increase in the EPSPs. (After Gustafsson et al., 1987.)

Figure 8.9 Properties of LTP at a CA1 pyramidal neuron receiving synaptic inputs from two independent sets of Schaffer collateral axons. (A) Strong activity initiates LTP at active synapses (pathway 1) without initiating LTP at nearby inactive synapses (pathway 2). (B) Weak stimulation of pathway 2 alone does not trigger LTP. However, when the same weak stimulus to pathway 2 is activated together with strong stimulation of pathway 1, both sets of synapses are strengthened.

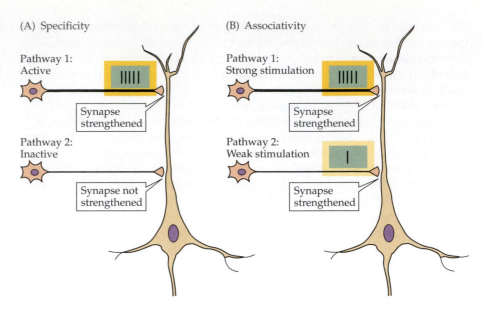

(A) Specificity

Pathway 1: Active — Synapse strengthened

Pathway 2: Inactive — Synapse not strengthened

(B) Associativity

Pathway 1: Strong stimulation — Synapse strengthened

Pathway 2: Weak stimulation — Synapse strengthened

fer collaterals—which would not normally elicit LTP—is paired with strong depolarization of the postsynaptic CA1 cell, the activated Schaffer collateral synapses undergo LTP. The increase occurs only if the paired activities of the presynaptic and postsynaptic cells are tightly linked in time, such that the strong postsynaptic depolarization occurs within about 100 ms of presynaptic transmitter release. Such a requirement for coincident presynaptic and postsynaptic activity is the central postulate of a theory of learning devised by Donald Hebb in 1949. Hebb proposed that coordinated activity of a presynaptic terminal and a postsynaptic neuron will strengthen the synaptic connection between them, precisely as is observed for LTP. Hebb's postulate has also been useful in thinking about the role of neuronal activity in other brain functions, most notably development of neural circuits (see Chapter 24).

A second property of LTP is *input specificity*: LTP induced by activation of one synapse does not occur in other, inactive synapses that contact the same neuron (see Figure 8.7). Thus, LTP is restricted to activated synapses rather than to all of the synapses on a given cell (Figure 8.9A). This feature of LTP is consistent with its involvement in memory formation (or at least the selective storage of information at synapses). If activation of one set of synapses led to all other synapses—even inactive ones—being potentiated, it would be difficult to selectively enhance particular sets of inputs, as is presumably required to store specific information.

Another important property of LTP is **associativity** (Figure 8.9B). As noted, weak stimulation of a pathway will not by itself trigger LTP. However, if one pathway is weakly activated at the same time that a neighboring pathway onto the same cell is strongly activated, both synaptic pathways undergo LTP. This selective enhancement of conjointly activated sets of synaptic inputs is often considered a cellular analog of associative or classical conditioning. More generally, associativity is expected in any network of neurons that links one set of information with another.

Although there is clearly a gap between understanding LTP of hippocampal synapses and understanding learning, memory, or other aspects of behavioral plasticity in mammals, this form of long-term synaptic plasticity provides a plausible neural mechanism for enduring changes in a part of the brain that is known to be involved in the formation of certain kinds of memories.

Molecular Mechanisms Underlying LTP

Despite the fact that LTP was discovered nearly 40 years ago, its molecular under-pinnings were not well understood until recently. A key advance in this effort occurred in the mid-1980s, when it was discovered that antagonists of the NMDA type of glutamate receptor prevent LTP, but have no effect on the synaptic response evoked by low-frequency stimulation of the Schaffer collaterals. At about the same time, the unique biophysical properties of the NMDA receptor were first appreciated. As described in Chapter 6, the NMDA receptor channel is permeable to Ca^{2+}, but is blocked by physiological concentrations of Mg^{2+}. This property provides a critical insight into how LTP is selectively induced by high-frequency activity. During low-frequency synaptic transmission, glutamate released by the Schaffer collaterals binds to both NMDA-type and AMPA/kainate-type glutamate receptors. While both types of receptors bind glutamate, if the postsynaptic neuron is at its normal resting membrane potential, the pore of the NMDA receptor channel will be blocked by Mg^{2+} ions and no current will flow (Figure 8.10, left). Under such conditions, the EPSP will be mediated entirely by the AMPA receptors. Because blockade of the NMDA receptor by Mg^{2+} is voltage-dependent, the function of the synapse changes markedly when the postsynaptic cell is depolarized. Thus, high-frequency stimulation (as in Figure 8.7) will cause summation of EPSPs, leading to a prolonged depolarization that expels Mg^{2+} from the NMDA channel pore (Figure 8.10, right). Removal of Mg^{2+} allows Ca^{2+} to enter the postsynaptic neuron and the resulting increase in Ca^{2+} concentration within the dendritic spines of the postsynaptic cell turns out to be the trigger for LTP. The NMDA receptor thus behaves like a molecular coincidence detector. The channel of this receptor opens (to induce LTP) only when two events occur simultaneously: glutamate is bound to the receptor and the postsynaptic cell is depolarized to relieve the Mg^{2+} block of the channel pore.

These properties of the NMDA receptor can account for many of the characteristics of LTP. The specificity of LTP (see Figure 8.9A) can be explained by the fact that NMDA channels will be opened only at synaptic inputs that are active and releasing glutamate, thereby confining LTP to these sites. With respect to associativity (see Figure 8.9B), a weakly stimulated input releases glutamate, but cannot sufficiently depolarize the postsynaptic cell to relieve the Mg^{2+} block. If

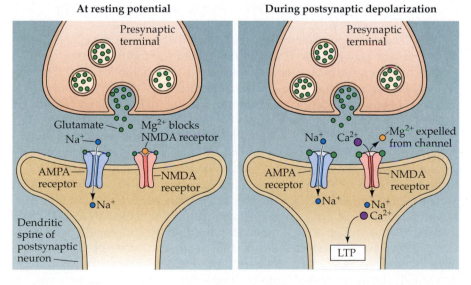

Figure 8.10 The NMDA receptor channel can open only during depolarization of the postsynaptic neuron from its normal resting level. Depolarization expels Mg^{2+} from the NMDA channel, allowing current to flow into the postsynaptic cell. This leads to Ca^{2+} entry, which in turn triggers LTP. (After Nicoll et al., 1988.)

Figure 8.11 Signaling mechanisms underlying LTP. During glutamate release, the NMDA channel opens only if the postsynaptic cell is sufficiently depolarized. Ca^{2+} enters the cell through the channel and activates postsynaptic protein kinases. These postsynaptic kinases trigger a series of reactions that leads to insertion of new AMPA receptors into the postsynaptic spine, thereby increasing the cell's sensitivity to glutamate.

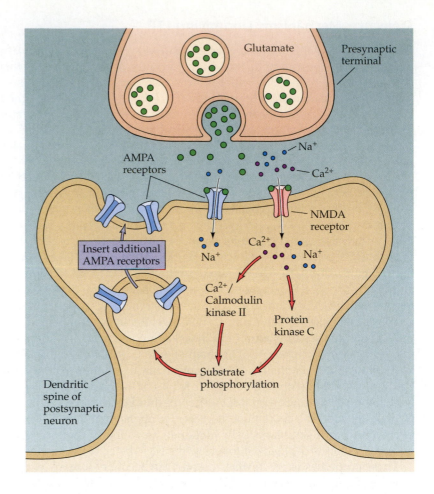

neighboring inputs are strongly stimulated, however, they provide the "associative" depolarization necessary to relieve the block. The state dependence of LTP, evident as the induction of LTP by the pairing of weak synaptic input with depolarization (see Figure 8.8), should work similarly—the synaptic input releases glutamate, while the coincident depolarization relieves the Mg^{2+} block of the NMDA receptor.

Several sorts of observations have confirmed that a rise in the concentration of Ca^{2+} in the postsynaptic CA1 neuron, due to Ca^{2+} ions entering through NMDA receptors, serves as a second messenger signal that induces LTP. Imaging studies, for instance, have shown that activation of NMDA receptors causes increases in postsynaptic Ca^{2+} levels. Furthermore, injection of Ca^{2+} chelators blocks LTP induction, whereas elevation of Ca^{2+} levels in postsynaptic neurons potentiates synaptic transmission. Ca^{2+} induces LTP by activating in the postsynaptic neuron complicated signal transduction cascades that include protein kinases. At least two Ca^{2+}-activated protein kinases have been implicated in LTP induction (Figure 8.11): Ca^{2+}/calmodulin-dependent protein kinase (CaMKII), and protein kinase C (PKC; see Chapter 7). CaMKII seems to play an especially important role; this enzyme is the most abundant postsynaptic protein at Schaffer collateral synapses, and pharmacological inhibition or genetic deletion of CaMKII prevents LTP. CaMKII also is able to phosphorylate itself, which has been proposed to cause a sustained activation of CaMKII that prolongs the duration of LTP. The downstream targets of these kinases are not yet fully known, but apparently include AMPA receptors and many other signaling proteins.

It appears that the strengthening of synaptic transmission during LTP mainly arises from an increase in the sensitivity of the postsynaptic cell to glutamate. Several recent observations indicate that excitatory synapses can dynamically regulate their postsynaptic glutamate receptors, and can even add new AMPA receptors to "silent" synapses that did not previously have postsynaptic AMPA receptors (Box 8B). The "expression" or maintenance of LTP apparently is due to such insertion of AMPA receptors into the postsynaptic membrane (as opposed to "induction" of LTP, which relies on activation of NMDA receptors). The resulting increase in the density of AMPA receptors in the postsynaptic spine increases the response of the postsynaptic cell to released glutamate (Figure 8.12A), yielding a strengthening of synaptic transmission that can last for as long as LTP is maintained (Figure 8.12B). At silent synapses, where synaptic activity generates no postsynaptic response at the normal resting potential, LTP adds AMPA receptors so that the synapse can produce postsynaptic responses (Figure 8.12C). Under some circumstances, LTP also can cause a sustained increase in the ability of presynaptic terminals to release glutamate. Because LTP clearly is triggered by the actions of Ca^{2+} within the postsynaptic neuron (see Figure 8.11), this presynaptic potentiation requires that a retrograde signal (perhaps NO) spread from the postsynaptic spine back to the presynaptic terminals.

The scheme described above can account for the changes in synaptic transmission that occur over the first hour or two after LTP is induced. However, there is also a later phase of LTP that depends on changes in gene expression and the synthesis of new proteins. The contributions of this late phase can be

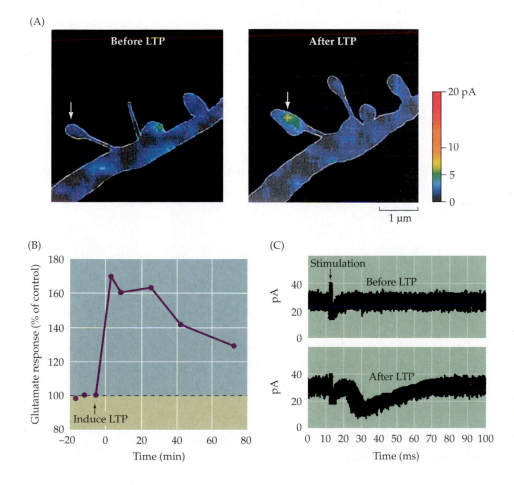

(A)

Figure 8.12 Addition of postsynaptic AMPA receptors during LTP. (A) Spatial maps of the glutamate sensitivity of a hippocampal neuron dendrite before (left) and 120 minutes after (right) induction of LTP. The color scale indicates amplitude of responses to highly localized glutamate application. LTP causes an increase in the glutamate response of a dendritic spine (arrow), due to an increase in the number of AMPA receptors on the spine membrane. (B) Time course of changes in glutamate sensitivity of dendritic spines during LTP. Induction of LTP (at time = 0) causes glutamate sensitivity to increase for more than 60 minutes. (C) LTP induces AMPA receptor responses at silent synapses in the hippocampus. Prior to inducing LTP, no EPSCs are elicited at −65 mV at this silent synapse (upper trace). After LTP induction, the same stimulus produces EPSCs that are mediated by AMPA receptors (lower trace). (A,B from Matsuzaki et al., 2004; C after Liao et al., 1995.)

BOX 8B   Silent Synapses

Several recent observations indicate that postsynaptic glutamate receptors are dynamically regulated at excitatory synapses. Early insight into this process came from the finding that stimulation of some glutamatergic synapses generates no postsynaptic electrical signal when the postsynaptic cell is at its normal resting membrane potential (Figure A). However, once the postsynaptic cell is depolarized, these "silent synapses" can transmit robust postsynaptic electrical responses. The fact that transmission at such synapses can be turned on or off in response to postsynaptic activity suggests an interesting and simple means of modifying neural circuitry.

Silent synapses are especially prevalent in development and have been found in many brain regions, including the hippocampus, cerebral cortex, and spinal cord. The silence of these synapses is evidently due to the voltage-dependent blockade of NMDA receptors by Mg^{2+} (see text and Chapter 6). At the normal resting membrane potential, presynaptic release of glutamate evokes no postsynaptic response at such synapses because their NMDA receptors are blocked by Mg^{2+}. However, depolarization of the postsynaptic neuron displaces the Mg^{2+}, allowing glutamate release to induce postsynaptic responses mediated by NMDA receptors.

Glutamate released at silent synapses evidently binds only to NMDA receptors. How, then, does glutamate release avoid activating AMPA receptors? One possibility is that glutamate released onto neighboring neurons diffuses to synapses on the neuron from which the electrical recording is being made. In this case, the diffusing glutamate may be present at concentrations sufficient to activate the high-affinity NMDA receptors, but not the low-affinity AMPA receptors. A second possibility is that a silent synapse

has both AMPA and NMDA receptors, but its AMPA receptors are somehow not functional. Finally, some excitatory synapses may have only NMDA receptors. Accumulating evidence supports the latter explanation. Most compelling are

immunocytochemical experiments demonstrating the presence of excitatory synapses that have only NMDA receptors (green spots in Figure B). Such NMDA receptor-only synapses are particularly abundant early in postnatal development

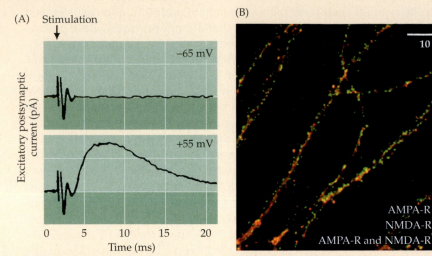

(A) Electrophysiological evidence for silent synapses. Stimulation of some axons fails to activate synapses when the postsynaptic cell is held at a negative potential (−65 mV, upper trace). However, when the postsynaptic cells is depolarized (+55 mV), stimulation produces a robust response (lower trace). (B) Immunofluorescent localization of NMDA receptors (green) and AMPA receptors (red) in a cultured hippocampal neuron. Many dendritic spines are positive for NMDA receptors but not AMPA receptors, indicating NMDA receptor-only synapses. (A after Liao et al., 1995; B courtesy of M. Ehlers.)

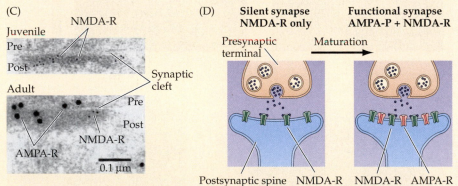

(C) Electron microscopy of excitatory synapses in CA1 stratum radiatum of the hippocampus from 10-day-old or 5-week-old (adult) rats double-labeled for AMPA receptors and NMDA receptors. The presynaptic terminal (pre), synaptic cleft, and postsynaptic spine (post) are indicated. AMPA receptors are abundant at the adult synapse, but absent from the younger synapse. (D) Diagram of glutamatergic synapse maturation. Early in postnatal development, many excitatory synapses contain only NMDA receptors. As synapses mature, AMPA receptors are recruited. (C from Petralia et al., 1999.)

and decrease in adults (Figure C). Thus, at least some silent synapses are not a separate class of excitatory synapses that lack AMPA receptors, but rather an early stage in the ongoing maturation of the glutamatergic synapse (Figure D). Evidently, AMPA and NMDA receptors are not inextricably linked at excitatory synapses, but are targeted via independent cellular mechanisms. Such synapse-specific glutamate receptor composition implies sophisticated mechanisms for regulating the localization of each type of receptor. Dynamic changes in the trafficking of AMPA and NMDA receptors can strengthen or weaken synaptic transmis-

sion and are important in LTP and LTD, as well as in the maturation of glutamatergic synapses.

Although silent synapses have begun to whisper their secrets, much remains to be learned about their physiological importance and the molecular mechanisms that mediate rapid recruitment or removal of synaptic AMPA receptors.

References

DERKACH, V. A., M. C. OH, E. S. GUIRE AND T. R. SODERLING (2007) Regulatory mechanisms of AMPA receptors in synaptic plasticity. *Nature Rev. Neurosci.* 8: 101–113.

GOMPERTS, S. N., A. RAO, A. M. CRAIG, R. C. MALENKA AND R. A. NICOLL (1998) Postsynaptically silent synapses in single neuron cultures. *Neuron* 21: 1443–1451.

LIAO, D., N. A. HESSLER AND R. MALINOW (1995) Activation of postsynaptically silent synapses during pairing-induced LTP in CA1 region of hippocampal slice. *Nature* 375: 400–404.

LUSCHER, C., R. A. NICOLL, R. C. MALENKA AND D. MULLER (2000) Synaptic plasticity and dynamic modulation of the postsynaptic membrane. *Nature Neurosci.* 3: 545–550.

PETRALIA, R. S. AND 6 OTHERS (1999) Selective acquisition of AMPA receptors over postnatal development suggests a molecular basis for silent synapses. *Nature Neurosci.* 2: 31–36.

observed by treating synapses with drugs that inhibit protein synthesis: blocking protein synthesis prevents LTP measured several hours after a stimulus but does not affect LTP measured at earlier times (Figure 8.13). The late phase of LTP appears to be initiated by protein kinase A, which goes on to activate transcription factors such as CREB, which stimulate the expression of other proteins. Although most of these newly synthesized proteins have not yet been identified, they include other transcriptional regulators, protein kinases, and AMPA receptors (Figure 8.14A). How these proteins contribute to the late phase of LTP is not yet known. There is evidence that the number and size of synaptic contacts increases during LTP (Figure 8.14B,C), so it is likely that some of the proteins newly synthesized during the late phase of LTP are involved in construction of new synaptic contacts that render LTP essentially permanent.

In conclusion, it appears that LTP in the mammalian hippocampus has many parallels to the long-term changes in synaptic transmission underlying behavioral sensitization *Aplysia*. Both consist of an early, transient phase that relies on protein kinases to produce posttranslational changes in membrane ion channels; and both have later, long-lasting phases that require changes in gene expression mediated by CREB. Both forms of long-term synaptic plasticity are likely to be involved in long-term storage of information, although the role of LTP in memory storage in the hippocampus is not firmly established.

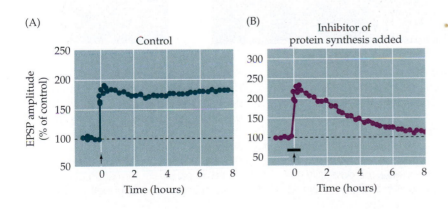

Figure 8.13 Role of protein synthesis for maintaining LTP. (A) Repetitive high-frequency stimulation (arrows) induces LTP that persists for many hours. (B) Treatment with anisomycin, an inhibitor of protein synthesis (at bar), causes LTP to decay within a few hours after the high-frequency stimulation (arrows). (After Frey and Morris, 1997.)

(A)   Short–term   Long–term

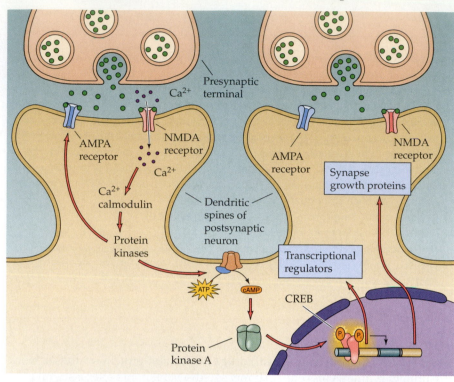

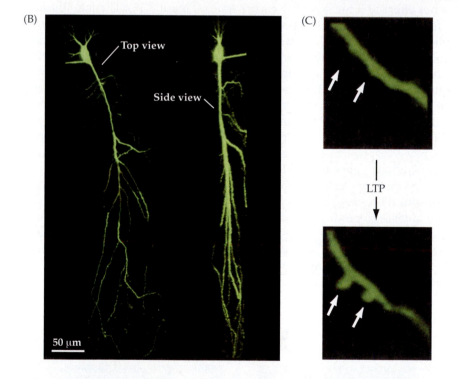

Figure 8.14   Mechanisms responsible for long-lasting changes in synaptic transmission during LTP. (A) The late component of LTP is due to PKA activating the transcriptional regulator CREB, which turns on expression of a number of genes that produce long-lasting changes in PKA activity and synapse structure. (B,C) Structural changes associated with LTP in the hippocampus. (B) The dendrites of a CA1 pyramidal neuron were visualized by filling the cell with a fluorescent dye. (C) New dendritic spines (white arrows) appear approximately 1 hour after a stimulus that induces LTP. The presence of novel spines raises the possibility that LTP may arise, in part, from formation of new synapses. (A after Squire and Kandel, 1999; B,C from Engert and Bonhoeffer, 1999.)

Long-Term Synaptic Depression

If synapses simply continued to increase in strength as a result of LTP, eventually they would reach some level of maximum efficacy, making it difficult to encode new information. Thus, to make synaptic strengthening useful, other processes must selectively weaken specific sets of synapses. Long-term depression (LTD) is such a process. In the late 1970s, LTD was found to occur at the synapses between the Schaffer collaterals and the CA1 pyramidal cells in the hippocampus. Whereas LTP at these synapses requires brief, high-frequency stimulation, LTD occurs when the Schaffer collaterals are stimulated at a low rate—about 1 Hz—for long periods (10–15 minutes). This pattern of activity depresses the EPSP for several hours and, like LTP, is specific to the activated synapses (Figure 8.15A,B). Moreover, LTD can erase the increase in EPSP size due to LTP, and, conversely, LTP can erase the decrease in EPSP size due to LTD. This complementarity suggests that LTD and LTP reversibly affect synaptic efficiency by acting at a common site.

(A)

(B)

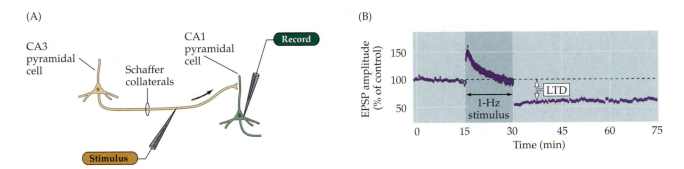

(C)

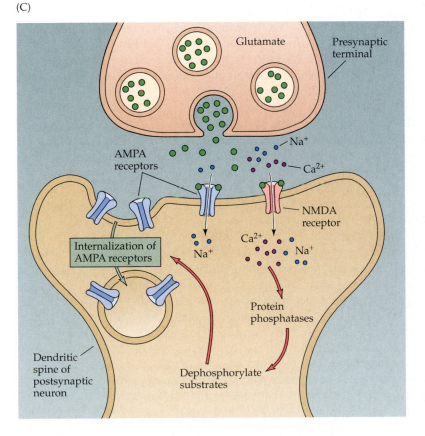

Figure 8.15 Long-term synaptic depression in the hippocampus. (A) Electrophysiological procedures used to monitor transmission at the Schaffer collateral synapses on to CA1 pyramidal neurons. (B) Low-frequency stimulation (1 per second) of the Schaffer collateral axons causes a long-lasting depression of synaptic transmission. (C) Mechanisms underlying LTD. A low-amplitude rise in Ca^{2+} concentration in the postsynaptic CA1 neuron activates postsynaptic protein phosphatases, which cause internalization of postsynaptic AMPA receptors, thereby decreasing the sensitivity to glutamate released from the Schaffer collateral terminals. (B after Mulkey et al., 1993.)

LTP and LTD at the Schaffer collateral-CA1 synapses actually share several key elements. Both require activation of NMDA-type glutamate receptors and the resulting entry of Ca^{2+} into the postsynaptic cell. The major determinant of whether LTP or LTD arises appears to be the nature of the Ca^{2+} signal in the postsynaptic cell: small and slow rises in Ca^{2+} lead to depression, whereas large and fast increases Ca^{2+} trigger potentiation. As noted above, LTP is at least partially due to activation of protein kinases, which phosphorylate their target proteins. LTD, on the other hand, appears to result from activation of Ca^{2+}-dependent phosphatases that cleave phosphate groups from these target molecules (see Chapter 7). Evidence in support of this idea is that phosphatase inhibitors prevent LTD, but have no effect on LTP. The different effects of Ca^{2+} during LTD and LTP may arise from the selective activation of protein phosphatases and kinases by the different types of Ca^{2+} signals. While the phosphatase substrates important for LTD have not yet been identified, it is possible that LTP and LTD phosphorylate and dephosphorylate the same set of regulatory proteins to control the efficacy of transmission at the Schaeffer collateral-CA1 synapse. Just as LTP at this synapse is associated with insertion of AMPA receptors, LTD is often associated with a loss of synaptic AMPA receptors. This loss probably arises from internalization of AMPA receptors into the postsynaptic cell (Figure 8.15C), due to the same sort of clathrin-dependent endocytosis mechanisms important for synaptic vesicle recycling in the presynaptic terminal (see Chapter 5).

A somewhat different form of LTD is observed in the cerebellum (see Chapter 19). LTD of synaptic inputs onto cerebellar Purkinje cells was first described by Masao Ito and colleagues in Japan in the early 1980s. Purkinje neurons in the cerebellum receive two distinct types of excitatory input: climbing fibers and parallel fibers (Figure 8.16A; also see Chapter 19). LTD reduces the strength of transmission at the parallel fiber synapse (Figure 8.16B) and has recently been found to depress transmission at the climbing fiber synapse as well. This form of LTD has been implicated in the motor learning that mediates the coordination, acquisition, and storage of complex movements within the cerebellum. Although the role of LTD in cerebellar motor learning remains controversial, it has nonetheless been a useful model system for understanding the cellular mechanisms of long-term synaptic plasticity.

Cerebellar LTD is associative in that it occurs only when climbing fibers and parallel fibers are activated at the same time (Figure 8.16C). The associativity arises from the combined actions of two distinct intracellular signal transduction pathways that are activated in the postsynaptic Purkinje cell due to the activity of climbing fiber and parallel fiber synapses. In the first pathway, glutamate released from the parallel fiber terminals activates two types of receptors, the AMPA-type and metabotropic glutamate receptors (see Chapter 6). Glutamate binding to the AMPA receptor results in membrane depolarization, whereas binding to the metabotropic receptor produces the second messengers inositol trisphosphate (IP_3) and diacylglycerol (DAG) (see Chapter 7). The second signal transduction pathway, initiated by climbing fiber activation, causes an influx of Ca^{2+} through voltage-gated channels and a subsequent increase in intracellular Ca^{2+} concentration. These second messengers work together to cause an amplified rise in intracellular Ca^{2+} concentration, due to IP_3 and Ca^{2+} acting together on IP_3-sensitive triggering release of Ca^{2+} from intracellular stores, and the synergistic activation of PKC by Ca^{2+} and DAG (Figure 8.16D). Thus, the associative property of cerebellar LTD appears to arise from both IP_3 receptors and PKC serving as coincidence detectors. While the downstream substrate proteins of PKC are still being determined, it appears that AMPA receptors are one of the proteins that is phosphorylated by PKC. The consequence of PKC activation is

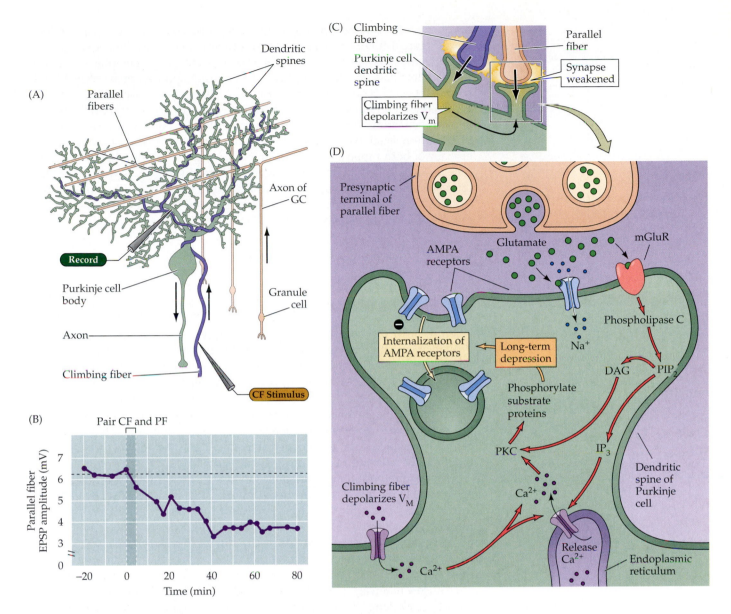

Figure 8.16   Long-term synaptic depression in the cerebellum. (A) Experimental arrangement. Synaptic responses were recorded from Purkinje cells following stimulation of parallel fibers and climbing fibers. (B) Pairing stimulation of climbing fibers (CF) and parallel fibers (PF) causes LTD that reduces the parallel fiber EPSP. (C) LTD requires depolarization of the Purkinje cell, produced by climbing fiber activation, as well as signals generated by active parallel fiber synapses. (D) Mechanism underlying cerebellar LTD. Glutamate released by parallel fibers activates both AMPA receptors and metabotropic glutamate receptors. The latter produces two second messengers, DAG and IP$_3$, which interact with Ca^{2+} that enters when climbing fiber activity opens voltage-gated Ca^{2+} channels. This leads to activation of PKC, which triggers clathrin-dependent internalization of postsynaptic AMPA receptors to weaken the parallel fiber synapse. (B after Sakurai, 1987.)

to cause an internalization of AMPA receptors via clathrin-dependent endocytosis (Figure 8.16D). This loss of AMPA receptors decreases the response of the postsynaptic Purkinje cell to glutamate release from the presynaptic terminals of the parallel fibers. Thus, in contrast to LTD in the hippocampus, cerebellar LTD requires the activity of a protein kinase, rather than a phosphatase, and does not involve Ca^{2+} entry through the NMDA type of glutamate receptor (which is not present in mature Purkinje cells). However, the net effect is the same in both cases: internalization of AMPA receptors is a common mechanism for decreased efficacy of both hippocampal and cerebellar synapses during LTD. As was the case for LTP at the hippocampal Schaffer collateral synapse, as well as for long-term synaptic plasticity in *Aplysia*, CREB appears to be required for a late phase of cerebellar LTD. It is not yet know which proteins are synthesized as a consequence of CREB activation.

Spike Timing-Dependent Plasticity

The preceding sections show that LTP and LTD are preferentially initiated by different rates of repetitive synaptic activity, with LTP requiring high-frequency activity and LTD being induced by low-frequency activity. However, it has recently been discovered that another important determinant of long-term synaptic plasticity is the temporal relationship between activity in the presynaptic and postsynaptic cells. At a given (low) frequency of synaptic activity, LTD will occur if presynaptic activity is preceded by a postsynaptic action potential, while LTP occurs if the postsynaptic action potential follows presynaptic activity (Figure 8.17A,B). The relationship between the time interval and the magnitude of the synaptic change is a very sensitive function of the time interval, with no changes observed if the presynaptic activity and postsynaptic activity are separated by 100 milliseconds or longer (Figure 8.17C).

This requirement for precise timing between presynaptic and postsynaptic activity for induction of these forms of long-lasting synaptic plasticity has led to them being called **spike timing-dependent plasticity** (**STDP**). Although the mechanisms involved are not yet understood, it appears that these properties of STDP arise from timing-dependent differences in postsynaptic Ca^{2+} signals. Specifically, if a postsynaptic action potential occurs after presynaptic activity, the resulting depolarization will relieve the block of NMDA receptors by Mg^{2+} and cause a relatively large amount of Ca^{2+} influx through postsynaptic NMDA receptors, yielding LTP. In contrast, if the postsynaptic action potential occurs *prior* to the presynaptic action potential, then the depolarization associated with the postsynaptic action potential will subside by the time that an EPSP occurs. This will reduce the amount of Ca^{2+} entry through the NMDA receptors, leading to LTD. It has been postulated that other signals, such as endocannabinoids (see Chapter 6), may also be required for LTD induction during STDP.

The requirement for a precise temporal relationship between pre- and postsynaptic activity means that STDP can perform several novel types of neuronal computation. STDP provides a means of encoding information about causality. For example, if a synapse generates a suprathreshold EPSP, the resulting postsynaptic action potential would rapidly follow presynaptic activity, and LTP would then encode the fact that the postsynaptic action potential resulted from the activity of that synapse. STDP could also serve as a mechanism for competition between synaptic inputs. Stronger inputs would be more likely to produce suprathreshold EPSPs and be reinforced by the resulting LTP, whereas weaker inputs would not generate postsynaptic action potentials that were correlated with presynaptic activity. There is evidence that STDP is important for neural

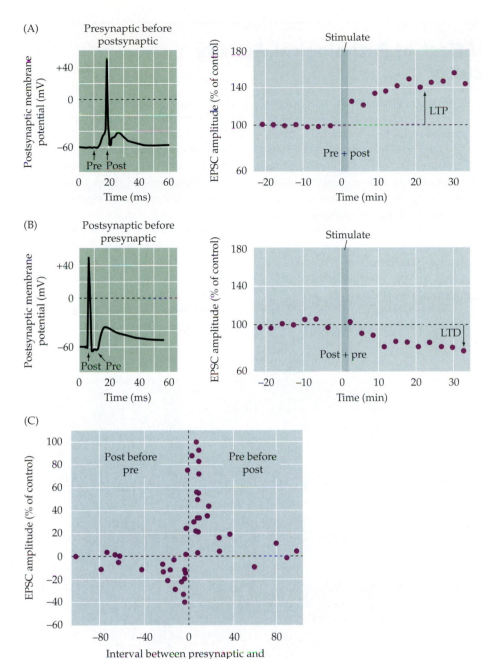

Figure 8.17 Spike timing-dependent synaptic plasticity in cultured hippocampal neurons. (A) The left-hand graph shows that stimulating a presynaptic neuron (Pre) causes an EPSP in the postsynaptic neuron; applying a subsequent stimulus to the postsynaptic neuron (Post) causes an action potential that is superimposed on the EPSP. At the right, repetitive application of the stimulus paradigm shown at left causes a long-term potentiation (LTP) of the EPSP. (B) Reversing the order of stimulation, so that the postsynaptic neuron is excited before the presynaptic neuron causes a long-term depression (LTD) of the EPSP. (C) Complex dependence of STDP on the interval between presynaptic and postsynaptic activity. If the presynaptic neuron is activated 40 ms or less before the postsynaptic neuron, LTP occurs. Conversely, if the postsynaptic neuron is activated 40 ms or less before the presynaptic neuron, LTD occurs. If the interval between the two events is longer than 40 ms, no STDP is observed. (After Bi and Poo, 1998.)

circuit function in vivo; for example, it appears to play a role in determining orientation maps in the visual system (see Chapter 12).

In summary, activity-dependent forms of synaptic plasticity cause changes in synaptic transmission that modify the functional connections within and among neural circuits. These changes in the efficacy and local geometry of synaptic connectivity can provide a basis for learning, memory, and other forms of brain plasticity. Activity-dependent changes in synaptic transmission may also be involved in some pathologies. Abnormal patterns of neuronal activity, such as those that occur in epilepsy, can stimulate abnormal changes in synaptic connections that may further increase the frequency and severity of seizures

BOX 8C Epilepsy: The Effect of Pathological Activity on Neural Circuitry

Epilepsy is a brain disorder characterized by periodic and unpredictable seizures mediated by the rhythmic firing of large groups of neurons. It seems likely that abnormal activity generates plastic changes in cortical circuitry that are critical to the pathogenesis of the disease.

The importance of neuronal plasticity in epilepsy is indicated most clearly by an animal model of seizure production called *kindling*. To induce kindling, a stimulating electrode is implanted in the brain, often in the amygdala (a component of the limbic system that makes and receives connections with the cortex, thalamus, and other limbic structures, including the hippocampus; see Chapter 29). At the beginning of such an experiment (laboratory rats or mice are typically used for such studies), weak electrical stimulation, in the form of a low-amplitude train of electrical pulses, has no discernible effect on the animal's behavior or on the pattern of electrical activity in the brain. As this weak stimulation is repeated once a day for several weeks, it begins to produce behavioral and electrical indications of seizures. By the end of the experiment, the same weak stimulus that initially had no effect now causes full-blown seizures. This phenomenon is essentially permanent; even after an interval of a year, the same weak stimulus will again trigger a seizure. Thus, repetitive weak activation produces long-lasting changes in the excitability of the brain that time cannot reverse. The word *kindling* is therefore quite appropriate: A single match can start a devastating fire.

The changes in the electrical patterns of brain activity detected in kindled animals resemble those in human epilepsy. The behavioral manifestations of epileptic seizures in human patients range from mild twitching of an extremity to loss of consciousness and uncontrollable convulsions. Although many highly accomplished people have suffered from epilepsy (Alexander the Great, Julius Caesar, Napoleon, Dostoyevsky, and van Gogh, to name a few), seizures of sufficient intensity and frequency can obviously interfere with many aspects of daily life. Moreover, uncontrolled convulsions can lead to excitotoxicity (see Box 6C). Up to 1 percent of the population is afflicted, making epilepsy one of the most common neurological problems.

Modern thinking about the causes (and possible cures) of epilepsy has focused on where seizures originate and the mechanisms that make the affected region hyperexcitable. Most of the evidence suggests that abnormal activity in small areas of the cerebral cortex (called foci) provide the triggers for a seizure that then spreads to other synaptically connected regions. For example, a seizure originating in the thumb area of

Electroencephalogram (EEG) recorded from a patient during a seizure. The traces show rhythmic activity that persisted much longer than the duration of this record. This abnormal pattern reflects the synchronous firing of large numbers of cortical neurons. (The designations are various positions of electrodes on the head; see Box 28C for additional information about EEG recordings.) (From Dyro, 1989.)

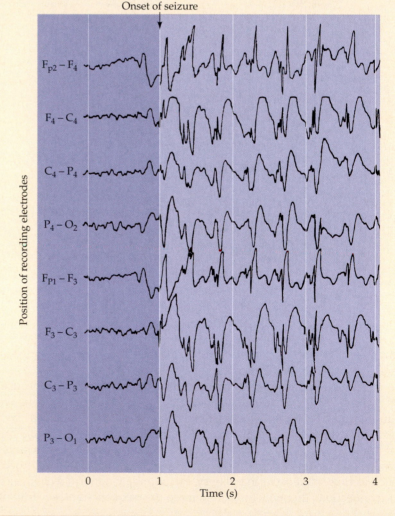

Onset of seizure

$F_{p2} - F_4$

$F_4 - C_4$

$C_4 - P_4$

$P_4 - O_2$

$F_{P1} - F_3$

$F_3 - C_3$

$C_3 - P_3$

$P_3 - O_1$

Position of recording electrodes

0   1   2   3   4

Time (s)

the right motor cortex will first be evident as uncontrolled movement of the left thumb that subsequently extends to other more proximal limb muscles, whereas a seizure originating in the visual association cortex of the right hemisphere may be heralded by complex hallucinations in the left visual field. The behavioral manifestations of seizures therefore provide important clues for the neurologist seeking to pinpoint the abnormal region of cerebral cortex.

Epileptic seizures can be caused by a variety of acquired or congenital factors, including cortical damage from trauma, stroke, tumors, congenital cortical dysgenesis (failure of the cortex to grow properly), and congenital vascular malformations. One rare form of epilepsy, Rasmussen's encephalitis, is an autoimmune disease in which the body's own immune system attacks the brain with humoral (antibodies) and cellular (lymphocytes and macrophages) agents that can destroy neurons. Some forms of

epilepsy are heritable, and more than a dozen distinct genes have been demonstrated to underlie unusual types of epilepsy. However, most forms of familial epilepsy (such as juvenile myoclonic epilepsy and petit mal epilepsy) are caused by the simultaneous inheritance of more than one mutant gene.

No effective prevention or cure exists for epilepsy. Pharmacological therapies that successfully inhibit seizures are based on two general strategies. One approach is to enhance the function of inhibitory synapses that use the neurotransmitter GABA; the other is to limit action potential firing by acting on voltage-gated Na^+ channels. Commonly used antiseizure medications include carbamazepine, phenobarbital, phenytoin (Dilantin®), and valproic acid. These agents, which must be taken daily, successfully inhibit seizures in 60–70 percent of patients. In a small fraction of patients, the epileptogenic region can be surgically excised. In extreme cases,

physicians resort to cutting the corpus callosum to prevent the spread of seizures (most of the "split-brain" subjects described in Chapter 26 were patients suffering from intractable epilepsy). One of the major reasons for controlling epileptic activity is to prevent the more permanent plastic changes that would ensue as a consequence of abnormal and excessive neural activity.

References

BERKOVIC, S. F., J. C. MULLEY, I. E. SCHEFFER AND S. PETROU (2006). Human epilepsies: Interaction of genetic and acquired factors. *Trends Neurosci.* 29: 391–397.

DYRO, F. M. (1989) *The EEG Handbook.* Boston: Little, Brown and Co.

ENGEL, J. JR. AND T. A. PEDLEY (1997) *Epilepsy: A Comprehensive Textbook.* Philadelphia: Lippincott-Raven Publishers.

MCNAMARA, J. O., Y. Z. HUANG AND A. S. LEONARD (2006) Molecular signaling mechanisms underlying epileptogenesis. *Science STKE* 356: re12.

(Box 8C). Despite the substantial advances in understanding the cellular and molecular bases of some forms of plasticity, how selective changes of synaptic strength encode memories or other complex behavioral modifications in the mammalian brain is simply not known.

Summary

Synapses exhibit many forms of plasticity that occur over a broad temporal range. At the shortest times (seconds to minutes), facilitation, augmentation, potentiation, and depression provide rapid but transient modifications in synaptic transmission. These forms of plasticity change the amount of neurotransmitter released from presynaptic terminals and are based on alterations in Ca^{2+} signaling and synaptic vesicle pools at recently active terminals. Longer lasting forms of synaptic plasticity such as LTP and LTD are also based on Ca^{2+} and other intracellular second messengers. At least some of the synaptic changes produced by these long-lasting forms of plasticity are postsynaptic, caused by changes in neurotransmitter receptor trafficking, although alterations in neurotransmitter release from the presynaptic terminal can also occur. In these more enduring forms of plasticity, protein phosphorylation and changes in gene expression greatly outlast the period of synaptic activity and can yield changes in synaptic strength that persist for hours, days, or even longer. Long-lasting synaptic plasticity can serve as a neural mechanism for many forms of brain plasticity, such as learning new behaviors or acquiring new memories.

Additional Reading

Reviews

BAILEY, C. H., E. R. KANDEL AND K. SI (2004) The persistence of long-term memory: A molecular approach to self-sustaining changes in learning-induced synaptic growth. *Neuron* 44: 49–57.

BLISS, T. V. P. AND G. L. COLLINGRIDGE (1993) A synaptic model of memory: Long-term potentiation in the hippocampus. *Nature* 361: 31–39.

BREDT, D. S. AND R A. NICOLL (2003) AMPA receptor trafficking at excitatory synapses. *Neuron* 40: 361–379.

DAN, Y. AND M. M. POO (2006) Spike timing-dependent plasticity: from synapse to perception. *Physiol Rev.* 86: 1033–1048.

HILFIKER, S., V. A. PIERIBONE, A. J. CZERNIK, H. T. KAO, G. J. AUGUSTINE AND P. GREENGARD (1999) Synapsins as regulators of neurotransmitter release. *Philos. Trans. Roy. Soc. Lond. B* 354: 269–279.

ITO, M. (2002) The molecular organization of cerebellar long-term depression. *Nature Rev. Neurosci.* 3: 896–902.

MALENKA, R. C. AND S. A. SIEGELBAUM (2001) Synaptic plasticity: Diverse targets and mechanisms for regulating synaptic efficacy. In *Synapses*. W. M. Cowan, T. C. Sudhof and C. F. Stevens (eds.). Baltimore: John Hopkins University Press, pp. 393–413.

MALINOW, R. AND R. C. MALENKA (2002) AMPA receptor trafficking and synaptic plasticity. *Annu. Rev. Neurosci.* 25: 103–126.

NICOLL, R. A. (2003) Expression mechanisms underlying long-term potentiation: A postsynaptic view. *Philos. Trans. Roy. Soc. Lond. B* 358: 721–726.

PITTENGER, C. AND E. R. KANDEL (2003) In search of general mechanisms for long-lasting plasticity: *Aplysia* and the hippocampus. *Philos. Trans. Roy. Soc. Lond. B* 358: 757–763.

ZUCKER, R. S. AND W. G. REGEHR (2002). Short-term synaptic plasticity. *Annu. Rev. Physiol.* 64: 355-405.

Important Original Papers

ABRAHAM, W. C., B. LOGAN, J. M. GREENWOOD AND M. DRAGUNOW (2002) Induction and experience-dependent consolidation of stable long-term potentiation lasting months in the hippocampus. *J. Neurosci.* 22: 9626–9634.

AHN, S., D. D. GINTY AND D. J. LINDEN (1999) A late phase of cerebellar long-term depression requires activation of CaMKIV and CREB. *Neuron* 23: 559–568.

BETZ, W. J. (1970) Depression of transmitter release at the neuromuscular junction of the frog. *J. Physiol.* (Lond.) 206: 629–644.

BI, G. Q. AND M. M. POO (1998) Synaptic modifications in cultured hippocampal neurons: dependence on spike timing, synaptic strength, and postsynaptic cell type. *J. Neurosci.* 18: 10464–10472.

BLISS, T. V. P. AND T. LOMO (1973) Long-lasting potentiation of synaptic transmission in the dentate area of the anaesthetized rabbit following stimulation of the perforant path. *J. Physiol.* 232: 331–356.

CHARLTON, M. P. AND G. D. BITTNER (1978) Presynaptic potentials and facilitation of transmitter release in the squid giant synapse. *J. Gen. Physiol.* 72: 487–511.

CHUNG, H. J., J. P. STEINBERG, R. L. HUGANIR AND D. J. LINDEN (2003) Requirement of AMPA receptor GluR2 phosphorylation for cerebellar long-term depression. *Science* 300: 1751–1755.

COLLINGRIDGE, G. L., S. J. KEHL AND H. MCLENNAN (1983) Excitatory amino acids in synaptic transmission in the Schaffer collateral-commissural pathway of the rat hippocampus. *J. Physiol.* 334: 33–46.

ENGERT, F. AND T. BONHOEFFER (1999) Dendritic spine changes associated with hippocampal long-term synaptic plasticity. *Nature* 399: 66–70.

FREY, U. AND R. G. MORRIS (1997) Synaptic tagging and long-term potentiation. *Nature* 385: 533–536.

GUSTAFSSON, B., H. WIGSTROM, W. C. ABRAHAM AND Y. Y. HUANG (1987) Long-term potentiation in the hippocampus using depolarizing current pulses as the conditioning stimulus to single volley synaptic potentials. *J. Neurosci.* 7: 774–780.

HAYASHI, Y., S. H. SHI, J. A. ESTEBAN, A. PICCINI, J. C. PONCER AND R. MALINOW (2000) Driving AMPA receptors into synapses by LTP and CaMKII: Requirement for GluR1 and PDZ domain interaction. *Science* 287: 2262–2267.

JUNGE, H. J. AND 7 OTHERS (2004) Calmodulin and munc13 form a Ca^{2+} sensor/effector complex that controls short-term synaptic plasticity. *Cell* 118: 389–401.

KATZ, B. AND R. MILEDI (1968) The role of calcium in neuromuscular facilitation. *J. Physiol.* (Lond.) 195: 481–492.

KAUER, J. A., R. C. MALENKA AND R. A. NICOLL (1988) A persistent postsynaptic modification mediates long-term potentiation in the hippocampus. *Neuron* 1: 911–917.

KONNERTH, A., J. DREESSEN AND G. J. AUGUSTINE (1992) Brief dendritic calcium signals initiate long-lasting synaptic depression in cerebellar Purkinje cells. *Proc. Natl. Acad. Sci. USA* 89: 7051–7055.

LEV-TOV, A., M. J. PINTER AND R. E. BURKE (1983) Posttetanic potentiation of group Ia EPSPs: Possible mechanisms for differential distribution among medial gastrocnemius motoneurons. *J. Neurophysiol.* 50: 379–398.

LIAO, D., N. A. HESSLER AND R. MALINOW (1995) Activation of postsynaptically silent synapses during pairing-induced LTP in CA1 region of hippocampal slice. *Nature* 375: 400–404.

MALINOW, R., H. SCHULMAN, AND R. W. TSIEN (1989) Inhibition of postsynaptic PKC or CaMKII blocks induction but not expression of LTP. *Science* 245: 862–866.

MATSUZAKI, M., N. HONKURA, G. C. ELLIS-DAVIES AND H. KASAI (2004) Structural basis of long-term potentiation in single dendritic spines. *Nature* 429: 761–766.

MIYATA, M. AND 9 OTHERS (2000) Local calcium release in dendritic spines required for long-term synaptic depression. *Neuron* 28: 233–244.

MULKEY, R. M., C. E. HERRON AND R. C. MALENKA (1993) An essential role for protein phosphatases in hippocampal long-term depression. *Science* 261: 1051–1055.

SAKURAI, M. (1987) Synaptic modification of parallel fibre-Purkinje cell transmission in *in vitro* guinea-pig cerebellar slices. *J. Physiol.* (Lond) 394: 463–480.

SILVA, A. J., R. PAYLOR, J. M. WEHNER AND S. TONEGAWA (1992) Impaired spatial learning in alpha-calcium-calmodulin kinase II mutant mice. *Science* 257: 206–211.

SWANDULLA, D., M. HANS, K. ZIPSER AND G. J. AUGUSTINE (1991) Role of residual calcium in synaptic depression and posttetanic potentiation: Fast and slow calcium signaling in nerve terminals. *Neuron* 7: 915–926.

ZAKHARENKO, S. S., L. ZABLOW AND S. A. SIEGELBAUM (2001) Visualization of changes in presynaptic function during long-term synaptic plasticity. *Nature Neurosci.* 4: 711–717.

Books

BAUDRY, M. AND J. D. DAVIS (1991) *Long-Term Potentiation: A Debate of Current Issues*. Cambridge, MA: MIT Press.

KATZ, B. (1966) *Nerve, Muscle, and Synapse*. New York: McGraw-Hill.

LANDFIELD, P. W. AND S. A. DEADWYLER (EDS.) (1988) *Long-Term Potentiation: From Biophysics to Behavior*. New York: A. R. Liss.

SQUIRE, L. R. AND E. R. KANDEL (1999) *Memory: From Mind to Molecules*. New York: Scientific American Library.

SENSATION and SENSORY PROCESSING

II

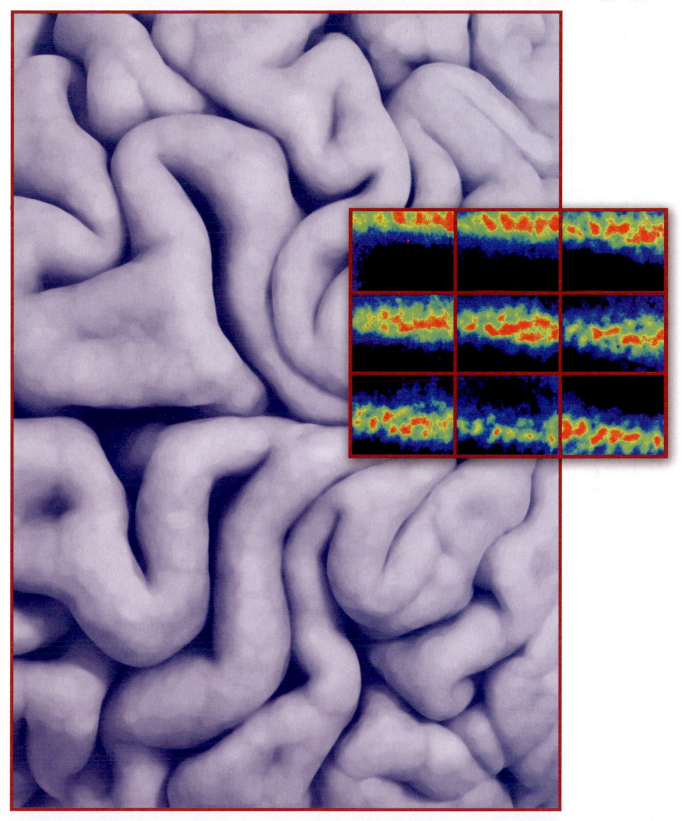

UNIT II

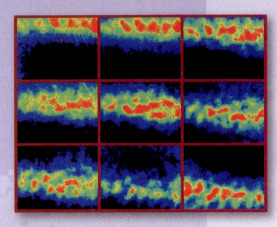

Surface view of the primary visual cortex illustrating patterns of neural activity visualized with intrinsic signal optical imaging techniques. Each panel illustrates the activity evoked by viewing a single thin vertical line. The smooth progression of the activated region from the upper left to the lower right panel illustrates the orderly mapping of visual space. The patchy appearance of the activated region in each panel reflects the columnar mapping of orientation preference. Red regions are the most active, black the least. (Courtesy of Bill Bosking, Justin Crowley, Tom Tucker, and David Fitzpatrick.)

Sensation entails the ability to transduce, encode, and ultimately perceive information generated by stimuli arising from both the external and internal environments. Much of the brain is devoted to these tasks. Although the basic senses—somatic sensation, vision, audition, vestibular sensation, and the chemical senses—are very different from one another, a few fundamental rules govern the way the nervous system deals with each of these diverse modalities. Highly specialized nerve cells called receptors convert the energy associated with mechanical forces, light, sound waves, odorant molecules, or ingested chemicals into neural signals—afferent sensory signals—that convey information about the stimulus to the brain. Afferent sensory signals activate central neurons capable of representing both the qualitative and quantitative aspects of the stimulus (what it is and how strong it is) and, in some modalities (somatic sensation, vision, and audition) the location of the stimulus in space (where it is).

The clinical evaluation of patients routinely requires an assessment of the sensory systems to infer the nature and location of potential neurological problems. Knowledge of where and how the different sensory modalities are transduced, relayed, represented, and further processed to generate appropriate behavioral responses is therefore essential to understanding and treating a wide variety of diseases. Accordingly, these chapters on the neurobiology of sensation also introduce some of the major structure/function relationships in the sensory components of the nervous system.

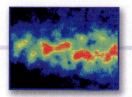

Chapter 9

The Somatic Sensory System: Touch and Proprioception

Overview

The somatic sensory system is arguably the most diverse of the sensory systems, mediating a range of sensations—touch, pressure, vibration, limb position, heat, cold, and pain—that are transduced by receptors within the skin or muscles and conveyed to a variety of central nervous system targets. Not surprisingly, this complex neurobiological machinery can be divided into functionally distinct subsystems with distinct sets of peripheral receptors and central pathways. One subsystem transmits information from cutaneous mechanoreceptors and mediates the sensation of fine touch, vibration, and pressure. Another originates in specialized receptors that are associated with muscles, tendons, and joints and is responsible for proprioception—our ability to sense the position of the limbs and other body parts in space. A third subsystem arises from receptors that supply information about painful stimuli and changes in temperature as well as coarse touch. This chapter focuses on the tactile and proprioceptive subsystems; the mechanisms responsible for sensations of pain, temperature, and coarse touch are considered in the following chapter.

Afferent Fibers Convey Somatic Sensory Information to the Central Nervous System

Somatic sensation originates from the activity of afferent nerve fibers whose peripheral processes ramify within the skin or in muscle (Figure 9.1A). The cell bodies of afferent fibers reside in a series of ganglia that lie alongside the spinal cord and the brainstem and are considered part of the peripheral nervous system. Neurons in the **dorsal root ganglia** and in the **cranial nerve ganglia** (for the body and head, respectively) are the critical links supplying central nervous system circuits with information about sensory events that occur in the periphery.

Action potentials generated in afferent fibers by events that occur in the skin or in muscle propagate along the fiber and past the location of the cell body in the ganglia until they reach the fibers' synaptic terminals, which are located in various target structures of the central nervous system (Figure 9.1B). Peripheral and central components of afferent fibers are continuous, attached to the cell body in the ganglia by a single process. For this reason, neurons in the dorsal root ganglia are often called **pseudounipolar**. Because of this configuration, conduction of electrical activity through the membrane of the cell body is not an obligatory step in conveying sensory information to central targets. Nevertheless, cell bodies of sensory afferents play a critical role in maintaining the cellular machinery that mediates transduction, conduction, and transmission by sensory afferent fibers.

The fundamental mechanism of **sensory transduction**—the process of converting the energy of a stimulus into an electrical signal—is similar in all

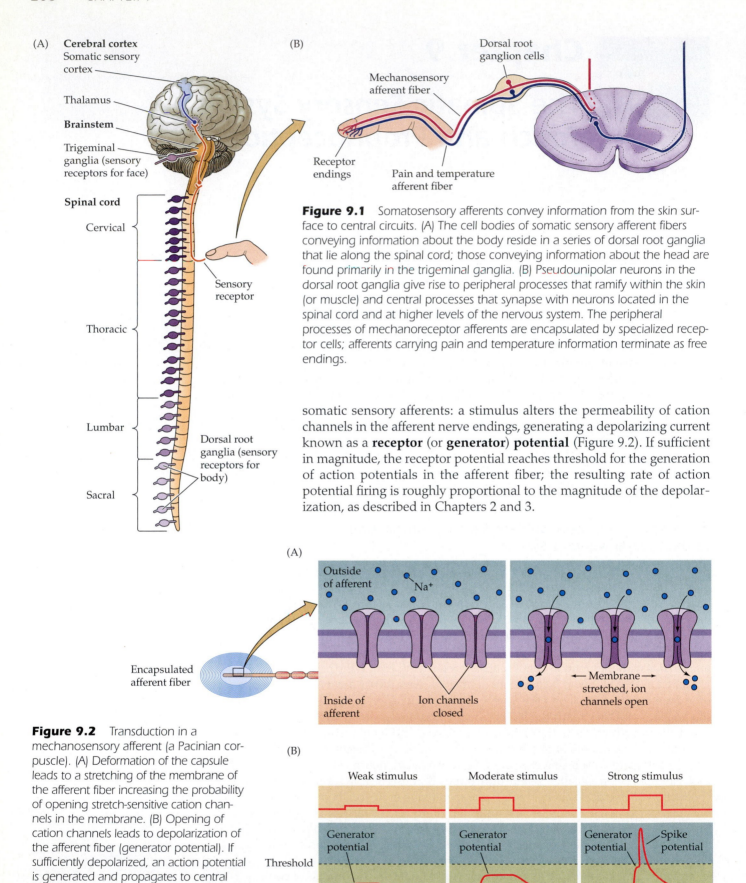

Figure 9.1 Somatosensory afferents convey information from the skin surface to central circuits. (A) The cell bodies of somatic sensory afferent fibers conveying information about the body reside in a series of dorsal root ganglia that lie along the spinal cord; those conveying information about the head are found primarily in the trigeminal ganglia. (B) Pseudounipolar neurons in the dorsal root ganglia give rise to peripheral processes that ramify within the skin (or muscle) and central processes that synapse with neurons located in the spinal cord and at higher levels of the nervous system. The peripheral processes of mechanoreceptor afferents are encapsulated by specialized receptor cells; afferents carrying pain and temperature information terminate as free endings.

somatic sensory afferents: a stimulus alters the permeability of cation channels in the afferent nerve endings, generating a depolarizing current known as a **receptor** (or **generator**) **potential** (Figure 9.2). If sufficient in magnitude, the receptor potential reaches threshold for the generation of action potentials in the afferent fiber; the resulting rate of action potential firing is roughly proportional to the magnitude of the depolarization, as described in Chapters 2 and 3.

Figure 9.2 Transduction in a mechanosensory afferent (a Pacinian corpuscle). (A) Deformation of the capsule leads to a stretching of the membrane of the afferent fiber increasing the probability of opening stretch-sensitive cation channels in the membrane. (B) Opening of cation channels leads to depolarization of the afferent fiber (generator potential). If sufficiently depolarized, an action potential is generated and propagates to central targets.

Afferent fibers are often encapsulated by specialized receptor cells (**mechanoreceptors**) that help tune the afferent fiber to particular features of somatic stimulation. Afferent fibers that lack specialized receptor cells are referred to as **free nerve endings** and are especially important in the sensation of pain (see Chapter 10). Afferents that have encapsulated endings generally have lower thresholds for action potential generation and are thus are more sensitive to sensory stimulation than free nerve endings.

BOX 9A   Dermatomes

Each dorsal root (sensory) ganglion and its associated spinal nerve arises from an iterated series of embryonic tissue masses called *somites*. This fact of development explains the overall segmental arrangement of somatic nerves (and the targets they innervate) in the adult. The territory innervated by each spinal nerve is called a **dermatome**. In humans, the cutaneous area of each dermatome has been defined in patients in whom specific dorsal roots were affected (as in herpes zoster, or "shingles") or after sur-

gical interruption (for relief of pain or other reasons). Such studies show that dermatomal maps vary among individuals. Moreover, dermatomes overlap substantially, so that injury to an individual dorsal root does not lead to complete loss of sensation in the relevant skin region. The overlap is more extensive for sensations of touch, pressure, and vibration than for pain and temperature. Thus, testing for pain sensation provides a more precise assessment of a segmental nerve injury than does testing for

responses to touch, pressure, or vibration. The segmental distribution of proprioceptors, however, does not follow the dermatomal map but is more closely allied with the pattern of muscle innervation. Despite these limitations, knowledge of dermatomes is essential in the clinical evaluation of neurological patients, particularly in determining the level of a spinal lesion.

The innervation arising from a single dorsal root ganglion and its spinal nerve is called a dermatome. The full set of sensory dermatomes is shown here for a typical adult. Knowledge of this arrangement is particularly important in defining the location of suspected spinal (and other) lesions. The numbers refer to the spinal segments by which each nerve is named. (After Rosenzweig et al., 2002.)

Somatic Sensory Afferents Exhibit Distinct Functional Properties

Somatic sensory afferents differ significantly in their response properties and these differences, taken together, define distinct classes of afferents, each of which makes unique contributions to somatic sensation. Axon diameter is one of the factors that differentiates classes of somatic sensory afferents (Table 9.1). The largest diameter sensory afferents (designated **Ia**) are those that supply the sensory receptors in the muscles. Most of the information subserving touch is conveyed by slightly smaller diameter fibers (**Aβ** afferents), and information about pain and temperature is conveyed by even smaller diameter fibers (**Aδ** and **C**). The diameter of the axon determines the action potential conduction speed and is well matched to the properties of the central circuits and the various behavior demands for which each type of sensory afferent is employed (see Chapter 16).

Another distinguishing feature of sensory afferents is the size of the **receptive field**—the area of the skin surface over which stimulation results in a significant change in the rate of action potentials (Figure 9.3A). A given region of the body surface is served by sensory afferents that vary significantly in the size of their receptive fields. The size of the receptive field is largely a function of the branching characteristics of the afferent within the skin; smaller arborizations result in smaller receptive fields. Moreover, there are systematic regional variations in the average size of afferent receptive fields that reflect the density of afferent fibers supplying the area. The receptive fields in regions with dense innervation (fingers, lips, toes) are relatively small compared to those in the forearm or back that are innervated by a smaller number of afferent fibers (Figure 9.3B).

Regional differences in receptive field size and innervation density are the major factors that limit the spatial accuracy with which tactile stimuli can be sensed. Thus measures of **two-point discrimination**—the minimum interstim-

---

TABLE 9.1 **Somatic Sensory Afferents that Link Receptors to the Central Nervous System**

Sensory function	Receptor type	Afferent axon type[a]	Axon diameter	Conduction velocity
Proprioception	Muscle spindle	Ia, II (Axon, Myelin)	13–20 μm	80–120 m/s
Touch	Merkel, Meissner, Pacinian, and Ruffini cells	Aβ	6–12 μm	35–75 m/s
Pain, temperature	Free nerve endings	Aδ	1–5 μm	5–30 m/s
Pain, temperature, itch	Free nerve endings	C	0.2–1.5 μm	0.5–2 m/s

[a]During the 1920s and 1930s, there was a virtual cottage industry classifying axons according to their conduction velocity. Three main categories were discerned, called A, B, and C. A comprises the largest and fastest axons, C the smallest and slowest. Mechanoreceptor axons generally fall into category A. The A group is further broken down into subgroups designated α (the fastest), β, and δ (the slowest). To make matters even more confusing, muscle afferent axons are usually classified into four additional groups—I (the fastest), II, III, and IV (the slowest)—with subgroups designated by lowercase roman letters!

(After Rosenzweig et al., 2005.)

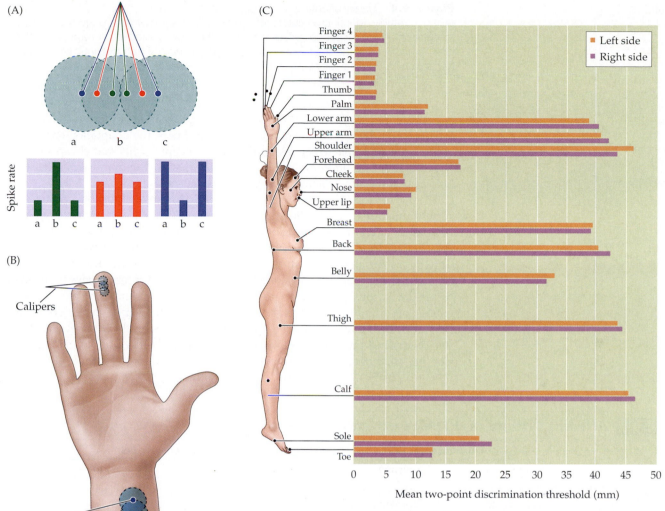

Figure 9.3   Receptive fields and two-point discrimination threshold. (A) Patterns of activity in three mechanosensory afferent fibers with overlapping receptive fields a, b, and c on the skin surface. When two-point discrimination stimuli are closely spaced (green dots and histogram), there is one focus of neural activity, with afferent b firing the most. As the stimuli are moved farther apart (red dots and histogram), the activity in afferents a and c increases and the activity in b decreases. At some separation distance (blue dots and histogram), the activity in a and c exceeds that in b to such an extent that two discrete foci of stimulation can be identified. This differential pattern of activity forms the basis for the two-point discrimination threshold. (B) The two-point discrimination threshold in the fingers is much finer than that in the wrist because of differences in the sizes of afferent receptive fields—that is, the separation distance necessary to produce two distinct foci of neural activity in the population of afferents innervating the wrist is much greater than that for the afferents innervating the fingertips. (C) Differences in the two-point discrimination threshold across the surface of the body. Somatic acuity is much higher in the fingers, toes, and face than in the arms, legs, or torso. (C after Weinstein, 1968.)

ulus distance required to perceive two simultaneously applied stimuli as distinct—vary dramatically across the skin surface (Figure 9.3C). On the fingertips, stimuli (the indentation points produced by the tips of a caliper, for example) are perceived as distinct if they are separated by roughly 2 millimeters, but the same stimuli applied to the forearm are not perceived as distinct until they are at least 40 millimeters apart!

Sensory afferents are further differentiated by the temporal dynamics of their response to sensory stimulation. Some afferents fire rapidly when a stimulus is first presented, then fall silent in the presence of continued stimulation; others generate a sustained discharge in the presence of an ongoing stimulus (Figure 9.4). **Rapidly adapting afferents** (those that become quiescent in the face of continued stimulation) are thought to be particularly effective in conveying information about changes in ongoing stimulation such as those produced by stimulus movement. In contrast, **slowly adapting afferents** are better suited

Stimulus

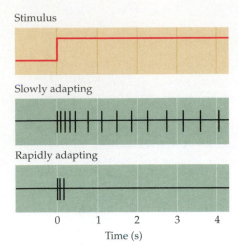

Slowly adapting

Rapidly adapting

Time (s)

Figure 9.4 Slowly adapting mechanoreceptors continue responding to a stimulus, whereas rapidly adapting receptors respond only at the onset (and often the offset) of stimulation. These functional differences allow the mechanoreceptors to provide information about both the static (via slowly adapting receptors) and dynamic (via rapidly adapting receptors) qualities of a stimulus.

to provide information about the spatial attributes of the stimulus, such as size and shape. At least for some classes of afferent fibers, the adaptation characteristics are attributable to the properties of the receptor cells that encapsulate it. Rapidly adapting afferents that are associated with Pacinian corpuscles (see below) become slowly adapting when the corpuscle is removed.

Finally, sensory afferents respond differently to the qualities of somatic sensory stimulation. Due to differences in the properties of the channels expressed in sensory afferents, and/or to the filter properties of the specialized receptor cells that encapsulate many sensory afferents, generator potentials are produced only by a restricted set of stimuli that impinge on a given afferent fiber. For example, the afferents encapsulated within specialized receptor cells in the skin respond vigorously to mechanical deformation of the skin surface, but not to changes in temperature or to the presence of mechanical forces or chemicals that are known to elicit painful sensations. The latter stimuli are especially effective in driving the responses of sensory afferents known as *nociceptors* (see Chapter 10) that terminate in the skin as free nerve endings. Further subtypes of

TABLE 9.2 Afferent Systems and Their Properties

	Small receptor field		Large receptor field	
	Merkel	**Meissner**	**Pacinian**	**Ruffini**
Location	Tip of epidermal sweat ridges	Dermal papillae (close to skin surface)	Dermis and deeper tissues	Dermis
Axon diameter	7–11 μm	6–12 μm	6–12 μm	6–12 μm
Conduction velocity	40–65 m/s	35–70 m/s	35–70 m/s	35–70 m/s
Sensory function	Form and texture perception	Motion detection; grip control	Perception of distant events through transmitted vibrations; tool use	Tangential force; hand shape; motion direction
Effective stimuli	Edges, points, corners, curvature	Skin motion	Vibration	Skin stretch
Receptive field area[a]	9 mm^2	22 mm^2	Entire finger or hand	60 mm^2
Innervation density (finger pad)	100/cm^2	150/cm^2	20/cm^2	10/cm^2
Spatial acuity	0.5 mm	3 mm	10+ mm	7+ mm
Response to sustained indentation	Sustained (slow adaptation)	None (rapid adaptation)	None (rapid adaptation)	Sustained (slow adaptation)
Frequency range	0–100 Hz	1–300 Hz	5–1000 Hz	0–? Hz
Peak sensitivity	5 Hz	50 Hz	200 Hz	0.5 Hz
Threshold for rapid indentation or vibration:				
Best	8 μm	2 μm	0.01 μm	40 μm
Mean	30 μm	6 μm	0.08 μm	300 μm

[a]Receptive field areas as measured with rapid 0.5-mm indentation.
(After K. O. Johnson, 2002.)

mechanoreceptors and nociceptors are identified on the basis of their distinct responses to somatic stimulation.

While a given sensory afferent can give rise to multiple peripheral branches, the transduction properties of all the branches of a single fiber are identical. As a result, somatic sensory afferents constitute **parallel pathways** that differ in conduction velocity, receptive field size, dynamics, and effective stimulus features. As will become apparent, these different pathways remain segregated through several stages of central processing and their activity contributes in unique ways to the extraction of somatosensory information that is necessary for the appropriate control of both goal-oriented and reflexive movements.

Mechanoreceptors Specialized to Receive Tactile Information

Our understanding of the contribution of distinct afferent pathways to cutaneous sensation is best developed for the glabrous (hairless) portions of the hand (i.e., the palm and the fingertips). These regions of the skin surface are specialized for providing a high-definition neural image of manipulated objects. Active touching, or **haptics**, involves the interpretation of complex spatiotemporal patterns of stimuli that are likely to activate many classes of mechanoreceptors. Indeed, manipulating an object with the hand can often provide enough information to identify the object, a capacity called **stereognosis**. By recording the responses of individual sensory afferents in the nerves of humans and non-human primates, it has been possible to characterize the responses of these afferents under controlled conditions and gain insights into their contribution to somatic sensation. Here we consider four distinct classes of mechanoreceptive afferents that innervate the glabrous skin of the hand (Figure 9.5 and Table 9.2).

Merkel cell afferents are slowly adapting fibers that account for about 25 percent of the mechanosensory afferents in the hand. They are especially enriched in the fingertips, and are the only afferents to sample information from receptor cells located in the epidermis. Merkel cell/neurite complexes lie in the tips of the primary epidermal ridges—extensions of the epidermis into the underlying dermis that coincide with the prominent ridges ("fingerprints") on the finger surface. The exact role of Merkel cells in somatosensory transduction remains unclear. Merkel cells are excitable cells that express voltage-sensitive calcium channels and molecules that are required for synaptic vesicle release. However, action potentials appear to arise from mechanosensitive ion channels in the membrane of the afferent fiber, suggesting that Merkel cells serve to modulate the activity of the afferents rather than as the site of transduction. Merkel cell afferents have the highest spatial resolution of all the sensory afferents—individual Merkel afferents can resolve spatial details of 0.5 mm. They are also highly sensitive to points, edges, and curvature, which makes them ideally suited for processing information about form and texture.

Meissner afferents are rapidly adapting fibers that innervate the skin even more densely than Merkel afferents, accounting for about 40 percent of the mechanosensory innervation of the human hand. Meissner corpuscles lie in the tips of the dermal papillae adjacent to the primary ridges and closest to the skin surface (see Figure 9.5). These elongated receptors are formed by a connective tissue capsule that comprises several lamellae of Schwann cells. The center of the capsule contains 2–6 afferent nerve fibers that terminate as disks between the Schwann cell laminae, a configuration thought to contribute to the transient response of these afferents to somatic stimulation. Due at least in part to their close proximity to the skin surface, Meissner afferents are more than four times as sensitive to skin deformation as Merkel afferents; however, their receptive

Figure 9.5 The skin harbors a variety of morphologically distinct mechanoreceptors. This diagram represents the smooth, hairless (also called glabrous) skin of the fingertip. The major characteristics of the various receptor types are summarized in Table 9.2. (After Johansson and Vallbo, 1983.)

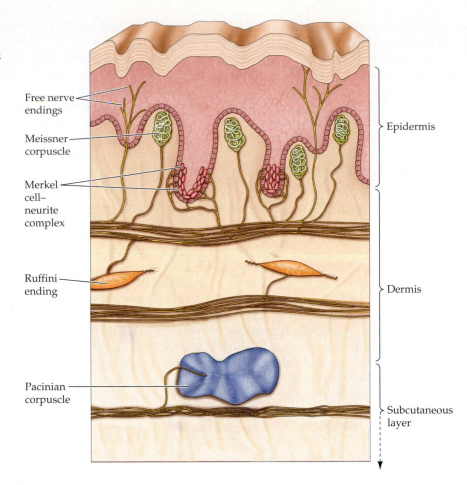

Free nerve endings

Meissner corpuscle

Merkel cell–neurite complex

Ruffini ending

Pacinian corpuscle

Epidermis

Dermis

Subcutaneous layer

fields are larger than those of Merkel afferents and thus they transmit signals with reduced spatial resolution (see Table 9.2).

Meissner corpuscles are particularly efficient in transducing information about the relatively low-frequency vibrations (3–40 Hz) that occur when textured objects are moved across the skin. Several lines of evidence suggest that the information conveyed by Meissner afferents is responsible for detecting slippage between the skin and an object held in the hand, essential feedback information for the efficient control of grip.

Pacinian afferents are rapidly adapting fibers that make up 10–15% of the mechanosensory innervation in the hand. Pacinian corpuscles are located deep in the dermis or in the subcutaneous tissue; their appearance resembles that of a small onion, with concentric layers of membranes surrounding a single afferent fiber (see Figure 9.5). This laminar capsule acts as a filter, allowing only transient disturbances at high frequencies (250–350 Hz) to activate the nerve endings. Pacinian corpuscles adapt more rapidly than Meissner corpuscles and have a lower response threshold. The most sensitive Pacinian afferents generate action potentials for displacements of the skin as small as 10 nanometers. Because they are so sensitive, the receptive fields of Pacinian afferents are often large and their boundaries are difficult to define. The properties of Pacinian afferents make them well suited to detect vibrations transmitted through objects that contact the hand or are being grasped in the hand—no doubt important for the skilled use of tools (e.g., using a wrench, cutting bread with a knife, writing).

Ruffini afferents are slowly adapting fibers and are the least understood of the cutaneous mechanoreceptors. Ruffini endings are elongated, spindle-shaped,

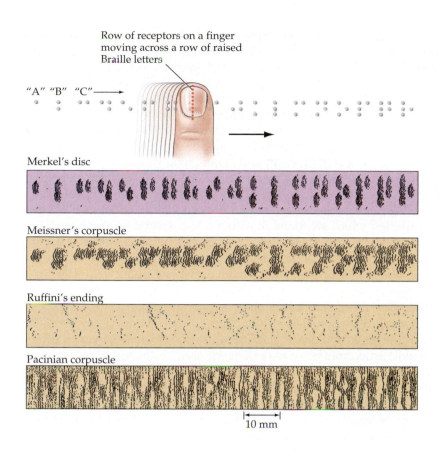

Row of receptors on a finger moving across a row of raised Braille letters

"A" "B" "C" ——→

Merkel's disc

Meissner's corpuscle

Ruffini's ending

Pacinian corpuscle

|← 10 mm →|

Figure 9.6 Simulation of the patterns of activity in different classes of mechanosensory afferents as a fingertip is moved across a row of Braille type. In the response records, each dot represents an action potential recorded from a single mechanosensory afferent fiber innervating the human finger. A horizontal line of dots in the raster plot represents the pattern of activity in the afferent as a result of moving the pattern from left to right across the finger. The position of the pattern (relative to the tip of the finger) was then displaced by a small distance, and the pattern was once again moved across the finger. Repeating this pattern multiple times produces a record that simulates the pattern of activity that would arise in a population of afferents whose receptive fields lie along a line in the finger tip (red dots). Only slowly adapting Merkel disc afferents (purple panel) provide a high-fidelity representation of the Braille pattern—that is, the individual Braille dots can be distinguished only in the pattern of Merkel afferent neural activity. (After Phillips et al., 1990.)

capsular specializations located deep in the skin, as well as in ligaments and tendons (see Figure 9.5). The long axis of the corpuscle is usually oriented parallel to the stretch lines in skin; thus, Ruffini corpuscles are particularly sensitive to the cutaneous stretching produced by digit or limb movements; they account for about 20% of the mechanoreceptors in the human hand. Although there is still some question as to their function, Ruffini corpuscles are thought to be especially responsive to internally generated stimuli, such as the movement of the fingers. Information supplied by Ruffini afferents contributes, along with muscle receptors, to providing an accurate representation of finger position and the conformation of the hand (see the section on proprioception, below).

The different kinds of information that sensory afferents convey to central structures is nicely illustrated in experiments conducted by Johnson and colleagues who compared the responses of different afferents as a fingertip was moved across a row of raised Braille letters (Figure 9.6). Clearly, all of the afferent types are activated by this stimulation, but the information supplied by each type varies enormously. The pattern of activity in the Merkel afferents is sufficient to recognize the details of the Braille pattern, and the Meissner afferents supply a slightly coarser version of this pattern. But these details are lost in the response of the Pacinian and Ruffini afferents; presumably these responses have more to do with tracking the movement and position of the finger than with the specific identity of the Braille characters.

Mechanoreceptors Specialized for Proprioception

While cutaneous mechanoreceptors provide information derived from external stimuli, another major class of receptors provides information about mechanical forces arising within the body itself, particularly from the musculoskeletal system. The purpose of these **proprioceptors** ("receptors for self") is primarily to give detailed and continuous information about the position of the limbs and other body parts in space. Low-threshold mechanoreceptors, including muscle spindles, Golgi tendon organs, and joint receptors, provide this kind of sensory information, which is essential to the accurate performance of complex movements. Information about the position and motion of the head is particularly important; in this case, proprioceptors are integrated with the highly specialized vestibular system, which is considered separately in Chapter 14. (Specialized proprioceptors also exist in the heart and major blood vessels to provide information about blood pressure, but these neurons are considered to be part of the visceral motor system; see Chapter 21.)

The most detailed knowledge about proprioception derives from studies of **muscle spindles**, which are found in all but a few striated (skeletal) muscles. Muscle spindles consist of four to eight specialized **intrafusal muscle fibers** surrounded by a capsule of connective tissue. The intrafusal fibers are distributed among and in a parallel arrangement with the **extrafusal fibers** of skeletal muscle, which are the true force-producing fibers (Figure 9.7A). Sensory afferents are coiled around the central part of the intrafusal spindle and, when the muscle is stretched, tension placed on the intrafusal fibers activates mechanically gated ion channels in the nerve endings, triggering action potentials. Innervation of the muscle spindle arises from two classes of fibers: primary and secondary endings. Primary endings arise from the largest myelinated sensory axons (**group Ia afferents**) and have rapidly adapting responses to changes in muscle length; in contrast, secondary endings (**group II afferents**) produce sustained responses to constant muscle lengths. Primary endings are thought to transmit information about limb dynamics—the velocity and direction of movement—whereas secondary endings provide information about the static position of limbs.

Changes in muscle length are not the only factors impacting the response of spindle afferents. The intrafusal fibers are themselves contractile muscle fibers and are controlled by a separate set of motor neurons (**γ motor neurons**) that lie in the ventral horn of the spinal cord. While intrafusal fibers do not add

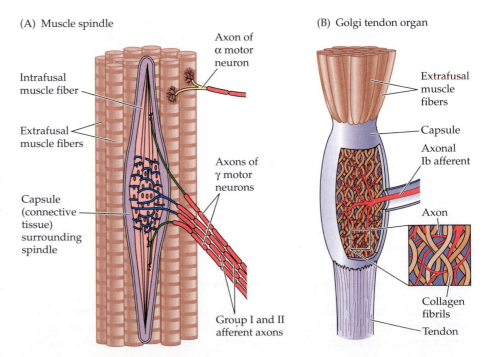

(A) Muscle spindle

(B) Golgi tendon organ

Intrafusal muscle fiber

Extrafusal muscle fibers

Capsule (connective tissue) surrounding spindle

Axon of α motor neuron

Axons of γ motor neurons

Group I and II afferent axons

Extrafusal muscle fibers

Capsule

Axonal Ib afferent

Axon

Collagen fibrils

Tendon

Figure 9.7   Proprioceptors in the musculoskeletal system provide information about the position of the limbs and other body parts in space. (A) A muscle spindle and several extrafusal muscle fibers. The specialized intrafusal muscle fibers of the spindle are surrounded by a capsule of connective tissue. (B) Golgi organs are low-threshold mechanoreceptors found in tendons; they provide information about changes in muscle tension. (A after Matthews, 1964.)

appreciably to the force of muscle contraction, changes in the tension of intra-fusal fibers have significant impact on the sensitivity of the spindle afferents to changes in muscle length. Thus, in order for central circuits to provide an accurate account of limb position and movement, the level of activity in the γ system must be taken into account. (A more detailed explanation of the interaction of the γ system and the activity of spindle afferents is given in Chapters 16 and 17.)

The density of spindles in human muscles varies. Large muscles that generate coarse movements have relatively few spindles; in contrast, extraocular muscles and the intrinsic muscles of the hand and neck are richly supplied with spindles, reflecting the importance of accurate eye movements, the need to manipulate objects with great finesse, and the continuous demand for precise positioning of the head. This relationship between receptor density and muscle size is consistent with the generalization that the sensory motor apparatus at all levels of the nervous system is much richer for the hands, head, speech organs, and other parts of the body that are used to perform especially important and demanding tasks. Spindles are lacking altogether in a few muscles, such as those of the middle ear, which do not require the kind of feedback that these receptors provide.

Whereas muscle spindles are specialized to signal changes in muscle length, low-threshold mechanoreceptors in tendons inform the central nervous system about changes in muscle tension. These mechanoreceptors, called **Golgi tendon organs**, are formed by branches of **group Ib afferents** distributed among the collagen fibers that form the tendons (Figure 9.7B). Each Golgi tendon organ is arranged in series with a small number (10–20) of extrafusal muscle fibers. Taken together, the population of Golgi tendon organs for a given muscle provides an accurate sample of the tension that exists in the muscle.

How each of these proprioceptive afferents contributes to the perception of limb position, movement, and force remains an area of active investigation. Experiments using vibrators to stimulate the spindles of specific muscles have provided compelling evidence that the activity of these afferents can give rise to vivid sensations of movement in immobilized limbs. For example, vibration of the biceps muscle leads to the illusion that the elbow is moving to an extended position, as if the biceps is being stretched. Similar illusions of motion have been evoked in postural and facial muscles. In some cases, the magnitude of the effect is so great that it produces a percept that is anatomically impossible; for example, when an extensor muscle of the wrist is vigorously vibrated, subjects report that the hand is hyperextended to the point where it is almost in contact with the back of the forearm. In all such cases, the illusion occurs only if the subject is blindfolded and cannot see the position of the limb, demonstrating that even though proprioceptive afferents alone can provide cues about limb position, under normal conditions both somatic and visual cues play important roles.

Prior to these studies, the primary source of proprioceptive information about limb position and movement was thought to arise from mechanoreceptors in and around joints. These **joint receptors** resemble many of the receptors found in the skin, including Ruffini endings and Pacinian corpuscles. However, individuals who have had artificial joint replacements were found to exhibit only minor deficits in judging the position or motion of limbs, and anesthetizing a joint such as the knee has no effect on judgments of the joint's position or movement. While they make little contribution to limb proprioception, joint receptors appear to be important for judging position of the fingers. Along with cutaneous signals from Ruffini afferents and inputs from muscle spindles that contribute to fine representation of finger position, joint receptors appear to play a protective role in signaling positions that lie near the limits of normal finger joint range of motion.

Central Pathways Conveying Tactile Information from the Body: The Dorsal Column–Medial Lemniscal System

The axons of cutaneous mechanosensory afferents enter the spinal cord through the dorsal roots and the majority ascend ipsilaterally through the **dorsal columns** (also called the **posterior funiculi**) of the cord to the lower medulla, where they synapse on neurons in the **dorsal column nuclei** (Figure 9.8A). The term column refers to the gross "columnar" appearance of these fibers as they run the length of the spinal cord. These **first-order neurons** in the pathway can have quite long axonal processes: neurons innervating the lower extremities, for example, have axons that extend through the entire length of the cord.

The dorsal columns of the spinal cord are topographically organized such that the fibers conveying information from lower limbs lie most medial and travel in a circumscribed bundle known as **fasciculus gracilis**, while those that convey information from the upper limbs, trunk, and neck lie in a more lateral bundle known as the **fasciculus cuneatus** (*fasciculus* is Latin for "bundle"). In turn, the fibers in these two tracts end in different subdivisions of the dorsal column nuclei: a medial subdivision, the **nucleus gracilis**; and a lateral subdivision, the **nucleus cuneatus**.

The **second-order neurons** in the dorsal column nuclei send their axons to the somatic sensory portion of the thalamus (see Figure 9.8A). The axons exiting from dorsal column nuclei are identified as the **internal arcuate fibers**. The internal arcuate fibers subsequently cross the midline and then form a dorsoventrally elongated tract known as the **medial lemniscus**. The word *lemniscus* means "ribbon"; the crossing of the internal arcuate fibers is called the *decussation* of the medial lemniscus, from the Roman numeral X, or *decem* (ten). In a cross-section through the medulla, such as the one shown in Figure 9.8A, the medial lemniscal axons carrying information from the lower limbs are located ventrally, whereas the axons related to the upper limbs are located dorsally. As the medial lemniscus ascends through the pons and midbrain, it rotates 90 degrees laterally, so that the fibers representing the upper body are eventually located in the medial portion of the tract and those representing the lower body are in the lateral portion. The axons of the medial lemniscus synapse with thalamic neurons located in the **ventral posterior lateral nucleus** (**VPL**).

Third-order neurons in the VPL send their axons via the **internal capsule** to terminate in the **postcentral gyrus** of the cerebral cortex, a region known as the **primary somatosensory cortex**, or **SI**. Neurons in the VPL also send axons to the **secondary somatosensory cortex** (**SII**), a smaller region that lies in the upper bank of the lateral sulcus.

Central Pathways Conveying Tactile Information from the Face: The Trigeminothalamic System

Cutaneous mechanoreceptor information from the face is conveyed centrally by a separate set of first-order neurons that are located in the **trigeminal (cranial nerve V) ganglion** (Figure 9.8B). The peripheral processes of these neurons form the three main subdivisions of the **trigeminal nerve** (the **ophthalmic**, **maxillary**, and **mandibular branches**). Each branch innervates a well-defined territory on the face and head, including the teeth and the mucosa of the oral and nasal cavities. The central processes of trigeminal ganglion cells form the sensory roots of the trigeminal nerve; they enter the brainstem at the level of the pons to terminate on neurons in the **trigeminal brainstem complex**.

The trigeminal complex has two major components: the **principal nucleus** and the **spinal nucleus** (a third component, the trigeminal mesencephalic

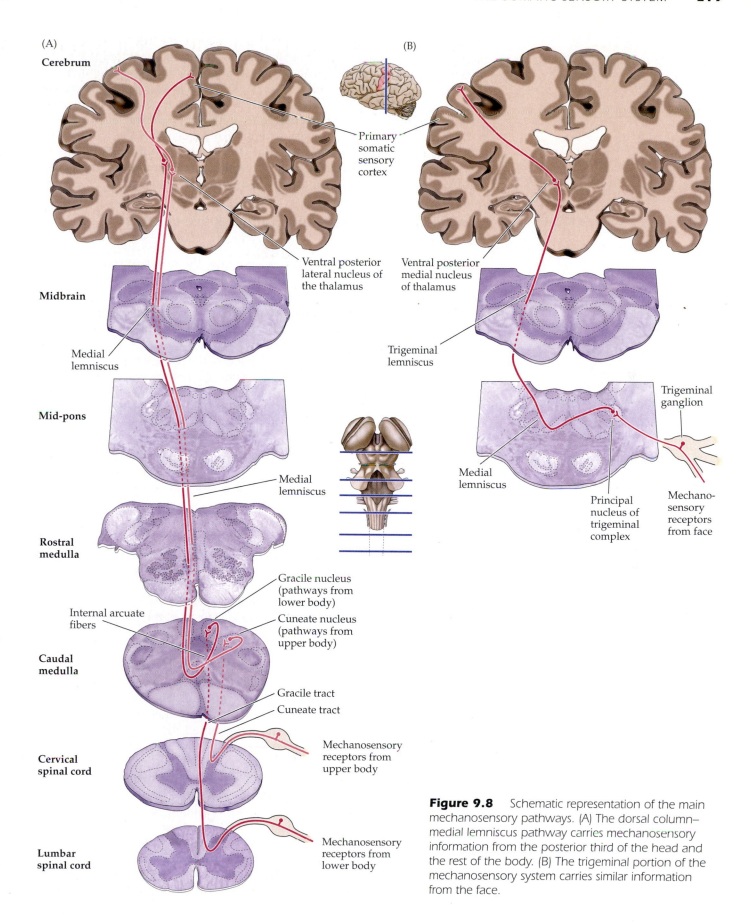

(A) Cerebrum

(B)

Primary somatic sensory cortex

Ventral posterior lateral nucleus of the thalamus

Ventral posterior medial nucleus of thalamus

Midbrain

Medial lemniscus

Trigeminal lemniscus

Mid-pons

Trigeminal ganglion

Medial lemniscus

Rostral medulla

Medial lemniscus

Principal nucleus of trigeminal complex

Mechano-sensory receptors from face

Gracile nucleus (pathways from lower body)

Internal arcuate fibers

Cuneate nucleus (pathways from upper body)

Caudal medulla

Gracile tract

Cuneate tract

Cervical spinal cord

Mechanosensory receptors from upper body

Lumbar spinal cord

Mechanosensory receptors from lower body

Figure 9.8 Schematic representation of the main mechanosensory pathways. (A) The dorsal column–medial lemniscus pathway carries mechanosensory information from the posterior third of the head and the rest of the body. (B) The trigeminal portion of the mechanosensory system carries similar information from the face.

nucleus, will be considered separately later). Most of the afferents conveying information from low-threshold cutaneous mechanoreceptors terminate in the principal nucleus. In effect, this nucleus corresponds to the dorsal column nuclei that relay mechanosensory information from the rest of the body. The spinal nucleus contains neurons that are sensitive to pain, temperature, and coarse touch and will be discussed more fully in Chapter 10. The second-order neurons of the trigeminal brainstem nuclei give off axons that cross the midline and ascend to the **ventral posterior medial (VPM) nucleus** of the thalamus by way of the **trigeminothalamic tract** (also called the **trigeminal lemniscus**). Neurons in the VPM send their axons to cortical areas SI and SII.

Central Pathways Conveying Proprioceptive Information from the Body

Like their counterparts for cutaneous sensation, the axons of proprioceptive afferents enter the spinal cord through the dorsal roots and, for much of their course, travel with the axons conveying cutaneous information. However, there are some differences in the spinal cord routes for proprioceptive pathways that reflect the important role proprioceptive information plays in the reflex regulation of motor control as well as perception.

First, when they enter the spinal cord, many of the fibers from proprioceptive afferents bifurcate into ascending and descending branches, which in turn send collateral branches to several spinal segments (Figure 9.9). Some collateral branches penetrate the dorsal horn of the spinal cord and synapse on neurons located there, as well as on neurons in the ventral horn. These synapses mediate, among other things, segmental reflexes such as the "knee-jerk" or myotatic reflex described in Chapter 1, and are further considered in Chapters 16 and 17.

Second, the information supplied by proprioceptive afferents is important not only for our ability to sense limb position; it is also essential for the functions of the cerebellum, a structure that regulates the timing of muscle contractions necessary for the performance of voluntary movements. As a consequence, proprioceptive information reaches higher cortical circuits as branches of pathways that are also targeting the cerebellum, and some of these axons run through spinal cord tracts whose names reflect their association with this structure.

The association with cerebellar pathways is especially clear for the route that conveys proprioceptive information for the lower part of the body to the dorsal column nuclei. First-order proprioceptive afferents that enter the spinal cord between the mid lumbar and thoracic levels (L2–T1) synapse on neurons in **Clark's nucleus**, a nucleus in the medial aspect of the dorsal horn (see Figure 9.9, red pathway). Those that enter below this level ascend through the dorsal column and then synapse with neurons in Clark's nucleus. Second-order neurons in Clark's nucleus send their axons into the ipsilateral posterior lateral column of the spinal cord where they travel up to the level of the medulla in the dorsal spinocerebellar tract. These axons continue into the cerebellum, but in their course, give off collaterals that synapse with neurons lying just outside the nucleus gracilus (for the present purpose, "proprioceptive neurons" of the dorsal column nuclei). Axons of these third-order neurons decussate and join the medial lemniscus, accompanying the fibers from cutaneous mechanoreceptors in their course to the VPL of the thalamus.

First-order proprioceptive afferents from the upper limbs have a course that is similar to cutaneous mechanoreceptors (see Figure 9.9, blue pathway). They enter the spinal cord and travel via the dorsal columns (fasciculus cuneatus) up to the level of the medulla where they synapse on proprioceptive neurons in the

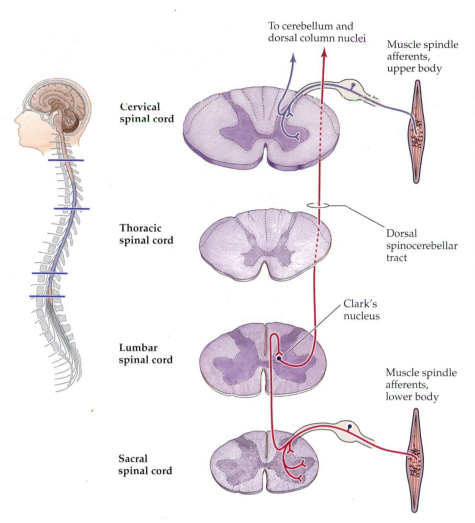

To cerebellum and
dorsal column nuclei

Cervical
spinal cord

Muscle spindle
afferents,
upper body

Thoracic
spinal cord

Dorsal
spinocerebellar
tract

Clark's
nucleus

Lumbar
spinal cord

Muscle spindle
afferents,
lower body

Sacral
spinal cord

Figure 9.9 *Proprioceptive pathways for the upper and lower body. Proprioceptive afferents for the lower part of the body synapse on neurons in the dorsal and ventral horn of the spinal cord and on neurons in Clark's nucleus. Neurons in Clark's nucleus send their axons via the dorsal spinocerebellar tract to the dorsal column nuclei and the cerebellum. Proprioceptive afferents for the upper body also have synapses in the dorsal and ventral horns, but then ascend via the dorsal column to the cerebellum and to the dorsal column nuclei. Proprioceptive target neurons in the dorsal column nuclei send their axons across the midline and ascend through the medial lemniscus to the ventral posterior nucleus (see Figure 9.8).*

dorsal column nuclei. Second-order neurons send their axons across the midline, where they join the medial lemniscus and ascend to the VPL of the thalamus.

Central Pathways Conveying Proprioceptive Information from the Face

Like the information from cutaneous mechanoreceptors, proprioceptive information from the face is conveyed through the trigeminal nerve. However, the cell bodies of the first-order proprioceptive neurons for the face have an unusual location: instead of residing in the trigeminal ganglia, they are found within the central nervous system, in the **mesencephalic trigeminal nucleus**, a well-defined cluster of neurons lying at the lateral extent of the central gray region of the midbrain. Like their counterparts in the trigeminal and dorsal root ganglia, these pseudounipolar neurons have peripheral processes that innervate muscle spindles and Golgi tendon organs associated with facial musculature (especially the jaw muscles) and central processes that include projections to brainstem nuclei responsible for reflex control of facial muscles. Although the exact route is not clear, information from proprioceptive afferents in the mesencephalic trigeminal nucleus also reaches the thalamus and is represented in somatic sensory cortex.

The Somatic Sensory Components of the Thalamus

Each of the several ascending somatic sensory pathways originating in the spinal cord and brainstem converge on the **ventral posterior complex** of the thalamus and terminate in an organized fashion (Figure 9.10). One of the organizational features of this complex instantiated by the pattern of afferent terminations is a complete and orderly somatotopic representation of the body and head. As already mentioned, the more laterally located ventral posterior lateral nucleus (VPL) receives projections from the medial lemniscus carrying somatosensory information from the body and posterior head whereas the more medially located ventral posterior medial nucleus (VPM) receives axons from the trigeminal lemniscus conveying somatosensory information from the face. In addition, inputs carrying different types of somatosensory information—for example, those that respond to different types of mechanoreceptors, to muscle spindle afferents or to Golgi tendon organs—terminate on separate populations of relay cells within the ventral posterior complex. Thus the information supplied by different somatosensory receptors remains segregated in its passage to cortical circuits.

Primary Somatic Sensory Cortex

The majority of the axons arising from neurons in the ventral posterior complex of the thalamus project to cortical neurons located in layer 4 of primary somatic sensory cortex (see Box 26A for a more detailed description of cortical lamination). The **primary somatic sensory cortex** in humans (also called **SI**) is located in the postcentral gyrus of the parietal lobe and comprises four distinct regions, or fields, known as **Brodmann's areas 3a, 3b, 1**, and **2** (Figure 9.11A). Mapping studies in humans and other primates show further that each of these four cortical areas contains a separate and complete representation of the body. In these **somatotopic maps**, the foot, leg, trunk, forelimbs, and face are represented in a medial to lateral arrangement, as shown in Figure 9.11B.

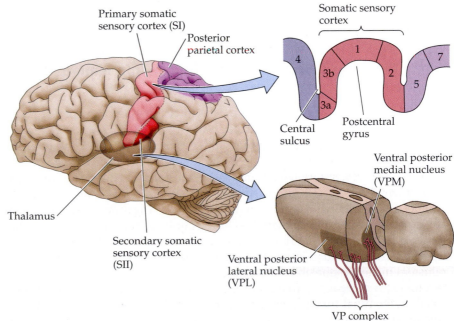

Figure 9.10   Diagram of the somatic sensory portions of the thalamus and their cortical targets in the postcentral gyrus. The ventral posterior nuclear complex comprises the VPM, which relays somatic sensory information carried by the trigeminal system from the face, and the VPL, which relays somatic sensory information from the rest of the body. The diagram at the upper right shows the organization of the primary somatosensory cortex in the postcentral gyrus, shown here in a section cutting across the gyrus from anterior to posterior. (After Brodal, 1992 and Jones et al., 1982.)

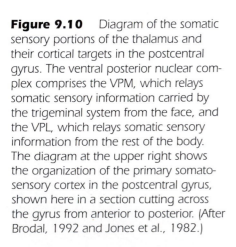

(A)

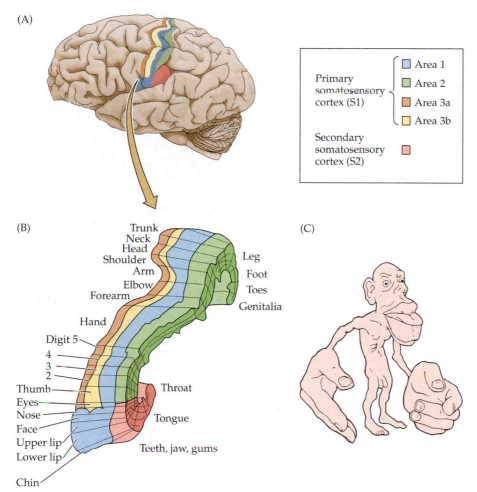

Primary somatosensory cortex (S1)	Area 1
	Area 2
	Area 3a
	Area 3b
Secondary somatosensory cortex (S2)	

(B)

Trunk
Neck
Head
Shoulder
Arm
Elbow
Forearm
Hand
Digit 5
4
3
2
Thumb
Eyes
Nose
Face
Upper lip
Lower lip
Chin

Leg
Foot
Toes
Genitalia

Throat

Tongue

Teeth, jaw, gums

(C)

Figure 9.11 Somatotopic order in the human primary somatic sensory cortex. (A) Diagram showing the region of the human cortex from which electrical activity is recorded following mechanosensory stimulation of different parts of the body. (The patients in the study were undergoing neurosurgical procedures for which such mapping was required.) Although modern imaging methods are now refining these classical data, the human somatotopic map defined in the 1930s has remained generally valid. (B) Diagram showing the somatotopic representation of body parts from medial to lateral. (C) Cartoon of the homunculus constructed on the basis of such mapping. Note that the amount of somatic sensory cortex devoted to the hands and face is much larger than the relative amount of body surface in these regions. A similar disproportion is apparent in the primary motor cortex, for much the same reasons (see Chapter 18). (After Penfield et al., 1953 and Corsi, 1991.)

A salient feature of somatotopic maps, recognized soon after their discovery, is their failure to represent the human body in its actual proportions. When neurosurgeons determined the representation of the human body in the primary sensory (and motor) cortex, the homunculus (literally, "little man") defined by such mapping procedures had a grossly enlarged face and hands compared to the torso and proximal limbs (Figure 9.11C). These anomalies arise because manipulation, facial expression, and speech are extraordinarily important for humans and require a great deal of circuitry, both central and peripheral, to govern them. Thus in humans, the cervical spinal cord is enlarged to accommodate the extra circuitry related to the hand and upper limb and, as stated earlier, the density of receptors is greater in regions such as the hands and lips.

Such distortions are also apparent when topographical maps are compared across species. In the rat brain, for example, an inordinate amount of the somatic sensory cortex is devoted to representing the large facial whiskers that are key components of the somatic sensory input for rats and mice (Box 9B), while raccoons overrepresent their paws and the platypus its bill. In short, the sensory input (or motor output) that is particularly significant to a given species gets relatively more cortical representation.

Although the topographic organization of the several somatic sensory areas is similar, the functional properties of the neurons in each region are distinct. Experiments carried out in non-human primates indicate that neurons in areas

BOX 9B Patterns of Organization within the Sensory Cortices: Brain Modules

Observations over the last 40 years have made it clear that there is an iterated substructure within the somatic sensory (and many other) cortical maps. This substructure takes the form of units called *modules*, each involving hundreds or thousands of nerve cells in repeating patterns. The advantages of these iterated patterns for brain function remain largely mysterious; for the neurobiologist, however, such iterated arrangements have provided important clues about cortical connectivity and the mechanisms by which neural activity influences brain development (see Chapters 23 and 24).

The observation that the somatic sensory cortex comprises elementary units of vertically linked cells was first noted in the 1920s by the Spanish neuroanatomist Rafael Lorente de Nó, based on his studies in the rat. The potential importance of cortical modularity remained largely unexplored until the 1950s, however, when electrophysiological experiments indicated an arrangement of repeating units in the brains of cats and, later, monkeys. Vernon Mountcastle, a neurophysiologist at Johns Hopkins, found that vertical microelectrode penetrations in the primary somatosensory cortex of these animals encountered cells that responded to the same sort of mechanical stimulus presented at the same location on the body surface. Soon

after Mountcastle's pioneering work, David Hubel and Torsten Wiesel discovered a similar arrangement in the cat primary visual cortex. These and other observations led Mountcastle to the general view that "the elementary pattern of organization of the cerebral cortex is a vertically oriented column or cylinder of cells capable of input-output functions of considerable complexity." Since these discoveries in the late 1950s and early 1960s, the view that modular circuits represent a fundamental feature of the mammalian cerebral cortex has gained wide acceptance, and many such entities have now been described in various cortical regions (see figure).

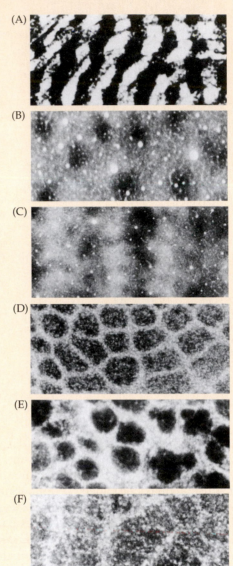

(A)
(B)
(C)
(D)
(E)
(F)

Examples of iterated, modular substructures in the mammalian brain. (A) Ocular dominance columns in layer IV in the primary visual cortex (V1) of a rhesus monkey. (B) Repeating units called "blobs" in layers II and III in V1 of a squirrel monkey. (C) Stripes in layers II and III in V2 of a squirrel monkey. (D) Barrels in layer IV in primary somatic sensory cortex of a rat. (E) Glomeruli in the olfactory bulb of a mouse. (F) Iterated units called "barreloids" in the thalamus of a rat. These and other examples indicate that modular organization is commonplace in the brain. These units are on the order of one hundred to several hundred microns across. (From Purves et al., 1992.)

3b and 1 respond primarily to cutaneous stimuli, whereas neurons in 3a respond mainly to stimulation of proprioceptors; area 2 neurons process both tactile and proprioceptive stimuli. These differences in response properties reflect, at least in part, parallel sets of inputs from functionally distinct classes of neurons in the ventral posterior complex. In addition, a rich pattern of corticocortical connections between SI areas contributes significantly to the elaboration of SI response properties. Area 3b receives the bulk of the input from the ventral posterior complex and provides a particularly dense projection to areas 1 and 2. This arrangement of connections establishes a functional hierarchy in which area 3b serves as an obligatory first step in cortical processing of somato-

This wealth of evidence for such patterned circuits has led many neuroscientists to conclude, like Mountcastle, that modules are a fundamental feature of the cerebral cortex, essential for perception, cognition, and perhaps even consciousness. Despite the prevalence of iterated modules, there are some problems with the view that modular units are universally important in cortical function. First, although modular circuits of a given class are readily seen in the brains of some species, they have not been found in the same brain regions of other, sometimes closely related, animals. Second, not all regions of the mammalian cortex are organized in a modular fashion. And third, no clear function of such modules has been discerned, much effort and speculation notwithstanding. This salient feature of the organization of the somatic sensory cortex and other cortical (and some subcortical) regions therefore remains a tantalizing puzzle.

References

HUBEL, D. H. (1988) *Eye, Brain, and Vision*. Scientific American Library. New York: W. H. Freeman.

LORENTE DE NÓ, R. (1949) The structure of the cerebral cortex. *Physiology of the Nervous System*, 3rd Ed. New York: Oxford University Press.

MOUNTCASTLE, V. B. (1957) Modality and topographic properties of single neurons of cat's somatic sensory cortex. *J. Neurophysiol.* 20: 408–434.

MOUNTCASTLE, V. B. (1998) *Perceptual Neuroscience: The Cerebral Cortex*. Cambridge: Harvard University Press.

PURVES, D., D. RIDDLE AND A. LAMANTIA (1992) Iterated patterns of brain circuitry (or how the cortex gets its spots). *Trends Neurosci.* 15: 362–369.

WOOLSEY, T. A. AND H. VAN DER LOOS (1970) The structural organization of layer IV in the somatosensory region (SI) of mouse cerebral cortex. The description of a cortical field composed of discrete cytoarchitectonic units. *Brain Res.* 17: 205–242.

sensory information (Figure 9.12). Consistent with this view, lesions of area 3b in non-human primates result in profound deficits in all forms of tactile sensations mediated by cutaneous mechanoreceptors, while lesions limited to areas 1 or 2 result in partial deficits and an inability to use tactile information to discriminate either the texture of objects (area 1 deficit) or the size and shape of objects (area 2 deficit).

Even finer parcellations of functionally distinct neuronal populations exist within single cortical areas. Based on his analysis of electrode penetrations in primary somatosensory cortex, Vernon Mountcastle was the first to suggest that neurons with similar response properties might be clustered together into functionally distinct "columns" that traversed the depth of the cortex. Subsequent studies of finely spaced electrode penetrations in area 3b provided strong evidence in support of this idea, demonstrating that neurons with responses to rapidly and slowly adapting mechanoreceptors were clustered into separate zones within the representation of a single digit (Figure 9.13). Similar clusters of neurons responding preferentially to different somatosensory afferents have

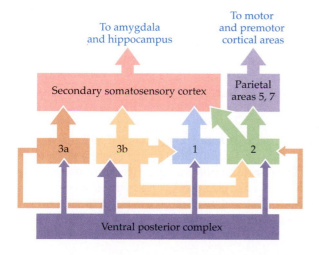

Figure 9.12   Connections within the somatosensory cortex establish functional hierarchies. Inputs from the ventral posterior nucleus of the thalamus terminate in areas 3a, 3b, 1, and 2, with the greatest density of projections in area 3b. Area 3b in turn projects heavily to areas 1 and 2, and the functions of these areas are dependent on the activity of area 3b. All subdivisions of primary somatosensory cortex project to secondary somatosensory cortex, and the functions of SII are dependent on the activity of SI.

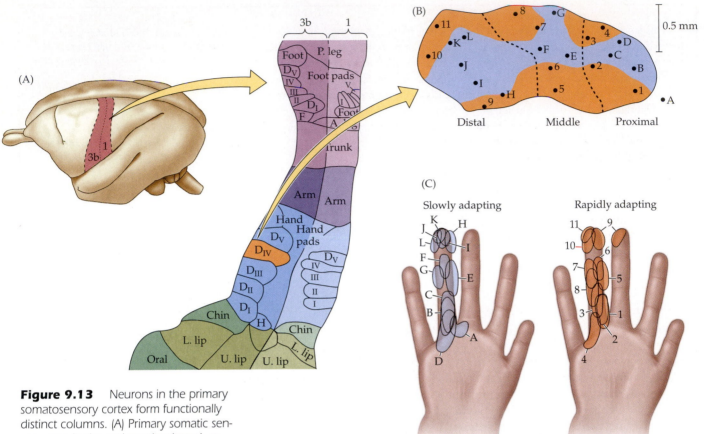

Figure 9.13 Neurons in the primary somatosensory cortex form functionally distinct columns. (A) Primary somatic sensory map in the owl monkey based, as for the human in Figure 9.11, on the electrical responsiveness of the cortex to peripheral stimulation. The enlargement on the right shows Brodmann's areas 3b and 1, which process most cutaneous mechanosensory information. The arrangement is generally similar to that determined in humans. Note the presence of regions that are devoted to the representation of individual digits. (B) Modular organization of responses within the representation of a single digit, showing the location of electrode penetrations that encountered rapidly adapting (orange) and slowly adapting (blue) responses within the representation of the fourth digit. (C) Distribution of slowly adapting and rapidly adapting receptive fields used to derive the plot in (B). Although the receptive fields of these different classes of afferents overlap on the skin surface, they are segregated within the cortical representation (A after Kaas, 1983; C after Sur et al., 1984.)

been described in areas 2 and 1. This modular organization of cortical areas is a fundamental feature of cortical organization, that is especially pronounced in visual cortical areas (see Chapter 11). While these patterns reflect specificity in the underlying patterns of thalamocortical and corticocortical connections, the functional significance of columns remains unclear (see Box 9B)

Beyond SI: Corticocortical and Descending Pathways

Somatic sensory information is distributed from the primary somatic sensory cortex to "higher order" cortical fields. One of these higher order cortical centers, the secondary somatosensory cortex (SII) lies in the upper bank of the lateral sulcus (see Figures 9.10 and 9.12). SII receives convergent projections from all subdivisions of SI, and these inputs are necessary for the function of SII; lesions of SI eliminate the somatosensory responses of SII neurons. Area SII sends projections in turn to limbic structures such as the amygdala and hippocampus (see Chapters 29 and 31). This latter pathway is believed to play an important role in tactile learning and memory.

Neurons in SI also project to parietal areas posterior to area 2, especially areas 5a and 7b. These areas receive direct projections from area 2 and, in turn, supply inputs to neurons in motor and premotor areas of the frontal lobe. This is a major route by which information derived from proprioceptive afferents signaling the current state of muscle contraction gains access to circuits that initiate voluntary movements. More generally, the projections from parietal cortex to motor cortex are critical for the integration of sensory and motor information (see Chapters 20 and 26, which consider the functions of "association" regions of the cerebral cortex).

Finally, a fundamental but often neglected feature of the somatic sensory system is the presence of massive descending projections. These pathways originate in sensory cortical fields and run to the thalamus, brainstem, and spinal cord. Indeed, descending projections from the somatic sensory cortex outnumber ascending somatic sensory pathways. Although their physiological role is not well understood, it is generally assumed (with some experimental support) that descending projections modulate the ascending flow of sensory information at the level of the thalamus and brainstem.

Plasticity in the Adult Cerebral Cortex

The analysis of maps of the body surface in primary somatic sensory cortex and the responses to altered patterns of activity in peripheral afferents has been instrumental in understanding the potential for the reorganization of cortical circuits in adults. Jon Kaas and Michael Merzenich were the first to explore this issue, by examining the impact of peripheral lesions (e.g., cutting a nerve that innervates the hand, or amputation of a digit) on the topographic maps in somatosensory cortex. Immediately after the lesion, the corresponding region of the cortex was found to be unresponsive. After a few weeks, however, the unresponsive area became responsive to stimulation of neighboring regions of the skin (Figure 9.14). For example, if the third digit was amputated, cortical neurons that formerly responded to stimulation of digit 3 now responded to stimulation of digits 2 or 4. Thus the central representation of the remaining digits had expanded to take over the cortical territory that had lost its main input. Such "functional re-mapping" also occurs in the somatic sensory nuclei in the thalamus and brainstem; indeed, some of the reorganization of cortical circuits may depend on this concurrent subcortical plasticity. This sort of adjustment in the somatic sensory system may contribute to the altered sensation of phantom limbs after amputation (see Box 10D). Similar plastic changes have been demonstrated in the visual, auditory, and motor cortices, suggesting that some ability to reorganize after peripheral deprivation or injury is a general property of the mature neocortex.

Appreciable changes in cortical representation also can occur in response to more physiological changes in sensory or motor experience. For instance, if a monkey is trained to use a specific digit for a particular task that is repeated many times, the functional representation of that digit determined by electrophysiological mapping can expand at the expense of the other digits (Figure 9.15). In fact, significant changes in receptive fields of somatic sensory neurons can be detected when a peripheral nerve is blocked temporarily by a local anesthetic. The transient loss of sensory input from a small area of skin induces a reversible reorganization of the receptive fields of both cortical and subcortical neurons. During this period, the neurons assume new receptive fields that respond to tactile stimulation of the

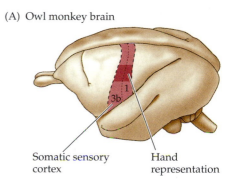

(A) Owl monkey brain

Somatic sensory cortex   Hand representation

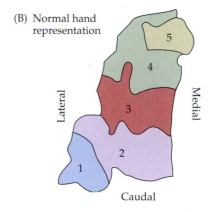

(B) Normal hand representation

Lateral   Medial

Caudal

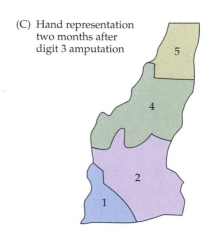

(C) Hand representation two months after digit 3 amputation

Figure 9.14 *Functional changes in the somatic sensory cortex of an owl monkey following amputation of a digit. (A) Diagram of the somatic sensory cortex in the owl monkey, showing the approximate location of the hand representation. (B) The hand representation in the animal before amputation; the numbers correspond to different digits. (C) The cortical map determined in the same animal two months after amputation of digit 3. The map has changed substantially; neurons in the area formerly responding to stimulation of digit 3 now respond to stimulation of digits 2 and 4. (After Merzenich et al., 1984.)*

Figure 9.15 *Functional expansion of a cortical representation by a repetitive behavioral task. (A) An owl monkey was trained in a task that required heavy usage of digits 2, 3, and occasionally 4. (B) The map of the digits in the primary somatic sensory cortex prior to training is shown. (C) After several months of "practice," a larger region of the cortex contained neurons activated by the digits used in the task. Note that the specific arrangements of the digit representations are somewhat different from the monkey shown in Figure 9.14, indicating the variability of the cortical representation in particular animals. (After Jenkins et al., 1990.)*

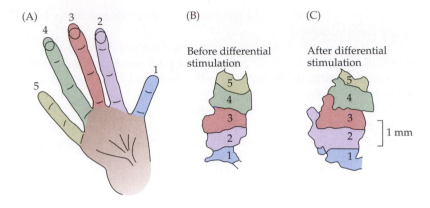

(A)

(B) Before differential stimulation

(C) After differential stimulation

1 mm

skin surrounding the anesthetized region. Once the effects of the local anesthetic subside, the receptive fields of cortical and subcortical neurons return to their usual size. The common experience of an anesthetized area of skin feeling disproportionately large—as experienced, for example, following dental anesthesia—may be a consequence of this temporary change.

Despite these intriguing observations, the mechanism, purpose, and significance of the reorganization of sensory and motor maps that occurs in adult cortex are not known. Clearly, changes in cortical circuitry occur in the adult brain; but, as centuries of clinical observations show, these changes appear to be of limited value for recovery of function following brain injury, and they may well lead to symptoms that detract from rather than enhance the quality of life following neural damage. Given their rapid and reversible character, most of these changes in cortical function probably reflect alterations in the strength of synapses already present. Indeed, finding ways to prevent or redirect the synaptic events that underlie injury-induced plasticity could reduce the long-term impact of acute brain damage.

Summary

The components of the somatic sensory system process information conveyed by mechanical stimuli that either impinge on the body surface (cutaneous mechanoreception) or are generated within the body itself (proprioception). Somatic sensory processing is performed by neurons distributed across several brain structures that are connected by both ascending and descending pathways. Transmission of afferent mechanosensory information from the periphery to the brain begins with a variety of receptor types that initiate action potentials. This activity is conveyed centrally via a chain of nerve cells, referred to as the first-, second-, and third-order neurons. First-order neurons are located in the dorsal root and cranial nerve ganglia. Second-order neurons are located in brainstem nuclei. Third-order neurons are found in the thalamus, from whence they project to the cerebral cortex. These pathways are topographically arranged throughout the system, the amount of cortical and subcortical space allocated to various body parts being proportional to the density of peripheral receptors. Studies of non-human primates show that specific cortical regions correspond to each functional submodality; area 3b, for example, processes information from low-threshold cutaneous receptors, while area 3a processes inputs from proprioceptors. Thus, at least two broad criteria operate in the organization of the somatic sensory system: modality and somatotopy. The end result of this complex interaction is the unified perceptual representation of the body and its ongoing interaction with the environment.

Additional Reading

Reviews

CHAPIN, J. K. (1987) Modulation of cutaneous sensory transmission during movement: Possible mechanisms and biological significance. In *Higher Brain Function: Recent Explorations of the Brain's Emergent Properties.* S. P. Wise (ed.). New York: John Wiley and Sons, pp. 181–209.

DARIAN-SMITH, I. (1982) Touch in primates. *Annu. Rev. Psychol.* 33: 155–194.

JOHANSSON, R. S. AND A. B. VALLBO (1983) Tactile sensory coding in the glabrous skin of the human. *Trends Neurosci.* 6: 27–32.

JOHNSON, K. O. (2002) Neural basis of haptic perception. In *Seven's Handbook of Experimental Psychology*, 3rd ed. Vol 1: *Sensation and Perception.* H. Pashler and S. Yantis (eds.). New York: Wiley, pp. 537–583.

KAAS, J. H. (1990) Somatosensory system. In *The Human Nervous System.* G Paxinos (ed.). San Diego: Academic Press, pp. 813–844.

KAAS, J. H. (1993) The functional organization of somatosensory cortex in primates. *Ann. Anat.* 175: 509–519.

KAAS, J. H. AND C. E. COLLINS (2003) The organization of somatosensory cortex in anthropoid primates. *Adv. Neurol.* 2003: 93: 57–67.

MOUNTCASTLE, V. B. (1975) The view from within: Pathways to the study of perception. *Johns Hopkins Med. J.* 136: 109–131.

NICOLELIS, M. A. AND E. E. FANSELOW (2002) Thalamocortical optimization of tactile processing according to behavioral state. *Nature Neurosci.* 5(6): 517–523.

PETERSEN, R. S., S. PANZERI AND M. E. DIAMOND (2002) Population coding in somatosensory cortex. *Curr. Opin. Neurobiol.* 12: 441–447.

WOOLSEY, C. (1958) Organization of somatic sensory and motor areas of the cerebral cortex. In *Biological and Biochemical Bases of Behavior.* H. F. Harlow and C. N. Woolsey (eds.). Madison, WI: University of Wisconsin Press, pp. 63–82.

Important Original Papers

ADRIAN, E. D. AND Y. ZOTTERMAN (1926) The impulses produced by sensory nerve endings. II. The response of a single end organ. *J. Physiol.* 61: 151–171.

FRIEDMAN, R. M., L. M. CHEN AND A. W. ROE (2004) Modality maps within primate somatosensory cortex. *Proc. Natl. Acad. Sci. USA* 101: 12724-12729.

JOHANSSON, R. S. (1978) Tactile sensibility of the human hand: Receptive field characteristics of mechanoreceptive units in the glabrous skin. *J. Physiol. (Lond.)* 281: 101–123.

JOHNSON, K. O. AND G. D. LAMB (1981) Neural mechanisms of spatial tactile discrimination: Neural patterns evoked by Braille-like dot patterns in the monkey. *J. Physiol. (Lond.)* 310: 117–144.

JONES, E. G. AND D. P. FRIEDMAN (1982) Projection pattern of functional components of thalamic ventrobasal complex on monkey somatosensory cortex. *J. Neurophysiol.* 48: 521–544.

JONES, E. G. AND T. P. S. POWELL (1969) Connexions of the somatic sensory cortex of the rhesus monkey. I. Ipsilateral connexions. *Brain* 92: 477–502.

LAMOTTE, R. H. AND M. A. SRINIVASAN (1987) Tactile discrimination of shape: Responses of rapidly adapting mechanoreceptive afferents to a step stroked across the monkey fingerpad. *J. Neurosci.* 7: 1672–1681.

LAUBACH, M., J. WESSBER AND M. A. L. NICOLELIS (2000) Cortical ensemble activity increasingly predicts behavior outcomes during learning of a motor task. *Nature* 405: 567–571.

MOORE, C. I. AND S. B. NELSON (1998) Spatiotemporal subthreshold receptive fields in the vibrissa representation of rat primary somatosensory cortex. *J. Neurophysiol.* 80: 2882– 2892.

MOORE, C. I., S. B. NELSON AND M. SUR (1999) Dynamics of neuronal processing in rat somatosensory cortex. *Trends Neurosci.* 22: 513–520.

NICOLELIS, M. A. L., L. A. BACCALA, R. C. S. LIN AND J. K. CHAPIN (1995) Sensorimotor encoding by synchronous neural ensemble activity at multiple levels of the somatosensory system. *Science* 268: 1353–1359.

SUR, M. (1980) Receptive fields of neurons in areas 3b and 1 of somatosensory cortex in monkeys. *Brain Res.* 198: 465–471.

WALL, P. D. AND W. NOORDENHOS (1977) Sensory functions which remain in man after complete transection of dorsal columns. *Brain* 100: 641–653.

ZHU, J. J. AND B. CONNORS (1999) Intrinsic firing patterns and whisker-evoked synaptic responses of neurons in the rat barrel cortex. *J. Neurophysiol.* 81: 1171–1183.

Chapter 10

Pain

Overview

A natural assumption is that the sensation of pain arises from excessive stimulation of the same receptors that generate other somatic sensations (i.e., those discussed in Chapter 9). This is not the case. Although similar in some ways to the sensory processing of ordinary mechanical stimulation, the perception of pain, called nociception, depends on specifically dedicated receptors and pathways. Since alerting the brain to the dangers implied by noxious stimuli differs substantially from informing it about innocuous somatic sensory stimuli, it makes good sense that a special subsystem be devoted to the perception of potentially threatening circumstances. The overriding importance of pain in clinical practice, as well as the many aspects of pain physiology and pharmacology that remain imperfectly understood, continue to make nociception an extremely active area of research.

Nociceptors

The relatively unspecialized nerve cell endings that initiate the sensation of pain are called **nociceptors** (from the Latin *nocere*, "to hurt"). Like other cutaneous and subcutaneous receptors, nociceptors transduce a variety of stimuli into receptor potentials, which in turn trigger afferent action potentials. Moreover, nociceptors, like other somatic sensory receptors, arise from cell bodies in dorsal root ganglia (or in the trigeminal ganglion) that send one axonal process to the periphery and the other into the spinal cord or brainstem (see Figure 9.1).

Because peripheral nociceptive axons terminate in unspecialized "free endings," it is conventional to categorize nociceptors according to the properties of the axons associated with them (see Table 9.1). As described in the previous chapter, the somatic sensory receptors responsible for the perception of innocuous mechanical stimuli are associated with myelinated axons that have relatively rapid conduction velocities. The axons associated with nociceptors, in contrast, conduct relatively slowly, being only lightly myelinated or, more commonly, unmyelinated. Accordingly, axons conveying information about pain fall into either the Aδ group of myelinated axons, which conduct at 5–30 m/s, or into the C fiber group of unmyelinated axons, which conduct at velocities generally less than 2 m/s. Thus, even though the conduction of all nociceptive information is relatively slow, there are fast and slow pain pathways.

In general, the faster-conducting Aδ nociceptors respond either to dangerously intense mechanical or to both intense mechanical and thermal stimuli. The majority of unmyelinated C-fiber nociceptors tend to respond to thermal, mechanical, and chemical stimuli, and are therefore said to be *polymodal*. In short, there are three major classes of nociceptive afferents supplying the skin:

(A)

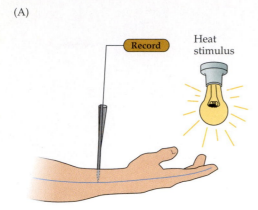

(B)

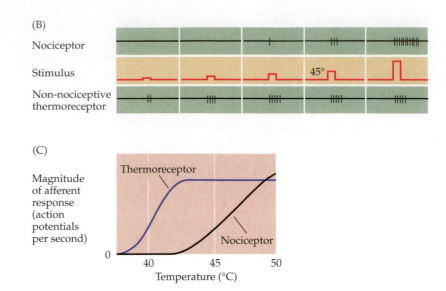

Figure 10.1 Experimental demonstration that nociception involves specialized neurons, not simply greater discharge of the neurons that respond to normal stimulus intensities. (A) Arrangement for transcutaneous nerve recording. (B) In the painful stimulus range, the axons of thermoreceptors fire action potentials at the same rate as at lower temperatures; the number and frequency of action potential discharge in the nociceptive axon, however, continues to increase. (Note that 45°C is the approximate threshold for pain.) (C) Summary of results. (After Fields, 1987.)

Aδ mechanosensitive nociceptors; Aδ mechanothermal nociceptors; and **polymodal nociceptors,** the latter being specifically associated with C fibers.

Studies carried out in both humans and experimental animals demonstrated some time ago that the rapidly conducting axons that subserve somatic sensory sensation are not involved in the transmission of pain. A typical experiment of this sort is illustrated in Figure 10.1. The peripheral axons responsive to nonpainful mechanical or thermal stimuli do not discharge at a greater rate when painful stimuli are delivered to the same region of the skin surface. The nociceptive axons, on the other hand, begin to discharge only when the strength of the stimulus (a thermal stimulus in the example in Figure 10.1) reaches high levels; at this same stimulus intensity, other thermoreceptors discharge at a rate no different from the maximum rate already achieved within the nonpainful temperature range, indicating that there are both nociceptive and nonnociceptive thermoreceptors. Equally important, direct stimulation of the large-diameter somatic sensory afferents at any frequency in humans does not produce sensations that are described as painful. In contrast, the smaller-diameter, more slowly conducting Aδ and C fibers are active when painful stimuli are delivered; and when stimulated electrically in human subjects, they produce pain.

How, then, do these different classes of nociceptors lead to the perception of pain? As mentioned, one way of determining the answer has been to stimulate different nociceptors in human volunteers while noting the sensations reported. In general, two categories of pain perception have been described: a sharp **first pain** and a more delayed, diffuse, and longer-lasting sensation that is generally called **second pain** (Figure 10.2A). Stimulation of the large, rapidly conducting Aα and Aβ axons in peripheral nerves does not elicit the sensation of pain. When the stimulus intensity is raised to a level that activates a subset of Aδ fibers, however, a tingling sensation or, if the stimulation is intense enough, a feeling of sharp pain is reported. If the stimulus intensity is increased still further, so that the small-diameter, slowly conducting C fiber axons are brought into play, then a duller, longer-lasting sensation of pain is experienced. It is also possible to selectively anesthetize C fibers and Aδ fibers; in general, these selective blocking experiments confirm that the Aδ fibers are responsible for first pain, and that C fibers are responsible for the duller, longer-lasting second pain (Figure 10.2B,C).

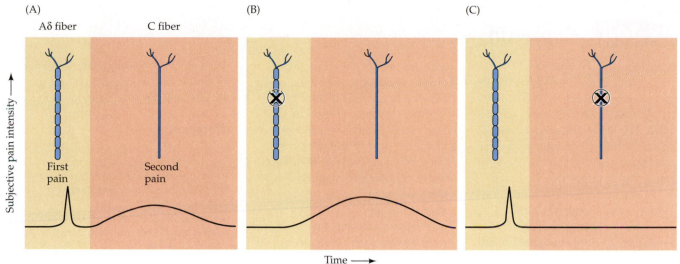

Figure 10.2 Pain can be separated into an early perception of sharp pain and a later sensation that is described as having a duller, burning quality. (A) First and second pain, as these sensations are called, are carried by different axons, as can be shown by (B) the selective blockade of the more rapidly conducting myelinated axons that carry the sensation of first pain, or (C) blockade of the more slowly conducting C fibers that carry the sensation of second pain. (After Fields, 1990.)

Transduction of Nociceptive Signals

Given the variety of stimuli (mechanical, thermal, and chemical) that can give rise to painful sensations, the transduction of nociceptive signals is a complex task. While many puzzles remain, some insights have come from the identification of specific receptors associated with nociceptive afferent endings. These receptors are sensitive to both heat and capsaicin, the ingredient in chili peppers that is responsible for the familiar tingling or burning sensation produced by spicy foods (Box 10A). The so-called vanilloid receptor (VR-1 or TRPV1) is found in C and Aδ fibers and is activated by moderate heat (45°C—a temperature that is perceived as uncomfortable) as well as by capsaicin. Another type of receptor (vanilloid-like receptor, VRL-1 or TRPV2) has a higher threshold response to heat (52°C), is not sensitive to capsaicin, and is found in Aδ fibers. Both are members of the larger family of *transient receptor potential* (TRP) channels, first identified in studies of the phototransduction pathway in fruit flies and now known to comprise a large number of receptors sensitive to different ranges of heat and cold. Structurally, TRP channels resemble voltage-gated potassium or cyclic nucleotide-gated channels, having six transmembrane domains with a pore between domains 5 and 6. Under resting conditions the pore of the channel is closed. In the open, activated state, these receptors allow an influx of sodium and calcium that initiates the generation of action potentials in the nociceptive fibers.

Since the same receptor is responsive to heat as well as capsaicin, it is not surprising that chili peppers seem "hot." A puzzle, however, is why the nervous system has evolved receptors that are sensitive to a chemical in chili peppers. As with the case of other plant compounds that selectively activate neural receptors (see the discussion of opiates below), it seems likely that TRPV1 receptors detect endogenous substances whose chemical structure resembles that of capsaicin. In fact, there is now some evidence that "endovanilloids" are produced by peripheral tissues in response to injury, and that these substances, along with other factors, contribute to the nociceptive response to injury.

Central Pain Pathways Are Distinct from Mechanosensory Pathways

Pathways responsible for pain originate with other sensory neurons in dorsal root ganglia, and, like other sensory nerve cells, the central axons of nociceptive nerve

BOX 10A   Capsaicin

Capsaicin, the principle ingredient responsible for the pungency of hot peppers, is eaten daily by over a third of the world's population. Capsaicin activates responses in a subset of nociceptive C fibers (polymodal nociceptors) by opening ligand-gated ion channels that permit the entry of Na^+ and Ca^{2+}. One of these channels, VR-1, has been cloned and has been found to be activated by capsaicin, acid, and anandamide (an endogeneous compound that also activates cannabanoid receptors), and by heating the tissue to about 43°C. It follows that anandamide and temperature are probably the endogenous activators of these channels. Mice whose VR-1 receptors have been knocked out drink capsaicin solutions as if they were water. Receptors for capsaicin have been found in polymodal nociceptors of all mam-

mals, but are not present in birds (leading to the production of squirrel-proof birdseed laced with capsaicin!).

When applied to the mucus membranes of the oral cavity, capsaicin acts as an irritant, producing protective reactions. When injected into skin, it produces a burning pain and elicits hyperalgesia to thermal and mechanical stimuli. Repeated applications of capsaicin also desensitize pain fibers and prevent neuromodulators such as substance P, VIP, and somatostatin from being released by peripheral and central nerve terminals. Consequently, capsaicin is used clinically as an analgesic and anti-inflammatory agent; it is usually applied topically in a cream (0.075%) to relieve the pain associated with arthritis, postherpetic neuralgia, mastectomy, and trigeminal neuralgia. Thus, this remarkable chemical

irritant not only gives gustatory pleasure on an enormous scale, but is also a useful pain reliever.

References

CATERINA, M. J., M. A. SCHUMACHER, M. TOMINAGA, T. A ROSEN, J. D. LEVINE AND D. JULIUS (1997) The capsaicin receptor: A heat-activated ion channel in the pain pathway. *Nature* 389: 816–766.

CATERINA, M. J. AND 8 OTHERS (2000) Impaired nociception and pain sensation in mice lacking the capsaicin receptor. *Science* 288: 306–313.

SZALLASI, A. AND P. M. BLUMBERG (1999) Vanilloid (capsaicin) receptors and mechanisms. *Pharm. Reviews* 51: 159–212.

TOMINAGA, M. AND 8 OTHERS (1998) The cloned capsaicin receptor integrates multiple pain-producing stimuli. *Neuron* 21: 531–543.

ZYGMUNT, P. M. AND 7 OTHERS (1999) Vanilloid receptors on sensory nerves mediate the vasodilator action of anandamide. *Nature* 400: 452–457.

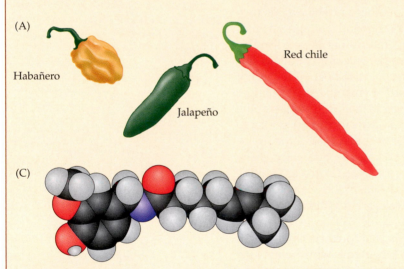

(A) Habañero Jalapeño Red chile

(B) Capsaicin
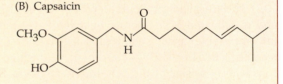

(D)
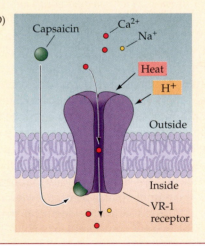

(A) Some popular peppers that contain capsaicin. (B) The chemical structure of capsaicin. (C) The capsaicin molecule. (D) Schematic of the VR-1/capsaicin receptor channel. This channel can be activated by capsaicin intracellularly, or by heat or protons (H^+) at the cell surface.

cells enter the spinal cord via the dorsal roots (Figure 10.3A). When these centrally projecting axons reach the dorsal horn of the spinal cord, they branch into ascending and descending collaterals, forming the **dorsolateral tract of Lissauer** (named after the German neurologist who first described this pathway in the late nineteenth century). Axons in Lissauer's tract typically run up and down for one

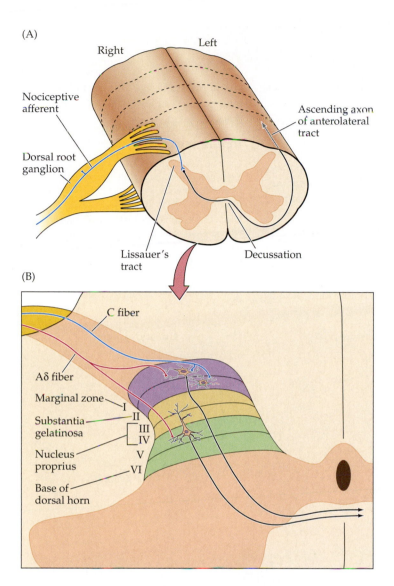

(A)

Right Left

Nociceptive afferent

Dorsal root ganglion

Ascending axon of anterolateral tract

Lissauer's tract

Decussation

(B)

C fiber

Aδ fiber

Marginal zone —I

Substantia —II
gelatinosa   III
              IV

Nucleus —V
proprius —VI

Base of dorsal horn

Figure 10.3 The anterolateral system. (A) Primary afferents in the dorsal root ganglia send their axons via the dorsal roots to terminate in the dorsal horn of the spinal cord. Afferents branch and course for several segments up and down the spinal cord in Lissauer's tract, giving rise to collateral branches that terminate in the dorsal horn. Second-order neurons in the dorsal horn send their axons (black) across the midline to ascend to higher levels in the anterolateral column of the spinal cord. (B) C-fiber afferents terminate in Rexed's laminae 1 and 2 of the dorsal horn, while Aδ fibers terminate in layers 1 and 5. The axons of second-order neurons in laminae 1 and 5 cross the midline and ascend to higher centers.

or two spinal cord segments before they penetrate the gray matter of the dorsal horn. Once within the dorsal horn, the axons give off branches that contact second-order neurons located in Rexed's laminae 1 and 5. (These laminae are the descriptive divisions of the spinal gray matter in cross section, again named after the neuroanatomist who described these details in the 1950s; see the Appendix.)

The axons of these second-order neurons in the dorsal horn of the spinal cord cross the midline and ascend to the brainstem and thalamus in the anterolateral (also called ventrolateral) quadrant of the contralateral half of the spinal cord (Figure 10.3B). For this reason, the neural pathway that conveys pain and temperature information to higher centers is often referred to as the **anterolateral system**, to distinguish it from the dorsal column–medial lemniscus system that conveys mechanosensory information (see Chapter 9).

The sites where axons conveying information for these two systems cross the midline are quite different, and this difference provides a clinically relevant sign that is useful for defining the locus of a spinal cord lesion. Axons of the first-order neurons for the dorsal column–medial lemniscal system enter the spinal cord, turn and ascend in the ipsilateral dorsal columns all the way to the

BOX 10B Referred Pain

Surprisingly, there are few, if any, neurons in the dorsal horn of the spinal cord that are specialized solely for the transmission of *visceral* pain. Obviously, we recognize such pain, but it is conveyed centrally via dorsal horn neurons that are also concerned with *cutaneous* pain. As a result of this economical arrangement, the disorder of an internal organ is sometimes perceived as cutaneous pain. A patient may therefore present to the physician with the complaint of pain at a site other than its actual source, a potentially confusing phenomenon called

referred pain. The most common clinical example is anginal pain (pain arising from heart muscle that is not being adequately perfused with blood) referred to the upper chest wall, with radiation into the left arm and hand. Other important examples are gallbladder pain referred to the scapular region, esophogeal pain referred to the chest wall, ureteral pain (e.g., from passing a kidney stone) referred to the lower abdominal wall, bladder pain referred to the perineum, and the pain from an inflamed appendix referred to the anterior abdominal wall

around the umbilicus. Understanding referred pain can lead to an astute diagnosis that might otherwise be missed.

References

CAPPS, J. A. AND G. H. COLEMAN (1932) *An Experimental and Clinical Study of Pain in the Pleura, Pericardium, and Peritoneum*. New York: Macmillan.

HEAD, H. (1893) On disturbances of sensation with special reference to the pain of visceral disease. *Brain* 16: 1–32.

KELLGREW, J. H. (1939–1942) On the distribution of pain arising from deep somatic structures with charts of segmental pain areas. *Clin. Sci.* 4: 35–46.

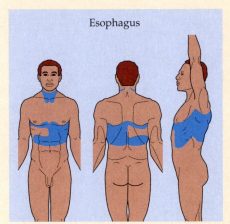
Esophagus

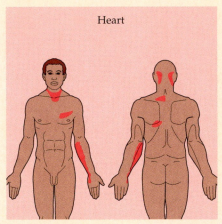
Heart

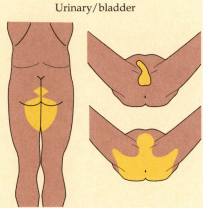
Urinary/bladder

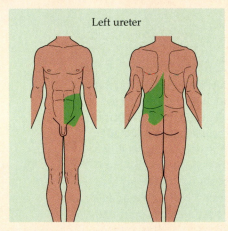
Left ureter

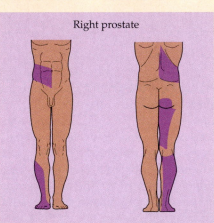
Right prostate

Examples of pain arising from a visceral disorder referred to a cutaneous region (color).

medulla, where they synapse on neurons in the dorsal column nuclei (Figure 10.4A). The axons of neurons in the dorsal column nuclei then cross the midline and ascend to the contralateral thalamus. In contrast, the crossing point for information conveyed by the anterolateral system lies within the spinal cord. First-order neurons contributing to the anterolateral system terminate in the

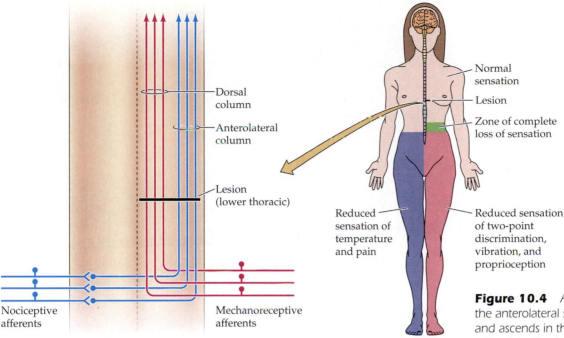

Figure 10.4   As diagrammed here, the anterolateral system (blue) crosses and ascends in the contralateral anterolateral column of the spinal cord, while the dorsal column-medial leminiscal system (magenta) ascends in the ipsilateral dorsal column. A lesion restricted to the left half of the spinal cord results in dissociated sensory loss and mechanosensory deficits on the left half of the body, with pain and temperature deficits experienced on the right.

dorsal horn, and second-order neurons in the dorsal horn send their axons across the midline and ascend on the contralateral side of the cord (in the anterolateral column) to their targets in the thalamus and brainstem.

Because of this anatomical difference in the site of decussation, a unilateral spinal cord lesion results in dorsal column symptoms (loss of touch, pressure, vibration, and proprioception) on the side of the body *ipsilateral* to the lesion, and anterolateral symptoms (deficits of pain and temperature) on the *contralateral* side of the body (Figure 10.4B). Because the deficits are due to the interruption of fibers ascending from lower levels of the cord, the deficits generally include all regions of the body (on either the contralateral or ipsilateral side) that are innervated by spinal cord segments that lie below the level of the lesion. This pattern of **dissociated sensory loss** (contralateral pain and temperature, ipsilateral touch and pressure) is a signature of spinal cord lesions and, together with local dermatomal signs (see Box 9A), can be used to define the level of the lesion.

Parallel Pain Pathways

Second-order fibers in the anterolateral system project to a number of different structures in the brainstem and forebrain, making it clear that pain is processed by a diverse and distributed network of neurons. While the full significance of this complex pattern of connections remains unclear, these central destinations are likely to mediate different aspects of the sensory and behavioral response to a painful stimulus.

One component of this system mediates the **sensory discriminative** aspects of pain: the location, intensity, and quality of the noxious stimulation. These aspects of pain are thought to depend on information that is relayed through the ventral posterior lateral nucleus (VPL) to neurons in the primary and secondary somatosensory cortex (Figures 10.5 and 10.6A). (The pathway for relay of information from the face to the ventral posterior medial nucleus, or VPM, is considered in the next section). Although axons from the anterolateral system overlap those from the dorsal column system in the ventral posterior nuclei, these axons

Figure 10.5 The anterolateral system supplies information to different parts of the brainstem and forebrain that contribute to different aspects of the experience of pain: those that are responsible for the sensory discrimination of pain, and those that are responsible for the affective and motivational responses to pain. See text for details.

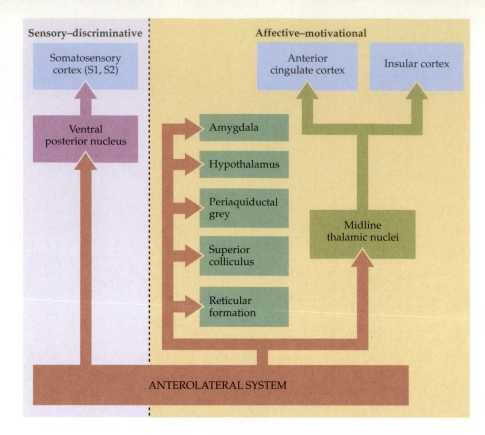

contact different classes of relay neurons, so that nociceptive information remains segregated up to the level of cortical circuits. Consistent with mediating the discriminative aspects of pain, electrophysiological recordings from nociceptive neurons in S1 show that these neurons have small, localized receptive fields—properties commensurate with behavioral measures of pain localization.

Other parts of the system convey information about the **affective–motivational** aspects of pain—the unpleasant feeling, the fear and anxiety, and the autonomic activation that accompany exposure to a noxious stimulus (the classic "fight-or-flight" reaction; see Chapter 21). Targets of these projections include several subdivisions of the reticular formation, the deep layers of the superior colliculus, the central gray, the hypothalamus, and the amygdala. In addition, a distinct set of thalamic nuclei that lie medial to the ventral posterior nucleus, which we group together here as the midline thalamic nuclei (see Figure 10.5), are thought to play an important role in transmitting nociceptive signals to the anterior cingulate cortex, and to the insula (the region of cortex that lies on the medial wall of the lateral fissure).

The view that the sensory–discriminative and affective–motivational aspects of pain are mediated by different brain regions is supported by evidence from functional imaging studies in humans. The presentation of a painful stimulus results in the activation of both primary somatosensory cortex and anterior cingulate cortex; however, by using hypnotic suggestion to selectively increase or decrease the unpleasantness of a painful stimulus, it has been possible to tease apart the neural response to changes in the intensity of a painful stimulus versus changes in its unpleasantness. Changes in intensity are accompanied by changes in the activity of neurons in somatosensory cortex, with little change in the activity of cingulate cortex; whereas changes in unpleasantness are highly correlated with changes in the activity of neurons in cingulate cortex.

From this description, it should be evident that the full experience of pain involves the cooperative action of an extensive network of brain regions whose properties are only beginning to be understood (Box 10C). The cortical represen-

BOX 10C A Dorsal Column Pathway for Visceral Pain

Chapters 9 and 10 have presented a framework for considering the central neural pathways that convey innocuous mechanosensory signals and painful signals from cutaneous and deep somatic sources. Considering just the signals derived from the body below the head, discriminitive mechanosensory and proprioceptive information travels to the ventral posterior thalamus via the dorsal column–medial lemniscal system (see Figure 9.6A), while nociceptive information travels to the same (and additional) thalamic relays via the anterolateral systems (see Figure 10.3A). But how do painful signals that arise in the visceral organs of the pelvis, abdomen, and thorax enter the central nervous system and ultimately reach consciousness?

The answer is via a newly discovered component of the dorsal column–medial lemniscal pathway that conveys visceral nociception. Although Chapter 21 will present more information on the systems that receive and process visceral sensory information, at this juncture it is worth considering this component of the pain pathways and how this particular pathway has begun to impact clinical medicine.

Primary visceral afferents from the pelvic and abdominal viscera enter the spinal cord and synapse on second-order neurons in the dorsal horn of the lumbar-sacral spinal cord. As discussed in Box 10A and Chapter 21, some of these second-order neurons are cells that give rise to the anterolateral systems and contribute to referred visceral pain patterns. However, other neurons—perhaps primarily those that give rise to nociceptive signals—synapse upon neurons in the intermediate gray region of the spinal cord near the central canal. These neurons, in turn, send their axons not through the anterolateral white matter of the spinal cord (as might be expected for a pain pathway) but through the dorsal columns in a position

very near the midline (see Figure A). Similarly, second-order neurons in the thoracic spinal cord that convey nociceptive signals from thoracic viscera send their axons rostrally through the dorsal columns along the dorsal intermediate septum, near the division of the gracile and cuneate fasciculi. These second order axons then synapse in the dorsal column nuclei of the caudal medulla, where neurons give rise to arcuate fibers that form the contralateral medial lemniscus and eventually synapse on thalamocortical projection neurons in the ventral-posterior thalamus.

This dorsal column visceral sensory projection now appears to be the principal pathway by which painful sensations arising in the viscera are detected and discriminated. Several observations support this conclusion: (1) neurons in the ventral posterior lateral nucleus, nucleus gracilis, and near the central canal of the spinal cord all respond to noxious visceral stimulation; (2) responses of neurons in the ventral posterior lateral nucleus and nucleus gracilis to such stimulation are greatly reduced by spinal lesions of the dorsal columns (Figure B), but not lesions of the anterolateral white matter; and (3) infusion of drugs that block nociceptive synaptic transmission into the intermediate gray region of the sacral spinal cord blocks the responses of neurons in the nucleus gracilis to noxious visceral stimulation, but not to innocuous cutaneous stimulation.

The discovery of this visceral sensory component in the dorsal column–medial lemniscal system has helped to explain why surgical transection of the axons that run in the medial part of the dorsal columns (a procedure termed *midline myelotomy*) generates significant relief from the debilitating pain that can result from visceral cancers in the abdomen and pelvis. Although the initial development of this surgical procedure preceded the elucidation of this visceral pain path-

way, these new discoveries have renewed interest in midline myelotomy as a palliative neurosurgical intervention for cancer patients whose pain is otherwise unmanageable. Indeed, precise knowledge of the visceral sensory pathway in the dorsal columns has led to further refinements that permit a minimally invasive ("punctate") surgical procedure that attempts to interrupt the second-order axons of this pathway within just a single spinal segment (typically, a mid- or lower-thoracic level; Figure C). In so doing, this procedure offers some hope to patients who struggle to maintain a reasonable quality of life in extraordinarily difficult circumstances.

References

AL-CHAER, E. D., N. B. LAWAND, K. N. WESTLUND AND W. D. WILLIS (1996) Visceral nociceptive input into the ventral posterolateral nucleus of the thalamus: a new function for the dorsal column pathway. *J. Neurophys.* 76: 2661–2674.

AL-CHAER, E. D., N. B. LAWAND, K. N. WESTLUND AND W. D. WILLIS (1996) Pelvic visceral input into the nucleus gracilis is largely mediated by the postsynaptic dorsal column pathway. *J. Neurophys.* 76: 2675–2690.

BECKER, R., S. GATSCHER, U. SURE AND H. BERTALANFFY (2001) The punctate midline myelotomy concept for visceral cancer pain control – case reort and review of the literature. *Acta Neurochir.* [Suppl.] 79: 77–78.

HITCHCOCK, E. R. (1970) Stereotactic cervical myelotomy. *J. Neurol. Neurosurg. Psychiatry* 33: 224–230.

KIM, Y. S. AND S. J. KWON (2000) High thoracic midline dorsal column myelotomy for severe visceral pain due to advanced stomach cancer. *Neurosurgery* 46:85–90.

NAUTA, H. AND 8 OTHERS (2000) Punctate midline myelotomy for the relief of visceral cancer pain. *J. Neurosurg.* (*Spine 2*) 92: 125–130.

WILLIS, W. D., E. D. AL-CHAER, M. J. QUAST AND K. N. WESTLUND (1999) A visceral pain pathway in the dorsal column of the spinal cord. *Proc. Natl. Acad. Sci. USA* 96: 7675–76710.

(Continued on next page)

BOX 10C *(Continued)*

(A)

Cerebrum

(B) Sham lesion   Dorsal column lesion

Before surgery

4 months after surgery

Ventral posterior nuclear complex of thalamus

Insular cortex

(C)

Needle

Dorsal columns

Dorsal horn

Midbrain

Gastrointestinal tract

Gracile nucleus

Cuneate nucleus

Medial lemniscus

Medulla

Dorsal root ganglion cells

Spinal cord

(A) A visceral pain pathway in the dorsal column–medial lemniscal system. For simplicity, only the pathways that mediate visceral pain from the pelvis and lower abdomen are illustrated. The mechanosensory component of this system for the discrimination of tactile stimuli and the anterolateral system for the detection of painful and thermal cutaneous stimuli are also shown for comparison (see also Figures 8.6A and 10.3A). (B) Empirical evidence supporting the existence of the visceral pain pathway shown in (A). Increased neural activity was observed with functional MRI techniques in the thalamus of monkeys that were subjected to noxious distention of the colon and rectum, indicating the processing of visceral pain. This activity was abolished by lesion of the dorsal columns at T10, but not by "sham" surgery. (C) Top, one method of punctate midline myelotomy for the relief of severe visceral pain. Bottom, myelin-stained section of the thoracic spinal cord (T10) from a patient who underwent midline myelotomy for the treatment of colon cancer pain that was not controlled by analgesics. After surgery, the patient experienced relief from pain during the remaining three months of his life. (B from Willis et al., 1999; C from Hirshberg et al., 1996; drawing after Nauta et al., 1997.)

tation of pain is the least well documented aspect of the central pathways for nociception, and further studies will be needed to elucidate the contribution of regions outside the somatosensory areas of the parietal lobe. Nevertheless, a prominent role for these areas in the perception of pain is suggested by the fact that ablations of the relevant regions of the parietal cortex do not generally alleviate chronic pain (although they impair contralateral mechanosensory perception, as expected).

Pain and Temperature Pathways for the Face

Information about noxious and thermal stimulation of the face originates from first-order neurons that are located in the trigeminal ganglion and from ganglia associated with cranial nerves VII, IX, and X (Figure 10.6B). After entering the pons, these small myelinated and unmyelinated trigeminal fibers descend to the medulla, forming the spinal trigeminal tract (or spinal tract of cranial nerve V) and terminate in two subdivisions of the spinal trigeminal complex: the pars interpolaris and pars caudalis. Axons from the second-order neurons in these two trigeminal nuclei cross the midline and terminate in a variety of targets in the brainstem and thalamus. Like their counterparts in the dorsal horn of the spinal cord, these targets can be grouped into those that mediate the discriminative aspects of pain, and those that mediate the affective/motivational aspects. The discriminative aspects of facial pain are thought to be mediated by projections to the contralateral ventral posterior medial nucleus (via the trigeminothalamic tract) and projections from the VPM to primary and secondary somatosensory cortex. The affective/motivational aspects are mediated by connections to various targets in the reticular formation and midbrain, as well as the midline nuclei of the thalamus, which supply the cingulate and insular regions of cortex.

Other Modalities Mediated by the Anterolateral System

While the anterolateral system plays a critical role in mediating nociception, it is also responsible for transmitting a variety of other innocuous information to higher centers. For example, in the absence of the dorsal column system, the anterolateral system appears to be capable of mediating what is commonly called "nondiscriminative touch" a form of tactile sensitivity that lacks the fine spatial resolution that can only be supplied by the dorsal column system. Thus following damage to the dorsal column–medial lemniscal system, a crude form of tactile sensation remains, one in which two point discrimination thresholds are increased, and the ability to identify objects by touch alone (stereognosis) is markedly impaired.

As already alluded to, the anterolateral system is responsible for mediating innocuous temperature sensation. The sensation of warmth and cold is thought to be subserved by separate sets of primary afferents: warm fibers that respond with increasing spike discharge rates to increases in temperature, and cold fibers that respond with increasing spike discharge to decreases in temperature. Neither of these afferents responds to mechanical stimulation, and they are distinct from afferents that respond to temperatures that are considered painful (noxious heat, above 42°C; or noxious cold, below 17°C). The recent identification of TRP channels with sensitivity to temperatures in the innocuous range—TRPV3 and TRPV4 that respond to warm temperatures and TRPM8 that responds to cold temperatures—raises the possibility of labeled lines for the transmission of warmth and cold, beginning at the level of transduction and continuing within central pathways. Consistent with this idea, the information

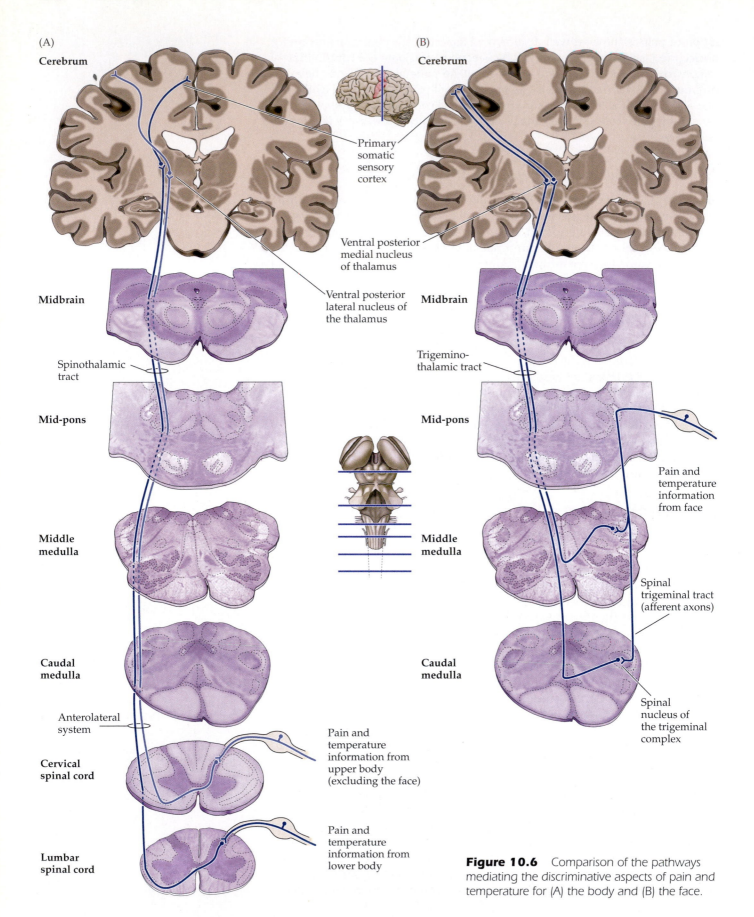

(A)

Cerebrum

Primary
somatic
sensory
cortex

Ventral posterior
medial nucleus
of thalamus

Ventral posterior
lateral nucleus of
the thalamus

Midbrain

Spinothalamic
tract

Mid-pons

Middle
medulla

Caudal
medulla

Anterolateral
system

Cervical
spinal cord

Pain and
temperature
information from
upper body
(excluding the face)

Lumbar
spinal cord

Pain and
temperature
information from
lower body

(B)

Cerebrum

Midbrain

Trigemino-
thalamic tract

Mid-pons

Pain and
temperature
information
from face

Middle
medulla

Spinal
trigeminal tract
(afferent axons)

Caudal
medulla

Spinal
nucleus of
the trigeminal
complex

Figure 10.6   Comparison of the pathways
mediating the discriminative aspects of pain and
temperature for (A) the body and (B) the face.

supplied by innocuous warm and cold afferents is relayed to higher centers by distinct classes of secondary neurons that reside in lamina 1 of the spinal cord.

Indeed, the emerging view of lamina 1 is that it consists of a number of distinct classes of modality-selective neurons that convey noxious and innocuous types of sensory information into the anterolateral system. These include individual classes of neurons that are sensitive to: sharp (first) pain, burning (second) pain; innocuous warmth; innocuous cold; histamine (mediating the sense of "itch"); slow mechanical stimulation ("sensual touch"); and a class of inputs that innervates muscles and senses lactic acid and other metabolites that are released during muscle contraction. The latter could contribute to the "burn" or ache that can accompany strenuous exercise.

Is lamina 1 merely an eclectic mixture of cells with different properties, or is there a unifying theme that might account for this diversity? It has been proposed that the lamina 1 system functions as the sensory input to a network that is responsible for representing the physiological condition of the body—a modality that has been called **interoception**, to distinguish it from **exteroception** (touch and pressure) and proprioception. These inputs drive the homeostatic mechanisms that maintain an optimal internal state. Some of these mechanisms are automatic and the changes necessary to maintain homeostasis can be mediated by reflexive adjustment of the autonomic nervous system (see Chapter 21). For example, changes in temperature evoke autonomic reflexes (sweating, or shivering) that counter a disturbance to the body's optimal temperature. Others cannot be mediated by autonomic reflexes alone and require behavioral adjustments (putting on or taking off a sweater) to restore balance. In this conception, the sensations associated with the activation of the lamina 1 system—whether pleasant or noxious—motivate the initiation of behaviors appropriate to maintaining the physiological homeostasis of the body.

Sensitization

Following a painful stimulus associated with tissue damage (e.g., cuts, scrapes, and bruises), stimuli in the area of the injury and the surrounding region that would ordinarily be perceived as slightly painful are perceived as significantly more so, a phenomenon referred to as **hyperalgesia**. A good example of hyperalgesia is the increased sensitivity to temperature that occurs after a sunburn. This effect is due to changes in neuronal sensitivity that occur at the level of peripheral receptors as well as their central targets.

Peripheral sensitization results from the interaction of nociceptors with the "inflammatory soup" of substances released when tissue is damaged. These products of tissue damage include extracellular protons, arachidonic acid and other lipid metabolites, bradykinin, histamine, serotonin, prostaglandins, nucleotides, and nerve growth factor (NGF), all of which can interact with receptors or ion channels of nociceptive fibers, augmenting their response (Figure 10.7). For example, the responses of the TRPV1 receptor to heat can be potentiated by direct interaction of the channel with extracellular protons or lipid metabolites. NGF and bradykinin also potentiate the activity of the TRPV1 receptors, but do so indirectly through the actions of separate cell-surface receptors (TrkA and bradykinin receptors, respectively) and their associated intracellular signaling pathways. The prostaglandins are thought to contribute to peripheral sensitization by binding to G-protein-coupled receptors that increase levels of cyclic AMP within nociceptors. Prostaglandins also reduce the threshold depolarization required for generating action potentials via phosphorylation of a specific class of TTX-resistant sodium channels that are expressed in nociceptors. In addition, electrical activity in the nociceptors causes them to release peptides and neuro-

Figure 10.7 *Inflammatory response to tissue damage. Substances released by damaged tissues augment the response of nociceptive fibers. In addition, electrical activation of nociceptors causes the release of peptides and neurotransmitters that further contribute to the inflammatory response.*

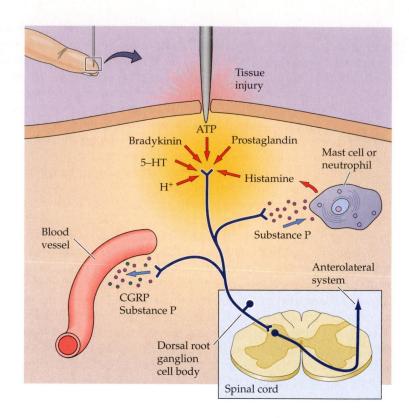

transmitters such as substance P, calcitonin gene-related peptide (CGRP), and ATP, all of which further contribute to the inflammatory response (vasodilation, swelling, and the release of histamine from mast cells). The presumed purpose of the complex chemical signaling cascade arising from local damage is not only to protect the injured area (as a result of the painful perceptions produced by ordinary stimuli close to the site of damage), but also to promote healing and guard against infection by means of local effects such as increased blood flow and the migration of white blood cells to the site. Identifying the components of the inflammatory soup and their mechanisms of action is a fertile area to explore in the search for potential analgesics (compounds that reduce pain's intensity). For example, so-called nonsteroidal anti-inflammatory drugs (NSAIDs), which include aspirin and ibuprofen, act by inhibiting cyclooxygenase (COX), an enzyme important in the biosynthesis of prostaglandins.

Central sensitization refers to an immediate-onset, activity-dependent increase in the excitability of neurons in the dorsal horn of the spinal cord following high levels of activity in the nociceptive afferents. As a result, activity levels in nociceptive afferents that were subthreshold prior to the sensitizing event become sufficient to generate action potentials in dorsal horn neurons, contributing to an increase in pain sensitivity. Although central sensitization is triggered in dorsal horn neurons by activity in nociceptors, the effects generalize to other inputs that arise from low-threshold mechanoreceptors. Thus, stimuli that under normal conditions would be innocuous (such as brushing the surface of the skin) activate second-order neurons in the dorsal horn that receive nociceptive inputs, giving rise to a sensation of pain. The induction of pain by a normally innocuous stimulus is referred to as **allodynia**. This phenomenon typically occurs immediately after the painful event and can outlast the pain of the original stimulus by several hours.

Like its peripheral counterpart, a number of different mechanisms contribute to central sensitization, and these can be divided broadly into transcription-independent and transcription-dependent processes. One form of transcription-independent central sensitization is called "windup" and involves a progressive increase in the discharge rate of dorsal horn neurons in response to repeated low-frequency activation of nociceptive afferents. A behavioral correlate of the windup phenomenon has been studied by examining the perceived intensity of pain in response to multiple presentations of a noxious stimulus. Although the intensity of the stimulation is constant, the perceived intensity increases with each stimulus presentation. Windup lasts only during the period of stimulation and arises from the summation of the slow synaptic potentials evoked in dorsal horn neurons by nociceptive inputs. The sustained depolarization of the dorsal horn neurons results in part from the activation of voltage-dependent L-type calcium channels, and from the removal of the Mg block of NMDA receptors. Removing the Mg block increases the sensitivity of the dorsal horn neuron to glutamate, the transmitter in nociceptive afferents

Other forms of central sensitization that last longer than the period of sensory stimulation (such as allodynia) are thought to involve an LTP-like enhancement of postsynaptic potentials (see Chapter 8). The longest lasting forms, resulting from transcription-dependent processes, can be elicited by changes in neuronal activity or by humoral signals. Those elicited by neuronal activity are localized to the site of the injury, while humoral activation can lead to more widespread changes. For example, cytokines released from microglia and from other sources promote the widespread transcription of COX-2 and the production of prostaglandins in dorsal horn neurons. As described for nociceptive afferents, increased levels of prostaglandins in CNS neurons augment neuronal excitability. Thus the analgesic effects of drugs that inhibit COX are due to actions in both the periphery and within the dorsal horn.

As injured tissue heals, the sensitization induced by peripheral and central mechanisms typically declines and the threshold for pain returns to pre-injury levels. However, when the afferent fibers or central pathways themselves are damaged—a frequent complication in pathological conditions including diabetes, shingles, AIDS, multiple sclerosis, and stroke—these processes can persist. The resulting condition is referred to as **neuropathic pain**: a chronic, intensely painful experience that is difficult to treat with conventional analgesic medications. (See Box 10D for a description of neuropathic pain associated with amputation of an extremity.) Neuropathic pain can arise spontaneously (i.e., without any stimulus), or it can be produced by mild stimuli that are common to everyday experience, such as the gentle touch and pressure of clothing, or warm and cool temperatures. Patients often describe their experience as a constant burning sensation interrupted by episodes of shooting, stabbing, or electric shocklike jolts. Because the disability and psychological stress associated with chronic neuropathic pain can be severe, much present research is being devoted to better understanding of the mechanisms of peripheral and central sensitization with the hope of developing more effective therapies for this debilitating syndrome.

Descending Control of Pain Perception

With respect to the *interpretation* of pain, observers have long commented on the difference between the objective reality of a painful stimulus and the subjective response to it. Modern studies of this discrepancy have provided considerable insight into how circumstances affect pain perception and, ultimately, into the anatomy and pharmacology of the pain system.

BOX 10D Phantom Limbs and Phantom Pain

Following the amputation of an extremity, nearly all patients have an illusion that the missing limb is still present. Although this illusion usually diminishes over time, it persists in some degree throughout the amputee's life and can often be reactivated by injury to the stump or other perturbations. Such phantom sensations are not limited to amputated limbs; phantom breasts following mastectomy, phantom genitalia following castration, and phantoms of the entire lower body following spinal cord transection have all been reported. Phantoms are also common after local nerve block for surgery. During recovery from brachial plexus anesthesia, for example, it is not unusual for the patient to experience a phantom arm, perceived as whole and intact, but displaced from the real arm. When the real arm is viewed, the phantom appears to "jump into" the arm and may emerge and reenter intermittently as the anesthesia wears off. These sensory phantoms demonstrate that the central machinery for processing somatic sensory information is not idle in the absence of peripheral stimuli; apparently, the central sensory processing apparatus continues to operate independently of the periphery, giving rise to these bizarre sensations.

Phantoms might simply be a curiosity—or a provocative clue about higher-order somatic sensory processing—were it not for the fact that a substantial number of amputees also develop phantom pain. This common problem is usually described as a tingling or burning sensation in the missing part. Sometimes, however, the sensation becomes a more serious pain that patients find increasingly debilitating. Phantom pain is, in fact, one of the more common causes of chronic pain syndromes and is extraordinarily difficult to treat. Because of the widespread nature of central pain processing, ablation of the spinothalamic tract, portions of the thalamus, or even primary sensory cortex does not generally relieve the discomfort felt by these patients.

Indeed, considerable functional reorganization of somatotopic maps in the primary somatosensory cortex occurs in amputees. This reorganization starts immediately after the amputation and tends to evolve for several years. One of the effects of this process is that neurons that have lost their original inputs (i.e., from the removed limb) respond to tactile stimulation of other body parts. A surprising consequence is that stimulation of the face, for example, can be experienced as if the missing limb had been touched.

Further evidence that the phenomenon of phantom limb is the result of a central representation is the experience of children born without limbs. Such individuals have rich phantom sensations, despite the fact that a limb never developed. This observation suggests that a full representation of the body exists independently of the peripheral elements that are mapped. Based on these results, Ronald Melzack proposed that the loss of a limb generates an internal mismatch between the brain's representation of the body and the pattern of peripheral tactile input that reaches the neocortex. The consequence would be an illusory sensation that the missing body part is still present and functional. With time, the brain may adapt to this loss and alter its intrinsic somatic representation to better accord with the new configuration of the body. This change could explain why the phantom sensation appears almost immediately after limb loss, but usually decreases in intensity over time.

References

MELZACK, R. (1989) Phantom limbs, the self, and the brain. The D.O. Hebb Memorial Lecture. *Canad. Psychol.* 30: 1–14.

MELZACK, R. (1990) Phantom limbs and the concept of a neuromatrix. *Trends Neurosci.* 13: 88–92.

NASHOLD, B. S., JR. (1991) Paraplegia and pain. In *Deafferentation Pain Syndromes: Pathophysiology and Treatment*, B. S. Nashold, Jr. and J. Ovelmen-Levitt (eds.). New York: Raven Press, pp. 301–319.

RAMACHANDRAN, V. S. AND S. BLAKESLEE (1998) *Phantoms in the Brain.* New York: William Morrow & Co.

SOLONEN, K. A. (1962) The phantom phenomenon in amputated Finnish war veterans. *Acta. Orthop. Scand. Suppl.* 54: 1–37.

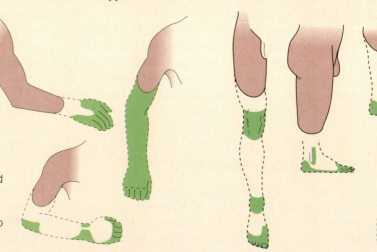

Drawings of phantom arms and legs, based on patients' reports. The phantom is indicated by a dashed line, with the colored regions showing the most vividly experienced parts. Note that some phantoms are telescoped into the stump. (After Solonen, 1962.)

During World War II, Henry Beecher and his colleagues at Harvard Medical School made a fundamental observation. In the first systematic study of its kind, they found that soldiers suffering from severe battle wounds often experienced little or no pain. Indeed, many of the wounded expressed surprise at this odd dissociation. Beecher, an anesthesiologist, concluded that the perception of pain depends on its context. For instance, the pain of an injured soldier on the battle-field would presumably be mitigated by the imagined benefits of being removed from danger, whereas a similar injury in a domestic setting would present quite a different set of circumstances that could exacerbate the pain (loss of work, financial liability, and so on). Such observations, together with the well-known placebo effect (discussed in the next section), make clear that the perception of pain is subject to central modulation; indeed, all sensations are subject to at least some degree of this kind of modification. This statement should not be taken as a vague acknowledgment about the importance of psychological or "top-down" influences on sensory experience. On the contrary, there has been a gradual realization among neuroscientists and neurologists that such "psychological" effects are as real and important as any other neural phenomenon. This appreciation has provided a much more rational view of psychosomatic problems in general, and pain in particular.

The Placebo Effect

The word *placebo* means "I will please," and the placebo effect has a long history of use (and abuse) in medicine. The placebo effect is defined as a physiological response following the administration of a pharmacologically inert "remedy"; its reality is undisputed. In one classic study, medical students were given one of two different pills, one said to be a sedative and the other a stimulant. In fact, both pills contained only inert ingredients. Of the students who received the "sedative," more than two-thirds reported that they felt drowsy, and students who took two such pills felt sleepier than those who had taken only one. Conversely, a large fraction of the students who took the "stimulant" reported that they felt less tired. Moreover, about a third of the entire group reported side effects ranging from headaches and dizziness to tingling extremities and a staggering gait! Only 3 of the 56 students studied reported that the pills they took had no appreciable effect.

In another study of this general sort, 75 percent of patients suffering from postoperative wound pain reported satisfactory relief after an injection of sterile saline. The researchers who carried out this work noted that the responders were indistinguishable from the non-responders, both in the apparent severity of their pain and psychological makeup. Most tellingly, this placebo effect in postoperative patients could be blocked by naloxone, a competitive antagonist of opiate receptors, indicating a substantial pharmacological basis for the pain relief experienced (see the next section).

A common misunderstanding about the placebo effect is the view that patients who respond to a therapeutically meaningless reagent are not suffering real pain, but only "imagining" it; this is certainly not the case. Among other things, the placebo effect probably explains the efficacy of acupuncture anesthesia and the analgesia that can sometimes be achieved by hypnosis. In China, surgery has often been carried out under the effect of a needle (often carrying a small electrical current) inserted at locations dictated by ancient acupuncture charts. Before the advent of modern anesthetic techniques, operations such as thyroidectomies for goiter were commonly done without extraordinary discomfort, particularly among populations where stoicism was the cultural norm.

The mechanisms of pain amelioration on the battlefield, in acupuncture anesthesia, and with hypnosis are presumably related. Although the mechanisms by which the brain affects the perception of pain are only beginning to be understood, the effect is neither magical nor a sign of a suggestible intellect. In short, the placebo effect is quite real.

The Physiological Basis of Pain Modulation

Understanding the central modulation of pain perception (on which the placebo effect is presumably based) was greatly advanced by the finding that electrical or pharmacological stimulation of certain regions of the midbrain produces relief of pain. This analgesic effect arises from activation of descending pain-modulating pathways that project to the dorsal horn of the spinal cord (as well as to the spinal trigeminal nucleus) and regulate the transmission of information to higher centers. One of the major brainstem regions that produce this effect is located in the periaqueductal gray of the midbrain. Electrical stimulation at this site in experimental animals not only produces analgesia by behavioral criteria, but also demonstrably inhibits the activity of nociceptive projection neurons in the dorsal horn of the spinal cord.

Further studies of descending pathways to the spinal cord that regulate the transmission of nociceptive information have shown that they arise from a number of brainstem sites, including the parabrachial nucleus, the dorsal raphe, and locus coeruleus and the medullary reticular formation (Figure 10.8A). The analgesic effects of stimulating the periaqueductal gray are mediated through these brainstem sites. These centers employ a wealth of different neurotransmitters (noradrenaline, serotonin, dopamine, histamine, acetylcholine) and can exert both facilitatory and inhibitory effects on the activity of neurons in the dorsal horn. The complexity of these interactions is made even greater by the fact that descending projections can exert their effects on a variety of sites within the dorsal horn, including the synaptic terminals of nociceptive afferents, excitatory and inhibitory interneurons, and the synaptic terminals of the other descending pathways, as well as by contacting the projection neurons themselves. Although these descending projections were originally viewed as a mechanism that served primarily to inhibit the transmission of nociceptive signals, it is now evident that these projections provide a balance of facilitatory and inhibitory influences that ultimately determines the efficacy of nociceptive transmission.

In addition to descending projections, local interactions between mechanoreceptive afferents and neural circuits within the dorsal horn can modulate the transmission of nociceptive information to higher centers (Figure 10.8B). These interactions are thought to explain the ability to reduce the sensation of sharp pain by activating low-threshold mechanoreceptors—for example, if you crack your shin or stub a toe, a natural (and effective) reaction is to vigorously rub the site of injury for a minute or two. Such observations, buttressed by experiments in animals, led Ronald Melzack and Patrick Wall to propose that the flow of nociceptive information through the spinal cord is modulated by concomitant activation of the large myelinated fibers associated with low-threshold mechanoreceptors. Even though further investigation led to modification of some of the original propositions in Melzack and Wall's **gate theory of pain**, the idea stimulated a great deal of work on pain modulation and has emphasized the importance of synaptic interactions within the dorsal horn for modulating the perception of pain intensity.

The most exciting advance in this long-standing effort to understand central mechanisms of pain regulation has been the discovery of **endogenous opioids**.

(A)

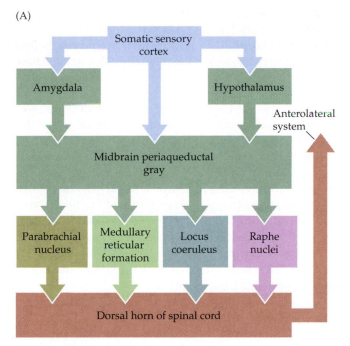

(B)

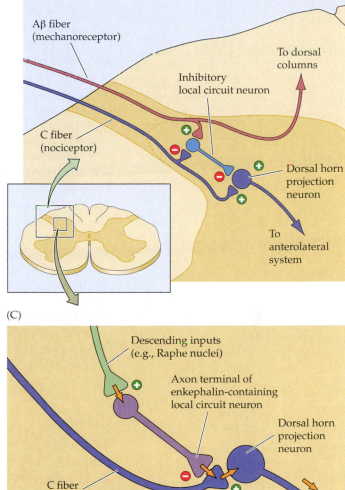

(C)

Figure 10.8 The descending systems that modulate the transmission of ascending pain signals. (A) These modulatory systems originate in the somatic sensory cortex, the hypothalamus, the periaqueductal gray matter of the midbrain, the raphe nuclei, and other nuclei of the rostral ventral medulla. Complex modulatory effects occur at each of these sites, as well as in the dorsal horn. (B) Gate theory of pain. Activation of mechanoreceptors modulates the transmission of nociceptive information to higher centers. (C) The role of enkephalin-containing local circuit neurons in the descending control of nociceptive signal transmission.

For centuries opium derivatives such as morphine have been known to be powerful analgesics—indeed, they remain a mainstay of analgesic therapy today. In the modern era, animal studies have shown that a variety of brain regions are susceptible to the action of opiate drugs, particularly—and significantly—the periaqueductal gray matter and other sources of descending projections. There are, in addition, opiate-sensitive neurons within the dorsal horn of the spinal cord. In other words, the areas that produce analgesia when stimulated are also responsive to exogenously administered opiates. It seems likely, then, that opiate drugs act at most or all of the sites shown in Figure 10.8 in producing their dramatic pain-relieving effects.

The analgesic action of opiates implied the existence of specific brain and spinal cord receptors for these drugs long before the receptors were actually found during the 1960s and 1970s. Since such receptors are unlikely to have evolved in response to the exogenous administration of opium and its derivatives, the conviction grew that *endogenous* opiate-like compounds must exist in order to explain the evolution of these receptors in the body (see Chapter 6). Several categories of endogenous opioids have now been isolated from the brain and intensively studied. These agents are found in the same regions that are involved in the modulation of nociceptive afferents, although each of the families of endogenous opioid peptides has a somewhat different distribution. All three of the major groups—**enkephalins**, **endorphins**, and **dynorphins** (see Table 6.2)—are present in the periaqueductal gray matter. The enkephalins and dynorphins have also been found in the rostral ventral medulla and in the spinal cord regions involved in the modulation of pain.

One of the most compelling examples of the mechanism by which endogenous opiates modulate transmission of nociceptive information occurs at the first synapse in the pain pathway between nociceptive afferents and projection neurons in the dorsal horn of the spinal cord (see Figure 10.8B). A class of enkephalin-containing local circuit neurons within the dorsal horn synapses with the axon terminals of nociceptive afferents, which in turn synapse with dorsal horn projection neurons. The release of enkephalin onto the nociceptive terminals inhibits their release of neurotransmitter onto the projection neuron, reducing the level of activity that is passed on to higher centers. Enkephalin-containing local circuit neurons are themselves the targets of descending projections, thus providing a powerful mechanism by which higher centers can decrease the activity relayed by nociceptive afferents.

A particularly impressive aspect of this story is the wedding of physiology, pharmacology, and clinical research to yield a much richer understanding of the intrinsic modulation of pain. This information has finally begun to explain the subjective variability of painful stimuli and the striking dependence of pain perception on the context of experience. Precisely how pain is modulated is now being explored in many laboratories, motivated by the tremendous clinical (and economic) benefits that would accrue from still deeper knowledge of the pain system and its molecular underpinnings.

Summary

Whether from a structural or functional perspective, pain is an extraordinarily complex sensory modality. Because of their importance in warning an animal about dangerous circumstances, the mechanisms and pathways that subserve nociception are widespread and redundant. A distinct set of pain afferents with membrane receptors known as nociceptors transduces noxious stimulation and conveys this information to neurons in the dorsal horn of the spinal cord. The major central pathway responsible for transmitting the discriminative aspects of pain (location, intensity, and quality) differs from the mechanosensory pathway primarily in that the central axons of dorsal root ganglion cells synapse on second-order neurons in the dorsal horn; the axons of the second-order neurons then cross the midline in the spinal cord and ascend to thalamic nuclei that relay information to the somatic sensory cortex of the postcentral gyrus. Additional pathways involving a number of centers in the brainstem, thalamus, and cortex mediate the affective and motivational responses to painful stimuli. Descending pathways interact with local circuits in the spinal cord to regulate the transmission of nociceptive signals to higher centers. Tremendous progress in understanding pain has been made in the last 25 years, and much more seems likely, given the importance of the problem. No patients are more distressed—or more difficult to treat—than those with chronic pain. Indeed, some aspects of pain seem much more destructive to the sufferer than required by any physiological purposes. Perhaps such seemingly excessive effects are a necessary but unfortunate by-product of the protective benefits of this vital sensory modality.

Additional Reading

Reviews

CATERINA, M. J. AND D. JULIUS (1999) Sense and specificity: A molecular identity for nociceptors. *Curr. Opin. Neurobiol.* 9: 525–530.

DI MARZO, V., P. M. BLUMBERG AND A. SZALLASI (2002) Endovanilloid signaling in pain. *Curr. Opin. Neurobiol.* 12: 372–3710.

DUBNER, R. AND M. S. GOLD (1999) The neurobiology of pain. *Proc. Natl. Acad. Sci. USA* 96: 7627–7630.

FIELDS, H. L. AND A. I. BASBAUM (1978) Brainstem control of spinal pain transmission neurons. *Annu. Rev. Physiol.* 40: 217–248.

HUNT, S. P. AND P. W. MANTYH (2001) The molecular dynamics of pain control. *Nature Rev. Neurosci.* 2: 83–91.

JI, R. R., T. KOHNO, K. A. MOORE AND C. J. WOOLF (2003) Central sensitization and LTP: Do pain and memory share similar mechanisms? *Trends Neurosci.* 26: 696–705.

JULIUS, D. AND A. I. BASBAUM (2001) Molecular mechanisms of nociception. *Nature* 413: 203–209.

MILLAN, M. J. (2002) Descending control of pain. *Prog. Neurobiol.* 66: 355–474.

PATAPOUTIAN, A., A. M. PEIER, G. M. STORY AND V. VISWANATH (2003) ThermoTRP channels and beyond: Mechanisms of temperature sensation. *Nature Rev. Neurosci.* 4: 529–539.

RAINVILLE, P. (2002) Brain mechanisms of pain affect and pain modulation. *Curr. Opin. Neurobiol.* 12: 195–204.

SCHOLZ, J. AND C. J. WOOLF (2002) Can we conquer pain? *Nature Rev. Neurosci.* 5 (Suppl): 1062–1067.

TREEDE, R. D., D. R. KENSHALO, R. H. GRACELY AND A. K. JONES (1999) The cortical representation of pain. *Pain* 79: 105–111.

Important Original Papers

BASBAUM, A. I. AND H. L. FIELDS (1979) The origin of descending pathways in the dorsolateral funiculus of the spinal cord of the cat and rat: Further studies on the anatomy of pain modulation. *J. Comp. Neurol.* 187: 513–522.

BEECHER, H. K. (1946) Pain in men wounded in battle. *Ann. Surg.* 123: 96.

BLACKWELL, B., S. S. BLOOMFIELD AND C. R. BUNCHER (1972) Demonstration to medical students of placebo response and non-drug factors. *Lancet* 1: 1279–1282.

CATERINA, M. J. AND 8 OTHERS (2000) Impaired nociception and pain sensation in mice lacking the capsaicin receptor. *Science* 288: 306–313.

CRAIG, A. D., E. M. REIMAN, A. EVANS AND M. C. BUSHNELL (1996) Functional imaging of an illusion of pain. *Nature* 384: 258–260.

LEVINE, J. D., H. L. FIELDS AND A. I. BASBAUM (1993) Peptides and the primary afferent nociceptor. *J. Neurosci.* 13: 2273–2286.

Books

FIELDS, H. L. (1987) *Pain.* New York: McGraw-Hill.

FIELDS, H. L. (ed.) (1990) *Pain Syndromes in Neurology.* London: Butterworths.

KOLB, L. C. (1954) *The Painful Phantom.* Springfield, IL: Charles C. Thomas.

SKRABANEK, P. AND J. MCCORMICK (1990) *Follies and Fallacies in Medicine.* New York: Prometheus Books.

WALL, P. D. AND R. MELZACK (1989) *Textbook of Pain.* New York: Churchill Livingstone.

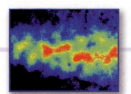

Chapter 11

Vision: The Eye

Overview

The human visual system is extraordinary in the quantity and quality of information it supplies about the world. A glance is sufficient to describe the location, size, shape, color, and texture of objects and, if the objects are moving, their direction and speed. Equally remarkable is the fact that visual information can be discerned over a wide range of stimulus intensities, from the faint light of stars at night to bright sunlight. The next two chapters describe the molecular, cellular, and higher order mechanisms that allow us to see. The first steps in the process of seeing involve transmission and refraction of light by the optics of the eye, the transduction of light energy into electrical signals by photoreceptors, and the refinement of these signals by synaptic interactions within the neural circuits of the retina.

Anatomy of the Eye

The eye is a fluid-filled sphere enclosed by three layers of tissue (Figure 11.1). Only the innermost layer of the eye, the **retina**, contains neurons that are sensitive to light and are capable of transmitting visual signals to central targets. The immediately adjacent layer of tissue includes three distinct but continuous structures collectively referred to as the **uveal tract**. The largest component of the uveal tract is the **choroid**, which is composed of a rich capillary bed (important for nourishing the photoreceptors of the retina) as well as a high concentration of the light absorbing pigment melanin. Extending from the choroid near the front of the eye is the **ciliary body**, a ring of tissue that encircles the **lens** and consists of a muscular component that is important for adjusting the refractive power of the lens, and a vascular component (the so-called ciliary processes) that produces the fluid that fills the front of the eye. The most anterior component of the uveal tract is the **iris**, the colored portion of the eye that can be seen through the cornea. It contains two sets of muscles with opposing actions, which allow the size of the **pupil** (the opening in its center) to be adjusted under neural control. The **sclera** forms the outermost tissue layer of the eye and is composed of a tough white fibrous tissue. At the front of the eye, however, this opaque outer layer is transformed into the **cornea**, a specialized transparent tissue that permits light rays to enter the eye.

Beyond the cornea, light rays pass through two distinct fluid environments before striking the retina. In the **anterior chamber**, just behind the cornea and in front of the lens, lies **aqueous humor**, a clear, watery liquid that supplies nutrients to both of these structures. Aqueous humor is produced by the ciliary processes in the **posterior chamber** (the region between the lens and the iris) and flows into the anterior chamber through the pupil. The amount of fluid produced by the ciliary processes is substantial: it is estimated that the entire volume of fluid in the anterior chamber is replaced 12 times a day. Thus the rates of

Figure 11.1 *Anatomy of the human eye.*

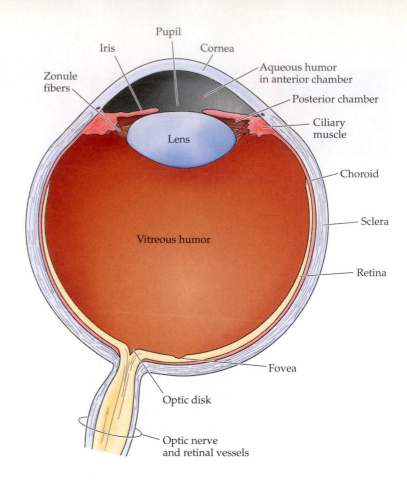

aqueous humor production must be balanced by comparable rates of drainage from the anterior chamber in order to ensure a constant intraocular pressure. A specialized meshwork of cells that lies at the junction of the iris and the cornea—a region called the **limbus**—is responsible for aqueous drainage. Failure of adequate drainage results in a disorder known as **glaucoma**, in which high levels of intraocular pressure can reduce the blood supply to the eye and eventually damage retinal neurons.

The space between the back of the lens and the surface of the retina is filled with a thick, gelatinous substance called the **vitreous humor**, which accounts for about 80% of the volume of the eye. In addition to maintaining the shape of the eye, the vitreous humor contains phagocytic cells that remove blood and other debris that might otherwise interfere with light transmission. The housekeeping abilities of the vitreous humor are limited, however, as a large number of middle-aged and elderly individuals with vitreal "floaters" will attest. Floaters are collections of debris too large for phagocytic consumption that therefore remain to cast annoying shadows on the retina; they typically arise when the aging vitreous membrane pulls away from the overly long eyeball of myopic individuals (Box 11A).

The Formation of Images on the Retina

Normal vision requires that the optical media of the eye be transparent, and both the cornea and the lens are remarkable examples of tissue specialization, achieving a level of transparency that rivals that found in inorganic materials such as glass. Not surprisingly, alterations in the composition of the cornea or the lens can significantly reduce their transparency and have serious consequences for visual perception. Indeed, opacities in the lens known as **cataracts** account for

BOX 11A   Myopia and Other Refractive Errors

Discrepancies of the various physical components of the eye cause a majority of the human population to have some form of refractive error, called **ametropia**. People who are unable to bring distant objects into clear focus are said to be nearsighted, or myopic (Figure B). **Myopia** can be caused by the corneal surface being too curved, or by the eyeball being too long. In either case, with the lens as flat as it can be, the image of distant objects focuses in front of, rather than on, the retina.

People who are unable to focus on near objects are said to be farsighted, or hyperopic. **Hyperopia** can be caused by the eyeball being too short or the refracting system too weak (Figure C). Even with the lens in its most rounded-up state, the image is out of focus on the retinal surface (focusing at some point behind it). Both myopia and hyperopia are correctable by appropriate lenses—concave (minus) and convex (plus), respectively—or by the increasingly popular technique of corneal surgery.

Myopia is by far the most common ametropia; an estimated 50 percent of the U.S. population is affected. Given the large number of people who need glasses, contact lenses, or surgery to correct this refractive error, one naturally wonders how nearsighted people coped before spectacles were invented only a few centuries ago. From what is now known about myopia, most people's vision may have been considerably better in ancient times. The basis for this assertion is the surprising finding that the growth of the eyeball is strongly influenced by focused light falling on the retina. This phenomenon was first described in 1977 by Torsten Wiesel and Elio Raviola at Harvard Medical School, who studied monkeys reared with their lids sutured (the same approach used to demonstrate the effects of visual deprivation on cortical connections in the visual system; see Chapter 24), a proce-

dure that deprives the eye of focused retinal images. They found that animals growing to maturity under these conditions show an elongation of the eyeball. The effect of focused light deprivation appears to be a local one, since the abnormal growth of the eye occurs in experimental animals even if the optic nerve is cut. Indeed, if only a portion of the retinal surface is deprived of focused light, then only that region of the eyeball grows abnormally.

Although the mechanism of light-mediated control of eye growth is not fully understood, many experts now believe that the modern prevalence of myopia may be due to some aspect of modern civilization—perhaps learning to read and write at an early age—that interferes with the normal feedback control of vision on eye development, leading to abnormal elongation of the eyeball. A corollary of this hypothesis is that if children (or, more likely, their parents) wanted to improve their vision, they might be able to do so by practicing far vision to counterbalance the near-work "overload." Practically, of course, most people would probably choose wearing glasses or contacts or having corneal surgery rather than indulging in the onerous daily practice that would presumably be required. Furthermore, not everyone agrees that such a remedy would be effective, and a number of investigators (and drug companies) are exploring the possibility of pharmacological intervention during the period of childhood when abnormal eye growth is presumed to occur. In any event, it is a remarkable fact that deprivation of focused light on the retina causes a compensatory growth of the eye and that this feedback loop is so easily perturbed.

Even people with normal (**emmetropic**) vision as young adults eventually experience difficulty focusing on near objects. One of the many consequences of aging is that the lens loses its elastic-

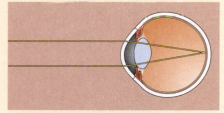

(A) Emmetropia (normal)

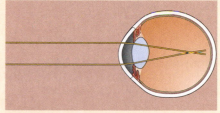

(B) Myopia (nearsighted)

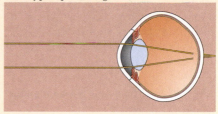

(C) Hyperopia (farsighted)

Refractive errors. (A) In the normal eye, with ciliary muscles relaxed, an image of a distant object is focused on the retina. (B) In myopia, light rays are focused in front of the retina. (C) In hyperopia, images are focused at a point beyond the retina.

ity; as a result, the maximum curvature the lens can achieve when the ciliary muscle contracts is gradually reduced. The near point (the closest point that can be brought into clear focus) thus recedes, and objects (such as this book) must be farther and farther away from the eye in order to focus them on the retina. At some point, usually during early middle age, the accommodative ability of the eye is so reduced that near vision tasks like reading become difficult or impossible (Figure D). This condition is referred to as **presbyopia**. Presbyopia can be corrected by convex lenses for near-vision tasks, or by bifocal lenses

(Continued on next page)

BOX 11A   *(Continued)*

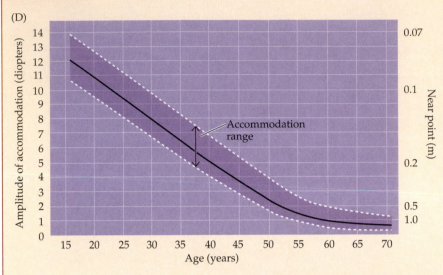

(D)

(D) Changes in the ability of the lens to round up (accommodate) with age. The graph also shows how the near point (the closest point to the eye that can be brought into focus) changes. Accommodation, which is an optical measurement of the refractive power of the lens, is given in diopters. (After Westheimer, 1974.)

if myopia is also present (which requires a negative correction).

Bifocal correction presents a particular problem for those who prefer contact lenses. Because contact lenses float on the surface of the cornea, having the dis-

tance correction above and the near correction below (as in conventional bifocal glasses) doesn't work (although "omni-focal" contact lenses have recently been used with some success). A surprisingly effective solution to this problem for

some contact lens wearers has been to put a near correcting lens in one eye and a distance correcting lens in the other! The success of this approach is another testament to the remarkable ability of the visual system to adjust to a wide variety of unusual demands.

References

Bock, G. and K. Widdows (1990) *Myopia and the Control of Eye Growth*. Ciba Foundation Symposium 155. Chichester: Wiley.

Coster, D. J. (1994) *Physics for Ophthalmologists*. Edinburgh: Churchill Livingston.

Kaufman, P. L. and A. Alm (eds.) (2002) *Adler's Physiology of the Eye: Clinical Application*, 10th Ed. St. Louis, MO: Mosby Year Book.

Sherman, S. M., T. T. Norton and V. A. Casagrande (1977) Myopia in the lid-sutured tree shrew. *Brain Res.* 124: 154–157.

Wallman, J., J. Turkel and J. Tractman (1978) Extreme myopia produced by modest changes in early visual experience. *Science* 201: 1249–1251.

Wallman, J. and J. Winawer (2004) Homeostasis of eye growth and the question of myopia. *Neuron* 43: 447–468.

Wiesel, T. N. and E. Raviola (1977) Myopia and eye enlargement after neonatal lid fusion in monkeys. *Nature* 266: 66–68.

roughly half the cases of blindness in the world, and almost everyone over the age of 70 will experience some loss of transparency in the lens that ultimately degrades the quality of visual experience. Fortunately, successful surgical treatments for cataracts can restore vision in most cases. Furthermore, the recognition that a major factor in the production of cataracts is exposure to ultraviolet (UV) solar radiation has heightened public awareness of the need to protect the lens (and the retina) by reducing UV exposure through the use of sunglasses.

Beyond efficiently transmitting light energy, the primary function of the optical components of the eye is to achieve a focused image on the surface of the retina. The cornea and lens are primarily responsible for the refraction (bending) of light necessary for the formation of focused images on the photoreceptors of the retina (Figure 11.2). The cornea contributes most of the necessary refraction, as can be appreciated by considering the hazy, out-of-focus images experienced when swimming underwater. Water, unlike air, has a refractive index close to that of the cornea; as a result, immersion in water virtually eliminates the refraction that normally occurs at the air/cornea interface; thus the image is no longer focused on the retina. The lens has considerably less refractive power than the cornea; however, the refraction supplied by the lens is adjustable, allowing objects at various distances from the observer to be brought into sharp focus.

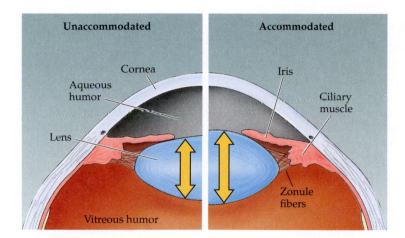

Figure 11.2 Diagram showing the anterior part of the human eye in the unaccommodated (left) and accommodated (right) state. Accommodation for focusing on near objects involves the contraction of the ciliary muscle, which reduces the tension in the zonule fibers and allows the elasticity of the lens to increase its curvature.

Dynamic changes in the refractive power of the lens are referred to as **accommodation**. When viewing distant objects, the lens is made relatively thin and flat and has the least refractive power. For near vision, the lens becomes thicker and rounder and has the most refractive power (see Figure 11.2). These changes result from the activity of the **ciliary muscle** that surrounds the lens. The lens is held in place by radially arranged connective tissue bands called **zonule fibers** that are attached to the ciliary muscle. The shape of the lens is thus determined by two opposing forces: the elasticity of the lens, which tends to keep it rounded up (removed from the eye, the lens becomes spheroidal); and the tension exerted by the zonule fibers, which tends to flatten it. When viewing distant objects, the force from the zonule fibers is greater than the elasticity of the lens, and the lens assumes the flatter shape appropriate for distance viewing. Focusing on closer objects requires relaxing the tension in the zonule fibers, allowing the inherent elasticity of the lens to increase its curvature. This relaxation is accomplished by the sphincter-like contraction of the ciliary muscle. Because the ciliary muscle forms a ring around the lens, when the muscle contracts, the attachment points of the zonule fibers move toward the central axis of the eye, thus reducing the tension on the lens. Unfortunately, changes in the shape of the lens are not always able to produce a focused image on the retina, in which case a sharp image can be focused only with the help of additional corrective lenses (see Box 11A).

Adjustments in the size of the pupil also contribute to the clarity of images formed on the retina. Like the images formed by other optical instruments, those generated by the eye are affected by spherical and chromatic aberrations, which tend to blur the retinal image. Since these aberrations are greatest for light rays that pass farthest from the center of the lens, narrowing the pupil reduces both spherical and chromatic aberration, just as closing the iris diaphragm on a camera lens improves the sharpness of a photographic image. Reducing the size of the pupil also increases the depth of field—that is, the distance within which objects are seen without blurring. However, a small pupil also limits the amount of light that reaches the retina, and, under conditions of dim illumination, visual acuity becomes limited by the number of available photons rather than by optical aberrations. An adjustable pupil thus provides an effective means of reducing optical aberrations, while maximizing depth of field to the extent that different levels of illumination permit. The size of the pupil is controlled by innervation from both sympathetic and parasympathetic divisions of the visceral motor system, which in turn are modulated by several brainstem centers (see Chapters 20 and 21).

The Surface of the Retina

Using an ophthalmoscope, the inner surface of the retina, or **fundus**, can be visualized through the pupil (Figure 11.3). Numerous blood vessels, both arteries and veins, fan out over the inner surface of the retina. These blood vessels arise from the ophthalmic artery and vein, which enter the eye through a whiteish circular area known as the **optic disk** or **optic papilla**. The optic disk is also the site where retinal axons leave the eye and travel through the optic nerve to reach target structures in the thalamus and midbrain. This region of the retina contains no photoreceptors and, because it is insensitive to light, produces the perceptual phenomena known as the blind spot (Box 11B). In addition to being a conspicuous retinal landmark, the appearance of the optic disk is a useful gauge of intracranial pressure. The subarachnoid space surrounding the optic nerve is continuous with that of the brain; as a result, increases in intracranial pressure—a sign of serious neurological problems such as space-occupying lesions—can be detected as a swelling of the optic disk.

Another prominent feature of the fundus is the **macula lutea**, an oval spot containing yellow pigment (xanthophyll), roughly 1.5 millimeters in diameter and located near the center of the retina. The macula is the region of the retina that supports high visual acuity (the ability to resolve fine details). Acuity is greatest at the center of the macula, a small depression or pit in the retina called the **fovea**. The pigment xanthophyll has a protective role, filtering ultraviolet wavelengths that could be harmful to the photoreceptors. Damage to this region of the retina, as occurs in age-related macular degeneration has a devastating impact on visual perception (Box 11C).

Retinal Circuitry

Despite its peripheral location, the retina, which is the neural portion of the eye, is actually part of the central nervous system. During development, the retina forms as an outpocketing of the diencephalon called the optic vesicle. The optic

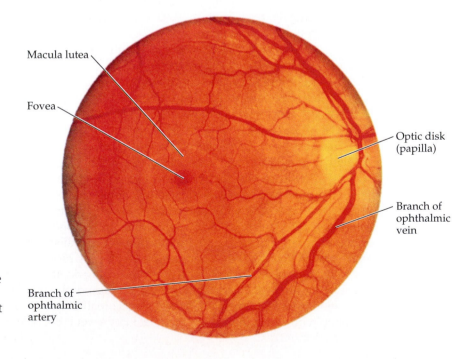

Figure 11.3 The inner surface of the retina, viewed with an ophthalmoscope. The optic disk is the region where the ganglion cell axons leave the retina to form the optic nerve; it is also characterized by the entrance and exit, respectively, of the ophthalmic arteries and veins that supply the retina. The macula lutea can be seen as a distinct area at the center of the optical axis (the optic disk lies nasally); the macula is the region of the retina that has the highest visual acuity. The fovea is a depression or pit about 1.5 mm in diameter that lies at the center of the macula.

Macula lutea

Fovea

Optic disk (papilla)

Branch of ophthalmic vein

Branch of ophthalmic artery

BOX 11B **The Blind Spot**

It is logical to suppose that a visual field defect (or **scotoma**, to use the medical term) arising from damage to the retina or central visual pathways would be obvious to the individual suffering from such pathology. When the deficit involves a peripheral region of the visual field, however, a scotoma often goes unnoticed until a car accident or some other mishap all too dramatically reveals the sensory loss. In fact, all of us have a physiological scotoma of which we are quite unaware, the so-called "blind spot." The blind spot is the substantial gap in each monocular visual field that corresponds to the location of the optic disk, the receptor-free region of the retina where the optic nerve leaves the eye (see Figure 11.1).

To find the "blind spot" of your right eye, hold this book 30–40 centimeters (about a foot) away, close your left eye, and fixate on the X shown in the figure. Take a pencil in your right hand and, without breaking fixation, move the tip slowly toward the X from the right side of the page. At some point, the tip of the pencil (indeed, the whole end of the pencil) will disappear; mark this point and continue to move the pencil to the left until it reappears; then make another mark. The borders of the blind spot along the vertical axis can be determined in the same way by moving the pencil up and down so that its path falls

between the two horizontal marks.

To prove that information from the region of visual space bounded by the marks is really not perceived, put a penny inside the demarcated area. When you fixate the X with both eyes and then close the left eye, the penny will disappear (a seemingly magical event that amazed the French royal court when it was first reported by the natural philosopher Edmé Mariotte in 1668).

How can we be unaware of such a large (typically about 5°–8°) defect in the visual field? The optic disk is located in the nasal retina of each eye. With both eyes open, information about the corresponding region of visual space is, of course, available from the temporal retina of the other eye. But this fact does not explain why the blind spot remains undetected with one eye closed. When the world is viewed monocularly, the visual system appears to "fill in" the missing part of the scene based on the information supplied by the regions surrounding the optic disk. To observe this phenomenon, notice what happens when a pencil or some other object lies *across* the optic disk representation. Remarkably, the pencil looks complete! Although electrophysiological recordings have shown that neurons in the visual cortex whose receptive fields lie in the optic disk representation can be activated by stimulating the regions that

surround the optic disk of the contralateral eye—a fact that suggests filling in the blind spot is based on cortical mechanisms that integrate information from different points in the visual field—the mechanism of this striking phenomenon is not clear. Herman von Helmholtz pointed out in the nineteenth century that it may just be that this part of the visual world is ignored, and the pencil is completed across the blind spot because the rest of the scene simply "collapses" around it.

References

FIORANI, M., M. G. P. ROSA, R. GATTASS AND C. E. ROCHA-MIRANDA (1992) Dynamic surrounds of receptive fields in striate cortex: A physiological basis for perceptual completion? *Proc. Natl. Acad. Sci. USA* 89: 8547–8551.

GILBERT, C. D. (1992) Horizontal integration and cortical dynamics. *Neuron* 9: 1–13.

RAMACHANDRAN, V. S. AND T. L. GREGORY (1991) Perceptual filling in of artificially induced scotomas in human vision. *Nature* 350: 699–702.

VON HELMHOLTZ, H. (1968). *Helmholtz's Treatise on Physiological Optics*, Vols. I–III (Translated from the Third German Ed. published in 1910). J. P. C. Southall (ed.). New York: Dover Publications. See pp. 204ff in Vol. III.

BOX 11C Macular Degeneration

An estimated 6 million people in the United States suffer from a condition known as **age-related macular degeneration (AMD)**, which causes a progressive loss of central vision. Because central vision is critical for sight, diseases that affect the macula (see Figure 11.1) severely limit a person's ability to perform visual tasks. Indeed, in the United States, AMD is the most common cause of vision loss in people over age 55, and its incidence is rising as the population ages.

The underlying problem, which remains poorly understood, is degeneration of the photoreceptors. Usually, patients first notice a blurring of central vision when performing tasks such as reading. Images may also appear distorted. A chart known as the Amsler grid is used as a simple test for early signs of AMD. By focusing on a marked spot in the middle of the grid (which resembles a piece of graph paper), the patient can assess whether the parallel and perpendicular lines on the grid appear blurred or distorted. Blurred central vision often progresses to having blind spots within central vision, and in most cases both eyes are eventually involved.

Although the risk of developing AMD clearly increases with age, the causes of the disease are not known. Various studies have implicated hereditary factors, cardiovascular disease, environmental factors such as smoking and light exposure, and nutritional causes. Indeed, it may be that all these contribute to the risk of developing AMD.

In descriptive terms, macular degeneration is broadly divided into two types. In the *exudative-neovascular form*, or "wet" AMD, which accounts for 10 percent of all cases, abnormal blood vessel growth occurs under the macula. These blood vessels leak fluid and blood into the retina and cause damage to the photoreceptors. Wet AMD tends to progress rapidly and can cause severe damage; rapid loss of central vision may occur

over just a few months. The treatment for this form of the disease is laser therapy. By transferring thermal energy, the laser beam destroys the leaky blood vessels under the macula, thus slowing the rate of vision loss. A disadvantage of this approach is that the high thermal energy delivered by the beam also destroys nearby healthy tissue. An improvement in the laser treatment of AMD involves a light-activated drug to target abnormal blood vessels. Once the drug is administered, relatively low energy laser pulses aimed at the abnormal blood vessels are delivered to stimulate the drug, which in turn destroys the abnormal blood vessels with minimal damage to the surrounding tissue.

The remaining 90 percent of AMD cases are the nonexudative, or "dry" form. In these patients there is a gradual disappearance of the retinal pigment epithelium, resulting in circumscribed areas of atrophy. Since photoreceptor loss follows the disappearance of the pigment epithelium, the affected retinal areas have little or no visual function. Vision loss from dry AMD occurs more gradually, typically over the course of many years. These patients usually retain some central vision, although the loss can be severe enough to compromise performance of tasks that require seeing details. Unfortunately, at the present time there is no treatment for dry AMD. A radical and quite fascinating new approach that offers some promise entails surgically repositioning the retina away from the abnormal area.

Occasionally, macular degeneration occurs in much younger individuals. Many of these cases are caused by a number of distinct genetic mutations, each with its own clinical manifestations. The most common form of juvenile macular degeneration is known as *Stargardt disease*, which is inherited as an autosomal recessive. Patients are usually diagnosed before they reach the age of

20. Although the progression of vision loss is variable, most of these patients are legally blind by age 50. Mutations that cause Stargardt disease have been identified in the *ABCR* gene, which codes for a protein that transports retinoids across the photoreceptor membrane. Thus, the visual cycle of photopigment regeneration (see Figure 11.6) may be disrupted, presumably by dysfunctional proteins encoded by the abnormal gene. Interestingly, the *ABCR* gene is expressed only in rods, suggesting that the cones may have their own visual cycle enzymes.

In sporadic forms of AMD, DNA sequence variation in two genes involved in the complement cascade—factor B and factor H—have been identified as risk factors, thereby implicating the complement cascade as a potential therapeutic target for some forms of AMD.

References

BRESSLER, N. M., S. B. BRESSLER, AND S. L. FINE (2006) Neovascular (exudative) age-related macular degeneration. In *Retina*, 4th Ed., Vol. 2: *Medical Retina*. S. J. Ryan (ed.-in-chief). Philadelphia: Elsevier Mosby, pp. 1074–1114.

BRESSLER, S. B., N. M. BRESSLER, S. H. SARKS AND J. P. SARKS (2006) Age-related macular degeneration: Nonneovascular early AMD, intermediate AMD, and geographic atrophy. In *Retina*, 4th Ed., Vol. 2: *Medical Retina*. S. J. Ryan (ed.-in-chief). Philadelphia: Elsevier Mosby, pp. 1041–1074.

DEUTMAN, A. F., C. B. HOYNG AND J. J. C. VAN LITH-VERHOEVEN (2006) Macular dystrophies. In *Retina*, 4th Ed., Vol. 2: *Medical Retina*. S. J. Ryan (ed.-in-chief). Philadelphia: Elsevier Mosby, pp. 1163–1210.

FINE, S. L., J. W. BERGER, M. G. MAGUIRE AND A. C. HO (2000) Drug therapy: Age-related macular degeneration. *New Eng. J. Med.* 342: 483–492.

THE FOUNDATION FIGHTING BLINDNESS of Hunt Valley, MD, maintains a web site that provides updated information about many forms of retinal degeneration: www.blindness.org

RETNET provides updated information, including references to original articles, on genes and mutations associated with retinal diseases: www.sph.uth.tmc.edu/RetNet

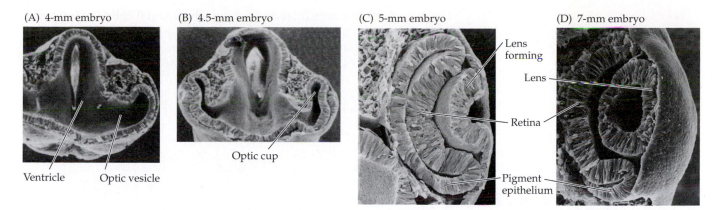

(A) 4-mm embryo   (B) 4.5-mm embryo   (C) 5-mm embryo   (D) 7-mm embryo

Lens forming

Lens

Retina

Pigment epithelium

Optic cup

Ventricle   Optic vesicle

Figure 11.4   *Development of the human eye. (A) The retina develops as an outpocketing from the neural tube, called the optic vesicle. (B) The optic vesicle invaginates to form the optic cup. (C,D) The inner wall of the optic cup becomes the neural retina, while the outer wall becomes the pigment epithelium. (A–C from Hilfer and Yang, 1980; D courtesy of K. Tosney.)*

vesicle undergoes invagination to form the optic cup (Figure 11.4; see also Chapter 22). The inner wall of the optic cup gives rise to the retina, while the outer wall gives rise to the **retinal pigment epithelium**. This epithelium is a thin, melanin-containing structure that reduces backscattering of the light that enters the eye and, as will become apparent, plays a critical role in maintaining the phototransduction machinery of retinal photoreceptors.

Consistent with its status as a full-fledged part of the central nervous system, the retina exhibits complex neural circuitry that converts the graded electrical activity of specialized photosensitive neurons—the photoreceptors—into action potentials that travel to central targets via axons in the optic nerve. Although it has the same types of functional elements and neurotransmitters found in other parts of the central nervous system, the retina comprises fewer broad classes of neurons, and these are arranged in a manner that has been less difficult to unravel than the circuits in other areas of the brain.

There are five basic classes of neurons in the retina: **photoreceptors**, **bipolar cells**, **ganglion cells**, **horizontal cells**, and **amacrine cells**. The cell bodies and processes of these neurons are stacked in alternating layers, with the cell bodies located in the inner nuclear, outer nuclear, and ganglion cell layers, and the processes and synaptic contacts located in the inner plexiform and outer plexiform layers (Figure 11.5A,B).

The retina contains two types of photoreceptors: **rods** and **cones** (Figure 11.5B,C). Both types have an outer segment (adjacent to the pigment epithelium) that is composed of membranous disks containing light-sensitive photopigment; and an inner segment that contains the cell nucleus and gives rise to synaptic terminals that contact bipolar or horizontal cells.

A three-neuron chain—photoreceptor cell to bipolar cell to ganglion cell—is the most direct pathway of information flow from photoreceptors to the optic nerve. Absorption of light by the photopigment in the outer segment of the photoreceptors initiates a cascade of events that changes the membrane potential of the receptor, and therefore the amount of neurotransmitter released by the photoreceptor terminals. (This process, called **phototransduction**, is discussed in detail later in the chapter.) The synapses between photoreceptor terminals and bipolar cells (and horizontal cells) occur in the outer plexiform layer; more specifically, the cell bodies of photoreceptors make up the outer nuclear layer, whereas the cell bodies of bipolar cells lie in the inner nuclear layer. The short axonal processes of bipolar cells make synaptic contacts in turn on the dendritic processes of ganglion cells in the inner plexiform layer. The much larger axons of the ganglion cells form the **optic nerve** and carry information about retinal stimulation to the rest of the central nervous system.

The two other types of neurons in the retina, **horizontal cells** and **amacrine cells**, have their cell bodies in the inner nuclear layer and have processes that are limited to the outer and inner plexiform layers respectively (see Figure 11.5B). The processes of horizontal cells enable lateral interactions between photoreceptors and bipolar cells that are thought to maintain the visual system's sensitivity to contrast, over a wide range of light intensities, or **luminance**. The processes of amacrine cells are postsynaptic to bipolar cell terminals

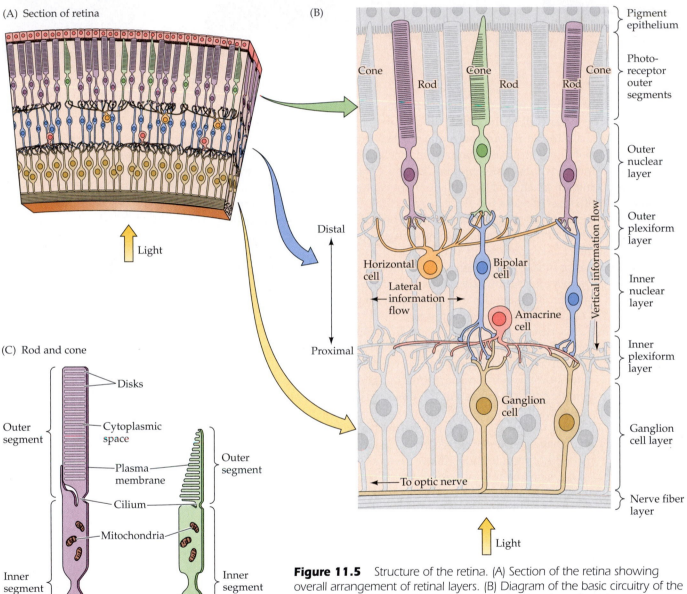

Figure 11.5 Structure of the retina. (A) Section of the retina showing overall arrangement of retinal layers. (B) Diagram of the basic circuitry of the retina. A three-neuron chain—photoreceptor, bipolar cell, and ganglion cell—provides the most direct route for transmitting visual information to the brain. Horizontal cells and amacrine cells mediate lateral interactions in the outer and inner plexiform layers, respectively. The terms *inner* and *outer* designate relative distances from the center of the eye (inner, near the center of the eye; outer, away from the center, or toward the pigment epithelium). (C) Structural differences between rods and cones. Although generally similar in structure, rods and cones differ in their size and shape, as well as in the arrangement of the membranous disks in their outer segments.

and presynaptic to the dendrites of ganglion cells. Different subclasses of amacrine cells are thought to make distinct contributions to visual function. For example, one amacrine cell type has an obligatory role in the pathway that transmits information from rod photoreceptors to retinal ganglion cells. Another type is believed to be critical for generating the direction-selective responses exhibited by a specialized subset of ganglion cells.

The variety of amacrine cell subtypes illustrates the more general rule that, even with only five basic classes of retinal neurons, there can be considerable diversity within a given cell class. This diversity is also a hallmark of retinal ganglion cells and the basis for pathways that convey different sorts of information to central targets in a parallel manner, a topic that is considered in more detail in Chapter 12.

Retinal Pigment Epithelium

The spatial arrangement of retinal layers at first seems counterintuitive: light rays must pass through various non-light-sensitive elements of the retina as well as the retinal vasculature (which branches extensively on the inner surface of the retina) before reaching the outer segments of the photoreceptors, which is where photons are absorbed (see Figure 11.5A,B). The reason for this curious feature of retinal organization lies in the special relationship that exists among the outer segments of the photoreceptors and the pigment epithelium. The cells that make up the retinal pigment epithelium have long processes that extend into the photoreceptor layer, surrounding the tips of the outer segments of each photoreceptor (Figure 11.6A).

Figure 11.6 Removal of photoreceptor disks by the pigment epithelium. (A) The tips of the outer segments of photoreceptors are embedded in pigment epithelium. Epithelial cell processes extend down between the outer segments. (B) The life span of photoreceptor disks is seen in the movement of radioactively labeled amino acids injected into the inner segment and incorporated into disks. The labeled disks migrate from the inner to the outer portion of the outer segment over a 12-day period. (C) Expended disks are shed from the outer segment and phagocytosed. The photopigment from the disks enters the pigment epithelium, where it will be biochemically cycled back to "newborn" photoreceptor disks. (A after Oyster, 1999; B,C after Young, 1971.)

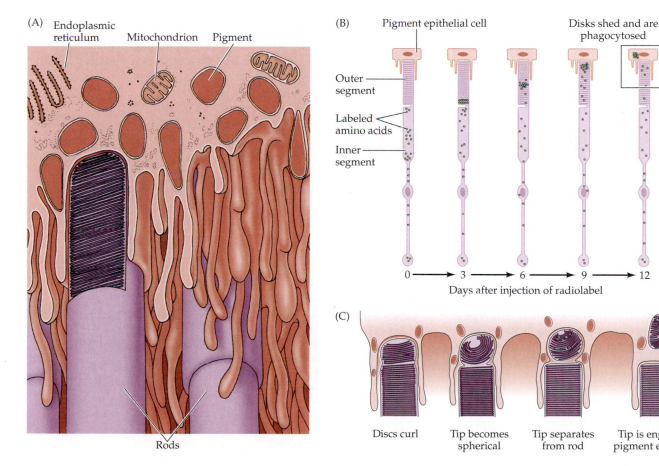

(A) Endoplasmic reticulum Mitochondrion Pigment

Rods

(B) Pigment epithelial cell

Outer segment

Labeled amino acids

Inner segment

Disks shed and are phagocytosed

0 → 3 → 6 → 9 → 12

Days after injection of radiolabel

(C)

Discs curl | Tip becomes spherical | Tip separates from rod | Tip is engulfed by pigment epithelium

The pigment epithelium plays two roles that are critical to the function of retinal photoreceptors. First, the membranous disks in the outer segment, which house the light-sensitive photopigment and other proteins involved in phototransduction, have a relatively limited life span of about 12 days. New outer segment disks are continuously being formed near the base of the outer segment, while the oldest disks are removed or "shed" at the tip of the outer segment (Figure 11.6B). During their life span, disks move progressively from the base of the outer segment to the tip, where the pigment epithelium plays an essential role in removing the expended receptor disks. Shedding involves the "pinching off" of a clump of receptor disks by the outer segment membrane of the photoreceptor. This enclosed clump of disks is then phagocytosed by the pigment epithelium (Figure 11.6C). The epithelium's second role, as we will emphasize in more detail in the next section, is to regenerate photopigment molecules after they have been exposed to light. Photopigment is cycled continuously between the outer segment of the photoreceptor and the pigment epithelium.

These considerations, along with the fact that the capillaries in the choroid underlying the pigment epithelium are the primary source of nourishment for retinal photoreceptors, presumably explain why rods and cones are found in the outermost rather than the innermost layer of the retina. Indeed, disruptions in this normal relationship between the pigment epithelium and retinal photoreceptors have severe consequences for vision (Box 11D).

Phototransduction

In most sensory systems, activation of a receptor by the appropriate stimulus causes the cell membrane to depolarize, ultimately stimulating an action potential and transmitter release onto the neurons it contacts. In the retina, however, photoreceptors do not exhibit action potentials; rather, light activation causes a graded change in membrane potential and a corresponding change in the rate of transmitter release onto postsynaptic neurons. Indeed, much of the processing within the retina is mediated by graded potentials, largely because action potentials are not required to transmit information over the relatively short distances involved.

Perhaps even more surprising is that shining light on a photoreceptor, either a rod or a cone, leads to membrane *hyperpolarization* rather than depolarization (Figure 11.7). In the dark, the receptor is in a depolarized state, with a membrane potential of roughly –40 mV (including those portions of the cell that release transmitters). Progressive increases in the intensity of illumination cause the potential across the receptor membrane to become more negative, a

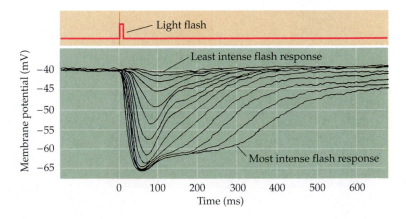

Figure 11.7 An intracellular recording from a single cone stimulated with different amounts of light (the cone has been taken from the turtle retina, which accounts for the relatively long time course of the response). Each trace represents the response to a brief flash that was varied in intensity. At the highest light levels, the response amplitude saturates (at about –65 mV). The hyperpolarizing response is characteristic of vertebrate photoreceptors; interestingly, some invertebrate photoreceptors depolarize in response to light. (After Schnapf and Baylor, 1987.)

BOX 11D Retinitis Pigmentosa

Retinitis pigmentosa (RP) refers to a heterogeneous group of hereditary eye disorders characterized by progressive vision loss due to a gradual degeneration of photoreceptors. An estimated 100,000 people in the United States have RP. In spite of the name, inflammation is not a prominent part of the disease process; instead the photoreceptor cells appear to die by apoptosis (determined by the presence of DNA fragmentation).

Classification of this group of disorders under one rubric is based on the clinical features commonly observed in these patients. The hallmarks of RP are night blindness, a reduction of peripheral vision, narrowing of the retinal vessels, and the migration of pigment from disrupted retinal pigment epithelium into the retina, forming clumps of various sizes, often next to retinal blood vessels.

Typically, patients first notice difficulty seeing at night due to the loss of rod photoreceptors; the remaining cone photoreceptors then become the mainstay of visual function. Over many years, the cones also degenerate, leading to a progressive loss of vision. In most RP patients, visual field defects begin in the midperiphery, between 30° and 50° from the point of foveal fixation. The defective regions gradually enlarge, leaving islands of vision in the periphery and a constricted central field—a condition known as tunnel vision. When the visual field contracts to 20° or less and/or central vision is 20/200 or worse, the patient is categorized as legally blind.

Inheritance patterns indicate that RP can be transmitted in an X-linked (XLRP), autosomal dominant (ADRP), or recessive (ARRP) manner. In the United States, the percentage of these genetic types is estimated to be 9%, 16%, and 41%, respectively. When only one member of a pedigree has RP, the case is classified as "simplex," which accounts for about a third of all cases.

Among the three genetic types of RP, ADRP is the mildest. These patients often retain good central vision until 60 years of age or older. In contrast, patients with the XLRP form of the disease are usually legally blind by 30 to 40 years of age. However, the severity and age of onset of the symptoms varies greatly among patients with the same type of RP, and even within the same family (when, presumably, all the affected members have the same genetic mutation).

To date, RP-inducing mutations of 30 genes have been identified. Many of these genes encode photoreceptor-specific proteins, several being associated with phototransduction in the rods. Among the latter are genes for rhodopsin, subunits of the cGMP phosphodiesterase, and the cGMP-gated channel. Multiple mutations have been found in each of these cloned genes. For example, in the case of the rhodopsin gene, 90 different mutations have been identified among ADRP patients.

The heterogeneity of RP at all levels, from genetic mutations to clinical symptoms, has important implications for understanding its pathogenesis and designing therapies. Given the complex molecular etiology of RP, it is unlikely that a single cellular mechanism will explain all cases. Regardless of the specific mutation or causal sequence, the vision loss that is most critical is the result of gradual degeneration of cones. But in many cases, the RP-causing mutation affects proteins that are not even expressed in the cones—the prime example is the rod-specific visual pigment rhodopsin—and the loss of cones may be an indirect result of a rod-specific mutation. Identifying the cellular mechanisms that directly cause cone degeneration should lead to a better understanding of this pathology.

References

RIVOLTA, C., D. SHARON, M. M. DEANGELIS AND T. P. DRYJA. (2002) Retinitis pigmentosa and allied diseases: Numerous diseases, genes and inheritance patterns. *Hum. Molec. Genet.* 11: 1219–1227.

WELEBER, R. G. AND K. GREGORY-EVANS (2001) Retinitis pigmentosa and allied disorders. In *Retina*, 4th Ed., Vol. 1: *Basic Science and Inherited Retinal Diseases.* S. J. Ryan (ed. in chief). Philadelphia: Elsevier Mosby, pp. 395–498.

THE FOUNDATION FIGHTING BLINDNESS of Hunt Valley, MD, maintains a web site that provides updated information about many forms of retinal degeneration: www.blindness.org

RETNET provides updated information, including references to original articles, on genes and mutations associated with retinal diseases: www.sph.uth.tmc.edu/RetNet

Characteristic appearance of the retina in patients with retinitis pigmentosa. Note the dark clumps of pigment that are the hallmark of this disorder.

response that saturates when the membrane potential reaches about −65 mV. Although the sign of the potential change may seem odd, the only logical requirement for subsequent visual processing is a consistent relationship between luminance changes and the rate of transmitter release from the photoreceptor terminals. As in other nerve cells, transmitter release from the synaptic terminals of the photoreceptor is dependent on voltage-sensitive Ca^{2+} channels in the terminal membrane. Thus, in the dark, when photoreceptors are relatively depolarized, the number of open Ca^{2+} channels in the synaptic terminal is high, and the rate of transmitter release is correspondingly great; in the light, when receptors are hyperpolarized, the number of open Ca^{2+} channels is reduced, and the rate of transmitter release is also reduced. The reason for this unusual arrangement compared to other sensory receptor cells is not known.

In the dark, cations (both Na^+ and Ca^{2+}) flow into the outer segment through membrane channels that are gated by the nucleotide **cyclic guanosine monophosphate** (cGMP) in a fashion similar to other second-messenger systems (see Chapter 7). This inward current is opposed by an outward current mediated by potassium-selective channels in the inner segment. Thus the depolarized state of the photoreceptor in the dark reflects the net contribution of Na^+ and Ca^{2+} influx, which acts to depolarize the cell, and K^+ efflux, which acts to hyperpolarize the cell (Figure 11.8A). Absorption of light by the photoreceptor reduces the concentration of cGMP in the outer segment, leading in turn to a closure of the cGMP-gated channels in the outer segment membrane and, consequently, a reduction in the inward flow of Na^+ and Ca^{2+}. As a result, positive charge (carried by K^+) flows out of the cell more rapidly than positive charge (carried by Na^+ and Ca^{2+}) flows in, and the cell becomes hyperpolarized (Figure 11.8B).

The series of biochemical changes that ultimately leads to a reduction in cGMP levels begins when a photon is absorbed by the photopigment in the

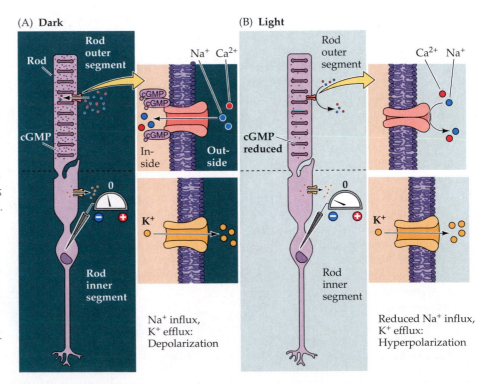

Figure 11.8 Cyclic GMP-gated channels in the outer segment membrane are responsible for the light-induced changes in the electrical activity of photoreceptors. A rod is shown in this simplified diagram, but the same scheme applies to cones. (A) In the dark, cGMP levels in the outer segment are high; cGMP binds to the Na^+-permeable channels in the membrane, keeping them open and allowing sodium and other cations to enter, thus depolarizing the cell. (B) Absorption of photons leads to a decrease in cGMP levels, closing the cation channels and resulting in receptor hyperpolarization.

receptor disks. The photopigment contains the light-absorbing chromophore **retinal** (an aldehyde of vitamin A) coupled to one of several possible proteins called **opsins**. The different opsins tune the molecule's absorption of light to a particular region of the spectrum; indeed, it is the differing protein components of the photopigments in rods and cones that allows the functional specialization of these two receptor types.

Most of what is known about the molecular events of phototransduction has been gleaned from experiments in rods, in which the photopigment is **rhodopsin**. The seven transmembrane domains of the opsin molecule traverse the membrane of the disks in the outer segment, forming a pocket in which the retinal molecule resides (Figure 11.9A). When retinal absorbs a photon of light, one of the double bonds between the carbon atoms in the retinal molecule

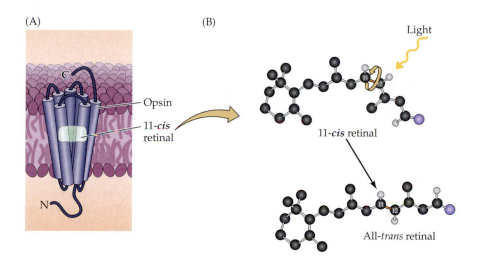

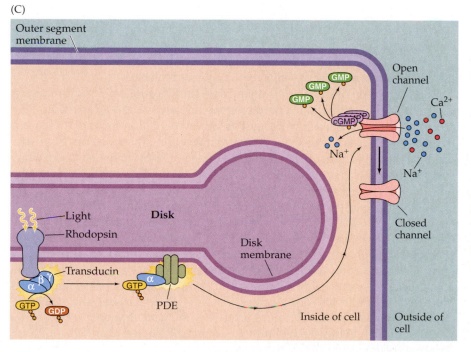

Figure 11.9 Details of phototransduction in rod photoreceptors. (A) Rhodopsin resides in the disk membrane of the photoreceptor outer segment. The seven transmembrane domains of the opsin molecule enclose the light-sensitive retinal molecule. (B) Absorption of a photon of light by retinal leads to a change in configuration from the 11-*cis* to the all-*trans* isomer. (C) The second messenger cascade of phototransduction. The change in the retinal isomer activates transducin, which in turn activates a phosphodiesterase (PDE). The phosphodiesterase then hydrolyzes cGMP, reducing its concentration in the outer segment and leading to the closure of channels in the outer segment membrane.

breaks and its configuration changes from the 11-*cis* isomer to all-*trans* retinal (Figure 11.9B); this change triggers a series of alterations in the opsin component of the molecule. The changes in opsin lead, in turn, to the activation of an intracellular messenger called **transducin**, which activates a phosphodiesterase that hydrolyzes cGMP. All of these events take place within the disk membrane. The hydrolysis by phosphodiesterase at the disk membrane lowers the concentration of cGMP throughout the outer segment, and thus reduces the number of cGMP molecules that are available for binding to the channels in the surface of the outer segment membrane, leading to channel closure (Figure 11.9C).

One of the important features of this complex biochemical cascade initiated by photon capture is that it provides enormous signal amplification. It has been estimated that a single light-activated rhodopsin molecule can activate 800 transducin molecules—roughly 8 percent of the transducin molecules on the disk surface. Although each transducin molecule activates only one phosphodiesterase molecule, each PDE is capable of catalyzing the breakdown of as many as 6 cGMP molecules. As a result, the absorption of a single photon by a rhodopsin molecule results in the closure of approximately 200 ion channels, or about 2 percent of the number of channels in each rod that are open in the dark. This number of channel closures causes a net change in the membrane potential of about 1 mV.

Once initiated, additional mechanisms limit the duration of this amplifying cascade and restore the various molecules to their inactivated states. Activated rhodopsin is rapidly phosphorylated by **rhodopsin kinase**, which permits the protein **arrestin** to bind to rhodopsin. Bound arrestin blocks the ability of activated rhodopsin to activate transducin, thus effectively truncating the phototransduction cascade.

The restoration of retinal to a form capable of signaling photon capture is a complex process known as the **retinoid cycle** (Figure 11.10A). The all-*trans* retinal dissociates from opsin and diffuses into the cytosol of the outer segment; there it is converted to all-*trans* retinol and transported into the pigment epithelium via a chaperone protein, **interphotoreceptor retinoid binding protein** (**IRBP**; see Figure 11.10A), where appropriate enzymes ultimately convert it to 11-*cis* retinal. After being transported back into the outer segment via IRBP, 11-*cis* retinal recombines with opsin in the receptor disks. The retinoid cycle is critically important for maintaining the light sensitivity of photoreceptors. Even under intense illumination, the rate of retinal regeneration is sufficient to maintain a significant number of active photopigment molecules.

The magnitude of the phototransduction amplification varies with the prevailing level of illumination, a phenomenon known as **light adaptation**. Photoreceptors are most sensitive to light at low levels of illumination. As levels of illumination increase, sensitivity decreases, preventing the receptors from saturating and thereby greatly extending the range of light intensities over which they operate. The concentration of Ca^{2+} in the outer segment appears to play a key role in the light-induced modulation of photoreceptor sensitivity. The cGMP-gated channels in the outer segment are permeable to both Na^+ and Ca^{2+} (Figure 11.10B); thus, light-induced closure of these channels leads to a net decrease in the internal Ca^{2+} concentration. This decrease triggers a number of changes in the phototransduction cascade, all of which tend to reduce the sensitivity of the receptor to light. For example, the decrease in Ca^{2+} increases the activity of guanylate cyclase, the cGMP-synthesizing enzyme, leading to increased cGMP levels. The decrease in Ca^{2+} also increases the activity of rhodopsin kinase, pemitting more arrestin to bind to rhodopsin. Finally, the decrease in Ca^{2+} increases the affinity of the cGMP-gated channels for cGMP, reducing the impact of the light-induced reduction of cGMP levels. The regula-

(A)

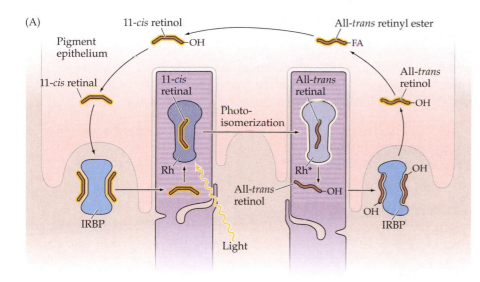

(B)

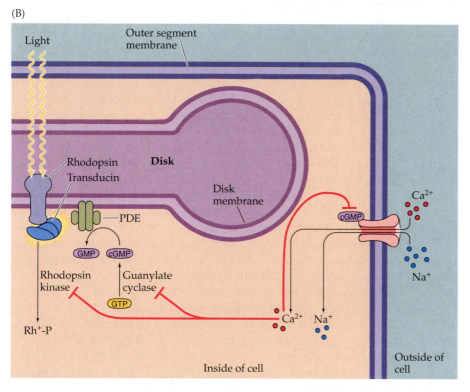

Figure 11.10 The retinoid cycle and photoadaptation. (A) Following photoisomerization, all-*trans* retinal is converted into all-*trans* retinol and is transported by the chaperone protein IRBP into the pigment epithelium. There, in a series of steps, it is converted to 11-*cis* retinal and transported back to the outer segment (again via IRBP), where it recombines with opsin. (B) Photoreceptor adaptation. Calcium in the outer segment inhibits the activity of guanylate cyclase and rhodopsin kinase, and reduces the affinity of cGMP-gated channels for cGMP. Light-induced closure of channels in the outer segment membrane leads to a reduction in Ca^{2+} concentration and a reduction in Ca^{2+}-mediated inhibition of these elements of the cascade. As a result, the photoreceptor's sensitivity to photon capture is reduced.

tory effects of Ca^{2+} on the phototransduction cascade are only one part of the mechanism that adapts retinal sensitivity to background levels of illumination; another important contribution comes from neural interactions between horizontal cells and photoreceptor terminals (discussed later in this chapter).

Functional Specialization of the Rod and Cone Systems

The two types of photoreceptors, rods and cones, are distinguished by shape (from which they derive their names), type of photopigment they contain, distribution across the retina, and pattern of synaptic connections. These properties reflect the fact that the rod and cone systems (i.e., the receptor cells and their connections within the retina) are specialized for different aspects of vision. The

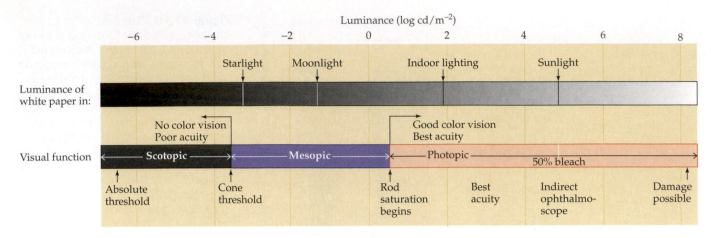

Figure 11.11 The range of luminance values over which the visual system operates. At the lowest levels of illumination, only rods are activated. Cones begin to contribute to perception at about the level of starlight and are the only receptors that function under relatively bright conditions.

rod system has very low spatial resolution but is extremely sensitive to light; it is therefore specialized for sensitivity at the expense of resolution. Conversely, the cone system has very high spatial resolution but is relatively insensitive to light; it is specialized for acuity at the expense of sensitivity. The properties of the cone system also allow humans and many other animals to see color.

The range of illumination over which the rods and cones operate is shown in Figure 11.11. At the lowest levels of illumination, only the rods are activated. Such rod-mediated perception is called **scotopic vision**. The difficulty of making fine visual discriminations under very low light conditions where only the rod system is active is a common experience. The problem is primarily the poor resolution of the rod system (and, to a lesser extent, the fact that there is no perception of color because in dim light there is no significant involvement of the cones). Although cones begin to contribute to visual perception at about the level of starlight, spatial discrimination at this light level is still very poor.

As illumination level increases, cones become more and more dominant in determining what is seen, and they are the major determinant of perception under conditions such as normal indoor lighting or sunlight. The contributions of rods to vision drops out nearly entirely in **photopic vision** because their response to light saturates—that is, the membrane potential of individual rods no longer varies as a function of illumination because all of the membrane channels are closed (see Figure 11.9). **Mesopic vision** occurs in levels of light at which both rods and cones contribute—at twilight, for example. From these considerations it should be clear that most of what we think of as normal "seeing" is mediated by the cone system, and that loss of cone function is devastating, as occurs in elderly individuals suffering from macular degeneration (see Box 11C). People who have lost cone function are legally blind, whereas those who have lost rod function only experience difficulty seeing at low levels of illumination (night blindness; see Box 11D).

Differences in the transduction mechanisms utilized by the two receptor types are a major factor in the ability of rods and cones to respond to different ranges of light intensity. For example, rods produce a reliable response to a single photon of light, whereas more than 100 photons are required to produce a comparable response in a cone. It is not, however, that cones fail to effectively capture photons. Rather, the change in current produced by single photon capture in cones is comparatively small and difficult to distinguish from background noise. Another difference is that the response of an individual cone does not saturate at high levels of steady illumination, as the rod response does. Although both rods and cones adapt to operate over a range of luminance val-

(A)

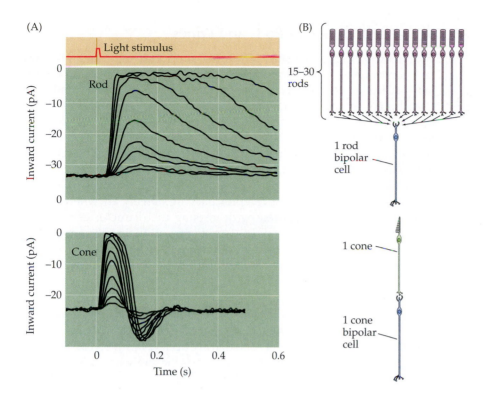

(B)

15–30 rods

1 rod bipolar cell

1 cone

1 cone bipolar cell

Figure 11.12 *Differential responses of primate rods and cones. (A) Suction electrode recordings of the reduction in inward current produced by flashes of successively higher light intensity. For moderate to long flashes, the rod response continues for more than 600 ms, whereas even for the brightest flashes tested, the cone response returns to baseline (with an overshoot) in roughly 200 ms. (B) Difference in the amount of convergence in the rod and cone pathway. Each rod bipolar cell receives synapses from 15–30 rods. Additional convergence occurs at downstream sites in the rod pathway (see text). In contrast, in the center of the fovea, each bipolar cell receives its input from a single cone, and synapses with a single ganglion cell. (A after Baylor, 1987.)*

ues, the adaptation mechanisms of the cones are more effective. This difference in adaptation is apparent in the time course of the response of rods and cones to light flashes. The response of a cone, even to a bright light flash that produces the maximum change in photoreceptor current, recovers in about 200 milliseconds, more than four times faster than rod recovery (Figure 11.12A).

The arrangement of the circuits that transmit rod and cone information to retinal ganglion cells also contributes to the different characteristics of scotopic and photopic vision. In most parts of the retina, rod and cone signals converge on the same ganglion cells; that is, individual ganglion cells respond to both rod and cone inputs, depending on the level of illumination. The early stages of the pathways that link rods and cones to ganglion cells, however, are largely independent. For example, the pathway from rods to ganglion cells involves a distinct class of bipolar cell, called rod bipolar cells, that, unlike cone bipolar cells, do not contact retinal ganglion cells. Instead, rod bipolar cells synapse with the dendritic processes of a specific class of amacrine cell that makes gap junctions and chemical synapses with the terminals of cone bipolars; these processes, in turn, make synaptic contacts on the dendrites of ganglion cells in the inner plexiform layer. Another dramatic difference between rod and cone circuitry is their degree of convergence (Figure 11.12B). Each rod bipolar cell is contacted by a number of rods, and many rod bipolar cells contact a given amacrine cell. In contrast, the cone system is much less convergent. Thus, each retinal ganglion cell that dominates central vision (called midget ganglion cells) receives input from only one cone bipolar cell, which, in turn, is contacted by a single cone. Convergence makes the rod system a better detector of light, because small signals from many rods are pooled to generate a large response in the bipolar cell. At the same time, convergence reduces the spatial resolution of the rod system, since the source of a signal in a rod bipolar cell or retinal ganglion cell could have come from anywhere within a relatively large area of the retinal surface. The one-to-one relationship of cones to bipolar and ganglion cells is just what is required to maximize acuity.

Anatomical Distribution of Rods and Cones

The distribution of rods and cones across the surface of the retina also has important consequences for vision. Despite the fact that perception in typical daytime light levels is dominated by cone-mediated vision, the total number of rods in the human retina (about 90 million) far exceeds the number of cones (roughly 4.5 million). As a result, the density of rods is much greater than cones throughout most of the retina (Figure 11.13A). However, this relationship changes dramatically in the **fovea**, the highly specialized region in the center of the macula that measures about 1.2 millimeters in diameter (see Figure 11.1). In the fovea (which literally means "pit"), cone density increases almost 200-fold, reaching, at its center, the highest receptor packing density anywhere in the retina. This high density is achieved by decreasing the diameter of the cone outer segments such that foveal cones resemble rods in their appearance. The increased density of cones in the fovea is accompanied by a sharp decline in the density of rods. In fact, the central 300 μm of the fovea, called the **foveola**, is totally rod-free (Figure 11.13B).

The extremely high density of cone receptors in the fovea, coupled with the one-to-one relationship with bipolar cells and retinal ganglion cells (see Figure 11.12), endows this component of the cone system with the capacity to mediate the highest levels of visual acuity. As cone density declines with eccentricity and the degree of convergence onto retinal ganglion cells increases, acuity is markedly

(A)

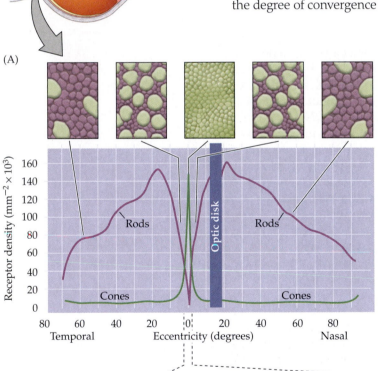

Figure 11.13 Distribution of photoreceptors in the human retina. (A) Cones are present at a low density throughout the retina, with a sharp peak in the center of the fovea (the foveola). Conversely, rods are present at high density throughout most of the retina, with a sharp decline in the fovea; rods are absent in the foveola. Boxes show face-on sections through the outer segments of the photoreceptors at different eccentricities. The increased density of cones in the fovea is accompanied by a striking reduction in the diameter of their outer segments. Note also the lack of receptors at the optic disk, where retinal ganglion cell axons gather to exit the retina. (B) Diagrammatic cross section through the fovea. The overlying cellular layers and blood vessels are displaced so that light is subjected to a minimum of scattering before photons strike the outer segments of the cones in the foveola.

(B)

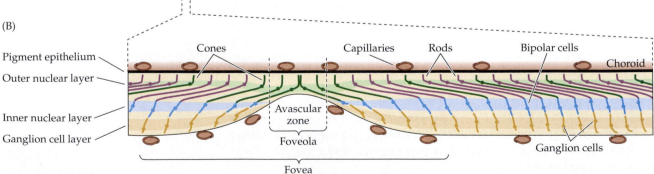

reduced. Just 6° eccentric to the line of sight, acuity is reduced by 75 percent, a fact that can be readily appreciated by trying to read the words on any line of this page beyond the word being fixated on. The restriction of highest acuity vision to such a small region of the retina is the main reason humans spend so much time moving their eyes (and heads) around—in effect directing the foveas of the two eyes to objects of interest (see Chapter 20). It is also the reason why disorders that affect the functioning of the fovea have such devastating effects on sight (see Box 11C). Conversely, the exclusion of rods from the fovea, and their presence in high density away from the fovea, explain why the threshold for detecting a light stimulus is lower outside the region of central vision. It is easier to see a dim object (such as a faint star) by looking slightly away from it, so that the stimulus falls on the region of the retina that is richest in rods (see Figure 11.13A).

Another anatomical feature of the fovea that contributes to the superior acuity of the cone system is the displacement of the inner layers of the retina such that photons are subjected to a minimum of scattering before they strike the photoreceptors. The fovea is also devoid of another source of optical distortion that lies in the light path to the receptors elsewhere in the retina—the retinal blood vessels (see Figure 11.13B). This avascular central region of the fovea is thus dependent on the underlying choroid and pigment epithelium for oxygenation and metabolic sustenance.

Cones and Color Vision

Perceiving color allows humans and many other animals to discriminate objects on the basis of the distribution of the wavelengths of light that they reflect to the eye. While differences in luminance are often sufficient to distinguish objects, color adds another perceptual dimension that is especially useful when differences in light intensity are subtle or nonexistent. Color obviously gives us a quite different way of perceiving and describing the world we live in, and our color vision is the result of special properties of the cone system.

Unlike rods, which contain a single photopigment, there are three types of cones that differ in the photopigment they contain. Each of the photopigments is differentially sensitive to light of different wavelengths, and for this reason cones are referred to as "blue," "green," and "red" or, more appropriately, short (S), medium (M), and long (L) wavelength cones—terms that more or less describe their spectral sensitivities (Figure 11.14A). This nomenclature can be a bit misleading in that it seems to imply that individual cones provide color information for the wavelength of light that excites them best. In fact, individual cones, like rods, are entirely color blind in that their response is simply a reflection of the number of photons they capture, regardless of the wavelength of the photon (or, more properly, its vibrational energy). It is impossible, therefore, to determine whether the change in the membrane potential of a particular cone has arisen from exposure to many photons at wavelengths to which the receptor is relatively insensitive, or fewer photons at wavelengths to which it is most sensitive. This ambiguity can only be resolved by *comparing* the activity in different classes of cones. Based on the responses of individual ganglion cells and cells at higher levels in the visual pathway (see Chapter 12), comparisons of this type are clearly involved in how the visual system extracts color information from spectral stimuli, although a full understanding of the neural mechanisms that underlie color perception has been elusive (Box 11E).

While it might seem natural to assume that the three cone types are present in roughly the same number, this is clearly not the case. S cones make up only about 5–10 percent of the cones in the retina, and they are virtually absent from the center of the fovea. While the M and L types are the predominant retinal

(A)

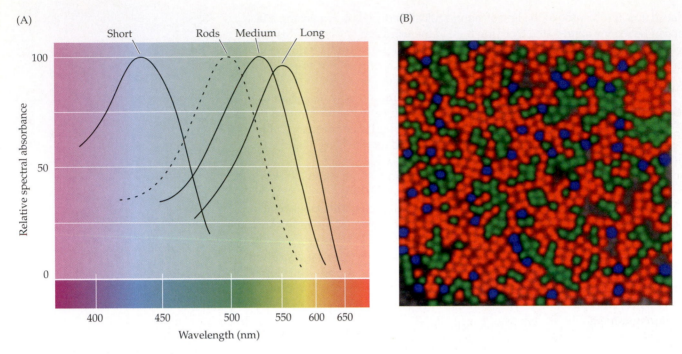

(B)

Figure 11.14 *Absorption spectra and distribution of cone opsins. (A) Light absorption spectra of the four photopigments in normal human retina. (Recall that* light *is defined as electromagnetic radiation having wavelengths between ~400 and 700 nm.* Absorbance *is defined as the log value of the intensity of incident light divided by intensity of transmitted light.) Solid curves represent the three cone opsins; the dashed curve shows rod rhodopsin for comparison. (B) Using a technology known as adaptive optics and clever "tricks" of light adaptation, it is possible to map with great precision the distribution of different cone types within the living retina. Pseudo-color has been used to identify the short- (blue), medium- (green), and long-wavelength (red) cones. (B from Hofer et al., 2005.)*

cones, the ratio of M to L varies considerably from individual to individual, as has been shown using optical techniques that permit visualization of identified cone types in the intact human retina (Figure 11.14B). Interestingly, large differences in the ratio of M and L cone types (from nearly 4:1 to 1:1) do not appear to have a significant impact on color perception.

Much information about color vision has come from studies of individuals with abnormal color-detecting abilities. Color vision deficiencies result either from the inherited failure to make one or more of the cone pigments, or from an alteration in the absorption spectra of cone pigments (or, rarely, from lesions in the central stations that process color information; see Chapter 12). Under normal conditions, most people can match any color in a test stimulus by adjusting the intensity of three superimposed light sources generating long, medium, and short wavelengths. The fact that only three such sources are needed to match nearly all the perceived colors is strong confirmation of the fact that color sensation is based on the relative levels of activity in three sets of cones with different absorption spectra.

That color vision is **trichromatic** was first recognized by Thomas Young at the beginning of the nineteenth century (thus, people with normal color vision are called *trichromats*). But for about 5–6 percent of the male population in the United States (and for a much smaller percentage of the female population), color vision is more limited. Only two bandwidths of light are needed to match all the colors that these individuals can perceive; the third color category is simply not seen. Such **dichromacy**, or "color blindness" as it is commonly called, is inherited as a recessive, sex-linked characteristic and exists in two forms: *protanopia*, in which all color matches can be achieved by using only green and blue light, and *deuteranopia*, in which all matches can be achieved by using only blue and red light (Figure 11.15). In another major class of color deficiencies, all three light sources (i.e., short, medium, and long wavelengths) are needed to make all possible color matches, but the matches are made using values that are significantly different from those used by most individuals. Some of these *anomalous trichromats* require more red than normal to match other colors (protanomalous trichromats); others require more green than normal (deuteranomalous trichromats).

BOX 11E The Importance of Context in Color Perception

Seeing color logically demands that retinal responses to different wavelengths in some way be *compared*. The discovery of the three human cone types and their different absorption spectra is correctly regarded, therefore, as the basis for human color vision. Nevertheless, how these human cone types and the higher-order neurons they contact (see Chapter 12) produce the sensations of color is still unclear. Indeed, this issue has been debated by some of the greatest minds in science (Hering, Helmholtz, Maxwell, Schroedinger, and Mach, to name only a few) since Thomas Young first proposed that humans must have three different receptive "particles"—i.e., the three cone types.

A fundamental problem has been that, although the relative activities of three cone types can more or less explain the colors perceived in color-matching experiments performed in the laboratory, the perception of color is strongly influenced by context. For example, a patch returning the exact same spectrum of wavelengths to the eye can appear quite different depending on its surround, a phenomenon called *color contrast* (Figure A). Moreover, test patches returning different spectra to the eye can appear to be the same color, an effect called *color constancy* (Figure B). Although these phenomena were well known in the nineteenth century, they were not accorded a central place in color vision theory until Edwin Land's work in the 1950s. In his most famous demonstration, Land (who among other

achievements founded the Polaroid company and became a billionaire) used a collage of colored papers that have been referred to as "the Land Mondrians" because of their similarity to the work of the Dutch artist Piet Mondrian.

Using a telemetric photometer and three adjustable illuminators generating short, middle, and long wavelength light, Land showed that two patches that in white light appeared quite different in color (e.g., green and brown) continued to look their respective colors even when the three illuminators were adjusted so that the light being returned from the "green" surfaces produced exactly the same readings on the three telephotometers as had previously come from the "brown" surface—a striking demonstration of color constancy.

The phenomena of color contrast and color constancy have led to a heated modern debate about how color percepts are generated that now spans sev-

eral decades. For Land, the answer lay in a series of ratiometric equations that could integrate the spectral returns of different regions over the entire scene. It was recognized even before Land's death in 1991, however, that his so-called retinex theory did not work in all circumstances and was in any event a description rather than an explanation. An alternative explanation of these contextual aspects of color vision is that color, like brightness, is generated empirically according to what spectral stimuli have typically signified in past experience (see Box 11F).

References

LAND, E. (1986) Recent advances in Retinex theory. *Vision Research* 26: 7–21.

PURVES, D. AND R. B. LOTTO (2003) *Why We See What We Do: An Empirical Theory of Vision*, Chapters 5 and 6. Sunderland MA: Sinauer Associates, pp. 89–138.

(A)

(B)

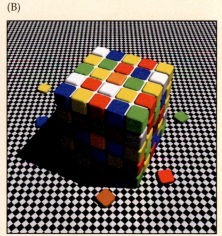

The genesis of contrast and constancy effects by exactly the same context. The two panels demonstrate the effects on apparent color when two *similarly* reflective target surfaces (A) or two *differently* reflective target surfaces (B) are presented in the *same* context in which all the information provided is consistent with illumination that differs only in intensity. The appearances of the relevant target surfaces in a neutral context are shown in the insets below. (From Purves and Lotto, 2003.)

Figure 11.15 *Simulation of the image of a flower as it would appear to an observer with normal color vision (A), an observer with protanopia (loss of long wavelength sensitive cones) (B) and an observer with deuteranopia (loss of medium wavelength sensitive cones) (C). Simulation of the image of a flower as it would appear to an observer with normal color vision (A), an observer with protanopia (loss of long wavelength sensitive cones) (B) and an observer with deuteranopia (loss of medium wavelength sensitive cones) (C). The graphs show absorption spectra of retinal cones in males with defective color vision. (Photographic color simulation courtesy of vischeck.com.)*

(A) Normal (trichromat)

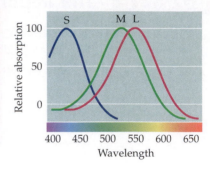

(B) Protanopia

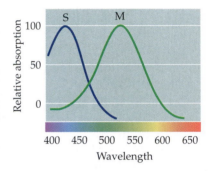

(C) Deuteranopia

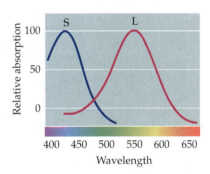

It is difficult for a normal trichromat to appreciate what the world looks like for those who do not possess the full complement of normal cone types. However, some sense of this can be gleaned from looking at color images that have been filtered to simulate the appearance of objects to color blind individuals (see Figure 11.15). Although in most cases abnormal color vision is not a hindrance to normal life, it can pose significant problems for performance on visual tasks that require chromatic discriminations.

Jeremy Nathans and his colleagues at Johns Hopkins University have provided a deeper understanding of color vision deficiencies by identifying and sequencing the genes that encode the three human cone pigments (Figure 11.16A). The genes that encode the red and green pigments show a high degree of sequence homology and lie adjacent to each other on the X chromosome,

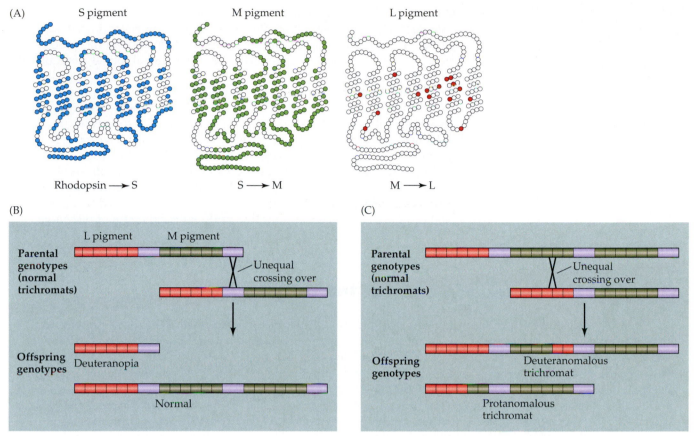

(A) S pigment   M pigment   L pigment

Rhodopsin ⟶ S   S ⟶ M   M ⟶ L

(B)

Parental genotypes (normal trichromats)

L pigment   M pigment

Unequal crossing over

Offspring genotypes

Deuteranopia

Normal

(C)

Parental genotypes (normal trichromats)

Unequal crossing over

Offspring genotypes

Deuteranomalous trichromat

Protanomalous trichromat

Figure 11.16 Genetics of the cone pigments. (A) In these representations of the amino acid sequences of human S-, M- and L-cone pigments, colored dots identify amino acid differences between each photopigment and a comparison pigment. There are substantial differences in the amino acid sequences of rhodopsin and the S-cone pigment, and between the S- and M-cone pigments; however, only a few amino acid differences separate the M- and L-cone pigment sequences. (B,C) Many deficiencies of color vision arise from alterations in the M- or L-cone pigment genes as a result of chromosomal crossing over during meiosis. Colored squares represent the six exons of the L and M genes. (B) Unequal recombination in the intergenic region results in loss of a gene (or duplication of a gene). Loss of a gene results in dichromatic color capabilities (protanopia or deuteranopia). (C) Intragenic recombination results in hybrid genes that code for photopigments with abnormal absorption spectra, consistent with the color vision capabilities of anomalous trichromats. (A after Nathans, 1987; B,C after Deeb, 2005.)

thus explaining the prevalence of color blindness in males. Normal trichromats have one gene for the red pigments and can have anywhere from 1 to 5 genes for green pigments. In contrast, the blue-sensitive pigment gene is found on chromosome 7 and is quite different in its amino acid sequence. These facts suggest that the red and green pigment genes evolved relatively recently, perhaps as a result of the duplication of a single ancestral gene; they also explain why most color vision abnormalities involve the red and green cone pigments.

Human dichromats lack one of the three cone pigments, either because the corresponding gene is missing or because it exists as a hybrid of the red and green pigment genes (Figure 11.16B,C). For example, some dichromats lack the green pigment gene altogether, while others have a hybrid gene that is thought to produce a red-like pigment in the "green" cones. Anomalous trichromats also possess hybrid genes, but these genes elaborate pigments whose spectral properties lie between those of the normal red and green pigments. Thus, although most anomalous trichromats have distinct sets of medium- and long-wavelength cones, there is more overlap in their absorption spectra than in normal trichromats, and thus less difference in how the two sets of cones respond to a given wavelength, with resulting anomalies in color perception.

Retinal Circuits for Detecting Luminance Change

Despite the aesthetic pleasure inherent in color vision, most of the information in visual scenes consists of spatial variations in light intensity; a black and white movie, for example, has most of the information a color version has, although it is deficient in some respects and usually is less fun to watch. How the spatial patterns of light and dark that fall on the photoreceptors are deciphered by central targets has been a vexing problem (Box 11F). To understand what is accomplished by the complex neural circuits within the retina during this process, it is useful to start by considering the responses of individual retinal ganglion cells to small spots of light.

Stephen Kuffler, working at Johns Hopkins University, pioneered this approach in the 1950s when he characterized the responses of single ganglion cells in the cat retina. He found that each ganglion cell responds to stimulation of a small circular patch of the retina, which defines the cell's receptive field (see Chapter 9 for a discussion of receptive fields). Based on these responses, Kuffler distinguished two classes of ganglion cells, "on"-center and "off"-center. Turning on a spot of light in the receptive field center of an **on-center ganglion cell** produces a burst of action potentials. The same stimulus applied to the receptive field center of an **off-center ganglion cell** reduces the rate of discharge, and

---

BOX 11F   The Perception of Light Intensity

Understanding the link between retinal stimulation and what we see (perception) is arguably the central problem in vision, and the relation of luminance (a physical measurement of light intensity) and brightness (the sensation elicited by light intensity) is probably the simplest place to consider this challenge.

As indicated in the text, how we see the brightness differences (i.e., contrast) between adjacent territories with distinct luminances depends in the first instance on the relative firing rate of retinal ganglion cells, modified by lateral interactions. However, there is a problem with the assumption that the central nervous system simply "reads out" these relative rates of ganglion cell activity to sense brightness. The difficulty, as in perceiving color, is that the brightness of a given target is markedly affected by its context in ways that are difficult or impossible to explain in terms of the retinal output as such. The accompanying figures, which illustrate two simultaneous brightness contrast illusions, help make this point. In Figure A, two photometrically identical (equiluminant) gray squares appear

differently bright as a function of the background in which they are presented.

A conventional interpretation of this phenomenon is that the receptive field properties illustrated in Figures 11.14 through 11.17 cause ganglion cells to fire differently depending on whether the surround of the equiluminant target is dark or light. The demonstration in Figure B, however, undermines this explanation, since in this case the target surrounded by more dark area actually looks *darker* than the same target surrounded by more light area.

An alternative interpretation of luminance perception that can account for these puzzling phenomena is that brightness percepts are generated on a statistical basis as a means of contending with the inherent ambiguity of lumi-

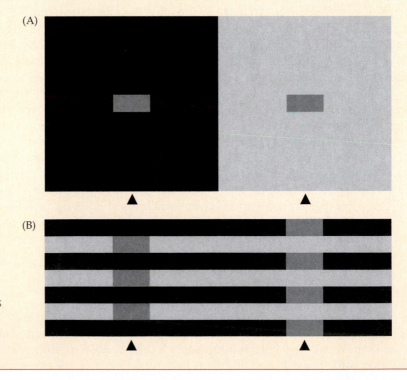

(A) Standard illusion of simultaneous brightness contrast. (B) Another illusion of simultaneous brightness contrast that is difficult to explain in conventional terms.

when the spot of light is turned off, the cell responds with a burst of action potentials (Figure 11.17A). Complementary patterns of activity are found for each cell type when a dark spot is placed in the receptive field center (Figure 11.17B). Thus, on-center cells increase their discharge rate to luminance *increments* in the receptive field center, whereas off-center cells increase their discharge rate to luminance *decrements* in the receptive field center.

On- and off-center ganglion cells are present in roughly equal numbers. The receptive fields have overlapping distributions, so that every point on the retinal surface (that is, every part of visual space) is analyzed by several on-center and several off-center ganglion cells. A rationale for having two distinct types of retinal ganglion cells was suggested by Peter Schiller and his colleagues at the Massachusetts Institute of Technology, who examined the effects of pharmacologically inactivating on-center ganglion cells on a monkey's ability to detect visual stimuli. After silencing on-center ganglion cells, the animals showed a

(C)

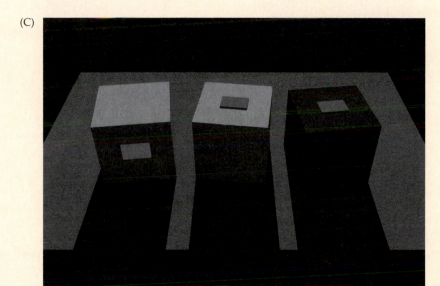

(C) Cartoons of some possible sources of the standard simultaneous brightness contrast illusion in (A). (Courtesy of R. B. Lotto and D. Purves.)

nance (i.e., the fact that a given value of luminance can be generated by many different combinations of illumination and surface reflectance properties). Since to be successful an observer has to respond to the real-world sources of luminance and not to light intensity as such, this ambiguity of the retinal stimulus presents a quandary. A plausible solution to the inherent uncertainty of the relationship between luminance values and their actual sources would be to generate the sensation of brightness elicited by a given luminance (e.g., in the brightness of the identical test patches in the figure) on the basis of what the

luminance of the test patches had typically turned out to be in past experience of human observers. To get the gist of this explanation consider Figure C, which illustrates the point that the two equiluminant target patches in Figure A could have been generated by two differently painted surfaces in different illuminants, as in a comparison of the target patches on the left and middle cubes, or two similarly reflecting surfaces in similar amounts of light, as in a comparison of the target patches on the middle and right cubes. An expedient—and perhaps the only—way the visual system can cope with this ambiguity is to generate

the perception of the stimulus in Figure A (and in Figure B) empirically, i.e., based on what the target patches typically turned out to signify in the past. Since the equiluminant targets will have arisen from a variety of possible sources, it makes sense to have the brightness elicited by the patches determined statistically by the relative frequency of occurrence of that luminance in the particular context in which it is presented. The advantage of seeing luminance according to the relative probabilities of the possible sources of the stimulus is that percepts generated in this way give the observer the best chance of making appropriate behavioral responses to profoundly ambiguous stimuli.

References

ADELSON, E. H. (1999) Light perception and lightness illusions. In *The Cognitive Neurosciences*, 2nd Ed. M. Gazzaniga (ed.). Cambridge, MA: MIT Press, pp. 339–351.

PURVES, D. AND R. B. LOTTO (2003) *Why We See What We Do: An Empirical Theory of Vision*, Chapters 3 and 4. Sunderland MA: Sinauer Associates, pp. 41–87.

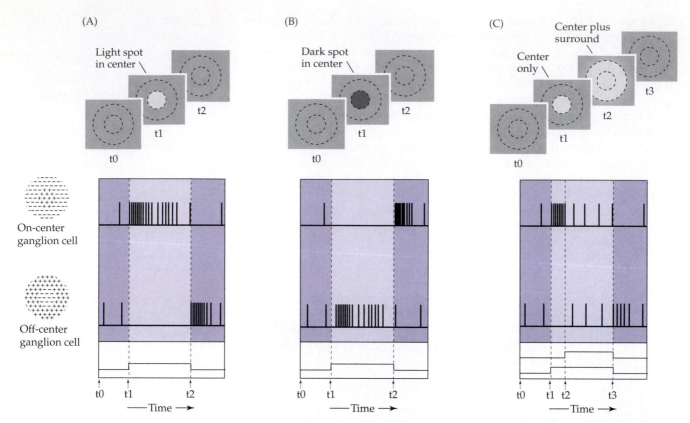

Figure 11.17 The responses of on-center and off-center retinal ganglion cells to stimulation of different regions of their receptive fields. Upper panels indicate the time sequence of stimulus changes; note the overlapping receptive fields. (A) Effects of light spot in the receptive field center. (B) Effects of dark spot in the receptive field center. (C) Effects of light spot in the center followed by the addition of light in the surround.

deficit in their ability to detect stimuli that were brighter than the background; however, they could still see objects that were darker than the background.

These observations imply that information about increases or decreases in luminance is carried separately to the brain by the axons of the two types of retinal ganglion cells. Having separate luminance "channels" means that changes in light intensity, whether increases or decreases, are always conveyed to the brain by an increased number of action potentials. Because ganglion cells rapidly adapt to changes in luminance, their "resting" discharge rate in constant illumination is relatively low. Although an increase in discharge rate above resting level serves as a reliable signal, a decrease in firing rate from an initially low rate of discharge might not be. Having luminance changes signaled by two classes of adaptable cells provides unambiguous information about both increments and decrements in luminance.

The functional differences between these two cell types can be understood in terms of both their anatomy and their physiological properties and relationships. On- and off-center ganglion cells have dendrites that arborize in separate strata of the inner plexiform layer, forming synapses selectively with the terminals of on- and off-center bipolar cells that respond to luminance increases and decreases, respectively (Figure 11.18A). As mentioned previously, the principal difference between ganglion cells and bipolar cells lies in the nature of their electrical response. Like most other cells in the retina, bipolar cells have graded potentials rather than action potentials. Graded depolarization of bipolar cells leads to an increase in transmitter release (glutamate) at their synapses and consequent depolarization of the on–center ganglion cells that they contact via AMPA, kainate, and NMDA receptors.

The selective response of on- and off-center bipolar cells to light increments and decrements is explained by the fact that they express different types of glu-

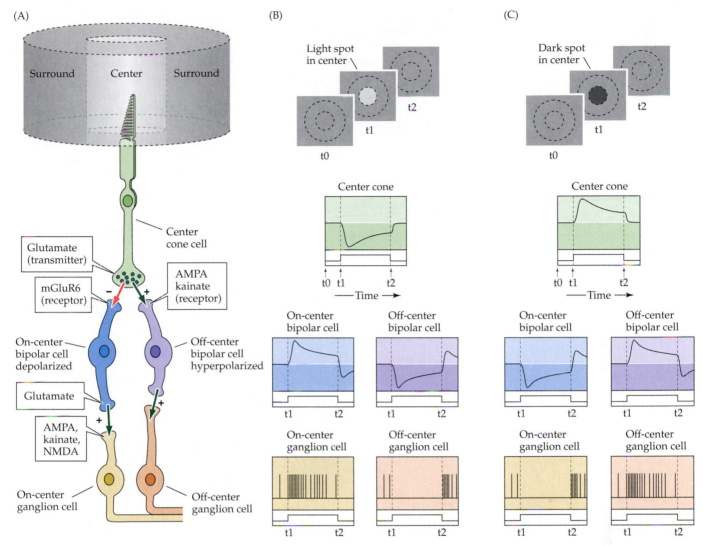

(A)

Surround   Center   Surround

Glutamate (transmitter)

Center cone cell

mGluR6 (receptor)

AMPA kainate (receptor)

On-center bipolar cell depolarized

Off-center bipolar cell hyperpolarized

Glutamate

AMPA, kainate, NMDA

On-center ganglion cell

Off-center ganglion cell

(B)

Light spot in center

t0   t1   t2

Center cone

t0   t1   t2
— Time →

On-center bipolar cell

Off-center bipolar cell

t1   t2   t1   t2

On-center ganglion cell

Off-center ganglion cell

t1   t2   t1   t2

(C)

Dark spot in center

t0   t1   t2

Center cone

t0   t1   t2
— Time →

On-center bipolar cell

Off-center bipolar cell

t1   t2   t1   t2

On-center ganglion cell

Off-center ganglion cell

t1   t2   t1   t2

Figure 11.18   Circuitry responsible for generating receptive field center responses of retinal ganglion cells. (A) Functional anatomy of cone inputs to the center of a ganglion cell receptive field. A plus indicates a sign-conserving synapse; a minus represents a sign-inverting synapse. (B) Responses of the various cell types to the presentation of a light spot in the center of the ganglion cell receptive field. (C) Responses of the various cell types to the presentation of a dark spot in the center of the ganglion cell receptive field.

tamate receptors (see Figure 11.18A). Off-center bipolar cells have ionotropic receptors (AMPA and kainate) that cause the cells to depolarize in response to glutamate released from photoreceptor terminals. In contrast, on-center bipolar cells express a G-protein-coupled metabotropic glutamate receptor (mGluR6). When bound to glutamate, these receptors activate an intracellular cascade that closes cGMP-gated Na$^+$ channels, reducing inward current and hyperpolarizing the cell. Thus, glutamate has opposite effects on these two classes of cells, depolarizing off-center bipolar cells and hyperpolarizing on-center cells. Photoreceptor synapses with off-center bipolar cells are called *sign-conserving*, since the sign of the change in membrane potential of the bipolar cell (depolarization or hyperpolarization) is the same as that in the photoreceptor. Photoreceptor synapses with on-center bipolar cells are called *sign-inverting* because the change in the membrane potential of the bipolar cell is the opposite of that in the photoreceptor.

In order to understand the response of on- and off-center bipolar cells to changes in light intensity, recall that photoreceptors hyperpolarize in response to light increments, decreasing their release of neurotransmitter (Figure 11.18B). Under these conditions, on-center bipolar cells contacted by the photoreceptors are freed from the hyperpolarizing influence of the photoreceptor's transmitter, and they depolarize. In contrast, for off-center cells, the reduction in glutamate represents the withdrawal of a depolarizing influence, and these cells hyperpolarize. Decrements in light intensity naturally have the opposite effect on these two classes of bipolar cells, hyperpolarizing on-center cells and depolarizing off-center ones (Figure 11.18C).

Kuffler's work also called attention to the fact that retinal ganglion cells do not act as simple photodetectors. Indeed, most ganglion cells are relatively poor at signaling differences in the level of diffuse illumination. Instead, they are sensitive to *differences* between the level of illumination that falls on the receptive field center and the level of illumination that falls on the surround—that is, to **luminance contrast**. The center of a ganglion cell receptive field is surrounded

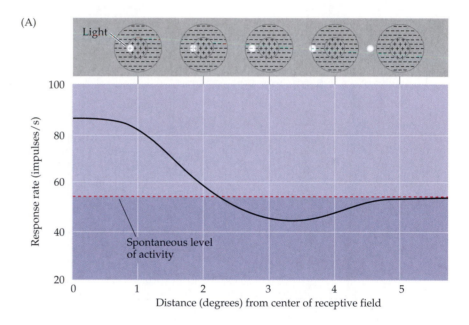

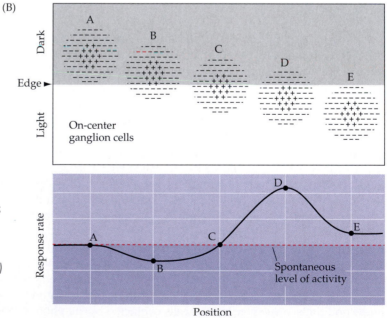

Figure 11.19   *Responses of on-center ganglion cells to different light conditions. (A) The rate of discharge of an on-center cell to a spot of light as a function of the distance of the spot from the receptive field center. Zero on the x axis corresponds to the center; at a distance of 5°, the spot falls outside the receptive field. (B) Responses of a hypothetical population of on-center ganglion cells whose receptive fields (A–E) are distributed across a light-dark edge. Those cells whose activity is most affected have receptive fields that lie along the light-dark edge.*

by a concentric region that, when stimulated, antagonizes the response to stimulation of the receptive field center (see Figure 11.17C). For example, as a spot of light is moved from the center of the receptive field of an on-center cell toward its periphery, the response of the cell to the spot of light decreases. When the spot falls completely outside the center (that is, in the surround), the response of the cell falls below its resting level; the cell is effectively inhibited until the distance from the center is so great that the spot no longer falls on the receptive field at all, in which case the cell returns to its resting level of firing (Figure 11.19A). Off-center cells exhibit a similar surround antagonism. Stimulation of the surround by light opposes the decrease in firing rate that occurs when the center is stimulated alone, and reduces the response to light decrements in the center.

Because of their antagonistic surrounds, most ganglion cells respond much more vigorously to small spots of light confined to their receptive field centers than to either large spots or to uniform illumination of the visual field. To appreciate how center-surround antagonism makes the ganglion cell sensitive to luminance contrast, consider the activity levels in a hypothetical population of on-center ganglion cells whose receptive fields are distributed across a retinal image of a light-dark edge (Figure 11.19B). The neurons whose firing rates are most affected by this stimulus—either increased (neuron D) or decreased (neuron B)—are those with receptive fields that lie along the light-dark border; those with receptive fields completely illuminated (or completely darkened) are less affected (neurons A and E). Thus, the information supplied by the retina to central visual stations for further processing does not give equal weight to all regions of the visual scene; rather, it emphasizes the regions where there are differences in luminance.

Contribution of Retinal Circuits to Light Adaptation

In addition to making ganglion cells especially sensitive to light-dark borders in the visual scene, center-surround mechanisms make a significant contribution to the process of **light adaptation**. In Figure 11.20, the response rate of an on-

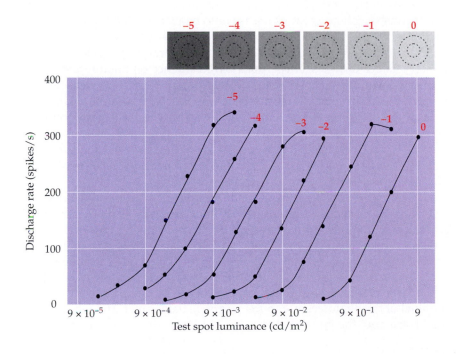

Figure 11.20   A series of curves illustrating the discharge rate of a single on-center ganglion cell to the onset of a small test spot of light in the center of its receptive field. Each curve represents the discharge rate evoked by spots of varying intensity at a constant background level of illumination, which is given by the red numbers at the top of each curve. The highest background level is 0, the lowest –5. Examples at top of figure depict the receptive field at different background levels of illumination. The response rate is proportional to stimulus intensity over a range of 1 log unit, but the operating range shifts to the right as the background level of illumination increases. (After Sakmann and Creutzfeldt, 1969.)

center ganglion cell to a small spot of light turned on in its receptive field center varies as a function of the spot's intensity. In fact, response rate is proportional to the spot's intensity over a range of about one log unit. However, the intensity of spot illumination required to evoke a given discharge rate is dependent on the background level of illumination. Increases in background level of illumination are accompanied by adaptive shifts in the cell's operating range such that greater stimulus intensities are required to achieve the same discharge rate. Thus, firing rate is not an absolute measure of light intensity, but rather signals the *difference from background level* of illumination.

Because the range of light intensities over which we can see is enormous compared to the narrow range of ganglion cell discharge rates (see Figure 11.11), adaptational mechanisms are essential. By scaling the ganglion cell's response to ambient levels of illumination, the entire dynamic range of a neuron's firing rate is used to encode information about intensity differences over the range of luminance values that are relevant for a given visual scene. Due to the antagonistic center-surround organization of retinal ganglion cells, the signal sent to the brain from the retina downplays the background level of illumination. This arrangement presumably explains why the relative brightness of objects remains much the same over a wide range of lighting conditions. The print on this page, for example, reflects considerably more light to the eye when seen in bright sunlight than when we read it in room light. In fact, the *print* reflects more light in sunlight than the *paper* reflects in room light; yet it continues to look black and the page white, indoors or out.

Like the mechanism responsible for generating the on- and off-center response, the antagonistic surround of ganglion cells is thought to be a product of interactions that occur at the early stages of retinal processing. Much of the antagonism is thought to arise via lateral connections established by horizontal cells and receptor terminals (Figure 11.21). Horizontal cells receive synaptic inputs from photoreceptor terminals and are linked via gap junctions with a vast network of other horizontal cells distributed over a wide area of the retinal surface. As a result, the activity in horizontal cells reflects levels of illumination over a broad area of the retina. Although the details of their actions are not entirely clear, horizontal cells are thought to exert their influence via the release of neurotransmitter directly onto photoreceptor terminals, regulating the amount of transmitter that the photoreceptors release onto bipolar cell dendrites.

Glutamate release from photoreceptor terminals has a depolarizing effect on horizontal cells (sign-conserving synapse), while the transmitter released from horizontal cells (GABA) has a hyperpolarizing influence on photoreceptor terminals (sign-inverting synapse) (Figure 11.21A). As a result, the net effect of inputs from the horizontal cell network is to oppose changes in the membrane potential of the photoreceptor that are induced by phototransduction events in the outer segment. How these events lead to surround suppression in an on-center ganglion cell is illustrated in Figure 11.21B. A small spot of light centered on a photoreceptor supplying input to the center of the ganglion cell's receptive field produces a strong hyperpolarizing response in the photoreceptor. Under these conditions, changes in the membrane potential of the horizontal cells that synapse with the photoreceptor terminal are relatively small, and the response of the photoreceptor to light is largely determined by its phototransduction cascade. With the addition of light to the surround, however, the impact of the horizontal network becomes significantly greater; the light-induced reduction in the release of glutamate from the photoreceptors in the surround leads to a strong hyperpolarization of the horizontal cells whose processes converge on the terminal of the photoreceptor in the receptive field center. The reduction in GABA release from the horizontal cells has a depolarizing effect on the mem-

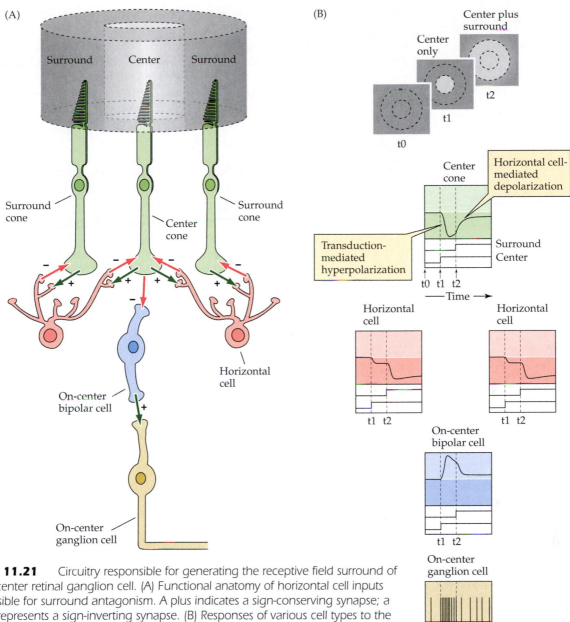

Figure 11.21 *Circuitry responsible for generating the receptive field surround of an on-center retinal ganglion cell. (A) Functional anatomy of horizontal cell inputs responsible for surround antagonism. A plus indicates a sign-conserving synapse; a minus represents a sign-inverting synapse. (B) Responses of various cell types to the presentation of a light spot in the center of the receptive field (t1) followed by the addition of light stimulation in the surround (t2). Light stimulation of the surround leads to hyperpolarization of the horizontal cells and a decrease in the release of inhibitory transmitter (GABA) onto the photoreceptor terminals. The net effect is to depolarize the center cone terminal, offsetting much of the hyperpolarization induced by the transduction cascade in the center cone's outer segment.*

brane potential of the central photoreceptor, reducing the light-evoked response and ultimately reducing the firing rate of the on-center ganglion cell.

Thus, even at the earliest stages in visual processing, neural signals do not represent the absolute numbers of photons that are captured by a receptor, but rather the relative intensity of stimulation—how much the current level of stimulation differs from ambient levels. While it may seem that the actions of horizontal cells decrease the sensitivity of the retina, they play a critical role in allowing the full range of the photoreceptor's electrical response (about 30 mV) to be applied to the limited range of stimulus intensities present at any given moment. The network mechanisms of adaptation described here function in conjunction with cellular mechanisms in the receptor outer segments that regulate the sensitivity of the phototransduction cascade at different light levels. Together, they allow retinal circuits to convey the most salient aspects of luminance changes to the central stages of the visual system described in the following chapter.

Summary

Light falling on photoreceptors is transformed by retinal circuitry into a pattern of action potentials that ganglion cell axons convey to the visual centers in the brain. This process begins with phototransduction, a biochemical cascade that ultimately regulates the opening and closing of ion channels in the membrane of the photoreceptor's outer segment, and thereby the amount of neurotransmitter the photoreceptor releases. Two systems of photoreceptors—rods and cones—allow the visual system to meet the conflicting demands of sensitivity and acuity, respectively. Retinal ganglion cells operate quite differently from the photoreceptor cells. The center-surround arrangement of ganglion cell receptive fields makes these neurons particularly sensitive to luminance contrast and relatively insensitive to the overall level of illumination. It also allows the retina to adapt, such that it can respond effectively over the enormous range of illuminant intensities in the world. The underlying organization is generated by the synaptic interactions between photoreceptors, horizontal cells, and bipolar cells in the outer plexiform layer. As a result, the signal sent to the visual centers in the brain is already highly processed when it leaves the retina, emphasizing those aspects of the visual scene that convey the most information.

Additional Reading

Reviews

ARSHAVSKY, V. Y., T. D. LAMB AND E. N. PUGH JR. (2002) G proteins and phototransduction. *Annu. Rev. Physiol.* 64: 153–187.

BURNS, M. E. AND D. A. BAYLOR (2001) Activation, deactivation, and adaptation in vertebrate photoreceptor cells. *Annu. Rev. Neurosci.* 24: 779–805.

DEEB, S. S. (2005) The molecular basis of variation in human color vision. *Clin. Genet.* 67: 369–377.

LAMB, T. D. AND E. N. PUGH JR. (2004) Dark adaptation and the retinoid cycle of vision. *Prog. Retin. Eye Res.* 23: 307–380.

NATHANS, J. (1987) Molecular biology of visual pigments. *Annu. Rev. Neurosci.* 10: 163–194.

SCHNAPF, J. L. AND D. A. BAYLOR (1987) How photoreceptor cells respond to light. *Sci. Amer.* 256 (April): 40–47.

STERLING, P. (1990) Retina. In *The Synaptic Organization of the Brain*, G. M. Shepherd (ed.). New York: Oxford University Press, pp. 170–213.

STRYER, L. (1986) Cyclic GMP cascade of vision. *Annu. Rev. Neurosci.* 9: 87–119.

WASSLE, H. (2004) Parallel processing in the mammalian retina. *Nature Rev. Neurosci.* 5: 747–757.

Important Original Papers

BAYLOR, D. A., M. G. F. FUORTES AND P. M. O'BRYAN (1971) Receptive fields of cones in the retina of the turtle. *J. Physiol. (Lond.)* 214: 265–294.

DOWLING, J. E. AND F. S. WERBLIN (1969) Organization of the retina of the mud puppy, *Necturus maculosus*. I. Synaptic structure. *J. Neurophysiol.* 32: 315–338.

ENROTH-CUGELL, C. AND R. M. SHAPLEY (1973) Adaptation and dynamics of cat retinal ganglion cells. *J. Physiol.* 233: 271–309.

FASENKO, E. E., S. S. KOLESNIKOV AND A. L. LYUBARSKY (1985) Induction by cyclic GMP of cationic conductance in plasma membrane of retinal rod outer segment. *Nature* 313: 310–313.

HOFER, H., J. CARROLL, J. NEITZ, M. NEITZ AND D. R. WILLIAMS (2005) Organization of the human trichromatic cone mosaic. *J. Neurosci.* 25: 9669–9679.

KUFFLER, S. W. (1953) Discharge patterns and functional organization of mammalian retina. *J. Neurophysiol.* 16: 37–68.

NATHANS, J., D. THOMAS AND D. S. HOGNESS (1986) Molecular genetics of human color vision: The genes encoding blue, green, and red pigments. *Science* 232: 193–202.

NATHANS, J., T. P. PIANTANIDA, R. EDDY, T. B. SHOWS AND D. S. HOGNESS (1986) Molecular genetics of inherited variation in human color vision. *Science* 232: 203–211.

ROORDA, A. AND D. R. WILLIAMS (1999) The arrangement of the three cone classes in the living human eye. *Nature* 397: 520–522.

SAKMANN, B. AND O. D. CREUTZFELDT (1969) Scotopic and mesopic light adaptation in the cat's retina. *Pflügers Arch.* 313: 168–185.

SCHILLER, P. H., J. H. SANDELL AND J. H. R. MAUNSELL (1986) Functions of the "on" and "off" channels of the visual system. *Nature* 322: 824–825.

WERBLIN, F. S. AND J. E. DOWLING (1969) Organization of the retina of the mud puppy, *Necturus maculosus*. II. Intracellular recording. *J. Neurophysiol.* 32: 339–354.

YOUNG, R. W. (1978) The daily rhythm of shedding and degradation of rod and cone outer segment membranes in the chick retina. *Invest. Ophthalmol. Vis. Sci.* 17: 105–116.

Books

BARLOW, H. B. AND J. D. MOLLON (1982) *The Senses*. London: Cambridge University Press.

DOWLING, J. E. (1987) *The Retina: An Approachable Part of the Brain*. Cambridge, MA: Belknap Press.

FAIN, G. L. (2003) *Sensory Transduction*. Sunderland, MA: Sinauer Associates.

HART, W. M. J. (ed.) (1992) *Adler's Physiology of the Eye: Clinical Application*, 9th Ed. St. Louis, MO: Mosby Year Book.

HELMHOLTZ, H. L. F. VON (1924) *Helmholtz's Treatise on Physiological Optics*, Vol. I–III.

Transl. from the Third German Edition by J. P. C. Southall. Menasha, WI: George Banta Publishing Company.

HOGAN, M. J., J. A. ALVARADO AND J. E. WEDDELL (1971) *Histology of the Human Eye: An Atlas and Textbook*. Philadelphia: Saunders.

HUBEL, D. H. (1988) *Eye, Brain, and Vision*, Scientific American Library Series. New York: W. H. Freeman.

HURVICH, L. (1981) *Color Vision*. Sunderland, MA: Sinauer Associates, pp. 180–194.

OGLE, K. N. (1964) *Researches in Binocular Vision*. Hafner: New York.

OYSTER, C. (1999) *The Human Eye: Structure and Function*. Sunderland, MA: Sinauer Associates.

POLYAK, S. (1957) *The Vertebrate Visual System*. Chicago: The University of Chicago Press.

RODIECK, R. W. (1973) *The Vertebrate Retina*. San Francisco: W. H. Freeman.

RODIECK, R. W. (1998) *First Steps in Seeing*. Sunderland, MA: Sinauer Associates.

WANDELL, B. A. (1995) *Foundations of Vision*. Sunderland, MA: Sinauer Associates.

Chapter 12

Central Visual Pathways

Overview

Information supplied by the retina initiates interactions between multiple subdivisions of the brain; these interactions eventually lead to conscious perception of the visual scene, at the same time stimulating more conventional reflexes such as adjusting the size of the pupil, directing the eyes to targets of interest, and regulating homeostatic behaviors that are tied to the day/night cycle. The pathways and structures mediating this broad range of functions are necessarily diverse. Of these, the primary visual pathway from the retina to the dorsal lateral geniculate nucleus in the thalamus and on to the primary visual cortex is the most important and certainly the most thoroughly studied component of the visual system. Different classes of neurons within this pathway encode the variety of visual information—luminance, spectral differences, orientation, and motion—that we ultimately "see." The parallel processing of different categories of visual information continues in cortical pathways that extend beyond the primary visual cortex, supplying visual areas in the occipital, parietal, and temporal lobes. Visual areas in the temporal lobe are primarily involved in object recognition, whereas those in the parietal lobe are concerned with motion. Normal vision depends on the integration of information in all these cortical areas. The processes underlying visual perception are not understood and remain one of the central challenges of modern neuroscience.

Central Projections of Retinal Ganglion Cells

As we saw in Chapter 11, ganglion cell axons exit the retina through a circular region in its nasal part called the **optic disk** (or optic papilla), where they bundle together to form the **optic nerve**. Axons in the optic nerve run a straight course to the **optic chiasm** at the base of the diencephalon (Figure 12.1). In humans, about 60 percent of these fibers cross in the chiasm; the other 40 percent continue toward thalamus and midbrain targets on the same side.

Once past the optic chiasm, the ganglion cell axons on each side form the **optic tract**. Thus the optic tract, unlike the optic nerve, contains fibers from *both* eyes. The partial crossing (*decussation*) of ganglion cell axons at the optic chiasm allows information from corresponding points on the two retinas to be processed by approximately the same cortical site in each hemisphere, an important feature that will be considered in the next section.

The ganglion cell axons in the optic tract reach a number of structures in the diencephalon and midbrain (see Figure 12.1). The major target in the diencephalon is the **dorsal lateral geniculate nucleus** of the thalamus. Neurons in the lateral geniculate nucleus, like their counterparts in the thalamic relays of other sensory systems, send their axons to the cerebral cortex via the internal capsule. These axons pass through a portion of the internal capsule called the

Figure 12.1 *Central projections of retinal ganglion cells. Ganglion cell axons terminate in the lateral geniculate nucleus of the thalamus, the superior colliculus, the pretectum, and the hypothalamus. For clarity, only the crossing axons of the right eye are shown (view is looking up at the inferior surface of the brain).*

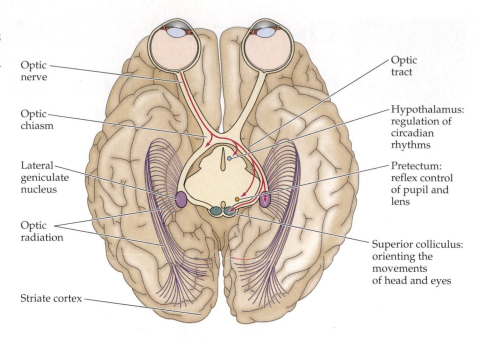

Optic nerve

Optic chiasm

Lateral geniculate nucleus

Optic radiation

Striate cortex

Optic tract

Hypothalamus: regulation of circadian rhythms

Pretectum: reflex control of pupil and lens

Superior colliculus: orienting the movements of head and eyes

optic radiation and terminate in the **primary visual cortex (V1)**, or **striate cortex** (also referred to as **Brodmann's area 17**), which lies largely along and within the calcarine fissure in the occipital lobe. The **retinogeniculostriate pathway**, or **primary visual pathway**, conveys information that is essential for most of what is thought of as seeing; damage anywhere along this route results in serious visual impairment.

A second major target of ganglion cell axons is a collection of neurons that lies between the thalamus and the midbrain in a region known as the **pretectum**. Although small in size compared to the lateral geniculate nucleus, the pretectum is particularly important as the coordinating center for the **pupillary light reflex** (i.e., the reduction in the diameter of the pupil that occurs when sufficient light falls on the retina; Figure 12.2). The initial component of the pupillary light reflex pathway is a bilateral projection from the retina to the pretectum.

Pretectal neurons, in turn, project to the **Edinger-Westphal nucleus**, a small group of nerve cells that lies close to the nucleus of the oculomotor nerve (cranial nerve III) in the midbrain. The Edinger-Westphal nucleus contains the preganglionic parasympathetic neurons that send their axons via the oculomotor nerve to terminate on neurons in the ciliary ganglion (see Chapter 20). Neurons in the ciliary ganglion innervate the constrictor muscle in the iris, which decreases the diameter of the pupil when activated. Shining light in the eye leads to an increase in the activity of pretectal neurons, which stimulates the Edinger-Westphal neurons and the ciliary ganglion neurons they innervate, thus constricting the pupil.

In addition to its normal role in regulating the amount of light that enters the eye, the pupillary reflex provides an important diagnostic tool that allows the physician to test the integrity of the visual sensory apparatus, the motor outflow to the pupillary muscles, and the central pathways that mediate the reflex. Under normal conditions, the pupils of both eyes respond identically, regardless of which eye is stimulated; that is, light in one eye produces constriction of both the stimulated eye (the direct response) and the unstimulated eye (the consensual response). Comparing the responses in the two eyes is often helpful in localizing a lesion. For example, a direct response in the left eye without a con-

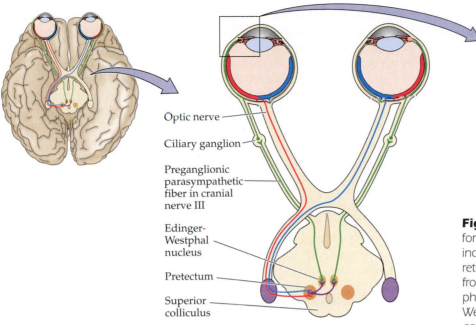

Optic nerve

Ciliary ganglion

Preganglionic parasympathetic fiber in cranial nerve III

Edinger-Westphal nucleus

Pretectum

Superior colliculus

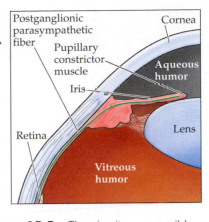

Figure 12.2 The circuitry responsible for the pupillary light reflex. This pathway includes bilateral projections from the retina to the pretectum and projections from the pretectum to the Edinger-Westphal nucleus. Neurons in the Edinger-Westphal nucleus terminate in the ciliary ganglion, and neurons in the ciliary ganglion innervate the pupillary constrictor muscles. Notice that the afferent axons activate both Edinger-Westphal nuclei via the neurons in the pretectum.

sensual response in the right eye suggests a problem with the visceral motor outflow to the right eye, possibly as a result of damage to the oculomotor nerve or Edinger-Westphal nucleus in the brainstem. Failure to elicit a response (either direct or indirect) to stimulation of the left eye if both eyes respond normally to stimulation of the right eye suggests damage to the sensory input from the left eye, possibly to the left retina or optic nerve.

There are several other important targets of retinal ganglion cell axons. One is the **suprachiasmatic nucleus** of the hypothalamus, a small group of neurons at the base of the diencephalon (see Box 21A). The **retinohypothalamic pathway** is the route by which variation in light levels influences the broad spectrum of visceral functions that are entrained to the day/night cycle (see Chapter 28). Another target is the **superior colliculus**, a prominent structure visible on the dorsal surface of the midbrain (see Figures 12.1 and 12.2). The superior colliculus coordinates head and eye movements to visual (as well as other) targets; its functions are considered in Chapter 20.

The type of visual information required to perform the functions of these different retinal targets is quite different. Reading the text on this page, for example, requires a high-resolution sampling of the retinal image, whereas regulating circadian rhythms and adjusting the pupil accordingly require only a measure of overall changes in light levels, and little or no information about the features of the image. It should come as no surprise, then, that there is a diversity of ganglion cell types that provide information appropriate to the functions of these different targets.

Projections to the lateral geniculate nucleus (which are described in more detail later) arise from at least three broad classes of ganglion cells, whose tuning properties are appropriate for mediating the richness of visual perception (high acuity, color, motion). In contrast, projections to the hypothalamus and the pretectum arise from ganglion cells that lack these properties and are highly suited for detecting luminance flux. The retinal specializations responsible for constructing these distinct classes of retinal ganglion cells are only beginning to be identified; they include not only differences in ganglion cell synaptic connections, but in the locus of the phototransduction event itself. Unlike the majority of ganglion cells, which depend on rods and cones for their sensitivity to light, the ganglion cells that project to the hypothalamus and pretectum express their own light-sensitive photopigment (*melanopsin*) and are capable of modulating their response to changes in light levels in the absence of signals from rods and cones. The presence of light sensitivity within this class of ganglion cells presumably explains why normal circadian rhythms are maintained in animals that have completely lost form vision as a result of degeneration of rod and cone photoreceptors.

The Retinotopic Representation of the Visual Field

The spatial relationships among the ganglion cells in the retina are maintained in most of their central targets as orderly representations or "maps" of visual space. Most of these structures receive information from both eyes, requiring that these inputs be integrated to form a coherent map of individual points in space. As a general rule, information from the left half of the visual world, whether it originates from the left or right eye, is represented in the right half of the brain, and vice versa.

Understanding the neural basis for the appropriate arrangement of inputs from the two eyes requires considering how images are projected onto the two retinas, and the central destination of the ganglion cells located in different parts of the retina. Each eye sees a part of visual space that defines its **visual field** (Figure 12.3A). For descriptive purposes, each retina and its corresponding visual field are divided into quadrants. In this scheme, the surface of the retina is subdivided by vertical and horizontal lines that intersect at the center of the fovea (Figure 12.3B). The vertical line divides the retina into **nasal** and **temporal divisions**, and the horizontal line divides the retina into **superior** and **inferior divisions**. Corresponding vertical and horizontal lines in visual space (also called *meridians*) intersect at the **point of fixation** (the point in visual space that falls on the fovea) and define the quadrants of the visual field. The crossing of light rays diverging from different points on an object at the pupil causes the images of objects in the visual field to be inverted and left-right reversed on the retinal surface. As a result, objects in the temporal part of the visual field are seen by the nasal part of the retina, and objects in the superior part of the visual field are seen by the inferior part of the retina. (It may help in understanding Figure 12.3B to imagine that you are looking at the back surfaces of the retinas, with the corresponding visual fields projected onto them.)

With both eyes open, the two foveas are normally aligned on a single target in visual space, causing the visual fields of both eyes to overlap extensively (see Figures 12.3B and 12.4). This **binocular field** of view consists of two symmetrical visual hemifields (left and right). The left binocular hemifield includes the nasal visual field of the right eye and the temporal visual field of the left eye; the right hemifield includes the temporal visual field of the right eye and the nasal visual field of the left eye. The temporal visual fields are more extensive than the nasal visual fields, reflecting the size of the nasal and temporal retinas respectively. As a result, vision in the periphery of the field of view is strictly monocular, mediated by the most medial portion of the nasal retina. Most of the rest of the field of view can be seen by both eyes; i.e., individual points in visual space lie in the nasal visual field of one eye and the temporal visual field of the other. It is worth noting, however, that the shape of the face and nose impact the extent of this region of binocular vision. In particular, the inferior nasal visual fields are less extensive than the superior nasal fields, and consequently the binocular field of view is smaller in the lower visual field than in the upper (see Figure 12.3B).

(A)

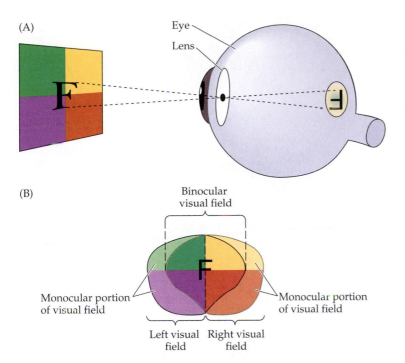

(B)

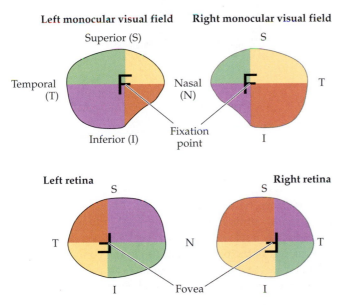

Figure 12.3   Projection of the visual fields onto the left and right retinas. (A) Projection of an image onto the surface of the retina. The passage of light rays through the pupil of the eye results in images that are inverted and left–right reversed on the retinal surface. (B) Retinal quadrants and their relation to the organization of monocular and binocular visual fields, as viewed from the back surface of the eyes. Vertical and horizontal lines drawn through the center of the fovea define retinal quadrants (bottom). Comparable lines drawn through the point of fixation define visual field quadrants (center). Color coding illustrates corresponding retinal and visual field quadrants. The overlap of the two monocular visual fields is shown at the top.

Ganglion cells that lie in the nasal division of each retina give rise to axons that cross in the optic chiasm, while those that lie in the temporal retina give rise to axons that remain on the same side (see Figure 12.4). The boundary, or line of decussation, between contralaterally and ipsilaterally projecting ganglion cells runs through the center of the fovea and defines the border between the nasal and temporal hemiretinas. Images of objects in the left visual hemifield (such as point B in Figure 12.4) fall on the nasal retina of the left eye and the temporal retina of the right eye, and the axons from ganglion cells in these regions of the two retinas project through the right optic tract. Objects in the right visual hemifield (such as point C in Figure 12.4) fall on the nasal retina of the right eye and the temporal retina of the left eye; the axons from ganglion cells in these regions

Figure 12.4 Projection of the binocular field of view onto the two retinas and its relation to the crossing of fibers in the optic chiasm. Points in the binocular portion of the left visual field (B) fall on the nasal retina of the left eye and the temporal retina of the right eye. Points in the binocular portion of the right visual field (C) fall on the nasal retina of the right eye and the temporal retina of the left eye. Points that lie in the monocular portions of the left and right visual fields (A and D) fall on the left and right nasal retinas, respectively. The axons of ganglion cells in the nasal retina cross in the optic chiasm, whereas those from the temporal retina do not. As a result, information from the left visual field is carried in the right optic tract, and information from the right visual field is carried in the left optic tract

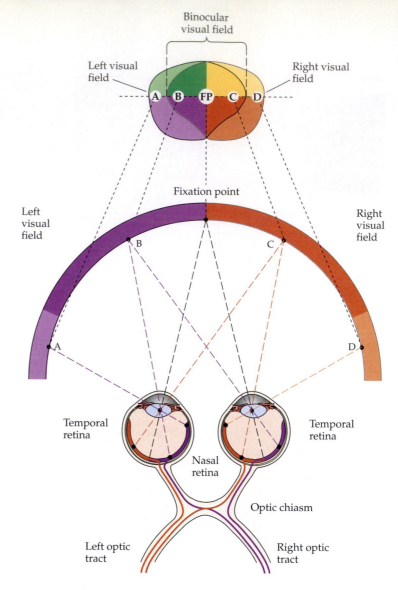

project through the left optic tract. As mentioned previously, objects in the monocular portions of the visual hemifields (points A and D in Figure 12.4) are seen only by the most peripheral nasal retina of each eye; the axons of ganglion cells in these regions (like the rest of the nasal retina) run in the contralateral optic tract. Thus, unlike the optic nerve, the optic tract contains the axons of ganglion cells that originate in both eyes and represent the contralateral field of view.

Optic tract axons terminate in an orderly fashion within their target structures, thus generating well-ordered maps of the contralateral hemifield. For the primary visual pathway, the map of the contralateral hemifield established in the lateral geniculate nucleus is maintained in the projections of the lateral geniculate nucleus to the striate cortex (Figure 12.5). Thus the fovea is represented in the posterior part of the striate cortex, whereas the more peripheral regions of the retina are represented in progressively more anterior parts of the striate cortex. The upper visual field is mapped below the calcarine sulcus, and the lower visual field is mapped above it. As in the somatic sensory system, the amount of cortical area devoted to each unit area of the sensory surface is not uniform, but reflects the density of receptors and sensory axons that supply the peripheral region. Like the representation of the hand region in the somatic sensory cortex, the representation of the macula is therefore disproportionately large, occupying most of the caudal pole of the occipital lobe.

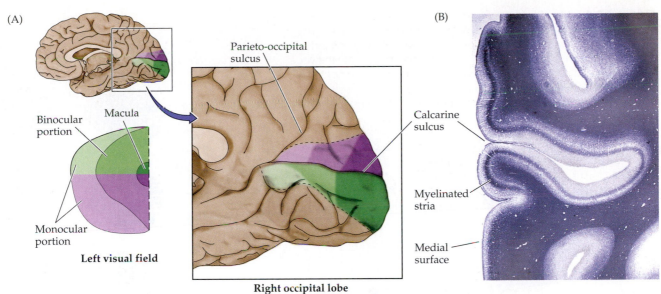

(A)

Parieto-occipital sulcus

Binocular portion

Macula

Monocular portion

Left visual field

Right occipital lobe

(B)

Calcarine sulcus

Myelinated stria

Medial surface

Figure 12.5 Visuotopic organization of the striate cortex in the right occipital lobe, as seen in midsagittal view. (A) The primary visual cortex occupies a large part of the occipital lobe. The area of central vision (the fovea) is represented over a disproportionately large part of the caudal portion of the lobe, whereas peripheral vision is represented more anteriorly. The upper visual field is represented below the calcarine sulcus, the lower field above the calcarine sulcus. (B) Coronal section of the human striate cortex, showing the characteristic myelinated band, or *stria*, that gives this region of the cortex its name. The calcarine sulcus on the medial surface of the occipital lobe is indicated. (B courtesy of T. Andrews and D. Purves.)

Visual Field Deficits

A variety of retinal or more central pathologies that involve the primary visual pathway can cause visual field deficits that are limited to particular regions of visual space. Because the spatial relationships in the retinas are maintained in central visual structures, a careful analysis of the visual fields can often indicate the site of neurological damage. Relatively large visual field deficits are called **anopsias**; smaller ones are called **scotomas**. The former term is combined with various prefixes to indicate the specific region of the visual field from which sight has been lost.

Damage to the retina or one of the optic nerves before it reaches the optic chiasm results in a loss of vision that is limited to the eye of origin (Figure 12.6A). In contrast, damage in the region of the chiasm—or more centrally—results in specific types of deficits that involve the visual fields of both eyes (Figure 12.6B–E). Damage to structures that are central to the optic chiasm, including the optic tract, lateral geniculate nucleus, optic radiation, and visual cortex, results in deficits that are limited to the contralateral visual hemifield. For example, interruption of the optic tract on the right (Figure 12.6C) results in a loss of sight in the left visual field (that is, blindness in the temporal visual field of the left eye and the nasal visual field of the right eye). Because such damage affects corresponding parts of the visual field in each eye, there is a complete loss of vision in the affected region of the binocular visual field, and the deficit is referred to as a **homonymous hemianopsia** (in this case, a left homonymous hemianopsia).

In contrast, damage to the optic chiasm results in visual field deficits that involve noncorresponding parts of the visual field of each eye. For example, damage to the middle portion of the optic chiasm (which is often the result of pituitary tumors) can affect the fibers that are crossing from the nasal retina of each eye, leaving the uncrossed fibers from the temporal retinas intact. The resulting loss of vision is confined to the temporal visual field of each eye and is known as **bitemporal hemianopsia** (see Figure 12.6B). It is also called **heteronomous hemianopsia** to emphasize that the parts of the visual field that are lost in each eye do not overlap. Individuals with this condition are able to see in both left and right visual fields, provided both eyes are open. However, all information from the most peripheral parts of visual fields (which are seen only by the nasal retinas) is lost.

Figure 12.6 Visual field deficits resulting from damage at different points along the primary visual pathway. The diagram on the left illustrates the basic organization of the primary visual pathway and indicates the location of various lesions. The right panels illustrate the visual field deficits associated with each lesion. (A) Loss of vision in right eye. (B) Bitemporal (heteronomous) hemianopsia. (C) Left homonymous hemianopsia. (D) Left superior quadrantanopsia. (E) Left homonymous hemianopsia with macular sparing.

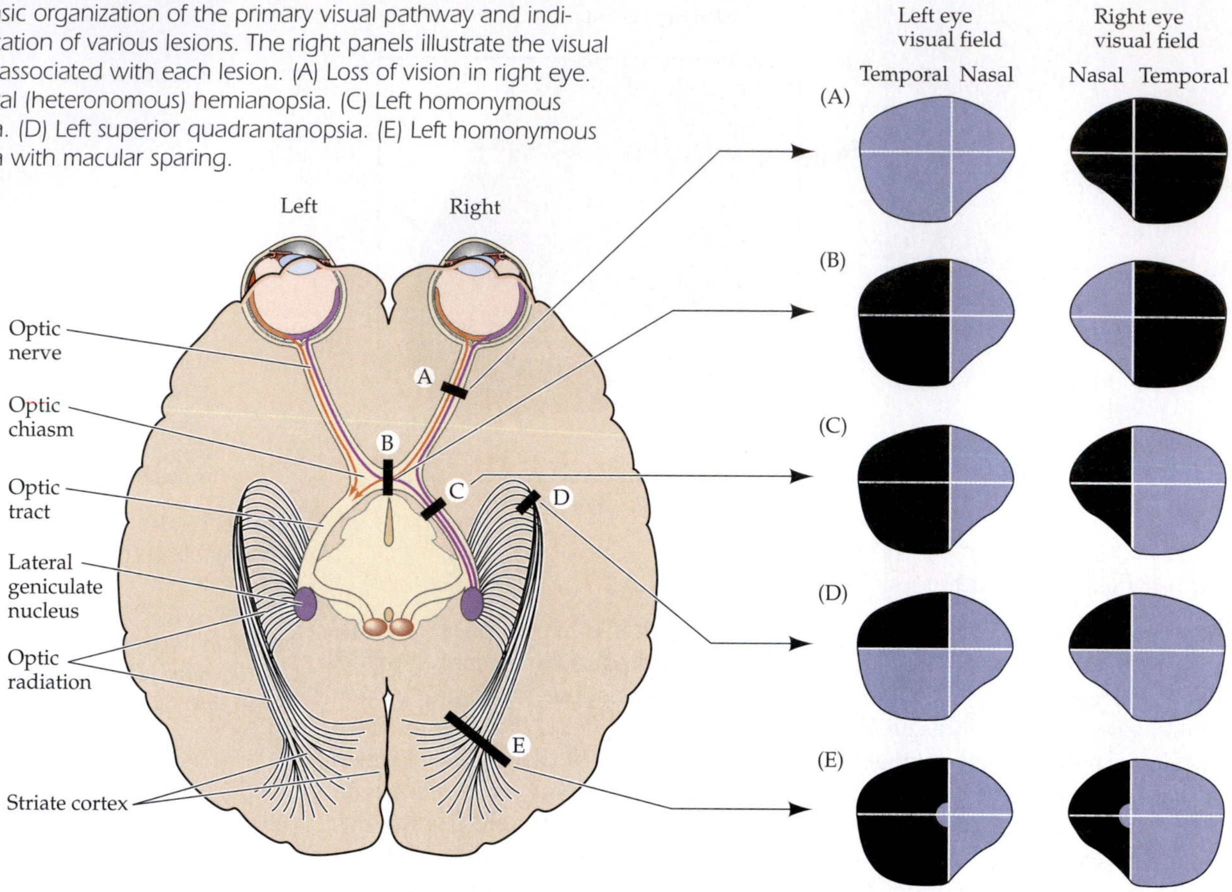

Damage to central visual structures is rarely complete. As a result, the deficits associated with damage to the chiasm, optic tract, optic radiation, or visual cortex are typically more limited than those shown in Figure 12.6. This is especially true for damage along the optic radiation, which fans out under the temporal and parietal lobes in its course from the lateral geniculate nucleus to the striate cortex. Some of the optic radiation axons run out into the temporal lobe on their route to the striate cortex, a branch called **Meyer's loop** (Figure 12.7). Meyer's loop carries information from the superior portion of the contralateral visual field. More medial parts of the optic radiation, which pass under the cortex of the parietal lobe, carry information from the inferior portion of the contralateral visual field. Damage to parts of the temporal lobe with involvement of Meyer's loop can thus result in a superior **homonymous quadrantanopsia** (see Figure 12.6D); damage to the optic radiation underlying the parietal cortex results in an inferior homonymous quadrantanopsia.

Injury to central visual structures can also lead to a phenomenon called *macular sparing*, i.e., the loss of vision throughout wide areas of the visual field, with the exception of foveal vision (see Figure 12.6E). Macular sparing is commonly found with damage to the cortex, but can be a feature of damage anywhere along the length of the visual pathway. Although several explanations for macular sparing have been offered, including overlap in the pattern of crossed and uncrossed ganglion cells supplying central vision, the basis for this selective preservation is not clear.

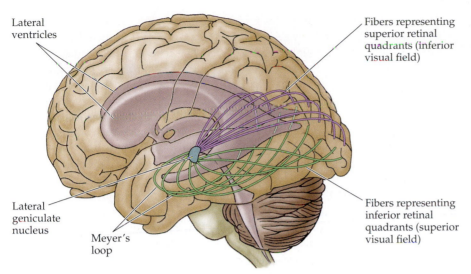

Figure 12.7 Course of the optic radiation to the striate cortex. Axons carrying information about the superior portion of the visual field sweep around the lateral horn of the ventricle in the temporal lobe (Meyer's loop) before reaching the occipital lobe. Those carrying information about the inferior portion of the visual field travel in the parietal lobe.

Spatiotemporal Tuning Properties of Neurons in Primary Visual Cortex

Much of our current understanding of the functional organization of visual cortex had its origin in the pioneering studies of David Hubel and Torsten Wiesel, who used microelectrode recordings in anesthetized animals to examine the responses of individual neurons in the lateral geniculate nucleus and the cortex to various patterns of retinal stimulation (Figure 12.8A). The responses of neurons in the lateral geniculate nucleus were found to be remarkably similar to those in the retina, with a center-surround receptive field organization and selectivity for luminance increases or decreases. However, the small spots of light that were so effective at stimulating neurons in the retina and lateral geniculate nucleus were largely ineffective in visual cortex. Instead, most corti-

Figure 12.8 Neurons in the primary visual cortex respond selectively to oriented edges. (A) An anesthetized animal is fitted with contact lenses to focus the eyes on a screen, where images can be projected; an extracellular electrode records the neuronal responses. (B) Neurons in the primary visual cortex typically respond vigorously to a bar of light oriented at a particular angle and less strongly—or not at all—to other orientations. (C) Orientation tuning curve for a neuron in primary visual cortex. In this example, the highest rate of action potential discharge occurs for vertical edges—the neuron's "preferred" orientation.

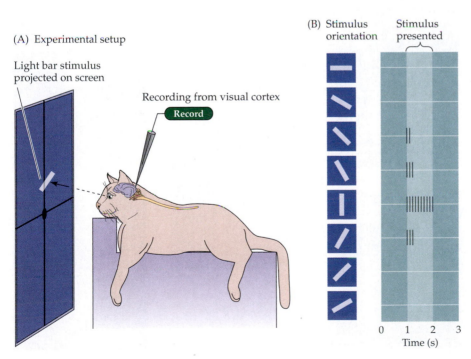

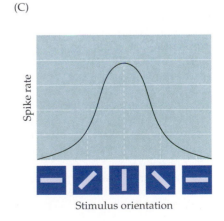

cal neurons in cats and monkeys responded vigorously to light–dark bars or edges, and only if the bars were presented at a particular range of orientations within the cell's receptive field (Figure 12.8B). The responses of cortical neurons are thus tuned to the orientation of edges, much like cone receptors are tuned to the wavelength of light; the peak in the tuning curve (the orientation to which a cell is most responsive) is referred to as the neuron's *preferred orientation* (Figure 12.8C). By sampling the responses of a large number of single cells, Hubel and Wiesel demonstrated that all edge orientations were roughly equally represented in visual cortex. As a result, a given orientation in a visual scene appears to be "encoded" in the activity of a distinct population of orientation-selective neurons.

To appreciate how the properties of an image might be represented by populations of neurons that are tuned to different orientations, an image can be decomposed into its frequency components (an analytical approach discovered by the French mathematician Joseph Fourier) and then filtered to create a set of images whose spectral composition simulate the information that would be conveyed by neurons tuned to different orientations (Figure 12.9). Each class of orientation-selective neuron transmits only a fraction of the information in the scene—the part that matches its filter properties—but the information from these different filters contains all the spatial information necessary to generate a faithful representation of the original image.

Orientation preference is only one of the qualities that defines the filter properties of neurons in primary visual cortex. A substantial fraction of cortical neurons are also tuned to the direction of stimulus motion, for example responding much more vigorously when a stimulus moves to the right than when it moves to the left. Neurons can also be characterized by their preference for spatial frequency (the coarseness or fineness of the variations in contrast that fall within their receptive fields) as well as temporal frequency (rate of change in contrast). Why should cortical neurons show selectivity for these particular stimulus dimensions? Computational analyses suggest that receptive

Figure 12.9 The representation of a visual image by neurons selective for different stimulus orientations. This simulation uses image mathematics (selective filtering of the two-dimensional Fourier transform of the image) to illustrate the attributes of a visual image (greyhound and fence) that would be represented in the responses of populations of cortical neurons tuned to different preferred orientations. The panels surrounding the image illustrate the components of the image that would be detected by neurons tuned to vertical, horizontal, and oblique orientations (blue boxes). In ways that are still not understood, the activity in these different populations of neurons is integrated to yield a coherent representation of the image features. (Photos courtesy of Steve Van Hooser and Elizabeth Johnson.)

fields with properties like these are well-matched to the statistical structure of natural scenes and serve as the basis for an efficient neural code that maximizes the amount of information encoded with a minimum of redundancy.

Primary Visual Cortex Architecture

Like all neocortex, visual cortex is a sheet of neurons approximately 2 millimeters thick that exhibits a conspicuous laminar structure in preparations stained to reveal the density and size of neuronal cell bodies (see Chapter 26 for general overview of cortical structure and cell types). By convention, neocortex is divided into six cellular layers (layers 1–6; Figure 12.10A). To accommodate the laminar complexity exhibited by primate visual cortex, the layers can be further subdivided using Latin and Greek lettering (e.g., layer 4Cβ).

Although the organization of intracortical circuits is complex and not fully understood, it is useful for didactic purposes to outline the basic input–output organization of visual cortex (Figure 12.10B). Lateral geniculate axons terminate primarily in cortical layer 4C, which is composed of a class of neuron (spiny stellate) whose axons convey the activity supplied by the lateral geniculate nucleus to other cortical layers. Pyramidal neurons (named for the cone shape of their cell bodies) in the superficial layers of visual cortex are the source of projections to extrastriate cortical areas, while those in the deeper cortical layers send their axons to subcortical targets, including the lateral geniculate nucleus and the superior colliculus. Thus the laminar organization of visual cortex, like that of other cortical areas, serves to segregate populations of neurons with distinct patterns of connections.

What cannot be discerned from a cursory examination of anatomical sections is that the cortex also exhibits a striking degree of organization in the radial dimension. Microelectrode penetrations perpendicular to the cortical surface encounter columns of neurons that have similar receptive field properties, responding, for example, to stimulation arising from the same region of visual space and exhibiting preferences for similar stimulus properties, such as edge ori-

Figure 12.10 Organization of primary visual (striate) cortex. Striate cortex is divided into six principal cellular layers that differ in cell packing density, cellular morphology, and connections. (A) Primary visual cortex visualized using a histological stain that reveals neuronal cell bodies. In primates, layer 4 has several subdivisions (4A, 4B, and 4C; see also Figure 12.5). (B) Pyramidal cells with prominent apical and basilar dendrites are the most numerous cell type in the neocortex; they are located in all layers except 4C. Layer 4C is dominated by spiny stellate neurons, whose dendrites are confined to this layer. (C) Laminar organization of inputs from the lateral geniculate nucleus (LGN). Lateral geniculate axons terminate most heavily in layers 4C and 4A, with less dense projections to layers 1, 2/3, and 6; the terminations in layer 2/3 are "patchy." (D) Laminar organization of major intracortical connections. Neurons in layer 4C give rise to axons that terminate in more superficial layers (4B and 2/3). Axons of layer 2/3 neurons terminate heavily in layer 5. Axons of layer 6 neurons terminate in layer 4C. (E) Laminar organization of neurons projecting to different targets. Connections with extrastriate cortex arise primarily from neurons in layers 2/3 and 4B (red). Descending projections to the lateral geniculate nucleus arise from layer 6 neurons (blue), while those projecting to the superior colliculus reside in layer 5 (green).

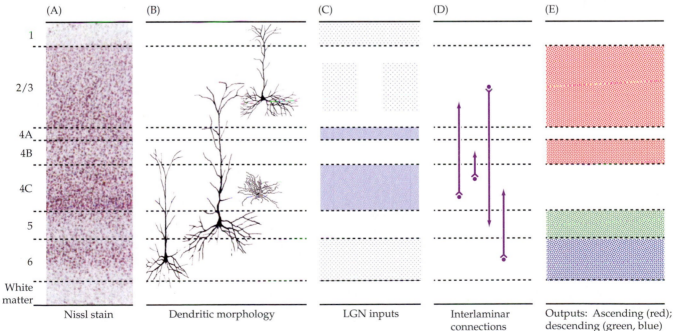

(A)	(B)	(C)	(D)	(E)
Nissl stain	Dendritic morphology	LGN inputs	Interlaminar connections	Outputs: Ascending (red); descending (green, blue)

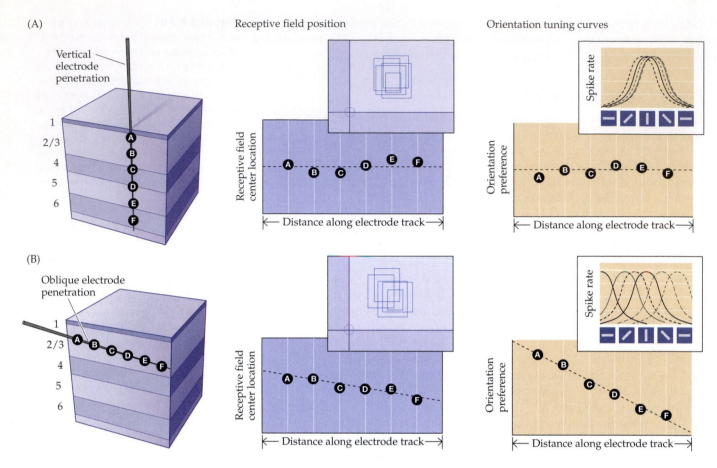

(A) Vertical electrode penetration

Receptive field position

Orientation tuning curves

(B) Oblique electrode penetration

Figure 12.11 Orderly progression of columnar response properties forms the basis of functional maps in primary visual cortex. (A) Neurons displaced along the radial axis of the cortex have receptive fields that are centered on the same region of visual space and exhibit similar orientation preferences. At left is a depiction of a vertical microelectrode penetration into primary visual cortex, Neuronal receptive fields encountered along the electrode track are located in the upper part of the right visual field (central panel, top; intersection of axes represents center of gaze). Note that there is little variation in the location of the receptive field centers (central panel, bottom). The orientation tuning curves (right panel, top) and preferred orientation (right panel, bottom) for neurons encountered along the electrode track show that there is little variation in the orientation preference of the neurons. (B) Neurons displaced along the tangential axis of the cortex exhibit an orderly progression of receptive field properties. Neurons encountered along the electrode penetration have receptive field centers (center panel) and orientation preferences (right panel) that shift in a progressive fashion.

entation and direction of motion (Figure 12.11). The uniformity in response along the radial axis raises the obvious question of how response properties change across adjacent columns. From the previous description of the mapping of visual space in primary visual cortex, it should come as no surprise that adjacent columns have similar but slightly shifted receptive field locations, such that tangential electrode penetrations encounter columns of neurons whose receptive field locations overlap significantly, but shift progressively in a fashion that is consistent with the global mapping of visual space.

(A)　　　　　　　　　　　　　(B)

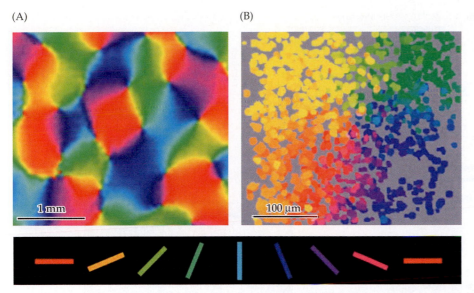

Figure 12.12 Functional imaging techniques reveal the orderly mapping of orientation preference in primary visual cortex. (A) Surface view of the visual cortex using intrinsic signal imaging techniques to visualize the map of preferred orientation. Colors indicate the average preferred orientation of columns at a given location; red indicates the location of columns that respond preferentially to horizontal orientations, blue those that respond preferentially to vertical orientations. The smooth progression of preferred orientations is interrupted by point discontinuities (pinwheel centers; circle). (B) Single-cell view of a "pinwheel" visualized using two-photon imaging of calcium signals. Note that adjacent cells have similar preferred orientations except at the very center, where nearby cells exhibit nearly orthogonal orientation preferences. (A courtesy of D. Fitzpatrick; B modified from Ohki et al., 2006.)

Like receptive field location, the orientation tuning curves of neurons in adjacent columns also overlap significantly, but tangential penetrations frequently reveal an orderly progression in orientation preference (see Figure 12.11B). The availability of functional imaging techniques has made it possible to visualize the two-dimensional layout of the map of orientation preference on the surface of visual cortex (Figure 12.12). Much of the map of orientation preference exhibits smooth progressive change, like that seen for the mapping of visual space. This smooth progression is interrupted periodically by point discontinuities, where neurons with disparate orientation preferences lie close to each other in a pattern resembling a child's pinwheel. The full range of orientation preferences (0–180 degrees) is replicated many times such that neurons with the same orientation preference are arrayed in an iterated fashion, repeating at approximately 1-millimeter intervals across the primary visual cortex. This iteration ensures that the full range of orientation values are represented for each region of visual space—that is, there are no "holes" in the capacity to perceive all stimulus orientations. Thus each point in visual space lies in the receptive fields of a large population of neurons that collectively occupy several millimeters of cortical surface area, an area that contains neurons with the full range of orientation preferences. As described in Box 9C, there are a number of other cortical regions that show a similar columnar arrangement of their processing circuitry.

Combining Inputs from Two Eyes

Unlike neurons at earlier stages in the primary visual pathway, most neurons in striate cortex are binocular, responding to stimulation of both the left and right eyes. Inputs from both eyes are present at the level of the lateral geniculate nucleus, but contralateral and ipsilateral retinal axons terminate in separate layers, so that individual geniculate neurons are strictly monocular, driven either by the left or right eye but not by both (Figure 12.13A–C). Activity arising from the left and right eyes that is conveyed by geniculate axons continues to be segregated at the earliest stages of cortical processing as the axons of geniculate neu-

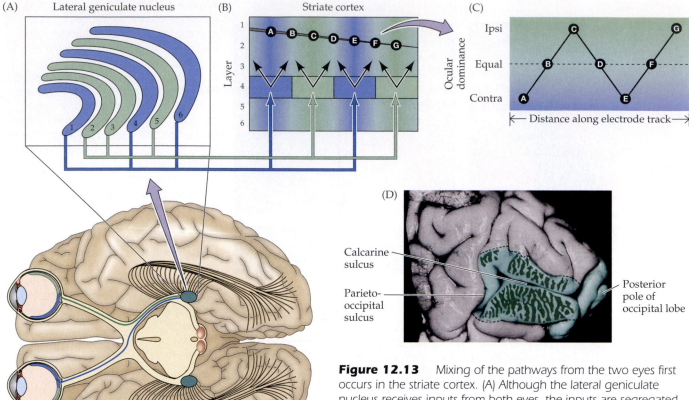

(A) Lateral geniculate nucleus

(B) Striate cortex

(C)

Ipsi

Equal

Contra

Ocular dominance

← Distance along electrode track →

Layer

(D)

Calcarine sulcus

Parieto-occipital sulcus

Posterior pole of occipital lobe

Figure 12.13 Mixing of the pathways from the two eyes first occurs in the striate cortex. (A) Although the lateral geniculate nucleus receives inputs from both eyes, the inputs are segregated in separate layers. (B) In many species, including most primates, inputs from the two eyes remain segregated in the ocular dominance columns of layer 4. Layer 4 neurons send their axons to other cortical layers; it is at this stage that the information from the two eyes converges onto individual neurons. (B,C) Physiological demonstration of columnar organization of ocular dominance in primary visual cortex. Cortical neurons vary in the strength of their response to the inputs from the two eyes, from complete domination by one eye to equal influence of the two eyes. Neurons encountered in a vertical electrode penetration (other than those neurons that lie in layer 4) tend to have similar ocular dominance. Tangential electrode penetration across the superficial cortical layers reveals a gradual shift in the strength of response to the inputs from the two eyes, from complete domination by one eye to equal influence of the two eyes. (D) Pattern of ocular dominance columns in human striate cortex. The alternating left and right eye columns in layer 4 have been reconstructed from tissue sections and projected onto a photograph of the medial wall of the occipital lobe. (D from Horton and Hedley-Whyte, 1984.)

rons terminate in alternating eye-specific **ocular dominance columns** within cortical layer 4 (Figure 12.13D). Beyond this point, however, signals from the two eyes converge as the axons from layer 4 neurons in adjacent monocular stripes synapse on individual neurons in other cortical layers. While most neurons outside of layer 4 are binocular, the relative strength of the inputs from the two eyes varies from neuron to neuron in a columnar fashion that reflects the pattern of ocular dominance stripes in layer 4. Thus neurons that are located over the centers of layer 4 ocular dominance columns respond almost exclusively to the left or right eye, while those that lie over the borders between ocular dominance columns in layer 4 respond equally well to stimulation of both eyes. Similar to the mapping of orientation preference, tangential electrode penetrations across the superficial layers reveal a gradual continuous shift in the ocular dominance of the recorded neurons (see Figure 12.13B,C). With the exception of layer 4, which is strictly monocular, vertical penetrations encounter neurons with similar ocular preferences.

Bringing together the inputs from the two eyes at the level of the striate cortex provides a basis for **stereopsis**, the sensation of depth that arises from viewing nearby objects with two eyes instead of one. Because the two eyes look at the world from slightly different angles, objects that lie in front of or behind the plane of fixation project to non-corresponding points on the two retinas. To convince yourself of this fact, hold your hand at arm's length and fixate on the tip of one finger. Maintain fixation on the finger as you hold a pencil in your other hand about half as far away. At this distance, the image of the pencil falls on non-corresponding points on the two retinas and will therefore be perceived as two separate pencils (a phenomenon called double vision, or *diplopia*). If the pencil is moved toward the finger (the point of fixation), the two images of the pencil fuse and a single pencil is seen in front of the finger. Thus, for a small distance on either side of the plane of fixation, where the disparity between the two views of the world remains modest, a single image is perceived; the disparity between the two eye views of objects nearer or farther than the point of fixation is interpreted as *depth* (Figure 12.14). While disparity cues normally arise from objects in the visual scene, disparity cues alone are powerful enough to give rise to the appearance of objects that cannot be seen by monocular viewing (Box 12A).

Some binocular neurons in the striate cortex and in other visual cortical areas have receptive field properties that make them good candidates for extracting information about binocular disparity. In these neurons, the receptive fields driven by the left and right eyes are slightly offset either in their position in visual space or in their internal organization so that the cell is maximally activated by stimuli that fall on non-corresponding parts of the retinas. Some of these neurons (so-called **far cells**) discharge to retinal disparities that arise from points beyond the plane of fixation, while others (**near cells**) respond to retinal disparities that arise from points in front of the plane of fixation. A third class of neurons (**tuned zero**) respond selectively to points that lie on the plane of fixation. The relative activity in these different classes of neurons is thought to mediate sensations of stereoscopic depth.

Interestingly, the presence of binocular responses in cortical neurons is contingent on normal patterns of activity from the two eyes during early postnatal life. Any factor that creates an imbalance in the activity of the two eyes—for example, the clouding of one lens or the abnormal alignment of the eyes during infancy (*strabismus*)—can permanently reduce the effectiveness of one eye in driving cortical neurons, thus impairing the ability to use binocular information as a cue for depth. Early detection and correction of visual problems is therefore essential for normal visual function in maturity (see Chapter 24).

Division of Labor within the Primary Visual Pathway

In addition to being specific for input from one eye or the other, the layers in the lateral geniculate are also distinguished on the basis of cell size. Two ventral layers are composed of large neurons and are referred to as the **magnocellular layers**, while more dorsal layers are composed of small neurons and are

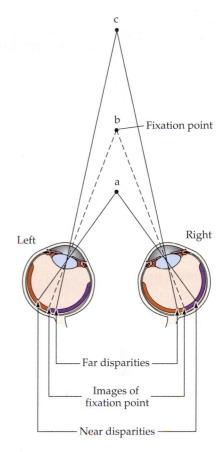

Figure 12.14 Binocular disparities are generally thought to be the basis of stereopsis. When the eyes are fixated on point b, points that lie beyond the plane of fixation (point c) or in front of the point of fixation (point a) project to non-corresponding points on the two retinas. When these disparities are small, the images are fused and the disparity is interpreted by the brain as small differences in depth. When the disparities are greater, double vision occurs (although this normal phenomenon is generally unnoticed).

BOX 12A Random Dot Stereograms and Related Amusements

An important advance in studies of stereopsis was made in 1959 when Bela Julesz, then working at the Bell Laboratories in Murray Hill, New Jersey, discovered an ingenious way of showing that stereoscopy depends on matching information seen by the two eyes without any prior recognition of what object(s) such matching might generate. Julesz, a Hungarian whose background was in engineering and physics, was working on the problem of how to "break" camouflage. He surmised that the brain's ability to fuse the slightly different views of the two eyes to bring out new information would be an aid in overcoming military camouflage. Julesz also realized that, if his hypothesis was correct, a hidden figure in a random pattern presented to the two eyes should emerge when a portion of the otherwise identical pattern was shifted horizontally in the view of one eye or the other. A horizontal shift in one direction would cause the hidden object to appear in front of the plane of the background, whereas a shift in the other direction would cause the hidden object to appear in back of the plane. Such a figure, called a *random dot stereogram*, and the method of its creation are shown in Figures A and B. The two images can be easily fused in a stereoscope (like the familiar ViewMaster®

toy), but can also be fused simply by allowing the eyes to diverge. Most people find it easiest to do this by imagining that they are looking "through" the figure; after some seconds, during which the brain tries to make sense of what it is presented with, the two images merge and the hidden figure appears (in this case, a square that occupies the middle portion of the figure). The random dot stereogram has been widely used in stereoscopic research for about 40 years, although how such stimuli elicit depth remains very much a matter of dispute.

An impressive—and extraordinarily popular—derivative of the random dot stereogram is the autostereogram (Figure C). The possibility of autostereo-

grams was first discerned by the nineteenth-century British physicist David Brewster. While staring at a Victorian wallpaper with an iterated but offset pattern, he noticed that when the patterns were fused, he perceived two different planes. The plethora of autostereograms that can be seen today in posters, books, and newspapers are close cousins of the random dot stereogram in that computers are used to shift patterns of iterated information with respect to each other. The result is that different planes emerge from what appears to be a meaningless array of visual information (or, depending on the taste of the creator, an apparently "normal" scene in which the iterated and displaced information is

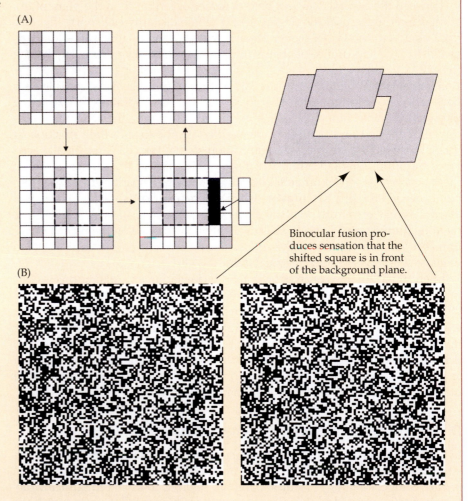

(A)

(B)

Binocular fusion produces sensation that the shifted square is in front of the background plane.

Random dot stereograms. (A) A random dot pattern is created to be observed by one eye. The stimulus for the other eye is created by copying the first image, displacing a particular region horizontally, and then filling in the gap with a random sample of dots. (B) When the right and left images are viewed simultaneously but independently by the two eyes (by using a stereoscope or fusing the images by converging or diverging the eyes), the shifted region (a square) appears to be in a different plane from the other dots. (A after Wandell, 1995.)

hidden). Some autostereograms are designed to reveal the hidden figure when the eyes diverge, and others when they converge. (Looking at a plane more distant than the plane of the surface causes divergence; looking at a plane in front of the picture causes the eyes to converge; see Figure 12.14.)

The elevation of the autostereogram to a popular art form should probably be attributed to Chris W. Tyler, a student of Julesz's and a visual psychophysicist, who was among the first to create commercial autostereograms. Numerous graphic artists—preeminently in Japan, where the popularity of the autostereogram has been enormous—have generated many of such images. As with the random dot stereogram, the task in viewing the autostereogram is not clear to the observer. Nonetheless, the hidden figure emerges, often after minutes of effort in which the brain automatically tries to make sense of the occult information.

(C) An autostereogram. The hidden figure (three geometrical forms) emerges by diverging the eyes in this case. (Courtesy of Jun Oi.)

References

JULESZ, B. (1971) *Foundations of Cyclopean Perception*. Chicago: The University of Chicago Press.

JULESZ, B. (1995) *Dialogues on Perception*. Cambridge, MA: MIT Press.

N. E. THING ENTERPRISES (1993) *Magic Eye: A New Way of Looking at the World*. Kansas City: Andrews and McMeel.

referred to as the **parvocellular layers**. The magno- and parvocellular layers receive inputs from distinct populations of ganglion cells that exhibit corresponding differences in cell size. M ganglion cells that terminate in the magnocellular layers have larger cell bodies, more extensive dendritic fields, and larger-diameter axons than the P ganglion cells that terminate in the parvocellular layers (Figure 12.15A,B). Moreover, the axons of relay cells in the magno- and parvocellular layers of the lateral geniculate nucleus terminate on distinct populations of neurons located in separate strata within layer 4C of primary visual cortex: magnocellular axons terminate in the upper part of layer 4C (4Cα), while parvocellular axons terminate in the lower part of layer 4C (4Cβ) (Figure 12.15C). Thus the retinogeniculate pathway is composed of parallel magnocellular and parvocellular pathways that convey distinct types of information to the initial stages of cortical processing.

The response properties of the M and P ganglion cells provide important clues about the contributions of the magno- and parvocellular pathways to visual perception. M ganglion cells have larger receptive fields than P cells, and their axons have faster conduction velocities. M and P ganglion cells also differ in ways that are not so obviously related to their morphology. M cells respond

Figure 12.15 Magno-, parvo-, and koniocellular pathways. (A) Tracings of M, P, and K ganglion cells as seen in flat mounts of the retina. M cells have large-diameter cell bodies and large dendritic fields. They supply the magnocellular layers of the lateral geniculate nucleus. P cells have smaller cell bodies and dendritic fields. They supply the parvocellular layers of the lateral geniculate nucleus. K cells have small cell bodies and intermediate size dendritic fields. They supply the koniocellular layers of the lateral geniculate nucleus. (B) The human lateral geniculate nucleus showing the magnocellular, parvocellular, and koniocellular layers. (C) Termination of lateral geniculate axons in striate cortex. Magnocellular layers terminate in layer 4Cα, parvocellular layers terminate in layer 4Cβ, and koniocellular layers terminate in a patchy pattern in layers 2 and 3. Inputs to other layers omitted for simplicity (see Figure 12.10). (A after Watanabe and Rodieck, 1989; B courtesy of T. Andrews and D. Purves.)

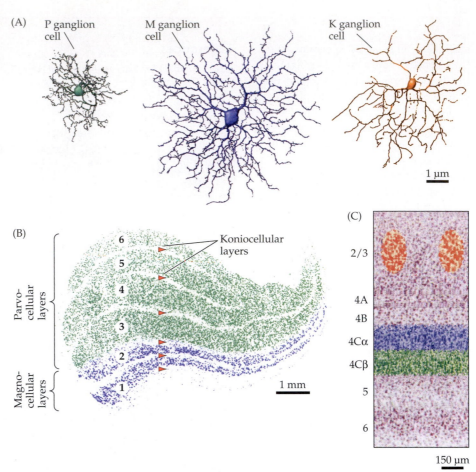

transiently to the presentation of visual stimuli, while P cells respond in a sustained fashion. Moreover, P ganglion cells can transmit information about color, whereas M cells cannot. P cells convey color information because their receptive field centers and surrounds are driven by different classes of cones (i.e., cones responding with greatest sensitivity to short-, medium-, or long-wavelength light). For example, some P ganglion cells have centers that receive inputs from long-wavelength-sensitive cones and surrounds that receive inputs from medium-wavelength cones. Others have centers that receive inputs medium wavelength-sensitive cones and surrounds from long wavelength-sensitive cones (see Chapter 11). As a result, P cells are sensitive to differences in the wavelengths of light striking their receptive field center and surround. Although M ganglion cells also receive inputs from cones, there is no difference in the type of cone input to the receptive field center and surround; the center and surround of each M cell receptive field is driven by all cone types. The absence of cone specificity to center-surround antagonism makes M cells largely insensitive to differences in the wavelengths of light that strike their receptive field centers and surrounds, and they are thus unable to transmit color information to their central targets.

The contribution of the magno- and parvocellular pathways to visual perception has been tested experimentally by examining the visual capabilities of monkeys after selectively damaging either the magno- or parvocellular layers of the lateral geniculate nucleus. Damage to the magnocellular layers has little effect on visual acuity or color vision but sharply reduces the ability to perceive rapidly changing stimuli. In contrast, damage to the parvocellular layers has no effect on motion perception but severely impairs visual acuity and color perception. These observations suggest that the visual information conveyed by the parvocellular pathway is particularly important for high spatial resolution—the detailed analysis of the shape, size, and color of objects. The magnocellular pathway, on the other hand, appears critical for tasks that require high temporal resolution, such as evaluating the location, speed, and direction of a rapidly moving object.

In addition to the magno- and parvocellular pathways, a third distinct anatomical pathway—the **koniocellular**, or **K-cell pathway**—has been identified within the lateral geniculate nucleus (see Figure 12.15). Neurons contributing to the K-cell pathway reside in the interlaminar zones that separate lateral geniculate layers; these neurons receive inputs from fine-caliber retinal axons and project in a patchy fashion to the superficial layers (layers 2 and 3) of primary visual cortex. Although the contribution of the K-cell pathway to perception is not understood, it appears that some aspects of color vision, especially information derived from short-wavelength-sensitive cones, may be transmitted via the K-cell rather than the P-cell pathway. Why short-wavelength-sensitive cone signals should be processed differently from middle- and long-wavelength information is not clear, but the distinction may reflect the earlier evolutionary origin of the K-cell pathway (see Chapter 11).

The Functional Organization of Extrastriate Visual Areas

Anatomical and electrophysiological studies in monkeys have led to the discovery of a multitude of areas in the occipital, parietal, and temporal lobes that are involved in processing visual information (Figure 12.16). Each of these areas contains a map of visual space, and each is largely dependent on the primary visual cortex for its activation. The response properties of the neurons in some of these regions suggest that they are specialized for different aspects of the visual scene. For example, the **middle temporal area** (**MT**) contains neurons that respond selectively to the direction of a moving edge without regard to its color. In contrast, area **V4** contains a high percentage of neurons that respond selectively to the color of a visual stimulus without regard to its direction of movement. These physiological findings are supported by behavioral evidence; thus, damage to area MT leads to a specific impairment in a monkey's ability to perceive the direction of motion in a stimulus pattern, while other aspects of visual perception remain intact.

Functional imaging studies have indicated a similar arrangement of visual areas within human extrastriate cortex. Using retinotopically restricted stimuli, it has been possible to localize at least ten separate representations of the visual field (Figure 12.17). One of these areas exhibits a large motion-selective signal, suggesting that it is the homologue of the motion-selective middle temporal area described in monkeys. Another area exhibits color-selective responses, suggesting that it may be similar to V4 in non-human primates. A role for these areas in the perception of motion and color, respectively, is further supported by evidence for increases in activity not only during the presentation of the relevant stimulus, but also during periods when subjects experience motion or color afterimages.

The clinical description of selective visual deficits after localized damage to various regions of extrastriate cortex also supports functional specialization of extrastriate visual areas in humans. For example, a well-studied patient who suffered a stroke that damaged the extrastriate region thought to be comparable to area MT in the monkey was unable to appreciate the motion of objects, a rare disorder called **cerebral akinetopsia**. The neurologist who treated her noted that she had difficulty in pouring tea into a cup because the fluid seemed to be

(A)

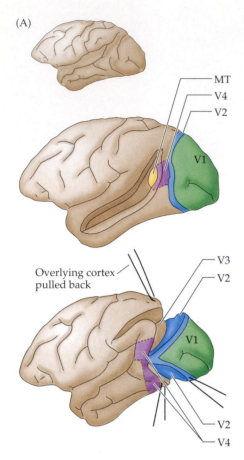

(B)

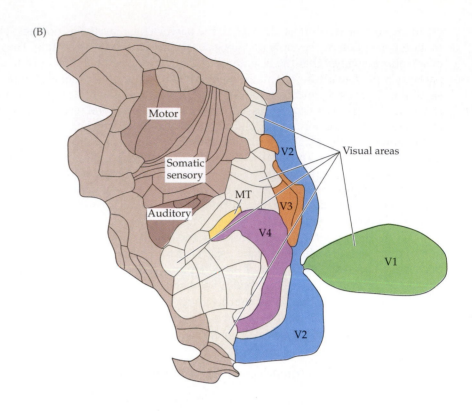

Figure 12.16 *Subdivisions of the extrastriate cortex in the macaque monkey. (A) Each of the subdivisions indicated in color contains neurons that respond to visual stimulation. Many are buried in sulci, and the overlying cortex must be removed in order to expose them. Some of the more extensively studied extrastriate areas are specifically identified (V2, V3, V4, and MT). V1 is the primary visual cortex; MT is the middle temporal area. (B) The arrangement of extrastriate and other areas of neocortex in a flattened view of the monkey neocortex. There are at least 25 areas that are predominantly or exclusively visual in function, plus 7 other areas suspected to play a role in visual processing. (A after Maunsell and Newsome, 1987; B after Felleman and Van Essen, 1991.)*

"frozen." In addition, she could not stop pouring at the right time because she was unable to perceive when the fluid level had risen to the brim. The patient also had trouble following a dialogue because she could not follow the movements of the speaker's mouth. Crossing the street was potentially terrifying because she couldn't judge the movement of approaching cars. As the patient related, "When I'm looking at the car first, it seems far away. But then, when I want to cross the road, suddenly the car is very near." Her ability to perceive other features of the visual scene, such as color and form, was intact.

Another example of a specific visual deficit as a result of damage to extrastriate cortex is **cerebral achromatopsia**. These patients lose the ability to see the world in color, although other aspects of vision remain in good working order. The normal colors of a visual scene are described as being replaced by "dirty" shades of gray, much like looking at a poor-quality black-and-white movie. Achromatopsic individuals know the normal colors of objects—that a school bus is yellow, an apple red—but can no longer see them. When asked to draw objects from memory, they have no difficulty with shapes but are unable to appropriately color the objects they have represented. It is important to distinguish this condition from the color blindness that arises from the congenital absence of one or more cone pigments in the retina (see Chapter 11). In achromatopsia, the three types of cones are functioning normally; it is damage to specific extrastriate cortical areas that renders the patient unable to use the information supplied by the retina.

Based on the anatomical connections between visual areas, differences in electrophysiological response properties, and the effects of cortical lesions, a consensus has emerged that extrastriate cortical areas are organized into two largely separate systems that eventually feed information into cortical association areas in the temporal and parietal lobes (see Chapter 26). One system, called the ventral stream, includes area V4 and leads from the striate cortex into the inferior part of the temporal lobe. This system is thought to be responsible

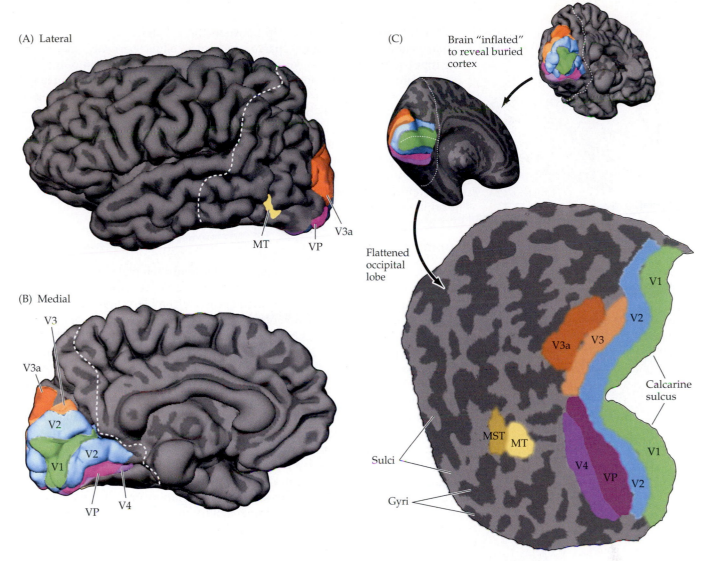

(A) Lateral

MT   VP   V3a

(B) Medial

V3
V3a
V2
V1
V2
VP   V4

(C) Brain "inflated" to reveal buried cortex

Flattened occipital lobe

V1
V2
V3a   V3
Calcarine sulcus
MST   MT
V1
Sulci
V4
VP   V2
Gyri

Figure 12.17 Localization of multiple visual areas in the human brain using fMRI. (A,B) Lateral and medial views (respectively) of the human brain, illustrating the location of primary visual cortex (V1) and additional visual areas V2, V3, VP (ventral posterior area), V4, MT (middle temporal area), and MST (medial superior temporal area). (C) Unfolded and flattened view of retinotopically defined visual areas in the occipital lobe. Dark grey areas correspond to cortical regions that were buried in sulci; light regions correspond to regions that were located on the surface of gyri. Visual areas in humans show a close resemblance to visual areas originally defined in monkeys (compare with Figure 12.16). (After Sereno et al., 1995.)

for high-resolution form vision and object recognition. The dorsal stream, which includes the middle temporal area, leads from striate cortex into the parietal lobe. This system is thought to be responsible for spatial aspects of vision, such as the analysis of motion, and positional relationships between objects in the visual scene (Figure 12.18).

The functional dichotomy between these two streams is supported by observations on the response properties of neurons and the effects of selective cortical lesions. Neurons in the ventral stream exhibit properties that are important for object recognition, such as selectivity for shape, color, and texture. At the highest levels in this pathway, neurons exhibit even greater selectivity, responding preferentially to faces and objects (see Chapter 26). In contrast, those in the dorsal stream are not tuned to these properties, but show selectivity for direction and speed of movement. Consistent with this interpretation, lesions of the parietal cortex severely impair an animal's ability to distinguish objects on the basis of their position, while having little effect on its ability to perform object recognition tasks. In contrast, lesions of the inferotemporal cortex produce profound impairments in the ability to perform recognition tasks but no impairment in spatial tasks. These effects are remarkably similar to the syndromes associated with damage to the parietal and temporal lobe in humans (see Chapters 26 and 27).

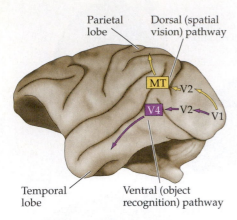

Parietal lobe

Dorsal (spatial vision) pathway

MT   V2

V4   V2

V1

Temporal lobe

Ventral (object recognition) pathway

Figure 12.18   *The visual areas beyond the striate cortex are broadly organized into two pathways: a ventral pathway that leads to the temporal lobe, and a dorsal pathway that leads to the parietal lobe. The ventral pathway plays an important role in object recognition, the dorsal pathway in spatial vision.*

What, then, is the relationship between these "higher order" extrastriate visual steams and the magno-, parvo- and koniocellular pathways that supply the input to primary visual cortex? Despite the initial segregation of magno-, parvo-, and koniocellular inputs in primary visual cortex, at subsequent stages of processing these inputs at least partially converge. The extrastriate areas in the ventral stream clearly have access to the information conveyed by all three pathways, and the dorsal stream, while dominated by inputs from the magnocellular pathway, also receives inputs from the parvo- and koniocellular pathways. Even within primary visual cortex there is ample evidence for the convergence of information from these different lateral geniculate pathways. Thus, it would appear that the functions of higher visual areas involve the integration of information derived from distinct geniculocortical pathways.

Summary

Distinct populations of retinal ganglion cells send their axons to a number of central visual structures that serve different functions. The most important projections are to the pretectum for mediating the pupillary light reflex; to the hypothalamus for the regulation of circadian rhythms; to the superior colliculus for the regulation of eye and head movements; and—most important of all—to the lateral geniculate nucleus for mediating vision and visual perception. The retinogeniculostriate projection (the primary visual pathway) is arranged topographically such that central visual structures contain an organized map of the contralateral visual field. Damage anywhere along the primary visual pathway, which includes the optic nerve, optic tract, lateral geniculate nucleus, optic radiation, and striate cortex, results in a loss of vision confined to a predictable region of visual space. Compared to retinal ganglion cells, neurons at higher levels of the visual pathway become increasingly selective in their stimulus requirements. Thus, most neurons in the striate cortex respond to light–dark edges only if they are presented at a certain orientation, or to movement of the edge in a specific direction. The neural circuitry in the striate cortex also brings together information from the two eyes; most cortical neurons (other than those in layer 4, which are segregated into eye-specific columns) have binocular responses. Binocular convergence is presumably essential for the detection of binocular disparity, an important component of depth perception. The primary visual pathway is composed of separate functional pathways that convey information from different types of retinal ganglion cells to the initial stages of cortical processing. The magnocellular pathway conveys information that is critical for the detection of rapidly changing stimuli, the parvocellular pathway mediates high acuity vision and appears to share responsibility for color vision with the koniocellular pathway. Finally, beyond striate cortex, parcellation of function continues in the ventral and dorsal streams that lead to the extrastriate and association areas in the temporal and parietal lobes, respectively. Areas in the inferotemporal cortex are especially important in object recognition, whereas areas in the parietal lobe are critical for understanding the spatial relations between objects in the visual field.

Additional Reading

Reviews

BERSON, D. M. (2003) Strange vision: Ganglion cells as circadian photoreceptors. *Trends Neurosci.* 26: 314–320.

CALLAWAY, E. M. (2005) Neural substrates within primary visual cortex for interactions between parallel visual pathways. *Prog. Brain Res.* 149: 59–64.

COURTNEY, S. M. AND L. G. UNGERLEIDER (1997) What fMRI has taught us about human vision. *Curr. Op. Neurobiol.* 7: 554–561.

FELLEMAN, D. J. AND D. C. VAN ESSEN (1991) Distributed hierarchical processing in primate cerebral cortex. *Cerebral Cortex* 1: 1–47.

FELSEN, G. AND Y. DAN (2005) A natural approach to studying vision. *Nature Neurosci.* 8: 1643–1646.

GRILL-SPECTOR, K. AND R. MALACH (2004) The human visual cortex. *Annu. Rev Neurosci.* 27: 649–677.

HENDRY, S. H. AND R. C. REID (2000) The koniocellular pathway in primate vision. *Annu. Rev. Neurosci.* 23: 127–153.

HORTON, J. C. (1992) The central visual pathways. In *Alder's Physiology of the Eye*. W. M. Hart (ed.). St. Louis: Mosby Yearbook.

HUBEL, D. H. AND T. N. WIESEL (1977) Functional architecture of macaque monkey visual cortex. *Proc. R. Soc. (Lond.) B:* 198: 1–59.

MAUNSELL, J. H. R. (1992) Functional visual streams. *Curr. Opin. Neurobiol.* 2: 506–510.

OLSHAUSEN, B. A. AND D. J. FIELD (2004) Sparse coding of sensory inputs. *Curr. Opin. Neurobiol.* 14: 481–487.

SCHILLER, P. H. AND N. K. LOGOTHETIS (1990) The color-opponent and broad-band channels of the primate visual system. *Trends Neurosci.* 13: 392–398.

SINCICH, L. C. AND J. C. HORTON (2005) The circuitry of V1 and V2: Integration of color, form, and motion. *Annu. Rev. Neurosci.* 28: 303–326.

TOOTELL, R. B., A. M. DALE, M. I. SERENO AND R. MALACH (1996) New images from human visual cortex. *Trends Neurosci.* 19: 481–489.

UNGERLEIDER, J. G. AND M. MISHKIN (1982) Two cortical visual systems. In *Analysis of Visual Behavior.* D. J. Ingle, M. A. Goodale and R. J. W. Mansfield (eds.). Cambridge, MA: MIT Press, pp. 549–586.

Important Original Papers

HATTAR, S., H. W. LIAO, M. TAKAO, D. M. BERSON AND K. W. YAU (2002) Melanopsin-containing retinal ganglion cells: Architecture, projections, and intrinsic photosensitivity. *Science* 295: 1065–1070.

HUBEL, D. H. AND T. N. WIESEL (1962) Receptive fields, binocular interaction and functional architecture in the cat's visual cortex. *J. Physiol.* (Lond.) 160: 106–154.

HUBEL, D. H. AND T. N. WIESEL (1968) Receptive fields and functional architecture of monkey striate cortex. *J. Physiol.* (Lond.) 195: 215–243.

OHKI, K., S. CHUNG, P. KARA, M. HUBENER, T. BONHOEFFER AND R. C. REID (2006) Highly ordered arrangement of single neurons in orientation pinwheels. *Nature* 442: 925–928.

SERENO, M. I. AND 7 OTHERS (1995) Borders of multiple visual areas in humans revealed by functional magnetic resonance imaging. *Science* 268: 889–893.

ZIHL, J., D. VON CRAMON AND N. MAI (1983) Selective disturbance of movement vision after bilateral brain damage. *Brain* 106: 313–340.

Books

CHALUPA, L. M. AND J. S. WERNER (EDS.) (2004) *The Visual Neurosciences.* Cambridge, MA: MIT Press.

HUBEL, D. H. (1988) *Eye, Brain, and Vision.* New York: Scientific American Library.

RODIECK, R. W. (1998) *The First Steps in Seeing.* Sunderland, MA: Sinauer Associates.

ZEKI, S. (1993) *A Vision of the Brain.* Oxford: Blackwell Scientific Publications.

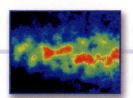

Chapter 13

The Auditory System

Overview

The auditory system is one of the engineering masterpieces of the human body. At the heart of the system is an array of miniature acoustical detectors packed into a space no larger than a pea. These detectors can faithfully transduce vibrations as small as the diameter of an atom, and they can respond a thousand times faster than visual photoreceptors. Such rapid auditory responses to acoustical cues facilitate the initial orientation of the head and body to novel stimuli, especially those that are not initially within the field of view. Although humans are highly visual creatures, much human communication is mediated by the auditory system; indeed, loss of hearing can be more socially debilitating than blindness. From a cultural perspective, the auditory system is essential not only to understanding speech, but also to music, one of the most aesthetically sophisticated forms of human expression. For these and other reasons, audition represents a fascinating and especially important mode of sensation.

Sound

In physical terms, *sound* refers to pressure waves generated by vibrating air molecules (somewhat confusingly, sound is used more casually to refer to an auditory percept). Sound waves are much like the ripples that radiate outward when a rock is thrown into a pool of water. However, instead of occurring across a two-dimensional surface, sound waves propagate in three dimensions, creating spherical shells of alternating compression and rarefaction. Like all wave phenomena, sound waves have four major features: **waveform**, **phase**, **amplitude** (usually expressed in log units known as decibels, abbreviated dB), and **frequency** (expressed in cycles per second or Hertz, abbreviated Hz). For human listeners, the amplitude and frequency of a sound pressure change at the ear roughly correspond to **loudness** and **pitch**, respectively.

The waveform of a sound stimulus is its amplitude plotted against time. It helps to begin by visualizing an acoustical waveform as a sine wave. At the same time, it must be kept in mind that sounds composed of single sine waves (i.e., pure tones) are extremely rare in nature; most sounds in speech, for example, consist of acoustically complex waveforms. Interestingly, such complex waveforms can often be modeled as the sum of sinusoidal waves of varying amplitudes, frequencies, and phases. In engineering applications, an algorithm called the Fourier transform decomposes a complex signal into its sinusoidal components. In the auditory system, as will be apparent later in the chapter, the inner ear acts as a sort of acoustical prism, decomposing complex sounds into a myriad of constituent tones.

Figure 13.1 diagrams the behavior of air molecules near a tuning fork that vibrates sinusoidally when struck. The vibrating tines of the tuning fork produce

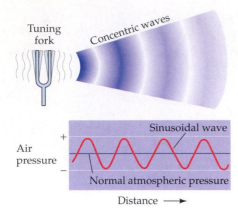

Figure 13.1 Diagram of the periodic condensation and rarefaction of air molecules produced by the vibrating tines of a tuning fork. The molecular disturbance of the air is pictured as if frozen at the instant the constituent molecules responded to the resultant pressure wave. Shown below is a plot of the air pressure versus distance from the fork. Note its sinusoidal quality.

local displacements of the surrounding molecules, such that when the tine moves in one direction, there is molecular condensation; when it moves in the other direction, there is rarefaction. These changes in density of the air molecules are equivalent to local changes in air pressure.

Such regular, sinusoidal cycles of compression and rarefaction can be thought of as a form of circular motion, with one complete cycle equivalent to one full revolution (360°). This point can be illustrated with two sinusoids of the same frequency projected onto a circle, a strategy that also makes it easier to understand the concept of phase (Figure 13.2). Imagine that two tuning forks, both of which resonate at the same frequency, are struck at slightly different times. At a given time $t = 0$, one wave is at position P and the other at position Q. By projecting P and Q onto the circle, their respective phase angles, θ_1 and θ_2, are apparent. The sine wave that starts at P reaches a particular point on the circle, say 180°, at time t_1, whereas the wave that starts at Q reaches 180° at time t_2. Thus, phase differences have corresponding time differences, a concept that is important in appreciating how the auditory system locates sounds in space.

The human ear is extraordinarily sensitive to sound pressure. At the threshold of hearing, air molecules are displaced an average of only 10 picometers (10^{-11} m), and the intensity of such a sound is about one-trillionth of a watt per square meter! This means a listener on an otherwise noiseless planet could hear a 1-watt, 3-kHz sound source located over 450 km away (consider that even a very dim light bulb consumes more than 1 watt of power). Even dangerously high sound pressure levels (>100 dB) have power at the eardrum that is only in the milliwatt range (Box 13A).

The Audible Spectrum

Humans can detect sounds in a frequency range from about 20 Hz to 20 kHz. Human infants can actually hear frequencies slightly higher than 20 kHz, but lose some high-frequency sensitivity as they mature; the upper limit in average adults is closer to 15–17 kHz. Not all mammalian species are sensitive to the same range of frequencies. Most small mammals are sensitive to very high frequencies, but not to low frequencies. For instance, some species of bats are sensitive to tones as high as 200 kHz, but their lower limit is around 20 kHz—the upper limit for young people with normal hearing.

One reason for these differences is that small objects, including the auditory structures of these small mammals, resonate at high frequencies, whereas large objects tend to resonate at low frequencies—which explains why the violin has a higher pitch than the cello. Different animal species tend to emphasize frequency bandwidths in both their vocalizations and their range of hearing. In general, vocalizations by virtue of their periodicity can be distinguished from the noise "barrier" created by environmental sounds, such as wind and rustling leaves. Animals that echolocate, such as bats and dolphins, rely on very high-frequency vocal sounds to maximally resolve spatial features of the target, while animals intent on avoiding predation have auditory systems "tuned" to the low-frequency vibrations that approaching predators transmit through the substrate. These behavioral differences are mirrored by a wealth of anatomical and functional specializations throughout the auditory system.

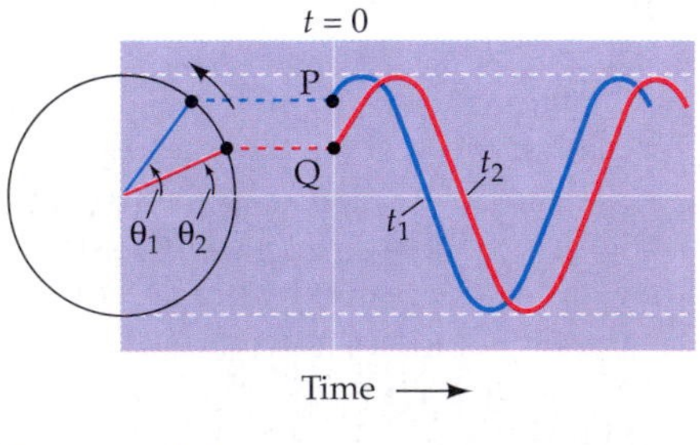

Figure 13.2 A sine wave and its projection as circular motion. The two sinusoids shown are at different phases, such that point P corresponds to phase angle θ_1 and point Q corresponds to phase angle θ_2.

A Synopsis of Auditory Function

The auditory system transforms sound waves into distinct patterns of neural activity, which are then integrated with information from other sensory systems to guide behavior, including orienting movements to acoustical stimuli and

BOX 13A Four Causes of Acquired Hearing Loss

Acquired hearing loss is an increasingly common sensory deficit that can lead to impaired oral communication and social isolation. Four major causes of acquired hearing loss are acoustical trauma, infection of the inner ear, ototoxic drugs, and presbyacusis (literally, "the hearing of the old").

The exquisite sensitivity of the receptor cells in the auditory periphery, combined with the direct mechanical linkage between these receptor cells and the acoustical stimulus, make the ear especially susceptible to acute or chronic acoustical trauma. Extremely loud, percussive sounds, such as those generated by explosives or gunfire, can rupture the eardrum and so severely distort the inner ear that the organ of Corti is torn. The resultant loss of hearing is abrupt and often quite severe.

Less well appreciated is the fact that repeated exposure to less dramatic but nonetheless loud sounds, including those produced by industrial or household machinery or by amplified musical instruments, can also damage the inner ear. Although these sounds leave the eardrum intact, specific damage is done to the hair bundle: the stereocilia of cochlear hair cells of animals exposed to loud sounds either shear off at their pivot points with the hair cell body, or fuse together in a platelike fashion that impedes movement. In humans, the mechanical resonance of the ear to stimulus frequencies centered about 3 kHz means that exposure to loud, broadband noises (such as those generated by jet engines) results in especially pronounced deficits near this resonant frequency.

Ototoxic drugs include aminoglycoside antibiotics such as gentamycin and kanamycin, which directly affect hair cells; and ethacrynic acid, which poisons the potassium-extruding cells of the stria vascularis. In the absence of these K^+-pumping cells, the endocochlear potential, which supplies the energy to drive the auditory signal transduction process, is lost. Although still a matter of some debate, the relatively nonselective transduction channel apparently affords a means of entry for aminoglycoside antibiotics, which then poison hair cells by disrupting phosphoinositide metabolism. In particular, outer hair cells and those inner hair cells that transduce high-frequency stimuli are more affected, simply because of their greater energy requirements.

Finally, presbyacusis, the hearing loss associated with aging, may stem in part from atherosclerotic damage to the especially fine microvasculature of the inner ear, as well as from genetic predispositions to hair-cell damage. Recent advances in understanding the genetic transmission of acquired hearing loss in both humans and mice point to mutations in myosin isoforms unique to hair cells as a likely culprit.

References

CHENG, A. G., L. L. CUNNINGHAM AND E. W. RUBEL (2005) Mechanisms of hair cell death and protection. *Curr. Opin. Otolaryngol Head Neck Surg.* 13: 343–348.

GATES, G. A. AND J. H. MILLS (2005) Presbycusis. *Lancet* 366: 1111–1120.

HANSON, D. R. AND R. W. FEARN (1975) Hearing acuity in young people exposed to pop music and other noise. *Lancet* 2: 203–205.

HOLT, J. R. AND D. P. COREY (1999) Ion channel defects in hereditary hearing loss. *Neuron* 22: 217–219.

KEATS, B. J. AND D. P. COREY (1999) The usher syndromes. *Amer. J. Med. Genet.* 89: 158–166.

PRIUSKA, E. M. AND J. SCHACT (1997) Mechanism and prevention of aminoglycoside ototoxicity: Outer hair cells as targets and tools. *Ear, Nose, Throat J.* 76: 164–171.

intraspecies communication. The first stage of this transformation occurs at the external and middle ears, which collect sound waves and amplify their pressure, so that the sound energy in the air can be successfully transmitted to the fluid-filled cochlea of the inner ear. In the inner ear, a series of biomechanical processes occur that break up the signal into simpler, sinusoidal components, with the result that the frequency, amplitude, and phase of the original signal are all faithfully transduced by the sensory **hair cells** and encoded by the electrical activity of the **auditory nerve fibers**. One product of this process of acoustical decomposition is the systematic representation of sound frequency along the length of the cochlea, referred to as **tonotopy**, which is an important organizational feature preserved throughout the central auditory pathways. The earliest stage of central processing occurs at the cochlear nucleus, where the peripheral auditory information diverges into a number of parallel central pathways. Accordingly, the output of the cochlear nucleus has several targets. One of these is the superior olivary complex, the first place that information from the two ears interacts and the site of the initial processing of the cues that allow lis-

BOX 13B Music

Even though everyone recognizes music when they hear it, the concept of music is a difficult one. The *Oxford English Dictionary* defines it as "The art or science of combining vocal or instrumental sounds with a view toward beauty or coherence of form and expression of emotion." In terms of the present chapter, music chiefly concerns the aspect of human audition that is experienced as *tones*. The stimuli that give rise to tonal percepts are periodic, meaning that they repeat systematically over time, like the sine wave in Figure 13.1. However, natural periodic stimuli do not occur as sine waves, but rather as complex repetitions involving a number of different frequencies; such stimuli give rise to a sense of harmony when sounded together in appropriate combinations, and a sense of melody when they occur sequentially.

Although we usually take for granted the way tone-evoking stimuli are heard, this aspect of audition presents some profound puzzles. The most obvious of these is that humans perceive periodic stimuli whose fundamental frequencies have a 2:1 ratio as highly similar, and, for the most part, musically interchangeable. Thus, in Western musical terminology, any two tones related by an interval of one or more octaves are given the same name (i.e., the notes A, B, C, … G) and are distinguished only by a qualifier that denotes their relative ordinal position

(e.g., C_1, C_2, C_3, etc.). As a result, music is characterized by repeating intervals (called *octaves*) defined by these more or less interchangeable tones. A key question, then, is why periodic sound stimuli whose fundamental frequencies have a 2:1 ratio are perceived as similar.

A second puzzling feature is that most, if not all, musical traditions subdivide octaves into a relatively small set of intervals for composition and performance, each interval being defined by its relationship to the lowest tone of the set. Such sets are called musical *scales*. The scales predominantly employed in all cultures over the centuries have used some (or occasionally all) of the 12 tonal intervals that in Western musical terminology are referred to as the *chromatic scale* (see figure). Moreover, some intervals of the chromatic scale—such as the fifth, the fourth, the major third, and the major sixth—are used more often than others in composition and performance. These form the majority of the intervals employed in the pentatonic and diatonic major scales, the two most frequently used scales in music worldwide. Again, there is no principled explanation of these preferences among all the possible intervals within the octave.

Perhaps the most fundamental question in music—and arguably the common denominator of all musical tonality—is why certain combinations of

Ten of the 12 tones in the chromatic scale, related to a piano keyboard. The function above the keyboard indicates that these tones correspond statistically to peaks of power in normalized human speech. (After Schwartz et al., 2003.)

tones are perceived as relatively consonant, or "harmonious," and others relatively dissonant, or "inharmonious." These perceived differences among the possible combinations of tones making up the chromatic scale are the basis for polytonal music in which the perception of relative harmony guides the composition of chords and melodic lines. The more compatible of these combinations

---

teners to localize sound in space. The cochlear nucleus also projects to the inferior colliculus of the midbrain, a major integrative center and the first place where auditory information can interact with the motor system. The inferior colliculus relays auditory information to the thalamus and cortex, where additional integrative aspects (such as harmonic and temporal combinations) of sound especially germane to speech and music, are processed (Box 13B). The large number of stations between the auditory periphery and the cortex far exceeds those in other sensory systems, providing a hint that the perception of communication and environmental sounds is an especially intensive neural process. Furthermore, both the peripheral and central auditory systems are "tuned" to conspecific communication vocalizations, pointing to the interdependent evolution of neural systems used for generating and perceiving these signals.

are typically used to convey "resolution" at the end of a musical phrase or piece, whereas less compatible combinations are used to indicate a transition, a lack of resolution, or to introduce a sense of tension in a chord or melodic sequence. Like octaves and scales, the reason for this phenomenology remains a mystery.

The classical approaches to rationalizing octaves, scales, and consonance have been based on the fact that the musical intervals corresponding to octaves, fifths, and fourths (in modern musical terminology) are produced by physical sources whose relative proportions (e.g., the relative lengths of two plucked strings or their fundamental frequencies) have ratios of 2:1, 3:2, or 4:3, respectively; these relationships were first described by Pythagoras. This coincidence of numerical simplicity and perceptual effect has been so impressive over the centuries that attempts to rationalize phenomena such as consonance and scale structure in terms of mathematical relationships have tended to dominate the thinking about these issues. This conceptual framework, however, fails to account for many perceptual observations, including, most famously, why people hear the pitch of the fundamental frequency of stimuli comprising only upper harmonics (called "hearing the missing fundamental") and why, when the frequencies of a set of harmonics are changed by a constant value

such that they lack a common divisor, the pitch heard corresponds to neither the fundamental frequency nor the frequency spacing between the harmonics (called the "pitch shift of the residue").

It seems likely that a better way to consider the problem is in terms of the biological rationale for evolving a sense of tonality in the first place rather than by pondering mathematical relationships. Since the auditory system evolved in the world of natural sounds, it is presumably important that the majority of periodic sounds humans were exposed to over the course of evolution were those made by the human vocal tract in the process of communication, initially pre-linguistic but, more recently, speech sounds (see Chapter 27). Thus, developing a sense of tonality would enable listeners to respond not only to the distinctions among the different speech sounds that are important for understanding spoken language, but also to information about the probable sex, age, and emotional state of the speaker. It may thus be that music reflects the advantage of facilitating a listener's ability to glean the linguistic intent and biological state of fellow humans through vocal utterances.

In keeping with this general idea, Michael Lewicki and his colleagues have argued that both music and speech are based on the demands of processing vocal sounds compared to the demands

of processing nonvocal environmental sounds. That auditory systems may have evolved to deal with different categories of natural sound stimuli provides a promising framework for ultimately rationalizing the phenomenology of music.

References

BURNS, E. M. (1999) Intervals, scales, and tuning. In *The Psychology of Music*, D. Deutsch (ed.). New York: Academic Press, pp. 215–264.

CARTERETTE, E. C. AND R. A. KENDALL (1999) Comparative music perception and cognition. In *The Psychology of Music*, D. Deutsch (ed.). New York: Academic Press.

LEWICKI, M. S. (2002) Efficient coding of natural sounds. *Nature Neurosci.* 5: 356–363.

PIERCE, J. R. (1983, 1992) *The Science of Musical Sound*. New York: W. H. Freeman and Co., Chapters 4–6.

PLOMP, R. AND W. J. LEVELT (1965) Tonal consonance and critical bandwidth. *J. Acoust. Soc. Amer.* 28: 548–560.

RASCH, R. AND R. PLOMP (1999) The perception of musical tones. In *The Psychology of Music*, D. Deutsch (ed.). New York: Academic Press, pp. 89–113.

SCHWARTZ, D. A., C. Q. HOWE AND D. PURVES (2003) The statistical structure of human speech sounds predicts musical universals. *J. Neurosci.* 23: 7160–7168.

SCHWARTZ, D. A. AND D. PURVES (2004) Pitch is determined by naturally occurring periodic sounds. *Hearing Research* 194: 31–46.

SMITH, E. AND M. S. LEWICKI (2006). Efficient auditory coding. *Nature* 439: 978–982.

TERHARDT, E. (1974) Pitch, consonance, and harmony. *J. Acoust. Soc. Amer.* 55: 1061–1069.

The External Ear

The external ear, which consists of the **pinna**, **concha**, and **auditory meatus**, gathers sound energy and focuses it on the eardrum, or **tympanic membrane** (Figure 13.3). One consequence of the configuration of the human auditory meatus is that it selectively boosts the sound pressure 30- to 100-fold for frequencies around 3 kHz via passive resonance effects. This amplification makes humans especially sensitive to frequencies in the range of 2–5 kHz—and also explains why they are particularly prone to hearing loss near this frequency following exposure to loud broadband noises, such as those generated by heavy machinery or high explosives (see Box 13A). The sensitivity to this frequency range in the human auditory system appears to be directly related to speech

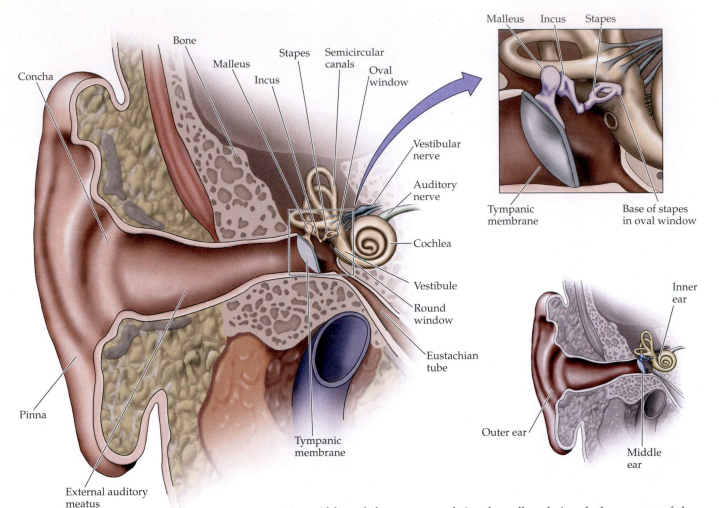

Figure 13.3   *The human ear. Note the large surface area of the tympanic membrane (eardrum) relative to the oval window. This feature, along with the lever action of the malleus, incus, and stapes, facilitates transmission of airborne sounds to the fluid-filled cochlea.*

perception. Although human speech is a broadband signal, the energy of the plosive consonants (e.g., *ba* and *pa*) that distinguish different phonemes (the elementary human speech sounds) is concentrated around 3 kHz (see Box 27A). Therefore, selective hearing loss in the 2–5 kHz range disproportionately degrades speech recognition.

A second important function of the pinna and concha is to selectively filter different sound frequencies in order to provide cues about the elevation of the sound source. The vertically asymmetrical convolutions of the pinna are shaped so that the external ear transmits more high-frequency components from an elevated source than from the same source at ear level. This effect can be demonstrated by recording sounds from different elevations after they have passed through an "artificial" external ear; when the recorded sounds are played back via earphones, so that the whole series is presented from a source at the same elevation relative to the listener, the recordings from higher elevations are perceived as coming from positions higher in space than the recordings from lower elevations.

The Middle Ear

Sounds impinging on the external ear are airborne; however, the environment within the inner ear, where the sound-induced vibrations are converted to neural impulses, is aqueous. The major function of the middle ear is to match relatively low-impedance airborne sounds to the higher-impedance fluid of the inner ear. The term "impedance" in this context describes a medium's resistance to movement. Normally, when sound waves travel from a low-impedance medium like air to a much higher-impedance medium like water, almost all (more than 99.9%) of the acoustical energy is reflected. The middle ear (see Figure 13.3) overcomes this problem and ensures transmission of the sound energy across the air–fluid boundary by boosting the pressure measured at the tympanic membrane almost 200-fold by the time it reaches the inner ear.

Two mechanical processes occur within the middle ear to achieve this large pressure gain. The first and major boost is achieved by focusing the force impinging on the relatively large-diameter tympanic membrane onto the much

---

BOX 13C **Sensorineural Hearing Loss and Cochlear Implants**

The same features that make the auditory periphery exquisitely sensitive to detecting airborne sounds also make it highly vulnerable to damage. By far the most common forms of hearing loss involve the peripheral auditory system—namely, those structures that transmit and transduce sounds into neural impulses. Monaural hearing deficits are the defining symptom of a peripheral hearing loss, because unilateral damage at or above the auditory brainstem results in a binaural deficit (due to the extensive bilateral organization of the central auditory system). Peripheral hearing insults can be further divided into conductive hearing losses, which involve damage to the outer or middle ear, and sensorineural hearing losses, which stem from damage to the inner ear, most typically the cochlear hair cells or the auditory nerve itself. Although both forms of peripheral hearing loss manifest themselves as a raised threshold for hearing on the affected side, their diagnoses and treatments differ.

Conductive hearing loss can be due to occlusion of the ear canal by wax or foreign objects, rupture of the tympanic membrane itself, or arthritic ossification of the middle ear bones. In contrast, sensorineural hearing loss usually is due to congenital or environmental insults that lead to hair cell death (see Box 13A) or damage to the auditory nerve. As hair cells are relatively few in number and do not regenerate in humans, their depletion leads to a diminished ability to

detect sounds. The Weber test, a simple test involving a tuning fork, can be used to distinguish between these two forms of hearing loss. If a resonating tuning fork (~256 Hz) is placed on the vertex, a patient with conductive hearing loss will report that the sound is louder in the affected ear. In the "plugged" state, sounds propagating through the skull do not dissipate so freely back out through the auditory meatus, and thus a greater amount of sound energy is transmitted to the cochlea on the blocked side. In contrast, a patient with a monaural sensorineural hearing loss will report that a Weber test sounds louder on the intact side, because even though the inner ear may vibrate equally on the two sides, the damaged side cannot transduce this vibration into a neural signal.

Treatment also differs for these two types of deafness. In conductive hearing losses, an external hearing aid is used to boost sounds to compensate for the reduced efficiency of the conductive apparatus. These miniature devices are inserted into the ear canal, and contain a microphone and speaker, as well as an amplifier. One limitation of hearing aids is that they often provide rather flat amplification curves, which can interfere with listening in noisy environments; moreover, they do not achieve a high degree of directionality. The use of digital signal-processing strategies partly overcomes these problems, and hearing aids obviously provide significant benefits to many people.

The treatment of sensorineural hearing loss is more complicated and invasive; conventional hearing aids are useless, because no amount of mechanical amplification can compensate for the inability to generate or convey a neural impulse from the cochlea. However, if the auditory nerve is intact, cochlear implants can be used to partially restore hearing. The cochlear implant consists of a peripherally mounted microphone and digital signal processor that transforms a sound into its spectral components, and additional electronics that use this information to activate different combinations of contacts on a threadlike multisite stimulating electrode array. The electrode is inserted into the cochlea through the round window (see figure) and positioned along the length of the tonotopically organized basilar membrane and auditory nerve endings. This placement enables electrical stimulation of the nerve in a manner that mimics some aspects of the spectral decomposition naturally performed by the cochlea.

Cochlear implants can be remarkably effective in restoring hearing to people with hair-cell damage, permitting them to engage in spoken communication. Despite such success in treating those who have lost their hearing *after* having learned to speak, whether cochlear implants can enable development of spoken language in the congenitally deaf is still a matter of debate. Although cochlear implants cannot help patients

(Continued on next page)

BOX 13C (Continued)

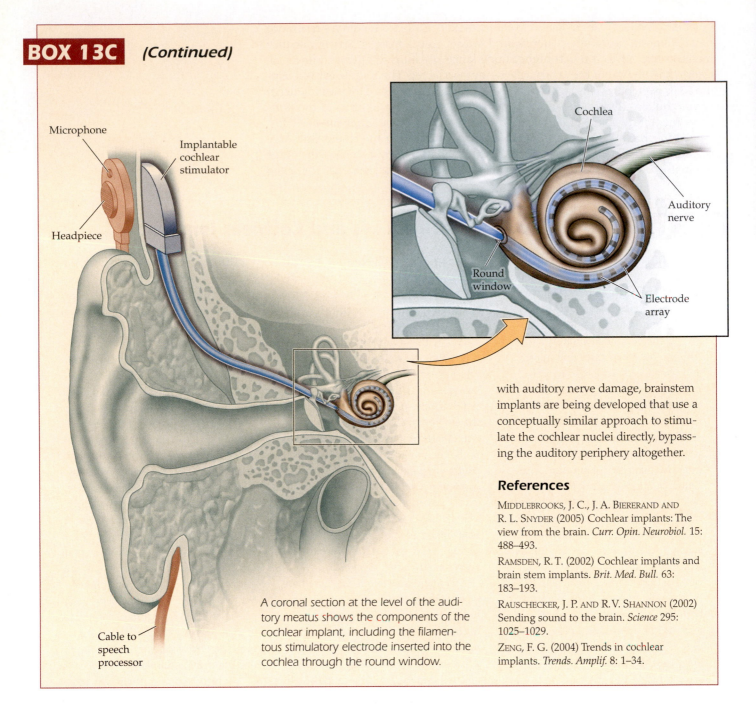

A coronal section at the level of the auditory meatus shows the components of the cochlear implant, including the filamentous stimulatory electrode inserted into the cochlea through the round window.

with auditory nerve damage, brainstem implants are being developed that use a conceptually similar approach to stimulate the cochlear nuclei directly, bypassing the auditory periphery altogether.

References

MIDDLEBROOKS, J. C., J. A. BIERERAND AND R. L. SNYDER (2005) Cochlear implants: The view from the brain. *Curr. Opin. Neurobiol.* 15: 488–493.

RAMSDEN, R. T. (2002) Cochlear implants and brain stem implants. *Brit. Med. Bull.* 63: 183–193.

RAUSCHECKER, J. P. AND R. V. SHANNON (2002) Sending sound to the brain. *Science* 295: 1025–1029.

ZENG, F. G. (2004) Trends in cochlear implants. *Trends. Amplif.* 8: 1–34.

smaller-diameter **oval window**, the site where the bones of the middle ear contact the inner ear. A second and related process relies on the mechanical advantage gained by the lever action of the three small interconnected middle ear bones, or **ossicles** (i.e., the malleus, incus, and stapes; see Figure 13.3), which connect the tympanic membrane to the oval window. **Conductive hearing losses**, which involve damage to the external or middle ear, lower the efficiency at which sound energy is transferred to the inner ear and can be partially overcome by artificially boosting sound pressure levels with an external hearing aid (Box 13C). In normal hearing, the efficiency of sound transmission to the inner ear also is regulated by two small muscles in the middle ear, the tensor tympani, innervated by cranial nerve V, and the stapedius, innervated by cranial nerve VII (see the Appendix). Flexion of these muscles, which is triggered automatically by loud noises or during self-generated vocalization, stiffens the ossicles and reduces the amount of sound energy transmitted to the cochlea, serving to protect the inner ear. Conversely, conditions that lead to flaccid paralysis of either of these muscles, such as Bell's palsy (nerve VII), can trigger a painful sensitivity to moderate or even low-intensity sounds known as **hyperacusis**.

Bony and soft tissues, including those surrounding the inner ear, have impedances close to that of water. Therefore, even without an intact tympanic membrane or middle-ear ossicles, acoustical vibrations can still be transferred directly through the bones and tissues of the head to the inner ear. In the clinic, bone conduction can be exploited using a simple test involving a tuning fork to determine whether hearing loss is due to conductive problems or is due to damage to the hair cells of the inner ear or to the auditory nerve itself (**sensorineural hearing loss**; see Boxes 13A and 13C)

The Inner Ear

The **cochlea** of the inner ear is arguably the most critical and fascinating structure in the auditory pathway, for it is there that the energy from sonically generated pressure waves is transformed into neural impulses. The cochlea not only amplifies sound waves and converts them into neural signals, but it also acts as a mechanical frequency analyzer, decomposing complex acoustical waveforms into simpler elements. Many features of auditory perception derive from aspects of the physical properties of the cochlea; hence, it is important to consider this structure in some detail.

The cochlea (from the Latin for "snail") is a small (about 10 mm wide) coiled structure, which, were it uncoiled, would form a tube about 35 mm long (Figures 13.4 and 13.5). Both the oval window and the **round window**, another region where the bone is absent surrounding the cochlea, are at the basal end of this tube. The cochlea is bisected from its basal end almost to its apical end by the cochlear partition, which is a flexible structure that supports the **basilar membrane** and the **tectorial membrane**. There are fluid-filled chambers on each side of the cochlear partition, named the **scala vestibuli** and the **scala tympani**; a distinct channel, the **scala media**, runs within the cochlear partition. The cochlear partition does not extend all the way to the apical end of the cochlea; instead, there is an opening, known as the **helicotrema**, that joins the scala vestibuli to the scala tympani, allowing their fluid, known as **perilymph**, to mix. One consequence of this structural arrangement is that inward movement of the oval window displaces the fluid of the inner ear, causing the round window to bulge out slightly and deforming the cochlear partition.

The manner in which the basilar membrane vibrates in response to sound is the key to understanding cochlear function. Measurements of the vibration of different parts of the basilar membrane, as well as the discharge rates of individual auditory nerve fibers that terminate along its length, show that both of these features are highly tuned; that is, they respond most intensely to sounds of a specific frequency. Frequency tuning within the inner ear is attributable in part to the geometry of the basilar membrane, which is wider and more flexible at the apical end and narrower and stiffer at the basal end. One feature of such a system is that, regardless of where energy is supplied to it, movement always begins at the stiff end (i.e., the base), and then propagates to the more flexible end (i.e., the apex). Georg von Békésy, working at Harvard University, showed that a membrane that varies systematically in its width and flexibility vibrates maximally at different positions as a function of the stimulus frequency (see Fig-

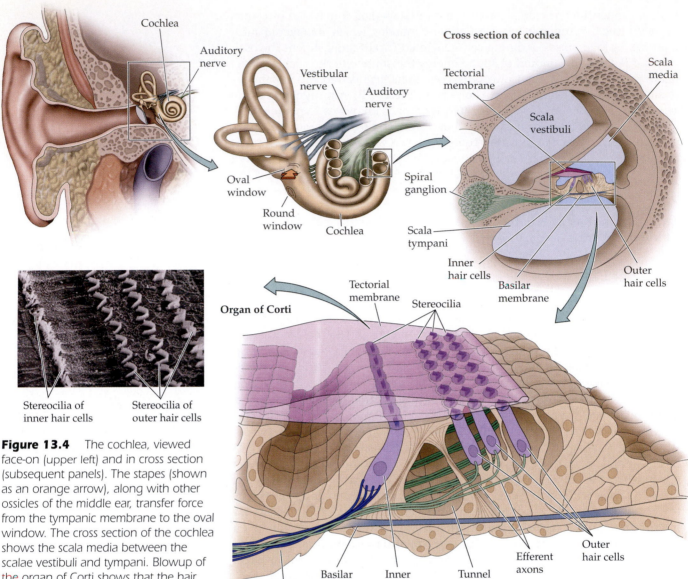

Figure 13.4 The cochlea, viewed face-on (upper left) and in cross section (subsequent panels). The stapes (shown as an orange arrow), along with other ossicles of the middle ear, transfer force from the tympanic membrane to the oval window. The cross section of the cochlea shows the scala media between the scalae vestibuli and tympani. Blowup of the organ of Corti shows that the hair cells are located between the basilar and tectorial membranes; the latter is rendered transparent in the line drawing and removed in the scanning electron micrograph. The hair cells are named for their tufts of stereocilia; inner hair cells receive afferents from cranial nerve VIII, whereas outer hair cells receive mostly efferent innervation. (Micrograph from Kessel and Kardon, 1979.)

ure 13.5). Using tubular models and human cochleas taken from cadavers, von Békésy found that an acoustical stimulus initiates a **traveling wave** of the same frequency in the cochlea, which propagates from the base toward the apex of the basilar membrane, growing in amplitude and slowing in velocity until a point of maximum displacement is reached. This point of maximal displacement is determined by the sound frequency. The points responding to high frequencies are at the base of the basilar membrane where it is stiffer, and the points responding to low frequencies are at the apex, giving rise to a topographical mapping of frequency (i.e., **tonotopy**). An important feature is that complex sounds cause a pattern of vibration equivalent to the superposition of the vibrations generated by the individual tones making up that complex sound, thus accounting for the decompositional aspects of cochlear function mentioned earlier. This process of spectral decomposition appears to be an important strategy for detecting the various harmonic combinations that distinguish different natural sounds, including speech.

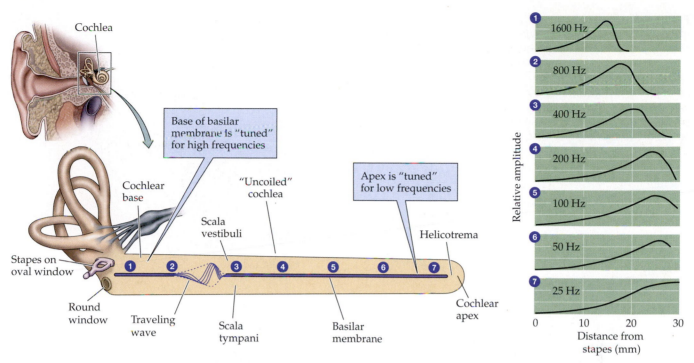

Figure 13.5 Traveling waves along the cochlea. A traveling wave is shown at a given instant along the cochlea, which has been uncoiled for clarity. The graphs on the right, profile the amplitude of the traveling wave along the basilar membrane for different frequencies. The position (i.e., 1–7) where the traveling wave reaches its maximum amplitude varies directly with the frequency of stimulation, with higher frequencies mapping to the base and lower frequencies mapping to the apex. (Drawing after Dallos, 1992; graphs after von Békésy, 1960.)

Von Békésy's model of cochlear mechanics was a passive one, resting on the premise that the basilar membrane acts like a series of linked resonators, much as a concatenated set of tuning forks. Each point on the basilar membrane was postulated to have a characteristic frequency at which it vibrated most efficiently; because it was physically linked to adjacent areas of the membrane, each point also vibrated (if somewhat less readily) at other frequencies, thus permitting propagation of the traveling wave. It is now clear, however, that the tuning of the auditory periphery, whether measured at the basilar membrane or recorded as the electrical activity of auditory nerve fibers, is too sharp to be explained by passive mechanics alone. At very low sound intensities, the basilar membrane vibrates 100-fold more than would be predicted by linear extrapolation from the motion measured at high intensities. Therefore, the ear's sensitivity arises from an active biomechanical process, as well as from its passive resonant properties (Box 13D). The outer hair cells, which together with the inner hair cells constitute the sensory cells of the inner ear, are the most likely candidates for driving this active process.

The motion of the traveling wave initiates sensory transduction by displacing the hair cells that sit atop the basilar membrane. Because these structures are anchored at different positions, the vertical component of the traveling wave is translated into a shearing motion between the basilar membrane and the overlying tectorial membrane (Figure 13.6). This motion bends the tiny processes, called **stereocilia**, that protrude from the apical ends of the hair cells, leading to voltage changes across the hair-cell membrane. How the bending of stereocilia leads to receptor potentials in hair cells is considered in the following section.

Hair Cells and the Mechanoelectrical Transduction of Sound Waves

The hair cell is an evolutionary triumph that solves the problem of transforming vibrational energy into an electrical signal. The scale at which the hair cell operates is truly amazing. At the limits of human hearing, hair cells can faithfully

Figure 13.6 Vertical movement of the basilar membrane is translated into a shearing force that bends the stereocilia of the hair cells. The pivot point of the basilar membrane is offset from the pivot point of the tectorial membrane, so that when the basilar membrane is displaced, the tectorial membrane moves across the tops of the hair cells, bending the stereocilia.

(A) Resting position

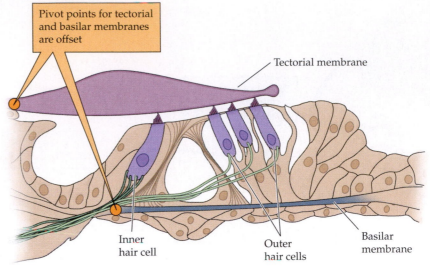

Pivot points for tectorial and basilar membranes are offset

Tectorial membrane

Inner hair cell

Outer hair cells

Basilar membrane

(B) Sound-induced vibration

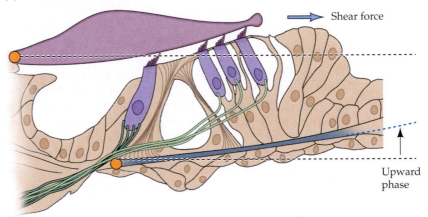

Shear force

Upward phase

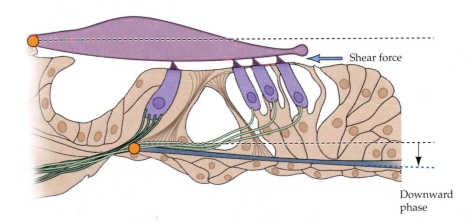

Shear force

Downward phase

detect movements of atomic dimensions and respond in tens of microseconds! Furthermore, hair cells can adapt rapidly to constant stimuli, thus allowing the listener to extract signals from a noisy background.

BOX 13D The Sweet Sound of Distortion

As early as the first half of the eighteenth century, musical composers such as G. Tartini and W. A. Sorge discovered that, upon playing pairs of tones, other tones not present in the original stimulus are also heard. These combination tones, fc, are mathematically related to the played tones, f_1 and f_2 ($f_2 > f_1$), by the formula

$$fc = mf_1 \pm nf_2$$

where m and n are positive integers. Combination tones have been used for a variety of compositional effects, as they can strengthen the harmonic texture of a chord. Furthermore, organ builders sometimes use the difference tone ($f_2 - f_1$) created by two smaller organ pipes to produce the extremely low tones that would otherwise require building one especially large pipe.

Modern experiments suggest that this distortion product is due at least in part to the nonlinear properties of the inner ear. M. Ruggero and his colleagues placed small glass beads (10–30 nm in diameter) on the basilar membrane of an anesthetized animal and then determined the velocity of the basilar membrane in response to different combinations of tones by measuring the Doppler shift of laser light reflected from the beads. When two tones were played into the ear, the basilar membrane vibrated not only at those two frequencies but also at other frequencies predicted by the above formula.

Related experiments on hair cells studied *in vitro* suggest that these nonlinearities result from the properties of the mechanical linkage of the transduction apparatus. By moving the hair bundle sinusoidally with a metal-coated glass fiber, A. J. Hudspeth and his coworkers found that the hair bundle exerts a force at the same frequency. However, when two sinusoids were applied simultaneously, the forces exerted by the hair bundle occurred not only at the primary frequencies but at several combination frequencies, as well. These distortion products are due to the transduction apparatus, because blocking the transduction channels causes the forces exerted at the combination frequencies to disappear, even though the forces at the primary frequencies remain unaffected. It seems that the tip links add a certain extra springiness to the hair bundle in the small range of motions over which the transduction channels are changing between closed and open states. If nonlinear distortions of basilar membrane vibrations arise from the properties of the hair bundle, then it is likely that hair cells can indeed influence basilar membrane motion, thereby accounting for the cochlea's extreme sensitivity. When we hear difference tones, we may be paying the price in distortion for an exquisitely fast and sensitive transduction mechanism.

References

JARAMILLO, F., V. S. MARKIN AND A. J. HUDSPETH (1993) Auditory illusions and the single hair cell. *Nature* 364: 527–529.

PLANCHART, A. E. (1960) A study of the theories of Giuseppe Tartini. *J. Music Theory* 4: 32–61.

ROBLES, L., M. A. RUGGERO AND N. C. RICH (1991) Two-tone distortion in the basilar membrane of the cochlea. *Nature* 439: 413–414.

The hair cell is a flask-shaped epithelial cell named for the bundle of hairlike processes that protrude from its apical end into the scala media. Each hair bundle contains anywhere from 30 to a few hundred stereocilia, with one taller **kinocilium** (Figure 13.7A). Despite their names, only the kinocilium is a true ciliary structure, with the characteristic two central tubules surrounded by nine doublet tubules that define cilia (Figure 13.7B). The function of the kinocilium is unclear, and in mammalian cochlear hair cells it actually disappears shortly after birth (Figure 13.7C). The stereocilia are simpler, containing only an actin cytoskeleton. Each stereocilium tapers where it inserts into the apical membrane, forming a hinge about which each stereocilium pivots. The stereocilia are graded in height and are arranged in a bilaterally symmetric fashion (see Figure 13.7C); in vestibular hair cells, this plane runs through the kinocilium (see Figure 13.7A). Fine filamentous structures, known as tip links, run in parallel to the plane of bilateral symmetry, connecting the tips of adjacent stereocilia (Figure 13.7D).

The tip links provide the means for rapidly translating hair bundle movement into a receptor potential. Displacement of the hair bundle parallel to the plane of bilateral symmetry in the direction of the tallest stereocilia stretches the tip links, directly opening cation-selective channels located at the end of the link

(A)

Kinocilium

Plane of cross section

(B) Kinocilium

(C)

(D)

Tip link

Figure 13.7 The structure and function of the hair bundle in cochlear and vestibular hair cells. The vestibular hair bundles shown here resemble those of cochlear hair cells, except for the presence of the kinocilium, which disappears in the mammalian cochlea shortly after birth. (A) The hair bundle of a guinea pig vestibular hair cell. This view shows the increasing height leading to the kinocilium. (B) Cross section through the vestibular hair bundle shows the 9 + 2 array of microtubules in the kinocilium at the top, which contrasts with the simpler actin filament structure of the stereocilia. (C) Scanning electron micrograph of a cochlear outer hair cell bundle viewed along the plane of mirror symmetry. Note the graded lengths of the stereocilia, and the absence of a kinocilium. (D) Tip links connect adjacent stereocilia and are believed to be mechanical linkages that open and close transduction channels. (A from Lindeman, 1973; B from Hudspeth, 1983; C courtesy of David Furness and Carole Hackney, Keele University, UK; D from Fain 2003.)

and depolarizing the hair cell (Figure 13.8). Movement in the opposite direction compresses the tip links, closing channels and hyperpolarizing the hair cell. As the linked stereocilia pivot back and forth, the tension on the tip link varies, modulating the ionic flow and resulting in a graded receptor potential that follows the movements of the stereocilia. The tip link model also explains why only deflections along the axis of the hair bundle activate transduction channels, because tip links join adjacent stereocilia along the axis directed toward the tallest stereocilia (Figure 13.9; see also Box 14B).

Mechanotransduction in hair cells is both fast and remarkably sensitive. The hair bundle movements at the threshold of hearing are approximately 0.3 nm—about the diameter of an atom of gold. Hair cells can convert the displacement of the stereociliary bundle into an electrical potential in as little as 10 μs; as described below, such speed is required for the accurate localization of the source of the sound. The need for microsecond resolution requires direct mechanical gating of the transduction channel, rather than the relatively slow second-messenger pathways used in visual and olfactory transduction (see Chapters 11 and 15). Although mechanotransduction is extremely fast, springiness of the tip link introduces distortion effects that, in some cases, are audible (see Box 13D). Moreover, the exquisite mechanical sensitivity of the stereocilia also presents substantial risks. High intensity sounds can shear off the hair bundle, resulting in profound hearing deficits. Because human stereocilia, unlike those in fishes and birds, do not regenerate, such damage is irreversible. The small number of hair cells (a total of about 30,000 in a human, or 15,000 per ear) further compounds the sensitivity of the inner ear to environmental and genetic insults. An important goal of current research is to identify stem cells and factors that could contribute to the regeneration of human hair cells, thus affording a possible therapy for some forms of sensorineural hearing loss.

(A)

Hyperpolarization ⟵

(B)

⟶ **Depolarization**

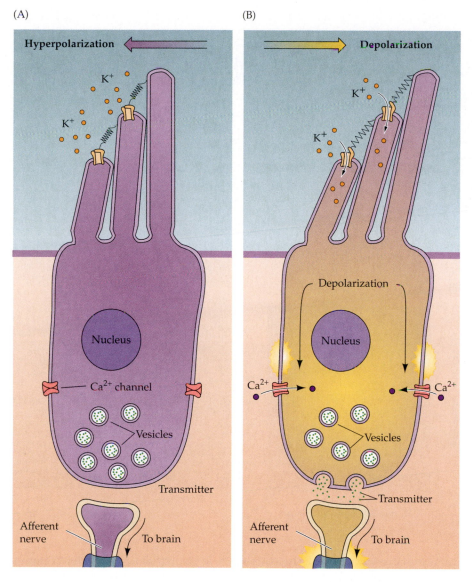

Figure 13.8 Mechanoelectrical transduction mediated by hair cells. (A,B) When the hair bundle is deflected toward the tallest stereocilium, cation-selective channels open near the tips of the stereocilia, allowing K⁺ to flow into the hair cell down their electrochemical gradient (see text for the explanation of this peculiar situation). The resulting depolarization of the hair cell opens voltage-gated Ca^{2+} channels in the cell soma, allowing calcium entry and release of neurotransmitter onto the nerve endings of the auditory nerve. (After Lewis and Hudspeth, 1983.)

The Ionic Basis of Mechanotransduction in Hair Cells

Understanding the ionic basis of hair-cell transduction has been greatly advanced by intracellular recordings made from these tiny structures. The hair cell has a resting potential between –45 and –60 mV relative to the fluid that bathes the basal end of the cell. At the resting potential, only a small fraction of the transduction channels are open. When the hair bundle is displaced in the direction of the tallest stereocilium, more transduction channels open, causing depolarization as K⁺ enter the cell (see Figure 13.9). Depolarization in turn opens voltage-gated calcium channels in the hair-cell membrane, and the resultant Ca^{2+} influx causes transmitter release from the basal end of the cell onto the auditory nerve endings (see Figure 13.8A,B). Such calcium-dependent exocytosis is similar to chemical neurotransmission elsewhere in the central and peripheral nervous systems (see Chapters 5 and 6); thus, the hair cell has become a useful model for studying calcium-dependent transmitter release. Because some transduction channels are open at rest, the receptor potential is biphasic: movement toward the tallest stereocilia depolarizes the cell, while

(A)

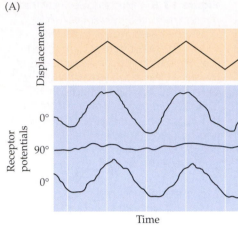

(C)

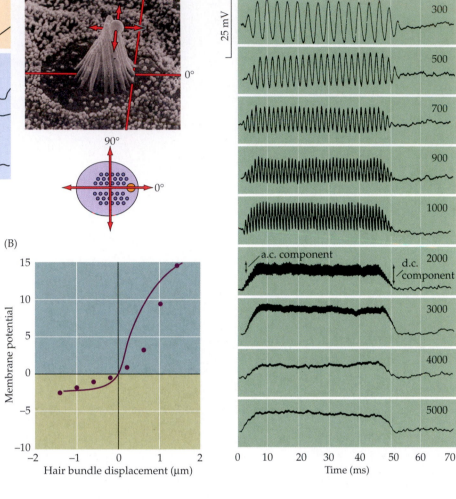

Figure 13.9 *Mechanoelectrical transduction mediated by vestibular hair cells. (A) Vestibular hair cell receptor potentials (bottom three traces; blue) measured in response to symmetrical displacement (top trace; yellow) of the hair bundle about the resting position, either parallel (0°) or orthogonal (90°) to the plane of bilateral symmetry. (B) The asymmetrical stimulus/response (x-axis/y-axis) function of the hair cell. Equivalent displacements of the hair bundle generate larger depolarizing responses than hyperpolarizing responses, because most transduction channels are closed "at rest" (i.e., 0 μm). (C) Receptor potentials generated by an individual hair cell in the cochlea in response to pure tones (indicated in Hz, right). Note that the hair-cell potential faithfully follows the waveform of the stimulating sinusoids for low frequencies (< 3 kHz) but still responds with a DC offset to higher frequencies due to the asymmetrical stimulus-response function and the electrical filtering properties of the hair cells. (A after Shotwell et al., 1981; B after Hudspeth and Corey, 1977; C after Palmer and Russell, 1986.)*

movement in the opposite direction leads to hyperpolarization. This situation allows the hair cell to generate a sinusoidal receptor potential in response to a sinusoidal stimulus, thus preserving the temporal information present in the original signal, up to frequencies of around 3 kHz (see Figure 13.9). Hair cells still can signal at frequencies above 3 kHz, although without preserving the exact temporal structure of the stimulus; the asymmetric displacement-receptor current function of the hair-cell bundle is filtered by the cell's membrane time-constant to produce a tonic depolarization of the soma, augmenting transmitter release and thus exciting auditory nerve terminals.

The high-speed demands of mechanoelectrical transduction have resulted in some impressive specializations of the ion fluxes within the inner ear. An unusual adaptation of the hair cell in this regard is that K^+ serves both to depolarize *and* repolarize the cell, enabling the hair cell's K^+ gradient to be largely maintained by passive ion movement alone. As with other epithelial cells, the basal and apical surfaces of the hair cell are separated by tight junctions, allowing separate extracellular ionic environments at these two surfaces. The apical end (including the stereocilia) protrudes into the scala media and is exposed to the K^+-rich, Na^+-poor **endolymph** produced by dedicated ion-pumping cells in the **stria vascularis** (Figure 13.10). The basal end of the hair-cell body is bathed in **perilymph**—the same fluid that fills the scala tympani. Perilymph resembles

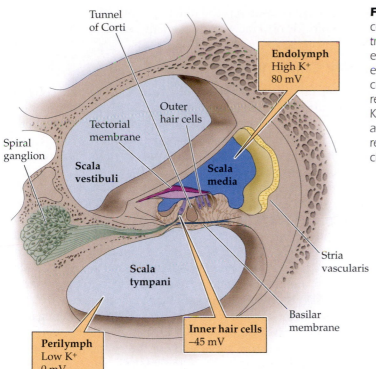

Figure 13.10 Depolarization and repolarization of hair cells is mediated by K+. The stereocilia of the hair cells protrude into the endolymph, which is high in K+ and has an electrical potential of +80 mV relative to the perilymph. This endocochlear potential drives K+ into open transduction channels located at the apical ends of the stereocilia; the resulting depolarization of the hair-cell body opens somatic K+ channels. The negative resting potential of the hair cell and the low K+ concentration in the surrounding perilymph results in an outward K+ current through these somatic K+ channels.

other extracellular fluids in that it is K+-poor and Na+-rich. However, the compartment containing endolymph is about 80 mV more positive than the perilymph compartment (this difference is known as the *endocochlear potential*), while the inside of the hair cell is about 45 mV more negative than the perilymph and about 125 mV more negative than the endolymph.

The resulting electrical gradient across the membrane of the stereocilia (about 125 mV, or the difference between the hair cell's resting potential and the endocochlear potential) drives K+ through open transduction channels into the hair cell, even though these cells already have a high internal K+ concentration. K+ entry via the transduction channels electrotonically depolarizes the hair cell, opening voltage-gated Ca²⁺ and K+ channels located in the membrane of the hair-cell soma (see Box 14B). The opening of *somatic* K+ channels favors K+ efflux, and thus repolarization; the efflux occurs because the perilymph surrounding the basal end is low in K+ relative to the cytosol, and because the equilibrium potential for K+ is more negative than the hair cell's resting potential ($E_{K_{basal}} = -85$ mV). Repolarization of the hair cell via K+ efflux is also facilitated by Ca²⁺ entry. In addition to modulating the release of neurotransmitter, Ca²⁺ entry opens Ca²⁺-dependent K+ channels, which provide another avenue for K+ to enter the perilymph. Indeed, the interaction of Ca²⁺ influx and Ca²⁺-dependent K+ efflux can lead to electrical resonances that enhance the tuning of response properties within the inner ear (also explained in Box 14B). In essence, the hair cell operates as two distinct compartments, each dominated by its own Nernst equilibrium potential for K+; this arrangement ensures that the hair cell's ionic gradient does not run down, even during prolonged stimulation. At the same time, rupture of Reissner's membrane (which normally separates the scalae media and vestibuli), or compounds such as ethacrynic acid that selectively poison the ion-pumping cells of the stria vascularis, can cause the endocochlear potential to dissipate, resulting in a sensorineural hearing deficit (see

Box 13A). In short, the hair cell exploits the different ionic milieus of its apical and basal surfaces to provide extremely fast and energy-efficient repolarization.

Two Kinds of Hair Cells in the Cochlea

The cochlear hair cells in humans consist of one row of **inner hair cells** and three rows of **outer hair cells** (see Figures 13.4 and 13.7). The inner hair cells are the actual sensory receptors, and 95 percent of the fibers of the auditory nerve that project to the brain arise from this subpopulation. The terminations on the outer hair cells are almost all from efferent axons that arise from cells in the superior olivary complex.

A clue to the significance of this efferent pathway was provided by the discovery that an active process within the cochlea, as mentioned, influences basilar membrane motion. First, it was found that the cochlea actually emits sound under certain conditions. These otoacoustical emissions, which can be detected by placing a sensitive microphone at the eardrum and monitoring the response after briefly presenting a tone or click, provide a useful means to assess cochlear function in the newborn, and this test is now done routinely to rule out congenital deafness. Such emissions can also occur spontaneously, especially in certain pathological states, and are thus one potential source of **tinnitus** (ringing in the ears). These observations clearly indicate that a process within the cochlea is capable of producing sound. Second, stimulation of the crossed olivocochlear bundle, which supplies efferent input to the outer hair cells, can broaden auditory nerve tuning curves. Third, the high sensitivity notch of auditory nerve tuning curves is lost when the outer hair cells are selectively inactivated. Finally, isolated outer hair cells contract and expand in response to small electrical currents, thus providing a potential source of energy to drive an active process within the cochlea. Thus, it seems likely that the outer hair cells sharpen the frequency-resolving power of the cochlea by actively contracting and relaxing, thus changing the stiffness of the tectorial membrane at particular locations. This active process contributes to the nonlinear vibration of the basilar membrane at low sound intensities.

Tuning and Timing in the Auditory Nerve

The rapid response time of the transduction apparatus allows the membrane potential of the hair cell to follow deflections of the hair bundle up to moderately high frequencies of oscillation. In humans, the receptor potentials of certain hair cells and the action potentials of their associated auditory nerve fibers can follow stimuli of up to about 3 kHz in a one-to-one fashion. Such real-time encoding of stimulus frequency by the pattern of action potentials in the auditory nerve is known as the "volley theory" of auditory information transfer. Even these extraordinarily rapid processes, however, fail to follow frequencies above 3 kHz (see Figure 13.9). Accordingly, some other mechanism must be used to transmit auditory information at higher frequencies. The tonotopically organized basilar membrane provides an alternative to temporal coding—namely, a "labeled-line" coding mechanism.

In labeled-line coding, frequency information is specified by preserving the tonotopy of the cochlea at higher levels in the auditory pathway. Because the auditory nerve fibers associate with the inner hair cells in approximately a one-to-one ratio (although several or more auditory nerve fibers synapse on a single hair cell), each auditory nerve fiber transmits information about only a small part of the audible frequency spectrum. As a result, auditory nerve fibers related to the apical end of the cochlea respond to low frequencies, and fibers that are related to the basal end respond to high frequencies (see Figure 13.5). The properties of specific fibers can be seen in electrophysiological recordings of responses to sound (Figure 13.11). These threshold functions are called **tuning curves**, and the lowest threshold of the tuning curve is called the **characteristic frequency**. Since the topographical order of the characteristic frequency of neurons is retained throughout the system, information about frequency is also preserved. Cochlear implants (see Box 13C) exploit the tonotopic organization of the cochlea, and particularly its auditory nerve afferents, to roughly recreate the patterns of auditory nerve activity elicited by sounds. In patients with damaged hair cells, such implants can effectively bypass the impaired transduction apparatus, and thus restore some degree of auditory function.

Figure 13.11 Response properties of auditory nerve fibers. (A) Frequency tuning curves of six different fibers in the auditory nerve. Each graph plots the minimum sound level required to increase the fiber's firing rate above its spontaneous firing level, across all frequencies to which the fiber responds. The lowest point in the plot is the weakest sound intensity to which the neuron will respond. The frequency at this point is called the neuron's characteristic frequency. (B) The frequency tuning curves of auditory nerve fibers superimposed and aligned with their approximate relative points of innervation along the basilar membrane. (In the side-view schematic, the basilar membrane is represented as a black line within the unrolled cochlea.) (C) Temporal response patterns of a low-frequency axon in the auditory nerve. The stimulus waveform is indicated beneath the histograms, which show the phase-locked responses to a 50-ms tone pulse of 260 Hz. Note that the spikes are all locked to the same phase of the sinusoidal stimulus. (A after Kiang and Moxon, 1972; C after Kiang, 1984.)

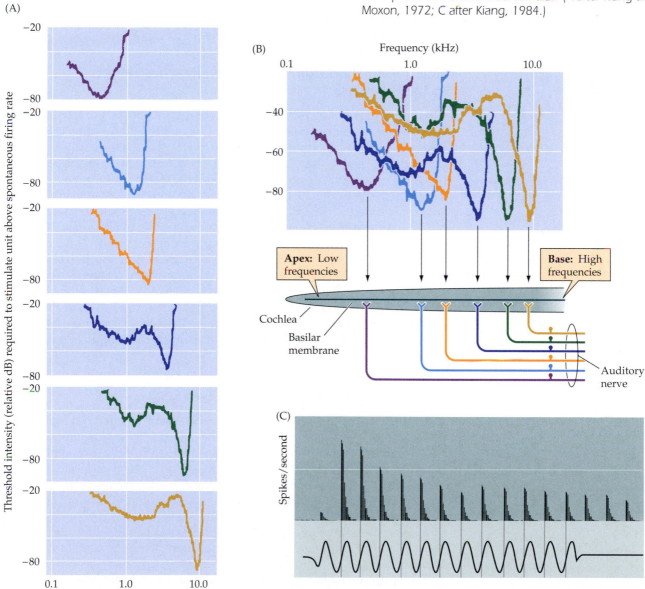

The other prominent feature of hair cells—their ability to follow the wave-form of low-frequency sounds—is also important in other, more subtle aspects of auditory coding. As mentioned earlier, hair cells have biphasic response properties. Because hair cells release transmitter only when depolarized, auditory nerve fibers fire only during the positive phases of low-frequency sounds (see Figure 13.11). The resultant "phase-locking" provides temporal information from the two ears to neural centers that compare interaural time differences. The evaluation of interaural time differences provides a critical cue for sound localization and the perception of auditory "space." That auditory space can be perceived is remarkable, given that the cochlea, unlike the retina, cannot represent space directly.

A final point is that auditory nerve activity patterns are not simply faithful neural replicas of the auditory stimulus itself. Indeed, William Bialek and his colleagues at Princeton University have shown that the auditory nerve in the bullfrog encodes conspecific mating calls more efficiently than artificial sounds with similar frequency and amplitude characteristics. This study suggests that the auditory periphery is optimized to transmit natural sounds, including species-typical vocal sounds, rather than simply transmitting all sounds equally well to central auditory areas.

How Information from the Cochlea Reaches Targets in the Brainstem

A hallmark of the ascending auditory system is its parallel organization. This arrangement becomes evident as soon as the auditory nerve enters the brainstem, where it branches to innervate the three divisions of the cochlear nucleus. The auditory nerve (which, along with the vestibular nerve, constitutes cranial nerve VIII) comprises the central processes of the bipolar spiral ganglion cells in the cochlea (see Figure 13.4); each of these cells sends a peripheral process to contact one inner hair cell and a central process to innervate the cochlear nucleus. Within the cochlear nucleus, each auditory nerve fiber branches, sending an ascending branch to the anteroventral cochlear nucleus, and a descending branch to the posteroventral cochlear nucleus and the dorsal cochlear nucleus (Figure 13.12). The tonotopic organization of the cochlea is maintained in the three parts of the cochlear nucleus, each of which contains different populations of cells with quite different properties. In addition, the patterns of termination of the auditory nerve axons differ in density and type; thus, there are several opportunities at this level for transformation of the information from the hair cells.

Integrating Information from the Two Ears

Just as the auditory nerve branches to innervate several different targets in the cochlear nuclei, the neurons in these nuclei give rise to several different pathways (see Figure 13.12). One clinically relevant feature of the ascending projections of the auditory brainstem is a high degree of bilateral connectivity, which means that damage to central auditory structures is almost never manifested as a monaural hearing loss. Indeed, a monaural hearing loss strongly implicates unilateral peripheral damage, either to the middle or inner ear, or to the auditory nerve itself (see Box 13C). Given the relatively byzantine organization already present at the level of the auditory brainstem, it is useful to consider these pathways in the context of their functions.

The best-understood function mediated by the auditory brainstem nuclei, and certainly the one most intensively studied, is sound localization. Humans

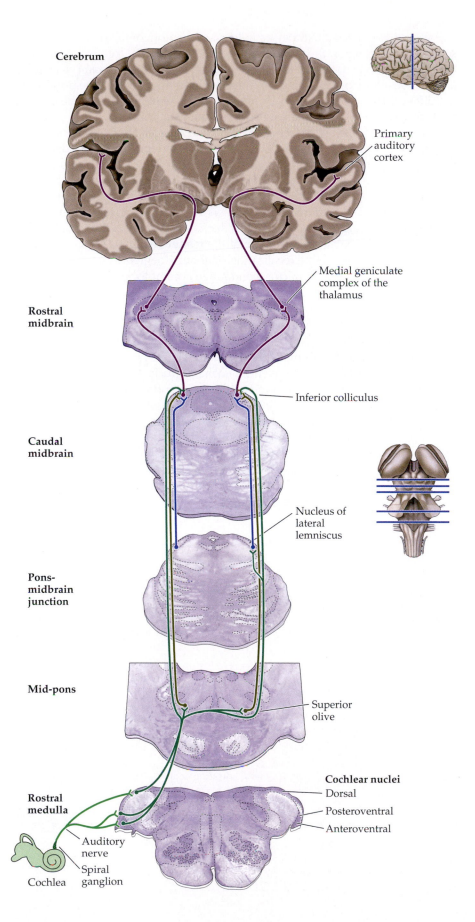

Cerebrum

Primary
auditory
cortex

Medial geniculate
complex of the
thalamus

**Rostral
midbrain**

Inferior colliculus

**Caudal
midbrain**

Nucleus of
lateral
lemniscus

**Pons-
midbrain
junction**

Mid-pons

Superior
olive

Cochlear nuclei
Dorsal

**Rostral
medulla**

Posteroventral

Anteroventral

Auditory
nerve

Spiral
ganglion

Cochlea

Figure 13.12 Diagram of the major auditory pathways. Although many details are missing from this simplified diagram, two important points are evident: (1) the auditory system entails several parallel pathways, and (2) information from each ear reaches both sides of the system, even at the level of the brainstem.

use at least two different strategies to localize the horizontal position of sound sources, depending on the frequencies in the stimulus. For frequencies below 3 kHz (which can be followed in a phase-locked manner), interaural *time* differences are used to localize the source; above these frequencies, interaural *intensity* differences are used as cues. Parallel pathways originating from the cochlear nucleus serve each of these strategies for sound localization.

The human ability to detect interaural time differences is remarkable. The longest interaural time differences, which are produced by sounds arising directly lateral to one ear, are on the order of only 700 microseconds (a value given by the width of the head divided by the speed of sound in air, about 340 meters per second). Psychophysical experiments show that humans can actually detect interaural time differences as small as 10 microseconds; two sounds presented through earphones separated by such small interaural time differences are perceived as arising from the side of the leading ear. This sensitivity translates into accuracy for sound localization of about 1 degree.

How is timing in the 10-microsecond range accomplished by neural components that operate in the milliseconds range? The neural circuitry that computes such tiny interaural time differences consists of binaural inputs to the **medial superior olive (MSO)** that arise from the right and left anteroventral cochlear nuclei (Figure 13.13; see also Figure 13.12). The MSO contains cells with bipolar dendrites that extend both medially and laterally. The lateral dendrites receive input from the ipsilateral anteroventral cochlear nucleus, and the medial dendrites receive input from the contralateral anteroventral cochlear nucleus (both inputs are excitatory). As might be expected, the cells of the MSO work as **coincidence detectors**, responding when both excitatory signals arrive at the same time. For a coincidence mechanism to be useful in localizing sound, different neurons must be maximally sensitive to different interaural time delays. The axons that project from the anteroventral cochlear nucleus evidently vary sys-

Figure 13.13 *Illustration of how the MSO computes the location of a sound by interaural time differences. A given MSO neuron responds most strongly when the two inputs arrive simultaneously, as occurs when the contralateral and ipsilateral inputs precisely compensate (via their different lengths) for differences in the time of arrival of a sound at the two ears. The systematic (and inverse) variation in the delay lengths of the two inputs creates a map of sound location: In this model, neuron E in the MSO would be most sensitive to sounds located to the left, and neuron A to sounds from the right; neuron C would respond best to sounds coming from directly in front of the listener. (After Jeffress, 1948.)*

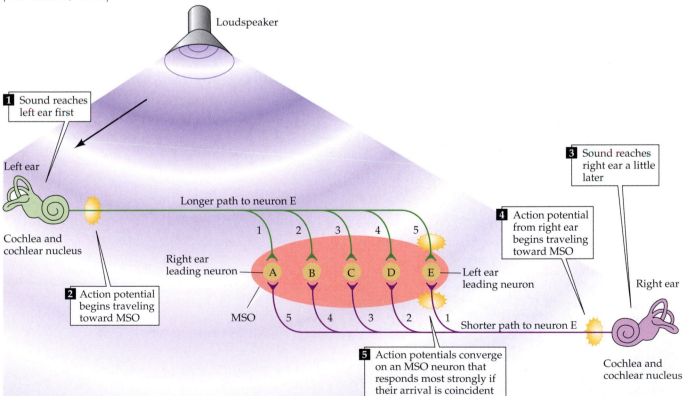

tematically in length to create delay lines. (Remember that the length of an axon divided by its conduction velocity equals the conduction time.) These anatomical differences compensate for sounds arriving at slightly different times at the two ears, so that the resultant neural impulses arrive at a particular MSO neuron simultaneously, making each cell especially sensitive to sound sources in a particular place. The mechanisms enabling MSO neurons to function as coincidence detectors at the microsecond level are still poorly understood, but certainly reflect one of the more impressive biophysical specializations in the nervous system.

Sound localization perceived on the basis of interaural time differences requires phase-locked information from the periphery, which, as already emphasized, is available to humans only for frequencies below 3 kHz. (In barn owls, the reigning champions of sound localization, phase-locking occurs at up to 9 kHz.) Therefore, a second mechanism must come into play at higher frequencies. One clue to the solution is that at frequencies higher than about 2 kHz, the human head begins to act as an acoustical obstacle, because the wavelengths of the sounds are too short to bend around it. As a result, when high-frequency sounds are directed toward one side of the head, an acoustical "shadow" of lower intensity is created at the far ear. These intensity differences provide a second cue about the location of a sound. The circuits that compute the position of a sound source on this basis are found in the **lateral superior olive (LSO)** and the **medial nucleus of the trapezoid body (MNTB)** (Figure 13.14). Excitatory axons project directly from the ipsilateral anteroventral cochlear nucleus to the LSO (as well as to the MSO; see Figure 13.13). Note that the LSO also receives inhibitory input from the contralateral ear, via an inhibitory neuron in the MNTB. This excitatory/inhibitory interaction results in a net excitation of the LSO on the same side of the body as the sound source. For sounds arising directly lateral to the listener, firing rates will be highest in the

Figure 13.14 LSO neurons encode sound location through interaural intensity differences. (A) LSO neurons receive direct excitation from the ipsilateral cochlear nucleus; input from the contralateral cochlear nucleus is relayed via inhibitory interneurons in the MNTB. (B) This arrangement of excitation–inhibition makes LSO neurons fire most strongly in response to sounds arising directly lateral to the listener on the same side as the LSO, because excitation from the ipsilateral input will be great and inhibition from the contralateral input will be small. In contrast, sounds arising from in front of the listener, or from the opposite side, will silence the LSO output, because excitation from the ipsilateral input will be minimal, but inhibition driven by the contralateral input will be great. Note that LSOs are paired and bilaterally symmetrical; each LSO only encodes the location of sounds arising from the ipsilateral hemifield.

(A)

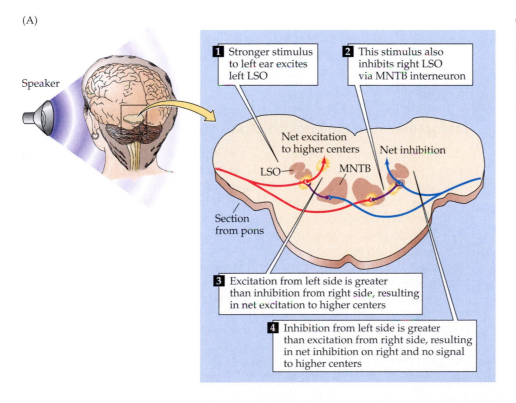

(B)

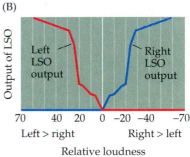

LSO on that side; in this circumstance, the excitation via the ipsilateral anteroventral cochlear nucleus will be maximal, and inhibition from the contralateral MNTB minimal. In contrast, sounds arising closer to the listener's midline will elicit lower firing rates in the ipsilateral LSO because of increased inhibition arising from the contralateral MNTB. For sounds arising at the midline, or from the other side, the increased inhibition arising from the MNTB is powerful enough to completely silence LSO activity. Note that each LSO only encodes sounds arising in the ipsilateral hemifield; therefore, it takes both LSOs to represent the full range of horizontal positions.

In summary, there are two separate pathways—and two separate mechanisms—for localizing sound along the azimuth. Interaural time differences are processed in the medial superior olive, and interaural intensity differences are processed in the lateral superior olive. These two pathways are eventually merged in the midbrain auditory centers. As mentioned earlier, the elevation of sound sources is determined by spectral filtering mediated by the external pinnae; recent experimental evidence suggests that these spectral "notches" created by the pinnae are detected by neurons in the dorsal cochlear nucleus. Thus, binaural cues play an important role in localizing the azimuthal position of sound sources, while spectral cues are used to localize the elevation of sound sources.

Monaural Pathways from the Cochlear Nucleus to the Lateral Lemniscus

The binaural pathways for sound localization are only part of the output of the cochlear nucleus. This fact is hardly surprising, given that auditory perception involves much more than locating the position of the sound source. A second major set of pathways from the cochlear nucleus bypasses the superior olive and terminates in the **nuclei of the lateral lemniscus** on the contralateral side of the brainstem (see Figure 13.12). These particular pathways respond to sound arriving at one ear only and are thus referred to as monaural. Some cells in the lateral lemniscus nuclei signal the onset of sound, regardless of its intensity or frequency. Other cells in the lateral lemniscus nuclei process other temporal aspects of sound, such as duration. The precise role of these pathways in processing temporal features of sound is not yet known. As with the outputs of the superior olivary nuclei, the pathways from the nuclei of the lateral lemniscus converge at the midbrain.

Integration in the Inferior Colliculus

Auditory pathways ascending via the olivary and lemniscal complexes, as well as other projections that arise directly from the cochlear nucleus, project to the midbrain auditory center, also known as the **inferior colliculus**. In examining how integration occurs in the inferior colliculus, it is again instructive to turn to the most completely analyzed auditory mechanism, the binaural system for localizing sound. As already noted, space is not mapped on the auditory receptor surface; thus, the perception of auditory space must somehow be synthesized by circuitry in the lower brainstem and midbrain. Experiments in the barn owl, an extraordinarily proficient animal at localizing sounds, show that the convergence of binaural inputs in the midbrain produces something entirely new, relative to the periphery—namely, a computed topographical representation of auditory space. Neurons within this **auditory space map** in the colliculus respond best to sounds originating in a specific region of space; thus, they have both a preferred elevation and a preferred horizontal location, or azimuth. Although the circuit mechanisms underlying sound localization in humans are unknown, humans have a clear perception of both the elevational and azimuthal components of a sound's location, suggesting that we have an auditory space map.

Another important property of the inferior colliculus is its ability to process sounds with complex temporal patterns. Many neurons in the inferior colliculus respond only to frequency-modulated sounds, while others respond only to sounds of specific durations. Such sounds are typical components of biologically relevant sounds, such as those made by predators, or intraspecific communication sounds, which, in humans, include speech.

The Auditory Thalamus

Despite the parallel pathways in the auditory stations of the brainstem and midbrain, the **medial geniculate complex** (**MGC**) in the thalamus is an obligatory relay for all ascending auditory information destined for the cortex (see Figure 13.12). Most input to the MGC arises from the inferior colliculus, although a few auditory axons from the lower brainstem bypass the inferior colliculus to reach the auditory thalamus directly. The MGC has several divisions, including the ventral division, which functions as the major thalamocortical relay, and the dorsal and medial divisions, which are organized like a belt around the ventral division.

In some mammals, the strictly maintained tonotopy of the lower brainstem areas is exploited by convergence onto MGC neurons, generating specific responses to certain spectral combinations. The original evidence for this statement came from research on the response properties of cells in the MGC of echolocating bats. Some cells in the so-called belt areas of the bat MGC respond only to combinations of widely spaced frequencies that are specific components of the bat's echolocation signal and of the echoes that are reflected from objects in the bat's environment. In the mustached bat, where this phenomenon has been most thoroughly studied, the echolocation pulse has a changing frequency (frequency-modulated, or FM) component that includes a fundamental frequency and one or more harmonics. The fundamental frequency (FM_1) has low intensity and sweeps from 30–20 kHz. The second harmonic (FM_2) is the most intense component and sweeps from 60–40 kHz. Note that these frequencies do not overlap. Most of the echoes are from the intense FM_2 sound, and virtually none arise from the weak FM_1, even though the emitted FM_1 is loud enough for the bat to hear. Apparently, the bat assesses the distance to an object by measuring the delay between the FM_1 emission and the FM_2 echo. Certain MGC neurons respond when FM_2 follows FM_1 by a specific delay, providing a mechanism for sensing such frequency combinations. Because each neuron responds best to a particular delay, the population of MGC neurons encodes a range of distances.

Bat sonar illustrates two important points about the function of the auditory thalamus. First, the MGC is the first station in the auditory pathway where pronounced selectivity for combinations of frequencies is found. The mechanism responsible for this selectivity is presumably the ultimate convergence of inputs from cochlear areas with different spectral sensitivities. Second, cells in the MGC are selective not only for frequency combinations, but also for specific time intervals between the two frequencies. The principle is the same as that described for binaural neurons in the medial superior olive, but in this instance, two monaural signals with different frequency sensitivities coincide, and the time difference is in the millisecond rather than the microsecond range.

In summary, neurons in the medial geniculate complex receive convergent inputs from spectrally and temporally separate pathways. This complex, by virtue of its convergent inputs, mediates the detection of specific spectral and temporal combinations of sounds. In many species, including humans, varying spectral and temporal cues are especially important features of communication sounds. It is not known whether cells in the human medial geniculate are selective to combinations of sounds, but the processing of speech certainly requires both spectral and temporal combination sensitivity.

The Auditory Cortex

The ultimate target of afferent auditory information is the auditory cortex. Although the auditory cortex has a number of subdivisions, a broad distinction can be made between a primary area and peripheral, or secondary, areas. The **primary auditory cortex**, or **A1**, is located on the superior temporal gyrus in the temporal lobe and receives point-to-point input from the ventral division of the medial geniculate complex; thus, it contains a precise tonotopic map. The **belt areas** of the auditory cortex receive more diffuse input from the belt areas of the medial geniculate complex, as well as input from the primary auditory cortex, and are less precise in their tonotopic organization.

The primary auditory cortex has a topographical map of the cochlea (Figure 13.15), just as the primary visual cortex (V1) and the primary somatic sensory cortex (S1) have topographical maps of their respective sensory epithelia. Unlike the visual and somatic sensory systems, however, the cochlea has already decomposed the acoustical stimulus so that it is arrayed tonotopically along the length of the basilar membrane. Thus, A1 is said to comprise a tonotopic map, as do most of the ascending auditory structures between the cochlea and the

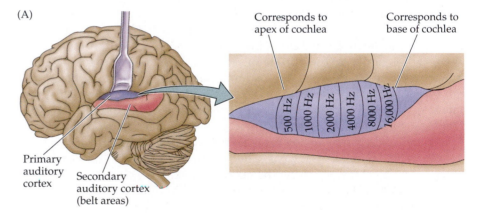

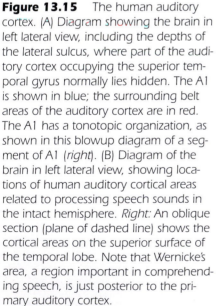

Figure 13.15 The human auditory cortex. (A) Diagram showing the brain in left lateral view, including the depths of the lateral sulcus, where part of the auditory cortex occupying the superior temporal gyrus normally lies hidden. The A1 is shown in blue; the surrounding belt areas of the auditory cortex are in red. The A1 has a tonotopic organization, as shown in this blowup diagram of a segment of A1 (*right*). (B) Diagram of the brain in left lateral view, showing locations of human auditory cortical areas related to processing speech sounds in the intact hemisphere. *Right:* An oblique section (plane of dashed line) shows the cortical areas on the superior surface of the temporal lobe. Note that Wernicke's area, a region important in comprehending speech, is just posterior to the primary auditory cortex.

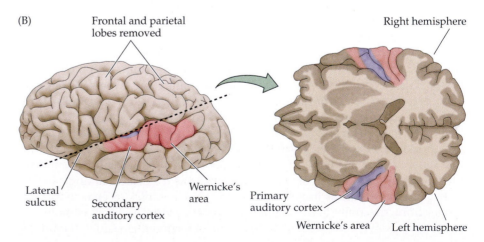

cortex. Orthogonal to the frequency axis of the tonotopic map are irregular patches of neurons that are excited by both ears (and are therefore called EE cells) interspersed with patches of cells that are excited by one ear and inhibited by the other ear (EI cells). The EE and EI stripes alternate, an arrangement that is reminiscent of the ocular dominance columns in V1 (see Chapter 12).

The auditory cortex obviously does much more than provide a tonotopic map and respond differentially to ipsi- and contralateral stimulation. Although the sorts of sensory processing that occur in the auditory cortex are not well understood, they are likely to be important to higher order processing of natural sounds, especially those used for communication (Box 13E; see also Chapter 27). One clue about such processing comes from work in marmosets, a small neotropical primate with a complex vocal repertoire. The A1 and belt areas of these animals are indeed organized tonotopically, but also contain neurons that are strongly responsive to spectral combinations that characterize certain vocalizations. The responses of these neurons to the tonal stimuli do not accurately predict their responses to the spectral combinations, suggesting that, in accordance with peripheral optimization, cortical processing is, in part, dedicated to detecting particular intraspecific vocalizations. Recent studies in marmosets and humans also implicate secondary regions of the auditory cortex in the perception of pitch. This percept is especially important to our musical sense and to vocal communication, because it enables us to hear two speech sounds as distinct even when they have overlapping spectral content and arise from the same location. A curious feature of pitch perception is that, for the harmonically complex sounds that typify speech and music, pitch corresponds to the fundamental frequency, even when it is absent from the actual stimulus. This synthetic aspect of pitch processing further underscores the integrative nature of sensory precessing in the auditory cortex.

Another clue about the role of the primary auditory cortex in the processing of intraspecific communication sounds comes from work in echolocating bats. Consistent with the essential role that echolocation plays in the survival of these crepuscular animals, certain regions of the bat A1, like those described in the medial geniculate complex, are tuned in a systematic manner to the delays between frequency-modulated pulses and their echoes, thus providing information about target distance and velocity. These delay-tuned neurons can exhibit highly specific responses to intraspecific communication calls, suggesting that the same cortical neurons can serve these two distinct auditory functions (see Box 13E). Evidently, the general ability of the mammalian auditory cortex to detect certain spectral and temporal combinations of natural sounds has been exploited in bats to serve sonar-mediated navigation, yielding these dual function neurons.

Many of the dually specialized neurons are categorized as "combination-sensitive" neurons—that is, neurons that show a nonlinear increase in their response magnitude when presented with a combination of tones and/or noise bands in comparison to the total magnitude of the response elicited by presenting each sound element separately. Combination-sensitive neurons are tuned to more than one frequency and are specialized to recognize complex species-specific sounds and extract information that is critical for survival. This sensitivity to combinations of simple sound elements appears to be a universal property of neurons for the perception of complex sounds by many animal species, such as frogs, birds, bats, and non-human primates. Therefore, combination-sensitive neurons most likely partake in the recognition of complex sounds in the human auditory cortex as well.

Sounds that are especially important for intraspecific communication often have a highly ordered temporal structure. In humans, the best example of such time-varying signals is speech, where different phonetic sequences are perceived as distinct syllables and words (see Box 27A). Behavioral studies in cats

BOX 13E   Representing Complex Sounds in the Brains of Bats and Humans

Most natural sounds are complex, meaning that they differ from the pure tones or clicks that are frequently used in neurophysiological studies of the auditory system. Rather, natural sounds are tonal: They have a fundamental frequency that largely determines the "pitch" of the sound, and one or more harmonics of different intensities that contribute to the quality or "timbre" of a sound. The frequency of a harmonic is, by definition, a multiple of the fundamental frequency, and both may be modulated over time. Such *frequency-modulated* (FM) sweeps can rise or fall in frequency, or change in a sinusoidal (or some other) fashion. Occasionally, multiple nonharmonic frequencies may be simultaneously present in some communication or musical sounds. In some sounds, a level of spectral splatter or "broadband noise" is embedded within tonal or frequency-modulated sounds. The variations in the sound spectrum are typically accompanied by a modulation of the amplitude envelope of the complex sound, as well. All of these features can be visualized by performing a spectrographic analysis.

How does the brain represent such complex natural sounds? Cognitive studies of complex sound perception provide some understanding of how a large but limited number of neurons in

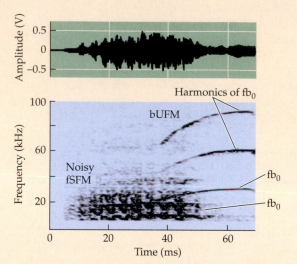

(A) Amplitude envelope (above) and spectrogram (below) of a composite syllable emitted by mustached bats for social communication. This composite consists of two simple syllables: a fixed Sinusoidal FM (fSFM), and a bent Upward FM (bUFM) that emerges from the fSFM after some overlap. Each syllable has its own fundamental (fa0 and fb0) and multiple harmonics. (Courtesy of Jagmeet Kanwal.)

the brain can dynamically represent an infinite variety of natural stimuli in the sensory environment of humans and other animals. In bats, specializations for processing complex sounds are apparent. Studies in echolocating bats show that both communication and echolocation sounds (Figure A) are processed not only within some of the same areas, but also within the same neurons in the auditory cortex. In humans, multiple modes of processing are also likely, given the large overlap within the superior and middle temporal gyri in the temporal lobe for the representation of different types of complex sounds.

Asymmetrical representation is another common principle of complex sound processing that results in lateralized (though largely overlapping) representations of natural stimuli. Thus, speech sounds that are important for communication are lateralized to the left in the belt regions of the auditory cortex, whereas environmental sounds that are important for reacting to and recognizing aspects of the auditory environment are represented in each hemisphere (Figure B). Musical sounds that can either motivate us to march in war or to relax and meditate when coping with physical and emotional stress are highly lateralized to the right in the belt

and monkeys show that the auditory cortex is especially important for processing temporal sequences of sound. If the auditory cortex is ablated in these animals, they lose the ability to discriminate between two complex sounds that have the same frequency components but which differ in temporal sequence. Thus, without the auditory cortex, monkeys cannot discriminate one conspecific communication sound from another. The physiological basis of such temporal sensitivity likely requires neurons that are sensitive to time-varying cues in communication sounds. Indeed, electrophysiological recordings from the primary auditory cortices of both marmosets and bats show that some neurons that respond to intraspecific communication sounds do not respond as strongly when the sounds are played in reverse, indicating sensitivity to the sounds' temporal features. Studies of human patients with bilateral damage to the auditory cortex also reveal severe problems in processing the temporal order of sounds.

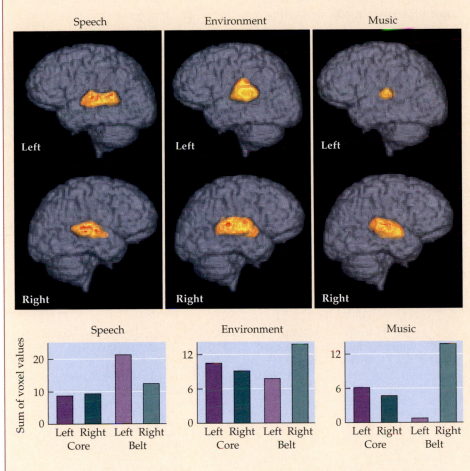

Speech Environment Music

Left Left Left

Right Right Right

(B) Reconstructed functional magnetic resonance images of BOLD contrast signal change (average for 8 subjects) showing significant (p < 0.001) activation elicited by speech, environmental, and musical sounds on surface views of the left versus the right side of the human brain. The bar graphs show the total significant activation to each category of complex sounds in the core and belt areas of the auditory cortex for the left versus the right side. (Courtesy of Jagmeet Kanwal.)

Speech — Environment — Music (Sum of voxel values; Left/Right Core, Left/Right Belt)

regions of the auditory cortex. The extent of lateralization for speech, and possibly music, may vary with sex, age, and training. In some species of bats, mice, and primates, processing of natural communication sounds appears to be lateralized to the left hemisphere. In summary, natural sounds are complex and their representation within the sensory cortex tends to be asymmetric across the two hemispheres.

References

EHRET, G. (1987) Left hemisphere advantage in the mouse brain for recognizing ultrasonic communication calls. *Nature* 325: 249–251.

ESSER, K.-H., C. J. CONDON, N. SUGA AND J. S. KANWAL (1997) Syntax processing by auditory cortical neurons in the FM-FM area of the mustached bat, *Pteronotus parnellii*. *Proc. Natl. Acad. Sci. USA* 94: 14019–14024.

HAUSER, M. D. AND K. ANDERSSON (1994) Left hemisphere dominance for processing vocalizations in adult, but not infant, rhesus monkeys: Field experiments. *Proc. Natl. Acad. Sci. USA* 91: 3946–3948.

KANWAL, J. S., J. KIM AND K. KAMADA (2000) Separate, distributed processing of environmental, speech and musical sounds in the cerebral hemispheres. *J. Cog. Neurosci.* (Supp.): 32.

KANWAL, J. S., J. S. MATSUMURA, K. OHLEMILLER AND N. SUGA (1994) Acoustic elements and syntax in communication sounds emitted by mustached bats. *J. Acous. Soc. Am.* 96: 1229–1254.

KANWAL, J. S. AND N. SUGA (1995) Hemispheric asymmetry in the processing of calls in the auditory cortex of the mustached bat. *Assoc. Res. Otolaryngol.* 18: 104.

It seems likely, therefore, that specific regions of the human auditory cortex are specialized for processing elementary speech sounds, as well as other temporally complex acoustical signals, such as music (see Box 13B). Thus, Wernicke's area, which is critical to the comprehension of human language, is contiguous with the secondary auditory area (see Figure 13.15 and Chapter 27).

Summary

Sound waves are transmitted via the external and middle ear to the cochlea of the inner ear, which exhibits a traveling wave when stimulated. For high-frequency sounds, the amplitude of the traveling wave reaches a maximum at the base of the cochlea; for low-frequency sounds, the traveling wave reaches a maximum at the apical end. The associated motions of the basilar membrane are transduced primarily by the inner hair cells, while the basilar membrane motion is itself actively modulated by the outer hair cells. Damage to the outer or middle ear results in conductive hearing loss, while hair cell damage results in a sensorineural hearing deficit. The tonotopic organization of the cochlea is retained at all levels of the central auditory system. Projections from the cochlea travel via the auditory nerve to the three main divisions of the cochlear nucleus. The targets of the cochlear nucleus neurons include the superior olivary complex and nuclei of the lateral lemniscus, where the binaural cues for sound localization are processed. The inferior colliculus, the target of nearly all of the auditory pathways in the lower brainstem, carries out important integrative functions, such as processing sound frequencies and integration of the cues for localizing sound in space. The primary auditory cortex, which is also organized tonotopically, is essential for basic auditory functions, such as frequency discrimination and sound localization, and also plays an important role in processing of intraspecific communication sounds. The belt areas of the auditory cortex have a less strict tonotopic organization and also process complex sounds, such as those that mediate communication. In the human brain, the major speech comprehension areas are located in the zone immediately adjacent to the auditory cortex.

Additional Reading

Reviews

COREY, D. P. (1999) Ion channel defects in hereditary hearing loss. *Neuron.* 22(2): 217–9.

GARCIA-ANOVEROS, J. AND D. P. COREY (1997) The molecules of mechanosensation. *Ann. Rev. Neurosci.* 20: 567–597.

HEFFNER, H. E. AND R. S. HEFFNER (1990) Role of primate auditory cortex in hearing. In *Comparative Perception, Volume II: Complex Signals.* W. C. Stebbins and M. A. Berkley (eds.). New York: John Wiley.

HUDSPETH, A. J. (2000) Hearing and deafness. *Neurobiol. Dis.* 7: 511–514.

HUDSPETH, A. J. (2001–2002) How the ear's works work: Mechanoelectrical transduction and amplification by hair cells of the internal ear. *Harvey Lect.* 97: 41–54.

HUDSPETH, A. J. AND M. KONISHI (2000) Auditory neuroscience: Development, transduction, and integration. Proc. Natl. Acad. Sci. USA 97: 11690–11691.

HUDSPETH, A. J., Y. CHOE, A. D. MEHTA AND P. MARTIN (2000) Putting ion channels to work: Mechanoelectrical transduction, adaptation, and amplification by hair cells. *Proc. Natl. Acad. Sci. USA* 97: 11765–11772.

KIANG, N. Y. S. (1984) Peripheral neural processing of auditory information. In *Handbook of Physiology,* Section 1: *The Nervous System,* Volume III. *Sensory Processes,* Part 2. J. M. Brookhart, V. B. Mountcastle, I. Darian-Smith and S. R. Geiger (eds.). Bethesda, MD: American Physiological Society.

LEMASURIER, M. AND P. G. GILLESPIE (2005) Hair-cell mechanotransduction and cochlear amplification. *Neuron* 48: 403–415.

LIN, S. Y. AND D. P. COREY (2005) TRP channels in mechanotransduction. *Curr. Opin. Neurobiol.* 15: 350–357.

NELKEN, I. (2002) Feature detection by the auditory cortex. In *Integrative Functions in the Mammalian Auditory Pathway, Springer Handbook of Auditory Research, Volume 15.* D. Oertel, R. Fay and A. N. Popper (eds.). New York: Springer-Verlag, pp. 358–416.

NELKEN, I. (2004) Processing of complex stimuli and natural scenes in the auditory cortex. *Curr. Opin. Neurobiol.* 14: 474–480.

SMITH, E. C. AND M. S. LEWICKI (2006) Efficient auditory coding. *Nature* 439: 978–982.

TRAMO, M. J., P. A. CARIANI, C. K. KOH, N. MAKRIS AND L. D. BRAIDA (2005) Neurophysiology and neuroanatomy of pitch perception: Auditory cortex. *Ann. N.Y. Acad. Sci.* 1060: 148–174.

Important Original Papers

BARBOUR, D. L. AND X. WANG (2005) The neuronal representation of pitch in primate auditory cortex. *Nature* 436: 1161–1165.

CRAWFORD, A. C. AND R. FETTIPLACE (1981) An electrical tuning mechanism in turtle cochlear hair cells. *J. Physiol.* 312: 377–413.

FITZPATRICK, D. C., J. S. KANWAL, J. A. BUTMAN AND N. SUGA (1993) Combination-sensitive neurons in the primary auditory cortex of the mustached bat. *J. Neurosci.* 13: 931–940.

COREY, D. P. AND A. J. HUDSPETH (1979) Ionic basis of the receptor potential in a vertebrate hair cell. *Nature* 281: 675–677.

MIDDLEBROOKS, J. C., A. E. CLOCK, L. XU AND D. M. GREEN (1994) A panoramic code for sound location by cortical neurons. *Science* 264: 842–844.

KNUDSEN, E. I. AND M. KONISHI (1978) A neural map of auditory space in the owl. *Science* 200: 795–797.

JEFFRESS, L. A. (1948) A place theory of sound localization. *J. Comp. Physiol. Psychol.* 41: 35–39.

NELKEN, I., Y. ROTMAN AND O. BAR YOSEF (1999) Responses of auditory-cortex neurons to structural features of natural sounds. *Nature* 397: 154–157.

RIEKE, F., D. A. BODNAR, AND W. BIALEK (1995) Naturalistic stimuli increase the rate and efficiency of information transfer by primary auditory afferents. *Proc. Biol. Sci.* 262(1365): 259–265.

SUGA, N., W. E. O'NEILL AND T. MANABE (1978) Cortical neurons sensitive to combinations of information-bearing elements of biosonar signals in the mustache bat. *Science* 200: 778–781.

VON BÉKÉSY, G. (1960) *Experiments in Hearing.* New York: McGraw-Hill. (A collection of von Békésy's original papers.)

Books

MOORE, B. C. J. (2003) *An Introduction to the Psychology of Hearing.* London: Academic Press.

PICKLES, J. O. (1988) *An Introduction to the Physiology of Hearing.* London: Academic Press.

YOST, W. A. AND G. GOUREVITCH (EDS.) (1987) *Directional Hearing.* Berlin: Springer Verlag.

YOST, W. A. AND D. W. NIELSEN (1985) *Fundamentals of Hearing.* Fort Worth: Holt, Rinehart and Winston.

Chapter 14

The Vestibular System

Overview

The vestibular system has important sensory functions, contributing to the perception of self-motion, head position, and spatial orientation relative to gravity. It also serves important motor functions, helping to stabilize gaze, head, and posture. The peripheral portion of the vestibular system includes inner ear structures that function as miniaturized accelerometers and inertial guidance devices, continually reporting information about the motions and position of the head and body to integrative centers in the brainstem, cerebellum, and somatic sensory cortices. The central portion of the system includes the vestibular nuclei, which make extensive connections with brainstem and cerebellar structures. The vestibular nuclei also directly innervate motor neurons controlling extraocular, cervical, and postural muscles. This motor output is especially important to stabilization of gaze, head orientation, and posture during movement. Although we are normally unaware of its functioning, the vestibular system is a key component in postural reflexes and eye movements. Balance, gaze stabilization during head movement, and sense of orientation in space are all adversely affected if the system is damaged. These manifestations of vestibular damage are especially important in the evaluation of brainstem injury. Because the circuitry of the vestibular system extends through a large part of the brainstem, simple clinical tests of vestibular function can be performed to determine brainstem involvement, even on comatose patients.

The Vestibular Labyrinth

The main peripheral component of the vestibular system is an elaborate set of interconnected chambers—the **labyrinth**—that has much in common, and is in fact continuous with, the cochlea (see Chapter 13). Like the cochlea, the labyrinth is derived from the otic placode of the embryo, and it uses the same specialized set of sensory cells—hair cells—to transduce physical motion into neural impulses. In the cochlea, the motion is due to airborne sounds; in the labyrinth, the motions transduced arise from head movements, inertial effects due to gravity, and ground-borne vibrations (Box 14A).

The labyrinth is buried deep in the temporal bone and consists of the two **otolith organs** (the **utricle** and **saccule**) and three **semicircular canals** (Figure 14.1). The elaborate and tortuous architecture of these components explains why this part of the vestibular system is called the labyrinth. The utricle and saccule are specialized primarily to respond to *linear accelerations* of the head and *static head position relative to the gravitational axis*, whereas the semicircular canals, as their shapes suggest, are specialized for responding to *rotational accelerations* of the head.

The intimate relationship between the cochlea and the labyrinth goes beyond their common embryonic origin. Indeed, the cochlear and vestibular spaces are

BOX 14A   A Primer on Vestibular Navigation

The function of the vestibular system can be simplified by remembering some basic terminology of classical mechanics. All bodies moving in three dimensions have six degrees of freedom; three of these are translational and three are rotational. The translational components may be given in terms of movements along x, y, and z axes of the head, which form a right-handed coordinate system for the head. The otoliths detect accelerations along these axes as well as tilts of the head relative to gravity that induce equivalent accelerations. Rotation of the head about these axes activates the semicircular canals, which are responsible for sensing rotational or angular accelerations of the head. Rotations about the x, y, and z axes are commonly referred to as *roll*, *pitch*, and *yaw*. The positive direction of head rotation follows a right-hand rule—that is, if the fingers of the right hand are curled in the direction of the arrows, the thumb points in the positive direction of the axis.

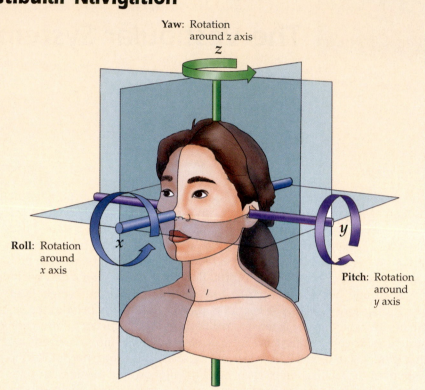

Yaw: Rotation around z axis

Roll: Rotation around x axis

Pitch: Rotation around y axis

actually joined (see Figure 14.1), and the specialized ionic environments of the vestibular end organ parallel those of the cochlea. The membranous sacs within the bone are filled with fluid (endolymph) and are collectively called the membranous labyrinth. The endolymph (like the cochlear endolymph) is similar to intracellular solutions in that it is high in K^+ and low in Na^+. Between the bony walls (the osseous labyrinth) and the membranous labyrinth is another fluid, the perilymph, which is similar in composition to cerebrospinal fluid (i.e., low in K^+ and high in Na^+; see Chapter 13).

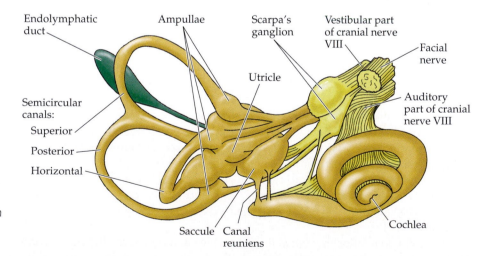

Figure 14.1 The labyrinth and its innervation. The vestibular and auditory portions of cranial nerve VIII are shown; the small connection from the vestibular nerve to the cochlea contains auditory efferent fibers. General orientation within the head is shown in Figure 13.3; see also Figure 14.8.

Endolymphatic duct

Ampullae

Scarpa's ganglion

Vestibular part of cranial nerve VIII

Facial nerve

Utricle

Auditory part of cranial nerve VIII

Semicircular canals:
Superior
Posterior
Horizontal

Saccule   Canal reuniens

Cochlea

The vestibular hair cells are located in the utricle and saccule and in three juglike swellings called **ampullae**, located at the base of the semicircular canals next to the utricle. Within each ampulla, the vestibular hair cells extend their hair bundles into the endolymph of the membranous labyrinth. As in the cochlea, tight junctions seal the apical surfaces of the vestibular hair cells, ensuring that endolymph selectively bathes the hair cell bundle while remaining separate from the perilymph surrounding the basal portion of the hair cell.

Vestibular Hair Cells

The vestibular hair cells, which like cochlear hair cells transduce minute displacements into behaviorally relevant receptor potentials, provide the basis for vestibular function. Vestibular and auditory hair cells are quite similar; a detailed description of hair cell structure and function has already been given in Chapter 13. As in the case of auditory hair cells, movement of the stereocilia toward the kinocilium in the vestibular end organs opens mechanically gated transduction channels located at the tips of the stereocilia, depolarizing the hair cell and causing neurotransmitter release onto (and excitation of) the vestibular nerve fibers (Figure 14.2). Movement of the stereocilia in the direction away from the kinocilium closes the channels, hyperpolarizing the hair cell and thus reducing vestibular nerve activity. The biphasic nature of the receptor potential means that some transduction channels are open in the absence of stimulation, with the result that hair cells tonically release transmitter, thereby generating consid-

Figure 14.2 The morphological polarization of vestibular hair cells and the polarization maps of the vestibular organs. (A) A cross section of hair cells shows that the kinocilia of a group of hair cells are all located on the same side of the hair cell. The arrow indicates the direction of deflection that depolarizes the hair cell. (B) View looking down on the hair bundles. (C) In the ampulla located at the base of each semicircular canal, the hair bundles are oriented in the same direction. In the sacculus and utricle, the striola divides the hair cells into populations with opposing hair bundle polarities.

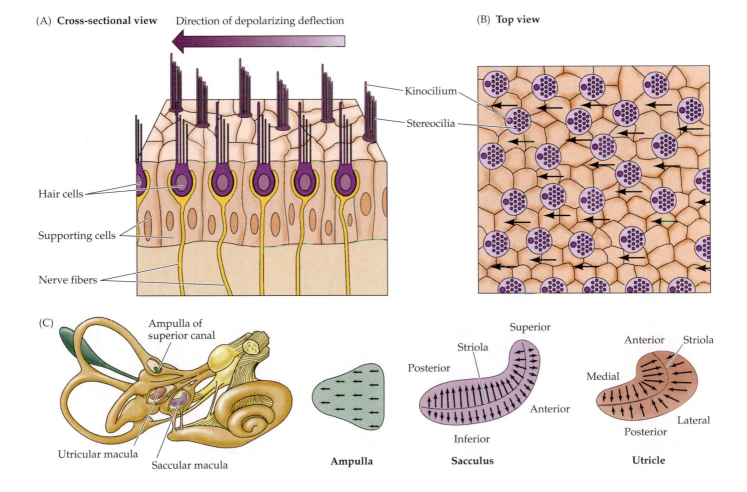

(A) **Cross-sectional view**   Direction of depolarizing deflection

(B) **Top view**

Kinocilium

Stereocilia

Hair cells

Supporting cells

Nerve fibers

(C)
Ampulla of superior canal

Utricular macula

Saccular macula

Ampulla

Superior

Striola

Posterior

Anterior

Inferior

Sacculus

Anterior   Striola

Medial

Lateral

Posterior

Utricle

BOX 14B Adaptation and Tuning of Vestibular Hair Cells

Hair Cell Adaptation

The minuscule movement of the hair bundle at sensory threshold has been compared to the displacement of the top of the Eiffel Tower by a thumb's breadth! Despite its great sensitivity, the hair cell can adapt quickly and continuously to static displacements of the hair bundle caused by large movements. Such adjustments are especially useful in the otolith organs, where adaptation permits hair cells to maintain sensitivity to small linear and angular accelerations of the head despite the constant input from gravitational forces that are over a million times greater.

In other receptor cells, such as photoreceptors, adaptation is accomplished by regulating the second messenger cascade induced by the initial transduction event. The hair cell has to depend on a different strategy, however, because there is no second messenger system between the initial transduction event and the subsequent receptor potential (as might be expected for receptors that respond so rapidly).

Adaptation occurs in both directions in which the hair bundle displacement generates a receptor potential, albeit at different rates for each direction. When the hair bundle is pushed toward the kinocilium, tension is initially increased in the gating spring. During adaptation, tension decreases back to the resting level, perhaps because one end of the gating spring repositions itself along the shank of the stereocilium. When the hair bundle is displaced in the opposite direction, away from the kinocilium, tension in the spring initially decreases; adaptation then involves an increase in spring tension. One theory is that a calcium-regulated motor such as a myosin ATPase climbs along actin filaments in the stereocilium and actively resets the tension in the transduction spring (Figure A). During sustained depolarization, some Ca^{2+} enters through the transduction channel, along with K^+. Ca^{2+} then causes the motor to spend a greater fraction of its time unbound from the actin, resulting in slippage of the spring down the side of the stereocilium. During sustained hyperpolarization, Ca^{2+} levels drop below normal resting levels and the

motor spends more of its time bound to the actin, thus climbing up the actin filaments and increasing the spring tension. As tension increases, some of the previously closed transduction channels open, admitting Ca^{2+} and thus slowing the motor's progress until a balance is struck between the climbing and slipping of the motor. In support of this model, when internal Ca^{2+} is reduced artificially, spring tension increases. This model of hair cell adaptation presents an elegant molecular solution to the regulation of a mechanical process.

Electrical Tuning

Although mechanical tuning plays an important role in generating frequency selectivity in the cochlea, there are other mechanisms that contribute to this process in vestibular and auditory nerve cells. These other tuning mechanisms are especially important in the otolith organs, where, unlike the cochlea, there are no obvious macromechanical resonances to selectively filter and/or

Adaptation is explained in the gating spring model by adjustment of the insertion point of tips links. Movement of the insertion point up or down the shank of the stereocilium, perhaps driven by a Ca^{2+}-dependent protein motor, can continually adjust the resting tension of the tip link. (After Hudspeth and Gillespie, 1994.)

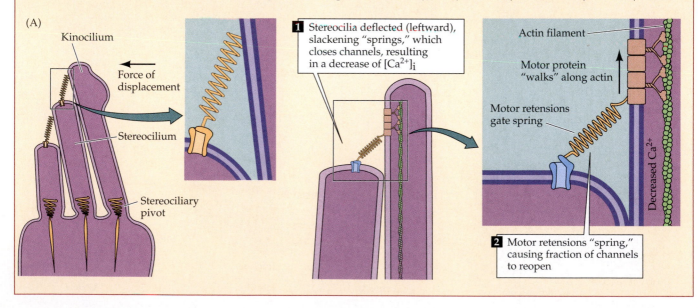

(A)

Kinocilium

Force of displacement

Stereocilium

Stereociliary pivot

1 Stereocilia deflected (leftward), slackening "springs," which closes channels, resulting in a decrease of $[Ca^{2+}]_i$

Actin filament

Motor protein "walks" along actin

Motor retensions gate spring

Decreased Ca^{2+}

2 Motor retensions "spring," causing fraction of channels to reopen

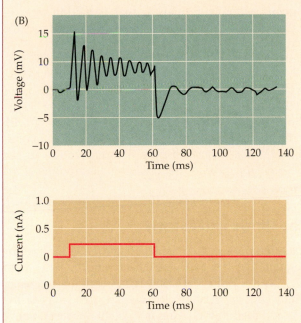

(B)

Voltage oscillations (upper trace) in an isolated hair cell in response to a depolarizing current injection (lower trace). (After Lewis and Hudspeth, 1983.)

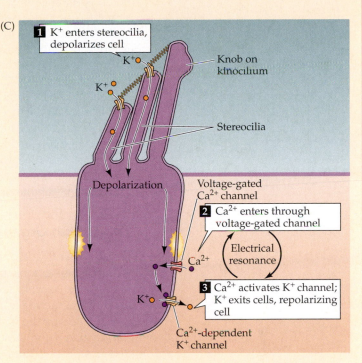

(C)

Proposed ionic basis for electrical resonance in hair cells. (After Hudspeth, 1985.)

enhance biologically relevant movements. One such mechanism is an electrical resonance displayed by hair cells in response to depolarization: the membrane potential of a hair cell undergoes damped sinusoidal oscillations at a specific frequency in response to the injection of depolarizing current pulses (Figure B).

The ionic mechanism of this process involves two major types of ion channels located in the membrane of the hair cell soma. The first of these is a voltage-activated Ca^{2+} conductance, which lets Ca^{2+} into the cell soma in response to depolarization, such as that generated by the transduction current. The second is a Ca^{2+}-activated K^+ conductance, which is triggered by the rise in internal Ca^{2+} concentration. These two currents produce an interplay of depolarization and repolarization that results in electrical resonance (Figure C). Activation of the hair cell's calcium-activated K^+ conductance occurs 10 to

100 times faster than that of similar currents in other cells. Such rapid kinetics allow this conductance to generate an electrical response that usually requires the fast properties of a voltage-gated channel.

Although a hair cell responds to hair bundle movement over a wide range of frequencies, the resultant receptor potential is largest at the frequency of electrical resonance. The resonance frequency represents the characteristic frequency of the hair cell, and transduction will be most efficient at that frequency. This electrical resonance has important implications for structures like the utricle and sacculus, which may encode a range of characteristic frequencies based on the different resonance frequencies of their constituent hair cells. Thus, electrical tuning in the otolith organs can generate enhanced tuning to biologically relevant frequencies of stimulation, even in the absence of macromechanical resonances within these structures.

References

Assad, J. A. and D. P. Corey (1992) An active motor model for adaptation by vertebrate hair cells. *J. Neurosci.* 12: 3291–3309.

Crawford, A. C. and R. Fettiplace (1981) An electrical tuning mechanism in turtle cochlear hair cells. *J. Physiol.* 312: 377–412.

Hudspeth, A. J. (1985) The cellular basis of hearing: The biophysics of hair cells. *Science* 230: 745–752.

Hudspeth, A. J. and P. G. Gillespie (1994) Pulling strings to tune transduction: Adaptation by hair cells. *Neuron* 12: 1–9.

Lewis, R. S. and A. J. Hudspeth (1988) A model for electrical resonance and frequency tuning in saccular hair cells of the bull-frog, *Rana catesbeiana. J. Physiol.* 400: 275–297.

Lewis, R. S. and A. J. Hudspeth (1983) Voltage- and ion-dependent conductances in solitary vertebrate hair cells. *Nature* 304: 538–541.

Shepherd, G. M. G. and D. P. Corey (1994) The extent of adaptation in bullfrog saccular hair cells. *J. Neurosci.* 14: 6217–6229.

erable spontaneous activity in vestibular nerve fibers (see Figure 14.6). One consequence of these spontaneous action potentials is that the firing rates of vestibular fibers can increase or decrease in a manner that faithfully mimics the receptor potentials produced by the hair cells (Box 14B).

Importantly, the hair cell bundles in each vestibular organ have specific orientations (Figure 14.2). As a result, the organ as a whole is responsive to displacements in all directions. In a given semicircular canal, the hair cells in the ampulla are all polarized in the same direction. In the utricle and saccule, a specialized area called the **striola** divides the hair cells into two populations with opposing polarities (Figure 14.2C; see also Figure 14.4C). The directional polarization of the receptor surfaces is a basic principle of organization in the vestibular system, as will become apparent in the following descriptions of the individual vestibular organs.

The Otolith Organs: The Utricle and Saccule

Displacements and linear accelerations of the head, such as those induced by tilting or translational movements (see Box 14A), are detected by the two otolith organs: the saccule and the utricle. Both of these organs contain a sensory epithelium, the **macula**, which consists of hair cells and associated supporting cells. Overlying the hair cells and their hair bundles is a gelatinous layer; above this layer is a fibrous structure, the **otolithic membrane**, in which are embedded crystals of calcium carbonate called **otoconia** (Figures 14.3 and 14.4A). The crystals give the otolith organs their name (*otolith* is Greek for "ear stones"). The otoconia make the otolithic membrane considerably heavier than the structures and fluids surrounding it; thus, when the head tilts, gravity causes the membrane to shift relative to the sensory epithelium (Figure 14.4B). The resulting shearing motion between the otolithic membrane and the macula displaces the hair bundles, which are embedded in the lower, gelatinous surface of the membrane. This displacement of the hair bundles generates a receptor potential in the hair cells. A shearing motion between the macula and the otolithic membrane also occurs when the head undergoes linear accelerations (see Figure 14.5); the greater relative mass of the otolithic membrane causes it to lag behind the macula temporarily, leading to transient displacement of the hair bundle.

The similar effects exerted on otolithic hair cells by certain head tilts and linear accelerations would be expected to render these different stimuli perceptually equivalent when visual feedback is absent, as occurs in the dark or when the eyes are closed. Nevertheless, evidence suggests that subjects can discriminate between these two stimulus categories, apparently through combined activity of the otolith organs and the semicircular canals.

As already mentioned, the orientation of the hair cell bundles is organized relative to the striola, which demarcates the overlying layer of otoconia (see Figure 14.4A). The striola forms an axis of mirror symmetry such that hair cells on opposite sides of the striola have opposing morphological polarizations. Thus, a tilt along the axis of the striola will excite the hair cells on one side while inhibiting the hair cells on the other side. The saccular macula is oriented vertically and the utricular macula horizontally, with a continuous variation in the morphological polarization of the hair cells located in each macula (as shown in Figure 14.4C, where the arrows indicate the direction of movement that produces excitation). Inspection of the excitatory orientations in the maculae indicates that the utricle responds to movements of the head in the horizontal plane, such as sideways head tilts and rapid lateral displacements, whereas the saccule responds to movements in the vertical plane (up–down and forward–backward movements in the sagittal plane).

Figure 14.3 Scanning electron micrograph of calcium carbonate crystals (otoconia) in the utricular macula of the cat. Each crystal is about 50 μm long. (From Lindeman, 1973.)

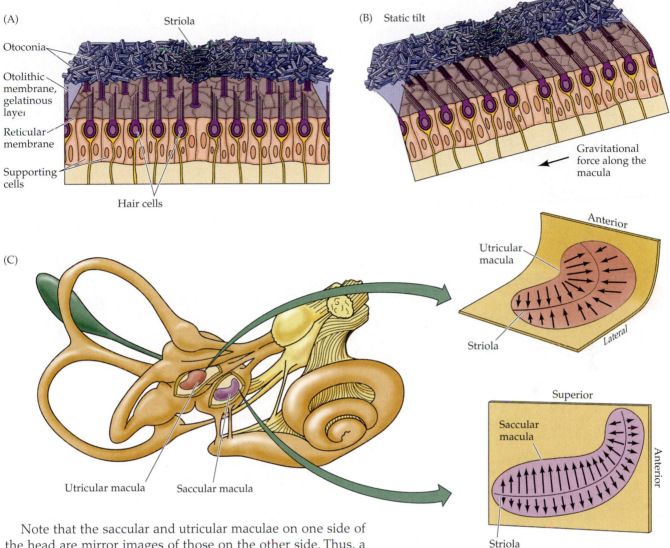

Figure 14.4   Morphological polarization of hair cells in the utricular and saccular maculae. (A) Cross section of the utricular macula showing hair bundles projecting into the gelatinous layer when the head is level. (B) Cross section of the utricular macula when the head is tilted. The hair cells are deflected by the otoconia in the direction of the gravitational force along the macular plane. An equivalent linear acceleration opposite to this force would induce the same deflection of the otoconia and is referred to as the *equivalent acceleration*. (C) Orientation of the utricular and saccular maculae in the head; arrows show orientation of the kinocilia, as in Figure 14.2. The *saccules* on either side are oriented more or less vertically, and the *utricles* more or less horizontally. The striola is a structural landmark consisting of small otoconia arranged in a narrow trench that divides each otolith organ. In the utricular macula, the kinocilia are directed toward the striola. In the saccular macula, the kinocilia point away from the striola. Note that, given the utricle and sacculus on both sides of the body, there is a continuous representation of all directions of body movement.

Note that the saccular and utricular maculae on one side of the head are mirror images of those on the other side. Thus, a tilt of the head to one side has opposite effects on corresponding hair cells of the two utricular maculae. This concept is important in understanding how the central connections of the vestibular periphery mediate the interaction of inputs from the two sides of the head.

How Otolith Neurons Sense Linear Accelerations of the Head

The structure of the otolith organs enables them to sense both static displacements, as would be caused by tilting the head relative to the gravitational axis, and transient displacements caused by translational movements of the head. Figure 14.5 illustrates some of the forces produced by head tilt and linear accelerations on the utricular macula.

The mass of the otolithic membrane relative to the surrounding endolymph, as well as the otolithic membrane's physical uncoupling from the underlying macula, means that hair bundle displacement will occur transiently in response to linear accelerations, and tonically in response to tilting of the head. This tran-

Upright

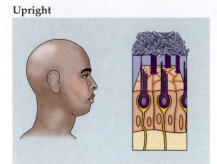

Figure 14.5 Forces acting on the head result in displacement of the otolithic membrane of the utricular macula. For each of the positions and accelerations due to translational movements, some set of hair cells will be maximally excited, whereas another set will be maximally inhibited. Note that head tilts produce displacements similar to certain accelerations.

Head tilt; sustained

Backward

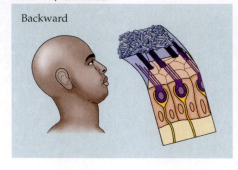

Forward

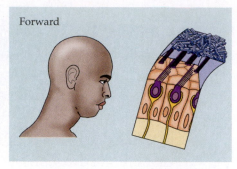

No head tilt; transient

Forward acceleration

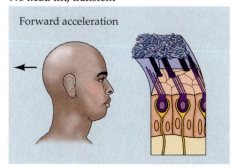

Deceleration

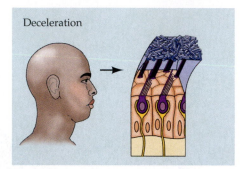

sient displacement is the result of components of the gravitational force acting along the macular plane (see Figure 14.4B).

These properties of hair cells are reflected in the responses of the vestibular nerve fibers that innervate the otolith organs. The nerve fibers have a steady and relatively high firing rate when the head is upright. The change in firing rate in response to a given movement can be either sustained or transient, thereby signaling either absolute head position or linear acceleration. An example of the sustained response of a vestibular nerve fiber innervating the utricle is shown in Figure 14.6. The responses were recorded from axons in a monkey seated in a chair that could be tilted for several seconds to produce a steady force. Prior to the tilt, the axon has a high firing rate, which increases or decreases depending on the direction of the tilt. Notice also that the response remains at a high level as long as the tilting force remains constant; thus, such neurons faithfully encode the static force being applied to the head (Figure 14.6A). When the head is returned to the original position, the firing level of the neurons returns to baseline value. Conversely, when the tilt is in the opposite direction, the neurons respond by decreasing their firing rate below the resting level (Figure 14.6B) and remain depressed as long as the static force continues. In a similar fashion, transient increases or decreases in firing rate from spontaneous levels signal the direction of linear accelerations of the head.

The range of orientations of hair bundles within the otolith organs enables them to transmit information about linear forces in every direction the body moves (see Figure 14.4C). The utricle, which is primarily concerned with motion in the horizontal plane, and the saccule, which is concerned with vertical motion, combine to effectively gauge the linear forces acting on the head at any instant in three dimensions. Tilts of the head off the horizontal plane and translational movements of the head in any direction stimulate a distinct subset of hair cells in the saccular and utricular maculae, while simultaneously suppressing the responses of other hair cells in these organs. Ultimately, variations in

(A)

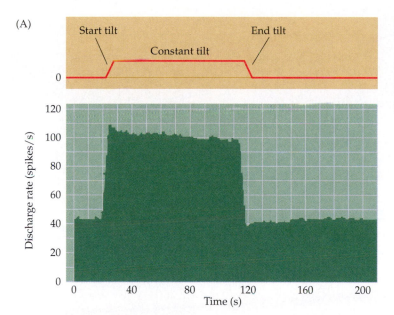

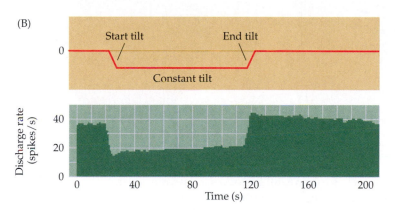

Figure 14.6 Response of a vestibular nerve axon from an otolith organ (the utricle in this example). (A) The stimulus (top) is a change in head tilt. The spike histogram shows the neuron's response to tilting in a particular direction. (B) A response of the same fiber to tilting in the opposite direction. (After Goldberg and Fernandez, 1976.)

(B)

hair cell polarity within the otolith organs produce patterns of vestibular nerve fiber activity that, at a population level, can unambiguously encode head position and the forces that influence it.

The Semicircular Canals

Whereas the otolith organs are primarily concerned with head translations and orientation with respect to gravity, the semicircular canals sense head *rotations*, arising either from self-induced movements or from angular accelerations of the head imparted by external forces. Each of the three semicircular canals has at its base a bulbous expansion called the **ampulla** (Figure 14.7), which houses the sensory epithelium, or **crista**, that contains the hair cells. The structure of the canals suggests how they detect the angular accelerations that arise through rotation of the head. The hair bundles extend out of the crista into a gelatinous mass, the **cupula**, that bridges the width of the ampulla, forming a viscous barrier through which endolymph cannot circulate. As a result, the relatively compliant cupula is distorted by movements of the endolymphatic fluid. When the head turns in the plane of one of the semicircular canals, the inertia of the endolymph produces a force across the cupula, distending it away from the direction of head movement and causing a displacement of the hair bundles

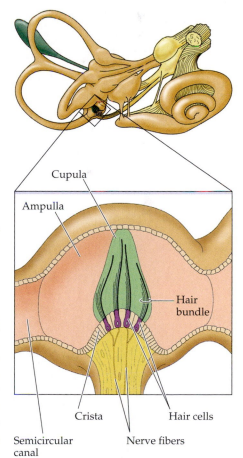

Figure 14.7 The ampulla of the posterior semicircular canal showing the crista, hair bundles, and cupula. The cupula is distorted by the fluid in the membranous canal when the head rotates.

Figure 14.8 Functional organization of the semicircular canals. (A) The position of the cupula without angular acceleration. (B) Distortion of the cupula during angular acceleration. When the head is rotated in the plane of the canal (arrow outside canal), the inertia of the endolymph creates a force (arrow inside canal) that displaces the cupula. (C) Arrangement of the canals in pairs. The two horizontal canals form a pair; the right anterior canal (AC) and the left posterior canal (PC) form a pair; and the left AC and the right PC form a pair.

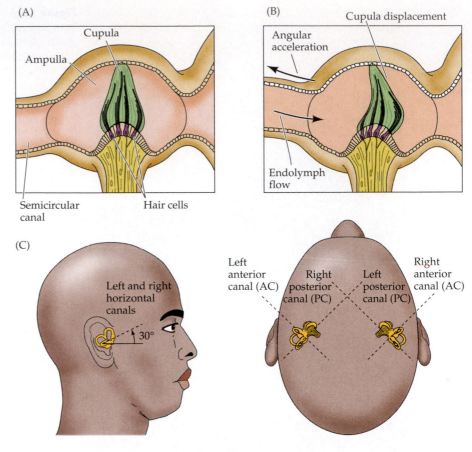

within the crista (Figure 14.8A,B). In contrast, linear accelerations of the head produce equal forces on the two sides of the cupula, so the hair bundles are not displaced.

Unlike the saccular and utricular maculae, all of the hair cells in the crista within each semicircular canal are organized with their kinocilia pointing in the same direction (see Figure 14.2C). Thus, when the cupula moves in the appropriate direction, the entire population of hair cells is depolarized and activity in all of the innervating axons increases. When the cupula moves in the opposite direction, the population is hyperpolarized and neuronal activity decreases. Deflections orthogonal to the excitatory–inhibitory direction produce little or no response.

Each semicircular canal works in concert with the partner located on the other side of the head that has its hair cells aligned oppositely. There are three such pairs: the two pairs of horizontal canals, and the superior canal on each side working with the posterior canal on the other side (Figure 14.8C). Head rotation deforms the cupula in opposing directions for the two partners, resulting in opposite changes in their firing rates. Thus, the orientation of the horizontal canals makes them selectively sensitive to rotation in the horizontal plane. More specifically, the hair cells in the canal toward which the head is turning are depolarized, while those on the other side are hyperpolarized.

For example, when the head accelerates to the left, the cupula is pushed toward the kinocilium in the left horizontal canal, and the firing rate of the relevant axons in the left vestibular nerve increases. In contrast, the cupula in the right horizontal canal is pushed away from the kinocilium, with a concomitant decrease in the firing rate of the related neurons. If the head movement is to the right, the result is just the opposite. This push–pull arrangement operates for all three pairs of canals; the pair whose activity is modulated is in the plane of the rotation, and the member of the pair whose activity is increased is on the side toward which the head is turning. The net result is a system that provides information about the rotation of the head in any direction.

How Semicircular Canal Neurons Sense Angular Accelerations

Like axons that innervate the otolith organs, the vestibular fibers that innervate the semicircular canals exhibit a high level of spontaneous activity. As a result, they can transmit information by either increasing or decreasing their firing rate, thus more effectively encoding head movements (see above). The bidirectional responses of fibers innervating the hair cells of the semicircular canal have been studied by recording the axonal firing rates in a monkey's vestibular nerve. Seated in a chair, the monkey was rotated continuously in one direction during three phases: an initial period of acceleration, then a period of several seconds at constant velocity, and finally a period of sudden deceleration to a stop (Figure 14.9). The maximum firing rates observed correspond to the period of acceleration, when the cupula is deflected; the maximum inhibition corresponds to the period of deceleration, when the cupula is deflected in the opposite direction. During the constant-velocity phase, firing rates return to a baseline level as the cupula returns to its non-deflected state over a time course that is related to the cupular elasticity and the viscosity of the endolymph. Note that the time it takes the cupula to return to its undistorted state (and for the hair bundles to return to their undeflected position) can occur while the head is still turning, as long as a constant angular velocity is maintained. Such constant forces are rare in nature, although they are encountered aboard ships, airplanes, and space vehicles, where prolonged acceleratory arcs are sometimes described.

An interesting aspect of the cupula–endolymph system dynamic is that it "smooths" the transduction of head accelerations into neural signals. For example, when the head is accelerated to a constant velocity rather rapidly (corresponding to high-frequency rotational movements of the head), vestibular units associated with the affected canal generate a velocity signal; note that the unit in Figure 14.9 rises linearly during the acceleration phase. However, when the head is moving at a constant angular velocity (i.e., low-frequency rotational movements), the unit decays to the spontaneous level (corresponding to an acceleration of zero). This transduction process results in a velocity signal at high frequencies and an acceleration signal at low frequencies, a behavior that can be clearly seen when applying sinusoidal stimuli over a wide frequency range.

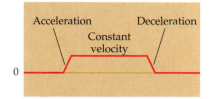

Figure 14.9 Response of a vestibular nerve axon from the semicircular canal to angular acceleration. The stimulus (top) is a rotation that first accelerates, then maintains constant velocity, and then decelerates the head. The stimulus-evoked change in firing rate of this vestibular unit (bottom) reflects the fact that the endolymph has viscosity and inertia and that the cupola has elasticity. Thus, during the initial acceleration, the deflection of the cupula causes the unit activity to rapidly increase. During constant angular velocity, the cupula returns to its non-deflected state over a time course related to its elasticity and the viscosity of the fluid, and the unit activity returns to the baseline rate. During deceleration, the cupula is deflected in the opposite direction, causing a transient decrease in the unit firing rate. This behavior can be thought of as the cupula–endolymph system dynamic; the inertia of the fluid plays a minor role in this dynamic, coming into play only at very high frequencies of head movement. (After Goldberg and Fernandez, 1971.)

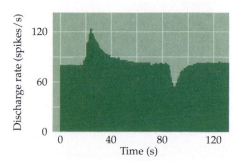

BOX 14C   Throwing Cold Water on the Vestibular System

Testing the integrity of the vestibular system can indicate much about the condition of the brainstem, particularly in comatose patients.

Normally, when the head is not being rotated, the output of the nerves from the right and left sides are equal; thus, no eye movements occur. When the head is rotated in the horizontal plane, the vestibular afferent fibers on the side toward the turning motion increase their firing rate, while the afferents on the opposite side decrease their firing rate (Figures A and B). The net difference in

firing rates then leads to slow movements of the eyes counter to the turning motion. This reflex response generates the slow component of a normal eye movement pattern called physiological nystagmus, which means "nodding" or oscillatory movements of the eyes (Figure B1). (The fast component is a saccade that resets the eye position; see Chapter 20.)

Pathological nystagmus can occur if there is unilateral damage to the vestibular system. In this case, the silencing of the spontaneous output from the damaged side results in an unphysiological difference in firing rate because the spontaneous discharge from the intact side remains (Figure B2). The difference in firing rates causes nystag-

mus, even though no head movements are being made.

Responses to vestibular stimulation are thus useful in assessing the integrity of the brainstem in unconscious patients. If the individual is placed on his or her back and the head is elevated to about 30° above horizontal, the horizontal semicircular canals lie in an almost vertical orientation. Irrigating one ear with cold water will then lead to spontaneous eye movements because convection currents in the canal and direct cooling of the nerve mimic rotatory head movements away from the irrigated ear (Figure C). In normal individuals, these eye movements consist of a slow movement toward the irrigated ear and a fast movement away from it. The fast move-

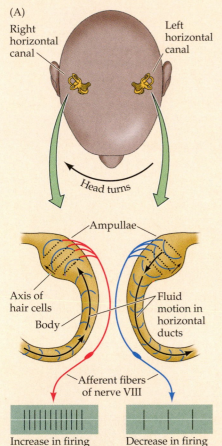

(A)

Right horizontal canal

Left horizontal canal

Head turns

Ampullae

Axis of hair cells

Body

Fluid motion in horizontal ducts

Afferent fibers of nerve VIII

Increase in firing   Decrease in firing

(A) View looking down on the top of a person's head illustrates the fluid motion generated in the left and right horizontal canals, and the changes in vestibular nerve firing rates when the head turns to the right.

(B)

(1) Physiological nystagmus

Head rotation

Slow eye movement

Fast eye movement

Right horizontal canal

Left horizontal canal

Primary vestibular afferents

Increased firing   Decreased firing

(2) Spontaneous nystagmus

Baseline firing   No firing

(B) In normal individuals, rotating the head elicits physiological nystagmus (1), which consists of a slow eye movement counter to the direction of head turning. The slow component of the eye movements is due to the net differences in left and right vestibular nerve firing rates acting via the central circuit diagrammed in Figure 14.10. Spontaneous nystagmus (2), where the eyes move rhythmically from side to side in the absence of any head movements, occurs when one of the canals is damaged. In this situation, net differences in vestibular nerve firing rates exist even when the head is stationary because the vestibular nerve innervating the intact canal fires steadily when at rest, in contrast to a lack of activity on the damaged side.

(C)

(C) Caloric testing of vestibular function is possible because irrigating an ear with water slightly warmer than body temperature generates convection currents in the canal that mimic the endolymph movement induced by turning the head to the irrigated side. Irrigation with cold water induces the opposite effect. These currents result in changes in the firing rate of the associated vestibular nerve, with an increased rate on the warmed side and a decreased rate on the chilled side. As in head rotation and spontaneous nystagmus, net differences in firing rates generate eye movements.

ment is most readily detected by the observer, and the significance of its direction can be kept in mind by using the mnemonic COWS ("Cold Opposite, Warm Same"). This same test can also be used in unconscious patients. In patients who are comatose due to dysfunction of both cerebral hemispheres but whose brainstem is intact, saccadic movements are no longer made and the response to cold water consists of only the slow movement component of the eyes to side of the irrigated ear (Figure D). In the presence of brainstem lesions involving either the vestibular nuclei themselves, the connections from the vestibular nuclei to oculomotor nuclei (the third, fourth, or sixth cranial nerves), or the peripheral nerves exiting these nuclei, vestibular responses are abolished (or altered, depending on the severity of the lesion).

(D) Caloric testing can be used to test the function of the brainstem in an unconscious patient. The figures show eye movements resulting from cold or warm water irrigation in one ear for (1) a normal subject, and in three different conditions in an unconscious patient: (2) with the brainstem intact; (3) with a lesion of the medial longitudinal fasciculus (MLF; note that irrigation in this case results in lateral movement of the eye only on the less active side); and (4) with a low brainstem lesion (see Figure 14.10).

(D)

Central Pathways for Stabilizing Gaze, Head, and Posture

The vestibular end organs communicate via the vestibular branch of cranial nerve VIII with targets in the brainstem and the cerebellum that process much of the information necessary to compute head position and motion. As with the cochlear nerve, the vestibular nerves arise from a population of bipolar neurons, the cell bodies of which in this instance reside in the **vestibular nerve ganglion** (also called **Scarpa's ganglion**; see Figure 14.1). The distal processes of these cells innervate the semicircular canals and the otolith organs, while the central processes project via the vestibular portion of cranial nerve VIII to the **vestibular nuclei** (and also directly to the cerebellum; Figure 14.10). The vestibular nuclei are important centers of integration, receiving input from the vestibular nuclei of the opposite side, as well as from the cerebellum and the visual and somatic sensory systems.

Although the canal and otolith afferents are largely segregated in the periphery, there is a large amount of canal–otolith convergence over a wide range of neurons in the vestibular nuclei. Because vestibular and auditory fibers run together in cranial nerve VIII, damage to this structure often results in both auditory and vestibular disturbances.

The central projections of the vestibular system participate in three major classes of reflexes: (1) helping to maintain equilibrium and gaze during movement; (2) maintaining posture; and (3) maintaining muscle tone. The first of these reflexes helps coordinate head and eye movements to keep gaze fixated on objects of interest during movements (other functions include protective or escape reactions; see Box 14D). The **vestibulo-ocular reflex** (**VOR**) in particular is a mechanism for producing eye movements that counter head movements, thus permitting the gaze to remain fixed on a particular point (Box 14C; see also Chapter 20). For example, activity in the left horizontal canal induced by leftward rotary acceleration of the head excites neurons in the left vestibular nucleus and results in compensatory eye movements to the right.

The circuits mediating this reflex are illustrated in Figure 14.10. Vestibular nerve fibers originating in the left horizontal semicircular canal project to the medial and lateral vestibular nuclei. Excitatory fibers from the medial vestibular nucleus cross to the contralateral abducens nucleus, which has two outputs. One of these is a motor pathway that causes the lateral rectus of the right eye to contract; the other is an excitatory projection that crosses the midline and ascends via the **medial longitudinal fasciculus** to the left oculomotor nucleus, where it activates neurons that cause the medial rectus of the left eye to contract. Finally, inhibitory neurons project from the medial vestibular nucleus to the left abducens nucleus, directly causing the motor drive on the lateral rectus of the left eye to decrease and also indirectly causing the right medial rectus to relax. The consequence of these several connections is that excitatory input from the horizontal canal on one side produces eye movements toward the opposite side. Therefore, turning the head to the left causes eye movements to the right.

In a similar fashion, head turns in other planes activate other semicircular canals, causing other appropriate compensatory eye movements. Thus, the VOR also plays an important role in vertical gaze stabilization in response to the linear vertical head oscillations that accompany locomotion, and in response to vertical angular accelerations of the head, as can occur when riding on a swing. The rostrocaudal set of cranial nerve nuclei involved in the VOR (i.e., the vestibular, abducens, and oculomotor nuclei), as well as the VOR's persistence in the unconscious state, make this reflex especially useful for detecting brainstem damage in the comatose patient (see Box 14C).

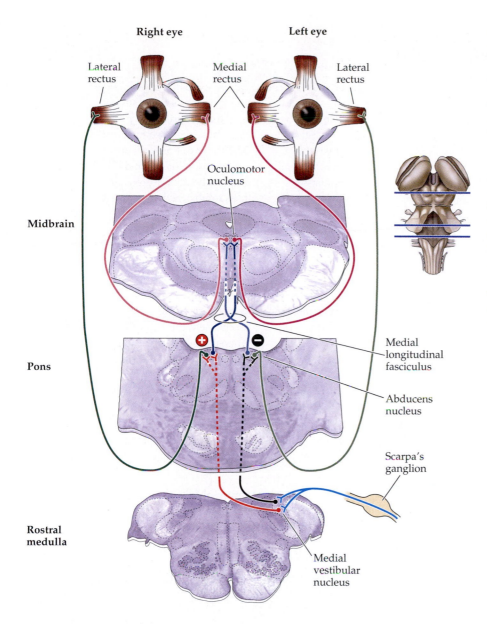

Right eye   **Left eye**

Lateral rectus   Medial rectus   Lateral rectus

Oculomotor nucleus

Midbrain

Pons

Medial longitudinal fasciculus

Abducens nucleus

Scarpa's ganglion

Rostral medulla

Medial vestibular nucleus

Figure 14.10   Connections underlying the vestibulo-ocular reflex. Projections of the vestibular nucleus to the nuclei of cranial nerves III (oculomotor) and VI (abducens). The connections to the oculomotor nucleus and to the contralateral abducens nucleus are excitatory (red), whereas the connections to ipsilateral abducens nucleus are inhibitory (black). There are connections from the oculomotor nucleus to the medial rectus of the left eye and from the adbucens nucleus to the lateral rectus of the right eye. This circuit moves the eyes to the right, that is, in the direction away from the left horizontal canal, when the head rotates to the left. Turning to the right, which causes increased activity in the right horizontal canal, has the opposite effect on eye movements. The projections from the right vestibular nucleus are omitted for clarity.

Loss of the VOR can have severe consequences. A patient with vestibular damage finds it difficult or impossible to fixate on visual targets while the head is moving, a condition called **oscillopsia** ("bouncing vision"). If the damage is unilateral, the patient usually recovers the ability to fixate objects during head movements. However, a patient with bilateral loss of vestibular function has the persistent and disturbing sense that the world is moving when the head moves. The underlying problem in such cases is that information about head and body movements normally generated by the vestibular organs is not available to the oculomotor centers, so that compensatory eye movements cannot be made.

Descending projections from the vestibular nuclei are essential for postural adjustments of the head, mediated by the vestibulo-cervical reflex (VCR), and body, mediated by the vestibulo-spinal reflex (VSR). As with the VOR, these postural reflexes are extremely fast, in part due to the small number of synapses interposed between the vestibular organ and the relevant motor neurons (Box 14D). Like the VOR, the VCR and the VSR are both compromised in patients with bilateral vestibular damage. Such patients exhibit diminished head and

BOX 14D   Mauthner Cells in Fish

A primary function of the vestibular system is to provide information about the direction and speed of ongoing movements, ultimately enabling rapid, coordinated reflexes to compensate for both self-induced and externally generated forces. One of the most impressive and speediest vestibular-mediated reflexes is the tail-flip escape behavior of fish (and larval amphibians), a stereotyped response that allows a potential prey to elude its predators (Figure A; tap on the side of a fish tank if you want to observe the reflex). In response to a perceived risk, fish flick their tail and are thus propelled laterally away from the approaching threat.

The circuitry underlying the tail-flip escape reflex includes a pair of giant medullary neurons called Mauthner cells, their vestibular inputs, and the spinal cord motor neurons to which the Mauthner cells project. (In most fish, there is one pair of Mauthner cells in a stereotypic location. Thus, these cells can be consistently visualized and studied from animal to animal.) Movements in the water, such as might be caused by an approaching predator, excite saccular hair cells in the vestibular labyrinth. These receptor potentials are transmitted via the central processes of vestibular ganglion cells in cranial nerve VIII to the two Mauthner cells in the brainstem. As in the vestibulo-spinal pathway in humans, the Mauthner cells project directly to spinal motor neurons. The small number of synapses intervening between the receptor cells and the motor neurons is one of the ways that this circuit has been optimized for speed by natural selection, an arrangement evident in humans as well. The large size of the Mauthner axons is another; the axons from these cells in a goldfish are about 50 μm in diameter.

The optimization for speed and direction in the escape reflex also is reflected in the synapses vestibular nerve afferents make on each Mauthner cell (Figure B). These connections are electrical synapses that allow rapid and faithful transmission of the vestibular signal.

An appropriate direction for escape is promoted by two features: (1) each Mauthner cell projects only to contralateral motor neurons; and (2) a local network of bilaterally projecting interneurons inhibits activity in the Mauthner cell away from the side on which the vestibular activity originates. In this way, the Mauthner cell on one side faithfully generates action potentials that command contractions of contralateral tail musculature, thus moving the fish out of the path of the oncoming predator. Conversely, the Mauthner cell on the opposite side is silenced by the local inhibitory network during the response (Figure C).

The Mauthner cells in fish are analogous to the reticulospinal and vestibulospinal pathways that control balance, posture, and orienting movements in mammals. The equivalent behavioral

(A) Bird's-eye view of the sequential body orientations of a fish engaging in a tail-flip escape behavior, with time progressing from left to right. This behavior is largely mediated by vestibular inputs to Mauthner cells.

(A)

postural stability, resulting in gait deviations; they also have difficulty balancing. These balance defects become more pronounced in low light or while walking on uneven surfaces, indicating that balance normally is the product of vestibular, visual, and proprioceptive inputs.

The anatomical substrate for the VCR involves the medial vestibular nucleus; axons from this nucleus descend in the medial longitudinal fasciculus to reach the upper cervical levels of the spinal cord (Figure 14.11). This pathway regulates head position by reflex activity of neck muscles in response to stimulation of the semicircular canals from rotational accelerations of the head. For example, during a downward pitch of the body (e.g., tripping), the superior canals are activated and the head muscles reflexively pull the head up. The dorsal flexion

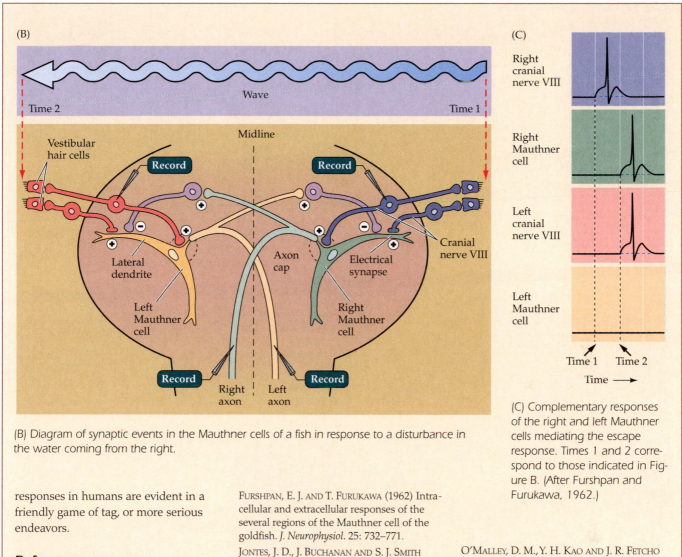

(B)

Wave

Time 2 — Time 1

Vestibular hair cells — Midline

Record — Record

Lateral dendrite — Axon cap — Electrical synapse — Cranial nerve VIII

Left Mauthner cell — Right Mauthner cell

Record — Right axon — Left axon — Record

(B) Diagram of synaptic events in the Mauthner cells of a fish in response to a disturbance in the water coming from the right.

(C)

Right cranial nerve VIII

Right Mauthner cell

Left cranial nerve VIII

Left Mauthner cell

Time 1 — Time 2

Time →

(C) Complementary responses of the right and left Mauthner cells mediating the escape response. Times 1 and 2 correspond to those indicated in Figure B. (After Furshpan and Furukawa, 1962.)

responses in humans are evident in a friendly game of tag, or more serious endeavors.

References

EATON, R. C., R. A. BOMBARDIERI AND D. L. MEYER (1977) The Mauthner-initiated startle response in teleost fish. *J. Exp. Biol.* 66: 65–81.

FURSHPAN, E. J. AND T. FURUKAWA (1962) Intracellular and extracellular responses of the several regions of the Mauthner cell of the goldfish. *J. Neurophysiol.* 25: 732–771.

JONTES, J. D., J. BUCHANAN AND S. J. SMITH (2000) Growth cone and dendrite dynamics in zebrafish embryos: Early events in synaptogenesis imaged *in vivo*. *Nature Neurosci.* 3: 231–237.

O'MALLEY, D. M., Y. H. KAO AND J. R. FETCHO (1996) Imaging the functional organization of zebrafish hindbrain segments during escape behaviors. *Neuron* 17: 1145–1155.

of the head initiates other reflexes, such as forelimb extension and hindlimb flexion, to stabilize the body and protect against a fall (see Chapter 17).

The VSR is mediated by a combination of pathways, including the lateral and medial vestibulospinal tracts and the reticulospinal tract. The inputs from the otolith organs project mainly to the lateral vestibular nucleus, which in turn sends axons in the lateral vestibulospinal tract to the spinal cord (see Figure 14.11). These axons terminate monosynaptically on extensor motor neurons, and they disynaptically inhibit flexor motor neurons; the net result is a powerful excitatory influence on the extensor (antigravity) muscles. When hair cells in the otolith organs are activated, signals reach the medial part of the ventral horn. By activating the ipsilateral pool of motor neurons innervating extensor muscles in

Figure 14.11 *Descending projections from the medial and lateral vestibular nuclei to the spinal cord underlie the VCR and VSR. The medial vestibular nuclei project bilaterally in the medial longitudinal fasciculus to reach the medial part of the ventral horns and mediate head reflexes in response to activation of semicircular canals. The lateral vestibular nucleus sends axons via the lateral vestibular tract to contact anterior horn cells innervating the axial and proximal limb muscles. Neurons in the lateral vestibular nucleus receive input from the cerebellum, allowing the cerebellum to influence posture and equilibrium.*

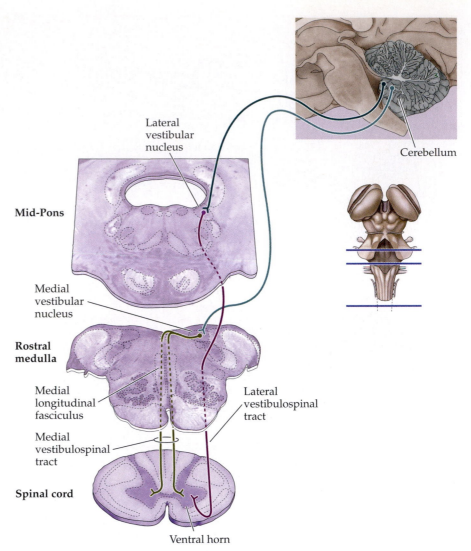

the trunk and limbs, this pathway mediates balance and the maintenance of upright posture.

Decerebrate rigidity, characterized by rigid extension of the limbs, arises when the brainstem is transected above the level of the vestibular nucleus. Decerebrate rigidity in experimental animals is relieved when the vestibular nuclei are lesioned, underscoring the importance of the vestibular system to the maintenance of muscle tone. The tonic activation of extensor muscles in decerebrate rigidity suggests further that the vestibulospinal pathway is normally suppressed by descending projections from higher levels of the brain, especially the cerebral cortex (see also Chapter 17).

Vestibular Pathways to the Thalamus and Cortex

In addition to these several descending projections, the superior and lateral vestibular nuclei send axons to the ventral posterior nuclear complex of the thalamus, which in turn projects to two cortical areas relevant to vestibular sensations (Figure 14.12). One of these cortical targets is just posterior to the primary somatosensory cortex, near the representation of the face; the other is at the transition between the somatic sensory cortex and the motor cortex (Brod-

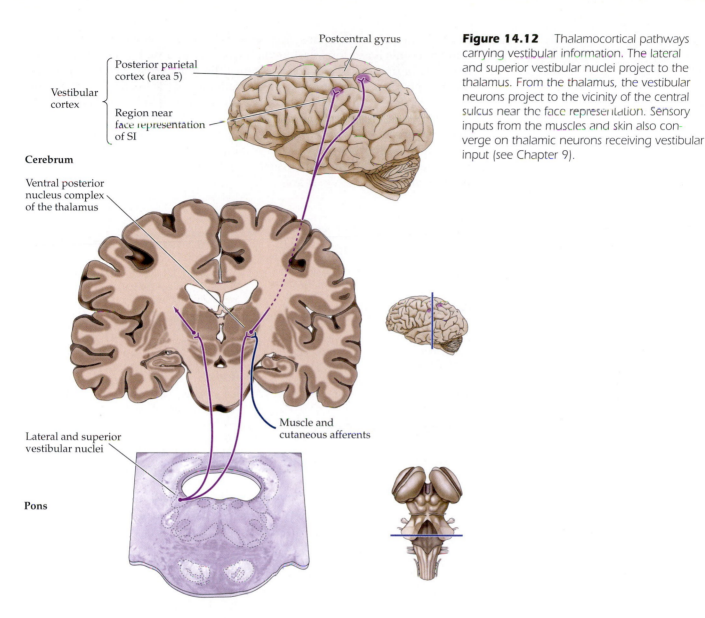

Figure 14.12 *Thalamocortical pathways carrying vestibular information. The lateral and superior vestibular nuclei project to the thalamus. From the thalamus, the vestibular neurons project to the vicinity of the central sulcus near the face representation. Sensory inputs from the muscles and skin also converge on thalamic neurons receiving vestibular input (see Chapter 9).*

mann's area 3a; see Chapter 9). Electrophysiological studies of individual neurons in these areas show that the relevant cells respond to proprioceptive and visual stimuli as well as to vestibular stimuli. Many of these neurons are activated by moving visual stimuli as well as by rotation of the body (even with the eyes closed), suggesting that these cortical regions are involved in the perception of body orientation in extrapersonal space. Consistent with this interpretation, patients with lesions of the right parietal cortex suffer altered perception of personal and extrapersonal space, as discussed in greater detail in Chapter 26.

Summary

The vestibular system provides information about the position, orientation, and motion of the head and body in space. The sensory receptor cells of the vestibular system are located in the otolith organs and the semicircular canals of the inner ear. The otolith organs provide information necessary for postural adjustments of the somatic musculature, particularly the axial musculature, when the head tilts in various directions or undergoes linear accelerations. This information represents linear forces acting on the head that arise through static effects of gravity or from translational movements. The semicircular canals, in contrast, provide information about rotational accelerations of the head. This latter information generates reflex movements that adjust the eyes, head, and body during motor activities. Among the best studied of these reflexes are eye movements that compensate for head movements, thereby stabilizing the visual scene when the head moves. Input from all the vestibular organs is integrated with input from the visual and somatic sensory systems to provide perceptions of body position and orientation in space.

Additional Reading

Reviews

BENSON, A. (1982) The vestibular sensory system. In *The Senses*, H. B. Barlow and J. D. Mollon (eds.). New York: Cambridge University Press.

BRANDT, T. (1991) Man in motion: Historical and clinical aspects of vestibular function. A review. *Brain* 114: 2159–2174.

FURMAN, J. M. AND R. W. BALOH (1992) Otolith-ocular testing in human subjects. *Ann. New York Acad. Sci.* 656: 431–451.

GOLDBERG, J. M. (1991) The vestibular end organs: Morphological and physiological diversity of afferents. *Curr. Opin. Neurobiol.* 1: 229–235.

GOLDBERG, J. M. AND C. FERNANDEZ (1984) The vestibular system. In *Handbook of Physiology*, Section 1: *The Nervous System*, Volume III: *Sensory Processes*, Part II, J. M. Brookhart, V. B. Mountcastle, I. Darian-Smith and S. R. Geiger (eds.). Bethesda, MD: American Physiological Society.

HESS, B. J. (2001) Vestibular signals in self-orientation and eye movement control. *News Physiolog. Sci.* 16: 234–238.

RAPHAN, T. AND B. COHEN. (2002) The vestibulo-ocular reflex in three dimensions. *Exp. Brain Res.* 145: 1–27.

Important Original Papers

GOLDBERG, J. M. AND C. FERNANDEZ (1971) Physiology of peripheral neurons innervat-ing semicircular canals of the squirrel monkey, Parts 1, 2, 3. *J. Neurophysiol.* 34: 635–684.

GOLDBERG, J. M. AND C. FERNANDEZ (1976) Physiology of peripheral neurons innervat-ing otolith organs of the squirrel monkey, Parts 1, 2, 3. *J. Neurophysiol.* 39: 970–1008.

LINDEMAN, H. H. (1973) Anatomy of the otolith organs. *Adv. Oto.-Rhino.-Laryng.* 20: 405–433.

Books

BALOH, R. W. AND V. HONRUBIA (2001) *Clinical Neurophysiology of the Vestibular System*, 3rd Ed. New York: Oxford University Press.

BALOH, R. W. (1998) *Dizziness, Hearing Loss, and Tinnitus*. Philadelphia: F. A. Davis Company.

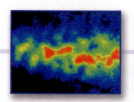

Chapter 15

The Chemical Senses

Overview

There are three sensory systems associated with the nose and mouth: the olfactory (smell), gustatory (taste), and trigeminal (chemosensory irritant) systems. Each of these systems is dedicated to the detection of chemicals in the environment. The olfactory system detects airborne molecules called odorants. In humans, odorants provide information about self and other people, animals, and plants, as well as helping to identify food and noxious or hazardous substances in the environment. Olfactory information thus influences social interactions, reproduction, defensive responses, and feeding behavior. The gustatory system detects ingested tastants, which are primarily water- or fat-soluble molecules. Tastants provide information about the quality, quantity, and safety of ingested food. Finally, the trigeminal chemosensory system provides information about irritating or noxious molecules that come into contact with skin or mucous membranes of the eyes, nose, and mouth. All three chemosensory systems rely on receptors that interact with relevant molecules in the environment. For smell and taste, essential aspects of sensory transduction rely on G-protein-coupled receptors and second messenger-mediated signaling. Information from primary sensory receptors in the nose, tongue, and other mucous membranes is relayed to regions of the central nervous system that guide the broad range of behaviors influenced by chemosensation. From an evolutionary perspective, the chemical senses—particularly olfaction—are deemed to be the "oldest" or "most primitive" sensory systems, yet they remain in many ways the least understood of the sensory modalities.

The Organization of the Olfactory System

The olfactory system is the most thoroughly studied component of the chemosensory triad, and processes information about the identity, concentration, and quality of a wide range of airborne, volatile chemical stimuli called **odorants**. Odorants interact with olfactory receptor neurons found in an epithelial sheet—the **olfactory epithelium**—that lines the interior of the nose (Figure 15.1A,B). The axons arising from the receptor cells project directly to neurons in the **olfactory bulb**, which in turn sends projections to the **pyriform cortex** in the temporal lobe, as well as to other structures in the forebrain via an axon pathway known as the **olfactory tract** (Figure 15.1C,D). The olfactory system is unique among the sensory systems in that it does not include a thalamic relay from primary receptors en route to a neocortical region that processes the sensory information. The pyriform cortex is three-layered archicortex. It is a specialized cortical region dedicated to olfaction and considered to be phylogenetically older than the six-layered neocortex. Although the initial pathway for olfactory information bypasses the thalamus, the thalamus does play an important role in subsequent stages. Olfactory information from pyriform cortex is

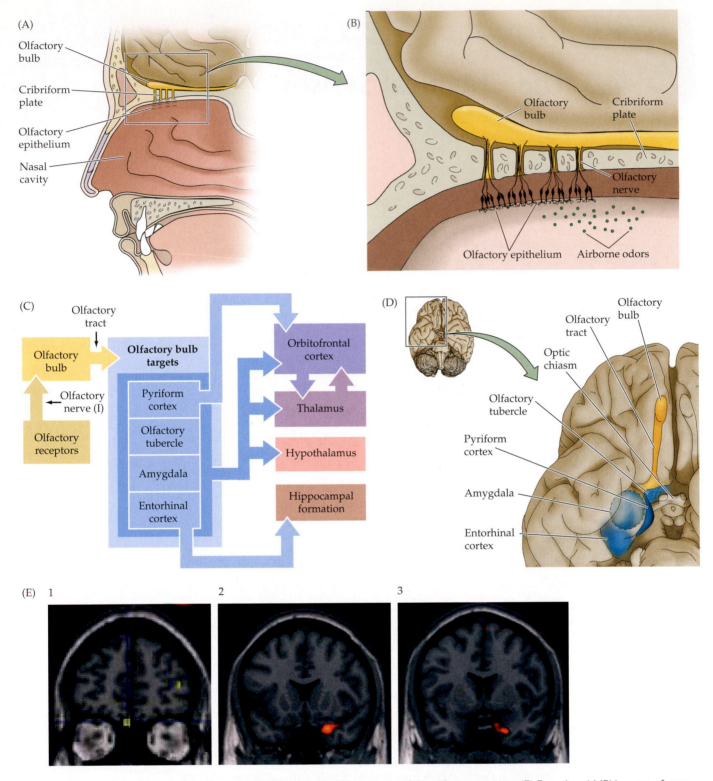

Figure 15.1 Organization of the human olfactory system. (A) Peripheral and central components of the primary olfactory pathway. (B) Enlargement of region boxed in (A) showing the relationship between the olfactory epithelium (which contains the olfactory receptor neurons) and the olfactory bulb (the central target of olfactory receptor neurons). (C) Diagram of the basic pathways for processing olfactory information. (D) Central components of the olfactory system. (E) Functional MRI images of coronal sections through the human brain at the level of the orbitofrontal cortex (1), the pyriform cortex and olfactory bulbs (2), and the amygdala (3). Maximal focal activation in response to odor presentation [in this case correlated either with pleasantness (1) or intensity (2, 3)] is seen in the orbitofrontal and pyriform cortices as well as in the amygdala. (E from Rolls et al., 2003.)

relayed to the thalamus en route to association areas in the neocortex, where further processing occurs. Together, the pyriform cortex and the olfactory association areas of the neocortex are thought to be essential for the conscious appreciation of odorants and the association of odors with other sensory characteristics of environmental stimuli. The olfactory tract also projects directly to a number of other forebrain targets, including the hypothalamus and amygdala (Figure 15.1E). The neural computations that occur in these regions influence motor, visceral, and emotional reactions to olfactory stimuli, particularly those that are relevant to feeding, reproduction, and aggressive behaviors.

Despite its phylogenetic age (the olfactory system is thought to represent the "primordial" sensory system in all animals) and the unusual route by which it relays information to the neocortex, the olfactory system abides by the same principle that governs other sensory modalities: interactions with stimuli—in this case, airborne chemical odorants—at the periphery are transduced and encoded by receptors into electrical signals, which are then relayed to higher order centers. Nevertheless, the central representation of olfactory information is less well understood than that of other sensory pathways. For example, the somatic sensory and visual cortices feature spatial maps of the relevant receptor surface, and the auditory cortex features maps that represent frequency and other distinct physical or computational aspects of sound. Whether analogous maps of specific odorants (e.g., rose or pine) or odorant attributes (e.g., sweet or acrid) exist in the olfactory bulb or pyriform cortex is not yet known. Indeed, until recently it has been difficult to imagine what sensory qualities would be represented in an olfactory map, or what features might be processed in parallel (like form versus motion in vision) as occurs in other sensory systems.

Olfactory Perception in Humans

In humans, olfaction is often considered the least acute of the senses, and a number of animals are obviously superior to humans in their olfactory abilities. This difference may reflect the larger number of olfactory receptor neurons and odorant receptor molecules in the olfactory epithelium of other species, as well as the proportionally larger area of the forebrain devoted to olfaction (Figure 15.2A,B). The proportional size of the olfactory bulb and related structures versus the neocortical regions dedicated to processing other sensory modalities is quite large in rodents and carnivores and correspondingly modest in humans. Humans are nonetheless quite good at detecting and identifying a number of airborne molecules present at very low concentrations in the environment (Figure 15.2C). For instance, the major aromatic constituent of bell pepper (2-isobutyl-3-methoxypyrazine) can be detected in room air at a concentration of 0.01 nM—or approximately one molecule per *billion*. The threshold concentrations for detecting and identifying odorants vary greatly; most cannot be detected until they reach far higher concentrations. (Ethanol, for example, cannot be identified until its concentration reaches approximately 2 mM.) The human olfactory system is also capable of making perceptual distinctions based on small changes in molecular structure; for example, the molecule D-carvone smells like caraway seeds, whereas L-carvone smells like spearmint.

Because the number of perceived odors is extremely large, there have been several attempts to classify them in groups. The most widely used scheme was developed in the 1950s by John Amoore, who categorized odors based on their perceived quality, molecular structure, and the fact that some people have difficulty smelling one or another group. Amoore classified odorants as *pungent, floral, musky, earthy, ethereal, camphor, peppermint, ether,* and *putrid* (Figure 15.2C). Although this classification scheme remains entirely empirical, it is still used to

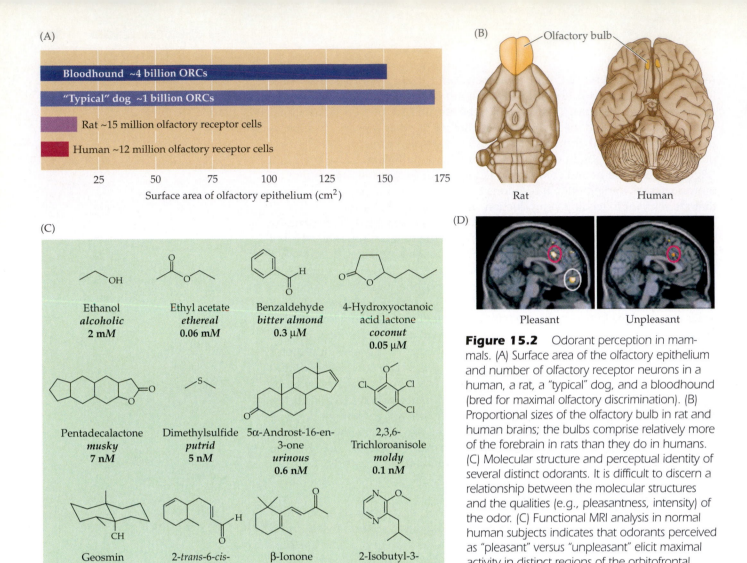

Figure 15.2 Odorant perception in mammals. (A) Surface area of the olfactory epithelium and number of olfactory receptor neurons in a human, a rat, a "typical" dog, and a bloodhound (bred for maximal olfactory discrimination). (B) Proportional sizes of the olfactory bulb in rat and human brains; the bulbs comprise relatively more of the forebrain in rats than they do in humans. (C) Molecular structure and perceptual identity of several distinct odorants. It is difficult to discern a relationship between the molecular structures and the qualities (e.g., pleasantness, intensity) of the odor. (C) Functional MRI analysis in normal human subjects indicates that odorants perceived as "pleasant" versus "unpleasant" elicit maximal activity in distinct regions of the orbitofrontal (white oval) and cingulate (red ovals) cortex. (A data from Shier et al., 2004; C after Pelosi, 1994; D from Rolls et al., 2003.)

describe odors, to study the cellular mechanisms of olfactory transduction, and to discuss the central representation of olfactory information.

A further complication in rationalizing the perception of odors is that their quality may change with odorant concentration. For example, at low concentrations the molecule indole has a "floral" odor, whereas at higher concentrations it smells "putrid." Despite these problems, the continued relevance of Amoore's scheme suggests that the olfactory system can identify odorant classes with shared perceptual qualities (e.g., rose and lilac both smell "floral") that evoke somewhat predictable responses. Indeed, humans perceive distinct odorant molecules as a particular, identifiable smell. Thus, coconuts, violets, cucumbers, and bell peppers each have a distinctive odor generated by a specific molecule (see Figure 15.2C), and most individuals report all four odors as having the same pleasant or appetizing qualities. These basic properties of olfaction—such as pleasant or unpleasant—are apparently represented in distinct cortical regions that mediate olfactory perception (Figure 15.2D). This suggests that "aesthetic" properties of odorant have distinct representations. Most naturally occurring odors, however, are blends of several odorant molecules, even though they are typically experienced as a single smell (such as the perceptions elicited

Figure 15.3 *Anosmia is the inability to identify common odors. When subjects are presented with seven common odors (a test frequently used by neurologists), the vast majority of "normal" individuals can identify all seven odors correctly (in this case, baby powder, chocolate, cinnamon, coffee, mothballs, peanut butter, and soap). Some people, however, have difficulty identifying even these common odors. In this example, individuals previously identified as anosmics were presented with the same battery of odors, only a few could identify all of the odors (less than 15%), and more than half could not identify any of the odors (After Cain and Gent, in Meiselman and Rivlin, 1986.)*

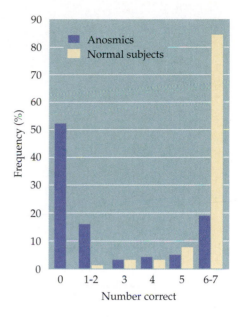

by perfumes or the bouquet of a wine). Thus, it remains to be determined if there is a "map" of odors based on single perceptual attributes.

Psychologists and neurologists have developed a variety of tests that measure a person's ability to detect common odors. Although most people are able to consistently identify a broad range of test odorants, some fail to identify one or more common smells (Figure 15.3). Such chemosensory deficits, called **anosmias**, are often restricted to a single odorant, suggesting that a specific element in the olfactory system—either an olfactory receptor gene (see below) or genes that control expression or function of specific odorant receptor genes—is inactivated. Genetic analysis of anosmic individuals has yet to confirm this possibility, however, and, unlike other sensory impairments (blindness and deafness), olfactory loss is difficult to classify as either peripheral or central in its origins.

Anosmias can be either congenital, or acquired due to chronic sinus infection or inflammation, traumatic head injuries, or the consequences of aging or disease. In most cases, the disruption or loss of human olfactory sensitivity is not a source of great concern (for example, the transient anosmia that occurs with a severe cold). Nevertheless, it can diminish the enjoyment of food, or if sustained, olfactory loss can influence appetite and cause weight loss and malnutrition (especially in aged individuals; see below). If an anosmia is particularly specific and severe, it can affect the ability to identify and respond appropriately to potentially dangerous odors such as spoiled food or smoke. Anosmias often target perception of distinct, noxious odorants. About 1 person in 1000 is insensitive to butyl mercaptan, the foul-smelling odorant released by skunks. More serious is the inability to detect hydrogen cyanide (1 person in 10), which can be lethal, or ethyl mercaptan, the chemical added to natural gas to enable people to detect gas leaks.

Like other sensory modalities, human olfactory capacity normally decreases with age. If otherwise healthy subjects are challenged to identify a large battery of common odorants, people between 20 and 40 years of age can typically identify 50–75 percent of the odors, whereas those between ages 50 and 70 correctly identify only about 30–45 percent (Figure 15.4A). These changes may reflect diminished peripheral sensitivity, or altered activity of CNS olfactory structures in otherwise healthy aged individuals (Figure 15.4B). A more radically diminished or distorted sense of smell often accompanies Alzheimer's and Parkinson's diseases. In fact, odor discrimination (usually measured by a standardized "scratch and sniff" test known as the University of Pennsylvania Smell Identification Test, or UPSIT) is often part of a battery of diagnostic tests administered at the early stages of age-related dementia and other neurodegenerative diseases. In addi-

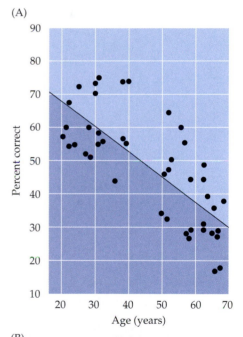

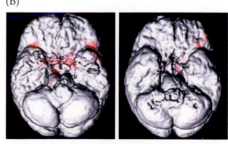

Figure 15.4 *Normal decline in olfactory sensitivity with age. (A) The ability to identify 80 common odorants declines markedly between 20 and 70 years of age. (B) Comparison of maximal activation (shown with red shading) of orbitofrontal and medial (pyriform cortex/amygdala) cerebral cortex by familiar odors in young and normal (i.e., without dementia) aged subjects. While the areas of focal activation remain similar, there is clearly diminished activity in the older subjects. (A after Murphy, 1986; B from Wang et al., 2005.)*

tion to normal and pathological age-related changes in olfaction, olfactory sensation and perception can be disrupted by eating disorders, psychotic disorders (especially schizophrenia), diabetes, and taking certain medications; the reasons for such disruption remain obscure. For example, olfactory hallucinations (perception of a stimulus that is not present in the environment) are among the earliest symptoms of psychosis in schizophrenic patients.

Physiological and Behavioral Responses to Odorants

In addition to olfactory perceptions, odorants can elicit a variety of physiological responses. Examples are the visceral motor responses to the aroma of appetizing food (salivation and increased gastric motility) or to a noxious smell (gagging and, in extreme cases, vomiting). Olfaction can also influence reproductive and endocrine functions. For instance, it has been documented that women living in single-sex dormitories tend to have synchronized menstrual cycles, a phenomenon that appears to be mediated by olfaction. Volunteers exposed to gauze pads from the underarms of women at different stages of their menstrual cycles also experience synchronized menses, and this synchronization can be disrupted by exposure to analogous gauze pads from men. These responses are thought to reflect in part detection of gender-specific odorants (see below). Olfaction also influences mother–child interactions. Infants recognize their mothers within hours after birth by smell, preferentially orienting toward their mothers' breasts and showing increased rates of suckling when fed by their mother compared to being fed by other lactating females, or when presented experimentally with their mother's odor versus that of an unrelated female. This recognition ability in infants is matched by that in mothers, who can reliably discriminate their own infant's odor when challenged with a range of odor stimuli from other infants of similar age.

In other animals, including many mammals, species-specific odorants called **pheromones** play important roles in behavior, influencing social, reproductive, and parenting behaviors. In rats and mice, odorants thought to be pheromones are detected by G-protein-coupled receptors located at the base of the nasal

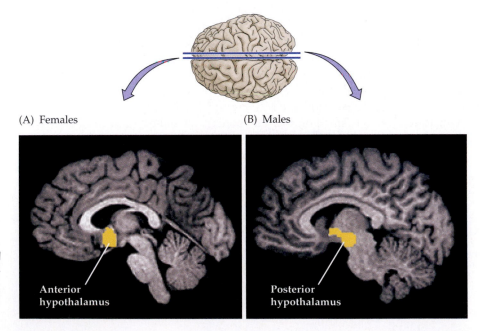

Figure 15.5 Differential patterns of activation in the hypothalamus of normal human female (right) and male (left) subjects after exposure to an estrogen- or androgen-containing odor mix. (From Savic et al., 2001.)

(A) Females

(B) Males

Anterior hypothalamus

Posterior hypothalamus

cavity in distinct, encapsulated chemosensory structures called **vomeronasal organs** (Box 15A). Although prominent in rodents and other animals, vomeronasal organs are found bilaterally in only 8 percent of adult humans, and there is no clear indication that these human structures have any significant function. Moreover, human genes that encode vomeronasal receptor proteins are pseudogenes and thus not expressed. It is unlikely that human pheromone perception, if it exists, is mediated by the vomeronasal system. Nevertheless, recent observations suggest that exposure to androgen- and estrogen-like compounds that may approximate potential pheromones for sexual attraction at concentrations below the level of conscious detection can elicit behavioral responses and different patterns of brain activation in adult female and male human subjects (see Chapter 30). The foci of activation by these odors include distinct regions of the hypothalamus (Figure 15.5) as well as the amygdala—areas believed to mediate reproductive, emotional, and social behaviors. Thus, olfactory related structures in the human brain can evidently detect signals that may affect reproductive and other homeostatic behaviors.

Olfactory Epithelium and Olfactory Receptor Neurons

The transduction of olfactory information occurs in the olfactory epithelium, the sheet of neurons and supporting cells that lines approximately half the nasal cavity surface (see Figure 15.1). The remaining surface is lined by respiratory epithelium similar to that in other airways, including the trachea and lungs. Respiratory epithelium primarily maintains appropriate moisture for inhaled air (which may also be important for the concentration and presentation of odorants) and provides an immune barrier that protects the nasal cavity from irritation and infection.

The olfactory epithelium includes several cell types (Figure 15.6A). The most important of these are the **olfactory receptor neurons** (**ORNs**). These bipolar cells give rise to small-diameter, unmyelinated axons at their basal surface that

Figure 15.6 Structure and function of the olfactory epithelium. (A) Diagram of the olfactory epithelium showing the major cell types: olfactory receptor neurons and their cilia, sustentacular (supporting) cells that detoxify potentially dangerous chemicals, and basal cells. Bowman's glands produce mucus. Bundles of unmyelinated axons and blood vessels run in the basal part of the mucosa (called the lamina propria). Olfactory receptor neurons are generated continuously from dividing stem cells maintained among the basal cells of the olfactory epithelium. (B) Generation of receptor potentials in response to odors takes place in the cilia of receptor neurons. Thus, odorants evoke a large inward (depolarizing) current when applied to the cilia (left), but only a small current when applied to the cell body (right). (A after Anholt, 1987; B after Firestein et al., 1991.)

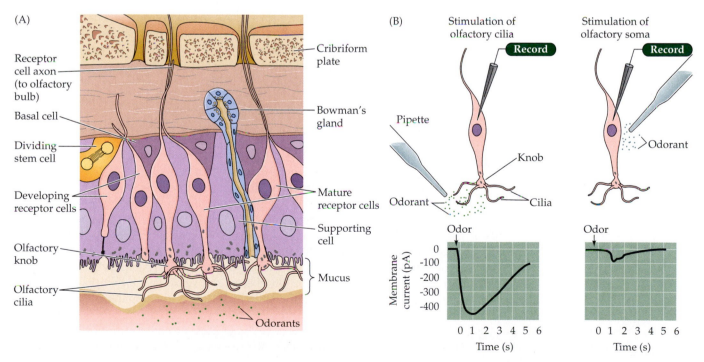

BOX 15A Pheromones, Reproduction, and the Vomeronasal System

Many dog owners (and the occasional brave cat owner) have noticed the conspicuous openings in the mucous membranes above the upper gum line of their companion animal—usually while trying to pry an especially chewable household item from a recalcitrant pet's jaws. These modest openings represent a second division of the olfactory system that is prominent in carnivores (including dogs and cats) and rodents, but less robust or absent in primates (especially humans). This **vomeronasal system** encompasses a distinct receptor population in a separate compartment of the nasal epithelium called the vomeronasal organ (VNO), as well as a separate region of the olfactory bulb—called the accessory olfactory bulb (AOB)—where axons from chemosensory receptor cells in the vomeronasal organ synapse (see figure).

The projections of the accessory olfactory bulb are distinct from those of the remainder of the olfactory bulb (referred to as the "main" olfactory bulb in rodents and carnivores) and include the hypothalamus and amygdala as their major target zones. This anatomical distinction provides an important clue to the primary function of the vomeronasal system: it is believed to encode and process information about odorants from conspecifics or predators and to mediate sexual, reproductive, and aggressive responses. The specific odorants detected and represented by the vomeronasal system are referred to as **pheromones**. The existence of pheromones, distinct from consciously perceived odors, remains a focus of research for a variety of purposes, including animal population control and assisted reproduction.

While the existence and behavioral significance of pheromones that influence human sexual or aggressive behavior remains unclear, it is well established that certain chemicals in secreted or excreted from rodents and carnivores—especially components in urine and feces—elicit vomeronasal responses. In the late 1990s, the distinct identity of the vomeronasal system was confirmed at the molecular level with the cloning of a family of vomeronasal receptors—the VRs—whose genomic identity and expression was specific for the chemosensory neurons of the vomeronasal organ. The VRs are seven-transmembrane G-protein-coupled receptors (see figure). They fall into two major classes, V1Rs and V2Rs, each of which uses a different G-protein-coupled cascade to activate signaling. In addition, the V2Rs are co-expressed (and thought to interact functionally with) members of the major histocompatibility (MHC) family of cell surface receptors. This is significant because many of the ligands produced by individuals and thought to be pheromones are associated with the MHC. Thus, although vomeronasal receptor cells look much like their counterparts in the olfactory epithelium (and share expression of some molecules), their G-protein-coupled receptors are genetically and structurally distinct. Moreover, signal transduction is accomplished via a completely different set of second messengers and cyclic nucleotide regulated ion channels: transient receptor potential (TRP) channels are found in vomeronasal receptor neurons, whereas a cyclic nucleotide-regulated ion channel is the primary molecular mediator of excitability in olfactory receptor neurons.

Genetic manipulation of the expression of VRs or downstream signaling molecules like TRP channels, as well as electrical recordings from single neurons in animals exposed to putative pheromones, provides functional evidence that the vomeronasal system is indeed a distinct entity, functioning in parallel with the "main" olfactory system. Elimination of the TRP channels that mediate vomeronasal signal transduction, (or elimination, replacement, or mutation of the receptors themselves) leads to changes in sexual or reproductive behavior, often in a sex-specific manner (that is, males and females are differentially compromised). Electrical recordings from

transmit olfactory information centrally. At the apical surface, an ORN gives rise to a single dendritic process that expands into a knoblike protrusion from which several microvilli, called **olfactory cilia**, extend into a thick layer of mucus. The mucus that lines the nasal cavity protects the exposed receptor neurons and supporting cells of the olfactory epithelium and controls the ionic milieu of the olfactory cilia. Mucus is produced by secretory specializations called **Bowman's glands** that are distributed throughout the olfactory epithelium. When the mucus layer becomes thicker, as during a cold, olfactory acuity decreases significantly. Two other cell classes, basal cells and sustentacular (supporting) cells, are also present in the olfactory epithelium. This entire apparatus—mucus layer and epithelium with neural and supporting cells—is called the **nasal mucosa**.

Olfactory receptor neurons have direct access to odorant molecules as air is inspired past the nasal mucosa into the lungs. A consequence of this direct

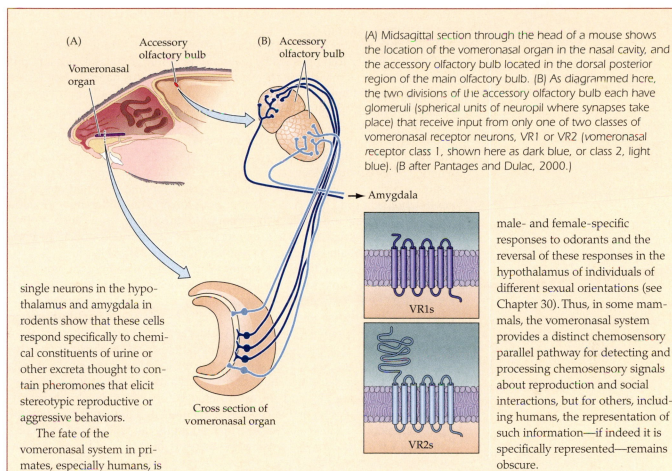

(A) Midsagittal section through the head of a mouse shows the location of the vomeronasal organ in the nasal cavity, and the accessory olfactory bulb located in the dorsal posterior region of the main olfactory bulb. (B) As diagrammed here, the two divisions of the accessory olfactory bulb each have glomeruli (spherical units of neuropil where synapses take place) that receive input from only one of two classes of vomeronasal receptor neurons, VR1 or VR2 (vomeronasal receptor class 1, shown here as dark blue, or class 2, light blue). (B after Pantages and Dulac, 2000.)

single neurons in the hypothalamus and amygdala in rodents show that these cells respond specifically to chemical constituents of urine or other excreta thought to contain pheromones that elicit stereotypic reproductive or aggressive behaviors.

The fate of the vomeronasal system in primates, especially humans, is mysterious. The vomeronasal organ is absent in most primates, as is a region of the olfactory bulb that corresponds to the accessory olfactory bulb. There are few recognizable VR genes in the human genome, and those that have some homology are pseudogenes—they are not expressed and do not appear to encode functional proteins. Nevertheless, primates, including humans, have behavioral responses that can be attributed to pheromone or pheromone-like stimuli, including the control of menstrual cycle in women exposed either to same-sex or opposite-sex individuals, and much more controversial studies of male- and female-specific responses to odorants and the reversal of these responses in the hypothalamus of individuals of different sexual orientations (see Chapter 30). Thus, in some mammals, the vomeronasal system provides a distinct chemosensory parallel pathway for detecting and processing chemosensory signals about reproduction and social interactions, but for others, including humans, the representation of such information—if indeed it is specifically represented—remains obscure.

References

DULAC, C. AND A. T. TORELLO (2003) Molecular detection of pheromone signals in mammals: From genes to behaviour. *Nature Rev. Neurosci.* 4: 551–562.

PANTAGES, E. AND C. DULAC (2000) A novel family of candidate pheromone receptors in mammals. *Neuron* 28: 835–845.

access, however, is that ORNs are exceptionally exposed to airborne pollutants, allergens, microorganisms, and other potentially harmful substances, subjecting these neurons to more or less continual damage. Several mechanisms help maintain the integrity of the olfactory epithelium in the face of this trauma. The mucus secreted by Bowman's glands traps and neutralizes potentially harmful agents. In both the respiratory and olfactory epithelium, immunoglobulins in this mucus provide an initial line of defense against harmful antigens. The sustentacular cells also contain enzymes (cytochrome P450s and others) that catabolize organic chemicals and other potentially damaging molecules. In addition, macrophages found throughout the nasal mucosa isolate and remove harmful material—as well as the remains of degenerating ORNs—since the ultimate solution to the vulnerability of olfactory receptor neurons is to replace these cells in a normal cycle of degeneration and regeneration, analagous to that in other exposed epithelia (such as the intestine and lung).

In rodents, the entire population of olfactory neurons is renewed every 6–8 weeks. This feat is accomplished by maintaining among the basal cells a population of stem cells that divide to give rise to new receptor neurons (see Figure 15.6A; see also Chapter 25). This naturally occurring regeneration provides an opportunity to investigate how neural stem cells can successfully produce new neurons and reconstitute function in the mature central nervous system, a topic of broad clinical interest. Many of the molecules that influence neuronal differentiation, axon outgrowth, and synapse formation during development elsewhere in the nervous system (see Chapters 22 and 23) are apparently retained to perform similar functions for regenerating olfactory receptor neurons in the adult. Understanding how new olfactory receptor neurons differentiate, extend axons to the brain, and reestablish appropriate functional connections is obviously relevant to stimulating the regeneration of functional connections elsewhere in the brain after injury or disease (see Chapter 25).

Odor transduction in the olfactory epithelium begins with odorant binding to specific odorant receptors proteins (discussed below) concentrated on the external surface of olfactory cilia. Prior to the identification of odorant receptor proteins, the compartmental sensitivity of the cilia to odors was demonstrated in physiological experiments (Figure 15.6B). Odorants presented to the *cilia* of an isolated olfactory receptor neuron elicit a robust electrical response; those presented to the *cell body* do not. Despite their external appearance, olfactory cilia do not have the cytoskeletal features of motile cilia (the so-called 9 + 2 arrangement of microtubules). Instead, the actin-rich olfactory cilia more closely resemble the microvilli of other epithelia (such as the lung and gut), and thus provide a greatly expanded cellular surface to which odorants can bind. Many molecules that are crucial for olfactory transduction are enriched or found exclusively in the cilia (see Figure 15.8).

Odorant Receptor Proteins

The central role of odorant receptor proteins in the encoding and transducing of olfactory information was acknowledged when the 2004 Nobel Prize for Physiology or Medicine was awarded to Richard Axel and Linda Buck for their discovery of the odorant receptor gene family. Olfactory receptor molecules are homologous to a large family of G-protein-linked receptors that includes β-adrenergic receptors and muscarinic acetylcholine receptors, as well as the photopigments rhodopsin and the cone opsins (see Chapters 6, 7, and 11). In all invertebrates and vertebrates examined thus far, odorant receptor proteins have seven membrane-spanning hydrophobic domains, potential odorant binding sites in the extracellular domain of the protein, and the ability to interact with G-proteins at the carboxyl terminal region of their cytoplasmic domain (Figure 15.7A). The amino acid sequences for these molecules show substantial variability in several of the membrane-spanning regions, as well as in the extracellular and cytoplasmic domains (Figure 15.7B). The specificity of odorant recognition and signal transduction is presumably the result of this molecular variety of the odorant receptor molecules present in the nasal epithelium; however, the molecular mechanism by which individual receptors bind specific odorants remains poorly understood.

The number of odorant receptor genes, though substantial in all species, varies widely. Nevertheless, in all mammals, odorant receptors are the largest known gene family, representing 3–5 percent of all mammalian genes. Analysis of the complete human genome sequence has identified approximately 950 odorant receptor genes, and the number is similar in other primates, including

(A)

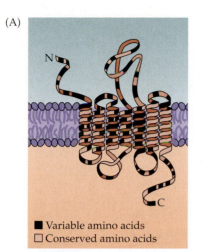

N

C

■ Variable amino acids
□ Conserved amino acids

Figure 15.7 Odorant receptor genes. (A) The generic structure of putative olfactory odorant receptors. These proteins have seven transmembrane domains, plus a variable cell surface region and a cytoplasmic tail that interacts with G-proteins. As many as a thousand genes encode proteins of similar inferred structure in several mammalian species, including humans. Each gene presumably encodes a receptor protein that detects a particular set of odorant molecules. (B) The seven transmembrane domains characteristic of G-protein-coupled receptors are shown as dark regions on maps of receptor genes in the nematode *C. elegans*, the fruit fly *Drosophila*, and mammals. The comparative size of each domain, as well as the size of the intervening cytoplasmic or cell surface domains, varies from species to species. In addition, splice sites (arrowheads) reflect introns in the genomic sequences of the two invertebrates; in contrast, genes for mammalian odorant receptors lack introns. The number of genes that encode odorant receptors in *C. elegans*, *Drosophila*, mice, and humans are indicated in the appropriate boxes (A adapted from Menini, 1999; B after Dryer, 2000.)

(B)

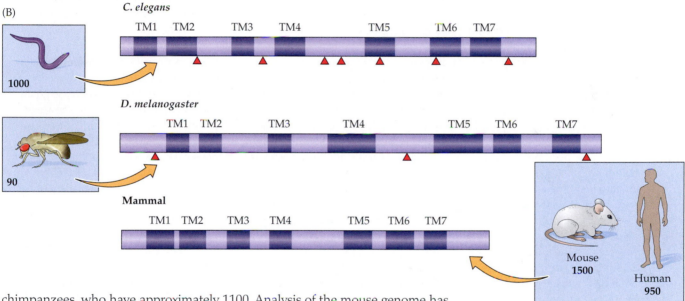

chimpanzees, who have approximately 1100. Analysis of the mouse genome has identified about 1500 different odorant receptor genes, and in certain dogs, including those noted for their olfactory abilities (Box 15B), the number is around 1200. Additional sequence analysis of apparent mammalian odorant receptor genes, however, suggests that many of these genes—around 60 percent in humans and chimps versus 15–20 percent in mice and dogs—are not transcribed. Thus, the number of functional odorant receptor proteins is estimated to be around 400 in humans and chimps versus about 1200 in mice and 1000 in dogs. In mammals, the number of expressed odorant receptors apparently is correlated with the olfactory capacity of different species. Similar analysis of complete genome sequences from the worm *C. elegans* and the fruit fly *D. melanogaster* indicate that there are approximately 1000 odorant receptor genes in the worm, but only about 60 in the fruit fly. The significance of these quite different numbers of odorant receptor genes is not known.

Expression in olfactory receptor neurons has been confirmed for only a limited subset of the huge number of odorant receptor genes. Messenger RNAs for different odorant receptor genes are expressed in subsets of olfactory neurons that occur in bilaterally symmetric zones of olfactory epithelium. Additional evidence for restricted patterns of odorant receptor gene expression in spatially restricted subsets of olfactory receptor neurons comes from molecular

BOX 15B The "Dogtor" Is In

Common wisdom holds that having a pet, particularly a dog, is good for your health. Most of us assume that the primary benefits come from the companionship, as well as the daily exercise, that one's dog provides. However, there may be more critical benefits of pet ownership that reflect the remarkable acuity of the canine olfactory system. The family dog may in fact be a reliable source of early diagnosis for several cancers—albeit a diagnostician that likes to chew shoes and has a wet nose.

In the late 1980s, anecdotal reports emerged that claimed family dogs could use smell to identify moles and other skin blemishes on their owners that turned out to be malignant. In recounting this seemingly strange capacity of several dogs, H. Williams, one of the original discoverers, reported "a patient whose dog constantly sniffed at a mole on her leg. On one occasion, the dog even tried to bite the lesion off. … [The] constant attention [of the dog] prompted her to seek medical advice. The lesion was excised and histology showed the lesion to be a malignant melanoma."

Subsequently, similar diagnoses by individual pets for their owners were reported, including a Labrador who detected a basal cell carcinoma that had developed from an eczema lesion on his master's skin. A slightly less anecdotal study relied on techniques used to train explosive-sniffing dogs for airport security. In this instance, George, a schnauzer, was trained to distinguish malignant melanomas in cell culture from their nonmalignant melanocyte counterparts. George was then introduced to a patient who had several moles. One mole caused George "to go crazy." Biopsy of that mole proved that it was indeed an early malignant melanoma.

Over the ensuing years, further anecdotal evidence suggested that dogs could recognize lung, breast, and bladder cancer using olfaction. These reports remained in the category of "shaggy (or sniffy) dog" stories until 2006, when a truly systematic analysis of this apparent diagnostic capacity was published. In this study, five ordinary adult dogs were trained to distinguish exhaled breath samples from lung and breast cancer patients versus healthy controls. The dogs were then tested for their ability to distinguish patients from controls in an entirely novel sample population. In this instance, the specificity and sensitivity of the dogs ability to detect lung cancer from early to late stages was 99 percent as accurate as that of conventional biopsy diagnosis. The accuracy of breast cancer detection was slightly lower—approximately 90 percent that of conventional methods.

A similar study challenged dogs to discriminate urine from patients with and without bladder cancer with parallel but somewhat less robust results. During the course of this study, however, the dogs consistently identified a presumed "control" sample as that from a cancer patient. Clinicians were sufficiently alerted to perform further diagnostic tests, and in fact discovered a kidney carcinoma in this individual.

Aside from writing a new chapter in the saga of the salutary relationship between humans and dogs, these intriguing observations have several implications for understanding the mechanisms and biological significance of olfactory acuity and selectivity. First, there is evidence that the concentration of alkanes as well as other volatile organic compounds is increased in air exhaled from patients with lung cancer. Thus, as indicated by preliminary studies of odorant receptor molecule sensitivity, seven transmembrane G-protein-coupled odorant receptors may be specialized to detect and discriminate a wide—and biologically significant—spectrum of volatile organic compounds at low concentrations. Second, the discrimination made between patients and controls, either by untrained individual dogs or the trained group of dogs, suggests that subtle distinctions in olfactory perception are clearly represented and can guide behavior. This apparent heightened olfactory ability in dogs may reflect a somewhat larger number of odorant receptors that increases specificity, a relatively larger olfactory periphery that allows increased sensitivity, or specialized circuitry in the olfactory bulb, pyriform cortex, or other brain regions that assign cognitive significance to distinct olfactory stimuli. Whether this ability has adaptive significance for dogs or is just the ultimate smart pet trick is unclear.

Does this mean the term "pet scan" will soon assume a new meaning in clinical medicine? Clearly, the complexity of making critical diagnoses and the potential lack of reliability of dogs—however well trained—render routine use of diagnostic dogs difficult to imagine. Nevertheless, the remarkable olfactory capacity of these animals provides a starting point to understand molecular specificity of odorant receptors, as well as processing capacity and representations of olfactory information in the central nervous system. Such understanding may not only illuminate the functional characteristics of the olfactory system; it may provide a natural guide to specific molecules associated with disease states, and the design of better diagnostic tools—or at least diagnoses that don't rely on cold, wet noses.

References

McCulloch, M., T. Jezierski, M. Broffman, A. Hubbard, K. Turner and T. Janecki (2006) Diagnostic accuracy of canine scent detection in early- and late-stage lung and breast cancers. *Integ. Cancer Therap.* 5: 30–39.

Willis, C. M. and 7 others (2004) Olfactory detection of human bladder cancer by dogs: Proof of principle study. *BMJ* 329: 712 (25 September 2004).

Phillips, M. and 7 others (2003) Detection of lung cancer with volatile markers in the breath. *Chest* 123: 2115–2123.

Church, J. and H. Williams (2001) Another sniffer dog for the clinic? *Lancet* 358: 930.

(A)   (B)   (C)   (D)

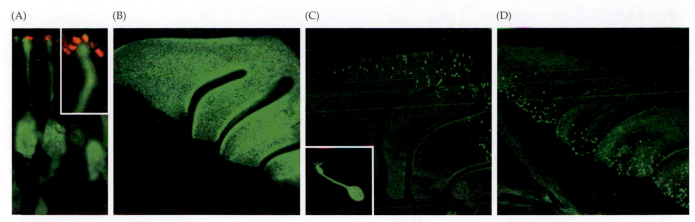

Figure 15.8 Odorant receptor gene expression. (A) Individual olfactory receptor neurons labeled immunohistochemically with the olfactory marker protein OMP (green; OMP is selective for *all* olfactory receptor neurons) and the olfactory receptor neuron-specific adenylyl cyclase III (red) that is limited to olfactory cilia (inset). The labels are in register with the segregation of signal transduction components to this domain. (B) The distribution of OMP-expressing olfactory receptor neurons throughout the entire nasal epithelium of an adult mouse, demonstrated with an OMP-GFP reporter transgene. The protuberances oriented diagonally from left to right represent individual turbinates within the olfactory epithelium. The remaining bony and soft-tissue structures of the nose have been dissected away. (C) The distribution of olfactory receptor neurons expressing the I7 odorant receptor. These cells are restricted to a distinct domain or zone in the epithelium. The inset shows that odorant receptor-expressing cells are indeed cilia-bearing olfactory receptor neurons. (D) Olfactory receptor neurons expressing the M71 odorant receptor are limited to a zone that is completely distinct from that of the I7 receptor. (A courtesy of A.-S. LaMantia; B–D from Bozza et al., 2002.)

genetic experiments (done primarily in mice and fruit flies) in which reporter proteins like β-galactosidase or GFP are inserted into odorant receptor gene loci (Figure 15.8).

Genetic as well as cell biological analyses have shown that each olfactory receptor neuron expresses only one, or at most a few, odorant receptor genes; furthermore, only one of the two copies of each odorant receptor gene is expressed in any particular olfactory receptor neuron. Thus, different odors must activate molecularly and spatially distinct subsets of olfactory receptor neurons, and one of the two alleles for each odorant receptor gene must be silenced in each olfactory receptor neuron. The molecular diversity of the odorant receptors and the accompanying cellular diversity for olfactory receptor neurons most certainly mediates, at least in part, the capacity of most olfactory systems to detect and encode a wide range of complex and novel odors in the environment.

The Transduction of Olfactory Signals

Once an odorant is bound to an odor receptor protein, several additional steps are required to generate a receptor potential that converts chemical information into electrical signals that can be interpreted by the brain. In mammals, the principal pathway for generating electrical activity in olfactory receptors involves cyclic nucleotide-gated ion channels similar to those found in rod photoreceptors (see Chapter 11). The olfactory receptor neurons express an olfactory-specific G-protein, G_{olf}, which in turn activates **adenyl cyclase III** (**ACIII**), an olfactory-specific adenylate cyclase (Figure 15.9A). Both of these proteins are restricted to the olfactory knob and cilia, consistent with the idea that odor transduction occurs in these domains of the olfactory receptor neuron (see Figure 15.6A). Stimulation of odorant receptor molecules leads to an increase in cyclic AMP (cAMP), which opens **cyclic nucleotide-gated channels** that permit the entry of Na^+ and Ca^{2+} (mostly Ca^{2+}), thus depolarizing the neuron. This depolarization, amplified by a Ca^{2+}-activated Cl^- current, is conducted passively from the cilia to the axon hillock region of the olfactory receptor neuron, where action potentials are generated via voltage-regulated Na^+ channels and transmitted to the olfactory bulb.

In genetically engineered mice, inactivation of any one of the major signal transduction elements (G_{olf}, ACIII, or the cyclic nucleotide-gated channel) results in a loss of receptor potential response to odorants in olfactory receptor neurons—neurons which otherwise appear normal in these animals—including continued expression of the olfactory marker protein that is uniquely expressed in differentiated ORNs. There is also complete loss of behavioral response to

Figure 15.9 Molecular mechanisms of odorant transduction. (A) Odorants in the mucus bind directly (or are shuttled via odorant binding proteins) to one of many receptor molecules located in the membranes of the cilia. This association activates an odorant-specific G-protein (G_{olf}) that, in turn, activates an adenylate cyclase, resulting in the generation of cyclic AMP (cAMP). One target of cAMP is a cation-selective channel that, when open, permits the influx of Na^+ and Ca^{2+} into the cilia, resulting in depolarization. The ensuing increase in intracellular Ca^{2+} opens Ca^{2+}-gated Cl^- channels that provide most of the depolarization of the olfactory receptor potential. The receptor potential is reduced in magnitude when cAMP is broken down by specific phosphodiesterases to reduce its concentration. At the same time, Ca^{2+} complexes with calmodulin (Ca^{2+}-CAM) and binds to the channel, reducing its affinity for cAMP. Finally, Ca^{2+} is extruded through the Ca^{2+}/Na^+ exchange pathway. (B) Consequences of inactivation of critical molecules in the odorant signal transduction cascade. The images of olfactory receptor neurons show expression of G_{olf}, ACIII, and the cyclic nucleotide-gated channel. The traces below show odorant-elicited electrical activity in the olfactory epithelium, measured extracellularly using the electro-olfactogram (EOG). In the wild-type, there is a robust response when either pleasant (citralva) or pungent (isomenthone) odors are presented. These responses are abolished by inactivating any of the major signal transduction molecules linked to the seven-transmembrane odorant receptors. (A adapted from Menini, 1999; B from Belluscio et al., 1998 [G_{olf}]; Wong et al., 2000 [ACIII]; and Brunet et al., 1996 [CNG].)

(A)

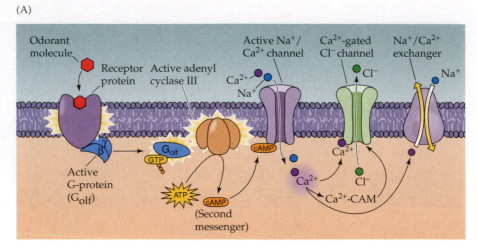

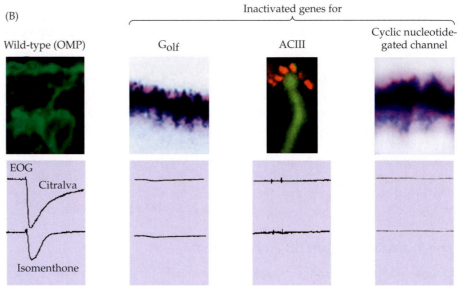

(B)

odorants; in other words, these mice are completely anosmic (Figure 15.9B). This common endpoint following loss of function of each molecule demonstrates that each step contributes to the transduction of odorants.

Like other sensory receptor cells, individual olfactory receptor neurons are sensitive to subsets of stimuli; thus, there is receptor specificity. There is a range of relationships between specific odorant stimuli and electrical responses of olfactory receptor neurons. Some olfactory receptor neurons exhibit marked selectivity to a single chemically defined odorant, whereas others are activated by a number of different odorant molecules (Figure 15.10). Presumably, this reflects the expression of a single odorant receptor gene in each olfactory receptor neuron. There is currently no chemical or physiological data that indicates a correspondence between high affinity binding of an odorant to a receptor molecule, electrical activation of the olfactory receptor neuron, and perception of a specific odor. Nevertheless, a selective relationship between classes of odorants and selective responses has been demonstrated in individual cells isolated from olfactory epithelium and labeled genetically to identify the odorant receptor molecule expressed in that cell. There is a response "signature" to different odorants for ORNs that express individual odorant receptors (Figure 15.11A). The different

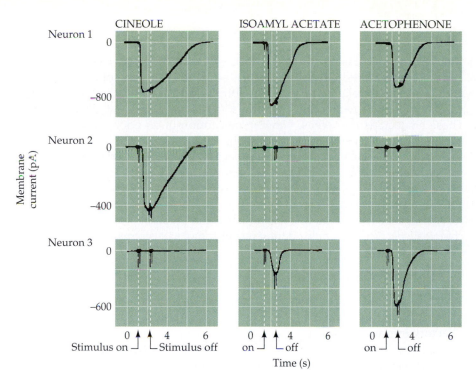

Figure 15.10 Responses of olfactory receptor neurons to selected odorants. Neuron 1 responds similarly to three different odorants. In contrast, neuron 2 responds to only one of these odorants. Neuron 3 responds to two of the three stimuli. The responses of these receptor neurons were recorded by whole-cell patch clamp recording; downward deflections represent inward currents measured at a holding potential of –55 mV. (After Firestein, 1992.)

responses seem to reflect chemical differences in subsets of odorants (for example, differences in carbon chain length of the molecular "backbone"), as well as the overall quality of the odorants in each group (Figure 15.11B). It is not known, however, whether the odorant molecules tested represent the "best" or most environmentally relevant odorants for any given receptor protein.

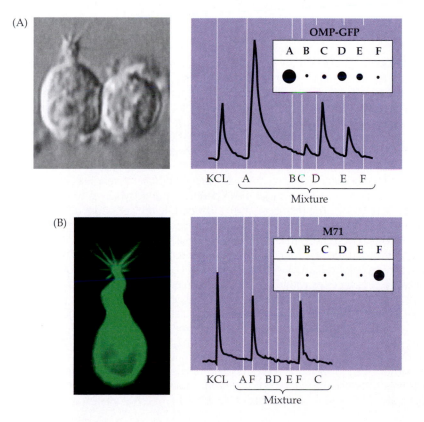

Figure 15.11 Odorant receptor protein selectivity demonstrated by responses to different combinations of molecularly identified odorants. (A) Randomly chosen odorant receptor neurons (ORNs) were tested for their responses to six different odorant molecule mixtures (indicated here as A–F) and the responses compared to those of an ORN known to express the M71 odorant receptor (see Figure 15.8C). (B) The M71R-expressing cell was isolated by linking its gene to green fluorescent protein. The size of the dots and magnitude of the spikes in the graphs indicate the strength of the electrical response to each odorant mixture. Randomly chosen ORNs respond to several of the six mixtures, while M71-selected ORNs respond to preferentially only to mixture F. (After Bozza et al., 2002; photographs courtesy of T. Bozza.)

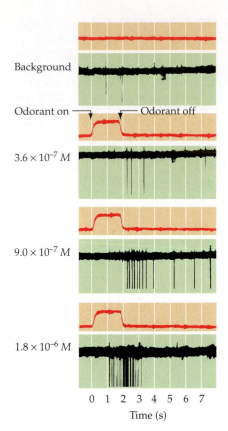

Background

Odorant on ——⌐   ⌐—— Odorant off

3.6×10^{-7} M

9.0×10^{-7} M

1.8×10^{-6} M

0 1 2 3 4 5 6 7
Time (s)

Figure 15.12 *Responses of a single olfactory receptor neuron to changes in the concentration of a single odorant, isoamyl acetate (banana). The upper trace in each panel (red) indicates the duration of the odorant stimulus; the lower trace the neuronal response. The frequency and number in each panel of action potentials increases as the odorant concentration increases. (After Getchell, 1986.)*

In addition to sensing different subsets of stimuli, olfactory receptor neurons exhibit different thresholds for a particular odorant. That is, receptor neurons that are inactive at a background concentration of a given odorant may be activated when exposed to higher concentrations of the same odorant (Figure 15.12). These characteristics suggest why the perception of an odor can change as a function of its concentration. In addition, these physiological properties most likely contribute both to the adaptive capacity of the olfactory system—a capacity easily appreciated by olfactory experiences like the decreased awareness of being in a "smoking" room at a hotel as more time is spent there. The relationship between these two aspects of physiological tuning of single olfactory receptor neurons (i.e., responses to the abundance and duration of a stimulus) and the chemical specificity of single odorant receptor molecules remains unclear.

The Olfactory Bulb

The transduction of odorants in the olfactory cilia and the subsequent changes in electrical activity in the olfactory receptor neuron are only the first steps in olfactory information processing. Unlike other primary sensory receptor cells (e.g., photoreceptors in the retina or hair cells in the cochlea), olfactory receptor neurons have axons, and these axons relay odorant information directly to the brain via action potentials. As the axons leave the olfactory epithelium, they coalesce to form a large number of bundles that together make up the **olfactory nerve** (cranial nerve I). Each olfactory nerve projects ipsilaterally to the **olfactory bulb**, which in humans lies on the ventral anterior aspect of the ipsilateral

Figure 15.13 *The organization of the mammalian olfactory bulb. (A) When the bulb is viewed from its dorsal surface (visualized here in a living mouse in which the overlying bone has been removed), olfactory glomeruli can be seen. The dense accumulation of dendrites and synapses that constitute glomeruli are stained here with a vital fluorescent dye that recognizes neuronal processes. The inset shows a similar arrangement of glomeruli in the mushroom body (the equivalent of the olfactory bulb) in Drosophila. (B) Among the major neuronal components of each glomerulus are the apical tufts of mitral cells, which project to the pyriform cortex and other bulb targets (see Figure 15.1C). In this image of a coronal section through the bulb, they have been labeled retrogradely by placing the lipophilic tracer Di-I in the lateral olfactory tract. (C) The cellular structure of the olfactory bulb, shown in a Nissl-stained coronal section. The five layers of the bulb are indicated. The glomerular layer includes the tufts of mitral cells, the axon terminals of olfactory receptor neurons, and periglomerular cells that define the margins of each glomerulus. The external plexiform layer is made up of lateral dendrites of mitral cells, cell bodies and lateral dendrites of tufted cells, and dendrites of granule cells that make dendrodendritic synapses with the other dendritic elements. The mitral cell layer is defined by the cell bodies of mitral cells, and mitral cell axons are found in the internal plexiform layer. Finally, granule cell bodies are densely packed into the granule cell layer. (D) Diagram of the laminar and circuit organization of the olfactory bulb, shown in a cutaway view from its medial surface. Olfactory receptor cell axons synapse with mitral cell apical dendritic tufts and periglomerular cell processes within glomeruli. Granule cells and mitral cell lateral dendrites constitute the major synaptic elements of the external plexiform layer. (E) Axons from olfactory receptor neurons that express a particular odorant receptor gene converge on a small subset of bilaterally symmetrical glomeruli. These glomeruli, indicated in the boxed area in the upper panel, are shown at higher magnification in the lower panel. The projections from the olfactory epithelium have been labeled by a reporter transgene inserted by homologous recombination ("knocked in") into the genetic locus that encodes the particular receptor. (A from LaMantia et al., 1992; B,C from Pomeroy et al., 1990; E from Mombaerts et al., 1996.)*

cerebral hemisphere. The most distinctive feature of the olfactory bulb is an array of more or less spherical accumulations of neuropil 100–200 μm in diameter called **glomeruli**, which lie just beneath the surface of the bulb and are the synaptic target of the primary olfactory axons (Figure 15.13A–D). In vertebrates, ORN axons make excitatory glutamatergic synapses within the glomeruli.

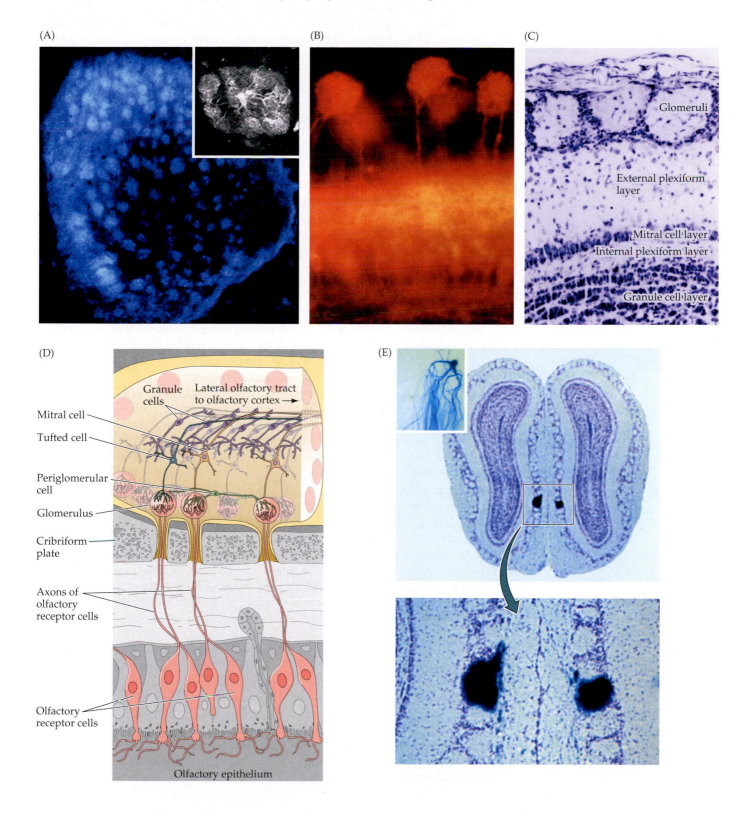

(A)

(B)

(C)

Glomeruli

External plexiform layer

Mitral cell layer
Internal plexiform layer

Granule cell layer

(D)

Granule cells

Lateral olfactory tract to olfactory cortex →

Mitral cell

Tufted cell

Periglomerular cell

Glomerulus

Cribriform plate

Axons of olfactory receptor cells

Olfactory receptor cells

Olfactory epithelium

(E)

Remarkably, this relationship between the olfactory periphery (the nose, or similar structures) and glomeruli in the central nervous system is maintained across the animal kingdom (Figure 15.13A, inset).

Within each glomerulus, the axons of the receptor neurons contact apical dendrites of **mitral cells**, which are the principal projection neurons of the olfactory bulb. The cell bodies of the mitral cells are located in a distinct layer of the olfactory bulb deep to the glomeruli. A mitral cell extends its primary dendrite into a single glomerulus, where the dendrite gives rise to an elaborate tuft of branches onto which the axons of olfactory receptor neurons synapse (Figure 15.13B,D). In the mouse, whose glomerular connectivity has been studied quantitatively, each glomerulus includes the apical dendrites of approximately 25 mitral cells, which in turn receive input from approximately 25,000 olfactory receptor axons. Remarkably, all 25,000 of these axons come from olfactory receptor neurons that express the same single odorant receptor gene (Figure 15.13E). This degree of convergence presumably increases the sensitivity of mitral cells to ensure maximal fidelity of odor detection. It may also maximize the signal strength from the convergent olfactory receptor neuron input by averaging out uncorrelated "background" noise. Each glomerulus also includes dendritic processes from two other classes of local circuit neurons: approximately 50 *tufted cells* and 25 *periglomerular cells* contribute to each glomerulus (Figure 15.13D). Although it is generally assumed that these neurons sharpen the sensitivity of individual glomeruli to specific odorants, their function is unclear.

Finally, *granule cells*, which constitute the innermost layer of the vertebrate olfactory bulb, synapse primarily on the basal dendrites of mitral cells within the external plexiform layer (Figure 15.13C,D). Granule cells lack an identifiable axon, and instead make reciprocal dendro-dendritic synapses with mitral cells. Granule cells are thought to establish local lateral inhibitory circuits with mitral cells as well as participating in synaptic plasticity in the olfactory bulb. Olfactory granule cells and periglomerular cells are among the few classes of neurons in the forebrain that can be replaced throughout life (see Chapter 25).

The relationship between olfactory receptor neurons expressing one odorant receptor and small subsets of glomeruli (Figure 15.13E) suggests that individual glomeruli respond specifically (or at least selectively) to distinct odorants. The selective (but not unique) responsiveness of subsets of glomeruli to particular odorants has been confirmed physiologically in invertebrates like *Drosophila*, as well as in mice, using single and multiunit recordings, metabolic mapping, voltage-sensitive dyes, genetically encoded sensors of electrical activity, or intrinsic signals that depend on blood oxygenation. Such studies show that increasing the odorant concentration increases the activity of individual glomeruli, as well as the number of glomeruli activated. In addition, different single odorants (Figure 15.14A), or odorants with distinct chemical structures (for example, length of the carbon chain in the backbone of the odorant molecule; Figure 15.14B) maximally activate one or a few glomeruli. It is still not clear how (or whether) odor identity and concentration is mapped across the array of glomeruli.

Given the response of small numbers of glomeruli to isolated odorants, one might expect that complex natural odorants such as coffee, fruits, cheeses, spices—each of which is composed of more than a hundred compounds— would activate a large number of olfactory glomeruli. Surprisingly, this is not the case. Natural odorants presented at their normal concentrations activate a relatively small number of glomeruli, each of which responds selectively to one or two molecules that characterize the complex odor. Thus, to solve the problem of representing complex odorants, the olfactory system appears to employ a sparse coding mechanism that cues in on a small number of dominant chemicals within a mixture (see Figure 15.11). One useful metaphor is to consider the

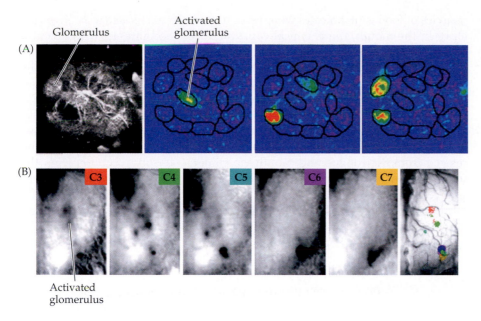

Glomerulus

Activated glomerulus

(A)

(B) C3 C4 C5 C6 C7

Activated glomerulus

Figure 15.14 Mapping responses of chemically distinct odorants in individual glomeruli. (A) At left, the array of glomeruli in the *Drosophila melanogaster* olfactory lobe (the equivalent of the mammalian olfactory bulb), is visualized with a fluorescent protein expressed under the genetic control of an olfactory lobe-specific gene. The subsequent panels show that three distinct odorants—1-octan-3-ol, an insect attractant; hexane, which has a chemical smell to humans; and isomyl acetate, the major molecular constituent of the aroma given off by bananas—activate different glomeruli. Red indicates maximal activity, and in each case activation is limited to one or two distinct glomeruli. (B) The surface of the mouse olfactory bulb, imaged for intrinsic electrical activity in response to the odorants of different carbon chain lengths (C3–C7, color-coded in the upper right-hand corners). Focal electrical activity shows up as darker spots on the light background, as indicated in the first panel. The spots correspond to individual glomeruli or a few glomeruli. The far-right panel summarizes the relationship between carbon chain length and location of activation on the dorsal surface of the olfactory bulb. Color-coding each focal spot with its corresponding odorant shows segregated activation of specific odorants. (A from Wang et al., 2003; B from Belluscio and Katz, 2001.)

sheet of glomeruli in the olfactory bulb as an array of lights on a movie marquee: the spatial distribution of active and inactive glomeruli provides a message that is unique for a given odorant at a particular concentration.

Central Projections of the Olfactory Bulb

Mitral cell axons provide the only relay for olfactory information to the rest of the brain. The mitral cell axons from each olfactory bulb form a bundle—the **lateral olfactory tract**—that projects to the accessory olfactory nuclei, the olfactory tubercle, the pyriform and entorhinal cortex, as well as portions of the amygdala (see Figure 15.1A). The major target of the olfactory tract is the three-layered **pyriform cortex** in the ventromedial aspect of the human temporal lobe, near the optic chiasm. Neurons in pyriform cortex respond to odors, and mitral cell inputs from glomeruli that receive odorant receptor-specific projections remain partially segregated. Most of projections of the lateral olfactory tract are ipsilateral; however, a subset of mitral cell axons cross the midline, presumably initiating bilateral processing of some aspects of olfactory information. The axons of pyramidal cells in the pyriform cortex project in turn to thalamic and hypothalamic nuclei as well as to the hippocampus and amygdala. Additional neurons in the pyriform cortex innervate directly a variety of areas in the neocortex including the orbitofrontal cortex, where multimodal responses to complex stimuli, particularly food, include an olfactory component. The representation of olfactory information in the pyriform cortex as well as the neocortex, however, is even less well understood than that for the olfactory bulb. Nevertheless, these pathways insure that information about odors reaches a variety of forebrain regions, allowing olfactory perception to influence cognitive, visceral, emotional, and homeostatic behaviors

The Organization of the Taste System

The second chemosensory system, the taste system, represents the chemical as well as physical qualities of ingested substances, including food. In concert with the olfactory and trigeminal systems, taste reflects the aesthetic and nutritive qualities of food, and indicates whether or not a food item is safe to be ingested. Once in the mouth, the chemical constituents of food interact with receptor proteins on **taste cells** located in epithelial specializations called **taste buds** in the tongue. Taste cells transduce these stimuli to encode information about the identity, concentration and pleasant, unpleasant, or potentially harmful qualities of the substance. This information also prepares the gastrointestinal system to receive and digest food by causing salivation and swallowing—or, if the substance is noxious, gagging and regurgitation. Information about the temperature and texture of food (including viscosity and fat content) is transduced and relayed from the tongue and mouth via somatic sensory receptors from the trigeminal and other sensory cranial nerves to the thalamus and somatic sensory cortices (see Chapters 9 and 10). Of course, food is not simply eaten for nutritional value or avoided because of unpleasant or potentially harmful qualities; "taste" also depends on cultural and psychological factors. How else can one explain why so many people enjoy consuming hot peppers or bitter-tasting liquids such as beer?

Like the olfactory system, the taste system is defined by its specialized peripheral receptors as well as a number of central pathways that relay and process taste information (Figure 15.15). Taste cells (the peripheral receptors) are found in taste buds distributed on the dorsal surface of the tongue, soft palate, pharynx, and the upper part of the esophagus (see Figures 15.16 and 15.17). Taste cells synapse with primary sensory axons that run in the chorda tympani and greater superior petrosal branches of the facial nerve (cranial nerve VII), the lingual branch of the glossopharyngeal nerve (cranial nerve IX), and the superior laryngeal branch of the vagus nerve (cranial nerve X) to innervate the taste buds in the tongue, palate, epiglottis, and esophagus, respectively. The central axons of these primary sensory neurons in the respective cranial nerve ganglia project to rostral and lateral regions of the **nucleus of the solitary tract** in the medulla (Figure 15.15A), which is also known as the **gustatory nucleus** of the solitary tract complex. (The posterior region of the solitary nucleus is the main target of afferent visceral sensory information related to the sympathetic and parasympathetic divisions of the visceral motor system; see Chapter 21.)

The distribution of the cranial nerves that innervate taste buds in the oral cavity is topographically represented along the rostral–caudal axis of the rostral portion of the gustatory nucleus; the terminations from the facial nerve are rostral, the glossopharyngeal are in the midregion, and those from the vagus nerve are more caudal in the nucleus (see Figure 15.15A). Integration of taste and visceral sensory information is presumably facilitated by this arrangement. The caudal part of the nucleus of the solitary tract also receives innervation from subdiaphragmatic branches of the vagus nerve, which control gastric motility. Interneurons connecting the rostral and caudal regions of the nucleus represent the first interaction between visceral and gustatory stimuli, and this circuit can be thought of as the sensory limb of a gustatory/visceral reflex arc. This close relationship between gustatory and visceral information makes good sense, since an animal must quickly recognize if it is eating something that is likely to make it sick and respond accordingly.

Axons from the rostral (gustatory) part of the solitary nucleus project to the ventral posterior complex of the thalamus, where they terminate in the medial half of the **ventral posterior medial nucleus**. This nucleus projects in turn to several regions of the neocortex, including the anterior insula in the temporal lobe (the insular taste cortex; see Figure 15.15A) and the operculum of the frontal lobe. There is also a secondary neocortical taste area in the caudolateral orbitofrontal cortex, where neurons respond to combinations of visual, somatic sensory, olfactory, and gustatory stimuli. Interestingly, when a given food is con-

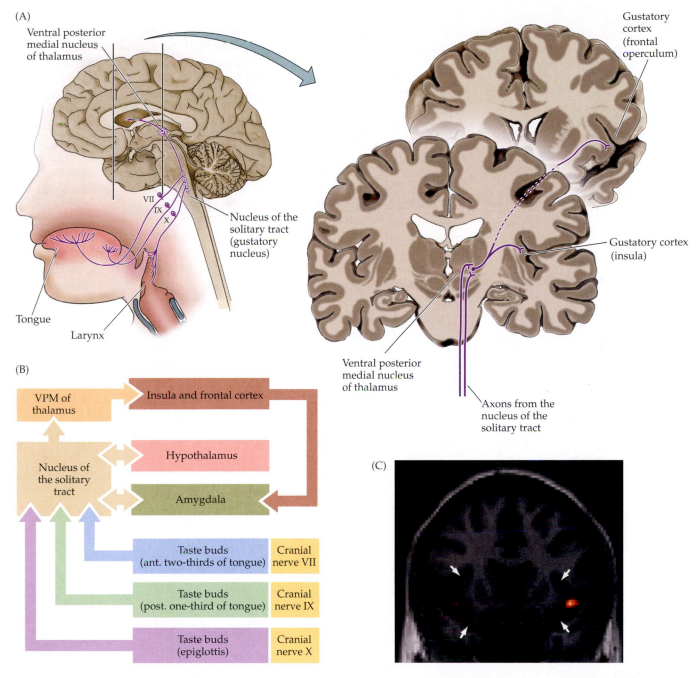

(A)

Ventral posterior medial nucleus of thalamus

VII
IX
X

Nucleus of the solitary tract (gustatory nucleus)

Tongue

Larynx

Gustatory cortex (frontal operculum)

Gustatory cortex (insula)

Ventral posterior medial nucleus of thalamus

Axons from the nucleus of the solitary tract

(B)

VPM of thalamus	Insula and frontal cortex

Hypothalamus

Amygdala

Nucleus of the solitary tract

Taste buds (ant. two-thirds of tongue)	Cranial nerve VII
Taste buds (post. one-third of tongue)	Cranial nerve IX
Taste buds (epiglottis)	Cranial nerve X

(C)

Figure 15.15 The human taste system. (A) The drawing shows the relationship between receptors in the mouth and upper alimentary canal, and the nucleus of the solitary tract in the medulla. The coronal section shows the VPM nucleus of the thalamus and its connection with gustatory regions of the cerebral cortex. (B) Basic pathways for processing taste information. (C) Functional MRI of a normal human subject consuming food. Note bilateral focal activation (red) in the insular cortex (arrows), with a bias for greater activation in the dominant hemisphere (the left in most humans). (C from Schoenfeld et al. 2004.)

sumed to the point of satiety, specific orbitofrontal neurons in the monkey diminish their activity to that tastant, suggesting that these neurons are involved in the conscious motivation to eat (or not to eat) particular foods. Finally, reciprocal projections connect the nucleus of the solitary tract via nuclei in the pons to the hypothalamus and amygdala. These projections presumably influence affective aspects (e.g., pleasurable vs. aversive experience of food; food-seeking behavior) of appetite, satiety, and other homeostatic responses associated with eating (recall that the hypothalamus is the major center governing homeostasis; see Chapter 21).

Taste Perception in Humans

The taste system encodes information about the quantity as well as the identity of stimuli. Most taste stimuli are nonvolatile, hydrophilic molecules that are soluble in saliva. In general, the perceived intensity of taste is directly proportional to the concentration of the taste stimulus. Threshold concentrations for most ingested tastants are quite high. For example, in humans, the threshold concentration for citric acid is about 2 mM; for salt (NaCl), 10 mM; and for sucrose, 20 mM. (In contrast, recall that the perceptual threshold for some odorants is as low as 0.01 nM.) Because the body requires substantial concentrations of salts and carbohydrates, taste cells may respond only to relatively high concentrations of these essential substances in order to promote an adequate intake. Clearly, it is advantageous for the taste system to detect potentially dangerous substances (e.g., bitter-tasting plant compounds, which may be noxious or poisonous) at much lower concentrations. Thus the threshold concentration for such tastants is relatively low: quinine is 0.008 mM, and for strychnine 0.0001 mM.

Tastants are detected over the full surface the tongue in receptive specializations called **taste papillae** (Figure 15.16A). Papillae are defined by multicellular protuberances surrounded by local invaginations in the tongue epithelium that form a trench to concentrate solubilized tastants. Taste buds, the sites of taste receptor cells, are distributed along the lateral surfaces of the papillar protuberance as well as the trench walls. There are three types of papillae: **fungiform** (which contain about 25 percent of the total number of taste buds), **circumvallate** (50 percent), and **foliate** (the remaining 25 percent) The three classes are distributed discontinuously on the surface of the tongue. Fungiform papillae are found only on the anterior two-thirds of the tongue; the highest density (about 30 per cm^2) is at the tip. Fungiform papillae have a mushroom-like structure (hence

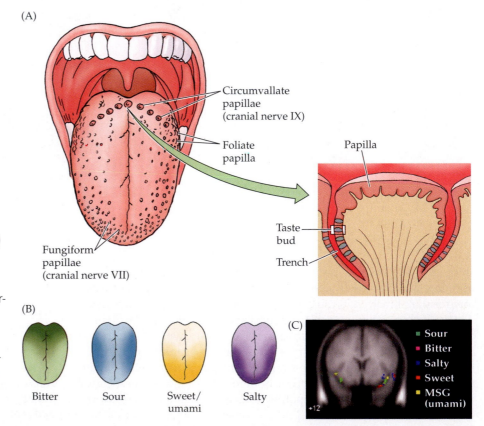

Figure 15.16 Taste buds and the peripheral innervation of the tongue. (A) Distribution of taste papillae on the dorsal surface of the tongue. The blow-up shows the location of individual taste buds on a circumvallate papilla. (B) Different responses to sweet, salty, sour, and bitter tastants recorded in the three cranial nerves that innervate the tongue and epiglottis. (C) Composite fMRI showing different locations of focal activation in the insular cortex in response to each of the five basic tastes encoded by taste receptors.

(A)

Circumvallate papillae (cranial nerve IX)

Foliate papilla

Papilla

Fungiform papillae (cranial nerve VII)

Taste bud

Trench

(B)

Bitter

Sour

Sweet/umami

Salty

(C)

- Sour
- Bitter
- Salty
- Sweet
- MSG (umami)

+12

their name) and typically have about three taste buds at their apical surface. There are nine circumvallate papillae arranged in a chevron at the rear of the tongue. Each consists of a circular trench containing about 250 taste buds along the trench walls. Two foliate papillae are present on the posterolateral tongue, each having about 20 parallel ridges with about 600 taste buds in their walls. Thus, chemical stimuli on the tongue first stimulate receptors in the fungiform papillae and then in the foliate and circumvallate papillae. Tastants subsequently stimulate scattered taste buds in the pharynx, larynx, and upper esophagus.

Based on general agreement across cultures, the taste system detects five perceptually distinct categories of tastants: **salt**, **sour**, **sweet**, **bitter**, and **umami** (from the Japanese word for delicious, *umami* refers to savory tastes, including monosodium glutamate and other amino acids that provide the flavor in cooked meat and other protein-rich foods). These five perceptual categories also have dietary and metabolic significance: salt tastes include NaCl, which is needed for electrolyte balance; essential amino acids such as glutamate are needed for protein synthesis; sugars such as glucose and other carbohydrates are needed for energy; sour tastes, associated with acidity and thus protons (H^+), indicate the palatability of various foods (for example the citric acid in oranges); bitter-tasting molecules, including plant alkaloids like atropine, quinine, and strychnine, indicate foods that may be poisonous.

There are obvious limitations to this classification. People experience a variety of gustatory or ingestive sensations in addition to these five, including astringent (cranberries and tea), pungent (hot peppers and ginger), fat, starchy, and various metallic tastes, to name only a few. In addition, mixtures of chemicals may elicit entirely new taste sensations. Finally, at low concentrations the protective response to aversive tastes can be overridden, leading to acquired tastes for a variety of sour and bitter tastes like lemons (sour) and quinine (bitter).

Although all tastes can be detected over the entire surface of the tongue, different regions of the tongue have different thresholds for various tastes (Figure 15.16B). These discontinuities in taste sensitivity may be related to the aesthetic, metabolic, and potentially toxic qualities detected by the taste receptors in the tongue. The tip of the tongue is most responsive to sweet, umami, and salty-tasting compounds, all of which produce pleasurable sensations at somewhat higher concentrations. Thus, tastes encountered by this region—the initial point of contact for most ingested foods—activate feeding behaviors such as mouth movements, salivary secretion, insulin release, and swallowing. The acquisition of foods high in carbohydrates and amino acids is beneficial (in moderation), and thus it is not surprising that the most exposed region of the tongue is especially sensitive to these tastes, thus promoting the intake of relevant foods.

Sour and bitter taste sensitivity is lowest toward the tip of the tongue, and are greatest on the sides and back of the tongue. It seems reasonable that, once analyzed for nutrient content, the receptor surface might next evaluate other aesthetic characteristics, like acidity and bitterness, that indicate lack of palatability (excessive sourness), or even toxicity (bitter). Sour-tasting compounds elicit grimaces, puckering responses, and massive salivary secretion to dilute the tastant. Activation of the rear of the tongue by bitter-tasting substances elicits protrusion of the tongue and other protective reactions (expectoration and gagging) that prevent ingestion.

Each of the five primary tastes represented over the surface of the tongue correspond to distinct classes of receptor molecules expressed in subsets of taste cells (see the next section). Thus, the categories of taste perception, and their representation in taste buds is closely linked to the molecular biology of taste transduction. These taste categories are also maintained in the representation of taste information in the central nervous system, including that in the insular

Figure 15.17 Taste buds, taste cells, and taste transduction. (A) Diagram and light micrograph of a taste bud, showing various types of taste cells and the associated gustatory nerves. The apical surface of the receptor cells have microvilli that are oriented toward the taste pore. (B) The basic components of sensory transduction in taste cells. Taste cells are polarized epithelial cells with an apical and a basal domain separated by tight junctions. Tastant-transducing channels (salt and sour) and G-protein-coupled receptors (sweet, amino acid, and bitter) are limited to the apical domain. Intracellular signaling components that are coupled to taste receptor molecules (G-proteins and various second messenger-related molecules) are also enriched in the apical domain. Voltage-regulated Na⁺, K⁺, and Ca²⁺ channels mediate release of neurotransmitter from presynaptic specializations at the base of the cell onto terminals of peripheral sensory afferents. These channels are limited to the basolateral domain, as is endoplasmic reticulum that also modulates intracellular Ca²⁺ concentration and contributes to the release of neurotransmitter. The neurotransmitter serotonin, among others, is found in taste cells, and serotonin receptors are found on the sensory afferents. Finally, the TRPM$_5$ channel, which facilitates G-protein-coupled receptor-mediated depolarization, is expressed in taste cells. Its localization to apical versus basal domains is not yet known. (Micrograph from Ross et al., 1995.)

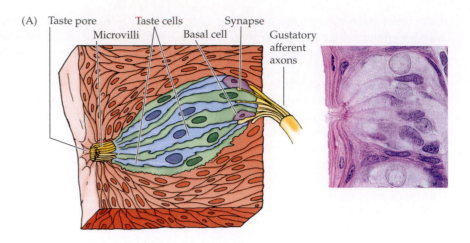

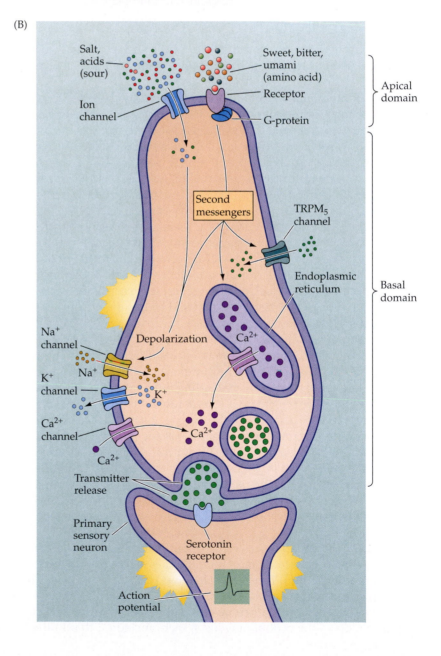

taste cortex (see Figure 15.15C). Mapping of responses to sweet, bitter, salty, sour, and umami in normal human subjects shows that each of these tastes elicits focal activity in the taste cortex, suggesting that information about each taste category remains somewhat segregated throughout the taste system.

As in olfaction, gustatory sensitivity declines with age. An obvious index of this decline is the tendency of adults to add more salt and spices to food than children do. The decreased sensitivity to salt can be problematic for older people with hypertension as well as electrolyte and/or fluid balance problems. Unfortunately, a safe and effective substitute for table salt (NaCl) has not yet been developed.

Taste Buds Taste Cells, Receptor Proteins, and Transduction

The initial transduction and encoding of taste information occurs in taste buds. In humans, approximately 4000 taste buds are distributed in papillae throughout the oral cavity and upper alimentary canal. Taste buds consist of specialized neuroepithelial receptor cells called **taste cells**, some supporting cells, and occasional **basal cells**. Taste buds, found primarily in the taste papillae, are distributed throughout the surface of the tongue (as well as the palate, epiglottis, and esophagus; see Figure 15.16). Taste cells are clustered around a 1-mm opening in the taste bud near the surface of the tongue called a **taste pore** (Figure 15.17A), and solubilized tastants are further concentrated and presented directly to the exposed sensory receptor cells in this relatively small space. Like olfactory receptor neurons (and presumably for the same reasons—exposure to infectious agents and environmental toxins), taste cells have a lifetime of only about two weeks. They are apparently regenerated from basal cells, which constitute a local stem cell population that is retained in the mature tongue.

Only the taste cell is specialized for sensory transduction, and its basic structure and function are uniform across all classes of papillae and their constituent taste buds. Taste cells in individual taste buds have distinct apical and basal domains, reflecting their epithelial character. Chemosensory transduction is initiated in the apical domain of the taste cells, and electrical signals, via graded receptor potentials (and corresponding secretion of neurotransmitters) are generated at the basal domain (Figure 15.17B). The specific neurotransmitters released by taste cells remain uncertain, but are thought to include serotonin and ATP. Taste receptor proteins and related signaling molecules, like those in olfactory receptor neurons, are concentrated on microvilli that emerge from the taste cell apical surface. These receptor-rich apical surfaces of the basal domain consist of secretory specializations that synapse with primary afferent axons from branches of three cranial nerves: the facial (VII), glossopharyngeal (IX), and vagus (X) nerves (see Figure 15.15A).

The major perceptual taste categories—salty, sour, sweet, umami, and bitter—are represented by five distinct classes of taste receptor molecules. These receptor molecules are thought to be concentrated primarily in the apical microvilli of taste cells. Salty and sour tastes are elicited by ionic stimuli such as the positively charged ions in salts (like Na^+ from NaCl), or the H^+ in acids (acetic acid, for example, which gives vinegar its sour taste). Thus, the ions in salty and sour tastants initiate sensory transduction via specific ion channels, most likely an amiloride-sensitive Na^+ channel for salty tastes (Figure 15.18A) and, for sour, an H^+-permeant, non-selective cation channel that is a member of the transient receptor potential (TRP) channel family (Figure 15.18B). The sour receptor channel is related to a similar channel protein that is mutated in polycystic kidney disease; thus the channel is referred to as PKD.

The sour receptor is expressed in a distinct subset of taste cells, similar to the segregated expression of receptor proteins for sweet, umami, and bitter (see

Figure 15.18 Molecular mechanisms of taste transduction via ion channels and G-protein-coupled receptors. (A) Cation selectivity of the amiloride-sensitive Na⁺ versus the H⁺-sensitive proton channel provides the basis for specificity of salty tastes. (B) Sour tastants are transduced by a proton-permeant, nonselective cation channel that is a member of the transient receptor potential (TRP) channel family. In both cases, positive current via the cation channel leads to depolarization of the cell. (C–E) For sweet, amino acid (umami), and bitter tastants, different classes of G-protein-coupled receptors mediate transduction. (C) For sweet tastants, heteromeric complexes of the T1R2 and T1R3 receptors transduce stimuli via a PLC$_{\beta2}$-mediated, IP$_3$-dependent mechanism that leads to activation of the TRPM$_5$ Ca^{2+} channel. (D) For amino acids, heteromeric complexes of T1R1 and T1R3 receptors transduce stimuli via the same PLC$_{\beta2}$/IP$_3$/TRPM$_5$-dependent mechanism. (E) Bitter tastes are transduced via a distinct set of G-protein-coupled receptors, the T2R receptor subtypes. The details of T2R receptors are less well established; however, they apparently associate with the taste cell-specific G-protein gustducin, which is not found in sweet or amino acid receptor-expressing taste cells. Nevertheless, stimulus-coupled depolarization for bitter tastes relies upon the same PLC$_{\beta2}$/IP$_3$/TRPM$_5$-dependent mechanism used for sweet and amino acid taste transduction.

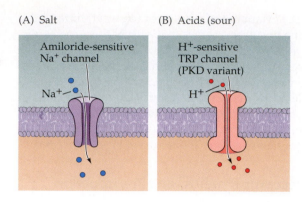

(A) Salt

(B) Acids (sour)

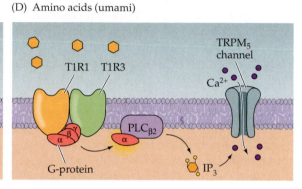

(C) Sweet

(D) Amino acids (umami)

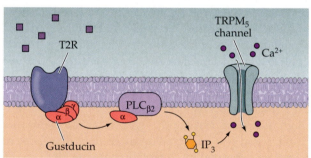

(E) Bitter

below). The receptor potentials generated by the positive inward current carried either by Na⁺ for salty or H⁺ for sour directly depolarize the relevant taste cell. The initial depolarization leads to the activation of voltage-gated Na⁺ channels in the basolateral aspect of the taste cell. This additional depolarization activates voltage-gated Ca^{2+} channels, leading to the release of neurotransmitter from the basal aspect of the taste cell and the activation of action potentials in ganglion cell axons (see Figure 15.17B).

In humans and other mammals, sweet and umami receptors are heterodimeric G-protein-coupled receptors that share a common seven-transmembrane receptor subunit called T1R3, paired with the T1R2 seven-transmembrane receptor for perception of sweet, or with the T1R1 receptor for amino acids (Figure 15.18C,D). The T1R2 and T1R1 receptors are expressed in different subsets of taste cells, indicating that there are, respectively, sweet- and amino acid-selective cells in the taste buds. Upon binding sugars or other sweet stimuli, the T1R2/T1R3 receptor heterodimer initiates a G-protein-mediated signal transduction cascade that leads to the activation of the phospholipase C isoform PLC$_{\beta2}$, leading in turn to increased concentrations of inositol triphosphate (IP$_3$) and to the opening of TRP channels (specifically the TRPM$_5$ channel), which depolarizes the taste cell via increased intracellular Ca^{2+}. Similarly, the T1R1/T1R3 receptor is broadly tuned to the 20 standard L-amino acids found in proteins (but not to their D-amino acid enantiomers). Transduction of amino acid stimuli via the T1R1/T1R3 receptor also reflects G-protein-coupled intracellular signaling leading to PLC$_{\beta2}$-mediated activation of the TRPM$_5$ channel and depolarization of the taste cell (Figure 15.18D).

Another family of G-protein-coupled receptors known as T2R receptors transduce bitter tastes. There are approximately 30 T2R subtypes encoded by 30 genes in humans and other mammals, and multiple T2R subtypes are expressed in single taste cells (see Figure 15.19). Indeed, in humans, a well known mutation for the perception of a specific bitter tastant, phenothylcarbamide (PTC), was originally discovered in the early 1930s and identified as a simple Mendelian trait shortly thereafter; this has proven to be a mutation of a human T2R gene. The observation of a selective single gene mutation for bitter taste indicates that this taste category is distinct and encoded specifically in taste receptor cells. The distribution of T2R receptors among taste cells supports this view. T2Rs are not expressed in the same taste cells as T1R1, 2, and 3 receptors. Thus the receptor cells for bitter tastants are presumably completely distinct from those for sweet and umami, which share at least one heterodimeric G-protein-coupled receptor subunit.

Although the transduction of bitter stimuli relies on a similar mechanism to that for sweet and amino acid tastes, a taste cell-specific G-protein, **gustducin**, is found primarily in T2R-expressing taste cells and apparently contributes to the transduction of bitter tastes (Figure 15.18E). The role of gustducin versus that of the other G-proteins for sweet and umami/amino acids remains unclear. The remaining steps in bitter transduction are similar to those for sweet and amino acids: PLC$_{\beta2}$-mediated activation of TRPM$_5$ channels depolarizes the taste cell, resulting in the release of neurotransmitter at the synapse between the taste cell and sensory ganglion cell axon.

Neural Coding in the Taste System

In the taste system, **neural coding** refers to the way that the identity, concentration, and "hedonic" (pleasurable or aversive) value of tastants is represented in the pattern of action potentials relayed to the brain from the taste buds. Neurons in the taste system (or in any other sensory system) might be specifically "tuned" to respond with a maximal change in electrical activity to a single taste stimulus. Such tuning might rely on specificity at the level of the receptor cells, as well as on the maintenance of separate channels for the relay of this information from the periphery to the brain. This sort of coding scheme is often referred to as a **labeled line code**, since responses in specific cells at multiple points in the pathway presumably correspond to distinct stimuli. The segregated expression of sour, sweet, amino acid, and bitter receptors in different taste cells (see Figure 15.18), and the maintenance of focal activation for each class of taste in the insular taste cortex (see Figure 15.15C) is consistent with labeled line coding.

Molecular genetic experiments in mice indicate that the five perceptual taste categories are established based on the identity of T1Rs and T2Rs (or amiloride/Na$^+$ and the PKD-TRP channels for salt and sour, respectively) that are expressed in individual taste cells. Initial support came from studies in which the genes that specify the sweet and amino acid heteromeric receptors (T1R2 and T1R1) were inactivated in mice. Such mice lack behavioral responses to a broad range of sweet or amino acid stimuli, depending on the gene that has been inactivated. Moreover, recordings of electrical activity in the relevant branches of cranial nerves VII, IX, or X showed that action potentials in response to sweet or amino acid stimuli were lost in parallel with the genetic mutation and behavioral change. Finally, these deficits in transduction and perception were unchanged at a broad range of concentrations, indicating that the molecular specificity of each receptor is quite rigid—the remaining receptors could not respond, even at high concentrations of sweet or amino acid stimuli.

These observations suggest that sweet and amino acid transduction and perception depend on labeled lines from the periphery. Bitter taste proved harder to analyze because of the larger number of T2R bitter receptors. To circumvent this challenge, Charles Zuker, Nicholas Ryba and colleagues took advantage of the shared aspects of intracellular signaling for sweet, amino acid, and bitter tastes. Thus, if the genes for either the $TRPM_5$ channel or $PLC_{\beta 2}$ are inactivated, behavioral and physiological responses to sweet, amino acid, and bitter stimuli are all abolished (Figure 15.19) while salty and sour perception—which do not rely on the G-protein-coupled, $PLC_{\beta 2}$-mediated transduction mechanism—remain. To evaluate whether taste cells expressing the T2R family of receptors provide a labeled line for bitter tastes, $PLC_{\beta 2}$ was selectively re-expressed in T2R-expressing taste cells in a $PLC_{\beta 2}$ mutant mouse. Thus, in these mice, only the taste cells that normally express the T2R subset of receptor genes can now tranduce taste signals. If these receptors specifically and uniquely encode bitter tastes, the "rescued" mice (i.e., those expressing $PLC_{\beta 2}$ in T2R cells) should regain their perceptual and physiological responses to bitter, but not sweet or amino acid tastes. This was indeed the result of the experiment—behavioral and physiological responses to bitter tastes, but not sweet or amino acid tastes, were restored to normal levels (see Figure 15.19). Evidently, receptor proteins uniquely expressed in subsets of taste cells encode sweet, amino acid, and bitter—judged by taste perception, action potential activity in peripheral nerves and behavioral responses. This specificity and segregation of receptor cells at the periphery can be considered to establish labeled lines that relay the information to the central nervous system, where information about the identity of the five primary taste categories remains segregated (see also Figure 15.15C). Given the clear distinctions one makes between sweet, sour, bitter, and umami tastes, it seems likely that this perceptual clarity established by peripheral receptors is

Figure 15.19 Specificity in peripheral taste coding supports the labeled line ▶ hypothesis. (A–C) Sweet (A), amino acid (B), and bitter (C) receptors are expressed in different subsets of taste cells. (D–E) The gene for the $TRPM_5$ channel can be inactivated, or "knocked out," in mice ($TRPM_5^{-/-}$) and behavioral responses measured with a taste preference test. The mouse is presented with two drinking spouts, one with water and the other with a tastant; behavioral responses are measured as the frequency of licking of the two spouts. For pleasant tastes like sweet (sucrose; D) or umami (glutamate; E) control mice lick the spout with the tastant more frequently, and higher concentrations of tastant leads to increased response (blue lines). In $TRPM_5^{-/-}$ mice, this behavioral response (i.e., a preference for the tastant versus water) is eliminated at all concentrations (red lines). (F) For an aversive tastant like bitter quinine, control mice prefer water. This behavioral response—which is initially low—is further diminished with higher quinine concentrations (blue line). Inactivation of $TRPM_5$ also eliminates this behavioral response, regardless of tastant concentration (red line). (G–I) When the $PLC_{\beta 2}$ gene is knocked out, behavioral response to (G) sucrose, (H) glutamate, and (I) quinine are eliminated (red lines). When $PLC_{\beta 2}$ is re-expressed only in T2R-expressing taste cells, behavioral responses to sucrose and glutamate are not rescued (dotted green lines in G and H); however, the behavioral response to quinine is restored to normal levels (compare the blue and dotted green lines in I). (After Zhang et al., 2003.)

maintained by central representations and used to guide specific ingestive (sweet, salty, umami) or aversive (sour, bitter) behaviors.

Trigeminal Chemoreception

The third chemosensory system, the trigeminal chemosensory system, consists of polymodal nociceptive neurons and their axons in the trigeminal nerve (cranial nerve V) and, to a lesser degree, nociceptive neurons whose axons run in the glossopharyngeal and vagus nerves (IX and X). Thus, this system can be considered a specialized component of the pain- and temperature-sensing somatosensory pathways in the head and neck (see Chapter 10), particularly the trigeminal pain system. The irritant-sensitive polymodal nociceptors of the

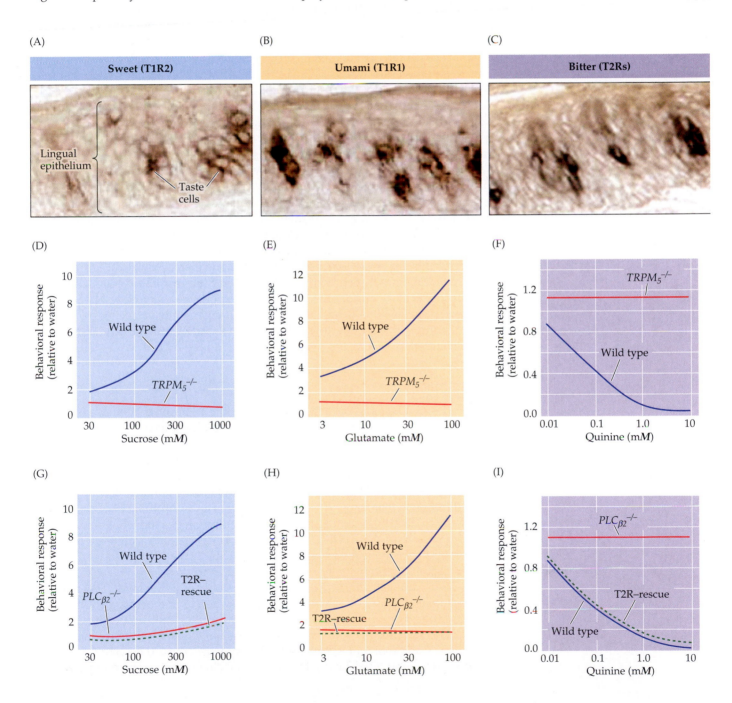

trigeminal system help alert the organism to potentially harmful chemical stimuli that have been respired, or have come in contact with the face at high environmental concentrations. The peripheral receptor neurons and their associated endings are typically activated by relatively high concentrations of irritating chemicals that come into direct contact with the mucous membranes of the head, including the mouth, nose, and eyes. Stimuli for the trigeminal chemosensory system include air pollutants such as sulfur dioxide, ammonia (smelling salts), ethanol (liquor), acetic acid (vinegar), carbon dioxide (in soft drinks), menthol (in various inhalant sensations; see Box 10A), and capsaicin (the compound in hot chili peppers that elicits the characteristic burning sensation). The receptors for irritants are primarily on the terminal branches of polymodal nociceptive neurons, as described for the pain and temperature systems in Chapter 10. With the exception of capsaicin and acidic stimuli, both of which activate cation-selective TRP channels, little is known about the transduction mechanisms for irritants, or their central processing. Each of these classes of compounds can also be detected by the olfactory and gustatory systems. The concentration thresholds that activate trigeminal chemosensory endings versus specialized olfactory and taste receptors, however, are significantly higher, and it is likely that the majority of these stimuli are transduced via distinct cellular and molecular mechanisms by the trigeminal chemosensory system.

Most trigeminal chemosensory information from the face, scalp, cornea, and mucous membranes of the oral and nasal cavities is relayed via the three major sensory branches of the trigeminal nerve: ophthalmic, maxillary, and mandibular. The central target of these afferent axons is the spinal component of the trigeminal nucleus, which relays this information to the ventral posterior medial nucleus of the thalamus and thence to the somatic sensory cortex and other cortical areas that process facial irritation and pain (see Chapter 10). A variety of physiological responses mediated by the trigeminal chemosensory system are triggered by exposure to irritants. These include increased salivation, vasodilation, tearing, nasal secretion, sweating, decreased respiratory rate, and bronchoconstriction. Consider, for instance, the experience that follows the ingestion of capsaicin (see Box 10A). These reactions are generally protective in that they dilute the stimulus (tearing, salivation, sweating) and prevent inhaling or ingesting more of it.

Summary

The chemical senses—olfaction, taste, and the trigeminal chemosensory system—all contribute to sensing airborne or soluble molecules from a variety of sources. Humans and other mammals rely on this information for behaviors as diverse as attraction, reproduction, feeding, and avoiding potentially dangerous circumstances. Receptor neurons in the olfactory epithelium transduce chemical stimuli into neuronal activity by stimulation of a large family of G-protein-coupled receptors that mediate second messengers to cation channels. These events generate olfactory receptor potentials, and ultimately action potentials, in the afferent axons of these cells. The large number of odorant receptor molecules in most species is believed to establish sensitivity to the myriad odors that most animals can discriminate. Taste receptor cells, in contrast, use a variety of mechanisms for transducing a more limited range of chemical stimuli. There are five perceptual categories of taste: salty, sour, sweet, amino acids (also known as umami), and bitter, and each is encoded by receptor cells that express distinct receptor proteins. Salts and protons (acids) directly activate two different ion channels, and there are specific sets of G-protein-coupled receptors for sweet, amino acid, and bitter tastes. The trigeminal chemosensory system transduces irritants via free nerve endings in facial mucous membranes that are in many ways similar to those that encode pain and temperature information in the skin. Olfaction, taste, and trigeminal chemosensation all are relayed via specific pathways in the central nervous system. Olfactory receptor neurons project directly to the olfactory bulb. In the taste system, information is initially relayed centrally by cranial sensory ganglion cells to the solitary nucleus in the brainstem. In the trigeminal chemosensory system, information is relayed via trigeminal ganglion cell projections to the spinal trigeminal nucleus in the brainstem. Each of these structures projects in turn to many sites in the brain, including (via thalamic relays) to specific neocortical areas in the frontal and temporal lobes that process chemosensory information in ways that give rise to some of the most sublime pleasures humans can experience.

Additional Reading

Reviews

AXEL, R. (2005) Scents and sensibility: A molecular logic of olfactory perception (Nobel lecture). *Angew Chem.*, Int. Ed. (English) 44(38): 6110–6127.

BUCK, L. B. (2000) The molecular architecture of odor and pheromone sensing in mammals. *Cell* 100: 611–618.

HERNESS, M. S. AND T. A. GILBERTSON (1999) Cellular mechanisms of taste transduction. *Annu. Rev. Physiol.* 61: 873–900.

HILDEBRAND, J. G. AND G. M. SHEPHERD (1997) Mechanisms of olfactory discrimination: Converging evidence for common principles across phyla. *Annu. Rev. Neurosci.* 20: 595–631.

LINDEMANN, B. (1996) Taste reception. *Physiol. Rev.* 76: 719–766.

MOMBAERTS, P. (2004) Genes and ligands for odorant, vomeronasal and taste receptors *Nat. Rev. Neurosci.* 5: 263–278.

SCOTT, K. (2004) The sweet and the bitter of mammalian taste. *Curr. Opin. Neurobiol.* 14: 423–427.

ZUFALL, F. AND T. LEINDERS-ZUFALL (2000) The cellular and molecular basis of odor adaptation. *Chem. Senses* 25: 473–481.

Important Original Papers

ADLER, E., M. A. HOON, K. L. MUELLER, J. CHRANDRASHEKAR, N. J. P. RYBA AND C. S. ZUCKER (2000) A novel family of mammalian taste receptors. *Cell* 100: 693–702.

ASTIC, L. AND D. SAUCIER (1986) Analysis of the topographical organization of olfactory epithelium projections in the rat. *Brain Res. Bull.* 16: 455–462.

AVANET, P. AND B. LINDEMANN (1988) Amiloride-blockable sodium currents in isolated taste receptor cells. *J. Memb. Biol.* 105: 245–255.

BOZZA, T., P. FEINSTEIN, C. ZHENG AND P. MOMBAERTS (2002) Odorant receptor expression defines functional units in the mouse olfactory system. *J. Neurosci.* 22: 3033–3043.

BUCK, L. AND R. AXEL (1991) A novel multigene family may encode odorant receptors: A molecular basis for odor recognition. *Cell* 65: 175–187.

CATERINA, M. J. AND 8 OTHERS (2000) Impaired nociception and pain sensation in mice lacking the capsaicin receptor. *Science* 288: 306–313.

CHAUDHARI, N., A. M. LANDIN AND S. D. ROPER (2000) A metabotropic glutamate receptor variant functions as a taste receptor. *Nature Neurosci.* 3: 113–119.

GRAZIADEI, P. P. C. AND G. A. MONTI-GRAZIADEI (1980) Neurogenesis and neuron regeneration in the olfactory system of mammals. III. Deafferentation and reinnervation of the olfactory bulb following section of the fila olfactoria in rat. *J. Neurocytol.* 9: 145–162.

KAY, L. M. AND G. LAURENT (2000) Odor- and context-dependent modulation of mitral cell activity in behaving rats. *Nature Neurosci.* 2: 1003–1009.

LIN, D.Y., S. D. SHEA AND L. D. KATZ (2006) Representation of natural stimuli in the rodent main olfactory bulb. *Neuron* 50: 937–949

MALNIC, B., J. HIRONO, T. SATO AND L. B. BUCK (1999) Combinatorial receptor codes for odors. *Cell* 96: 713–723.

MOMBAERTS, P. AND 7 OTHERS (1996) Visualizing an olfactory sensory map. *Cell* 87: 675–686.

NELSON, G., M. A. HOON, J. CHANDRASHEKAR, Y. ZHANG, N. J. P. RYBA AND C. S. ZUKER (2001) Mammalian sweet taste receptors. *Cell* 106: 381–390.

NELSON, G. AND 6 OTHERS. (2002) An amino-acid taste receptor. *Nature* 416: 199–202.

VASSAR, R., S. K. CHAO, R. SITCHERAN, J. M. NUNEZ, L. B. VOSSHALL AND R. AXEL (1994) Topographic organization of sensory projections to the olfactory bulb. *Cell* 79: 981–991.

WONG, G. T., K. S. GANNON AND R. F. MARGOLSKEE (1996) Transduction of bitter and sweet taste by gustducin. *Nature* 381: 796–800.

ZHANG, Y. AND 7 OTHERS. (2003) Coding of sweet, bitter, and umami tastes: Different receptor cells sharing similar signaling pathways. *Cell* 112: 293–301.

ZHAO, G. Q. AND 6 OTHERS (2003) The receptors for mammalian sweet and umami taste. *Cell* 115: 255–266.

Books

BARLOW, H. B. AND J. D. MOLLON (1989) *The Senses.* Cambridge: Cambridge University Press, Chapters 17–19.

DOTY, R. L. (ed.) (1995) *Handbook of Olfaction and Gustation.* New York: Marcel Dekker.

FARBMAN, A. I. (1992) *Cell Biology of Olfaction.* New York: Cambridge University Press.

GETCHELL, T. V., L. M. BARTOSHUK, R. L. DOTY AND J. B. SNOW, JR. (1991) *Smell and Taste in Health and Disease.* New York: Raven Press.

MOVEMENT AND ITS CENTRAL CONTROL

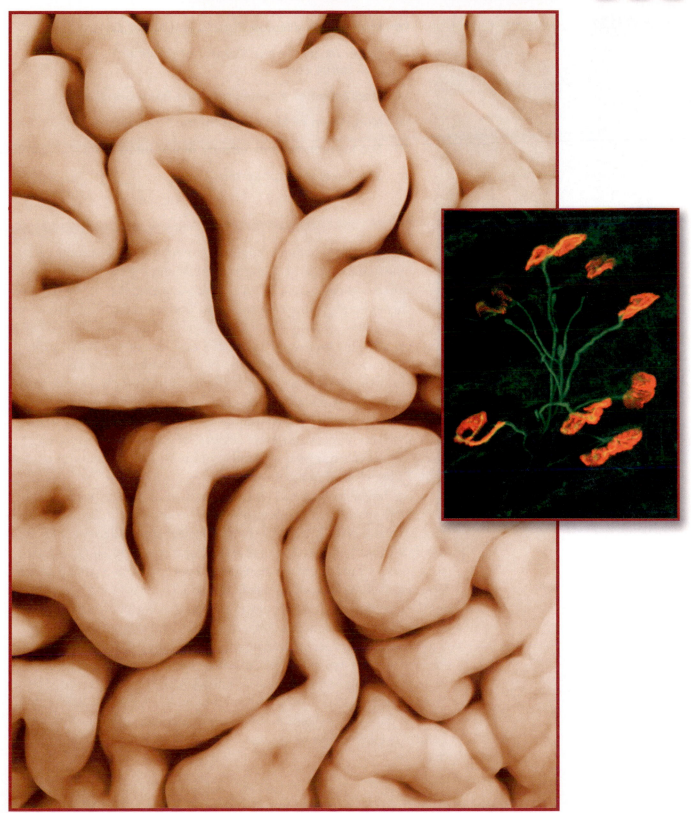

UNIT III

MOVEMENT AND ITS CENTRAL CONTROL

Fluorescence photomicrograph showing motor axons (green) and neuromuscular synapses (orange) in transgenic mice that have been genetically engineered to express fluorescent proteins. (Courtesy of Bill Snider and Jeff Lichtman.)

Movements, whether voluntary or involuntary, are produced by spatial and temporal patterns of muscular contractions orchestrated by neural circuits in the brain and spinal cord. Analysis of these circuits is fundamental to an understanding of both normal behavior and the etiology of a variety of neurological disorders. This unit considers the brainstem and spinal cord circuitry that make elementary reflex movements possible, as well as the circuits that organize the intricate patterns of neural activity responsible for more complex motor acts. Ultimately, all movements produced by the skeletal musculature are initiated by "lower" motor neurons in the spinal cord and brainstem that directly innervate skeletal muscles; the innervation of visceral smooth muscles is separately organized by the autonomic divisions of the visceral motor system.

The lower motor neurons are controlled directly by local circuits within the spinal cord and brainstem that coordinate individual muscle groups, and indirectly by "upper" motor neurons in higher centers that regulate those local circuits, thus enabling and coordinating complex sequences of movements and ensuring appropriate autonomic activity in support of the behavioral needs at hand. Circuits in the basal ganglia and cerebellum regulate upper motor neurons, facilitating the initiation and performance of movement with spatial and temporal precision.

Specific disorders of movement often signify damage to a particular brain region. For example, clinically important neurodegenerative disorders, such as amyotrophic lateral sclerosis, Parkinson's disease, and Huntington's disease, result from pathological changes in different parts of the motor system. Knowledge of the various levels of motor control is essential for understanding, diagnosing, and treating these diseases.

Chapter 16

Lower Motor Neuron Circuits and Motor Control

Overview

Skeletal (striated) muscle contraction is initiated by "lower" motor neurons in the spinal cord and brainstem. The cell bodies of the lower neurons are located in the ventral horn of the spinal cord gray matter and in the motor nuclei of the cranial nerves in the brainstem. These neurons (also called α motor neurons) send axons directly to skeletal muscles via the ventral roots and spinal peripheral nerves, or via cranial nerves, in the case of brainstem motor nuclei. The spatial and temporal patterns of activation of lower motor neurons are determined primarily by local circuits located within the spinal cord and brainstem. The local circuit neurons receive direct input from sensory neurons and mediate sensory-motor reflexes; they also maintain precise interconnections that enable the coordination of a rich repertoire of rhythmical and stereotyped behavior. Descending pathways from higher centers comprise the axons of "upper" motor neurons that modulate the activity of lower motor neurons by influencing this local circuitry. The cell bodies of upper motor neurons are located in brainstem centers, such as the vestibular nucleus, superior colliculus, and reticular formation, as well as in the cerebral cortex, which governs the execution of volitional movement. The axons of the upper motor neurons typically contact the local circuit neurons in the brainstem and spinal cord, which, via relatively short axons, in turn, contact the appropriate combinations of lower motor neurons. Lower motor neurons, therefore, are the final common pathway for transmitting information from a variety of sources to the skeletal muscles. Comparable circuits of interneurons and lower motor neurons may be recognized within the divisions of the visceral motor system, but consideration of these motor circuits will be reserved for Chapter 21; otherwise, the principal context for our exploration of the central control of movement will be those movements that are executed by musculoskeletal systems.

Neural Centers Responsible for Movement

The neural circuits responsible for the control of movement can be divided into four distinct but highly interactive subsystems, each of which makes a unique contribution to motor control (Figure 16.1). The first of these subsystems is the local circuitry within the gray matter of the spinal cord and the tegmentum of the brainstem. The relevant cells include the **lower motor neurons** (which send their axons out of the brainstem and spinal cord to innervate the skeletal muscles of the head and body, respectively) and the **local circuit neurons** (which are the major source of synaptic input to the lower motor neurons). All commands for movement, whether reflexive or voluntary, are ultimately conveyed to the muscles by the activity of the lower motor neurons; thus, these neurons comprise, in the words of the great British neurophysiologist, Charles Sherring-

Figure 16.1 Overall organization of neural structures involved in the control of movement. Four systems—local spinal cord and brainstem circuits, descending modulatory pathways, the cerebellum, and the basal ganglia—make essential and distinct contributions to motor control.

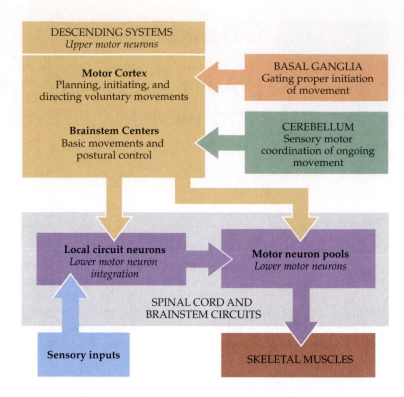

ton, the "final common path" for initiating movement. The local circuit neurons receive sensory inputs as well as descending projections from higher centers. Thus, the circuits they form provide much of the coordination between different muscle groups that is essential for organized movement. Even after the spinal cord is disconnected from the brain in an experimental animal, such as a cat, appropriate stimulation of local spinal circuits elicits involuntary but highly coordinated limb movements that resemble walking.

The second motor subsystem consists of the **upper motor neurons**, whose cell bodies lie in the brainstem or cerebral cortex and whose axons descend to synapse with the local circuit neurons or (more rarely) with the lower motor neurons directly. The upper motor neuron pathways that arise in the cortex are essential for the initiation of voluntary movements and for complex spatiotemporal sequences of skilled movements. In particular, descending projections from cortical areas in the frontal lobe, including Brodmann's area 4 (the **primary motor cortex**) and several divisions of **premotor cortex** in Brodmann's area 6, are essential for planning, initiating, and directing sequences of voluntary movements involving the limbs. The frontal lobe also contains cortical areas that play a similar role in the control of eye movements. In addition, cortical areas in the anterior cingulate gyrus (Brodmann's area 24) govern the expression of emotions, especially with respect to the facial musculature. Upper motor neurons originating in the brainstem are responsible for regulating muscle tone and for orienting the eyes, head, and body with respect to vestibular, somatic, auditory, and visual sensory information. Their contributions are thus critical for basic navigational movements and for the control of posture.

The third and fourth subsystems are complex circuits with output pathways that have no direct access to either the local circuit neurons or the lower motor neurons; instead, they control movement by regulating the activity of the upper motor neurons. The larger of these subsystems, the **cerebellum**, overlies the pons and fourth ventricle in the posterior cranium (see the Appendix). The cerebellum acts via its efferent pathways to the upper motor neurons as a servomechanism, detecting the difference, or "motor error," between an intended movement and the movement actually performed (see Chapter 19). The cerebellum uses this information about discrepancies to mediate both real-time and long-term reductions in these inevitable motor errors (the latter being a form of motor learning). As might be expected from this account, patients with cerebellar damage exhibit persistent errors in ongoing movement. The final subsystem, embedded in the depths of the forebrain, consists of a group of structures collectively referred to as the **basal ganglia**. The basal ganglia suppress unwanted movements and prepare (or "prime") upper motor neuron circuits for the initiation of movements. The problems associated with disorders of basal ganglia, such as Parkinson's disease and Huntington's disease, attest to the importance of this complex in the initiation of voluntary movements (see Chapter 18).

Despite much effort, the sequence of events that leads from volitional thought and emotion to movement is still poorly understood. The picture is clearest, however, at the level of control of the skeletal muscles themselves. It therefore makes sense to begin an account of motor behavior by considering the anatomical and physiological relationships between lower motor neurons and the striated muscle fibers they innervate.

Motor Neuron–Muscle Relationships

An orderly relationship between the location of the motor neuron pools and the muscles they innervate is evident both along the length of the spinal cord and across the medial-to-lateral dimension of the cord, an arrangement that, in effect, provides a spatial map of the body's musculature. This map can be demonstrated in animal experiments by injecting individual muscle groups with visible tracers that are transported by the axons of the lower motor neurons back to their cell bodies; the lower motor neurons that innervate each of the body's skeletal muscles can then be seen in histological sections of the ventral horns of the spinal cord. Each lower motor neuron innervates muscle fibers within a single muscle, and all the motor neurons innervating a single muscle (called the **motor neuron pool** for that muscle) are grouped together into a rod-shaped cluster that runs parallel to the long axis of the spinal cord for one or more spinal cord segments (Figure 16.2). For example, the motor neuron pools that innervate the arm are located in the cervical enlargement of the cord and those that innervate the leg are located in the lumbar enlargement (see the Appendix). The mapping, or *topography*, of motor neuron pools in the medial-to-lateral dimension can be appreciated in a cross section through the cervical enlargement (illustrated in Figure 16.3). Thus, neurons that innervate the axial musculature (i.e., the postural muscles of the trunk) are located medially in the spinal cord. Lateral to these cell groups are motor neuron pools innervating muscles located progressively more laterally in the body. Neurons that innervate the muscles of the shoulders (or pelvis, if one were to look at a similar section in the lumbar enlargement; see Figure 16.2) are the next most lateral group, whereas those that innervate the proximal muscles of the arm (or leg) are located laterally to these. The motor neuron pools that innervate the distal parts of the extremities, the fingers or toes, lie farthest from the midline.

This spatial organization of motor neuron pools in the ventral horn provides a framework for understanding the control of the body's musculature in posture

(A)

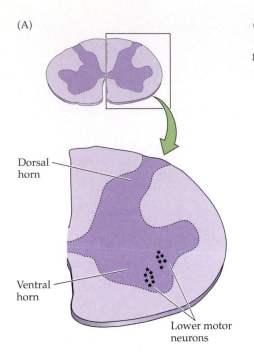

Dorsal horn

Ventral horn

Lower motor neurons

Figure 16.2 Spatial distribution of lower motor neurons in the ventral horn of the spinal cord demonstrated by labeling their cell bodies following injection of a retrograde tracer in individual muscles. Neurons were identified by placing a retrograde tracer into the medial gastrocnemius (or soleus muscle) of the cat. (A) Transverse section through the lumbar level of the spinal cord showing the distribution of labeled cell bodies. Lower motor neurons form distinct clusters (motor pools) in the ipsilateral ventral horn. Spinal cord cross sections (B) and a reconstruction seen from the dorsal surface (C) illustrate the distribution of motor neurons innervating individual skeletal muscles in both axes of the cord. The cylindrical shape and distinct distribution of different pools are especially evident in the dorsal view of the reconstructed cord. The dashed lines in (C) represent individual lumbar and sacral spinal cord segments. (After Burke et al., 1977.)

(B)

Medial gastrocnemius injection

Soleus injection

(C)

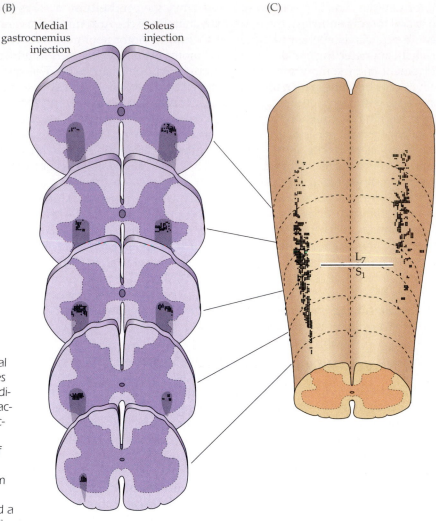

L_7
S_1

and movement, as well as how descending projections of upper motor neurons and intersegmental spinal cord circuits are organized to modulate motor behavior. Thus, medial motor neuronal pools that govern postural control and the maintenance of balance receive input from upper motor neurons via long projection systems that run in the medial and anterior (ventral) white matter of the spinal cord. The more lateral motor

Figure 16.3 Somatotopic organization of lower motor neurons in a cross section of the ventral horn at the cervical level of the spinal cord. Motor neurons innervating axial musculature are located medially, whereas those innervating the distal musculature are located more laterally.

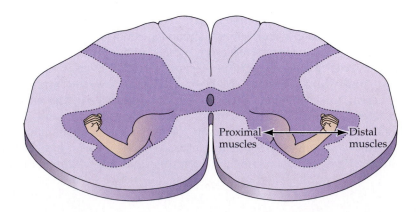

Proximal muscles

Distal muscles

neuronal pools that innervate the distal extremities are often concerned with the execution of skilled behavior, especially the lateral motor neurons of the cervical enlargement that innervate muscles of the forearm and hand in primates. These laterally placed lower motor neurons are governed by projections from motor divisions of the cerebral cortex that, in primates, run through the lateral white matter of the spinal cord. This same somatotopic plan is reflected in the connections made by intrinsic spinal cord circuits that interconnect neurons in the longitudinal axis of the spinal cord (Figure 16.4). Thus, the patterns of connections made by such interneurons in the medial region of the intermediate zone are different from the patterns made by those in the lateral region, and these differences are related to their respective functions. The medial interneurons, which supply the lower motor neurons in the medial ventral horn, have axons that project to many spinal cord segments. Indeed, some projections run between the cervical and lumbar enlargements and participate in the coordination of rhythmic movements of the upper and lower limbs (see the upcoming section on "Spinal Cord Circuitry and Locomotion"), while other axons terminate along the entire length of the cord. Moreover, many of these neurons have axonal branches that cross the midline in the commissure of the spinal cord to innervate lower motor neurons in the medial part of the contralateral hemicord. This arrangement ensures that groups of axial muscles on both sides of the body act in concert to maintain and adjust posture. In contrast, local circuit neurons in the lateral region of the intermediate zone have shorter axons that typically extend fewer than five segments and are predominantly ipsilateral. This more restricted pattern of connectivity provides the finer and more differentiated control that is exerted over the muscles of the distal extremities, such as that required for the independent movement of individual fingers during manipulative tasks.

Two types of lower motor neuron are found in the motor neuronal pools of the ventral horn. Small **γ motor neurons** innervate specialized muscle fibers that, in combination with the nerve fibers that innervate them, are actually sensory receptors called muscle spindles (see Chapter 9). The muscle spindles are embedded within connective tissue capsules in the muscle and are thus referred to as intrafusal muscle fibers (*fusal* means capsular). The intrafusal muscle fibers are also innervated by sensory axons that send information to the spinal cord and brainstem about the length of the muscle. The function of the γ motor neurons is to regulate this sensory input by setting the intrafusal muscle fibers to an appropriate length (see the next section). The second type of lower motor neuron, called **α motor neurons**, innervates the extrafusal muscle fibers, which are the striated muscle fibers that actually generate the forces needed for posture and movement.

Although the following discussion focuses on the lower motor neurons in the spinal cord, comparable sets of motor neurons responsible for the control of muscles in the head and neck are located in the brainstem. The latter neurons are distributed in the eight somatic and branchial motor nuclei of the cranial nerves in the medulla, pons, and midbrain (see the Appendix). Somewhat confusingly, but quite appropriately, these motor neurons in the brainstem are also called lower motor neurons.

The Motor Unit

Most extrafusal skeletal muscle fibers in mature mammals are innervated by only a single α motor neuron (developing muscle fibers are innervated by several α motor neurons; see Chapter 23). Since there are by far more muscle fibers than motor neurons, individual motor axons branch within muscles to synapse on many different fibers that are typically distributed over a relatively wide area

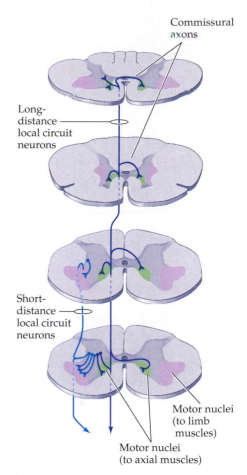

Figure 16.4 Local circuit neurons that supply the medial region of the ventral horn are situated medially in the intermediate zone of the spinal cord gray matter and have axons that extend over a number of spinal cord segments and terminate bilaterally. In contrast, local circuit neurons that supply the lateral parts of the ventral horn are located more laterally, have axons that extend over a few spinal cord segments, and terminate only on the same side of the cord. Pathways that contact the medial parts of the spinal cord gray matter are involved primarily in the control of posture; those that contact the lateral parts are involved in the fine control of the distal extremities.

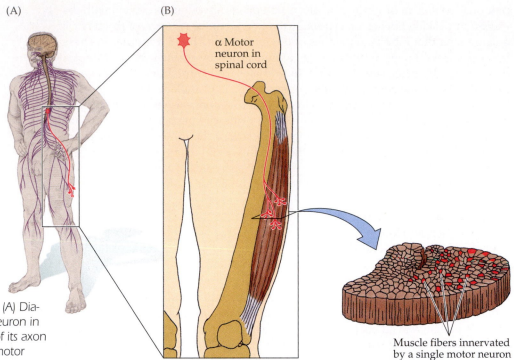

(A)   (B)

α Motor
neuron in
spinal cord

Muscle fibers innervated
by a single motor neuron

Figure 16.5   The motor unit. (A) Diagram showing a lower motor neuron in the spinal cord and the course of its axon to its target muscle. (B) Each α motor neuron synapses with multiple fibers within the muscle. The α motor neuron and the muscle fibers it contacts define the motor unit. Cross section through the muscle shows the relatively diffuse distribution of muscle fibers (red) contacted by a single a motor neuron.

within the muscle, presumably to ensure that the contractile force is spread evenly (Figure 16.5). In addition, this arrangement reduces the chance that damage to one or a few α motor neurons will significantly alter a muscle's action. Because an action potential generated by a motor neuron normally brings to threshold all of the muscle fibers it contacts, a single α motor neuron and its associated muscle fibers together constitute the smallest unit of force that can be activated to produce movement. Sherrington was again the first to recognize this fundamental relationship between an α motor neuron and the muscle fibers it innervates, for which he coined the term **motor unit**.

Both motor units and the α motor neurons themselves vary in size. Small α motor neurons innervate relatively few muscle fibers to form motor units that generate small forces, whereas large motor neurons innervate larger, more powerful motor units. Motor units also differ in the types of muscle fibers that they innervate. In most skeletal muscles, the smaller motor units comprise small "red" muscle fibers that contract slowly and generate relatively small forces; but, because of their rich myoglobin content, plentiful mitochondria, and rich capillary beds, such small red fibers are resistant to fatigue (these units are also innervated by relatively small α motor neurons). These small units are called **slow (S) motor units** and are especially important for activities that require sustained muscular contraction, such as the maintenance of an upright posture. Larger α motor neurons innervate larger, pale muscle fibers that generate more force; however, these fibers have sparse mitochondria and are therefore easily fatigued. These units are called **fast fatigable (FF) motor units** and are especially important for brief exertions that require large forces, such as running or jumping. A third class of motor units has properties that lie between those of the other two. These **fast fatigue-resistant (FR) motor units** are of intermediate size and are not quite as fast as FF motor units. They generate about twice the force of a slow motor unit and, as the name implies, are resistant to fatigue (Figure 16.6).

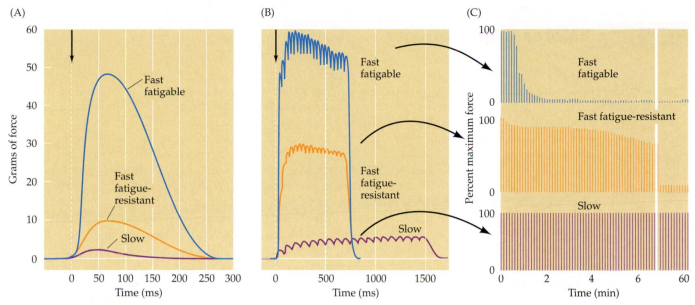

Figure 16.6 Comparison of the force and fatigability of the three different types of motor units. In each case, the response reflects stimulation of a single α motor neuron. (A) Change in muscle tension in response to a single action potential. (B) Tension in response to repetitive stimulation of each type of motor unit. (C) Response to repeated stimulation at a level that initially evokes maximum tension. The ordinate represents the force generated by each stimulus. Note the strikingly different fatigue rates. (After Burke et al., 1974.)

These distinctions among different types of motor units indicate how the nervous system produces movements appropriate for different circumstances. In most muscles, small, S motor units have lower thresholds for activation than the larger units and are tonically active during motor acts that require sustained effort (standing, for instance). The thresholds for the large, fast motor units are reached only when rapid movements requiring great force are made, such as jumping.

The functional distinctions between the various classes of motor units also explain some structural differences among muscle groups. For example, a motor unit in the soleus (a muscle important for posture that comprises mostly small motor units) has an average innervation ratio of 180 muscle fibers for each motor neuron. In contrast, the gastrocnemius, a muscle that comprises both small and larger motor units, has an innervation ratio of ~1000–2000 muscle fibers per motor neuron, and can generate forces needed for sudden changes in body position. Other differences are related to the highly specialized functions of particular muscles. For instance, the eyes require rapid, precise movements but little strength; in consequence, extraocular muscle motor units are extremely small (with an average innervation ratio of only 3!) and have a very high proportion of muscle fibers capable of contracting with maximal velocity. More subtle variations are present in athletes on different training regimens; indeed, both the myofibril and neuronal properties of motor units are subject to use-dependent plasticity. This potential for change, in part, underlies neuromuscular adaptations to physical exercise (Box 16A). Thus, muscle biopsies show that sprinters have a larger proportion of powerful, but rapidly fatiguing, pale fibers in their leg muscles than do marathoners.

The Regulation of Muscle Force

Increasing or decreasing the number of motor units active at any one time changes the amount of force produced by a muscle. In the 1960s, Elwood Henneman and his colleagues at Harvard Medical School found that progressive increases in muscle tension could be produced by progressively increasing the activity of axons that provide input to the relevant pool of lower motor neurons.

BOX 16A Motor Unit Plasticity

Organisms with complex nervous systems demonstrate an astounding ability to acquire new motor skills and modify the strength and endurance of motor behavior. The neural basis of these abilities depends heavily on the operations of supraspinal motor centers (i.e., neural centers above the spinal cord) whose functions in volitional motor behavior and motor learning are described in Chapters 17–19. But what role—if any—do the motor units themselves play in the functional changes that underlie such abilities? Are motor units subject to use-dependent plasticity and, if so, to what extent are the anatomical and physiological properties of motor units changeable? To address these questions, it is necessary to consider in more detail the range of phenotypes expressed by motor units.

When considering the structure and function of skeletal muscle, it is convenient to classify the constituent motor units into one of three categories: slow (S), fast fatigable (FF), or fast fatigue-resistant (FR) (see Figure 16.6). However, with increasingly sophisticated means of characterizing the intrinsic architecture, biochemistry, and physiology of muscle fibers, it has become clear that most skeletal muscles possess a broad spectrum of fiber phenotypes that vary in speed of contraction, tension

generation, oxidative capacity, and endurance. These variations among muscle fibers combine with corresponding variations in the morphological and biophysical properties of α motor neurons to determine the size and physiological function of motor units (Figure A). Thus the features of α motor neurons that serve small motor units explain why such neurons are easily depolarized to firing threshold, but typically maintain only slower and steady rates of firing—properties that are well suited for the control of slow muscle fibers that mediate postural stability. On the other hand, α motor neurons that serve large motor units are more difficult to depolarize to threshold but are capable of achieving high rates of firing upon activation—properties consistent with the force-generating potential of FF muscle fibers that are recruited for production of maximal tension. (Not surprisingly, muscle fibers that are intermediate in their functional properties are supplied by α motor neurons whose phenotype is midway between these extremes.)

An early answer to the question of motor unit plasticity came from a classic series of "cross-innervation" experiments performed by the great Australian physiologist J. C. Eccles and his colleagues, most notably A. J. Buller. The results demonstrated that the physio-

logical properties of slow and fast muscle fibers could be reversed when the innervation to these fibers was surgically altered so that slow muscle fibers were innervated by a nerve that normally supplies fast fibers, and vice versa. Subsequent studies by other investigators demonstrated that the actual pattern of neural activity in a motor nerve, in addition to—or perhaps instead of—the identity of the innervating motor neurons themselves, provides an instructive signal that can influence the expression of muscle fiber phenotype. For example, chronic electrical nerve stimulation transforms the metabolic and contractile properties of FF fibers to those consistent with S fibers (Figure B). Corresponding changes were also observed in the biophysical properties of the α motor neurons whose axons were stimulated. Although the effects were more subtle, stimulated α motor neurons were modified toward slow, fatigue-resistant motor units, with increased excitability, lengthened after-hyperpolarizations, and short-term depression of EPSP amplitudes following high-frequency activation.

It is much more difficult to control and interpret studies of exercising organisms; nevertheless, the same general principles of motor unit plasticity that were derived from nerve stimulation studies apply to neuromuscular adaptation in more naturalistic contexts, including resistance and endurance training. Thus, the nature and degree of adaptation following exercise is a function of the tensions exerted by muscle fibers and the duration of increased muscle activity. Most commonly, exercise regimes can "slow" the contractile properties of motor units while increasing the endurance and strength of muscle fibers. However, the impact of exercise is distributed proportionately to motor units in

WITH INCREASED MOTOR UNIT SIZE, α MOTOR NEURONS EXHIBIT:	
INCREASED Cell body size Dendritic complexity Short-term EPSP potentiation with repeated activation Axonal diameter (i.e., faster conduction) Number of axonal branches (i.e., more muscle fibers innervated)	**DECREASED** Input resistance Excitability Ia EPSP amplitude PSP decay constant Duration of after-hyperpolarization

(A) Morphological and biophysical properties of α motor neurons that scale proportionally with the size of motor units.

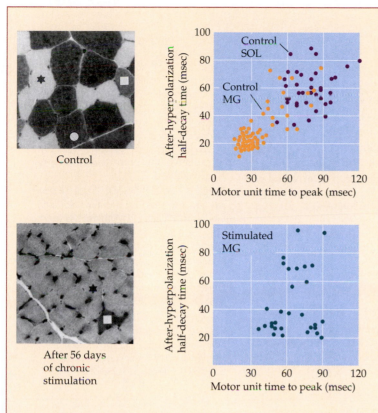

(B) (Left) Photomicrographs of muscle fibers in cat medial gastrocnemius (MG) stained to demonstrate the presence of myosin ATPase activity in alkaline conditions. In control muscle, fast fatigable fibers (circle) and fast fatique-resistant fibers (square) stain darkly, but slow oxidative fibers (star) stain very lightly. Following 56 days of chronic electrical nerve stimulation, nearly all fibers acquired the histochemical phenotype of slow oxidative fibers. (Right) The electrophysiological properties of the α motor neurons supplying the stimulated nerve also shifted toward those more characteristic of the slower motor units of the soleus muscle (SOL). The upper scatterplot shows control data in which the faster motor units of the medial gastrocnemius are differentiated from the slower motor units of the soleus muscle by shorter neuronal after-hyperpolarizations and time-to-peak tension in the supplied muscle fibers. The lower graph shows the impact of chronic stimulation, which shifts the properties of MG motor neurons toward those seen in SOL motor neurons. (Micrographs from Gordon et al., 1997; plots from Munson et al. 1997.)

order of their recruitment during training activities, with S motor units being affected most at low exertion levels, and FR and FF motor units being affected only if recruited by higher intensities of exercise.

Interestingly, neural contributions to exercise-induced changes in performance are not limited to alterations of motor unit phenotype. Indeed, the increase in strength achieved in the early phases of resistance training often exceeds what can be attributable to changes in the structure and function of muscle fibers, implying the operation of spinal and/or supraspinal neural mechanisms that mediate increased motor function. At the motor unit level, these neural adaptations include an increase in the

(Continued on next page)

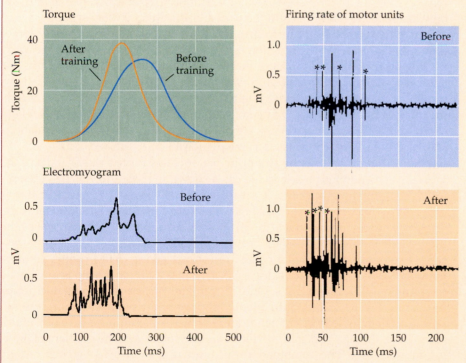

(C) (*Left*) Comparison of torque and electromyogram (EMG) activity during ballistic contractions of ankle dorsiflexor muscles in humans before and after dynamic training. Note the increased rate of tension development after training and the accompanying increase in rectified, surface EMG activity in the early phase of contraction. (*Right*) These changes are associated with an increase in the instantaneous firing rate of motor units recorded from intramuscular electrodes; asterisks mark repetitive discharges of the same motor unit. (After Van Cutsem et al., 1998.)

instantaneous discharge rate and a marked decrease in interspike interval at the onset of contraction, which facilitate the rapid generation of tension (Figure C). Furthermore, studies of unilateral exercise (e.g., training one arm and not the other) have shown appreciable gains in the non-exercised limb, indicating the recruitment and adaptation of central neural circuits that have access to contralateral motor units. There are even documented gains in muscle strength with *imagined* exercise—a provocative result that may eventually have profound implications for athletic training and rehabilitation science.

There is still much to be learned about how motor units respond to use alterations in strength and endurance training, and scientists are only beginning to probe the neurobiological and neuromuscular mechanisms that underlie skill acquisition. Pursuit of these aims will surely lead to a better understanding of how to maximize motor performance in exercising human (and non-human) subjects, as well as in the rehabilitation of patients coping with neurological or neuromuscular impairments and physical disability.

References

BULLER, A. J., J. C. ECCLES AND R. M. ECCLES (1960a) Differentiation of fast and slow muscles in the cat hind limb. *J. Physiol.* 150: 399–416.

BULLER, A. J., J. C. ECCLES AND R. M. ECCLES (1960b) Interactions between motoneurones and muscles in respect of the characteristic speeds of their responses. *J. Physiol.* 150: 417–439.

CLOSE, R. (1965) Effects of cross-union of motor nerves to fast and slow skeletal muscles. *Nature* 206: 831–832.

GORDON, T., N. TYREMAN, V. F. RAFUSE AND J. B. MUNSON (1997) Fast-to-slow conversion following chronic low-frequency activation of medial gastrocnemius muscle in cats. I. Muscle and motor unit properties. *J. Neurophysiol.* 77: 2585–2604.

LIEBER, R. L. (2002) *Skeletal Muscle Structure, Function, and Plasticity*, 2nd Ed. Baltimore: Lippincott Williams & Wilkins.

MUNSON, J. B., R. C. FOEHRING, L. M. MENDELL AND T. GORDON (1997) Fast-to-slow conversion following chronic low-frequency activation of medial gastrocnemius muscle in cats. II. Motoneuron properties. *J. Neurophysiol.* 77: 2605–2615.

VAN CUTSEM, M., J. DUCHATEAU AND K. HAINAUT (1998) Changes in single motor unit behaviour contribute to the increase in contraction speed after dynamic training in humans. *J. Physiol.* 513: 295–305.

This gradual increase in tension results from the recruitment of motor units in a fixed order, according to their size. By stimulating either sensory nerves or upper motor pathways that project to a lower motor neuron pool while measuring the tension changes in the muscle, Henneman found that, in experimental animals, only the smallest motor units in the pool are activated by weak synaptic stimulation. When synaptic input to a motor pool increases, progressively larger motor units that generate larger forces, are recruited. Thus, as the synaptic activity driving a motor neuron pool increases, low threshold S motor units are recruited first, then FR motor units, and finally, at the highest levels of activity, the FF motor units. Since these original experiments, evidence for the orderly recruitment of motor units has been found in a variety of voluntary and reflexive movements, including exercise activities. As a result, this systematic relationship has come to be known as the **size principle**.

An illustration of how the size principle operates for the motor units of the medial gastrocnemius muscle in the cat is shown in Figure 16.7. When the animal is standing quietly, the force measured directly from the muscle tendon is only a small fraction (about 5%) of the total force that the muscle can generate. The force is provided by the S motor units, which make up about 25% of the motor units in this muscle. When the cat begins to walk, larger forces are necessary. Locomotor activities that range from slow walking to fast running require up to 25% of the muscle's total force capacity. This additional need is met by the recruitment of FR motor units. Only movements such as galloping and jumping, which are performed infrequently and for short periods, require the full power of the muscle; such demands are met by the recruitment of the FF motor units. Thus, the size principle provides a simple solution to the problem of grading muscle force. The combination of motor units activated by such orderly

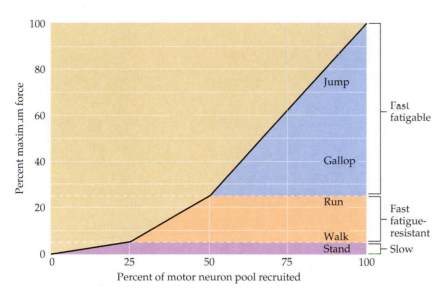

Figure 16.7 The recruitment of motor neurons in the cat medial gastrocnemius muscle under different behavioral conditions. Slow (S) motor units provide the tension required for standing. Fast fatigue-resistant (FR) motor units provide the additional force needed for walking and running. Fast fatigable (FF) motor units are recruited for the most strenuous activities, such as jumping. (After Walmsley et al., 1978.)

recruitment optimally matches the physiological properties of different motor unit types with the range of forces required to perform different motor tasks.

The frequency of the action potentials generated by motor neurons also contributes to the regulation of muscle tension. The increase in force that occurs with increased firing rate reflects the summation of successive muscle contractions. The muscle fibers are activated by the next action potential before they have time to completely relax, and the forces generated by the temporally overlapping contractions are summed (Figure 16.8). The lowest firing rates during a voluntary movement are on the order of 8 per second (Figure 16.9). As the firing rate of individual units rises to a maximum of about 20–25 per second in the muscle being studied here, the amount of force produced increases.

At the highest firing rates, individual muscle fibers are in a state of "fused tetanus"—that is, the tension produced in individual motor units no longer has peaks and troughs that correspond to the individual twitches evoked by the motor neuron's action potentials. Under normal conditions, the maximum firing rate of motor neurons is less than that required for fused tetanus (see Figure 16.9). However, the asynchronous firing of different lower motor neurons provides a steady level of input to the muscle, which causes the contraction of a relatively constant number of motor units, and averages out the changes in tension due to contractions and relaxations of individual motor units. All this allows the resulting movements to be executed smoothly.

Figure 16.8 The effect of stimulation rate on muscle tension. (A) At low frequencies of stimulation, each action potential in the motor neuron results in a single twitch of the related muscle fibers. (B) At higher frequencies, the twitches sum to produce a force greater than that produced by single twitches. (C) At a still higher frequency of stimulation, the force produced is greater, but individual twitches are still apparent. This response is referred to as unfused tetanus. (D) At the highest rates of motor neuron activation, individual twitches are no longer apparent (a condition called fused tetanus).

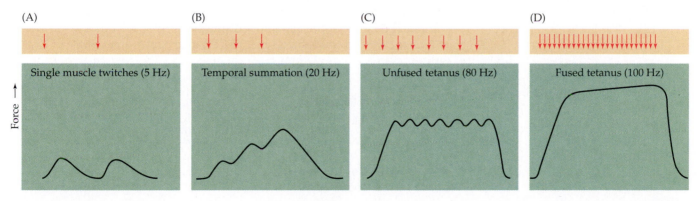

Figure 16.9 Motor units (represented by the lines between the dots) recorded transcutaneously in a muscle of the human hand as the amount of voluntary force produced is progressively increased. The lowest threshold motor units generate the least amount of voluntary force and are recruited first. As the subject generates more and more force, both the number and the rate of firing of the active motor units increase; note that all motor units fire initially at about 8 Hz. (After Monster and Chan, 1977.)

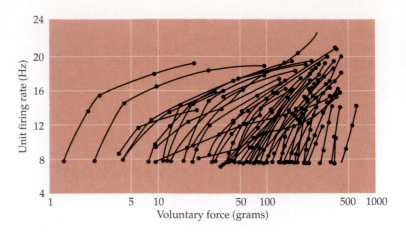

The Spinal Cord Circuitry Underlying Muscle Stretch Reflexes

The local circuitry within the spinal cord mediates a number of sensory motor reflex actions. The simplest of these reflex arcs entails a sensory response to muscle stretch, which provides direct excitatory feedback to the motor neurons innervating the muscle that has been stretched (Figure 16.10). As already mentioned, the sensory signal for the **stretch reflex** originates in **muscle spindles**, the sensory receptors embedded within most muscles. The spindles comprise eight to ten intrafusal fibers arranged in parallel with the extrafusal fibers that make up the bulk of the muscle (Figure 16.10A).

As described in Chapter 9, there are two structural and functional classes of intrafusal fibers: the nuclear bag fibers and the nuclear chain fibers. These fibers differ in the position of their nuclei (giving rise to their nomenclature), the intrinsic architecture of their myofibrils, and their dynamic sensitivity to stretch. Large-diameter sensory axons (group Ia and group II afferents; see Table 9.1) are coiled around the central part of the intrafusal fibers. These afferents are the largest axons in peripheral nerves and, because action potential conduction velocity is a direct function of axon diameter (see Chapters 2 and 3), they mediate very rapid reflex adjustments when the muscle is stretched. The stretch imposed on the muscle deforms the intrafusal muscle fibers, which in turn initiates action potentials by activating mechanically gated ion channels in the afferent axons coiled around the spindle.

Group Ia afferents, which preferentially innervate nuclear bag fibers, respond phasically to small stretches, while group II afferents, which innervate both fiber types, signal the level of sustained stretch by firing tonically in proportion to the degree of stretch. The centrally projecting branch of the sensory neuron forms monosynaptic excitatory connections with the α motor neurons in the ventral horn of the spinal cord that innervate the same (homonymous) muscle and, via local circuit neurons, forms inhibitory connections with the α motor neurons of antagonistic (heteronymous) muscles. This arrangement is an example of what is

Figure 16.10 Stretch reflex circuitry. (A) Diagram of a muscle spindle, the sensory ▶ receptor that initiates the stretch reflex. (B) Stretching a muscle spindle leads to increased activity in Ia afferents and an increase in the activity of α motor neurons that innervate the same muscle. Ia afferents also excite the motor neurons that innervate synergistic muscles, and inhibit indirectly the motor neurons that innervate antagonists (see Figures 1.7–1.9). (C) The stretch reflex operates as a negative feedback loop to regulate muscle length.

(A) Muscle spindle

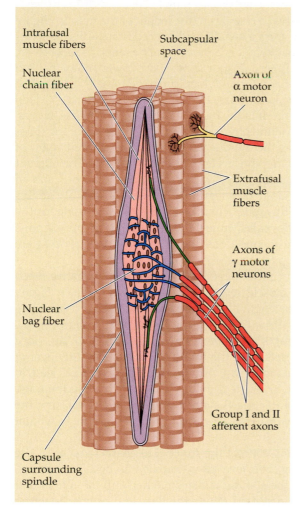

Intrafusal muscle fibers

Subcapsular space

Nuclear chain fiber

Axon of α motor neuron

Extrafusal muscle fibers

Axons of γ motor neurons

Nuclear bag fiber

Group I and II afferent axons

Capsule surrounding spindle

(B)

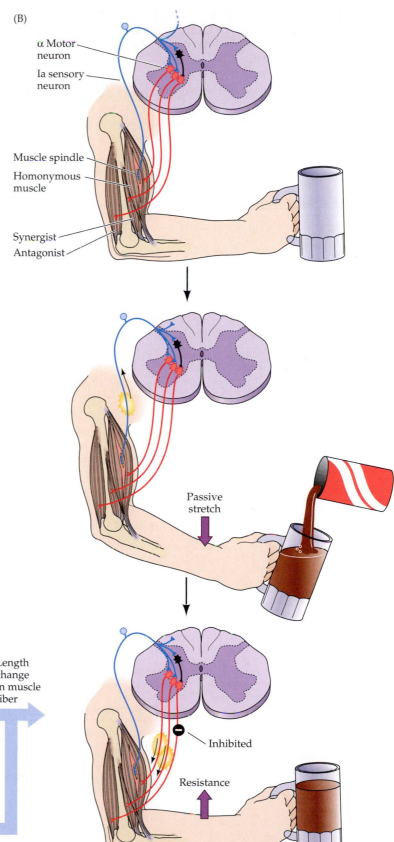

α Motor neuron

Ia sensory neuron

Muscle spindle

Homonymous muscle

Synergist

Antagonist

Passive stretch

Inhibited

Resistance

(C)

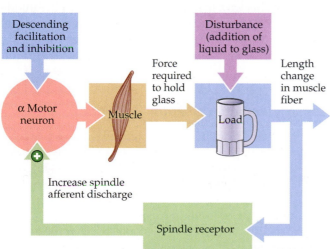

Descending facilitation and inhibition

Disturbance (addition of liquid to glass)

Force required to hold glass

Length change in muscle fiber

α Motor neuron

Muscle

Load

Increase spindle afferent discharge

Spindle receptor

called *reciprocal innervation* and results in rapid contraction of the stretched muscle and simultaneous relaxation of the antagonist muscle. All of this leads to especially rapid and efficient responses to changes in the length of the muscle (Figure 16.10B). The excitatory pathway from a spindle to the α motor neurons innervating the same muscle is unusual in that it is a monosynaptic reflex; in most cases, sensory neurons from the periphery do not contact the lower motor neuron directly but instead exert their effects through local circuit neurons.

This monosynaptic reflex arc is variously referred to as the "stretch," "deep tendon," or "myotatic" reflex, and it is the basis of the knee, ankle, jaw, biceps, or triceps response tested in a routine physical examination. The tap of the reflex hammer on the tendon stretches the muscle, which evokes an afferent volley of activity in the Ia sensory axons that innervate the muscle spindles. The afferent volley is relayed to the α motor neurons in the brainstem or spinal cord, and an efferent volley returns to the muscle (see Figure 1.7). Since muscles are always under some degree of stretch, this reflex circuit is normally responsible for the steady level of tension in muscles called **muscle tone**. Changes in muscle tone occur in a variety of pathological conditions, and it is these changes that are assessed by examination of deep tendon reflexes (see Box 17E).

In terms of engineering principles, the stretch reflex arc is a negative feedback loop used to maintain muscle length at a desired value (see Figure 16.10C). The appropriate muscle length is specified by the activity of descending upper motor neuron pathways that influence the lower motor neuron pool. Deviations from the desired length are detected by the muscle spindles, since increases or decreases in the stretch of the intrafusal fibers alter the level of activity in the sensory axons that innervate the spindles. These changes lead, in turn, to adjustments in the activity of the α motor neurons, returning the muscle to the desired length by contracting the stretched muscle and relaxing the opposing muscle group, and by restoring the level of spindle activity to what it was before.

The smaller γ motor neurons control the functional characteristics of the muscle spindles by modulating their level of excitability. As described earlier, when the muscle is stretched, the spindle is also stretched and the rate of discharge in the afferent fibers is increased. When the muscle shortens, however, the spindle is relieved of tension, or "unloaded," and the sensory axons that innervate the spindle might therefore be expected to fall silent during contraction. However, they remain active. The γ motor neurons terminate on the contractile poles of the intrafusal fibers, and the activation of these neurons causes intrafusal fiber contraction—in this way, maintaining the tension on the middle (or equatorial region) of the intrafusal fibers where the sensory axons terminate. Thus, *co-activation* of the α and γ motor neurons allows spindles to function (i.e., send information centrally) at all muscle lengths during movements and postural adjustments.

The Influence of Sensory Activity on Motor Behavior

The level of γ motor neuron activity is often referred to as γ bias, or *gain*, and can be adjusted by upper motor neuron pathways as well as by local reflex circuitry. The gain of the myotatic reflex refers to the amount of muscle force generated in response to a given stretch of the intrafusal fibers. If the gain of the reflex is high, then a small amount of stretch applied to the intrafusal fibers will produce a large increase in the number of α motor neurons recruited and a large increase in their firing rates; this, in turn, leads to a large increase in the amount of tension produced by the extrafusal fibers. If the gain is low, a greater stretch is required to generate the same amount of tension in the extrafusal muscle fibers.

In fact, the gain of the stretch reflex is continuously adjusted to meet different functional requirements. For example, while standing in a moving bus, the gain of the stretch reflex can be modulated by upper motor neuron pathways to compensate for the variable changes that occur as the bus stops and starts or progresses relatively smoothly. During voluntary stretching, such as when warming up for athletic performance, the gain of myotatic reflexes must be reduced to facilitate the lengthening of muscle fibers and other elastic elements of the musculotendinous system that is desirable under these temporary circumstances. Thus, under the various demands of voluntary (and involuntary) movement, α and γ motor neurons are often co-activated by higher centers to prevent muscle spindles from being unloaded (Figure 16.11).

In addition, the level of γ motor neuron activity can be modulated independently of α motor neuron activity if the context of a movement requires it. In general, the baseline activity level of γ motor neurons is high if a movement is relatively difficult and demands rapid and precise execution. For example, recordings from cat hindlimb muscles show that γ motor neuron activity is high when the animal has to perform a difficult movement, such as walking across a narrow beam. Unpredictable conditions, as when the animal is picked up or handled, also lead to marked increases in γ motor neuron activity and greatly increased spindle responsiveness.

Gamma motor neuron activity, however, is not the only factor that sets the gain of the stretch reflex. The gain also depends on the level of excitability of the α motor neurons that serve as the efferent side of this reflex loop. Thus, in addition to the influence of descending upper motor neuron projections, other local circuits in the spinal cord can change the gain of the stretch reflex by excitation

Figure 16.11   The role of γ motor neuron activity in regulating the responses of muscle spindles. (A) When α motor neurons are stimulated without activation of γ motor neurons, the response of the Ia fiber decreases as the muscle contracts. (B) When both α and γ motor neurons are activated, there is no decrease in Ia firing during muscle shortening. Thus, the γ motor neurons can regulate the gain of muscle spindles so they can operate efficiently at any length of the parent muscle. (After Hunt and Kuffler, 1951.)

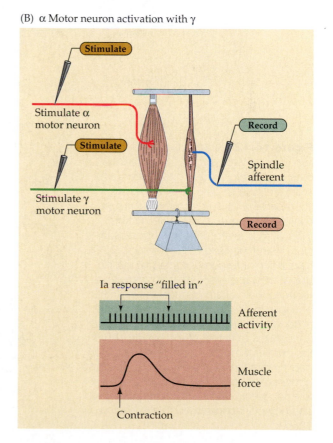

or inhibition of either α or γ motor neurons. There are also inhibitory interneurons that maintain axo-axonal synapses on the terminals of Ia afferents and are thus positioned to suppress the transfer of excitatory drive to motor neurons. The activities of local circuits in the spinal cord are themselves influenced by the projections of upper motor neurons in the brainstem and cerebral cortex, as well as by neuromodulatory systems that originate in the brainstem reticular forma-

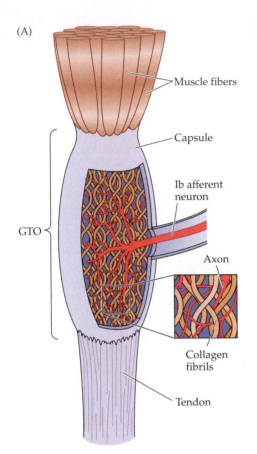

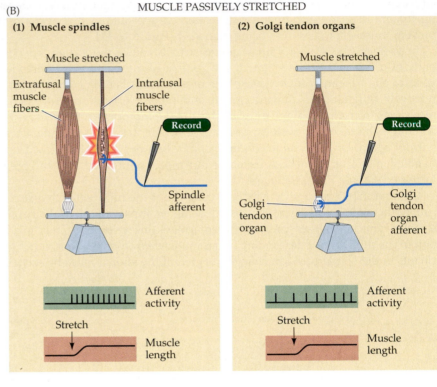

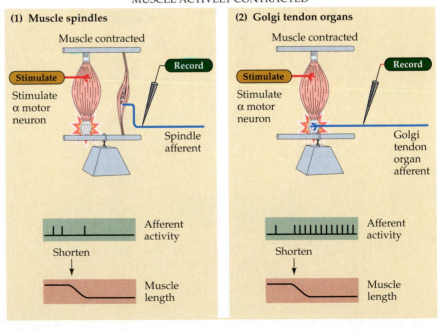

Figure 16.12 Comparison of the function of muscle spindles and Golgi tendon organs. (A) Golgi tendon organs are arranged in series with extrafusal muscle fibers because of their location at the junction of muscle and tendon. (B) The two types of muscle receptors, the muscle spindles (1) and the Golgi tendon organs (2), have different responses to passive muscle stretch (*top*) and active muscle contraction (*bottom*). Both afferents discharge in response to passively stretching the muscle, although the Golgi tendon organ discharge is much less than that of the spindle. When the extrafusal muscle fibers are made to contract by stimulation of their α motor neurons, however, the spindle is unloaded and therefore falls silent, whereas the rate of Golgi tendon organ firing increases. (B after Patton, 1965.)

tion (see Chapter 17). Many of these neuromodulatory projections release biogenic amine neurotransmitters that bind to G-protein-coupled receptors and mediate long-lasting effects on the gain of segmental circuits in the spinal cord.

Other Sensory Feedback That Affects Motor Performance

Another sensory receptor that is important in the reflex regulation of motor unit activity is the **Golgi tendon organ**. Golgi tendon organs are encapsulated afferent nerve endings located at the junction of a muscle and tendon (Figure 16.12A). Each tendon organ is innervated by a single group Ib sensory axon (Ib axons are slightly smaller than the Ia axons that innervate the muscle spindles; see Table 9.1). In contrast to the parallel arrangement of extrafusal muscle fibers and spindles, Golgi tendon organs are in series with the extrafusal muscle fibers. When a muscle is passively stretched, most of the change in length occurs in the muscle fibers, since they are more elastic than the fibrils of the tendon. When a muscle actively contracts, however, the force acts directly on the tendon, leading to an increase in the tension of the collagen fibrils in the tendon organ and compression of the intertwined sensory receptors. As a result, Golgi tendon organs are exquisitely sensitive to increases in muscle *tension* that arise from muscle contraction but, unlike spindles, are relatively insensitive to *passive stretch* (Figure 16.12B).

The Ib axons from Golgi tendon organs contact inhibitory local circuit neurons in the spinal cord (called Ib inhibitory interneurons) that synapse, in turn, with the α motor neurons that innervate the same muscle. The Golgi tendon circuit is thus a negative feedback system that regulates muscle tension; it decreases the activation of a muscle when exceptionally large forces are generated and, in this way, it protects the muscle. This reflex circuit also operates at reduced levels of muscle force, counteracting small changes in muscle tension by increasing or decreasing the inhibition of α motor neurons. Under these conditions, the Golgi tendon system tends to maintain a steady level of force, counteracting effects that diminish muscle force (such as fatigue). In short, the muscle spindle system is a feedback system that monitors and maintains muscle *length*, and the Golgi tendon system is a feedback system that monitors and maintains muscle *force*.

Like the muscle spindle system, the Golgi tendon organ system is not a closed loop. The Ib inhibitory interneurons also receive synaptic inputs from a variety of other sources, including upper motor neurons, cutaneous receptors, muscle spindles, and joint receptors, with the latter comprising several types of receptors resembling Ruffini's and Pacinian corpuscles located in joint capsules. Joint receptors signal hyperextension or hyperflexion of the joint, thereby contributing to the protective functions mediated by Ib inhibitory interneurons (Figure 16.13). Acting in concert, these diverse inputs regulate the responsiveness of Ib interneurons to activity arising in Golgi tendon organs.

Flexion Reflex Pathways

So far, this discussion has focused on reflexes driven by sensory receptors located within muscles or tendons. Other reflex circuitry mediates the withdrawal of a limb from a painful stimulus, such as a pinprick or the heat of a flame. Contrary to what might be imagined given the speed with which we are able to withdraw from a painful stimulus, this **flexion reflex** involves slowly conducting afferent axons and several synaptic links (Figure 16.14). As a result of activity in this circuitry, stimulation of nociceptive sensory fibers leads to withdrawal of the limb from the source of pain by excitation of ipsilateral flexor

Figure 16.13   Negative feedback regulation of muscle tension by Golgi tendon organs. The Ib afferents from tendon organs contact inhibitory interneurons that decrease the activity of α motor neurons innervating the same muscle. The Ib inhibitory interneurons also receive input from other sensory fibers (not illustrated), as well as from descending pathways. This arrangement prevents muscles from generating excessive tension.

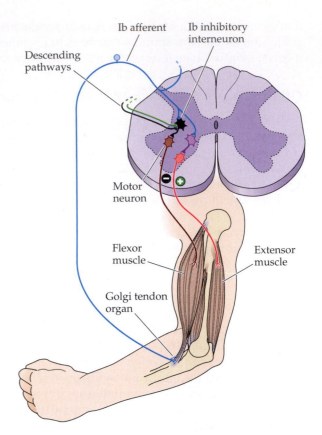

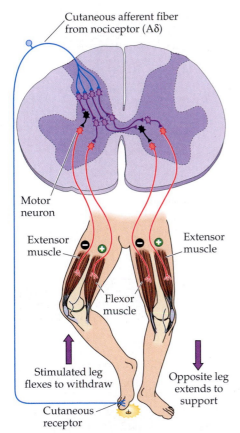

Figure 16.14   Spinal cord circuitry responsible for the flexion-crossed extension reflex. Stimulation of cutaneous receptors in the foot (by stepping on a tack, in this example) leads to activation of spinal cord local circuits that serve to withdraw (flex) the stimulated extremity and extend the other extremity to provide compensatory support.

muscles and reciprocal inhibition of ipsilateral extensor muscles. Flexion of the stimulated limb is also accompanied by an opposite reaction in the contralateral limb (i.e., the contralateral extensor muscles are excited while flexor muscles are inhibited). This **crossed extension reflex** provides postural support during withdrawal of the affected limb from the painful stimulus.

Like the other reflex pathways, local circuit neurons in the flexion reflex pathway receive converging inputs from several different sources, including other spinal cord interneurons and upper motor neuron pathways. Although the functional significance of this complex pattern of connectivity is unclear, changes in the character of the reflex, following damage to descending pathways, provide some insight. Under normal conditions, a noxious stimulus is required to evoke the flexion reflex; following damage to descending pathways, however, other types of stimulation, such as squeezing a limb, can sometimes produce the same response. Alternatively, under some conditions, descending pathways can suppress the reflex withdrawal from a painful stimulus. These observations suggest that the descending projections to the spinal cord modulate the responsiveness of the local circuitry to a variety of sensory inputs.

Spinal Cord Circuitry and Locomotion

The contribution of local circuitry to motor control is not, of course, limited to reflexive responses to sensory inputs. Studies of rhythmic movements, such as locomotion and swimming in animal models (Box 16B), have demonstrated that local circuits in the spinal cord, called **central pattern generators**, are fully capable of controlling the timing and coordination of such complex patterns of movement, and of adjusting them in response to altered circumstances (Box 16C).

A good example is locomotion (walking, running, etc.). The movement of a single limb during locomotion can be thought of as a cycle consisting of two phases: a *stance phase*, during which the limb is extended and placed in contact with the ground to propel humans or other bipeds forward; and a *swing phase*, during which the limb is flexed to leave the ground and then brought forward to begin the next stance phase (Figure 16.15A). Increases in the speed of locomotion reduce the amount of time it takes to complete a cycle, and most of the change in cycle time is due to shortening of the stance phase; the swing phase remains relatively constant over a wide range of locomotor speeds.

In quadrupeds, changes in locomotor speed are also accompanied by changes in the sequence of limb movements. At low speeds, for example, there is a back-

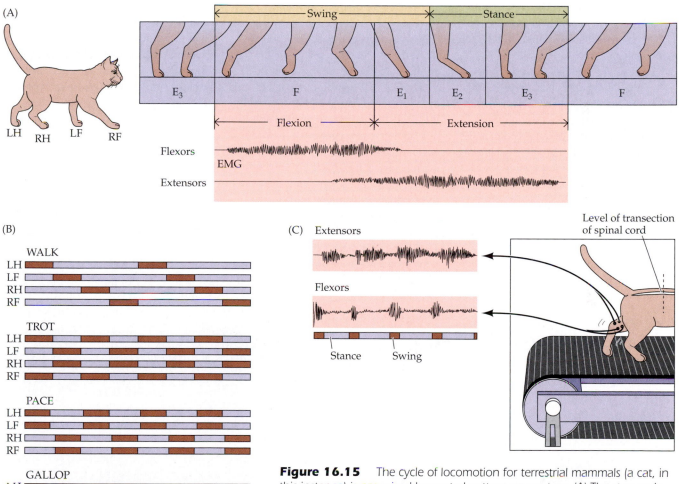

Figure 16.15 The cycle of locomotion for terrestrial mammals (a cat, in this instance) is organized by central pattern generators. (A) The step cycle, showing leg flexion and extension and their relation to the swing and stance phases of locomotion. EMG indicates electromyographic recordings. (B) Comparison of the stepping movements for different gaits. Brown bars, foot lifted (swing phase); gray bars, foot planted (stance phase). (C) Transection of the spinal cord at the thoracic level isolates the hindlimb segments of the cord. The hindlimbs are still able to walk on a treadmill after recovery from surgery, and reciprocal bursts of electrical activity can be recorded from flexors during the swing phase and from extensors during the stance phase of walking. (After Pearson, 1976.)

BOX 16B Locomotion in the Leech and the Lamprey

All animals must coordinate body movements so they can navigate successfully in their environment. All vertebrates, including mammals, use local circuits in the spinal cord (central pattern generators) to control the coordinated movements associated with locomotion. The cellular basis of organized locomotor activity, however, has been most thoroughly studied in an invertebrate, the leech, and in a simple vertebrate, the lamprey.

Both the leech and the lamprey lack peripheral appendages for locomotion possessed by many vertebrates (limbs, flippers, fins, or their equivalent). Furthermore, their bodies comprise repeating muscle segments (as well as repeating skeletal elements in the lamprey). Thus, in order to move through the water, both animals must coordinate the movement of each segment. They do this by orchestrating a sinusoidal displacement of each body segment in sequence, so that the animal is propelled forward through the water.

The leech is particularly well-suited for studying the circuit basis of coordinated movement. The nervous system in the leech consists of a series of interconnected segmental ganglia, each with motor neurons that innervate the corresponding segmental muscles (Figure A). These segmental ganglia facilitate electrophysiological studies, because there is a limited number of neurons in each, and each neuron has a distinct identity. The neurons can thus be recognized and studied from animal to animal, and their electrical activity correlated with the sinusoidal swimming movements.

A central pattern generator circuit coordinates this undulating motion. In the leech, the relevant neural circuit is an ensemble of sensory neurons, interneurons, and motor neurons repeated in each segmental ganglion that controls the local sequence of contraction and relaxation in each segment of the body wall musculature (Figure B). The sensory neurons detect the stretching and contraction of the body wall associated with the sequential swimming movements. Dorsal and ventral motor neurons in the circuit provide innervation to dorsal and ventral muscles, whose phasic contractions propel the leech forward. Sensory information and motor neuron signals are coordinated by interneurons that fire rhythmically, setting up phasic patterns of activity in the dorsal and ventral cells that lead to sinusoidal movement. The intrinsic swimming rhythm is established by a variety of membrane conductances that mediate periodic bursts of suprathreshold action potentials followed by well-defined periods of hyperpolarization.

The lamprey, one of the simplest vertebrates, is distinguished by its clearly segmented musculature and by its lack of bilateral fins or other appendages. In order to move through the water, the lamprey contracts and relaxes each mus-

(A) The leech propels itself through the water by sequential contraction and relaxation of the body wall musculature of each segment. The segmental ganglia in the ventral midline coordinate swimming, with each ganglion containing a population of identified neurons. (B) Electrical recordings from the ventral (EMG_V) and dorsal (EMG_D) muscles in the leech and the corresponding motor neurons show a reciprocal pattern of excitation for the dorsal and ventral muscles of a given segment.

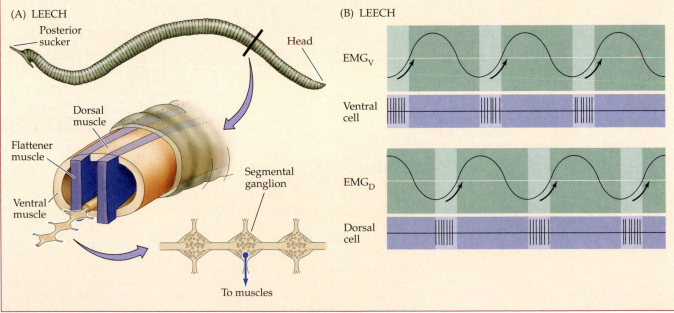

(A) LEECH

Posterior sucker

Head

Dorsal muscle

Flattener muscle

Ventral muscle

Segmental ganglion

To muscles

(B) LEECH

EMG_V

Ventral cell

EMG_D

Dorsal cell

cle segment in sequence (Figure C), which produces a sinusoidal motion, much like that of the leech. Again, a central pattern generator coordinates this sinusoidal movement.

Unlike the leech with its segmental ganglia, the lamprey has a continuous spinal cord that innervates its muscle segments. The lamprey spinal cord is simpler than that of other vertebrates, and several classes of identified neurons occupy stereotyped positions. This orderly arrangement again facilitates the identification and analysis of neurons that constitute the central pattern generator circuit.

In the lamprey spinal cord, the intrinsic firing pattern of a set of interconnected sensory neurons, interneurons and motor neurons, establishes the pattern of undulating muscle contractions that underlie swimming (Figure D). The patterns of connectivity between neurons, the neurotransmitters used by each class of cell, and the physiological properties of the elements in the lamprey pattern generator, are now known. Different neurons in the circuit fire with distinct rhythmicity, thus controlling specific aspects of the swim cycle (Figure E). Particularly important are reciprocal inhibitory connections across the midline that coordinate the pattern-generating circuitry on each side of the spinal cord. This circuitry in the lamprey thus provides a basis for understanding the circuits that control locomotion in more complex vertebrates.

These observations on pattern-generating circuits for locomotion in relatively simple animals have stimulated parallel studies of terrestrial mammals in which central pattern generators in the spinal cord also coordinate locomotion. Although different in detail, terrestrial locomotion ultimately relies on sequential movements similar to those that propel the leech and the lamprey through aquatic environments, as well as intrinsic physiological properties of spinal cord neurons that establish rythmicity for coordinated movement.

References

GRILLNER, S., D. PARKER AND A. EL MANIRA (1998) Vertebrate locomotion: A lamprey perspective. *Ann. N.Y. Acad. Sci.* 860: 1–18.

KRISTAN, JR., W. B., R. L. CALABRESE AND W. O. FRIESEN (2005) Neuronal control of leech behavior. *Prog. Neurobiol.* 76: 279–327.

MARDER, E. AND R. L. CALABRESE (1996) Principles of rhythmic motor pattern generation. *Physiol. Rev.* 76: 687–717.

STENT, G. S., W. B. KRISTAN, W. O. FRIESEN, C. A. ORT, M. POON AND R. L. CALABRESE (1978) Neural generation of the leech swimming movement. *Science* 200: 1348–1357.

(C) LAMPREY

Duration of EMG activity in each segmental muscle

Anterior

Posterior

1 swim cycle

(D) LAMPREY

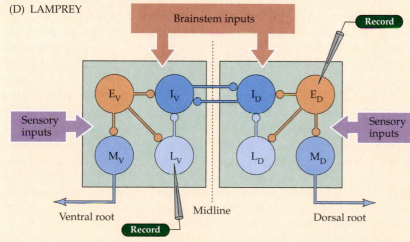

(E) LAMPREY

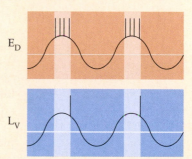

(C) In the lamprey, the pattern of activity across segments is also highly coordinated. (D) The elements of the central pattern generator in the lamprey have been worked out in detail, providing a guide to understanding homologous circuitry in more complex spinal cords. (E) As in the leech, different patterns of electrical activity in lamprey spinal neurons (neurons E_D and L_V, in this example) correspond to distinct periods in the sequence of muscle contractions related to the swim cycle.

BOX 16 C The Autonomy of Central Pattern Generators: Evidence from the Lobster Stomatogastric Ganglion

A principle that has emerged from studies of central pattern generators is that rhythmic patterns of firing elicit complex motor responses without need of ongoing sensory stimulation. A good example is the behavior mediated by a small group of nerve cells, called the stomatogastric ganglion (STG), that controls the muscles of the gut in lobsters and other crustaceans (Figure A). This ensemble of 30 motor neurons and interneurons in the lobster is perhaps the most completely characterized neural circuit known. Of the 30 cells, defined subsets are essential for two distinct rhythmic movements: gastric mill movements that mediate grinding of food by "teeth" in the lobster's foregut, and pyloric movements that propel food into the hindgut. Phasic firing patterns of the motor neurons and interneurons of the STG are directly correlated with these two rhythmic movements. Each of the relevant cells has now been identified based on its position in the ganglion,

and its electrophysiological and neuropharmacological properties characterized (Figures B and C).

Patterned activity in the motor neurons and interneurons of the ganglion begins only if the appropriate neuro-

modulatory input is provided by sensory axons that originate in other ganglia. Depending upon the activity of the sensory axons, neuronal ensembles in the STG produce one of several characteristic rhythmic firing patterns. Once

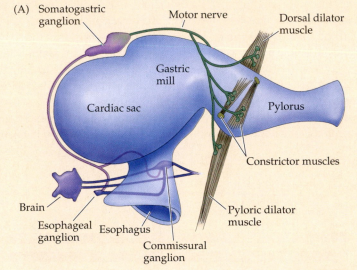

(A) Location of the lobster stomatogastric ganglion in relation to the gut.

to-front progression of leg movements, first on one side and then on the other. As the speed increases to a trot, the movements of the right forelimb and left hindlimb are synchronized (as are the movements of the left forelimb and right hindlimb). At the highest speeds (a gallop), the movements of the two front legs are synchronized, as are the movements of the two hindlimbs (Figure 16.15B).

Given the precise timing of the movement of individual limbs and the coordination among limbs that are required in this process, it is natural to assume that locomotion is accomplished by higher centers that organize the spatial and temporal activity patterns of the individual limbs. Indeed, centers in the brainstem, such as the mesencephalic locomotor region, can trigger locomotion and change the speed of the movement by changing their amount of input to the spinal cord. However, following transection of the spinal cord at the thoracic level, a cat's hindlimbs will still make coordinated locomotor movements if the animal is supported and placed on a moving treadmill (Figure 16.15C). Under these conditions, the speed of locomotor movements is determined by the speed of the treadmill, suggesting that the movement is nothing more than a reflexive response to stretching the limb muscles. This possibility is ruled out, however, by experiments in which the dorsal roots are also sectioned. In this condition, locomotion can be induced by the activation of local circuits triggered by the act of transecting the spinal cord, or by the intravenous injection of L-DOPA (a

(B)

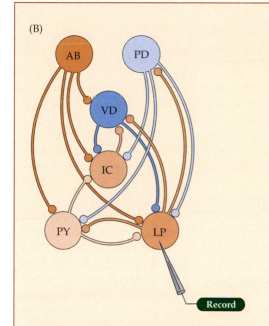

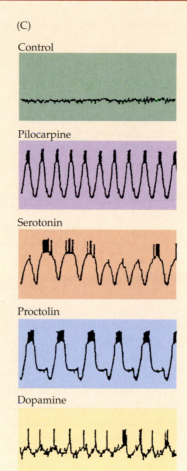

(C)

Control

Pilocarpine

Serotonin

Proctolin

Dopamine

(B) Subset of identified neurons in the stomatogastric ganglion that generates gastric mill and pyloric activity. The abbreviations indicate individual identified neurons, all of which project to different pyloric muscles (except the AB neuron, which is an interneuron). (C) Recording from one of the neurons—the lateral pyloric or LP neuron, in this circuit—showing the different patterns of activity elicited by several neuromodulators known to be involved in the normal synaptic interactions in this ganglion

activated, however, the intrinsic membrane properties of identified cells within the ensemble sustain the rhythmicity of the circuit in the absence of further sensory input.

Another key fact that has emerged from this work is that the same neurons can participate in different programmed motor activities, as circumstances demand. For example, the subset of neurons producing gastric mill activity overlaps the subset that generates pyloric activity. This economic use of neuronal subsets has not yet been described in the central pattern generators of mammals, but seems likely to be a feature of all such circuits.

References

HARTLINE, D. K. AND D. M. MAYNARD (1975) Motor patterns in the stomatogastric ganglion of the lobster, *Panulirus argus. J. Exp. Biol.* 62: 405–420.

MARDER, E. AND R. M. CALABRESE (1996) Principles of rhythmic motor pattern generation. *Physiol. Rev.* 76: 687–717.

SELVERSTON, A. I. (2005) A neural infrastructure for rhythmic motor patterns. *Cell. Mol. Neurobiol.* 25: 223–244.

dopamine precursor), which may serve to release neurotransmitter from the axon terminals of the now-transected upper motor neuron pathways. Although the speed of walking is slowed and the movements are less coordinated than under normal conditions, appropriate locomotor movements are still observed.

These and other observations in experimental animals show that the rhythmic patterns of limb movement during locomotion are not dependent on sensory input, nor are they dependent on input from descending projections from higher centers. Rather, each limb appears to have its own central pattern generator responsible for the alternating flexion and extension of the limb during locomotion (see Box 16C). Under normal conditions, the central pattern generators for the limbs are variably coupled to each other by additional local circuits in order to achieve the different sequences of movements that occur at different speeds.

Although some locomotor movements can also be elicited in humans following damage to descending pathways, these are considerably less effective than the movements seen in the cat. The reduced ability of the transected spinal cord to mediate rhythmic stepping movements in humans presumably reflects an increased dependence of local circuitry on upper motor neuron pathways. Perhaps bipedal locomotion carries with it requirements for postural control greater than can be accommodated by spinal cord circuitry alone. Whatever the explanation, the basic oscillatory circuits that control such rhythmic behaviors as flying, walking, and swimming in many animals also play an important part in human locomotion.

The Lower Motor Neuron Syndrome

The complex of signs and symptoms that arise from damage to the lower motor neurons of the brainstem and spinal cord is referred to as the "lower motor neuron syndrome." In clinical neurology, this constellation of problems must be distinguished from the "upper motor neuron syndrome" that results from damage to the descending upper motor neuron pathways. (See Chapter 17 for a discussion of the signs and symptoms associated with damage to upper motor neurons.)

Damage to lower motor neuron cell bodies, or their peripheral axons, results in paralysis (loss of movement) or paresis (weakness) of the affected muscles, depending on the extent of the damage. In addition to paralysis and/or paresis, the lower motor neuron syndrome includes a loss of reflexes (areflexia) due to interruption of the efferent (motor) limb of the sensory motor reflex arcs. Damage to lower motor neurons also entails a loss of muscle tone, since tone is, in part, dependent on the monosynaptic reflex arc that links the muscle spindles to the lower motor neurons (see Box 17E). A somewhat later effect is atrophy of the affected muscles due to denervation and disuse. The muscles involved may also exhibit fibrillations and fasciculations, which are spontaneous twitches characteristic of single denervated muscle fibers or motor units, respectively. These phenomena arise from changes in the excitability of denervated muscle fibers in the case of fibrillation, and from abnormal activity of injured α motor neurons in the case of fasciculations. These spontaneous contractions can be readily recognized in an electromyogram, providing an especially helpful clinical tool in diagnosing lower motor neuron disorders (Box 16D).

Summary

Four distinct but highly interactive motor subsystems—local circuits in the spinal cord and brainstem, descending upper motor neuron pathways that control these circuits, the basal ganglia, and the cerebellum—all make essential contributions to motor control. Alpha motor neurons located in the spinal cord and in the cranial nerve nuclei in the brainstem directly link the nervous system and muscles, with each motor neuron and its associated muscle fibers constituting a functional entity called the motor unit. Motor units vary in size, amount of tension produced, speed of contraction, and degree of fatigability. Graded increases in muscle tension are mediated by both the orderly recruitment of different types of motor units and an increase in lower motor neuron firing frequency. Local circuitry involving sensory inputs, local circuit neurons, and α and γ motor neurons are especially important in the reflexive control of muscle activity. The stretch reflex is a monosynaptic circuit with connections between sensory fibers arising from muscle spindles and the α motor neurons that innervate the same, or synergistic, muscles. Gamma motor neurons regulate the gain of the stretch reflex by adjusting the level of tension in the intrafusal muscle fibers of the muscle spindle. This mechanism sets the baseline level of activity in α motor neurons and helps to regulate muscle length and tone. Other reflex circuits provide feedback control of muscle tension and mediate essential functions, such as the rapid withdrawal of limbs from painful stimuli. Much of the spatial coordination and timing of muscle activation required for complex

BOX 16D Amyotrophic Lateral Sclerosis

Amyotrophic lateral sclerosis (ALS) is a neurodegenerative disease that affects an estimated 0.05 percent of the population in the United States. It is also called Lou Gehrig's disease, after the New York Yankees baseball player who died of the disorder in 1936. ALS is characterized by the slow but inexorable degeneration of α motor neurons in the ventral horn of the spinal cord and brainstem (*lower* motor neurons), and of neurons in the motor cortex (*upper* motor neurons). Affected individuals show progressive weakness due to upper and/or lower motor neuron involvement, wasting of skeletal muscles due to lower motor neuron involvement, and usually die within 5 years of onset. Sadly, these patients are condemned to watch their own demise, since the intellect remains intact. No available therapy effectively prevents the progression of this disease.

Approximately 10 percent of ALS cases are familial, and several distinct familial forms have been identified. An autosomal dominant form of familial ALS (FALS) is caused by mutations of the gene that encodes the cytosolic antioxidant enzyme copper/zinc superoxide dismutase (SOD1). Mutations of *SOD1* account for roughly 20 percent of families with FALS. A rare, autosomal recessive, juvenile-onset form is caused by mutations in a protein called alsin, a putative GTPase regulator. Another rare type of FALS consists of a slowly progressive, autosomal dominant, lower motor neuron disease without sensory symptoms, with onset in early adult-

hood; this form is caused by mutations of the protein dynactin, which binds to microtubules.

How these mutant genes lead to the phenotype of motor neuron disease is uncertain. Defects of axonal transport have long been hypothesized to cause ALS, perhaps because both upper and lower motor neurons give rise to some of the longest axonal projections in the nervous system and may be at greatest risk for injury secondary to impairments of intrinsic axonal structure and/or transport mechanisms. Evidence for this cause is that transgenic mice with mutant *SOD1* exhibit defects in slow axonal transport early in the course of the disease, and that mutant dynactin may modify fast axonal transport along microtubules. However, whether defective axonal transport is the cellular mechanism by which these mutant proteins lead to motor neuron disease remains to be clearly established. Recent studies have explored a variety of other pathogenic factors that may also play a role in the majority of ALS cases that are sporadic (i.e., non-familial). Among the plausible pathological mechanisms are glutamate excitotoxicity, the activity of reactive oxygen species, the induction of apoptotic pathways, pro-inflammatory interactions between neurons and microglia, mitochondrial dysfunction, and deregulation of calcium homeostasis. Given that motor neurons are exceptionally vulnerable to disruption of mitochondrial function and that they tend to be weak in their capacity to buffer intracellular calcium, these last

two factors are likely to contribute to the selective vulnerability of motor neurons in ALS.

Despite these pathogenic uncertainties and the growing list of candidate mechanisms of neurodegeneration in sporadic cases of ALS, demonstration that specific mutations can cause familial ALS has given scientists valuable clues about the molecular pathogenesis of at least some forms of this tragic disorder.

References

ADAMS, R. D. AND M. VICTOR (2005) *Principles of Neurology,* 8th Ed. New York: McGraw-Hill, pp. 938–944.

BOILLEE, S., C. VANDE VELDE AND D. W. CLEVELAND (2006) ALS: A disease of motor neurons and their nonneuronal neighbors. *Neuron* 52: 39–59.

HADANO, S. AND 20 OTHERS (2001) A gene encoding a putative GTPase regulator is mutated in familial amyotrophic lateral sclerosis 2. *Nature Gen.* 29: 166–173.

PULS, I. AND 13 OTHERS (2003) Mutant dynactin in motor neuron disease. *Nature Gen.* 33: 455–456.

ROSEN, D. R. AND 32 OTHERS (1993) Mutations in Cu/Zn superoxide dismutase gene are associated with familial amyotrophic lateral sclerosis. *Nature* 362: 59–62.

VON LEWINSKI, F. AND B. U. KELLER (2005) Ca²⁺, mitochondria, and selective motoneuron vulnerability: Implications for ALS. *Trends Neurosci.* 28: 494–500.

YANG, Y. AND 16 OTHERS (2001) The gene encoding alsin, a protein with three guanine-nucleotide exchange factor domains, is mutated in a form of recessive amyotrophic lateral sclerosis. *Nature Gen.* 29: 160–165.

rhythmic movements, such as locomotion, are provided by specialized local circuits called central pattern generators. Because of their essential role in all of these circuits, damage to lower motor neurons leads to paralysis of the associated muscle and to other changes, including the loss of reflex activity, loss of muscle tone, and eventually, muscle atrophy.

Additional Reading

Reviews

BURKE, R. E. (1981) Motor units: Anatomy, physiology and functional organization. In *Handbook of Physiology*, V. B. Brooks (ed.). Section 1: *The Nervous System*. Volume 1, Part 1. Bethesda, MD: American Physiological Society, pp. 345–422.

BURKE, R. E. (1990) Spinal cord: Ventral horn. In *The Synaptic Organization of the Brain*, 3rd Ed. G. M. Shepherd (ed.). New York: Oxford University Press, pp. 88–132.

GRILLNER, S. AND P. WALLEN (1985) Central pattern generators for locomotion, with special reference to vertebrates. *Annu. Rev. Neurosci.* 8: 233–261.

HENNEMAN, E. (1990) Comments on the logical basis of muscle control. In *The Segmental Motor System*, M. C. Binder and L. M. Mendell (eds.). New York: Oxford University Press, pp. 7–10.

HENNEMAN, E. AND L. M. MENDELL (1981) Functional organization of the motoneuron pool and its inputs. In *Handbook of Physiology*, V. B. Brooks (ed.). Section 1: *The Nervous System*. Volume 1, Part 1. Bethesda, MD: American Physiological Society, pp. 423–507.

LUNDBERG, A. (1975) Control of spinal mechanisms from the brain. In *The Nervous System*, Volume 1: *The Basic Neurosciences*. D. B. Tower (ed.). New York: Raven Press, pp. 253–265.

NISTRI, A., K. OSTOUMOV, E. SHARIFULLINA AND G. TACCOLA (2006) Tuning and playing a motor rhythm: How metabotropic glutamate receptors orchestrate generation of motor patterns in the mammalian central nervous system. *J. Physiol.* 572: 323–334.

PATTON, H. D. (1965) Reflex regulation of movement and posture. In *Physiology and Biophysics*, 19th Ed., T. C. Rugh and H. D. Patton (eds.). Philadelphia: Saunders, pp. 181–206.

PEARSON, K. (1976) The control of walking. *Sci. Amer.* 235: 72–86.

PROCHAZKA, A., M. HULLIGER, P. TREND AND N. DURMULLER (1988) Dynamic and static fusimotor set in various behavioral contexts. In *Mechanoreceptors: Development, Structure, and Function*. P. Hnik, T. Soulup, R. Vejsada and J. Zelena (eds.). New York: Plenum, pp. 417–430.

SCHMIDT, R. F. (1983) Motor systems. In *Human Physiology*. R. F. Schmidt and G. Thews (eds.). Berlin: Springer Verlag, pp. 81–110.

Important Original Papers

BURKE, R. E., D. N. LEVINE, M. SALCMAN AND P. TSAIRES (1974) Motor units in cat soleus muscle: Physiological, histochemical, and morphological characteristics. *J. Physiol.* 238: 503–514.

BURKE, R. E., P. L. STRICK, K. KANDA, C. C. KIM AND B. WALMSLEY (1977) Anatomy of medial gastrocnemius and soleus motor nuclei in cat spinal cord. *J. Neurophysiol.* 40: 667–680.

HENNEMAN, E., E. SOMJEN, AND D. O. CARPENTER (1965) Excitability and inhibitability of motoneurons of different sizes. *J. Neurophysiol.* 28: 599–620.

HUNT, C. C. AND S. W. KUFFLER (1951) Stretch receptor discharges during muscle contraction. *J. Physiol.* 113: 298–315.

LIDDELL, E. G. T. AND C. S. SHERRINGTON (1925) Recruitment and some other factors of reflex inhibition. *Proc. R. Soc. London* 97: 488–518.

LLOYD, D. P. C. (1946) Integrative pattern of excitation and inhibition in two-neuron reflex arcs. *J. Neurophysiol.* 9: 439–444.

MONSTER, A. W. AND H. CHAN (1977) Isometric force production by motor units of extensor digitorum communis muscle in man. *J. Neurophysiol.* 40: 1432–1443.

WALMSLEY, B., J. A. HODGSON AND R. E. BURKE (1978) Forces produced by medial gastrocnemius and soleus muscles during locomotion in freely moving cats. *J. Neurophysiol.* 41: 1203–1215.

Books

BRODAL, A. (1981) *Neurological Anatomy in Relation to Clinical Medicine*, 3rd Ed. New York: Oxford University Press.

LIEBER, R. L. (2002) *Skeletal Muscle Structure, Function, and Plasticity*, 2nd Ed. Baltimore: Lippincott Williams & Wilkins.

SHERRINGTON, C. (1947) *The Integrative Action of the Nervous System*, 2nd Ed. New Haven: Yale University Press.

Chapter 17

Upper Motor Neuron Control of the Brainstem and Spinal Cord

Overview

The axons of upper motor neurons descend from higher centers to influence the local circuits in the brainstem and spinal cord that organize movements by coordinating the activity of the lower motor neurons described in the previous chapter. The sources of these upper motor neuron pathways include several brainstem centers and a number of cortical areas in the frontal lobe. The motor control centers in the brainstem are especially important in ongoing postural control, orienting toward sensory stimuli, locomotion, and orofacial behavior, with each center having a distinct influence. The mesencephalic locomotor area controls locomotion. Two other centers, the vestibular nuclear complex and the reticular formation, have widespread effects on body position. The reticular formation also contributes to a variety of somatic and visceral motor circuits that govern the expression of autonomic and stereotyped somatic motor behavior. Also in the brainstem, the superior colliculus contains upper motor neurons that initiate orienting movements of the head and eyes. The primary motor cortex and a mosaic of "premotor" areas in the frontal lobe, in contrast, are responsible for the planning and precise control of complex sequences of voluntary movements, as well as motivating the somatic expression of emotional states. Most upper motor neurons, regardless of their source, influence the generation of movements by modulating the activity of the local circuits in the brainstem and spinal cord. Upper motor neurons in the cortex also control movement indirectly, via pathways that project to motor control centers in the brainstem, which, in turn, project to the local organizing circuits in the brainstem and spinal cord. A major function of these indirect pathways is to maintain the body's posture during cortically initiated voluntary movements.

Organization of Descending Motor Control

Some insight into the functions of different sets of upper motor neurons is provided by the way the lower motor neurons and local circuit neurons—the ultimate targets of the upper motor neurons—are arranged within the spinal cord. As described in Chapter 16, lower motor neurons in the ventral horn of the spinal cord are organized in a somatotopic fashion: the most medial part of the ventral horn contains lower motor neuron pools that innervate axial muscles or proximal muscles of the limbs, whereas the more lateral parts contain lower motor neurons that innervate the distal muscles of the limbs (Figure 17.1). The local circuit neurons, which lie primarily in the intermediate zone of the spinal cord and supply much of the direct input to the lower motor neurons, are also topographically arranged. Thus, the medial region of the intermediate zone of the spinal cord gray matter contains the local circuit neurons that synapse with lower motor neurons in the medial part of the ventral horn, whereas the lateral

Figure 17.1 Overview of descending motor control. (A) Somatotopic organization of the ventral horn in the cervical enlargement. The locations of descending projections from the motor cortex in the lateral white matter and from the brainstem in the anterior-medial white matter are shown. (B) Schematic illustration of the major pathways for descending motor control. The medial ventral horn contains lower motor neurons that govern posture, balance, and orienting movements of the head and neck during shifts of visual gaze. These medial motor neurons receive descending input from pathways that originate mainly in the brainstem, course through the anterior-medial white matter of the spinal cord, and then terminate bilaterally. The lateral ventral horn contains lower motor neurons that mediate the expression of skilled voluntary movements of the distal extremities. These lateral motor neurons receive a major descending projection from the contralateral motor cortex via the main (lateral) division of the corticospinal tract, which runs in the lateral white matter of the spinal cord. For simplicity, only one side of the brainstem, motor cortex, and lateral ventral horn is shown, and the minor anterior corticospinal tract is not illustrated.

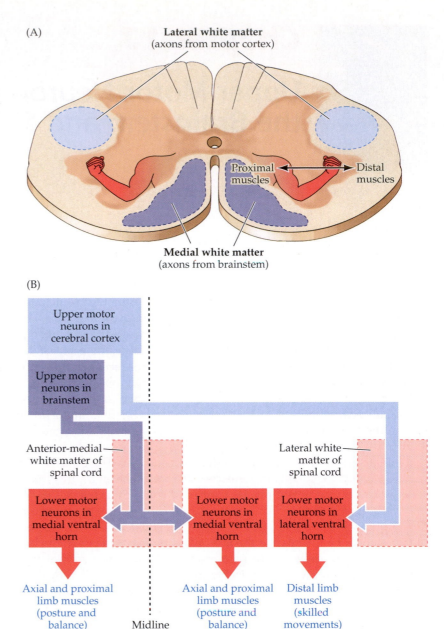

regions of the intermediate zone contain local neurons that synapse primarily with lower motor neurons in the lateral ventral horn. As emphasized in Chapter 16, the somatotopic organization of the ventral horn provides an important framework for understanding the control of the body's musculature in posture and movement, as well as the how descending projections of upper motor neurons are organized to influence motor behavior.

Differences in the way upper motor neuron pathways from the cortex and brainstem terminate in the spinal cord conform to functional distinctions between the local circuits that organize the activity of axial and distal muscle groups. Thus, most upper motor neurons that project to the medial part of the ventral horn also project to the medial region of the intermediate zone. The axons of these upper motor neurons course through the anterior-medial white matter of the spinal cord and give rise to collateral branches that terminate over many spinal cord segments among medial cell groups on both sides of the spinal cord. The sources of these projections are located primarily in the brainstem and, as their terminal zones in the medial spinal cord gray matter suggest, they are concerned primarily with posture, balance and orienting mechanisms (see Figure 17.1B). In contrast, the large majority of axons that project from the motor cortex to the spinal cord course through the lateral white matter of the spinal cord and terminate in lateral parts of the ventral horn, with terminal fields that are restricted to only a few spinal cord segments. The major component of this corticospinal pathway is concerned with the voluntary expression of precise, skilled movements involving more distal parts of the limbs.

Motor Control Centers in the Brainstem: Upper Motor Neurons That Maintain Balance, Govern Posture, and Orient Gaze

There are several important structures in the brainstem that contain circuits of upper motor neurons whose activities serve to organize a variety of somatic movements involving the axial musculature and the muscles of the proximal limbs. These movements include the maintenance of balance, the regulation of posture and the orienting of visual gaze, and they are governed by upper motor neurons in the nuclei of the vestibular complex, the reticular formation, and the superior colliculus. Such movements are usually necessary to support the expression of skilled motor behaviors involving the more distal parts of the extremities or, in the case of visual gaze, when attention is directed toward a particular sensory stimulus. Indeed, the relevant brainstem circuits are competent to direct motor behavior without supervision by higher motor centers in the cerebral cortex. However, these brainstem motor centers usually work in concert with divisions of the motor cortex that organize volitional movements, which always entail both skilled (voluntary) and supporting (reflexive) motor activities.

As described in Chapter 14, the vestibular nuclei are the major destination of the axons that form the vestibular division of the eighth cranial nerve; as such, they receive sensory information from the semicircular canals and the otolith organs that specifies the position and the angular and linear acceleration of the head. Many of the cells in the vestibular nuclei that receive this information are upper motor neurons with descending axons that terminate in the medial region of the spinal cord gray matter, although some extend more laterally to contact the neurons that control the proximal muscles of the limbs. The projections from the vestibular nuclei that control axial muscles and those that influence proximal limb muscles originate from different cells and take somewhat different routes to the spinal cord (Figure 17.2). Neurons in the medial vestibular nucleus give rise to a **medial vestibulospinal tract** that terminates bilaterally in the medial ventral horn of the cervical cord, where it regulates head position by reflex activation of neck muscles in response to the stimulation of the semicircular canals resulting from rotational accelerations of the head. Neurons in the lateral vestibular nucleus are the source of the **lateral vestibulospinal tract**, which courses through the anterior white matter of the spinal cord in a slightly more lateral position, relative to the medial vestibulospinal tract. Despite the modifier in its name, the lateral vestibulospinal tract terminates among medial lower motor neuronal pools that govern proximal muscles of the limbs. As discussed in more detail in Chapter 14, this tract facilitates the activation of limb extensor (antigravity) muscles when the otolith organs signal deviations from stable balance and upright posture. Other upper motor neurons in the vestibular nuclei project to lower motor neurons in the cranial nerve nuclei that control eye movements (the third, fourth, and sixth cranial nerve nuclei). This pathway produces the eye movements that maintain fixation while the head is moving (the vestibulo-ocular reflex; see Chapters 14 and 20).

(A) LATERAL AND MEDIAL
VESTIBULOSPINAL TRACTS

(B) RETICULOSPINAL TRACT

(C) COLLICULOSPINAL TRACT

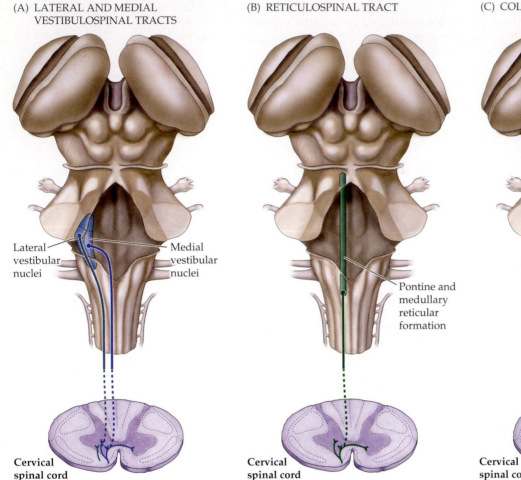
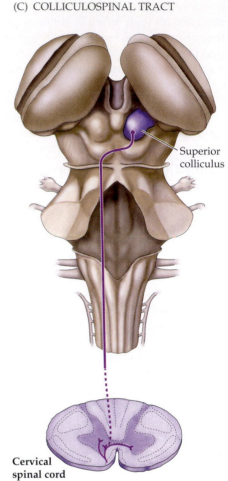

Figure 17.2 *Descending projections from the brainstem to the spinal cord. Pathways that influence motor neurons in the medial part of the ventral horn originate in the vestibular nuclei (A), reticular formation (B), and superior colliculus (C).*

The **reticular formation** is a complicated network of circuits in the core of the brainstem that extends from the rostral midbrain to the caudal medulla; it is similar in structure and function to circuitry in the intermediate gray matter of the spinal cord (Figure 17.3 and Box 17A). Unlike the well-defined sensory and motor nuclei of the cranial nerves, the reticular formation comprises clusters of neurons scattered among a welter of interdigitating axon bundles; it is therefore difficult to subdivide anatomically. The neurons within the reticular formation have a variety of functions, including cardiovascular and respiratory control (see Chapter 21), governance of myriad sensory motor reflexes (see Chapters 16 and 21), the coordination of eye movements (see Chapter 20), regulation of sleep and wakefulness (see Chapter 28), and, most important for present purposes, the temporal and spatial coordination of limb and trunk movements. The descending motor control pathways from the reticular formation to the spinal cord are similar to those of the vestibular nuclei; they terminate primarily in the medial parts of the gray matter where they influence the local circuit neurons that coordinate axial and proximal limb muscles (see Figure 17.2B).

Both the vestibular nuclei and the reticular formation provide information to the spinal cord that maintains posture in response to environmental (or self-induced) disturbances of body position and stability. Direct projections from the vestibular nuclei to the spinal cord ensure a rapid compensatory response to

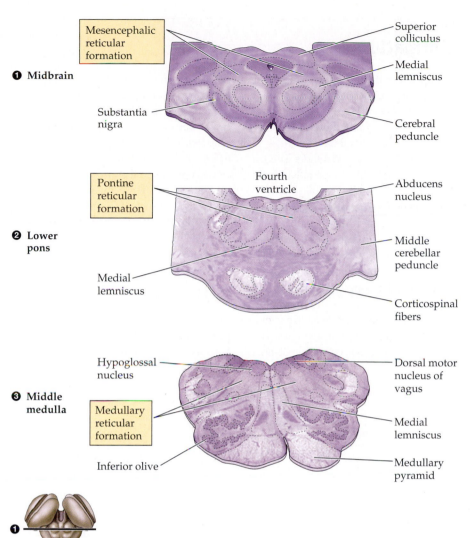

❶ **Midbrain**

Mesencephalic reticular formation

Superior colliculus

Medial lemniscus

Substantia nigra

Cerebral peduncle

Pontine reticular formation

Fourth ventricle

Abducens nucleus

❷ **Lower pons**

Middle cerebellar peduncle

Medial lemniscus

Corticospinal fibers

Hypoglossal nucleus

Dorsal motor nucleus of vagus

❸ **Middle medulla**

Medullary reticular formation

Medial lemniscus

Inferior olive

Medullary pyramid

Figure 17.3 The location of the reticular formation in relation to some other major landmarks at different levels of the brainstem. Neurons in the reticular formation are scattered among the axon bundles that course through the medial portion of the midbrain, pons, and medulla (see Box 17A).

any postural instability detected by the vestibular labyrinth (see Chapter 14). In contrast, the motor centers in the reticular formation are controlled largely by other motor centers in the cerebral cortex, hypothalamus, or brainstem. The relevant neurons in the reticular formation initiate adjustments that stabilize posture during ongoing movements.

The way neurons of the reticular formation maintain posture can be appreciated by analyzing their activity during voluntary movements. Even the simplest movements are accompanied by the activation of muscles that at first glance seem to have little to do with the primary purpose of the movement. For exam-

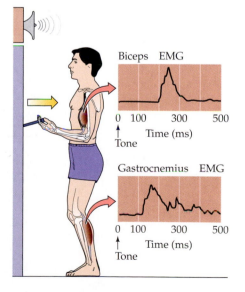

Biceps EMG

0   100   300   500
Time (ms)
↑
Tone

Gastrocnemius EMG

0   100   300   500
Time (ms)
↑
Tone

Figure 17.4 Anticipatory maintenance of body posture. At the onset of a tone, the subject pulls on a handle, contracting the biceps muscle. To ensure postural stability, contraction of the gastrocnemius muscle precedes that of the biceps. EMG refers to the electromyographic recording of muscle activity.

BOX 17A   **The Reticular Formation**

If one were to exclude from the structure of the brainstem the cranial nerve nuclei, the nuclei that provide input to the cerebellum, the long ascending and descending tracts that convey explicit sensory and motor signals, and the structures that lie dorsal and lateral to the ventricular system, what would be left is a central core region known as the *tegmentum* (Latin for "covering structure"), so named because it "covers" the ventral part of the brainstem. Scattered among the diffuse fibers that course through the tegmentum are small clusters of neurons that are collectively known as the reticular formation. With few exceptions, these clusters of neurons are difficult to recognize as distinct nuclei in standard histological preparations. Indeed, the modifying term *reticular* (net-like) was applied to this loose collection of neuronal clusters because the early neurohistologists envisioned these neurons as part of a sparse network of diffusely connected cells that extends from the intermediate gray regions of the cervical spinal cord to the lateral regions of the hypothalamus and certain nuclei along the midline of the thalamus.

These early anatomical concepts were influenced by lesion experiments in animals and clinical observations in human patients made in the 1930s and 1940s. These studies showed that damage to the upper brainstem tegmentum produced coma, suggesting the existence of a neural system in the midbrain and rostral pons that supported normal conscious brain states and transitions between sleep and wakefulness. These ideas were articulated most influentially by G. Moruzzi and H. Magoun when they proposed a "reticular activating system" to account for these functions and the critical role of the brainstem reticular formation. Current evidence generally supports the notion of an activating function of the rostral reticular formation; however, neuroscientists now recognize the complex interplay of a variety of neurochemical systems (with diverse postsynaptic effects) comprising distinct cell clusters in the rostral tegmentum, and a myriad of other functions performed by neuronal clusters in more caudal parts of the reticular formation. Thus, with the advent of more precise means of demonstrating anatomical connections, as well as more sophisticated means of identifying neurotransmitters and the activity patterns of individual neurons, the concept of a "sparse network" engaged in a common function is now obsolete.

Nevertheless, the term *reticular formation* remains, as does the daunting challenge of understanding the anatomical complexity and functional heterogeneity of this complex brain region. Fortunately, two simplfying generaliza-

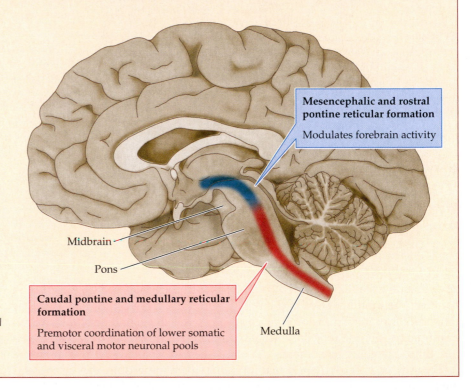

Midsagittal view of the brain showing the longitudinal extent of the reticular formation and highlighting the broad functional roles performed by neuronal clusters in its rostral (blue) and caudal (red) sectors.

Mesencephalic and rostral pontine reticular formation

Modulates forebrain activity

Midbrain

Pons

Caudal pontine and medullary reticular formation

Premotor coordination of lower somatic and visceral motor neuronal pools

Medulla

ple, Figure 17.4 shows the pattern of muscle activity that occurs as a subject uses his arm to pull on a handle in response to an auditory tone. Activity in the biceps muscle begins about 200 ms after the tone. However, as the records show, the contraction of the biceps is accompanied by a significant increase in the activity of a proximal leg muscle, the gastrocnemius (as well as many other muscles not monitored in the experiment). In fact, contraction of the gastrocnemius muscle begins well before contraction of the biceps.

tions can be made. First, the functions of the different clusters of neurons in the reticular formation can be grouped into two broad categories: *modulatory functions* and *premotor functions*. Second, the modulatory functions are primarily found in the rostral sector of the reticular formation, whereas the premotor functions are localized in more caudal regions.

Several clusters of large ("magnocellular") neurons in the midbrain and rostral pontine reticular formation participate—together with certain diencephalic nuclei—in the modulation of conscious states (see Chapter 28). These effects are accomplished by long-range, diencephalic projections of cholinergic neurons near the superior cerebellar peduncle, as well as the more widespread forebrain projections of noradrenergic neurons in the locus coeruleus and serotonergic neurons in the raphe nuclei. Generally speaking, these biogenic amine neurotransmitters function as neuromodulators (see Chapter 6) that alter the membrane potential and thus the firing patterns of thalamocortical and cortical neurons (the details of these effects are explained in Chapter 28). Also included in this category are the dopaminergic systems of the ventral midbrain that modulate cortico-striatal interactions in the basal ganglia (see Chapter 18) and the responsiveness of neurons in the prefrontal cortex and limbic forebrain (see Chapter 29). However, not all modulatory projections from the rostral reticular formation are directed toward the forebrain. Although not always considered part of the reticular formation, it is helpful to include in this functional group certain neuronal columns in the periaqueductal gray (surrounding the cerebral aqueduct) that project to the dorsal horn of the spinal cord and modulate the transmission of nociceptive signals (see Chapter 10).

Reticular formation neurons in the caudal pons and medulla oblongata generally serve a premotor function in the sense that they integrate feedback sensory signals with executive commands from upper motor neurons and deep cerebellar nuclei and, in turn, organize the efferent activities of lower visceral motor and certain somatic motor neurons in the brainstem and spinal cord. Examples of this functional category include the smaller (parvocellular) neurons that coordinate a broad range of motor activities, including the gaze centers discussed in Chapter 20 and local circuit neurons near the somatic motor and branchiomotor nuclei that organize mastication, facial expressions, and a variety of reflexive orofacial behaviors such as sneezing, hiccuping, yawning, and swallowing. In addition, there are "autonomic centers" that organize the efferent activities of specific pools of primary visceral motor neurons. Included in this subgroup are distinct clusters of neurons in the ventral-lateral medulla that generate respiratory rhythms, and others that regulate the cardioinhibitory output of neurons in the nucleus ambiguus and the dorsal motor nucleus of the vagus nerve. Still other clusters organize more complex activities that require the coordination of both somatic motor and visceral motor outflow, such as gagging and vomiting, and even laughing and crying.

One set of neuronal clusters that does not fit easily into this rostral-caudal framework is the set of neurons that give rise to the reticulospinal projections. As described in the text, these neurons are distributed in both rostral and caudal sectors of the reticular formation and they give rise to long-range projections that innervate lower motor neuronal pools in the medial ventral horn of the spinal cord. The reticulospinal inputs serve to modulate the gain of segmental reflexes involving the muscles of the trunk and proximal limbs and to initiate certain stereotypical patterns of limb movement.

In summary, the reticular formation is best viewed as a heterogeneous collection of distinct neuronal clusters in the brainstem tegmentum that either modulate the excitability of distant neurons in the forebrain and spinal cord or coordinate the firing patterns of more local lower motor neuronal pools engaged in reflexive or stereotypical somatic motor and visceral motor behavior.

References

BLESSING, W. W. (1997) Inadequate frameworks for understanding bodily homeostasis. *Trends Neurosci.* 20: 235–239

HOLSTEGE, G., R. BANDLER AND C. B. SAPER (EDS.) (1996) *Progress in Brain Research*, Vol. 107. Amsterdam: Elsevier.

LOEWY, A. D. AND K. M. SPYER (EDS.) (1990) *Central Regulation of Autonomic Functions*. New York: Oxford.

MASON, P. (2001) Contributions of the medullary raphe and ventromedial reticular region to pain modulation and other homeostatic functions. *Annu. Rev. Neurosci.* 24: 737–777.

MORUZZI, G. AND H. W. MAGOUN (1949) Brain stem reticular formation and activation of the EEG. *EEG Clin. Neurophys.* 1: 455–476.

These observations show that postural control entails an anticipatory, or feedforward, mechanism (Figure 17.5). As part of the motor plan for moving the arm, the effect of the impending movement on body stability is "evaluated" and used to generate a change in the activity of the gastrocnemius muscle. This change actually precedes and provides preparatory postural support for the movement of the arm. In the example given in Figure 17.4, contraction of the biceps would tend to pull the entire body forward, an action that is opposed by

Figure 17.5 Feedforward and feedback mechanisms of postural control. Feedforward postural responses are "preprogrammed" and typically precede the onset of limb movement (see Figure 17.4). Feedback responses are initiated by sensory inputs that detect postural instability.

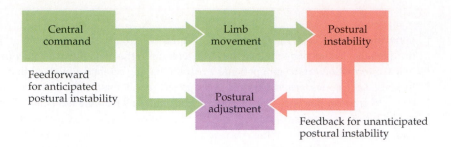

the contraction of the gastrocnemius muscle. In short, this feedforward mechanism predicts the resulting disturbance in body stability and generates an appropriate stabilizing response.

The importance of the reticular formation for feedforward mechanisms of postural control has been explored in more detail in cats trained to use a forepaw to strike an object. As expected, the forepaw movement is accompanied by feedforward postural adjustments in the other legs to maintain the animal upright. These adjustments shift the animal's weight from an even distribution over all four feet to a diagonal pattern, in which the weight is carried mostly by the contralateral, nonreaching forelimb and the ipsilateral hindlimb. Lifting of the forepaw and postural adjustments in the other limbs can also be induced in an alert cat by electrical stimulation of the motor cortex. After pharmacological inactivation of the reticular formation, however, electrical stimulation of the motor cortex evokes only the forepaw movement, without the feedforward postural adjustments that normally accompany them.

The results of this experiment can be understood in terms of the fact that the upper motor neurons in the motor cortex influence the spinal cord circuits by two routes: direct projections to the spinal cord (considered in more detail below) and indirect projections to brainstem centers that in turn project to the spinal cord. The reticular formation is one of the major destinations of these latter projections from the motor cortex; thus, cortical upper motor neurons initiate both the reaching movement of the forepaw and also the postural adjustments in the other limbs necessary to maintain body stability. The forepaw movement is initiated by the direct pathway from the cortex to the spinal cord, whereas the postural adjustments are mediated via pathways from the motor cortex that reach the spinal cord indirectly, after an intervening relay in the reticular formation (the cortico-reticulospinal pathway; Figure 17.6).

Further evidence for the contrasting functions of the direct and indirect pathways from the motor cortex to the spinal cord comes from experiments carried out by the Dutch neurobiologist Hans Kuypers, who examined the behavior of rhesus monkeys that had the direct pathway to the spinal cord transected at the level of the medulla, leaving the indirect pathways to the spinal cord via the brainstem centers intact. Immediately after the surgery, the animals were able to use axial and proximal muscles to stand, walk, run, and climb, but they had great difficulty using the distal parts of their limbs (especially their hands) independently of other body movements. For example, the monkeys could cling to the cage but were unable to reach toward and pick up food with their fingers; rather, they used the entire arm to sweep the food toward them. After several weeks, the animals recovered some independent use of their hands and were again able to pick up objects of interest, but this action involved the concerted closure of all of the fingers. The ability to make independent, fractionated movements of the fingers, as in opposing the movements of the fingers and thumb to pick up an object, never returned.

Figure 17.6 *Indirect pathways from the motor cortex to the spinal cord. Neurons in the motor cortex that supply the lateral part of the ventral horn to initiate movements of the distal limbs (see Figure 17.9) also terminate on neurons in the reticular formation to mediate postural adjustments that support the movement. The reticulospinal pathway terminates in the more medial parts of the ventral horn, where lower motor neurons that innervate axial and proximal muscles are located. Thus, the motor cortex has both direct and indirect routes by which it can influence the activity of spinal cord neurons.*

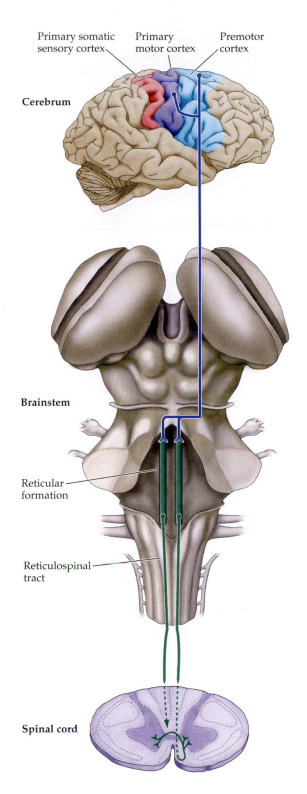

These observations show that following damage to the direct projections from the motor cortex to the spinal cord at the level of the medulla, the indirect projections from the motor cortex via the brainstem centers (or from brainstem centers alone) are capable of sustaining motor behavior that involves primarily the use of proximal muscles. In contrast, the direct projections from the motor cortex to the spinal cord provide the speed and agility of movements, and enable a higher degree of precision in fractionated finger movements than is possible using the indirect pathways alone (see below).

Two additional brainstem structures, the **superior colliculus** and, in non-humans, the **red nucleus**, also contribute upper motor neuron pathways to the spinal cord. The axons arising from neurons in deep layers of the superior colliculus project via the **colliculospinal tract** to medial cell groups in the cervical cord, where they influence the lower motor neuron circuits that control axial musculature in the neck (see Figure 17.2C). These projections are particularly important in generating orienting movements of the head (the role of the superior colliculus in the generation of head and eye movements is covered in detail in Chapter 20). In non-human primates and other mammals, a large nucleus in the tegmentum of the midbrain, termed the red nucleus, projects via the **rubrospinal tract** to the cervical level of the spinal cord. (*Rubro* is Latin for red; the adjective is derived from the reddish color of this nucleus in fresh tissue, presumably due to the enrichment of its neurons with iron–protein complexes.)

Unlike the other projections from the brainstem to the spinal cord discussed thus far, the rubrospinal tract is located in the lateral white matter of the spinal cord and its axons terminate in lateral regions of the ventral horn and intermediate zone, where circuits of lower motor neurons governing the distal musculature of the upper extremities reside. Presumably, this projection participates together with the direct pathway from the motor cortex in the control of the arms (or forepaws). The limited distribution of rubrospinal projections may seem surprising, given the large size of the red nucleus in most mammals. However, the rubrospinal tract arises from especially large (magnocellular) neurons in the caudal pole of the red nucleus, which account for a relatively small fraction of the total number of neurons in the nucleus. In the human midbrain, there are very few—if any—large neurons in the red nucleus; thus, if the rubrospinal tract exists in humans (which may not be the case), its significance for motor control is dubious. Indeed, nearly all of the neurons in the red nucleus in humans are small (parvocellular) and do not project to the spinal cord at all; many of these neurons relay information to the inferior olive, an important source of learning signals for the cerebellum (see Chapter 19).

(A) Lateral view

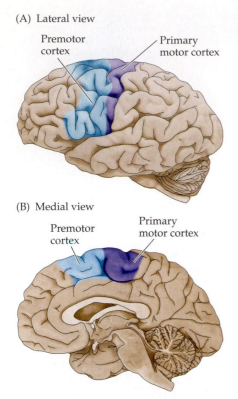

(B) Medial view

Figure 17.7   The primary motor cortex and the premotor area in the human cerebral cortex as seen in lateral (A) and medial (B) views. The primary motor cortex is located in the precentral gyrus; the mosaic of premotor areas is more rostral.

The Corticospinal and Corticobulbar Pathways: Projections from Upper Motor Neurons That Initiate Complex Voluntary Movements

The upper motor neurons in the cerebral cortex reside in several adjacent and highly interconnected areas in the posterior frontal lobe, which together mediate the planning and initiation of complex temporal sequences of voluntary movements. These cortical areas all receive regulatory input from the basal ganglia and cerebellum via relays in the ventrolateral thalamus (see Chapters 18 and 19), as well as inputs from sensory regions of the parietal lobe (see Chapter 9). Although the phrase "motor cortex" is sometimes used to refer to these frontal areas collectively, the term is more commonly restricted to the **primary motor cortex** located in the precentral gyrus and the paracentral lobule (Figure 17.7). The primary motor cortex can be distinguished from a complex mosaic of adjacent "premotor" areas both cytoarchitectonically (it is area 4 in Brodmann's nomenclature; see Chapter 26) and by the low intensity of current necessary to elicit movements by electrical stimulation in this region. The low threshold for eliciting movements is an indicator of a relatively large and direct pathway from the primary area to the lower motor neurons of the brainstem and spinal cord. This section and the next focus on the organization and functions of the primary motor cortex and its descending pathways. The subsequent section addresses the contributions of the adjacent premotor areas.

The pyramidal cells of cortical layer 5 are the upper motor neurons of the primary motor cortex. Among these neurons are the conspicuous Betz cells, which are the largest neurons (by soma size) in the human central nervous system (Figure 17.8). Although it is often assumed that Betz cells are the principal upper motor neurons of the motor cortex, there are far too few of them to account for the direct projection from the motor cortex to the lower motor neurons in the brainstem and spinal cord; indeed, in the human central nervous

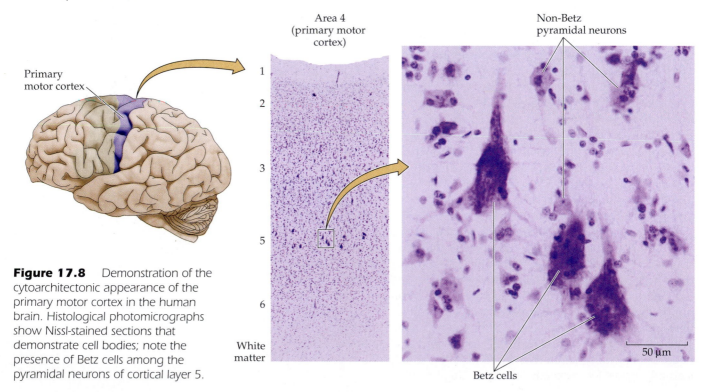

Figure 17.8   Demonstration of the cytoarchitectonic appearance of the primary motor cortex in the human brain. Histological photomicrographs show Nissl-stained sections that demonstrate cell bodies; note the presence of Betz cells among the pyramidal neurons of cortical layer 5.

BOX 17B Patterns of Facial Weakness and Their Importance for Localizing Neurological Injury

The signs and symptoms pertinent to the cranial nerves and their nuclei are of special importance to clinicians seeking to pinpoint the neurological lesions that produce motor deficits. An especially instructive example is provided by the muscles of facial expression. It has long been recognized that the distribution of facial weakness provides important localizing clues indicating whether the underlying injury involves lower motor neurons in the facial motor nucleus (and/or their axons in the facial nerve) or the inputs that govern these neurons, which arise from upper motor neurons in the cerebral cortex. Damage to the facial motor nucleus or its nerve affects all the muscles of facial expression on the side of the lesion (lesion C in the figure); this is expected given the intimate anatomical and functional linkage between lower motor neurons and skeletal muscles. A pattern of impairment that is more difficult to explain accompanies unilateral injury to the motor areas in the lateral frontal lobe (primary motor and premotor cortex), as occurs with strokes that involve the middle cerebral artery (lesion A in the figure). Most patients with such injuries have difficulty controlling the contralateral muscles around the mouth but retain the ability to symmetrically raise their eyebrows, wrinkle their forehead, and squint.

Until recently, it was assumed that this pattern of inferior facial paresis with superior facial sparing could be attributed to (presumed) bilateral projections from the face representation in the *primary motor cortex* to the facial motor nucleus; in this conception, the intact ipsilateral corticobulbar projec-

Organization of projections from cerebral cortex to the facial motor nucleus and the effects of upper and lower motor neuron lesions.

tions were considered sufficient to motivate the contractions of the superior muscles of the face. However, recent tract-tracing studies in non-human primates have suggested a different explanation. These studies demonstrate two important facts that clarify the relations among the face representations in the cerebral cortex and the facial motor nucleus. First, the corticobulbar projections of the primary motor cortex are directed predominantly toward the lateral cell columns in the contralateral

facial motor nucleus, which control the movements of the perioral musculature. Thus, the more dorsal cell columns in the facial motor nucleus that innervate superior facial muscles do not receive significant input from the primary motor cortex. Second, these dorsal cell columns are governed by premotor areas in the anterior cingulate gyrus, a cortical region that is associated with emotional processing (see Chapter 29). Therefore, a better interpretation is that strokes

(Continued on next page)

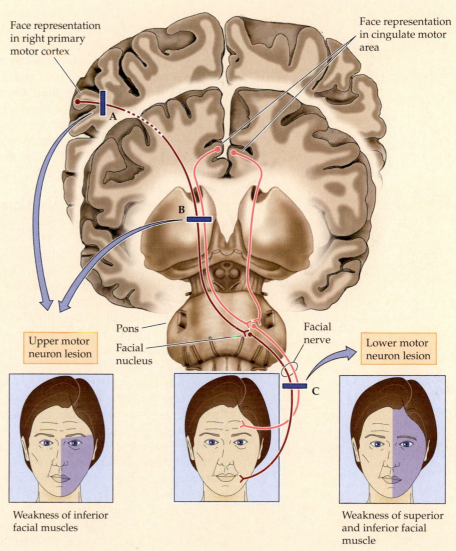

Face representation in right primary motor cortex

Face representation in cingulate motor area

A

B

Pons

Facial nucleus

Upper motor neuron lesion

Facial nerve

Lower motor neuron lesion

C

Weakness of inferior facial muscles

Weakness of superior and inferior facial muscle

BOX 17B (Continued)

involving the middle cerebral artery spare the superior aspect of the face because the relevant upper motor neurons are in the cingulum, which is supplied by the anterior cerebral artery.

An additional puzzle has also been resolved by these studies. Strokes involving the anterior cerebral artery or subcortical lesions that interrupt the corticobulbar projection (lesion B in the figure) seldom produce significant paresis of the superior facial muscles. Superior facial sparing in these situations may arise

because these *cingulate motor areas* (see Figure 17.14) send descending projections through the corticobulbar pathway that bifurcate and innervate dorsal facial motor cell columns on both sides of the brainstem. Thus, the superior muscles of facial expression are controlled by symmetrical inputs from the cingulate motor areas in both hemispheres.

References

JENNY, A. B. AND C. B. SAPER (1987) Organization of the facial nucleus and corticofacial projection in the monkey: A reconsideration of the upper motor neuron facial palsy. *Neurology* 37: 930–939.

KUYPERS, H. G. J. M. (1958) Corticobulbar connexions to the pons and lower brainstem in man. *Brain* 81: 364–489.

MORECRAFT, R. J., J. L. LOUIE, J. L. HERRICK AND K. S. STILWELL-MORECRAFT (2001) Cortical innervation of the facial nucleus in the non-human primate: A new interpretation of the effects of stroke and related subtotal brain trauma on the muscles of facial expression. *Brain* 124: 176–208.

system, they could account for no more than 5 percent of the projection to the spinal cord. The remaining upper motor neurons are non-Betz pyramidal neurons of layer 5 that are found in the primary motor cortex and in each division of the premotor cortex. The axons of these upper motor neurons descend in the **corticobulbar** and **corticospinal tracts**, terms that are used to distinguish axons that terminate in the brainstem ("bulbar" refers to brainstem nuclei) or spinal cord. Along their course, these axons pass through the posterior limb of the internal capsule in the forebrain to enter the cerebral peduncle at the base of the midbrain (Figure 17.9). They then run through the base of the pons, where they are scattered among the transverse pontine fibers and nuclei of the basal pontine gray matter, coalescing again on the ventral surface of the medulla where they form the **medullary pyramids**. The components of this upper motor neuron pathway that innervate cranial nerve nuclei, the reticular formation, the red nucleus, and the basilar pontine nuclei (that is, the corticobulbar tract) leave the pathway at the appropriate levels of the brainstem (see Figure 17.9 and Box 17B). At the caudal end of the medulla, a large majority—about 90 percent—of the axons in the pyramidal tract cross the midline (*decussate*) to enter the lateral columns of the spinal cord on the opposite side, where they form the **lateral corticospinal tract**. The remaining 10 percent enter the spinal cord without crossing; these axons, which constitute the **ventral (anterior) corticospinal tract**, terminate either ipsilaterally or bilaterally (with collateral branches crossing the midline via the ventral white commissure of the spinal cord). The ventral corticospinal pathway arises primarily from dorsal and medial regions of the motor cortex that serve axial and proximal limb muscles, the same divisions of the motor cortex that give rise to projections to the reticular formation (see Figure 17.6).

The lateral corticospinal tract forms a direct pathway from the cortex to the spinal cord and terminates primarily in the lateral portions of the ventral horn and intermediate gray matter, with some axons synapsing directly on α motor neurons that govern the distal extremities (see Figures 17.1 and 17.9). However, this privileged synaptic contact on lower motor neurons is restricted to a subset of α motor neurons that supply the muscles of the forearm and hand; most axons of the lateral corticospinal tract terminate among pools of local circuit neurons that coordinate the activities of the lower motor neurons in the lateral

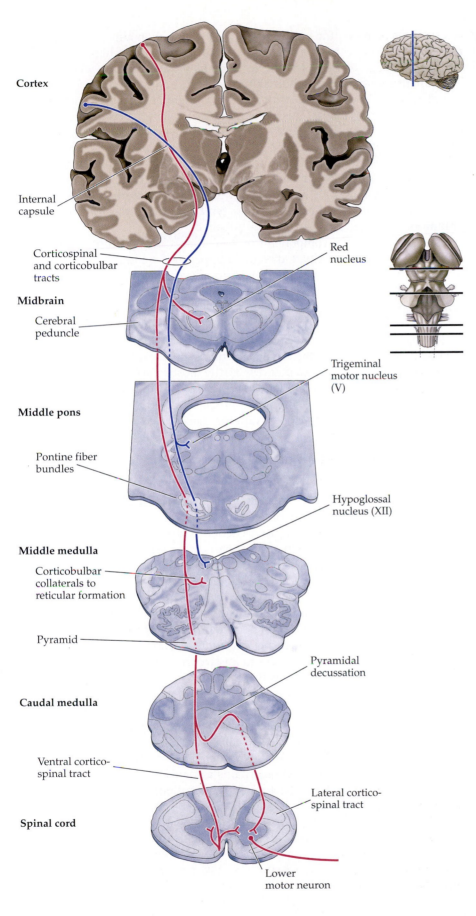

Cortex

Internal capsule

Corticospinal and corticobulbar tracts

Midbrain

Cerebral peduncle

Middle pons

Pontine fiber bundles

Middle medulla

Corticobulbar collaterals to reticular formation

Pyramid

Caudal medulla

Ventral cortico-spinal tract

Spinal cord

Red nucleus

Trigeminal motor nucleus (V)

Hypoglossal nucleus (XII)

Pyramidal decussation

Lateral cortico-spinal tract

Lower motor neuron

Figure 17.9 The corticospinal and corticobulbar tracts. Neurons in the motor cortex give rise to axons that travel through the internal capsule and coalesce on the ventral surface of the midbrain, within the cerebral peduncle. These axons continue through the pons and come to lie on the ventral surface of the medulla, giving rise to the pyramids. As they course through the brainstem, some axons give rise to collaterals that innervate brainstem nuclei. Most of these pyramidal fibers cross in the caudal part of the medulla to form the lateral corticospinal tract in the spinal cord. Those axons that do not cross form the ventral corticospinal tract.

cell columns of the ventral horn. This terminal distribution implies an important role for the lateral cortical spinal tract in the control of the hands. Although selective damage to this pathway in humans is rarely seen, evidence from experimental studies in non-human primates indicates that direct projections from the motor cortex to the spinal cord are essential for the performance of discrete finger movements. This evidence helps explain the limited recovery in humans after damage to the motor cortex or some component of this pathway. Immediately after such an injury, such patients are typically paralyzed. With time, however, some ability to perform voluntary movements reappears. These movements, which are presumably mediated by motor centers in the brainstem, are crude for the most part, and the ability to perform fractionated finger movements, such as those required for writing, typing, playing a musical instrument, or buttoning clothes, typically remains impaired.

Finally, there are also components of the corticobulbar and corticospinal projections that do not participate directly in upper motor control of lower motor neurons. These components are derived from layer 5 neurons in somatic sensory regions of the anterior parietal lobe and terminate among local circuit neurons near the trigeminal nulcei of the brainstem and in the dorsal horn of the spinal cord. They are likely involved in modulating the transmission of proprioceptive signals and other mechanosensory inputs that are relevant for monitoring bodily movements. Interestingly, the corticospinal projection to the ventral horn is greatest in vertebrates that display the most complex repertoire of fractionated movements with their hands or forepaws. In animals that display little ability to execute skilled movements with their forepaws, the corticospinal projection is predominantly directed toward the dorsal horn.

Functional Organization of the Primary Motor Cortex

Clinical observations and experimental work dating back a hundred years or more provide a foundation for understanding the functional organization of the motor cortex. By the end of the nineteenth century, experimental work in animals by the German physiologists G. Theodor Fritsch and Eduard Hitzig had shown that electrical stimulation of the motor cortex elicits contractions of muscles on the contralateral side of the body. At about the same time, the British neurologist John Hughlings Jackson surmised that the motor cortex contains a complete spatial representation, or map, of the body's musculature. Jackson reached this conclusion from his observation that the abnormal movements during some types of epileptic seizures "march" systematically from one part of the body to another. For instance, partial motor seizures may start with abnormal movements of a finger, progress to involve the entire hand, then the forearm, the arm, the shoulder, and, finally, the face.

This early evidence for motor maps in the cortex was confirmed shortly after the turn of the nineteenth century when Charles Sherrington published his classic maps of the organization of the motor cortex in great apes, created using focal electrical stimulation. During the 1930s, one of Sherrington's students, the American neurosurgeon Wilder Penfield, extended this work by demonstrating that the human motor cortex also contains a spatial map of the contralateral body. By correlating the location of muscle contractions with the site of electrical stimulation on the surface of the motor cortex (the same method used by Sherrington), Penfield mapped the motor representation in the precentral gyrus in over 400 neurosurgical patients (Figure 17.10). He found that this motor map shows the same general disproportions observed in the somatic sensory maps in the postcentral gyrus (see Chapter 9). Thus, the musculature used in tasks requiring fine motor control (such as movements of the face and hands) is rep-

(A)

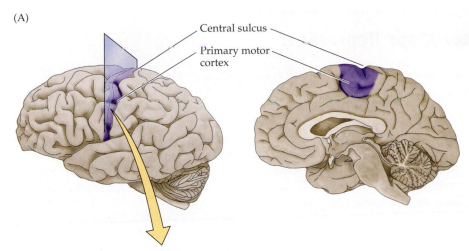

Central sulcus

Primary motor cortex

Figure 17.10 Topographic map of movement in the primary motor cortex. (A) Location of primary motor cortex in the precentral gyrus. (B) Section along the precentral gyrus, illustrating the coarse somatotopic organization of the motor cortex. In contrast to the precise and detailed representation of the contralateral body in the primary somatic sensory cortex (see Figure 9.11), the somatotopy of the primary motor cortex is much more coarse.

(B)

Corticospinal tract

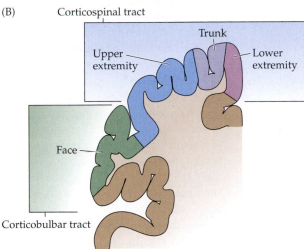

Trunk

Upper extremity

Lower extremity

Face

Corticobulbar tract

resented by a greater area of motor cortex than is the musculature requiring less precise motor control (such as that of the trunk). The behavioral implications of cortical motor maps are considered in Boxes 17C and 17D.

The introduction in the 1960s of intracortical microstimulation (a more refined method of cortical activation) allowed a more detailed understanding of motor maps. Microstimulation entails the delivery of brief electrical currents an order of magnitude smaller than those used by Sherrington and Penfield. By passing the current through the sharpened tip of a metal microelectrode inserted into the cortex, the upper motor neurons in layer 5 that project to lower motor neuron circuitry can be stimulated focally. Although intracortical stimulation generally confirmed Penfield's spatial map in the motor cortex, it also showed that the finer organization of the map is rather different than most neuroscientists imagined.

For example, when microstimulation was combined with recordings of muscle electrical activity, even the smallest currents capable of eliciting a response initiated the excitation of several muscles (and the simultaneous suppression of others), suggesting that organized movements rather than individual muscles are represented in the map (see Box 17C). Furthermore, within major subdivisions of the map (e.g., forearm or face regions), a particular movement could be elicited by stimulation of widely separated sites, supporting the argument that neurons in nearby regions are linked by local circuits in the cortex and spinal

BOX 17C What Do Motor Maps Represent?

Electrical stimulation studies carried out by the neurosurgeon Wilder Penfield and his colleagues in human patients (and by Sherrington and later Clinton Woolsey and his colleagues in experimental animals) clearly demonstrated a systematic motor map in the precentral gyrus (see text). The fine structure of this map and the nature of its representation, however, have been a continuing source of controversy. Is the map in the motor cortex a map of musculature that operates like a "piano keyboard" for the control of individual muscles, or is it a map of movements, in which specific sites control multiple muscle groups that contribute to the generation of particular actions?

Initial experiments implied that the map in the motor cortex is a fine-scale representation of individual muscles. Thus, stimulation of small regions of the map activated single muscles, suggesting that vertical columns of cells in the motor cortex were responsible for controlling the actions of particular muscles, much as columns in the somatic sensory map are thought to analyze stimulus information from particular locations in the body (see Chapter 9).

More recent studies using anatomical and physiological techniques, however, have shown that the map in the motor cortex is far more complex than a columnar representation of particular muscles. Individual pyramidal tract axons are now known to terminate on sets of spinal motor neurons that innervate different muscles. This relationship is evident even for neurons in the hand representation of the motor cortex, the region that controls the most discrete, fractionated movements.

Furthermore, cortical microstimulation experiments have shown that contraction of a single muscle can be evoked by stimulation over a wide region of the motor cortex (about 2–3 mm in macaque monkeys) in a complex, mosaic fashion. It seems likely that horizontal connections within the motor cortex and local circuits in the spinal cord create ensembles of neurons that coordinate the pattern of firing in the population of ventral horn cells that ultimately generate a given movement.

Thus, while the somatotopic maps in the motor cortex generated by early studies are correct in their overall topo-

graphy, the fine structure of the map is far more abstract. As discussed in the text, it is now widely accepted that the functional maps in the primary motor and premotor cortex are maps of movement. Although coarse somatotopy provides one means of understanding the organization of these motor maps (see Figure 17.10), Michael S. A. Graziano and his colleagues at Princeton University proposed another scheme. Their microstimulation studies of awake, behaving monkeys suggest that the topographic representations of movement in the motor cortex are organized around ethologically relevant categories of motor behavior.

For example, microstimulation of sites in the arm region of the primary motor cortex often invoked movements of the arm that brought the monkey's hand into central space where the animal might visually inspect and manipulate a held object (see Figure 17.12C). Stimulation of more lateral regions (toward the face representation) often lead to hand-to-mouth motions and mouth opening, whereas more medial stimulation sites (toward the trunk and

cord to organize specific movements. This interpretation has been supported by the observation that the regions responsible for initiating different movements overlap substantially. The conclusion that movements rather than muscles are encoded in the cortex also applies to the motor areas of the frontal cortex that control eye movements (see Chapter 20).

About the same time that these studies were being undertaken, Ed Evarts and his colleagues at the National Institutes of Health were pioneering a technique in which implanted microelectrodes were used to record the electrical activity of individual motor neurons in awake, behaving monkeys. The monkeys were trained to perform a variety of motor tasks, thus providing a means of correlating neuronal activity with voluntary movements. Evarts and his group found that the force generated by contracting muscles changed as a function of the firing rate of upper motor neurons. Moreover, the firing rates of the active neurons often changed *prior* to movements involving very small forces. Evarts therefore proposed that the primary motor cortex contributes to the initial phase of recruitment of lower motor neurons involved in the generation of finely controlled movements. Additional experiments showed that the activity of primary motor neurons is correlated not only with the magnitude, but also

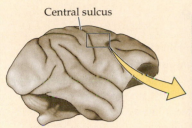

Central sulcus

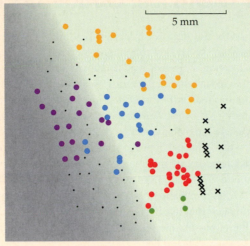

5 mm

- ● Hand-to-mouth
- ● Defensive
- ● Central space/manipulation
- ● Reach
- · Other outward arm movements
- ● Climbing/leaping
- × No movement

Topographic distribution of microstimulation sites that evoke behaviorally relevant movements in a macaque monkey. Rectangular region on brain (left) shows the portion of the motor cortex under investigation. The shaded region in the map of stimulation sites indicates cortex folded into the anterior bank of the central sulcus. (After Graziano et al., 2005.)

GRAZIANO, M. S. A., T. N. S. AFLALO AND D. F. COOKE (2005) Arm movements evoked by electrical stimulation in the motor cortex of monkeys. *J. Neurophysiol.* 94: 4209–4223.

LEMON, R. (1988) The output map of the primate motor cortex. *Trends Neurosci.* 11: 501–506.

PENFIELD, W. AND E. BOLDREY (1937) Somatic motor and sensory representation in the cerebral cortex of man studied by electrical stimulation. *Brain* 60: 389–443.

SCHIEBER, M. H. AND L. S. HIBBARD (1993) How somatotopic is the motor cortex hand area? *Science* 261: 489–491.

WOOLSEY, C. N. (1958) Organization of somatic sensory and motor areas of the cerebral cortex. In *Biological and Biochemical Bases of Behavior*, H. F. Harlow and C. N. Woolsey (eds.). Madison, WI: University of Wisconsin Press, pp. 63–81.

leg representations) evoked climbing- or leaping-like postures. These observations suggests that the posterior regions of the motor cortex, including the primary motor cortex, are most concerned with manual and oral behaviors that occur in central, personal space. Just anterior to this cortical region (in the premotor cortex) are sites that when stimulated elicit reaching motions and other outward arm movements directed away from the body. There are other anterior sites from which coordinated, defensive postures may be evoked, perhaps reflecting the integration of threatening sensory signals derived from extrapersonal space.

Unraveling the details of what is represented in motor maps still holds the key to understanding how patterns of activity in the primate motor cortex generate the rich repertoire of volitional movement.

References

BARINAGA, M. (1995) Remapping the motor cortex. *Science* 268: 1696–1698.

with the direction of the force produced by muscles. Thus, some neurons show progressively less activity as the vector of the movement deviates from the neuron's "preferred direction."

A further advance was made in the mid 1970s with the introduction of *spike-triggered averaging* (Figure 17.11). By correlating the timing of the cortical neuron's discharges with the onset times of the contractions generated by the various muscles used in a movement, this method provides a way of measuring the influence of a single cortical motor neuron on a population of lower motor neurons in the spinal cord. Recording such activity from different muscles as monkeys performed wrist flexion or extension demonstrated that the activity of a number of different muscles is directly facilitated by the discharges of a given upper motor neuron. This peripheral muscle group is referred to as the *muscle field* of the upper motor neuron. On average, the size of the muscle field in the wrist region is 2–3 muscles per upper motor neuron. These observations confirmed that single upper motor neurons contact several lower motor neuron pools, and are consistent with the general conclusion that *movements*, rather than individual muscles, are controlled by the activity of the upper motor neurons in the cortex (see Box 17C).

BOX 17D Sensory Motor Talents and Cortical Space

Are special sensory motor talents, such as the exceptional speed and coordination displayed by talented athletes, ballet dancers, or concert musicians visible in the structure of the nervous system? The widespread use of noninvasive brain imaging techniques (see Box 1A) has generated a spate of studies that have tried to answer this and related questions. Most of these studies have sought to link particular sensory motor skills to the amount of brain space devoted to such talents. For example, a study of professional violinists, cellists, and classical guitarists purported to show that representations of the "fingering" digits of the left hand in the right primary somatic sensory cortex are larger than the corresponding representations in the opposite hemisphere.

Although such studies in humans remain controversial (the techniques are only semiquantitative), the idea that greater motor talents (or any other ability) will be reflected in a greater amount of brain space devoted to that task makes good sense. In particular, comparisons across species show that special talents are invariably based on commensurately sophisticated brain circuitry, which means more neurons, more synaptic contacts between neurons, and more supporting glial cells—all of which occupy more space within the brain. The size and proportion of bodily representations in the primary somatic sensory and motor cortices of various animals reflects species-specific nuances of mechanosensory discrimination and motor control. Thus, the representations of the paws are disproportionately large in the sensorimotor cortex of raccoons; rats and mice devote a great deal of cortical space to representations of their prominent facial whiskers; and a large fraction of the sensorimotor cortex of the star-nosed mole is given over to representing the elaborate nasal appendages that provide critical mechanosensory informa-

tion for this burrowing species. The link between behavioral competence and the allocation of space is equally apparent in animals in which a particular ability has diminished, or has never developed fully, during the course of evolution. Nevertheless, it remains uncertain how—or if—this principle applies to variations in behavior among members of the same species, including humans.

One promising domain of human behavior that has been a fruitful paradigm for exploring the relationship between skill and brain structure is handedness. Investigators have reasoned that some size asymmetry might exist in the somatic sensory and motor systems simply because 90 percent of humans prefer to use the right hand when they perform challenging manual tasks. Several *in vivo* and post mortem studies have now examined the morphometry of the central sulcus in the two hemispheres and, despite some conflicting results, there is a degree of consensus favoring a deeper central sulcus on the left side of right-handers. This asymmetry in gyral size could reflect an underlying difference in the extent of the relevant Brodmann's areas in the two hemispheres (areas 3, 1, and 2 in the postcentral gyrus; areas 6 and 4 in the precentral gyrus), but initial cytoarchitectonic studies have not shown a corresponding asymmetry in the overall size of areas 3 and 4. However, microanatomical studies have shown that there is more tissue volume occupied by cellular processes and synapses in area 4 of the left hemisphere, and that the precentral portion of the corticospinal/corticobulbar tract is larger on the left side than on the right. Although the neurobiological basis of human handedness remains an unresolved issue, such studies suggest that it may be possible to understand the relation between skilled behavior and cortical space if the relevant brain structures

are identified and their functional connections assessed histologically.

It seems likely that individual sensory motor talents among humans will be reflected in the allocation of an appreciably different amount of space to those behaviors, but this issue is just beginning to be explored with quantitative methods that are adequate to the challenge.

References

AMUNTS, K., G. SCHLAUG, A. SCHLEICHER, H. STEINMETZ, A. DABRINGHAUS, P. E. ROLAND AND K. ZILLES (1996) Asymmetry in the human motor cortex and handedness. *NeuroImage* 4: 216–222.

AMUNTS, K., L. JÄNCKE, H. MOHLBERG, H. STEINMETZ AND K. ZILLES (2000) Interhemispheric asymmetry of the human motor cortex related to handedness and gender. *Neuropsychologia* 38: 304–312.

CATANIA, K. C. AND J. H. KAAS (1995) Organization of the somatosensory cortex of the star-nosed mole. *J. Comp. Neurol.* 351: 549–567.

ELBERT, T., C. PANTEV, C. WIENBRUCH, B. ROCKSTROH AND E. TAUB (1995) Increased cortical representation of the fingers of the left hand in string players. *Science* 270: 305–307.

RADEMACHER, J., U. BÜRGEL, S. GEYER, T. SCHORMANN, A. SCHLEICHER, H.-J. FREUND AND K. ZILLES (2001) Variability and asymmetry in the human precentral motor system. *Brain* 124: 2232–2258.

WELKER, W. I. AND S. SEIDENSTEIN (1959) Somatic sensory representation in the cerebral cortex of the raccoon (*Procyon lotos*). *J. Comp. Neurol.* 111: 469–501.

WHITE, L. E., T. J. ANDREWS, C. HULETTE, A. RICHARDS, M. GROELLE, J. PAYDARFAR AND D. PURVES (1997) Structure of the human sensorimotor system. II. Lateral symmetry. *Cereb. Cortex* 7: 31–47.

WOOLSEY, T. A. AND H. VAN DER LOOS (1970) The structural organization of layer IV in the somatosensory region (SI) of mouse cerebral cortex. The description of a cortical field composed of discrete cytoarchitectonic units. *Brain Res.* 17: 205–242.

(A) Detection of postspike facilitation

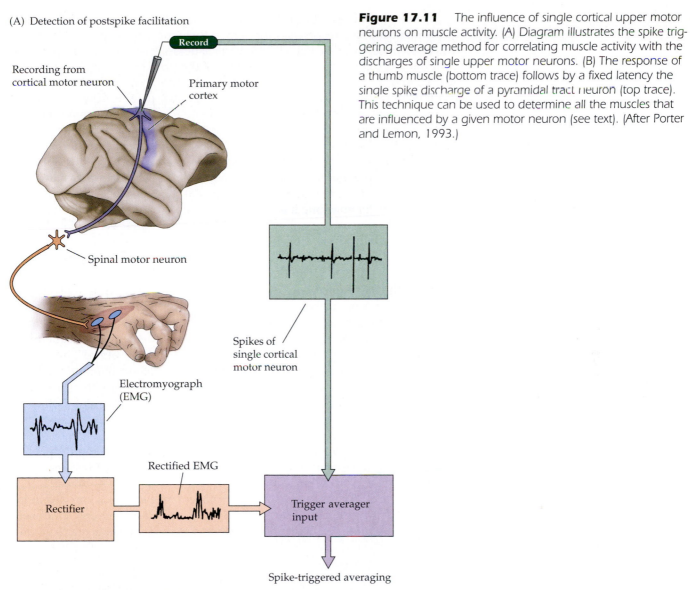

Record

Recording from cortical motor neuron

Primary motor cortex

Spinal motor neuron

Spikes of single cortical motor neuron

Electromyograph (EMG)

Rectified EMG

Rectifier

Trigger averager input

Spike-triggered averaging

Figure 17.11 The influence of single cortical upper motor neurons on muscle activity. (A) Diagram illustrates the spike triggering average method for correlating muscle activity with the discharges of single upper motor neurons. (B) The response of a thumb muscle (bottom trace) follows by a fixed latency the single spike discharge of a pyramidal tract neuron (top trace). This technique can be used to determine all the muscles that are influenced by a given motor neuron (see text). (After Porter and Lemon, 1993.)

(B) Postspike facilitation by cortical motor neuron

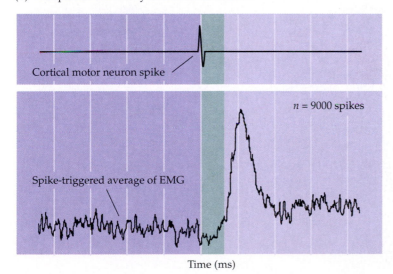

Cortical motor neuron spike

$n = 9000$ spikes

Spike-triggered average of EMG

Time (ms)

For the several reasons discussed above, the motor map in the precentral gyrus is much less precise than the somatotopic map in the postcentral gyrus, where the receptive fields of adjacent cortical neurons overlap in a smooth and continuous progression across the surface of the primary somatic sensory cortex. Indeed, it is problematic to represent the motor map in the form of a homunculus cartoon that would be analogous to the somatic sensory homunculus in the postcentral gyrus (see Figure 9.11), since the representation of muscle movement is not organized at the level of individual muscles or body parts, and the distribution of muscle fields among neighboring cortical neurons is neither spatially continuous or temporally fixed. However, this apparent imprecision in the motor map does not indicate a degenerate representation of the body's musculature in the motor cortex. Rather, it suggests a dynamic and flexible means for encoding higher order movement parameters that entails the coordinated activation of multiple muscle groups across several joints to perform behaviorally useful actions.

This principle of upper motor neural control has been demonstrated by Michael S. A. Graziano and his colleagues at Princeton University, who extended the technique of cortical microstimulation in behaving monkeys to a timescale that more closely corresponds to the duration of volitional movements (from hundreds of milliseconds to several seconds). When such stimuli are applied to the precentral gyrus, the resulting movements are sequentially distributed across multiple joints and are strikingly purposeful (Figure 17.12). Examples of motor patterns frequently elicited with prolonged microstimulation of the precentral gyrus are movements of the hand to the mouth as if to feed, movements that bring the hand to central space as if to inspect an object of interest, and defensive postures as if to protect the body from an impending collision. These findings reinforce the current view that purposeful movements are mapped in the primary motor cortex and that their somatotopic organization is best understood in the context of ethologically relevant behaviors (see Box 17C and the next section).

Finally, the ability to command precise movement patterns is encoded in the integrated activity of a large population of upper motor neurons. One well-studied paradigm for exploring the nature of this neural code involves visually

Figure 17.12 *Purposeful movements of the contralateral arm and hand with prolonged microstimulation of the primary motor cortex in a macaque monkey. Stimulation of sites near the middle of the precentral gyrus elicit coordinated movements of the hand and mouth (A) or movements of the arm that bring the hand toward central space, as if to visually inspect and manipulate a held object (B). The starting positions of the contralateral hand are indicated by the blue crosses, the elicited movements are illustrated with the curved black lines, and the final positions of the hand at the end of microstimulation are indicated by the red dots. (After Graziano et al., 2005.)*

(A)   (B)

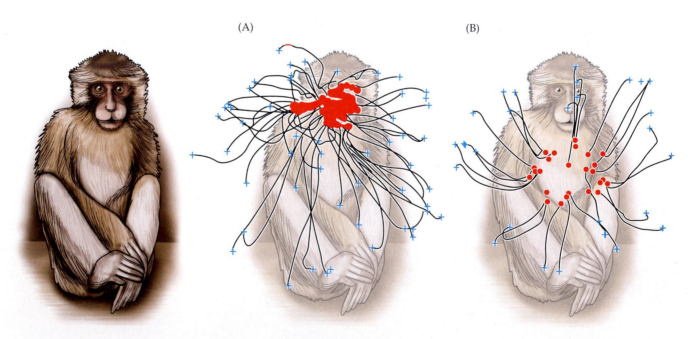

guided reaching movements of the arm and hand. Using this paradigm, the direction of arm movements in monkeys could be predicted by calculating a "neuronal population vector" derived simultaneously from the discharges of upper motor neurons that are "broadly tuned" in the sense that they discharge prior to movements in many directions (Figure 17.13). These observations showed that the discharges of individual upper motor neurons cannot specify the direction of an arm movement, simply because they are tuned too broadly; rather, each arm movement must be encoded by the concurrent discharges of a

(A)

(B)

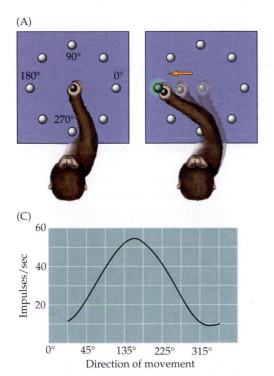

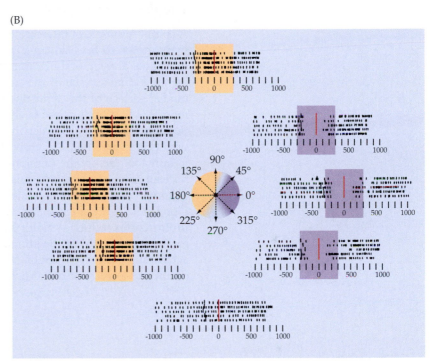

(C)

(D)

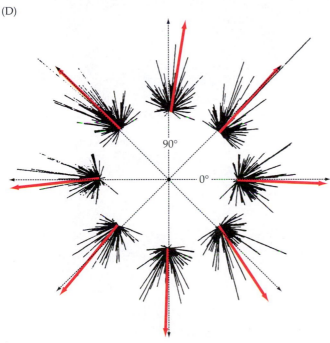

Figure 17.13   Directional tuning of an upper motor neuron in the primary motor cortex. (A) A monkey is trained to move a joystick in the direction indicated by a light. (B) The activity of a single neuron was recorded during arm movements in each of eight different directions (0 indicates the time of movement onset; each short vertical line in this raster plot represents an action potential). The activity of the neuron increased before movements between 90° and 225° (yellow zone), but decreased in anticipation of movements between 45° and 315° (purple zone). (C) Plot showing that the neuron's discharge rate was greatest before movements in a particular direction, which defines the neuron's "preferred direction" in this experimental paradigm. (D) The black lines indicate the discharge rates of individual upper motor neurons prior to each direction of movement. By combining the responses of all the neurons, a "population vector" (red arrow) can be derived that represents the movement direction encoded by the simultaneous activity of the entire population. (After Georgeopoulos et al., 1986.)

large population of such neurons. The fact that the same site in the primary motor cortex can encode different trajectories of motion depending upon the starting position of the limb (see Figure 17.12) suggests that multiple parameters of movement may be selected by the relevant ensemble of upper motor neurons to achieve a behaviorally useful action. Thus, taken together, the microstimulation experiments described above utilize exogenous electrical currents to engage populations of upper motor neurons whose output encodes not simply the trajectory of arm motion, but also the final position of the hand.

The Premotor Cortex

A complex mosaic of interconnected frontal lobe areas that lie rostral to the primary motor cortex also contributes to motor functions (see Figure 17.7). This functional division of the motor cortex includes Brodmann's areas 6, 8, and 44/45 on the lateral surface of the frontal lobe and parts of areas 23 and 24 on the medial surface of the hemisphere. Although the organization of this premotor mosaic is best understood in the macaque monkey (Figure 17.14), recent functional brain imaging studies in human subjects, as well as studies in patients with frontal lobe injuries, supports a comparable distribution of premotor areas in humans. Each of the divisions of the premotor cortex receives extensive multisensory input from regions of the inferior and superior parietal lobules, as well as more complex signals related to motivation and intention from the rostral ("prefrontal") divisions of the frontal lobe. The upper motor neurons in this **premotor cortex** influence motor behavior both indirectly through extensive reciprocal connections with the primary motor cortex, and directly via axons that project through the corticobulbar and corticospinal pathways to influence local circuit and lower motor neurons of the brainstem and spinal cord.

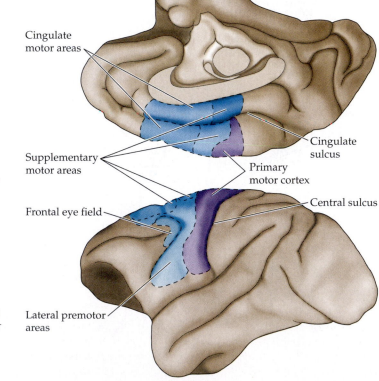

Figure 17.14   Divisions of the motor cortex in the macaque monkey brain. As in humans, the primary motor cortex resides in the anterior bank of the central sulcus. Anterior to this region is a complex mosaic of premotor areas that extends from the frontal operculum on the lateral surface of the frontal lobe to the cingulate gyrus on the medial hemispheric surface. The lateral premotor and supplementary motor areas are involved in selecting and organizing purposeful movements of the limbs and face; the frontal eye fields organize voluntary gaze shifts (see Chapter 20) and the cingulate motor areas are involved in the expression of emotional somatic behavior (see Chapter 29). Current evidence supports the existence of comparable premotor areas in the human motor cortex. (After Geyer et al., 2000.)

Indeed, over 30 percent of the axons in the corticospinal tract arise from neurons in the premotor cortex. Thus, past arguments that the premotor cortex occupies a higher position in a cortical hierarchy of motor control operating through feedforward signals to the primary motor cortex are no longer tenable. Rather, a variety of experiments indicate that the premotor cortex uses information from other cortical regions to select movements appropriate to the context of the action (see Chapter 26). The principal difference between the premotor cortex and primary motor cortex lies in the strength of their connections to lower motor neurons, with more upper motor neurons in the primary motor cortex making monosynaptic connections to α-motor neurons, especially those in the ventral horn of the cervical spinal cord that control the distal upper extremities. Recent evidence suggests that other differences may reflect the mapping of purposeful movements relative to personal and extrapersonal space and the nature of the signals that lead to the initiation of motor commands (see Box 17C).

The functions of the premotor cortex may be considered in terms of differences between the lateral and medial components of this region. As many as 65 percent of the neurons in the lateral premotor cortex have responses that are linked in time to the occurrence of movements; as in the primary motor area, many of these cells fire most strongly in association with movements made in a specific direction. However, these neurons are especially important in conditional ("closed-loop") motor tasks. That is, in contrast to the neurons in the primary motor area, when a monkey is trained to reach in different directions depending on the nature of a visual cue, the appropriately tuned lateral premotor neurons begin to fire at the appearance of the cue, well before the monkey receives a signal to actually make the movement. As the animal learns to associate a new visual cue with the movement, appropriately tuned neurons begin to increase their rate of discharge in the interval between the cue and the onset of the signal to perform the movement. Rather than directly commanding the initiation of a movement, these neurons appear to encode the monkey's *intention* to perform a particular movement; thus, they seem to be particularly involved in the *selection* of movements based on external events.

One particular division in the ventrolateral portion of the premotor cortex has received considerable attention in recent years, after the discovery that a subset of its neurons responds not just in preparation for the execution of particular movements, such as a precision grip to retrieve a morsel of food, but also when the same action is *observed*, being performed by another (monkey or human). For example, these premotor neurons fire action potentials when a monkey observes the hand of a human trainer engaging in the same or similar action that would activate these neurons during self-initiated movements (Figure 17.15). However, these so-called **mirror motor neurons** respond much less well when the same actions are pantomimed without the presence of a behavioral goal, such as an object to be grasped. Furthermore, they respond during the observation of goal-directed behavior even when the final stage of the action is hidden from view—for example, the grasping of an object known to the monkey to have been placed behind a small barrier. These findings suggest that the mirror motor system is involved in encoding the intentions and behaviorally relevant actions of others and may participate in an extended network of parietal and frontal regions that subserve imitation learning. The functions of the mirror motor system are among the most actively studied and debated domains of motor and cognitive neuroscience, but the full scope of this system's contributions to motor control, motor learning, and to complex brain functions such as social communication, language, and empathy, remain to be elucidated.

Further evidence that the lateral premotor area is concerned with movement selection comes from the effects of cortical damage on motor behavior. Lesions in

Figure 17.15 Demonstration of the activity of a mirror motor neuron in a ventral-anterior sector of the lateral premotor cortex. In each panel, the upper graphics illustrate the monkey's view of the hand of the trainer placing a food morsel on a tray and the monkey's own hand extending to retrieve the morsel. The middle graphics illustrate raster plots that show the firing of the neuron relative to the observed and executed movements (each tick mark indicates an action potential and each row represents one trial). The lower graphs are peristimulus response histograms aligned to the overlying raster plots. The mirror motor neuron fires during the passive observation of a human hand placing the morsel of food on the tray (A), as well as during the execution of a similar action to retrieve the food (the vertical line in the raster plots indicates the time at which the food was placed on the tray; 1–2 seconds later, the monkey reaches to retrieve the morsel). The same neuron does not respond when the food is placed with the aid of pliers (B), but it does fire during the monkey's reaching and retrieval movements when the monkey is allowed to observe its reach (B) and when the behavior is executed behind a barrier (C). These findings suggest that this division of the premotor cortex plays a role in encoding the actions of the others. (After Rizzolatti et al., 1996.)

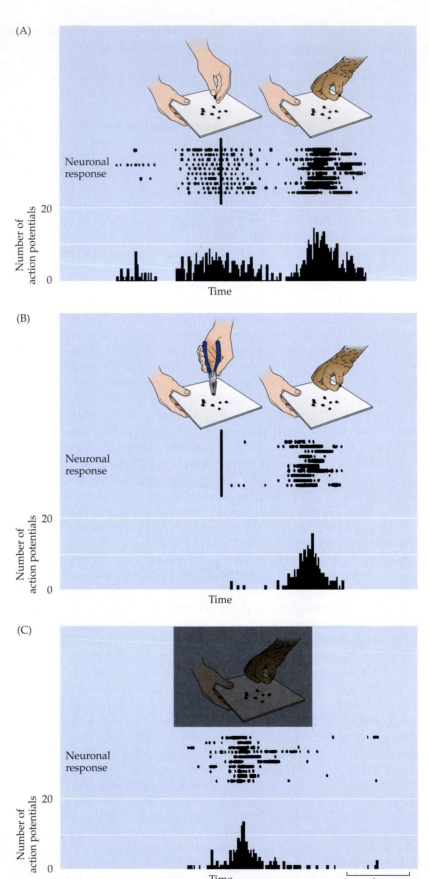

this region severely impair the ability of monkeys to perform visually cued conditional tasks, even though they can still respond to the visual stimulus and can perform the same movement in a different setting. Similarly, patients with frontal lobe damage have difficulty learning to select a particular movement to be performed in response to a visual cue, even though they understand the instructions and can perform the movements. Individuals with lesions in the premotor cortex may also have difficulty performing movements in response to verbal commands.

Finally, a rostral division of the lateral premotor cortex in the human brain, especially in the left hemisphere, has evolved to play a special role in organizing vocal tract articulators that are involved in the production of speech sounds. This region, which corresponds to **Broca's area** (Brodmann's areas 44 and 45), is critical for the production of speech and will be considered in detail in Chapter 27. The evolution of this premotor division in primates and its functional relation to semantic processing regions in the parietal and temporal lobe remains an area of active investigation.

The medial premotor cortex, like the lateral area, mediates the selection of movements. However, this region appears to be specialized for initiating movements specified by *internal* rather than *external* cues ("open-loop" conditions). In contrast to lesions in the lateral premotor area, removal of the medial premotor area in a monkey reduces the number of self-initiated or "spontaneous" movements the animal makes, whereas the ability to execute movements in response to external cues remains largely intact. Imaging studies suggest that this cortical region in humans functions in much the same way. For example, PET scans show that the medial region of the premotor cortex is activated when the subjects perform motor sequences from memory (i.e., without relying on an external instruction). In accord with this evidence, single unit recordings in monkeys indicate that many neurons in the medial premotor cortex begin to discharge one or two seconds before the onset of a self-initiated movement. Among the areas of the medial premotor cortex are two divisions that will be considered in more detail elsewhere: a frontal eye field (see Figure 17.14) involved in directing visual gaze toward a location of interest (see also Chapter 20); and a set of areas in the depths of the cingulate sulcus (see Figure 17.14 and Box 17B) that plays a role in the expression of emotional behavior (see also Chapter 29).

In summary, both the lateral and medial areas of the premotor cortex are intimately involved in selecting a specific movement or sequence of movements from the repertoire of possible behaviorally relevant actions. The functions of the areas differ, however, in the relative contributions of external and internal cues to the selection process.

Damage to Descending Motor Pathways: The Upper Motor Neuron Syndrome

Injury to upper motor neurons is common because of the large amount of cortex occupied by the motor areas, and because their pathways extend all the way from the cerebral cortex to the lower end of the spinal cord. Damage to the descending motor pathways anywhere along this trajectory gives rise to a set of symptoms called **upper motor neuron syndrome**.

This clinical picture differs markedly from the lower motor neuron syndrome described in Chapter 16 and entails a characteristic set of motor deficits (Table 17.1). Damage to the motor cortex or the descending upper motor axons in the internal capsule causes an immediate flaccidity of the muscles on the contralateral side of the body and lower face. Given the topographical arrangement of the motor system, identifying the specific parts of the body that are affected helps localize the site of the injury. The acute manifestations tend to be most severe in

TABLE 17.1	Signs and Symptoms of Upper and Lower Motor Neuron Lesions	
Upper Motor Neuron Syndrome	**Lower Motor Neuron Syndrome**	
Weakness	Weakness or paralysis	
Spasticity	Decreased superficial reflexes	
Increased tone	Hypoactive deep reflexes	
Hyperactive deep reflexes	Decreased tone	
Clonus	Fasciculations and fibrillations	
Babinski's sign	Severe muscle atrophy	
Loss of fine voluntary movements		

the arms and legs: if the affected limb is elevated and released, it drops passively, and all reflex activity on the affected side is abolished. In contrast, control of trunk muscles is usually preserved, either by the remaining brainstem pathways or because of the bilateral projections of the corticospinal pathway to local circuits that control midline musculature. This initial period of "hypotonia" after upper motor neuron injury is called **spinal shock**, and reflects the decreased activity of spinal circuits suddenly deprived of input from the motor cortex and brainstem.

After several days, however, the spinal cord circuits regain much of their function for reasons that are not fully understood. Thereafter, a consistent pattern of motor signs and symptoms emerges, including:

- *The Babinski sign.* The normal response in an adult to stroking the sole of the foot is flexion of the big toe, and often the other toes. Following damage to descending upper motor neuron pathways, however, this stimulus elicits extension of the big toe and a fanning of the other toes (Figure 17.16). A similar response occurs in human infants before the maturation of the corticospinal pathway and presumably indicates incomplete upper motor neuron control of local motor neuron circuitry.

- *Spasticity.* Spasticity is increased muscle tone (Box 17E), hyperactive stretch reflexes, and clonus (oscillatory contractions and relaxations of muscles in response to muscle stretching). Extensive upper motor neuron lesions may be accompanied by rigidity of the extensor muscles of the leg and the flexor muscles of the arm (called *decerebrate rigidity*; see below). Spasticity is probably caused by the removal of inhibitory influences exerted by the cortex on the postural centers of the vestibular nuclei and reticular formation. In experimental animals, for instance, lesions of the vestibular nuclei ameliorate the spasticity that follows damage to the corticospinal tract. Spasticity is also eliminated by sectioning the dorsal roots, suggesting that it represents an abnormal increase in the *gain* of the spinal cord stretch reflexes due to loss of descending inhibition (see Chapter 16). This increased gain is also thought to explain clonus.

- *A loss of the ability to perform fine movements.* If the lesion involves the descending pathways that control the lower motor neurons to the upper limbs, the ability to execute fine movements (such as independent movements of the fingers) is lost.

Although these upper motor neuron signs and symptoms may arise from damage anywhere along the descending pathways, the spasticity that follows damage to descending pathways in the spinal cord is less marked than the spasticity that follows damage to the cortex or internal capsule.

(A) Normal plantar response

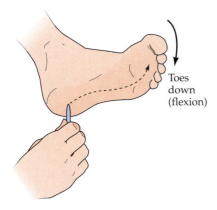

Toes down (flexion)

(B) Extensor plantar response (Babinski sign)

Up

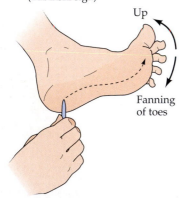

Fanning of toes

Figure 17.16 The Babinski sign. Following damage to descending corticospinal pathways, stroking the sole of the foot causes an abnormal fanning of the toes and the extension of the big toe.

BOX 17E   Muscle Tone

Muscle tone is the resting level of tension in a muscle. In general, maintaining an appropriate level of muscle tone allows a muscle to make an optimal response to voluntary or reflexive commands in a given context. Tone in the extensor muscles of the legs, for example, helps maintain posture while standing. By keeping the muscles in a state of readiness to resist stretch, tone in the leg muscles prevents the amount of sway that normally occurs while standing from becoming too large. During activities such as walking or running, the "background" level of tension in leg muscles also helps to store mechanical energy, in effect enhancing the muscle tissue's springlike qualities. Muscle tone depends on the resting level of discharge of α motor neurons. Activity in the Ia spindle afferents—the neurons responsible for the stretch reflex—is the major contributor to this tonic level of firing. As described in Chapter 16, the γ efferent system (by its action on intrafusal muscle fibers) regulates the resting level of activity in the Ia afferents and establishes the baseline level of α motor neuron activity in the absence of muscle stretch.

Clinically, muscle tone is assessed by judging the resistance of a patient's limb to passive stretch. Damage to either the α motor neurons or the Ia afferents carrying sensory information to the α motor neurons results in a decrease in muscle tone, called *hypotonia*. In general, damage to descending pathways that terminate in the spinal cord has the opposite effect, leading to an increase in muscle tone, or *hypertonia* (except during the initial phase of spinal shock; see text). The neural changes responsible for hypertonia following damage to higher centers are not well understood; however, at least part of this change is due to an increase in the responsiveness of α motor neurons to Ia sensory inputs. Thus, in experimental animals in which descending inputs have been severed, the resulting hypertonia can be eliminated by sectioning the dorsal roots.

Increased resistance to passive movement following damage to higher centers is called spasticity, and is associated with two other characteristic signs: the clasp-knife phenomenon and clonus. When first stretched, a spastic muscle provides a high level of resistance to the stretch and then suddenly yields, much

like the blade of a pocket knife (or clasp knife, in old-fashioned terminology). Hyperactivity of the stretch reflex loop is the reason for the increased resistance to stretch in the clasp-knife phenomenon. The physiological basis for the inhibition that causes the sudden collapse of the stretch reflex (and loss of muscle tone) may involve the activation of Golgi tendon organs (see Chapter 16).

Clonus refers to a rhythmic pattern of contractions (3–7 per second) due to the alternate stretching and unloading of the muscle spindles in a spastic muscle. Clonus can be demonstrated in the flexor muscles of the leg by pushing up on the sole of patient's foot to dorsiflex the ankle. If there is damage to descending upper motor neuron pathways, holding the ankle loosely in this position generates rhythmic contractions of both the gastrocnemius and soleus muscles. Both the increase in muscle tone and the pathological oscillations seen after damage to descending pathways are very different from the tremor at rest and cogwheel rigidity present in basal ganglia disorders such as Parkinson's disease, phenomena discussed in Chapters 18 and 19.

For example, the spastic extensor muscles in the legs of a patient with spinal cord damage cannot support the individual's body weight, whereas those of a patient with damage at the cortical level often can. On the other hand, lesions that interrupt the descending pathways in the brainstem above the level of the vestibular nuclei but below the level of the red nucleus cause even greater extensor tone than that which occurs after damage to higher regions. Sherrington, who first described this phenomenon, called the increased tone **decerebrate rigidity**. In the cat, the extensor tone in all four limbs is so great after lesions that spare the vestibulospinal tracts that the animal can stand without support. Patients with severe brainstem injury at the level of the pons may exhibit similar signs of decerebration: arms and legs stiffly extended, jaw clenched, and neck retracted. The relatively greater hypertonia following damage to the nervous system above the level of the spinal cord is presumably explained by the remaining activity of the intact descending pathways from the vestibular nuclei and reticular formation, which have a net excitatory influence on these stretch reflexes.

Summary

Two sets of upper motor neuron pathways make distinct contributions to the control of the local circuitry in the brainstem and spinal cord. One set originates from neurons in brainstem centers—primarily the reticular formation and the vestibular nuclei—and is responsible for postural regulation. The reticular formation is especially important in *feedforward* control of posture (that is, movements that occur in anticipation of changes in body stability). In contrast, the neurons in the vestibular nuclei that project to the spinal cord are especially important in *feedback* postural mechanisms (i.e., in producing movements that are generated in response to sensory signals that indicate an existing postural disturbance). The other major upper motor neuron pathway originates from the frontal lobe and includes projections from the primary motor cortex and the nearby premotor areas. The premotor cortices are responsible for planning and selecting movements, especially movements that are triggered by sensory cues or internal motivations, whereas the primary motor cortex is especially involved with the execution of skilled movements of the limb and facial musculature. The motor cortex influences movements *directly* by contacting lower motor neurons and local circuit neurons in the spinal cord and brainstem, and *indirectly* by innervating neurons in brainstem centers (mainly the reticular formation) that in turn project to lower motor neurons and circuits. Although the brainstem pathways can independently organize gross motor control, direct projections from the motor cortex to local circuit neurons in the brainstem and spinal cord are essential for the fine, fractionated movements of the distal parts of the limbs, tongue, and face that are especially important in our daily lives.

Additional Reading

Reviews

DUM, R. P. AND P. L. STRICK (2002) Motor areas in the frontal lobe of the primate. *Physiol. Behav.* 77: 677–682.

GAHERY, Y. AND J. MASSION (1981) Coordination between posture and movement. *Trends Neurosci.* 4: 199–202.

GEORGEOPOULOS, A. P., M. TAIRA AND A. LUKASHIN (1993) Cognitive neurophysiology of the motor cortex. *Science* 260: 47–52.

GEYER, S., M. MATELLI AND G. LUPPINO (2000) Functional neuroanatomy of the primate isocortical motor system. *Anat. Embryol.* 202: 443–474.

GRAZIANO, M. S. A. (2006) The organization of behavioral repertoire in motor cortex. *Annu. Rev. Neurosci.* 29: 105–134.

KUYPERS, H. G. J. M. (1981) Anatomy of the descending pathways. In *Handbook of Physiology*, Section 1: *The Nervous System*, Volume II, *Motor Control*, Part 1, V. B. Brooks (ed.). Bethesda, MD: American Physiological Society.

NASHNER, L. M. (1979) Organization and programming of motor activity during posture control. In *Reflex Control of Posture and Movement*, R. Granit and O. Pompeiano (eds.). *Prog. Brain Res.* 50: 177–184.

NASHNER, L. M. (1982) Adaptation of human movement to altered environments. *Trends Neurosci.* 5: 358–361.

RIZZOLATTI, G. AND L. CRAIGHERO (2004) The mirror-neuron system. *Annu. Rev. Neurosci.* 27: 169–192.

SHERRINGTON, C. AND S. F. GRUNBAUM (1901) Observations on the physiology of the cerebral cortex of some of the higher apes. *Proc. Roy. Soc.* 69: 206–209.

Important Original Papers

EVARTS, E. V. (1981) Functional studies of the motor cortex. In *The Organization of the Cerebral Cortex*, F. O. Schmitt, F. G. Worden, G. Adelman and S. G. Dennis (eds.). Cambridge, MA: MIT Press, pp. 199–236.

FETZ, E. E. AND P. D. CHENEY (1978) Muscle fields of primate corticomotoneuronal cells. *J. Physiol. (Paris)* 74: 239–245.

FETZ, E. E. AND P. D. CHENEY (1980) Postspike facilitation of forelimb muscle activity by primate corticomotoneuronal cells. *J. Neurophysiol.* 44: 751–772.

GEORGEOPOULOS, A. P., A. B. SWARTZ AND R. E. KETTER (1986) Neuronal population coding of movement direction. *Science* 233: 1416–1419.

GRAZIANO, M. S. A., T. N. S. AFLALO AND D. F. COOKE (2005) Arm movements evoked by electrical stimulation in the motor cortex of monkeys. *J. Neurophysiol.* 94: 4209–4223.

LAWRENCE, D. G. AND H. G. J. M. KUYPERS (1968) The functional organization of the motor system in the monkey. I. The effects of bilateral pyramidal lesions. *Brain* 91: 1–14.

MITZ, A. R., M. GODSCHALK AND S. P. WISE (1991) Learning-dependent neuronal activity in the premotor cortex: Activity during the acquisition of conditional motor associations. *J. Neurosci.* 11: 1855–1872.

RIZZOLATTI, G., L. FADIGA, V. GALLESE AND L. FOGASSI (1996) Premotor cortex and the recognition of motor actions. *Cogn. Brain Res.* 3: 131–141.

ROLAND, P. E., B. LARSEN, N. A. LASSEN AND E. SKINHOF (1980) Supplementary motor area and other cortical areas in organization of voluntary movements in man. *J. Neurophysiol.* 43: 118–136.

SANES, J. N. AND W. TRUCCOLO (2003) Motor "binding": Do functional assemblies in primary motor cortex have a role? *Neuron* 38: 115–125.

SCHIEBER, M. H. AND L. S. HIBBARD (1993) How somatotopic is the motor cortex hand area? *Science* 261: 489–492.

Books

ASANUMA, H. (1989) *The Motor Cortex.* New York: Raven Press.

BRODAL, A. (1981) *Neurological Anatomy in Relation to Clinical Medicine*, 3rd Ed. New York: Oxford University Press.

BROOKS, V. B. (1986) *The Neural Basis of Motor Control*. New York: Oxford University Press.

PASSINGHAM, R. (1993) *The Frontal Lobes and Voluntary Action*. Oxford: Oxford University Press.

PENFIELD, W. AND T. RASMUSSEN (1950) *The Cerebral Cortex of Man: A Clinical Study of Localization of Function*. New York: Macmillan.

PHILLIPS, C. G. AND R. PORTER (1977) *Corticospinal Neurons: Their Role in Movement*. London: Academic Press.

PORTER, R. AND R. LEMON (1993) *Corticospinal Function and Voluntary Movement*. Oxford: Oxford University Press.

SHERRINGTON, C. (1947) *The Integrative Action of the Nervous System*, 2nd Ed. New Haven: Yale University Press.

SJÖLUND, B. AND A. BJÖRKLUND (1982) *Brainstem Control of Spinal Mechanisms*. Amsterdam: Elsevier.

Chapter 18

Modulation of Movement by the Basal Ganglia

Overview

As described in the preceding chapter, motor regions of the cerebral cortex and brainstem contain upper motor neurons that initiate movement by controlling the activity of local circuit and lower motor neurons in the brainstem and spinal cord. This chapter and the next discuss two additional regions of the brain that are important in motor control: the basal ganglia and the cerebellum. In contrast to the components of the motor system that harbor upper motor neurons, the basal ganglia and cerebellum do not project directly to either the local circuit or lower motor neurons; instead, they influence movement by regulating the activity of upper motor neurons. The term *basal ganglia* refers to a large and functionally diverse set of nuclei that lies deep within the cerebral hemispheres. The subset of these nuclei relevant to this account of motor function includes the caudate, the putamen, and the globus pallidus. Two additional structures, the substantia nigra in the base of the midbrain and the subthalamic nucleus in the ventral thalamus, are closely associated with the motor functions of these basal ganglia nuclei. The motor components of the basal ganglia, together with the substantia nigra and the subthalamic nucleus, effectively make a subcortical loop that links most areas of the cerebral cortex with upper motor neurons in the primary motor and premotor cortex and in the brainstem. The neurons in this loop modulate their activity in anticipation of and during movements, and their effects on upper motor neurons are required for the normal course of voluntary movements. When one of these components of the basal ganglia or associated structures is compromised, the motor systems cannot switch smoothly between commands that initiate a movement and those that terminate the movement. The disordered movements that result can be understood as a consequence of abnormal upper motor neuron activity in the absence of the regulatory control normally provided by the basal ganglia.

Projections to the Basal Ganglia

The motor nuclei of the basal ganglia are divided into several functionally distinct groups. The first and larger of these groups is called the **corpus striatum**, which includes two nuclei, the **caudate** and the **putamen** (Figure 18.1). The name *corpus striatum*, which means "striped body," reflects the fact that caudate and the dorsal part of the putamen are joined by slender bridges of gray matter that extend through the internal capsule and confer a striped appearance in parasagittal sections through this area. The two subdivisions of the corpus striatum comprise the *input* zone of the basal ganglia, their neurons being the destinations of most of the pathways that reach this complex from other parts of the brain (Figure 18.2). The destinations of the incoming axons from the cerebral cortex are the dendrites of a class of cells in the corpus striatum called **medium**

Figure 18.1 Motor components of the human basal ganglia. The basal ganglia comprise a set of gray matter structures, most of which are are buried deep in the telencephalon but some are found in the diencephalon and midbrain. The major components that receive and process movement related signals are the corpus striatum (caudate and putamen) and the pallidum (globus pallidus and substantia nigra pars reticulata); these structures border the internal capsule in the forebrain and midbrain (the cerebral peduncle is a caudal extension of the internal capsule). Smaller but functionally significant components of the basal ganglia system are the substantia nigra pars compacta and the subthalamic nucleus, which provide input to the corpus striatum and pallidum, respectively. For the control of limb movements, output from the basal ganglia arises in the internal segment of the globus pallidus and is sent to the ventral anterior and ventral lateral nuclei (VA/VL complex) of the thalamus, which interact directly with circuits of upper motor neurons in frontal cortex. The substantia nigra pars reticulata projects to the superior colliculus and controls orienting movements of the eyes and head.

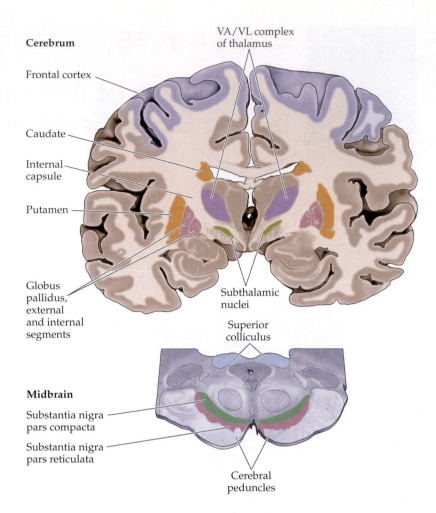

spiny neurons (Figure 18.3). The large dendritic trees of these neurons allow them to integrate inputs from a variety of cortical, thalamic, and brainstem structures. The axons arising from the medium spiny neurons converge on neurons in the **pallidum**, which includes the **globus pallidus** and the **substantia nigra pars reticulata**. The globus pallidus and substantia nigra pars reticulata are the main sources of *output* from the basal ganglia complex (see Figure 18.5).

Nearly all regions of the cerebral cortex project directly to the corpus striatum, making the cortex the source of the largest input to the basal ganglia. Indeed, the only cortical areas that do not project to the corpus striatum are the primary visual and primary auditory cortices (Figure 18.4). Of the cortical areas that do innervate the striatum, the heaviest projections are from association areas in the frontal and parietal lobes, but substantial contributions also arise from the temporal, insular, and cingulate cortices. All of these projections, referred to collectively as the **corticostriatal pathway**, travel through the subcortical white matter on their way to the caudate and putamen (see Figure 18.2).

The cortical inputs to the caudate and putamen are not equivalent, however, and the differences in input reflect functional differences between these two nuclei. The caudate receives cortical projections primarily from multimodal association cortices, and from motor areas in the frontal lobe that control eye movements. As the name implies, the association cortices do not process any one type of sensory information; rather, they receive inputs from a number of primary and secondary sensory cortices and associated thalamic nuclei (see Chapter 26). The putamen, on the other hand, receives input from the primary

Cerebrum

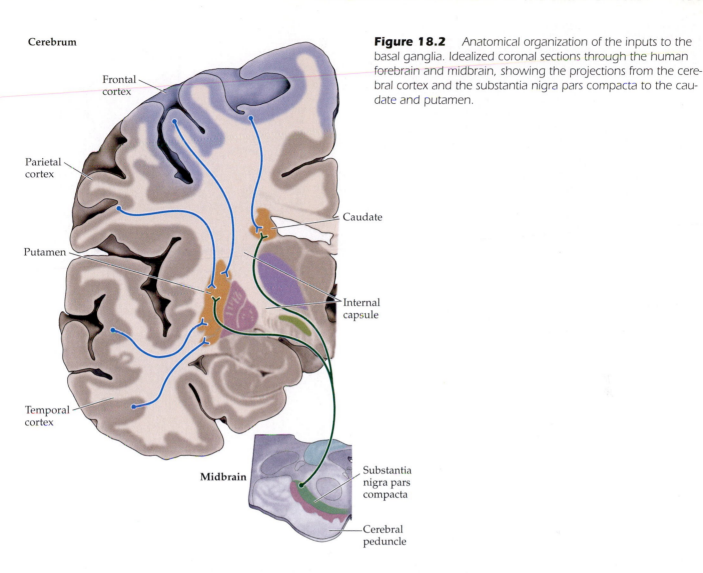

Figure 18.2 *Anatomical organization of the inputs to the basal ganglia. Idealized coronal sections through the human forebrain and midbrain, showing the projections from the cerebral cortex and the substantia nigra pars compacta to the caudate and putamen.*

Frontal cortex

Parietal cortex

Putamen

Temporal cortex

Caudate

Internal capsule

Midbrain

Substantia nigra pars compacta

Cerebral peduncle

and secondary somatic sensory cortices in the parietal lobe, the secondary (extrastriate) visual cortices in the occipital and temporal lobes, the premotor and motor cortices in the frontal lobe, and the auditory association areas in the temporal lobe. The fact that different cortical areas project to different regions of the striatum implies that the corticostriatal pathway consists of multiple parallel pathways serving different functions. This interpretation is supported by the observation that the segregation is maintained in the structures that receive projections from the striatum, and in the pathways that project from the basal ganglia to other brain regions.

There are other indications that the corpus striatum is functionally subdivided according to its inputs. For example, visual and somatic sensory cortical projections are topographically mapped within different regions of the putamen. Moreover, the cortical areas that are functionally interconnected at the level of the cortex give rise to projections that overlap extensively in the striatum. Anatomical studies by Ann Graybiel and her colleagues at the Massachusetts Institute of Technology have shown that regions of different cortical areas concerned with the hand (see Chapter 9) converge in specific rostrocaudal bands within the striatum; conversely, regions in the same cortical areas concerned with the leg converge in other striatal bands. These rostrocaudal bands therefore appear to be functional units concerned with the movement of partic-

(A)

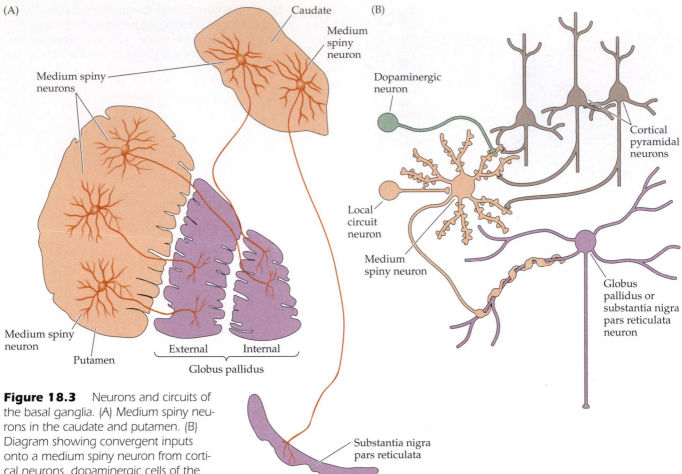

Medium spiny neurons

Caudate

Medium spiny neuron

Medium spiny neuron

Putamen

External   Internal

Globus pallidus

(B)

Dopaminergic neuron

Local circuit neuron

Medium spiny neuron

Cortical pyramidal neurons

Globus pallidus or substantia nigra pars reticulata neuron

Substantia nigra pars reticulata

Figure 18.3 *Neurons and circuits of the basal ganglia. (A) Medium spiny neurons in the caudate and putamen. (B) Diagram showing convergent inputs onto a medium spiny neuron from cortical neurons, dopaminergic cells of the substantia nigra, and local circuit neurons within the corpus striatum. The arrangement of these synapses indicates that the response of the medium spiny neurons to their principal input, derived from the cerebral cortex, can be modulated by dopamine and the inputs of local circuit neurons. The primary output of the medium spiny cells is to pallidal neurons in the globus pallidus and substantia nigra pars reticulata.*

ular body parts. Another study by the same group showed that the more extensively cortical areas are interconnected by corticocortical pathways, the greater the overlap in their projections to the striatum. Thus, the specialization of functional units within the corpus striatum reflects the specialization of the cortical areas that provide input.

A further indication of functional subdivision within the corpus striatum is evident when tissue sections obtained postmortem are stained for the presence of different neurotransmitters and their related enzymes. For example, when the striatum is stained for the enzyme acetylcholinesterase, which inactivates acetylcholine (see Chapter 6), a compartmental organization is revealed within the striatum. The compartments are defined by lightly stained regions, called "patches" or "striosomes," surrounded by densely stained tissue, called "matrix." Subsequent studies of the distributions of other neurochemicals, including peptide neurotransmitters, have catalogued a variety of neuroactive substances that localize to patch or matrix compartments. Tract-tracing experiments in animals have likewise shown differences between these striatal compartments in the sources of their inputs from the cortex and in the destinations of their projections to other parts of the basal ganglia. For example, the matrix comprises the bulk of the corpus striatum; it receives input from most areas of the cerebral cortex and sends projections to the globus pallidus and the substantia nigra pars reticulata. The patches in the caudate receive most of their input from the pre-

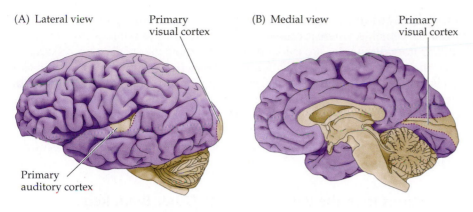

(A) Lateral view — Primary visual cortex — Primary auditory cortex

(B) Medial view — Primary visual cortex

Figure 18.4 Regions of the cerebral cortex (shown in purple) that project to the corpus striatum. The caudate, putamen, and ventral striatum receive cortical projections primarily from the association areas of the frontal, parietal, and temporal lobes (see Box 18C).

frontal cortex (see Chapter 26) and project preferentially to a different subdivision of the substantia nigra (see below). These differences between the connectivity of medium spiny neurons in the patches and matrix further support the conclusion that functionally distinct pathways project in parallel from the cerebral cortex to the corpus striatum.

The nature of the signals transmitted to the caudate and putamen from the cerebral cortex is not understood. It is known, however, that collateral axons of corticocortical, corticothalamic, and corticospinal pathways all form excitatory glutamatergic synapses on the dendritic spines of medium spiny neurons (see Figure 18.3B). The arrangement of these cortical synapses is such that the number of contacts established between an individual cortical axon and a single medium spiny cell is very small, whereas the number of spiny neurons contacted by a single axon is extremely large. This divergence of the input from corticostriatal axons allows a single medium spiny neuron to integrate the influences of thousands of cortical cells.

The medium spiny cells also receive inputs from several other sources, besides the cerebral cortex, including other medium spiny neurons via their local axon collaterals, intrinsic interneurons of the corpus striatum, neurons in the midline and intralaminar nuclei of the thalamus, and neurons in several aminergic nuclei of the brainstem. In contrast to the cortical inputs to the dendritic spines, the local circuit neuron and thalamic synapses are made on the dendritic shafts and close to the cell soma, where they can modulate the effectiveness of cortical synaptic activation arriving from the more distal dendrites. The main aminergic inputs are dopaminergic, and they originate in a nigral subdivision called the **substantia nigra pars compacta** because of its densely packed cells. (The striatum also receives serotonergic inputs from the raphe nuclei; see Chapter 6.) The dopaminergic synapses are located on the base of the spine, in close proximity to the cortical synapses, where they more directly and selectively modulate cortical input (see Figure 18.3B). As a result, inputs from both the cortex and the substantia nigra pars compacta are relatively far from the initial segment of the medium spiny neuron axon, where the nerve impulse is generated. Furthermore, medium spiny neurons express inward-rectifier potassium conductances that tend to remain open near resting membrane potentials, but close with depolarization. Accordingly, these neurons exhibit very little spontaneous activity and must simultaneously receive many excitatory inputs from cortical and nigral neurons to overcome the stabilizing influence of this potassium conductance.

When the medium spiny neurons do become active, their firing is associated with the occurrence of a movement. Extracellular recordings show that these neurons typically increase their rate of discharge before an impending movement. Neurons in the putamen tend to discharge in anticipation of limb and trunk movements, whereas caudate neurons fire prior to eye movements. These anticipatory discharges are evidently part of a movement selection process; in fact, they can precede the initiation of movement by as much as several seconds. Similar recordings have also shown that the discharges of some striatal neurons vary according to the location in space of the *destination* of a movement, rather than with the starting position of the limb relative to the destination. Thus the activity of these cells may encode the *decision to move* toward a goal rather than the direction and amplitude of the actual movement necessary to reach the goal.

Projections from the Basal Ganglia to Other Brain Regions

The medium spiny neurons of the caudate and putamen give rise to inhibitory GABAergic projections that terminate in the pallidal nuclei in the basal ganglia complex: the globus pallidus and the substantia nigra pars reticulata (Figure 18.5). *Globus pallidus* means "pale body," a name which reflects its appearance due to the large number of myelinated axons in this nucleus; pars reticulata is

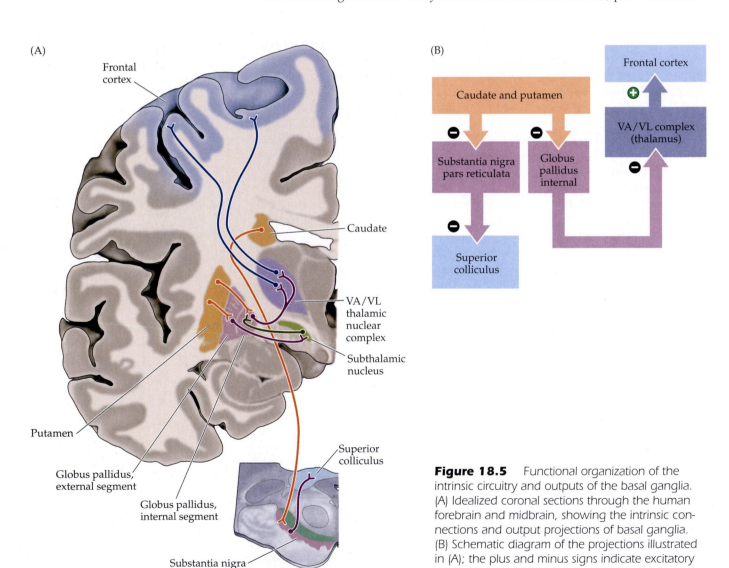

Figure 18.5 Functional organization of the intrinsic circuitry and outputs of the basal ganglia. (A) Idealized coronal sections through the human forebrain and midbrain, showing the intrinsic connections and output projections of basal ganglia. (B) Schematic diagram of the projections illustrated in (A); the plus and minus signs indicate excitatory and inhibitory projections, respectively.

so named because, unlike the pars compacta, axons passing through give it a netlike, or reticulate, appearance.

The globus pallidus and substantia nigra pars reticulata contain the same types of neurons and perform comparable functions, albeit on the different types of signals they receive from the parallel streams of processing that flow through the basal ganglia. In fact, the pars reticulata may be understood as being a part of the globus pallidus that, during early brain development, became displaced to the midbrain by the formation of the internal capsule and cerebral peduncle. The striatal projections to these two nuclei resemble the corticostriatal pathways in that they terminate in rostrocaudal bands, the locations of which vary with the locations of sources in the striatum. A striking feature of these projections is the degree of convergence from the medium spiny neurons to the neurons of the globus pallidus and substantia nigra pars reticulata. In humans, for example, the corpus striatum contains approximately 100 million neurons, about 75 percent of which are medium spiny neurons. In contrast, the main destination of their axons, the globus pallidus, comprises only about 700,000 cells. Thus, on average, more than 100 medium spiny neurons innervate each cell in the globus pallidus. However, despite this impressive degree of convergence, individual axons from the corpus striatum sparsely contact many pallidal neurons before terminating densely on the dendrites of a particular neuron. Consequently, ensembles of medium spiny neurons exert a broad but functionally weak influence over many pallidal neurons, while at the same time strongly influencing a subset of neurons in the globus pallidus or substantia nigra pars reticulata. This pattern of innervation is important for understanding the role of the corpus striatum in the selection and initiation of intended motor programs, as will be described below.

The efferent neurons of the globus pallidus and substantia nigra pars reticulata together give rise to the major pathways that allow the basal ganglia to influence the activity of upper motor neurons located in the motor cortex and in the brainstem (see Figure 18.5). The pathway to the cortex arises primarily in the medial division of the globus pallidus, called the **internal segment**, and reaches the motor cortex via a relay in the **ventral anterior** and **ventral lateral nuclei** of the dorsal thalamus. These thalamic nuclei project directly to motor areas of the cerebral cortex, thus completing a vast loop of neural circuitry that originates in multiple areas of the cortex and terminates in the motor areas of the frontal lobe, after processing in the basal ganglia and thalamus. In contrast, many efferent axons from substantia nigra pars reticulata have more direct access to upper motor neurons by synapsing on neurons in the superior colliculus that command head and eye movements, without an intervening relay in the thalamus. This difference between the globus pallidus and substantia nigra pars reticulata is not absolute, however, since many reticulata axons also project to the thalamus (mediodorsal and ventral anterior nuclei), where they contact relay neurons that project to the frontal eye fields of the premotor cortex (see Chapter 20).

Because the efferent cells of both the globus pallidus and substantia nigra pars reticulata are GABAergic, the main output of the basal ganglia is *inhibitory*. In contrast to the quiescent medium spiny neurons, the neurons in both these output zones have high levels of spontaneous activity that prevents unwanted movement by tonically inhibiting cells in the thalamus and superior colliculus. Because the medium spiny neurons of the striatum also are GABAergic and inhibitory, the net effect of the excitatory inputs that reach the striatum from the cortex is to inhibit the tonically active inhibitory cells of the globus pallidus and substantia nigra pars reticulata (Figure 18.6). For example, in the absence of volitional body movements, the globus pallidus neurons provide tonic inhibition

Figure 18.6 A chain of nerve cells arranged in a disinhibitory circuit. *Top:* Diagram of the connections between inhibitory neurons A and B and an excitatory neuron, C. *Bottom:* Pattern of the action potential activity of cells A, B, and C when A is at rest, and when neuron A fires transiently as a result of its excitatory inputs. Such circuits are central to the gating operations of the basal ganglia.

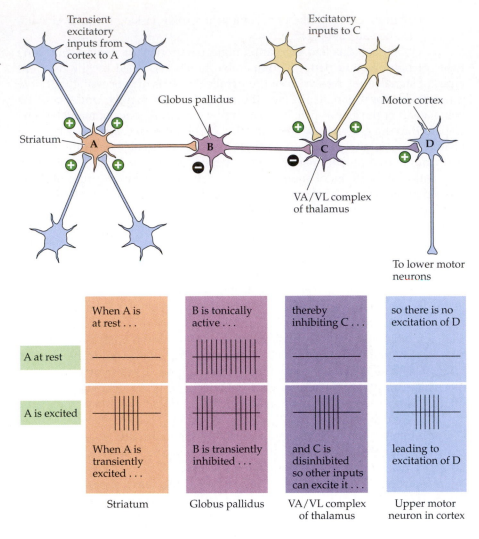

to the relay cells in the ventral lateral and anterior nuclei of the thalamus. When the pallidal cells are inhibited by activity of the medium spiny neurons, the thalamic neurons are *disinhibited* and can relay signals from other sources to the upper motor neurons in the cortex. This disinhibition is what normally allows the upper motor neurons to send commands to local circuit and lower motor neurons that initiate movement.

Evidence from Studies of Eye Movements

The permissive role of the basal ganglia in the initiation of movement is perhaps most clearly demonstrated by studies of eye movements carried out by Okihide Hikosaka and Robert Wurtz at the National Institutes of Health (Figure 18.7). As described in the previous section, the substantia nigra pars reticulata is part of the output circuitry of the basal ganglia. Instead of projecting to the thalamus, however, it sends axons mainly to the deep layers of the superior colliculus. The upper motor neurons in these layers command the rapid, orienting movements of the eyes called *saccades* (see Chapter 20). When the eyes are fixating a visual target, these upper motor neurons are tonically inhibited by the spontaneously active reticulata cells, thus preventing unwanted saccades. Shortly before the

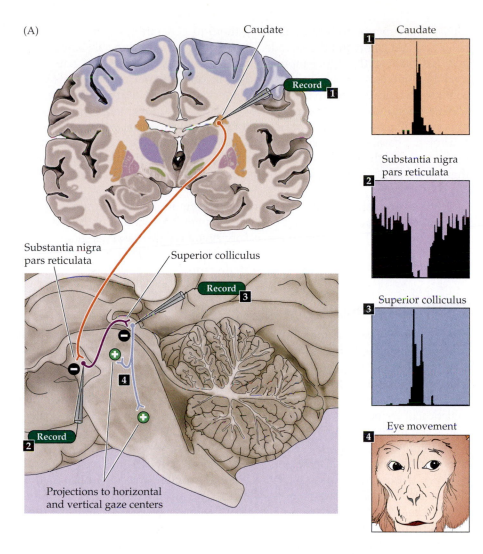

(A)

Caudate

Record 1

Substantia nigra
pars reticulata

Superior colliculus

Record 3

Record 2

Projections to horizontal
and vertical gaze centers

Caudate
1

Substantia nigra
pars reticulata
2

Superior colliculus
3

Eye movement
4

Figure 18.7 The role of basal ganglia disinhibition in the generation of saccadic eye movements. (A) Medium spiny cells in the caudate respond with a transient burst of action potentials to an excitatory input from the cerebral cortex (1). The spiny cells inhibit the tonically active GABAergic cells in substantia nigra pars reticulata (2). As a result, the upper motor neurons in the deep layers of the superior colliculus are no longer tonically inhibited and can generate the bursts of action potentials that command a saccade (3, 4). (B) The temporal relationship between inhibition in substantia nigra pars reticulata (purple) and disinhibition in the superior colliculus (light blue) preceding a saccade to a visual target. (After Hikosaka and Wurtz, 1989.)

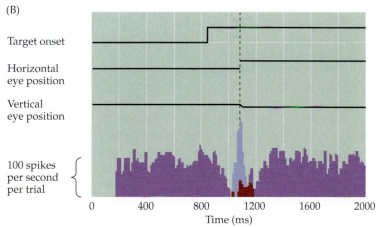

(B)

Target onset

Horizontal
eye position

Vertical
eye position

100 spikes
per second
per trial

0   400   800   1200   1600   2000

Time (ms)

onset of a saccade, the tonic discharge rate of the reticulata neurons is sharply reduced by input from the GABAergic medium spiny neurons of the caudate, which have been activated by signals from the cortex. The subsequent reduction in the tonic discharge from reticulata neurons disinhibits the upper motor neurons of the superior colliculus, allowing them to generate the bursts of action potentials that command the saccade. Thus, the projections from substantia nigra pars reticulata to the upper motor neurons act as a physiological "gate" that must be "opened" to allow either sensory or other higher order signals from cognitive centers to activate the upper motor neurons and initiate a saccade.

This brief account of the genesis of saccadic eye movements provides an important illustration of the principal functions of the basal ganglia in motor control: the basal ganglia facilitate the *initiation* of motor programs that express movement and the *suppression* of competing or non-synergistic motor programs that would otherwise interfere with the expression of sensory-driven or goal-directed behavior.

A more complete account of sensorimotor integration and the origins of eye movements is provided in Chapter 20; the remaining sections of this chapter will explain how the intrinsic and accessory circuits of the basal ganglia accomplish the principal functions in motor control, and why disease afflicting elements of these circuits can lead to devastating movement disorders.

Circuits within the Basal Ganglia System

The projections from the medium spiny neurons of the caudate and putamen to the internal segment of the globus pallidus constitute the so-called "direct pathway" through the basal ganglia and, as illustrated in Figure 18.6, serve to release from tonic inhibition the thalamic neurons that drive upper motor neurons. Thus, this direct pathway provides a means for the basal ganglia to facilate the initiation of volitional movement. The functional organization of the direct pathway is summarized in Figure 18.8A.

To reinforce the suppression of inappropriate movements, there are additional circuits of the basal ganglia that constitute a so-called "indirect pathway" between the corpus striatum and the internal segment of the globus pallidus (Figure 18.8B). This second pathway serves to increase the level of tonic inhibition mediated by the projection neurons of the internal segment (and the substantia nigra pars reticulata). In the indirect pathway, a distinct population of medium spiny neurons projects to the lateral division of the globus pallidus, called the **external segment**. The external segment of the globus pallidus sends projections both to the adjacent internal segment and to the **subthalamic nucleus** of the ventral thalamus (see Figure 18.1). The subthalamic nucleus also receives excitatory projections from neurons in the cerebral cortex that work synergistically with the projection from the external segment of the globus pallidus. In turn, the subthalamic nucleus projects diffusely back to the internal segment of the globus pallidus and to the substantia nigra pars reticulata. Thus, the indirect pathway feeds back onto the output nuclei that provide the means by which the basal ganglia gain access to upper motor neurons. But as will become clear in the following discussion, *the indirect pathway antagonizes the activity of the direct pathway;* together, they function to open or shut the physiological gates that initiate and terminate movements.

The indirect pathway through the basal ganglia modulates the disinhibitory actions of the direct pathway. The subthalamic nucleus neurons that project to the internal segment of the globus pallidus and substantia nigra pars reticulata use glutamate as their neurotransmitter and are excitatory. Normally, when the indirect pathway is activated by signals from the cortex, the medium spiny neurons discharge and inhibit the tonically active GABAergic neurons of the external globus pallidus. As a result of the removal of this tonic inhibition and the simultaneous arrival of excitatory inputs from the cerebral cortex, the subthalamic cells become more active and, by virtue of their excitatory synapses with

(A) Direct pathway

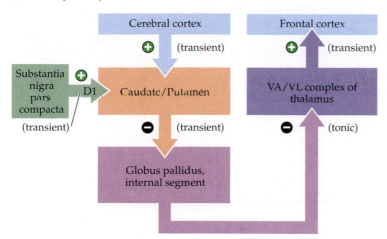

(B) Indirect and direct pathways

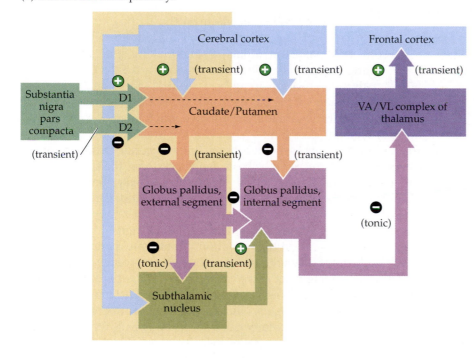

Figure 18.8 Disinhibition in the direct and indirect pathways through the basal ganglia. (A) In the direct pathway, transiently inhibitory projections from the caudate and putamen project to tonically active inhibitory neurons in the *internal* segment of the globus pallidus, which project in turn to the VA/VL complex of the thalamus. Transiently excitatory inputs to the caudate and putamen from the cortex and substantia nigra are also shown, as is the transiently excitatory input from the thalamus back to the cortex. (B) In the indirect pathway (shaded yellow), transiently active inhibitory neurons from the caudate and putamen project to tonically active inhibitory neurons of the *external* segment of the globus pallidus. Note that the influence of nigral dopaminergic input to neurons in the indirect pathway is inhibitory. The globus pallidus (external segment) neurons project to the subthalamic nucleus, which also receives a strong excitatory input from the cortex. The subthalamic nucleus in turn projects to the globus pallidus (internal segment), where its transiently excitatory drive acts to oppose the disinhibitory action of the direct pathway. In this way, the indirect pathway modulates the effects of the direct pathway.

cells of the internal segment of the globus pallidus and substantia nigra pars reticulata, they increase the inhibitory outflow of the basal ganglia. In contrast to the direct pathway, which when activated releases thalamocortical and collicular circuits from tonic inhibition, the net effect of activity in the indirect pathway is to increase the inhibitory influences of the basal ganglia. The balance of activity mediated by the direct and indirect pathways is the principal determinant of when output from the pallidum to the thalamus or superior colliculus will facilitate the expression of the intended motor program.

These circuits not only facilitate the selection of a motor program, they also suppress competing motor programs that could interfere with the timely expression of sensory-driven or goal-oriented behavior. One concept that has guided understanding of this antagonistic interaction is called *focused selection*. According to this concept, the direct and indirect pathways are functionally

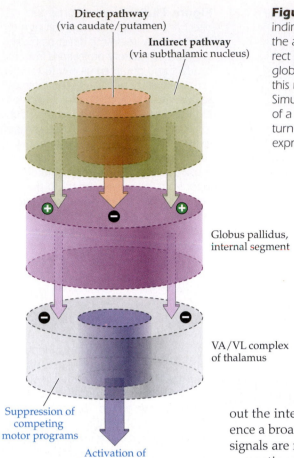

Direct pathway
(via caudate/putamen)

Indirect pathway
(via subthalamic nucleus)

Globus pallidus,
internal segment

VA/VL complex
of thalamus

Suppression of
competing
motor programs

Activation of
intended motor
programs

Figure 18.9 *Center-surround functional organization of the direct and indirect pathways. Integration of cortical input by the corpus striatum leads to the activation of the direct and indirect pathways. With activation of the indirect pathway, neurons in a "surround" region of the internal segment of the globus pallidus are driven by excitatory inputs from the subthalamic nucleus; this reinforces the suppression of a broad set of competing motor programs. Simultaneously, activation of the direct pathway leads to the focal inhibition of a more restricted "center" cluster of neurons in the internal segment; this in turn results in the disinhibition (open arrow) of the VA/VL complex and the expression of the intended motor program.*

organized in a center-surround fashion within the output nuclei of the basal ganglia (Figure 18.9). The influence of the direct pathway is tightly focused on particular functional units in the internal segment of globus pallidus (and substantia nigra pars reticulata), whereas the influence of the indirect pathway is much more diffuse, covering a broader range of functional units. Recall that individual axons from the corpus striatum to the internal segment of the globus pallidus tend to synapse densely on single pallidal neurons (despite making sparse contacts on numerous pallidal cells); this provides a means for the direct pathway to focus its input on a "central" functional unit at the output stage of the basal ganglia. In contrast, afferents from the subthalamic nucleus are distributed much more evenly throughout the internal segment, providing a means for the indirect pathway to influence a broader "surrounding" set of functional units. Accordingly, when cortical signals are received and processed by basal ganglia systems, the suppression of competing motor programs is reinforced and, simultaneously, the activation of the particular thalamocortical (or collicular) circuits that underlie the intended movement is facilitated.

Precisely how the these complex circuits of the basal ganglia interact to assist upper motor neuron systems in the execution of volitional behavior remains poorly understood, and this simplified description will undoubtedly be subject to revision as further anatomical and physiological details become available. Nevertheless, this account serves as a useful model for understanding the architecture and function of neural systems that achieve fine control of their output by an interplay between neural excitation and inhibition (recall, for example, the "center-surround" antagonism of ganglion cell receptive fields in the retina; see Chapter 11). Furthermore, this model provides an instructive framework for understanding disorders of movement that result from injury or disease that afflicts one or more components of the basal ganglia system (see below).

Dopamine Modulates Basal Ganglia Circuits

Another circuit within the basal ganglia system involves the dopaminergic cells in the pars compacta subdivision of the substantia nigra. Although this circuit derives from a relatively small pool of neurons, it exerts a profound influence over the integration of cortical input in the corpus striatum. The medium spiny neurons of the corpus striatum project directly to substantia nigra pars compacta, which in turn sends widespread dopaminergic projections back to the medium spiny neurons. The effects of dopamine on the spiny neurons are complex; they illustrate the principal that the action of a neurotransmitter is determined by the types of receptors expressed in postsynaptic neurons, and by the downstream signaling pathways to which the receptors are linked (see Chapter 6). In this case, the same nigral neurons can provide excitatory inputs to the spiny cells that project to the internal globus pallidus (the direct pathway) and inhibitory inputs to the spiny cells that project to the external globus pallidus (the indirect pathway). This duality is achieved by the differential expression of two types of dopamine receptors—types D1 and D2—by the medium spiny neurons.

Both D1- and D2-type dopamine receptors are members of the 7-transmembrane, G-protein-coupled family of cell surface receptors; the major difference between them is that the D1 receptors mediate the activation of G-proteins that *stimulate* cAMP, while D2 receptors act through different G-proteins that *inhibit* cAMP. For either type, the dopaminergic synapses on medium spiny neurons tend to be located on the shafts of the spines that receive synaptic input from the cerebral cortex. This arrangement suggests that dopamine exerts its effects on the spiny neurons by modulating their responses to cortical input, with D1 receptors positioned to enhance the excitatory input from cortex and D2 receptors positioned to negate this excitation. Since the actions of the direct and indirect pathways on the output of the basal ganglia are antagonistic, these different influences of dopamine on medium spiny neurons have the same effects—a decrease in the inhibitory outflow of the basal ganglia, and the consequent release of thalamocortical or collicular circuits.

This dopaminergic input to the corpus striatum may contribute to reward-related modulation of behavior. For example, in monkeys, the latencies of saccades toward a target are shorter when the goal of the movement is associated with a larger reward. This effect is eliminated by caudate injections of the dopamine D1 receptor antagonist and enhanced by injections in the same site of the D2 receptor antagonist. These results suggest that the influence of motivation on motor performance may be modulated by circuits in the basal ganglia. The role of dopamine in motivated behavior and the deleterious impact of addictive drugs of abuse on dopaminergic modulation of basal ganglia function are discussed in more detail in Chapter 29.

Hypokinetic and Hyperkinetic Movement Disorders

The modulatory influences of this dopaminergic circuit may also help explain many of the manifestations of basal ganglia disorders. For example, **Parkinson's disease** (Box 18A) is caused by the loss of the nigrostriatal dopaminergic neurons (Figure 18.10A). As described above, activation of the nigrostriatal projection leads to opposite but synergistic effects on the direct and indirect pathways: the release of dopamine in the corpus striatum increases the responsiveness of the direct pathway to corticostriatal input (a D1 affect) while decreasing the responsiveness of the indirect pathway (a D2 affect). Normally, both of these dopaminergic effects serve to decrease the inhibitory outflow of the basal ganglia and thus to increase the excitability of upper motor neurons. In contrast, when the dopaminergic cells of the pars compacta are destroyed, as occurs in

(A) Parkinson's disease

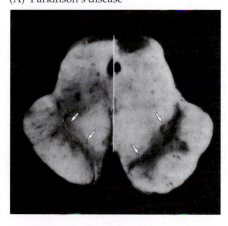

(B) Huntington's disease

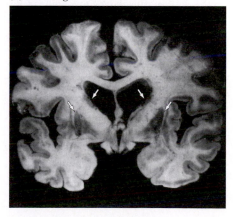

Figure 18.10   *The pathological changes in certain neurological diseases provide insights about the function of the basal ganglia. (A) Left: The midbrain from a patient with Parkinson's disease. The substantia nigra (pigmented area) is largely absent in the region above the cerebral peduncles (arrows). Right: The midbrain from a normal subject, showing intact substantia nigra (arrows). (B) The size of the caudate and putamen (the corpus striatum; arrows) is dramatically reduced in patients with Huntington's disease. (From Bradley et al., 1991.)*

BOX 18A Parkinson's Disease: An Opportunity for Novel Therapeutic Approaches

Parkinson's disease is the second most common degenerative disease of the nervous system (Alzheimer's disease being the leader; see Chapter 31). Described by James Parkinson in 1817, this disorder is characterized by tremor at rest, slowness of movement (*bradykinesia*), rigidity of the extremities and neck, and minimal facial expressions. Walking entails short steps, stooped posture, and a paucity of associated movements such as arm swinging. In some patients these abnormalities of motor function are associated with dementia. Following a gradual onset between the ages of 50 and 70, the disease progresses slowly and culminates in death 10 to 20 years later.

The defects in motor function are due to the progressive loss of dopaminergic neurons in the substantia nigra pars compacta, a population that projects to and innervates neurons in the caudate and putamen (see Figure 18.10A). Although the cause of the progressive deterioration of these dopaminergic neurons is not known, genetic investigations are providing clues to the etiology and pathogenesis. Whereas the majority of cases of Parkinson's disease are sporadic, there may be specific forms of susceptibility genes that confer increased risk of acquiring the disease, just as the *apoE4* allele increases the risk of Alzheimer's disease. Familial forms of the disease caused by single gene mutations account for less than 10 percent of all cases; identification of these rare genes, however, is likely to give some insight into molecular pathways that may underlie the disease. Mutations of three distinct genes—*α-synuclein*, *Parkin*, and *DJ-1*—have been implicated in rare forms of Parkinson's disease. Identification of these genes provides an opportunity to generate mutant mice carrying the mutant form of the human gene, potentially providing a useful animal model in which the pathogenesis can be elucidated and therapies can be tested.

In contrast to other neurodegenerative diseases (such as Alzheimer's disease or amyotrophic lateral sclerosis), in Parkinson's disease the spatial distribution of the degenerating neurons is largely restricted to the substantia nigra pars compacta. This spatial restriction, combined with the defined and relatively homogeneous phenotype of the degenerating neurons (i.e., dopaminergic neurons), has provided an opportunity for novel therapeutic approaches to this disorder.

One strategy is so-called gene therapy. *Gene therapy* refers to the correction of a disease phenotype through the introduction of new genetic information into the affected organism. Although still in its infancy, this approach has the potential to revolutionize treatment of human disease. One therapy for Parkinson's disease would be to enhance release of dopamine in the caudate and putamen. In principle, this could be accomplished by implanting cells genetically modified to express tyrosine hydroxylase, the enzyme that converts tyrosine to L-DOPA, which in turn is converted by a nearly ubiquitous decarboxylase into the neurotransmitter dopamine.

An alternative strategy to treating Parkinsonian patients involves "neural grafts" using stem cells. Stem cells are self-renewing, multipotent progenitors with broad developmental potential (see Chapters 22 and 25). Instead of isolating mature dopaminergic neurons from the fetal midbrain for transplantation, this approach isolates neuronal progenitors at earlier stages of development, when these cells are actively proliferating. Critical to this approach is to prospectively identify and isolate stem cells that are multipotent and self-renewing, and to identify the growth factors needed to promote differentiation into the desired phenotype (e.g., dopaminergic neurons). The prospective identification and isolation of multipotent mammalian stem cells has already been accomplished, and several factors likely to be important in differentiation of midbrain precursors into dopamine neurons have been identified. Establishing the efficacy of this approach for Parkinsonian patients would increase the possibility of its application to other neurodegenerative diseases.

These therapeutic strategies remain experimental, while other novel approaches are now entering clinical practice (see Box 18C); with ongoing basic science investigations in animal models and clinical studies in human patients, it is likely that some of them will succeed.

References

BJÖRKLUND, A. AND U. STENEVI (1979) Reconstruction of the nigrostriatal dopamine pathway by intracerebral nigral transplants. *Brain Res.* 177: 555–560.

DAUER, W. AND S. PRZEDBORSKI (2003) Parkinson's disease: Mechanisms and models. *Neuron* 39: 889–909.

DAWSON, T. M. AND V. L. DAWSON (2003) Rare genetic mutations shed light on the pathogenesis of Parkinson disease. *J. Clin. Invest.* 111: 145–151.

LEE, V. M. AND J. Q. TROJANOWSKI (2006) Mechanisms of Parkinson's disease linked to pathological a-synuclein: New targets for drug discovery. *Neuron* 52: 33–38.

Parkinson's disease, the inhibitory outflow of the basal ganglia is abnormally high, and timely thalamic activation of upper motor neurons in the motor cortex is therefore less likely (Figure 18.11A).

In fact, many of the symptoms seen in Parkinson's disease and other *hypokinetic* movement disorders reflect a failure of the disinhibition normally mediated by the basal ganglia. Thus, Parkinson's patients tend to have diminished facial expressions and lack associated movements, such as arm-swinging during walking. Indeed, any movement is difficult to initiate and, once initiated, is often difficult to terminate. Disruption of the same circuits also increases the discharge rate of the inhibitory cells in substantia nigra pars reticulata. The resulting increase in tonic inhibition reduces the excitability of the upper motor neurons in the superior colliculus and causes saccades to be reduced in both frequency and amplitude.

(A) Parkinson's disease (hypokinetic)

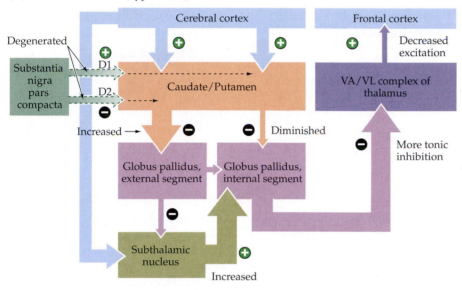

(B) Huntington's disease (hyperkinetic)

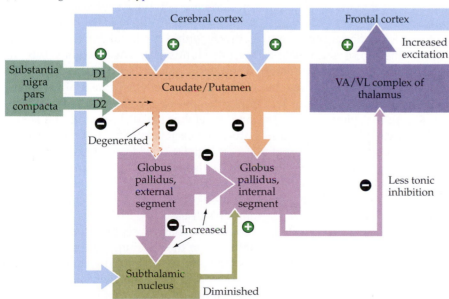

Figure 18.11 In both hypokinetic disorders such as Parkinson's disease and hyperkinetic disorders like Huntington's disease, the balance of inhibitory signals in the direct and indirect pathways is altered, leading to a diminished ability of the basal ganglia to control the thalamic output to the cortex. (A) Parkinson's disease. The dopaminergic inputs provided by the substantia nigra pars compacta are diminished (dashed arrows), making it more difficult to generate the transient inhibition from the caudate and putamen. The result of this change in the direct pathway is to sustain the tonic inhibition from the globus pallidus (internal segment) to the thalamus, making thalamic excitation of the motor cortex less likely (thinner arrow from thalamus to cortex). (B) Huntington's disease. The projection from the caudate and putamen to the external segment of the globus pallidus is diminished (dashed arrow). This effect increases the tonic inhibition from the globus pallidus to the subthalamic nucleus (larger arrow), making the excitatory subthalamic nucleus less effective in opposing the action of the direct pathway (thinner arrow). Thus, thalamic excitation of the cortex is increased (larger arrow), leading to greater and often inappropriate motor activity. (After DeLong, 1990.)

BOX 18B   Huntington's Disease

In 1872, a physician named George Huntington described a group of patients seen by his father and grandfather in their practice in East Hampton, Long Island. The disease he defined, which became known as Huntington's disease (HD), is characterized by the gradual onset of defects in behavior, cognition, and movement beginning in the fourth and fifth decades of life. The disorder is inexorably progressive, resulting in death within 10 to 20 years. HD is inherited in an autosomal dominant pattern, a feature that has led to a much better understanding of its cause in molecular terms.

One of the more common inherited neurodegenerative diseases, HD usually presents as an alteration in mood (especially depression) or a change in personality that often takes the form of in-creased irritability, suspiciousness, and impulsive or eccentric behavior. Defects of memory and attention may also occur. The hallmark of the disease, however, is a movement disorder consisting of rapid, jerky motions with no clear purpose; these choreiform movements may be confined to a finger or may involve a whole extremity, the facial musculature, or even the vocal apparatus. The movements themselves are involuntary, but the patient often incorporates them into apparently deliberate actions, presumably in an effort to obscure the problem. There is no weakness, ataxia, or deficit of sensory function. Occasionally the disease begins in childhood or adolescence. The clinical manifestations in juveniles include rigidity, seizures, more marked dementia, and a rapidly progressive course.

A distinctive neuropathology is associated with these clinical manifestations: a profound but selective atrophy of the caudate and putamen, with some associated degeneration of the frontal and temporal cortices (see Figure 18.10B). This pattern of destruction is thought to explain the disorders of movement, cognition, and behavior, as well as the sparing of other neurological functions.

The availability of extensive HD pedigrees has allowed geneticists to decipher the molecular cause of this disease. HD was one of the first human diseases in which DNA polymorphisms were used to localize the mutant gene, which in 1983 was mapped to the short arm of chromosome 4. This discovery led to an intensive effort to identify the HD gene within this region by positional cloning. Ten years later, these

---

Support for this explanation of hypokinetic movement disorders like Parkinson's disease comes from studies of monkeys in which degeneration of the dopaminergic cells of substantia nigra has been induced by the neurotoxin 1-methyl-4-phenyl-1,2,3,6-tetrahydropyridine (MPTP). Monkeys (or humans) exposed to MPTP develop symptoms that are very similar to those of patients with Parkinson's disease. Furthermore, a second lesion placed in the subthalamic nucleus results in significant improvement in the ability of these animals to initiate movements, as would be expected based on the circuitry of the indirect pathway (see Figure 18.11B).

Knowledge about the indirect pathway within the basal ganglia also helps explain the motor abnormalities seen in Huntington's disease (Box 18B), an exemplary *hyperkinetic* movement disorder. In patients with Huntington's disease, medium spiny neurons that project to the external segment of the globus pallidus degenerate (Figure 18.10B). In the absence of their normal inhibitory input from the spiny neurons, the external globus pallidus cells become abnormally active; this activity reduces in turn the excitatory output of the subthalamic nucleus to the internal segment of the globus pallidus (Figure 18.11B), and the inhibitory outflow of the basal ganglia is reduced. Without the restraining influence of the basal ganglia, upper motor neurons can be activated by inappropriate signals, resulting in the undesired ballistic and choreiform (dancelike) movements that characterize Huntington's disease.

Similarly, imbalances in the fine control mechanism represented by the convergence of the direct and indirect pathways in the pallidum are apparent in diseases that affect primarily the subthalamic nucleus. These disorders remove a source of excitatory input to the internal segment of the globus pallidus and pars reticulata, thus abnormally reducing the inhibitory outflow of the basal ganglia.

efforts culminated in identification of the gene (named *Huntingtin*) responsible for the disease. In contrast to previously recognized forms of mutations such as point mutations, deletions, or insertions, the mutation of *Huntingtin* is an unstable triplet repeat. In normal individuals, *Huntingtin* contains between 15 and 34 repeats, whereas the gene in HD patients contains from 42 to over 66 repeats.

HD is one of a growing number of diseases attributed to unstable DNA segments. Other examples are fragile X syndrome, myotonic dystrophy, spinal and bulbar muscular atrophy, and spinocerebellar ataxia type 1. In the latter two and HD, the repeats consist of a DNA segment (CAG) that codes for the amino acid glutamine and is present within the coding region of the gene.

The mechanism by which the increased number of polyglutamine repeats injures neurons is not clear. The leading hypothesis is that the increased numbers of glutamines alter protein folding, which somehow triggers a cascade of molecular events culminating in dysfunction and neuronal death. Interestingly, although *Huntingtin* is expressed predominantly in the expected neurons in the basal ganglia, it is also present in regions of the brain that are not affected in HD. Indeed, the gene is expressed in many organs outside the nervous system. How and why the mutant *Huntingtin* uniquely injures striatal neurons is unclear. Continuing to elucidate this molecular pathogenesis will no doubt provide further insight into this and other triplet repeat diseases.

References

ADAMS, R. D. AND M. VICTOR (2005) *Principles of Neurology*, 8th Ed. New York: McGraw-Hill, pp. 910–913.

CATTANEO, E., ZUCCATO, C. AND M. TARTARI (2005) Normal huntingtin function: An alternative approach to Huntington's disease. *Nature Rev. Neurosci.* 6: 919–930.

GUSELLA, J. F. AND 13 OTHERS (1983) A polymorphic DNA marker genetically linked to Huntington's disease. *Nature* 306: 234–238.

HUNTINGTON, G. (1872) On chorea. *Med. Surg. Reporter* 26: 317.

HUNTINGTON'S DISEASE COLLABORATIVE RESEARCH GROUP (1993) A novel gene containing a trinucleotide repeat that is expanded and unstable on Huntington's disease chromosomes. *Cell* 72: 971–983.

WEXLER, A. (1995) *Mapping Fate: A Memoir of Family, Risk, and Genetic Research.* New York: Times Books.

YOUNG, A. B. (2003) Huntingtin in health and disease. *J. Clin. Invest.* 111: 299–302.

A basal ganglia syndrome called **hemiballismus**, which is characterized by violent, involuntary movements of the limbs, is the result of damage to the subthalamic nucleus. As in Huntington's disease, the involuntary movements of hemiballismus are initiated by the abnormal discharges of upper motor neurons that are receiving less than adequate tonic inhibition from the basal ganglia.

The hyperkinetic movement disorders emphasize the importance of the subthalamic nucleus in modulating the output of the basal ganglia. Indeed, this nucleus has become an important target of novel clinical inerventions aimed at restablishing permissive patterns of neural activity in basal ganglia circuits in human patients (Box 18C).

As predicted by this account of hypokinetic and hyperkinetic movement disorders, GABA agonists and antagonists applied to substantia nigra pars reticulata of monkeys produce symptoms similar to those seen in human basal ganglia disease. For example, intranigral injection of bicuculline, which blocks the GABAergic inputs from the striatal medium spiny neurons to the reticulata cells, increases the amount of tonic inhibition on the upper motor neurons in the deep collicular layers. These animals exhibit fewer and slower saccades, reminiscent of patients with Parkinson's disease. In contrast, injection of the GABA agonist muscimol into substantia nigra pars reticulata decreases the tonic GABAergic inhibition of the upper motor neurons in the superior colliculus, with the result that the injected monkeys generate spontaneous, irrepressible saccades that resemble the involuntary movements characteristic of basal ganglia diseases such as hemiballismus and Huntington's disease (Figure 18.12).

The central theme of this unit is "Movement and Its Central Control," and the focus of this account of the basal ganglia has been its role in the modulation of movement. However, there are several parallel streams of processing that run

BOX 18C   Deep Brain Stimulation

Since the seminal demonstration by G. Fritsch and E. Hitzig in the nineteenth century that bodily movements could be induced when electrical currents were applied to cerebral tissue (in this case, motor cortex), clinicians have considered the possibility that certain neurological disorders affecting volitional movement could be treated with the application of acute or chronic electrical stimulation to key motor centers in the brain. However, it was not until well after the internally implanted cardiac pacemaker was introduced in the 1960s that technological advances made possible the implantation of a comparable device for the focal stimulation of brain structures. Such devices were introduced in the 1990s for the treatment of movement disorders, with their targets being components of the basal ganglia and thalamus located deep in the forebrain; hence the common term for this intervention, "deep brain stimulation."

Prior to this time, the treatment options for patients with movement disorders were limited to pharmacological intervention (e.g., using drugs designed to increase dopamine levels in the corpus striatum to treat Parkinson's disease; see Box 18A), physical therapy and, in the most intractable cases, neurosurgical ablation of sites in the basal ganglia and thalamus that gate the initiation of movement. For the latter patients, the introduction of deep brain stimulation provided an obviously welcome alternative to the permanent destruction of brain circuitry, as well as to the slow pace of progress attending the development of other experimental strategies.

Deep brain stimulation entails the implantation of battery-powered generator units, usually near the clavicles. These units produce electrical discharges that are passed through subcutaneous wires to electrodes implanted bilaterally into the brain (Figure A). (Recall that,

with the exception of medial divisions of the premotor cortex, all of the cortical and subcortical neural circuitry in the forebrain that governs the activity of upper motor neurons is organized unilaterally; therefore, bilateral deep brain stimulation is necessary to achieve symmetrical results.) The placement of the electrodes requires careful stereotaxic surgery combined with radiological imaging of the patient's brain and electrophysiological recordings of spontaneous and movement-related neuronal activity. The neuronal recordings are essential so that the neurosurgical team can recognize by sight and sound the characteristic discharge patterns of neurons in different nuclei of the basal ganglia and thalamus, as action potential waveforms are displayed on oscilloscopes and audio and computer monitors (see Figure 18.6).

Once the target structures are localized, stimulation is tested to determine if the desired clinical effect can be observed. After recovery from the implantation procedure, the generator units are activated and the parameters of stimulation are fine-tuned as various combinations of pulse widths, current amplitudes, and temporal patterns of pulse trains are employed and modified as needed.

A careful consideration of Figure 18.11 suggests several possible target sites of deep brain stimulation for patients with hypokinetic or hyperkinetic movement disorders. In both diagnostic categories, neural activity in motor nuclei of the thalamus is abnormal; accordingly, the VA/VL complex of the thalamus is an appropriate target. However, abnormal activity in the thalamus is often the consequence of disturbances within the basal ganglia itself, and it is arguably more desirable to manipulate the influence of the basal ganglia on thalamocortical circuits rather than directly altering neural activ-

ity in upper motor neuronal circuits with exogenous electrical stimulation. Thus, the two most common sites for deep brain stimulation in patients with movement disorders are the internal segment of the globus pallidus and the subthalamic nucleus. Regardless of whether the neurological disorder to be rectified is hypokinetic (difficulty expressing movement) or hyperkinetic (expressing unwanted movement), deep brain stimulation can be used to override intrinsic, pathological discharge patterns with stable and highly structured patterns of neural activity that better facilitate the initiation and termination of volitional movement (Figure B).

Given the complexity of even a small volume of neural tissue (such as might be present near the tip of a stimulating electrode in the globus pallidus or subthalamic nucleus), it is not surprising that the exogenous induction of electrical currents can lead to complex patterns of activity and inactivity in the affected neural elements. Different intensities of deep brain stimulation may lead to the local release of neurotransmitters and neuromodulators. They may also lead to the generation of action potentials in afferent axons, neuronal cell bodies, efferent axons, and "fibers of passage" that originate elsewhere. However, the effects of electrical stimulation on certain intrinsic membrane properties, including voltage-dependent ionic conductances, may block the generation of action potentials, thus silencing the affected neurons. Ideally, the net effect of these diverse changes is the amelioration of abnormal network activity that hinders the normal operation of upper motor neurons.

Despite ongoing uncertainties regarding the mechanisms of action, deep brain stimulation has offered hope to thousands of patients who suffer neurological dysfunction ranging from the movement disorders discussed here to related disorders of non-motor basal ganglia loops, such as Tourette's syndrome, depression, and obsessive compulsive disorder (see Box 18D). The fact that stimulation protocols are adjustable provides clinicians with an unprecedented ability to manipulate the activities and functions of basal ganglia circuits whose operations are crucial to the normal expression of thought, emotion, and motor behavior.

References

GARCIA, L., G. D'ALESSANDRO, B. BIOULAC AND C. HAMMOND (2005) High-frequency stimulation in Parkinson's disease: More or less? *Trends Neurosci.* 28: 209–216.

HASHIMOTO, T., C. M. ELDER, M. S. OKUN, S. K. PATRICK AND J. L. VITEK (2003) Stimulation of the subthalamic nucleus changes the firing pattern of pallidal neurons. *J. Neurosci.* 23: 1916–1923.

KLEINER-FISMAN, G. AND 7 OTHERS (2006) Subthalamic nucleus deep brain stimulation: Summary and meta-analysis of outcomes. *Movement Disord.* 21: S290–S304.

PERLMUTTER, J. S. AND J. W. MINK (2006) Deep brain stimulation. *Annu. Rev. Neurosci.* 29: 229–257.

WICHMANN, T. AND M. R. DELONG (2006) Deep brain stimulation for neurologic and neuropsychiatric disorders. *Neuron* 52: 197–204.

(A)

Electrode brain implant

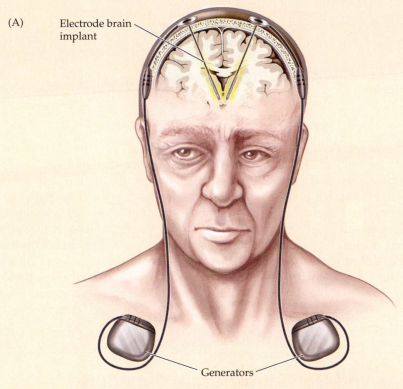

Generators

(A) Illustration of a patient following implantation of a device for deep brain stimulation. (Courtesy of L. Kibiuk.)

(B) Raster plots of action potentials recorded from a neuron in the internal segment of the globus pallidus in an awake rhesus monkey rendered Parkinsonian by the systemic administration of MPTP; each row lasts one second. The endogenous pattern of discharge is marked by irregular clusters of burst activity (*top panel*). Within seconds of the onset of subthalamic nucleus stimulation, the Parkinsonian symptoms abated and the discharge of the pallidal neuron became much more regular (*middle panel*). (From Hashimoto et al., 2003.)

(B)

Pre-stimulation

During stimulation

Poststimulation

Figure 18.12 After the tonically active cells of substantia nigra pars reticulata are inactivated by an intranigral injection of muscimol (A), the upper motor neurons in the deep layers of the superior colliculus are disinhibited and the monkey generates spontaneous irrepressible saccades (B). The cells in both substantia nigra pars reticulata and the deep layers of the superior colliculus are arranged in spatially organized motor maps of saccade vectors (see Chapter 20), and so the direction of the involuntary saccades—in this case toward the upper left quadrant of the visual field—depends on the precise location of the injection site in the substantia nigra.

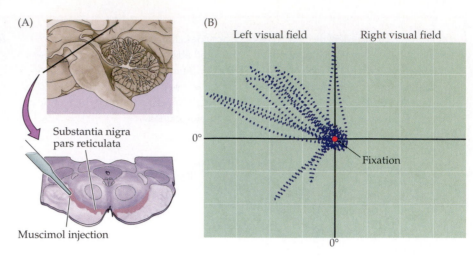

through different sectors of the basal ganglia, including functional loops that modulate the expression of cognitive and affective behavior (Box 18D). Studies of the anatomical and physiological organization of the better understood motor and oculomotor loops has provided the foundation for investigations of anterior and ventral circuitry of the basal ganglia that subserves a variety of non-motor functions (see Chapter 29). Thus, each functional loop through the basal ganglia is likely to exert a similar influence on the selection, initiation, and suppression of motor or non-motor programs, with equally significant clinical implications should injury, disease, or neurochemical imbalance impair the function of one or more component of the diverse basal ganglia systems.

BOX 18D Basal Ganglia Loops and Non-Motor Brain Functions

The basal ganglia traditionally have been regarded as motor structures that regulate the initiation of voluntary movements, such as those involving the limbs and eyes. However, the basal ganglia are also central structures in anatomical circuits or loops that are involved in modulating non-motor aspects of behavior. These parallel loops originate in different regions of the cerebral cortex, engage specific subdivisions of the basal ganglia and thalamus, and ultimately affect areas of the frontal lobe outside of the primary motor and premotor cortices. The most prominent of these non-motor loops are a dorsolateral prefrontal loop, involving the dorsolateral sector of the prefrontal cortex and the head of the caudate (see Chapter 26); and a "limbic" loop that originates

in the orbitomedial prefrontal cortex, amygdala, and hippocampal formation and runs through ventral divisions of the corpus striatum (see Chapter 29).

The anatomical similarity of these loops to the better understood motor loops suggests that the non-motor regulatory functions of the basal ganglia may be generally the same as the roles of basal ganglia in regulating the initiation of movement. For example, the prefrontal loop may regulate the initiation and termination of cognitive processes such as planning, working memory, and attention. By the same token, the limbic loop may regulate emotional and motivated behavior, as well as the transitions from one mood state to another. Indeed, the deterioration of cognitive and emotional function in both Parkinson's and

Huntington's diseases (see Boxes 18A and 18B) could be the result of the disruption of these non-motor loops.

In fact, a variety of other disorders are now thought to be caused, at least in part, by damage to non-motor components of the basal ganglia. For example, patients with Tourette's syndrome produce inappropriate utterances and obscenities as well as unwanted vocal-motor "tics" and repetitive grunts. These manifestations may be a result of excessive activity in basal ganglia loops that regulate the cognitive circuitry of the prefrontal speech areas. Another example is schizophrenia, which some investigators have argued is associated with aberrant activity within the limbic and prefrontal loops, resulting in hallucinations, delusions, disordered thoughts, and loss of

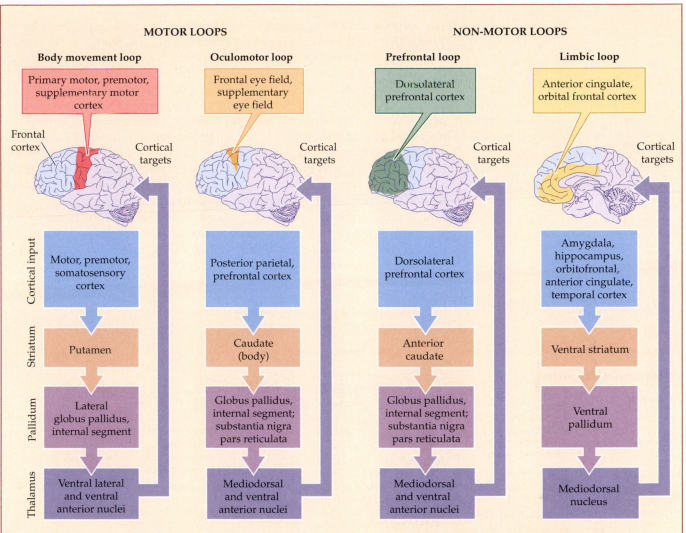

MOTOR LOOPS

Body movement loop

Primary motor, premotor, supplementary motor cortex

Frontal cortex

Cortical targets

Oculomotor loop

Frontal eye field, supplementary eye field

Cortical targets

NON-MOTOR LOOPS

Prefrontal loop

Dorsolateral prefrontal cortex

Cortical targets

Limbic loop

Anterior cingulate, orbital frontal cortex

Cortical targets

Cortical input:
- Motor, premotor, somatosensory cortex
- Posterior parietal, prefrontal cortex
- Dorsolateral prefrontal cortex
- Amygdala, hippocampus, orbitofrontal, anterior cingulate, temporal cortex

Striatum:
- Putamen
- Caudate (body)
- Anterior caudate
- Ventral striatum

Pallidum:
- Lateral globus pallidus, internal segment
- Globus pallidus, internal segment; substantia nigra pars reticulata
- Globus pallidus, internal segment; substantia nigra pars reticulata
- Ventral pallidum

Thalamus:
- Ventral lateral and ventral anterior nuclei
- Mediodorsal and ventral anterior nuclei
- Mediodorsal and ventral anterior nuclei
- Mediodorsal nucleus

Comparison of motor and non-motor basal ganglia loops.

emotional expression. In support of the argument for a basal ganglia contribution to schizophrenia, antipsychotic drugs are known to act on dopaminergic receptors, which are found in high concentrations in the corpus striatum.

Still other psychiatric disorders, including obsessive-compulsive disorder, depression, and chronic anxiety, may also involve dysfunctions of the limbic loop. Indeed, one particular component of the limbic loop in a ventral division of the corpus striatum is the nucleus accumbens. This structure is implicated in both the neuropharmocology of addiction to drugs of abuse and to the

expression of addictive reward-seeking behavior (see Chapter 29). A challenge for future research is to understand more fully the relationships between these clinical problems and the functions of the basal ganglia.

References

ALEXANDER, G. E., M. R. DELONG AND P. L. STRICK (1986) Parallel organization of functionally segregated circuits linking basal ganglia and cortex. *Annu. Rev. Neurosci.* 9: 357–381.

BHATIA, K. P. AND C. D. MARSDEN (1994) The behavioral and motor consequences of focal lesions of the basal ganglia in man. *Brain* 117: 859–876.

BLUMENFELD, H. (2002) *Neuroanatomy through Clinical Cases.* Sunderland, MA: Sinauer Associates.

DREVETS, W. C. AND 6 OTHERS (1997) Subgenual prefrontal cortex abnormalities in mood disorders. *Nature* 386: 824–827.

GRAYBIEL, A. M. (1997) The basal ganglia and cognitive pattern generators. *Schiz. Bull.* 23: 459–469.

JENIKE, M. A., L. BAER AND W. E. MINICHIELLO (1990) *Obsessive Compulsive Disorders: Theory and Management.* Chicago: Year Book Medical Publishers, Inc.

MARTIN, J. H. (1996) *Neuroanatomy: Text and Atlas.* New York: McGraw-Hill.

MIDDLETON, F. A. AND P. L. STRICK (2000) Basal ganglia output and cognition: Evidence from anatomical, behavioral, and clinical studies. *Brain Cogn.* 42: 183–200.

Summary

The contribution of the basal ganglia to motor control is apparent from the deficits that result from damage to the component nuclei. Such lesions compromise the initiation and performance of voluntary movements, as exemplified by the paucity of movement in Parkinson's disease and in the inappropriate "release" of movements in Huntington's disease. The organization of the basic circuitry of the basal ganglia indicates how this constellation of nuclei modulates movement. With respect to motor function, the system forms a loop that originates in almost every area of the cerebral cortex and eventually terminates, after enormous convergence within the basal ganglia, on the upper motor neurons in the motor and premotor areas of the frontal lobe and in the superior colliculus. The efferent neurons of the basal ganglia influence the upper motor neurons in the cortex by gating the flow of information through relays in the ventral nuclei of the thalamus. The upper motor neurons in the superior colliculus that initiate saccadic eye movements are controlled by monosynaptic projections from substantia nigra pars reticulata. In each case, the basal ganglia loop regulates movement by a process of disinhibition that results from the serial interaction within the basal ganglia circuitry of two GABAergic neurons. Internal circuits within the basal ganglia system modulate the amplification of the signals that are transmitted through the loop.

Additional Reading

Reviews

ALEXANDER, G. E. AND M. D. CRUTCHER (1990) Functional architecture of basal ganglia circuits: Neural substrates of parallel processing. *Trends Neurosci.* 13: 266–271.

DELONG, M. R. (1990) Primate models of movement disorders of basal ganglia origin. *Trends Neurosci.* 13: 281–285.

GERFEN, C. R. AND C. J. WILSON (1996) The basal ganglia. In *Handbook of Chemical Neuroanatomy*, Vol. 12: *Integrated Systems of the CNS*, Part III. L. W. Swanson, A. Björklund and T. Hokfelt (eds.). New York: Elsevier Science Publishers, pp. 371–468.

GOLDMAN-RAKIC, P. S. AND L. D. SELEMON (1990) New frontiers in basal ganglia research. *Trends Neurosci.* 13: 241–244.

GRAYBIEL, A. M. AND C. W. RAGSDALE (1983) Biochemical anatomy of the striatum. In *Chemical Neuroanatomy*, P. C. Emson (ed.). New York: Raven Press, pp. 427–504.

GRILLNER, S., J. HELLGREN, A. MÉNARD, K. SAITOH AND M. A. WIKSTRÖM (2005) Mechanisms for selection of basic motor programs: Roles for the striatum and pallidum. *Trends Neurosci.* 28: 364–370.

HIKOSAKA, O. AND R. H. WURTZ (1989) The basal ganglia. In *The Neurobiology of Eye Movements*, R. H. Wurtz and M. E. Goldberg (eds.). New York: Elsevier Science Publishers, pp. 257–281.

KAJI, R. (2001) Basal ganglia as a sensory gating devise for motor control. *J. Med. Invest.* 48: 142–146.

MINK, J. W. AND W. T. THACH (1993) Basal ganglia intrinsic circuits and their role in behavior. *Curr. Opin. Neurobiol.* 3: 950–957.

POLLACK, A. E. (2001) Anatomy, physiology, and pharmacology of the basal ganglia. *Neurol. Clin.* 19: 523–534.

SLAGHT, S. J, T. PAZ, S. MAHON, N. MAURICE, S. CHARPIER AND J. M. DENIAU (2002) Functional organization of the circuits connecting the cerebral cortex and the basal ganglia. Implications for the role of the basal ganglia in epilepsy. *Epileptic Disord.* Suppl 3: S9–S22.

WILSON, C. J. (1990) Basal ganglia. In *Synaptic Organization of the Brain*. G. M. Shepherd (ed.). Oxford: Oxford University Press, Chapter 9.

Important Original Papers

ANDEN, N.-E., A. DAHLSTROM, K. FUXE, K. LARSSON, K. OLSON AND U. UNGERSTEDT (1966) Ascending monoamine neurons to the telencephalon and diencephalon. *Acta Physiol. Scand.* 67: 313–326.

BRODAL, P. (1978) The corticopontine projection in the rhesus monkey: Origin and principles of organization. *Brain* 101: 251–283.

CRUTCHER, M. D. AND M. R. DELONG (1984) Single cell studies of the primate putamen. *Exp. Brain Res.* 53: 233–243.

DELONG, M. R. AND P. L. STRICK (1974) Relation of basal ganglia, cerebellum, and motor cortex units to ramp and ballistic movements. *Brain Res.* 71: 327–335.

DIFIGLIA, M., P. PASIK AND T. PASIK (1976) A Golgi study of neuronal types in the neostriatum of monkeys. *Brain Res.* 114: 245–256.

KEMP, J. M. AND T. P. S. POWELL (1970) The cortico-striate projection in the monkey. *Brain* 93: 525–546.

KIM, R., K. NAKANO, A. JAYARAMAN AND M. B. CARPENTER (1976) Projections of the globus pallidus and adjacent structures: An autoradiographic study in the monkey. *J. Comp. Neurol.* 169: 217–228.

NAKAMURA, K. AND O. HIKOSAKA (2006) Role of dopamine in the primate caudate nucleus in reward modulation of saccades. *J. Neurosci.* 26: 5360–5369.

KOCSIS, J. D., M. SUGIMORI AND S. T. KITAI (1977) Convergence of excitatory synaptic inputs to caudate spiny neurons. Brain Res. 124: 403–413.

MINK, J. W. (1996) The basal ganglia: Focused selection and inhibition of competing motor programs. *Prog. Neurobiol.* 50: 381–425.

SMITH, Y., M. D. BEVAN, E. SHINK AND J. P. BOLAM (1998) Microcircuitry of the direct and indirect pathways of the basal ganglia. *Neuroscience* 86: 353–387.

Books

BRADLEY, W. G., R. B. DAROFF, G. M. FENICHEL AND C. D. MARSDEN (EDS.). (1991) *Neurology in Clinical Practice*. Boston: Butterworth-Heinemann, Chapters 29 and 77.

KLAWANS, H. L. (1989) *Toscanini's Fumble and Other Tales of Clinical Neurology*. New York: Bantam, Chapters 7 and 10.

Chapter 19

Modulation of Movement by the Cerebellum

Overview

In contrast to the upper motor neurons described in Chapter 17, the efferent cells of the cerebellum do not project directly either to the local circuits of the brainstem and spinal cord that organize movement, or to the lower motor neurons that innervate muscles. Instead—like the basal ganglia—the cerebellum influences movements primarily by modifying the activity patterns of the upper motor neurons. In fact, the cerebellum sends prominent projections to virtually all circuits that govern upper motor neurons. Anatomically, the cerebellum has two main gray matter structures: a laminated cerebellar cortex on its surface and clusters of cells buried deep in the white matter of the cerebellum referred to collectively as the deep cerebellar nuclei. Pathways that reach the cerebellum from other brain regions (in humans, predominantly from the cerebral cortex) project to both components; thus, the afferent axons send branches to both the deep nuclei and the cerebellar cortex. Neurons in the deep cerebellar nuclei are also the main source of the output from the cerebellum; however, these cells do not simply return the same signals they receive. Rather, their patterns of efferent activity are sculpted by descending input from the overlying cerebellar cortex. In this way, input signals are modified as they are processed by the cerebellum before returning to circuits of upper motor neurons in the cerebral cortex, by way of thalamic relays, and in the brainstem. The primary function of the cerebellum is evidently to detect the difference, or "motor error," between an intended movement and the actual movement, and, through its influence over upper motor neurons, to reduce the error. These corrections can be made both during the course of the movement and as a form of motor learning when the correction is stored. When this feedback loop is damaged, as occurs in many cerebellar diseases, the afflicted individual makes persistent errors when executing movement. The specific character of the errors depends on the location of the damage.

Organization of the Cerebellum

The **cerebellar hemispheres** can be subdivided into three main parts based on differences in their sources of input (Figure 19.1; Table 19.1). By far the largest subdivision in humans is the **cerebrocerebellum**. It occupies most of the lateral cerebellar hemisphere and receives input indirectly from many areas of the cerebral cortex. This region of the cerebellum is especially well developed in primates and is particularly prominent in humans. The cerebrocerebellum is concerned with the regulation of highly skilled movements, especially the planning and execution of complex spatial and temporal sequences of movement (including speech). Just medial to the cerebrocerebellum is the **spinocerebellum**. The spinocerebellum occupies the median and paramedian zone of the cerebellar

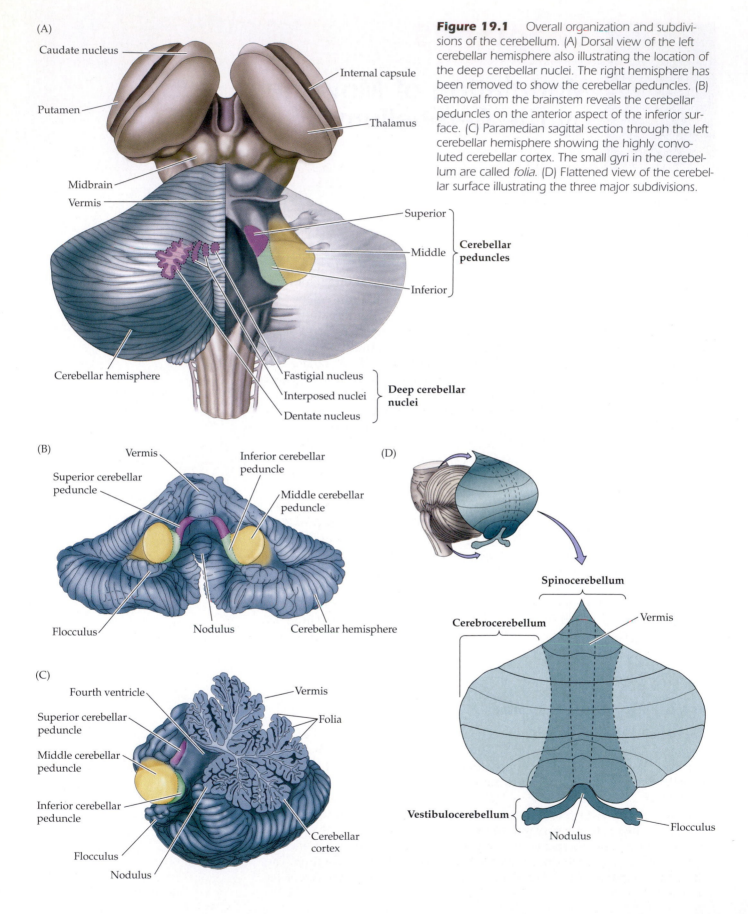

(A)

Caudate nucleus

Internal capsule

Putamen

Thalamus

Midbrain

Vermis

Superior

Middle — **Cerebellar peduncles**

Inferior

Cerebellar hemisphere

Fastigial nucleus

Interposed nuclei — **Deep cerebellar nuclei**

Dentate nucleus

Figure 19.1 *Overall organization and subdivisions of the cerebellum. (A) Dorsal view of the left cerebellar hemisphere also illustrating the location of the deep cerebellar nuclei. The right hemisphere has been removed to show the cerebellar peduncles. (B) Removal from the brainstem reveals the cerebellar peduncles on the anterior aspect of the inferior surface. (C) Paramedian sagittal section through the left cerebellar hemisphere showing the highly convoluted cerebellar cortex. The small gyri in the cerebellum are called* folia. *(D) Flattened view of the cerebellar surface illustrating the three major subdivisions.*

(B)

Vermis

Superior cerebellar peduncle

Inferior cerebellar peduncle

Middle cerebellar peduncle

Flocculus

Nodulus

Cerebellar hemisphere

(D)

Spinocerebellum

Cerebrocerebellum

Vermis

Vestibulocerebellum

Nodulus

Flocculus

(C)

Fourth ventricle

Vermis

Superior cerebellar peduncle

Folia

Middle cerebellar peduncle

Inferior cerebellar peduncle

Flocculus

Nodulus

Cerebellar cortex

hemispheres and is the only part that receives input directly from the spinal cord. The more lateral (paramedian) part of the spinocerebellum is primarily concerned with movements of distal muscles. The median strip of cerebellar hemisphere along the midline, called the **vermis**, is primarily concerned with movements of proximal muscles; it also regulates certain forms of eye movements (see Chapter 20). The last of the major subdivisions is the **vestibulocerebellum**, which is also the phylogenetically oldest part of the cerebellum. This portion comprises the caudal-inferior lobes of the cerebellum and includes the **flocculus** and **nodulus** (see Figure 19.1). As its name suggests, the vestibulocerebellum receives input from the vestibular nuclei in the brainstem and is primarily concerned with the regulation of movements underlying posture and equilibrium, as well as the vestibulo-ocular reflex (see Chapter 14).

The connections between the cerebellum and other parts of the nervous system occur by way of three large pathways called **cerebellar peduncles** (Figures 19.1 to 19.3). The **superior cerebellar peduncle** (or **brachium conjunctivum**) is almost entirely an efferent pathway. The neurons that give rise to this pathway are in the deep cerebellar nuclei, and their axons project to upper motor neurons in the deep layers of the superior colliculus, and, after a relay in the dorsal thalamus, the primary motor and pre-motor areas of the cortex (see Chapter 17). In nonhuman species, they also provide input to upper motor neurons in the caudal part of the red nucleus. The **middle cerebellar peduncle** (or **brachium pontis**) is an afferent pathway to the cerebellum; most of the cell bodies that give rise to this pathway are in the base of the contralateral pons, where they form the **pontine nuclei** (Figure 19.2). The pontine nuclei receive input from a wide variety of sources, including almost all areas of the cerebral cortex and the superior colliculus. The axons of the cells in the pontine nuclei, called **transverse pontine fibers**, cross the midline and enter the cerebellum via the middle cerebellar peduncle (Figure 19.3). Each of the two middle cerebellar peduncles contain over 20 million axons, making this one of the largest pathways in the brain. In comparison, the optic and pyramidal tracts each contain about a million axons. Thus, the prominence of the cerebral peducles in the ventral portion of the human midbrain (each of which also contains about 20 million axons) primarily reflects the magnitude of the projection from the cere-

TABLE 19.1	Major Components of the Cerebellum

Cerebellar cortex
  Cerebrocerebellum
  Spinocerebellum
  Vestibulocerebellum

Deep cerebellar nuclei
  Dentate nucleus
  Interposed nuclei
  Fastigial nucleus

Cerebellar peduncles
  Superior peduncle
  Middle peduncle
  Inferior peduncle

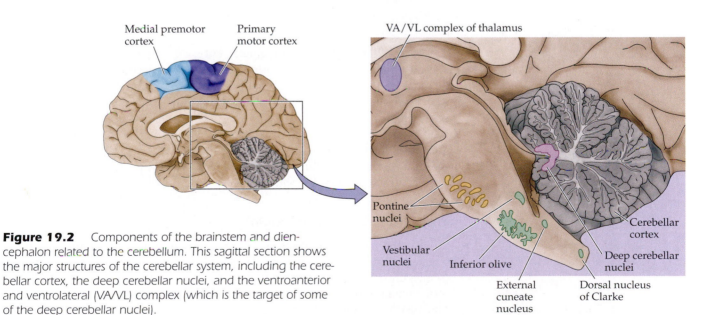

Figure 19.2 *Components of the brainstem and diencephalon related to the cerebellum. This sagittal section shows the major structures of the cerebellar system, including the cerebellar cortex, the deep cerebellar nuclei, and the ventroanterior and ventrolateral (VA/VL) complex (which is the target of some of the deep cerebellar nuclei).*

(A)

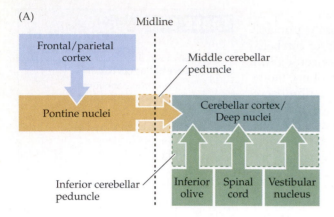

(B)

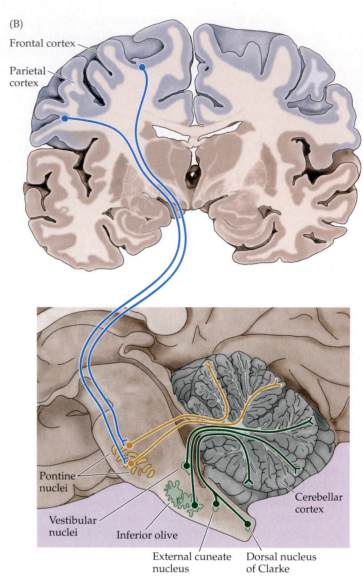

Figure 19.3 Functional organization of the inputs to the cerebellum. (A) Diagram of the major inputs. (B) Idealized coronal and sagittal sections through the human brainstem and cerebrum, showing inputs to the cerebellum from the cortex, vestibular system, spinal cord, and brainstem. The cortical projections to the cerebellum are made via relay neurons in the pons. These pontine axons then cross the midline within the pons and project to the cerebellum via the middle cerebellar peduncle. Axons from the inferior olive, spinal cord, and vestibular nuclei enter via the inferior cerebellar peduncle.

bral cortex that, via the pontine nuclei, enters the cerebellum. In contrast, the corticospinal projection (which is often assumed to account for the bulk of the cerebral peduncles) comprises a mere 5 percent of the total number of axons in each cerebral peduncle. Finally, the **inferior cerebellar peduncle** (or **restiform body**) is the smallest but most complex of the cerebellar peduncles, containing multiple afferent and efferent pathways. Afferent pathways in this peduncle include axons from the vestibular nuclei, the spinal cord, and several regions of the brainstem tegmentum, while efferent pathways project to the vestibular nuclei and the reticular formation.

Projections to the Cerebellum

The cerebral cortex is by far the largest source of input signals that reach the cerebellum, and the major destination of this input is the cerebrocerebellum (see Figure 19.3). These afferent pathways arise from a somewhat more circumscribed area of the cortex than do those to the basal ganglia (see Chapter 18). The majority originate in the primary motor and premotor cortices of the frontal

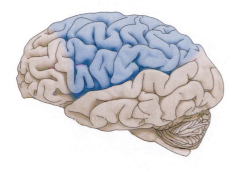

Figure 19.4 Regions of the cerebral cortex that project to the cerebellum (shown in blue). The cortical projections to the cerebellum are mainly from the sensory association cortex of the parietal lobe and motor association areas of the frontal lobe.

lobe, the primary and secondary somatic sensory cortices of the anterior parietal lobe, and the higher order visual regions of the posterior parietal lobe (Figure 19.4). The visual input to the cerebellum originates mostly in association areas concerned with processing moving visual stimuli and visuospatial awareness (see Chapter 12). Indeed, regulating the visual guidance of ongoing movement is one of the major tasks carried out by the cerebrocerebellum. As outlined above, these cortical axons do not project directly into the cerebellum. Rather, they synapse on neurons in the pontine nuclei that are located on the same side of the brainstem as their hemisphere of origin. The pontine nuclei in turn give rise to the transverse projections that cross the midline and form the middle cerebellar peduncle, thus relaying these cortical signals to the contralateral cerebellar hemisphere. Consequently, signals derived from one *cerebral* hemisphere are received and processed by neural circuits in the opposite *cerebellar* hemisphere (see Figure 19.3A).

Sensory pathways also project to the cerebellum (see Figure 19.3). Vestibular axons from the eighth cranial nerve and axons from the vestibular nuclei in the medulla project to the vestibulocerebellum. In addition, relay neurons in the **dorsal nucleus of Clarke** in the spinal cord and the **external** (or accessory) **cuneate nucleus** of the caudal medulla send their axons to the spinocerebellum (recall that these nuclei comprise groups of relay neurons innervated by proprioceptive axons from the lower and upper parts of the body, respectively; see Chapter 9). Proprioceptive signals from the face are likewise relayed via the **mesencephalic nucleus** of the trigeminal complex to the spinocerebellum, although the relevant pathway is not clear. The vestibular, spinal, and trigeminal inputs provide the cerebellum with information from the labyrinth in the ear, muscle spindles, and other mechanoreceptors that monitor the position and motion of the body. Finally, visual and auditory signals are relayed via brainstem nuclei to the vermis; presumably, they provide the cerebellum with additional sensory signals that compliment the proprioceptive information regarding body position and motion.

The somatic sensory input remains topographically mapped in the spinocerebellum such that there are orderly representations of the body surface within the cerebellum (Figure 19.5). These maps are "fractured," however: That is, fine-grain electrophysiological analysis indicates that each small area of the body is represented multiple times by spatially separated clusters of cells, rather than by a specific site within a single continuous somatotopic map. The vestibular and spinal inputs remain ipsilateral from their point of entry in the brainstem, traveling in the inferior cerebellar peduncle (see Figure 19.3A). This arrangement ensures that the right cerebellum is concerned with the right half of the body and the left cerebellum with the left half. Thus, while the cerebrum is generally concerned with *contralateral* representation (of body and external space), the cerebellum is concerned with *ipsilateral* representation. The following section will detail how signals leave the cerebellum to interact with circuits of upper motor neurons on the appropriate side of the midline.

Figure 19.5 Somatotopic maps of the body surface in the cerebellum. The spinocerebellum contains at least two maps of the body.

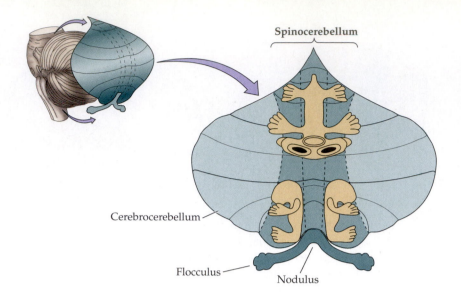

Spinocerebellum

Cerebrocerebellum

Flocculus

Nodulus

Finally, the entire cerebellum receives modulatory inputs from the **inferior olive** and the locus ceruleus in the brainstem. These nuclei participate in the learning and memory functions served by cerebellar circuitry.

Projections from the Cerebellum

The efferent neurons of the cerebellar cortex project to the deep cerebellar nuclei and to the vestibular complex; these structures project, in turn, to upper motor neurons in the brainstem and to thalamic nuclei that innervate upper motor neurons in the motor cortex. In each cerebellar hemisphere, there are four major deep nuclei: the **dentate nucleus** (by far the largest in humans), two **interposed nuclei**, and the **fastigial nucleus**. Each receives input from a different region of the cerebellar cortex. Although the borders are not distinct, in general the cerebrocerebellum projects primarily to the dentate nucleus, the spinocerebellum to the interposed and fastigial nuclei, and the vestibulocerebellum to the vestibular complex (For this reason, parts of the vestibular complex may be considered a functional and anatomical component of the deep cerebellar nuclei; Figure 19.6.)

Figure 19.6 Functional organization of cerebellar ouputs. The three major functional divisions of the cerebellar hemispheres project to corresponding deep cerebellar nuclei and the vestibular nuclei, which in turn provide input to neural circuits that govern different aspects of motor control.

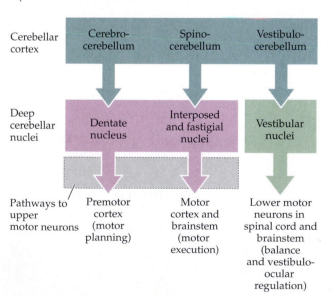

Cerebrocerebellar pathways from the dentate nucleus are mainly destined for the premotor and associational cortices of the frontal lobe, which function in planning volitional movements. They reach these cortical areas after a relay in the ventral nuclear complex in the thalamus (Figure 19.7A). Since each cerebellar hemisphere is concerned with the ispsilateral side of the body, this pathway must cross the midline if the motor cortex in each hemisphere, which is concerned with contralateral musculature, is to receive information from the appropriate cerebellar hemisphere. Consequently, the dentate axons exit the cerebellum via the superior cerebellar peduncle, cross at the **decussation of the superior cerebellar peduncle** in the caudal midbrain, and then ascend to the thalamus. Along the way toward the thalamus, this pathway sends collaterals to the parvocellular (small-celled) division of the **red nucleus** in the midbrain (which accounts for virtually the entire red nucleus in the human midbrain) (Figure 19.7B). This division of the red nucleus projects, in turn, to the inferior olive, thus providing a means for cerebellar output to feed

(A)

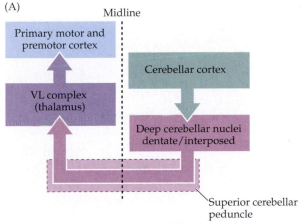

Superior cerebellar
peduncle

Figure 19.7 Functional organization of the major ascending outputs from the cerebellum. (A) Diagram of major outputs that affect upper motor neurons in the cerebral cortex. The axons of the deep cerebellar nuclei cross in the midbrain in the decussation of the superior cerebellar peduncle before reaching the thalamus. (B) Idealized coronal and sagittal sections through the cerebrum and brainstem, showing the location of the structures and pathways diagrammed in (A) and a feedback circuit by which cerebellar output is directed to the inferior olive via the red nucleus.

(B)

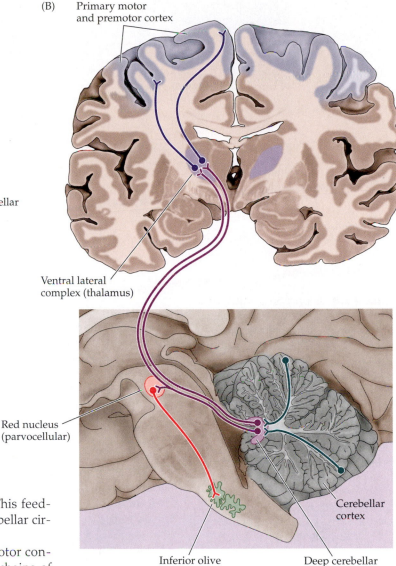

back upon a major source of cerebellar input. This feedback is crucial for the adaptive functions of cerebellar circuits (see below).

In addition to projections concerned with motor control, anatomical studies using viruses to trace chains of connections between nerve cells have shown that large parts of the cerebrocerebellum send information back to non-motor areas of the cortex to form "closed loops." That is, a region of the cerebellum sends projections back to the same cortical areas (via thalamic projections) from which its input signals originated. Such closed loops, for example, characterize cerebellar pathways that modulate cognitive programs organized in the prefrontal cortex. These cerebellar circuits may influence the coordination of non-motor programs (such as mental problem-solving) in a manner that is analogous to their operations on movement-related signals. These closed loops run in parallel to "open loops" that receive input from multiple cortical areas and funnel output back to upper motor neurons in specific regions of the motor and pre-motor cortices.

Spinocerebellar pathways are directed toward circuits of upper motor neurons that govern the execution of movement (see Figure 19.6). The somatotopic organization of this division of the cerebellum is reflected in the organization of its efferent projections, both of which conform to the medial–lateral organization of motor control in the spinal cord (see Chapter 16). Thus the fastigial nuclei (which underlie the vermis near the midline of the cerebellum) project via the inferior cerebellar peduncle to nuclei of the reticular formation and vestibular

(A)

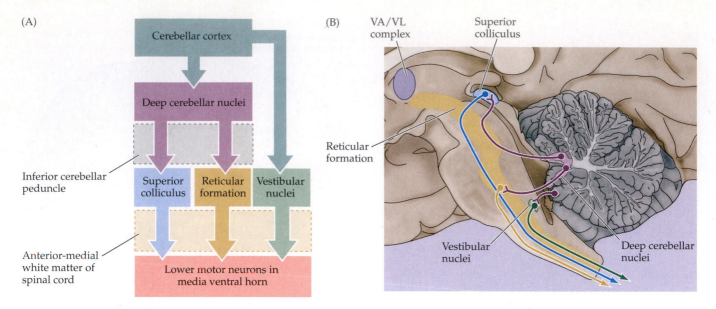

(B)

Figure 19.8   *Functional organization of the major descending outputs from the cerebellum. (A) Diagram of major outputs that affect upper motor neurons in the brainstem. The axons of the deep cerebellar nuclei and the vestibulocerebellar cortex project to upper motor neurons that contribute to the control of axial and proximal limb musculature in the medial ventral horn of the spinal cord. (B) Idealized sagittal section through the brainstem, showing the location of the structures diagrammed in (A). In addition to the connection illustrated in (A), the dentate and interposed nuclei also project to the contralateral superior colliculus, via the superior cerebellar peduncle.*

complex that give rise to medial tracts governing the axial and proximal limb musculature (Figure 19.8). The more laterally positioned interposed nuclei (which underlie the paramedian subdivision of the spinocerebellum) project via the superior cerebellar peduncle to thalamic circuits that interact with motor regions in the frontal lobe concerned with volitional movements of the limbs (see Figure 19.7). In non-human primates, projections from the interposed nuclei also send collaterals to a magnocellular (large-celled) division of the red nucleus that gives rise to the rubrospinal tract, a lateral tract of the spinal cord that functions synergistically with the lateral corticospinal tract. (As discussed in Chapter 17, this division of the red nucleus and its spinal projection is vestigial in humans relative to other primates and is unlikely to be of any functional significance.)

Most of the projection from the cerebellum to the upper motor neurons in the superior colliculus arises in the dentate and interposed nuclei, which receive their input from the lateral portions of the cerebellar cortex. The output pathway travels in the superior cerebellar peduncle and crosses the midline to terminate in the superior colliculus on the contralateral side. So, for example, the right cerebellar hemisphere projects to the left superior colliculus, which in turn controls saccades toward the right half of the visual field (see Chapter 20).

The thalamic nuclei that receive projections from the cerebrocerebellum (dentate nuclei) and spinocerebellum (interposed nuclei) are segregated in two distinct subdivisions of the ventral lateral nuclear complex: the oral, or anterior, part of the posterolateral segment, and a region simply called "area X." Both of these thalamic relays project directly to primary motor and premotor association cortices. Thus, the cerebellum has access to the upper motor neurons that organize the sequence of muscular contractions underlying complex voluntary movements, as well as the associational circuits that exert executive control over movement (see Chapter 17).

Lastly, vestibulocerebellum projections course through the inferior cerebellar peduncle and terminate in nuclei of the vestibular complex that govern the movements of the eyes, head, and neck, compensating for linear and rotational accelerations of the head (see Figure 19.6).

Circuits within the Cerebellum

The ultimate destination of the afferent pathways to the cerebellar cortex is a distinctive cell type called the **Purkinje cell**. However, the input from the cere-

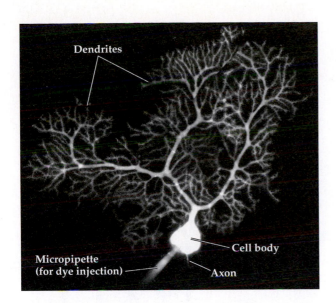

Figure 19.9 Photomicrograph of a cerebellar Purkinje neuron in a living slice from mouse cerebellum. The neuron has been visualized by insertion into the cell body of a micropipette containing a fluorescent dye that indicates Ca^{2+} concentrations. (Courtesy of K. Tanaka and G. Augustine.)

bral cortex to the Purkinje cells is indirect. Neurons in the pontine nuclei receive a massive projection from the cerebral cortex and relay the information to the contralateral cerebellar cortex. The axons from the pontine nuclei—and most other sources in the brainstem and spinal cord—are called **mossy fibers** because of the appearance of their synaptic terminals. Mossy fibers synapse on neurons in the deep cerebellar nuclei and on **granule cells** in the granule cell layer of the cerebellar cortex. Cerebellar granule cells are widely held to be the most abundant class of neurons in the human brain; they give rise to axons called **parallel fibers** that ascend to the **molecular layer** of the cerebellar cortex. The parallel fibers bifurcate in the molecular layer to form T-shaped branches that relay information via excitatory synapses onto the dendritic spines of Purkinje cells.

The Purkinje cells present the most striking histological feature of the cerebellum (Figure 19.9). Elaborate dendrites extend into the molecular layer from a single subjacent layer of these giant nerve cell bodies (called the Purkinje layer). Once in the molecular layer, the Purkinje cell dendrites branch extensively in a plane at right angles to the trajectory of the parallel fibers (Figure 19.10A). In this way, each Purkinje cell is in a position to receive input from a large number of parallel fibers (about 200,000), and each parallel fiber can contact a very large number of Purkinje cells (on the order of tens of thousands). The Purkinje cells also receive a direct modulatory input on their dendritic shafts from the **climbing fibers**, all of which arise in the inferior olive (Figure 19.10B). Each Purkinje cell receives numerous synaptic contacts from a single climbing fiber. In most models of cerebellum function, the climbing fibers provide a "training" signal that modulates the effectiveness of the mossy-parallel fiber connection with the Purkinje cells (see below).

The Purkinje cells project in turn to the deep cerebellar nuclei. They are the only output cells of the cerebellar cortex. Since Purkinje cells are GABAergic, the output of the cerebellar cortex is wholly inhibitory. However, the neurons in the deep cerebellar nuclei receive excitatory input from the collaterals of the mossy and climbing fibers. The inhibitory projections of Purkinje cells serve to shape the discharge patterns that deep nuclei neurons generate in response to this direct mossy and climbing fiber input (Figure 19.11).

Inputs from local circuit neurons modulate the inhibitory activity of Purkinje cells. The most powerful of these local inputs are inhibitory nests of synapses made with the Purkinje cell bodies by **basket cells** (see Figure 19.10). Another type of local circuit neuron, the **stellate cell**, receives input from the parallel

(A)

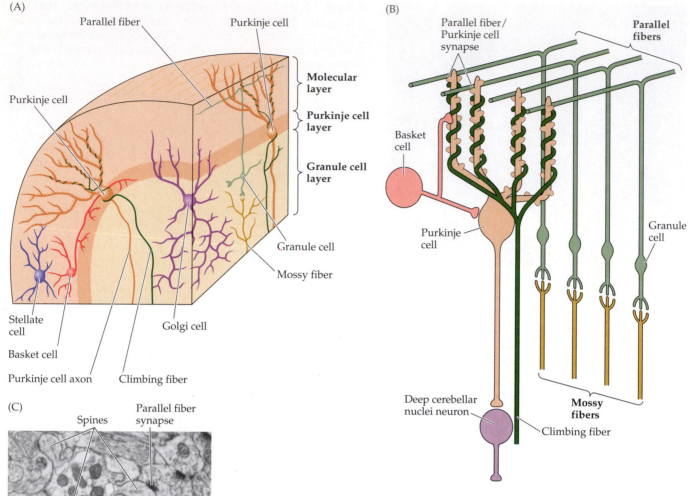

(C)

Figure 19.10 Neurons and circuits of the cerebellum. (A) Neuronal types in the cerebellar cortex. Note that the various neuron classes are found in distinct layers. (B) Diagram showing convergent inputs onto the Purkinje cell from parallel fibers and local circuit neurons [the boxed region is shown at higher magnification in (C)]. The output of the Purkinje cells is to the deep cerebellar nuclei. (C) Electron micrograph showing Purkinje cell dendritic shaft with three spines contacted by synapses from a trio of parallel fibers. (C courtesy of A.-S. La Mantia and P. Rakic.)

fibers and provides an inhibitory input to the Purkinje cell dendrites. Finally, the molecular layer contains the apical dendrites of a cell type called **Golgi cells**; these neurons have their cell bodies in the granular cell layer. Golgi cells receive input from the parallel fibers and provide an inhibitory feedback to the cells of origin of the parallel fibers (the granule cells).

This basic circuit is repeated over and over throughout every subdivision of the cerebellum in all mammals; it is the fundamental functional module of the cerebellum. Transformation of signal flow through these modules provides the basis for both the real-time regulation of movement and the long-term changes in regulation that underlie motor learning. Description of the flow of signals through this admittedly complex intrinsic circuitry may be simplified by distinguishing the two basic stages of cerebellar processing, beginning with the deep cerebellar nuclei. Mossy fiber and climbing fiber collaterals drive the activation of neurons in the deep cerebellar nuclei; this constitutes an *excitatory loop* where input signals converge on the output stage of cerebellar processing. However, as suggested above, the spatiotemporal patterns of the output activity are not simply faithful replications of the input patterns. The activity patterns of the deep cerebellar nuclei are sculpted by the descending inhibitory inputs of Purkinje cells, which are driven by these same two pathways (i.e., the mossy and climbing fiber projections to the cerebellar cortex). For their part, the Purkinje cells integrate these principal inputs and invert their "sign" by responding to excitatory drive with an inhibitory output (see Figure 19.11). Thus, the Purkinje cells convey the product of computations performed by an *inhibitory loop* that com-

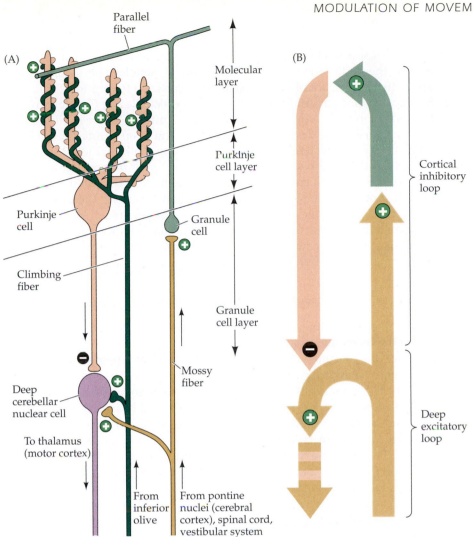

(A)

Parallel fiber

Molecular layer

Purkinje cell layer

Purkinje cell

Granule cell

Granule cell layer

Climbing fiber

Mossy fiber

Deep cerebellar nuclear cell

To thalamus (motor cortex)

From inferior olive

From pontine nuclei (cerebral cortex), spinal cord, vestibular system

(B)

Cortical inhibitory loop

Deep excitatory loop

Figure 19.11   Excitatory and inhibitory connections in the cerebellar cortex and deep cerebellar nuclei. (A) The excitatory input from mossy fibers and climbing fibers to Purkinje cells and deep nuclear cells is basically the same. Additional convergent input onto the Purkinje cell from local circuit neurons (basket and stellate cells) and other Purkinje cells (not illustrated) establishes a basis for the comparison of ongoing movement and sensory feedback derived from it. The output of the Purkinje cell onto the deep cerebellar nuclear cell is inhibitory. (B) Conceptual diagram of the circuitry illustrated in (A). The deep cerebellar nuclei and their excitatory afferents constitute a "deep excitatory loop" whose ouput is shaped by a "cortical inhibitory loop" that inverts the sign of the input signals. The Purkinje neuron output to the deep cerebellar nuclear cell thus generates an error correction signal that can modify movements. The climbing fibers modify the efficacy of the parallel fiber–Purkinje cell connection, producing long-term changes in cerebellar output. (A after Stein, 1986.)

prises the circuitry of the cerebellar cortex, including the interneurons of the granule and molecular layers, as well as the Purkinje cells themselves. The Golgi, stellate, and basket cells control the flow of information through the cerebellar cortex. For example, the Golgi cells form an inhibitory feedback circuit that controls the gain of the granule cell input to the Purkinje cells, whereas the basket cells provide lateral inhibition that may focus the spatial distribution of Purkinje cell activity.

The modulation of cerebellar output by the cerebellar cortex may be responsible for the motor learning aspect of cerebellar function. According to a model proposed by Masao Ito and his colleagues at Tokyo University, the climbing fibers relay the message of a motor error to the Purkinje cells. This message is derived from inputs that the inferior olive receives from multiple structures (including the cerebral cortex and spinal cord) as well as from feedback signals from the cerebellum via the red nucleus, as discussed above. The thousand or so synapses made by a single climbing fiber with the proximal dendrites of a single Purkinje cell constitutes one of the most powerful excitatory connections in the entire central nervous system. The strength of this input is further enhanced by gap junctions that electronically join and synchronize the activity of the neurons in the inferior olive. Thus, ensembles of olivary neurons can simultaneously both drive the activation of cerebellar circuits and promote adaptive plasticity in the inhibitory output of the cerebellar cortex. This plasticity results from long-term reductions in the

Purkinje cell responses to parallel fiber inputs through a complex chain of events leading to the endocytosis of AMPA receptors at parallel fiber–Purkinje cell synapses. (For an account of the cellular mechanism for this long-term reduction in the efficacy of the parallel fiber synapse on Purkinje cells, see Chapter 8.)

The reduction in the efficacy of parallel fiber input to Purkinje cells has the effect of increasing the response of neurons in the deep cerebellar nuclei to afferent activity (by weakening the influence of the inhibitory loop). Thus the signals returned from the cerebellum to circuits of upper motor neurons in the motor cortex and brainstem are altered as a consequence of climbing fiber activation. It is not yet understood how this alteration mediates a "correction" of movement error. Nevertheless, it is clear from studies of animal models and of human patients with damage to the inferior olive that both short-term sensorimotor adaptation (error correction) and long-term motor learning require the modulation of cerebellar processing by climbing fiber activation.

Cerebellar Circuitry and the Coordination of Ongoing Movement

As expected for a structure that monitors and adjusts motor behavior, neuronal activity in the cerebellum changes continually during the course of a movement. For instance, the execution of a relatively simple task like flipping the wrist back and forth elicits a dynamic pattern of activity in both the Purkinje cells and the deep cerebellar nuclear cells that closely follows the ongoing movement (Figure 19.12). Both types of cells are tonically active at rest and change their frequency of firing as movements occur. The neurons respond selectively to various aspects of movement, including relaxation or contraction of specific muscles, the position of the joints, and the direction of the next movement that will occur. All this information is encoded by changes in the discharge pattern of Purkinje cells, which modulates the activity of the deep cerebellar nuclear cells.

As these neuronal response properties predict, cerebellar lesions and disease tend to disrupt the modulation and coordination of ongoing movements (Box

Figure 19.12   *Activity of Purkinje cells (A) and deep cerebellar nuclear cells (B) at rest (upper traces) and during movement of the wrist (lower traces). The lines below the action potential records show changes in muscle tension, recorded by electromyography. The durations of the wrist movements are indicated by the colored blocks. Both classes of cells are tonically active at rest. Rapid alternating movements result in the transient inhibition of the tonic activity of both cell types. (After Thach, 1968.)*

(A) PURKINJE CELL

At rest

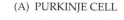

During alternating movement

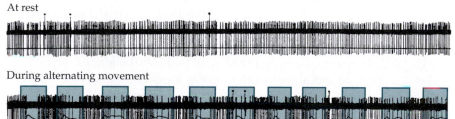

(B) DEEP NUCLEAR CELL

At rest

During alternating movement

19A). Thus the hallmark of patients with cerebellar damage is difficulty producing smooth, well-coordinated movements. Instead, movements tend to be jerky and imprecise, a condition referred to as **cerebellar ataxia**. Many of these difficulties in performing movements can be explained as disruption of the cerebellum's role in correcting errors in ongoing movements, since the cerebellar error correction mechanism normally ensures that movements are modified to cope with changing circumstances. As described earlier, the Purkinje cells and deep cerebellar nuclear cells recognize potential errors by comparing patterns of convergent activity that are concurrently available to both cell types; the deep nuclear cells then send corrective signals to the upper motor neurons in order to maintain or improve the accuracy of the movement.

As in the case of the basal ganglia, studies of the oculomotor system (saccades in particular) have contributed greatly to understanding the contribution that the cerebellum makes to motor error reduction. For example, cutting part of the tendon to the lateral rectus muscles in one eye of a monkey weakens horizontal eye movements by that eye (Figure 19.13). When a patch is then placed over the normal eye to force the animal to use its weak eye, the saccades per-

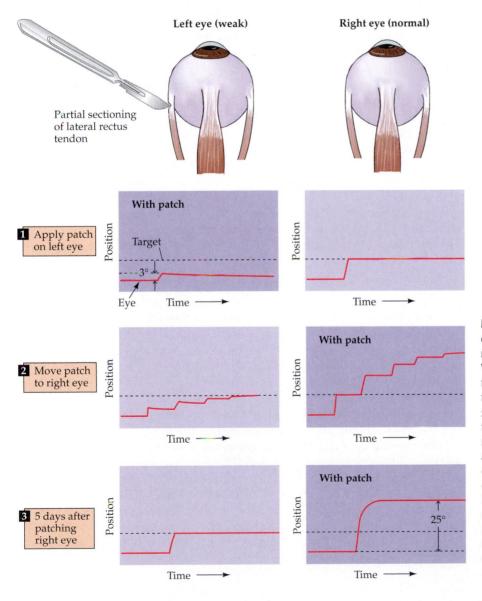

Figure 19.13   Contribution of the cerebellum to the experience-dependent modification of saccadic eye movements. Weakening of the lateral rectus muscle of the left eye causes the eye to undershoot the target (1). When the experimental subject (in this case a monkey) is forced to use this eye by patching the right eye, multiple saccades must be generated to acquire the target (2). After 5 days of experience with the weak eye, the gain of the saccadic system has been increased and a single saccade is now used to fixate the target. (3) This adjustment of the gain of the saccadic eye movement system depends on an intact cerebellum. (After Optican and Robinson, 1980.)

BOX 19A   Prion Diseases

Creutzfeldt-Jakob disease (CJD) is a rare but devastating neurological disorder characterized by cerebellar ataxia, myoclonic jerks, seizures, and the fulminant progression of dementia. The onset is usually in middle age, and death typically follows within a year. The distinctive histopathology of the disease, termed *spongiform degeneration*, consists of neuron loss and extensive glial proliferation, mainly in the cortex of the cerebellum and cerebrum; the peculiar spongiform pattern is due to vacuoles in the cytoplasm of neurons and glia. CJD is the only human disease known to be transmitted by inoculation (either orally or into the bloodstream), or to be inherited through the germline. In contrast to other transmissible diseases mediated by microorganisms such as viruses or bacteria, the agent in this case is a protein called a prion.

Observations dating back some 30 years suggested that CJD was infective. The major clue came from scrapie, a once-obscure disease of sheep that is also characterized by cerebellar ataxia, wasting, and intense itching. The ability to transmit scrapie from one sheep to another strongly suggested an infectious

agent. Another clue came from the work of Carlton Gajdusek, a neurologist studying a peculiar human disease called kuru that occurred specifically in a group of New Guinea natives known to practice ritual cannibalism. Like CJD, kuru is a neurodegenerative disease characterized by devastating cerebellar ataxia and subsequent dementia, usually leading to death within a year. The striking similarities in the distinctive histopathology of scrapie and kuru—namely, spongiform degeneration—suggested a common pathogenesis and led to the successful transmission of kuru to apes and chimpanzees in the 1960s, confirming that CJD was indeed infectious. The prolonged period (months to years) between inoculation and disease onset led Gajdusek to suggest that the transmissible agent was what he called a "slow virus."

These extraordinary findings spurred an intensive search for the infectious agent. The transmission of scrapie from sheep to hamsters by Stanley Prusiner at the University of California at San Francisco permitted biochemical characterization of partially purified fractions of scrapie agent from hamster brain.

Oddly, he found that the infectivity was extraordinarily resistant to ultraviolet irradiation or nucleases—both treatments that degrade nucleic acids. It therefore seemed unlikely that a virus could be the causal agent. Conversely, procedures that modified or degraded proteins markedly diminished infectivity. In 1982, Prusiner coined the term *prion* to refer to the agent causing these transmissible spongiform encephalopathies. He chose the term to emphasize that the agent was a proteinaceous infectious particle (and made the abbreviation a little more euphonious in the process). Since Prusiner's discoveries, a half-dozen more animal diseases—including the widely publicized bovine spongiform encephalopathy (BSE), or "mad cow disease"—and four more human diseases have been shown to be caused by prions.

Whether prions contain undetected nucleic acids or are really proteins remained controversial for some years. Prusiner strongly advocated a "protein-only" hypothesis, a revolutionary concept with respect to transmissible diseases. He proposed that the prion is a protein consisting of a modified (scrapie)

formed by the weak eye are initially *hypometric*—they fall short of visual targets. Over the next few days, the amplitude of the saccades gradually increases until they again become accurate. If the patch is then switched to cover the weakened eye, the saccades performed by the normal eye are now *hypermetric*. In other words, over a period of a few days the nervous system corrects the errors in the saccades made by the weak eye by increasing the gain in the saccade motor system (see Chapter 20). Lesions in the vermis of the spinocerebellum (see Figure 19.1) eliminate this ability to reduce the motor error.

Similar evidence of the cerebellar contribution to movement has come from studies of the vestibulo-ocular reflex (VOR) in monkeys and humans. The VOR keeps the eyes trained on a visual target during head movements (see Chapter 14). The relative simplicity of this reflex has made it possible to analyze some of the mechanisms that enable motor learning as a process of error reduction. When a visual image on the retina shifts its position as a result of head movement, the eyes must move at the same velocity in the opposite direction to

form (PrP^Sc) of the normal host protein (PrP^C, for "prion protein control"), the propagation of which occurs by a conformational change of endogenous PrP^C to PrP^Sc autocatalyzed by PrP^Sc. That is, the modified form of the protein (PrP^Sc) transforms the normal form (PrP^C) into the modified form, much as crystals form in supersaturated solutions. Differences in the secondary structures of PrP^C and PrP^Sc seen with optical spectroscopy supported this idea. An alternative hypothesis, however, was that the agent is simply an unconventional nucleic acid-containing virus, and that the accumulation of PrP^Sc is an incidental consequence of infection and cell death.

In the past decade, a compelling body of evidence in support of the "protein-only" hypothesis has emerged. First, PrP^Sc and scrapie infectivity copurify by a number of procedures, including affinity chromatography using an anti-PrP monoclonal antibody; no nucleic acid has been detected in highly purified preparations, despite intensive efforts. Second, spongiform encephalopathies can be inherited in humans, and the cause is now known to be a mutation (or mutations) in the gene coding for PrP. Third, transgenic mice carrying a mutant *PrP* gene equivalent to one of

the mutations of inherited human prion disease develop a spongiform encephalopathy. Thus, a defective protein is sufficient to account for the disease. Finally, transgenic mice carrying a null mutation for *PrP* do not develop spongiform encephalopathy when inoculated with scrapie agent, whereas wild-type mice do. These results argue convincingly that PrP^Sc must indeed interact with endogenous PrP^C to convert PrP^C to PrP^Sc, propagating the disease in the process. The protein is highly conserved across mammalian species, suggesting that it serves some essential function, although mice carrying a null mutation of *PrP* exhibit no detectable abnormalities.

These advances notwithstanding, many questions remain. What is the mechanism by which the conformational transformation of PrP^C to PrP^Sc occurs? How do mutations at different sites of the same protein culminate in the distinct phenotypes evident in diverse prion diseases of humans? Are conformational changes of proteins a common mechanism of other neurodegenerative diseases? And do these findings suggest a therapy for the dreadful manifestations of spongiform encephalopathies?

Despite these unanswered questions, this work represents one of the most exciting chapters in modern neurological research, and won Nobel Prizes in Physiology or Medicine for both Gajdusek (in 1976) and Prusiner (in 1997).

References

BUELER, H. AND 6 OTHERS (1993) Mice devoid of PrP are resistant to scrapie. *Cell* 73: 1339–1347.

GAJDUSEK, D. C. (1977) Unconventional viruses and the origin and disappearance of kuru. *Science* 197: 943–960.

GIBBS, C. J., D. C. GAJDUSEK, D. M. ASHER AND M. P. ALPERS (1968) Creutzfeldt-Jakob disease (spongiform encephalopathy): Transmission to the chimpanzee. *Science* 161: 388–389.

HARRIS, D. A. AND H. L. TRUE (2006) New insights into prion structure and toxicity. *Neuron* 50: 353–357.

MALLUCCI, G. AND J. COLLINGE (2005). Rational targeting for prion therapeutics. *Nature Rev. Neurosci.* 6: 23–34.

PRUSINER, S. B. (1982) Novel proteinaceous infectious particles cause scrapie. *Science* 216: 136–144.

PRUSINER, S. V., M. R. SCOTT, S. J. DEARMOND AND G. E. COHEN (1998) Prion protein biology. *Cell* 93: 337–348.

RHODES, R. (1997) *Deadly Feasts: Tracking the Secrets of a Terrifying New Plague.* New York: Simon and Schuster.

maintain a stable percept. In these studies, the adaptability of the VOR to changes in the nature of incoming sensory information is challenged by fitting subjects (either monkeys or humans) with magnifying or minifying spectacles (Figure 19.14). Because the glasses alter the size of the visual image on the retina, the compensatory eye movements, which would normally have maintained a stable image of an object on the retina, are either too large or too small. Over time, subjects (whether monkeys or humans) learn to adjust the distance the eyes must move in response to head movements to accord with the artificially altered size of the visual field. Moreover, this change is retained for significant periods after the spectacles are removed and can be detected electrophysiologically in recordings from Purkinje cells and neurons in the deep cerebellar nuclei. Once again, if the cerebellum is damaged or removed, the ability of the VOR to adapt to the new conditions is lost. These observations support the conclusion that the cerebellum is critically important in error reduction during motor learning.

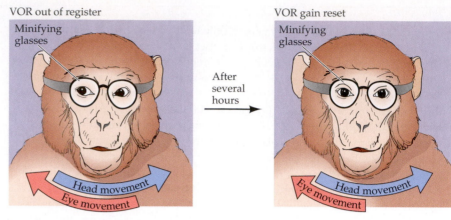

Normal vestibulo-ocular reflex (VOR)

VOR out of register
Minifying glasses

After several hours →

VOR gain reset
Minifying glasses

Head movement
Eye movement

Head movement
Eye movement

Eye movement
Head movement

Head and eyes move in a coordinated manner to keep image on retina

Eyes move too far in relation to image movement on the retina when the head moves

Eyes move smaller distances in relation to head movement to compensate

Figure 19.14 *Learned changes in the vestibulo-ocular reflex in monkeys. Normally, this reflex operates to move the eyes as the head moves, so that the retinal image remains stable. When the animal observes the world through minifying spectacles, the eyes initially move too far with respect to the "slippage" of the visual image on the retina. After some experience, however, the gain of the VOR is reset and the eyes move an appropriate distance in relation to head movement, thus compensating for the altered size of the visual image.*

Further Consequences of Cerebellar Lesions

We have seen that patients with cerebellar damage, regardless of the cause or location, exhibit persistent errors in movement. These movement errors are always on the same side of the body as the damage to the cerebellum, reflecting the cerebellum's unusual status as a brain structure in which sensory and motor information is represented ipsilaterally rather than contralaterally. Furthermore, somatic, visual, and other inputs are represented topographically within the cerebellum; as a result, the movement deficits following circumscribed cerebellar damage may be quite specific. For example, one of the most common cerebellar syndromes is caused by degeneration in the anterior portion of the cerebellar cortex in patients with a long history of alcohol abuse (Figure 19.15). Such damage specifically affects movement in the lower limbs, which are represented in the anterior spinocerebellum (see Figure 19.5). The consequences include a wide and staggering gait, but with little impairment of arm or hand movements.

Figure 19.15 *The pathological changes in a variety of neurological diseases provide insights about the function of the cerebellum. In this example, chronic alcohol abuse has caused degeneration of the anterior vermis (arrows), while leaving other cerebellar regions intact. The patient had difficulty walking but little impairment of arm movements or speech. The orientation of this paramedian sagittal section is the same as Figure 19.1C. (From Victor et al., 1959.)*

Thus, the topographical organization of the cerebellum allows cerebellar damage to disrupt the coordination of movements performed by some muscle groups but not others.

The implication of these pathologies is that the cerebellum is normally capable of integrating the moment-to-moment actions of muscles and joints throughout the body to ensure the smooth execution of a full range of motor behaviors. Thus cerebellar lesions lead first and foremost to a lack of coordination of ongoing movements (Box 19B). For example, damage to the vestibulocerebellum impairs the ability to stand upright and maintain the direction of gaze. The eyes have difficulty maintaining fixation; they drift from the target and then jump back to it with a corrective saccade, a phenomenon called **nystagmus**. Disruption of the pathways to the vestibular nuclei may also result in a reduction of muscle tone. In contrast, patients with damage to the spinocerebellum have difficulty controlling walking movements; they have a wide-based gait with small shuffling movements, which represents the inappropriate operation of groups of muscles that normally rely on sensory feedback to produce smooth, concerted actions. The patients also have difficulty performing rapid alternating movements, a sign referred to as **dysdiadochokinesia**. Over- and underreaching, or **dysmetria**, may also occur. Tremors—called **action** or **intention tremors**—accompany over- and undershooting of the movement due to disruption of the mechanism for detecting and correcting movement errors. Finally, lesions of the cerebrocerebellum produce impairments in highly skilled sequences of learned movements, such as speech or playing a musical instrument. The common denominator of all of these signs, regardless of the site of the lesion, is the inability to perform smooth, directed movements.

Summary

The cerebellum receives input from regions of the cerebral cortex that plan and initiate complex and highly skilled movements; it also receives innervation from sensory systems that monitor the course of movements. This arrangement enables a comparison of an intended movement with the actual movement and a reduction in the difference, or "motor error." The corrections of motor error produced by the cerebellum occur both in real time and over longer periods, as motor learning. For example, the vestibulo-ocular reflex allows an observer to maintain fixation on an object of interest while the head moves. Lenses that change image size produce a long-term change in the gain of this reflex that depends on an intact cerebellum. Knowledge of cerebellar circuitry suggests that motor learning is mediated by climbing fibers that ascend from the inferior olive to contact the dendrites of the Purkinje cells in the cerebellar cortex. Information provided by the climbing fibers modulates the effectiveness of the second major input to the Purkinje cells, which arrives via the parallel fibers from the granule cells. The granule cells receive information about the intended movement from the vast number of mossy fibers that enter the cerebellum from multiple sources, including the cortico-ponto-cerebellar pathway. As might be expected, the output of the cerebellum from the deep cerebellar nuclei projects to circuits that govern all the major sources of upper motor neurons described in Chapter 17. The effects of cerebellar disease provide strong support for the idea that the cerebellum regulates the performance of movements. Thus, patients with cerebellar disorders show severe ataxias in which the site of the lesion determines the particular movements affected.

BOX 19B Genetic Analysis of Cerebellar Function

Since the early 1950s, investigators interested in motor behavior have identified and studied strains of mutant mice in which movement is compromised. These mutant mice are easy to spot: following induced or spontaneous mutagenesis, the "screen" is simply to look for animals that have difficulty moving.

Genetic analysis suggested that some of these abnormal behaviors could be explained by single autosomal recessive or semidominant mutations, in which homozygotes are most severely affected. The strains were given names like *reeler*, *weaver*, *lurcher*, *staggerer*, and *leaner* that reflected the nature of the motor dysfunction they exhibited (see table). The relatively large number of mutations that compromise movement suggested it might be possible to understand some

aspects of motor circuits and function at the genetic level.

A common feature of the mutants is ataxia resembling that associated with cerebellar dysfunction in humans. Indeed, all the mutations are associated with some form of cerebellar pathology. The pathologies associated with the *reeler* and *weaver* mutations are particularly striking. In the *reeler* cerebellum, Purkinje cells, granule cells, and interneurons are all displaced from their usual laminar positions, and there are fewer granule cells than normal. In *weaver*, most of the granule cells are lost prior to their migration from the external granule layer (a proliferative region where cerebellar granule cells are generated during development), leaving only Purkinje cells and local interneurons to

carry on the work of the cerebellum. Thus, these mutations causing deficits in motor behavior impair the development and final disposition of the neurons that comprise the major processing circuits of the cerebellum (see Figure 19.8).

Efforts to characterize the cellular mechanisms underlying these motor deficits were unsuccessful, and the molecular identity of the affected genes remained obscure until recently. In the past few years, however, both the *reeler* and *weaver* genes have been identified and cloned.

The *reeler* gene was cloned through a combination of good luck and careful observation. In the course of making transgenic mice by inserting DNA fragments in the mouse genome, investigators in Tom Curran's laboratory created a

Motor Mutations in Mice

Mutation	Inheritance	Chromosome affected	Behavioral and morphological characteristics
reeler (*rl*)	Autosomal recessive	5	Reeling ataxia of gait, dystonic postures, and tremors. Systematic malposition of neuron classes in the forebrain and cerebellum. Small cerebellum, reduced number of granule cells.
weaver (*wv*)	Autosomal recessive	?	Ataxia, hypotonia, and tremor. Cerebellar cortex reduced in volume. Most cells of external granular layer degenerate prior to migration.
leaner (*tg1a*)	Autosomal recessive	8	Ataxia and hypotonia. Degeneration of granule cells, particularly in the anterior and nodular lobes of the cerebellum. Degeneration of a few Purkinje cells.
lurcher (*lr*)	Autosomal semidominant	6	Homozygote dies. Heterozygote is ataxic with hesitant, lurching gait and has seizures. Cerebellum half normal size; Purkinje cells degenerate; granule cells reduced in number.
nervous (*nr*)	Autosomal recessive	8	Hyperactivity and ataxia. Ninety percent of Purkinje cells die between 3 and 6 weeks of age.
Purkinje cell degeneration (*pcd*)	Autosomal recessive	13	Moderate ataxia. All Purkinje cells degenerate between the fifteenth embryonic day and third month of age.
staggerer (*sg*)	Autosomal recessive	9	Ataxia with tremors. Dendritic arbors of Purkinje cells are simple (few spines). No synapses of Purkinje cells with parallel fibers. Granule cells eventually degenerate.

Adapted from Caviness and Rakic, 1978.

new strain of mice that behaved much like *reeler* mice and had similar cerebellar pathology. This "synthetic" *reeler* mutation was identified by finding the position of the novel DNA fragment—which turned out to be on the same chromosome as the original *reeler* mutation. Further analysis showed that the same gene had indeed been mutated, and the *reeler* gene was subsequently identified. Remarkably, the protein encoded by this gene is homologous to known extracellular matrix proteins such as tenascin, laminin, and fibronectin (see Chapter 23). This finding makes good sense, since the pathophysiology of the *reeler* mutation entails altered cell migration, resulting in misplaced neurons in the cerebellar cortex as well as the cerebral cortex and hippocampus.

Molecular genetic techniques have also led to cloning the *weaver* gene. Using linkage analysis and the ability to clone and sequence large pieces of mammalian chromosomes, Andy Peterson and his colleagues "walked" (i.e.,

sequentially cloned) several kilobases of DNA in the chromosomal region to find where the *weaver* gene mapped. By comparing normal and mutant sequences within this region, they determined *weaver* to be a mutation in an inward rectifier K+ channel (see Chapter 4). How this particular molecule influences the development of granule cells or causes their death in the mutants is not yet clear.

The story of the proteins encoded by the *reeler* and *weaver* genes indicates both the promise and the challenge of a genetic approach to understanding cerebellar function. Identifying motor mutants and their pathology is reasonably straightforward, but understanding their molecular genetic basis depends on hard work and good luck.

References

CAVINESS, V. S. JR. AND P. RAKIC (1978) Mechanisms of cortical development: A view from mutations in mice. *Annu. Rev. Neurosci.* 1. 297–326.

D'ARCANGELO, G., G. G. MIAO, S. C. CHEN, H. D. SOARES, J. I. MORGAN AND T. CURRAN (1995) A protein related to extracellular matrix proteins deleted in the mouse mutation *reeler*. *Nature* 374: 719–723.

PATIL, N., D. R. COX, D. BHAT, M. FAHAM, R. M. MEYERS AND A. PETERSON (1995) A potassium channel mutation in *weaver* mice implicates membrane excitability in granule cell differentiation. *Nature Genetics* 11: 126–129.

RAKIC, P. AND V. S. CAVINESS JR. (1995) Cortical development: A view from neurological mutants two decades later. *Neuron* 14: 1101–1104.

The cerebellar cortex is disrupted in both the *reeler* and *weaver* mutations. (A) The cerebellar cortex in homozygous *reeler* mice. The *reeler* mutation causes the major cell types of the cerebellar cortex to be displaced from their normal laminar positions. Despite the disorganization of the cerebellar cortex in *reeler* mutants, the major inputs—mossy fibers and climbing fibers—find appropriate targets. (B) The cerebellar cortex in homozygous *weaver* mice. The granule cells are missing, and the major cerebellar inputs synapse inappropriately on the remaining neurons. (After Rakic, 1977.)

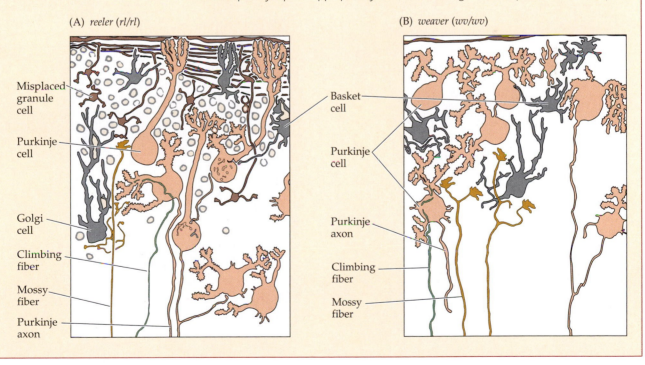

(A) *reeler (rl/rl)*

Misplaced granule cell
Purkinje cell
Golgi cell
Climbing fiber
Mossy fiber
Purkinje axon

(B) *weaver (wv/wv)*

Basket cell
Purkinje cell
Purkinje axon
Climbing fiber
Mossy fiber

Additional Reading

Reviews

ALLEN, G. AND N. TSUKAHARA (1974) Cerebrocerebellar communication systems. *Physiol. Rev.* 54: 957–1006.

GLICKSTEIN, M. AND C. YEO (1990) The cerebellum and motor learning. *J. Cog. Neurosci.* 2: 69–80.

LISBERGER, S. G. (1988) The neural basis for learning of simple motor skills. *Science* 242: 728–735.

OHYAMA, T., W. L. NORES, M. MURPHY AND M. D. MAUK (2003) What the cerebellum computes. *Trends Neurosci.* 26: 222–227.

ROBINSON, F. R. AND A. F. FUCHS (2001) The role of the cerebellum in voluntary eye movements. *Annu. Rev. Neurosci.* 24: 981–1004.

STEIN, J. F. (1986) Role of the cerebellum in the visual guidance of movement. *Nature* 323: 217–221.

THACH, W. T., H. P. GOODKIN AND J. G. KEATING (1992) The cerebellum and adaptive coordination of movement. *Annu. Rev. Neurosci.* 15: 403–442.

Important Original Papers

ASANUMA, C., W. T. THACH AND E. G. JONES (1983) Distribution of cerebellar terminals and their relation to other afferent terminations in the ventral lateral thalamic region of the monkey. *Brain Res. Rev.* 5: 237–265.

BRODAL, P. (1978) The corticopontine projection in the rhesus monkey: Origin and principles of organization. *Brain* 101: 251–283.

DELONG, M. R. AND P. L. STRICK (1974) Relation of basal ganglia, cerebellum, and motor cortex units to ramp and ballistic movements. *Brain Res.* 71: 327–335.

ECCLES, J. C. (1967) Circuits in the cerebellar control of movement. *Proc. Natl. Acad. Sci. USA* 58: 336–343.

McCORMICK, D. A., G. A. CLARK, D. G. LAVOND AND R. F. THOMPSON (1982) Initial localization of the memory trace for a basic form of learning. *Proc. Natl. Acad. Sci. USA* 79: 2731–2735.

THACH, W. T. (1968) Discharge of Purkinje and cerebellar nuclear neurons during rapidly alternating arm movements in the monkey. *J. Neurophysiol.* 31: 785–797.

THACH, W. T. (1978) Correlation of neural discharge with pattern and force of muscular activity, joint position, and direction of intended next movement in motor cortex and cerebellum. *J. Neurophysiol.* 41: 654–676.

VICTOR, M., R. D. ADAMS AND E. L. MANCALL (1959) A restricted form of cerebellar cortical degeneration occurring in alcoholic patients. *Arch. Neurol.* 1: 579–688.

Books

BRADLEY, W. G., R. B. DAROFF, G. M. FENICHEL AND C. D. MARSDEN (EDS.) (1991) *Neurology in Clinical Practice*. Boston: Butterworth-Heinemann, Chapters 29 and 77.

ITO, M. (1984) *The Cerebellum and Neural Control*. New York: Raven Press.

KLAWANS, H. L. (1989) *Toscanini's Fumble and Other Tales of Clinical Neurology*. New York: Bantam, Chapters 7 and 10.

Chapter 20

Eye Movements and Sensory Motor Integration

Overview

Eye movements are in many ways easier to study than movements of other parts of the body. This fact arises in part from the relative simplicity of muscle actions on the eyeball. There are only six extraocular muscles, each of which has a specific role in adjusting eye position. Moreover, there is a limited set of stereotyped eye movements, each with its own control circuitry. Eye movements have therefore been a useful model for understanding the mechanisms of motor control. Indeed, much of what is known about the regulation of movements by the cerebellum, basal ganglia, and vestibular system has come from the study of eye movements (see Chapters 14, 18, and 19). In this chapter, the major features of eye movement control are used to illustrate principles of sensory motor integration that also apply to more complex motor behaviors.

What Eye Movements Accomplish

Eye movements are important in humans because high visual acuity is restricted to the fovea, the small circular region (about 1.2 mm in diameter) in the central retina that is densely packed with cone photoreceptors (see Chapter 11). Eye movements can direct the fovea to new objects of interest (a process called "foveation") or compensate for disturbances that cause the fovea to be displaced from a target already being foveated.

Several decades ago, the Russian physiologist Alfred Yarbus demonstrated that eye movements reveal a good deal about the strategies used to inspect a scene. Yarbus used contact lenses with small mirrors on them to document (by the position of a reflected beam) the pattern of eye movements made while subjects examined a variety of objects and scenes. Figure 20.1 shows the changes in the direction of a subject's gaze while viewing a photograph (of the famous bust of Queen Nefertiti in this experiment). The thin, straight lines represent the quick, ballistic eye movements (**saccades**) used to align the foveas with particular parts of the scene. Little or no visual perception occurs during a saccade, which occupies only a few tens of milliseconds. The denser spots along these lines represent points of fixation where the observer paused for a variable period to take in visual information. The results obtained by Yarbus, and subsequently many others, showed that vision is an active process in which eye movements typically shift the view several times each second to selected parts of the scene to examine especially interesting features. The spatial distribution of the fixation points indicates that much more time is spent scrutinizing Nefertiti's eye, nose, mouth, and ear than examining the middle of her cheek or neck. Thus, eye movements allow us to scan the visual field, pausing to focus attention on the portions of the scene that convey the most significant information. As is apparent in Figure 20.1, tracking eye movements can be used to determine

Figure 20.1 Eye movements of a subject viewing a photograph of the bust of Queen Nefertiti. The photograph at the top is what the subject saw; the diagram on the bottom shows the subject's eye movements over a 2-minute viewing period. (From Yarbus, 1967.)

which aspects of a scene are particularly arresting; in fact today's corporate advertisers can use modern versions of Yarbus's method to determine which pictures and scene arrangements will best sell their products.

The importance of eye movements for visual perception has also been demonstrated by experiments in which a visual image is stabilized on the retina, either by paralyzing the extraocular eye muscles or by moving a scene in exact register with eye movements so that the different features of the image always fall on exactly the same parts of the retina (Box 20A). Stabilized visual images rapidly disappear, for reasons that remain poorly understood. Nonetheless, these observations on motionless images make it plain that eye movements are essential for normal visual perception.

The Actions and Innervation of Extraocular Muscles

Three antagonistic pairs of muscles control eye movements: the **lateral** and **medial rectus muscles**, the **superior** and **inferior rectus muscles**, and the **superior** and **inferior oblique muscles**. These muscles are responsible for movements of the eye along three different axes: *horizontal*, either toward the nose (adduction) or away from the nose (abduction); *vertical*, either elevation or depression; and *torsional*, movements that bring the top of the eye toward the nose (intorsion) or away from the nose (extorsion). Horizontal movements are controlled entirely by the medial and lateral rectus muscles; the medial rectus muscle is responsible for adduction, the lateral rectus muscle for abduction (Figure 20.2). Vertical movements require the coordinated action of the superior and inferior rectus muscles, as well as the oblique muscles. The relative contribution of the rectus and oblique groups depends on the horizontal position of the eye. In the primary position (eyes straight ahead), both of these groups contribute to vertical movements. Elevation is due to the action of the superior rectus and inferior oblique muscles, while depression is due to the action of the inferior rectus and superior oblique muscles. When the eye is abducted, the rectus muscles are the prime vertical movers. Elevation is due to the action of the superior rectus, and depression is due to the action of the inferior rectus. When the eye is adducted, the oblique muscles are the prime vertical movers. Elevation is due to the action of the inferior oblique muscle, while depression is due to the action of the superior oblique muscle. The oblique muscles are also primarily responsible for torsional movements.

Figure 20.2 The contributions of the six pairs of extraocular muscles to vertical and horizontal eye movements. Horizontal movements are mediated by the medial and lateral rectus muscles, while vertical movements are mediated by the superior and inferior rectus and the superior and inferior oblique muscle groups.

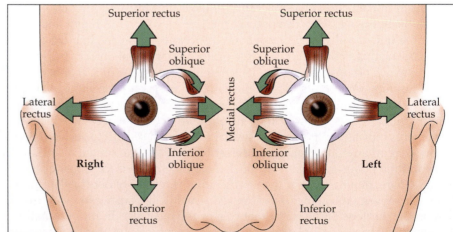

BOX 20A   The Perception of Stabilized Retinal Images

Visual perception depends critically on frequent changes of scene. Normally, our view of the world is changed by saccades, and tiny saccades that continue to move the eyes abruptly over a fraction of a degree of visual arc occur even when the observer stares intently at an object of interest. Moreover, continual drift of the eyes during fixation progressively shifts the image onto a nearby but different set of photoreceptors. As a consequence of these several sorts of eye movements (Figure A), our point of view changes more or less continually.

The importance of a continually changing scene for normal vision is dramatically revealed when the retinal image is stabilized. If a small mirror is attached to the eye by means of a contact lens and an image reflected off the mirror onto a screen, then the subject necessarily sees the same thing, whatever the position of the eye: every time the eye moves, the projected image moves by exactly the same amount (Figure B). Under these circumstances, the stabilized image actually disappears from perception within a few seconds!

A simple way to demonstrate the rapid disappearance of a stabilized retinal image is to visualize one's own retinal blood vessels. The blood vessels, which lie in front of the photoreceptor layer, cast a shadow on the underlying

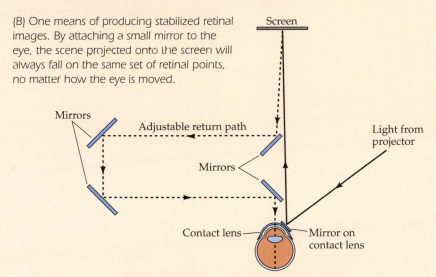

(B) One means of producing stabilized retinal images. By attaching a small mirror to the eye, the scene projected onto the screen will always fall on the same set of retinal points, no matter how the eye is moved.

receptors. Although normally invisible, the vascular shadows can be seen by moving a source of light across the eye, a phenomenon first noted by J. E. Purkinje more than 150 years ago. This perception can be elicited with an ordinary penlight pressed gently against the lateral side of the closed eyelid. When the light is wiggled vigorously, a rich network of black blood vessel shadows appears against an orange background. (The vessels appear black because they are shadows.) By starting and stopping the movement, it is readily apparent that the image of the

blood vessel shadows disappears within a fraction of a second after the light source is stilled.

The conventional interpretation of the rapid disappearance of stabilized images is retinal adaptation. In fact, the phenomenon is at least partly of central origin. Stabilizing the retinal image in one eye, for example, diminishes perception through the other eye, an effect known as interocular transfer. Although the explanation of these remarkable effects is not entirely clear, they emphasize the point that the visual system is designed to deal with novelty.

References

BARLOW, H. B. (1963) Slippage of contact lenses and other artifacts in relation to fading and regeneration of supposedly stable retinal images. *Q. J. Exp. Psychol.* 15: 36–51.

COPPOLA, D. AND D. PURVES (1996) The extraordinarily rapid disappearance of entopic images. *Proc. Natl. Acad. Sci. USA* 96: 8001–8003.

HECKENMUELLER, E. G. (1965) Stabilization of the retinal image: A review of method, effects and theory. *Psychol. Bull.* 63: 157–169.

KRAUSKOPF, J. AND L. A. RIGGS (1959) Interocular transfer in the disappearance of stabilized images. *Amer. J. Psychol.* 72: 248–252.

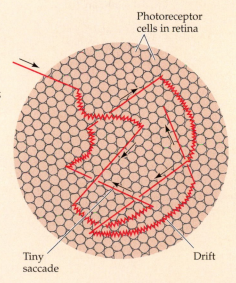

(A) Diagram of the types of eye movements that continually change the retinal stimulus during fixation. The straight lines indicate microsaccades and the zig-zag lines drift; the structures in the background are photoreceptor cells drawn approximately to scale. The normal scanning movements of the eyes (saccades) are much too large to be shown here, but obviously contribute to the changes of view that we continually experience, as do slow tracking eye movements (although the fovea tracks a particular object, the scene nonetheless changes). (After Pritchard, 1961.)

The extraocular muscles are innervated by lower motor neurons that form three cranial nerves: the abducens, the trochlear, and the oculomotor (Figure 20.3). The **abducens nerve** (cranial nerve VI) exits the brainstem from the pons–medullary junction and innervates the lateral rectus muscle. The **trochlear nerve** (cranial nerve IV) exits from the caudal portion of the midbrain and supplies the superior oblique muscle. In distinction to all other cranial nerves, the trochlear nerve exits from the dorsal surface of the brainstem and crosses the midline to innervate the superior oblique muscle on the contralateral side. The **oculomotor nerve** (cranial nerve III), which exits from the rostral midbrain near the cerebral peduncle, supplies all the rest of the extraocular muscles. Although the oculomotor nerve governs several different muscles, each receives its innervation from a separate group of lower motor neurons within the nuclear complex that supplies the third nerve.

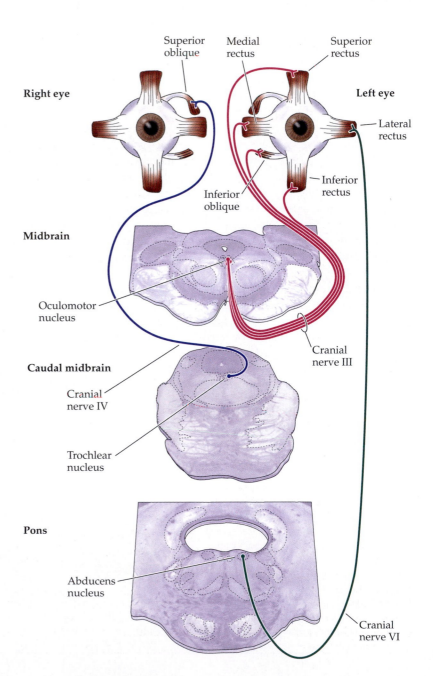

Figure 20.3   Organization of the cranial nerve nuclei that govern eye movements, showing their innervation of the extraocular muscles. The abducens nucleus innervates the ipsilateral lateral rectus muscle; the trochlear nucleus innervates the contralateral superior oblique muscle; and the oculomotor nucleus innervates all the rest of the ipsilateral extraocular muscles (the medial rectus, inferior rectus, superior rectus, and inferior oblique).

In addition to supplying the extraocular muscles, a distinct cell group within the oculomotor complex innervates the levator muscles of the eyelid; the axons from these neurons also travel in the third nerve. Finally, the third nerve carries parasympathetic axons that are responsible for pupillary constriction (see Chapter 12) from the nearby Edinger-Westphal nucleus. Thus, damage to the third nerve results in three characteristic deficits: impairment of eye movements, drooping of the eyelid (ptosis), and pupillary dilation.

Types of Eye Movements and Their Functions

There are five basic types of eye movements that can be grouped into two functional categories: those that serve to *shift* the direction of gaze when new targets are foveated or when foveated targets move, and those that serve to *stabilize* gaze when the head moves or when there are large-scale movements of the visual field. Thus, saccades, smooth pursuit movements, and vergence movements shift the direction of gaze, and vestibulo-ocular and optokinetic movements stabilize gaze. The functions of each type of eye movement are introduced here; in subsequent sections, the neural circuitry responsible for movements that shift the direction of gaze is presented in more detail (see Chapters 14 and 19 for further discussion of neural circuitry underlying gaze-stabilizing movements).

As noted earlier, saccades are rapid, ballistic movements of the eyes that abruptly change the direction of fixation. They range in amplitude from the small movements made while reading to the much larger movements made while gazing around a room. Saccades can be elicited voluntarily, but occur reflexively whenever the eyes are open, even when fixated on a target (see Box 20A). The rapid eye movements (REM) that occur during an important phase of sleep are saccades (see Chapter 28).

The time course of a saccadic eye movement is shown in Figure 20.4. After the onset of a target for a saccade (in this example, the stimulus was the movement of an already fixated target), it takes about 200 milliseconds for eye movement to begin. During this delay, the position of the target with respect to the fovea is computed (that is, how far the eye has to move), and the difference between the initial and intended position is converted into a motor command that activates the extraocular muscles to move the eyes the correct distance in the appropriate direction. Saccadic eye movements are said to be ballistic because the saccade-generating system usually does not respond to subsequent changes in the position of the target during the course of the eye movement. If the target moves again during this time (which is on the order of 15–100 ms), the saccade will miss the target, and a second saccade must be made to correct the error.

Smooth pursuit movements are much slower tracking movements of the eyes designed to keep a moving stimulus on the fovea once foveation is achieved. Such movements are under voluntary control in the sense that the observer can choose whether or not to track a moving stimulus (Figure 20.5). (Saccades can be voluntary, but are also made unconsciously.) Surprisingly, however, only highly trained observers can make a smooth pursuit movement in the absence of a moving target. Most people who try to move their eyes in a smooth fashion without a moving target simply make a saccade.

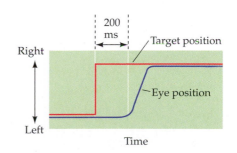

Figure 20.4   *The metrics of a saccadic eye movement. The red line indicates the position of a fixation target and the blue line the position of the fovea. When the target moves suddenly to the right, there is a delay of about 200 ms before the eye begins to move to the new target position. (After Fuchs, 1967.)*

Figure 20.5 *The metrics of smooth pursuit eye movements. These traces show eye movements (blue lines) tracking a stimulus moving at one of three different velocities (red lines). After a quick saccade to capture the target, the eye movement attains a velocity that matches the velocity of the target. (After Fuchs, 1967.)*

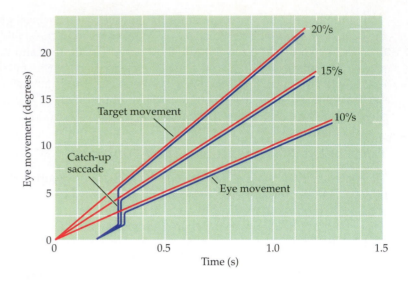

Vergence movements align the fovea of each eye with targets located at different distances from the observer. Although vergence movements are required to track a visual target that may be moving closer or further away, they are more commonly employed when abruptly shifting gaze, for example from a near object to one that is more distant. Unlike other types of eye movements, in which the two eyes move in the same direction (**conjugate eye movements**), vergence movements are **disconjugate** (or **disjunctive**); they involve either a convergence or divergence of the lines of sight of each eye to see an object that is nearer or farther away. Convergence is one of the three reflexive visual responses elicited by interest in a near object. The other components of the so-called **near reflex triad** are accommodation of the lens, which brings the object into focus, and pupillary constriction, which increases the depth of field and sharpens the image on the retina (see Chapter 11).

Vestibulo-ocular movements and **optokinetic eye movements** operate together to stabilize the eyes relative to the external world, thus compensating for head movements. These reflex responses prevent visual images from "slipping" on the surface of the retina as head position varies and, more rarely, when confronted with large-scale movements of the visual scene (such as a flowing river or a passing train).

The action of vestibulo-ocular movements can be appreciated by fixating an object and moving the head from side to side; the eyes automatically compensate for the head movement by moving the same distance and at the same velocity but in the opposite direction, thus keeping the image of the object at more or less the same place on the retina. The vestibular system detects brief, transient changes in head position and produces rapid corrective eye movements (see Chapter 14). Sensory information from the semicircular canals directs the eyes to move in a direction opposite to the head movement. Although the vestibular system operates effectively to counteract rapid movements of the head, it is relatively insensitive to slow movements (below 1 Hz) or to persistent rotation of the head. For example, if the vestibulo-ocular reflex is tested with continuous rotation and without visual cues about the movement of the image (i.e., with eyes closed or in the dark), the compensatory eye movements cease after only about 30 seconds of rotation. However, if the same test is performed with visual cues, eye movements persist. The compensatory eye movements in this case are due to the activation of another system that relies not on vestibular information, but on visual cues indicating motion of the visual field. This optokinetic system is especially sensitive to

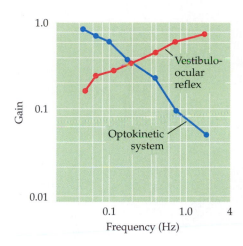

Figure 20.6 Comparison of the operational range of the vestibulo-ocular and optokinetic systems. Functions of the vestibulo-ocular and optokinetic systems were assessed independently in rabbits by rotating the animals with their eyes closed (to isolate the vestibulo-ocular reflex) or following recovery from bilateral labyrinthectomy (to isolate the optokinetic system). At low frequencies of movement (below 1 Hz or one back-and-forth cycle of stimulation per second), the gain of the vestibulo-ocular reflex (the ratio of eye movement to head movement) diminishes below unity. However, the gain of the optokinetic system (the ratio of eye movement to retinal slip) approaches unity at such low frequencies of stimulation. Thus the vestibulo-ocular and optokinetic systems act in a complementary, frequency-dependent fashion to stabilize gaze over a broad range of stimulation frequencies (After Baarsma and Collewijn, 1974.)

slow movements (below 1 Hz) of large areas of the visual field, and its response builds up slowly. These features complement the properties of the vestibulo-ocular reflex, especially as head movements slow down and vestibular signals decay (Figure 20.6). Thus, should a visual image slowly "slip" across the retina, the optokinetic system will respond by inducing compensatory movements of the eyes that restore foveation of a visual target in the image.

The optokinetic system can be tested by placing a subject inside a rotating cylinder with vertical stripes. (In practice, this is usually done by simply seating the subject in front of a screen on which a series of horizontally moving vertical bars is presented.) The eyes automatically track the stripes until they reach the end of their excursion. Then there is a quick saccade in the direction opposite to the movement, followed once again by smooth pursuit of the stripes. This alternating slow and fast movement of the eyes in response to such stimuli is called **optokinetic nystagmus**. Optokinetic nystagmus is a normal reflexive response of the visual and oculomotor systems in response to large-scale movements of the visual scene, and should not be confused with the pathological nystagmus that can result from certain kinds of brain injury (for example, damage to the vestibular system or the cerebellum; see Chapter 19).

Neural Control of Saccadic Eye Movements

The problem of moving the eyes to fixate a new target in space (or indeed any other movement) entails two separate issues: controlling the *amplitude* of movement (how far), and controlling the *direction* of the movement (which way). The amplitude of a saccadic eye movement is encoded by the duration of neuronal activity in the lower motor neurons of the oculomotor nuclei. For instance, as shown in Figure 20.7, neurons in the abducens nucleus fire a burst of action potentials just prior to abducting the eye (by causing the lateral rectus muscle to contract) and are silent when the eye is adducted. The amplitude of the movement is correlated with the duration of the burst of action potentials in the abducens neuron. Following each saccade, the abducens neurons reach a new baseline level of discharge that is correlated with the position of the eye in the orbit. The steady baseline level of firing holds the eye in its new position.

The direction of the movement is determined by which eye muscles are activated. Although in principle any given direction of movement could be specified by independently adjusting the activity of individual eye muscles, the complexity of the task would be overwhelming. Instead, the direction of eye movement is controlled by the local circuit neurons in two **gaze centers** in the reticular for-

Figure 20.7   Motor neuron activity in relation to saccadic eye movements. The experimental setup is shown on the right. In this example, an abducens lower motor neuron fires a burst of activity (upper trace) that precedes and extends throughout the movement (solid line). An increase in the tonic level of firing is associated with more lateral displacement of the eye. Note also the decline in firing rate during a saccade in the opposite direction. (After Fuchs and Luschei, 1970.)

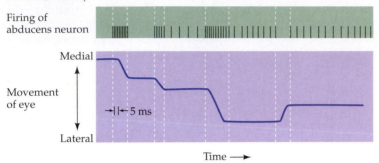

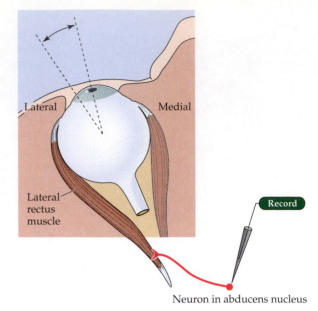

mation (see Box 17A), each of which is responsible for generating movements along a particular axis. The **paramedian pontine reticular formation** (**PPRF**), or **horizontal gaze center**, is a collection of local circuit neurons near the midline in the pons and is responsible for generating horizontal eye movements. The **rostral interstitial nucleus**, or **vertical gaze center**, is located in the rostral part of the midbrain reticular formation and is responsible for vertical movements. Activation of each gaze center separately results in movements of the eyes along a single axis, either horizontal or vertical. Activation of the gaze centers in concert results in oblique movements whose trajectories are specified by the relative contribution of each center.

An example of how the PPRF works with the abducens and oculomotor nuclei to generate a horizontal saccade to the right is shown in Figure 20.8. Neurons in the PPRF innervate cells in the abducens nucleus on the same side of the brain. There are two types of neurons in the abducens nucleus. One type comprises the lower motor neurons that innervate the lateral rectus muscle on the same side. The other type, called internuclear neurons, send their axons across the midline. These axons ascend in a fiber tract called the **medial longitudinal fasciculus**, terminating in the portion of the oculomotor nucleus that contains lower motor neurons that innervate the medial rectus muscle. As a result of this arrangement, activation of PPRF neurons on the right side of the brainstem causes horizontal movements of both eyes to the right; the converse is true for the PPRF neurons in the left half of the brainstem.

Neurons in the PPRF also send axons to the medullary reticular formation, where they contact inhibitory local circuit neurons. These local circuit neurons, in turn, project to the contralateral abducens nucleus, where they terminate on lower motor neurons and internuclear neurons. In consequence, activation of neurons in the PPRF on the right results in a reduction in the activity of the lower motor neurons in the left abducens nucleus, whose muscles would oppose movements of the eyes to the right (see Figure 20.7). This inhibition of antagonists resembles the strategy used by local circuit neurons in the spinal cord to control limb muscle antagonists (see Chapter 16).

Although saccades can occur in complete darkness, they are often elicited when something attracts attention and the observer directs the foveas toward the stimulus. How then is sensory information about the location of a target in

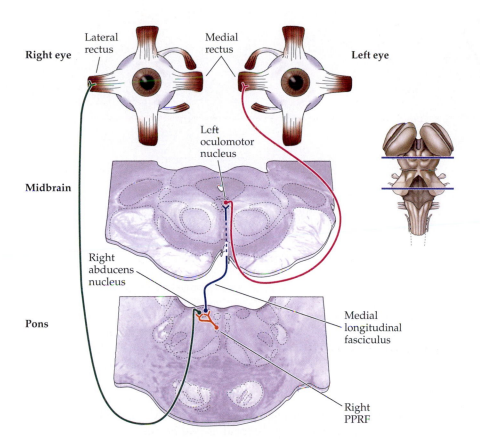

Figure 20.8 Simplified diagram of synaptic circuitry responsible for horizontal movements of the eyes to the right. Activation of local circuit neurons in the right horizontal gaze center (the PPRF; orange) leads to increased activity of lower motor neurons (red and green) and internuclear neurons (blue) in the right abducens nucleus. The lower motor neurons innervate the lateral rectus muscle of the right eye. The internuclear neurons project via the medial longitudinal fasciculus to the contralateral oculomotor nucleus, where they activate lower motor neurons that in turn innervate the medial rectus muscle of the left eye.

space transformed into an appropriate pattern of activity in the horizontal and vertical gaze centers? Two regions of the brain that project to the gaze centers are demonstrably important for the initiation and accurate targeting of saccadic eye movements: the **superior colliculus** of the midbrain (called the **optic tectum** in non-mammalian vertebrates), and several areas in the frontal and parietal cortex. Especially well studied is a region of the frontal lobe that lies in a rostral portion of the premotor cortex, known as the **frontal eye field** (**Brodmann's area 8**). Upper motor neurons in both the superior colliculus and frontal eye fields, each of which contains a topographical map of eye movement vectors, discharge immediately prior to saccades. Thus, activation of a particular site in the superior colliculus or in the frontal eye field produces saccadic eye movements in a specified direction and for a specified distance. This movement is independent of the initial position of the eyes in the orbit; the direction and distance are always the same for a given site, but direction and distance of movement change systematically when different sites are activated.

Both the superior colliculus and the frontal eye field also contain cells that respond to visual stimuli; however, the relation between the sensory and motor responses of individual cells is better understood for the superior colliculus. An orderly map of visual space is established by the termination of retinal axons within the superior colliculus, which are supplemented by cortical inputs from visual areas that participate in the "dorsal spatial vision pathway" (see Figure 12.18). This sensory map is in register with the motor map that generates eye movements. Thus, neurons in a particular region of the superior colliculus are activated by the presentation of visual stimuli in a limited region of visual space. This activation leads to the generation of a saccade that moves the eye by an amount just sufficient to align the foveas with the region of visual space that provided the stimulation (Figure 20.9).

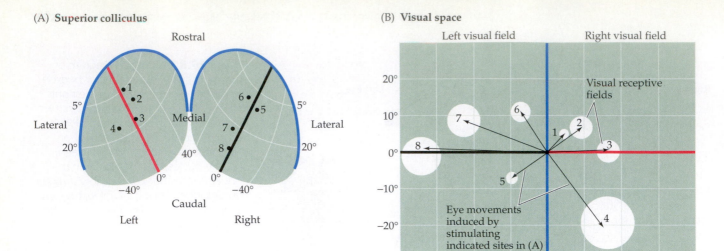

(A) Superior colliculus

Rostral

5°
Lateral
20°

1
2
3
4

Medial

6
5
7
8

5°
Lateral
20°

−40°
0°

0°
−40°

Caudal

Left

Right

(B) Visual space

Left visual field   Right visual field

20°

10°

0°

−10°

−20°

Visual receptive fields

7
6
1 2
8
3
5
4

Eye movements induced by stimulating indicated sites in (A)

30° 20° 10° 0° 10° 20° 30°

Figure 20.9   Evidence for the registration of sensory and motor maps obtained from electrical recording and stimulation in the superior colliculus. (A) Surface views of the superior colliculus illustrating the location of eight separate electrode recording and stimulation sites. (B) Map of visual space showing the visual receptive field location of the sites in (A) (white circles), and the amplitude and direction of the eye movements elicited by stimulating these sites electrically (arrows). In each case, electrical stimulation results in eye movements that align the fovea with a region of visual space that corresponds to the visual receptive fields of neurons at that site. (After Schiller and Stryker, 1972.)

Figure 20.10   Saccades are encoded in movement coordinates, not retinotopic coordinates. (A) Map of visual space illustrating experimental design. Monkeys were trained to fixate a central location (F, in black) and then perform a saccade to a remembered target location cued by a brief flash at a location 10° above the starting position (black T). After cueing, but before expression of the cued saccade, an electrical stimulus was applied to a site in the superior colliculus that induced a saccade down and to the left (to the location marked by the red F). If the cued saccade was encoded in retinotopic coordinates, the monkey should move its eyes 10° above the stimulus-induced position of foveation (red F) to a location marked by the dash-encircled T. If the saccade was encoded in movement coordinates, then a compensatory saccade to the cued target location (black T) would be expected. (B) Consistent with the encoding of saccades in movement coordinates, the monkeys performed compensatory saccades upward and to the right, toward the location of the cued target. Dots represent eye movements sampled at 500 Hz. Colored axes are as in Figure 20.9. (After Sparks and Mays, 1983.)

Neurons in the superior colliculus also respond to auditory and somatic sensory stimuli. Indeed, location in space for these other modalities is mapped in register with the motor map in the colliculus. Topographically organized maps of auditory space and of the body surface in the superior colliculus can therefore orient the eyes (and the head, via the colliculospinal tract and inputs from the superior colliculus to neurons that give rise to the reticulospinal tract; see Chapter 17) in response to a variety of different sensory stimuli. This registration of the sensory and motor maps in the colliculus illustrates an important principle of topographical maps in the motor system: namely, they provide an efficient mechanism for the transformation of sensory signals into movement parameters that are appropriate for the governed motor apparatus (in this case, the extraocular muscles and muscles of the posterior head and neck) (Box 20B). However, the motor map in the deep layers of the superior colliculus is not simply organized in the framework established by the spatial distribution of the sensory

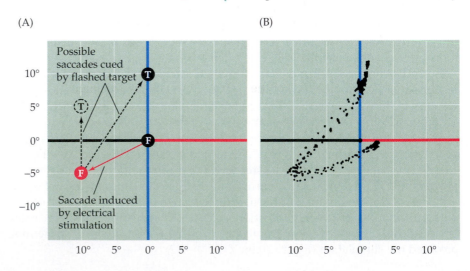

(A)

Possible saccades cued by flashed target

10°
5°
0°
−5°
−10°

T
T
F
F

Saccade induced by electrical stimulation

10° 5° 0° 5° 10°

(B)

10° 5° 0° 5° 10°

inputs. Rather, the input signals must also be encoded in movement coordinates so that sensory cues and cognitive signals (see below) can activate the motor responses required to move the eyes to the intended position in the orbits. Thus, motor circuitry of the superior colliculus specifies motor goals rather than movements to fixed positions in external space or on the body surface.

The organizing framework of this motor map was demonstrated in an ingenious series of studies performed by David Sparks and his colleagues at the University of Alabama. They showed that retinal error signals (i.e., the distance and direction of the retinal projection of the target from the fovea) in retinotopic coordinates are often not sufficient to localize saccade targets. Using trained monkeys, the investigators cued a voluntary saccade with a brief flash of light, but before the saccade could be initiated, they stimulated a site in the deep layers of the superior colliculus that induced a saccade away from the point of fixation. They recorded eye movements to determine whether the change in eye position induced by the stimulation had an impact on the direction and distance of the cued saccade (Figure 20.10). If the saccade vectors were determined simply by the retinotopic coordinates of the target, then the monkey would be expected to make a saccade of the cued direction and distance; but because of the deviated starting position, the saccade should systematically miss the target position by the amount of the stimulation-induced deviation (indicated by the red arrow in Figure 20.10A). The results consistently showed, however, that this was not the case. The animals compensated for the stimulation-induced shift by performing a compensatory saccade (a saccade indicated by the black dashed arrow to T—the actual target location—in Figure 20.10A). This compensatory action was based on stored information about location of the retinal image and current information about the position of the eyes in the orbit. The upper motor neurons that initiate the compensatory saccade are located at the expected site in the motor map of saccade vectors, but their activation depends on information in addition to the retinotopic location of the target. This information may be provided by circuits in the cerebral cortex and sent to the site in the superior colliculus that initiates the compensatory saccade. Thus, the monkeys performed saccades that brought the foveas into alignment with the location of the cued target (Figure 20.10B).

This study and several that followed showed that signals from different sensory modalities are integrated and transformed into a common motor frame of reference that encodes the direction and distance of the eye movements necessary to foveate an intended target. This "place code" for intended eye position generated in the upper motor neurons of the superior colliculus is then translated into a "rate code" by downstream gaze centers in the reticular formation that can then direct the activity of lower motor neurons in the ocular motor nuclei (Box 20C).

The eye movement regions of the cerebral cortex also participate with the superior colliculus in controlling saccades. Thus, the frontal eye field projects to the superior colliculus, and the superior colliculus projects to the PPRF on the contralateral side (Figure 20.11). (It

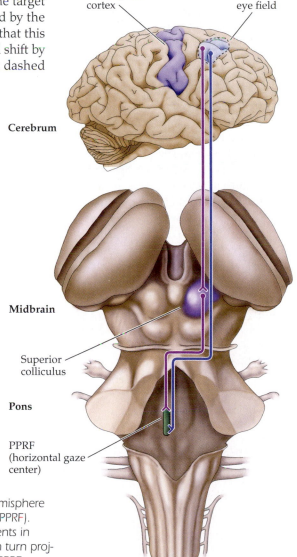

Primary motor cortex

Frontal eye field

Cerebrum

Midbrain

Superior colliculus

Pons

PPRF (horizontal gaze center)

Figure 20.11 Projections from the frontal eye field in the right cerebral hemisphere (Brodmann's area 8) to the superior colliculus and the horizontal gaze center (PPRF). There are two routes by which the frontal eye field can influence eye movements in humans: indirectly, by projections to the ipsilateral superior colliculus, which in turn projects to the contralateral PPRF; and directly, by projections to the contralateral PPRF.

BOX 20B   Sensory Motor Integration in the Superior Colliculus

The superior colliculus is a laminated structure in which differences between the layers provide clues about how sensory and motor maps interact to produce appropriate movements. As discussed in the text, the superficial or "visual" layer of the colliculus receives input from retinal axons that form a topographic map. Thus, each site in the superficial layer is activated maximally by the presence of a stimulus at a particular point of visual space. In contrast, neurons in the deeper or "motor" layers generate bursts of action potentials that command saccades, effectively generating a motor map; thus, activation of different sites generates saccades having different vectors (see Figure 20.9). The visual and motor maps are *in register*, so that visual cells responding to a stimulus in a specific region of visual space are located directly above the motor cells that command eye movements toward that same region (Figure A).

The registration of the visual and motor maps suggests a simple strategy for how the eyes might be guided toward an object of interest in the visual field. When an object appears at a particular location in the visual field, it will activate neurons in the corresponding part of the visual map. As a result, bursts of action potentials are generated by the subjacent motor cells to command a saccade that rotates the two eyes just the right amount to direct the foveas toward that same location in the visual field. This behavior is called "visual grasp" because successful sensory motor integration results in the accurate foveation of a visual target.

This seemingly simple model, formulated in the early 1970s when the collicular maps were first found, assumes point-to-point connections between the visual and motor maps. In practice, however, these connections have been difficult to demonstrate. Neither the anatomical nor the physiological methods available at the time were sufficiently precise to establish these postulated synaptic connections. At about the same time, motor neurons were found to command saccades to nonvisual stimuli; moreover, spontaneous saccades occur in the dark. Thus, it was clear that visual layer activity is not always necessary for saccades. To confuse matters further, animals could be trained *not* to make a saccade when an object appeared in the visual field, showing that the activation of visual neurons is sometimes insufficient to command saccades. The fact that activity of neurons in the visual map is *neither necessary nor sufficient* for eliciting saccades led investigators away from the simple model of direct connections between corresponding regions of the two maps, toward models that linked the layers indirectly through pathways that detoured through the cortex.

Eventually, however, new and better methods resolved this uncertainty. Techniques for filling single cells with axonal tracers showed an overlap between descending visual layer axons and ascending motor layer dendrites, in accord with direct anatomical connections between corresponding regions of the maps. At the same time, in vitro whole-cell patch clamp recording (see Box 4A) permitted more discriminating functional studies that distinguished exci-

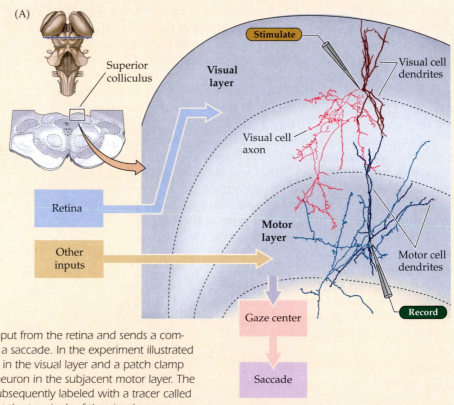

(A)

(A) The superior colliculus receives visual input from the retina and sends a command signal to the gaze centers to initiate a saccade. In the experiment illustrated here, a stimulating electrode activates cells in the visual layer and a patch clamp pipette records the response evoked in a neuron in the subjacent motor layer. The cells in the visual and motor layers were subsequently labeled with a tracer called biocytin. This experiment demonstrates that the terminals of the visual neuron are located in the same region as the dendrites of the motor neuron.

tatory and inhibitory inputs to the motor cells. These experiments showed that the visual and motor layers do indeed have the functional connections required to initiate the command for a visually guided saccadic eye movement. A single brief electrical stimulus delivered to the superficial layer generates a prolonged burst of action potentials that resembles the command bursts that normally occur just before a saccade (Figures B,C).

These direct connections presumably provide the substrate for the very short latency reflex-like "express saccades" that are unaffected by destruction of the frontal eye fields. Other visual and non-visual inputs to the deep layers probably explain why activation of the retina is neither necessary nor sufficient for the production of saccades.

References

LEE, P. H., M. C. HELMS, G. J. AUGUSTINE AND W. C. HALL (1997) Role of intrinsic synaptic circuitry in collicular sensorimotor integration. *Proc. Natl. Acad. Sci. USA* 94: 13299–13304.

OZEN, G., G. J. AUGUSTINE AND W. C. HALL (2000) Contribution of superficial layer neurons to premotor bursts in the superior colliculus. *J. Neurophysiol.* 84: 460–471.

SPARKS, D. L. AND J. S. NELSON (1987) Sensory and motor maps in the mammalian superior colliculus. *Trends Neurosci.* 10: 312–317.

WURTZ, R. H. AND J. E. ALBANO (1980) Visual-motor function of the primate superior colliculus. *Annu. Rev. Neurosci.* 3: 189–226.

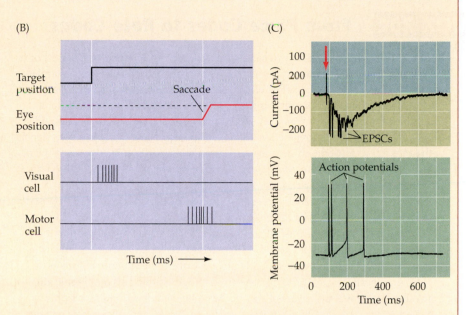

(B) The onset of a target in the visual field (top trace) is followed after a short interval by a saccade to foveate the target (second trace). In the superior colliculus, the visual cell responds shortly after the onset of the target, while the motor cell responds later, just before the onset of the saccade. (C) Bursts of excitatory postsynaptic currents (EPSCs) recorded from a motor layer neuron in response to a brief (0.5 ms) current stimulus applied via a steel wire electrode in the visual layer (top; see arrow). These synaptic currents generate bursts of action potentials in the same cell (bottom). (B after Wurtz and Albano, 1980; C after Ozen et al., 2000.

---

also projects to the vertical gaze center, but for simplicity the discussion here is limited to the PPRF.) The frontal eye field can thus control eye movements by activating selected populations of superior colliculus neurons. This cortical area also projects directly to the contralateral PPRF; as a result, the frontal eye field can also control eye movements independently of the superior colliculus. The parallel inputs to the PPRF from the frontal eye field and superior colliculus are reflected in the deficits that result from damage to these structures. Injury to the frontal eye field results in an inability to make saccades to the contralateral side and a deviation of the eyes to the side of the lesion. These effects are transient, however; in monkeys with experimentally induced lesions of this cortical region, recovery is virtually complete in two to four weeks. Lesions of the superior colliculus change the latency, accuracy, frequency, and velocity of saccades; yet saccades still occur, and the deficits also improve with time. These results suggest that the frontal eye fields and the superior colliculus provide complementary pathways for the control of saccades. Moreover, one of these structures appears to be able to compensate (at least partially) for the loss of the other. In support of this interpretation, combined lesions of the frontal eye field and the superior colliculus produce a dramatic and permanent loss in the ability to make saccadic eye movements.

BOX 20C From Place Codes to Rate Codes

How does the pattern of activity in the superior colliculus get translated into a motor command that can be delivered to muscle fibers? Recall that neurons in the superior colliculus have "movement fields": they discharge in conjunction with saccadic eye movements of a particular direction and amplitude. Movement fields are conceptually similar to the receptive fields that occur in various sensory areas of the brain. Across the entire population of collicular neurons, all possible saccade vectors are represented (Figure A). Because the movement fields are topographically organized, the superior colliculus forms a *motor map* of saccade vectors (or motor goals; see text).

The direction and amplitude of eye movements is encoded quite differently by the extraocular muscles (Figure B). *Direction* is controlled by the ratio of activation of the different muscles, and *amplitude* is controlled by the magnitude of the activation of those muscles. In other words, to make a saccade go farther, the muscle pulling the eye must pull harder and longer than it would for a shorter saccade. Amplitude is therefore a *monotonic* function of muscle activation.

The pattern of activity must be transformed from a code in which collicular neurons are tuned for particular saccade amplitudes to a code in which most or all α motor neurons respond, regardless of saccade amplitude, but the level and/or duration of their activity varies monotonically with saccade amplitude. This transformation occurs before signals from the superior colliculus reach the α motor neurons that activate the extraocular muscles.

Various models have been proposed to explain this transformation. The basic idea, shared by all models, is that the saccade vector, as signaled by the locus of activity in the superior colliculus, is decomposed into two monotonic amplitude signals corresponding roughly to the horizontal and vertical components of the saccade vector. The weights of the projections from the superior colliculus to the horizontal and vertical gaze control centers are thought to be tailored to accomplish this. For example, a site in the superior colliculus where the movement fields encode 5° rightward movements would project to the rightward horizontal gaze control center with a modest strength. A site encoding 10° rightward saccades would send a stronger projection to that center. A site encoding an oblique saccade with a 10° horizontal and a 5° vertical component would project to both the horizontal and vertical centers, with weights proportional to the required contribution along each direction (Figure C).

This model is too simple to account for all the relevant experimental findings. However, it serves to give a general idea of how the brain might convert information encoded in one kind of format into another. This kind of transformation is a likely requirement of sensory motor integration in many behavioral contexts where sensory cues guide movement.

References

Fuchs, A. F. and E. S. Luschei (1970) Firing patterns of abducens neurons of alert monkeys in relationship to horizontal eye movement. *J. Neurophysiol.* 33: 382–392.

Groh, J. M. (2001) Converting neural signals from place codes to rate codes. *Biol. Cybern.* 85: 159–165.

Sparks, D. L. (1975) Response properties of eye movement-related neurons in the monkey superior colliculus. *Brain Res.* 90: 147–152.

(A) Direction tuning of three neurons recorded from the deep layers of the superior colliculus in macaque monkeys. Each neuron is broadly tuned, but responds best to a particular direction (and amplitude) of saccadic eye movement. (B) Relation of firing frequency to steady eye position in two neurons in the abducens nucleus of a macaque.

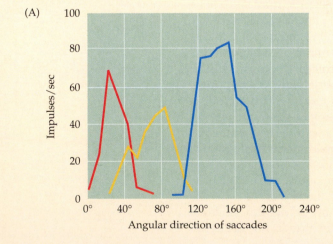

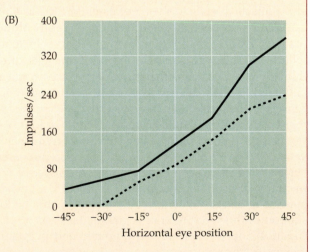

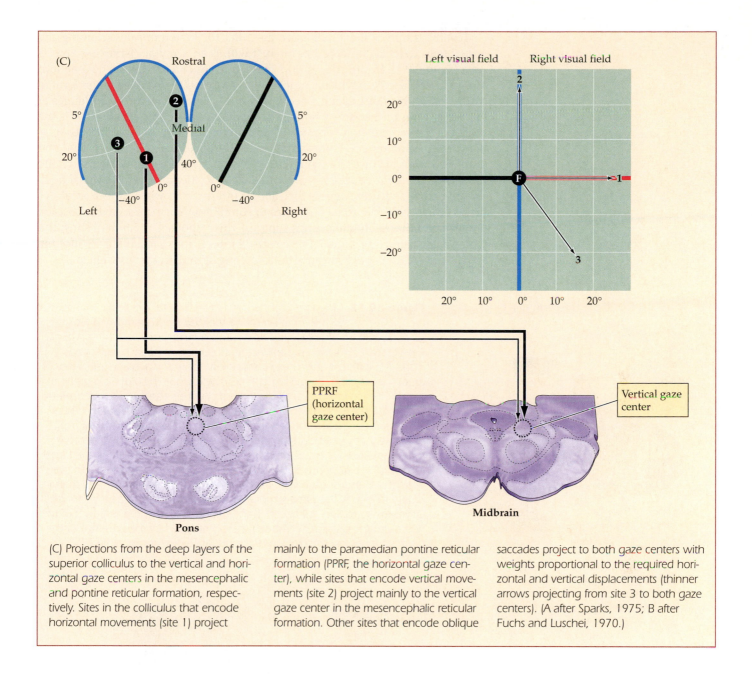

(C) Projections from the deep layers of the superior colliculus to the vertical and horizontal gaze centers in the mesencephalic and pontine reticular formation, respectively. Sites in the colliculus that encode horizontal movements (site 1) project mainly to the paramedian pontine reticular formation (PPRF, the horizontal gaze center), while sites that encode vertical movements (site 2) project mainly to the vertical gaze center in the mesencephalic reticular formation. Other sites that encode oblique saccades project to both gaze centers with weights proportional to the required horizontal and vertical displacements (thinner arrows projecting from site 3 to both gaze centers). (A after Sparks, 1975; B after Fuchs and Luschei, 1970.)

These observations do not, however, imply that the frontal eye fields and the superior colliculus have the same functions. Superior colliculus lesions produce a permanent deficit in the ability to perform very short latency, reflex-like eye movements called "express saccades." Express saccades are evidently mediated by direct pathways to the superior colliculus from the retina or visual cortex that can access the upper motor neurons in the colliculus without extensive, and more time-consuming, processing in the frontal cortex (see Box 20B). In contrast, frontal eye field lesions produce permanent deficits in the ability to make saccades that are not guided by an external target. For example, patients (or monkeys) with a lesion in the frontal eye fields cannot voluntarily direct their eyes *away* from a stimulus in the visual field, a type of eye movement called an "anti-saccade." Such lesions also eliminate the ability to make a saccade to the remembered location of a target that is no longer visible.

Finally, the frontal eye fields are essential for systematically scanning the visual field to locate an object of interest within an array of distracting objects (see Figure 20.1). Figure 20.12 shows the responses of a frontal eye field neuron during a visual task in which a monkey was required to foveate a target located within an array of distracting objects. This frontal eye field neuron discharges at different levels to the same stimulus, depending on whether the stimulus is the target of the saccade or a "distractor," and on the location of the distractor relative to the actual target. For example, the differences between the middle and the left and right traces in Figure 20.12 demonstrate that the response to the distractor is much reduced if it is located close to the target in the visual field. Results such as these suggest that lateral interactions within the frontal eye fields enhance the neuronal responses to stimuli that will be selected as saccade targets, and that such interactions suppress the responses to uninteresting and potentially distracting stimuli. These sorts of interactions presumably reduce the occurrence of unwanted saccades to distracting stimuli in the visual field.

Figure 20.12 *Responses of neurons in the frontal eye fields. (A) Locus of the left frontal eye field on a lateral view of the rhesus monkey brain. (B) Activation of a frontal eye field neuron during visual search for a target. The vertical tickmarks represent action potentials, and each row of tick marks is a different trial. The graphs below show the average frequency of action potentials as a function of time. The change in color from green to purple in each row indicates the time of onset of a saccade toward the target. In the left trace (1), the target (red square) is in the part of the visual field "seen" by the neuron, and the response to the target is similar to the response that would be generated by the neuron even if no distractors (green squares) were present (not shown). In the right trace (3), the target is far from the response field of the neuron. The neuron responds to the distractor in its response field. However, it responds at a lower rate than it would to exactly the same stimulus if the square were not a distractor but a target for a saccade (left trace). In the middle trace (2), the response of the neuron to the distractor has been sharply reduced by the presence of the target in a neighboring region of the visual field. (After Schall, 1995.)*

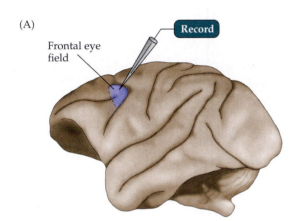

(A) Frontal eye field — Record

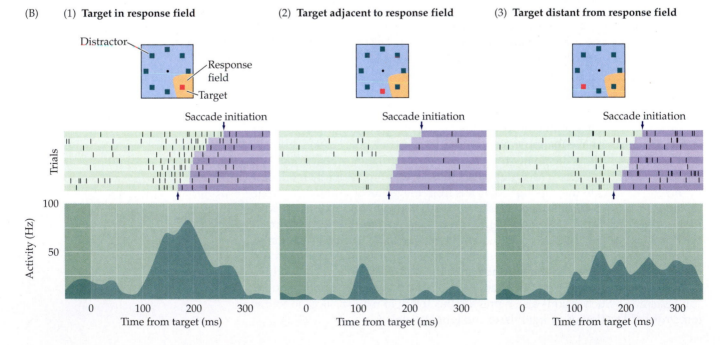

(B)

(1) **Target in response field**   (2) **Target adjacent to response field**   (3) **Target distant from response field**

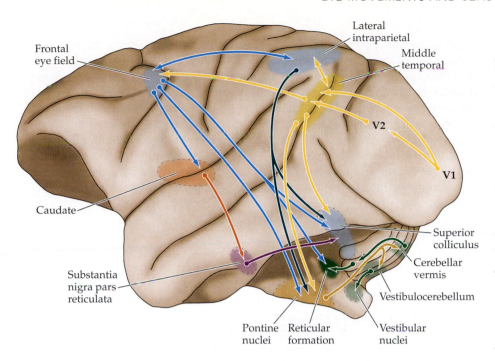

Frontal
eye field

Lateral
intraparietal

Middle
temporal

V2

V1

Caudate

Superior
colliculus

Cerebellar
vermis

Substantia
nigra pars
reticulata

Vestibulocerebellum

Pontine
nuclei

Reticular
formation

Vestibular
nuclei

Figure 20.13   *Summary of the sensory and motor structures and their connections that govern saccadic and pursuit eye movements, as studied in the rhesus monkey brain. Although these two types of eye movements were previously thought to be controlled by separate circuits in the forebrain and brainstem, it is now recognized that they are governed by similar networks of cortical and subcortical structures. Visual signals are processed by the dorsal spatial vision pathway, including the middle temporal and lateral intraparietal areas. Sensory and attentional signals then guide motor planning areas in the frontal eye field. These cortical areas interact with subcortical structures, including basal ganglia (caudate and substantia nigra, pars reticulata) and pontine-cerebellar structures (pontine nuclei, cerebellar vermis, and vestibulocerebellum) that modulate the initiation and coordination of eye movements by the superior colliculus and downstream oculomotor centers in the reticular formation and vestibular nuclei. The eye movements regulated by this complex circuitry are guided by a variety of sensory and cognitive signals, including perception, attention, memory, and reward expectation. (After Krauzlis, 2005.)*

Neural Control of Smooth Pursuit Movements

Traditionally, smooth pursuit and saccades were considered to be mediated by different structures, but recent studies such as those carried out by Richard Krauzlis at the Salk Institute for Biological Studies indicate that these two types of eye movements involve many of the same structures. Not only are smooth pursuit movements mediated by neurons in the PPRF, they also are under the influence of motor control centers in the rostral superior colliculus and subareas within the frontal eye fields, both of which receive sensory input from the "dorsal spatial vision pathway" in the parietal and temporal lobes. The exact routes by which visual information reaches the PPRF to generate smooth pursuit movements is not known, but pathways from the cortex to the superior colliculus and PPRF similar to those that mediate saccades may play a role; an indirect pathway through the cerebellum also has been suggested (Figure 20.13). It is clear, however, that neurons in the striate and extrastriate visual areas provide sensory information that is essential for the initiation and accurate guidance of smooth pursuit movements. In monkeys, neurons in the middle temporal area (which is largely concerned with the perception of moving stimuli; see Chapter 12) respond selectively to targets moving in a specific direction. Moreover, damage to this area disrupts smooth pursuit movements. In humans, damage of comparable areas in the parietal and occipital lobes also results in abnormalities of smooth pursuit movements.

Neural Control of Vergence Movements

When a person wishes to look from one object to another object that is located at a different distance from the eyes, a saccade is made that shifts the direction of gaze toward the new object and the eyes either diverge or converge until the object falls on the fovea of each eye. The structures and pathways responsible for mediating such vergence movements are not well understood, but appear to include several extrastriate areas in the occipital lobe. Information about the location of retinal activity is relayed through the two lateral geniculate nuclei to the cortex, where the information from the two eyes is integrated. The appropriate command to diverge or converge the eyes, which is based largely on information from the two eyes about the amount of binocular disparity (see Chapter 12), is then sent from the occipital cortex to "vergence centers" in the brainstem. One such center is a population of local circuit neurons located in the midbrain near the oculomotor nucleus. These neurons generate a burst of action potentials. The onset of the burst is the command to generate a vergence movement, and the frequency of the burst determines its velocity. There is a division of labor within the vergence center, so that some neurons command convergence movements while others command divergence movements. These neurons also coordinate vergence movements of the eyes with accommodation of the lens and pupillary constriction to maximize the clarity of images formed on the retina, as discussed in Chapter 11.

Summary

Despite their specialized function, the systems that control eye movements have much in common with the motor systems that govern movements of other parts of the body. Just as the spinal cord provides the basic circuitry for coordinating the actions of muscles around a joint, the reticular formation of the pons and midbrain provides the basic circuitry that mediates movements of the eyes. Descending projections from higher order centers in the superior colliculus and the frontal eye field innervate the brainstem gaze centers, providing a basis for integrating eye movements with a variety of sensory information that indicates the location of objects in space. The superior colliculus and the frontal eye field are organized in a parallel as well as a hierarchical fashion, enabling one of these structures to compensate for the loss of the other. Eye movements, like other movements, are also under the control of the basal ganglia and cerebellum; this control ensures the proper initiation and successful execution of these relatively simple motor behaviors, thus allowing observers to interact efficiently with the universe of things that can be seen.

Additional Reading

Reviews

FUCHS, A. F., C. R. S. KANEKO AND C. A. SCUDDER (1985) Brainstem control of eye movements. *Annu. Rev. Neurosci.* 8: 307–337.

HIKOSAKA, O AND R. H. WURTZ (1989) The basal ganglia. In *The Neurobiology of Saccadic Eye Movements: Reviews of Oculomotor Research*, Volume 3. R. H. Wurtz and M. E. Goldberg (eds.). Amsterdam: Elsevier, pp. 257–281.

KRAUZLIS, R. J. (2005) The control of voluntary eye movements: New perspectives. *Neuroscientist* 11: 124–137.

MAY, P. J. (2006) The mammalian superior colliculus: Laminar structure and connections. *Prog. Brain Res.* 151: 321–378.

ROBINSON, D. A. (1981) Control of eye movements. In *Handbook of Physiology*, Section 1: *The Nervous System*, Volume II: *Motor Control*, Part 2. V. B. Brooks (ed.). Bethesda, MD: American Physiological Society, pp. 1275–1320.

SCHALL, J. D. (1995) Neural basis of target selection. *Reviews in the Neurosciences* 6: 63–85.

SPARKS, D. L. AND L. E. MAYS (1990) Signal transformations required for the generation of saccadic eye movements. *Annu. Rev. Neurosci.* 13: 309–336.

ZEE, D. S. AND L. M. OPTICAN (1985) Studies of adaption in human oculomotor disorders. In *Adaptive Mechanisms in Gaze Control: Facts and Theories*. A Berthoz and G. Melvill Jones (eds.). Amsterdam: Elsevier, pp. 165–176.

Important Original Papers

BAARSMA, E. AND H. COLLEWIJN (1974) Vestibulo-ocular and optokinetic reactions to rotation and their interaction in the rabbit. *J. Physiol.* 238: 603-625.

FUCHS, A. F. AND E. S. LUSCHEI (1970) Firing patterns of abducens neurons of alert monkeys in relationship to horizontal eye movements. *J. Neurophysiol.* 33: 382–392.

OPTICAN, L. M. AND D. A. ROBINSON (1980) Cerebellar-dependent adaptive control of primate saccadic system. *J. Neurophysiol.* 44: 1058–1076.

SCHILLER, P. H. AND M. STRYKER (1972) Single unit recording and stimulation in superior colliculus of the alert rhesus monkey. *J. Neurophysiol.* 35: 915–924.

SCHILLER, P. H., S. D. TRUE AND J. L. CONWAY (1980) Deficits in eye movements following frontal eye-field and superior colliculus ablations. *J. Neurophysiol.* 44: 1175–1189.

SPARKS, D. L. AND L. E. MAYS (1983) Spatial localization of saccade targets. I. Compensation for stimulation-induced perturbations in eye position. *J. Neurophysiol.* 49: 45–63.

Books

HALL, W. C. AND A. MOSCHOVAKIS (EDS.) (2004) *The Superior Colliculus: New Approaches for Studying Sensorimotor Integration.* Methods and New Frontiers in Neuroscience Series. New York: CRC Press.

LEIGH, R. J. AND D. S. ZEE (1983) *The Neurology of Eye Movements.* Contemporary Neurology Series. Philadelphia: Davis.

SCHOR, C. M. AND K. J. CIUFFREDA (EDS.) (1983) *Vergence Eye Movements: Basic and Clinical Aspects.* Boston: Butterworth.

YARBUS, A. L. (1967) *Eye Movements and Vision.* Basil Haigh (trans.). New York: Plenum Press.

Chapter 21

The Visceral Motor System

Overview

The visceral, or autonomic, motor system controls involuntary functions mediated by the activity of smooth muscle fibers, cardiac muscle fibers, and glands. The system comprises two major divisions, the sympathetic and parasympathetic subsystems; the specialized innervation of the gut provides a further, semi-independent division that is usually referred to as the enteric nervous system. Although these divisions are always active at some level, the sympathetic division mobilizes the body's resources for dealing with challenges of one sort or another. Conversely, parasympathetic activity predominates during states of relative quiescence, so that energy sources previously expended can be restored. This continuous neural regulation of the expenditure and replenishment of the body's resources is crucial for the overall physiological balance of bodily functions called homeostasis. Whereas the major controlling centers for somatic motor activity are the primary and premotor motor cortices in the frontal lobes and a variety of related subcortical nuclei, the major locus of central control in the visceral motor system is the hypothalamus and the ill-defined circuitry that it controls in the reticular formation of the brainstem and in the spinal cord. The status of both principal divisions of the visceral motor system is modulated by descending pathways from these centers to preganglionic neurons in the brainstem and spinal cord, which in turn determine the activity of the primary, or "lower," visceral motor neurons in autonomic ganglia located outside the central nervous system. The autonomic regulation of several organ systems of particular importance in clinical practice (including cardiovascular function, control of the bladder, and the governance of the reproductive organs) is considered in more detail as specific examples of visceral motor control.

Early Studies of the Visceral Motor System

Although humans must always have been aware of involuntary motor reactions to stimuli in the environment (e.g., narrowing of the pupil in response to bright light, constriction of superficial blood vessels in response to cold or fear, increased heart rate in response to exertion), it was not until the late nineteenth century that the neural control of these and other visceral functions came to be understood in modern terms. The researchers who first rationalized the workings of the **visceral motor system** were Walter Gaskell and John Langley, British physiologists at Cambridge University. Gaskell's work preceded that of Langley and established the overall anatomy of the system as he carried out early physiological experiments that demonstrated some of its salient functional characteristics (e.g., that the heartbeat of an experimental animal is accelerated by stimulating the outflow of the upper thoracic spinal cord segments). Based on these and other observations, Gaskell concluded in 1866 that "every tissue is

TABLE 21.1 Summary of the Major Functions of the Visceral Motor System

Sympathetic Division			
Target organ	**Location of preganglionic neurons**	**Location of ganglionic neurons**	**Actions**
Eye	Upper thoracic spinal cord (C8–T7)	Superior cervical ganglion	Pupillary dilation
Lacrimal gland			Tearing
Submandibular and sublingual glands			Vasoconstriction
Parotid gland			Vasoconstriction
Head, neck (blood vessels, sweat glands, piloerector muscles)			Sweat secretion, vasoconstriction, piloerection
Upper extremity	T3–T6	Stellate and upper thoracic ganglia	Sweat secretion, vasoconstriction, piloerection
Heart	Middle thoracic spinal cord (T1–T5)	Superior cervical and upper thoracic ganglia	Increased heart rate and stroke volume, dilation of coronary arteries
Bronchi, lungs		Upper thoracic ganglia	Vasodilation, bronchial dilation
Stomach	Lower thoracic spinal cord (T6–T10)	Celiac ganglion	Inhibition of peristaltic movement and gastric secretion, vasoconstriction
Pancreas		Celiac ganglion	Vasoconstriction, inhibition of insulin secretion
Ascending small intestine, transverse large intestine		Celiac, superior, and inferior mesenteric ganglia	Inhibition of peristaltic movement and secretion
Descending large intestine, sigmoid, rectum		Inferior mesenteric hypogastric and pelvic plexus	Inhibition of peristaltic movement and secretion
Adrenal gland	T9–L2	Cells of gland are modified neurons	Catecholamine secretion
Ureter, bladder	T11–L2	Hypogastric and pelvic plexus	Relaxation of bladder wall muscle and contraction of internal sphincter
Lower extremity	T10–L2	Lower lumbar and upper sacral ganglia	Sweat secretion, vasoconstriction, piloerection

TABLE 21.1

Parasympathetic Division			
Target organ	**Location of preganglionic neurons**	**Location of ganglionic neurons**	**Actions**
Eye	Edinger-Westphal nucleus	Ciliary ganglion	Pupillary constriction, accommodation
Lacrimal gland	Superior salivatory nucleus	Pterygopalatine ganglion	Secretion of tears
Submandibular and sublingual glands	Superior salivatory nucleus	Submandibular ganglion	Secretion of saliva, vasodilation
Parotid gland	Inferior salivatory nucleus	Otic ganglion	Secretion of saliva, vasodilation
Head, neck (blood vessels, sweat glands, piloerector muscles)	None	None	None
Upper extremity	None	None	None
Heart	Nucleus ambiguus and dorsal motor nucleus of the vagus nerve	Cardiac plexus	Reduced heart rate
Bronchi, lungs	Dorsal motor nucleus of the vagus nerve	Pulmonary plexus	Bronchial constriction and secretion
Stomach	Dorsal motor nucleus of the vagus nerve	Myenteric and submucosal plexus	Peristaltic movement and secretion
Pancreas	Dorsal motor nucleus of the vagus nerve	Pancreatic plexus	Secretion of insulin and digestive enzymes
Ascending small intestine, transverse large intestine	Dorsal motor nucleus of the vagus nerve	Ganglia in the myenteric and submucosal plexus	Peristaltic movement and secretion
Descending large intestine, sigmoid, rectum	S3–S4	Ganglia in the myenteric and submucosal plexus	Peristaltic movement and secretion
Adrenal gland	None	None	None
Ureter, bladder	S2–S4	Pelvic plexus	Contraction of bladder wall and inhibition of internal sphincter
Lower extremity	None	None	None

innervated by two sets of nerve fibers of opposite characters," and he further surmised that these actions showed "the characteristic signs of opposite chemical processes."

Using similar electrical stimulation techniques in experimental animals, Langley went on to establish the function of **autonomic ganglia** (which harbor the lower visceral motor neurons), defined the terms "preganglionic" and "postganglionic" (see below), and coined the phrase **autonomic nervous system**, which is commonly used as a synonym for visceral motor system (although certain somatic motor activities may also be considered "autonomic"; see Chapter 29). Langley's work on the pharmacology of the autonomic system initiated the classic studies indicating the roles of acetylcholine and the catecholamines in visceral motor function, and in neurotransmitter function more generally (see Chapter 6). In short, Langley's ingenious physiological and anatomical experiments established in detail the general proposition put forward by Gaskell on more circumstantial grounds.

The third major figure in the pioneering studies of the visceral motor system was Walter Cannon at Harvard Medical School, who during the early to mid-1900s devoted his career to understanding visceral motor functions in relation to homeostatic mechanisms and to the emotions and higher brain functions (see Chapter 29). Like Gaskell and Langley before him, his work was based primarily on electrical stimulation in experimental animals, but included activation of brainstem and other brain regions as well as the peripheral components of the system. He also established the effects of denervation in the visceral motor system, laying some of the basis for further work on what is now referred to as "neuronal plasticity" (see Chapter 8).

Distinctive Features of the Visceral Motor System

Chapters 16 and 17 discussed in detail the organization of lower motor neurons in the central nervous system, their relationships to striated muscle fibers, and the means by which their activities are governed by higher motor centers. With respect to the efferent systems that govern the actions of smooth muscle fibers, cardiac muscle fibers, and glands, it is instructive to recognize the anatomical and functional features of the visceral motor system that distinguish it from the somatic motor system.

First, although it is useful to recognize medial (postural control) and lateral (distal extremity control) components of the somatic motor system, the anatomical and functional distinctions that justify this division of the somatic motor system are not nearly so great as they are for the subsystems that comprise the visceral motor system (Figure 21.1).

Second, the lower motor neurons of the visceral motor system are located outside of the central nervous system. The cell bodies of these primary visceral motor neurons are found in autonomic ganglia that are either close to the spinal cord (sympathetic division) or embedded in a neural plexus (*plexus* means "network") very near or in the target organ (parasympathetic and enteric divisions).

Third, the contacts between visceral motor neurons and the viscera are much less differentiated than the neuromuscular junctions of the somatic motor system. Visceral motor axons tend to be highly branched and give rise to many synaptic terminals at varicosities (swellings) along the length of the terminal axonal branch. Moreover, the surfaces of the visceral muscle usually lack the highly ordered structure of the motor endplates that characterizes postsynaptic target sites on striated muscle fibers. As a consequence, the neurotransmitters released by visceral motor terminals often diffuse for hundreds of microns

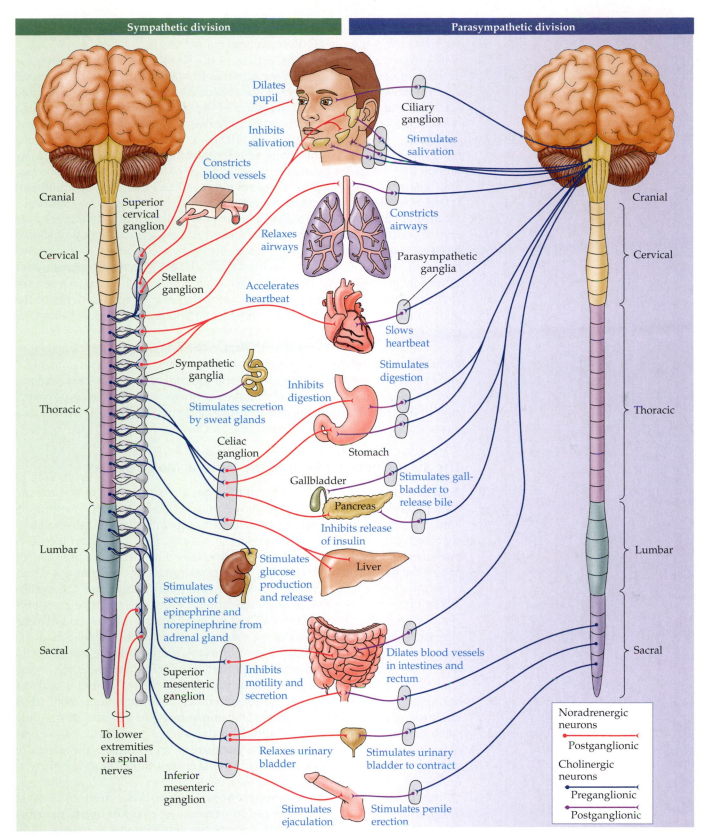

Figure 21.1 Overview of the sympathetic (left) and parasympathetic (right) divisions of the visceral motor system.

before binding to postsynaptic receptors—a far greater distance than at the synaptic cleft of the somatic neuromuscular junction.

Fourth, whereas the principal actions of the somatic motor system are governed by motor cortical areas in the posterior frontal lobe (discussed in Chapter 17), the activities of the visceral motor system are coordinated by a distributed set of cortical and subcortical structures in the ventral and medial parts of the forebrain; collectively, these structures comprise a central autonomic network.

Finally, visceral motor terminals release a variety of neurotransmitters, including primary small-molecule neurotransmitters (which differ depending on whether the motor neuron in question is sympathetic or parasympathetic), and one or more of a variety of co-neurotransmitters that may be a different small-molecule neurotransmitter or a neuropeptide (see Chapter 6). These neurotransmitters interact with a diverse set of postsynaptic receptors that mediate myriad postsynaptic effects in smooth and cardiac muscle and glands. It should be clear, then, that while the major effect of somatic motor activation on striated muscle is nearly the same throughout the body, the effects of visceral motor activation are remarkably varied. This fact should come as no surprise, given the challenge of maintaining homeostasis across the many organ systems of the body in the face of variable environmental conditions and dynamic behavioral contingencies.

In the remaining sections of this chapter, the sympathetic and parasympathetic divisions and the enteric nervous system are separately considered. General principles of visceral motor control and the central and reflexive coordination of visceral motor and somatic motor activity are illustrated in more detail later in the chapter, in a discussion of specific autonomic reflexes related to cardiovascular control, urination, and sexual functioning.

The Sympathetic Division of the Visceral Motor System

Activity of the neurons that make up the sympathetic division of the visceral motor system ultimately prepares individuals for "flight or fight," as Cannon famously put it. Cannon meant that, in extreme circumstances, heightened levels of sympathetic neural activity allow the body to make maximum use of its resources (particularly its metabolic resources), thereby increasing the chances of survival or success in threatening or otherwise challenging situations. Thus, during high levels of sympathetic activity, the pupils dilate and the eyelids retract (allowing more light to reach the retina and the eyes to move more efficiently); the blood vessels of the skin and gut constrict (rerouting blood to muscles, thus allowing them to extract a maximum of available energy); the hairs stand on end (making our hairier ancestors look more fearsome); the bronchi dilate (increasing oxygenation); the heart rate accelerates and the force of cardiac contraction is enhanced (maximally perfusing skeletal muscles and the brain); and digestive and other vegetative functions become quiescent (thus diminishing activities that are temporarily unnecessary) (see Figure 21.1). At the same time, sympathetic activity stimulates the adrenal medulla to release epinephrine and norepinephrine into the bloodstream and mediates the release of glucagon from the pancreas, further enhancing energy-mobilizing (or catabolic) functions.

The neurons in the central nervous system that drive these effects are located in the spinal cord. They are arranged in a column of **preganglionic neurons** that extends from the uppermost thoracic to the upper lumbar segments (T1 to L2 or L3; see Table 21.1) in a region of the spinal cord gray matter called the **intermediolateral column** or **lateral horn** (Figure 21.2). The preganglionic neurons that control sympathetic outflow to the organs in the head and thorax

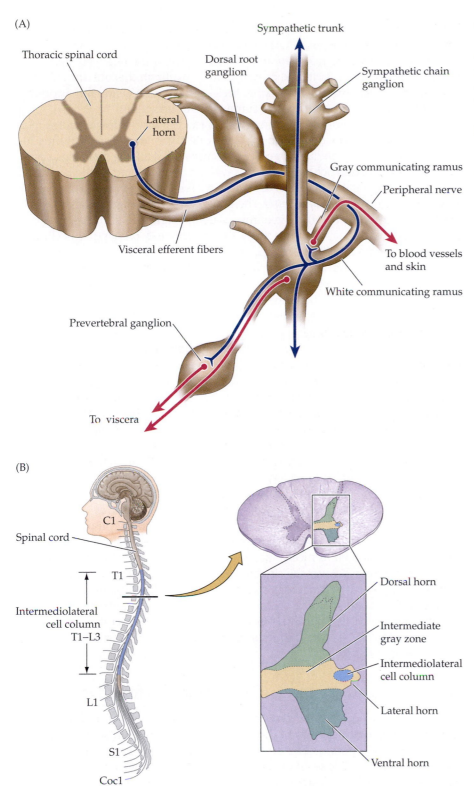

(A)

Thoracic spinal cord

Sympathetic trunk

Dorsal root ganglion

Sympathetic chain ganglion

Lateral horn

Gray communicating ramus

Peripheral nerve

Visceral efferent fibers

To blood vessels and skin

White communicating ramus

Prevertebral ganglion

To viscera

(B)

C1

Spinal cord

T1

Intermediolateral cell column
T1–L3

L1

S1

Coc1

Dorsal horn

Intermediate gray zone

Intermediolateral cell column

Lateral horn

Ventral horn

Figure 21.2 Organization of the preganglionic spinal outflow to sympathetic ganglia. (A) General organization of the sympathetic division of the visceral motor system in the spinal cord and the preganglionic outflow to the sympathetic ganglia that contain the primary visceral motor neurons. (B) Cross section of thoracic spinal cord at the level indicated, showing location of the sympathetic preganglionic neurons in the intermediolateral cell column of the lateral horn.

are in the upper and middle thoracic segments, whereas those that control the abdominal and pelvic organs and targets in the lower extremities are in the lower thoracic and upper lumbar segments. The axons that arise from these spinal preganglionic neurons typically extend only a short distance, terminating in a series of paravertebral or sympathetic chain ganglia, which, as the name implies, are arranged in a chain that extends along most of the length of the vertebral column (see Figure 21.1). These preganglionic pathways to the ganglia are known as the *white communicating rami* because of the relatively light color imparted to the rami (singular, *ramus*) by the myelinated axons they contain (see Figure 21.2A). Roughly speaking, these preganglionic spinal neurons are comparable to somatic motor interneurons (see Chapter 16).

The neurons in **sympathetic ganglia** are the primary or lower motor neurons of the sympathetic division in that they directly innervate smooth muscles, cardiac muscle, and glands. The **postganglionic axons** arising from these **paravertebral sympathetic chain** neurons travel to various targets in the body wall, joining the segmental spinal nerves of the corresponding spinal segments by way of the *gray communicating rami*. The gray rami are another set of short linking nerves, so named because the unmyelinated postganglionic axons give them a somewhat darker appearance than the myelinated preganglionic linking nerves (see Figure 21.2A).

In addition to innervating the sympathetic chain ganglia, the preganglionic axons that govern the viscera extend a longer distance from the spinal cord in the splanchnic nerves (nerves that innervate thoracic and abdominal viscera) to reach sympathetic ganglia that lie in the chest, abdomen, and pelvis. These **prevertebral ganglia** include sympathetic ganglia in the cardiac plexus, the celiac ganglion, the superior and inferior mesenteric ganglia, and sympathetic ganglia in the pelvic plexus. The postganglionic axons arising from the prevertebral ganglia provide sympathetic innervation to the heart, lungs, gut, kidneys, pancreas, liver, bladder, and reproductive organs; many of these organs also receive some postganglionic innervation from neurons in the sympathetic chain ganglia. Finally, a subset of thoracic preganglionic fibers in the splanchnic (visceral) nerves innervate the adrenal medulla, which is generally regarded as a sympathetic ganglion modified for a specific endocrine function—namely, the release of catecholamines into the circulation to enhance a widespread sympathetic response to stress. In summary, sympathetic axons contribute to virtually all peripheral nerves, carrying innervation to an enormous range of target organs (see Table 21.1).

Cannon's memorable truism that the sympathetic activity prepares the animal for "fight or flight" notwithstanding, the sympathetic division of the visceral motor system is tonically active to maintain sympathetic function at appropriate levels whatever the circumstances. Nor should the sympathetic system be thought of as responding in an all-or-none fashion; many specific sympathetic reflexes operate more or less independently, as might be expected from the obvious need to specifically control various organ functions (e.g., the heart during exercise, the bladder during urination, and the reproductive organs during sexual intercourse).

The Parasympathetic Division of the Visceral Motor System

In contrast to the sympathetic division, the preganglionic outflow from the central nervous system to the ganglia of the parasympathetic division stems from neurons whose distribution is limited to the brainstem and the sacral part of the spinal cord (Figure 21.3; see also Figure 21.1). The cranial preganglionic innervation arising from the brainstem, which is analogous to the preganglionic sympathetic outflow from the spinal cord, includes the **Edinger-Westphal nucleus** in the midbrain (which innervates the ciliary ganglion via the oculomotor nerve and mediates the diameter of the pupil in response to light; see Chapter 12); the **superior** and **inferior salivatory nuclei** in the pons and medulla (which innervate the salivary glands and tear glands, mediating salivary secretion and the production of tears); a visceral motor division of the **nucleus ambiguus** in the

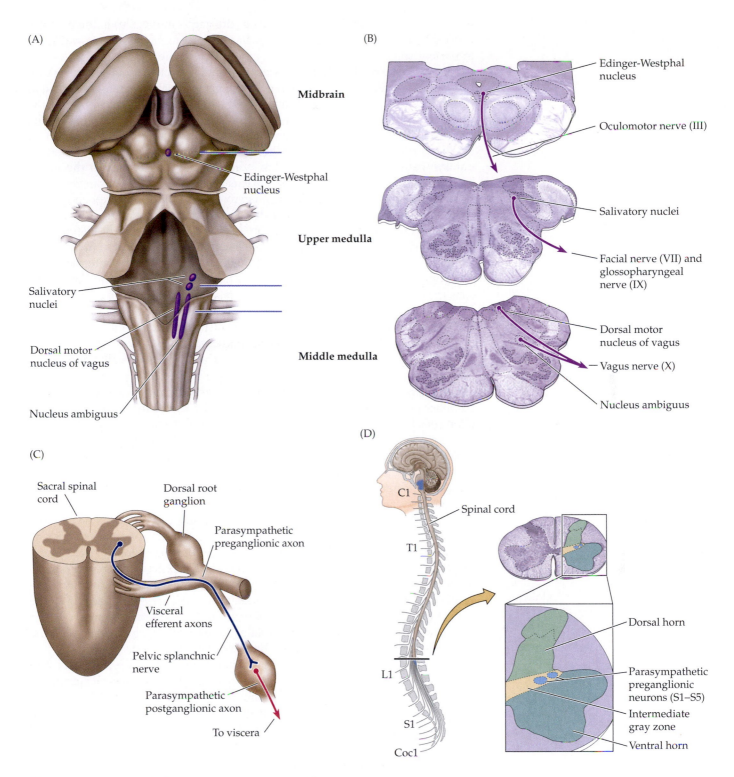

Figure 21.3 Organization of the preganglionic outflow to parasympathetic ganglia. (A) Dorsal view of brainstem showing the location of the nuclei of the cranial part of the parasympathetic division of the visceral motor system. (B) Cross section of the brainstem at the relevant levels [indicated by horizontal lines in (A)] showing location of these parasympathetic nuclei. (C) Main features of the parasympathetic preganglionics in the sacral segments of the spinal cord. (D) Cross section of the sacral spinal cord showing location of sacral preganglionic neurons.

medulla; and the **dorsal motor nucleus of the vagus nerve**, also in the medulla. Neurons in the ventral-lateral part of the nucleus ambiguus also provide an important source of cardio-inhibitory innervation to the cardiac ganglia via the vagus nerve. The more dorsal part of the dorsal motor nucleus of the vagus nerve primarily governs glandular secretion via the parasympathetic ganglia located in the viscera of the thorax and abdomen, whereas the more ventral part of the nucleus controls the motor responses of the heart, lungs, and gut elicited by the vagus nerve (e.g., slowing of the heart rate and constricting the bronchioles). In addition, other preganglionic neurons in the nucleus ambiguus innervate parasympathetic ganglia in the submandibular salivary glands and the medi-astinum (a different division of the nucleus ambiguus provides branchiomotor innervation of striated muscle in the pharynx and larynx; see the Appendix). The location of the parasympathetic brainstem nuclei is shown in Figure 21.3.

The sacral preganglionic innervation arises from neurons in the lateral gray matter of the sacral segments of the spinal cord, which are located in much the same position as the sympathetic preganglionic neurons in the intermediolateral column of the thoracic cord (Figure 21.3C,D). The axons from these neurons travel in the splanchnic nerves to innervate parasympathetic ganglia in the lower third of the colon, rectum, bladder, and reproductive organs.

The **parasympathetic ganglia** innervated by preganglionic outflow from both cranial and sacral levels are in or near the end organs they serve. In this way they differ from the ganglionic targets of the sympathetic system (recall that both the paravertebral chain and prevertebral ganglia are located relatively far from their target organs; see Figure 21.1). An important anatomical difference between sympathetic and parasympathetic ganglia at the cellular level is that sympathetic ganglion cells tend to have extensive dendritic arbors and are, as might be expected from this arrangement, innervated by a large number of preganglionic fibers. Parasympathetic ganglion cells have few if any dendrites and consequently are each innervated by only one or a few preganglionic axons. This arrangement implies a greater diversity of converging influences upon sympathetic ganglion neurons compared to parasympathetic ganglion neurons.

The overall function of the parasympathetic system, as Gaskell, Langley, and Cannon demonstrated, is generally opposite to that of the sympathetic system, serving to increase metabolic and other resources during periods when the animal's circumstances allow it to "rest and digest." In contrast to the sympathetic functions enumerated earlier, the activity of the parasympathetic system constricts the pupils, slows the heart rate, and increases the peristaltic activity of the gut. At the same time, diminished activity in the sympathetic system allows the blood vessels of the skin and gut to dilate, the piloerector muscles to relax, and the outflow of catecholamines from the adrenal medulla to decrease.

Although most organs receive innervation from *both* the sympathetic and parasympathetic divisions of the visceral motor system (as Gaskell surmised), some receive only sympathetic innervation. These exceptional targets include the sweat glands, the adrenal medulla, the piloerector muscles of the skin, and most arterial blood vessels (see Table 21.1).

The Enteric Nervous System

An enormous number of neurons are specifically associated with the gastrointestinal tract to control its many functions; indeed, more neurons are said to reside in the human gut than in the entire spinal cord. As already noted, the activity of the gut is modulated by both the sympathetic and the parasympathetic divisions of the visceral motor system. However, the gut also has an extensive system of nerve cells in its wall (as do its accessory organs such as the pancreas and gallbladder) that do not fit neatly into the sympathetic or parasympathetic divisions of the visceral motor system (Figure 21.4A). To a surprising degree, these neurons and the complex enteric networks in which they are found operate more or less independently according to their own reflex rules; as a result, many gut functions continue perfectly well without sympathetic or parasympathetic supervision (peristalsis, for example, occurs in isolated gut segments in vitro). Thus, most investigators prefer to classify the enteric nervous system as a separate component of the visceral motor system.

The neurons in the gut wall include local and centrally projecting sensory neurons that monitor mechanical and chemical conditions in the gut, local cir-

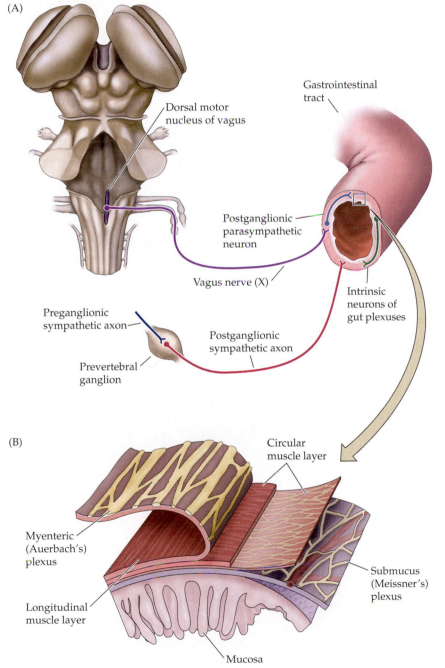

Figure 21.4 Organization of the enteric component of the visceral motor system. (A) Sympathetic and parasympathetic innervation of the enteric nervous system, and the intrinsic neurons of the gut. (B) Detailed organization of nerve cell plexuses in the gut wall. The neurons of the submucus plexus (Meissner's plexus) are concerned with the secretory aspects of gut function, and the myenteric plexus (Auerbach's plexus) with the motor aspects of gut function (e.g., peristalsis).

cuit neurons that integrate this information, and motor neurons that influence the activity of the smooth muscles in the wall of the gut and glandular secretions (e.g., of digestive enzymes, mucus, stomach acid, and bile). This complex arrangement of nerve cells intrinsic to the gut is organized into (1) the myenteric (or Auerbach's) plexus, which is specifically concerned with regulating the musculature of the gut; and (2) the submucus (or Meissner's) plexus, which is located, as the name implies, just beneath the mucus membranes of the gut and is concerned with chemical monitoring and glandular secretion (Figure 21.4B).

As already mentioned, the preganglionic parasympathetic neurons that influence the gut are primarily in the dorsal motor nucleus of vagus nerve in the brainstem and the intermediate gray zone in the sacral spinal cord segments. The preganglionic sympathetic innervation that modulates the action of the gut plexuses derives from the thoracolumbar cord, primarily by way of the celiac, superior, and inferior mesenteric ganglia.

Sensory Components of the Visceral Motor System

Although the focus of this unit is "movement and its central control," it is important to understand the sources of visceral sensory information and the means by which this input becomes integrated with visceral motor networks in the central nervous system. Generally speaking, afferent activity arising from the viscera serves two important functions: (1) it provides feedback input to local reflexes that modulate moment-to-moment visceral motor activity within individual organs; and (2) it serves to inform higher integrative centers of more complex patterns of stimulation that may signal potentially threatening conditions and/or require the coordination of more widespread visceral motor, somatic motor, neuroendocrine, and behavioral activities (Figure 21.5). The **nucleus of the solitary tract** in the medulla is the central structure in the brain

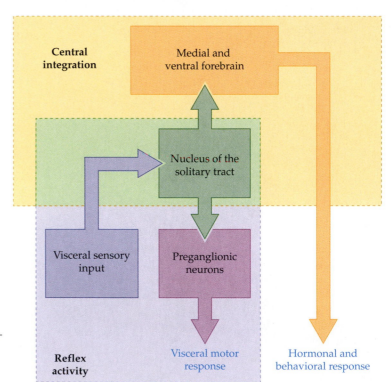

Figure 21.5 *Distribution of visceral sensory information by the nucleus of the solitary tract to serve either local reflex responses or more complex hormonal and behavioral responses via integration within a central autonomic network. As illustrated in Figure 21.7, forebrain centers also provide input to visceral motor effector systems in the brainstem and spinal cord.*

that receives visceral sensory information and distributes it accordingly to serve both purposes.

The afferent fibers that provide this visceral sensory input arise from cell bodies in the dorsal root ganglia (as is the case of somatic sensory modalities; see Chapters 9 and 10) and the sensory ganglia associated with the glossopharyngeal and vagus cranial nerves. However, there are far fewer visceral sensory neurons (by a factor of about 10) in comparison to mechanosensory neurons that innervate the skin and deeper somatic structures. This relative sparseness of peripheral visceral sensory innervation accounts in part for why most visceral sensations are diffuse and difficult to localize precisely.

The spinal visceral sensory neurons in the dorsal root ganglia send axons peripherally, through sympathetic nerves, ending in sensory receptor specializations such as nerve endings that are sensitive to pressure or stretch (in the walls of the heart, bladder, and gastrointestinal tract); endings that innervate specialized chemosensitive cells (oxygen-sensitive cells in the carotid bodies); or nociceptive endings that respond to damaging stretch, ischemia, or the presence of irritating chemicals. The central axonal processes of these dorsal root ganglion neurons terminate on second-order neurons and local interneurons in the dorsal horn and in intermediate gray regions of the spinal cord. Some primary visceral sensory axons terminate near the lateral horn, where the preganglionic neurons of sympathetic and parasympathetic divisions are located; these terminals mediate visceral reflex activity in a manner not unlike the segmental somatic motor reflexes described in Chapter 16.

In the dorsal horn, many of the second-order neurons that receive visceral sensory inputs are actually neurons of the anterolateral system, which also receive nociceptive and/or crude mechanosensory input from more superficial sources (see Chapter 10). As described in Box 10B, this is one means by which painful visceral sensations may be "referred" to more superficial somatic territories. Axons of these second-order visceral sensory neurons travel rostrally in the ventrolateral white matter of the spinal cord and the lateral sector of the brainstem, and eventually reach the ventral posterior complex of the thalamus. However, the axons of other second-order visceral sensory neurons terminate before reaching the thalamus; the principal target of these axons is the nucleus of the solitary tract (Figure 21.6). Other brainstem targets of second-order visceral sensory axons are visceral motor centers in the medullary **reticular formation** (see Box 17A).

In the last decade, it has become clear that visceral sensory information, especially that related to painful visceral sensations, also ascends the central nervous system by another spinal pathway. Second-order neurons whose cell bodies are located near the central canal of the spinal cord send their axons through the dorsal columns to terminate in the dorsal column nuclei, where third-order neurons relay visceral nociceptive signals to the ventral posterior thalamus. Although the existence of this visceral pain pathway in the dorsal columns complicates the simplistic view of the dorsal column–medial lemniscal pathway as a discriminative mechanosensory projection and the anterolateral system as a pain pathway, mounting empirical and clinical evidence highlights the importance of this newly discovered *dorsal column pain pathway* in the central transmission of visceral nociception (see Box 10C).

Figure 21.6 *Organization of sensory input to the visceral motor system. Afferent input from the cranial nerves relevant to visceral sensation (as well as afferent input ascending from second-order visceral afferents in the spinal cord) converges on the caudal division of the nucleus of the solitary tract (the rostral division is a gustatory relay; see Chapter 15).*

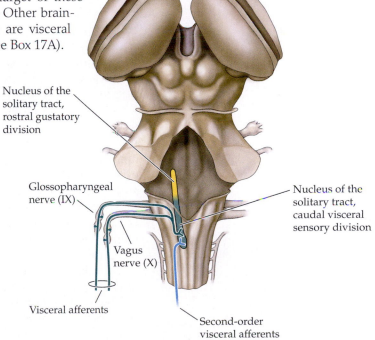

Nucleus of the solitary tract, rostral gustatory division

Glossopharyngeal nerve (IX)

Vagus nerve (X)

Visceral afferents

Nucleus of the solitary tract, caudal visceral sensory division

Second-order visceral afferents

In addition to these spinal visceral afferents, general visceral sensory inputs from thoracic and upper abdominal organs, as well as from viscera in the head and neck, enter the brainstem directly via the glossopharyngeal and vagus cranial nerves (see Figure 21.6). These glossopharyngeal and vagal visceral afferents terminate in the nucleus of the solitary tract. This nucleus, as described in the next section, integrates a wide range of visceral sensory information and transmits this information directly (and indirectly) to relevant visceral motor nuclei, to the brainstem reticular formation, and to several regions in the medial and ventral forebrain that coordinate visceral motor activity (see Figure 21.5).

Finally, unlike the somatic sensory system (where virtually all sensory signals gain access to conscious neural processing), sensory fibers related to the viscera convey only limited information to consciousness. For example, most of us are completely unaware of the subtle changes in peripheral vascular resistance that raise or lower our mean arterial blood pressure, yet such covert visceral afferent information is essential for the functioning of autonomic reflexes and the maintenance of homeostasis. Typically, only painful visceral sensations enter conscious awareness (see Chapter 29).

Central Control of Visceral Motor Functions

The nucleus of the solitary tract—and in particular the caudal part of this nucleus—is a key integrative center for reflexive control of visceral motor function and an important relay for visceral sensory information that reaches other brainstem nuclei and forebrain structures (Figure 21.7; see also Figure 21.5). The rostral part of this nucleus, as described in Chapter 15, is a gustatory relay, receiving input from primary taste afferents (cranial nerves VII, IX, and X) and sending projections to the gustatory nucleus in the ventroposterior thalamus. The caudal visceral sensory part of the nucleus of the solitary tract provides input to primary visceral motor nuclei, such as the dorsal motor nucleus of the vagus nerve and the nucleus ambiguus. It also projects to "premotor" autonomic centers in the medullary reticular formation, and to higher integrative centers in the amygdala (specifically, the central group of amygdaloid nuclei; see Box 29B) and hypothalamus (see below). In addition, the nucleus of the solitary tract projects to the **parabrachial nucleus** (so named because it envelopes the superior cerebellar peduncle, also known by the Latin name *brachium conjunctivum*). The parabrachial nucleus, in turn, relays visceral sensory information to the hypothalamus, amygdala, thalamus, and medial prefrontal and insular cortex (see Figure 21.7; for clarity, the cortical projections of the parabrachial nucleus are omitted).

Although one might argue that the posterior insular cortex is the primary visceral sensory area and the medial prefrontal cortex is the primary visceral motor area, it is more useful to emphasize the interactions among these cortical areas and related subcortical structures: taken together, they constitute a **central autonomic network**. This network accounts for the integration of visceral sensory information with input from other sensory modalities and from higher cognitive centers that process semantic and emotional experiences. Involuntary visceral reactions such as blushing in response to consciously embarrassing stimuli, vasoconstriction and pallor in response to fear, and autonomic responses to sexual situations are examples of the integrated activity of this network. Indeed, autonomic function is intimately related to emotional processing, as emphasized in Chapter 29.

The **hypothalamus** is a key component of this central autonomic network that deserves special consideration. The hypothalamus is a heterogeneous collection of nuclei in the base of the diencephalon that plays an important role in

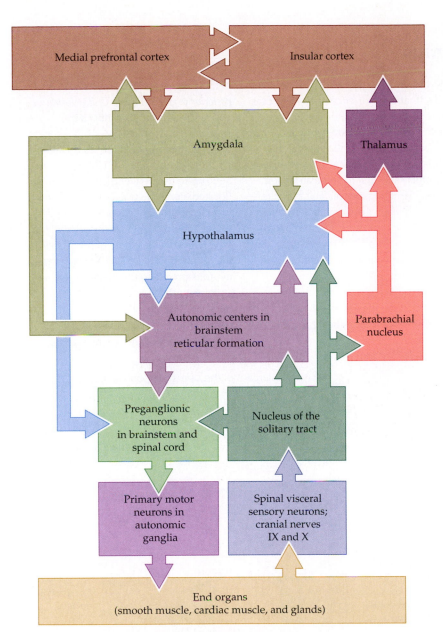

Figure 21.7 A central autonomic network for the control of visceral motor function. Overview of connections within the central autonomic network. The distribution of visceral sensory information within this network is illustrated on the right side of the figure and the generation of visceral motor commands is shown on the left. However, extensive interconnections among autonomic centers in the forebrain (between the amygdala and associated cortical regions or hypothalamus, for example) militate against a strict parsing of this network into afferent and efferent limbs. The hypothalamus is a key structure in this network that integrates visceral sensory input and higher order visceral motor signals (see Box 21A).

the coordination and expression of visceral motor activity (Box 21A). The major outflow from the relevant hypothalamic nuclei is directed toward "autonomic centers" in the reticular formation; these centers can be thought of as dedicated premotor circuits that coordinate the efferent activity of preganglionic visceral motor neurons. They organize specific visceral functions such as cardiac reflexes, bladder-control reflexes, sexual function reflexes, and critical reflexes underlying respiration and vomiting (see Box 17A).

In addition to these important projections to the reticular formation, hypothalamic control of visceral motor function is also exerted more directly by projections to the cranial nerve nuclei that contain parasympathetic preganglionic neurons, and to the sympathetic and parasympathetic preganglionic neurons in the spinal cord. Nevertheless, the autonomic centers of the reticular formation and the preganglionic visceral motor neurons that they control are competent to function autonomously should disease or injury impede the ability of the hypo-

BOX 21A   The Hypothalamus

The hypothalamus is located at the base of the forebrain, bounded by the optic chiasm rostrally and the midbrain tegmentum caudally. It forms the floor and ventral walls of the third ventricle and is continuous through the infundibular stalk with the posterior pituitary gland, as illustrated in Figure A. Given its central position in the brain and its proximity to the pituitary, it is not surprising that the hypothalamus integrates information from the forebrain, brainstem, spinal cord, and various intrinsic chemosensitive neurons.

What is surprising about this structure is the remarkable diversity of homeostatic functions that are governed by this relatively small region of the forebrain. The diverse functions in which hypothalamic involvement is at least partially understood include: *the control of blood flow* (by promoting adjustments in cardiac output, vasomotor tone, blood osmolality, and renal clearance, and by motivating drinking and salt consumption); the *regulation of energy metabolism* (by monitoring blood glucose levels and regulating feeding behavior, digestive functions, metabolic rate, and temperature); the *regulation of reproductive activity* (by influencing gender identity, sexual orientation and mating behavior and, in females, by governing menstrual cycles, pregnancy, and lactation); and the *coordination of responses to threatening conditions* (by governing the release of stress hormones, modulating the balance between sympathetic and parasympathetic tone, and influencing the regional distribution of blood flow).

Despite the impressive scope of hypothalamic control, the individual components of the hypothalamus utilize similar physiological mechanisms to exert their influence over these many functions (Figure B). Thus, hypothalamic circuits receive sensory and contextual information, compare that information with biological set points, and activate

relevant visceral motor, neuroendocrine, and somatic motor effector systems that restore homeostasis and/or elicit appropriate behavioral responses.

Like the overlying thalamus—and consistent with the scope of hypothalamic functions—the hypothalamus comprises a large number of distinct nuclei, each with its own specific pattern of connections and functions. The nuclei, most of which are intricately interconnected, can be grouped in three longitudinal regions referred to as *periventricular, medial,* and *lateral*. They can also be grouped along the anterior–posterior dimension, the groups being those nuclei

in the *anterior* (or preoptic), *tuberal,* and *posterior* regions (Figure C). The anterior-periventricular group contains the suprachiasmatic nucleus, which receives direct retinal input and drives circadian rhythms (see Chapter 28). More scattered neurons in the periventricular region (located along the wall of the third ventricle) manufacture peptides known as releasing or inhibiting factors that control the secretion of a variety of hormones by the anterior pituitary. The axons of these neurons project to the median eminence, a region at the junction of the hypothalamus and pituitary stalk, where the peptides are secreted into the portal circulation that supplies the anterior pituitary.

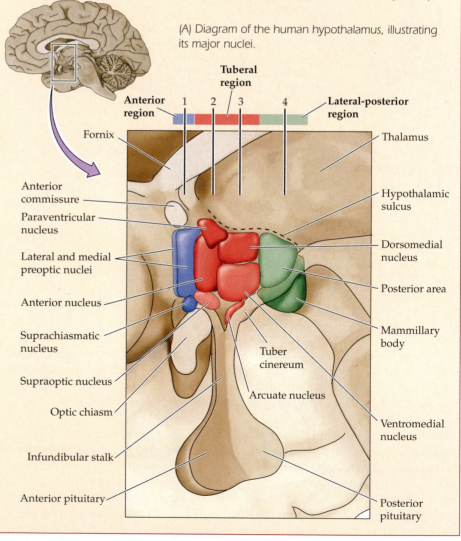

(A) Diagram of the human hypothalamus, illustrating its major nuclei.

(B) Physiological mechanisms underlying hypothalamic function.

Contextual information
(Cerebral cortex, amygdala, hippocampal formation)

Hypothalamus
(Compares input to biological set points)

Sensory inputs
(Visceral and somatic sensory pathways, chemosensory and humoral signals)

Visceral motor, somatic motor, neuroendocrine, behavioral responses

the neurosecretory neurons whose axons extend into the posterior pituitary. With appropriate stimulation, these neurons secrete oxytocin or vasopressin (antidiuretic hormone) directly into the bloodstream. Other neurons in the paraven-

tricular nucleus project to autonomic centers in the reticular formation, as well as preganglionic neurons of the sympathetic and parasympathetic divisions in the brainstem and spinal cord; these cells are thought to exert hypothalamic control over the visceral motor system. The paraventricular nucleus receives inputs from other hypothalamic zones, which are in turn related to the cerebral cortex, hippocampus, amygdala, and other central structures that are all capable of influencing visceral motor function.

(Continued on next page)

The medioal-tuberal region nuclei ("tuberal" refers to the *tuber cinereum*, the anatomical name given to the middle portion of the inferior surface of the hypothalamus) include the paraventricular and supraoptic nuclei, which contain

(C) Coronal sections through the human hypothalamus (see Figure A for location of sections 1–4). Color coding of the nuclei illustrates the two dimensions by which hypothalamic nuclei are subdivided (see text). Blue, red, and green illustrate nuclei in the anterior, tuberal, and posterior regions, respectively. The relative shading of these hues illustrates the three mediolateral zones: Lighter shading represents nuclei in the periventricular zone, whereas darker shades represent medial zone nuclei. Nuclei in the lateral zone are stippled. (1) Section through the anterior region illustrating the preoptic and suprachiasmatic nuclei. (2) Rostral tuberal region. (3) Caudal tuberal region. (4) Section through the posterior region illustrating the mammillary bodies.

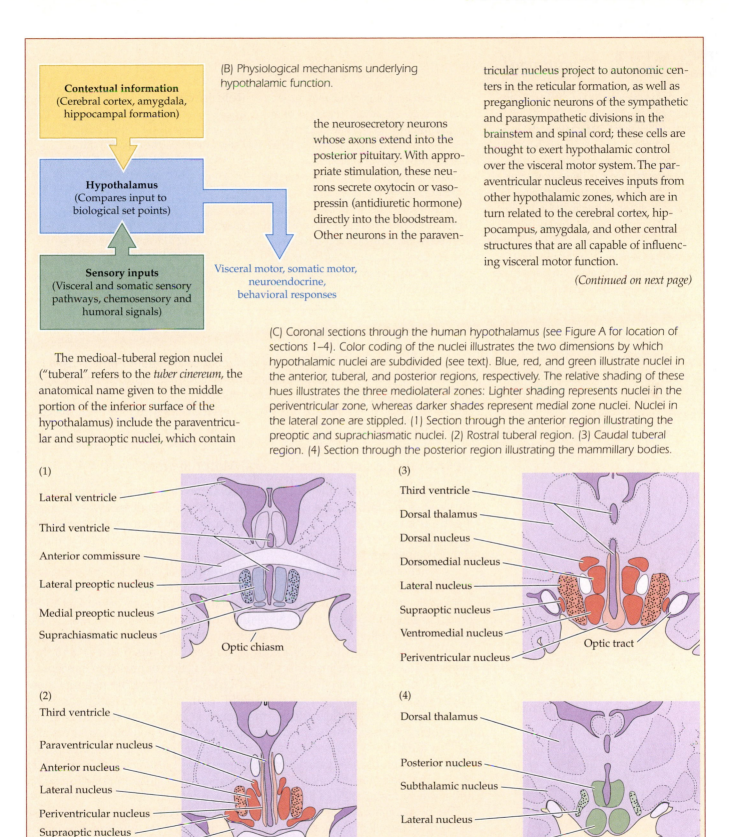

(1)
Lateral ventricle
Third ventricle
Anterior commissure
Lateral preoptic nucleus
Medial preoptic nucleus
Suprachiasmatic nucleus
Optic chiasm

(2)
Third ventricle
Paraventricular nucleus
Anterior nucleus
Lateral nucleus
Periventricular nucleus
Supraoptic nucleus
Optic tract
Optic chiasm

(3)
Third ventricle
Dorsal thalamus
Dorsal nucleus
Dorsomedial nucleus
Lateral nucleus
Supraoptic nucleus
Ventromedial nucleus
Periventricular nucleus
Optic tract

(4)
Dorsal thalamus
Posterior nucleus
Subthalamic nucleus
Lateral nucleus
Mammillary body

Also in this region of the hypothalamus are the dorsomedial and ventromedial nuclei, which are involved in feeding, reproductive and parenting behavior, thermoregulation, and water balance. These nuclei receive inputs from structures of the limbic system, as well as from visceral sensory nuclei in the brainstem (e.g., the nucleus of the solitary tract).

Finally, the lateral region of the hypothalamus is really a rostral continuation of the midbrain reticular formation (see Box 17A). Thus, the neurons of the lateral region are not grouped into nuclei, but are scattered among the fibers of the medial forebrain bundle, which runs through the lateral hypothalamus. These cells control behavioral arousal and shifts of attention, especially as related to reproductive activities.

In summary, the hypothalamus regulates an enormous range of physiological and behavioral activities and serves as the key controlling center for visceral motor activity and for homeostatic functions generally.

References

SAPER, C. B. (1990) Hypothalamus. In *The Human Nervous System*. G. Paxinos (ed.). San Diego: Academic Press, pp. 389–414.

SWANSON, L. W. (1987) The hypothalamus. In *Handbook of Chemical Neuroanatomy*, Vol. 5: *Integrated Systems of the CNS*, Part I: *Hypothalamus, Hippocampus, Amygdala, Retina*. A. Björklund and T. Hokfelt (eds.). Amsterdam: Elsevier, pp. 1–124.

SWANSON, L. W. AND P. E. SAWCHENKO (1983) Hypothalamic integration: Organization of the paraventricular and supraoptic nuclei. *Annu. Rev. Neurosci.* 6: 269–324.

thalamus to govern the many bodily systems that maintain homeostasis. The general organization of this central autonomic control is summarized in Figure 21.7; some important clinical manifestations of damage to this descending system are illustrated in Box 21B; Box 21C shows the relevance of this central control to obesity.

Neurotransmission in the Visceral Motor System

The neurotransmitters utilized by the visceral motor system are of enormous importance in clinical practice, and drugs that act on the autonomic system are among the most important in the clinical armamentarium. Moreover, autonomic transmitters have played a major role in the history of efforts to understand synaptic function.

Acetylcholine is the primary neurotransmitter of both sympathetic and parasympathetic preganglionic neurons. Nicotinic receptors on autonomic ganglion cells are ligand-gated ion channels that mediate a so-called fast EPSP (much like nicotinic receptors at the neuromuscular junction). In contrast, muscarinic acetylcholine receptors on ganglion cells are members of the 7-transmembrane G-protein-linked receptor family (see Chapters 6 and 7), and they mediate slower synaptic responses. The primary action of muscarinic receptors in autonomic ganglion cells is to close K^+ channels, making the neurons more excitable and generating a prolonged EPSP. Acting in concert with the muscarinic activities are neuropeptides that serve as co-neurotransmitters at the ganglionic synapses. As described in Chapter 6, peptide neurotransmitters also tend to exert slowly developing but long-lasting effects on postsynaptic neurons. As a result of these two acetylcholine receptor types and a rich repertoire of neuropeptide transmitters, ganglionic synapses mediate both rapid excitation and a slower modulation of autonomic ganglion cell activity.

The postganglionic effects of autonomic ganglion cells on their smooth muscle, cardiac muscle, or glandular targets are mediated by two primary neurotransmitters: norepinephrine (NE) and acetylcholine (ACh). For the most part, sympathetic ganglion cells release norepinephrine onto their targets (a notable exception is the cholinergic sympathetic innervation of sweat glands), whereas parasympathetic ganglion cells typically release acetylcholine. As expected from

BOX 21B Horner's Syndrome

The characteristic clinical presentation of damage to the pathway that controls the sympathetic division of the visceral motor system to the head and neck is called Horner's syndrome, after the Swiss ophthalmologist who first described this clinical picture in the mid-nineteenth century. The main features, as illustrated in Figure A, are decreased diameter of the pupil on the side of the lesion (*miosis*), a droopy eyelid (*ptosis*), and a sunken appearance of the affected eye (*enophthalmos*). Less obvious signs are decreased sweating, increased skin temperature, and flushing of the skin on the same side of the face and neck.

All these signs are explained by a loss of sympathetic tone due to damage somewhere along the pathway that connects visceral motor centers in the hypothalamus and reticular formation with sympathetic preganglionic neurons in the intermediolateral cell column of the thoracic spinal cord (Figure B). Lesions that interrupt these fibers often spare the descending parasympathetic pathways, which are located more medially in the brainstem and are more diffuse. The sympathetic preganglionic targets that are affected by such lesions include the neurons in the intermediolateral column in spinal segments T1–T3 that control the dilator muscle of the iris and the tone in smooth muscles of the eyelid and globe, the paralysis of which leads to miosis, ptosis, and enophthalmos. The flushing and decreased sweat-

ing are likewise the result of diminished sympathetic tone, in this case governed by intermediolateral column neurons in somewhat lower thoracic segments (~T3–T8). Damage to the descending sympathetic pathway in the brainstem will, of course, affect sweating and vascular tone in the rest of the body on the side of the lesion. However, if the damage is to the upper thoracic outflow (as

is more typical), the upper thoracic chain, or the superior cervical ganglion, then the manifestations of Horner's syndrome will be limited to the head and neck. Typical causes in these sites are stab or gunshot wounds or other traumatic injuries to the head and neck, and tumors of the apex of the lung, thyroid, or cervical lymph nodes.

(A)

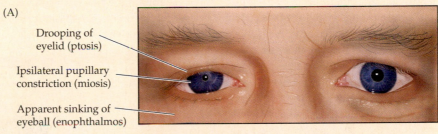

Drooping of eyelid (ptosis)

Ipsilateral pupillary constriction (miosis)

Apparent sinking of eyeball (enophthalmos)

(B)

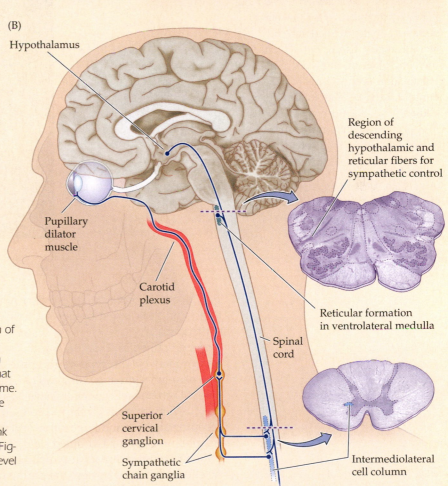

Hypothalamus

Region of descending hypothalamic and reticular fibers for sympathetic control

Pupillary dilator muscle

Carotid plexus

Reticular formation in ventrolateral medulla

Spinal cord

Superior cervical ganglion

Sympathetic chain ganglia

Intermediolateral cell column

(A) Major features of the clinical presentation of Horner's syndrome. (B) Diagram of the descending sympathetic pathways arising in the hypothalamus and reticular formation that can be interrupted to cause Horner's syndrome. Damage to the preganglionic neurons in the upper thoracic cord, to the superior cervical ganglion, or to the cervical sympathetic trunk can also cause Horner's syndrome (see also Figure 21.1). The transverse lines indicate the level of the sections shown at right.

BOX 21C   Obesity and the Brain

Obesity and its relationship to a broad range of diseases—including diabetes, cardiovascular disease and cancer—has become a major public health concern in most developed countries, particularly the United States. Whereas the signature of obesity is obviously an excess of body fat, the underlying cause or causes are generally thought to lie in abnormal regulation by the brain circuits that control appetite and satiety. This fact makes weight loss particularly difficult for many obese individuals. Thus understanding of the central nervous system mechanisms that regulate food intake and metabolism is essential for developing effective strategies to combat this serious health problem.

The brain regulates appetite and satiety (the feeling of fullness following a meal) via the neural activity that is modulated by chemical signals that are secreted into the circulation by fat storing adipose tissues throughout the body. Since this feedback loop entails some of the central components of the visceral motor system, in addition to endocrine mechanisms via insulin and growth hormone, it is discussed here. The peptide **ghrelin** is secreted by the stomach prior to feeding, presumably as a signal of hunger; adipocytes (the cells that concentrate lipid in fatty tissues) increase their secretion of **leptin** into the circulation following feeding, presumably one of several signals for satiety. The receptors for these peptides are concentrated in small groups of neurons in the ventrolateral and anterior hypothalamus (see Box 21A), which contact additional hypothalamic neurons in the arcuate region. These grehlin- and leptin-sensitive cells modulate the activity of neurons expressing the opiomelanocortin propeptide (POMC) and the subsequent secretion of α-melanocyte secreting hormone (α-MSH), one of the peptides encoded by the POMC transcript. This hormone evidently regulates appetite and satiety by acting on specific receptors (particularly the melanocortin receptor subtype called MCR-4) located on additional populations of hypothalamic and brainstem neurons (particularly those in the nucleus of the solitary tract), as well as by endocrine mechanisms that remain poorly understood.

The interactions of leptin, grehlin, α-MSH, and MCR-4 were determined in animal models. Two recessive mutations in mice—the obese (*ob/ob*) and the misnamed diabetic (*db/db*) mice—were identified based on excessive body weight

(A)

(B)

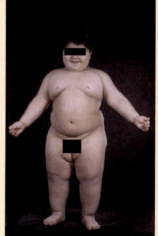 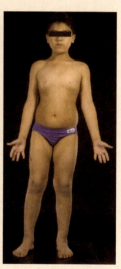

(A) A *POMC* knockout mouse (left) and a wild-type littermate (right). (B) The effect of leptin treatment in a human. At age 3 years, the subject weighed 42 kg (left); at age 7 years, following treatment, the same child weighed 32 kg (right). (A from Yaswen et al., 1999, B from O'Rahilly et al., 2003.)

the foregoing account, these two neurotransmitters usually have opposing effects on their target tissue—contraction versus relaxation of smooth muscle, for example.

As described in Chapters 6 and 7, the specific effects of either ACh or NE are determined by the type of receptor expressed in the target tissue, and the downstream signaling pathways to which these receptors are linked. Peripheral sympathetic targets generally have two subclasses of noradrenergic receptors in their cell membranes, referred to as α and β receptors. Like muscarinic ACh receptors, both α and β receptors and their subtypes belong to the seven-transmembrane G-protein-coupled class of cell surface receptors. The different distribution of these receptors in sympathetic targets allows for a variety of postsynaptic effects mediated by norepinephrine released from postganglionic sympathetic nerve endings (Table 21.2).

and failure to regulate food intake. When each mutation was cloned, the mutant gene in *ob* mice turned out to be the gene for leptin, and the *db* gene that for the leptin receptor. Mutations in the POMC (Figure A) and *MCR4* genes also lead to obesity in mice. The results of inactivation of the *ghrelin* gene are less clear; however, pharmacological and physiological studies associate changes in ghrelin levels with altered feeding and weight loss. Studies in mice have thus provided a solid framework for examining the physiological mechanisms regulating food intake in humans. Nonetheless, their relevance to morbid human obesity remained unclear, until recently.

Genetic analysis of individuals in human pedigrees with extreme obesity (measured body mass indices and weight/height ratios) revealed mutations in one or more of the leptin, leptin receptor, or *MCR4* genes. As a result, these individuals have little sense of satiety after eating, and thus fail to regulate food intake based on signals other than gastric distension, pain and plasma osmolality. How this pathophysiology is related to less extreme degrees of obesity is not yet known, but is being intensely studied because of its implications for normal weight control.

The emerging understanding of body weight regulation by hypothalamic circuits that are modulated by feedback from by hormonal signals from fat tissues has provided new ways of thinking about pharmacological therapies for weight control. While leptin mimetics have proven generally ineffective, leptin administration in human subjects with leptin deficiencies does reduce food intake and obesity (Figure B). Currently, there is great interest in drugs that modulate α-MSH signaling via MCR-4. Although no effective pharmacological therapies presently exist, there is hope that such drugs, when combined with behavioral changes in dietary practices, will effectively combat this often intractable and increasingly common health problem.

References

HORVATH, T. L. AND S. DIANO (2004) The floating blueprint of hypothalamic feeding circuits. *Nature Rev. Neurosci.* 5: 662–667.

O'RAHILLY, S., I. S. FAROOQI, G. S. H. YEO AND B. G. CHALLIS (2003) Human obesity—lessons from monogenic disorders. *Endocrinology* 144: 3757–3764.

SCHWARTZ, M. W., S. C. WOODE, D. PORTE, R. J. SEELY AND D. G. BASKIN (2000) Central nervous system control of food intake. *Nature* 404: 661–671.

SAPER, C. B., T. C. CHOU AND J. K. ELMQUIST (2002) The need to feed: Homeostatic and hedonic control of eating. *Neuron* 36: 199–201.

TABLE 21.2 Summary of Adrenergic Receptor Types and Some of Their Effects in Sympathetic Targets

Receptor	G-protein	Tissue	Response
α_1	G_q	Smooth muscle of blood vessels, iris, ureter, hairs, uterus, bladder	Contraction of smooth muscle
		Heart muscle	Positive inotropic effect ($\beta_1 \gg \alpha_1$)
		Salivary gland	Secretion
		Adipose tissue	Glycogenolysis, gluconeogenesis
		Sweat glands	Secretion
		Kidney	Na$^+$ reabsorbed
α_2	G_i	Adipose tissue	Inhibition of lipolysis
		Pancreas	Inhibition of insulin release
		Smooth muscle of blood vessels	Contraction
β_1	G_s	Heart muscle	Positive inotropic effect; positive chronotropic effect
		Adipose tissue	Lipolysis
		Kidney	Renin release
β_2	G_s	Liver	Glycogenolysis, gluconeogenesis
		Skeletal muscle	Glycogenolysis, lactate release
		Smooth muscle of bronchi, uterus, gut, blood vessels	Relaxation
		Pancreas	Inhibits insulin secretion
		Salivary glands	Thickened secretions
β_3	G_s	Adipose tissue	Lipolysis
		Smooth muscle of gut	Modulation of intestinal mobility

TABLE 21.3	Summary of Cholinergic Receptor Types and Some of Their Effects in Parasympathetic Targets			
Receptor	G-protein	Tissue	Response
Nicotinic	—	Most parasympathetic targets (and all autonomic ganglion cells)	Relatively fast postsynaptic response
Muscarinic (M1)	G_q	Smooth muscles and glands of the gut	Smooth muscle contraction and glandular secretion (relatively slow response)
Muscarinic (M2)	G_i	Smooth and cardiac muscle of cardiovascular system	Reduction in heart rate; smooth muscle contraction
Muscarinic (M3)	G_q	Smooth muscles and glands of all targets	Smooth muscle contraction, glandular secretion

The effects of acetylcholine released by parasympathetic ganglion cells onto smooth muscles, cardiac muscle, and glandular cells also vary according to the subtypes of muscarinic cholinergic receptors found in the peripheral target (Table 21.3). The two major subtypes are known as M1 and M2 receptors, M1 receptors being found primarily in the gut and M2 receptors in the cardiovascular system. (Another subclass of muscarinic receptors, M3, occurs in both smooth muscle and glandular tissues.) Muscarinic receptors are coupled to a variety of intracellular signal transduction mechanisms that modify K^+ and Ca^{2+} channel conductances. They can also activate nitric oxide synthase, which promotes the local release of NO in some parasympathetic target tissues (see, for example, the section below on autonomic control of sexual function).

In contrast to the relatively restricted responses generated by norepinephrine and acetylcholine released by sympathetic and parasympathetic ganglion cells, respectively, neurons of the enteric nervous system achieve an enormous diversity of effects by virtue of many different neurotransmitters, most of which are neuropeptides associated with specific cell groups in either the myenteric or submucus plexuses mentioned earlier. The details of these agents and their actions are beyond the scope of this introductory account.

Many examples of specific autonomic functions could be used to illustrate in more detail how the visceral motor system operates. The three outlined here—control of cardiovascular function, control of the bladder, and control of sexual function—have been chosen primarily because of their importance in human physiology and clinical practice.

Autonomic Regulation of Cardiovascular Function

The cardiovascular system is subject to precise reflex regulation so that an appropriate supply of oxygenated blood can reliably be provided to different body tissues under a wide range of circumstances. The sensory monitoring for this critical homeostatic process entails primarily mechanical (*barosensory*) information about pressure in the arterial system and, secondarily, chemical (*chemosensory*) information about the levels of oxygen and carbon dioxide in the blood. The parasympathetic and sympathetic activity relevant to cardiovascular control is determined by the information supplied by these sensors.

The mechanoreceptors, called baroreceptors, are located in the heart and major blood vessels; the chemoreceptors are located primarily in the carotid bodies, which are small, highly specialized organs located at the bifurcation of the common carotid arteries (some chemosensory tissue is also found in the aorta). The nerve endings in baroreceptors are activated by deformation as the elastic elements of the vessel walls expand and contract. The chemoreceptors in the carotid bodies and aorta respond directly to the partial pressure of oxygen and carbon dioxide in the blood. Visceral afferents from the aortic arch and carotid bifurcation reach the brainstem via the vagus nerve and glossopharyngeal nerve, respectively. Both afferent systems convey their signals to the nucleus of the solitary tract, which relays this information to the hypothalamus and the relevant autonomic centers in the reticular formation (Figure 21.8).

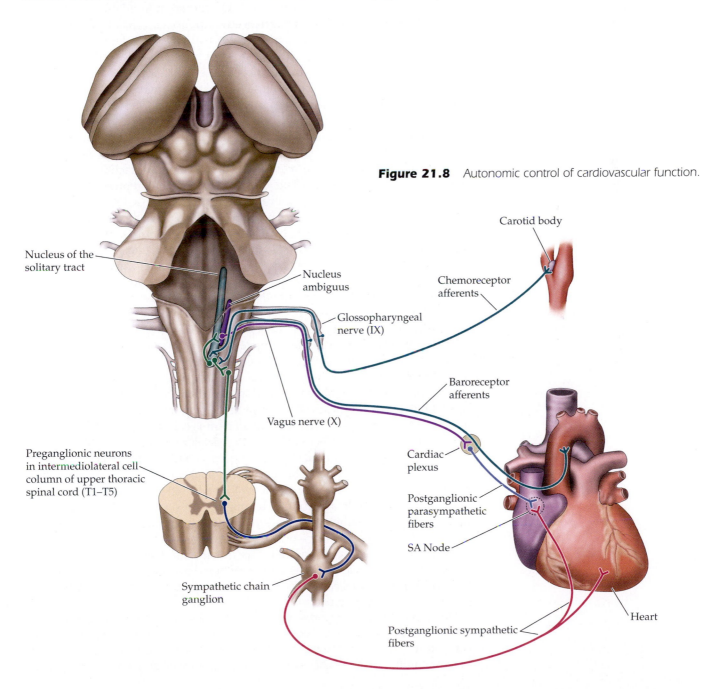

Figure 21.8 Autonomic control of cardiovascular function.

The afferent information derived from changes in arterial pressure and blood gas levels reflexively modulates the activity of the relevant visceral motor pathways and, ultimately, smooth and cardiac muscles and other more specialized structures. For example, a rise in blood pressure activates baroreceptors that, via the pathway illustrated in Figure 21.8, inhibit the tonic activity of sympathetic preganglionic neurons in the spinal cord. In parallel, the pressure increase stimulates the activity of the parasympathetic preganglionic neurons in the nucleus ambiguus and the dorsal motor nucleus of the vagus that influence heart rate. The carotid chemoreceptors also have some influence, but less than that stemming from the baroreceptors.

As a result of this shift in the balance of sympathetic and parasympathetic activity, the stimulatory noradrenergic effects of postganglionic sympathetic innervation on the cardiac pacemaker and cardiac musculature is reduced. These effects are abetted by the decreased output of catecholamines from the adrenal medulla and the decreased vasoconstrictive effects of sympathetic innervation on the peripheral blood vessels. At the same time, activation of the cholinergic parasympathetic innervation of the heart decreases the discharge rate of the cardiac pacemaker in the sinoatrial node and slows the ventricular conduction system. These parasympathetic influences are mediated by an extensive series of parasympathetic ganglia in and near the heart, which release acetylcholine onto cardiac pacemaker cells and cardiac muscle fibers. As a result of this combination of sympathetic and parasympathetic effects, heart rate and the effectiveness of atrial and ventricular myocardial contraction are reduced and the peripheral arterioles dilate, thus lowering the blood pressure.

In contrast to this sequence of events in response to raised blood pressure, a fall in blood pressure (as might occur from blood loss) has the opposite effect—it inhibits parasympathetic activity while increasing sympathetic activity. As a result, norepinephrine is released from sympathetic postganglionic terminals, increasing the rate of cardiac pacemaker activity and enhancing cardiac contractility, at the same time increasing release of catecholamines from the adrenal medulla (which further augments these and many other sympathetic effects that enhance the response to this threatening situation). Norepinephrine released from the terminals of sympathetic ganglion cells also acts on the smooth muscles of the arterioles to increase the tone of the peripheral vessels, particularly those in the skin, subcutaneous tissues, and muscles, thus shunting blood away from these tissues to those organs where oxygen and metabolites are urgently needed to maintain function (e.g., brain, heart, and kidneys in the case of blood loss). If these reflex sympathetic responses fail to raise the blood pressure sufficiently (in which case the patient is said to be in shock), the vital functions of these organs begin to fail, often catastrophically.

A more mundane circumstance that requires a reflex autonomic response to a fall in blood pressure is standing up. Rising quickly from a prone position produces a shift of some 300–800 milliliters of blood from the thorax and abdomen to the legs, resulting in a sharp (approximately 40%) decrease in the output of the heart. The adjustment to this normally occurring drop in blood pressure (called *orthostatic hypotension*) must be rapid and effective, as evidenced by the dizziness sometimes experienced in this situation. Indeed, normal individuals can briefly lose consciousness as a result of blood pooling in the lower extremities, which is the usual cause of fainting among healthy individuals who stand still for abnormally long periods.

The sympathetic innervation of the heart arises from the preganglionic neurons in the intermediolateral column of the spinal cord, extending from roughly the first through fifth thoracic segments (see Table 21.1). The primary visceral motor neurons are in the adjacent thoracic paravertebral and prevertebral ganglia of the cardiac plexus. The parasympathetic preganglionics, as already mentioned, are in the nucleus ambiguus and the dorsal motor nucleus of the vagus nerve and project to parasympathetic ganglia in and around the heart and great vessels.

Autonomic Regulation of the Bladder

The autonomic regulation of the bladder provides a good example of the interplay between components of the somatic motor system that are subject to volitional control (we normally have voluntary control over urination), and the sympathetic and parasympathetic divisions of the visceral motor system, which operate involuntarily. This should not be surprising given that, for many mammals, the act of urination (and of defecation) places the individual at increased risk for attack, since the capacity for immediate fight or flight is reduced. In addition, for many mammals, urine contains chemical signals that mediate complex social behaviors. The neural control of bladder function therefore involves the coordination of relevant autonomic, somatic motor and cognitive faculties that inhibit or promote urination.

The arrangement of afferent and efferent innervation of the bladder is shown in Figure 21.9A. The parasympathetic control of the bladder musculature, the contraction of which causes bladder emptying, originates with neurons in the sacral spinal cord segments (S2–S4) that innervate visceral motor neurons in parasympathetic ganglia in or near the bladder wall. The sympathetic innerva-

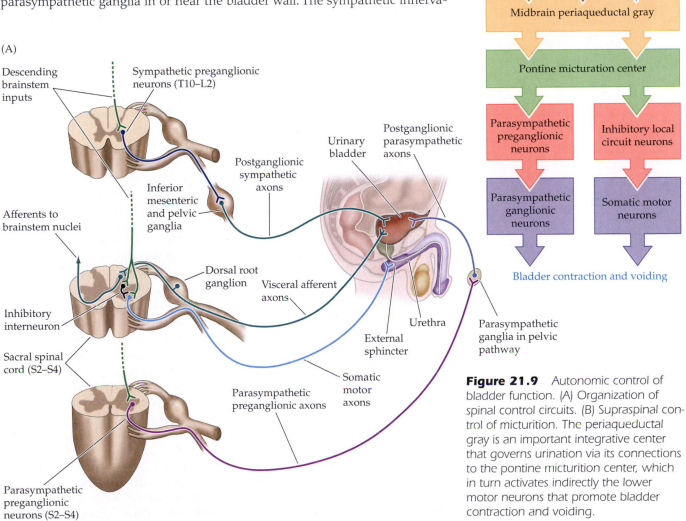

Figure 21.9 Autonomic control of bladder function. (A) Organization of spinal control circuits. (B) Supraspinal control of micturition. The periaqueductal gray is an important integrative center that governs urination via its connections to the pontine micturition center, which in turn activates indirectly the lower motor neurons that promote bladder contraction and voiding.

tion of the bladder originates in the lower thoracic and upper lumbar spinal cord segments (T10–L2), with preganglionic axons running to primary sympathetic neurons in the inferior mesenteric ganglion and the ganglia of the pelvic plexus. The postganglionic fibers from these ganglia travel in the hypogastric and pelvic nerves to the bladder, where sympathetic activity causes the internal urethral sphincter to close (postganglionic sympathetic fibers also innervate the blood vessels of the bladder). Stimulation of this sympathetic pathway in response to a modest increase in bladder pressure from the accumulation of urine thus closes the internal sphincter and inhibits the contraction of the bladder wall musculature, allowing the bladder to fill. At the same time, moderate distension of the bladder inhibits parasympathetic activity (which would otherwise contract the bladder and allow the internal sphincter to open).

The afferent limb of this reflexive circuit is supplied by mechanoreceptors in the bladder wall that convey visceral afferent information to second order neurons in the dorsal horn of the spinal cord. In addition to local connections within spinal cord circuitry, these neurons project to higher integrative centers in the periaqueductal gray of the midbrain. This midbrain region (which is also involved in the descending control of nociception; see Chapter 10) receives input from the hypothalamus, amygdala, and orbital and medial prefrontal cortex. These forebrain structures participate in limbic networks that evaluate risk and the emotional significance of contextual cues (see Chapter 29); in the context of bladder filling, they signal when it is safe and socially appropriate to urinate.

When the bladder is full, parasympathetic activity increases and sympathetic activity decreases, causing the bladder to contract and the internal sphincter muscle to relax. However, urine is held in check by the voluntary somatic motor innervation of the external urethral sphincter muscle. The voluntary control of the external sphincter is mediated by motor neurons of the ventral horn in sacral spinal cord segments (S2–S4), which cause the striated muscle fibers of the sphincter to contract. During bladder filling (and subsequently, until circumstances permit urination) these neurons are active, keeping the external sphincter closed and preventing bladder emptying (or *voiding*, as clinicians often call this process). During urination, this tonic activity is temporarily inhibited, leading to relaxation in the external sphincter muscle. Normally, this is only possible when integrative signals derived from the periaqueductal gray activate a collection of premotor neurons in the pontine reticular formation, known as the "pontine micturition center" (or Barrington's nucleus). The pontine micturition center (*micturition* is "medicalese" for urination) projects to parasympathetic preganglionic neurons and inhibitory local circuit neurons in the sacral spinal cord; the net result is increased parasympathetic outflow (leading to stronger contraction of the bladder wall) and inhibition of the somatic lower motor neurons that innervate the external sphincter muscle (allowing for voiding; see Figure 21.9B). Thus, urination results from the coordinated activation of sacral parasympathetic neurons and temporary inactivation of motor neurons of the somatic motor system; this coordination is ultimately governed by the integration of visceral sensory, emotional, social, and contextual cues.

Importantly, paraplegic patients, or patients who have otherwise lost descending control of the sacral spinal cord, continue to exhibit reflexive, autonomic regulation of bladder function. Unfortunately, this reflex is not fully efficient in the absence of descending motor control, resulting in a variety of problems in paraplegics and others with diminished or defective central control of bladder function. The major difficulty in these cases is incomplete bladder emptying, which often leads to chronic urinary tract infections from the culture medium provided by retained urine, and thus the need for an indwelling catheter to ensure adequate drainage. Indeed, urinary tract morbidity is recognized as the second leading cause of death in patients with spinal cord injury. In other individuals with urge-incontinence and overactive bladder disorders, leakage of urine and an "absence of warning" is the problem. Mounting evidence obtained from functional and structural studies of the brains of such individuals implicates lesions or dysfunction that impairs the integrative activity of the periaqueductal gray and its control over the pontine micturition center.

Autonomic Regulation of Sexual Function

Much like control of the bladder, sexual responses are mediated by the coordinated activity of sympathetic, parasympathetic, and somatic innervation, all of which is governed by complex cognitive, emotional, and contextual cues processed in the limbic forebrain. Although these reflexes differ in detail in males and females, basic similarities, not only in humans but in mammals generally, allow the autonomic sexual responses of the two sexes to be considered together. These similarities include: (1) the mediation of vascular dilation, which causes penile or clitoral erection; (2) stimulation of prostatic or vaginal secretions; (3) smooth muscle contraction of the vas deferens during ejaculation in males or rhythmic vaginal contractions during orgasm in females; and (4) contractions of the somatic pelvic muscles that accompany orgasm in both sexes.

Like the urinary tract, the reproductive organs receive preganglionic parasympathetic innervation from the sacral spinal cord, preganglionic sympathetic innervation from the outflow of the lower thoracic and upper lumbar spinal cord segments, and somatic motor innervation from α motor neurons in the ventral horn of the lower spinal cord segments (Figure 21.10). The sacral

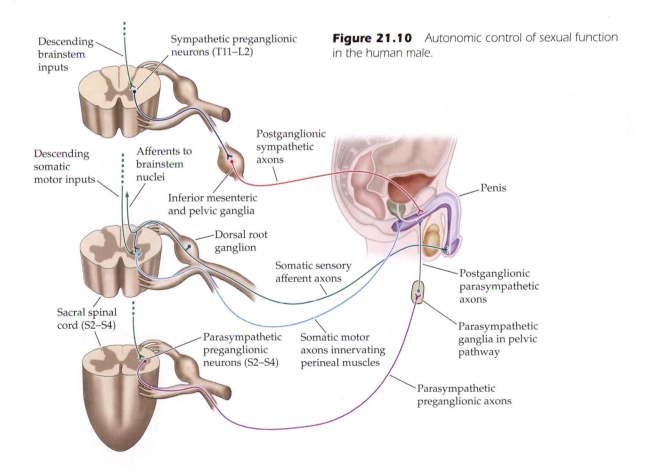

Figure 21.10 Autonomic control of sexual function in the human male.

Descending brainstem inputs

Sympathetic preganglionic neurons (T11–L2)

Descending somatic motor inputs

Afferents to brainstem nuclei

Postganglionic sympathetic axons

Inferior mesenteric and pelvic ganglia

Penis

Dorsal root ganglion

Somatic sensory afferent axons

Postganglionic parasympathetic axons

Sacral spinal cord (S2–S4)

Parasympathetic preganglionic neurons (S2–S4)

Somatic motor axons innervating perineal muscles

Parasympathetic ganglia in pelvic pathway

Parasympathetic preganglionic axons

parasympathetic pathway controlling the sexual organs in both males and females originates in the sacral segments S2–S4 and reaches the target organs via the pelvic nerves. Activity of the postganglionic neurons in the relevant parasympathetic ganglia causes dilation of penile or clitoral arteries, and a corresponding relaxation of the smooth muscles of the venous (cavernous) sinusoids, which leads to expansion of the sinusoidal spaces. As a result, the amount of blood in the tissue is increased, leading to a sharp rise in the pressure and an expansion of the cavernous spaces (i.e., erection). The mediator of the smooth muscle relaxation leading to erection is not acetylcholine (as in most postganglionic parasympathetic actions), but nitric oxide (see Chapter 7). The drug sildenafil (Viagra®), for instance, acts by inhibiting PDE-5, the predominant phosphodiesterase (PDE) expressed in erectile tissue, which leads to an increase in the intracellular concentration of cyclic GMP. This second messenger mediates the activity of endogenous NO; thus, PDE-5 inhibitors enhance the relaxation of the venous sinusoids and promote erection in males with erectile dysfunction. Parasympathetic activity also provides excitatory input to the vas deferens, seminal vesicles, and prostate in males, or vaginal glands in females.

In contrast, sympathetic activity causes vasoconstriction and loss of erection. The lumbar sympathetic pathway to the sexual organs originates in the thoracolumbar segments (T1–L2) and reaches the target organs via the corresponding sympathetic chain ganglia and the inferior mesenteric and pelvic ganglia, as in the case of the autonomic bladder control.

The afferent effects of genital stimulation are conveyed centrally from somatic sensory endings via the dorsal roots of S2–S4, eventually reaching the somatic sensory cortex (reflex sexual excitation may also occur by local stimulation, as is evident in paraplegics). The reflex effects of such stimulation are increased parasympathetic activity, which, as noted, causes relaxation of the smooth muscles in the wall of the sinusoids and subsequent erection.

Finally, the somatic component of reflex sexual function arises from α motor neurons in the lumbar and sacral spinal cord segments. These neurons provide excitatory innervation to the bulbocavernosus and ischiocavernosus muscles, which are active during ejaculation in males and mediate the contractions of the perineal (pelvic floor) muscles that accompany orgasm in both males and females.

Sexual functions are governed centrally by the anteromedial and mediotuberal zones of the hypothalamus, which contain a variety of nuclei pertinent to visceral motor control and reproductive behavior (see Box 21A). Although they remain poorly understood, these nuclei appear to act as integrative centers for sexual responses and are also thought to be involved in more complex aspects of sexuality, such as sexual preference and gender identity (see Chapter 30). The relevant hypothalamic nuclei receive inputs from several areas of the brain, including—as one might imagine—the cortical and subcortical structures concerned with emotion, hedonic reward, and memory (see Chapters 29 and 31).

Summary

Sympathetic and parasympathetic ganglia, which contain the visceral lower motor neurons that innervate smooth muscles, cardiac muscle, and glands, are controlled by preganglionic neurons in the spinal cord and brainstem. The sympathetic preganglionic neurons that govern ganglion cells in the sympathetic division of the visceral motor system arise from neurons in the thoracic and upper lumbar segments of the spinal cord; parasympathetic preganglionic neurons, in contrast, are located in the brainstem and sacral spinal cord. Sympathetic ganglion cells are distributed in the sympathetic chain (paravertebral) and prevertebral ganglia, whereas the parasympathetic motor neurons are more widely distributed in ganglia that lie within or near the organs they control. Most visceral structures receive inputs from both the sympathetic and parasympathetic systems, which act in a generally antagonistic fashion. The diversity of autonomic functions is achieved primarily by different types of receptors for the two primary classes of postganglionic autonomic neurotransmitters, norepinephrine in the case of the sympathetic division and acetylcholine in the parasympathetic division. The visceral motor system is regulated by sensory feedback provided by dorsal root and cranial nerve sensory ganglion cells that make local reflex connections in the spinal cord or brainstem and project to the nucleus of the solitary tract in the brainstem, and by descending pathways from the hypothalamus and brainstem reticular formation, the major control centers of the visceral motor system (and of homeostasis more generally). The importance of the visceral motor control of organs such as the heart, bladder, and reproductive organs—and the many pharmacological means of modulating autonomic function—have made visceral motor control a central theme in clinical medicine.

Additional Reading

Reviews

ANDERSSON, K.-E. AND G. WAGNER (1995) Physiology of penile erections. *Physiol. Rev.* 75: 191–236.

BROWN, D. A. AND 8 OTHERS (1997) Muscarinic mechanisms in nerve cells. *Life Sciences* 60(13–14): 1137–1144.

COSTA, M. AND S. J. H. BROOKES (1994) The enteric nervous system. *Am. J. Gastroenterol.* 89: S129–S137.

DAMPNEY, R. A. L. (1994) Functional organization of central pathways regulating the cardiovascular system. *Physiol. Rev.* 74: 323–364.

GERSHON, M. D. (1981) The enteric nervous system. *Annu. Rev. Neurosci.* 4: 227–272.

HOLSTEGE, G. (2005) Micturition and the soul. *J. Comp. Neurol.* 493: 15–21.

MUNDY, A. R. (1999) Structure and function of the lower urinary tract. In *Scientific Basis of Urology*, A. R. Mundy, J. M. Fitzpatrick, D. E. Neal, and N. J. R. George (eds.). Oxford: Isis Medical Media Ltd., pp. 217–242.

PATTON, H. D. (1989) The autonomic nervous system. In *Textbook of Physiology: Excitable Cells and Neurophysiology*, Vol. 1, Section VII: *Emotive Responses and Internal Milieu*, H. D. Patton, A. F. Fuchs, B. Hille, A. M. Scher, and R. Steiner (eds.). Philadelphia: Saunders, pp. 737–758.

PRYOR, J. P. (1999) Male sexual function. In *Scientific Basis of Urology*, A. R. Mundy, J. M. Fitzpatrick, D. E. Neal and N. J. R. George (eds). Oxford: Isis Medical Media, pp. 243–255.

Important Original Papers

JANSEN, A. S. P., X. V. NGUYEN, V. KARPITSKIY, T. C. METTENLEITER AND A. D. LOEWY (1995) Central command neurons of the sympathetic nervous system: Basis of the fight or flight response. *Science* 270: 644–646.

LANGLEY, J. N. (1894) The arrangement of the sympathetic nervous system chiefly on observations upon pilo-erector nerves. *J. Physiol. (Lond.)* 15: 176–244.

LANGLEY, J. N. (1905) On the reaction of nerve cells and nerve endings to certain poisons chiefly as regards the reaction of striated muscle to nicotine and to curare. *J. Physiol. (Lond.)* 33: 374–473.

LICHTMAN, J. W., D. PURVES AND J. W. YIP (1980) Innervation of sympathetic neurones in the guinea-pig thoracic chain. *J. Physiol.* 298: 285–299.

RUBIN, E. AND D. PURVES (1980) Segmental organization of sympathetic preganglionic neurons in the mammalian spinal cord. *J. Comp. Neurol.* 192: 163–174.

Books

APPENZELLER, O. (1997) The Autonomic Nervous System: An Introduction to Basic and Clinical Concepts, 5th Ed. Amsterdam: Elsevier Biomedical Press.

BLESSING, W. W. (1997) *The Lower Brainstem and Bodily Homeostasis.* New York: Oxford University Press.

BRADING, A. (1999) *The Autonomic Nervous System and Its Effectors.* Oxford: Blackwell Science.

BURNSTOCK, G. AND C. H. V. HOYLE (1995) *The Autonomic Nervous System*, Vol. 1: *Autonomic Neuroeffector Mechanism.* London: Harwood Academic.

CANNON, W. B. (1932) *The Wisdom of the Body.* New York: Norton.

FURNESS, J. B. AND M. COSTA (1987) *The Enteric Nervous System.* Edinburgh: Churchill Livingstone.

GABELLA, G. (1976) *Structure of the Autonomic Nervous System.* London: Chapman and Hall.

LANGLEY, J. N. (1921) *The Autonomic Nervous System.* Cambridge, England: Heffer & Sons.

LOEWY, A. D. AND K. M. SPYER (eds.) (1990) *Central Regulation of Autonomic Functions.* New York: Oxford.

PICK, J. (1970) *The Autonomic Nervous System: Morphological, Comparative, Clinical and Surgical Aspects.* Philadelphia: J.B. Lippincott Company.

RANDALL, W. C. (ed.) (1984) *Nervous Control of Cardiovascular Function.* New York: Oxford University Press.

THE CHANGING BRAIN

IV

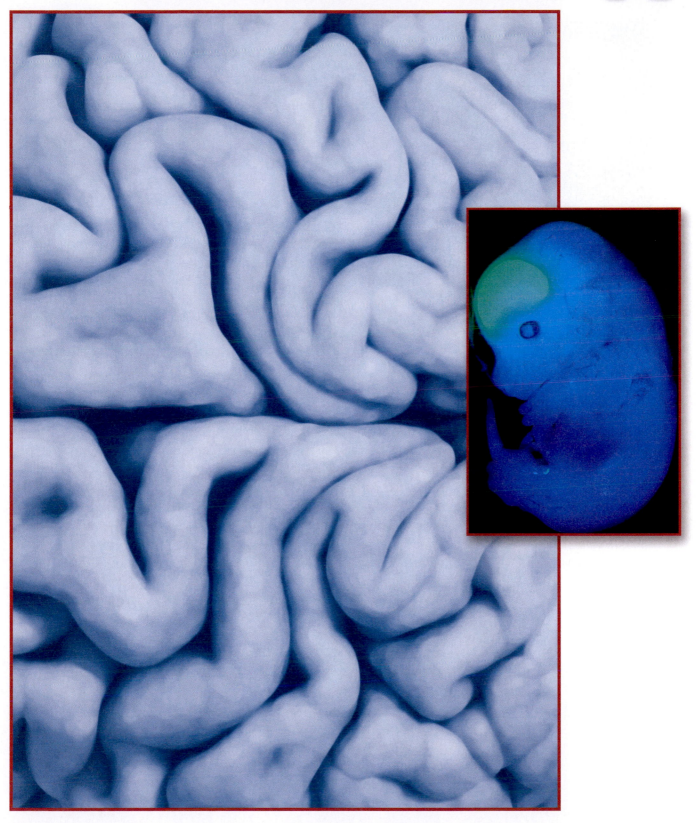

UNIT IV THE CHANGING BRAIN

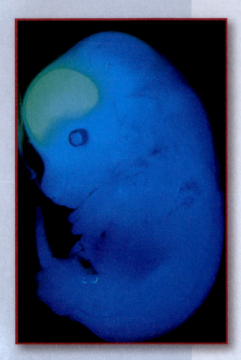

Mouse embryo illustrating the expression of Sox2, a transcription factor found only in neural precursor (stem) cells in the developing mammalian brain. Here a reporter gene for GFP (green fluorescent protein) has been "knocked in" to the animal's *Sox2* locus, and at this stage of development neural precursor cells in the forebrain are prominently labeled by fluorescence. (Courtesy of S. Hutton and L. Pevny.)

Although we think of ourselves as the same person throughout life, the structural and functional state of the brain changes dramatically over the human lifespan. The initial development of the nervous system entails the generation and differentiation of neurons, the formation of axonal pathways, and the elaboration of vast numbers of synapses. Each of these events relies upon the interplay of secreted signals, their receptors, and transcriptional regulators, as well as adhesion and recognition molecules that determine appropriate identity, positions, and connections for developing neurons. The circuits that emerge from these processes mediate an increasingly complex array of behaviors. Subsequent experience during postnatal life—and the activity-dependent molecular mechanisms that translate experience into changes in gene expression and neuronal growth—continues to shape neural circuits, the related behavioral repertoires, and ultimately cognitive abilities. These changes are most pronounced during developmental windows in early life called critical periods. Even in maturity, however, synaptic connections can be modified as new skills and memories are acquired and older ones are forgotten; even some new neurons can be generated in a few specialized brain regions. Some of the mechanisms used during early development are evidently retained and adapted to mediate these ongoing changes in the mature brain.

Finally, like any other organ, the brain is subject to disease and traumatic insults. Some of these processes activate repair mechanisms; however, the capacity of the mature brain for repair or regeneration is limited. Diseases like amyotrophic lateral sclerosis, Parkinson's disease, and Alzheimer's disease all reflect pathologies of processes that normally contribute to neuronal development and to the subsequent maintenance and modification of neural circuitry.

Chapter 22

Early Brain Development

Overview

The elaborate architecture of the adult brain is the product of genetic instructions, cell-to-cell signals, and, eventually, interactions between the growing child and the external world. The early development of the nervous system is dominated by events that occur prior to the formation of synapses and are therefore mostly independent of electrical activity. These early events include the establishment of the primordial nervous system in the embryo, the initial generation of neurons from undifferentiated precursor cells, the formation of the major brain regions, and the migration of neurons from sites of generation to their final positions. These processes set the stage for the subsequent formation of axon pathways and synaptic connections. When any of these processes goes awry—because of genetic mutation, disease, or exposure to drugs or chemicals—the consequences can be disastrous. Indeed, most well-studied congenital brain defects result from interference with the normal mechanisms of activity-independent neuronal development. With the aid of cell biological, molecular, and genetic tools, the multifaceted machinery underlying these extraordinarily complex events is beginning to be understood.

The Initial Formation of the Nervous System: Gastrulation and Neurulation

The cells that will generate the nervous system become distinct quite early in the generation of a vertebrate embryo, concurrent with the establishment of basic body axes: anterior–posterior, dorsal–ventral, and medial–lateral. Critical to this early framework is the process of **gastrulation**. Gastrulation begins as the local invagination of a subset of cells in the developing embryo (which starts out as a single sheet of cells). When this invagination is complete, the embryo consists of three primitive cell layers or **germ layers**: an outer **ectoderm**; a middle **mesoderm**; and an inner **endoderm**. Based on the position of the invaginating mesoderm and endoderm, gastrulation defines the midline, anterior–posterior, and dorsal–ventral axes of all vertebrate embryos.

The formation of the **notochord** is a central event in gastrulation and is essential for the subsequent development of the nervous system. The notochord is a distinct cylinder of mesodermal cells that condenses at the midline of the mesoderm and extends from the mid-anterior to the posterior aspect of the embryo. The notochord is generated by a surface indentation called the **primitive pit**, which subsequently elongates to form the **primitive streak**. As a result of these cell movements during gastrulation, the notochord comes to define the embryonic midline, and thus the axis of symmetry for the entire body. The ectoderm that lies immediately above the notochord is called **neuroectoderm** and gives rise to the entire nervous system. The notochord itself, however, is a transient structure that disappears once early development is complete.

In addition to specifying the basic topography of the embryo and determining the position of the nervous system, the notochord is required for subsequent early neural differentiation (Figure 22.1). Thus, the notochord (along with the primitive pit) sends **inductive signals** to the overlying ectoderm that cause a subset of neuroectodermal cells to differentiate into neural precursor cells. During this process, called **neurulation**, the midline ectoderm that contains these cells thickens into a distinct columnar epithelium called the **neural plate**. The lateral margins of the neural plate then fold inward, eventually transforming the neural plate into a tube. This **neural tube** subsequently gives rise to the brain and spinal cord.

The progenitor cells of the neural tube are known as **neural precursor cells**. These precursors are dividing **neural stem cells** that produce more precursors, all with the capacity to give rise to neurons, astrocytes, and oligodendroglial cells

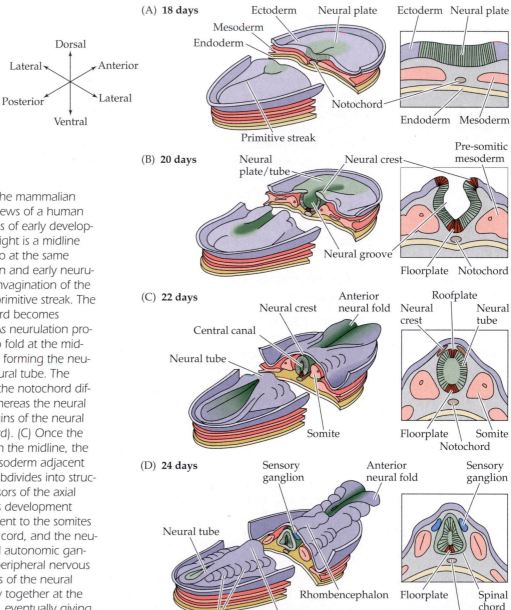

Figure 22.1 Neurulation in the mammalian embryo. On the left are dorsal views of a human embryo at several different stages of early development; each boxed view on the right is a midline cross section through the embryo at the same stage. (A) During late gastrulation and early neurulation, the notochord forms by invagination of the mesoderm in the region of the primitive streak. The ectoderm overlying the notochord becomes defined as the neural plate. (B) As neurulation proceeds, the neural plate begins to fold at the midline (adjacent to the notochord), forming the neural groove and ultimately the neural tube. The neural plate immediately above the notochord differentiates into the floorplate, whereas the neural crest emerges at the lateral margins of the neural plate (farthest from the notochord). (C) Once the edges of the neural plate meet in the midline, the neural tube is complete. The mesoderm adjacent to the tube then thickens and subdivides into structures called somites—the precursors of the axial musculature and skeleton. (D) As development continues, the neural tube adjacent to the somites becomes the rudimentary spinal cord, and the neural crest gives rise to sensory and autonomic ganglia (the major elements of the peripheral nervous system). Finally, the anterior ends of the neural plate (anterior neural folds) grow together at the midline and continue to expand, eventually giving rise to the brain.

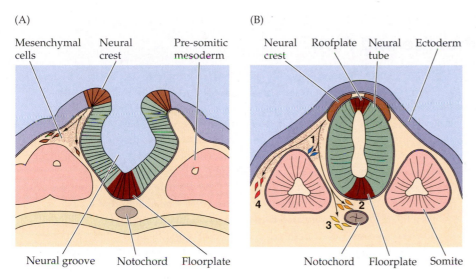

(A)

Mesenchymal cells — Neural crest — Pre-somitic mesoderm

Neural groove — Notochord — Floorplate

(B)

Neural crest — Roofplate — Neural tube — Ectoderm

Notochord — Floorplate — Somite

Figure 22.2 The neural crest. (A) Cross section through a developing mammalian embryo at a stage similar to that in Figure 22.1B. The neural crest cells are established based on their position at the boundary of the embryonic epidermis and neuroectoderm. Arrows indicate the initial migratory route of undifferentiated neural crest cells. (B) Four distinct migratory paths lead to differentiation of neural crest cells into specific cell types and structures. Cells that follow pathways (1) and (2) give rise to sensory and autonomic ganglia, respectively. The precursors of adrenal neurosecretory cells migrate along pathway (3) and eventually aggregate around the dorsal portion of the kidney. Cells destined to become non-neural tissues (for example, melanocytes) migrate along pathway (4). Each pathway permits the migrating cells to interact with different kinds of cellular environments, from which they receive inductive signals (see Figure 22.12). (After Sanes, 1988.)

(Box 22A). Eventually subsets of these neural precursor cells will generate non-dividing **neuroblasts** that differentiate into neurons. Not all cells in the neural tube, however, are neural precursors. The cells at the ventral midline of the neural tube differentiate into a special strip of epithelial-like cells called the **floorplate** (reflecting their proximity to the notochord). The floorplate provides molecular signals that specify the neuroblast cells. The position of the floorplate at the ventral midline defines the dorsoventral polarity of the neural tube and further influences the differentiation of neural precursor cells. Inductive signals from both the notochord and floorplate lead to differentiation of cells in the ventral portion of the neural tube that eventually give rise to spinal and hindbrain motor neurons (which are thus closest to the ventral midline). Precursor cells farther away from the ventral midline give rise to sensory relay neurons within the spinal cord and hindbrain. The differentiation of these dorsal cell groups is also facilitated by a narrow strip of neuroepithelium at the dorsal midline of the neural tube referred to as the **roofplate**. Like the notochord, the floorplate and roofplate are transient structures that disappear once initial development is complete.

At the dorsalmost limit of the neural tube, a third population of cells emerges in the region where the edges of the folded neural plate join together. Because of their location, this set of precursors is called the **neural crest** (Figure 22.2). The neural crest cells migrate away from the neural tube through loosely packed mesenchymal cells that fill the spaces between the neural tube, embryonic epidermis, and somites. Subsets of neural crest cells follow specific pathways where they are exposed to additional inductive signals that influence their differentiation. Thus neural crest cells give rise to a variety of progeny, including the neurons and glia of the sensory and visceral motor (autonomic) ganglia, the neurosecretory cells of the adrenal gland, and the neurons of the enteric nervous system. Neural crest cells also contribute to variety of non-neural structures such as pigment cells, cartilage, and bone, particularly in the face and skull.

The Molecular Basis of Neural Induction

During the first half of the twentieth century, gastrulation and neurulation were defined using a variety of classic experiments that relied on the removal or transfer of embryonic tissues to assess the potential or fate of undifferentiated cells in distinct embryonic structures. In the past two decades, molecular and genetic approaches have demonstrated that the generation of cell identity and

BOX 22A Stem Cells: Promise and Peril

One of the most highly publicized issues in biology over the past several years has been the potential use of stem cells in the treatment of a variety of neurodegenerative conditions, including Parkinson's, Huntington's, and Alzheimer's diseases. Amidst the social, political, and ethical debate set off by the promise of stem cell therapies, an issue that tends to get lost is what, exactly, is a stem cell?

Neural stem cells are an example of a broader class of somatic stem cells. **Somatic stem cells** are found in various tissues, both during development and in the adult. All somatic stem cells share two fundamental characteristics: they are self-renewing, and upon their terminal division and differentiation they can give rise to the full range of cell classes within the relevant tissue.

Thus, a neural stem cell can give rise to another neural stem cell, or to any of the differentiated cell types found in the central and peripheral nervous systems (i.e., inhibitory and excitatory neurons, astrocytes, and oligodendrocytes. A neural stem cell is distinct from a neural

progenitor cell, which is incapable of continuing self-renewal and usually has the capacity to give rise to only one class of differentiated progeny. An oligodendroglial progenitor cell, for example, continues to give rise to oligodendrocytes until its mitotic capacity is exhausted; a neural stem cell, in contrast, can generate more neural stem cells as well as a full range of differentiated neural cell classes, presumably indefinitely.

Neural stem cells, and indeed all classes of somatic stem cells, are distinct from *embryonic* stem cells. **Embryonic stem cells** (**ES cells**) are derived from pre-gastrula embryos. Like somatic stem cells, ES cells have the potential for infinite self-renewal. However, ES cells can give rise to *all* tissue and cell types of the organism—including the germ cells that undergo meiosis and generate haploid gametes as well as neural and other somatic stem cells (Figure A). Somatic stem cells, on the other hand, mitotically generate only diploid, tissue-specific cell types. Some experiments with hematopoietic (blood-forming) and neural

stem cells indicate that these cells can give rise to appropriately differentiated cells in other tissues; however, some of these experiments have not been replicated, and there is debate about the capacity of somatic stem cells to assume embryonic stem cell properties.

The ultimate therapeutic promise of stem cells—neural or other types—is their ability to generate newly differentiated cells and tissues to replace those that may have been lost due to disease

(A) Embryonic stem cells differentiate into various neuronal cell types. (i) Colonies of ES cells prior to differentiation. (ii, iii) After exposure to neuralizing signals, individual stem cell colonies express markers associated with different neural precursor cells. Cells in this colony express both Sox2 (green), a marker of early neural precursors, and nestin (red), a marker for later neural progenitor cells. (iv) After several days in culture, both neurons (red, labeled for neuron-specific tubulin) and astrocytes (green, labeled for glial fibrillary protein) have been generated from ES cells. (Photographs courtesy of L. Pevny.)

(A) (i)   (ii)   (iii)   (iv)

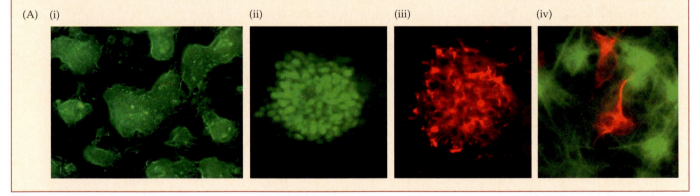

diversity—of which neural induction is but one mechanism—results from the spatial and temporal control of different sets of genes by endogenous signaling molecules provided by one distinct embryonic cell class or tissue to another adjacent cell class or tissue. These inducing signals—including those from the primitive pit and notochord—are, not surprisingly, molecules that can modify gene expression.

The embryonic structures that are critical for the initial induction and patterning of the central nervous system, including the notochord, floorplate,

or injury. Such therapies have been imagined for some forms of diabetes (replacement of islet cells that secrete insulin) and some hematopoietic diseases. In the nervous system, stem cell therapies have been suggested for replacement of dopaminergic cells lost to Parkinson's disease and replacing lost neurons in other degenerative disorders.

While intriguing, this projected use of stem cell technology raises some significant perils. These include ensuring the controlled division of stem cells when introduced into mature tissue, and identifying the appropriate molecular instructions to achieve differentiation of the desired cell class. Clearly, the latter challenge will need to be met with a fuller understanding of the signaling and transcriptional regulatory steps used during development to guide differentiation of relevant neuron classes in the embryo.

At present, there is no clinically validated use of stem cells for human therapeutic applications in the nervous system. Nevertheless, some promising work in mice and other experimental animals indicates that both somatic and ES cells can acquire distinct identities if given appropriate instructions in vitro (i.e., prior to introduction into the host), and if delivered into a supportive host environment. For example, ES cells grown in the presence of platelet-derived growth factor, which biases progenitors toward glial fates, have generated oligodendroglial cells that can myelinate axons in myelin-deficient rats. Similarly, ES cells pretreated with retinoic acid matured into motor neurons when introduced into the developing spinal cord (Figure

B). While such experiments suggest that a combination of proper instruction and correct placement can lead to appropriate differentiation of embryonic or somatic stem cells, there are still many issues to be resolved before the promise becomes reality.

References

BRAZELTON, T. R., F. M. V. ROSSI, G. I. KESHET AND H. M. BLAU (2000) From marrow to brain: Expression of neuronal phenotypes in adult mice. *Science* 290: 1776–1779.

BRUSTLE, O. AND 7 OTHERS (1999) Embryonic stem cell derived glial precursors: A source of myelinating transplants. *Science* 285: 754–756.

CASTRO, R. F., K. A. JACKSON, M. A. GOODELL, C. S. ROBERTSON, H. LIU AND H. D. SHINE (2002) Failure of bone marrow cells to transdifferentiate into neural cells in vivo. *Science* 297: 1299.

MEZEY, E., K. J. CHANDROSS, G. HARTA, R. A. MAKI AND S. R. MCKERCHER (2000) Turning blood into brain: Cells bearing neuronal antigens generated in vivo from bone marrow. *Science* 290: 1779–1782.

SEABERG, R. M. AND D. VAN DER KUOY (2003) Stem and progenitor cells: The premature desertion of rigorous definition. *Trends Neurosci.* 26: 125–131.

WICHTERLE, H., I. LIEBERAM, J. A. PORTER AND T. M. JESSELL (2002) Directed differentiation of embryonic stem cells into motor neurons. *Cell* 110: 385–397.

(B)   Injection of stem cells into spinal cord

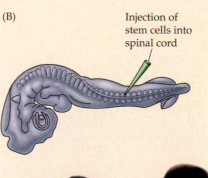

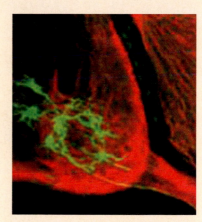

(B) Left: Injection of fluorescently labeled embryonic stem cells into the spinal cord of a host chicken embryo shows that ES cells integrate into the host spinal cord and apparently extend axons. Right: the progeny of the grafted ES cells are seen in the ventral horn of the spinal cord. They have motor neuron-like morphologies, and their axons extend into the ventral root. (From Wichterle et al., 2002.)

roofplate, and neural ectoderm itself, as well as adjacent tissues such as somites, all produce these signals (Figure 22.3A). Distinct classes of receptors transduce these signals within the neuroectoderm to drive further cellular differentiation. In some cases, the signals have graded effects based upon the distance of target cells from the source. These effects may represent a diffusion gradient of the signal, or graded activity due to distribution of receptors or other signaling components. Other signals are more specific in their action, being most effective at the boundaries between distinct cell populations. The

(A)

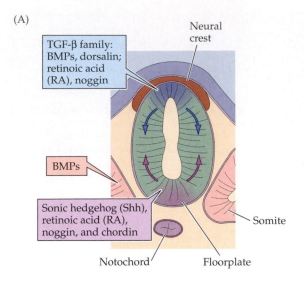

TGF-β family: BMPs, dorsalin; retinoic acid (RA), noggin

Neural crest

BMPs

Sonic hedgehog (Shh), retinoic acid (RA), noggin, and chordin

Somite

Notochord

Floorplate

(B) Retinoic acid (RA)

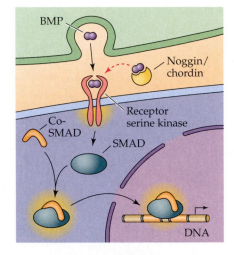

RA

RA binding protein ?

RA receptor

RA   Co-A

DNA

(C) Fibroblast growth factor (FGF)

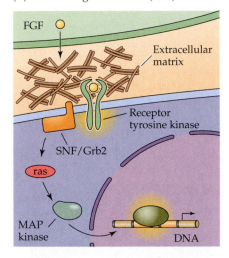

FGF

Extracellular matrix

Receptor tyrosine kinase

SNF/Grb2

ras

MAP kinase

DNA

(D) Bone morphogenetic protein (BMP)

BMP

Noggin/ chordin

Co-SMAD

Receptor serine kinase

SMAD

DNA

(E) Sonic hedgehog (Shh)

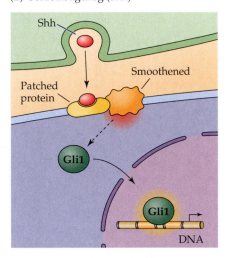

Shh

Patched protein

Smoothened

Gli1

Gli1

DNA

(F) Wnt

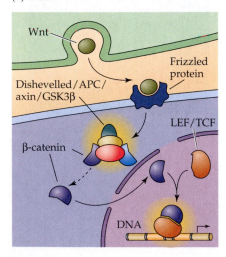

Wnt

Dishevelled/APC/ axin/GSK3β

Frizzled protein

β-catenin

LEF/TCF

DNA

Figure 22.3 Major inductive signaling pathways in vertebrate embryos. (A) The embryonic notochord, floorplate, and neural ectoderm, as well as adjacent tissues such as somites, produce the molecular signals that induce cell and tissue differentiation in the vertebrate embryo. (B–F) Schematics of ligands, receptors, and primary intracellular signaling molecules for retinoic acid (RA); members of the FGF and TGF-β (BMP) superfamily of peptide hormones; Sonic hedgehog (Shh); and the Wnt family of signals. Each of these pathways contributes to the initial establishment of the neural ectoderm, as well as to the subsequent differentiation of distinct classes of neurons and glia throughout the brain.

results of inductive signaling include changes in gene expression, shape, and motility in the target cells.

One of the first of these inductive signals to be identified was **retinoic acid**, a derivative of vitamin A and a member of the steroid/thyroid superfamily of hormones (Figure 22.3B and Box 22B). Retinoic acid activates a unique class of **transcription factors**—the **retinoid receptors**—that modulate the expression of a number of target genes. Peptide hormones provide another class of inductive signals, including those that belong to the **fibroblast growth factor (FGF)** and **transforming growth factor** (**TGF**) families (Figure 22.3C,D). Within the TGF family, the **bone morphogenetic proteins** (**BMPs**) are particularly important for a variety of events in neural induction and differentiation, including initial specification of the neural plate, and subsequent differentiation of the dorsal part of the spinal cord and hindbrain. Another peptide hormone essential for induction in the developing nervous system is **Sonic hedgehog** (**Shh**) (Figure 22.3E). Sonic hedgehog is thought to be particularly important for establishing identity of neurons—particularly motor neurons—in the ventral portion of the spinal cord and hindbrain. Finally, members of the **Wnt** family of secreted signals (vertebrate homologues of the *wingless* gene of *Drosophila*; Figure 22.3F) can modulate several aspects of neural induction and differentiation including some aspects of neural crest differentiation. The receptors for inductive signals, their locations, and their mode of action are clearly essential elements in determining how embryonic induction leads to cellular differentiation.

The receptors for the FGF and BMP families of peptide signals are transmembrane proteins with extracellular binding domains for their respective ligands, and intracellular protein kinase domains that initiate an intracellular signaling cascade; thus they are referred to as receptor-kinases. FGF receptors are tyrosine kinases (they catalyze the phosphorylation of tyrosine residues in target proteins) that bind FGF with the cooperation of extracellular matrix components, including heparan sulfate proteoglycan. Upon binding, activation of the intracellular kinase domains of the FGF receptors leads to activation of the RAS/MAP kinase pathway (see Chapter 7). This signaling pathway can modify cytoskeletal and cytoplasmic components and thus alter the shape or motility of a cell, or it can regulate gene expression, particularly genes that influence cell proliferation. BMP receptors are serine/threonine kinases (i.e., serine and threonine are the target amino acids) that phosphorylate a group of cytoplasmic proteins called SMADs. Upon phosphorylation, SMAD multimers translocate to the nucleus and interact with other DNA-binding proteins, thus modulating gene expression in response to the BMP signal.

Some inductive signals take more indirect routes. For example, the transduction of signals via Sonic hedgehog depends on the cooperative binding of two surface receptor proteins called patched and smoothened (the names are based on the appearances of their respective *Drosophila* mutants). In the absence of Shh, an inhibitory protein complex assembles that modulates a family of transcriptional regulators Gli1, 2, and 3 (originally discovered as oncogenes in gliomas). When this inhibitory complex is active, only Gli3, which represses transcription of target genes, is active. When Shh is present, it binds to patched and promotes the accumulation of smoothened on the cell surface. This event causes the disassembly of the inhibitory complex and allows translocation of Gli1 (or Gli2) into the nucleus to positively regulate gene expression (see Figure 22.3E). The transduction of Wnt signals has a similarly circuitous route, leading ultimately to the nucleus. Wnt receptors, including a family of proteins with the fanciful name "frizzled," initiate a cascade of events after Wnt binding that leads to the degradation of a cytoplasmic protein complex that normally prevents the translocation of β-catenin from the cytoplasm to the nucleus. Once freed from this inhibition, β-catenin enters the nucleus and influences expression of a number of downstream genes via interactions with a DNA-binding transcription factor called LEF/TCF (see Figure 22.3F).

A particularly distinctive aspect of neural induction is the mechanism by which the BMPs influence neural differentiation. As the name suggests, these peptide hormones, which are members of the TGF-β family, elicit osteogenesis from mesodermal cells. If ectodermal cells are exposed to BMPs, they assume an epidermal fate. How, then, does the ectoderm manage to become neuralized,

BOX 22B   Retinoic Acid: Teratogen and Inductive Signal

In the early 1930s, investigators noticed that vitamin A deficiency during pregnancy in animals led to a variety of fetal malformations. The most severe abnormalities affected the developing brain, which was often grossly malformed. At about the same time, experimental studies yielded the surprising finding that *excess* vitamin A caused similar defects. These observations suggested that a family of compounds—metabolic precursors or derivatives of vitamin A called retinoids—are teratogenic. (*Teratogenesis* is the term for birth defects induced by exogenous agents.) The retinoids include the alcohol, aldehyde, and acid forms of vitamin A (retinol, retinal, and retinoic acid, respectively).

The disastrous consequences of exposure to exogenous retinoids during

(A) At left, retinoic acid activates gene expression in a subset of cells in the normal developing forebrain of a mid-gestation mouse embryo (blue areas indicate β-galactosidase reaction product, an indicator of gene expression in this experiment). At right, after maternal ingestion of a small quantity of retinoic acid (0.00025 mg/g of maternal weight), gene expression is ectopically activated throughout the forebrain. (B) At left, the brain of a normal mouse at term; at right, the grossly abnormal brain of a mouse whose mother ingested this same amount of retinoic acid at mid-gestation. (A from Anchan et al., 1997; B from Linney and LaMantia, 1994.)

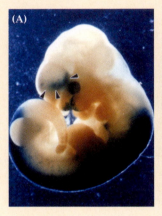

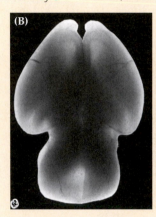
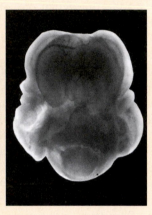

given the fact that BMPs are produced by the somites and surrounding mesodermal tissue? All of these structures are in position to send signals to the neuroectoderm, and thereby convert it to epidermis. This fate is evidently avoided in the neural plate by the local activity of other inductive signaling molecules such as noggin and chordin—two members of a broad class of endogenous antagonists that modulate signaling via the TGF-β family (including the BMPs). Both of these molecules bind directly to the BMPs and thus prevent their binding to BMP receptors. In this way, the neuroectoderm is "rescued" from becoming epidermis. Such negative regulation has reinforced the speculation that becoming a neuron is actually the "default" fate for embryonic ectodermal cells.

The local availability of retinoic acid, BMPs, FGFs, Shh, and Wnts are essential for establishing the basic axes of the neural tube (dorsal–ventral and medial–lateral) and for determining the identity of the neural crest (which will migrate away from the neural tube to form the peripheral nervous system and other structures). In addition to this initial function in determining pattern and cell identity in the nervous system, all of these signals have been implicated in determining the fates of specific classes of cells in the developing nervous system.

TGF-β family signals (including the BMPs) are important for the establishment of dorsal cells in the spinal cord—as well as the neural crest—and can influence other neuron classes in dorsal positions throughout the brain. Sonic hedgehog is essential for the differentiation of motor neurons in the ventral spinal cord, as well as some classes of neurons and glia in the hindbrain, mid-

human pregnancy were underscored in the early 1980s when the drug Accutane® (the trade name for isoretinoin, or 13-*cis*-retinoic acid) was introduced as a treatment for severe acne. Women who used this drug during pregnancy had an increased number of spontaneous abortions and children born with a range of birth defects. Despite the importance of these findings, the reasons for the adverse effects of retinoids on fetal development remained obscure well into the late twentieth century.

An important insight into teratogenic potential of retinoids came when embryologists working on limb development in chicks found that retinoic acid mimics the inductive ability of tissues in the limb bud. Still the mystery remained as to just what retinoic acid (or its absence) was doing to influence or compromise development. An essential clue came in the mid-1980s, when the receptors for retinoic acid were discovered. These receptors are members of the steroid/thyroid hormone receptor superfamily. When they bind retinoic acid or similar ligands, the receptors act as transcription factors to activate specific genes. Careful biochemical analysis showed that retinoic acid was synthesized by embryonic tissues, and subsequent studies showed that retinoic acid activates gene expression at several sites in the embryo, including the developing brain (see figure). Among the most important targets for retinoic acid regulation are genes for other inductive signals, including *Sonic hedgehog* and Hox genes (see Box 22C). Thus, an excess or deficiency of retinoic acid can disrupt normal development by eliciting inappropriate patterns of retinoid-induced gene expression.

The role of retinoic acid as both a teratogen and an endogenous signaling molecule implies that the retinoids cause birth defects by mimicking the normal signals that influence gene expression. The story provides a good example of how teratogenic, clinical, cellular, and molecular observations can be combined to explain seemingly bizarre developmental pathology.

References

EVANS, R. M. (1988) The steroid and thyroid hormone receptor superfamily. *Science* 240: 889–895.

JOHNSON R. L. AND C. J. TABIN (1997) Molecular models for vertebrate limb development. *Cell* 90: 979–990.

LaMANTIA, A.-S., M. C. COLBERT AND E. LINNEY (1993) Retinoic acid induction and regional differentiation prefigure olfactory pathway formation in the mammalian forebrain. *Neuron* 10: 1035–1048.

LAMMER, E. J. AND 11 OTHERS (1985) Retinoic acid embryopathy. *N. Engl. J. Med.* 313: 837–841.

SCHARDEIN, J. L. (1993) *Chemically Induced Birth Defects*, 2nd Ed. New York: Marcel Dekker.

THALLER, C. AND G. EICHELE (1987) Identification and spatial distribution of retinoids in the developing chick limb bud. *Nature* 327: 625–628.

TICKLE, C., B. ALBERTS, L. WOLPERT AND J. LEE (1982) Local application of retinoic acid to the limb bud mimics the action of the polarizing region. *Nature* 296: 564–565.

WARKANY, J. AND E. SCHRAFFENBERGER (1946) Congenital malformations induced in rats by maternal vitamin A deficiency. *Arch. Ophthalmol.* 35: 150–169.

brain, and forebrain. Retinoic acid influences both dorsal and ventral cell identity in the spinal cord and has multiple roles in neuronal differentiation throughout the rest of the developing nervous system. The Wnt signals also are essential for the differentiation of neural crest, cerebellar granule cells, and forebrain neurons. In most cases, these signals lead to the expression or activation of subsets of transcription factors that together define initial cell identity based on their coincident expression in differentiating populations of neurons. This process is best understood for the developing spinal cord (Figure 22.4).

In the spinal cord, the combined activities of retinoic acid, Shh, BMPs, and noggin/chordin specify a mosaic of transcription factors. These transcription factors identify subsets of precursor cells in the dorsal, intermediate, and ventral spinal cord. These molecularly distinct precursors then give rise to sensory relay neurons, interneurons, and motor neurons, respectively. Similar mechanisms are now thought to operate throughout the developing central and peripheral nervous system. Thus, inductive signals, their receptors, and the resulting regulation of gene expression (particularly in locally expressed transcription factors), can specifiy cell identity as well as influencing other aspects of neural development.

Awareness of the molecules involved in neural induction has led to a much more informed way of thinking about the etiology and prevention of a number of congenital disorders of the nervous system. Anomalies like **spina bifida** (failure of the posterior neural tube to close completely), **anencephaly** (failure of the anterior neural tube to close at all), **holoprosencephaly** (disrupted regional dif-

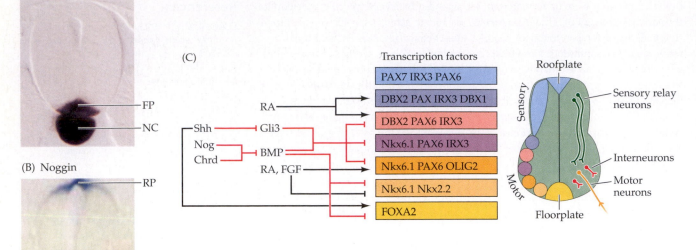

(A) Sonic hedgehog (Shh)

FP
NC

(B) Noggin

RP

FP
NC

(C)

Transcription factors

Roofplate

Sensory relay neurons

| PAX7 IRX3 PAX6 |
| DBX2 PAX IRX3 DBX1 |
| DBX2 PAX6 IRX3 |
| Nkx6.1 PAX6 IRX3 |
| Nkx6.1 PAX6 OLIG2 |
| Nkx6.1 Nkx2.2 |
| FOXA2 |

RA

Shh — Gli3
Nog
Chrd — BMP
RA, FGF

Sensory

Motor

Interneurons

Motor neurons

Floorplate

Figure 22.4 An integrated network of local signals from the ventral and dorsal spinal cord specifies sensory relay neurons, interneurons, and motor neurons. (A) A section through the embryonic chick spinal cord shows the distribution of the Sonic hedgehog signal (dark purple label) in the notochord and floorplate. (B) The BMP antagonist Noggin (light blue labeling), which helps preserve the neural identity of ectoderm by preventing BMP signaling, is available from the roofplate as well as the floor plate and notochord. This image is a section through the embryonic mouse spinal cord; the mouse notochord is a somewhat smaller structure than the chick's. (C) Interactions between Shh (via Gli3 repression), Noggin/Chordin, BMP, RA, and FGF lead to either expression (black) or repression (red) of a set of transcription factors that distinguish different precursors. These distinct precursors, based on their dorsal-to-ventral position in the spinal cord, will go on to become sensory relay neurons (dorsal), interneurons (intermediate), or motor neurons (ventral).

ferentiation of the forebrain), and other brain malformations (often accompanied by mental retardation) can result from environmental insults that disrupt inductive signaling, or from the mutation of genes that participate in this process. For example, excessive intake of vitamin A, the metabolic precursor of retinoic acid, can impede neural tube closure and differentiation or disrupt later aspects of neuronal differentiation due to an excess of ectopic retinoic acid signaling (see Box 22B). Embryonic exposure to a variety of other drugs—alcohol and thalidomide are good examples—can also elicit pathological differentiation of the embryonic nervous system by providing inductive signals at inappropriate times or places. Altered cholesterol metabolism can compromise Sonic hedgehog signaling due to the role of cholesterol in modulating the interaction of Shh with the patched receptor. Such metabolic disruptions, as well as rare mutations in the human *SHH* gene, are associated with holoprosencephaly (these disruptions and mutations, however, account for only a small proportion of the total number of holoprosencephaly cases seen). In addition, mutations in Shh and other proteins in the Shh signaling pathway—particularly the patched receptor—are associated with the most prevalent childhood brain tumor, **medulloblastoma** (Box 22C). Finally, dietary insufficiency of substances such as folic acid can disrupt neural tube formation by compromising cellular mechanisms essential for DNA replication, normal cell division, and motility. Because the consequences of disordered neural induction are so severe, pregnant women are well advised to avoid virtually all drugs and dietary supplements except those specifically prescribed by a physician, especially during the first trimester of pregnancy.

Formation of the Major Brain Subdivisions

Inductive signaling events and the establishment of cellular diversity occur in parallel with the formation of the basic anatomical structures that will define the gross subdivisions of the brain: the spinal cord, brainstem, midbrain, and forebrain. Soon after neural tube formation, the forerunners of the major brain regions become apparent as a result of morphogenetic movements that bend, fold, and constrict the neural tube. Initially, the anterior end of the tube forms a crook, giving it the shape of a cane handle (Figure 22.5A). The end of the cane nearest the sharpest bend, the **cephalic flexure**, balloons out to form the **prosencephalon**, which will give rise to the **forebrain**. The **midbrain**, or **mesencephalon**, forms as a bulge above the cephalic flexure. The **hindbrain**, or

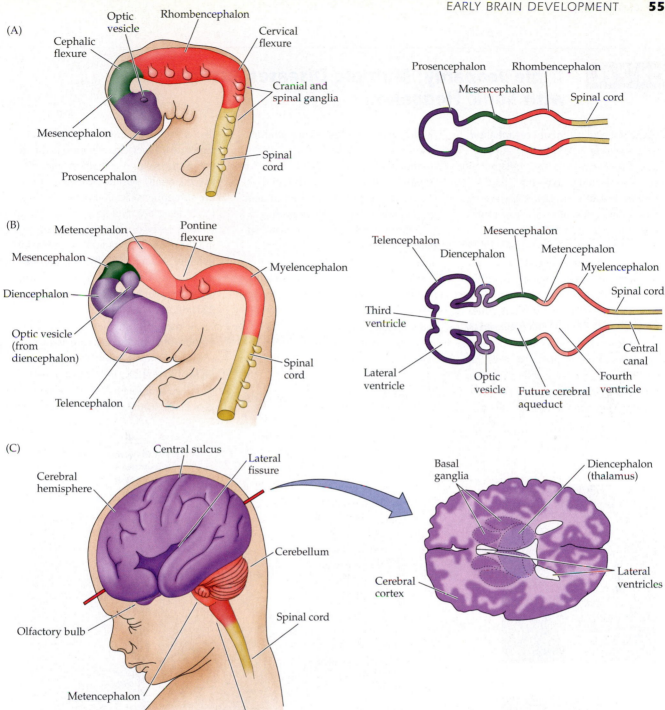

Figure 22.5   Regional specification of the developing brain. (A) Early in gestation the neural tube becomes subdivided into the prosencephalon (at the anterior end of the embryo), mesencephalon, and rhombencephalon. The spinal cord differentiates from the more posterior region of the neural tube. The initial bending of the neural tube at its anterior end leads to a cane shape. At right is a longitudinal section of the neural tube at this stage, showing the position of the major brain regions. (B) Further development distinguishes the telencephalon and diencephalon from the prosencephalon; two other subdivisions—the metencephalon and myelencephalon—derive from the rhombencephalon. These subregions give rise to the rudiments of the major functional subdivisions of the brain, while the spaces they enclose eventually form the ventricles of the mature brain. At right is a longitudinal section of the embryo at the developmental stage shown in (B). (C) The fetal brain and spinal cord are clearly differentiated by the end of the second trimester. Several major subdivisions, including the cerebral cortex and cerebellum, are clearly seen from the lateral surfaces. At right is a cross section through the forebrain at the level indicated showing the nascent sulci and gyri of the cerebral cortex, as well as the differentiation of the basal ganglia and thalamic nuclei.

BOX 22C Triple Jeopardy: Multiple Diseases Associated with Sonic Hedgehog

Many molecules that are essential for early nervous system patterning and morphogenesis—like the signaling molecule Sonic hedgehog—have odd names, and, at first glance, even more arcane functions. Nevertheless, mutations in *SHH* and related signaling genes are associated with at least three serious human disorders. One of these diseases, holoprosencephaly, disrupts the initial morphogenesis of two distinct cerebral hemispheres (Figure A). The second, medulloblastoma is caused by the cancerous transformation of cerebellar granule neuron precursor cells (Figure B). The third, basal cell carcinoma, is the most common cancer of the skin; it is commonly seen in fair-skinned adults

of middle age or older. While all three disorders have distinct cellular targets (neural plate cells, cerebellar precursors, or basal epidermal cells), each of them can be caused by mutations in human genes that encode Sonic hedgehog (*SHH*) or the related receptors and signaling proteins patched (*PTC*) and smoothened (*SMO*).*

Holoprosencephaly occurs in approximately 1 in every 16,000 live births and is the most common malformation of the forebrain known. It compromises the normal separation of the two hemispheres of the forebrain and can also

*Note that, per convention, gene names are italicized; the names for the corresponding proteins are in standard type.

disrupt development of midline facial structures; especially common is the failure of the eye primordium to split into two bilaterally symmetrical fields and the subsequent development of a single (cyclopean) eye. Holoprosencephaly has a broad range of phenotypes, from mildly affected individuals to embryonic lethality accompanied by early spontaneous abortion or stillbirth (in fact, this disorder accounts for approximately 1 in 250 stillbirths).

A small but consistent proportion of holoprosencephaly cases are associated with deletions or missense mutations in *SHH* on chromosome 7. Most of these cases are sporadic (i.e., they are not genetically inherited from a parent carry-

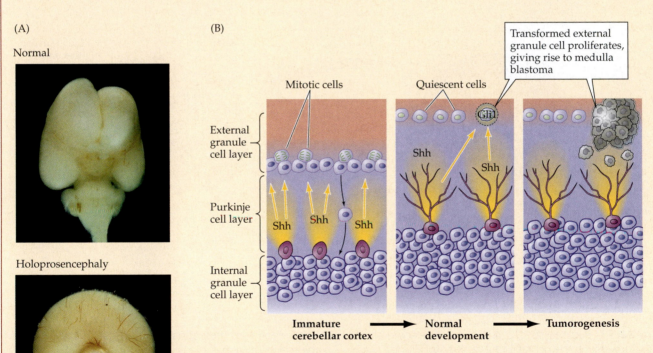

(A) A normal late-gestation human brain, and a brain from a fetus with holoprosencephaly. (B) The normal histogenesis of the cerebellum involves the migration of granule cell precursors (blue) to the external aspect of the cerebellum. The precursors divide in this location, and their postmitotic progeny cells migrate back into the cerebellum, where they differentiate into mature granule cells in the internal granule cell layer. In medulloblastoma, a subset of precursors transforms and divides uncontrollably (gray cells) due to a lack of Shh-mediated regulation of the Gli1 transcription factor. (Photographs courtesy of C. A. Walsh.)

ing the same mutation). Many cases reflect either microdeletions in the chromosomal region that includes *SHH,* or point mutations that cause missense transcripts that disrupt Sonic hedgehog protein structure and function. Significant support for the association of the *SHH* genomic lesion with holoprosencephaly in humans comes from studies of mice in which the *Shh* gene has been inactivated; and from zebrafish, in which a number of mutant alleles of *Shh* have been identified. In each instance, loss of Shh function results in the failure of forebrain hemisphere formation and results in midline facial defects, including the cyclopean eye that characterizes the most severe forms of holoprosencephaly. Thus, one of the initial obligate functions of Sonic hedgehog—whose absence is not easily compensated for when the gene is inactivated—appears to be guiding the formation of the two cerebral hemispheres as well as bilaterally symmetrical facial structures.

Medulloblastoma, the second disorder associated with altered *SHH* function, is the most common childhood brain tumor; even so, it is fortunately quite rare, with an estimated population frequency between 1 in 50,000 and 1 in 100,000 births. There is a 60 percent survival rate; however, these children are seriously compromised by the surgical and medical treatment necessary to prevent the growth of the tumor.

The pathogenesis of medulloblastomas subverts normal neurogenesis and cell migration in the cerebellum. Usually, a large number of cerebellar granule neurons are generated by precursors that migrate to the outside surface of the developing cerebellum, generating postmitotic neuroblasts that then migrate back past the Purkinje cells to their adult location (Figure B). Purkinje cells normally produce Sonic hedgehog, which acts as a mitogen to drive granule cell precursor division. This basic mechanism of granule cell genesis, along with the molecular pathology of medulloblas-

toma, led to the hypothesis that this devastating childhood tumor reflects altered SHH signaling. Most medulloblastoma cells have elevated levels of Gli1 (an oncogene product so named because it is found at elevated levels in various *gliomas,* or glial cell tumors), and Gli1 levels are regulated by Shh signaling. Usually, Shh binding to its receptor PTC (with the help of another Shh regulated protein called SUFU) results in maintaining low Gli1 levels, thus preventing the transcription of Gli1-regulated genes that drive cell proliferation. At least 9 percent of medulloblastoma patients have loss-of-function mutations in *PTC*, while an additional 9 percent have mutations in *SUFU*. The likely contribution of both of these genes to the pathogenesis of medulloblastomas was once again confirmed by studies in mutant mice in which the *Ptc* or *Sufu* gene was inactivated. These mutations, particularly when accompanied by mutations in other major tumor-suppressor genes, result in mice with medulloblastoma with varying frequency, depending upon the mutation.

The final *SHH*-associated disorder, basal cell carcinomas of the skin, are by far the most common malady associated with this multifaceted developmental signaling pathway. In the United States alone, there are at least 750,000 new basal cell carcinomas each year. A subset of basal cell carcinomas (which, because they are local neoplasms, are studied for somatic mutations in the tumor cells themselves rather than for heritable mutations in the individual) have mutations in either *SHH* (rare), *SMO* (rare), or *PTC* (very common). Once again, the use of mouse mutants confirmed the likely contribution of these genes to basal cell carcinoma pathogenesis. Moreover, basal carcinoma cells, when cultured, are responsive to manipulation of Shh signaling. Finally, and perhaps most intriguingly, a rare autosomal dominant syndrome called nevoid basal carcinoma, or Gorlin syndrome, is caused by loss-

of-function mutations in the *PTC* gene. Aside from a high incidence of basal cell carcinoma, these patients also have a significantly increased incidence of medulloblastoma.

Taken together, these observations indicate the central contributions of a seemingly obscure developmental signaling pathway to a number of disorders that, at first glance, might seem completely unrelated. The contrast between the morphogenetic effects of loss of Shh function (holoprosencephaly) versus disregulation of cell proliferation or differentiation (medulloblastoma and basal cell carcinoma) indicates how different tissue contexts can result in very different functions for the same molecules, and different pathologies when those functions are disrupted. Perhaps most importantly, the association of Shh signaling with these three diseases has led to some new therapeutic approaches, especially for medulloblastoma and basal cell carcinomas. In particular, the development of small-molecule inhibitors of SMO, which normally promotes GLI1 stability and nuclear translocation unless inhibited by PTC in the absence of Shh, have some promise as therapeutic agents.

References

DEYA-GROSJEAN, L. AND S. COUVE-PRIVAT (2005) Sonic hedgehog signaling in basal cell carcinomas. *Cancer Lett.* 225: 181–192.

MARINO, S. (2005) Medulloblastoma: Developmental mechanisms out of control. *Trends Molec. Med.* 11: 17–22.

MUENKE, M. AND P. A. BEACHY (2000) Genetics of ventral brain development and holoprosencephaly. *Curr. Op. Genet. Devel.* 10: 262–269.

ROYMER, J. AND T. CURRAN (2005) Targeting medulloblastoma: Small molecule inhibitors of the Sonic hedgehog pathway as potential cancer therapeutics. *Cancer Res.* 65: 4975–4978.

rhombencephalon, forms in the long, relatively straight stretch between the cephalic flexure and the more caudal **cervical flexure**. Caudal to the cervical flexure, the neural tube forms the precursor of the spinal cord. This bending and folding constricts or enlarges the lumen enclosed by the developing neural tube. These lumenal spaces eventually become the ventricles of the mature brain (Figure 22.5B; also see the Appendix).

Once the primitive brain regions are established, they undergo at least two more rounds of partitioning, each of which elaborates these developing brain regions into the forerunners of adult structures (Figure 22.5C). Thus, the lateral aspects of the rostral prosencephalon form the **telencephalon**. The two bilaterally symmetric telencephalic vesicles include dorsal and ventral territories. The dorsal territory will give rise to the rudiments of the cerebral cortex and hippocampus, while the ventral territory gives rise to the basal ganglia (derived from embryonic structures called the **ganglionic eminences**), basal forebrain nuclei, and olfactory bulb. The more caudal portion of the prosencephalon forms the **diencephalon**, which contains the rudiments of the thalamus and hypothalamus, as well as a pair of lateral outpocketings (the **optic cups**) from which the neural portion of the retina will form. The dorsal portion of the mesencephalon

---

BOX 22D   Homeotic Genes and Human Brain Development

The notion that particular genes influence the establishment of distinct regions in a developing embryo arose from efforts to catalog single-gene mutations that affect development of the fruit fly *Drosophila*. In the 1960s and 1970s, E. B. Lewis at the California Institute of Technology reported a number of mutations that resulted in either the duplication of a distinct body segment or the appearance of an inappropriate structure at an ectopic location in the fly. These genes were called **homeotic genes** because they were able to convert segments of one sort to those of another (*homeo* is Greek for "similar"). Subsequently, studies by C. Nusslein-Volhard and E. Wieschaus demonstrated the existence of numerous such "master control" genes, each part of a cascade of gene expression that leads to the distinctive segmentation of the developing embryo. Lewis, Nusslein-Volhard, and Wieschaus shared a 1995 Nobel Prize for these discoveries.

Homeotic genes code for DNA-binding proteins—that is, transcription factors—that bind to a particular sequence of genomic DNA called the **homeobox**. Similar genes have been found in most species, including humans. Using an approach known as cloning by homology, at least four "clusters" of homeobox genes have been identified in virtually all vertebrates that have been examined. The genes in each cluster are closely, but not consecutively, spaced on a single chromosome. Other motifs identified in *Drosophila* have led to the discovery of additional families of DNA-binding proteins, again found in a variety of species.

Importantly, a number of developmental anomalies in mice and humans have been associated with mutations in the homeotic or other developmental control genes initially identified in the fly. Relatively rare diseases like aniridia, Waardenburg syndrome, and Greig cephalopolysyndactyly syndrome (all disorders that disrupt the nervous system and peripheral structures like the iris or the digits; see text) have been associated with human genes that are homologues of *Drosophila* developmental control genes. In addition, several other developmental disorders, including autism and various forms of mental retardation, can be associated with mutations or polymorphisms of homeobox genes (see text). Thus, the initial insights into the molecular control of development gleaned from genetic studies of *Drosophila* have opened new avenues for exploring the molecular basis of developmental disorders in humans.

References

ENGELKAMP, D. AND V. VAN HEYNINGEN (1996) Transcription factors in disease. *Curr. Opin. Genet. Dev.* 6: 334–42.

GEHRING, W. J. (1993) Exploring the homeobox. *Gene* 135: 215–222.

GRUSS, P. AND C. WALTHER (1992) *Pax* in development. *Cell* 69: 719–722.

LEWIS, E. B. (1978) A gene complex controlling segmentation in *Drosophila*. *Nature* 276: 565–570.

NUSSLEIN-VOLHARD, C. AND E. WIESCHAUS (1980) Mutations affecting segment number and polarity in *Drosophila*. *Nature* 287: 795–801.

READ, A. P. AND V. E. NEWTON (1997) Waardenburg syndrome. *J. Med. Genet.* 34: 656–665.

SHIN, S. H., P. KOGERMAN, E. LINDSTROM, R. TOFTGARD AND L. G. BIESECKER (1999) Gli3 mutations in human disorders mimic *Drosophila* cubitus interruptus protein functions and localization. *Proc. Natl. Acad. Sci. USA* 96: 2880–2884.

gives rise to the superior and inferior colliculi, while the ventral portion gives rise to a collection of nuclei known as the midbrain tegmentum. The rostral part of the rhombencephalon becomes the **metencephalon** and gives rise to the adult cerebellum and pons. Finally, the caudal part of the rhombencephalon becomes the **myelencephalon** and gives rise to the adult medulla.

How can a simple tube of neuronal precursor cells produce such a variety of brain structures? At least part of the answer comes from the observation made early in the twentieth century that much of the neural tube is organized into repeating units called **neuromeres**. This discovery led to the idea that the process of **segmentation**—used by all animal embryos at the earliest stages of development to establish regional identity in the body by dividing the embryo into repeated units or segments—might also establish regional identity in the developing brain. Enthusiasm for this hypothesis was stimulated by observations of the development of the body plan of the fruit fly *Drosophila*. In the fly, early expression of a class of genes called **homeotic** or **homeobox genes** guides the differentiation of the embryo into distinct segments that give rise to the head, thorax, and abdomen (Figure 22.6A,B). *Drosophila* homeobox genes code for DNA-binding proteins that can modulate the expression of other genes (Box 22D). Similar genes in mammals (referred to as **Hox genes**) have also been identified. In some cases their pattern of expression coincides with, or even precedes, the formation of morphological features such as the various bends, folds, and constrictions that signify the progressive regionalization of the developing neural tube, particularly in the hindbrain and spinal cord (Box 22E; Figure 22.6C).

In all vertebrates, including humans, Hox gene expression does not extend into the midbrain or forebrain; however, regional differences in expression of other transcription factors are seen in these brain subdivisions prior to and during the morphogenetic events that define them anatomically. The genes seen in the midbrain and forebrain are also homologues of transcription factors that influence the development of major body structures in *Drosophila*, like appendages, head and mouthparts, and sensory organs. The patterned expression of Hox genes, as well as other developmentally regulated transcription factors and signaling molecules, does not by itself determine the fate of a group of embryonic neural precursors. Instead, this aspect of regionally distinct transcription factor expression during early brain development contributes to a broader series of genetic and cellular processes that eventually produce fully differentiated brain regions with appropriate classes of neurons and glia.

Genetic Abnormalities and Altered Development of Brain Regions

The recent explosion of information about molecules that influence brain development provides a basis for reevaluating the causes of a number of congenital brain malformations, as well as various forms of mental retardation. For instance, some forms of **hydrocephalus** (a condition in which impeded flow of cerebrospinal fluid increases pressure and results in enlarged ventricles and eventually cortical atrophy as a result of compression) can be traced to mutations of genes on the X chromosome, especially those in the L1 cell adhesion molecule (see Chapter 23). Similarly, **fragile-X syndrome**, the most common form of congenital mental retardation in males, is associated with triplet repeats in a subset of genes on the X chromosome, particularly the fragile-X protein, which is involved in stabilizing dendritic processes and synapses.

Beyond these X-linked abnormalities, some genetic disorders that compromise the nervous system reflect single gene mutations in homeobox-like tran-

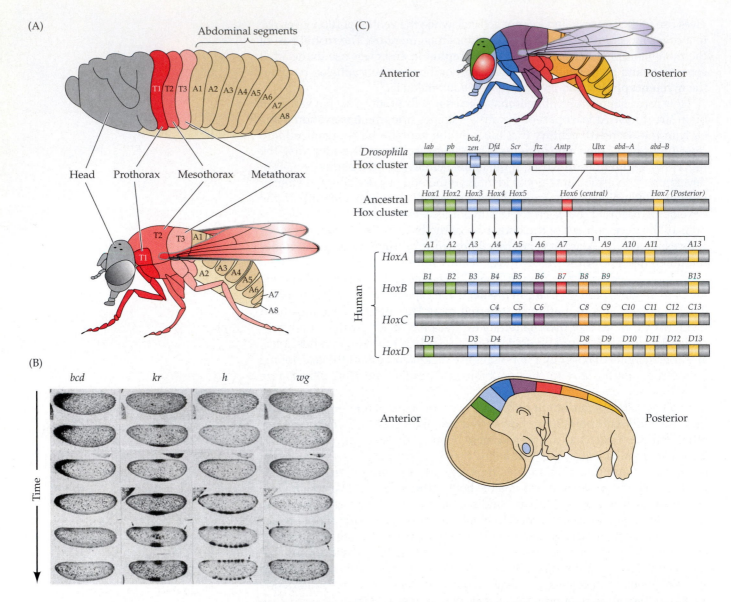

Figure 22.6 Sequential gene expression divides the embryo into regions and segments. (A) The relationship of the embryonic segments in the *Drosophila* larva, defined by sequential gene expression, to the body plan of the mature fruit fly. (B) Temporal pattern of expression of four genes that influence the establishment of the body plan in *Drosophila*. A series of sections through the anterior–posterior midline of the embryo are shown from early to later stages of development (top to bottom in each row). Initially, expression of the gene *bicoid* (*bcd*) helps define the anterior pole of the embryo. Next, *krüppel* (*kr*) is expressed in the middle and then at the posterior end of the embryo, defining the anterior–posterior axis. Then *hairy* (*h*) expression helps to delineate the domains that will eventually form the mature segmented body of the fly. Finally, the *wingless* (*wg*) gene is expressed, further refining the organization of individual segments. (C) Parallels between *Drosophila* segmental genes (the inferred "ancestral" homeobox genes from which invertebrate and vertebrate segmental genes evolved) and human Hox genes. Human Hox genes (and those of most mammals) have apparently been duplicated twice, leading to four independent groups, each on a distinct human chromosome. The anterior-to-posterior pattern of Hox gene expression in both flies and mammals (including humans) follows the 3'-to-5' orientation of these genes on their respective chromosomes. (A after Gilbert, 1994, and Lawrence, 1992; B from Ingham, 1988; C after Veraksa and McGinnis, 2000.)

BOX 22E   Rhombomeres

An interesting parallel between early embryonic segmentation and early brain development was noticed shortly after the dawn of the twentieth century. Several embryologists reported repeating units in the early neural plate and neural tube, which they called *neuromeres*. In the late 1980s, A. Lumsden, R. Keynes and their colleagues, as well as R. Krumlauf, R. Wilkinson and colleagues, noticed further that combinations of homeobox (Hox) and related genes (see Box 22C) are expressed in banded patterns in the developing chick nervous system, especially in the hindbrain (the common name for the rhombencephalon and its derivatives). These expression domains

defined **rhombomeres**, which in the chick (as well as in most mammals) are a series of seven transient bulges in the developing rhombencephalon corresponding to the neuromeres described earlier (Figure A). Rhombomeres are sites of differential cell proliferation (cells at rhombomere boundaries divide faster than cells in the rest of the rhombomere), differential cell mobility (cells from any one rhombomere cannot easily cross into adjacent rhombomeres), and differential cell adhesion (cells prefer to stick to those of their own rhombomere).

Later in development, the pattern of axon outgrowth from the cranial motor nerves also correlates with the earlier

rhombomeric pattern (Figure B). Cranial motor nerves (see the Appendix) originate either from a single rhombomere or from specific pairs of neighboring rhombomeres (transplantation experiments indicate that rhombomeres are in fact specified in pairs). Thus, Hox gene expression probably represents an early step in the formation of cranial nerves in the developing brain. Mutation or ectopic activation of Hox genes in mice alters the position of specific cranial nerves, or prevents their formation. Mutation of the *HoxA-1* gene by homologous recombination—the so-called "knockout" strategy for targeting muta-

(Continued on next page)

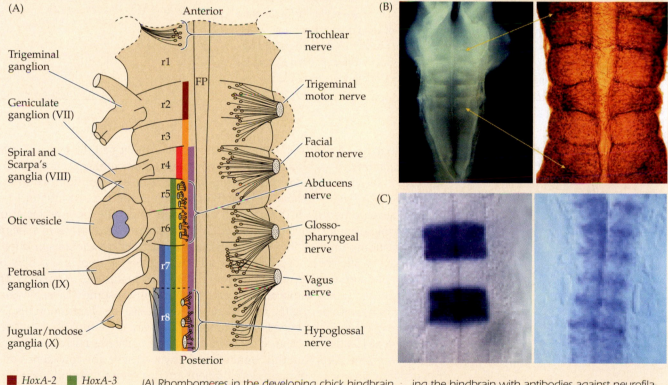

Legend colors:
- HoxA-2
- HoxB-2
- HoxB-1
- HoxA-1
- HoxB-3
- HoxA-3
- HoxB-4
- HoxA-4
- HoxC-4

(A) Rhombomeres in the developing chick hindbrain and their relationship to the pattern of Hox gene expression (color) and the differentiation of the cranial sensory ganglia and nerves (left) as well as cranial motor or mixed nerves (right). (B) In the chick, rhombomeres are a distinct set of ridges and valleys in the developing hindbrain (left). Developing axons follow rhombomere boundaries, visualized by label-

ing the hindbrain with antibodies against neurofilament proteins (right). (C) In the zebrafish, the expression of rhombomere-specific genes—in this case *Krox 20*, a Hox-related transcription factor—are limited to rhombomere boundaries (left). Other molecules, including the transcription factor mariposa, distinguish rhombomere boundaries (right). (B from Lumsden and Keynes, 1989; C courtesy of C. Moens.)

BOX 22E   (Continued)

tions to specific genes—prevents normal formation of rhombomeres. In these animals, development of the external, middle, and inner ear is also compromised, and cranial nerve ganglia are fused and located incorrectly. Conversely, when the *HoxA-1* gene is expressed in a rhombomere where it is usually not seen, the ectopic expression causes changes in rhombomere identity and subsequent differentiation. It is likely that problems in rhombomere formation are the underlying cause of congenital nervous system defects involving cranial nerves, ganglia, and peripheral structures derived from the cranial neural crest (the part of the neural crest that arises from the hindbrain).

The exact relationship between early patterns of rhombomere-specific gene transcription and subsequent cranial nerve development remains a puzzle. Nevertheless, the correspondence between these repeating units in the

embryonic brain and similar iterated units in the development of the insect body (see Figure 22.5) suggests that differential expression of transcription factors in specific regions is essential for the normal development of many species. In a wide variety of animals, spatially and temporally distinct patterns of transcription factor expression coincide with spatially and temporally distinct patterns of differentiation, including the differentiation of the nervous system. This has been elegantly demonstrated in zebrafish, where gene expression patterns clearly define rhombomeres (Figure C). Mutations in these genes disrupt region-specific neural differentiation in the hindbrain.

The idea that the bulges and folds in the neural tube are segments defined by patterns of gene expression provides an attractive framework for understanding the molecular basis of pattern formation in the developing vertebrate brain.

References

CARPENTER, E. M., J. M. GODDARD, O. CHISAKA, N. R. MANLEY AND M. CAPECCHI (1993) Loss of *HoxA-1* (*Hox-1.6*) function results in the reorganization of the murine hindbrain. *Development* 118: 1063–1075.

GUTHRIE, S. (1996) Patterning the hindbrain. *Curr. Opin. Neurobiol.* 6: 41–48.

LUMSDEN, A. AND R. KEYNES (1989) Segmental patterns of neuronal development in the chick hindbrain. *Nature* 337: 424–428.

VON KUPFFER, K. (1906) Die morphogenie des central nerven systems. In *Handbuch der vergleichende und experiementelle Entwicklungslehreder Wirbeltiere*, Vol. 2, 3: 1–272. Fischer Verlag, Jena.

WILKINSON, D. G. AND R. KRUMLAUF (1990) Molecular approaches to the segmentation of the hindbrain. *Trends Neurosci.* 13: 335–339.

ZHANG, M. AND 9 OTHERS (1993) Ectopic HoxA-1 induces rhombomere transformation in mouse hindbrain. *Development* 120: 2431–2442.

scription factors. **Aniridia** (characterized by loss of the iris in the eye and mild mental retardation) and **Waardenburg syndrome** (characterized by craniofacial abnormalities, spina bifida, and hearing loss) are caused by mutations in the *PAX6* and *PAX3* genes, respectively, both of which encode transcription factors. Finally, developmental disorders such as **autism** and other severe social or learning impairments have been linked in some cases to mutations in specific genes (including some of the Wnt family), as well as to microdeletions or duplications of specific chromosomal regions. Perhaps the best known example of this class of neurodevelopmental disorders is Down's syndrome, or **trisomy 21**, which is caused by the duplication of part or all of chromosome 21, usually due to failure of meiosis during the final stages of oogenesis. This duplication results in three copies of all the genes on chromosome 21; an as-yet unknown subset of these genes leads to increased levels of the relevant proteins and altered neural development. Although the connections between these aberrant genes and the resulting anomalies of brain development are not yet understood, such correlations provide a starting point for exploring the molecular pathogenesis of many congenital disorders of the nervous system.

The Initial Differentiation of Neurons and Glia

Once the neural tube has developed into a rudimentary brain and spinal cord, the generation and differentiation of the permanent cellular elements of the brain—neurons and glia—begins in earnest. The mature human brain contains about 100 *billion* neurons and many more glial cells, all generated over the course of only a few months from a small population of precursor cells. Except for a few specialized cases (see Chapter 25), the entire neuronal complement of the adult brain is produced during a time window that closes before birth; thereafter, precursor cells disappear, and few if any new neurons can be added to replace those lost by age or injury in most brain regions. The precursor cells are located in the **ventricular zone**, the innermost cell layer surrounding the lumen of the neural tube, and a region of extraordinary mitotic activity. It has been estimated that in humans about 250,000 new neurons are generated each minute during the peak of cell proliferation during gestation.

The dividing precursor cells in the ventricular zone undergo a stereotyped pattern of cell movements as they progress through the mitotic cycle, leading to the formation of either new stem cells or postmitotic **neuroblasts** (immature nerve cells) that differentiate into neurons (Figure 22.7). The distinction between different modes of cell division that give rise to either new stem cells or to neuroblasts is an essential aspect of the process of neurogenesis. New stem cells arise from *symmetrical* divisions of neuroectodermal cells. These cells divide relatively slowly and can renew themselves indefinitely. Surprisingly, neural stem cells seem to acquire and retain many of the molecular characteristics of glial cells. Thus, in the developing brain, some multipotent neural precursors are indistinguishable from the radial glial cells that also act as a substrate for migration of postmitotic neurons in the cerebral cortex (see below). In contrast, neuroblasts are generated from cells that divide *asymmetrically*: one of the two

Figure 22.7   Dividing precursor cells in the vertebrate neuroepithelium (neural plate and neural tube stages) are attached both to the pial (outside) surface of the neural tube and to its ventricular (lumenal) surface. The nucleus of the cell translocates between these two limits within a narrow cylinder of cytoplasm. When cells are closest to the outer surface of the neural tube, they enter a phase of DNA synthesis (the S stage); after the nucleus moves back to the ventricular surface (the G2 stage), the precursor cells lose their connection to the outer surface and enter mitosis (the M stage). When mitosis is complete, the two daughter cells extend processes back to the outer surface of the neural tube, and the new precursor cells enter a resting (G1) phase of the cell cycle. At some point a precursor cell generates either another progenitor cell that will go on dividing and a daughter cell—a neuroblast—that will not divide further, or two postmitotic daughter cells.

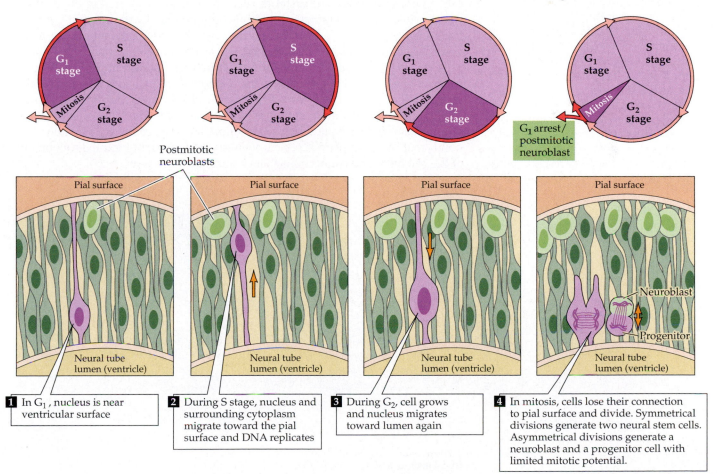

Postmitotic neuroblasts

G_1 arrest/ postmitotic neuroblast

Neuroblast

Progenitor

Pial surface

Neural tube lumen (ventricle)

1 In G_1, nucleus is near ventricular surface

2 During S stage, nucleus and surrounding cytoplasm migrate toward the pial surface and DNA replicates

3 During G_2, cell grows and nucleus migrates toward lumen again

4 In mitosis, cells lose their connection to pial surface and divide. Symmetrical divisions generate two neural stem cells. Asymmetrical divisions generate a neuroblast and a progenitor cell with limited mitotic potential.

daughters becomes a postmitotic neuroblast while the other re-enters the cell cycle to give rise to yet another postmitotic progeny via an asymmetric division. Such asymmetrically dividing progenitors tend to divide more rapidly, have a limited capacity for division over time, and are molecularly distinct from the slowly dividing precursors. These cells are also known as *transit amplifying cells*, because they are a transitional form between stem cells and differentiated neurons, and they account for the amplification in numbers of differentiated cells due to their rapid mitotic kinetics and serial asymmetric divisions.

Asymmetric cell divisions are thought to be mediated by the asymmetric distribution of molecular determinants, including proteins like **numb**, whose activity is modulated by and modulates signaling through **notch** and related proteins—which are master regulators of division and fate in all cells (see below). As cells become postmitotic, they leave the ventricular zone and migrate to their final positions in the developing brain. Knowing when the neurons destined to populate a given brain region are "born"—that is, when they become postmitotic (determined by performing birthdating studies; Box 22F)—has given considerable insight into how different regions of the brain are constructed.

Different populations of spinal cord neurons as well as nuclei of the brainstem and thalamus are distinguished by the times when their component neurons are generated, and some of these distinctions are influenced by local differences in signaling molecules and transcription factors that characterize the precursors (see Figure 22.4). In the cerebral cortex, most neurons of the six cortical layers are generated in an inside-out manner so that each layer consists of a cohort of cells "born" at a distinct time during development (see Box 22G for an intriguing exception to this rule). The firstborn cells are eventually located in the deepest layers, while later generations of neurons migrate radially from the site of their final division in the ventricular zone through the older cells and come to lie superficial to them (Figure 22.8). Indeed, in most regions of the brain where neurons are arranged into layered or laminar structures (hippocampus, cerebellum, superior colliculus), there is a systematic relationship between the layers and the time of cell origin. In these brain regions, neuroblasts from the ventricular zone migrate radially outward, and thus establish a systematic relationship between time of last cell division and laminar position. The implication of this phenomenon is that common periods of neurogenesis are important for the development of the cell types and connections that characterize each layer.

Figure 22.8 Generation of cortical neurons during the gestation of a rhesus monkey (a span of about 165 days). The final cell divisions of the neuronal precursors, determined by maximal incorporation of radioactive thymidine administered to the pregnant mother (see Box 22E), occur primarily during the first half of pregnancy and are complete on or about embryonic day 105. Each short horizontal line represents the position of a neuron heavily labeled by maternal injection of radiolabeled thymidine at the time indicated by the corresponding vertical line. The numerals on the left designate the cortical layers. The earliest generated cells are found in a transient layer called the subplate (a few of these cells survive in the white matter) and in cortical layer 1 (the Cajal-Retzius cells). (After Rakic, 1974.)

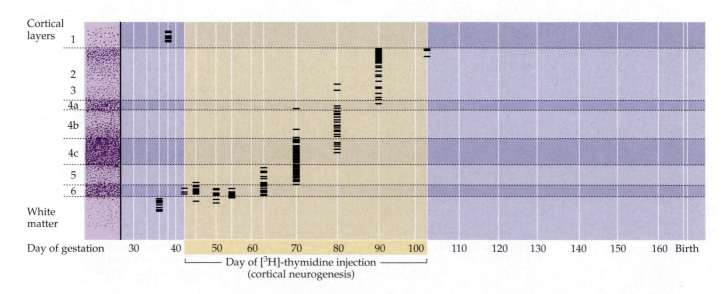

BOX 22F   Neurogenesis and Neuronal Birthdating

The process by which neurons are generated is generally referred to as neurogenesis. The time at which neurogenesis occurs for any particular neuron is called its neuronal "birthdate." At some point in development, neural stem cells (see Box 22A) undergo asymmetrical divisions that produce both another stem cell and a neuronal precursor (called a neuroblast) that will never again undergo cell division. Because neurons are generally unable to reenter the cell cycle once they have left it, the point at which a neuronal precursor leaves the cycle defines the birthdate of the resulting neuron.

In animals with extraordinarily simple nervous systems, such as the worm *Caenorhabditis elegans*, it is possible to directly monitor with a microscope each embryonic stem cell as it undergoes its characteristic series of cell divisions, and thereby to determine when a specific neuron is born. In the vastly more complex vertebrate brain, however, this approach is not feasible. Instead, neurobiologists rely on the characteristics of the cell cycle itself to label cells according to their date of birth. When cells are actively replicating DNA, they take up nucleotides—the building blocks of DNA (see Figure 22.8). Cell birthdating

studies use a labeled nucleotide that can be incorporated only into newly synthesized DNA—usually tritium-labeled thymidine or a chemically distinctive analog of thymidine (the DNA-specific nucleotide) such as bromodeoxyuridine (BrDU)—at a known time in the organism's developmental history. All stem cells that are actively synthesizing DNA incorporate the labeled tag and pass it on to their descendants. Because the labeled probe is only available for minutes to hours after being injected, if a stem cell continues to divide, the levels of labeled probe in the cell's DNA are quickly diluted. However, if a cell undergoes only a single division after incorporating the label and produces a postmitotic neuroblast, that neuron retains high levels of labeled DNA indefinitely. Once the animal has matured, histological sections prepared from the brain show the labeled neurons. The most heavily labeled cells are those that incorporated the tag just before their final division; they are therefore said to have been "born" at the time of injection.

One of the earliest insights obtained from this approach was that the layers of the cerebral cortex develop in an "inside-out" fashion (see Figure 22.8). In certain mutant mice, such as *reeler*

(see Box 19B), birthdating studies show that the oldest cells end up erroneously in the most superficial layers and the most recently generated cells in the deepest as a result of defective migration. Although neuronal birthdates do not, in themselves, tell the lineage of cells, or when they acquire specific phenotypic or molecular features, they mark a major transition in the genetic programs that dictate when and how nerve cells differentiate.

References

ANGEVINE, J. B. JR. AND R. L. SIDMAN (1961) Autoradiographic study of cell migration during histogenesis of the cerebral cortex in the mouse. *Nature* 192: 766–768.

CAVINESS, V. S. JR. AND R. L. SIDMAN (1973) Time of origin of corresponding cell classes in the cerebral cortex of normal and *reeler* mutant mice: An autoradiographic analysis. *J. Comp. Neurol.* 148: 141–151.

GRATZNER, H. G. (1982) Monoclonal antibody to 5-bromo and 5-iododeoxyuridine: A new reagent for the detection of DNA replication. *Science* 218: 474–475.

MILLER, M. W. AND R. S. NOWAKOWSKI (1988) Use of bromodeoxyuridine immunohistochemistry to examine the proliferation, migration, and time of origin of cells in the central nervous system. *Brain Res.* 457: 44–52.

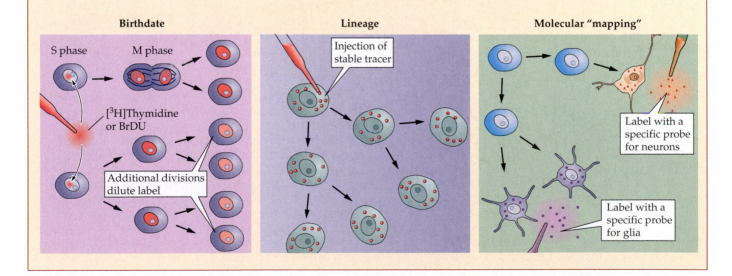

The Generation of Neuronal Diversity

In many ways, the neuronal precursor cells in the ventricular zone of the embryonic brain look and act more or less the same. Yet these precursors ultimately give rise to postmitotic cells that are enormously diverse in form and function. The spinal cord, cerebellum, cerebral cortex, and subcortical nuclei (including the basal ganglia and thalamus) each contain several neuronal cell types distinguished by morphology, neurotransmitter synthesis, cell surface molecules, and the types of synapses they make and receive. On an even more basic level, the stem cells of the ventricular zone produce both neurons and glia—cells with markedly different properties and functions. How and when are these different cell types determined?

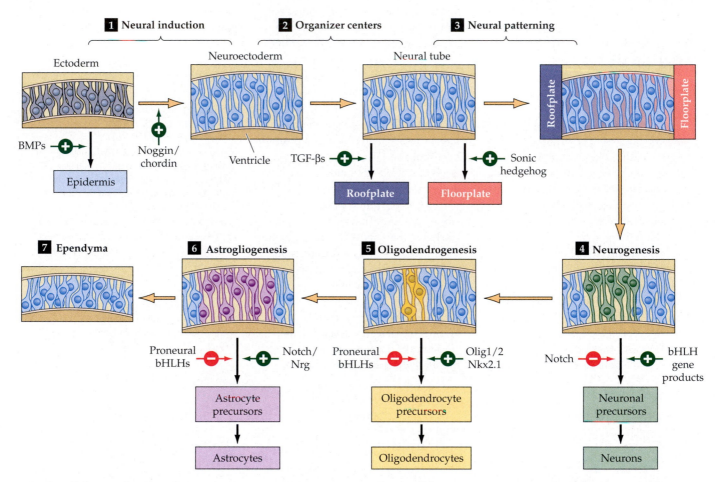

Figure 22.9 Essential molecular and cellular mechanisms that guide neuronal and glial differentiation in the neural ectoderm. (1–3) The steps by which ectoderm acquires its identity as neural ectoderm. Generation of neural precursors, or stem cells, relies first on the balance of BMP and its endogenous antagonists (e.g., noggin and chordin) in the developing embryo. Next, local sources of inductive signals, including TFG-β family members and Sonic hedgehog, establish gradients that influence subsequent neural precursor identities, as well as identifying local "organizers" (such as the floorplate and roofplate) that define the cellular identity of the inductive signaling centers. (4–7) Steps thought to define neurons, oligodendroglia, and astrocytes from multipotent neuronal precursors. Balanced signaling activity of notch and transcriptional control of the *bHLH* proneural genes (named based on their ability to bias neural progenitor cells toward a differentiated neural fate) influence neurogenesis. Similarly, antagonistic transcriptional regulation via either the products of the *bHLH* genes or three additional transcription factors—Olig1, Olig2, and Nkx2.2—influence the generation of oligodendroglia. Continued antagonism between bHLH proteins, Notch signaling proteins, and the signaling molecule neureglin (Nrg) is thought to influence the generation of mature astrocytes. Finally, in the adult brain, cells adjacent to the ventricles (which apparently have avoided becoming differentiated) remain as ependymal cells. These may include a subpopulation of neural stem cells (see Box 22A). (After Kintner, 2002.)

The bulk of the evidence favors the view that neuronal differentiation is based primarily on local cell–cell interactions followed by distinct histories of transcriptional regulation via a "code" of transcription factors expressed in each cell (Figure 22.9). Historically, experimental approaches to this issue relied on transplantation strategies, such as moving bits of a particular developing brain region to a different location in the brain of a host animal to determine whether the transplanted cells acquire the host phenotype or retain their original fate during subsequent development. In general, when very young precursor cells are transplanted, they tend to acquire the host region phenotype. Transplanted cells of increasingly older ages, however, usually retain the phenotype that reflects their region of origin.

The use of genetic approaches, particularly in simple, so-called "model" organisms such as fruit flies and the nematode *Caenorhabditis elegans*, has made clear the fact that cell lineages are influenced by local cell–cell interactions. Such work has also revealed some of the molecules that mediate these processes in neural fate determination. In the fruit fly eye, the position and identity of a variety of photoreceptor cells with distinct visual functions relies on signaling mediated by the notch protein and by additional cell-surface ligands (including one called boss) on one class of cells, and specific receptor kinases (e.g., sevenless) on adjacent cells (Figure 22.10). The balance between symmetric and asymmetric cell division, and its consequences for development of the nervous system, is particularly clear in *C. elegans*, where the complete lineage of the nervous system (as well as the entire worm) is known. For each class of neurons in *C. elegans*, a series of asymmetric cell divisions results in the full complement of relevant nerve cells. For example, the determination of midline neurons reflects their lineage, the proper functioning of genes involved in cell–cell signaling, transcriptional regulation, and whether subsets of precursors survive or die during apoptosis (programmed cell death). Genetic variants can be identified with great specificity, and analysis of mutant *C. elegans* has led to the identification of a number of genes that regulate lineage, division, and the survival or death of developing nerve cells in the nematode, providing a model for analysis of neural and non-neural cell lineages in other organisms.

A similar balance between cell lineage and cell–cell interactions has been invoked to explain differentiation of a number of neuronal and glial classes in the developing vertebrate brain. Perhaps not surprisingly, many of the signal

Figure 22.10 Development of the compound eye of the fruit fly *Drosophila* provides an example of how cell–cell interactions can determine cell fate. (A) Scanning electron micrograph of the eye in *Drosophila*. (B) Diagram of the structure of the fly eye. The eye consists of an array of identical ommatidia, each comprising an array of eight photoreceptors. (C) Arrangement of photoreceptors (R1–7) within each ommatidium and the cell–cell signaling that determines their fate. A membrane-bound ligand on R8 (the *boss* gene product) binds to a receptor (encoded by the *sevenless* gene, *sev*) on the R7 cell. These interactions eventually lead to the changes in gene expression that determine the fate of an R7 cell. The arrows between R8 and the remaining receptor cells indicate interactions necessary for determining the fates of R1–R6. (A courtesy of T. Venkatesh; B,C after Rubin, 1989.)

(A)

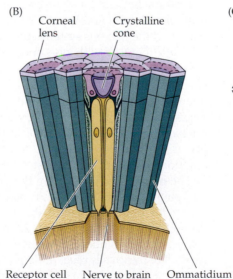

(B)

Corneal lens   Crystalline cone

Receptor cell   Nerve to brain   Ommatidium

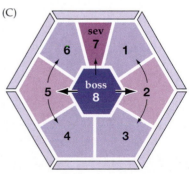

(C)

molecules that are essential for the initial steps of neural induction and region-alization—BMPs, Sonic hedgehog, and Wnts—all influence the genesis of specific classes of neurons and glia via local cell–cell interactions (see Figure 22.9). Additional signaling molecules that contribute to these processes in the vertebrate brain include the notch family of cell surface ligands and their delta receptors. Signaling through the notch–delta pathway tends to maintain progenitor cells in an undifferentiated state.

Among the targets of all of these pathways, a subset of transcription factor genes known as the **bHLH genes** (named for a shared *basic helix-loop-helix* amino acid motif that defines the DNA binding domain) has emerged as central to subsequent differentiation of distinct neural or glial fates. Some of these bHLH genes are named as homologues of similar genes originally discovered in the developing fly, while others have been identified based on their predicted amino acid sequences inferred from genomic sequences. These molecular details of cell signaling and subsequent lineage relationships provide an outline of how general cell classes are established. Nevertheless, there is presently no clear and complete explanation for how any specific neuronal class achieves its identity. This gap in knowledge presents a problem in using neural stem cells to generate replacements for specific cell classes lost in neurodegenerative diseases or after brain injury (see Box 22A and Chapter 25), as well as efforts to understand how neural precursors can become transformed into tumorogenic cells, like those that cause medulloblastoma and other cancers of the nervous system (see Box 22C).

Neuronal Migration in the Peripheral Nervous System

Cell migration, a ubiquitous feature in the embryo, brings distinct classes of cells into appropriate spatial relationships within differentiating tissues. In the nervous system, migration brings different classes of neurons together so that they can interact during development, and it insures the final position of many postmitotic neurons. The capacity of a neural precursor or an immature nerve cell to move and its transit through a changing cellular environment is essential for subsequent differentiation. The final location of a postmitotic nerve cell is presumably especially critical, since neural function depends on precise connections made by neurons and their targets. The developing neuron must be in the right place at the right time to be properly integrated into a functional circuit that can mediate behavior.

Neuronal migration involves much more than the mechanics of moving cells from one place to another. As is the case for inductive events during initial formation of the nervous system, stereotyped movements bring different classes of cells into contact with one another (often transiently), thereby providing a means of constraining cell–cell signaling to specific times and places. Such effects are most thoroughly documented for the migration and differentiation of neural crest cells from the neural tube to the periphery of the embryo, where migratory paths are influenced by the initial position of neural crest cells at distinct anterior–posterior locations in the neural tube. This initial positional identity is reflected in the final locations of neural crest cells from various anterior–posterior levels in distinct parts of the body.

The neural crest arises from the dorsal neural tube along the entire length of the spinal cord and hindbrain. Thus, as neural crest cells begin their journeys, they carry with them information about their point of origin, including expression of distinct Hox genes (see Figure 22.6C and Box 22D) that are limited to various spinal cord and hindbrain domains. Regardless of where neural crest cells originate, all of these cells must undergo an essential transition in order to begin their migration. They all begin as neuroepithelial cells, and thus have all of the intercel-

BOX 22G   *Mixing It Up: Long-Distance Neuronal Migration*

For many years, developmental neurobiologists assumed that "position was destiny" in the developing brain. For example, if a neuron was found in the thalamus, cerebellum, or cerebral cortex in the adult brain, it most likely came from a neural progenitor cell in the embryonic brain region that gave rise to the thalamus, cerebellum, or cerebral cortex. The identification of rhombomeres and subsequent evidence that these domains are compartments between which little mixing of cells occurs reinforced this notion. Nevertheless, a few observations hinted that all neurogenesis might not be local, and eventually led to a new idea of how neuronal classes in a variety of brain regions are integrated into mature structures and circuits.

The initial indication of this tendency for subsets of neurons to wander came in the late 1960s with a report that neurons in the pulvinar, a thalamic nucleus assumed to be derived from the dien-cephalon, were actually generated in the telencephalon. This observation received little notice until the mid 1980s, when a series of experiments using chick-quail chimeras suggested that a major portion of granule cells in the cerebellum (small local circuit neurons) were actually generated outside the rhombencephalon (the embryonic region associated with the generation of the cerebellum). Most of these extrinsic cells were thought to migrate from the mesencephalon (associated with the generation of the superior and inferior colliculus in the adult brain) into the external granule cell layer of the cerebellum. Together, these findings implied that adult brain structures might be derived from a broad range of embryonic brain subdivisions.

Around the same time, several investigators noticed a small but consistent proportion of cells in the cerebral cortex whose migratory route was apparently tangential rather than radial (via radial glial guides; see Figure 22.12). These observations were the focus of a lively debate that nevertheless failed to explain the significance of the apparent "escapees" from the radial migration framework in the developing cortex. Moreover, lineage analysis suggested that cortical projection neurons, interneurons, astrocytes, and oligodendroglia were probably not derived from the same precursor pools (see Figure 22.9). There was little consensus about these disparate observations until the mid 1990s when several groups realized that there was a massive migration of cells from the ventral forebrain—the region of the ganglionic eminence that gives rise to the caudate, putamen, and globus pallidus—to the cerebral cortex (Figure A). Moreover, these ventrally derived cells were not just any cell types; they constituted distinct classes of GABAergic interneurons in the cortex and olfactory bulb, as well as oligodendroglia throughout the entire forebrain. Newly generated granule and periglomular interneurons in the mature olfactory bulb are derived from a remnant of the ganglionic eminence called the anterior subventricular zone, which persists at the surface of the mature lateral ventricles.

A mosaic of transcriptional regulators whose expression and activity is restricted to various domains in the ventral forebrain orchestrates this long distance migration of distinct cell types (Figure B). When subsets of these transcription factors are mutated, migration of cells from the ventral forebrain to the

(Continued on next page)

(A)

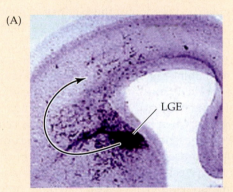

LGE

(C)

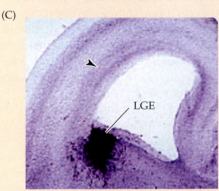

LGE

(B)

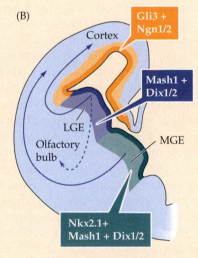

Cortex

Gli3 + Ngn1/2

Mash1 + Dix1/2

LGE

Olfactory bulb

MGE

Nkx2.1+ Mash1 + Dix1/2

(A) Migration of cells from the ventral forebrain to the neocortex during late gestation in the mouse. A tracer has been placed in the lateral ganglionic eminence, and labeled cells can be seen streaming toward the cortex, as well as in residence in the developing cortical plate. (B) Schematic of transcriptional regulators associated with the primary divisions of the ganglionic eminence (lateral and medial, LGE and MGE), and the basic migratory routes taken by cells in each division (arrows). (C) Diminished migration of cells from the ventral forebrain in mice with null mutations of both *Dlx1* and *Dlx2* (expressed throughout the lateral and medial ganglionic eminence).

cortex is dramatically diminished (Figure C), and the numbers of GABAergic interneurons is similarly reduced. The mechanisms for specifying cell identity, migration, and destination remain unknown; nevertheless, this regional diversity is apparently a consistent feature of neurogenesis in mammalian brains.

Not all neurons participate in this long-distance migration. In the human brain, for example, some GABAergic interneurons are generated locally in the cortical rudiment, in addition to those that migrate from the ventral forebrain. These locally generated interneurons apparently use the same radial glial migratory route as their glutamatergic neighbors.

The developmental and functional significance of this mixing of offspring from progenitors in various embryonic brain regions remains unknown. Perhaps the range and number of cell–cell interactions necessary to generate functionally distinct cell types is so large that the appropriate population can only be determined by exposing subsets of cells to a variety of environments, and then having those cells act as messengers to deliver additional molecular signals at a new location. Regardless of the purpose of this arduous journey, its consequence is the orderly establishment of cellular diversity in a number of brain regions.

References

ANDERSON, S. A., D. D. EISENSTAT, L. SHI AND J. L. RUBENSTEIN (1997) Interneuron migration from basal forebrain to neocortex: Dependence on *Dlx* genes. *Science* 278: 474–476.

HE, W., C. INGRAHAM, L. RISING, S. GODERIE AND S. TEMPLE (2001) Multipotent stem cells from the mouse basal forebrain contribute GABAergic neurons and oligodendrocytes to the cerebral cortex during embryogenesis. *J. Neurosci.* 21: 8854–8862.

MARTINEZ, S. AND R. M. ALVARADO-MALLART (1989) Rostral cerebellum originates from the caudal portion of the so-called "mesencephalic" vesicle: A study using chick/quail chimeras. *Eur. J. Neurosci.* 6: 549–560.

PARNAVELAS, J. G., J. A. BARFIELD, E. FRANKE AND M. B. LUSKIN (1991) Separate progenitor cells give rise to pyramidal and nonpyramidal neurons in the rat telencephalon. *Cereb. Cortex* 1: 463–468.

RAKIC, P. AND R. L. SIDMAN (1969) Telencephalic origin of pulvinar neurons in the fetal human brain. *Z. Anat. Entwicklungsgesch.* 129: 53–82.

WICHTERLE, H., D. H. TURNBULL, S. NERY, G. FISHELL AND A. ALVAREZ-BUYLLA (2001) *In utero* fate mapping reveals distinct migratory pathways and fates of neurons born in the mammalian basal forebrain. *Development* 128: 3759–3771.

lular junctions and adhesive interactions that keep epithelial cells in place. To move, neural crest cells must downregulate expression of these adhesive genes and undergo an epithelial-to-mesenchymal transition (epithelial cells being bound in a sheet, mesenchymal cells being more loosely packed and tending to migrate freely). Thus, presumptive neural crest cells express several transcription factors, including the bHLH family members Snail1 and Snail2 (see Figure 22.11B), which repress expression of intercellular junctional proteins and epithelial adhesion molecules. This process of delamination and migration in the neural crest is similar to the process of metastasis when transformed oncogenic cells escape their epithelial confines, assume a mesenchymal character, and become freely migratory (with grave pathological consequences). Indeed, some of the genes that modulate delamination in neural crest—particularly the *Snail* genes—can also be oncogenes if mutated in mature epithelial tissues. When the now-motile neural crest cells reach their final destination, they cease to express Snail and other transcription factors that favor the mesenchymal, migratory state. This change is thought to reflect the integration of a number of signals that neural crest cells encounter along their migratory route. (Understanding this normal event has implications for cancer biology: inducing transformed migratory cancerous cells to revert to a stationary state would have clear therapeutic value.)

Neural crest cells are largely guided along distinct migratory pathways provided by non-neural peripheral structures like somites (which eventually form axial muscles) and other rudimentary musculoskeletal or visceral tissues. The signals along these pathways can be secreted molecules (including some of the peptide hormones used at earlier times for neural induction), cell surface ligands and receptors (adhesion molecules and other signals), or extracellular matrix molecules; most of these molecules also are used at later stages of devel-

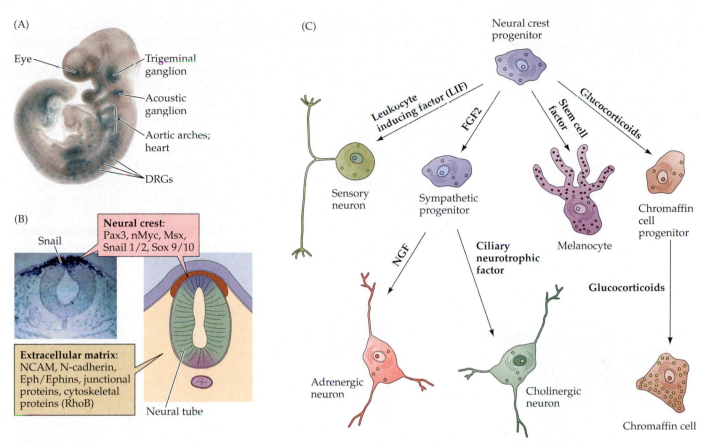

Figure 22.11 Migration and differentiation of the neural crest relies on regulation of gene expression and cell-cell signaling. (A) The migrating neural crest is visualized in a mouse embryo in which a reporter has been inserted into a gene that is normally expressed in the neural crest. Individual neural crest cells (blue) can be seen accumulating around the eye (where they will contribute to the pigment epithelium of the retina), the trigeminal (V) and acoustic (VIII) ganglia, the aortic arches and heart, and in the immature dorsal root ganglia (DRGs). (B) In a section through the spinal cord (left), neural crest cells labeled for the bHLH transcription factor Snail exit the dorsal spinal cord and stream toward the periphery. The schematic (right) indicates sites of action of molecules that favor the mesenchymal, migratory (for the neural crest), or neuroepithelial state. (C) Cell signaling during migration of neural crest cells influences progenitor identity and terminal differentiation. Each signal is available along a specific migratory route taken by subsets of neural crest cells (see Figure 22.2). (A courtesy of A. S. LaMantia; B courtesy of M. A. Nieto.)

opment to guide axonal growth and targeting (see Chapter 23). Thus, the surfaces of cells in the embryonic periphery display specialized adhesion molecules in the extracellular matrix like laminin or fibronectin, or adhesion molecules like neural cadherin (N-cadherin), neural cell adhesion molecule (NCAM), or the eph-receptors and their ligands, the ephrins (Figure 22.11A). In addition, secreted signals including neurotrophic molecules may influence the direction and trajectory of neural crest migration. Of particular significance is the fact that specific peptide hormone growth factors available in particular peripheral targets cause neural crest cells to differentiate into distinct phenotypes (Figure 22.11C; see also Figure 22.2). One of the consequences of these cues for generating peripheral neurons is the expression and activity of *bHLH* genes (see Figure 12.7). Clearly, these position-dependent cues, and their consequences for subsequent gene expression and differentiation are not restricted to the peripheral nervous system; in the cerebellum, for example, migratory neural precursors that form the external granule layer respond to similar cues. These cues modulate the expression of *bHLH* genes during the transition from migratory precursor to postmitotic neuroblast. Thus, the balance of migratory capacity, instructive cues, and modification of gene expression seen during the transit of the neural crest from the neural tube to the periphery illustrates the influence of migration on the establishment of neuronal identity.

Neuronal Migration in the Central Nervous System

Neuronal migration is not limited to the periphery. Neurons generated in several locations in the central nervous system must also move from the site of their initial genesis to a distant site, where they differentiate and are integrated into mature neural circuits. The mechanisms of central neuronal migration are diverse, and the successful completion of migration is essential for many aspects of normal brain function.

A minority of nerve and glial cells in the central nervous system (and a few in the periphery) use existing axon pathways as migratory guides. These include subsets of cranial nerve nuclei in the hindbrain; nuclei in the pons that project to the cerebellum; Schwann cells that myelinate peripheral axons; and a small population of neurons that migrate from the olfactory epithelium in the nose to the hypothalamus, where they secrete gonadotropin-releasing hormone (GnRH), which is essential for regulating reproductive functions in the mature animal.

The most prominent form of neuroblast migration within the central nervous system, however, is that guided by a particular type of glial cell. Many neuroblasts that migrate long distances within the central nervous system—including those in the cerebral cortex, cerebellum, and hippocampus—are guided to their final destinations by following the long processes of **radial glia** (Figure 22.12). Radial glial cells have multiple functions in the developing brain. In addition to

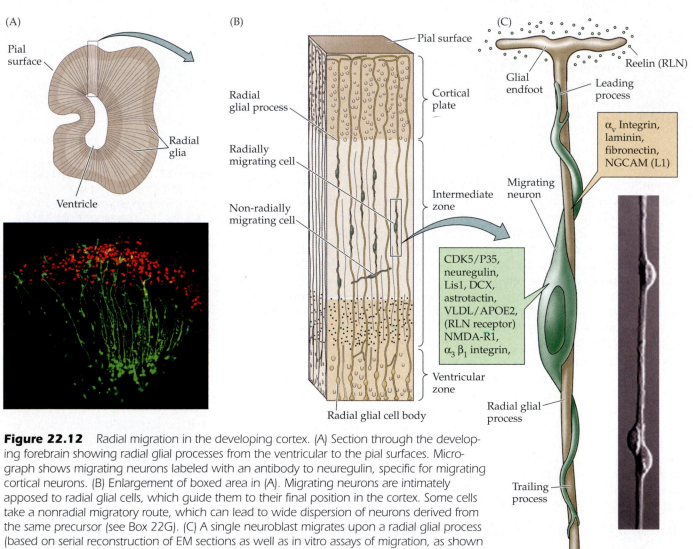

Figure 22.12 Radial migration in the developing cortex. (A) Section through the developing forebrain showing radial glial processes from the ventricular to the pial surfaces. Micrograph shows migrating neurons labeled with an antibody to neuregulin, specific for migrating cortical neurons. (B) Enlargement of boxed area in (A). Migrating neurons are intimately apposed to radial glial cells, which guide them to their final position in the cortex. Some cells take a nonradial migratory route, which can lead to wide dispersion of neurons derived from the same precursor (see Box 22G). (C) A single neuroblast migrates upon a radial glial process (based on serial reconstruction of EM sections as well as in vitro assays of migration, as shown in the accompanying micrograph). Cell adhesion and other signaling molecules or receptors found on the surface of either the neuron (green) or the radial glial process (tan) are indicated in the respective boxes. (After Rakic, 1974; micrographs courtesy of E. S. Anton and P. Rakic.)

acting as migratory guides, they are currently thought to be neuronal progenitor cells in the developing central nervous system (see also Chapter 24).

Histological observations of embryonic brains made by Wilhelm His and Ramón y Cajal during the nineteenth and early twentieth centuries suggested that neuroblasts in the developing cerebral cortical hemispheres followed glial guides to their final locations (Figure 22.12A). These light microscopic observations were supported by analyses of electron microscopic images of fixed tissue in the 1960s and 1970s (Figure 22.12B,C) as well as molecular labeling that identified the radial glial cells and migrating neurons as distinct cell classes. This apparent glial scaffold for radial movement of postmitotic neurons fit well with the orderly relationship between birthdates and final position of distinct cell types in the cerebral cortex as well as the cerebellum (see Figure 22.8 and Box 22F). By adhering to the radial glial process, a cell can move past already differentiating neurons in lower cortical layers (for example layers 5 and 6) and move toward the cortical surface, where they contact the endfeet of radial glial cells and disengage from the glial surface.

Innovations in cell culture techniques and light microscopy have made it possible to observe the process of radial glial-guided migration directly. When radial glial cells and immature neurons are isolated from the developing cerebellum or cerebral cortex and mixed together in vitro, the neurons attach to the glial cells, assume the characteristic shape of migrating cells seen in vivo, and begin moving along the glial processes. Indeed, the membrane constituents of glial cells, when coated onto thin glass fibers, support this sort of migration. Several cell surface adhesion molecules, extracellular matrix adhesion molecules, and associated signal transduction molecules apparently mediate this process. Many of these molecules are also essential for subsequent steps in neural development, such as axon growth, axon guidance, and synapse formation (see Chapters 8 and 23).

These observations suggest that the process of migration, particularly in the cortex, may be vulnerable to the effects of genetic mutations that disrupt either the ability of the nerve cell to move, the ability of radial glial cells to support migration, or both. A particularly compelling demonstration of the genetic bases of cortical migration has come from analysis of human patients with a variety of brain malformations that can be visualized using magnetic resonance imaging (Figure 22.13). Genetic analysis of these patients has identified mutations in several genes that cause disrupted cortical migration. In some cases (like that of the molecule Reelin, which is available near the glial endfeet and is thought to influence the detachment of neurons from the radial glia; see Figure 22.12C), mutations in humans result in a cortical phenotype (Figure 22.13B) very similar to that seen in mice in which the analogous gene is experimentally deleted. In patients with lissencephaly—"smooth brain," a condition in which the cortex has no sulci or gyri (Figure 22.13C)—novel genes have been identified, includ-

Figure 22.13 Mutations in genes that influence neuronal migration cause malformations of the human cerebral cortex. Yellow arrows point to the lateral ventricle, green arrows indicate the subcortical white matter in the internal capsule (a "thoroughfare" for axons entering and exiting the cerebral cortex), and red arrows highlight the normal appearance of sulci and gyri. (A) MRI image of normal cerebral cortex. (B) A patient with a mutation of the gene encoding Reelin, a protein that influences radial neuronal migration in the cortex. The lateral ventricles are enlarged, subcortical white matter is diminished, and the pattern of sulci and gyri is disrupted. (C) In a patient with a mutation in the *DCX* gene, the ventricles are dramatically enlarged, subcortical white matter is nearly absent, and sulci and gyri are completely absent. This dramatic cortical malformation is known clinically as lissencephaly ("smooth brain"). (Courtesy of C. A. Walsh.)

(A) Normal

(B) *Reelin* mutation

(C) Lissencephaly (*DCX*)

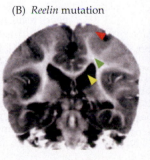
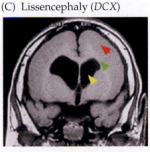

ing *doublecortin,* or *DCX.* DCX protein interacts with microtubules in migrating neurons and is thought to influence the integrity of the cytoskeleton in these cells as they move along glial guides.

Neuropathological observations as well as molecular and genetic studies indicate that some forms of mental retardation, epilepsy, and other neurological problems arise from the abnormal migration of cerebral cortical neurons. Mutations in additional genes that encode proteins associated with both neuronal migration and axon growth in the developing brain. In particular, neuregulin (a secreted signal), the neural cell adhesion molecule NCAM, and DISC1 (a gene that can be mutated or deleted in a small number of cases of schizophrenia) are highly associated with risk for psychiatric diseases. Thus, disrupted cell migration (and its consequences for subsequent development of brain circuits) may underlie the pathology of several serious brain disorders.

Summary

The initial development of the nervous system depends on an intricate interplay of inductive signals and cellular movements. In addition to the early establishment of regional identity and cellular position in the brain, substantial migration of neuronal precursors is necessary for the subsequent differentiation of distinct classes of neurons, as well as for the eventual formation of specialized patterns of synaptic connections (see Chapters 8 and 23). The fate of individual precursor cells is not determined simply by their mitotic history; rather, the information required for differentiation arises largely from interactions between the developing cells and the subsequent activity of distinct transcriptional regulators. All of these events depend upon the same categories of molecular and cellular phenomena: cell–cell signaling, changes in motility and adhesion, transcriptional regulation, and, ultimately, cell-specific changes in gene expression. The molecules that participate in signaling during early brain development are the same as the signals used by mature cells: hormones, transcription factors, and other second messengers (see Chapter 7), as well as cell adhesion molecules. The identification and characterization of these molecules in the developing brain has begun to explain a variety of congenital neurological defects as well as providing initial insight into the genetic and cellular basis of developmental disorders like autism psychiatric diseases. These associations with brain diseases most likely reflect the vulnerability of signaling and transcriptional regulation during early neural development to the effects of genetic mutations, and to the actions of the many drugs and toxins that can compromise the elaboration of a normal nervous system.

Additional Reading

Reviews

ANDERSON, D. J. (1993) Molecular control of cell fate in the neural crest: The sympathoadrenal lineage. *Annu. Rev. Neurosci.* 16: 129–158.

CAVINESS, V. S. JR. AND P. RAKIC (1978) Mechanisms of cortical development: A view from mutations in mice. *Annu. Rev. Neurosci.* 1: 297–326.

FRANCIS, N. J. AND S. C. LANDIS (1999) Cellular and molecular determinants of sympathetic neuron development. *Annu. Rev. Neurosci.* 22: 541–566.

HATTEN, M. E. (1993) The role of migration in central nervous system neuronal development. *Curr. Opin. Neurobiol.* 3: 38–44.

INGHAM, P. (1988) The molecular genetics of embryonic pattern formation in *Drosophila*. *Nature* 335: 25–34.

JESSELL, T. M. AND D. A. MELTON (1992) Diffusible factors in vertebrate embryonic induction. *Cell* 68: 257–270.

KESSLER, D. S. AND D. A. MELTON (1994) Vertebrate embryonic induction: Mesodermal and neural patterning. *Science* 266: 596–604.

KEYNES, R. AND R. KRUMLAUF (1994) Hox genes and regionalization of the nervous system. *Annu. Rev. Neurosci.* 17: 109–132.

KINTNER C. (2002) Neurogenesis in embryos and in adult neural stem cells. *J. Neurosci.* 22: 639–643.

LEWIS, E. M. (1992) The 1991 Albert Lasker Medical Awards. Clusters of master control genes regulate the development of higher organisms. *JAMA* 267: 1524–1531.

LINNEY, E. AND A. S. LAMANTIA (1994) Retinoid signaling in mouse embryos. *Adv. Dev. Biol.* 3: 73–114.

RICE, D. S. AND T. CURRAN (1999) Mutant mice with scrambled brains: Understanding the signaling pathways that control cell positioning in the CNS. *Genes Dev.* 13: 2758–2773.

RUBENSTEIN, J. L. R. AND P. RAKIC (1999) Genetic control of cortical development. *Cerebral Cortex* 9: 521–523.

SANES, J. R. (1989) Extracellular matrix molecules that influence neural development. *Annu. Rev. Neurosci.* 12: 491–516.

SELLECK, M. A., T. Y. SCHERSON AND M. BRONNER-FRASER (1993) Origins of neural crest cell diversity. *Dev. Biol.* 159: 1–11.

ZIPURSKY, S. L. AND G. M. RUBIN (1994) Determination of neuronal cell fate: Lessons from the R7 neuron of *Drosophila. Annu. Rev. Neurosci.* 17: 373–397.

Important Original Papers

ANCHAN, R. M., D. P. DRAKE, C. F. HAINES, E. A. GERWE AND A. S. LaMANTIA (1997) Disruption of local retinoid-mediated gene expression accompanies abnormal development in the mammalian olfactory pathway. *J. Comp. Neurol.* 379: 171–184.

ANGEVINE, J. B. AND R. L. SIDMAN (1961) Autoradiographic study of cell migration during histogenesis of cerebral cortex in the mouse. *Nature* 192: 766–768.

BULFONE, A., L. PUELLES, M. H. PORTEUS, M. A. FROHMAN, G. R. MARTIN AND J. L. RUBENSTEIN (1993) Spatially restricted expression of *Dlx-1, Dlx-2 (Tes-1), Gbx-2,* and *Wnt-3* in the embryonic day 12.5 mouse forebrain defines potential transverse and longitudinal segmental boundaries. *J. Neurosci.* 13: 3155–3172.

EKSIOGLU, Y. Z. AND 12 OTHERS (1996) Periventricular heterotopia: An X-linked dominant epilepsy locus causing aberrant cerebral cortical development. *Neuron* 16: 77–87.

ERICSON, J., S. MORTON, A. KAWAKAMI, H. ROELINK AND T. M. JESSELL (1996) Two critical periods of sonic hedgehog signaling required for the specification of motor neuron identity. *Cell* 87: 661–673.

GALILEO, D. S., G. E. GRAY, G. C. OWENS, J. MAJORS AND J. R. SANES (1990) Neurons and glia arise from a common progenitor in chicken optic tectum: Demonstration with two retroviruses and cell type-specific antibodies. *Proc. Natl. Acad. Sci. USA* 87: 458–462.

GRAY, G. E. AND J. R. SANES (1991) Migratory paths and phenotypic choices of clonally related cells in the avian optic tectum. *Neuron* 6: 211–225.

HAFEN, E., K. BASLER, J. E. EDSTROEM AND G. M. RUBIN (1987) *Sevenless,* a cell-specific homeotic gene of *Drosophila,* encodes a putative transmembrane receptor with a tyrosine kinase domain. *Science* 236: 55–63.

HEMMATI-BRIVANLOU, A. AND D. A. MELTON (1994) Inhibition of activin receptor signaling promotes neuralization in *Xenopus. Cell* 77: 273–281.

KRAMER, H., R. L. CAGAN AND S. L. ZIPURSKY (1991) Interaction of bride of sevenless membrane-bound ligand and the sevenless tyrosine-kinase receptor. *Nature* 352: 207–212.

LANDIS, S. C. AND D. L. KEEFE (1983) Evidence for transmitter plasticity in vivo: Developmental changes in properties of cholinergic sympathetic neruons. *Dev. Biol.* 98: 349–372.

LIEM, K. F. JR., G. TREMML AND T. M. JESSELL (1997) A role for the roof plate and its resident TGFβ-related proteins in neuronal patterning in the dorsal spinal cord. *Cell* 91: 127–138.

McMAHON, A. P. AND A. BRADLEY (1990) The *wnt-1 (int-1)* protooncogene is required for the development of a large region of the mouse brain. *Cell* 62: 1073–1085.

NODEN, D. M. (1975) Analysis of migratory behavior of avian cephalic neural crest cells. *Dev. Biol.* 42: 106–130.

PATTERSON, P. H. AND L. L. Y. CHUN (1977) The induction of acetylcholine synthesis in primary cultures of dissociated rat sympathetic neurons. *Dev. Biol.* 56: 263–280.

RAKIC, P. (1971) Neuron-glia relationship during granule cell migration in developing cerebral cortex: A Golgi and electronmicroscopic study in *Macacus rhesus. J. Comp. Neurol.* 141: 283–312.

RAKIC, P. (1974) Neurons in rhesus monkey visual cortex: Systematic relation between time of origin and eventual disposition. *Science* 183: 425–427.

SAUER, F. C. (1935) Mitosis in the neural tube. *J. Comp. Neurol.* 62: 377–405.

SPEMANN, H. AND H. MANGOLD (1924) Induction of embryonic primordia by implantation of organizers from a different species. Transl. V. Hamburger and reprinted in *Foundations of Experimental Embryology,* B. H. Willier and J. M. Oppenheimer (eds.) (1974). New York: Hafner Press.

STEMPLE, D. L. AND D. J. ANDERSON (1992) Isolation of a stem cell for neurons and glia from the mammalian neural crest. *Cell* 71: 973–985.

WALSH, C. AND C. L. CEPKO (1992) Widespread dispersion of neuronal clones across functional regions of the cerebral cortex. *Science* 255: 434–440.

YAMADA, T., M. PLACZEK, H. TANAKA, J. DODD AND T. M. JESSELL (1991) Control of cell pattern in the developing nervous system. Polarizing activity of the floor plate and notochord. *Cell* 64: 635–647.

ZIMMERMAN, L. B, J. M. DE JESUS-ESCOBAR AND R. M. HARLAND (1996) The Spemann organizer signal noggin binds and inactivates bone morphogenetic protein 4. *Cell* 86: 599–606.

Books

LAWRENCE, P. A. (1992) *The Making of a Fly: The Genetics of Animal Design.* Oxford: Blackwell Scientific.

MOORE, K. L. (1988) *The Developing Human: Clinically Oriented Embryology,* 4th Ed. Philadelphia: W. B. Saunders Company.

Chapter 23

Construction of Neural Circuits

Overview

Two central features of neural circuits emerge after neurons are generated and have migrated to their final positions. First, nerve cells in different regions become linked together via axon pathways. Second, orderly synaptic connections are made among appropriate pre- and postsynaptic partners. The cellular mechanisms that mediate axon growth and synapse formation are the major determinants of the orderly representation and processing of information in mature circuits throughout the peripheral and central nervous system. These events are critical for the genesis of nervous systems that can appropriately control behavior. The directed growth of axons and the recognition of appropriate synaptic targets depend critically upon the functions of a specialization at the tip of each growing axon called the growth cone. Growth cones detect and respond to signaling molecules that identify correct pathways, prohibit incorrect trajectories, and ultimately facilitate functional synaptic partnerships. These include cell surface adhesion molecules, extracellular matrix molecules, and diffusible signals that either attract or repel growing axons. In addition, specialized secreted neural growth factors influence axon growth and synapse formation and regulate appropriate numbers of connections between axons and their targets. These signals influence the growth cone by initiating signaling cascades that result either in local modifications of the cytoskeleton, growth cone surface, or gene expression. As in other instances of intercellular communication, a variety of receptors and second messenger molecules transduce the signals provided to the growth cone. These cell–cell signals initiate intracellular events that underlie directed growth of the axon, the conversion of the growth cone into a presynaptic specialization, and the elaboration of a distinct postsynaptic site. The dynamic interactions between growth cones and the embryonic environment, both in the periphery and within the brain, yield well-defined peripheral and central axon pathways, influence the establishment of topographic maps and other orderly representations of information in various neural systems, and insure the construction of complex neural circuits that allow animals to behave in ever more sophisticated ways as they mature.

The Axon Growth Cone

Among the many extraordinary features of nervous system development, one of the most fascinating is the ability of growing axons to navigate over millimeters or even centimeters, through complex embryonic terrain, to find appropriate synaptic partners. In 1910, Ross G. Harrison first observed axons extending in a living tadpole. He noticed that "the growing fibers are clearly endowed with considerable energy and have the power to make their way through the solid or semi-solid protoplasm of the cells of the neural tube. But we are at present in the dark with regard to the conditions which guide them to specific points." Subsequent efforts using increasingly sophisticated vital dyes and optical techniques with improved resolution have confirmed Harrison's initial description of the growing axons and their remarkable progress through the embryo (Figure 23.1A). Moreover, dendrites, especially primary dendrites like those of cortical pyramidal cells or cerebellar Purkinje cells must also extend over relatively long distances, and the manner in which they do so is very similar to that of axons.

Harrison recognized two fundamental features of axonal growth that continue to motivate efforts to understand the assembly of neural circuits in the embryonic nervous system. First, the "considerable energy … and power" of growing axons reflect the cellular properties of the **growth cone**, a specialized structure at the tip of the extending axon. Growth cones are highly motile structures that explore the extracellular environment, determine the direction of growth, and then guide the extension of the axon in that direction. The primary morphological characteristic of a growth cone is a sheetlike expansion of the growing axon at its tip called a **lamellipodium**. When growth cones are examined in vitro, lamellipodia can be seen clearly, as can numerous fine processes called **filopodia** that extend from lamellipodia (Figure 23.1B). Filopodia rapidly form and disappear from the terminal expansion, like fingers reaching out to

Figure 23.1 Growth cones guide axons in the developing nervous system. (A) Mauthner neurons (arrows) in the hindbrain of a zebrafish embryo, adjacent to the otic vesicles (OtV) which give rise to the sensory neurons of the inner ear. The inset shows a higher magnification of the hindbrain. The Mauthner neuron axons can be seen crossing the midline and extending down to the spinal cord. The right panel shows a single Mauthner axon labeled with a fluorescent dye in a living zebrafish embryo, led through the spinal cord by a relatively simple growth cone. Over 35 minutes, the axon advances approximately 50 μm. (B) A single dorsal root ganglion cell isolated in culture extends many processes called neurites, the tissue culture equivalent of axons and dendrites. Each neurite has a long shaft in which microtubules (green) predominate, tipped by a growth cone in which actin (red) is the major molecular constituent. (C) Growth cone shape varies at "decision" regions. In this example, from the drawings of Santiago Ramón y Cajal, growth cones from sensory relay neurons in the dorsal spinal cord are relatively simple near their origin ("C" in this drawing). They change their shape as they approach and cross the ventral midline (a major decision point) to form the spinothalamic tract ("A" and "B"). Once across the midline, they reassume the simpler morphology. (A from Takahashi et al., 2002; B courtesy of F. Zhou and W. D.Snider; C after S. Ramón y Cajal; courtesy of C. A. Mason.)

(A)

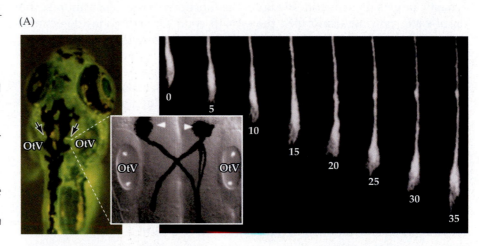

(B)   (C)

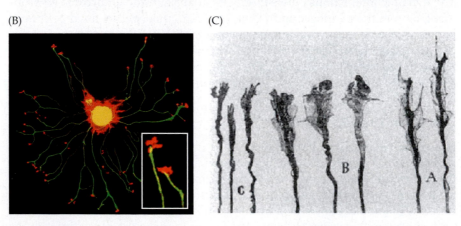

sense the environment. The lamellipodium and filopodia are distinguished from the axon shaft by different cytoskeletal molecules (Figure 23.1B; see the next section). Thus, the growth cone is a distinct, if transient, neuronal specialization whose activity is key for the construction of pathways and circuits in the developing brain. Indeed, once a growth cone reaches and recognizes an appropriate target it is gradually transformed into either a presynaptic ending for an axon, or the terminal domain of a dendrite.

Santiago Ramón y Cajal, Harrison's contemporary, observed in sections of fixed specimens that growth cones in established axon pathways presumably "pioneered" by other axons tend to be simple in shape. In contrast, when a "pioneer" axon (i.e., the first axon to extend through a given region, thus establishing that terrain as hospitable to axon growth) first extends in a new direction through a territory not previously innervated, or if it reaches a region where a choice must be made about the direction to take, the structure (and presumably the motility) of its growth cone changes dramatically (Figure 23.1C). The lamellipodium of an individual growth cone expands as it encounters a potential target and extends numerous filopodia; these actions suggest an active search for appropriate cues to direct subsequent growth. These changes of growth cone shape at "decision points" have been observed in both the peripheral and central nervous systems. In the periphery, the growth cones of both motor neurons and sensory neurons undergo shape changes as they enter the primordia of muscles in immature limbs, presumably facilitating the selection of appropriate targets in the developing musculature. In the central nervous system, growth cones in the cerebral commissures, olfactory nerve, and optic chiasm change shape when they reach critical points in their trajectories (Box 23A).

The Molecular Basis of Growth Cone Motility

Growth cone motility reflects rapid, controlled rearrangement of cytoskeletal elements. These elements include molecules related to the **actin cytoskeleton**, which regulates changes in lamellipodial and filopodial shape for directed growth, as well as the **microtubule cytoskeleton**, which is responsible for the elongation of the axon itself (Figure 23.2A,B). The molecular composition of both the actin and microtubule cytoskeleton change in distinct regions of the growing axon, suggesting a great deal of dynamism within growing neural processes (Figure 23.2B,C). Thus, the ways in which the actin and microtubule cytoskeleton are modified in growing axons are key for understanding how growth cones and axons extend.

Actin is the primary molecular constituent of a network of cellular filaments that are found in lamellipodia and filopodia. Tubulin is the primary molecular constituent of the microtubules that run parallel to the axis of the axon. Actin and tubulin are found in two basic forms in the growth cone and axon: freely soluble monomers in the cytoplasm; and polymers that form filaments (actin) or microtubules (tubulin). The dynamic polymerization and depolymerization of actin at the membrane of the lamellipodium, as well as within the filopodium sets the direction of growth cone movement, in part by generating local forces that orient the growth cone toward or away from local substrates. Similarly, the polymerization and depolymerization of tubulin into microtubules consolidates the direction of movement of the growth cone by stabilizing the axon shaft. The interface between the actin and microtubule cytoskeleton is particularly important for regulating the balance of active growth versus stability in a growing axon.

Figure 23.2 The structure and action of growth cones. (A) Distinct types of actin and tubulin are seen in discrete regions of the growth cone. Filamentous actin (F-actin, red) is seen in the lamellipodium and filopodia. Tyrosinated microtubules are the primary tubular constituents of the lamellar region (green), and acetylated microtubules are restricted to the axon itself (blue). (B) Dynamics of actin cytoskeleton (yellow, red) in a single growth cone imaged over an 8-hour interval. The distribution of filamentous actin, labeled here with a fluorescent actin-binding protein, changes in the region of the lamellipodium as well as in the filopodium. (C) The distribution and dynamics of cytoskeletal elements in the growth cone. Globular actin (G-actin) can be incorporated into F-actin at the leading edge of the filopodium in response to attractive cues. Repulsive cues support disassembly and retrograde flow of G-actin toward the lamellipodium. Organized microtubules make up the cytoskeletal core of the axon, while more broadly dispersed microtubule subunits are found at the transition between the axon shaft and the lamellipodium. Actin- and tubulin-binding proteins regulate the assembly and disassembly of subunits to filaments or tubules. This process is influenced by changes in intracellular Ca^{2+} via voltage regulated Ca^{2+} channels as well as transient receptor (TRP) channels. (D) Growth cones show rapid changes in Ca^{2+} concentration. In this example, a single filopodium (white arrowheads) undergoes a rapid increase in Ca^{2+}. (A courtesy of E. Dent and F. Gertler; B from Dent and Kalil, 2001; D from Gomez and Zheng, 2006.)

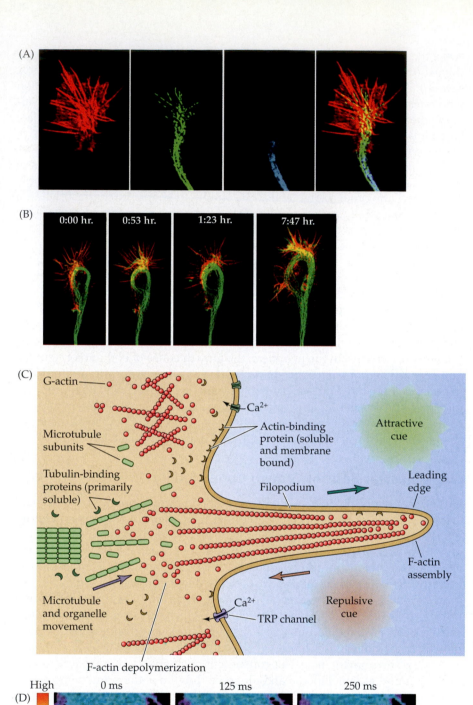

Several proteins regulate the polymerization and depolymerization of actin and tubulin by binding to these molecules and catalyzing posttranslational modifications, or by recruiting other enzymes that modify the primary molecular elements of the cytoskeleton. Actin-binding proteins are found throughout the growth cone cytoplasm. Most either bind actin directly, or modify actin monomers by phosphorylation and other posttranslational modifications. These molecules are particularly enriched at the inner surface of the growth cone

BOX 23A   *Choosing Sides: Axon Guidance at the Optic Chiasm*

The functional requirement that a subset of axons from retinal ganglion cells in each eye must cross while the remaining axons project to the ipsilateral side of the brain was predicted based on optical principles—most notably by Sir Isaac Newton in the seventeenth century—and confirmed (much later) by neuroanatomists and neurophysiologists (see Chapter 12). The partial crossing, or *decussation*, of retinal axons is most striking in primates, including humans, where approximately half of the axons cross and the other half does not. Although all other mammals also have crossed and uncrossed retinal projections, the percentage of uncrossed axons diminishes from 20–30 percent in carnivores to less than 5 percent in most rodents. The frequency of uncrossed axons decreases even more in other vertebrates; thus in amphibians, fish, and birds most or all of the retinal projection is crossed. For both functional and evolutionary reasons, the partial decussation of the retinal pathways and its variable extent in different species has engaged the imagination of biologists and others interested in vision.

For developmental neurobiologists, this phenomenon raises an obvious question: How do retinal ganglion cells "choose sides" such that some project contralaterally and others ipsilaterally? This question is central to understanding how the peripheral visual projection is

organized to construct two accurate visual hemifield maps that superimpose points of space seen jointly by the two eyes (see Chapter 12). It also speaks to the more general issue in neural development of how axons distinguish between ipsilateral and contralateral targets.

It is clear that the laterality of retinal axons is determined by initial cell identity and axon guidance mechanisms rather than by regressive processes that subsequently select or sculpt these projections. Thus, the distinction between the nasal and temporal retinal regions that project ipsilaterally and contralaterally is already apparent in the retina—as well as in axon trajectories at the midline and in the developing optic tract—long before the axons reach their targets. In the retina, this specificity is seen as a "line of decussation," or border, between ipsilaterally and contralaterally projecting retinal ganglion cells.

The line of decussation can be detected experimentally by injecting a retrograde tracer into the nascent optic tract of very young embryos. In the retinas of such embryos there is a distinct boundary between the population of retinal ganglion cells projecting ipsilaterally in one eye (found in the temporal retina), and a complementary boundary for contralaterally projecting cells in the other eye (see figure). A molecu-

lar basis for this specificity was initially suggested by studies of albino mammals, including mice and humans. In albinos, where single gene mutation disrupts melanin synthesis throughout the animal, including in the pigment epithelium of the retina, the ipsilateral component of the retinal projection from each eye is dramatically reduced, the line of decussation in the retina is disrupted, and the distribution of glia and other cells in the vicinity of the optic chiasm is altered. These and other observations suggested that identity of retinal axons with respect to decussation is established in the retina, and further reinforced by axonal "choices" influenced by cues provided by cells within the optic chiasm.

Analysis of growth cone cell morphologies has shown that the chiasm is indeed a region where growth cones explore the molecular environment in a

(Continued on next page)

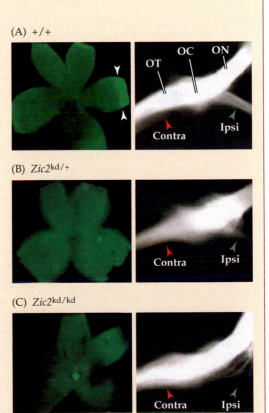

(A) +/+

(B) *Zic2*kd/+

(C) *Zic2*kd/kd

(A) A small population of *Zic2*-expressing retinal ganglion cells (arrowheads) is seen in the ventrotemporal region of the normal retina (at left, mounted flat by making several radial cuts). At right, the normal projection of one eye via the optic nerve (ON), through the optic chiasm (OC), and into the optic tract (OT) has been traced using a lipophilic dye placed in one eye. After the chiasm, labeled axons can be seen both in the contralateral (contra) as well as the ipsilateral (ipsi) optic tract. (B) When *Zic2* function is diminished in a mouse heterozygous for a *Zic2* "knockdown" mutation (in which expression of Zic2 protein is diminished, but not eliminated), the number of ipsilateral axons in the optic tract is similarly diminished. (C) When *Zic2* function is further diminished in homozygous *Zic2* knockdown mice, the ipsilateral projection can no longer be detected in the optic tract; thus, each of the optic tracts consist of contralateral axons. (From Herrera et al., 2003.)

BOX 23A   (Continued)

particularly detailed way, presumably to make choices pertinent to directed growth. Furthermore, molecular analysis reveals specialized neuroepithelial cells in and around the chiasm that express a number of cell adhesion molecules associated with axon guidance. Interestingly, some of these molecules—particularly netrins, slits, and their robo receptors—do not influence decussation in the chiasm as they do at other regions of the nervous system. Instead, they are expressed in cells where the chiasm forms, apparently constraining its location on the ventral surface of the diencephalon. The establishment of ipsilateral versus contralateral identity is evidently more dependent on the zinc finger transcription factor *Zic2*, as well as cell adhesion molecules of the ephrin family. *Zic2*, which is expressed specifically in the temporal retina, is associated with the expression of a distinct receptor, EphB1, in the axons arising from tempo-

ral retinal ganglion cells. The ephrin B2 ligand, which is recognized as a repellent of EphB1 axons, is found in midline glial cells in the optic chiasm. In support of the functional importance of these molecules, disrupting *Zic2*, *EphB1*, or *ephrin B2* gene function diminishes the degree of ipsilateral projection in developing mice; in accord with this finding, neither the *Zic2* nor the *ephrin B2* gene is expressed in vertebrate species that lack ipsilateral projections.

These observations provide a molecular framework for the identification of retinal ganglion cells and the sorting of their projections at the optic chiasm. How this sorting is related to the topography of tectal, thalamic, and cortical representations is not yet known. Most observations suggest that retinal topography is not faithfully preserved among axons in the optic tracts. The identity and position of axons from nasal and temporal retinas whose retinal ganglion cells

"see" a common point in the binocular hemifield must therefore be restored in the thalamus, and subsequently retained or re-established in the thalamic projections to cortex. Choosing sides at the chiasm is only a first step in establishing maps of visual space.

References

GUILLERY, R. W. (1974) Visual pathways in albinos. *Sci. Amer.* 230(5): 44–54.

GUILLERY, R. W., C. A. MASON AND J. S. TAYLOR (1995) Developmental determinants at the mammalian optic chiasm. *J. Neurosci.* 15: 4727–4737.

HERRERA, E. AND 8 OTHERS (2003) Zic2 patterns binocular vision by specifying the uncrossed retinal projection. *Cell* 114: 545–557.

RASBAND, K., M. HARDYV AND C. B. CHIEN (2003) Generating X: Formation of the optic chiasm. *Neuron* 39: 885–888.

WILLIAMS, S. E. AND 9 OTHERS (2003) Ephrin-B2 and EphB1 mediate retinal axon divergence at the optic chiasm. *Neuron* 39: 919–935.

plasma membrane, presumably to mediate assembly and membrane anchoring of actin filaments. The local anchoring of actin filaments is essential to generate the forces that create membrane extensions and direct the movement of the lamellipodium or filopodia (Figure 23.2C). Microtubule binding proteins are more concentrated in the axon shaft, and they modulate posttranslational modifications of monomeric and polymerized tubulin. The interaction of these proteins with tubulin is central for the axon to remain stable in the face of further exploration and growth of the growth cone.

The constant flux between monomeric actin and tubulin, and polymerized actin filaments and microtubules, is regulated via the binding proteins in response to signals from the environment. These signals are transduced by receptors and channels on the growth cone membrane surface, structures that influence the levels of intracellular messengers, particularly that of Ca^{2+} (Figure 23.2D). Indeed, the regulation of intracellular Ca^{2+} levels, either through voltage-regulated Ca^{2+} channels, transient receptor potential (TRP) channels activated by second messengers, or second-messenger pathways that mobilize intracellular Ca^{2+} stores, are thought to be a central mediator of actin and microtubule dynamics in the growing axon. Thus, the conditions that Ross Harrison suggested "guide [the growth cones] to specific points" are now understood to be changes in the cytoskeleton of the growth cone and axon, mediated through intracellular signaling cascades initiated by a broad range of adhesion molecules and diffusible signals available in the embryonic terrain through which the growth cone extends.

Non-Diffusible Signals for Axon Guidance

The complex behavior of growth cones during axonal extension suggests the presence of specific cues that cause the growth cone to move in a particular direction. In addition, the growth cone itself must have a specialized array of receptors and transduction mechanisms that respond to these cues. The identity of the cues themselves remained elusive for almost a century after Harrison's and Cajal's initial observations, but the identities of many of the relevant molecules have been established over the past 40 years. These signals comprise a large group of molecules associated with cell adhesion and cell–cell recognition throughout the organism, as well as with directed axon or growth cone motility in the developing nervous system. All of these molecules initiate intracellular signaling cascades that ultimately can change the actin or microtubule cytoskeleton, the complement of receptors and channels on the cell surface, or gene expression. The association of specific cell adhesion molecules with axon growth is based on experiments either in vitro, where addition or removal of a particular molecule results in modifying the relevant behavior of growing axons; or in vivo, where genetic mutation, deletion, or manipulation disrupts the growth, guidance, or targeting of a particular axon projection (see Box 23A).

Despite their daunting numbers, molecules that are known to influence axon growth and guidance can be grouped into families of ligands and their receptors (see Figure 23.2). The extracellular matrix molecules and their integrin receptors, the Ca^{2+}-independent cell adhesion molecules (CAMs), the Ca^{2+}-dependent cell adhesion molecules (cadherins), and the ephrins and eph receptors (see below) are the major classes of non-diffusible axon guidance molecules.

The **extracellular matrix cell adhesion molecules** were the first to be associated with axon growth. The most prominent members of this group are the **laminins**, the **collagens**, and **fibronectin**. As their family name indicates, all three of these signal types are found in the macromolecular complex or matrix outside of the cell (Figure 23.3A). The matrix components can be secreted by the cell itself or by its neighbors; however, rather than diffusing away from the cell after secretion, these molecules form polymers and create a more durable local extracellular substance. A broad class of receptors, known collectively as **integrins**, binds specifically to these molecules (see Figure 23.3A). Integrins themselves do not have kinase activity or any other direct signaling capacity. Instead, the binding of laminin, collagen, or fibronectin to integrins triggers a cascade of events—perhaps via interactions between the cytoplasmic domains of integrins with kinases and other signaling molecules—that generally stimulates axon growth and elongation.

The role of extracellular matrix molecules in axon guidance is particularly clear in the embryonic periphery. Axons extending through peripheral tissues grow through loosely arrayed mesenchymal cells that fill the interstices of the embryo, and the spaces between these mesenchymal cells are rich in extracellular matrix molecules. Peripheral axons also grow along the interface of mesenchyme and epithelial tissues, including the boundary between the neural tube and mesenchyme and the boundary between the mesenchyme and the epidermis, where organized sheets of extracellular matrix components called **basal lamina** provides a supportive substrate. In tissue culture as well as in the periphery of the embryo, the different extracellular matrix molecules have different capacities to stimulate axon growth. The role of these molecules in the central nervous system is less clear. Some of the same molecules are present in the extracellular spaces of the central nervous system, but are not organized into orderly substrates like the basal lamina in the periphery, and have therefore been harder to study.

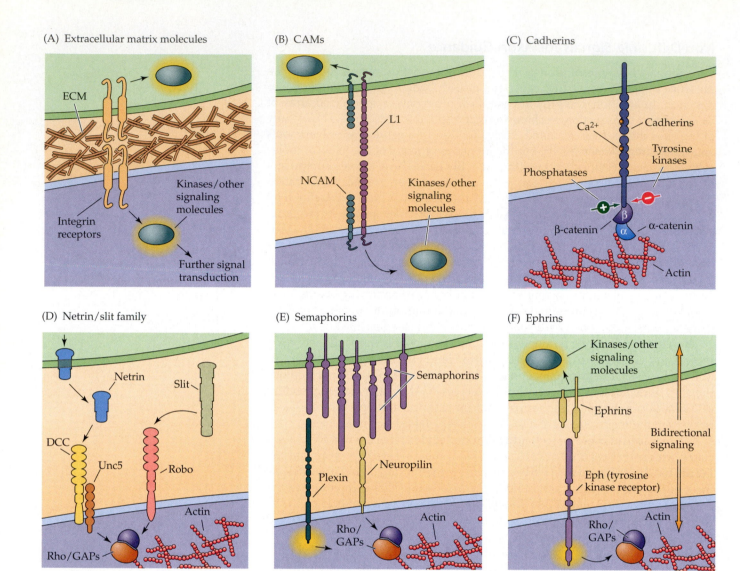

(A) Extracellular matrix molecules

ECM

Integrin receptors

Kinases/other signaling molecules

Further signal transduction

(B) CAMs

L1

NCAM

Kinases/other signaling molecules

(C) Cadherins

Ca^{2+} — Cadherins

Tyrosine kinases

Phosphatases

β-catenin

α-catenin

Actin

(D) Netrin/slit family

Netrin

Slit

DCC

Unc5

Robo

Actin

Rho/GAPs

(E) Semaphorins

Semaphorins

Plexin

Neuropilin

Actin

Rho/GAPs

(F) Ephrins

Kinases/other signaling molecules

Ephrins

Bidirectional signaling

Eph (tyrosine kinase receptor)

Actin

Rho/GAPs

Figure 23.3 *Several families of ligands and receptors constitute the major classes of axon guidance molecules. These ligand–receptor pairs can be either attractive or repulsive, depending on the identity of the molecules and the context in which they signal the growth cone. (A) Extracellular matrix molecules serve as the ligands for multiple integrin receptors. (B) Homophilic, Ca^{2+}-independent cell adhesion molecules (CAMs) are at once ligands and receptors. (C) Ca^{2+}-dependent adhesion molecules, or cadherins, are also capable of homophilic binding. (D) The netrin/slit family of attractive and repulsive secreted signals acts through two distinct receptors, DCC ("deleted in colorectal cancer"), which binds netrin, and robo, the receptor for slit. (E) Semaphorins are primarily repulsive cues that can either be bound to the cell surface or secreted. Their receptors (the plexins and neuropilin) are found on growth cones. (F) Ephrins, which can be transmembrane or membrane-associated, signal via the Eph receptors, which are receptor tyrosine kinases.*

The CAMs and cadherins are distinguished by their presence on growing axons and growth cones as well as surrounding cells or targets (Figure 23.3B,C). Moreover, both CAMs and cadherins have dual functions as ligands and receptors, usually via homophilic ("like with like") binding. Some of the CAMs, especially the L1 CAM, have been associated with the bundling, or *fasiculation*, of groups of axons as they grow. Cadherins have been suggested as important determinants of final target selection in the transition from growing axon to synapse (see below). For both CAMs and cadherins, the unique ability of each class to function as both ligand and receptor (for example, L1 is its own receptor) may be important for recognition between specific sets of axons and targets. Both CAMs and cadherins rely on a somewhat indirect route of signal transduction. The Ca^{2+}-independent CAMs interact with cytoplasmic kinases to initiate cellular responses, while the cadherins engage the APC/β-catenin pathway (also activated by Wnts; see Chapter 22).

The importance of adhesive interactions in axon growth and guidance is underscored by the pathogenesis of several inherited human developmental or neurological disorders. These syndromes include X-linked hydrocephalus, MASA (an acronym for *m*ental retardation, *a*phasia, *s*huffling gait, and *a*dducted thumbs), Kalman's syndrome (which compromises reproductive and chemosensory function) and X-linked spastic paraplegia. All are consequences of mutations in genes encoding the Ca^{2+}-independent CAMs. These mutations can also lead to the absence of the corpus callosum (referred to callosal agenesis), which connects the two cerebral hemispheres, and of the corticospinal tract, which carries cortical information to the spinal cord. These congenital anomalies (which are fortunately rare) are now understood to arise from errors in the signaling mechanisms normally responsible for axon navigation via cell surface adhesion molecules.

Diffusible Signals for Axon Guidance: Chemoattraction and Repulsion

Two major challenges in establishing appropriate patterns of connectivity are attracting axons to distant targets and insuring that the axons do not stray into inappropriate regions en route. With remarkable foresight, Cajal proposed early in the twentieth century that target-derived signals selectively influenced axonal growth cones, thereby attracting them to appropriate destinations. In addition to this *chemoattraction* predicted by Cajal, it was long supposed that there might also be *chemorepellent* signals that discouraged axon growth toward a particular region (Figure 23.4A).

Despite the clear importance of chemoattraction and repulsion in constructing pathways and circuits, the identity of the signals themselves remained uncertain until the last 15 years or so. One problem was the vanishingly small amounts of such factors expressed in the developing embryo. Another was that of distinguishing **tropic** molecules (which *guide* growing axons toward a source) from **trophic** molecules (which *support* the survival and growth of neurons and their processes once an appropriate target has been contacted). These problems were solved by laborious biochemical purification and analysis of attractive or repulsive activities from vertebrate (chick) embryos, and genetic analysis of axon growth in both *Drosophila* and *C. elegans*; this work eventually led to the identification of several genes that code for chemotropic factors. Remarkably, the identity and function of chemoattractants and chemorepellents across phyla is highly conserved.

The best characterized class of chemoattractant molecules is the **netrins** (Sanskrit, "to guide"; Figure 23.3D). In chick embryos, the netrins were identified as proteins with chemoattractant activity following biochemical purification. In *C. elegans*, netrins were first recognized as the product of a gene that influenced axon growth and guidance. The first such gene to be isolated was christened *Unc* for "uncoordinated," which describes the behavioral phenotype of the mutant worms; the cause is misrouted axons as a result of the absence of netrin. The netrins themselves have high homology to extracellular matrix molecules like laminin, and in some cases may actually interact with the extracellular matrix to influence directed axon growth.

Netrin signals are transduced by specific receptors including the molecule DCC (*d*eleted in *c*olorectal *c*ancer) as well as other co-receptors. Like many cell surface adhesion molecules, netrin receptors have repeated amino acid motifs in their extracellular domain, a transmembrane domain, and an intracellular domain with no known enzymatic activity. Thus, in order for netrin binding to cause changes in the target cell, other proteins with catalytic activity must interact with the cytoplasmic domain of DCC. Candidate enzymes include the focal adhesion kinase or FAK, which is known to modulate the actin cytoskeleton, and the tyro-

Figure 23.4 *Signals provided by the embryonic environment and their primary effects on growth cones and growing axons. (A) Diagram illustrating the basic signal classes. Chemoattractant, or tropic, signals (green pluses) can operate from a distance and reorient growth toward the source of the cue, often by acting on a pioneer growth cone that sets out a course distinct from the fasiculated followers. Similar cues acting on or near the surface of the axon shaft help maintain groups of axons as fasicles, which is essential for the formation of coherent nerves and tracts. Chemorepulsive signals (red minuses) can also act from a distance, or they act at regions where axons must defasiculate from a nascent nerve in order to change their trajectory or avoid an inappropriate target. Trophic signals (orange), support the further growth and differentiation of the axon and its parent nerve cell. (B) Calcium signaling is a key mediator of chemoattractive signaling. Here, the secreted chemoattractant netrin is applied to a growth cone in tissue culture at time 0. Within 1 minute, changes in Ca^{2+} concentrations are seen in the lamellipodium (white arrowheads); within 10 minutes, these changes intensify to the entire growth cone. (C) A growth cone in tissue culture collapses in the 30 minutes following application of semaphorin 3A, a chemorepellant molecule. This axon retracts each of its three lamellate growth cones and corresponding branches. (A after Huber et al., 2003; B from Gomez and Zheng, 2006; C from Dontchev and Letourneau, 2002.)*

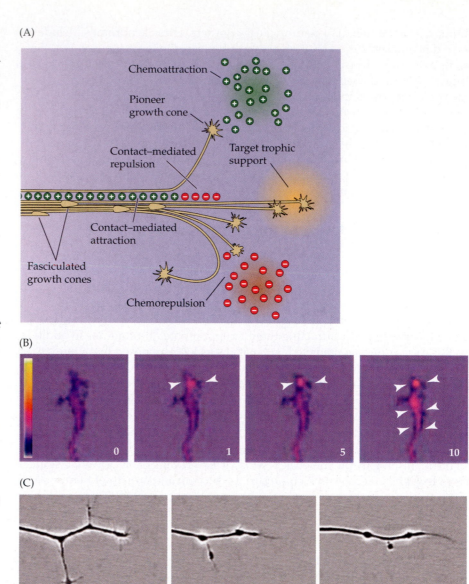

(A)

(B)

(C)

0 min   15 min after Sema3A   30 min after Sema3A

sine kinases SRC and Fyn, which are generally associated with cell adhesion signaling. The Rho/GAP family of signaling proteins, all of which modulate second messenger-mediated cytoskeletal modification, are thought to provide a final step for netrin signaling. Netrins are often found at the midline in the developing nervous system. Indeed, their initial characterization was guided by the observation that there seemed to be a chemoattractive signal in the spinal cord that influenced the growth of spinothalamic axons from the dorsal horn toward the ventral midline (Figure 23.5). After their initial purification and cloning, netrins were localized to the floorplate, which defines the ventral midline in the developing spinal cord. Consistent with their expression and in vitro activity, mutation of the *netrin-1* gene disrupts the development of axon pathways that cross the midline. These include the spinal cord anterior commissure as well as the corpus callosum, anterior commissure, and hippocampal commissure in the forebrain.

Since the midline is a point where there is overall symmetry in levels of netrin and other secreted factors, there has to be a mechanism to insure that

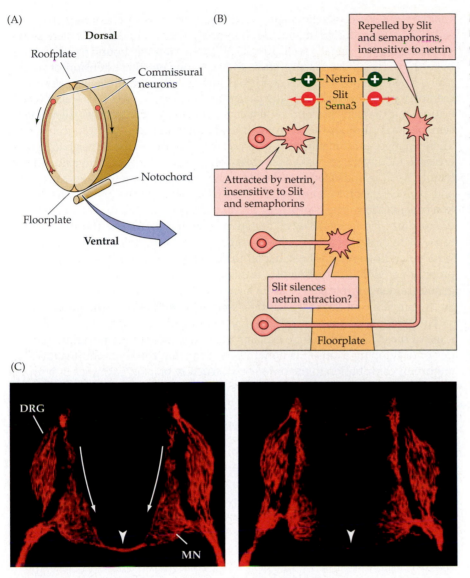

(A)

Dorsal

Roofplate

Commissural neurons

Notochord

Floorplate

Ventral

(B)

Repelled by Slit and semaphorins, insensitive to netrin

Netrin

Slit
Sema3

Attracted by netrin, insensitive to Slit and semaphorins

Slit silences netrin attraction?

Floorplate

(C)

DRG

MN

Figure 23.5 Netrins chemotropically regulate pathway formation in the developing spinal cord. (A) Commissural neurons send axons to the ventral region of the spinal cord, including the floorplate (see Figure 22.2). (B) Opposing activities of netrin and slit at the ventral midline of the spinal cord. This molecular guidance system ensures that the axons relaying pain and temperature via the anterolateral pathway cross the midline at appropriate levels of the spinal cord and remain on the contralateral side until they reach their targets in the thalamus. (C) At left, labeled commissural axons (red) descend through the spinal cord, pass the motor column (MC), and cross the midline into the anterior (ventral) commissure of the spinal cord. At right, the netrin gene of a mouse has been homozygously inactivated, and the commissural axons do not fasiculate, nor do they cross at the ventral midline (arrowheads). (A after Serafini et al., 1994; B after Dickinson, 2002; C from Serafini et al., 1996.)

once netrin directs axons across the midline, they do not cross back due to the ambiguous quantitative nature of signals at the midline. The secreted factor "slit" and its receptor "robo" (named for the phenotypes of *Drosophilia* mutants in which these genes were first identified) are important for preventing an axon from straying back over the midline once it has crossed initially in response to netrin. Slit and Robo are available immediately off the midline, and are thought to terminate the growth cone's sensitivity to netrin once it has crossed from one side of the CNS to the other. Thus slit, robo—and most likely several other molecules—and the signaling pathways they activate orchestrate the unidirectional crossing of axons at the midline. The successful crossing of axons is essential for the construction of some aspect of all major sensory, motor, and associational pathways in the mammalian brain (see also Box 23A).

Constructing the nervous system also entails telling axons where *not* to grow. Two additional broad classes of chemorepellent molecules beyond slit/robo (which serve a specific repulsive function) have been described. The first is associated with central nervous system myelin; molecules of this class are particularly important for axon regrowth following injury in the mature nervous system (see

Chapter 25). Molecules belonging to the second class of chemorepellents are active primarily during neural development. These molecules, called **semaphorins** (Greek, "signal"; see Figure 23.3E), are eventually bound to cell surfaces or to the extracellular matrix, where they can prevent the extension of nearby axons. Their receptors, like those for cell surface adhesion molecules, are transmembrane proteins (including the plexins and a protein called neuropilin) whose cytoplasmic domains have no known catalytic activity. Nevertheless, signaling by semaphorins through their receptors leads to changes in Ca^{2+} concentration. These changes presumably activate intercellular kinases and other signaling molecules to modify the growth cone cytoskeleton. Indeed, studies with cultured vertebrate neurons indicated that the semaphorins cause growth cones to collapse and axon extension to cease (see Figure 23.4C). Much of the initial characterization of semaphorin chemorepellent activity emerged from studies of invertebrates (particularly *Drosophila*) where mutation or manipulation of these genes can cause axons to grow abnormally. The activity of these molecules in developing vertebrates, however, has been harder to demonstrate in vivo. For example, inappropriate growth or targeting of axons is not apparent when single semaphorin genes are deleted in mice; nevertheless, local introduction of semaphorins can lead to altered axon trajectories as the axons avoid exogenous semaphorin.

The semaphorins represent the largest family of chemorepellants. Although none of these molecules alone explains the initial choices and resulting trajectories of developing axons, it is nonetheless clear that the semaphorins make an important contribution to the orderly construction of axon pathways in both the periphery and in the central nervous system.

The Formation of Topographic Maps

In the somatic sensory, visual, and motor systems, neuronal connections are arranged such that neighboring points in the periphery are represented at similarly adjacent locations in the appropriate regions of the central nervous system (see Chapters 9, 11, and 16). In other systems (e.g., the auditory and olfactory systems), there are also orderly representations of various stimulus attributes like frequency or receptor identity. How do growing axons distribute themselves with such fidelity once they reach appropriate target regions in the brain?

In the early 1960s, Roger Sperry, who later did pioneering work on the functional specialization of the cerebral hemispheres (see Chapter 27), articulated the **chemoaffinity hypothesis**, based primarily on work in the visual system of frogs and goldfish. In these animals, the terminals of retinal ganglion cells form a precise topographic map in the optic tectum (the tectum is homologous to the mammalian superior colliculus; Figure 23.6A). When Sperry crushed the optic nerve and allowed it to regenerate (fish and amphibians, unlike mammals, can regenerate axonal tracts in their central nervous system; see Chapter 25), he found that retinal axons reestablished the same pattern of connections in the tectum. Even if the eye was rotated 180°, the regenerating axons grew back to their original tectal destinations (causing some behavioral confusion for the frog; Figure 23.6B). Accordingly, Sperry proposed that each tectal cell carries a chemical "identification tag"; he further supposed that the growing terminals of retinal ganglion cells have complementary tags, such that they seek out a specific location in the tectum. In modern parlance, these "chemical tags" are cell adhesion or recognition molecules, and the "affinity" that they engender is a selective binding of receptor molecules on the growth cone to corresponding molecules on the tectal cells that signal their relative positions.

Further experiments in the amphibian and avian visual systems made the strictest form of the chemoaffinity hypothesis—labeling of each tectal location

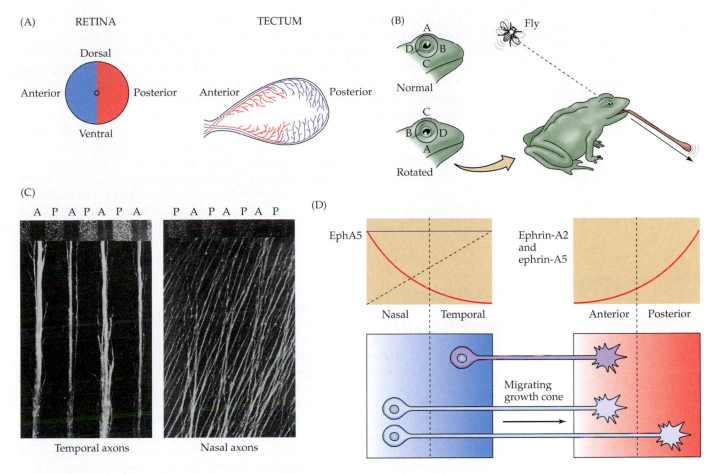

Figure 23.6 *Mechanisms of topographic mapping in the vertebrate visual system. (A) Posterior retinal axons project to the anterior tectum and anterior retinal axons to the posterior tectum. When the optic nerve of a frog is surgically interrupted, the axons regenerate with the appropriate specificity. (B) Even if the eye is rotated after severing the optic nerve, the axons regenerate to their original position in the tectum. This topographic constancy is evident from the frog's behavior: When a fly is presented above, the frog consistently strikes downward, and vice versa. (C) In vitro assay for cell surface molecules that contribute to topographic specificity in the optic tectum. A set of alternating, 90-µm wide stripes of membranes from anterior (A) and posterior (P) optic tectum of chicks was laid down on a glass coverslip. The posterior membranes have fluorescent particles added to make the boundaries of the stripes apparent (top of panels). Explants of retina from either nasal or temporal retina were placed on the stripes. Temporal axons prefer to grow on anterior membranes and are repulsed by posterior membranes. In contrast, nasal retinal axons grow equally well on both stripes. (D) Complementary gradients of Eph receptors (in afferent cells and their growth cones) and ephrins (in the target cells) lead to differential affinities and topograpic mapping. In this model, retinal growth cones with a high concentration of Eph receptors (for a mouse retina, EphA5) would be more likely to recognize a lower concentration of ligand (for the mouse superior colliculus, ephrins A2 and A5), whereas a growth cone with low Eph receptor concentration would recognize a higher concentration of ligand. (A,B after Sperry, 1963; C from Walter et al., 1987; D after Wilkinson, 2001.)*

by a different recognition molecule—untenable. Rather than displaying a precise "lock and key" affinity, the behavior of growing axons suggested that there are gradients of cell surface molecules to which growing axons respond to establish the basic axes of the retinotopic map. Normally, axons from the temporal region of the retina innervate the anterior pole of the tectum and avoid the posterior pole. Embryological experiments in which temporal and nasal regions of the retina or anterior and posterior regions of the tectum were reversed in their position suggested that there was some specificity. This specificity, however, was not absolute—if only posterior tectum was available to temporal retina axons, the axons would innervate the normally inhospitable target. Subsequent in vitro analysis showed that the specificity was generated by a comparison between different substrates. Temporal retinal axons, when presented with a choice of cell membranes derived from anterior or posterior tectal regions as a substrate (presented as an array of stripes affixed to a tissue culture substrate), grow exclusively on anterior membranes, avoiding membranes derived from the "wrong" region of the tectum (Figure 23.6C). The positive interactions probably are due to increased adhesion of the growth cones to the substrate, whereas the failure to grow into inappropriate regions may result from repulsive interactions that tend to collapse the growth cones.

A likely candidate for the negative guidance signal for temporal axons in the posterior tectum was subsequently purified and its gene cloned. The protein—initially called RAGS (repulsive *a*xon *g*uidance *s*ignal) and later re-named ephrin-A5—belongs to a family of **ephrin ligands** and **Eph receptors** (see Figure 23.3F). Ephrins and their receptors are found throughout the organism; however, in the developing nervous system, they seem to be especially important for development of topographic connections. In the eye and the tectum, ephrins and their receptors are distributed in complementary gradients so that similar levels of ligand and receptor are matched, thus leading to a topographic mapping of the nasal and temporal retina along the anterior–posterior axis of the tectum. Subsequent work has associated several members of this molecular family with topographic mapping in the visual system, as well as with the formation of axon pathways like the anterior commissure and with the migration of subpopulations of neural crest cells.

Ephrin ligands are cell adhesion-like molecules that can be either transmembrane or membrane-associated proteins. Eph receptors belong to the single transmembrane domain tyrosine receptor kinase family, and thus can directly transduce a signal from an Eph ligand. Subsequent work has also suggested that, upon binding with Eph receptors, ephrin ligands can generate intercellular signals in the cell expressing the ligand (therefore referred to as retrograde signalling) via interactions with cytoplasmic kinases and related molecules. Disruption of the genes for the ephrin ligands or their receptors results in subtle disruptions in the topographic organization of the retinocollicular or retinothalamic projection. These observations accord with the idea that chemoaffinity operates by a system of gradients in the retina and tectum that gives axons and their targets markers of general position within a system of coordinates (like N, S, E, and W on a map), rather than a one-to-one, lock-and-key sort of recognition. The Eph receptors and their ligands provide a model of how graded molecular information can organize topographic axonal growth in the visual system and other regions of the developing brain.

Selective Synapse Formation

Once they reach their correct target or target region, axons must make further, local determinations as to which particular cells to innervate among a variety of potential synaptic partners. The choices available to an axon include: establish synaptic contacts; retract and regrow to another target; or fail to form stable connections (a choice that can result in the death of the parent neuron). Because of the complexity of brain circuitry, this issue has been studied most thoroughly in the peripheral nervous system, particularly in the innervation of muscle fibers (Box 23B) and autonomic ganglion cells by spinal cord motor neurons.

The British physiologist John Langley first explored and defined synapse specificity at the end of the nineteenth century (see also Box 25A). Preganglionic sympathetic neurons located at different levels of the spinal cord inner-

BOX 23B Molecular Signals That Promote Synapse Formation

Synapses require a precise organization of presynaptic and postsynaptic elements in order to function properly (see Chapters 4–8). At the neuromuscular junction, for example, synaptic vesicles and the related release machinery are located at sites in the nerve terminal called active zones; and, in the postsynaptic muscle cell, acetylcholine receptors and other synapse-specific molecules are localized in high density in a region (the "postsynaptic density"; see Figure 23.7) exactly subjacent to the presynaptic active zones. During the past 25 years, a number of investigators have identified some of the molecular cues that guide the formation of these carefully apposed elements. Their efforts have met with the greatest success at the neuromuscular junction, where a molecule called agrin is now known to be responsible for initiating some of the events that lead to the formation of a fully functional synapse.

Agrin was originally identified as a result of its influence in the reinnervation of frog neuromuscular junctions following damage to the motor nerve. In mature skeletal muscle, each fiber typically receives a single synaptic con-

tact at a highly specialized region called the end plate (see Chapter 5). U. J. McMahan, Josh Sanes, and their colleagues at Harvard and later Stanford and Washington Universities found that regenerating axons reinnervate the original end plate site precisely. In seeking to determine the molecular signals underlying this phenomenon, they took advantage of the fact that each muscle fiber is surrounded by a sheath of extracellular matrix called the basal lamina. When muscle fibers degenerate, they leave the basal lamina behind (as do degenerating axons); moreover, a specific infolding of the basal lamina at the former end plate site allows its continued identification. Remarkably, presy-

naptic nerve terminals differentiate at these original sites even when the associated muscle fibers are absent. Equally remarkable is that regenerating muscle fibers form postsynaptic specializations—such as densely packed acetylcholine receptors—at precisely these same basal lamina locations in the absence of nerve fibers! These findings show that the signal(s) guiding synapse formation remain in the extracellular environment after removal of either nerve or muscle, presumably in the basal lamina "ghost" that surrounds each muscle fiber.

Using a bioassay based on the aggregation of acetylcholine receptors to ana-

(Continued on next page)

Development of neuromuscular junctions in agrin-deficient mice. Diaphragm muscles from control (left) and agrin-deficient (right) mice at embryonic day 15, 16, and 18 were double-stained for acetylcholine receptors and axons, then drawn with a camera lucida. The developing muscle fibers run vertically. In both control and mutant muscles, an intramuscular nerve (black) and aggregates of AChRs (red) are present by embryonic day 15. In controls, axonal branches and AChR clusters are confined to a band at the central end plate at all stages. Mutant AChR aggregates are smaller, less dense, and less numerous; axons form fewer branches and their synaptic relationships are disorganized. (From Gautam et al., 1996.)

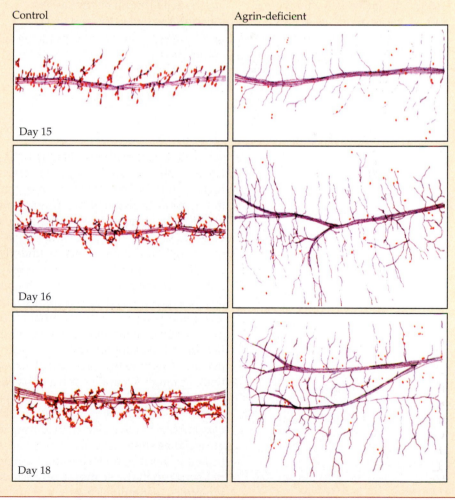

Control   Agrin-deficient

Day 15

Day 16

Day 18

BOX 23B *(Continued)*

lyze the constituents of the basal lamina, McMahan and colleagues isolated and purified agrin. Agrin is a proteoglycan found in both mammalian motor neurons and muscle fibers; it is also abundant in brain tissue. The neuronal form of agrin is synthesized by motor neurons, transported down their axons, and released from growing nerve fibers. Agrin binds to a postsynaptic receptor whose activation leads to a clustering of acetylcholine receptors and, evidently, to subsequent events in synaptogenesis. Support for the role of agrin as an organizer of synaptic differentiation is the finding by Sanes and his collaborators that genetically engineered mice lacking the gene for agrin develop in utero with few neuromuscular junctions (see figure). Importantly, mice lacking only neural agrin were as severely impaired as mice lacking both nerve and muscle agrin. Animals missing the agrin receptor also fail to develop neuromus-

cular junctions and die at birth. Agrin is therefore one of the first examples of a presynaptically derived molecule that promotes postsynaptic differentiation in target cells.

Because synapse formation requires an ongoing dialogue between pre- and postsynaptic partners, it is likely that postsynaptically derived organizers of presynaptic differentiation also exist. Based on the studies of basal lamina mentioned above, Sanes and his collaborators identified one such group of molecules, the β2-laminins (originally called S-laminin). Mice lacking β2-laminin show deficits in differentiation of motor nerve terminals and, unexpectedly, of terminal-associated glial (Schwann) cells. However, presynaptic defects in β2-laminin mutants are considerably less severe than postsynaptic defects in agrin mutants, suggesting that additional important retrograde signals remain to be identified.

References

BURGESS, R. W., Q. T. NGUYEN, Y.-J. SON, J. W. LICHTMAN AND J. R. SANES (1999) Alternatively spliced isoforms of nerve- and muscle-derived agrin: Their roles at the neuromuscular junction. *Neuron* 23: 33–44.

DECHIARA, T. M. AND 14 OTHERS (1996) The receptor tyrosine kinase MuSK is required for neuromuscular junction formation in vivo. *Cell* 85: 501–512.

GAUTAM, M. AND 6 OTHERS (1996) Defective neuromuscular synaptogenesis in agrin-deficient mutant mice. *Cell* 85: 525–535.

MCMAHAN, U. J. (1990) The agrin hypothesis. *Cold Spring Harbor Symp. Quant. Biol.* 50: 407–418.

NOAKES, P. G., M. GAUTAM, J. MUDD, J. R. SANES AND J. P. MERLIE (1995) Aberrant differentiation of neuromuscular junctions in mice lacking s-laminin/laminin β2. *Nature* 374: 258–262.

PATTON, B. L., A. Y. CHIU AND J. R. SANES (1998) Synaptic laminin prevents glial entry into the synaptic cleft. *Nature* 393: 698–701.

SANES, J. R., L. M. MARSHALL AND U. J. MCMAHAN (1978) Reinnervation of muscle fiber basal lamina after removal of myofibers. *J. Cell Biol.* 78: 176–198.

vate cells in sympathetic chain ganglia in a stereotyped and selective manner (see Chapter 21). In the superior cervical ganglion, for example, cells from the highest thoracic level (T1) innervate ganglion cells that project in turn to targets in the eye, whereas neurons from a somewhat lower level (T4) innervate ganglion cells that cause constriction of the blood vessels of the ear. Since the axons of all these neurons run together in the cervical sympathetic trunk to arrive at the ganglion, the mechanisms underlying the differential innervation of the ganglion cells must occur at the level of synapse formation rather than axon guidance to the general vicinity of target cells. Anticipating Sperry by more than 50 years (albeit in a different context), Langley concluded that selective synapse formation is based on differential affinities of the pre- and postsynaptic elements. Subsequent studies based on intracellular recordings from individual neurons in the superior cervical ganglion have shown, however, that the selective affinities between pre- and postsynaptic neurons are not especially restrictive: although synaptic connections to ganglion cells made by preganglionic neurons of a particular spinal level are preferred, synaptic contacts from neurons at other levels are not excluded (much like the rules that govern axon guidance). Despite this relative selectivity during synapse formation, a quite different line of work has shown that *where* a synapse forms on the target cell is tightly controlled by a set of molecules that are now understood in some detail. Perhaps not surprisingly, these molecules include variants of several of the cell adhesion molecules that influence growth cone behavior.

There are some absolute restrictions to synaptic associations. Thus, neurons do not innervate nearby glial or connective tissue cells, and many instances have been described in which various nerve and target cell types show little or no inclination to establish connections with one another. When synaptogenesis does proceed, however, neurons and their targets in both the central and peripheral nervous systems appear to associate according to a continuously variable system of preferences—as in the old song "if you can't be with the one you love, love the one you're with." Such biases guide the pattern of innervation that arises in development (or reinnervation during regeneration; see Chapter 25) without limiting it in an absolute way. The target cells residing in muscles, autonomic ganglia, or in the brain are certainly not equivalent, but neither are they unique with respect to the innervation they can receive. This relative promiscuity can cause problems following neural injury, since regenerated patterns of peripheral innervation are not always appropriate. It most likely reflects the fact that potential pre- and postsynaptic sites share many molecules that identify them as potential locations for making a connection. Thus, if there is a not a strong specific affinity, it is highly likely that generic recognition events will occur and anomalous connections will be made.

Much of this imprecision may reflect the overlapping subset of molecules that regulate synapse formation as well as axonal (and dendritic) growth. Several observations show that many of the same adhesion molecules that participate in axon guidance contribute to the identification and stabilization of a synaptic site on target cells, as well as to the ability of a growing axon to recognize specific sites as optimal (Figure 23.7). Thus, in the first stages of synapse formation, the cadherin family of Ca^{2+}-dependent adhesion molecules is thought to be essential for the recognition of a suitable synaptic site on a dendrite, cell body, or other appropriate target (i.e., muscle fibers; see Box 23B) by a nascent presynaptic process (derived by the conversion of a growth cone). In the next step, several additional molecules are thought to establish the clustering of synaptic vesicles and differentiation of an active zone as well as the construction of a postsynaptic density.

The **postsynaptic density** is a specialized intercellular junction that localizes neurotransmitter receptors and other relevant channels and signaling proteins to facilitate the postsynaptic response to neurotransmitter. Two molecules are particularly central to these events: **neurexin**, an adhesion molecule found in the presynaptic membrane; and its binding partner **neuroligin**, an adhesion molecule found in the postsynaptic membrane. In addition to the ability of neurexin and neuroligin to bind one another and promote adhesion between the pre- and postsynaptic membranes, neurexin has a specialized transmembrane domain that helps localize synaptic vesicles, docking proteins, and fusion molecules contributed by active zone vesicles in the presynaptic terminal. Neuroligin has a similar function for the postsynaptic site; it interacts with specialized postsynaptic proteins to promote the clustering of receptors and channels of the postsynaptic density as the synapse matures (Figure 23.7C).

The molecules depicted in Figure 23.7 are shared by all developing synapses (perhaps helping to explain the "love the one your with" phenomenon of synapse formation discussed above); however, there is diversity among developing synapses. Several candidate molecules have been proposed for the genesis of this specificity. Ephrins and their Eph receptors have been suggested to contribute to this process, as have cadherins. In both cases the diversity of these adhesion molecule families (there are at least 9 ephrin ligands and 15 Eph receptors, as well as a large number of cadherin family members) makes these adhesion molecules attractive candidates. Alternative splicing and posttranslational modification of several synaptic proteins may provide further distinctions between individual synapses.

Figure 23.7 Molecular mechanisms involved in synapse formation. (A) Initiation of a synapse depends centrally upon local recognition between the presumptive pre- and postsynaptic membranes mediated by members of the cadherin and protocadherin family of Ca²⁺ cell adhesion molecules (cadherins and protocadherins). This local recognition is accompanied by the initial accumulation of synaptic vesicles as well as transport vesicles that contain molecular components that contribute to the presynaptic active zone. (B) Once the initial specialization is established, additional adhesion molecules are recruited including Synaptic Cell Adhesion Molecule (SynCAM), a member of the Ca²⁺-independent, homophilic binding adhesion molecule family (like NCAM; see Figure 23.3B), neurexin and neuroligin (see panel C), and the ephrinB ligands and their EphB receptors. Adhesive signaling between these molecules initiates differentiation of the presynaptic active zone and the postsynaptic density. (C) The interaction of neurexin (a presynaptic transmembrane adhesion protein) with neuroligin (a postsynaptic adhesion protein) is central for recruiting and retaining cytoskeletal elements that localize synaptic vesicles to the presynaptic terminal and mediate their fusion. In addition, neurexin is important for localizing voltage-gated Ca²⁺ channels to insure local vesicle release. Neuroligin, upon binding neurexin, is essential for localizing neurotransmitter receptors and postsynaptic proteins to the postsynaptic specialization. (A,B after Waites et al, 2005; C after Dean and Dreshbach, 2006.)

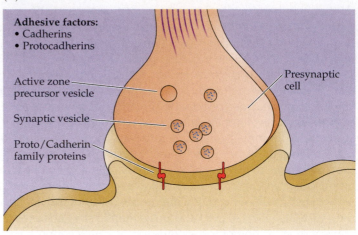

(A)

Adhesive factors:
• Cadherins
• Protocadherins

Active zone precursor vesicle

Synaptic vesicle

Proto/Cadherin family proteins

Presynaptic cell

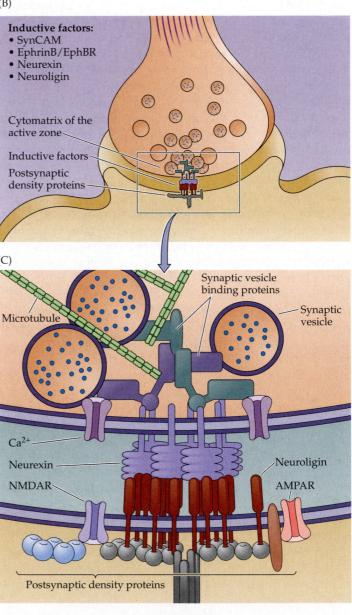

(B)

Inductive factors:
• SynCAM
• EphrinB/EphBR
• Neurexin
• Neuroligin

Cytomatrix of the active zone

Inductive factors

Postsynaptic density proteins

(C)

Synaptic vesicle binding proteins

Microtubule

Synaptic vesicle

Ca²⁺

Neurexin

Neuroligin

NMDAR

AMPAR

Postsynaptic density proteins

(A) DSCAM

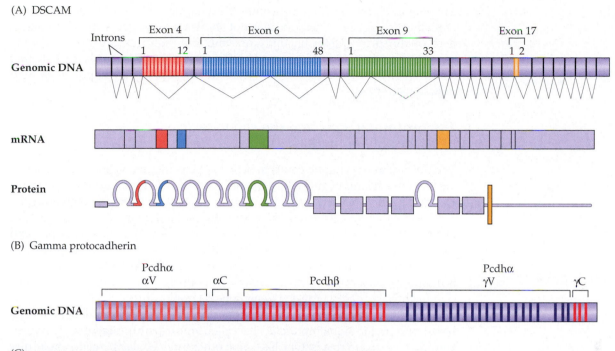

(B) Gamma protocadherin

(C)

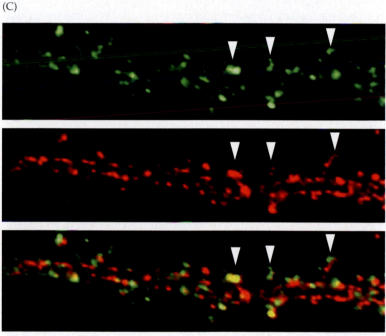

Figure 23.8   Potential molecular mediators of synapse identity. (A) Organization of the *DSCAM* gene in *Drosophila*. Each of four multiple-exon regions (4, 6, 9, and 17) has several alternative splice variants, and different combinations of these four regions yields a potential 38,000 isoforms of the DSCAM protein that can be expressed at distinct synaptic sites in the fly's developing nervous system. (B) Similar variability of multiple alternative exons is seen in the mammalian gene for γ-protocadherin. (C) Distinct γ-protocadherin isoforms (green and red) are expressed at subsets of synaptic contacts on dendrites of hippocampal neurons in culture, suggesting that different synaptic sites may have different complements of adhesion molecules perhaps conferring specify to those synaptic junctions. (A after Schmucker et al., 2000; B after Wang et al., 2002; C from Phillips et al., 2003.)

Speculation about adhesion molecule specificity and synapse formation has an intriguing parallel in *Drosophila*. In the fly, the gene for the cell adhesion molecule DSCAM (the fly ortholog of the mammalian *d*own *s*yndrome *c*ell *a*dhesion *m*olecule, the gene for which is located on chromosome 21, the chromosome that is duplicated in Down syndrome) has approximately 38,000 isoforms based upon the numbers of exons in the gene and predicted splicing (Figure 23.8A). In the fly, DSCAM is expressed at synaptic sites in the developing nervous system. It is not yet clear whether or not individual splice isoforms are differentially expressed at distinct synaptic sites; however, if this is the case, the genomic diversity may contribute to synaptic diversity. While the mammalian orthologue of DSCAM does not show a similar diversity, some members of the protocadherin family do, and the isoforms are not uniformly expressed at neighboring synapses (see Figure 23.8B,C). Thus, it is possible that splice isoforms of molecules like protocadherins invest synaptic sites with unique identities.

Trophic Interactions and the Ultimate Size of Neuronal Populations

The formation of synaptic contacts between growing axons and their synaptic partners marks the beginning of a new stage of development. Once synaptic contacts are established, neurons become dependent in some degree on the presence of their targets for continued survival and differentiation; in the absence of synaptic targets, the axons and dendrites of developing neurons atrophy and the innervating nerve cells may eventually die. This long-term dependency between neurons and their targets is referred to as **trophic interaction**. *Trophic* is from the Greek *trophé*, meaning, roughly, "nourishment." Despite this nomenclature, the sustenance provided to neurons by trophic interactions is not the sort derived from metabolites such as glucose or ATP. Rather, the dependence is based on specific signaling molecules called **neurotrophic factors**, or **neurotrophins**. Neurotrophic factors, like some other intercellular signaling molecules (mitogens that promote cell proliferation and cytokines that regulate inflammation and immune responses, for example), originate from target tissues and regulate differentiation, growth, and ultimately survival in nearby cells. In the nervous system, neurotrophins are unique in that, unlike inductive signaling molecules and cell adhesion molecules, their expression is limited primarily to neurons as well as some non-neural targets like muscles, and they are first detected after the initial populations of postmitotic neurons have been generated in the nascent central and peripheral nervous systems.

Why should neurons depend so strongly on their targets, and what specific cellular and molecular interactions mediate this dependence? The answer to the first part of this question lies in the changing scale of the developing nervous system and the body it serves, and the related need to precisely match the number of neurons in particular populations with the size of their targets. The basic mechanisms by which neurons are initially generated have already been considered in Chapter 22. There is, however, one more issue in establishing the final complement of neurons. A general—and surprising—strategy in the development of vertebrates is the production of an initial surplus of nerve cells (on the order of two- or threefold); the final population is subsequently established by the death of those neurons that fail to interact successfully with their intended targets (see below). The elimination of supernumerary neurons, particularly the initiation of **apoptosis**, the highly regulated processes that result in cell death, is mediated by neurotrophins (see below and Chapter 25).

An ongoing series of studies dating from the start of the twentieth century showed that targets play a major role in determining the size of the neuronal populations that innervate them, presumably based upon their production of neurotrophic factors. The seminal observation was that the removal of a limb bud from a chick embryo results, at later embryonic stages, in a striking reduction in the number of nerve cells (α motor neurons) in the corresponding portions of the spinal cord (Figure 23.9A,B). Apparently, there are signals in the limb (the target of spinal cord motor neuron axons) that ensure the survival of the motor neurons. Furthermore, in normal embryos, there is an apparent surplus of motor neurons that disappears by the end of early development. These

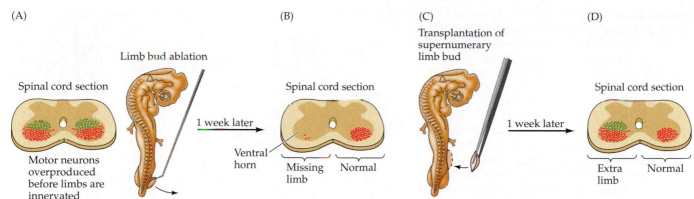

Figure 23.9 Target-derived trophic support regulates survival of related neurons. (A) The chicken spinal cord generates an excess of neurons (green) prior to the differentiation and innervation of the limb. Normally, some of these neurons are lost once the appropriate level of innervation is established in the developing limb bud. Limb bud amputation in a chick embryo at the appropriate stage of development (about 2.5 days of incubation) further depletes the pool of motor neurons that would have innervated the missing extremity. (B) Cross section of the lumbar spinal cord in an embryo that underwent this surgery about a week earlier. The motor neurons (dots) in the ventral horn that would have innervated the hindlimb degenerate almost completely after embryonic amputation; a normal complement of motor neurons is present on the other side; most of the normally supernumerary neurons have been lost. (C) Adding an extra limb bud before the normal period of cell death rescues early-generated neurons that normally would have died. (D) Such augmentation leads to an abnormally large number of limb motor neurons on the side related to the extra limb, and these neurons are recruited from the pool of cells overproduced at an earlier stage of development (green dots) rather than generated de novo through cell proliferation elicited by the added target. (After Hamburger, 1958, 1977, and Hollyday and Hamburger, 1976.)

supernumerary neurons are generated prior to the growth and innervation of the limb, and die due to a lack of trophic support as the motor pool becomes matched to the needs of the developing limb musculature.

A further interpretation of these observations is that neurons (in the spinal cord in this case) compete with one another for a resource present in the target (the developing limb) that is available in limited supply. In support of this idea, many neurons that would normally die can be rescued by augmenting the amount of target available, thereby providing extra trophic support—in this example, by adding another limb that can be innervated by the same spinal segments that innervate the normal limb (Figure 23.9C,D). Careful monitoring of cell proliferation versus cell death shows that the extra cells are not generated de novo (i.e., they do not arise from precursors in response to a mitogenic signal from the extra target). Instead, it is clear that they are "rescued" from a cell population that is overproduced in early development and normally winnowed based on the limited trophic support from the targets. Thus, the size of nerve cell populations in the adult is not fully determined by a rigid genetic program of cell proliferation followed by highly specified innervation of the target. The connections between nerve cells and their targets can be modified by idiosyncratic neuron–target interactions in each developing individual.

Further Competitive Interactions in the Formation of Neuronal Connections

Once the size of a neuronal population is established by trophic regulation, trophic interactions continue to modulate the formation of synaptic connections, beginning in embryonic life and extending far beyond birth. Among the problems that must be solved during the establishment of innervation is ensuring that the right number of remaining axons innervates each target cell, and that each axon innervates the right number of target cells. Getting these numbers right is another major achievement of trophic interactions between developing neurons and target cells, and is necessary for establishing appropriate circuits to support specific functional demands of each individual organism.

Studying synaptic refinement in the complex circuitry of the cerebral cortex or other regions of the central nervous system is a formidable challenge. As a result, many basic ideas about the ongoing modification of developing brain circuitry have come from simpler, more accessible systems, most notably the vertebrate neuromuscular junction and autonomic ganglion cells (Figure 23.10). Adult skeletal muscle fibers and neurons in some classes of autonomic ganglia (parasympathetic neurons) are each innervated by a single axon. Initially, however, each of these target cells is innervated by axons from several neurons, a

Figure 23.10 The number and pattern of synapses is adjusted during the first few weeks of postnatal life in the mammalian peripheral nervous system. In muscles (A) and in peripheral ganglia whose neurons have no dendrites (B), each axon innervates a larger number of target cells at birth than in maturity. Most of this rudimentary multiple innervation is eliminated shortly after birth. In both muscles and ganglia, however, the size and complexity of the terminal arbor that remains on each mature target cell increases. Thus, each axon elaborates more and more terminal branches and synaptic endings on the target cells it will innervate in maturity. The common denominator of this process is not a net loss of synapses, but the removal of immature contacts from all but one or a few axons on each target, and the focus on fewer target cells by a progressively increasing amount of synaptic machinery for each axon that remains. (After Purves and Lichtman, 1980.)

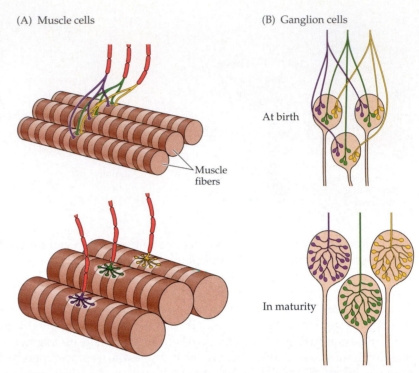

(A) Muscle cells

Muscle fibers

(B) Ganglion cells

At birth

In maturity

condition termed **polyneuronal innervation**. In such cases, inputs are gradually lost during early postnatal development until only one remains. This process of loss is generally referred to as **synapse elimination**, although the elimination actually refers to a reduction in the number of different axonal *inputs* to the target cells, not to a reduction in the overall number of synapses made on the postsynaptic cells.

In fact, the overall number of synapses (defined as the individual specialized junctions between pre- and postsynaptic cells) in the peripheral nervous system increases steadily during the course of development, as is the case throughout the brain. A variety of experiments have shown that the elimination of some initial inputs to muscle and ganglion cells is a process in which synapses originating from different neurons compete with one another for "ownership" of an individual target cell (see Box 23B). Patterns of electrical activity in the pre- and postsynaptic partners are thought to mediate this competition for target space. For example, if acetylcholine receptors at the neuromuscular junction are blocked by administering the potent AChR antagonist curare (see Chapter 6), polyneuronal innervation persists. Blocking presynaptic action potentials in the motor neuron axons (by silencing the nerve with tetrodotoxin, a sodium channel blocker) also prevents the reduction of polyneuronal innervation. Blocking neural activity therefore reduces or delays competitive interactions and the associated synaptic rearrangements.

Many useful insights into the nature of synaptic rearrangement during development have come from direct observations of this process. Using different colored fluorescent dyes that stain either the presynaptic terminal or the postsynaptic receptors, Jeff Lichtman and his colleagues followed the same neuromuscular junction over days, weeks, or longer (Figure 23.11). Their observations yielded some unexpected results. Competition between synapses arising from different motor neurons does not involve the active displacement of the "losing" input by the eventual "winner." Instead, it appears that the inputs of

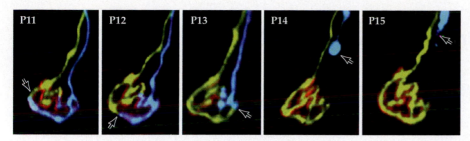

Figure 23.11 The elimination of multiple innervation at the neuromuscular junction. In this series of images, the same neuromuscular junction from a neonatal mouse has been imaged repeatedly beginning on the eleventh postnatal day (P11). Initially, two axons (green and blue) innervate the muscle fiber (the local clustering of postsynaptic ACh receptors is shown in red). The arrow indicates the boundary between the postsynaptic territory of the green and blue axon. By P12, the proportion of territory occupied by the green and blue axon has begun to shift, with the blue axon terminal apparently loosing postsynaptic space while the green axon terminal expands. This process continues at P13, and by P14 the blue axon has fully retreated, it's synaptic terminal transformed into a large retraction bulb (arrow at P14). Within an additional day the retracting axon is almost fully withdrawn from the synaptic site.

the two competitors initially occupy a subregion of the nascent postsynaptic specialization, but gradually segregate further. The "losing" axon eventually atrophies and retracts from the synaptic site; this is accomplished by a loss of the corresponding postsynaptic specializations associated with the "loser." Neurotransmitter receptors beneath the terminal branches that eventually will be eliminated are also lost. This receptor loss occurs before the nerve terminal has withdrawn and presumably reduces the synaptic strength of the input, which causes a further loss of postsynaptic receptors, leading to further reduction in the strength of the input. The downward spiral of synaptic efficacy presumably results in withdrawal of the presynaptic terminal. The remaining terminals continue to enlarge and strengthen in place as the end plate region expands during postnatal muscle growth.

A similar reorganization of synapse number and distribution is evident in a variety of other peripheral and central nervous system regions. In the peripheral nervous system, the numbers of presynaptic axons innervating target neurons also decrease, as demonstrated by studies of autonomic ganglia. A similar process has been described in the central nervous system. In the cerebellum, each adult Purkinje cell is innervated by a single climbing fiber (see Chapter 18); however, during early development, each Purkinje cell receives multiple climbing fiber inputs. Finally, in the visual cortex, initial binocular innervation of cells is eliminated to establish segregated, molecularly driven inputs (see Chapter 24).

Thus the pattern of synaptic connections that emerges in the adult is not simply a consequence of the biochemical identities of synaptic partners or other determinate developmental rules. Rather, the mature wiring plan is the result of a much more flexible process in which neuronal connections are formed, removed, and remodeled according to local circumstances that reflect molecular constraints, the detailed structure and size of the target, and ongoing electrical activity. These interactions guarantee that every target cell is innervated—and continues to be innervated—by the right number of inputs and synapses, and that every innervating axon contacts the right number of target cells with an appropriate number of synaptic endings. This regulation of **convergence** (the number of inputs to a target cell) and **divergence** (the number of connections made by a neuron) in the developing nervous system is another key consequence of trophic interactions among neurons and their targets. The regulation of convergence and divergence by neurotrophic interactions is also influenced by the shape and size of neurons, particularly the elaboration of dendrites (Box 23C); neurotrophins also control this aspect of circuit differentiation (see below).

Trophic interactions thus regulate three essential steps in the formation of mature neural circuits: they match numbers of neurons to the available target space; they regulate the degree of innervation of individual afferents and their postsynaptic partners; and they modulate the growth and shape of axonal and dendritic branches.

BOX 23C   Why Do Neurons Have Dendrites?

Perhaps the most striking feature of neurons is their diverse morphology. Some classes of neurons have no dendrites at all; others have a modest dendritic arborization; still others have an arborization that rivals the complex branching of a fully mature tree (see Figures 1.2 and 1.6). Why should this be? Although there are many reasons for this diversity, neuronal geometry influences the number of different inputs that a target neuron receives by modulating competitive interactions among the innervating axons.

Evidence that the number of inputs a neuron receives depends on its geometry has come from studies of the peripheral autonomic system, where it is possible to stimulate the full complement of axons innervating an autonomic ganglion and its constituent neurons. This approach is not usually feasible in the central nervous system because of the anatomical complexity of most central circuits. Since individual postsynaptic neurons can also be labeled via an intracellular recording electrode, electrophysiological measure-

ments of the number of different axons innervating a neuron can routinely be correlated with target cell shape. In both parasympathetic and sympathetic ganglia, the degree of preganglionic convergence onto a neuron is proportional to its dendritic complexity. Thus, neurons that lack dendrites altogether are generally innervated by a single input, whereas neurons with increasingly complex dendritic arborizations are innervated by a proportionally greater number of different axons (see figure). This correlation of

(A)

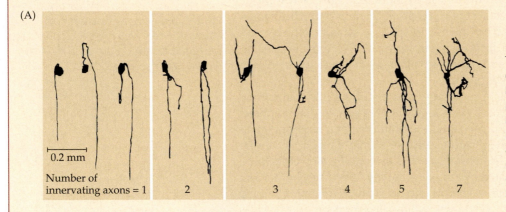

0.2 mm

Number of
innervating axons = 1   2   3   4   5   7

The number of axons innervating ciliary ganglion cells in adult rabbits. (A) Neurons studied electrophysiologically and then labeled by intracellular injection of a marker enzyme have been arranged in order of increasing dendritic complexity. The number of axons innervating each neuron is indicated.

Molecular Basis of Trophic Interactions

The three major functions of neurotrophic signaling—survival of a subset of neurons from a considerably larger population, subsequent formation and maintenance of appropriate numbers of connections, and the elaboration of axonal and dendritic branches to support these connections—can be rationalized in part by the supply and availability of trophic factors. These rules entail several general assumptions about neurons and their targets (which may be other neurons, or muscles and other peripheral structures). First, neurons depend on the availability of some minimum amount of trophic factor for survival, and subsequently for the persistence of appropriate numbers of target connections. Second, target tissues synthesize and make available to developing neurons appropriate trophic factors. Third, targets produce trophic factors in limited amounts; in consequence, the survival, persistence, and differentiation of developing neurons depends on neuronal competition for the available factor. One much-studied trophic protein, **nerve growth factor** (**NGF**), has provided support for these assumptions. Although the story of nerve growth factor does not explain all aspects of trophic interactions, it has been a useful paradigm for understanding in more detail the manner in which neural targets influence the survival and connections of the nerve cells that innervate them.

Rita Levi-Montalcini and Viktor Hamburger, working at Washington University in St. Louis, discovered NGF in the early 1950s. Based on experiments

neuronal geometry and input number holds within a single ganglion, among different ganglia in a single species, and among homologous ganglia across a range of species. Since ganglion cells that have few or no dendrites are initially innervated by several different inputs (see text), confining inputs to the limited arena of the developing cell soma evi-

dently enhances competition between them, whereas the addition of dendrites to a neuron allows multiple inputs to persist in peaceful coexistence. Importantly, the dendritic complexity of at least some classes of autonomic ganglion cells is influenced by neurotrophins.

A neuron innervated by a single axon will clearly be more limited in the scope

of its responses than a neuron innervated by 100,000 inputs (1 to 100,000 is the approximate range of convergence in the mammalian brain). By regulating the number of inputs that neurons receive, dendritic form greatly influences function.

References

HUME, R. I. AND D. PURVES (1981) Geometry of neonatal neurons and the regulation of synapse elimination. *Nature* 293: 469–471.

PURVES, D. AND R. I. HUME (1981) The relation of postsynaptic geometry to the number of presynaptic axons that innervate autonomic ganglion cells. *J. Neurosci.* 1: 441–452.

PURVES, D. AND J. W. LICHTMAN (1985) Geometrical differences among homologous neurons in mammals. *Science* 228: 298–302.

PURVES, D., E. RUBIN, W. D. SNIDER AND J. W. LICHTMAN (1986) Relation of animal size to convergence, divergence and neuronal number in peripheral sympathetic pathways. *J. Neurosci.* 6: 158–163.

SNIDER, W. D. (1988) Nerve growth factor promotes dendritic arborization of sympathetic ganglion cells in developing mammals. *J. Neurosci.* 8: 2628–2634.

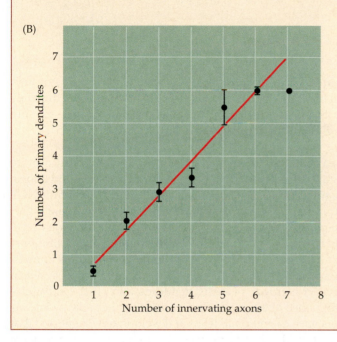

(B)

(B) This graph summarizes observations on a large number of cells. There is a strong correlation between dendritic geometry and input number. (After Purves and Hume, 1981.)

involving the survival of motor neurons after removal of developing limb buds (see Figure 23.8), the two researchers made an informed guess that target tissues provided some sort of signal to the relevant neurons. They reasoned further that limited amounts of such an agent from target tissue most likely explained the apparently competitive nature of nerve cell death. They then undertook a series of experiments to explore the source and nature of the postulated signal, focusing on dorsal root and sympathetic ganglion neurons rather than on spinal cord neurons.

Earlier, a former student of Hamburger's had removed a limb from a chick embryo and replaced it with a piece of mouse tumor. This experiment's surprising outcome was that the tumor apparently furnished an even more potent neuronal stimulus than the limb itself, causing an enlargement of the sensory and sympathetic ganglia that normally innervate the appendage. In further experiments, Levi-Montalcini and Hamburger provided evidence that the tumor (a mouse sarcoma—a tumor derived from muscle cells) secreted a soluble factor that stimulated the survival and growth of both sensory and sympathetic ganglion cells. Levi-Montalcini then devised a bioassay for the presumed agent and, in collaboration with Stanley Cohen, isolated and characterized the molecule—which had by then been named "nerve growth factor" for its ability to induce the massive outgrowth of neurites from explanted ganglia (Figure 23.12A). (The term *neurite* is used to describe neuronal branches when it is not

(A)   (B)   (C)   (D)

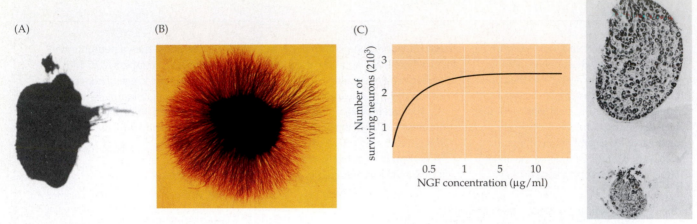

Figure 23.12   Effect of the neurotrophin NGF on the outgrowth of neurites and survival of neurons. (A) A chick sensory ganglion taken from an 8-day-old embryo and grown in organ culture for 24 hours in the absence of NGF. Few, if any, neuronal branches grow out into the medium in which the explant is embedded. (B) A similar ganglion in identical culture conditions 24 hours after the addition of NGF to the medium. NGF stimulates a halo of neurite outgrowth from the ganglion cells. (C) NGF influences the survival of newborn rat sympathetic ganglion cells grown in culture for 30 days. Dose-response curves confirm the dependence of these neurons on the availability of NGF. (D) Cross section of a superior cervical ganglion from a normal 9-day-old mouse (top) compared to a similar section from a littermate that was injected daily since birth with NGF antiserum (bottom). The ganglion of the treated mouse shows marked atrophy, with obvious loss of nerve cells. (A,B from Purves and Lichtman, 1985, courtesy of R. Levi-Montalcini; C after Chun and Patterson, 1977; D from Levi-Montalcini, 1972.)

known whether they are axons or dendrites.) NGF was identified as a protein and was purified from a rich biological source, the salivary glands of the male mouse. Subsequently, its amino acid sequence and three-dimensional structure were determined, and cDNAs encoding NGF have been cloned in several species.

Support for the idea that NGF is important for neuronal survival in more physiological circumstances emerged from a number of further observations. Depriving developing mice of NGF (by the chronic administration of an NGF antiserum or other strategies) resulted in adult mice who lacked most NGF-dependent neurons (Figure 23.12B). Conversely, injection of exogenous NGF into newborn rodents caused enlargement of sympathetic ganglia, an effect opposite to that of NGF deprivation. Neurons in ganglia in treated animals were both more numerous and larger; there was also more neuropil between cell bodies, suggesting an overgrowth of axons, dendrites, and other cellular elements. The dramatic influence of NGF on cell survival, together with what was known about the significance of neuronal death in development, suggested that NGF is indeed a target-derived signal that serves to match the number of nerve cells to the number of target cells.

The ability of NGF to support neuronal survival (and of NGF deprivation to enhance cell death) is not in itself unassailable proof of a physiological role for this factor in development. In particular, these observations provided no direct evidence for NGF synthesis by (and uptake from) neuronal targets. This gap was filled by another series of ingenious experiments in several laboratories that showed NGF to be present in sympathetic targets, and to be quantitatively correlated with the density of sympathetic innervation. Furthermore, messenger RNA for NGF was demonstrated in targets innervated by sympathetic and sensory ganglia, but not in the ganglia themselves or in targets innervated by other types of nerve cells. As might be expected from such specificity, the NGF-sensitive neurons were also shown to have receptor molecules for the trophic factor, although the precise molecular identity of those receptors remained unclear for several decades (see the next sections). Importantly, the NGF message appears only after ingrowing axons have reached their targets; this fact makes it unlikely that secreted NGF acts in vivo as a chemotropic (guidance) molecule in the way that netrins and other cell adhesion molecules discussed earlier do. Finally, the great majority of sympathetic neurons are lacking in mice in which the gene encoding NGF has been deleted.

In sum, several decades of work in a number of laboratories demonstrated that NGF mediates cell survival among two specific neuronal populations: sympathetic neurons and a subpopulation of sensory ganglion cells. This conclusion is supported by observations such as the death of the relevant neurons in the absence of NGF; the survival of a surplus of neurons in the presence of augmented levels of the factor; the presence and production of NGF in neuronal targets; and the existence of receptors for NGF in innervating nerve terminals. Indeed, these observations define the criteria that must be satisfied in order to conclude that a given molecule is indeed a neurotrophic factor.

Although NGF remains the most thoroughly studied neurotrophic factor, it was apparent from the outset that only certain classes of nerve cells respond to NGF. Furthermore, diversity of trophic factors most certainly would facilitate greater specificity between pre- and postsynaptic partners during development (or in regeneration; see Chapter 25). Work from a number of laboratories in the 1980s and 1990s showed that NGF is only one member of a family of related trophic molecules, the **neurotrophins**. Neurotrophins are similar to a broader class of signalling molecules found throughout the organism, referred to generically as "growth" or "trophic" factors. The expression and activity of the neurotrophins, however, is limited to neurons and their targets. At present, there are three well-characterized members of the neurotrophin family in addition to NGF: **brain-derived neurotrophic factor** (**BDNF**), **neurotrophin-3** (**NT-3**), and **neurotrophin 4/5** (**NT-4/5**) (Box 23D). Although several neurotrophins are homologous in amino acid sequence and structure, they are encoded by distinct genes, and are very different in their specificity (Figure 23.13A). For example, NGF supports the survival of (and neurite outgrowth from) sympathetic neurons, while another family member—BDNF—cannot. Conversely, BDNF, but not NGF, can support the survival of certain sensory ganglion neurons, which have a different embryonic origin. NT-3 supports both of these populations. Given the diverse systems whose growth and connectivity must be coordinated during neural development, this specificity makes good sense. Indeed, different neurotrophins are selectively available in different targets. For example the different receptor specializations in the skin that transduce somatosensory information express different neurotrophins, and this specificity is matched by expression of neurotrophin receptors (see the next section) that distinguish the sensory neurons that innervate each distinct structure (Figure 23.13B).

Aside from the neurotrophins, there are other secreted molecules (i.e., growth factors) with neurotrophic influences. These include the ciliary neurotrophic factor (CNTF), which is considered a cytokine because of its role in inflammation and immune responses beyond neurotrophic interactions, leukemia inhibitory factor (LIF), also a cytokine, and glial-derived growth factor (GGF), which is actually a neuregulin, a secreted signal that influences a variety of adhesive processes in the developing nervous system (see Figure 22.12). Thus, a wide variety of factors—some specific to the nervous system, others used for purposes beyond neural development—influence the survival and growth of developing nerve cells and therefore the elaboration of neural circuits.

Neurotrophin Signaling

All of the biological observations of neurotrophic interactions suggest that signaling via the neurotrophins will activate at least three different kinds of responses: cell survival/death; synapse stabilization/elimination; and neural process growth/retraction. The specific role of NGF in axon growth was initially demonstrated by experiments that presented NGF from a specific source and elicited directed growth toward the source, as well as experiments that isolated NGF availability to subsets of neural processes without exposing the cell body

BOX 23D The Discovery of BDNF and the Neurotrophin Family

During the 30 years or so of research that showed nerve growth factor to be a molecule fulfilling all the criteria for a target-derived neurotrophic factor, it also became clear that NGF affected only a few specific populations of peripheral neurons. It was therefore presumed that other neurotrophic factors must exist to support the survival and growth of other classes of neurons. In particular, whereas NGF was shown to be secreted by the peripheral targets of primary sensory and sympathetic neurons, other factors were presumably produced by target neurons in the brain and spinal cord that supported the central projections of sensory neurons.

The serendipity of the mouse salivary gland and its extraordinary levels of NGF was not repeated for these additional factors, however, and the hunt for the neurotrophic factors presumed to act in the central nervous system proved to be a long and arduous one. Indeed, it was not until the 1980s that the pioneering work of Yves Barde, Hans Thoenen, and their colleagues succeeded in identifying and purifying a factor from the brain that they named brain-derived neurotrophic factor (BDNF). As with NGF, this factor was purified on the basis of its ability to promote the survival and neurite outgrowth of sensory neurons. However, BDNF is expressed at such vanishingly small levels that over a million-fold purification was necessary

before the protein could be identified!

Thereafter, microsequencing and recombinant DNA technology allowed rapid progress even from the scant amounts of purified BDNF protein that were available. By 1989, Barde's group had succeeded in cloning the cDNA for BDNF. Surprisingly—despite its entirely different origin and distinct neuronal specificity—BDNF turned out to be a close relative of NGF. Based on the homologies between the primary structures of NGF and BDNF, the following year six independent laboratories (including Barde's) reported the cloning of a third member of the neurotrophin family, neurotrophin-3 (NT-3). At present, four members of the neurotrophin family have been reported in a variety of vertebrate species (see text).

Experiments on BDNF and other members of the neurotrophin family over last decade support the conclusion that, in both the periphery and the CNS, survival and growth of different neuronal populations depend on different neurotrophins, and that these relationships are mediated by expression of membrane receptors specific for each neurotrophin (see figure). However, the dramatic relationship between the survival of neuronal populations and neurotrophins has not been found in the CNS, which is where BDNF, NT-3, and NT-4/5, along with their receptors, are

primarily expressed. The most striking demonstration of this difference has been in "knockout" mice in which individual genes encoding neurotrophins or Trk receptors have been deleted. While these genetic deletions have led to predictable deficits in the PNS (see text), generally they have had minimal impact on CNS structure and function.

Thus the role of neurotrophins in the central nervous system remains uncertain. One possibility is that, in the CNS, neurotrophins are more involved in regulating neuronal differentiation and phenotype than in supporting neuronal survival per se. In this regard, the expression of neurotrophins is tightly regulated by electrical and synaptic activity, suggesting that they may also influence experience-dependent processes during the formation of central nervous system circuits.

References

HOFER, M. M. AND Y.-A. BARDE (1988) Brain-derived neurotrophic factor prevents neuronal death *in vivo. Nature* 331: 261–262.

HOHN, A., J. LEIBROCK, K. BAILEY AND Y.-A. BARDE (1990) Identification and characterization of a novel member of the nerve growth factor/brain-derived neurotrophic factor family. *Nature* 344: 339–341.

HORCH, H. W., A. KRUITTGEN, S. D. PORTBURY AND L. C. KATZ (1999) Destabilization of cortical dendrites and spines by BDNF. *Neuron* 23: 353–364.

LEIBROCK, J. AND 7 OTHERS (1989) Molecular cloning and expression of brain-derived neurotrophic factor. *Nature* 341: 149–152

SNIDER, W. D. (1994) Functions of the neurotrophins during nervous system development: What the knockouts are teaching us. *Cell* 77: 627–638.

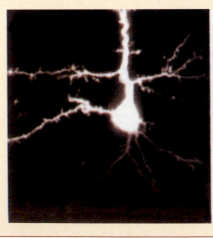
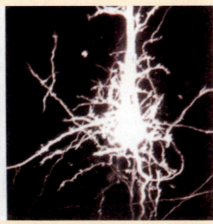

Neurotrophins influence dendritic arbors in the developing cerebral cortex. The cell on the left was transfected with the GFP gene alone, while the one on the right was transfected with GFP plus the gene encoding BDNF. Within a day, the BDNF-transfected neurons grew elaborate dendritic branches reminiscent of the NGF-induced halo in peripheral ganglia (see Figure 23.12B). (From Horch et al., 1999.)

to the factor (Figure 23.14A). The result of these experiments indicated that NGF could act locally to stimulate neurite growth—even while other processes of the same cell, deprived of NGF, are retracting. Apparently, general availability of NGF to an individual neuron does not support growth of all of that neuron's processes. Similar observations have been made for other neurotrophins. For example, BDNF can influence neurite growth by altering local Ca^{2+} signaling in

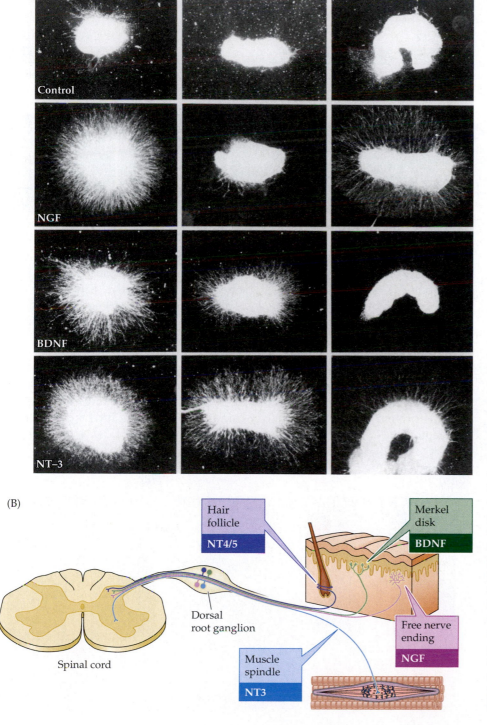

Figure 23.13 Neurotrophins have distinct effects on different target neurons. (A) Effect of NGF, BDNF, and NT-3 on the outgrowth of neurites from explanted, dorsal root ganglia (left column), nodose ganglia (middle column), and sympathetic ganglia (right column). The specificities of these several neurotrophins are evident in the ability of NGF to induce neurite outgrowth from sympathetic and dorsal root ganglia, but not from nodose ganglia (which are cranial nerve sensor ganglia that have a different embryological origin from dorsal root ganglia); of BDNF to induce neurite outgrowth from dorsal root and nodose ganglia, but not from sympathetic ganglia; and of NT-3 to induce neurite outgrowth from all three types of ganglia. (B) Specific influence of neurotrophins in vivo. Distinct classes of peripheral somatosensory receptors and the dorsal root ganglion cells that give rise to these sensory endings depend on different trophic factors in specific target tissues. (A from Maisonpierre et al., 1990; B after Bibel and Barde, 2000.)

Figure 23.14 Neurotrophins influence neurite growth by local action. (A) Three compartments of a culture dish (1, 2, and 3) are separated from one another by a plastic divider sealed to the bottom of the dish with grease. A magnified view looking down on the compartments is shown below. Isolated rat sympathetic ganglion cells plated in compartment 1 can grow through the grease seal and into compartments 2 and 3. Growth into a lateral well occurs as long as the compartment contains an adequate concentration of NGF, and this local application does not influence the neurites in the other lateral well. Subsequent removal of NGF from a well causes a local regression of neurites without affecting the survival of cells or neurites in the other compartments. (B) Imaging of Ca^{2+} signaling within a growth cone from a developing from spinal cord neuron in tissue culture. The color scale at left indicates the correspondence between colors and Ca^{2+} concentration. Within 10 minutes Ca^{2+} increases, and continues to increase over the subsequent hour. These observations show that neuritic growth can be locally controlled by neurotrophins. (A after Campenot, 1981; B from Li et al., 2005.)

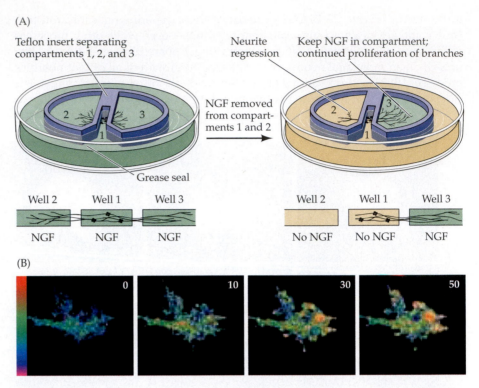

(A)

Teflon insert separating compartments 1, 2, and 3

Neurite regression

Keep NGF in compartment; continued proliferation of branches

NGF removed from compartments 1 and 2

Grease seal

Well 2   Well 1   Well 3
NGF   NGF   NGF

Well 2   Well 1   Well 3
No NGF   No NGF   NGF

(B)

0   10   30   50

Figure 23.15 Neurotrophin receptors and their specificity for the neurotrophins. (A) The Trk family of receptor tyrosine kinases for the neurotrophins. TrkA is primarily a receptor for NGF, TrkB a receptor for BDNF and NT-4/5, and TrkC a receptor for NT-3. Because of the high degree of structural homology among both the neurotrophins and the Trk receptors, there is some degree of cross-activation between factors and receptors. For example, NT-3 can bind to and activate TrkB under some conditions, as indicated by the dashed arrow. These distinct receptors allow various neurons to respond selectively to the different neurotrophins. (B) The p75 low-affinity neurotrophin receptor binds all neurotrophins at low affinities (as its name implies). This receptor confers the ability to respond to a broad range of neurotrophins upon fairly broadly distributed classes of neurons in the peripheral and central nervous systems.

growth cones (Figure 23.14B). In addition, physiological experiments indicate that NGF and other neurotrophins influence synaptic activity and plasticity (see Chapter 24), again independent of their effects on cell survival. Thus, there is a high degree of selectivity of neurotrophin actions, depending on the neurotrophic factor available and the stage of differentiation of the responding neuron, as well as the cellular domains where neurotrophins are available and neurotrophic signaling takes place.

The selective actions of the neurotrophins reflect their interactions with two classes of neurotrophin receptors: the **tyrosine kinase (Trk)** receptors and the **p75** receptor. There are three Trk receptors, each of which is a single transmembrane protein with a cytoplasmic tyrosine kinase domain. TrkA is primarily a receptor for NGF, TrkB a receptor for BDNF and NT-4/5, and TrkC a receptor for

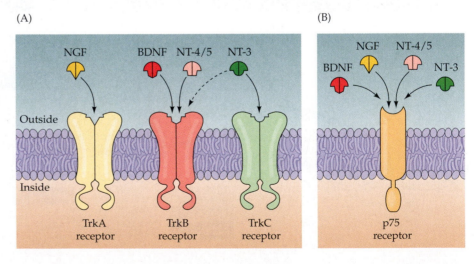

(A)

NGF   BDNF NT-4/5   NT-3

Outside

Inside

TrkA receptor   TrkB receptor   TrkC receptor

(B)

NGF   NT-4/5
BDNF   NT-3

p75 receptor

NT-3 (Figure 23.15). In addition, all neurotrophins can activate the p75 receptor protein. The interactions between neurotrophins and p75 demonstrate another level of selectivity and specificity of neurotrophin signaling. All neurotrophins are secreted in an unprocessed form that undergoes subsequent proteolytic cleavage. The p75 receptor has high affinity for unprocessed neurotrophins but low affinity for the processed ligands, while the Trk receptors have high affinity for processed ligands only. The expression of a particular Trk receptor subtype or p75 therefore confers the capacity to respond to the corresponding neurotrophin. Since neurotrophins, Trk receptors, and p75 are expressed only in subsets of neurons, the selective binding between ligand and receptor accounts for the specificity of the relevant neurotrophic interactions.

Signaling via either the Trk receptors or the p75 receptor can lead to changes in the three processes that are sensitive to neurotrophic signaling: cell survival/death, cell and process growth/differentiation, and activity-dependent synaptic stabilization or elimination. Each receptor class (Trk or p75) can engage distinct intracellular signaling cascades that lead to changes in cell state (motility, adhesion, etc.) or gene expression (Figure 23.16). Thus, understanding the specific effects of neurotrophic interactions for any cell relies on at least three pieces of information: the neurotrophins locally available; the combination of receptors on the relevant neuron; and the intracellular signaling pathways expressed by that neuron. The subtlety and diversity of neuronal circuits is set during development by different combinations of neurotrophins, their receptors, and signal transduction mechanisms. These mechanisms in concert determine the numbers of neurons, their shape, and their patterns of connections. Presumably, disruption of these neurotrophin-dependent processes, either during development or in the adult brain, can result in neurodegenerative conditions in which neu-

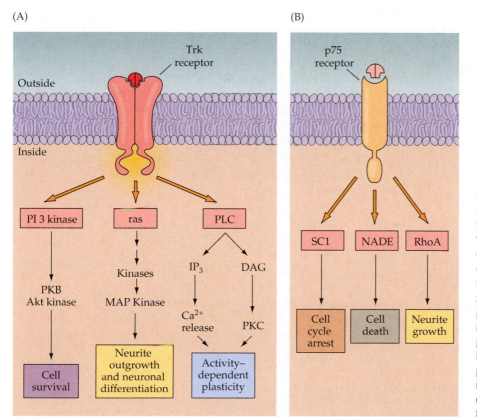

Figure 23.16 Signaling through the neurotrophins and their receptors. (A) Signaling via Trk dimers can lead to a variety of cellular responses, depending on the intracellular signaling cascade engaged by the receptor after binding to the ligand. The possibilities include cell survival (via the protein kinase C/AKT pathway); neurite growth (via the MAPKinase pathway); and activity-dependent plasticity (via the Ca^{2+}/calmodulin and PKC pathways). (B) Signaling via the p75 pathway can lead to neurite growth via interaction with Rho kinases, or to cell cycle arrest and cell death via other distinct intracellular signaling cascades.

rons die due to lack of appropriate trophic support or fail to make and maintain appropriate connections, with devastating consequences for the circuits that the cells define, and for the behaviors controlled by those circuits. Indeed, the pathogenic mechanisms of developmental disorders like schizophrenia and autism as well as neurodegenerative diseases as diverse as amyotrophic lateral sclerosis (ALS) and Parkinson's, Huntington's, and Alzheimer's diseases may all reflect deficiencies of neurotrophic regulation.

Summary

Neurons in the developing brain must integrate a variety of signals to determine where to send their axons, whether to live or die, what cells to form synapses upon, and how many synapses to make and retain. A remarkable transient cellular specialization, the growth cone, is responsible for axon growth and guidance. Growth cones explore the embryonic environment and determine the direction of axon growth as well as recognition of appropriate targets. The special motile properties that allow growth cones to approach, select, or avoid a target reflect modulation of the actin and microtubule cytoskeleton via a number of signaling mechanisms; many involve changes in intracellular Ca^{2+} signaling. The instructions that elicit these growth cone responses come from adhesive, chemotropic, chemorepulsive, and trophic molecules. These molecules are organized into an extracellular matrix found on cell surfaces, or are secreted to diffuse in extracellular spaces. Molecular cues ensure that coherent axon pathways form from one structure to another and prevent inappropriate connections. They can also establish a pattern of innervation that is the basis for topographic maps. Adhesive, attractive, and repulsive molecules also influence the transformation of the growth cone to a nascent synapse. Specific cell adhesion molecules organize the rudimentary molecular specializations that define a synapse. The earliest effects of trophic molecules—particularly neurotrophins that are made exclusively by neuronal targets—are on cell survival and differentiation. Once the appropriate number of neurons is established, trophic signals continue to govern the establishment of neural connections, particularly the extent of axonal and dendritic arborizations. Understanding the molecular basis of axon guidance, synapse formation, and trophic signaling began a century ago and has burgeoned into a broad effort that continues to identify new factors and signaling pathways and to illuminate their varied roles in both the developing and adult brain. Defects in the early guidance of axons are responsible for a variety of congenital neurological syndromes and developmental disorders. Neurotrophic dysfunction may also underlie degenerative diseases such as ALS and Parkinson's disease.

Additional Reading

Reviews

CULOTTI, J. G. AND D. C. MERZ (1998) DCC and netrins. *Curr. Opin. Cell Biol.* 10: 609–613.

HUBER, A. B., A. L. KOLODKIN, D. D. GINTY AND J. F. CLOUTIER (2003) Signaling at the growth cone: Ligand-receptor complexes and the control of axon growth and guidance. *Annu. Rev. Neurosci.* 26: 509–563.

LEVI-MONTALCINI, R. (1987) The nerve growth factor 35 years later. *Science* 237: 1154–1162.

LEWIN, G. R. AND Y. A. BARDE (1996) Physiology of the neurotrophins. *Annu. Rev. Neurosci.* 19: 289–317.

LICHTMAN, J. W. AND H. COLEMAN (2000) Synapse elimination and indelible memory. *Neuron* 25: 269–278.

PURVES, D. AND J. W. LICHTMAN (1978) Formation and maintenance of synaptic con-

nections in autonomic ganglia. *Physiol. Rev.* 58: 821–862.

PURVES, D. AND J. W. LICHTMAN (1980) Elimination of synapses in the developing nervous system. *Science* 210: 153–157.

PURVES, D., W. D. SNIDER AND J. T. VOYVODIC (1988) Trophic regulation of nerve cell morphology and innervation in the autonomic nervous system. *Nature* 336: 123–128.

RAPER, J. A. (2000) Semaphorins and their receptors in vertebrates and invertebrates. *Curr. Opin. Neurobiol.* 10: 88–94.

REICHARDT, L. F. AND K. J. TOMASELLI (1991) Extracellular matrix molecules and their receptors: Functions in neural development. *Annu. Rev. Neurosci.* 14: 531–570.

RUTISHAUSER, U. (1993) Adhesion molecules of the nervous system. *Curr. Opin. Neurobiol.* 3: 709–715.

SANES, J. R. AND J. W. LICHTMAN (1999) Development of the vertebrate neuro-muscular junction. *Annu. Rev. Neurosci.* 22: 389–442.

SCHWAB, M. E., J. P. KAPFHAMMER AND C. E. BANDTLOW (1993) Inhibitors of neurite growth. *Annu. Rev. Neurosci.* 16: 565–595.

SEGAL, R. A. AND M. E. GREENBERG (1996) Intracellular signaling pathways activated by neurotrophic factors. *Annu. Rev. Neurosci.* 19: 463–489.

SILOS-SANTIAGO, I., L. J. GREENLUND, E. M. JOHNSON JR. AND W. D. SNIDER (1995) Molecular genetics of neuronal survival. *Curr. Opin. Neurobiol.* 5: 42–49.

TEAR, G. (1999) Neuronal guidance: A genetic perspective. *Trends Genet.* 15: 113–118.

Important Original Papers

BAIER, H. AND F. BONHOEFFER (1992) Axon guidance by gradients of a target-derived component. *Science* 255: 472–475.

BALICE-GORDON, R. J. AND J. W. LICHTMAN (1994) Long-term synapse loss induced by focal blockade of postsynaptic receptors. *Nature* 372: 519–524.

BALICE-GORDON, R. J., C. K. CHUA, C. C. NELSON AND J. W. LICHTMAN (1993) Gradual loss of synaptic cartels precedes axon withdrawal at developing neuromuscular junctions. *Neuron* 11: 801–815.

BROWN, M. C., J. K. S. JANSEN AND D. VAN ESSEN (1976) Polyneuronal innervation of skeletal muscle in new-born rats and its elimination during maturation. *J. Physiol. (Lond.)* 261: 387–422.

CAMPENOT, R. B. (1977) Local control of neurite development by nerve growth factor. *Proc. Natl. Acad. Sci. USA* 74: 4516–4519.

DONTCHEV, V. D. AND P. C. LETOURNEAU (2002) Nerve growth factor and semaphorin 3A signaling pathways interact in regulating sensory neuronal growth cone motility. *J. Neurosci.* 22: 6659–6669.

DRESCHER, U., C. KREMOSER, C. HANDWERKER, J. LOSCHINGER, M. NODA AND F. BONHOEFFER (1995) In vitro guidance of retinal ganglion cell axons by RAGS, a 25 kDa tectal protein related to ligands for Eph receptor tyrosine kinases. *Cell* 82: 359–370.

FREDETTE, B. J. AND B. RANSCHT (1994) T-cadherin expression delineates specific regions of the developing motor axon-hindlimb projection pathway. *J. Neurosci.* 14: 7331–7346.

FARINAS, I., K. R. JONES, C. BACKUS, X.Y. WANG AND L. F. REICHARDT (1994) Severe sensory and sympathetic deficits in mice lacking neurotrophin-3. *Nature* 369: 658–661.

KAPLAN, D. R., D. MARTIN-ZANCA AND L. F. PARADA (1991) Tyrosine phosphorylation and tyrosine kinase activity of the *trk* proto-oncogene product induced by NGF. *Nature* 350: 158–160.

KENNEDY, T. E., T. SERAFINI, J. R. DE LA TORRE AND M. TESSIER-LAVIGNE (1994) Netrins are diffusible chemotropic factors for commissural axons in the embryonic spinal cord. *Cell* 78: 425–435.

KOLODKIN, A. L., D. J. MATTHES AND C. S. GOODMAN (1993) The *semaphorin* genes encode a family of transmembrane and secreted growth cone guidance molecules. *Cell* 75: 1389–1399.

LANGLEY, J. N. (1895) Note on regeneration of pre-ganglionic fibres of the sympathetic. *J. Physiol. (Lond.)* 18: 280–284.

LEVI-MONTALCINI, R. AND S. COHEN (1956) In vitro and in vivo effects of a nerve growth-stimulating agent isolated from snake venom. *Proc. Natl. Acad. Sci. USA* 42: 695–699.

LICHTMAN, J. W. (1977) The reorganization of synaptic connexions in the rat submandibular ganglion during postnatal development. *J. Physiol. (Lond.)* 273: 155–177.

LICHTMAN, J. W., L. MAGRASSI AND D. PURVES (1987) Visualization of neuromuscular junctions over periods of several months in living mice. *J. Neurosci.* 7: 1215-1222.

LUO, Y., D. RAIBLE AND J. A. RAPER (1993) Collapsin: A protein in brain that induces the collapse and paralysis of neuronal growth cones. *Cell* 75: 217–227.

MESSERSMITH, E. K., E. D. LEONARDO, C. J. SHATZ, M. TESSIER-LAVIGNE, C. S. GOODMAN AND A. L. KOLODKIN (1995) Semaphorin III can function as a selective chemorepellent to pattern sensory projections in the spinal cord. *Neuron* 14: 949–959.

OPPENHEIM, R. W., D. PREVETTE AND S. HOMMA (1990) Naturally occurring and induced neuronal death in the chick embryo in vivo requires protein and RNA synthesis: Evidence for the role of cell death genes. *Dev. Biol.* 138: 104–113.

SERAFINI, T. AND 6 OTHERS (1996) Netrin-1 is required for commissural axon guidance in the developing vertebrate nervous system. *Cell* 87: 1001–1014.

SPERRY, R. W. (1963) Chemoaffinity in the orderly growth of nerve fiber patterns and connections. *Proc. Natl. Acad. Sci. USA* 50: 703–710.

WALTER, J., S. HENKE-FAHLE AND F. BONHOEFFER (1987) Avoidance of posterior tectal membranes by temporal retinal axons. *Development* 101: 909–913.

Books

LETOURNEAU, P. C., S. B. KATER AND E. R. MACAGNO (EDS.) (1991) *The Nerve Growth Cone.* New York: Raven Press.

LOUGHLIN, S. E. AND J. H. FALLON (EDS.) (1993) *Neurotrophic Factors.* San Diego, CA: Academic Press.

PURVES, D. (1988) *Body and Brain: A Trophic Theory of Neural Connections.* Cambridge, MA: Harvard University Press.

RAMÓN Y CAJAL, S. (1928) *Degeneration and Regeneration of the Nervous System.* R. M. May (ed.). New York: Hafner Publishing.

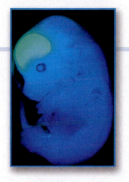

Chapter 24

Modification of Brain Circuits as a Result of Experience

Overview

The rich diversity of human personalities, abilities, and behavior is undoubtedly generated by the uniqueness of individual human brains. These fascinating neurobiological differences among humans derive from both genetic and environmental influences, and many of the most essential distinctions are established during the initial stages of life. The first steps in the construction of the brain's circuitry—the establishment of distinct brain regions, the generation of neurons, the formation of major axon tracts, the guidance of growing axons to appropriate targets, and the initiation of synaptogenesis—rely largely on the intrinsic cellular and molecular processes described in the previous chapters. Once the basic patterns of brain connections are established, however, patterns of neuronal activity (including those that are elicited by experience) modify the synaptic circuitry of the developing brain. Neuronal activity generated by interactions with the outside world in postnatal life thus provides a mechanism by which the environment can influence brain structure and function. Many of the effects of activity are transduced via signaling pathways that modify levels of intracellular Ca^{2+} and thus influence local gene expression, neurotransmission, and neurotrophic interactions (see Chapters 7 and 23). This activity-mediated influence on the developing brain is most consequential during temporal windows called critical periods. As humans (and other mammals) mature, the brain becomes increasingly refractory to the lessons of experience, and the cellular mechanisms that modify neural connectivity become less effective.

Critical Periods

The cellular and molecular mechanisms outlined in Chapters 22 and 23 construct a nervous system of impressive anatomical complexity. These embryonic mechanisms and their developmental consequences are sufficient to create some remarkably sophisticated innate or "instinctual" behaviors (see Box 30A). For most animals, the behavioral repertoire, including foraging, fighting, and mating strategies, largely relies on patterns of connectivity established by intrinsic developmental mechanisms. However, the nervous systems of animals with increasingly complex repertoires of behaviors, including humans, clearly adapt to and are influenced by the particular circumstances of an individual's environment. These environmental factors are especially influential in early life, during temporal windows called **critical periods**. Some behaviors, such as parental imprinting in hatchling birds, are expressed only if animals have certain specific experiences during a sharply restricted time in early postnatal (or posthatching) development (Box 24A). On the other hand, critical periods for sensorimotor skills and complex behaviors last longer and end far less abruptly. In some cases, such as the acquisition of language in humans, detailed instruc-

BOX 24A Built-In Behaviors

The idea that animals already possess a set of behaviors appropriate for a world not yet experienced has always been difficult to accept. However, the preeminence of instinctual responses is obvious to any biologist who looks at what animals actually do. Perhaps the most thoroughly studied examples occur in young birds. Hatchlings emerge from the egg with an elaborate set of innate behaviors. First, of course, is the complex behavior that allows the chick to escape from the egg. Having emerged, a variety of additional abilities indicate how much early behavior is "preprogrammed".

In a series of seminal observations based on his work with geese, Konrad Lorenz showed that goslings follow the first large, moving object they see and hear during their first day of life. Although this object is normally the mother goose, Lorenz found that goslings can imprint on a wide range of animate and inanimate objects presented during this period, including Lorenz himself. The window for imprinting in goslings is less than a day; if animals are not exposed to an appropriate stimulus during this time, they will never form the appropriate parental relationship. Once imprinting occurs, however, it is irreversible, and geese will continue to follow inappropriate objects (male conspecifics, people, or even inanimate objects).

In many mammals, auditory and visual systems are poorly developed at birth, and maternal imprinting relies on olfactory and/or gustatory cues. For example, during the first week of life (but not later), infant rats develop a lifelong preference for odors associated with their mother's nipples. As in birds, this filial imprinting plays a role in their social development and later sexual preferences.

Imprinting is a two-way street, with parents (especially mothers) rapidly forming exclusive bonds with their offspring. This phenomenon is especially important in animals like sheep that live in large groups or herds in which all the females produce offspring at about the same time of year. Ewes have a critical period 2–4 hours after giving birth during which they imprint on the scent of their own lamb. After about 4 hours, they rebuff approaches by other lambs.

The relevance of this work to primates was underscored in the 1950s by Harry Harlow and his colleagues at the University of Wisconsin. Harlow isolated monkeys within a few hours of birth and raised them in the absence of either a natural mother or a human substitute. In the best known of these experiments, the baby monkeys had one of two maternal surrogates: a "mother" constructed of a wooden frame covered with wire mesh that supported a nursing bottle, or a similarly shaped object covered with soft terrycloth but without any source of nourishment for the young monkey. When presented with this

Konrad Lorenz, followed by imprinted geese. (Photograph courtesy of H. Kacher.)

choice, the baby monkeys preferred the terrycloth mother and spent much of their time clinging to it, even though the feeding bottle was with the wire mother. Harlow took this to mean that newborn monkeys have a built-in need for maternal care and have at least some innate idea of what a mother should feel like. A number of other endogenous behaviors have been studied in infant monkeys, including a naïve monkey's fear reaction to the presentation of certain objects (e.g., a snake) and the "looming" response (fear elicited by the rapid approach of any formidable object). Most of these built-in behaviors have analogs in human infants.

Taken together, these observations make plain that many complicated behaviors, emotional responses, and other predilections are well established in the nervous system prior to any significant experience, and that the need for certain kinds of early experience for normal development is predetermined. These built-in behaviors and their neural substrates have presumably evolved to give newborns a better chance of surviving in a predictably dangerous world.

References

HARLOW, H. F. (1959) Love in infant monkeys. *Sci. Amer.* 2 (September): 68–74.

HARLOW, H. F. AND R. R. ZIMMERMAN (1959) Affectional responses in the infant monkey. *Science* 130: 421–432.

LORENZ, K. (1970) *Studies in Animal and Human Behaviour*. Translated by R. Martin. Cambridge, MA: Harvard University Press.

MACFARLANE, A. J. (1975) Olfaction in the development of social preferences in the human neonate. *Ciba Found. Symp.* 33: 103–117.

SCHAAL, B. E., H. MONTAGNER, E. HERTLING, D. BOLZONI, A. MOYSE AND R. QUICHON (1980) Les stimulations olfactives dans les relations entre l'enfant et la mère. *Reprod. Nutr. Dev.* 20(3b): 843–858.

TINBERGEN, N. (1953) *Curious Naturalists*. Garden City, NY: Doubleday.

tive influences from the environment are required for the normal development of the behavior (i.e., exposure to the individual's native language). The availability of these influences, as well as the neural capacity to respond to them, is key for successful completion of the critical period. These instructive influences are paralleled in some non-human species in which complex communication is important for territorial and reproductive behaviors. In some species of songbirds, male birds acquire the capacity to produce species-specific song by mimicking tutor birds. If this essential instruction is witheld or disrupted, the birds are not effective in using communication to define their territories and compete for mates.

Despite the fact that critical periods vary widely in both the behaviors affected and their duration, they all share some basic properties. A critical period is defined as the time during which a given behavior is especially susceptible to—indeed, requires—specific environmental influences in order to develop normally. These influences can be as subtle as the ongoing stimuli of light or sound that are encountered by an infant, or the precise instruction in one's native (or a foreign) language required to achieve fluent speech and accurate comprehension. Once a critical period ends, the behavior is largely unaffected by subsequent experience (or even by the complete absence of the relevant experience). Conversely, failure to be exposed to appropriate stimuli during the critical period is difficult or in some cases impossible to remedy subsequently. While psychologists and ethologists (biologists who study the natural behavior of animals) have long recognized that early postnatal or posthatching life is a period of special sensitivity to environmental influences, their studies of critical periods focused on the relationship between the acquisition of behaviors, or the ability to modify behavior, with the age of the individual.

Work in the last few decades has examined age-dependent capacities for modification in a fairly large number of brain circuits and systems, thus providing a biological substrate for the behavioral phenomenon. Several cellular and molecular mechanisms have been associated with circuit changes during critical periods. Neural systems that have critical periods tend to have complex synaptic connections, often have distinct topographic maps or patterns of modular circuits (see Box 9B), and subserve a functional repertoire that is built incrementally. In some cases, like that of neuromuscular control, increasingly fine movement is thought to depend on the proper size of motor units based on single innervation of muscle fibers by distinct motor neurons. For the neuromuscular junction, the underlying cellular mechanisms are thought to rely primarily on factors that influence the processes of synapse formation, multiple innervation, and competitive elimination of supernumerary innervation discussed in Chapters 8 and 23. For others, including the visual, auditory, somatic sensory, and olfactory systems, the known molecular mediators are primarily factors that mediate synaptic communication, including several neurotransmitters, neurotransmitter-processing enzymes, receptors, or second messenger-related signaling molecules (Table 24.1). This does not mean that other molecules do not contribute to critical period phenomena; rather, it indicates the primary role of synaptic activity in transducing final steps that lead to changes in neural circuit structure and function during a critical period. Other more complex behavioral phenomena, including cognitive and emotional functions like stress and anxiety, also exhibit critical periods. The significance of this sort of malleability is clear for understanding the consequences of early experience on a range of behavioral disturbances and psychiatric diseases later in life. The cellular and molecular mechanisms, however, remain undefined.

TABLE 24.1 Critical Periods and Molecular Regulators for Some Neural Systems

System	Species[a]	Critical period (postnatal)[b]	Confirmed molecular regulators[c]
Neuromuscular junction	Mouse	Prior to day 12	ACh
Cerebellum	Mouse	Day 15–16	NMDA, MGluR1, G_q, PLCβ, PKCγ
Lateral geniculate nucleus layers	Mouse, ferret, cat	Prior to day 10	ACh, cAMP, MAOA, NO, MHC1, CREB
Ocular dominance	Cat, rat, mouse, ferret	3 weeks–months	GABA, NMDA, PKA, ERK, CaMKII, CREB, BDNF, tPA, protein synthesis, NE, ACh
Orientation bias	Cat, mouse	Prior to day 28	NR1, NR2A, PSD95
Somatosensory map	Mouse, rat	Prior to day 7–16	NR1, MAOA, $5HT_{1B}$, cAMP, mGluR5, PLCβ, FGF8
Tonotopic map (cortex)	Rat	Days 16–50	ACh
Absolute pitch	Human	Before 7 years	Unknown
Taste, olfaction	Mouse	None	GABA, mGLuR2, NO, neurogenesis
Imprinting	Chick	14–42 hours	Catecholamines
Stress, anxiety	Rat, mouse	Prior to day 21	Hormones, $5HT_{1A}$
Slow-wave sleep	Cat, mouse	Day 40–60	NMDA
Sound localization	Barn owl	Prior to day 200	GABA, NMDA
Birdsong	Zebra finch	Prior to day 100	GABA, hormones, neurogenesis
Language	Human	0–12 years	Unknown

[a]Primary research species for elucidation of molecular mechanisms.
[b]Although the details vary from system to system and species to species, all critical periods are limited to a definite window of time during early postnatal (or posthatching) life and are complete before the onset of sexual maturation.
[c]The molecules known to regulate critical periods include neurotransmitters, their receptors, and related signaling proteins.
After Hensch, 2004.

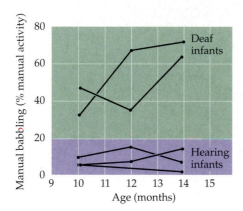

Figure 24.1 Manual "babbling" in two deaf infants raised by deaf, signing parents compared to manual babble in three hearing infants. Babbling was judged by scoring hand positions and shapes that showed some resemblance to the components of American Sign Language. In deaf infants with deaf signing parents, meaningful hand shapes increase as a percentage of manual activity between ages 10 and 14 months. Hearing children raised by hearing, speaking parents do not produce similar hand shapes. (After Petito and Marentette, 1991.)

The Development of Language: A Critical Period for Human Behavior

Many animals communicate by means of sound, and some (humans and songbirds are examples) learn these vocalizations during a critical period. In fact, there are provocative similarities in the development of human language and birdsong (Box 24B). Many animal vocalizations, like alarm calls in mammals and birds, are innate, and require no experience to be correctly produced. For example, quail raised in isolation or deafened at birth so that they never hear conspecifics nonetheless produce the full repertoire of species-specific vocalizations. The same may be true for isolation cries and other basic vocalizations in human infants. In contrast, humans obviously require extensive postnatal experience to produce and decode speech sounds that are the basis of language.

The various forms of early language exposure, including the "baby talk" that parents and other adults often use to communicate with infants and small children, may actually serve to emphasize important perceptual distinctions that facilitate proper language production and comprehension. This linguistic experience, to be effective, must occur in early life. The requirement for perceiving and practicing language (as opposed to specific auditory, visual, or motor skills) during a critical period is apparent in studies of language acquisition in congenitally deaf children, whose language acquisition relies on seeing and moving the hands and fingers rather than on listening and moving the lips, tongue, and larynx. Whereas most hearing and speaking babies begin producing speechlike sounds, or "babbling," at about 7 months, congenitally deaf infants show obvious deficits in their early vocalizations, and such individuals fail to develop language if not provided with an alternative form of symbolic expression such as sign language (see Chapter 27). If, however, deaf children are exposed to sign language at an early age (from approximately 6 months onward, which is particularly likely for the children of deaf, signing parents), they begin to "babble" with their hands just as a hearing infant babbles audibly (Figure 24.1). This manual babbling suggests that, regardless of the modality, early experience shapes language behavior. Children who have acquired speech but lose their hearing before puberty also suffer a substantial decline in spoken language, presumably because they are unable to hear themselves or others talk and thus lose the opportunity to refine their speech by auditory feedback during the final stages of the critical period for language.

Examples of pathological situations in which normal children were deprived of a significant amount of language experience make the same point. In one well-documented case, a girl raised by deranged parents until the age of 13 had almost no exposure to standard human language. Despite intense subsequent training, she never achieved more than a rudimentary level of verbal communication. This and other examples of so-called "feral" children starkly define the importance of early experience for language development, as well as other aspects of social communication and personality.

In contrast to the devastating effects of language deprivation on children, adults retain their ability to speak and comprehend language even if decades pass without exposure to human communication (a fictional example would be Robinson Crusoe). In short, the normal acquisition of human language is subject to a critical period: the process is sensitive to experience or deprivation during a restricted period of life (before puberty), and is relatively refractory to similar experience or deprivation in adulthood. Changes in the patterns of activity in language regions of the brain in children versus adults suggest that the relevant neural circuits may undergo functional or structural modifications during the critical period for language (Figure 24.2A).

On a more subtle level, the phonetic structure of the language an individual hears during early life shapes both the perception and production of speech. Many of the thousands of human languages and dialects use appreciably different speech elements called *phonemes* to produce spoken words (examples are the phonemes *ba* and *pa* in English; see Chapter 27). Very young human infants can perceive and discriminate between differences in *all* human speech sounds, and are not innately biased towards phonemes characteristic of any particular language. However, this universal perceptual capacity does not persist. For example, adult Japanese speakers cannot reliably distinguish between the *r* and *l* sounds in English, presumably because this phonemic distinction is not made in Japanese and thus not reinforced by experience during the critical period. Nonetheless, 4-month-old Japanese infants can make this discrimination as reliably as 4-month-olds raised in English-speaking households (as indicated by increased suckling frequency or head turning in the presence of a novel stimulus). By 6 months of age, however, infants begin to show preferences for phonemes in their native language over those in foreign languages, much as deaf infants do for moving digits that suggest signs. By the end of their first year they no longer respond robustly to phonetic elements peculiar to non-native languages.

The ability to perceive, learn, and produce phonemic contrasts, as well as the ability to acquire a sense of the rules of grammar and usage in a language (referred to as the semantic aspects of language) evidently persists for several more years, as evidenced by the fact that children can usually learn to speak a second language without accent and with fluent grammar until about age 7 or 8. After this age, however, performance gradually declines no matter what the extent of practice or exposure (Figure 24.2B).

A number of changes in the developing brain could explain these observations. One possibility is that experience acts selectively to preserve the circuits in the

(A)

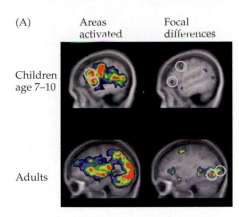

(B)

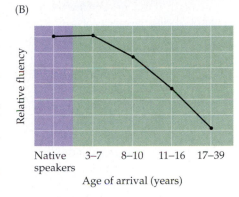

Figure 24.2 *Learning language.* (A) Maps derived from fMRI in adults and children performing visual word processing tasks. Images are sagittal sections with the front of the brain toward the left. The top row shows the range of active areas (left) and foci of activity based on group averages (right) for children ages 7–10. The bottom row shows analogous results for adults performing the same task. The differences in regions of maximal activation (shown in red in the images at left; highlighted by white circles in the right-hand images) indicate changes in either the circuitry or the mode of processing and performing the same task in children versus adults. (B) A critical period for learning language is shown by the decline in language ability (fluency) of non-native speakers of English as a function of their age upon arrival in the United States. The ability to score well on tests of English grammar and vocabulary declines from approximately age 7 onward. (A after Schlaggar et al., 2002; B after Johnson and Newport, 1989.)

BOX 24B Birdsong

Anyone witnessing a child learning to talk cannot help but be amazed at how quickly language development takes place. This facility contrasts with the adult acquisition of a new language, which can be a painfully slow process and rarely, if ever, produces complete fluency. In fact, many learned behaviors are acquired during a period in early life when experience exerts an especially potent influence on subsequent behavior. Particularly well characterized is the sensitive period for learning courtship songs by oscine songbirds such as canaries and finches. In these species, the quality of early sensory exposure is the major determinant of subsequent perceptual and behavioral capabilities. Furthermore, developmental periods for learning these and other behaviors are restricted during postnatal life, suggesting that the nervous system changes in some manner to become refractory to further experience. Understanding how critical periods are regulated has many implications, not least the possibility of reactivating this enhanced learning capacity in adults. Nonetheless,

such periods are often highly specialized for the acquisition of species-typical behaviors and are not merely times of general enhanced learning.

Avian song learning illustrates the interactions between intrinsic and environmental factors in this developmental process. Many birds sing to attract mates, but oscine songbirds are special in that their courtship songs are dependent on auditory and vocal experience. The sensitive period for song learning comprises an initial stage of sensory acquisition, when the juvenile bird listens to and memorizes the song of a nearby adult male tutor (usually of its own species), and a subsequent stage of vocal learning, when the young bird matches its own song to the now-memorized tutor model via auditory feedback. This sensory motor learning stage ends with the onset of sexual maturity, when songs become acoustically stable, or crystallized. In all species studied to date, young songbirds are especially impressionable during the first two months after hatching and become refractory to further exposure to

tutor song as they age. The impact of this early experience is profound, and the memory it generates can remain intact for months, perhaps years, before the onset of the vocal practice phase. Even constant exposure to other songs after sensory acquisition during the sensitive period ends does not affect this memory. The songs heard during sensory acquisition, but not later, are those that the bird vocally mimics. Early auditory experience is crucial to the bird's Darwinian success. In the absence of a tutor, or if raised only in the presence of another species, birds produce highly abnormal "isolate" songs, or songs of the foster species, neither of which succeeds in attracting females of their own kind.

Two other features of song learning indicate an intrinsic predisposition for this specialized form of vocal learning. First, juveniles often need to hear the tutor song only 10 or 20 times in order to vocally mimic it many months later. Second, when presented with a variety of songs played from tape recordings that include their own and other species'

brain that perceive phonemes and phonetic distinctions. The absence of exposure to non-native phonemes would then result in a gradual atrophy of the connections representing those sounds, accompanied by a declining ability to distinguish between them. In this formulation, circuits that are used are retained, whereas those that are unused get weaker (and presumably disappear, as described by the maxim "use it or lose it"). Alternatively, experience could promote the growth of rudimentary circuitry pertinent to the experienced sounds. Recent comparisons of patterns of activity in children aged 7–10 with the patterns of adults performing the same very specific word processing tasks suggest that different brain regions are activated for the same task in children versus adults (see Figure 24.2A). While the significance of such differences is not clear—they may reflect anatomical plasticity associated with critical periods, or distinct modes of performing language tasks in children versus adults—there is nevertheless an indication that brain circuits change to accommodate language function during early life.

Critical Periods in Visual System Development

Although critical periods for language and other distinctively human behaviors are in some ways the most compelling examples of this phenomenon, it is difficult

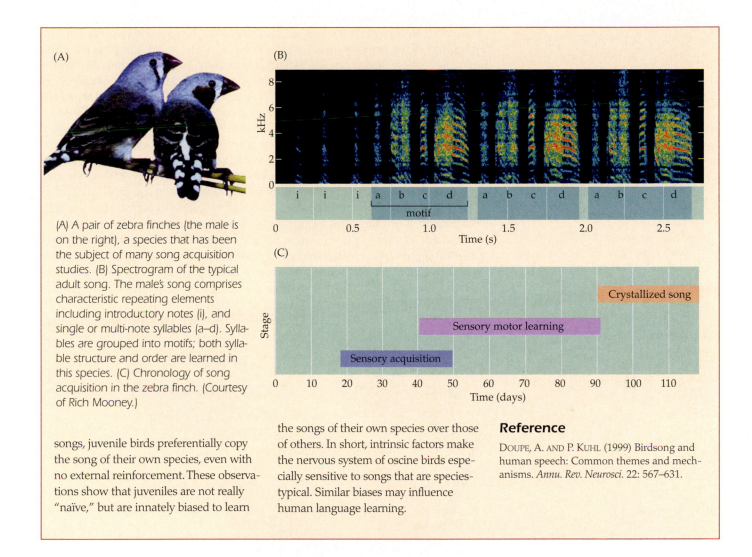

(A) A pair of zebra finches (the male is on the right), a species that has been the subject of many song acquisition studies. (B) Spectrogram of the typical adult song. The male's song comprises characteristic repeating elements including introductory notes (i), and single or multi-note syllables (a–d). Syllables are grouped into motifs; both syllable structure and order are learned in this species. (C) Chronology of song acquisition in the zebra finch. (Courtesy of Rich Mooney.)

songs, juvenile birds preferentially copy the song of their own species, even with no external reinforcement. These observations show that juveniles are not really "naïve," but are innately biased to learn the songs of their own species over those of others. In short, intrinsic factors make the nervous system of oscine birds especially sensitive to songs that are species-typical. Similar biases may influence human language learning.

Reference

DOUPE, A. AND P. KUHL (1999) Birdsong and human speech: Common themes and mechanisms. *Annu. Rev. Neurosci.* 22: 567–631.

if not impossible to study the underlying changes in the human brain. A much clearer understanding of how changes in connectivity might contribute to critical periods has come from studies of the developing visual system in experimental animals with highly developed visual abilities, particularly cats and monkeys. The visual system is extremely amenable to the sorts of experimental manipulations necessary to test the relationship between experience, activity, and circuitry. It is relatively easy to either deprive or augment visual experience in an experimental animal—eyes can be sutured shut, animals can be reared in all sorts of illumination conditions from total darkness to maximal light. Such control of sensory experience is almost impossible in any other modality; it is much harder to deprive an animal of auditory, somatosensory, olfactory, or taste stimuli.

In an extraordinarily influential series of experiments, David Hubel and Torsten Wiesel found that depriving animals of normal visual experience during a restricted period of early postnatal life irreversibly alters neuronal connections (and functions) in the visual cortex. These observations provided the first evidence that the brain translates the effects of early experience (that is, patterns of neural activity) into more or less permanently altered wiring. To understand these experiments and their implications, it is important to review the organization and development of the mammalian visual system.

Recall from Chapters 11 and 12 that information from the two eyes is first integrated in the primary visual (striate) cortex, where most afferents from the lateral geniculate nucleus of the thalamus terminate. In some mammals—carnivores, anthropoid primates, and humans—the afferent terminals form an alternating series of eye-specific domains in cortical layer 4 called **ocular dominance columns** (Figure 24.3). Ocular dominance columns can be visualized by injecting tracers such as radioactive amino acids into one eye; the tracer is then transported along the visual pathway to specifically label the geniculocortical terminals (i.e., synaptic terminals in the visual cortex) corresponding to that eye (Box 24C). In the adult macaque monkey, the domains representing the two eyes are stripes of about equal width (0.5 mm) that occupy roughly equal areas of layer 4 of the primary visual cortex. Electrical recordings confirm that the cells within layer 4 of macaques respond strongly or exclusively to stimulation of either the left or the right eye, while neurons in layers above and below layer 4 integrate inputs from the left and right eyes and respond to visual stimuli presented to either eye. Ocular dominance is thus apparent in two related phenomena: the degree to which individual cortical neurons are driven by stimulation of one eye or the other, and domains (stripes) in cortical layer 4 in which the majority of neurons are driven exclusively by one eye or the other. The clarity of these patterns of connectivity and the precision by which experience via

Figure 24.3 Ocular dominance columns (which in most anthropoid primates are really stripes or bands) in layer 4 of the primary visual cortex of an adult macaque monkey. The diagram indicates the labeling procedure (see also Box 24C); following transynaptic transport, the distribution of ipsilateral (I) versus contralateral (C) retinal ganglion cell axon terminals is seen in the LGN (lower left). Moreover, due to the transynaptic transport of the label, geniculocortical terminals (from the layers labeled in LGN shown in the left hand panel) related to the injected eye are visible as a pattern of bright stripes in this autoradiogram of a section through layer 4 in the plane of the cortex (that is, as if looking down on the cortical surface). The dark areas are the zones occupied by geniculocortical terminals related to the other eye. (LGN micrograph courtesy P. Rakic; ocular dominance columns from LeVay, Wiesel, and Hubel, 1980.)

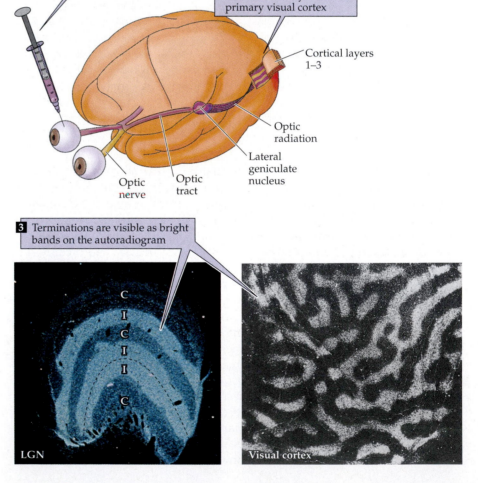

1 Radioactive amino acids injected in eye

2 Transynaptic transport through the LGN terminates in layer 4 of the primary visual cortex

Cortical layers 1–3

Optic radiation

Lateral geniculate nucleus

Optic nerve

Optic tract

3 Terminations are visible as bright bands on the autoradiogram

LGN

Visual cortex

BOX 24C   Transneuronal Labeling with Radioactive Amino Acids

Unlike many brain structures, ocular dominance columns are not easily visible by means of conventional histology. Thus, the striking cortical patterns evident in cats and monkeys were not seen until the early 1970s, when the technique of anterograde tracing using radioactive amino acids was introduced. In this approach, an amino acid commonly found in proteins (usually proline) is radioactively tagged and injected into the area of interest. Neurons in the vicinity take up the label from the extracellular space and incorporate it into newly made proteins. Some of these proteins are involved in the maintenance and function of the neuron's synaptic terminals; thus, they are shipped via anterograde axonal transport from the cell body to nerve terminals, where they accumulate. After a suitable interval, the tissue is fixed, and sections are made, placed on glass slides, and coated with a sensitive photographic emulsion. The radioactive decay of the labeled amino acids in the proteins causes silver grains to form in the emulsion. After several months of exposure, a heavy concentration of silver grains accumulates over the regions that contain synapses originating from the injected site. For example,

injections into the eye will heavily label the terminal fields of retinal ganglion cells in the lateral geniculate nucleus.

Transneuronal transport takes this process a step further. After tagged proteins injected into living animals reach the axon terminals, a fraction is actually released into the extracellular space, where the proteins are degraded into amino acids or small peptides that retain their radioactivity. An even smaller fraction of this pool of labeled amino acids is taken up by the postsynaptic neurons, incorporated again into proteins, and transported to synaptic terminals of the second set of neurons. Because the label passes from the presynaptic terminals of one set of cells to the postsynaptic target cells, the process is called transneuronal transport. By such transneuronal labeling, the chain of connections originating from a particular structure can be visualized. In the case of the visual system, proline injections into one eye label appropriate layers of the lateral geniculate nucleus (as well as other retinal gan-

glion cell targets such as the superior colliculus), and subsequently the terminals in the visual cortex of the geniculate neurons receiving inputs from that eye. Thus, when sections of the visual cortex are viewed with dark-field illumination to make the silver grains glow a brilliant white against the unlabeled background, ocular dominance columns in layer 4 are easily seen (see Figure 24.3).

References

COWAN, W. M., D. I. GOTTLIEB, A. HENDRICKSON, J. L. PRICE AND T. A. WOOLSEY (1972) The autoradiographic demonstration of axonal connections in the central nervous system. *Brain Res.* 37: 21–51

GRAFSTEIN, B. (1971) Transneuronal transfer of radioactivity in the central nervous system. *Science* 172: 177–179.

GRAFSTEIN, B. (1975) Principles of anterograde axonal transport in relation to studies of neuronal connectivity. In *The Use of Axonal Transport for Studies in Neuronal Connectivity*, W. M. Cowan and M. Cuénod (eds.). Amsterdam: Elsevier, pp. 47–68.

Transneuronal transport. A neuron in the retina is shown taking up a radioactive amino acid, incorporating it into proteins, and moving the proteins down the axons and across the extracellular space between neurons. This process is repeated in the thalamus, and eventually label accumulates in the thalamocortical terminals in layer 4 of the primary visual cortex.

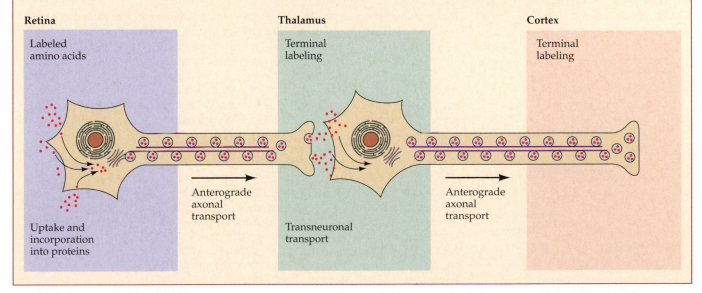

Retina
Labeled amino acids

Uptake and incorporation into proteins

Anterograde axonal transport

Thalamus
Terminal labeling

Transneuronal transport

Anterograde axonal transport

Cortex
Terminal labeling

the two eyes can be manipulated led to the series of experiments described in the following section. These experiments greatly clarified the neurobiological processes underlying critical periods, and integrated changes in functional innervation with change in patterns of connectivity. Thus, using a combination of electrical recordings and pathway tracing in the visual system, it was possible to generate an account of how experience during a critical period changes the way that the brain is wired, and how individual neurons respond to stimuli in the outside world.

Effects of Visual Deprivation on Ocular Dominance

As described in Chapter 12, if an electrode is passed at a shallow angle through the cortex while the responses of individual neurons to stimulation of one or the other eye are being recorded, detailed assessment of ocular dominance can be made at the level of individual cells (see Figure 12.11). In their original studies, Hubel and Wiesel assigned neurons to one of seven ocular dominance categories or groups, and this classification scheme has become standard in the field. Group 1 cells are defined as being driven only by stimulation of the contralateral eye; group 7 cells are driven entirely by the ipsilateral eye. Neurons driven equally well by either eye are assigned to group 4. Using this approach, they found that the ocular dominance distribution in all layers but layer 4 in primary visual cortex is roughly Gaussian in a normal adult (cats were used in these experiments). Most cells were activated to some degree by both eyes (distributed around a mean defined by "group 4" cells), and about a quarter were more activated by either the contralateral or ipsilateral eye (Figure 24.4A).

Hubel and Wiesel then asked whether this normal distribution of ocular dominance at the level of single cortical neurons could be altered by visual experience. When they sutured closed one eye of a kitten early in life and let the animal mature to adulthood (which takes about 6 months), a remarkable change was observed. Once the sutures were removed and the eyelid opened, electrophysiological recordings showed that very few cortical cells could be

Figure 24.4   Effect of early eyelid closure (deprivation of patterned light) of one ▶ eye on the distribution of cortical neurons driven by stimulation of both eyes. The histograms plot the number of cells that fall into one of the seven ocular dominance categories, defined based on the frequency of action potential activity elicited from visual cortical neurons following illumination in the relevant eye. In all three cases, the experimental readings were performed on cats 38 months after birth, at which time a penlight was shined into both wide-open eyes to elicit responses in the visual cortex. (A) Ocular dominance distribution of single-unit recordings from a large number of neurons in the primary visual cortex of normal adult cats. Cells in group 1 were activated exclusively by the contralateral eye, cells in group 7 by the ipsilateral eye. There were no cells that were not responsive (NR) to light stimulation in the retina. (B) One eye of a newborn kitten was closed from 1 week after birth until 2.5 months of age. After 2.5 months, the eye was opened and the kitten matured normally to 38 months. Note that the deprivation was relatively brief—the sutured eye had been open for 35.5 months of the cat's life. Even so, light presented to the open but transiently deprived eye elicited no electrical responses in visual cortical neurons. The only visually responsive cells responded to the ipsilateral (non-deprived) eye. (C) A much longer period of monocular deprivation in an adult cat showed little effect on ocular dominance, although overall cortical activity is diminished. Most of the responsive cells were driven by both eyes. In addition, some cells (groups 1 and 2) are uniquely or mostly responsive to the deprived eye. (A after Hubel and Wiesel, 1962; B after Wiesel and Hubel, 1963; C after Hubel and Wiesel, 1970.)

driven from the deprived (previously sutured) eye. The ocular dominance distribution had shifted; the eye that remained open was uniquely able to drive most cortical cells (Figure 24.4B). Recordings from the retina and lateral geniculate layers in response to direct electrical stimulation in the deprived eye (as opposed to light-elicited electrical responses) indicated that these more peripheral stations in the visual pathway worked quite normally. Thus, the absence of cortical cells that responded to stimulation of the closed eye was not a result of retinal degeneration or a loss of retinal connections to the thalamus. Rather, the deprived eye had been functionally disconnected from the visual cortex. Consequently, such animals are behaviorally blind in the deprived eye. This "cortical blindness," or amblyopia, is permanent (see next section). Even if the formerly deprived eye is unsutured and remains open, little or no recovery occurs.

Remarkably, the same manipulation—closing one eye—has no effect on the responses of cells in the visual cortex of an adult cat (Figure 24.4C). If one eye of a mature cat was closed for a year or more, both the ocular dominance distribution measured across all cortical layers and the animal's visual behavior were indistinguishable from normal when tested through the reopened eye. Thus, sometime between the time a kitten's eyes open (about a week after birth) and 1 year of age, visual experience determines how the visual cortex is wired with respect to eye dominance. After this time, deprivation or manipulation has little or no permanent, detectable effect. In fact, further experiments showed that eye closure is effective only if the deprivation occurs during the first 3 months of a kitten's life.

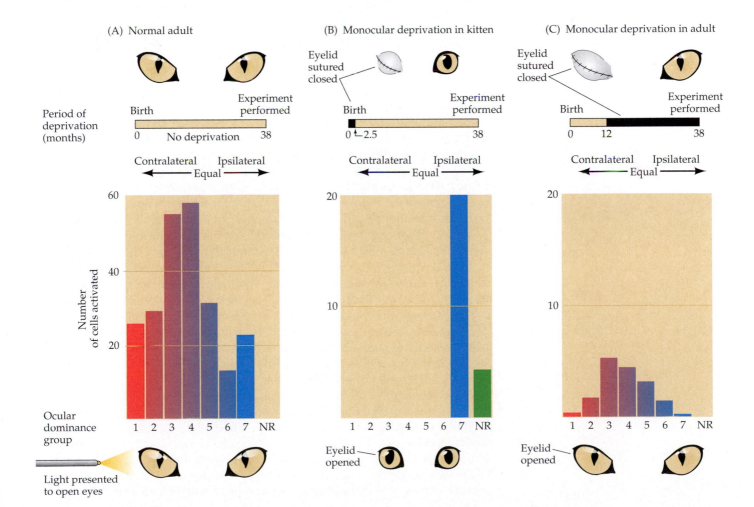

Figure 24.5 The consequences of a short period of monocular deprivation at the height of the critical period in the cat. (A) Just 3 days of deprivation produced a significant shift of cortical activation in favor of the non-deprived eye. (B) Six days of deprivation produced a shift of cortical activation in favor of the non-deprived eye that is almost as complete as that elicited by 2.5 months of deprivation (see Figure 24.4B). (After Hubel and Wiesel, 1970.)

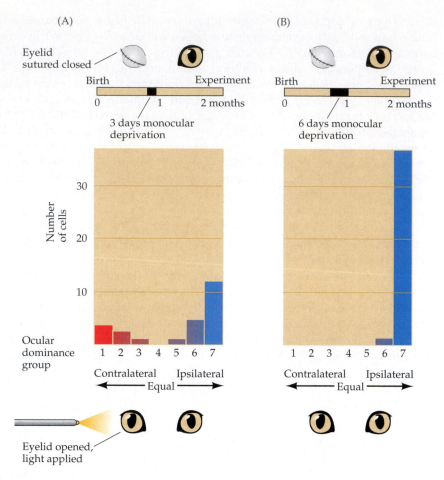

In keeping with the ethological observations described earlier in the chapter, Hubel and Wiesel called this period of susceptibility to visual deprivation the critical period for ocular dominance development. During the height of this critical period (about 4 weeks of age in the cat), as little as 3 to 4 days of eye closure profoundly alters the ocular dominance profile of the striate cortex (Figure 24.5). Similar experiments in monkeys have shown that the same phenomenon occurs in primates, although the critical period is longer (up to about 6 months of age).

The key advance arising from Hubel and Wiesel's early work was to show that visual deprivation must cause changes in cortical connectivity that influence the functional response properties of individual neurons. Moreover, their physiological studies showed that during the critical period, rapid and dramatic changes in responses could be elicited by very brief changes in the experience allotted one or the other eye. In contrast, when the same manipulations were done in maturity, after the close of the critical period, such changes were not observed, even when experience was altered for fairly long periods. Subsequent anatomical studies amplified the implications of physiological changes seen as result of experience. These studies established that the physiological changes were due to changes in patterns of connections. In monkeys, the stripe-like pattern of geniculocortical axon terminals in layer 4 that defines ocular dominance columns is already present at birth, and this pattern reflects the functional segregation of inputs from the two eyes (Figure 24.6A). Indeed, the early formation of this pattern in layer 4 reflects a significant amount of segregation of the lateral geniculate nucleus axons that relay information from one eye or the other, and this segregation occurs even in the absence of meaningful visual experience. Subsequent

(A)

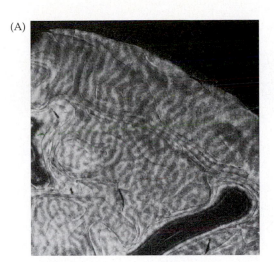

(B)

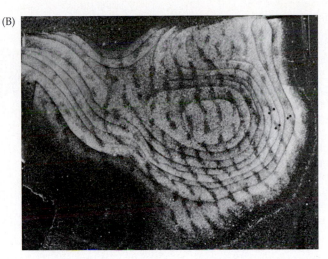

Figure 24.6 Effect of monocular deprivation on the pattern of ocular dominance columns in the macaque monkey. (A) In normal monkeys, ocular dominance columns are seen as alternating stripes of roughly equal width. (B) The picture is quite different after monocular deprivation. This dark-field autoradiograph shows a reconstruction of several sections through layer 4 of the primary visual cortex of a monkey whose right eye was sutured shut from 2 weeks to 18 months of age, when the animal was sacrificed. Two weeks before death, the normal (left) eye was injected with radiolabeled amino acids (see Box 24C). The columns related to the non-deprived eye (white stripes) are much wider than normal; those related to the deprived eye are shrunken. (A from Horton and Hocking, 1999; B from Hubel et al., 1977.)

observations have confirmed this initial experience-independent segregation, and there is some indication that specific molecular signals may distinguish LGN cells innervated by one eye or the other. Thus, the visual cortex is clearly not a blank slate on which the effects of experience are inscribed. Nevertheless, animals deprived from birth of vision in one eye develop abnormal patterns of ocular dominance stripes in the visual cortex (Figure 24.6B). The stripes related to the open eye are substantially wider than normal, and the stripes representing the deprived eye are correspondingly diminished. The absence of cortical neurons that respond to the deprived eye in electrophysiological studies is not simply a result of the relatively inactive inputs withering away. If this were the case, one would expect to see areas of layer 4 devoid of any thalamic innervation. Instead, inputs from the active (open) eye take over some—but not all—of the territory that formerly belonged to the inactive (closed) eye. These inputs then dominate the physiological responses of the target cortical neurons.

Hubel and Wiesel interpreted these results as demonstrating a **competitive interaction** between the two eyes during the critical period (see Chapter 22). In summary, the cortical representation of both eyes starts out equal. In normal animals, this balance is retained (and sharpened in terms of segregation of ocular dominance stripes in layer 4 of the cortex) if both eyes experience roughly comparable levels of visual stimulation. However, when an imbalance in visual experience is induced by monocular deprivation, the active eye gains a competitive advantage and replaces many of the synaptic inputs from the closed eye. In this case, even though LGN axons that arise from neurons innervated by the closed eye are retained in the cortex (albeit with much less extensive terminals), few if any neurons respond by firing action potentials when light is presented to the deprived eye. These observations in experimental animals have important implications for children with birth defects or ocular injuries that result in an imbalance of inputs from the two eyes. Unless the imbalance is corrected during the critical period, the child may ultimately have poor binocular fusion, diminished depth perception, and degraded acuity; in other words, the child's vision may be permanently impaired (see the next section).

The idea that a competitive imbalance underlies the altered distribution of inputs after deprivation has been confirmed by experimentally closing *both* eyes shortly after birth, thereby equally depriving all visual cortical neurons of normal experience during the critical period. The arrangement of ocular dominance recorded some months later is, by either electrophysiological or anatomical cri-

Figure 24.7 Terminal arborizations of lateral geniculate nucleus axons in the visual cortex can change rapidly in response to monocular deprivation during the critical period. (A) After only a week of monocular deprivation, axons terminating in layer 4 of the primary visual cortex from LGN neurons driven by the deprived eye have greatly reduced numbers of branches compared with those from the open eye. (B) Deprivation for longer periods does not result in appreciably larger changes in the arborization of geniculate axons. (After Antonini and Stryker, 1993.)

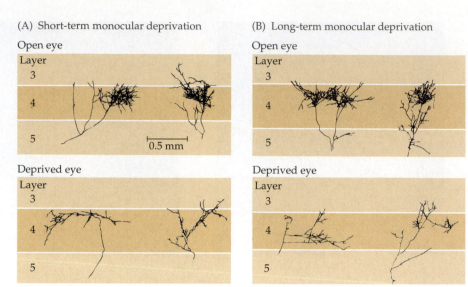

(A) Short-term monocular deprivation

(B) Long-term monocular deprivation

teria, much closer to normal than if just one eye is closed. Thus, the balance of inputs, not the absolute level of activity, is a key feature for shaping the normal pattern of connections. Although several peculiarities in the response properties of cortical cells deprived of normal light-dependent vision are apparent, roughly normal proportions of neurons responsive to the two eyes are present. Because there is no imbalance in the visual activity of the two eyes (both sets of related cortical inputs being deprived), both eyes retain their territory in the cortex. If disuse atrophy of the closed-eye inputs were the main effect of deprivation, then binocular deprivation during the critical period would cause the visual cortex to be largely unresponsive.

Experiments using techniques that label individual axons from distinct layers in the lateral geniculate nucleus have shown in greater detail what happens to the arborizations of individual LGN neurons in layer 4 of visual cortex after visual deprivation (Figure 24.7). As noted, monocular deprivation causes a loss of cortical territory related to the deprived eye, with a concomitant expansion of the open eye's territory. At the level of single axons, these changes are reflected in an increased extent and complexity of the arborizations related to the open eye (thus those arising from LGN cells innervated by the open eye), and a decrease in the size and complexity of the arborizations of LGN neurons related to the deprived eye. Individual neuronal arborizations can be substantially altered after as little as one week of deprivation, and perhaps even less. This latter finding highlights the ability of developing thalamic and cortical neurons to rapidly remodel their connections (presumably making and breaking synapses) in response to environmental circumstances.

Visual Deprivation and Amblyopia in Humans

Developmental phenomena in the visual systems of experimental animals accord with clinical problems in children who have experienced similar deprivation. The loss of acuity, diminished stereopsis, and problems with fusion that arise from early deficiencies of visual experience is called **amblyopia** (Greek, "dim sight"). These functional difficulties are all believed to reflect the essential contribution made by normal binocular input and competition for defining, in an experience-dependent manner, the cortical circuitry necessary for binocular vision and depth perception.

In humans, amblyopia is most often the result of **strabismus**—a misalignment of the two eyes due to improper control of the direction of gaze by the eye muscles (and referred to colloquially as "lazy eye"). Depending on the muscles affected, the misalignment can produce convergent strabismus, called **esotropia** ("cross-eyed"); or divergent strabismus, called **exotropia**. These alignment errors, which produce double vision, are surprisingly common, affecting about 5 percent of children. The response of the visual system in some of these individuals is to suppress input from one eye by mechanisms that are not completely understood, but which are thought to reflect competitive interactions during the critical period. Presumably, the inputs to the LGN from the properly aligned eye are competitively advantaged, and thus the corresponding LGN inputs are accorded more territory in the visual cortex. Functionally, the suppressed eye eventually comes to have very low acuity and may render the affected individual effectively blind in that eye. Early surgical correction of ocular misalignment (by adjusting the lengths of extraocular muscles) has become an essential treatment for strabismic children.

Another cause of visual deprivation in humans is cataracts. Cataracts, which can be caused by several congenital conditions, render the lens opaque. Diseases such as onchocerciasis ("river blindness," the result of infection by the parasitic nematode *Onchocerca volvulus*) and trachoma (caused by the parasitic bacterium *Chlamydia trachomatis*) affect millions of people in underdeveloped tropical regions, often inducing corneal opacity in one or both eyes. A cataract in one eye is functionally equivalent to monocular deprivation in experimental animals. Left untreated in children, this defect also results in an irreversible effect on the visual acuity of the deprived eye. If either the cataract or corneal opacity is removed before about 4 months of age, however, the consequences of monocular deprivation are largely avoided. As expected from Hubel and Wiesel's work, bilateral cataracts, which are similar to binocular deprivation in experimental animals, produce less dramatic deficits even if treatment is delayed. Apparently, unequal competition during the critical period for normal vision is more deleterious than the complete disruption of visual input that occurs with binocular deprivation.

In keeping with the findings in experimental animals, the visual abilities of individuals monocularly deprived of vision as adults (for example, by cataracts or corneal scarring) are much less compromised, even after decades of deprivation, when vision is restored (although there may be important psychological consequences of restoring sight after prolonged binocular blindness, as engagingly described by the neurologist Oliver Sacks, among others). Nor is there any evidence of anatomical change in this circumstance. For instance, a patient whose eye was surgically removed in adulthood showed normal ocular dominance columns when his brain was examined postmortem many years later. Thus, one can detect evidence of critical period phenomena for visual cortical development and behavior in the visual system of humans based upon careful examination of patients with opthalmic disease or other lesions.

Mechanisms by which Neuronal Activity Affects the Development of Neural Circuits

How, then, are differences in patterns of neural activity translated into changes in neural circuitry? In 1949, the psychologist D. O. Hebb hypothesized that the coordinated activity of a presynaptic terminal and a postsynaptic neuron strengthens the synaptic connection between them. *Hebb's postulate*, as it has come to be known, was originally formulated to explain the cellular basis of learning and memory, but the general concept has been widely applied to situ-

Figure 24.8 Representation of Hebb's postulate as it might operate during development of the visual system. The cell represents a postsynaptic neuron in layer 4 of the primary visual cortex. Early in development, inputs from the two eyes converge on single postsynaptic cells. The two sets of presynaptic inputs, however, have different patterns of electrical activity. Each pattern is more correlated with terminals driven by the same eye than that driven by the opposite eye. Activity patterns, corresponding to action potential frequency, are represented by the short vertical bars. In the example here, the three left eye inputs are better able to activate the postsynaptic cell. These inputs cause the postsynaptic cell to fire a pattern of action potentials that follows the pattern seen in the input. As a result, the activity of the presynaptic terminals and the postsynaptic neuron is highly correlated. According to Hebb's postulate, these synapses are therefore strengthened. The inputs from the right eye carry a different pattern of activity that is less well correlated with the majority of the activity elicited in the postsynaptic cell. These synapses gradually weaken and are eventually eliminated (right-hand side of figure), while the correlated inputs form additional synapses.

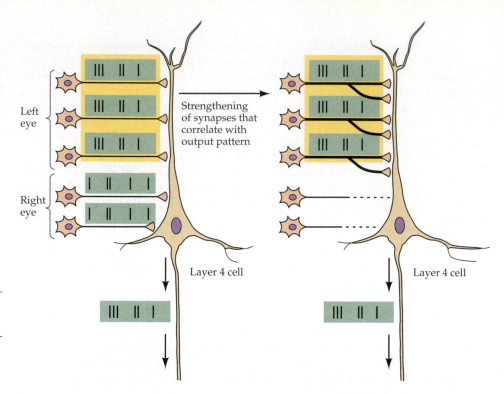

ations involving long-term modifications in synaptic strength, including those that occur during development of neural circuits. In this context, Hebb's postulate implies that synaptic terminals strengthened by correlated activity will be retained or sprout new branches, whereas those terminals that are persistently weakened by uncorrelated activity will eventually lose their hold on the postsynaptic cell (Figure 24.8; see also Chapter 22).

In the visual system, the action potentials of the thalamocortical inputs related to one eye are presumably better correlated with each other than with the activity related to the other eye—at least in layer 4. If sets of correlated inputs tend to dominate the activity of groups of locally connected postsynaptic cells, this relationship would exclude uncorrelated inputs (Box 24D describes a particularly striking example of this principle). Thus, patches of cortex occupied exclusively by inputs representing one eye or the other could arise. In this scenario, ocular dominance column rearrangements in layer 4 are generated by *cooperation* between inputs carrying similar patterns of activity, and by *competition* between inputs carrying dissimilar patterns.

Monocular deprivation, which dramatically changes ocular dominance columns, clearly alters both the levels and patterns of neural activity between the two eyes. To specifically test the role of correlated activity in driving the competitive postnatal rearrangement of cortical connections, it is necessary to create a situation in which activity levels in each eye remain the same but the correlations between the two eyes are altered. This circumstance can be created in experimental animals by cutting one of the extraocular muscles in one eye. As already mentioned, this condition, in which the two eyes can no longer be aligned, is called strabismus. The major consequence of strabismus is that objects in the same location in visual space no longer stimulate corresponding points on the two retinas at the same time. As a result, differences in the visually evoked patterns of activity between the two eyes are far greater than normal. Unlike monocular deprivation, however, the overall amount of activity in each eye remains roughly the same; only the correlation of activity arising from corresponding retinal points is changed.

BOX 24D   Correlation as Causation: Lessons from a Three-Eyed Frog

Any good statistician will tell an eager student that "correlation is not causation." In other words, it is not wise to draw strong conclusions solely from the coincidence of two events occurring in the same place at the same time. While this is wise advice for cautiously interpreting a link between the increase in urban ice cream sales and street crime, it is less useful for considering the implications of how correlated neural activity leads to segregated patterns of connectivity. The causal role for correlation of neural activity with the generation of periodic segregated patterns of inputs—like ocular dominance columns in the visual cortex—was given compelling support by creating, experimentally, a situation in which two sets of highly correlated inputs were forced to innervate a target territory that normally receives input from only one source. This sort of "augmentation" experiment is the equivalent of "gain-of-function" experiments used by geneticists to confirm their interpretations of mutant animals, or "loss of function" phenotypes that reflect inactivation of the same gene.

Thus, while most investigators agreed that the consequences of visual deprivation or, more dramatically, eye removal (the "loss of function" experiment) pointed toward a primary role for correlated activity shaping connections in the visual system, the definitive "gain of function" experiment—creating such patterns by introducing two independently correlated sets of inputs where previously there had only been one—seemed necessary. This experiment eventually was performed in a species that normally has no binocular input segregation: the frog *Rana pipiens*. The challenge was to create a circumstance in which two eyes had similar levels of activity and the molecular cues necessary to make connections in a shared target (insuring equal competitive opportunity for each), but adequate differences in

correlated activity between the each eye to facilitate competition, and, potentially, segregation of connections.

In the early 1970s, Martha Constantine-Paton and her colleagues, then at Princeton University, took advantage of the accessibility of frogs for embryonic manipulation to create a very odd, but very informative variant: a three-eyed *Rana pipiens*. By transplanting retinal primordium from a donor embryo to a position somewhere between the two immature eyes of a host tadpole, the research team generated an adult frog with three functional eyes. Remarkably, the third eye was able to differentiate somewhat normally. It generated all of the usual retinal cell types, was sensitive to light, and made neuronal connections with the central target structure of the two normal eyes—the optic tectum (the equivalent of the superior colliculus in mammals). This experimental oddity provided the long sought "augmentation" experiment. Moreover, the experiment demonstrated clearly that correlated activity was a primary mechanism for segregating inputs.

The three-eyed frogs would have been unremarkable—as well as somewhat unattractive—if it were not for the dramatic change that occured when the third eye innervated one of two normally singly innervated tecta. In a two-eyed frog, each eye exclusively innervates the

(Continued on next page)

(A)

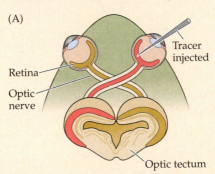

Retina
Optic nerve
Tracer injected
Optic tectum

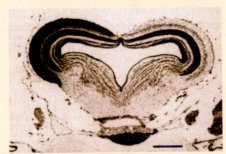

(B)

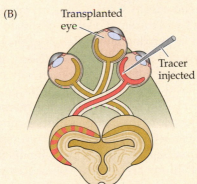

Transplanted eye
Tracer injected

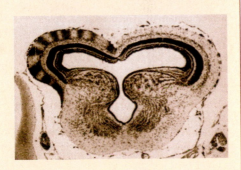

(A) The pattern of connections of a single eye in a normal frog. A micrograph of the frog tectum shows the continuous band of axon terminals seen when one eye has been injected with a tracer. (B) Arrangement of the eyes in a three-eyed frog, and the consequences of dual innervation of one tectum for patterned connections. In the micrograph, the sharp, periodic segregation of inputs from the two eyes is seen to create a pattern of eye-specific stripes—reminiscent of the mammalian optical dominance column—over the entire doubly innervated tectum. (After Katz and Crowley, 2002; micrographs from Constantine-Paton and Law, 1978.)

BOX 24D   (Continued)

optic tectum on the contralateral side. There is no partial decussation that leads to binocular innervation, and the input from each eye is distributed in one continuous band of axon terminals in the retinal-recipient layer of the tectum. When input from a third eye must be accommodated, however, a dramatic change occurs: the third eye innervates one of the two tecta selectively, and its axon terminals become segregated from that of the normal eye into periodic bands. These bands are of the same size and pattern as ocular dominance columns in the mammalian visual cortex: their boundaries are quite sharp, and the dendrites of tectal cells within each band tend not to transgress the territory innervated by the other band. Moreover, the development of this pattern relies on normal electrical activity in both the eyes that doubly innervate the tectum. The cellular basis of this apparently activity-dependent segregation is mediated by NMDA-dependent mechanisms that are thought generally to underlie Hebbian

phenomena in which correlated activity leads to segregated patterns of synaptic input.

This visual augmentation experiment thus confirms the interpretation of visual deprivation experiments in mammals. When a correlated, visually driven set of inputs is added to another set of similar inputs that normally do not segregate, the higher degree of correlation of activity within each set of inputs (versus between the inputs from the two independent eyes) leads to a periodic segregation of connections. Conversely, when one set of inputs is removed from a system where there is normal innervation by two eyes, the segregated pattern is disrupted and the remaining input changes to a continuous, or dramatically expanded pattern of connectivity driven by the intact eye. Here, correlation is indeed causation: the correlated activity of two distinct sets of visual inputs from two independent eyes causes segregation of connectivity in the target region. This segregation presumably reflects the

Hebbian mechanisms that are thought to govern competitive synapse formation and elimination throughout the developing (and mature) brain.

References

CLINE, H. T., E. A. DEBSKI AND M. CONSTANTINE-PATON (1987) N-Methyl-D-aspartate receptor antagonist desegregates eye-specific stripes. *Proc. Natl. Acad. Sci. USA* 84: 4342–4345.

CONSTANTINE-PATON, M. AND M. I. LAW (1978) Eye-specific termination bands in tecta of three-eyed frogs. *Science* 202: 639–641.

KATZ, L. C. AND M. CONSTANTINE-PATON (1988) Relationships between segregated afferents and postsynaptic neurones in the optic tectum of three-eyed frogs. *J. Neurosci.* 8: 3160–3180.

KATZ, L. C. AND J. C. CROWLEY (2002) Development of cortical circuits: Lessons from ocular dominance columns. *Nature Rev. Neurosci.* 3: 34–42.

REH, T. A. AND M. CONSTANTINE-PATON (1985) Eye-specific segregation requires neural activity in three-eyed *Rana pipiens*. *J. Neurosci.* 5: 1132–1143.

The effects of strabismus in experimental animals provide an illustration of the basic validity of Hebb's postulate. Recall that misalignment of the two eyes can, depending on the details of the situation, lead to suppression of the input from one eye and eventual loss of the related cortical connections. In other instances, however, input from both eyes is retained. The anatomical pattern of ocular dominance columns in layer 4 of cats in which input from both eyes remains (but is asynchronous) is sharper than normal. Apparently, the total independence of the two eyes further enhances ipsilateral- versus contralateral-correlated activity. This enhanced, eye-specific correlation actually accentuates the normal separation of cortical inputs from the two eyes. In addition, the ocular asynchrony prevents the binocular convergence that normally occurs in cells above and below layer 4; ocular dominance histograms from such animals show that cells in *all* layers are driven exclusively by one eye *or* the other (Figure 24.9). Evidently, strabismus not only accentuates the competition between the two sets of thalamic inputs in layer 4, but also prevents binocular interactions in the other layers, which are mediated by local connections originating from cells in layer 4.

Even before visual experience exerts these effects, innate mechanisms have ensured that the basic outlines of a functional system are present. These intrinsic mechanisms (including those that constrain axon growth and initiate the formation of topographic maps; see Chapter 23) establish the general circuitry required for vision, but allow slight modifications to accommodate the individ-

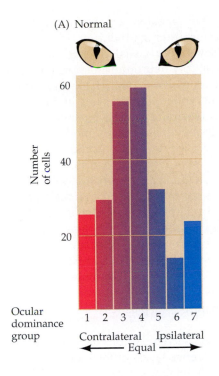

(A) Normal

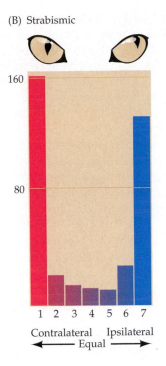

(B) Strabismic

Figure 24.9   Ocular dominance histograms obtained by electrophysiological recordings in (A) normal adult cats and (B) adult cats in which strabismus was induced during the critical period. The data in (A) is the same as that shown in Figure 24.3A. The number of binocularly driven cells (groups 3, 4, and 5) is sharply decreased as a consequence of strabismus; most of the cells are driven exclusively by stimulation of one eye or the other. This enhanced segregation of the inputs presumably results from the greater discrepancy in the patterns of activity between the two eyes as a result of surgically interfering with normal conjugate vision. This pathological state is thought to enhance the relative degree of correlation within inputs from each eye, and decrease the possibility of correlation between eye inputs. (After Hubel and Wiesel, 1965.)

ual requirements that occur with changes in head size or eye alignment. Normal visual experience evidently validates the initial wiring and preserves, augments, or adjusts the normal arrangement. In the case of abnormal experience, such as monocular deprivation, the mechanisms allowing these adjustments result in more dramatic anatomical (and ultimately behavioral) changes, such as those that occur in amblyopia. Presumably, these changes guarantee the maximal adaptive function of the system in the face of the altered capacity of the periphery to encode and relay information. The eventual decline of this capacity to remodel cortical (and subcortical) connections is most likely the cellular basis of critical periods in a variety of neural systems, including the development of language and other higher brain functions. By the same token, these differences in plasticity as a function of age presumably provide a neurobiological basis for the general observation that human behavior is much more susceptible to normal or pathological modification early in development than later on—a concept with obvious educational, psychiatric, and social implications. Although many cellular and molecular mechanisms have been proposed to explain these effects, the specific mechanisms responsible for creating and eventually terminating critical periods (as opposed to those that permit synaptic stabilization or rearrangement; see the next section) remain largely unknown.

Cellular and Molecular Correlates of Activity-Dependent Plasticity during Critical Periods

A basic question for understanding how experience changes neural circuits during critical periods is how patterns of activity are transduced to modify connections and to make these changes permanent. Clearly, the steps that initiate these processes must rely on signals generated by the synaptic activity associated with sensory experience or motor performance—the basic neural processes by which experience is represented. Neurotransmitters and a number of other signaling molecules, including neurotrophic factors, are obvious candidates for initiating changes that occur with correlated or repeated activity. Indeed, mice that lack genes for a number of neurotransmitter-synthetic or -degrading enzymes exhibit changes in experience-dependent visual cortical plasticity. Without doubt, the effects of correlated excitatory neurotransmission, particularly that via glutamate transduced by NMDA receptors (which are capable of detecting local correlated synaptic activity and modifying signaling in the postsynaptic cell; see Chapter 8), are key for understanding activity-dependent competition and plasticity. Similarly, the local and global effects of modified neurotrophin signaling—particularly for brain-derived neurotrophic factor (BDNF) and its receptors—include altered synaptic plasticity during critical periods. Thus, when these signaling molecules are manipulated experimentally, or when the genes encoding the ligands or receptors are mutated, some critical period phenomena are altered. One major target of all of these molecular signaling processes is apparently the network of local inhibitory connections made by GABAergic neurons. Regulation of the number and placement of local inhibitory synapses, as well as expression of GABA receptors at postsynaptic sites, all seems to be exquisitely sensitive to critical period manipulations. Thus the regulation of inhibitory connectivity has become a major focus for understanding the molecular and cellular mechanisms of plasticity during critical periods.

Intriguingly, neurotransmitters and other secreted molecules (like neurotrophins) are all thought to ultimately influence levels of intracellular Ca^{2+}, particularly in postsynaptic cells (Figure 24.10). Increased Ca^{2+} concentration in the affected cells can activate a number of kinases, including Ca^{2+}/calmodulin kinase (CaMK) II or IV, leading to phosphorylation-dependent modifications of the cytoskeleton and changes in dendritic and axonal branching. In addition, changes in Ca^{2+} can activate other kinases that translocate to the nucleus, including the extracellular signal-regulated kinases (ERKs; see Chapter 7). Once in the nucleus, ERKs can activate transcription factors like CREB (*c*yclic nucleotide *r*esponse *e*lement *b*inding transcription factor) via phosphorylation. In addition, particularly during critical periods, these kinases may facilitate modification of histone-binding proteins mediating chromatin conformations that favor gene expression. As the critical period draws to an end, it is possible

Figure 24.10   Transduction of electrical activity into cellular change via Ca^{2+} signaling. (A) A target neuron, showing two possible sites of action—the cell soma and the distal dendrites—for activity-dependent increases in Ca^{2+} signaling. (B) Correlated or sustained activity leads to increased Ca^{2+} conductances and increased intracellular Ca^{2+} concentration, which results in activation of Ca^{2+}/calmodulin kinase II or IV (CaMKII, CaMKIV) as well as ERK, and their subsequent translocation to the nucleus. ERK and CaMKII/IV then activate Ca^{2+}-regulated transcription factors such as CREB, as well as other chromatin binding proteins (not shown). The target genes for activated CREB may include neurotrophic signals like BDNF, which when secreted by a cell may help stabilize or promote the growth of active synapses on that cell. (C) Local increases in Ca^{2+} signaling in distal dendrites due to correlated or sustained activity may lead to local increases in Ca^{2+} concentration that modify cytoskeletal elements (actin- or tubulin-based structures), perhaps through the activity of kinases like CaMKII/IV operating in the cytoplasm rather than the nucleus. Changes in these cytoskeletal elements lead to local changes in dendritic structure. In addition, increased local Ca^{2+} concentration may influence local translation of transcripts in the endoplasmic reticulum, including transcripts for neurotransmitter receptors and other modulators of postsynaptic responses. Increased Ca^{2+} may also influence the trafficking of these proteins, their insertion into the postsynaptic membrane, and their interaction with local scaffolds for cytoplasmic proteins (see Figure 23.7). (After Wong and Ghosh, 2002.)

that such modification of chromatin-binding protein is minimal, thus making altered transcription—and corresponding plastic changes—more difficult. But when activation occurs during the critical period, these DNA-binding proteins, along with CREB, can influence gene expression, and thus alter the transcriptional state of the neuron to reflect experience-driven functional changes. Such changes may include (but are unlikely to be limited to) transcription of neurotrophin genes like *BDNF*.

Local increase of BDNF, especially if its secretion is limited to synapses from "stabilized" inputs, may lead to greater local elaboration of synaptic endings, additional dendritic growth, and the net increase of connectivity between the pre- and postsynaptic partners. GABAergic interneurons are particularly sensitive to altered BDNF signaling. This is consistent with the presumed role for local inhibitory circuits in consolidating the connections that are sorted out by activity-dependent competitive interactions during the critical period. It is also possible that activity-dependent mechanisms influence the expression and extracellular distribution of matrix components, including chondroitin sulfate proteoglycans. These matrix molecules might further influence the location and stability of subsets of synapses. Whether this sequence is correct or complete remains uncertain, but it provides a plausible scenario for the molecular and cellular events underlying activity-dependent plasticity.

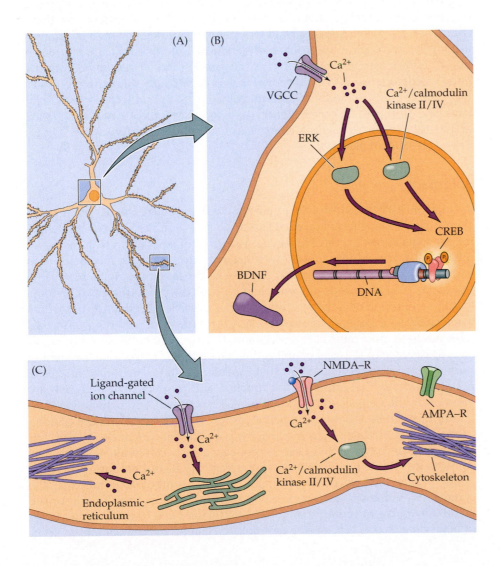

Evidence for Critical Periods in Other Sensory Systems

Although the neural basis of critical periods has been most thoroughly studied in the mammalian visual system, similar phenomena exist in a number of sensory systems, including the auditory, somatic sensory, and olfactory systems. Experiments on the role of auditory experience and neural activity in owls (who use auditory information to localize prey) indicate that neural circuits for auditory localization are similarly shaped by experience (see Table 24.1). Thus, deafening an owl or altering neural activity during early postnatal development compromises the bird's ability to localize sounds and can alter the neural circuits that mediate this capacity. The development of song in many species of birds provides another example in the auditory system. In the somatic sensory system, cortical maps can be changed by experience during a critical period of postnatal development. In mice or rats, for instance, the anatomical patterns of "whisker barrels" in the somatic sensory cortex (see Chapter 9) can be altered by abnormal sensory experience (or by removing subsets of sensory receptors like the whiskers) during a narrow window in early postnatal life. In the olfactory system, behavioral studies (outlined in Chapter 15) indicate that exposure to maternal odors for a limited period can alter the ability to respond to such odorants, a change that can persist throughout life.

Clearly, the phenomenon of the critical period is general to the development of sensory perceptual abilities and motor skills. The primary evidence for these critical periods comes from deprivation experiments, complemented by analysis using pharmacological approaches or genetically modified animals in which major neurotransmitter synthetic pathways are disabled, essential neurotransmitter receptors like NMDA receptors are lost, or major signaling molecules (e.g., the calcium/calmodulin kinases, BDNF, or the neurotrophin receptors) have been disrupted. In each instance, these modifications of synaptic signaling and its consequences result in changes in the duration or efficiency of critical period-dependent plasticity.

Summary

An individual animal's history of interaction with the environment—its "experience"—helps shape its neural circuitry and thus determines subsequent behavior. In some cases, experience functions primarily as a switch that activates innate behaviors. More often, however, experience during a specific time in early life—referred to as a critical period—helps shape the adult behavioral repertoire. Critical periods influence behaviors as diverse as maternal bonding and the acquisition of language. Although it is possible to define the behavioral consequences of critical periods for these complex functions, their biological basis has been more difficult to understand. The most accessible and thoroughly studied example of a critical period is the one pertinent to the establishment of normal vision. These studies show that experience is translated into patterns of neuronal activity that influence the function and connectivity of the relevant neurons. In the visual system, and other systems as well, competition between inputs with different patterns of activity is an important determinant of adult connectivity. Correlated patterns of activity in afferent axons tend to stabilize connections and conversely a lack of correlated activity can weaken or eliminate connections. When normal patterns of activity are disturbed during a critical period in early life (experimentally in animals or by pathology in humans), the connectivity in the visual cortex is altered, as is visual function. If not reversed before the end of the critical period, these structural and functional alterations of brain circuitry are difficult or impossible to change. The cellular and molecular mechanisms implicated in critical period include, not surprisingly, many neurotransmitters, receptors, and intracellular signaling cascades that can modify gene expression in response to changes in synaptic activity in a target cell. Genes for neurotrophins like BDNF, for extracellular matrix components, and for some classes of neurotransmitter receptors may all be targets for altered expression in response to synaptic activity during the critical period. Regardless of the specific molecular mechanisms, the influence of activity on neural connectivity during the critical period presumably enables the maturing brain to store the vast amounts of information, thus encoding early experiences that shape the abilities and idiosyncrasies of each individual.

Additional Reading

Reviews

KATZ, L. C. AND C. J. SHATZ (1996) Synaptic activity and the construction of cortical circuits. *Science* 274: 1133–1138.

KNUDSEN, E. I. (1995) Mechanisms of experience-dependent plasticity in the auditory localization pathway of the barn owl. *J. Comp. Physiol.* 184(A): 305–321.

SHERMAN, S. M. AND P. D. SPEAR (1982) Organization of visual pathways in normal and visually deprived cats. *Physiol. Rev.* 62: 738–855.

WIESEL, T. N. (1982) Postnatal development of the visual cortex and the influence of environment. *Nature* 299: 583–591.

WONG, W. O. AND A. GHOSH (2002) Activity-dependent regulation of dendritic growth and patterning. *Nat. Rev. Neurosci.* 10: 803–812.

HENSCH, T. K. (2004) Critical period regulation. *Annu. Rev. Neurosci.* 27: 549–579.

Important Original Papers

ANTONINI, A. AND M. P. STRYKER (1993) Rapid remodeling of axonal arbors in the visual cortex. *Science* 260: 1819–1821.

CABELLI, R. J., A. HOHN AND C. J. SHATZ (1995) Inhibition of ocular dominance column formation by infusion of NT-4/5 or BDNF. *Science* 267: 1662–1666.

HUANG, Z. J., A. KIRKWOOD, T. PIZZORUSSO, V. PORCIATTI, B. MORALES, M. F. BEAR, L. MAFFEI AND S. TONEGAWA (1999) BDNF regulates the maturation of inhibition and the critical period of plasticity in mouse visual cortex. *Cell* 98: 739–755.

HORTON, J. C. AND D. R. HOCKING (1999) An adult-like pattern of ocular dominance columns in striate cortex of newborn monkeys prior to visual experience. *J. Neurosci.* 16: 1791–1807.

HUBEL, D. H. AND T. N. WIESEL (1965) Binocular interaction in striate cortex of kittens reared with artificial squint. *J. Neurophysiol.* 28: 1041–1059.

HUBEL, D. H. AND T. N. WIESEL (1970) The period of susceptibility to the physiological effects of unilateral eye closure in kittens. *J. Physiol.* 206: 419–436.

HUBEL, D. H., T. N. WIESEL AND S. LeVAY (1977) Plasticity of ocular dominance columns in monkey striate cortex. *Phil. Trans. R. Soc. Lond. B* 278: 377–409.

KUHL, P. K., K. A. WILLIAMS, F. LACERDA, K. N. STEVENS AND B. LINDBLOM (1992) Linguistic experience alters phonetic perception in infants by 6 months of age. *Science* 255: 606–608.

LeVAY, S., T. N. WIESEL AND D. H. HUBEL (1980) The development of ocular dominance columns in normal and visually deprived monkeys. *J. Comp. Neurol.* 191: 1–51.

RAKIC, P. (1977) Prenatal development of the visual system in the rhesus monkey. *Phil. Trans. R. Soc. Lond. B* 278: 245–260.

STRYKER, M. P. AND W. HARRIS (1986) Binocular impulse blockade prevents the formation of ocular dominance columns in cat visual cortex. *J. Neurosci.* 6: 2117–2133.

WIESEL, T. N. AND D. H. HUBEL (1965) Comparison of the effects of unilateral and bilateral eye closure on cortical unit responses in kittens. *J. Neurophysiol.* 28: 1029–1040.

Books

CURTISS, S. (1977) *Genie: A Psycholinguistic Study of a Modern-Day "Wild Child."* New York: Academic Press.

HUBEL, D. H. (1988) *Eye, Brain, and Vision.* Scientific American Library Series. New York: W. H. Freeman.

PURVES, D. (1994) *Neural Activity and the Growth of the Brain.* Cambridge: Cambridge University Press.

Chapter 25

Repair and Regeneration in the Nervous System

Overview

The ability of brain tissue to alter, renew, or repair itself (beyond the molecular or cellular changes associated with synaptic plasticity) is limited. Unlike many other organs—notably the lungs, intestine, and liver—that generate new cells constitutively or after damage, human brains do not produce large numbers of new neurons once the initial complement is established from mid-gestation through early postnatal life. Moreover, neurons have only limited ability to replace damaged axons and dendrites; few adult neurons in the central nervous system can grow a new axon once the original axon has been severed or injured, nor can neurons replace dendrites lost due to local tissue damage or degenerative diseases. The fortunate exception is that peripheral axons can regrow through vacated peripheral nerve sheaths and eventually reinnervate sensory specializations in the skin or synaptic sites on muscles. In peripheral nerves following injury, Schwann cells produce a number of developmentally regulated molecules that promote axon growth and synapse formation. In the brain, however, three barriers restrain regenerative capacity. First, the consequences of local injury to brain tissue often lead to neuronal cell death. Second, several other cell classes, particularly glial cells, actively inhibit the growth of axons over long distances. And third, although neural stem cells (cells with the capacity to give rise to all of the cell classes in nervous tissue) are retained in the adult brain, most of these stem cells are constrained in their ability to divide, migrate, and differentiate. There are a few examples, however, where these impediments are circumvented. In the visual systems of some lower vertebrates and in the olfactory pathway and hippocampus in mammals, there is constitutive addition or replacement of neurons throughout life, as well as continual growth of newly generated axons through adult brain tissue. Efforts to understand these exceptions provide a foundation for ongoing research into potential therapies for brain repair following traumatic injury or degeneration due to pathologies such as Parkinson's, Huntington's, or Alzheimer's diseases.

The Damaged Brain

Many organs are capable of a great deal of repair and regeneration. The epithelial cells of the epidermis and intestinal lining are constantly lost and replaced, as are blood cells. Broken bones mend and wounds heal. Recent transplant surgeries have demonstrated that the adult liver has previously unsuspected regenerative abilities. However, the brain, especially in mammals, is generally refractory to repair. This deficiency was noted clinically at the dawn of medical history, millennia ago in ancient Egypt (Figure 25.1). Deeper understanding of the structure and function of nervous tissue gained in the intervening centuries only reinforced this sense of hopelessness. The realization that nervous tissue is made up

Figure 25.1 An ancient Egyptian papyrus acknowledges the difficulties of repairing the brain and spinal cord after serious injury. The symbols in brown translate to: *When you examine a man with a dislocation of a vertebra of his neck, and you find him unable to move his arms, and his legs…Then you have to say: a disease one cannot treat.* (From Case and Tessier-Lavigne, 2005.)

of many classes of highly branched, interconnected nerve cells that communicate via electrical impulses made it clear that repairing nervous tissue presents a far greater challenge than regenerating the liver. Moreover, as postmortem analysis and histological study of the nervous system progressed, it became clear that specifically localized lesions accompany the behavioral deficits seen after brain damage, and that these lesions remain visible even many years after the injury, suggesting there is little repair of the damaged tissue. In such patients, the injured region is characterized either by a fluid-filled cyst or by "scar tissue" that lacks the histological integrity of neighboring intact brain regions; thus we are led to conclude that adaptive changes in brain function are unlikely to reflect wholesale replacement or remodeling of the cellular constituents of the brain.

Many early students of the nervous system took the position that the brain, if damaged, could never be restored to its previous state. Although still widely held, this view has been challenged over the past century and continues to be vigorously debated. As described below, some of the most promising new evidence for the possibility of limited brain repair has come from a better understanding of the developmental mechanisms used to construct neural tissue in the first place.

Three Types of Brain Repair

The possibilities and limits of brain regeneration rest on three types of repair that can occur when nervous tissue is damaged. The first type of repair is the regrowth of axons from nerve cells in peripheral ganglia or those in the central nervous system whose peripherally projecting axons are severed (Figure 25.2A). This scenario for repair requires both a reactivation of the developmental processes for axon growth and guidance as well as initial synapse formation. In addition, such repair may require activity-dependent competitive mechanisms to insure proper quantitative matching of newly regrown afferents to temporarily denervated targets. This first type of repair is seen primarily when sensory or motor nerves are damaged in the periphery, leaving the nerve cell bodies in the relevant sensory and autonomic ganglia or the spinal cord intact. It is the most easily accomplished type of repair in the nervous system, and the most clinically successful.

The second type of repair is restoration of damaged nerve cells in the central nervous system that, while injured, nevertheless survive (Figure 25.2B). This response requires first that nerve cells be capable of restoring their damaged processes and connections to some level of functional integrity. Whereas this type of repair shares many of the requirements that are met in peripheral nerve repair, it also requires the cooperative regrowth of existing neuronal and glial elements in a more complex environment. Such repair is less easily accomplished than peripheral axonal repair, and often is accompanied by altered growth of glial cells at the expense of the local neurons. One possible reason for this failure is the delicate balance of immune system-mediated clearance of damaged tissue and the local inflammatory responses engendered by this process. Thus, loss of trophic support due to damage to axons and dendrites, plus the action of inflammatory cytokines (released by macrophages and other immune cells in response to tissue damage) may suppress reactivation of cellular mechanisms for axonal and dendritic regrowth and synapse formation.

The third type of brain repair is the wholesale genesis of new neurons to replace those that have been lost, whether through normal wear and tear or as a result of traumatic damage and subsequent neuronal death (Figure 25.2C). This last type of regeneration occurs rarely and its mechanisms are controversial. For such repair to occur, several criteria must be met. First, nervous tissue must retain

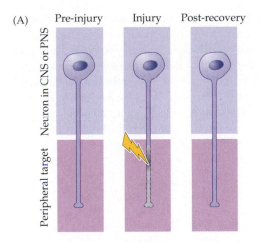

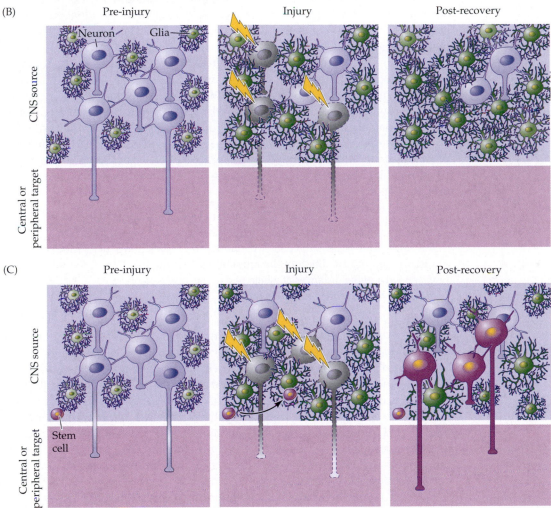

Figure 25.2 Three types of nervous system repair or regeneration. (A) In peripheral nerve regeneration, when peripheral axons are severed, the neuron, whether in a peripheral ganglion or in the central nervous system, regenerates the distal portion of the axon. (B) Repair of existing neurons at and around a site of injury in the central nervous system. Prior to injury, glial cells (dark green) are quiescent. Immediately following the injury, the glial cells grow, axons and dendrites degenerate and connections are lost. Following recovery, some modest axon and dendrite growth may be seen, but the hypertrophic glial cells remain to form a "scar" at the site of the tissue damage. (C) Neuronal replacement depends on the maintenance of a neural stem cell (dark purple). Following injury, this stem cell proliferates and gives rise to new neuroblasts that then differentiate and integrate into the damaged tissue. These new neurons (dark purple) make connections with existing cells.

a population of *multipotent neural stem cells* able to give rise to all of the cell types found in the mature part of the brain (see Box 22A). Second, these neural stem cells must be present in a distinct region or "niche" that retains an appropriate environment for the genesis and differentiation of new nerve cells and glia. Third, the regenerating tissue must preserve the capacity to recapitulate the migration, process outgrowth, and synapse formation necessary to reconstitute local functional networks of connections as well as long distance connections.

The remainder of this chapter considers each these three types of brain repair and regeneration. For the most part, the extent of neural repair possible appears to be most limited in the mammalian brain. Some other vertebrate species (and many more invertebrate species) exhibit the capacity for axon regrowth, neuron replacement, and tissue regeneration. These examples in other animals suggest to some investigators that a better understanding of cellular and molecular biology could possibly lead to treatments that promote the repair of damaged neural tissue in mammals, including humans.

Peripheral Nerve Regeneration

In the early 1900s, the British neurologist Henry Head provided a particularly dramatic account of repair in the peripheral nervous system. By this time it had become clear that damage to a peripheral nerve resulted in a gradual but usually incomplete restoration of sensory and motor function. The speed and precision of this recovery could be facilitated by the surgical reapposition of the two ends of the severed nerve, suggesting that if regeneration occurred in an environment that retained continuity of the proximal and distal ends of the severed nerve, functional recovery would be far better. Head's interest in this possibility culminated in a somewhat idiosyncratic approach to documenting the extent of regeneration of damaged peripheral sensory and motor nerves. Rather than continue evaluating recovery in patients with traumatic injuries that were variable in location and extent of tissue damage, he performed a nerve transection and reapposition experiment on himself, documenting the results as a personal narrative. In his 1905 paper, Head wrote:

> On April 25, 1903, the radial (ramus cutaneus radialis) and external cutaneous nerves were divided (cut) in the neighborhood of my elbow, and after small portions had been excised, the ends were united with silk sutures. Before the operation the sensory condition of the arm and back of the hand had been minutely examined and the distances at which two points of the compass could be discriminate had been everywhere measured.
>
> Head, Rivers, and Sherren, 1905, *Brain* 28: 99–115

Head went on to monitor the return of sensation and movement to the parts of his hand that had been rendered insensitive and paralyzed by the lesion (Figure 25.3). His observations emphasized several important aspects of peripheral nerve regeneration in this one-of-a-kind experiment. The first indication of recovery was a difference in the return of general sensitivity to pressure and touch that was not well localized (a sensitivity he called "protopathic"), beginning at approximately 6 weeks and lasting about 13 weeks. Thus Head noted that "the protopathic system regenerates more rapidly and with greater ease. It can triumph over want of apposition and the many disadvantages that are liable to follow traumatic division of a nerve."

Head also experienced a set of sensations that recovered more slowly and with less restoration to a normal sensory state. These phenomena included sensitivity to light touch, temperature discrimination, pinprick, and two-point discrimination, as well as fine motor control (which he referred to as "epicritic"). In fact, these faculties had not fully recovered over the 2 years between Head's surgical injury and the paper he wrote. He suggested that "the fibers of this system are more easily injured, and regenerate more slowly, than those of the protopathic system. They are evidently more highly developed and approach more nearly to the motor fibres that supply voluntary muscle in the time required for their regeneration."

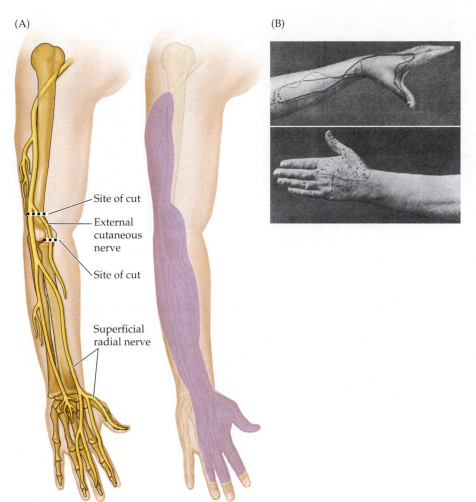

(A)

Site of cut

External cutaneous nerve

Site of cut

Superficial radial nerve

(B)

Figure 25.3 Henry Head's peripheral nerve regeneration experiment. (A) A diagram of the human arm showing the location of the radial nerve (left), which was severed and the proximal and distal ends surgically re-apposed in Head's experiment on his own arm; and the territory normally innervated by the radial nerve (purple, right). (B) Two photographs taken from Head's 1905 paper on his recovery from the peripheral nerve cut. The top panel shows the outlines of regions of Head's lower arm and hand that were insensitive to painful stimuli like pin prick, and the dotted line shows regions that were insensitive to light touch, like that from a cotton wisp. The lower panel shows the region of Head's hand and thumb that had regained sensation after an initial period of recovery (2–6 months). The various marks within the re-sensitized region indicate "hot" and "cold" spots that were more or less sensitive to stimulation (B from Head, Rivers, and Sherren, 1905.)

These remarkable observations were the first to articulate the distinction between the recuperative abilities of various classes of dorsal root ganglion cells and spinal motor neurons during the process of peripheral reinnervation. The distinction is presumably related to the initial specificity between different classes of dorsal root ganglion cell and motor axons and their targets during development—a specificity that relies on a variety of molecular signals, including several different neurotrophins (see Figure 23.15). The importance of this spectrum of cues in regeneration is reviewed in the following section.

The Cellular and Molecular Basis of Peripheral Nerve Repair

The cellular basis of peripheral nerve regeneration provides perhaps the clearest example of the relationship between the mechanisms used to promote initial axon growth and synapse formation during development and those that serve similar functions in later repair of neural damage. The major cellular elements that contribute to peripheral axon regrowth and the reinnervation of targets are *Schwann cells*, the peripheral glial cells that myelinate peripheral axons; and *macrophages*, immune system cells that clear the degenerating remains of severed axons. In addition to their respective roles in supporting intact axons and removing debris, both of these adult cell types secrete molecules that are essential for successful regeneration. In this process, the Schwann cell plays a domi-

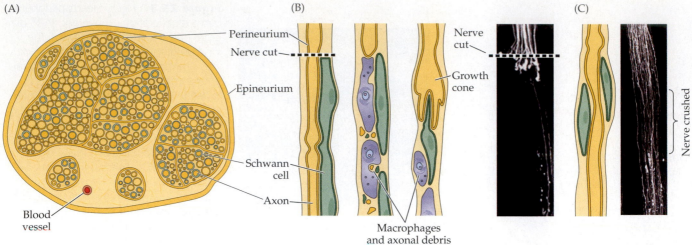

Figure 25.4 Regeneration in peripheral nerves. (A) Cross section through a peripheral nerve showing the connective tissue sheath of the epineurium and the extracellular matrix-rich perineurium that immediately surrounds the axons and Schwann cells. (B) Degeneration and regeneration in an idealized single peripheral nerve "tube" of perineurium/basal lamina. Once the axon is cut, the distal portion degenerates and is phagocytosed by macrophages. After the debris is mostly cleared, the proximal axon stump transforms into a growth cone, and this growth cone interacts with the adjacent Schwann cells. The image next to the drawing shows this step in peripheral nerve regeneration imaged in a living mouse after peripheral nerve damage. The nascent growth cones can be seen at the site of the cut, and in a few cases growing beyond it into the distal stump. Finally, the axon has regrown past the site of the cut, which is now repaired. (C) Regeneration is more efficient after crushing versus cutting a nerve. This panel shows a nerve adjacent to that which was cut in the same animal shown in (B). The remaining axons have recovered far more rapidly, and there is extensive regeneration across the site of the crush. (Photos from Pan et al., 2003.)

nant role in insuring the appropriate cellular and molecular milieu for regeneration following injury.

When a peripheral axon is severed, the axon segment distal to the site of the cut degenerates; when the axon is crushed, however, more rapid recovery occurs because the damaged distal segments provide a helpful guide to the regenerating proximal axons (Figure 25.4). In the case of a completely severed axon, only Schwann cells in the distal stump of the nerve and the basal lamina components secreted by the Schwann cell are available to stimulate and guide regeneration (Figure 25.5). The relevant molecules include laminin, fibronectin, and several others embedded in an extracellular matrix. The extracellular matrix, within the spaces defined by Schwann cell processes, provides a conduit of sorts for the regenerating axons. These remaining components in the distal peripheral nerve, referred to as bands of Bungner, are composed of a relatively orderly array of Schwann cells, matrix components, immune cells, and connective tissue. The extracellular matrix associated with axon fascicles (also referred to as the basal lamina, although a peripheral nerve is not really an epithelium) in the nerve before damage is more or less continuous to the axon's target. Accordingly, precise reapposition of distal and proximal nerve segments facilitates precise regeneration and better recovery of function, especially of fine touch and movement (Henry Head's "epicritic" abilities; see above).

Indeed, this sort of surgical reapposition, now done with microscopic technique, remains the primary therapeutic approach to peripheral nerve injury. The regenerating axons express integrins (see Figure 23.3) to mediate recognition of the matrix and subsequent intracellular signaling that facilitates growth. If an injury is so extensive that there is no apposable distal nerve left, new Schwann cells can be generated by Schwann cell precursors that remain in the damaged proximal nerves. These new cells can then provide an appropriate environment to support the extension of a newly generated growth cone from the axon stump. Nevertheless, the absence of a nearby distal stump makes regeneration limited and imprecise, and dramatically diminishes the recovery of function.

In addition to extracellular matrix molecules, the factors produced by Schwann cells that facilitate regeneration include many of the molecules thought to mediate axon guidance and growth during early development. Schwann cells increase the amount of cell adhesion molecules like N-CAM, L1, and N-cadherin (see Figure 23.3) on their surface in response to axon injury. Regenerating axons must therefore express complementary cell adhesion mole-

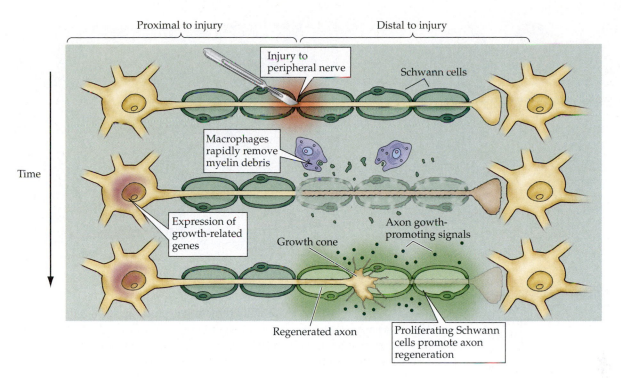

Proximal to injury   Distal to injury

Time

Injury to peripheral nerve

Schwann cells

Macrophages rapidly remove myelin debris

Expression of growth-related genes

Axon gowth-promoting signals

Growth cone

Regenerated axon

Proliferating Schwann cells promote axon regeneration

Figure 25.5 Molecular and cellular responses that promote peripheral nerve regeneration. The Schwann cell is essential for this process. Once the macrophages have cleared the debris from the degenerating peripheral stump, the Schwann cells proliferate, express adhesion molecules on their surface, and secrete neurotrophins and other growth-promoting signaling molecules. In parallel, the parent neuron of the regenerating axon expresses genes that restore it to a growth state. The gene products are often receptors, or signal transduction molecules, that allow the cell to respond to the factors provided by the Schwann cell.

cules on their surfaces. Schwann cells near the injury increase expression and secretion of a number of neurotrophins such as BDNF (which is thought to be especially crucial for motor axon growth). Expression of Trk as well as P75 neurotrophin receptors is elevated following injury on the newly generated growth cones of the regenerating peripheral axons. It is likely that this local availability of neurotrophins acts both to promote a "growth" state (i.e., trophic effects) for the damaged axons as well as defining a local target distal to the site of damage to grow toward (i.e., tropic effects). Further, the dynamics of the actin and microtubule cytoskeleton must be restored to a "growth" state so that growth cone navigation and axon extension can occur, and changes in gene expression enable this. The genes that are switched on include several associated with axon growth during development. A good example is growth-associated protein-43 (GAP43). GAP43 is normally found in growing axons in the embryo, but the protein is also seen at high levels following axotomy and the initiation of regrowth in the periphery as a result of gene activation. In addition, many of the signaling mechanisms that modulate the cytoskeleton and directed axon extension in development must be reactivated to accommodate regeneration.

Schwann cells and the milieu they create are the dominant factors in promoting peripheral axon regeneration. Accordingly, it seems possible that the peripheral nerve environment might have growth-promoting capacity for central nervous system axons. If axons in the optic nerve or spinal cord are severed (recall that the optic nerve and retina, although physically in the periphery, are actually part of the central nervous system) and the proximal stumps reapposed to a peripheral nerve graft that offers the Schwann cell-basal lamina-connective tissue components that normally support peripheral nerve regeneration, central axons grow readily through the peripheral nerve graft. Some of these axons can even make synapses in the target territory to which the distal end of the graft is connected (Figure 25.6). This sort of observation shows that Schwann cells define an environment in the peripheral nerve sheath that is particularly well adapted to initiate and support the regrowth of damaged adult axons, whether

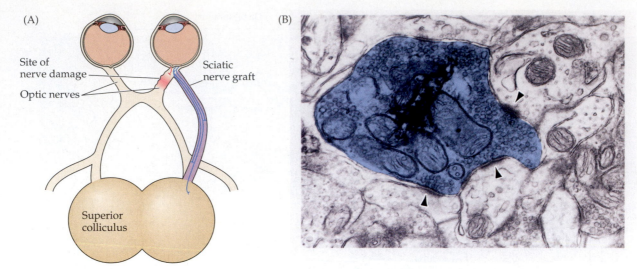

Figure 25.6   The growth-promoting properties of peripheral nerve sheaths and Schwann cells facilitate growth of damaged axons within the CNS. (A) Severed axons from the optic nerve are apposed to a peripheral nerve graft. The axons, which would normally not regenerate through the optic nerve (recall that the optic nerve, while physically in the periphery, is wholly within the central nervous system), now grow through the peripheral nerve graft to reach the superior colliculus, a normal target for retinal ganglion cells. (B) Regenerated axons make synapses with targets in the superior colliculus. The dark material is an electron-dense, intracellularly transported tracer that identifies specific synaptic terminals (arrowheads) as emanating from a regenerated retinal axon. (A after So and Aguyao, 1985; B from Bray et al., 1991.)

they project to the periphery or normally remain within the central nervous system. Why this same set of highly useful events does not normally occur in the central nervous system is an essential and challenging question that will be considered later in this chapter.

Regeneration of Peripheral Synapses

The extension of axons from mature sensory, autonomic, or motor neurons is only the first step in peripheral nerve regeneration. The next essential event in successful recovery of function is reinnervation of appropriate target tissues and re-establishment of synaptic connections. This process must occur for all three classes of peripheral axons (sensory, motor, and autonomic); however, it has been most thoroughly characterized at the neuromuscular junction and in the peripheral autonomic system (Box 25A). Because of the relative ease of identifying synaptic sites on muscle fibers, the regenerating neuromuscular junction has been studied in great detail (Figure 25.7A,B). The ability to define major molecular constituents—synaptic extracellular matrix, postsynaptic receptors, and related proteins—gives insight into the stability of a denervated synaptic site after damage, as well as the changes that accompany reinnveration.

When skeletal muscle fibers are denervated, the original neuromuscular synaptic sites remain intact for weeks. At these or nearby sites, many secreted signaling molecules are either increased or decreased in both the muscle cells and the Schwann cells near the denervated endplate sites (Figure 25.7C). Presumably, the neurotrophins whose expression is increased at the denervated neuromuscular junction—NGF and BDNF—enhance the tropic and trophic signaling necessary to recapitulate target recognition and synaptogenesis. Those that are diminished (NT3 and NT4) may be more important for the maintenance of established synapses. Their decline may insure that denervated synaptic sites are not refractory to innervation by a newly arrived regenerated axon.

The clustering of acetylcholine receptors that defines the postsynaptic membrane specialization also remains, as does the local scaffolding of proteins that retain the ACh receptors at the synaptic specialization on the muscle fiber. The secreted factor neuregulin, which is involved in the initiation of receptor clustering (see Chapter 23), is also maintained at the denervated synapse. In addition, the extracellular matrix components that distinguish the synaptic portion of the muscle basal lamina are maintained when mature muscle fibers are den-

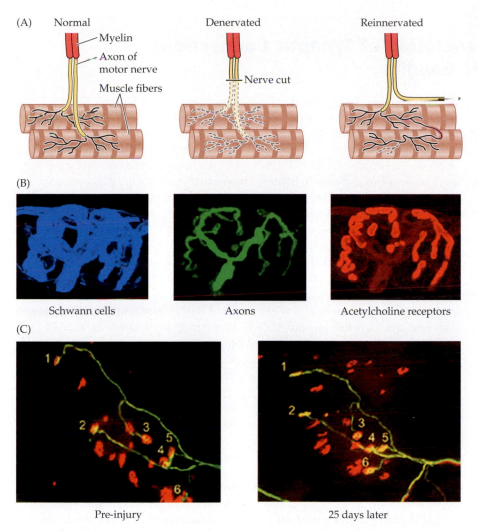

(A)

Normal | Denervated | Reinnervated

- Myelin
- Axon of motor nerve
- Muscle fibers
- Nerve cut

(B)

Schwann cells | Axons | Acetylcholine receptors

(C)

Pre-injury | 25 days later

Figure 25.7 Reinnveration of muscles following peripheral motor nerve damage. (A) Schematic showing the degeneration of a distal motor axon, its regrowth, and the maintenance of the postsynaptic specialization on the muscle surface during the period of denervation. (B) The cellular components usually found at the neuromuscular junction. When the axon degenerates, the Schwann cells and acetylcholine receptors (AChR) remain in place. (C) The pattern of motor innervation in an isolated muscle before an injury and 25 days after, imaged in a single living mouse. The postsynaptic specializations of the neuromuscular junctions on each individual muscle fiber are labeled in red (for AChR). The axon fluoresces green. This axon makes six synapses on six muscle fibers. Twenty-five days later, the axon has reinnervated all six sites, and the basic pattern is similar to that prior to the injury. (B from Pitts et al., 2006; C from Nguyen et al., 2002.)

ervated. This matrix includes specialized forms of laminin that are normally found at the neuromuscular synapse (synaptic, or S laminin), as well as molecules that are normally bound to the matrix (including acetylcholinesterase, the enzyme that normally degrades ACh at the neuromuscular synapse). It is also likely that the extracellular matrix localizes and concentrates the secreted growth factors produced at the synapse by specialized perisynaptic Schwann cells upon denervation. Accordingly, this highly organized region of the basement membrane, in concert with the postsynaptic receptor proteins and Schwann cells, defines the site of reinnervation. In fact, even if the muscle fiber is eliminated, motor axons recognize these specialized basal lamina sites as optimal locations for reinnervation.

Molecular specificity, however, provides only one part of the instructions for re-establishing synapses after peripheral nerve regeneration. Activity-dependent processes similar to those that eliminate polyneuronal innervation at neuromuscular synaptic sites during development (described in Chapter 24) are also essential for restoring function after peripheral nerve damage. There is a fair degree of imprecision in the reinnervation of specific targets, as is evident in Henry Head's description of the slowness and imprecision of his recovery of fine motor and sensory function. This result has been confirmed by more recent studies in which reinnervation is observed over time in adult animals with

BOX 25A · Specific Regeneration of Synaptic Connections in Autonomic Ganglia

A part of the vertebrate nervous system that has been studied in detail for more than a century with respect to neural regeneration is the peripheral autonomic system, where accessibility and the regenerative properties of peripheral axons allow a variety of investigations to be done with relative ease. Most of these studies have been done in the mammalian sympathetic system. Preganglionic sympathetic fibers, like other peripheral axons, regenerate when they are severed. Towards the end of the nineteenth century, the English physiologist John Langley working at Cambridge University found that sympathetic end organ responses (e.g., blood vessel constriction, piloerection, and pupillary dilation) recovered a few weeks after cutting the preganglionic nerve to the superior cervical ganglion. As indicated in Figure A, the normal innervation of this and other sympathetic ganglia is selectively organized in that preganglionic axons arising from different spinal segments innervate particular functional classes of cells in the ganglion. Langley found that after reinnervation the end organ responses were organized much as before; thus stimulation of T1 elicited

its particular constellation of largely nonoverlapping end organ effects compared to stimulation of the preganglionic axons arising from T4.

Modern experiments confirmed Langley's observations and showed further that the normal pattern of innervation observed with intracellular recording is indeed reestablished following regeneration of the preganglionic axons. Selective reinnervation also occurs in parasympathetic ganglia. In the chick ciliary ganglion, there are two functionally and anatomically distinct populations of ganglion cells, the ciliary cells and the choroidal cells. Because these ganglion cell types can be separately identified and are in turn innervated by preganglionic axons with different conduction velocities, one can ask whether these two populations are reinnervated by the same classes of axons that contacted them in the first place. As in the mammalian sympathetic system, appropriate contacts are reestablished during reinnervation.

The accurate reinnervation of different classes of sympathetic neurons is especially remarkable because the ganglion cells innervated by a particular

spinal segment (and that innervate a particular target) are distributed more or less randomly through the ganglion. This arrangement implies recognition of the pre- and postsynaptic elements must occur at the level of the target cells. A way to explore the implication that ganglion cells bear some more or less permanent identity is to transplant different sympathetic chain ganglia from a donor animal to a host where the ganglia can be exposed to the same segmental set of preganglionic axons during reinnervation. One can then ask whether two different ganglia, normally innervated by different sets of axons, are selectively reinnervated by axons arising from different spinal segments.

As seen in Figure B, different sympathetic chain ganglia (in this case, the superior cervical ganglion and the fifth thoracic ganglion) are indeed distinguished by the preganglionic axons in the host cervical sympathetic trunk. A donor superior cervical ganglion transplanted to the cervical sympathetic trunk is reinnervated in a manner that approximates its original segmental innervation; the fifth thoracic ganglion transplanted to that position, on the other

experimentally damaged peripheral motor nerves (see Figure 25.7B). The subsequent regeneration can either be fairly faithful to the original parrern or fairly imprecise. Imprecision is due not only to inappropriate target matching, but to the return of polyneuronal innervation at neuromuscular synapses during regeneration and reinnervation. Much of this innervation is eventually eliminated, presumably via the same activity-dependent mechanisms that operate during the postnatal period, when supernumerary axons are eliminated from the developing synapse (see Figure 23.10). If electrical activity is blocked during regeneration, either in the muscle fiber or in the afferent nerve, multiple innervation of the endplate sites remains.

Regeneration after Damage to the Central Nervous System

The second type of regeneration that occurs following damage to the adult brain involves the stabilization and regrowth of neurons and processes within an injured region of the central nervous system (spinal cord or brain). With the

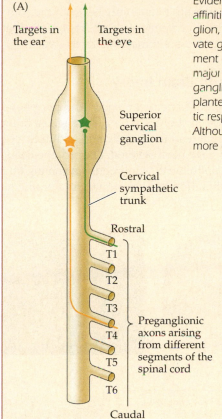

(A)

Targets in the ear

Targets in the eye

Superior cervical ganglion

Cervical sympathetic trunk

Rostral

T1
T2
T3
T4
T5
T6

Preganglionic axons arising from different segments of the spinal cord

Caudal

Evidence that synaptic connections between mammalian neurons form according to specific affinities between different classes of pre- and postsynaptic cells. (A) In the superior cervical ganglion, preganglionic neurons located in particular spinal cord segments (T1, for example) innervate ganglion cells that project to particular peripheral targets (the eye, for example). Establishment of these preferential synaptic relationships indicates that selective neuronal affinities are a major determinant of neural connectivity. (B) In a transplantation experiment, guinea pig donor ganglia C8 (superior cervical ganglion; control) and T5 (fifth thoracic ganglion) were transplanted into the superior cervical bed of a host animal. The graph shows the average postsynaptic response of neurons in the transplanted ganglia to stimulation of different spinal segments. Although there is overlap, the neurons in transplanted T5 ganglia are clearly reinnervated by a more caudal set of segments than are the transplanted C8 neurons. (B after Purves et al., 1981.)

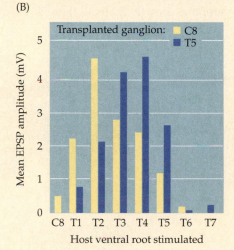

(B)

Transplanted ganglion: ☐ C8 ■ T5

Mean EPSP amplitude (mV)

Host ventral root stimulated

confirmation of Langley's original concept of "chemoaffinity" as a basis for the selectivity of target cell innervation (see Chapter 23).

References

LANGLEY, J. N. (1897) On the regeneration of pre-ganglionic and post-ganglionic visceral nerve fibres. *J. Physiol. (Lond.)* 22: 215–230.

LANDMESSER, L. AND G. PILAR (1970) Selective reinnervation of two cell populations in the adult pigeon ciliary ganglion. *J. Physiol. (Lond.)* 211: 203–216.

PURVES, D. AND J. W. LICHTMAN (1983) Specific connections between nerve cells. *Annu. Rev. Physiol.* 45: 553–565.

PURVES, D., W. THOMPSON AND J. W. YIP (1981) Re-innervation of ganglia transplanted to the neck from different levels of the guinea-pig sympathetic chain. *J. Physiol. (Lond.)* 313: 49–63.

hand, is reinnervated by an overlapping but caudally shifted subset of the thoracic spinal cord segments that normally contribute to the cervical sympathetic trunk. This more caudal innervation approximates the original segmental innervation of the fifth thoracic ganglion.

These results indicate that ganglion cells carry with them a property that biases the innervation they receive, in

exception of spinal and brainstem motor neurons, whose axons project into the periphery and therefore have access to instructions for the relatively successful peripheral regeneration described above, there is far less long distance axon growth and re-establishment of functional connections within the central nervous system following injury. The limited regrowth of damaged CNS axons whose cell bodies remain intact largely accounts for the relatively poor prognosis following brain or spinal cord injury.

Damage to the central nervous system usually occurs in one of three ways. First, the brain or spinal cord can be injured acutely by external physical **trauma** (e.g., auto accidents or gunshot wounds). A second type of damage is caused by **hypoxia**, a lack of local oxygen usually created by diminished blood flow (ischemia) due to local vascular occlusions (e.g., in strokes; see the Appendix) or a global deprivation of oxygen (e.g., due to drowning or cardiac arrest). The third type of damage arises from **neurodegenerative diseases** (e.g., Alzheimer's disease or amyotrophic lateral sclerosis). All three sources result in some neuronal death, either immediate or slowly progressive. In cases of physical damage or severe hypoxia, neurons die quite rapidly. When the insult is less severe, some neurons can survive and some local growth occurs. Because central axons have little ability to regenerate, the key to recovery from brain injury lies with the complex cellular events pertinent to the survival of neurons (see Chapter 23) that have not been killed outright, and whose processes remain relatively intact.

Cellular and Molecular Responses to Brain Injury

There are two major reasons for the differences between successful peripheral regeneration and the limited regeneration in the CNS. First, damage to brain tissue tends to engage the mechanisms that lead to necrotic and apoptotic cell death for nearby neurons whose process have been severed. Second, the cellular changes at the site of injury do not recapitulate developmental signaling that supports growth. Instead, there is a combination of glial growth and prolifera-

(A)

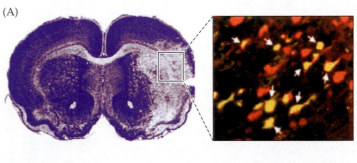

(B)

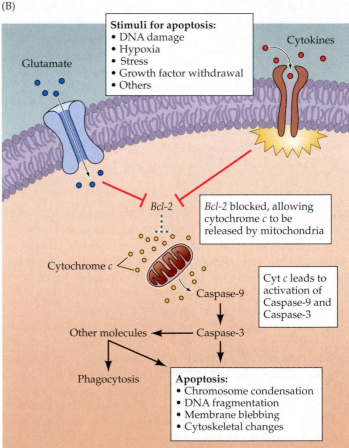

Figure 25.8 Consequences of hypoxia/ischemia in the mammalian brain. (A) Section through the brain of a 7-day-old mouse in which the carotid artery was transiently constricted. Nissl stain (see Chapter 1) was used to visualize cell bodies. The lighter region (i.e., little or no staining) shows the extent of cell damage and loss caused by this brief deprivation of oxygen. Cells in the higher magnification image were stained for the neuronal marker Neu-N (red) and for activated caspase-3, indicative of neurons undergoing apoptosis. (B) Model of the primary mechanism for neuronal apoptosis after injury. Apoptosis can be elicited by excitotoxicity via excess glutamate, and by the binding of inflammatory cytokines to receptors in the neuronal membrane. In addition, loss of neuronal connections to a target and resultant deprivation of trophic support can initiate apoptosis. Any or all of these stimuli, once present, result in the removal of the anti-apoptotic gene *Bcl-2*. Cytochrome *c* is then released from mitochondria, activating caspase-3 and obligating the cell to apoptotic death as caspase-3 stimulates destructive changes in downstream molecules. (A from Back et al., 2002.)

tion, and microglial activity (microglia have immune functions that lead to local inflammation) that actively inhibits growth as well as upregulation of growth-inhibiting molecules related to the chemorepellent factors that influence axon trajectories during development. Again, why these inhibitory phenomena occur is a central question for understanding the lack of successful regeneration in the brain.

One of the most striking differences in the consequences of CNS versus peripheral nerve cell damage is the extent of cell death that occurs after direct damage to the brain (Figure 25.8A). Neuronal cell death in the CNS is seen regardless of the type of damage (traumatic, hypoxic, or degenerative). It has been studied most extensively in brains where hypoxia has occurred due to vascular occlusion (e.g., stroke, vascular accidents, and asphyxiation). In such cases, there is a clear loss of cells in the hypoxic region. Where such cell loss has occurred, there is also enhanced activation of **caspase-3**, an enzyme that, when activated, obligates a cell to die via **apoptosis** (Figure 25.8B). This genetically regulated mechanism can be elicited by growth factor deprivation, hypoxia, or by DNA damage and other cellular stress. The transection of axons presumably can lead to growth factor deprivation by removing the target source for the parent neurons. DNA damage and cellular stress (including changes in oxidative metabolism) have been suggested to underlie some neurodegenerative diseases.

A major source of cellular stress is glutamatergic overstimulation caused by bursts of abnormal activity arising after local brain damage. Such overstimulation can also arise from epileptogenic foci and the seizure activity generated at these sites. This elevated activity and its consequences are referred to as **excitotoxicity**, and if unchecked it can lead to neuronal cell death (see Box 6D). Following injury or seizure, excessive amounts of neurotransmitters are released. This enhanced signaling modifies the effectiveness of members of the Bcl-2 family of anti-apoptotic molecules that normally oppose changes in mitochondrial function that reflect oxidative stress. Diminished *Bcl-2* activity allows cytochrome *c* to be liberated from mitochondria. Once in the cytoplasm, cytochrome *c* facilitates cleavage of caspase-3, activating this enzyme. Activated caspase-3 can then cause the fragmentation of nuclear DNA, membrane and cytoskeletal changes, and ultimately cell death (Figure 25.8B). Thus, one of the key determinants of the long-term effects of damage to adult neural tissue is the extent to which the damage activates apoptosis.

As might be expected from events in the periphery, glial cells found at the site of injury contribute to the degenerative and regenerative processes that occur after brain damage. All three glial classes—astrocytes, oligodendroglia, and microglia—display altered properties following brain injury (Figure 25.9). In addition, glia seem less susceptible to the stimuli that result in apoptosis. Most brain lesions cause limited proliferation of otherwise quiescent glial precursors, as well as extensive growth of existing glial cells within or around the site of injury. These reactions lead to the glial "scarring" referred to earlier, accompanied by increased secretion of a number of signals including transforming growth factor (TGF), fibroblast growth factor (FGF), tissue necrosis factor alpha (TNF-α), interleukins, interferon-γ, and insulin-like growth factor-1 (IGF-1). Depending on the cellular target (neuronal or glial), these signals (some of which are **cytokines**, most of which serve signaling roles in the immune system) can either promote cell death and phagocytosis (see Figure 25.8B), or provide protective signals for remaining nerve cells. Depending on the type of injury and the time elapsed since its occurrence, the **reactive glia** alter the relationships between remaining nerve cells and the glial environment, or they become the dominant cell type in the region of damage by replacing lost neurons and degenerated processes, resulting in the long-term glial scar.

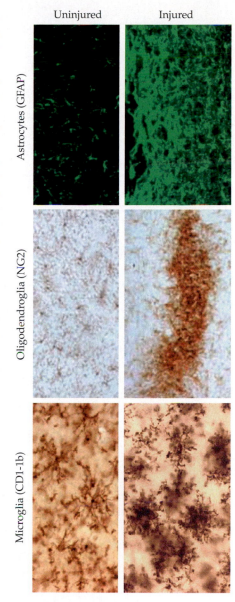

Uninjured   Injured

Astrocytes (GFAP)

Oligodendroglia (NG2)

Microglia (CD1-1b)

Figure 25.9 The reaction of the three major classes of glia in the central nervous system to local tissue damage. In each case, there is growth and change in expression of molecules normally associated with each cell class. (Top) Astrocytes labeled to visualize glial fibrillary acidic protein (GFAP) both before and after injury. (Center) The molecule NG2, notably present in glial scar tissue, is visualized here in oligodendroglial precursors and immature oligodendrocytes. (Bottom) CD1-1b, a marker for microglia. (Top from McGraw et al., 2001; center from Tan et al., 2005; bottom from Ladeby et al., 2005.)

Axon Growth after Brain Injury

One unfortunate consequence of glial scarring is an impediment to axon growth beyond the site of the lesion. The reasons for this barrier action (beyond the obvious physical interference) are not well understood, but it is clear that astroctyes within the glial scar produce several molecules that can inhibit axon growth. These include semaphorin 3A, several ephrins, and slit (see Figures 23.4 and 23.5). The receptors for each of these molecules are upregulated on the growth cones of axons that approach the glial scar, resulting in bulbous axonal abnormalities. In addition, matrix components that inhibit axon growth (particularly tenascin and chondroitin sulfate proteoglycan) are enriched in the extracellular spaces within the glial scar. Thus, the proliferation and hypertrophy of glial cells, and their expression of chemorepellent or growth inhibiting molecules dominates the tissue reaction to focal brain injury (Figure 25.10). There is considerable speculation about why the CNS response to injury inhibits regeneration—for example, the mature brain may place a premium on stable circuits and thus does not easily accommodate wholesale growth and change. There is no easy way, however, to confirm or refute such hypotheses.

One of the major unsolved issues is in the cell and molecular biology of central nervous system response to injury is how oligodendroglial cells (which are included in glial scars; see Figure 25.9) influence the failure of central regeneration following damage to brain tissue (this is somewhat distinct from the changes seen in oligodendroglia during phases of demyelination and remyelination in autoimmune diseases like multiple sclerosis). Several observations suggest that brain myelin, which is produced by oligodendroglia, inhibits axon growth, including the diminished ability of axons to grow on substrates enriched in myelin-associated proteins such as myelin-associated glycoprotein (MAG). This is something of a puzzle, since MAG is also produced by Schwann cells, but does not appear to be an impediment for peripheral regeneration.

The most unusual factor, however, is a molecule known as **NogoA**. NoGoA is selectively expressed in oligodendroglial cells, and it is not found in Schwann cells. In the early 1990s, Martin Schwab and his colleagues raised a polyclonal antibody to myelin-associated proteins from glial scars, and found that this antibody when

Figure 25.10   Cellular response to injury in the central nervous system. In the absence of a glial scar, which are particularly prominent in long axon pathways in the brain, there is a series of local cellular changes at or near an injured site. These include the local degeneration of myelin as well as other cellular elements, the clearing of this debris by microglia that act as phagocytic cells in the CNS, and the local production of inhibitory factors by reactive astrocytes, oligodendroglia, and microglia.

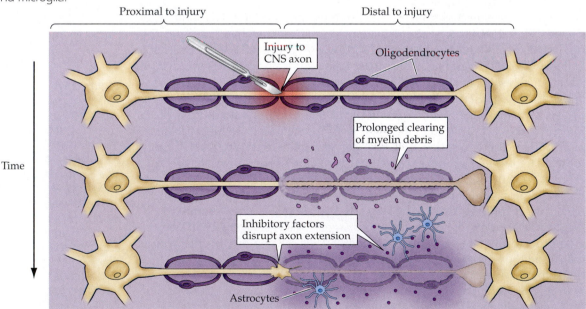

applied locally to a damaged area could promote axon growth through the glial scar. This antibody also recognizes NogoA. Given the promise of identifying a major inhibitory factor that if blocked might lead to improved regeneration of long axon pathways in the CNS, NogoA became a major focus of research on brain repair. These efforts resulted in a thorough characterization of NogoA and its receptor, but little resolution of the role of this unusual molecule in inhibiting regeneration, or as a target for therapies to enhance regeneration. In experiments in which the NogoA gene or its receptor was inactivated in mice, there was no clear indication of significantly enhanced regeneration across the site of injury.

Other therapeutic approaches for treating spinal cord injury have had limited efficacy at best. There is some indication that the peripheral glial cells that support ongoing growth of axons from newly generated olfactory receptor neurons may provide a more supportive substrate for the regrowth of damaged CNS axons. However, this approach, like most others based on cellular or molecular observations in animal models, has yet to be validated clinically in humans.

Generation and Replacement of Neurons in the Adult Brain

Few issues in modern neuroscience have engendered as much controversy or confusion as the ability of the adult nervous system to generate new neurons, especially in response to acute or degenerative damage to neural tissue. Until the early 1960s, with the advent of neuronal birthdating techniques that took advantage of incorporation of labeled thymidine analogs into the nuclei of dividing cells (see Box 22F), there had been no reliable way to assess the extent of cell proliferation and mitotic activity in the mature brain. Moreover, clinical experience, buttressed by decades of studies in animals with brain lesions, indicated that the brain was unlikely to undergo significant regeneration of nerve cells that could re-establish connections. This was particularly true for the mammalian brain; however, there was a sense that several lower vertebrates (including frogs and birds) had a greater capacity for neurogenesis in response to brain injury, as well as the ability to support extensive reintegration of new or regrown axons and dendrites into existing circuits. This was recognized clearly by the early 1950s, when several behavioral and anatomical studies evaluated the reinnervation and mapping of the optic tectum in frogs by eyes in which the optic nerve had been damaged (see Chapter 23). The issue of whether this sort of repair was accompanied by the addition, differentiation, and maintenance of new neurons remained mostly unaddressed by these studies.

The ability to label cells undergoing mitosis and to track their progeny led to a clear—albeit discouraging—assessment of the potential for new neurons to be added to the adult mammalian brain. Quite simply, there did not seem to be extensive addition of neurons to the brains of adult mammals after the completion of pre- and postnatal development. While this conclusion had extensive support, some reservations were raised as early as the mid 1960s. Joseph Altman and his colleagues, then at the Massachusetts Institute of Technology, found that small numbers of granule neurons in the hippocampus and olfactory bulb in guinea pigs and rats could be heavily labeled with tritiated thymidine injected at adult ages. Their work suggested that these inhibitory interneurons might be added to the brain during adulthood, either replacing or augmenting the cohort generated during development. At the time, however, the lack of additional markers that identified these cells as neurons led other investigators to conclude that many, if not all, of these cells were newly generated glia rather than neurons.

Modern approaches have led to a much clearer understanding of the identity of the progeny of mitotically active cells in the adult brain of many vertebrates, including mammals. A low level of glial cell proliferation continues throughout life. It is clear that existing, differentiated neurons do not de-differentiate and divide. Instead, at distinct locations in the brains of several species—including humans—there are regions where neural stem cells are maintained. A neural stem cell can give rise to the full complement of cell classes found in neural tissue—i.e., neurons, astrocytes, and oligodendroglia (see Box 22A), as well as more stem cells. Apparently, mature central nervous systems in several vertebrates can provide an environment that supports the maintenance of neural precursors. The extent to which these stem cells give rise to neurons that replace or augment existing populations varies depending on species, brain region, and the conditions (e.g., growth, seasonal change, injury) that influence neurogenesis in the adult brain.

Adult Neurogenesis in Non-Mammalian Vertebrates

Observations in several non-mammalian vertebrate species, particularly teleost fish like the goldfish and songbirds like the canary and zebra finch, make a strong case for the capacity of adult vertebrate brains to add new neurons and incorporate them into functional circuits that guide behavior. One of the first thoroughly characterized examples of ongoing vertebrate adult neurogenesis is from the goldfish. Goldfish, like many other fish, continue to grow throughout their lifetimes. This body growth is matched by growth of sensory structures in the periphery, particularly the eye. By the early 1970s, several investigators had recognized that this growth in the eye was accompanied by the generation of new retinal neurons. Subsequent work showed that these neurons are generated from a subset of precursor or stem cells that form a ring around the entire margin of the goldfish retina (Figure 25.11A). These cells are capable of generating all of the cell classes in the goldfish retina with the exception of rod photoreceptors, which are regenerated by a distinct precursor cell (see Chapter 11 for a review of retinal cell classes). The new neurons integrate into the existing retina in rings, or annulae, between the precursor cells at the periphery and the existing differentiated retina—much like the yearly growth rings that are added to tree trunks. The axons of the new retinal ganglion cells enter and grow through the optic nerve and tract to reinnervate the optic tectum. Most of this axon growth happens along the extracellular matrix that is deposited at the glial limiting membrane of the optic nerve—the equivalent of the basal lamina conduits made by Schwann cells in the periphery. Regeneration of all retinal cell classes can also happen in response to a local injury of existing retinal tissue; however the details of how this repair is completed are somewhat different than that for ongoing addition of cells. As surprising as these remarkable capacities are, it is perhaps equally surprising that the optic tectum in the brain adds new neurons to accommodate the quantitative expansion of the periperhal retinal projection. These cells are not added in complimentary rings; instead, populations of new neurons are added as crescents to the back of the tectum. Accordingly, the geometry of adult neurogenesis in the periphery and brain are mismatched. This divergent geometry requires that new retinal inputs be constantly remapped along with existing retinal projections. Thus, there must be a great deal of dynamism in synaptic connections in the adult goldfish tectum. Connections must be made, broken, and remade to maintain the integrity of the retinotopic map as the fish grows throughout its lifetime.

An equally striking example of adult neurogenesis has been found in several species of songbirds, including the canary and the zebra finch. This ongoing neurogenesis occurs in several regions of the avian brain; however, it has been most thoroughly studied in the structures that control vocalization and perception of song (Figure 25.11B; see also Box 24B). In most male songbirds, there is continual

(A)

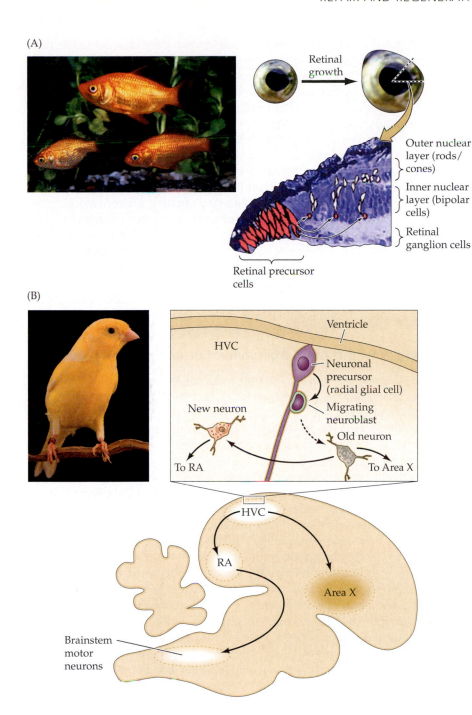

Retinal growth

Outer nuclear layer (rods/cones)

Inner nuclear layer (bipolar cells)

Retinal ganglion cells

Retinal precursor cells

(B)

Ventricle

HVC

Neuronal precursor (radial glial cell)

Migrating neuroblast

New neuron

Old neuron

To RA

To Area X

HVC

RA

Area X

Brainstem motor neurons

Figure 25.11 Adult neurogenesis in non-mammalian vertebrates. (A) Given favorable environmental conditions, teleost fishes such as goldfish grow throughout their entire adult lives; the growth of the fish's body is matched by the growth of its eyes and brain. The retina grows by adding new neurons generated from a population of stem cells distributed in a ring at the very margin of the retina (red). These stem cells give rise to all retinal cell types except the rods (which are regenerated from precursors found in the existing differentiated region of the retina). (B) In a process of ongoing adult neurogenesis, male songbirds like the canary lose and replace significant numbers of neurons in forebrain nuclei that control the production and perception of song. These song-control centers include the HVC (higher vocal center), RA (robustus archistriatus), and "area X," which is the equivalent of the caudate nucleus in the mammalian brain. In HVC, a population of radial stem cells is maintained. The cell bodies are adjacent to the ventricular space, and their processes extend into the neuropil of the nucleus. A subset of neurons is retained in the nucleus as new ones are added. Neuroblasts migrate from the ventricular zone along the radial processes of the precursor cells and then integrate into circuits with existing neurons. (A after Otteson and Hitchcock, 2003, photo © Juniors Bildarchiv/Alamy; B after Goldman, 1998, photo © Eric Isselée/istockphoto.com.)

loss and addition of neurons in these regions. In some, the cycle of loss and regeneration follows mating seasons, and is under the control of gonadal steroids (see Chapter 30), while in others it occurs constantly. Although it is tempting to speculate that the new neurons are a substrate for flexible acquisition or production of songs, no definitive evidence confirms this speculation. Moreover, many birds that have significant amounts of adult neurogenesis in song-control regions show very little flexibility in their song after the critical period for song learning is complete, even though new neurons have been added.

Regardless of the kinetics and behavioral consequences, it is estimated that birds replace most of the neurons in several song control centers of their brain several times over a lifetime. The new neurons are generated from precursors found in a limited region of the neural tissue immediately adjacent to the fore-brain lateral ventricles. Precursor or stem cell bodies are found in this zone, and their radial processes extend into the song control centers (Figure 25.11B). These cells function as both precursors, generating new neurons via asymmetric divisions, and as migration guides that constrain the translocation of new neurons from the ventricular zone to the song control nuclei. Thus, the precursors of newly generated adult neurons in the bird forebrain resemble the radial glia which serve a similar function during initial forebrain neurogenesis (see Chapter 22). Many of the new neurons integrate themselves into existing circuits and have functional properties that are consistent with a contribution to song production or perception. But a significant number also die before they can fully differentiate, suggesting that there may be limits to the capacity of newly generated neurons to establish sufficient trophic support and activity-dependent validation to survive. Such effects would be similar to the processes that limit neuron numbers during initial development in many species. A key feature of adult neurogenesis in the avian brain is that there is always a balance of existing, long-lived neurons and newly generated neurons. Thus, even this compelling example of adult neurogenesis occurs in the context of significant stability in the mature brain.

Neurogenesis in the Adult Mammalian Brain

Over the past decade, questions about the production of new neurons in the adult mammalian brain have been examined (or re-examined) in mice, rats, monkeys, and humans. Clearly, defining the capacity for ongoing regeneration in specific CNS regions would provide a model for how regeneration might be elicited following brain injury or to combat neurodegenerative disease. The results of these many examinations now seem quite clear. New nerve cells in the central nervous system are seen reliably in just two regions: (1) The olfactory bulb; and (2) the hippocampus (Figure 25.12). In addition to these sites, there is clear evidence that olfactory receptor neurons in the periphery are continually replaced.

In the central nervous system, new nerve cells are primarily interneurons: granule cells and periglomerular cells in the olfactory bulb (see Chapter 15), or granule cells in the dentate gyrus of the hippocampus (see Chapters 8 and 31). New neurons with long-distance projections have not been observed in either the bulb or hippocampus. These newly generated olfactory or hippocampal interneurons are apparently the progeny of precursor or stem cells located close to the surface of the lateral ventricles, relatively near to either the bulb or hippocampus. (In each case, however, some translocation from the site of final mitosis is required; see below.) At least some of these new nerve cells become integrated into functional synaptic circuits; however, most new neurons generated in the adult brain die before being integrated into existing circuitry. No one has yet explained the functional significance of restricting neurogenesis to just these few regions in the adult brain, and the ultimate behavioral consequences for the addition of such cells in mammals or other animals remains unclear. Moreover, the death of most of newly generated neurons suggests that there may be a premium placed on stability in the mammalian brain, thus limiting opportunities for new neurons to join existing circuits. Nevertheless, the fact that new neurons can be generated in at least a few adult brain regions shows that this phenomenon can occur in the mammalian CNS.

Cellular and Molecular Mechanisms of Adult Neurogenesis

In regenerating tissues like the intestine or lung, stem cells are found in a distinct location, or **niche**. Presumably there is a local environment within the stem cell

(A)

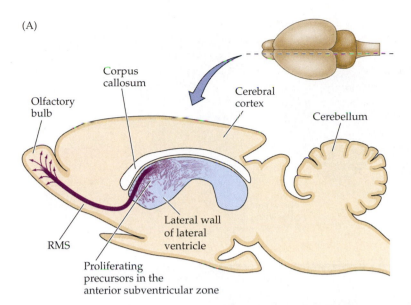

Figure 25.12 Neurogenesis in the adult mammalian brain. (A) Neural precursors in the epithelial lining of the anterior lateral ventricles in the forebrain (a region called the anterior subventricular zone, or SVZ) give rise to postmitotic neuroblasts that migrate to the olfactory bulb via a distinctive pathway known as the rostral migratory stream or RMS. Neuroblasts that migrate to the bulb via the RMS become either olfactory bulb granule cells or periglomerular cells; both cell types function as interneurons in the bulb. (B) In the mature hippocampus, a population of neural precursors is resident in the basal aspect of the granule cell layer of the dentate gyrus. These precursors give rise to postmitotic neuroblasts that translocate from the basal aspect of the granule cell layer to more apical levels. In addition, some of these neuroblasts elaborate dendrites and a local axonal process and apparently become GABAergic interneurons within the dentate gyrus. (After Gage, 2000.)

(B)

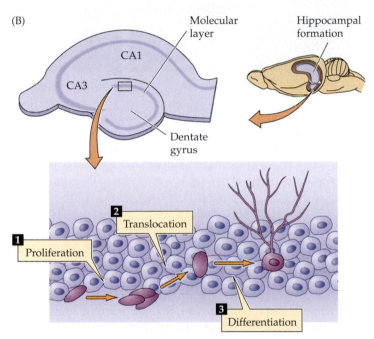

niche, established by secreted signals as well as cell surface molecules, that is conducive to the maintenance of stem cells as well as to the division and initial differentiation of cells that will reconstitute the adult tissue. In the adult mammal brain, new neurons are generated from regions referred to as **subventricular zones**, cell-dense regions adjacent to the ventricular space found in the cortical hemispheres. These zones in the adult mammalian brain are quite similar to the neurogenic zones in seen in lower vertebrates (see Figure 25.11). Adult stem cells can be isolated not only from the subventricular zone near the olfactory bulb (the anterior subventricular zone) and dentate gyrus (see Figure 25.12), but from the subventricular region of the cerebellum, midbrain, and spinal cord. When isolated in cell culture, these stem cells can give rise to neurons and glia. However, in the brain, at sites beyond the bulb and hippocampus, stem cells do not appear to produce new neurons that are integrated into adjacent CNS regions.

Figure 25.13 The forebrain anterior subventricular zone constitutes a stem cell niche. (A) Coronal section through the mouse brain indicating the location of the anterior subventricular zone in the anterior part of the lateral ventricles. The schematic shows the arrangement of cells at the ventricular surface. The ciliated ependymal cells form a tight epithelial boundary separating cerebrospinal fluid from brain tissue. Immediately adjacent are neural stem cells, whose processes are intercalated between the ependyma and all other cell types. These stem cells are also often seen in proximity to blood vessels. Transit amplifying cells are also found in this domain, and their progeny—the neuroblasts—are often clustered close by prior to adhering to glia that guide them into the rostral migratory stream (see Figure 25.14). (B) Isolation and propagation in vitro of neural stem cells from the anterior subventricular zone. At left is a "neurosphere," a ball of neural stem and transit amplifying cells that has been generated clonally from a single founder dissociated from the adult subventricular zone epithelium. The adjacent three panels show differentiated cell types generated from the neurosphere. From left to right: oligodendroglia, neurons, and astrocytes. (A after Alvarez-Buylla and Lim, 2004; B from Councill et al., 2006.)

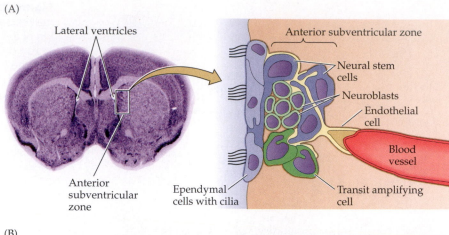

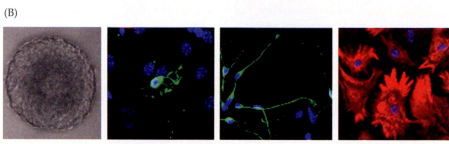

Three cell classes are found in mammalian brain subventricular zones: **neural stem cells**, **transit amplifying cells**, and **neuroblasts** (Figure 25.13). A fourth cell class, the **ependymal cell**, forms an epithelial barrier at the ventricular surface. Neural stem cells (also known as *multipotent neural precursors*) have the properties that distinguish stem cells in all tissues: they divide slowly and symmetrically; and they can give rise to all other classes of cells in the tissue—in this case, neurons, astrocytes, and oligodendroglial cells. Moreover, like stem cells in other tissues, neural stem cells can be isolated and propagated in vitro and, under appropriate cell culture conditions, will generate each major cell class found in the tissue (Figure 25.13B).

In mammalian subventricular zones, neural stem cells have many characteristics similar to astrocytes, including expressing multiple molecules that are also seen in astrocytes. Thus, similar to the developing brain, the multipotent neural stem cell has an apparent glial rather than neuronal identity (see Chapter 22). Furthermore, they are often found in close proximity to blood vessels, suggesting that they may be regulated by circulating as well as local signaling molecules. Although the route of delivery is different, this proximity to signals suggests that these precursors, like their developing counterparts, rely on cell–cell signaling to guide proliferation and differentiation. It is also possible that circulating signals relay information about the overall physiological state of the animal, thus linking broader homeostatic mechanisms with neurogenesis in the adult brain.

In order to generate differentiated neurons and glia, the neural stem cell must give rise to an intermediate precursor cell class, generally referred to as a **transit amplifying cell**. These cells retain the ability to divide; however, their cell cycles are much faster than those of stem cells, and they divide asymmetrically. After each cell division, a transit amplifying cell gives rise to one postmitotic neuroblast or glioblast, plus another transit amplifying cell that re-enters the cell cycle for an additional round of asymmetric division. Transit amplifying

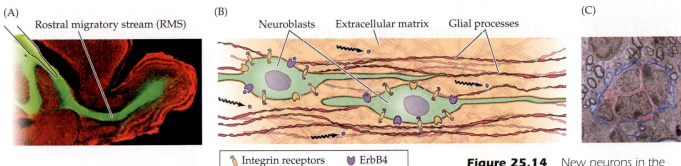

(A) Rostral migratory stream (RMS)

(B) Neuroblasts   Extracellular matrix   Glial processes

🖉 Integrin receptors   🟣 ErbB4

• Neuregulin   🟡 N-CAM

(C)

Figure 25.14 New neurons in the adult brain migrate via a specific pathway. (A) The rostral migratory stream can be demonstrated by injecting a tracer into the lateral ventricle. The cells in the SVZ take up the tracer (see Figure 25.13), and the labeled cells enter the forebrain tissue itself in a "stream" of migrating neurons. (B) A schematic of the RMS. Glial processes (color) form conduits for migrating neurons. The extracellular matrix (ECM) associated with these processes influences migration, mediated by integrin receptors for ECM components found on migrating neurons. Secreted neuregulin also influences motility of the migrating neurons in the RMS, via the ErbB4 neuregulin receptor. Finally, polysialyated N-CAM on the surfaces of newly generated neurons facilitates migration through the RMS. (C) Electron micrograph showing the glial conduit (blue) that encloses migrating newly generated neurons in the RMS (pink). (A,B after Ghashghaei et al., 2007; C from Alvarez-Buylla and Lim, 2004.)

cells are limited in their number of divisions, and eventually their potential for generating postmitotic blast cells is exhausted. Thus, for neurogenesis to proceed, the transit amplifying cells must be replenished from the stem cell population. The blast cells are no longer competent to divide, and usually they acquire the capacity to move away from the subventricular zone into brain regions where mature neurons or glia are found. In the hippocampus, this distance is relatively small, and the cells undergo a modest displacement to reach their final position. For newly generated cells destined for the olfactory bulb, however, the distance from the anterior subventricular zone to their final position in the bulb is considerable. A specific migratory route, defined by a distinct subset of glial cells, facilitates migration of newly generated neurons from the anterior subventricular zone to the bulb. This route is referred to as the **rostral migratory stream** (Figure 25.14), and within it, neuroblasts move along channels that are defined by the surfaces of elongated glial cells. In this migratory pathway, as at other sites of cell migration or axon growth, an extracellular matrix, presumably secreted by the glial cells, facilitates migration. In addition, the migrating cells express the polysialyated form of the neural cell adhesion molecule N-CAM, which promotes cell–cell interactions that facilitate migration. Secreted signaling molecules also influence migration in the rostral migratory stream. In this case, neuregulin, which also influences axon guidance and synapse formation in the periphery, facilitates motility via interaction with the Erb4 neuregulin receptor. Thus, in the rostral migratory stream, a number of developmentally regulated adhesion molecules and secreted signals mediate migration of new neurons through otherwise mature brain tissue.

Identification of molecular mediators of adult neurogenesis remains a major focus of current research into adult brain regeneration and repair. The most attractive hypothesis, and one that has been mostly supported by the available data, is that signaling molecules and transcriptional regulators that are used to define neural stem cells early in development are either retained or reactivated to facilitate neurogenesis in the adult. Accordingly, many of the inductive signaling molecules introduced in Chapter 22 as mediators of initial specification of neural precursors and their progeny in the neural plate and neural tube are also active in adult subventricular zones. These include Sonic hedgehog, members of the fibroblast growth factor family, TGF-β family members including the bone morphogenetic proteins or BMPs, and retinoic acid. Similarly, many of the adhesion molecules that influence cell migration during brain development (see Figures 23.3–23.5) also influence the migration of newly generated neurons, especially those destined for the olfactory bulb that must move through the rostral migratory stream.

BOX 25B Nuclear Weapons and Neurogenesis

The presence of stockpiles of nuclear weapons in an ever-increasing number of nations has always been a heavy burden for world affairs. It therefore may come as a surprise that the nuclear weapons testing carried out during the height of the Cold War (from the early 1950s through 1963) might play a positive, if unanticipated, role in resolving a major conflict in neuroscience.

In the late 1990s, a consensus emerged that the hippocampus and olfactory bulb were sites of gradual, limited addition of new neurons in the adult brain of all mammalian species, including humans. This consensus, however, did not extend to the question of whether new neurons are added to the cerebral cortex in adulthood. If adults do indeed add new neurons in significant numbers, such a mechanism would demand revision of current notions of plasticity, learning, and memory; it would also offer

new avenues for treating traumatic, hypoxic, and neurodegenerative cortical damage. Several reports in the mid 1990s, including some from experiments on non-human primates, suggested that there might be substantial addition of neurons to the adult cortex. Despite the provocative nature and

exciting implications of these findings, several other laboratories had difficulty replicating the surprising and controver-

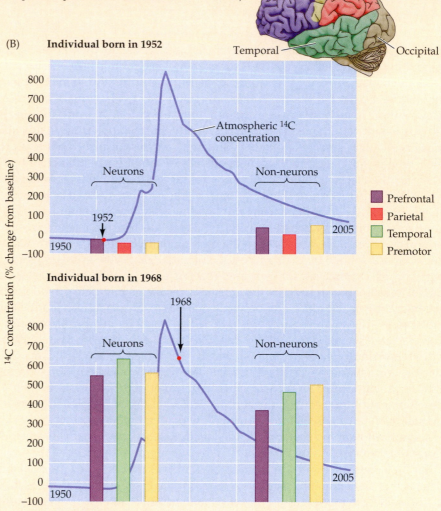

(A) Changes in atmospheric ^{14}C levels (right) and the availability of the isotope for incorporation into mitotic neurons at different times between 1955 (the start of frequent nuclear testing) and 1963 (when the Atomic Test Ban Treaty was put in place). Neurons generated between 1963 and 1970, whether in adults or in individuals who underwent gestation and birth during that timefame, would incorporate

significant amounts of ^{14}C into their nuclear DNA. (B) Autopsy results for the individual born in 1952 showed no cortical neurons with elevated ^{14}C levels; thus no neurons had been generated in the adult cortices. In contrast, the individual born in 1968 had significant numbers of ^{14}C-labeled neurons. Furthermore, in both cases, the neuronal level of ^{14}C corresponded to the atmospheric level around

the time of their gestation and birth. The slightly elevated levels for non-neural cells in the 1952 individual, and the lower levels in the 1968 individual, indicate that these cells turn over, and their ^{14}C content is altered by subsequent rounds of DNA synthesis and cell division. (A from Au and Fishell, 2006; B after Bhardwaj et al., 2006.)

(C)

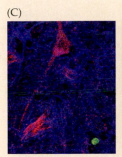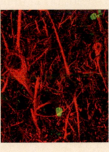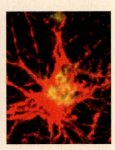

(C) The cortices from patients receiving BrdU during their lifetime were processed for BrdU histochemistry combined with neuronal and glial markers postmortem. Left: A BrdU nucleus (green) is distinct from cells labeled with Neu N (red), a neuronal marker. Center: A similar distinction between neurons labeled for neurofilaments (red) and BrdU-labeled cells (green). Right: Cells labeled for GFAP are coincident with the BrdU-labeled nuclei, suggesting that only glia are generated in the adult brain. (From Bhardwaj et al., 2006.)

sial results. The disparities engendered a polarized debate with no easy resolution. Clearly, what was needed was an independent means of assessing neurogenesis in the adult cortex, preferably in humans.

In a unique approach—and in one of the more successful searches for evidence of weapons of mass destruction in the last several years—Jonas Friesen and colleagues at the Karolinska Institute in Stockholm took advantage of fluctuations in environmental exposure to radioisotopes from nuclear weapons testing to evaluate when cortical neurons are indeed generated over an individual's lifetime (see Bhardwaj 2006). Their method relied on the recognition that the normally steady state of the isotope carbon-14 (^{14}C) in Earth's atmosphere had been dramatically altered for a brief period between the mid 1950s and the early 1960s. During this time, many countries conducted multiple tests of nuclear weapons, introducing large amounts of ionizing radiation into the atmosphere and nearly doubling the atmospheric concentration of ^{14}C. The Atomic Test Ban treaty of 1963 (which was adhered to by most countries until fairly recently) put a fairly abrupt end to this frightening period in human history, and atmospheric ^{14}C levels declined exponentially (Figure A).

This change in atmospheric ^{14}C concentration provided a natural version of experimental birthdating techniques described in Box 22E. Rather than injecting a bolus of tritiated thymidine or BrdU into the subject, people of varying ages had been naturally exposed to a "bolus"

of ^{14}C that was incorporated into DNA being synthesized at the time. Thus, regardless of the age of the subject, cortical neurons generated between 1955 and 1963 should have a higher concentration of ^{14}C in their nucleus than those generated before or after that time frame.

To assess neurogenesis in this ingenious way, the researchers obtained autopsy samples from the cerebral cortices of seven individuals born between 1933 and 1973. The logic was that those born before 1955 would have significant numbers of ^{14}C labeled neurons due to exposure as adults *only if there was indeed adult cortical neurogenesis*. If there was no adult neurogenesis, only those individuals born during (or shortly after) 1955–1963 should have ^{14}C-containing cortical neurons. To insure that only cortical neurons were assessed, dissociated cortical cells were labeled fluorescently with a neuron-specific marker and then sorted so that a comparison could be made between fluorescent-tagged neurons and the non-fluorescent glia and supporting cells (non-neurons). ^{14}C levels were then measured using accelerator mass spectroscopy.

The results were very clear: individuals born before 1955 had no cortical neurons with elevated ^{14}C levels; thus no neurons had been generated in their adult cortices (Figure B, top). In contrast, individuals born after 1955, but before the return of ^{14}C levels to baseline, had significant numbers of ^{14}C-labeled neurons; furthermore, the neuronal level of ^{14}C corresponded to the atmospheric level around the time of their gestation and birth (Figure B, bottom). The differ-

ent values for the non-neural cells indicate that these cells turn over, so their ^{14}C content declines because of dilution by subsequent rounds of DNA synthesis and cell division.

To amplify this result, Friesen and colleagues studied a group of patients who had skin cancers and whose treatment included injections of thymidine analogs like BrdU. They then examined cells heavily labeled with BrdU in the cortex postmortem. BrdU does not label neurons (recognized by staining for Neu-N, a neural marker) or neuron-specific neurofilaments; however, BrdU does label glial cells, recognized with antibodies against glial fibrillary protein (Figure C). This observation, especially when taken together with the ^{14}C study, argues strongly against significant neurogenesis in the adult cerebral cortex.

References

Au, E. and G. Fishell (2006) Adult cortical neurogenesis: Nuanced, negligible, or non-existent? *Nature Neurosci.* 9: 1086–1088.

Bhardwaj, R. D. and 10 others (2006) Neocortical neurogenesis in humans is restricted to development. *Proc. Natl. Acad. Sci. USA* 103: 12564–12568.

Gould, E, A. J. Reeves, M. S. Graziano and C. G. Gross (1999) Neurogenesis in the neocortex of adult primates. *Science* 286: 548–552.

Koketsu, D., A. Mikami, Y. Miyamoto and T. Hisatsune (2003) Nonrenewal of neurons in the cerebral neocortex of adult macaque monkeys. *J. Neurosci.* 23: 937–942.

Kornack, D. R. and P. Rakic (2001) Cell proliferation without neurogenesis in adult primate neocortex. *Science* 294: 2127–2130.

Rakic, P. (2006) No more cortical neurons for you. *Science* 313: 928–929.

Adult Neurogenesis, Stem Cells, and Brain Repair in Humans

These examples of addition or replacement of neurons in an adult brain—in fish, birds, rats, and mice—have provided clear examples of how new neurons may be integrated into existing circuits, presumably preserving, replacing, or augmenting function. In most cases, neuron replacement is gradual and likely related to ongoing, low-level neurogenesis rather than wholesale reconstitution of brain tissue in response to injury. Even in songbirds, neuron loss that accompanies neurogenesis is relatively gradual, and some neurons remain in the song control regions of the brain, presumably serving as a scaffold to guide where new cells will be added in the next season. The notable exceptions to this rule are the goldfish retina discussed earlier (where new neurons are added as part of the ongoing growth of the eye and brain), and the olfactory epithelium of most vertebrates (where major damage is repaired by reconstituting large numbers of neurons and connections).

Even the limited capacity of humans to replace neurons in an adult brain has offered some promise that, under the right conditions, neuron replacement might be used to repair the injured brain. In humans, the entire subventricular zone provides a supportive environment for neural stem cells. However, there is no evidence at present that adult neurogenesis occurs in the human brain outside the hippocampus and olfactory bulb. In fact, the available evidence suggests that there is no neurogenesis in the adult cerebral cortex (Box 25B)—an observation that establishes a much higher threshold for the possibility of stem cell-mediated repair of cortical circuitry damaged by trauma, hypoxia, or neuro-degenerative disease. Cell replacement therapies have been attempted in a relatively small number of patients with Parkinson's disease (see Box 18A); however, the overall effectiveness of such treatments has been limited. Furthermore, the extent to which truly undifferentiated human neural stem cells can be made to acquire characteristics of dopaminergic neurons of the substantia nigra (or any other distinctive neuronal identity) as well as an appropriate pattern of synaptic connections is unclear. Thus, although there is some promise in understanding ongoing neural replacement in the adult brain for repairing the damaged brain, the fulfillment of this promise may be extremely challenging.

Summary

There are three types of regeneration or repair in the adult nervous system. The first, and most effective, is the regrowth of severed peripheral axons, usually via the peripheral nerve sheaths once occupied by their forerunners, and the re-establishment of sensory and motor synapses on muscles or other targets. During this regeneration, mature Schwann cells and other supporting cells provide many of the molecules that mediate initial peripheral axon guidance and synaptogenesis during development. The second type of repair—local sprouting or longer extension of axons and dendrites at sites of traumatic damage or degenerative pathology in the brain or spinal cord—is far more limited. The major impediments to such local repair include death of the damaged neurons due to trophic deprivation or other stress; inhibition of growth by cytokines released during the immune response to brain tissue damage; and the formation of a glial scar by extensive hypertrophy of existing glial cells plus limited proliferation of glia at the site of the injury. The pathologic assembly of glial cell bodies and processes into a scar erects an impenetrable barrier to axon growth, in part due to secretion by the glia of many of molecules that inhibit axon growth and synapse formation during development. The third type of repair is the ongoing generation of new neurons in the adult brain. Although there is no evidence for wholesale replacement of neurons and circuits in most vertebrate brains, the capacity for more limited neuronal replacement exists in many species, including humans. In general, the olfactory bulb and the hippocampus are the sites of adult neurogenesis in most mammals. In both brain regions, new neurons are generated by neural stem cells retained in specific restricted locations, or niches, in the adult brain. Many of the molecules that regulate the maintenance, proliferation, and differentiation of adult neural stem cells and their progeny are molecules used for similar purposes during early development. The limited prospects of developing this capacity as a strategy for repair following brain injury or degenerative disease continue to captivate the imagination of patients and their physicians and motivate the efforts of many neuroscientists.

Additional Reading

Reviews

ALVAREZ-BUYLLA, A. AND D. A. LIM (2004) For the long run: Maintaining germinal niches in the adult brain. *Neuron* 41: 683–686.

BOYD, J. G. AND T. GORDON (2003) Neurotrophic factors and their receptors in axonal regeneration and functional recovery after peripheral nerve injury. *Mol. Neurobiol.* 27: 277–324.

CASE, L. C. AND M. TESSIER-LAVIGNE (2005) Regeneration of the adult central nervous system. *Curr. Biol.*15: 749–753.

DESHMUKH, M. AND E. M. JOHNSON JR. (1997) Programmed cell death in neurons: Focus on the pathway of nerve growth factor deprivation-induced death of sympathetic neurons. *Mol. Pharmacol.* 51: 897–906.

GAGE, F. H. (2000) Mammalian neural stem cells. *Science* 287: 1433–1488.

GOLDMAN, S. A. (1998) Adult neurogenesis: From canaries to the clinic. *J. Neurobiol.* 36: 267–286.

JOHNSON, E. O., A. B. ZOUBOS AND P. N. SOUCACOS (2005). Regeneration and repair of peripheral nerves. *Injury* 36S: S24–S29.

OTTESON, D. C. AND P. F. HITCHCOCK (2003) Stem cells in the teleost retina: Persistent neurogenesis and injury-induced regeneration. *Vision Res.* 43: 927–936.

SILVER, J. AND J. H. MILLER (2004) Regeneration beyond the glial scar. *Nature Rev. Neurosci.* 5: 146–156.

SONG, Y., J. A. PANZER, R. M. WYATT AND R. J. BALICE-GORDON (2006) Formation and plasticity of neuromuscular synaptic connections. *Int. Anesthesiol. Clin. Spring* 44: 145–178.

TERENGHI, G. (1999). Peripheral nerve regeneration and neurotrophic factors. *J. Anat.* 194: 1–14.

WIELOCH, T. AND K. NIKOLICH (2006) Mechanisms of neural plasticity following brain injury. *Curr. Opin. Neurobiol.* 16: 258–264.

Important Original Papers

ALTMAN, J. (1969) Autoradiographic and histological studies of postnatal neurogenesis. IV. Cell proliferation and migration in the anterior forebrain, with special reference to persisting neurogenesis in the olfactory bulb. *J. Comp. Neurol.* 136: 269–293.

ALTMAN, J. AND G. D. DAS (1967) Postnatal neurogenesis in the guinea-pig. *Nature* 214: 1098–1101.

BREGMAN, B. S., E. KUNKEL-BAGDEN, L. SCHNELL, H. N. DAI, D. GAO AND M. E. SCHWAB (1995) Recovery from spinal cord injury mediated by antibodies to neurite growth inhibitors. *Nature* 378: 498–501.

DAVID, S. AND A. J. AGUAYO (1981) Axonal elongation into peripheral nervous system "bridges" after central nervous system injury in adult rats. *Science* 214: 931–933.

EASTER, S. S. JR. AND C. A. STUERMER (1984) An evaluation of the hypothesis of shifting terminals in goldfish optic tectum. *J. Neurosci.* 4: 1052–1063.

ERIKSSON, P. S. AND 6 OTHERS (1998) Neurogenesis in the adult human hippocampus. *Nature Med.* 4: 1313–1317.

GOLDMAN, S. A. AND F. NOTTEBOHM (1983) Neuronal production, migration, and differentiation in a vocal control nucleus of the adult female canary brain. *Proc. Natl. Acad. Sci. USA* 80: 2390–2394.

GRAZIADEI, G. A. AND P. P. GRAZIADEI (1979) Neurogenesis and neuron regeneration in the olfactory system of mammals. II. Degeneration and reconstitution of the olfactory sensory neurons after axotomy. *J. Neurocytol.* 8: 197–213.

HEAD, H. AND W. H. R. RIVERS (1905) The afferent nervous system from a new aspect. *Brain* 28: 99–111.

KIM, J. E., S. LI, T. GRANDPRE, D. QIU AND S. M. STRITTMATTER (2003) Axon regeneration in young adult mice lacking Nogo-A/B. *Neuron* 38: 187–199.

LOIS, C., J. M. GARCIA-VERDUGO AND A. ALVAREZ-BUYLLA (1996) Chain migration of neuronal precursors. *Science* 271: 978–981.

LUSKIN, M. B. (1993) Restricted proliferation and migration of postnatally generated neurons derived from the forebrain subventricular zone. *Neuron* 11:173–189.

MARSHALL, L. M., J. R. SANES AND U. J. MCMAHAN (1977) Reinnervation of original synaptic sites on muscle fiber basement-membrane after disruption of the muscle cells. *Proc. Natl. Acad. Sci. USA* 74: 3073–3077.

NGUYEN, Q. T., J. R. SANES AND J. W. LICHTMAN (2002) Pre-existing pathways promote precise projection patterns. *Nature Neurosci.* 5: 861–867.

SANAI, N. AND 11 OTHERS (2004) Unique astrocyte ribbon in adult human brain contains neural stem cells but lacks chain migration. *Nature* 427: 740–744.

SKENE, J. H. AND M. WILLARD (1981) Axonally transported proteins associated with axon growth in rabbit central and peripheral nervous systems. *J. Cell Biol.* 89: 96–103.

SUHONEN, J. O., D. A. PETERSON, J. RAY AND F. H. GAGE (1996) Differentiation of adult hippocampus-derived progenitors into olfactory neurons in vivo. *Nature* 383: 624–627.

ZHENG, B., C. HO, S. LI, H. KEIRSTEAD, O. STEWARD AND M. TESSIER-LAVIGNE (2003) Lack of enhanced spinal regeneration in Nogo-deficient mice. *Neuron* 38: 213–224.

COMPLEX BRAIN FUNCTIONS

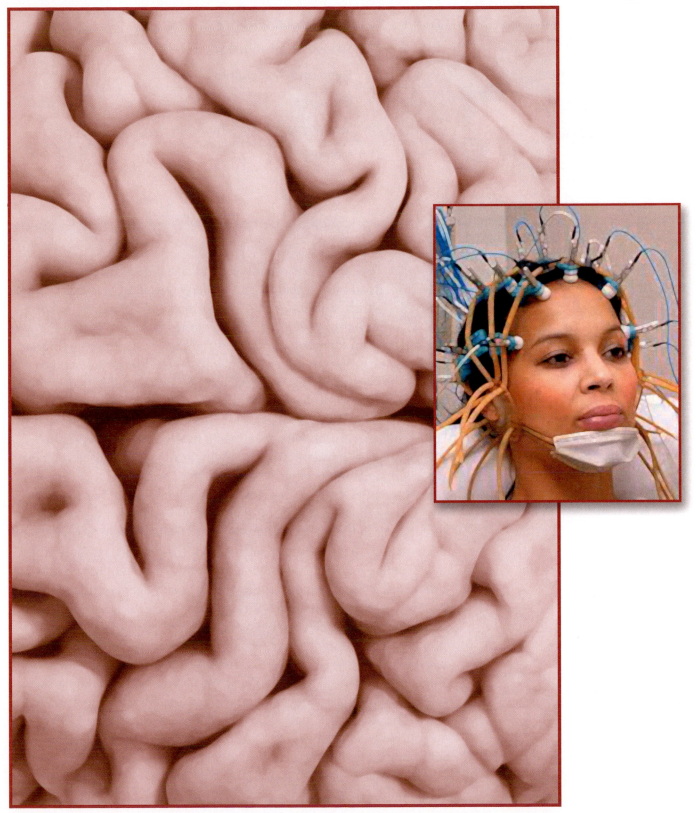

UNIT V COMPLEX BRAIN FUNCTIONS

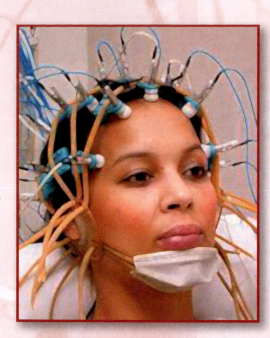

Subject with scalp leads in place for encephalographic (EEG) recording, one of several techniques that are widely used to assess brain activity during the performance of cognitive tasks. (Photograph © AJPhoto/Photo Researchers, Inc.)

The awareness of physical and social circumstances, the ability to have thoughts and feelings (emotions), to be sexually attracted to others, to express these things to our fellow humans by language, and to store such information in memory certainly rank among the most intriguing functions of the human brain. Given their importance in daily life—and for human culture generally—it is not surprising that much of the human brain is devoted to these and other complex mental functions. The intrinsic interest of these aspects of human behavior is unfortunately equaled by the difficulty—both technical and conceptual—involved in unraveling their neurobiological underpinnings. Nonetheless, a good deal of progress has been made in deciphering the structural and functional organization of the relevant brain regions. Especially important has been the steady accumulation of case studies during the last century or more that, by the signs and symptoms resulting from damage to specific brain regions, have indicated much about the primary location of various complex brain functions. More recently, the advent of noninvasive brain imaging techniques has provided a much deeper understanding of some of these abilities in normal human subjects as well as in neurological patients. Finally, complementary electrophysiological experiments in non-human primates and other experimental animals have begun to elucidate the cellular correlates of many of these functions. Taken together, these observations have established a rapidly growing body of knowledge about the more complex aspects of the human brain. This general domain of investigation has come to be called "cognitive neuroscience," a field that promises to offer greater insight into particularly human brain functions.

Chapter 26

The Association Cortices

Overview

The association cortices include most of the cerebral surface of the human brain and are largely responsible for the complex processing that goes on between the arrival of input to the primary sensory cortices and the generation of behavior. The diverse functions of the association cortices are loosely referred to as *cognition*, which literally means the process by which we come to know the world. ("Cognition" is perhaps not the best word to indicate this wide range of neural functions, but it has become part of the working vocabulary of neurologists and neuroscientists.) More specifically, cognition refers to the ability to attend to external stimuli or internal motivation; to identify the significance of such stimuli; and to make appropriate responses. Given the complexity of these tasks, it is not surprising that the association cortices receive and integrate information from a variety of sources, and that they influence a broad range of other cortical and subcortical targets. Inputs to the association cortices include projections from the primary and secondary sensory and motor cortices, the thalamus, and the brainstem. Outputs from the association cortices reach the hippocampus, the basal ganglia and cerebellum, the thalamus, and other cortical areas. Initial insight into the function of these cortical regions came primarily from observations of human patients with damage to one or another of these areas. Subsequent noninvasive brain imaging of normal subjects, functional mapping at neurosurgery, and electrophysiological analysis of comparable brain regions in non-human animals have generally confirmed clinical deductions. Together, these studies indicate that, among other functions, the parietal association cortex is especially important for attending to stimuli in the external and internal environment, that the temporal association cortex is especially important for identifying the nature of such stimuli, and that the frontal association cortex is especially important for selecting and planning appropriate behavioral responses.

The Association Cortices

The preceding chapters have considered in some detail the parts of the brain responsible for encoding sensory information and commanding movements (i.e., the primary sensory and motor cortices). But these regions account for only a fraction (perhaps a fifth) of the cerebral cortex. The consensus has long been that much of the remaining cortex is concerned with attending to complex stimuli, identifying the relevant features of such stimuli, recognizing the related objects, and planning appropriate responses (as well as storing aspects of this information). Collectively, these integrative abilities are referred to as **cognition**, and it is evidently the association cortices in the parietal, temporal, and frontal lobes that make cognition possible. (The extrastriate cortex of the occipital lobe

Figure 26.1 *Lateral and medial views of the human brain, showing the extent of the association cortices in blue. The primary sensory and motor regions of the neocortex are shaded in yellow. Notice that the primary cortices occupy a relatively small fraction of the total area of the cortical mantle. The remainder of the neocortex—defined by exclusion as the association cortices—is the seat of human cognitive ability. The term* association *refers to the fact that these regions of the cortex integrate (associate) information derived from other brain regions.*

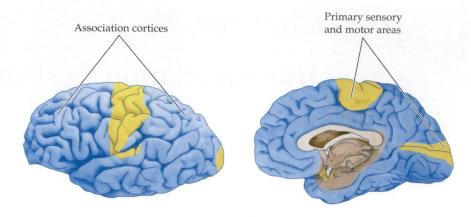

Association cortices

Primary sensory and motor areas

is equally important in cognition; its functions, however, are largely concerned with vision, and much of what is known about these areas has been discussed in Chapter 12.) These areas of the cerebral cortex are referred to collectively as the **association cortices** (Figure 26.1).

An Overview of Cortical Structure

Before delving into a more detailed account of the functions of the different regions of the association cortices, it is important to have a general understanding of cortical structure and the organization of its canonical circuitry. Most of the cortex that covers the cerebral hemispheres is **neocortex**, defined as cortex that has six cellular layers, or laminae. Each layer comprises more or less distinctive populations of cells based on their different densities, sizes, shapes, inputs, and outputs. The laminar organization and basic connectivity of the human cerebral cortex are summarized in Figure 26.2A and Table 26.1. Despite an overall uniformity, regional differences based on these laminar features have long been apparent (Box 26A), allowing investigators to identify numerous subdivisions of the cerebral cortex (Figure 26.2B). These histologically defined subdivisions are referred to as **cytoarchitectonic areas**, and, over the years, a zealous band of neuroanatomists has painstakingly mapped these areas in humans and in many of the more widely used laboratory animals.

Early in the twentieth century, cytoarchitectonically distinct regions were identified with little or no knowledge of their functional significance. Eventually, studies of patients in whom one or more of these cortical areas had been damaged, supplemented by electrophysiological mapping in both laboratory animals and neurosurgical patients, supplied this information. This work showed that many of

TABLE 26.1	**The Major Connections of the Neocortex**
Sources of cortical input	**Targets of cortical output**
Other cortical regions	Other cortical regions
Hippocampal formation	Hippocampal formation
Amygdala	Amygdala
Thalamus	Thalamus
Brainstem modulatory systems	Caudate and putamen (striatum)
	Brainstem
	Spinal cord

(A)

1
2
3
4
5
6

Pyramidal
cell

Local axon
collateral (local
circuitry)

Stellate
cell

Dendrites

Descending axon
(output)

White
matter

(B)

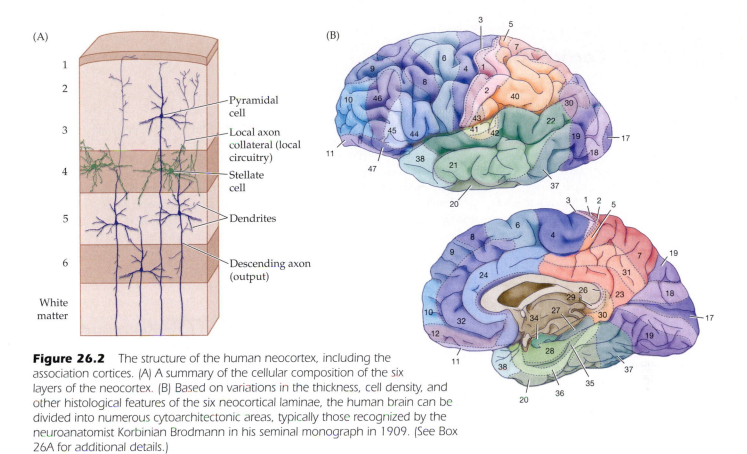

Figure 26.2   The structure of the human neocortex, including the association cortices. (A) A summary of the cellular composition of the six layers of the neocortex. (B) Based on variations in the thickness, cell density, and other histological features of the six neocortical laminae, the human brain can be divided into numerous cytoarchitectonic areas, typically those recognized by the neuroanatomist Korbinian Brodmann in his seminal monograph in 1909. (See Box 26A for additional details.)

the regions neuroanatomists had distinguished on histological grounds are also functionally distinct. Thus, cytoarchitectonic areas can sometimes be identified by the physiological response properties of their constituent cells, and often by their patterns of local and long-distance connections.

Despite significant variations among different cytoarchitectonic areas, the circuitry of all cortical regions has some common features (Figure 26.3). First, each cortical layer has a primary source of inputs and a primary output target. Second, each area has connections in the vertical axis (called *columnar* or *radial* connections) and connections in the horizontal axis (called *lateral* or *horizontal* connections). Third, cells with similar functions tend to be arrayed in radially aligned groups that span all the cortical layers and receive inputs that are often segregated into radial bands or columns. Finally, interneurons within specific cortical layers give rise to extensive local axons that extend horizontally in the cortex, often linking functionally similar groups of cells.

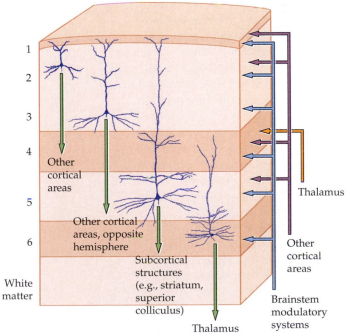

Figure 26.3   Canonical neocortical circuitry. Green arrows indicate outputs to the major targets of each of the neocortical layers in humans; orange arrow indicates thalamic input (primarily to layer 4); purple arrows indicate input from other cortical areas; and blue arrows indicate input from the brainstem modulatory systems to each layer.

BOX 26A Cortical Lamination

Much knowledge about the cerebral cortex is based on descriptions of differences in cell number and density throughout the cortical mantle. Nerve cell bodies, because of their high metabolic rate, are rich in basophilic substances (RNA, for instance), and therefore tend to stain darkly with reagents such as cresyl violet acetate. These *Nissl stains* (named after Franz Nissl, who first described this technique when he was a medical student in nineteenth-century Germany) provide a dramatic picture of brain structure at the histological level. The most striking feature revealed in this way is the distinctive lamination of the cortex in humans and other mammals, as seen in the figure. In humans, there are from three to six cortical layers, depending on the area of cortex. These layers, or laminae, are designated with the numerals 1–6 (or, often, with Roman numerals I–VI). Laminar subdivisions are indicated with letters (layers 4a, 4b, and 4c of the visual cortex, for example).

Each of the cortical laminae in the so-called *neocortex* (which covers the bulk of the cerebral hemispheres and is defined by six layers) has characteristic functional and anatomical features (see Figures 26.2 and 26.3). For example, cortical layer 4 is typically rich in stellate neurons with locally ramifying axons; in the primary sensory cortices, these neurons receive input from the thalamus, the major sensory relay from the periphery. Layer 5, and to a lesser degree layer 6, contain pyramidal neurons whose axons typically leave the cortex. The generally smaller pyramidal neurons in layers 2 and 3 (which are not as distinct as their numerical assignments suggest) have primarily corticocortical connections, and layer 1 contains mainly neuropil. Korbinian Brodmann, who early in the twentieth century devoted his career to an analysis of brain regions distinguished in this way, described about 50 distinct cortical regions, or cytoarchitec-

tonic areas (see Figure 26.2B). These structural features of the cerebral cortex continue to figure importantly in discussions of the brain, particularly in structural/functional correlation of intensely studied regions such as the primary sensory and motor cortices.

Not all cortex is six-layered neocortex. The hippocampal cortex, which lies deep in the temporal lobe and has been implicated in acquisition of declarative memories (see Chapter 31), has only three or four laminae. The hippocampal cortex is regarded as evolutionarily more primitive, and is therefore called *archicortex* to distinguish it from the six-layered *neocortex*. Another, presumably more primitive, type

of cortex, called *paleocortex* (*paleo*, "ancient"), generally has three layers and is found on the ventral surface of the cerebral hemispheres and along the parahippocampal gyrus in the medial temporal lobe.

The functional significance of different numbers of laminae in neocortex, archicortex, and paleocortex is not known, although it seems likely that the greater number of layers in neocortex reflects more complex information processing. The general similarity of neocortical structure across the entire cerebrum suggests a common denominator of cortical operation, although no one has yet deciphered what it is.

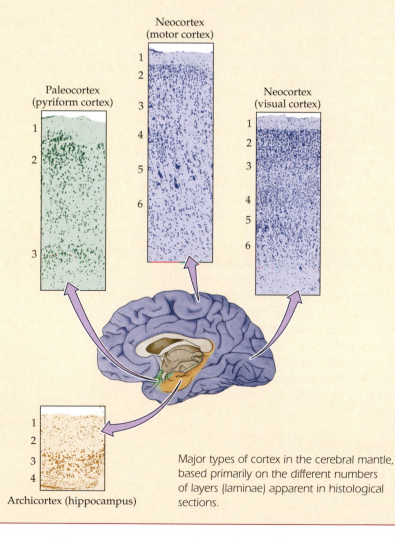

Major types of cortex in the cerebral mantle, based primarily on the different numbers of layers (laminae) apparent in histological sections.

The particular circuitry of any cortical region is a variation on this canonical pattern of inputs, outputs, and vertical and horizontal patterns of connectivity.

Specific Features of the Association Cortices

The connectivity of the association cortices is appreciably different from primary and secondary sensory and motor cortices, particularly with respect to inputs and outputs. For instance, two thalamic nuclei that are not involved in relaying primary motor or sensory information provide much of the subcortical input to the association cortices: the **pulvinar** projects to the parietal association cortex, while the **medial dorsal nuclei** project to the frontal association cortex. Several other thalamic nuclei, including the anterior and ventral anterior nuclei, innervate the association cortices as well.

Unlike the thalamic nuclei that receive peripheral sensory information and project to primary sensory cortices, the input to these association cortex-projecting nuclei comes from other regions of the cortex. In consequence, the signals coming into the association cortices via the thalamus reflect sensory and motor information that has *already* been processed in the primary sensory and motor areas of the cerebral cortex, and is being fed back to the association regions. The primary sensory cortices, in contrast, receive thalamic information that is more directly related to peripheral sense organs (see, for example, Chapter 9). Similarly, much of the thalamic input to primary motor cortex is derived from the thalamic nuclei related to the basal ganglia and cerebellum rather than to other cortical regions (see Unit III).

A second major difference in the sources of innervation to the association cortices is their enrichment in projections from other cortical areas, called **cortico-cortical connections** (see Figure 26.3). Indeed, these connections form the majority of the input to the association cortices. Ipsilateral corticocortical connections arise from primary and secondary sensory and motor cortices, and from other association cortices within the same hemisphere. Corticocortical connections also arise from both corresponding and noncorresponding cortical regions in the opposite hemisphere via the corpus callosum and anterior commissure, which together are referred to as **interhemispheric connections**. In the association cortices of humans and other primates, corticocortical connections often form segregated radial bands or columns in which interhemispheric projection bands are interdigitated with bands of ipsilateral corticocortical projections.

Another important source of innervation to the association areas is subcortical, arising from the dopaminergic nuclei in the midbrain, the noradrenergic and serotonergic nuclei in the brainstem reticular formation, and cholinergic nuclei in the brainstem and basal forebrain. These diffuse inputs project to different cortical layers and, among other functions, contribute to learning, motivation, and arousal (the continuum of mental states that ranges from deep sleep to high alert; see Chapter 28). A variety of behavioral and psychiatric disorders, such as addiction, depression, and attention deficit disorder are associated with dysfunction in one or more of these neuromodulatory circuits. Current pharmacological therapies for these diseases rely on manipulation of signaling via modulatory inputs to the association cortices.

The general wiring plan for the association cortices is summarized in Figure 26.4. Despite this degree of interconnectivity, the extensive inputs and outputs of the association cortices should not be taken to imply that everything is simply connected to everything else in these regions. On the contrary, each association area is defined by a distinct, if overlapping, subset of thalamic, corticocortical, and subcortical connections. It is nonetheless difficult to conclude much about the role of these different cortical areas based solely on connectivity. This

Figure 26.4 *Summary of the overall connectivity of the association cortices.*

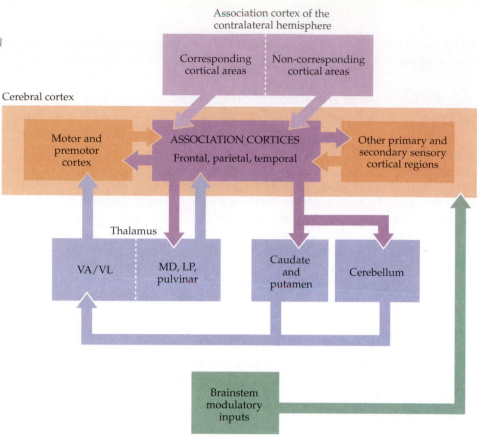

information is, in any event, quite limited for the human association cortices; most of the evidence comes from anatomical tracing studies in non-human primates, supplemented by the limited pathway tracing that can be done in human brain tissue postmortem. As a result, inferences about the function of human association areas continue to depend critically on observations of patients with cortical lesions. Damage to the association cortices in the parietal, temporal, and frontal lobes, respectively, results in specific cognitive deficits that indicate much about the operations and purposes of each of these regions. These deductions have largely been corroborated by patterns of neural activity observed in the homologous regions of the brains of experimental animals, as well as in humans using noninvasive imaging techniques.

Lesions of the Parietal Association Cortex: Deficits of Attention

In 1941, the British neurologist W. R. Brain reported three patients with unilateral parietal lobe lesions in whom the primary problem was varying degrees of attentional difficulty. Brain described their peculiar deficiency in the following way:

> Though not suffering from a loss of topographical memory or an inability to describe familiar routes, they nevertheless got lost in going from one room to another in their own homes, always making the same error of choosing a right turning instead of a left, or a door on the right instead of one on the left. In each case there was a massive lesion in the right parieto-occipital region, and it is suggested that this … resulted in an inattention to or neglect of the left half of external space.
>
> The patient who is thus cut off from the sensations which are necessary for the construction of a body scheme may react to the situation in several different

(A) "Draw a house"

Model   Patient's copy

(C) "Cancel the line"

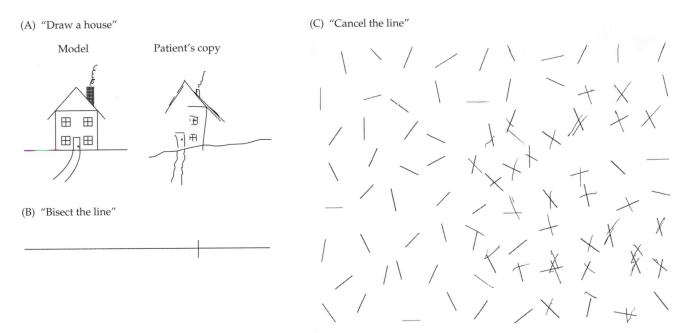

(B) "Bisect the line"

Figure 26.5   Characteristic performance on visuospatial tasks by individuals suffering from contralateral neglect syndrome. In (A), the patient was asked to draw a house by copying the figure on the left; on the right is the subject's imitation. In (B), the patient was asked to draw a vertical line through the center of (i.e., bisect) a horizontal line. In (C), the patient was asked to cross out each of the lines presented on the page. (A, B adapted from Posner and Raichle, 1994; C from Blumenfeld, 2002.)

ways. He may remember that the limbs on his left side are still there, or he may periodically forget them until reminded of their presence. He may have an illusion of their absence, i.e. they may 'feel absent' although he knows that they are there; he may believe that they are absent but allow himself to be convinced by evidence to the contrary; or, finally, his belief in their absence may be una- menable to reason and evidence to the contrary and so constitute a delusion.

W. R. Brain, 1941 (*Brain* 64: pp. 257 and 264)

This description is generally considered to be the first account of the link between parietal lobe lesions and deficits in attention or perceptual awareness. Based on a large number of patients studied since Brain's pioneering work, these deficits are now referred to as **contralateral neglect syndrome**.

The hallmark of contralateral neglect is an inability to attend to objects, or even one's own body, in a portion of space, despite the fact that visual acuity, somatic sensation, and motor ability remain intact. Affected individuals fail to report, respond to, or even orient to stimuli presented to the side of the body (or visual space) opposite the lesion (Figure 26.5). They may also have difficulty performing complex motor tasks on the neglected side, including dressing themselves, reaching for objects, writing, drawing, and, to a lesser extent, ori- enting to sounds. These motor deficits are called *apraxias*. The signs of the syn- drome can be as subtle as a temporary lack of contralateral attention that rap- idly improves as the patient recovers, or as profound as permanent denial of the existence of the side of the body and extrapersonal space opposite the lesion. Since Brain's original description of contralateral neglect and its relationship to lesions of the parietal lobe, it has been generally accepted that the parietal cor- tex, particularly the inferior parietal lobe, is the primary cortical region (but not the only region) governing attention (Figure 26.6A).

Importantly, contralateral neglect syndrome is typically associated with dam- age to the *right* parietal cortex. The unequal distribution of this particular cogni- tive function between the hemispheres is thought to arise because the right parietal cortex mediates attention to both left and right halves of the body and extrapersonal space, whereas the left hemisphere mediates attention primarily to the right (Figure 26.6B). Thus, left parietal lesions tend to be compensated by

(A)

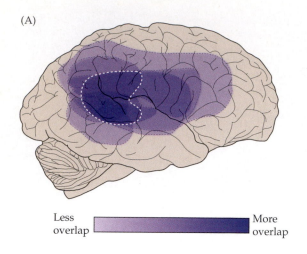

Less overlap ▮▮▮▮ More overlap

Figure 26.6 *Neuroanatomy of neglect syndromes. (A) Composite of the location of the underlying lesions in eight patients diagnosed with contralateral neglect syndrome. The site of damage was ascertained from CT scans (see Box 1B). While the lesions include parietal cortical areas, frontal areas, and the temporal lobe of the right hemisphere, the region of the right parietal lobe indicated by the dashed line is most often affected. (B) Schematic illustration of hemispheric asymmetry in attentional control inferred from neglect patients. In normal subjects, the right parietal cortex dominates the control of attention, as indicated by the thicker rays. A right parietal lesion (purple) results in severe left neglect, whereas a left parietal lesion leads to only minimal right neglect due to preserved attention within the right hemisphere. Bilateral parietal lesions cause right neglect due to diminished attentional processing in both hemispheres. (A after Heilman and Valenstein, 1985; B after Blumenfeld, 2002.)*

(B)

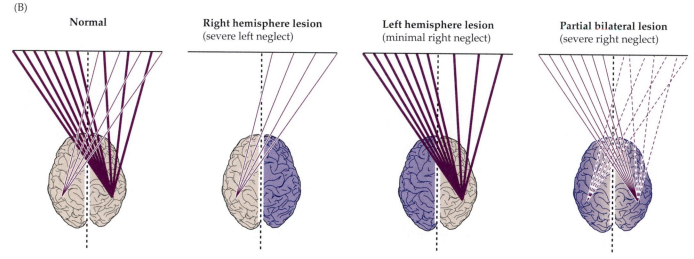

| Normal | Right hemisphere lesion (severe left neglect) | Left hemisphere lesion (minimal right neglect) | Partial bilateral lesion (severe right neglect) |

the intact right hemisphere. In contrast, when the right parietal cortex is damaged, there is little or no compensatory capacity in the left hemisphere to mediate attention to the left side of the body or extrapersonal space.

This interpretation has been confirmed by noninvasive imaging of parietal lobe activity during specific attention tasks carried out by normal subjects. Such studies show that neural activity is increased in *both* the right and left parietal cortices when subjects are asked to perform tasks in the *right* visual field requiring selective attention to distinct aspects of a visual stimulus such as its shape, velocity, or color. However, when a similar challenge is presented in the *left* visual field, only the *right* parietal cortex is activated (Figure 26.7). There is also evidence of increased activity in the right frontal cortex during such tasks (see also Figure 26.6A). This latter observation suggests that regions outside the parietal lobe also contribute to attentive behavior, and perhaps to some aspects of the pathology of neglect syndromes. Overall, however, metabolic mapping is consistent with the clinical fact that contralateral neglect typically arises from a right parietal lesion, and endorses the broader idea of hemispheric specialization for attention, in keeping with hemispheric specialization for a number of other cognitive functions (see below and Chapter 27).

Interestingly, patients with contralateral neglect are not simply deficient in their attentiveness to the left visual field, but to the left sides of objects generally. For example, when asked to cross out lines distributed throughout the

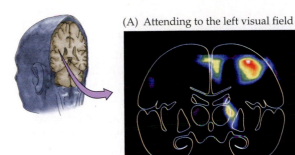

(A) Attending to the left visual field

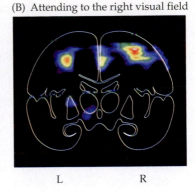

(B) Attending to the right visual field

L   R   L   R

Figure 26.7   Confirming the clinical impressions derived from neurological patients with parietal lobe damage, brain imaging (PET in this example) shows the right parietal cortex of normal subjects to be highly active during tasks requiring attention. (A) A subject has been asked to attend to objects in the left visual field; only the right parietal cortex is active. (B) When attention is shifted from the left visual field to the right, the right parietal cortex remains active, but activity is apparent in the left parietal cortex as well. This arrangement implies that damage to the left parietal lobe does not generate right-sided hemineglect because the right parietal lobe also serves this function. (After Posner and Raichle, 1994.)

visual field, contralateral neglect patients, as expected, tend to eliminate more lines on the right side of the field than on the left, consistent with a disruption in attentiveness to the left visual field (see Figure 26.5C). The lines they draw, however, tend to be biased towards the right side of each nonvertical line, no matter where the line happens to fall in the visual field. These observations suggest that attentiveness relies on a frame of reference anchored to the locations of objects and their relative dimensions.

Disruptions in spatial frames of reference are also associated with lesions of the parietal cortex that are more dorsal and medial than those typically associated with classical neglect. Such damage often presents as a triad of visuospatial deficits known as *Balint's syndrome* (named after the Austro-Hungarian neurologist who first described this condition). The three signs are an inability to perceive parts of a complex visual scene as a whole (called *simultanagnosia*), deficits in visually guided reaching (*optic ataxia*), and difficulty in voluntary scanning of visual scenes (*ocular apraxia*). In contrast to classical neglect, optic ataxia and ocular apraxia typically remit when movements are guided by nonvisual cues. These observations suggest that the parietal cortex participates in the construction of spatial representations that can guide both attention and movement.

Lesions of the Temporal Association Cortex: Deficits of Recognition

Clinical evidence from patients with lesions of the association cortex in the temporal lobe indicates that one of the major functions of this part of the brain is the recognition and identification of stimuli that are attended to, particularly complex stimuli. Thus, damage to either temporal lobe can result in difficulty recognizing, identifying, and naming different categories of objects. These disorders, collectively called **agnosias** (Greek for "not knowing"), are quite different from the neglect syndromes. As noted, patients with right parietal lobe damage often deny awareness of sensory information in the left visual field (and are less attentive to the left sides of objects generally), despite the fact that the sensory systems are intact (an individual with contralateral neglect syndrome typically withdraws his left arm in response to a pinprick, even though he may not admit the arm's existence). Patients with agnosia, on the other hand, acknowledge the presence of a stimulus, but are unable to report what it is. Agnosias have both a lexical aspect (a mismatching of verbal or other cognitive symbols with sensory stimuli; see Chapter 27) and a mnemonic aspect (a failure to recall stimuli when confronted with them again; see Chapter 31).

One of the most thoroughly studied agnosias following damage to the temporal association cortex in humans is the inability to recognize and identify faces. This disorder, called **prosopagnosia** (*prosopo*, "face" or "person"), was recognized by neurologists in the late nineteenth century and remains an area of intense investigation. After damage to the inferior temporal cortex, typically on the right, patients are often unable to identify familiar individuals by their facial characteristics, and in some cases cannot recognize a face at all. Nonetheless, such individuals are perfectly aware that some sort of visual stimulus is present and can describe particular aspects or elements of it without difficulty.

An example is the case of L.H., a patient described by the neuropsychologist N. L. Etcoff and colleagues. L.H. (the use of initials to identify neurological patients in published reports is standard practice) was a 40-year-old minister and social worker who had sustained a severe head injury as the result of an automobile accident when he was 18. After recovery, he could not recognize familiar faces, report that they were familiar, or answer questions about faces from memory. He could still identify other common objects, could discriminate subtle shape differences, and could recognize the sex, age, and even the "like-ability" of faces. Moreover, he could identify particular people by non-facial cues such as voice, body shape, and gait. The only other category of visual stimuli he had trouble recognizing was animals and their expressions, though these impairments were not as severe as for human faces, and he was able to lead a fairly normal and productive life. Noninvasive brain imaging showed that L.H.'s prosopagnosia was the result of damage to the right temporal lobe.

More recently, imaging studies in normal subjects have confirmed that the inferior temporal cortex mediates face recognition and that nearby regions are responsible for categorically different recognition functions (Figure 26.8). In general, lesions of the right temporal cortex lead to agnosia for faces and objects, whereas lesions of the corresponding regions of the left temporal cortex tend to result in difficulties with language-related material. (Recall that the primary auditory cortex is on the superior aspect of the temporal lobe; as described in the following chapter, the cortex adjacent to the auditory cortex in the left temporal lobe is specifically concerned with language.) The lesions that typically cause recognition deficits are in the inferior temporal cortex, in or near the fusiform gyrus; those that cause language-related problems in the left temporal

Figure 26.8 *Functional brain imaging of temporal lobe during face recognition. (A) Face stimulus presented to a normal subject at time indicated by arrow. Graph shows activity change in the relevant area of the right temporal lobe. (B) Location of fMRI activity in the right inferior temporal lobe. (Courtesy of Greg McCarthy.)*

(A)

(B)

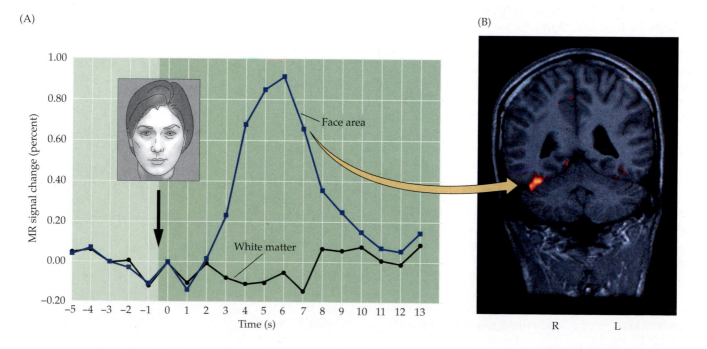

lobe tend to be on the lateral surface of the cortex. Consistent with these conclusions, direct cortical stimulation in subjects whose temporal lobes are being mapped for neurosurgery (typically removal of an epileptic focus) may induce transient prosopagnosia as a consequence of this abnormal activation of the relevant regions of the right temporal cortex.

Prosopagnosia and related agnosias involving objects are specific instances of a broad range of functional deficits that have as their hallmark the inability to recognize a complex sensory stimulus as familiar, and to identify and name that stimulus as a meaningful entity in the environment. Depending on the laterality, location, and size of the lesion in temporal cortex, agnosias can be as specific as for human faces, or as general as an inability to name most familiar objects.

Lesions of the Frontal Association Cortex: Deficits of Planning

The functional deficits that result from damage to the human frontal lobe are diverse and devastating, particularly if both hemispheres are involved. This broad range of clinical effects stems from the fact that the frontal cortex has a wider repertoire of functions than any other neocortical region (consistent with the fact that the frontal lobes in humans and other primates are the largest of the brain's lobes and comprises a greater number of cytoarchitectonic areas).

The particularly devastating nature of the behavioral deficits after frontal lobe damage reflects the role of this part of the brain in maintaining what is normally thought of as an individual's "personality." The frontal cortex integrates complex information from sensory and motor cortices, as well as from the parietal and temporal association cortices. The result is an appreciation of self in relation to the world that allows behaviors to be planned and executed normally. When this ability is compromised, the afflicted individual often has difficulty carrying out complex behaviors that are appropriate to the circumstances. These deficiencies in the normal ability to match ongoing behavior to present or future demands are, not surprisingly, interpreted as a change in the patient's "character."

The case that first called attention to the consequences of frontal lobe damage was that of Phineas Gage, a laborer on the Rutland and Burlington Railroad in mid-nineteenth-century Vermont. In that era, the conventional way of blasting rock was to tamp powder into a hole with a heavy metal rod. Gage, a popular and respected crew foreman, was undertaking this procedure one day in 1848 when his tamping rod sparked the powder, setting off an explosion that drove the rod—which was about a meter long and 4 or 5 centimeters in diameter—through his left orbit (eye socket), destroying much of the frontal part of his brain in the process. Gage, who never lost consciousness, was promptly taken to a local doctor who treated his wound. An infection set in, presumably destroying additional frontal lobe tissue, and Gage was an invalid for several months. Eventually he recovered and was to outward appearances well again. Those who knew Gage, however, were profoundly aware that he was not the "same" individual that he had been before. A temperate, hardworking, and altogether decent person had, by virtue of this accident, been turned into an inconsiderate, intemperate lout who could no longer cope with normal social intercourse or the kind of practical planning that had allowed Gage the social and economic success he enjoyed before.

The physician who looked after Gage until his death in 1863 summarized his impressions of Gage's personality as follows:

> [Gage is] fitful, irreverent, indulging at times in the grossest profanity (which was not previously his custom), manifesting but little deference for his fellows,

impatient of restraint or advice when it conflicts with his desires, at times pertinaciously obstinate, yet capricious and vacillating, devising many plans of future operations, which are no sooner arranged than they are abandoned in turn for others appearing more feasible. A child in his intellectual capacity and manifestations, he has the animal passions of a strong man. Previous to his injury, although untrained in the schools, he possessed a well-balanced mind, and was looked upon by those who knew him as a shrewd, smart businessman, very energetic and persistent in executing all his plans of operation. In this regard his mind was radically changed, so decidedly that his friends and acquaintances said he was 'no longer Gage.'

J. M. Harlow, 1868 (*Publications of the Massachusetts Medical Society* 2: 339–340)

Another classic case of frontal lobe deficits was that of a patient followed throughout the 1920s and 1930s by the neurologist R. M. Brickner. Joe A., as Brickner referred to his patient, was a stockbroker who at age 39 underwent bilateral frontal lobe resection because of a large tumor. After the operation, Joe A. had no obvious sensory or motor deficits; he could speak and understand verbal communication and was aware of people, objects, and temporal order in his environment. He acknowledged his illness and retained a high degree of intellectual power, as judged from an ongoing ability to play an expert game of checkers. Nonetheless, Joe A.'s personality underwent a dramatic change. This formerly restrained, modest man became boastful of professional, physical, and sexual prowess, showed little restraint in conversation, and was unable to match the appropriateness of what he said to his audience. Like Gage, his ability to plan for the future was largely lost, as was much of his earlier initiative and creativity. Even though he retained the ability to learn complex procedures, he was unable to return to work and had to rely on his family for support and care.

The effects of widespread frontal lobe damage documented by these case studies encompass a wide range of cognitive disabilities, including impaired restraint, disordered thought, perseveration (repetition of the same behavior), and the inability to plan appropriate action. Recent studies of patients with focal damage to particular regions of the frontal lobe suggest that some of the processes underlying these deficits may be localized. Working memory functions (see Chapter 31) are situated more dorsolaterally, and planning and social restraint functions are located more ventromedially. Some of these functions can be clinically assessed using standardized tests such as the Wisconsin Card Sorting Task for planning (see Box 26C), the delayed response task for working memory, and the "go–nogo" task for inhibition of inappropriate responses. All these observations are consistent with the idea that the common denominator of the cognitive functions subserved by the frontal cortex is the selection, planning, and execution of appropriate behavior, particularly in social contexts.

Sadly, the effects of damage to the frontal lobes have also been documented by the many thousands of frontal lobotomies ("leukotomies") performed in the 1930s and 1940s as a means of treating mental illness (Box 26B). The rise and fall of this "psychosurgery" provides a compelling example of the frailty of human judgment in medical practice, and of the conflicting approaches of neurologists, neurosurgeons, and psychiatrists in that era to the treatment of mental disease.

"Attention Neurons" in the Monkey Parietal Cortex

These clinical and pathological observations clearly indicate distinct cognitive functions for the parietal, temporal, and frontal lobes. They do not, however, provide much insight into how the nervous system represents this information

BOX 26B Psychosurgery

The consequences of frontal lobe destruction have been all too well documented by a disturbing yet fascinating episode in twentieth-century medical practice. From 1935 through the 1940s, neurosurgical destruction of the frontal lobe (frontal lobotomy, or leukotomy) was a popular treatment for certain mental disorders. More than 20,000 of these procedures were performed, mostly in the United States.

Enthusiasm for this approach to mental disease grew from the work of Egas Moniz, a respected Portuguese neurologist, who among other accomplishments did pioneering work on cerebral angiography before becoming the leading advocate of psychosurgery. Moniz recognized that the frontal lobes were important in personality structure and behavior, and concluded that interfering with frontal lobe function might alter the course of mental diseases such as schizophrenia and other chronic psychiatric disorders. He also recognized that destroying the frontal lobe would be relatively easy to do and, with the help of Almeida Lima, a neurosurgical colleague, introduced a simple surgical procedure for indiscriminately destroying most of the connections between the frontal lobe and the rest of the brain (see figure).

In the United States, the neurologist Walter Freeman, in collaboration with neurosurgeon James Watts, became an equally strong advocate of this approach and devoted his life to treating a wide variety of mentally disturbed patients in this way. He popularized a form of the procedure that could be carried out under local anesthesia and traveled widely across the country to demonstrate this technique and encourage its use.

Although it is easy in retrospect to be critical of such zealotry in the absence of either evidence or sound theory, it is important to remember that effective psychotropic drugs were not then available, and patients suffering from many of the disorders for which leukotomies were done were confined under custodial conditions that were at best dismal, and at worst brutal. Rendering a patient relatively tractable, albeit permanently altered in personality, no doubt seemed the most humane of the difficult choices that faced psychiatrists and others dealing with such patients during that period.

With the advent of increasingly effective psychotropic drugs in the late 1940s and the early 1950s, frontal lobotomy as a psychotherapeutic strategy rapidly disappeared, but not before Moniz was awarded the Nobel Prize for Physiology or Medicine in 1949. The history of this instructive episode in modern medicine has been compellingly told by Eliot Valenstein, and his book on the rise and fall of psychosurgery should be read by anyone contemplating a career in neurology, neurosurgery, or psychiatry.

References

BRICKNER, R. M. (1932) An interpretation of function based on the study of a case of bilateral frontal lobectomy. *Proceedings of the Association for Research in Nervous and Mental Disorders* 13: 259–351.

BRICKNER, R. M. (1952) Brain of patient A after bilateral frontal lobectomy: Status of frontal lobe problem. *Arch. Neurol. Psychiatry* 68: 293–313.

FREEMAN, W. AND J. WATTS (1942) *Psychosurgery: Intelligence, Emotion and Social Behavior Following Prefrontal Lobotomy for Mental Disorders.* Springfield, IL: Charles C. Thomas.

MONIZ, E. (1937) Prefrontal leukotomy in the treatment of mental disorders. *Amer. J. Psychiatry* 93: 1379–1385

VALENSTEIN, E. S. (1986) *Great and Desperate Cures: The Rise and Decline of Psychosurgery and Other Radical Treatments for Mental Illness.* New York: Basic Books.

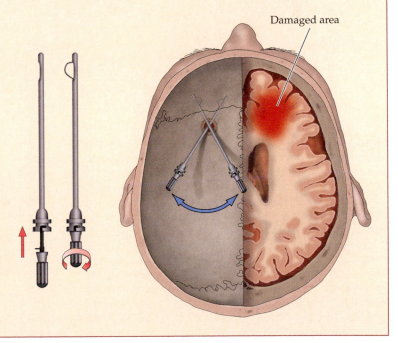

Damaged area

The surgical technique for frontal leukotomy under local anesthesia described and advocated by Egas Moniz and Almeida Lima. The "leukotome" was inserted into the brain at approximately the angles shown. When the leukotome was in place, a wire "knife" was extended and the handle rotated. The right side of the figure depicts a horizontal cutaway of the brain (parallel to the top of the skull) indicating Moniz's estimate of the extent of the damage done by the procedure. (After Moniz, 1937.)

in nerve cells and their interconnections. The apparent functions of the association cortices implied by clinical observations stimulated a number of informative electrophysiological studies in non-human primates, particularly macaque (usually rhesus) monkeys.

As in humans, a wide range of cognitive abilities in monkeys are mediated by the association cortices of the parietal, temporal, and frontal lobes (Figure 26.9A). Moreover, these functions can be tested using behavioral paradigms that assess attention, identification, and planning capabilities—the broad functions respectively assigned to the parietal, temporal, and frontal association cortices in humans. Using implanted electrodes, recordings from single neurons in the brains of awake, behaving monkeys are used to assess the activity of individual cells in the association cortices as various cognitive tasks are performed (Figure 26.9B).

Neurons in the parietal cortex of monkeys have been studied using this approach. The activity of parietal neurons is related to directing attention to objects, places, and other stimuli. The studies take advantage of the fact that monkeys can be trained to selectively attend to particular objects or events and report their experience in a variety of nonverbal ways, typically by looking at a response target (thus allowing their eye movements to be monitored) or manipulating a joystick. Attention-sensitive neurons can be identified by recording electrophysiological changes in neuronal activity associated with simultaneous changes in the attentive behavior of the animal. As might be expected from the

(A)

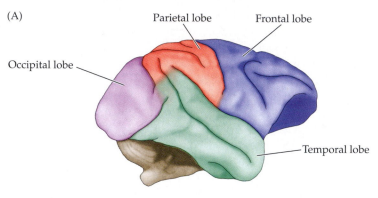

(B)

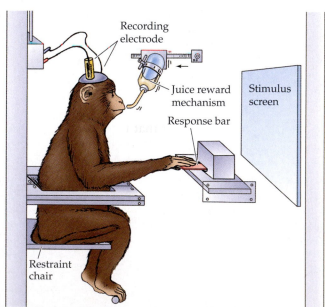

Figure 26.9 *Recording from single neurons in the brain of an awake, behaving rhesus monkey. (A) Lateral view of the rhesus monkey brain showing the parietal (red), temporal (green), and frontal (blue) cortices. The occipital cortex is shaded purple. (B) The animal is seated in a chair and gently restrained. Several weeks before data collection begins, a recording well is placed through the skull using a sterile surgical technique. For electrophysiological recording experiments, a tungsten microelectrode is inserted through the dura and arachnoid, and into the cortex. The screen and the response bar in front of the monkey are for behavioral testing. In this way, individual neurons can be monitored while the monkey performs specific cognitive tasks to gain a fruit juice reward.*

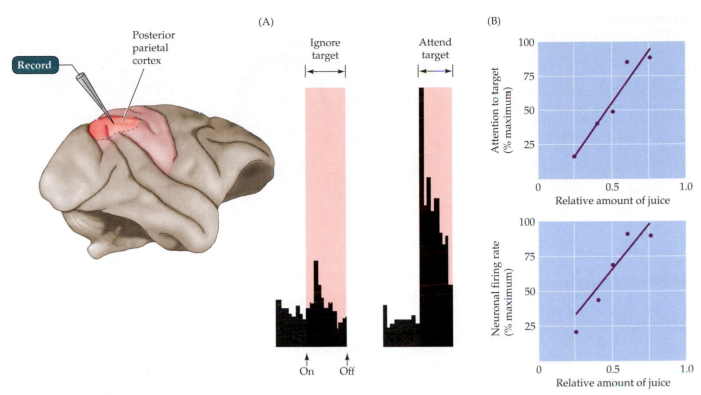

clinical evidence in humans, neurons in specific regions of the parietal cortex of the rhesus monkey are activated when the animal attends to a target, but not when the same stimulus is ignored (Figure 26.10A).

In another study, monkeys were rewarded with different amounts of fruit juice (a highly desirable treat) for attending to each of a pair of simultaneously illuminated targets (Figure 26.10B). Not surprisingly, the frequency with which monkeys attended to each target varied with the amount of juice they learned to expect for doing so. Moreover, the activity of some neurons in parietal cortex varied systematically as a function of the amount of juice associated with each target, and therefore the amount of attention paid by the monkey to the target. Thus the primate parietal cortex contains neurons that respond specifically when the animal attends to a behaviorally meaningful stimulus, and the vigor of the response reflects the amount of attention paid to the stimulus.

Intuitively, paying attention to a stimulus—whether it is a face in a crowd or a conversation at a crowded party—makes it easier to perceive that stimulus and facilitates appropriate behavioral responses. Indeed, behavioral studies in people and animals indicate that attended stimuli are responded to more quickly and more accurately than unattended stimuli. Such improvements in perception and action suggest that the underlying neuronal response to the attended stimulus is somehow improved as well. Recent neuroimaging and neurophysiological studies support this hypothesis. For example, fMRI scans of human subjects asked to pay attention to a visual stimulus in the lower visual hemifield reveal enhanced neural responses to those stimuli in the corresponding retinotopic portion of the contralateral occipital cortex (Figure 26.11A; see also Chapter 12). Furthermore, electrophysiological recordings from single neurons in extrastriate visual cortex in monkeys reveal enhanced responses to visual stimuli when they are attended compared with when they are ignored (Figure 26.11B). Such attention-related enhancement of neuronal responses in visual cortex seems likely to be responsible for the improvement in the speed and accuracy of behavioral responses to attended stimuli. Similar attention-related

Figure 26.10   Selective activation of neurons in the parietal cortex of a rhesus monkey as a function of attention (in this case, attention is directed to a light associated with a fruit juice reward). (A) Although the baseline level of activity of the neuron being studied here remains unchanged when the monkey ignores a visual target (left), firing rate increases dramatically when the monkey attends to the same stimulus (right). The histograms indicate action potential frequency per unit time. (B) When given a choice of where to attend, the monkey pays increasing attention to a particular visual target when more fruit juice reward can be expected for doing so (left), and the firing rate of the parietal neuron under study increases accordingly (A after Lynch et al., 1977; B after Platt and Glimcher, 1999.)

Figure 26.11 *Attending to a stimulus enhances neuronal responses in visual cortex. (A) When human subjects are instructed to attend to visual stimuli such as numbers and letters in the left visual field, neural activity in the contralateral (right) visual cortex increases. In contrast, when subjects attend to visual stimuli on the right, neural activity in the left visual cortex increases. (B) Responses of a single neuron in extrastriate visual cortex to a bar flashed at different orientations on a screen in front of a monkey. When the monkey attends to the bar the neuron responds with greater vigor, especially for bars at the preferred orientation. These data suggest that attention improves perceptual performance by selectively enhancing neuronal responses in the relevant sensory cortex. (A from Woldorff et al., 1997; B after McAdams and Maunsell, 1999.)*

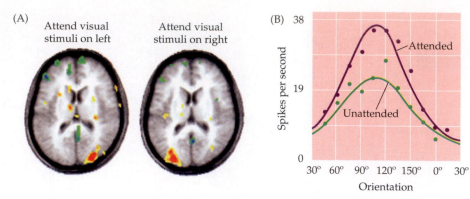

improvements in neuronal sensitivity have been found in auditory cortex for sounds and somatosensory cortex for tactile stimuli. The increased responsiveness of neurons in the sensory cortex to attended stimuli seems likely to be driven by inputs from attention-directing neurons in the parietal and prefrontal cortex.

"Recognition Neurons" in the Monkey Temporal Cortex

In keeping with human deficits of recognition following temporal lobe lesions, neurons with responses that correlate with the recognition of specific stimuli are present in the temporal cortex of rhesus monkeys (Figure 26.12). The behavior of these neurons is generally consistent with one of the major functions ascribed to the human temporal cortex, namely, the recognition and identification of complex stimuli. For example, some neurons in the inferior temporal gyrus of the rhesus monkey cortex respond specifically to the presentation of a monkey face. These cells are often quite selective: some respond only to the frontal view of a face and others only to profiles (Figure 26.12B,C). Furthermore, the cells are not easily deceived. When parts of faces or generally similar objects are presented, such cells typically fail to respond; in fact, the only things that confuse face-selective neurons are round or fuzzy objects such as apples, clock faces, or toilet brushes—all of which are vaguely facelike in appearance.

In principle, it is unlikely that such "face cells" are tuned to specific faces or objects, and no cells have so far been found that are selective for a particular face (sometimes referred to as a "grandmother" cell). However, it is not hard to imagine that populations of neurons differently responsive to various features of faces or other objects could act in concert to enable the recognition of such complex sensory stimuli. The notion of such "population coding" of objects is supported by the recent observation that face-selective neuronal responses in the temporal cortex of monkeys vary in intensity with respect to an average face. Both monkeys and humans are better at recognizing faces with extreme features—caricatures—than they are at recognizing less distinctive faces, suggesting that faces are identified by comparison with a mental standard or norm. Similarly, neurons in the inferior temporal cortex of monkeys respond much more strongly to caricatures of human faces than to an "average" human face, which is represented by the average activity of the neuronal population. Such norm-based tuning has also been reported for neuronal responses to shapes in the inferior temporal cortex.

Recent studies suggest that such complex response properties may be based on a columnar anatomical arrangement similar to that in the primary visual cortex (see Chapter 12). Each column has been taken to represent different

(A)

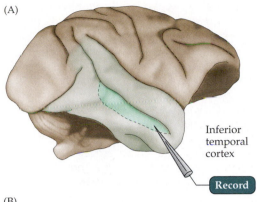

Inferior
temporal
cortex

Record

Figure 26.12   Selective activation of face cells in the inferior temporal cortex of a rhesus monkey. (A) Region of recording. (B) The neuron being recorded from in this case responds selectively to faces seen from the front. Scrambled parts of faces (stimulus 2) or faces with parts omitted (stimulus 3) do not elicit a maximal response. The cell responds best to different monkey faces, as long as they are complete and viewed from the front (stimulus 4); the cell also responds to a bearded human face (stimulus 5), although not quite as robustly. An irrelevant stimulus (a hand; stimulus 6) does not elicit a response. (C) In this example, the neuron being recorded from responds to profiles of faces. A face viewed from the front (stimulus 1), 30° (stimulus 2), or 60° (stimulus 3) is not as effective as a true profile (stimulus 4). The cell responds to profiles of different monkeys (stimulus 5), but is unresponsive to an irrelevant stimulus (a brush; stimulus 6). (After Desimone et al., 1984.)

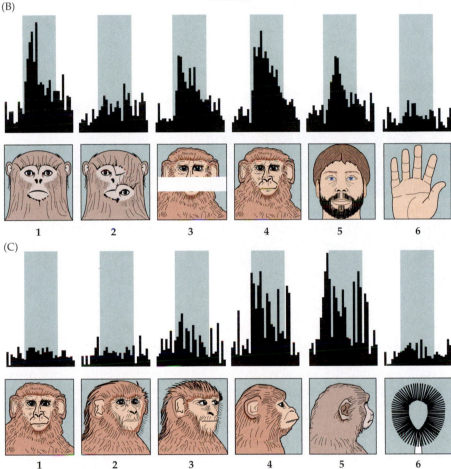

arrangements of complex features making up an object, the overall spatial pattern of neuronal activity representing the object in view. In keeping with this general idea, optical imaging of the surface of the temporal cortex shows that large populations of neurons are activated when monkeys view an object comprising several different geometric features. The locus of this activity in the upper layers of the cortex shifts systematically when object features, such as the orientation of a face, are systematically altered (Figure 26.13). Taken together, these further observations suggest that object identification relies on graded signals carried by a population of neurons rather than on the specific output of one or a few cells that are selective for a particular object.

(A)

Figure 26.13 *Possible scheme of object representation. (A) Suggested columnar organization of object representations in the inferotemporal cortex. Each cortical column is thought to signal a particular object class or point of view, with relatively smooth transitions between object features across columns. (B) Systematic movement of the active region of inferotemporal cortex with rotation of the face. Intrinsic signal optical images (below) were obtained for the views of five different positions of the face. Contours circumscribing significant cortical activation by these five different views are shown on the right. (A after Tanaka, 2001; B after Wang et al., 1996.)*

(B)

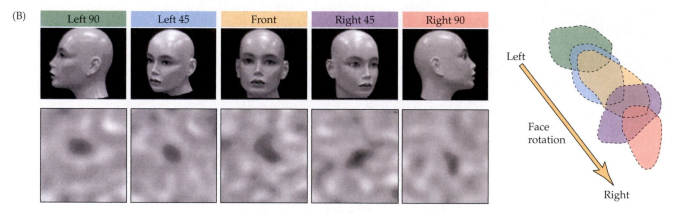

"Planning Neurons" in the Monkey Frontal Cortex

In confirmation of the clinical evidence about the function of the frontal association cortices in neurological patients, neurons that appear to be specifically involved in planning have been identified in the frontal cortices of rhesus monkeys.

One behavioral test used to study cells in the monkey frontal cortex is called the **delayed response task** (Figure 26.14A). Variants of this task are used to assess frontal lobe function in a variety of situations, including the clinical evaluation of frontal lobe function in humans (Box 26C). In the simplest version of the delayed response task, the monkey watches an experimenter place a food morsel in one of two wells; both wells are then covered. Subsequently, a screen is lowered for an interval of a few seconds to several minutes (the delay). When the screen is raised, the monkey gets only one chance to uncover the well containing food and retrieve the reward. Thus, the animal must decide that he wants the food, remember where it is placed, recall that the cover must be removed to obtain the food, and keep all this information available during the delay so that it can be used to get the reward. The monkey's ability to carry out this working memory task is diminished or abolished if the area anterior to the motor region of the frontal cortex—called the prefrontal cortex—is destroyed bilaterally (which is in accord with clinical findings in human patients).

(A)

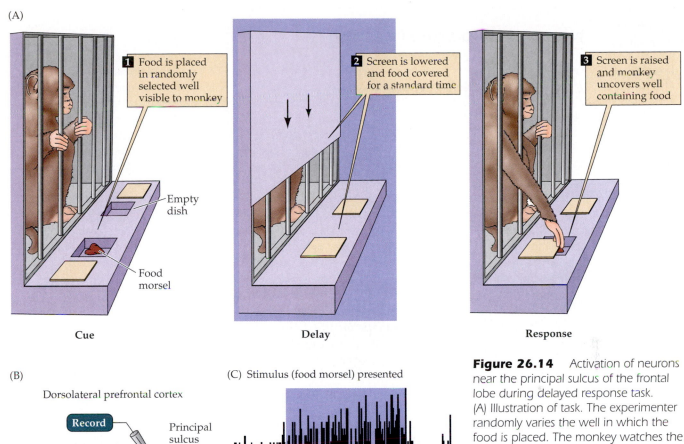

1 Food is placed in randomly selected well visible to monkey

Empty dish

Food morsel

Cue

2 Screen is lowered and food covered for a standard time

Delay

3 Screen is raised and monkey uncovers well containing food

Response

(B)

Dorsolateral prefrontal cortex

Record

Principal sulcus

(C) Stimulus (food morsel) presented

(D) No stimulus presented

Cue   Delay   Response

Figure 26.14 Activation of neurons near the principal sulcus of the frontal lobe during delayed response task. (A) Illustration of task. The experimenter randomly varies the well in which the food is placed. The monkey watches the morsel being covered, and then the screen is lowered for a standard time. When the screen is raised, the monkey is allowed to uncover only one well to retrieve the food. Normal monkeys learn this task quickly, usually performing at a level of 90 percent correct after less than 500 training trials, whereas monkeys with frontal lesions perform poorly. (B) Region of recording. (C) Activity of a delay-specific neuron in the prefrontal cortex of a rhesus monkey recorded during the delayed response task shown in (A). The histograms show the number of action potentials during the cue, delay, and response periods. The neuron begins firing when the screen is lowered and remains active throughout the delay period. (D) When the screen is lowered and raised but no food is presented, the same neuron is less active. (After Goldman-Rakic, 1987.)

Some neurons in the prefrontal cortex, particularly those in and around the principal sulcus (Figure 26.14B), are activated when monkeys perform the delayed response task, and they are maximally active during the period of the delay, as if their firing represented information about the location of the food morsel maintained from the presentation part of the trial (i.e., the cognitive information needed to guide behavior when the screen is raised; Figure 26.14C,D). Such neurons return to a low level of activity during the actual motor phase of the task, suggesting that they represent working memory and planning (see Chapter 31) rather than the actual movement itself. Delay-specific neurons in the prefrontal cortex are also active in monkeys that have been trained to perform a variant of the delayed response task in which well-learned movements are produced in the absence of any cue. Evidently, these neurons are equally capable of using stored information to guide behavior. Thus, if a

BOX 26C   Neuropsychological Testing

Long before PET scanning and functional MRI were available to evaluate normal and abnormal cognitive function, several "low-tech" methods provided reliable means of assessing these abilities in human subjects. From the late 1940s onward, psychologists and neurologists developed a battery of behavioral tests—generally called neuropsychological tests—to evaluate the integrity of cognitive function and to help localize lesions.

One of the most frequently used measures is the Wisconsin Card Sorting

Task illustrated here. In this test, the examiner places four cards with symbols that differ in number, shape, or color before the subject, who is given a set of response cards with similar symbols on them. The subject is then asked to place an appropriate response card in front of the stimulus card based on a sorting rule established, but not stated, by the examiner (i.e., sort by color, number, or shape). The examiner then indicates whether the response is "right" or "wrong." After 10 consecutive correct responses, the examiner changes the sorting rule simply by saying "wrong." The subject must then ascertain the new sorting rule and perform 10 correct trials. The sorting rule is then changed again, until six cycles have been completed.

In 1963, the neuropsychologist Brenda Milner at the Montreal Neurological Institute showed that patients with frontal lobe lesions have consistently poor performance in the Wisconsin Card Sorting Task. By comparing patients with known brain lesions as a result of surgery for epilepsy or tumor, Milner was able to demonstrate that this impairment is fairly specific for frontal

lobe damage. Particularly striking is the inability of frontal lobe patients to use previous information to guide subsequent behavior. A widely accepted explanation for the sensitivity of the Wisconsin Card Sorting Task to frontal lobe deficits is the "planning" aspect of this test. To respond correctly, the subject must retain information about the previous trial, which is then used to guide behavior on future trials. Processing this sort of information is characteristic of normal frontal lobe function.

A variety of other neuropsychological tests have been devised to evaluate the functional integrity of other cognitive functions. These include tasks in which a patient is asked to identify familiar faces in a series of pictures, and others in which "distractors" interfere with the patient's ability to attend to salient stimulus features. An example of the latter is the Stroop Test, in which patients are asked to read the names of colors presented in color-conflicting print (for example, the word "green" printed in red ink). This sort of challenge evaluates both attention and identification abilities.

Sort by color

Sort by shape

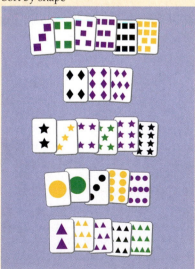

Sort by number

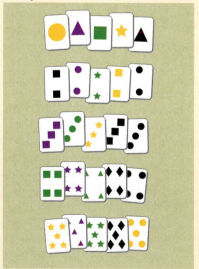

The simplicity, economy, and accumulated experience with such tests continue to make them a valuable means of evaluating cognitive functions.

References

BERG, E. A. (1948) A simple objective technique for measuring flexibility in thinking. *J. Gen. Psychol.* 39: 15–22.

LEZAK, M. D. (1995) *Neuropsychological Assessment*, 3rd Ed. New York: Oxford University Press.

MILNER, B. (1963) Effects of different brain lesions on card sorting. *Arch. Neurol.* 9: 90–100.

MILNER, B. AND M. PETRIDES (1984) Behavioural effects of frontal-lobe lesions in man. *Trends Neurosci.* 4: 403–407.

monkey is trained to associate looking at a particular target with a delayed reward, the delay-associated neurons in the prefrontal cortex will fire during the delay, even if the monkey shifts his gaze to the appropriate region of the visual field in the absence of the target.

In addition to maintaining cognitive information during short delays, some neurons in prefrontal cortex also appear to participate directly in longer range planning of sequences of movements. When monkeys are trained to perform a motor sequence, such as turning a joystick to the left, then right, then left again, some neurons in prefrontal cortex fire at a particular point in the sequence (such as the third response), regardless of which movement (e.g., left or right) is made. Prefrontal neurons have also been found that are selective for each position in a learned motor sequence, thus ruling out the possibility that these neurons merely encode task difficulty or proximity to reward as the monkey nears the end of the series of responses. When these regions of prefrontal cortex are inactivated pharmacologically, monkeys lose the ability to execute sequences of movements from memory. These observations endorse the notion, first inferred from studies of brain-damaged individuals like Phineas Gage, that the frontal lobe contributes specifically to the cognitive functions that use stored information to plan and guide appropriate sequences of behavior.

In short, the existence of planning-specific neurons in the frontal cortex of rhesus monkeys, as well as attention-specific cells in the parietal cortex and recognition-specific cells in the temporal cortex, supports the functions of these cortical areas inferred from clinical evidence in humans. Nonetheless, functional localization, whether inferred by examining human patients or by recording single neurons in monkeys, is an imprecise business. The observations summarized here are merely a rudimentary guide to thinking about how complex cognitive information is represented and processed in the brain, as well as how the relevant brain areas and their constituent neurons contribute to such important but still ill-defined qualities as personality, intelligence (Box 26D), and other cognitive functions that define what it means to be human.

Summary

The majority of the human cerebral cortex is devoted to tasks that transcend encoding primary sensations or commanding motor actions. Collectively, the association cortices mediate these cognitive functions of the brain—broadly defined as the ability to attend to, identify, and act meaningfully in response to complex external or internal stimuli. Descriptions of patients with cortical lesions, functional brain imaging of normal subjects, and behavioral and electrophysiological studies of non-human primates have established the general purpose of the major association areas. Thus, parietal association cortex is involved in attention and awareness of the body and the stimuli that act on it; temporal association cortex is involved in the recognition and identification of highly processed sensory

BOX 26D Brain Size and Intelligence

The fact that so much of the brain is occupied by the association cortices raises a fundamental question: Does having more of it provide individuals with greater cognitive ability? Humans and other animals obviously vary in their talents and predispositions for a wide range of cognitive behaviors. Does a particular talent imply a greater amount of neural space in the service of that function?

Historically, the most popular approach to the issue of brain size and behavior in humans has been to relate the overall size of the brain to a broad index of performance, conventionally measured in humans by "intelligence tests." This way of studying the relationship between brain and behavior has caused considerable trouble. In general terms, the idea that the size of brains from different species reflects intelligence represents a simple and apparently valid idea (see figure). The ratio of brain weight to body weight for fish is 1:5000; for reptiles it is about 1:1500; for birds, 1:220; for most mammals, 1:180; and for humans, it is 1:50. If intelligence is defined as the full spectrum of cognitive performance, surely no one would dispute that a human is more intelligent than a mouse, or that this difference is explained in large part by the 3000-fold difference in the size of the brains of these species.

Does it follow, however, that relatively small differences in the size of the brain among related species, strains, genders, or individuals—differences that often persist even after correcting for body size—are also a valid measure of cognitive abilities? Certainly no issue in neuroscience has provoked more heated debate than the notion that alleged differences in brain size among races (or the demonstrable differences in brain size between men and women) reflect differences in performance. The passion attending this controversy has been generated not only by the scientific issues involved, but also by the specters of racism and misogyny.

Nineteenth-century enthusiasm for brain size as a simple measure of human performance was championed by some remarkably astute scientists (including Darwin's cousin Francis Galton and the French neurologist Paul Broca), as well as others whose motives and methods are now suspect (see Gould, 1978, 1981 for a fascinating and authoritative commentary). Broca, one of the great neurologists of his day and a gifted observer, not only thought that brain size reflected intelligence, but was of the opinion (as was just about every other nineteenth-century male scientist) that white European males had larger and better developed brains than anyone else. Based on what was known about the human brain in the late nineteenth century, it was perhaps reasonable for Broca to consider it, like the liver or the lung, as an organ having a largely homogeneous function. Ironically, it was Broca himself who laid the groundwork for the modern view that the brain is a heterogeneous collection of highly interconnected but functionally discrete systems (see Chapter 27). Nonetheless, the simplistic nineteenth-century approach to brain size and intelligence has persisted in some quarters well beyond its time.

There are at least two reasons why measures such as brain weight or cranial capacity are not easily interpretable indices of intelligence, even though small observed differences may be statistically valid. First is the obvious difficulty of defining and accurately measuring intelligence, particularly among humans with different educational and cultural backgrounds. Second is the functional diversity and connectional complexity of the brain. Imagine assessing the relationship between body size and athletic ability, which might be considered the somatic analogue of intelligence. Body weight, or any other global measure of

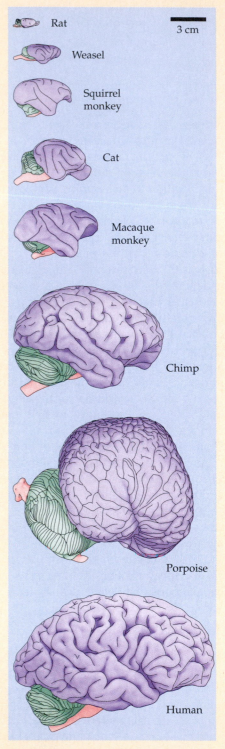

Typical size of the adult brain in several different species.

somatic phenotype, would be a woefully inadequate index of athletic ability. Although the evidence would presumably indicate that bigger is better in the context of sumo wrestling or basketball, more subtle somatic features would no doubt be correlated with extraordinary ability in ping-pong, gymnastics, or figure skating. The diversity of somatic function vis-à-vis athletic ability confounds the interpretation of any simple measure such as body size.

The implications of this analogy for the brain are straightforward. Any program that seeks to relate brain weight, cranial capacity, or some other measure of overall brain size to individual performance ignores the reality of the brain's functional diversity. Thus, quite

apart from the political or ethical probity of attempts to measure "intelligence" by brain size, by the yardstick of modern neuroscience (or simple common sense), this approach will inevitably generate more heat than light. A more rational approach to the issue has become feasible in the last few years, which is to relate the size of measurable regions of known function (the primary visual cortex, for example) to the corresponding functions (visual performance), as well as to cellular features such as synaptic density and dendritic arborization. These correlations have greater promise for exploring the sensible idea that better performance will be based on greater amounts of the underlying neural machinery.

References

BROCA, P. (1861) Sur le volume et la forme du cerveau suivant les individus et suivant les races. *Bull. Soc. Anthrop.* 2: 139–207, 301–321.

GALTON, F. (1883) *Inquiries into Human Faculty and Its Development*. London: Macmillan.

GOULD, S. J. (1978) Morton's ranking of races by cranial capacity. *Science* 200: 503–509.

GOULD, S. J. (1981) *The Mismeasure of Man*. New York: W. W. Norton and Company.

GROSS, B. R. (1990) The case of Phillipe Rushton. *Acad. Quest.* 3: 35–46.

SPITZKA, E. A. (1907) A study of the brains of six eminent scientists and scholars belonging to the American Anthropometric Society, together with a description of the skull of Professor E. D. Cope. *Trans. Amer. Phil. Soc.* 21: 175–308.

WALLER, A. D. (1891) *Human Physiology*. London: Longmans, Green.

information; and frontal association cortex is importantly involved in guiding complex behavior by planning responses to ongoing stimulation (or remembered information), matching such behaviors to the demands of a particular situation. The especially extensive association areas in our species compared to other primates support the cognitive processes that define human culture.

Additional Reading

Reviews

BEHRMANN, M. (1999) Spatial frames of reference and hemispatial neglect. In *The Cognitive Neurosciences*, 2nd Ed. M. Gazzaniga (ed.). Cambridge, MA: MIT Press, pp. 651–666.

DAMASIO, A. R. (1985) The frontal lobes. In *Clinical Neuropsychology*, 2nd Ed. K. H. Heilman and E. Valenstein (eds.). New York: Oxford University Press, pp. 409–460.

DAMASIO, A. R., H. DAMASIO AND G. W. VAN HOESEN (1982) Prosopagnosia: Anatomic basis and behavioral mechanisms. *Neurology* 32: 331–341.

DESIMONE, R. (1991) Face-selective cells in the temporal cortex of monkeys. *J. Cog. Neurosci.* 3: 1–8.

FILLEY, C. M. (1995) *Neurobehavioral Anatomy*. Chapter 8, Right hemisphere syndromes. Boulder: University of Colorado Press, pp. 113–130.

GOLDMAN-RAKIC, P. S. (1987) Circuitry of the prefrontal cortex and the regulation of behavior by representational memory. In *Handbook of Physiology*. Section 1, *The Nervous System*. Vol. 5, Higher Functions of the Brain, Part I. F. Plum (ed.). Bethesda: American Physiological Society, pp. 373–417.

HALLIGAN, P. W. AND J. C. MARSHALL (1994) Toward a principled explanation of unilateral neglect. *Cog. Neuropsych.* 11(2): 167–206.

LÁDAVAS, E., A. PETRONIO AND C. UMILTA (1990) The deployment of visual attention in the intact field of hemineglect patients. *Cortex* 26: 307–317.

MACRAE, D. AND E. TROLLE (1956) The defect of function is visual agnosia. *Brain* 77: 94–110.

POSNER, M. I. AND S. E. PETERSEN (1990) The attention system of the human brain. *Annu. Rev. Neurosci.* 13: 25–42.

VALLAR, G. (1998) Spatial hemineglect in humans. *Trends Cog. Sci.* 2(3): 87–96.

Important Original Papers

BRAIN, W. R. (1941) Visual disorientation with special reference to lesions of the right cerebral hemisphere. *Brain* 64: 224–272.

COLBY C. L., J. R. DUHAMEL AND M. E. GOLDBERG (1996) Visual, presaccadic, and cognitive activation of single neurons in monkey lateral intraparietal area. *J. Neurophysiol.* 76: 2841–2852.

DESIMONE, R., T. D. ALBRIGHT, C. G. GROSS AND C. BRUCE (1984) Stimulus-selective properties of inferior temporal neurons in the macaque. *J. Neurosci.* 4: 2051–2062.

ETCOFF, N. L., R. FREEMAN AND K. R. CAVE (1991) Can we lose memories of faces? Content specificity and awareness in a prosopagnosic. *J. Cog. Neurosci.* 3: 25–41.

FUNAHASHI, S., M. V. CHAFEE AND P. S. GOLDMAN-RAKIC (1993) Prefrontal neuronal activity in rhesus monkeys performing a delayed anti-saccade task. *Nature* 365: 753–756.

FUSTER, J. M. (1973) Unit activity in pre-frontal cortex during delayed-response performance: Neuronal correlates of transient memory. *J. Neurophysiol.* 36: 61–78.

GESCHWIND, N. (1965) Disconnexion syndromes in animals and man. Parts I and II. *Brain* 88: 237–294.

HARLOW, J. M. (1868) Recovery from the passage of an iron bar through the head. *Publications of the Massachusetts Medical Society* 2: 327–347.

LEOPOLD, D. A., I. V. BONDAR, AND M. A. GIESE (2006) Norm-based encoding by single neurons in the monkey inferotemporal cortex. *Nature* 442: 572–575.

MOUNTCASTLE, V. B., J. C. LYNCH, A. GEORGOPOULOUS, H. SAKATA AND C. ACUNA (1975) Posterior parietal association cortex of the monkey: Command function from operations within extrapersonal space. *J. Neurophys.* 38: 871–908.

PLATT, M. L. AND P. W. GLIMCHER (1999) Neural correlates of decision variables in parietal cortex. *Nature* 400: 233–238.

TANJI, J. AND K. SHIMA (1994) Role for supplementary motor area cells in planning several movements ahead. *Nature* 371: 413–416.

WANG, G., K. TANAKA AND M. TANIFUJI (1996) Optical imaging of functional organization in the monkey inferotemporal cortex. *Science* 272: 1665–1668.

Books

BRICKNER, R. M. (1936) *The Intellectual Functions of the Frontal Lobes.* New York: Macmillan.

DAMASIO, A. R. (1994) *Descartes' Error: Emotion, Reason and the Human Brain.* New York: Grosset/Putnam.

DEFELIPE, J. AND E. G. JONES (1988) *Cajal on the Cerebral Cortex: An Annotated Translation of the Complete Writings.* New York: Oxford University Press.

GAREY, L. J. (1994) *Brodmann's "Localisation in the Cerebral Cortex."* London: Smith-Gordon. (Translation of K. Brodmann's 1909 book. Leipzig: Verlag von Johann Ambrosius Barth.)

GLIMCHER, P. W. (2003) *Decisions, Uncertainty, and the Brain: The Science of Neuroeconomics.* Cambridge, MA: MIT Press.

HEILMAN, H. AND E. VALENSTEIN (1985) *Clinical Neuropsychology,* 2nd Ed, Chapters 8, 10, and 12. New York: Oxford University Press.

KLAWANS, H. L. (1988) *Toscanini's Fumble, and Other Tales of Clinical Neurology.* Chicago: Contemporary Books.

KLAWANS, H. L. (1991) *Newton's Madness.* New York: Harper Perennial Library.

POSNER, M. I. AND M. E. RAICHLE (1994) *Images of Mind.* New York: Scientific American Library.

SACKS, O. (1987) *The Man Who Mistook His Wife for a Hat.* New York: Harper Perennial Library.

SACKS, O. (1995) *An Anthropologist on Mars.* New York: Alfred A. Knopf.

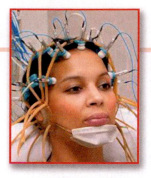

Chapter 27

Speech and Language

Overview

One of the most remarkable cortical functions in humans is the ability to associate arbitrary symbols with specific meanings to express thoughts and emotions to ourselves and others by means of spoken and in some cultures written language. Language is defined as the speech of a group of people, although in clinical medicine the word tends to be used to refer to speech production and comprehension by the brain. The achievements of human culture rest largely upon this kind of communication, and a person who for one reason or another fails to develop a facility for language as a child is severely incapacitated. Studies both of patients with damage to specific cortical regions and of normal subjects studied by electrophysiological methods or functional brain imaging indicate that the linguistic abilities of humans depend on the integrity of a number of specialized areas of the association cortices in the temporal and frontal lobes. Understanding functional localization and hemispheric lateralization of language is especially important in clinical practice. The loss of language is such a devastating blow that neurologists and neurosurgeons make every effort to identify and preserve those cortical areas involved in its comprehension and production. The need to map language functions in patients for the purpose of sparing these brain regions has provided another rich source of information about the neural organization of this critical human attribute. In the vast majority of people, the primary language functions for explicitly semantic processing are located in the left hemisphere: the linkages between speech sounds and their meanings are mainly represented in the left temporal cortex, and the circuitry for the motor commands that organize the production of meaningful speech is mainly found in the left frontal cortex. Despite this left-sided predominance of the lexical, grammatical, and syntactical aspects of language, the emotional (affective) content of speech is governed largely by the right hemisphere. Studies of congenitally deaf individuals have shown further that the cortical areas devoted to sign language are generally the same as those that organize spoken and heard communication. The regions of the brain devoted to language are therefore specialized for symbolic representation and communication rather than for heard and spoken language as such. A variety of evidence indicates that some aspects of these abilities and their neural substrates are not unique to humans.

Language Is Both Localized and Lateralized

It has been known for more than a century that two regions in the frontal and temporal association cortices of the left cerebral hemisphere are especially important for the explicitly verbal aspects of human language. That language abilities are both localized and lateralized is not surprising; ample evidence of the localization and lateralization of other cognitive functions was reviewed in Chapter 26. The unequal representation of language functions in the two cerebral hemispheres, however, provides an especially compelling example of this phenomenon.

Although the concept of lateralization has already been introduced in describing the unequal functions of the parietal lobes in attention and of the temporal lobes in recognizing different categories of objects, it is in language that this idea has been most thoroughly documented. Because language is so important to humans, its lateralization has given rise to the misleading idea that one hemisphere is actually "dominant" over the other—namely, the hemisphere in which the major capacity for language resides. But the true significance of lateralization for language (or any other cognitive ability) lies in the efficient subdivision of complex functions between the hemispheres rather than

BOX 27A   Speech

The organs that produce speech include the lungs, which serve as a reservoir of air; the larynx, which is the source of the periodic stimulus quality of "voiced" sounds; and the pharynx, oral, and nasal cavities and their included structures (e.g., tongue, teeth, and lips), which modify (or filter) the speech sounds that eventually emanate from the speaker. The fundamentally correct idea that the larynx is the "source" of speech sounds and the rest of the vocal tract acts as a filter that modulates the sound energy of the source is an old one, having been proposed by Johannes Mueller in the nineteenth century.

Although the physiological details are complex, the general operation of the vocal apparatus is simple. Air expelled from the lungs accelerates as it passes through a constricted opening between the **vocal folds** ("vocal cords") called the glottis, thus decreasing the pressure in the air stream according Bernoulli's principle. As a result, the vocal folds come together until the pressure buildup in the lungs forces them open again. The ongoing repetition of this process results in an oscillation of sound wave pressure, the frequency of which is determined primarily by the muscles that control the tension on the vocal cords. The fundamental frequencies of these oscillations—which are the basis of the power in voiced speech sounds—range from about 100 to about 400 Hz, depending

on the gender, size, and age of the speaker.

The larynx has many other consequential effects on the speech signal and these create additional speech sounds. For instance, the vocal folds can open suddenly to produce what is called a *glottal stop* (as in the beginning of the exclamation "Idiot!"). Alternatively, the vocal folds can hold an intermediate position for the production of consonants such as *h*, or they can be com-

pletely open for "unvoiced" consonants such as *s* or *f* (i.e., speech sounds that don't have the periodic quality derived from vocal fold oscillations). In short, the larynx is important in the production of virtually all vocal sounds.

The vocal system can be thought of as a sort of musical instrument capable of extraordinary subtlety and exquisite modulation. As in the sound produced by a musical instrument, however, the primary source of oscillation (e.g., the reed of a clarinet or the vocal folds in speech) is hardly the whole story. The entire pathway between the vocal folds and the lips (and nostrils) is equally criti-

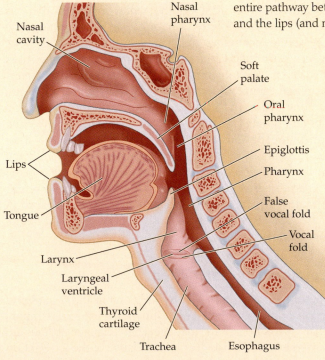

in any superiority of one hemisphere over the other. Indeed, pop psychological dogmas about cortical redundancy notwithstanding, it is a safe presumption that every region of the brain is doing *something* important.

A first step in the proper consideration of these issues is recognizing that the cortical representation of language is distinct from, although related to, the circuitry concerned with the motor planning and control of the larynx, pharynx, mouth, and tongue—the structures that produce speech sounds (Box 27A). The cortical representation of language is also distinct from, although clearly related to, the circuits underlying the auditory perception of spoken words and the visual perception of written words in the primary auditory and visual cortices, respectively (Figure 27.1). Whereas the neural substrates for language as such depend

cal in determining speech sounds, as is the structure of a musical instrument. The key determinants of the sound that emanates from an instrument are its natural resonances, which shape or filter the sound pressure oscillation. For the vocal tract, the natural resonances that modulate the air stream generated by the larynx are called *formants*. The resonance frequency of the major formant arises from the fact that the approximate length of the vocal tract in its relaxed state is 17 cm, which is the quarter wavelength of a 68-cm sound wave; quarter wavelengths determine the resonances of pipes open at one end, which essentially describes the vocal tract. Since the speed of sound is about 33,500 cm/sec, the lowest resonance frequency of an open tube or pipe of this length will be 33,500/68 or about 500 Hz; additional major resonant frequencies occur at about 1500 Hz and 2500 Hz. As a result, power in the laryngeal source at these formant frequencies will be reinforced, and power at other frequencies will, in varying degrees, be filtered out.

In any given language, the basic speech sounds are called *phones* and the percepts they elicit *phonemes*; different phones are produced as the muscles of the vocal tract change the tension on the vocal fold and the shape of the resonant cavities above the folds. Phones make up syllables in speech, which are used in turn to make up words, which are then strung together to create sentences. There are about 40 phonemes in Eng-

lish, and these are about equally divided between vowel and consonant speech sound percepts. Vowel sounds are by and large the voiced (periodic) elements of speech (i.e., the elemental sounds in any language generated by the oscillation of the vocal cords). In contrast, consonant sounds involve rapid changes in the sound signal and are more complex. In English, consonants begin and/or end syllables, each of which entails a vowel sound. Consonant sounds are categorized according to the site in the vocal tract that determines them (the *place of articulation*), or the physical way they are generated (the *manner of articulation*). With respect to place, there are labial consonants (such as *p* and *b*), dental consonants (*f* and *v*), palatal consonants (*sh*), and glottal consonants (*h*) (among many others). With respect to manner, there are plosive, fricative, nasal, liquid, and semi-vowel consonants. Plosives are produced by blocking the flow of air somewhere in the vocal tract, fricatives by producing turbulence, nasals by directing the flow of air through the nose, and so on. A further variation on the use of consonants is found in the "click languages" of southern Africa, of which about 30 survive today. Each of these languages has 4–5 different click sounds that are double consonants made by sucking the tongue down from the roof of the mouth.

It should be obvious then that speech sound stimuli are enormously complex (there are more than 200

phonemes in the approximately 6000 human languages in the world today). To make matters worse, Alvin Liberman, working at the Haskins Laboratory at Yale University, showed that there is no one-to-one correspondence between phonemes (as defined above) and the specific acoustic elements in speech. Because speech sounds changes continuously, they cannot be split up into discrete segments, as the concepts of phones and phonemes implies. This fact is now recognized as a fundamental problem that undermines any strictly phonemic or phonetic approach to language. Moreover, the formants for different vowels overlap in natural speech of men, women, and children. Evidence from studies of illiterates suggests that phones and phonemes are probably more related to learning how to read and spell than to actually hearing speech. Given this complexity, it is remarkable that we can communicate so readily.

References

LIBERMAN, A. M. (1996) *Speech: A Special Code.* Cambridge, MA: MIT Press.

MILLER, G. A. (1991) *The Science of Words*, Chapter 4, "The spoken word." New York: Scientific American Library.

PLOMP, R. (2002) *The Intelligent Ear: On the Nature of Sound Perception.* Mahwah, NJ: Erlbaum.

WARREN, R. M. (1999) *Auditory Perception: A New Analysis and Synthesis*, Chapter 7, "Speech." Cambridge: Cambridge University Press.

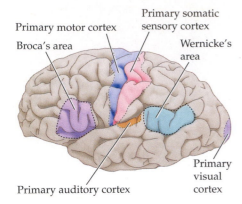

Figure 27.1 Diagram of the major brain areas involved in the comprehension and production of language. The primary sensory, auditory, visual, and motor cortices are indicated to show the relation of Broca's and Wernicke's language areas to these other areas that are necessarily involved in the comprehension and production of speech, albeit in a less specialized way.

on these essential motor and sensory functions, the regions of the brain that are specifically devoted to language transcend these more basic elements. The main concern of the areas of cortex that represent language is the usage of a system of symbols for purposes of communication—spoken and heard, written and read, or, in the case of sign language, gestured and seen (see below). Thus, the essential function of the cortical language areas, and indeed of language, is symbolic representation. Obedience to a set of rules for using these symbols (called grammar), ordering them to generate useful meanings (called syntax), and giving utterance to the appropriate emotional valence by varying intensity and pitch (called prosody) are all important and readily recognized elements of communication, regardless of the particular mode of representation and expression.

Given the profound biological and social importance of communication among the members of a species, it is not surprising that other animals communicate in ways that, while grossly impoverished compared to human language, nonetheless suggest the sorts of communicative skills and interactions from which human language evolved in the brains of our hominid and prehominid ancestors (Box 27B).

Aphasias

The distinction between language and the related sensory and motor capacities on which it depends was first apparent in patients with damage to specific brain regions. Clinical evidence from such cases showed that the ability to move the muscles of the larynx, pharynx, mouth, and tongue can be compromised without abolishing the ability to use spoken language to communicate (even though a motor deficit may make communication difficult). Similarly, damage to the auditory pathways can impede the ability to hear without interfering with language functions per se (as is obvious in individuals who have become partially or wholly deaf later in life). Damage to specific brain regions, however, can compromise essential language functions while leaving the sensory and motor infrastructure of verbal communication intact. These syndromes, collectively referred to as **aphasias**, diminish or abolish the ability to comprehend and/or to produce language as a vehicle for communicating meaningful statements, while sparing the ability to perceive the relevant stimuli and to produce intelligible words. Missing in these patients is the capacity to recognize or employ the symbolic value of words correctly, thus depriving such individuals of the linguistic understanding, grammatical and syntactical organization, and/or appropriate intonation that distinguishes language from nonsense (Box 27C).

Figure 27.2 The relationship of the major language areas to the classical cytoarchitectonic map of the cerebral cortex. As discussed in Chapter 26, about 50 histologically distinct regions (cytoarchitectonic areas) have been described in the human cerebral cortex. The language functions described by Broca and Wernicke are associated with at least three of the cytoarchitectonic areas defined by Brodmann (area 22, at the junction of the parietal and temporal lobes [Wernicke's area]; and areas 44 and 45, in the ventral and posterior region of the frontal lobe [Broca's area]), but are not coextensive with any of them.

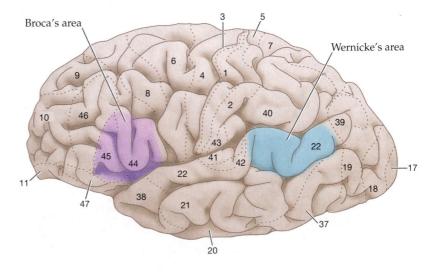

The first evidence for the localization of language function to a specific region (and to a hemisphere) of the cerebrum is usually attributed to the French neurologist Paul Broca and the German neurologist Carl Wernicke, who made a variety of seminal observations in the late 1800s. Both Broca and Wernicke examined the brains of individuals who had become aphasic and later died. Based on correlations of the clinical picture and the location of the brain damage seen in autopsy, Broca suggested that language abilities were localized in the ventroposterior region of the frontal lobe (Figures 27.1 and 27.2). More importantly, he observed that the loss of the ability to produce meaningful language—as opposed to the ability to move the mouth and produce words—was usually associated with damage to the left hemisphere. "*On parle avec l'hemisphere gauche,*" Broca concluded. The preponderance of aphasic syndromes

BOX 27B Do Other Animals Have Language?

Over the centuries, theologians, natural philosophers, and a good many modern neuroscientists have argued that language is uniquely human, this extraordinary behavior being seen as setting us qualitatively apart from our fellow animals. However, the gradual accumulation of evidence during the last 75 years demonstrating highly sophisticated systems of communication in species as diverse as bees, birds, monkeys, and whales (see Box 24B) has made this point of view increasingly untenable, at least in a broad sense. Until recently, however, human language *has* appeared unique in the ability to associate specific meanings with arbitrary symbols, ad infinitum. In the dance of the honeybee described so beautifully by Karl von Frisch, for example, each symbolic movement made by a foraging bee that returns to the hive encodes only a single meaning, whose expression and appreciation has been hardwired into the nervous systems of the actor and the respondents.

A series of highly controversial studies in great apes, however, have claimed that the rudiments of the human symbolic communication are evident in the behavior of our closest relatives. While

techniques have varied, most psychologists who study chimps have used some form of manipulable symbols that can be arranged to express ideas in an interpretable manner. For example, chimps can be trained to manipulate tiles or other symbols (such as the gestures of sign language) to represent words and syntactical constructs, allowing them to communicate simple demands, questions, and even spontaneous expressions. The most remarkable results have come from increasingly sophisticated work with chimps using keyboards with a variety of symbols (Figure A). With

appropriate training, chimps can choose from as many as 400 different symbols to construct expressions, allowing the researchers to have something resembling a rudimentary conversation with their charges. The more accomplished of these animals are alleged to have "vocabularies" of several thousand words or phrases (how they use these words compared to a child, however, is dramatically less impressive).

Given the challenge this work presents to some long-held beliefs about the uniqueness of human language, it is not

(*Continued on next page*)

Section of keyboard showing lexical symbols used to study symbolic communication in great apes. (From Savage-Rumbaugh et al., 1998.)

BOX 27B (Continued)

surprising that these claims continue to stir up debate. Nonetheless, the issues raised certainly deserve careful consideration by anyone interested in human language abilities and how our remarkable symbolic skills may have evolved from the communicative capabilities of our ancestors. The pressure for the evolution of some form of symbolic communication in great apes seems clear enough. Ethologists studying chimpanzees in the wild have described extensive social communication based on gestures, the manipulation of objects,

and facial expressions. In addition, studies of monkeys have shown that some species normally use a variety of vocalizations in socially meaningful ways, and that these vocalizations may activate regions in the frontal and temporal lobes that are homologous to Broca's and Wernicke's areas in humans (Figure B). This intricate social intercourse by gestural and vocal intercourse in non-human primates is likely to be the antecedent of human language; one need only think of the importance of gestures and facial expressions and non-verbal human vocal sounds as ancillary aspects of our own speech to appreciate this point. In the end, it may turn out to be that human language, for all its complexity, is based on the same general scheme of inherent and acquired neural associations between tokens and meanings that appears to be the basis of any animal communication.

References

GHAZANFAR, A. A. AND M. D. HAUSER (2001) The auditory behavior of primates: A neuroethological perspective. *Curr. Opin. Biol.* 16: 712–720.

GIL-DA-COSTA, R., A. MARTIN, M. A. LOPES, M. MUÑOZ, J. B. FRITZ AND A. R. BRAUN (2006) Species-specific calls activate homologs of Broca's and Wernicke's areas in the macaque. *Nature Neuroscience* 9: 1064–1070.

GOODALL, J. (1990) *Through a Window: My Thirty Years with the Chimpanzees of Gombe.* Boston: Houghton Mifflin Company.

GRIFFIN, D. R. (1992) *Animal Minds*. Chicago: The University of Chicago Press.

HAUSER, M .D. (1996) *The Evolution of Communication.* Cambridge, MA: Bradford/ MIT Press.

SAVAGE-RUMBAUGH, S., S. G. SHANKER, AND T. J. TAYLOR (1998) *Apes, Language, and the Human Mind*. New York: Oxford University Press.

SEFARTH, R. M. AND D. L. CHENEY (1984) The natural vocalizations of non-human primates. *Trends Neurosci.* 7: 66–73.

TERRACE, H. S. (1983) Apes who "talk": Language or projection of language by their teachers? In *Language in Primates: Perspectives and Implications*, J. de Luce and H. T. Wilder (eds.). New York: Springer-Verlag, pp. 19–42.

WHITEN, A., AND 8 OTHERS (1999) Cultures in chimpanzees. *Nature* 399: 682–685.

VON FRISCH, K. (1993) *The Dance Language and Orientation of Bees* (Transl. by Leigh E. Chadwick). Cambridge, MA: Harvard University Press.

(B)

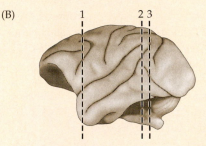

| Subject A | Subject B | Subject C |

Activation of areas in the frontal and temporal lobes of three rhesus monkeys responding to conspecific vocal calls. The areas activated are arguably similar to the major language areas in the human brain. (From Gil-da-Costa et al., 2006.)

BOX 27C   *Words and Meaning*

When Samuel Johnson (Figure A) compiled his *Dictionary of English Language* in 1755 under the sponsorship of Oxford University, he defined only 43,500 entries. The current *Oxford English Dictionary*, a lineal descendant of Johnson's seminal work and most recently revised in the 1980s, contains over 500,000 definitions! This quantitative difference is not the result of an increase in the number of English words since the eighteenth century, but rather is an indication of the difficulty collecting the enormous number of words we use in daily communication; the average educated speaker of English is said to have a working vocabulary of about 10,000 words and may know as many as 100,000.

Using words appropriately is made even more difficult by the fact that word meanings are continually changing, and by the enormous ambiguity of the words we do use. There is far more to a lexicon—be it a dictionary or a region of the left temporal cortex—than simply attaching meanings to words. Even when the meaning of a word is known, it must be understood in a particular context (Figure B) and used according to the rules of grammar and syntax in order to produce effective communication.

From the points of view of both neuroscience and linguistics, two related questions about words and grammar (i.e., the rules for putting words together) are especially germane in relation to this chapter. First, what is the nature of the neural machinery that allows us to learn language? And second, why do humans have such a profound drive to learn language? The major twentieth-century figure who has grappled with these questions is linguist Noam Chomsky, working at the Massachusetts Institute of Technology. Chomsky, while not interested in brain structure, argued that the complexity of language is such that it cannot simply be learned. He therefore proposed

that language must be predicated on a "universal grammar" laid down in the evolution of our species.

Although this argument is undoubtedly correct in one sense (the basic neural machinery for language, like all aspects of brain circuitry that support adult behavior, is indeed constructed during the normal development of each individual, primarily as a result of inheritance; see Chapters 23 and 24), Chomsky's eschewing of neurobiology avoids the central question of how, in evolutionary or developmental terms, this machinery comes to be and how it encodes words and strings them together into meaningful sentences. Whatever the mechanisms eventually prove to be, much of the language we use is obviously learned by making neuronal associations between arbitrary symbols and the objects, concepts, and interrelationships they signify in the real world. As such, human language provides a rich source for understanding how the relevant parts of the human cortex and their constituent neurons produce an enormous facility for making associations,

(A)

Samuel Johnson

which appears to be the fundamental characteristic of many cortical functions.

References

CHOMSKY, N. (1975) *Reflections on Language*. New York: Pantheon/Random House.

CHOMSKY, N. (1981) Knowledge of language: Its elements and origins. *Philos. Trans. Roy. Soc. Lond. B* 295: 223–234.

MILLER, G. A. (1991) *The Science of Words*. New York: Scientific American Library.

PINKER, S. (1994) *The Language Instinct*. New York: W. Morrow and Co.

WINCHESTER, S. (2003) *The Meaning of Everything: The Story of the Oxford English Dictionary*. Oxford: Oxford University Press.

(B)

The importance of context. When a person says "I'm going to our house on the lake," the meaning of the expression obviously depends on usage and context, rather than on the literal structure of the sentence uttered. This example indicates the daunting complexity of the task we all accomplish routinely. How this is done, even in principle, remains a central puzzle in language. (From Miller, 1991.)

associated with damage to the left hemisphere has supported his claim that one speaks primarily with the left hemisphere, a conclusion amply confirmed by a variety of modern studies using functional imaging (albeit with some important caveats, discussed later in the chapter).

Although Broca was basically correct, he failed to grasp the limitations of thinking about language as a unitary function localized in a single cortical region. This issue was better appreciated by Wernicke, who distinguished between patients who had lost the ability to comprehend language and those who could no longer produce language. Wernicke recognized that some aphasic patients retain the ability to produce utterances with reasonable grammatical and syntactical fluency, but without meaningful content. He concluded that lesions of the posterior and superior temporal lobe on the left side tend to result in deficits of this sort. In contrast, other patients continue to comprehend language but lack the ability to organize or control the content of their speech, even though it is clear that they understand what they are trying to say. Thus they repeat syllables and words, utter grammatically incomprehensible phrases, and repeat phrases—even though the meaning eventually gets through (see examples below). These deficits are associated with damage to the posterior and inferior region of the left frontal lobe, the area that Broca emphasized as an important substrate for language (see Figures 27.1 and 27.2).

As a consequence of these early observations, two rules about the localization of language have been taught ever since. The first is that lesions of the left frontal lobe in a region referred to as **Broca's area** affect the ability to *produce* language efficiently. This deficiency is called **motor** or **expressive aphasia**, and is also known as **Broca's aphasia**. (Such aphasias must be specifically distinguished from *dysarthria*, which is the inability to move the muscles of the face and tongue that mediate speaking.) The deficient motor-planning aspects of expressive aphasias accord with the complex motor functions of the posterior frontal lobe and its proximity to the primary motor cortex already discussed (see Chapters 16 and 26).

The second rule is that damage to the left temporal lobe causes difficulty *understanding* spoken language, a deficiency referred to as **sensory** or **receptive aphasia**, also known as **Wernicke's aphasia**. (Deficits of reading and writing— *alexias* and *agraphias*—are separate disorders that can arise from damage to related but different brain areas; most aphasics, however, also have difficulty with these closely linked abilities.) Receptive aphasia generally reflects damage to the auditory association cortices in the posterior temporal lobe, a region referred to as **Wernicke's area**.

A final broad category of language deficiency syndromes is **conduction aphasia**. These disorders arise from lesions to the pathways connecting the relevant temporal and frontal regions, such as the arcuate fasciculus in the subcortical white matter that links Broca's and Wernicke's areas. Interruption of this pathway may result in an inability to produce appropriate responses to heard communication, even though the communication is understood.

In a classic Broca's aphasia, the patient cannot express himself fluently because the organizational aspects of language (its grammar and syntax) have been disrupted, as shown in the following example reported by Howard Gardner (who is the interlocutor). The patient was a 39-year-old Coast Guard radio operator named Ford who had suffered a stroke that affected his left posterior frontal lobe.

'I am a sig…no…man…uh, well,…again.' These words were emitted slowly, and with great effort. The sounds were not clearly articulated; each syllable as uttered harshly, explosively, in a throaty voice. With practice, it was possible to understand him, but at first I encountered considerable difficulty in this. 'Let me help you,' I interjected. 'You were a signal…''A sig-nal man…right,' Ford completed my phrase triumphantly. 'Were you in the Coast Guard?''No, er, yes, yes, ...ship...Massachu...chusetts...Coastguard ...years.' He raised his hands twice, indicating the number nineteen.'Oh, you were in the Coast Guard for nineteen years.''Oh...boy...right...right,' he replied.'Why are you in the hospital, Mr. Ford?' Ford looked at me strangely, as if to say, Isn't it patently obvious? He pointed to his paralyzed arm and said,'Arm no good,' then to his mouth and said,'Speech...can't say...talk, you see.'

Howard Gardner, 1974
(The Shattered Mind: The Person after Brain Damage, pp. 60–61)

In contrast, the major difficulty in Wernicke's aphasia is putting together objects or ideas and the words that signify them. Thus, in a Wernicke's aphasia, speech is fluent and well structured, but makes little or no sense because words and meanings are not correctly linked, as is apparent in the following example (again from Gardner). The patient in this case was a 72-year-old retired butcher who had suffered a stroke affecting his left posterior temporal lobe.

Boy, I'm sweating, I'm awful nervous, you know, once in a while I get caught up, I can't get caught up, I can't mention the tarripoi, a month ago, quite a little, I've done a lot well, I impose a lot, while, on the other hand, you know what I mean, I have to run around, look it over, trebbin and all that sort of stuff. Oh sure, go ahead, any old think you want. If I could I would. Oh, I'm taking the word the wrong way to say, all of the barbers here whenever they stop you it's going around and around, if you know what I mean, that is tying and tying for repucer, repuccration, well, we were trying the best that we could while another time it was with the beds over there the same thing...

Ibid., p. 68

The major differences between these two classical aphasias are summarized in Table 27.1.

Despite the validity of Broca's and Wernicke's original observations, the classification of language disorders is considerably more complex in practice. An effort to refine the nineteenth-century categorization of aphasias was undertaken by the American neurologist Norman Geschwind during the 1950s and early 1960s. Based on clinical and anatomical data from a large number of patients and on the better understanding of cortical connectivity gleaned by that time from animal studies, Geschwind concluded correctly that several other regions of the parietal, temporal, and frontal cortices are critically involved in human linguistic capacities. Basically, he showed that damage to these additional areas results in identifiable, if more subtle, language deficits. His clarification of the definitions of language disorders has been largely confirmed by

TABLE 27.1 Characteristics of Broca's and Wernicke's Aphasias

Broca's aphasia[a]	Wernicke's aphasia[b]
Halting speech	Fluent speech
Tendency to repeat phrases or words (perseveration)	Little spontaneous repetition
Disordered syntax	Syntax adequate
Disordered grammar	Grammar adequate
Disordered structure of individual words	Contrived or inappropriate words
Comprehension intact	Comprehension not intact

[a] Also called motor, expressive, or production aphasia.
[b] Also called sensory or receptive aphasia.

functional brain imaging in normal subjects and remains the basis for much contemporary clinical work on language and aphasias.

Confirmation of Language Lateralization and Other Insights

Until the 1960s, observations about language localization and lateralization were based primarily on patients with brain lesions of varying severity, location, and etiology. Up until that time, the inevitable uncertainties of clinical findings had allowed skeptics to argue that language and other complex cognitive functions might not be lateralized (or even localized) in the brain. Definitive evidence supporting the inferences from neurological observations came from studies of patients whose corpus callosum and anterior commissure were severed as a treatment for medically intractable epileptic seizures. (Recall that a certain fraction of severe epileptics are refractory to medical treatment, and that interrupting the connection between the two hemispheres remains an effective way of treating epilepsy in highly selected patients; see Box 8C). In such patients, investigators could assess the function of the two cerebral hemispheres *independently*, since the major axon tracts that connect them had been interrupted. The first studies of these so-called **split-brain patients** were carried out by Roger Sperry and his colleagues at the California Institute of Technology in the 1960s and 1970s, and established the hemispheric lateralization of language beyond any doubt. This work also demonstrated many other functional differences between the left and right hemispheres (Figure 27.3) and stands as an extraordinary contribution to the understanding of brain organization.

To evaluate the functional capacity of each hemisphere in split-brain patients, it is essential to provide information to one side of the brain only. Sperry, Michael Gazzaniga (a key collaborator in this work), and others devised several simple ways to do this, the most straightforward of which was to ask the subject to use each hand independently to identify objects without any visual assistance (Figure 27.3A). Recall from Chapter 9 that somatic sensory information from the right hand is processed by the left hemisphere, and vice versa. By asking the subject to describe an item being manipulated by one hand or the other, the language capacity of the relevant hemisphere could be examined. Such testing showed clearly that the two hemispheres differ in their language ability (as expected from the postmortem correlations described earlier).

Using the left hemisphere, split-brain patients were able to name objects held in the right hand without difficulty. In contrast, and quite remarkably, an object held in the left hand could not be named. Using the right hemisphere, most split-brain patients could produce only an indirect description of the object that relied on rudimentary words and phrases rather than the precise lexical symbol for the object (for instance, "a round thing" instead of "a ball"); some could not provide any verbal account of what they held in their left hand. Observations using techniques to present visual information to the hemispheres independently (a method called *tachistoscopic presentation*; Figure 27.3B) showed further that the left hemisphere responds to written commands, whereas the right hemisphere typically responds only to nonverbal stimuli (e.g., pictorial instructions or, in some cases, rudimentary written commands). These distinctions reflect broader hemispheric differences summarized by the statement that the left hemisphere in most humans is specialized for (among other things) the verbal and symbolic processing important in communication, whereas the right hemisphere is specialized for (among other things) visuospatial and emotional processing.

The ingenious work of Sperry and his colleagues on split-brain patients put an end to the century-long controversy about language lateralization. In most individuals, the left hemisphere is unequivocally the seat of the explicitly verbal

(A)

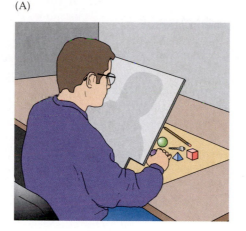

(C)

Left hemisphere functions	Right hemisphere functions
Analysis of right visual field	Analysis of left visual field
Stereognosis (right hand)	Stereognosis (left hand)
Lexical and syntactic language	Emotional coloring of language
Writing	Spatial abilities
Speech	Rudimentary speech

(B)

Normal individual

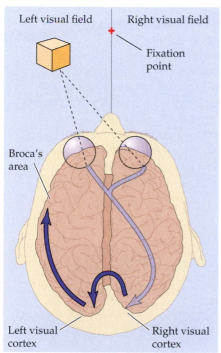

Split-brain individual

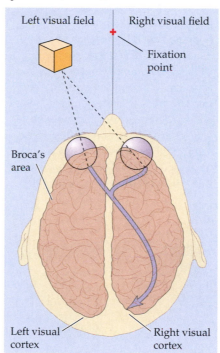

Split-brain individual

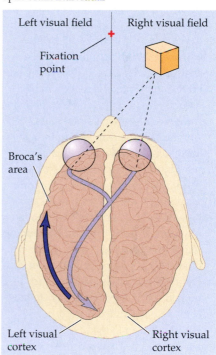

Figure 27.3 Confirmation of hemispheric specialization for language obtained by studying individuals in whom the connections between the right and left hemispheres have been surgically divided. (A) Single-handed, vision-independent stereognosis can be used to evaluate the language capabilities of each hemisphere in split-brain patients. Objects held in the right hand, which provides somatic sensory information to the left hemisphere, are easily named; objects held in the left hand, however, are not readily named by these patients. (B) Visual stimuli or simple instructions can be given independently to the right or left hemisphere in normal and split-brain individuals. Since the left visual field is perceived by the right hemisphere (and vice versa; see Chapter 12), a briefly presented (*tachistoscopic*) instruction in the left visual field is appreciated only by the right brain (assuming that the individual maintains fixation on a mark in the center of the viewing screen). In normal subjects, activation of the right visual cortex leads to hemispheric transfer of visual information via the corpus callosum to the left hemisphere. In split-brain patients, information presented to the left visual field cannot reach the left hemisphere, and patients are unable to produce a verbal report regarding the stimuli. However, such patients *are* able to provide a verbal report of stimuli presented to the right visual field. A wide range of hemispheric functions can be evaluated using this tachistoscopic method, even in normal subjects. (C) The list enumerates some of the different functional abilities of the left and right hemispheres, as deduced from a variety of behavioral tests in split-brain patients.

BOX 27D   Language and Handedness

Approximately 9 out of 10 people are right-handed, a proportion that appears to have been stable over thousands of years and across all cultures in which handedness has been examined. In addition, handedness, or its equivalent, is not peculiar to humans; many studies have demonstrated paw preference in animals ranging from mice to monkeys that is, at least in some ways, similar to human handedness.

Handedness is usually assessed by having individuals answer a series of questions about preferred manual behaviors, such as "Which hand do you use to write?"; "Which hand do you use to throw a ball?"; or "Which hand do you use to brush your teeth?" Each answer is given a value, depending on the preference indicated, providing a quantitative measure of the inclination toward right- or left-handedness. Anthropologists have determined the incidence of handedness in ancient cultures by examining artifacts—the shape of a flint ax, for example, can indicate whether it was made by a right- or left-handed individual. Handedness in antiquity has also been assessed by noting the incidence of people in artistic representations who are using one hand or the other. Based on this evidence, the human species appears always to have been a right-handed one.

Whether an individual is right- or left-handed has a number of interesting consequences. As will be obvious to left-handers, the world of human artifacts is in many respects a right-handed one (Figure A). Implements such as can openers, scissors, and power tools are constructed for the right-handed majority. Books and magazines are also designed for right-handers (compare turning this page with your left and right hands), as are golf clubs and guitars. By the same token, the challenge of penmanship is different for left- and right-handers by virtue of writing from left to right (Figure B). Perhaps as a consequence of such biases, the accident rate for left-handers in all categories (work, home, sports) is higher than for right-handers, including the rate of traffic fatalities. However, there are also some advantages to being left-handed. For example, an inordinate number of international fencing champions have been left-handed. The reason for this fact is simply that the majority of any individual's opponents will be right-handed; therefore, the average fencer, whether right- or left-handed, is less practiced at parrying thrusts from left-handers.

Hotly debated over the years have been the related questions of whether being left-handed is in any sense "pathological," and whether being left-handed entails a diminished life expectancy. No one disputes the fact that there is currently a surprisingly small number of left-handers among the elderly (Figure C). These data have come from studies of the general population and have been supported by information gleaned from *The Baseball Encyclopedia* (in which longevity and other characteristics of a large number of healthy left- and right-handers have been recorded because of interest in the U.S. national pastime).

Two explanations of this peculiar finding have been put forward. Stanley Coren and his collaborators at the University of British Columbia have argued that these statistics reflect a higher mortality rate among left-handers, partly as a result of increased accidents, but also

(A)

Right-handed

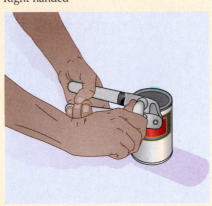

Left-handed

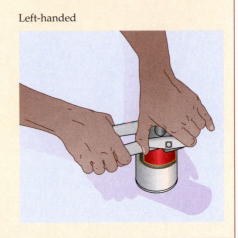

A simple manual can opener is one example of the many common objects designed for use by the right-handed majority.

language functions. There is significant variation in the degree of lateralization among individuals, however (Box 27D), and it would be wrong to suppose that the right hemisphere has no language capacity. As noted, in some individuals the right hemisphere can produce rudimentary words and phrases, a few individuals have fully right-sided verbal functions, and even for the majority with strongly left-lateralized language semantic abilities, the right hemisphere is normally the

(B)

Right-handed writing

Left-handed writing

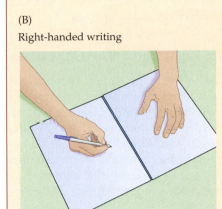

Writing techniques for right- and left-handed individuals.

(C)

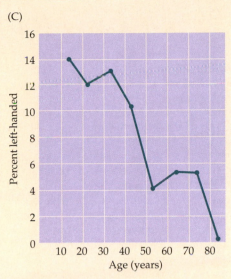

Age (years)

Percentage of left-handers in the population as a function of age (based on more than 5000 individuals). Taken at face value, these data indicate that right-handers live longer than left-handers. A more likely possibility, however, is that the paucity of elderly left-handers simply reflects changes in the social pressures on children to become right-handed. (After Coren, 1992.)

because other data show left-handedness to be associated with a variety of pathologies (there is, for instance, a higher incidence of left-handedness among individuals classified as mentally retarded). Coren and others have suggested that left-handedness may arise because of developmental problems in the pre- and/or perinatal period. If true, then a rationale for decreased longevity would have been identified that might combine with greater proclivity to accidents in a right-hander's world.

An alternative explanation, however, is that the diminished number of left-handers among the elderly at present is primarily a reflection of sociological factors—namely, a greater acceptance of left-handed children today compared to the first half of the twentieth century. In this view, there are fewer older left-handers because in earlier generations parents, teachers, and other authority figures encouraged (and sometimes insisted on) right-handedness. The weight of the evidence favors the sociological explanation.

The relationship between handedness and other lateralized functions—language in particular—has long been a source of confusion. It is unlikely that there is any direct relationship between language and handedness, despite much speculation to the contrary. The most straightforward evidence on this point comes from the results of the Wada test described in the text. The large number of such tests carried out for clinical purposes indicate that about 97% of humans, including the majority of left-handers, have explicitly verbal language functions in the left hemisphere (although it should be noted that right hemispheric dominance for language is much more common among left-handers). Since most left-handers have language function on the side of the brain opposite the control of their preferred hand, it is hard to argue for any strict relationship between these two lateralized functions. In all likelihood, handedness, like language, is first and foremost an example of the advantage of having any specialized function on a

single side of the brain or the other to make maximum use of the available neural circuitry in a brain of limited size.

References

BAKAN, P. (1975) Are left-handers brain damaged? *New Scientist* 67: 200–202.

COREN, S. (1992) *The Left-Hander Syndrome: The Causes and Consequence of Left-Handedness.* New York: The Free Press.

DAVIDSON, R. J. AND K. HUGDAHL (EDS.) (1995) *Brain Asymmetry.* Cambridge, MA: MIT Press.

SALIVE, M. E., J. M. GURALNIK AND R. J. GLYNN (1993) Left-handedness and mortality. *Am. J. Pub. Health* 83: 265–267.

source of our emotional coloring of language (see below and Chapter 29). Moreover, the right hemisphere in many split-brain patients understands language to a modest degree, since these patients can respond to simple visual commands presented tachistoscopically in the left visual field. Consequently, Broca's conclusion that we speak with our left brain is not strictly correct; it would be more accurate to say that most people understand language and speak very much better with the left hemisphere than with the right, and that the contributions of the two hemispheres to the overall goals of communication are different.

Anatomical Differences between the Right and Left Hemispheres

The differences in language function between the left and right hemispheres have naturally inspired neurologists and neuropsychologists to find a structural correlate of this behavioral lateralization. One hemispheric difference that has received much attention over the years was identified in the late 1960s by Norman Geschwind and his colleagues at Harvard Medical School, who found an asymmetry in the superior aspect of the temporal lobe known as the **planum temporale** (Figure 27.4). This area is significantly larger on the left side in about two-thirds of human subjects studied postmortem; a similar difference has been found in higher apes, but not in other primates.

Because the planum temporale is near (although certainly not congruent with) the regions of the temporal lobe that contain cortical areas essential to language (i.e., Wernicke's area and other auditory association areas), it was initially suggested that this leftward asymmetry reflected the greater involvement of the left hemisphere in language. However, these anatomical differences in the two hemispheres of the brain, which are recognizable at birth, are unlikely to be an anatomical correlate of the lateralization of language functions. The fact that a detectable planum asymmetry is present in only 67 percent of human brains, whereas the preeminence of language in the left hemisphere is evident in 97 percent of the population, argues that this association has some other cause. The structural correlate of the functional left–right differences in hemispheric lan-

Figure 27.4 *Asymmetry of the right and left human temporal lobes. (A) The superior portion of the brain has been removed as indicated to reveal the dorsal surface of the temporal lobes in the right-hand diagram (which presents a dorsal view of the horizontal plane). A region of the surface of the temporal lobe called the planum temporale is significantly larger in the left hemisphere of most (but far from all) individuals. (B) Measurements of the planum temporale in adult and infant brains. The mean size of the planum temporale is expressed in arbitrary planimetric units to get around the difficulty of measuring the curvature of the gyri within the planum. The asymmetry is evident at birth and persists in adults at roughly the same magnitude (on average, the left planum is about 50% larger than the right). (C) A magnetic resonance image in the frontal plane, showing this asymmetry (arrows) in a normal adult subject.*

(A)

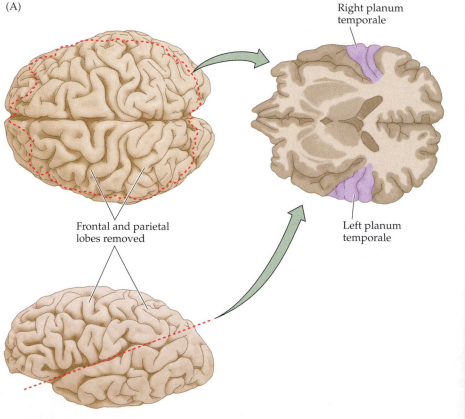

Right planum temporale

Frontal and parietal lobes removed

Left planum temporale

(B)

Planum temporale measurements of 100 adult and 100 infant brains		
	Left hemisphere	Right hemisphere
Infant	20.7	11.7
Adult	37.0	18.4

(C)   Right side   Left side

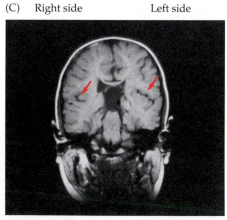

guage abilities, if indeed there is one at a gross anatomical level, is simply not clear, as is also the case for the lateralized hemispheric functions described in Chapter 26.

Mapping Language Functions

The pioneering work of Broca and Wernicke, and later Geschwind and Sperry, clearly established differences in hemispheric function. Several techniques have since been developed that allow hemispheric attributes to be assessed in neurological patients with an intact corpus callosum, and in normal subjects.

One method that has long been used for the clinical assessment of language lateralization was devised in the 1960s by Juhn Wada at the Montreal Neurological Institute. In the so-called Wada test, a short-acting anesthetic (e.g., sodium amytal) is injected into the left carotid artery; this procedure transiently "anesthetizes" the left hemisphere and thus tests the functional capabilities of the affected half of the brain. If the left hemisphere is indeed "dominant" for language, then the patient becomes transiently aphasic while carrying out an ongoing verbal task like counting. The anesthetic is rapidly diluted by the circulation, but not before its local effects on the hemisphere on the side of the injection can be observed. Since this test is potentially dangerous, its use is limited to neurological and neurosurgical patients.

Less invasive (but less definitive) ways to test the cognitive abilities of the two hemispheres in normal subjects include positron emission tomography, functional magnetic resonance imaging (see Box 1A), transcranial magnetic stimulation, and the sort of tachistoscopic presentation used so effectively by Sperry and his colleagues (even when the hemispheres are normally connected, subjects show delayed verbal responses and other differences when the right hemisphere receives the instruction). Application of these various techniques has amply confirmed the hemispheric lateralization of language functions. More importantly, such studies have provided valuable diagnostic tools to determine, in preparation for neurosurgery, which hemisphere is "eloquent": although most individuals have the major language functions in the left hemisphere, a few—about 3 percent of the population—do not (the latter are much more often left-handed; see Box 27D).

Once the appropriate hemisphere is known by these means, neurosurgeons typically map language functions more precisely by electrical stimulation of the cortex during the surgery to further refine their approach to the problem at hand. By the 1930s, the neurosurgeon Wilder Penfield and his colleagues at the Montreal Neurological Institute had already carried out a detailed localization of cortical capacities in a large number of patients (see Chapter 9). Penfield used electrical mapping techniques adapted from neurophysiological work in animals to delineate the language areas of the cortex prior to removing brain tissue in the treatment of tumors or epilepsy (Figure 27.5A). Such intraoperative mapping guaranteed that the cure would not be worse than the disease and has been widely used ever since, with increasingly sophisticated stimulation and recording methods. As a result, a wealth of more detailed information about language localization has emerged.

Penfield's observations, together with more recent studies performed by George Ojemann and his group at the University of Washington, have further advanced the conclusions inferred from postmortem correlations and other approaches (Figure 27.5B). As expected, these intraoperative studies using electrophysiological recording methods have shown that a large region of the perisylvian cortex of the left hemisphere is clearly involved in language production and comprehension. A surprise, however, has been the variability in language

Figure 27.5 Evidence for the variability of language representation among individuals, determined by electrical stimulation during neurosurgery. (A) Diagram from Penfield's original study illustrating sites in the left hemisphere at which electrical stimulation interfered with speech. (B) Diagrams summarizing data from 117 patients whose language areas were mapped by electrical recording at the time of surgery. The number in each red circle indicates the (quite variable) percentage of patients who showed interference with language in response to stimulation at that site. Note also that many of the sites that elicited interference fall outside the classic language areas (Broca's area, shown in purple; Wernicke's area, shown in blue). (A after Penfield and Roberts, 1959; B after Ojemann et al., 1989.)

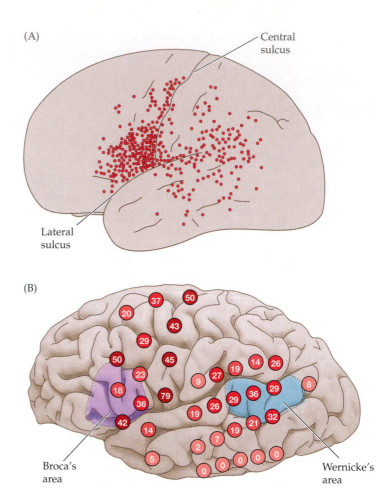

localization from patient to patient. Ojemann found that the brain regions involved in language are only approximately those indicated by older textbook treatments, and that their exact locations differ unpredictably among individuals. Equally unexpected, bilingual patients do not necessarily use the same bit of cortex for storing the names of the same objects in two different languages. Moreover, although neurons in the temporal cortex in and around Wernicke's area respond preferentially to spoken words, they do not show preferences for a particular word. Rather, a wide range of words can elicit a response in any given recording site.

Despite these advances, neurosurgical studies are complicated by their intrinsic difficulty, the risk involved, and to some extent by the fact that the brains of the patients in whom they are carried out are not normal. The advent of positron emission tomography in the 1980s, and more recently functional magnetic resonance imaging, has allowed the investigation of language regions in normal subjects by noninvasive brain imaging (Figure 27.6). Recall that these techniques reveal the areas of the brain that are active during a particular task because the related electrical activity increases local metabolic activity and therefore local blood flow. Much like Ojemann's studies in neurosurgical patients, the results of this approach, particularly in the hands of Marc Raichle, Steve Petersen, and their colleagues at Washington University in St. Louis, have challenged excessively rigid views of the localization and lateralization of linguistic function. Although high levels of activity occur in the expected regions, large areas of both hemispheres are activated in word recognition or production tasks.

Passively viewing words

Listening to words

Speaking words

Generating word associations

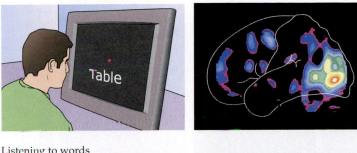

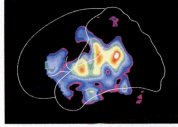

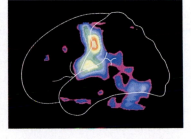

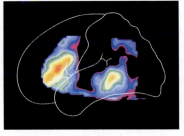

Figure 27.6 Language-related regions of the left hemisphere mapped by positron emission tomography (PET) in a normal human subject. Subjects reclined within the PET scanner and followed instructions on a special display (these details are not illustrated). The left panels indicate the task being practiced prior to scanning. The PET scan images are shown on the right. Language tasks such as listening to words and generating word associations elicit activity in Broca's and Wernicke's areas, as expected. However, there is also activity in primary and association sensory and motor areas for both active and passive language tasks. These observations indicate that language processing involves many cortical regions in addition to the classic language areas. (From Posner and Raichle, 1994.)

Finally, a number of investigators including Hanna Damasio and her colleagues at the University of Iowa and Alex Martin and his collaborators at the National Institutes of Mental Health, have shown that distinct regions of the temporal cortex are activated by tasks in which subjects carried out naming, viewing or matching tasks that involved faces, animals, or tools (Figure 27.7). This observation helps explain the clinical finding that when a relatively limited region of the temporal lobe is damaged (usually but by no means always on the left side), language deficits are sometimes restricted to a particular category of objects. These studies are also consistent with Ojemann's electrophysiological studies, which indicate that some aspects of language are organized according to categories of meaning rather than individual words. Taken together, such studies are rapidly augmenting the information available about how language is represented in the brain.

Figure 27.7 *Different regions in the temporal lobe are activated by different word categories using PET imaging. Dotted lines show location of the relevant temporal regions in these horizontal views. Note the different patterns of activity in the temporal lobe in response to each stimulus catagory. (From Damasio et al., 1996.)*

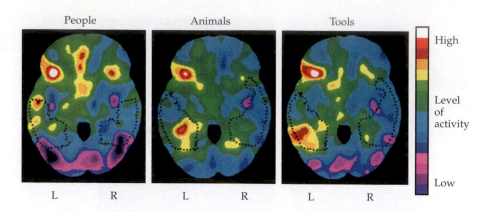

The Role of the Right Hemisphere in Language

Because the same gross anatomical and cytoarchitectonic areas exist in the cortex of both hemispheres, a puzzling issue remains. What do the comparable areas in the right hemisphere actually do? In fact, language deficits often *do* occur following damage to the right hemisphere. The most obvious effect of such lesions is an absence of the normal emotional and tonal components of language—called **prosodic elements**—that impart additional meaning to verbal communication. This "coloring" of speech is critical to the message conveyed; indeed, in some languages, such as Mandarin Chinese prosody is used to change the semantic meaning of the word uttered. Deficiencies in this ability, referred to as **aprosodias**, are associated with right-hemisphere damage to the cortical regions that correspond to Broca's and Wernicke's areas and associated regions in the left hemisphere. Aprosodias emphasize that although the left hemisphere (or, better put, specialized cortical regions within that hemisphere) figures prominently in the comprehension and production of language for most humans, other regions, including corresponding (and other) areas in the right hemisphere, are needed to generate the full richness of everyday speech.

In summary, whereas the classically defined regions of the left hemisphere operate more or less as advertised, a variety of more recent studies have shown that other left- and right-hemisphere areas clearly make a significant contribution to generation and comprehension of language.

Sign Language

The implication of several aspects of the foregoing account is that the cortical organization of language does not simply reflect specializations for hearing and speaking; the language regions of the brain appear to be more broadly organized for processing symbols pertinent to social communication. Strong support for this conclusion has come from studies of sign language in individuals deaf from birth.

American Sign Language has all the components (grammar, syntax, and emotional tone) of spoken and heard language. Based on this knowledge, Ursula Bellugi and her colleagues at the Salk Institute examined the cortical localization of sign language abilities in patients who had suffered lesions of either the left or right hemisphere. All these deaf individuals never learned language, had been signing throughout their lives, had deaf spouses, were members of the deaf community, and were right-handed. The patients with left-hemisphere lesions, which in each case involved the language areas of the frontal and/or temporal lobes, had measurable deficits in sign production and comprehension when compared to normal signers of similar age (Figure 27.8). In contrast, the patients with lesions in approximately the same areas in the right hemisphere did not have such signing "aphasias." Instead, as predicted from other hearing patients with similar lesions, right hemisphere abilities such as visuospatial processing, emotional processing and the emotional tone evident in signing were impaired. Although the number of subjects studied was necessarily small (deaf signers with lesions of the language areas are understandably difficult to find), the capacity for signed and seen communication is evidently represented predominantly in the left hemisphere, in the same areas as spoken language. This evidence accords with the idea that the language regions of the brain are specialized for the representation of social communication by means of symbols, rather than for heard and spoken language per se.

The capacity for seen and signed communication, like its heard and spoken counterpart, emerges in early infancy. Careful observation of babbling in hearing (and, eventually, speaking) infants shows the production of a predictable pattern of sounds related to the ultimate acquisition of spoken language. Thus, babbling prefigures mature language, and indicates that an innate capacity for language imitation is a key part of the process by which a full-blown language is ultimately acquired. The offspring of deaf, signing parents "babble" with their hands in gestures that are apparently the forerunners of mature signs (see Figure 24.1). Like verbal babbling, the amount of manual babbling increases with age until the child begins to form accurate, meaningful signs. These observations indicate that the strategy for acquiring the rudiments of symbolic communication from parental or other cues—regardless of the means of expression—is similar.

Patient with signing deficit:

Arrive   Stay   There

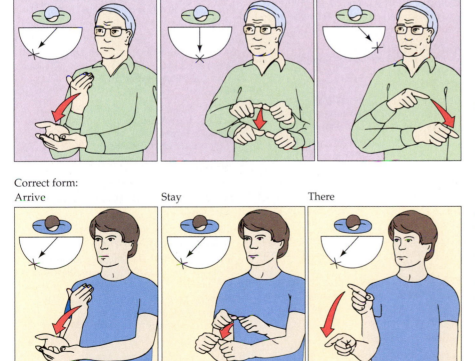

Correct form:

Arrive   Stay   There

Figure 27.8 Signing deficits in congenitally deaf individuals who had learned sign language from birth and later suffered lesions of the language areas in the left hemisphere. Left-hemisphere damage produced signing problems in these patients analogous to the aphasias seen after comparable lesions in hearing, speaking patients. In this example, the patient (upper panels) is expressing the sentence "We arrived in Jerusalem and stayed there." Compared to a normal control (lower panels), he cannot properly control the spatial orientation of the signs. The direction of the correct signs and the aberrant direction of the "aphasic" signs are indicated in the upper left-hand corner of each panel. (After Bellugi et al., 1989.)

Summary

A variety of methods have been used to understand the organization of language in the human brain. This effort began in the nineteenth century by correlating clinical signs and symptoms with the location of brain lesions determined postmortem. In the twentieth century, additional clinical observations together with studies of split-brain patients, mapping at neurosurgery, transient anesthesia of a single hemisphere, and noninvasive imaging techniques such as PET and *f*MRI have greatly extended knowledge about the neural substrates of language. Together, these various approaches show that the perisylvian cortices of the left hemisphere are especially important for normal language in the vast majority of humans. The right hemisphere also contributes importantly to language, most obviously by giving it emotional tone. The similarity of the deficits after comparable brain lesions in congenitally deaf individuals and their speaking counterparts have shown further that the cortical representation of language is independent of the means of its expression or perception (spoken and heard versus gestured and seen). The specialized language areas that have been identified are evidently the major components of a widely distributed set of brain regions that allow humans to communicate effectively by means of symbols that can be attached to objects, concepts, and feelings.

Additional Reading

Reviews

BELLUGI, U., H. POIZNER AND E. S. KLIMA (1989) Language, modality, and the brain. *Trends Neurosci.* 12: 380–388.

DAMASIO, A. R. (1992) Aphasia. *New Eng. J. Med.* 326: 531–539.

DAMASIO, A. R. AND H. DAMASIO (1992) Brain and language. *Sci. Amer.* 267 (Sept.): 89–95.

DAMASIO, A. R. AND N. GESCHWIND (1984) The neural basis of language. *Annu. Rev. Neurosci.* 7: 127–147.

ETCOFF, N. L. (1986) The neurophysiology of emotional expression. In *Advances in Clinical Neuropsychology,* Volume 3, G. Goldstein and R. E. Tarter (eds.). New York: Quantum, pp. 127–179.

LENNEBERG, E. H. (1967) Language in the context of growth and maturation. In *Biological Foundations of Language.* New York: John Wiley and Sons, pp. 125–395.

OJEMANN, G. A. (1983) The intrahemispheric organization of human language, derived with electrical stimulation techniques. *Trends Neurosci.* 4: 184–189.

OJEMANN, G. A. (1991) Cortical organization of language. *J. Neurosci.* 11: 2281–2287.

SPERRY, R. W. (1982) Some effects of disconnecting the cerebral hemispheres. *Science* 217: 1223–1227.

Important Original Papers

CHAO, L. L., J. V. HAXBY AND A. MARTIN (1999) Attribute-based neural substrates in temporal cortex for perceiving and knowing about objects. *Nature Neuroscience* 2: 913–919.

CREUTZFELDT, O., G. OJEMANN AND E. LETTICH (1989) Neuronal activity in the human temporal lobe. I. Response to speech. *Exp. Brain Res.* 77: 451–475.

CARAMAZZA, A. AND A. E. HILLIS (1991) Lexical organization of nouns and verbs in the brain. *Nature* 349: 788–790.

DAMASIO, H., T. J. GRABOWSKI, D. TRANEL, R. D. HICHWA AND A. DAMASIO (1996) A neural basis for lexical retrieval. *Nature* 380: 499–505.

EIMAS, P. D., E. R. SIQUELAND, P. JUSCZYK AND J. VIGORITO (1971) Speech perception in infants. *Science* 171: 303–306.

GAZZANIGA, M. S. (1998) The split brain revisited. *Sci. Amer.* 279 (July): 50–55.

GAZZANIGA, M. S., R. B. LURY AND G. R. MANGUN (1998) Ch. 8, Language and the Brain. In *Cognitive Neuroscience: The Biology of the Mind.* New York: W. W. Norton and Co., pp. 289–321.

GAZZANIGA, M. S. AND R. W. SPERRY (1967) Language after section of the cerebral commissures. *Brain* 90: 131–147.

GESCHWIND, N. AND W. LEVITSKY (1968) Human brain: Left-right asymmetries in temporal speech region. *Science* 161: 186–187.

OJEMANN, G. A. AND H. A. WHITAKER (1978) The bilingual brain. *Arch. Neurol.* 35: 409–412.

PETERSEN, S. E., P. T. FOX, M. I. POSNER, M. MINTUN AND M. E. RAICHLE (1988) Positron emission tomographic studies of the cortical anatomy of single-word processing. *Nature* 331: 585–589.

PETTITO, L. A. AND P. F. MARENTETTE (1991) Babbling in the manual mode: Evidence for the ontogeny of language. *Science* 251: 1493–1496.

WADA, J. A., R. CLARKE AND A. HAMM (1975) Cerebral hemispheric asymmetry in humans: Cortical speech zones in 100 adult and 100 infant brains. *Arch. Neurol.* 32: 239–246.

WESTBURY, C. F., R. J. ZATORRE AND A. C. EVANS (1999) Quantifying variability in the planum temporale: A probability map. *Cerebral Cortex* 9: 392–405.

Books

GARDNER, H. (1974) *The Shattered Mind: The Person After Brain Damage.* New York: Vintage.

LENNEBERG, E. (1967) *The Biological Foundations of Language.* New York: Wiley.

PINKER, S. (1994) *The Language Instinct: How the Mind Creates Language.* New York: William Morrow and Company.

POSNER, M. I. AND M. E. RAICHLE (1994) *Images of Mind.* New York: Scientific American Library.

Chapter 28

Sleep and Wakefulness

Overview

Sleep—which is defined behaviorally by the normal suspension of consciousness and electrophysiologically by specific brain wave criteria—consumes fully one-third of our lives. Sleep occurs in all mammals, and probably in all vertebrates. We crave sleep when deprived of it and, to judge from animal studies, continued sleep deprivation can ultimately be fatal. Shakespeare called sleep "nature's soft nurse," emphasizing (as have many others) the restorative nature of sleep. Surprisingly, however, sleep is not the result of a simple diminution of brain activity. Indeed, in REM, or "rapid eye movement" sleep, the brain is about as active as it is when people are awake. Rather, sleep is a series of precisely controlled physiological states, the sequence of which is governed by a group of brainstem nuclei that projects widely throughout the brain and spinal cord. The reason for the high levels of brain activity during REM sleep, the significance of dreaming, and the basis of the restorative effect of sleep are all topics that remain poorly understood. The clinical importance of sleep is obvious from the prevalence of sleep disorders. In any given year about 40 million Americans suffer from chronic sleep disorders, and an additional 30 million experience occasional (at least a few days each month) sleeping problems that are severe enough to interfere with their daily activities. In short, the phenomenology of sleep presents major challenges in both basic neurobiology and clinical medicine.

Why Do Humans (and Many Other Animals) Sleep?

To feel rested and refreshed upon awakening, most adults require 7–8 hours of sleep, although this number varies among individuals (Figure 28.1A). As a result, a substantial fraction of our lives is spent in this mysterious state. For infants, the requirement is much higher (~17 hours a day or more), and teenagers need on average about 9 hours of sleep. As people age, they tend to sleep more lightly and for shorter times (Figure 28.1B). Older adults often "make up" for shorter and lighter nightly sleep periods by napping during the day. Getting too little sleep creates a "sleep debt" that must be repaid in the following days. In the meantime, judgment, reaction time, and other functions are in varying degrees impaired. Poor sleep therefore has a price, sometimes with tragic consequences. In the United States, fatigue is estimated to contribute to more than 100,000 highway accidents each year, resulting in some 70,000 injuries and 1,500 deaths.

Sleep (or at least a physiological period of quiescence) is a highly conserved behavior found in animals ranging from fruit flies to humans (Box 28A). Despite this prevalence, *why* we sleep is not well understood. Because an animal is particularly vulnerable while sleeping, there must be evolutionary advantages that

(A)

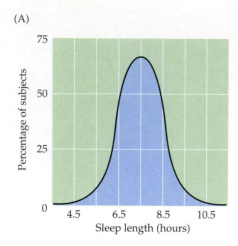

(B)

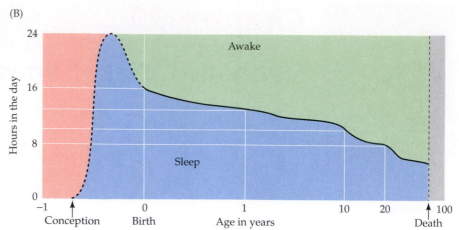

Figure 28.1 The duration of sleep. (A) The duration of sleep each night in adults is normally distributed with a mean of 7.5 hours and a standard deviation of about 1.25 hours. Thus, each night about two-thirds of the population sleeps between 6.25 and 8.75 hours. (B) The duration of daily sleep as a function of age. (After Hobson, 1989.)

outweigh this considerable disadvantage. From a perspective of energy conservation, one function of sleep is to replenish brain glycogen levels, which fall during the waking hours. In addition, since it is generally colder at night, more energy would have to be expended to keep warm were we nocturnally active. Human body temperature has a 24-hour cycle (as do many other indices of activity and stress; Figure 28.2), reaching a minimum at night and thus reducing heat loss. As might be expected, metabolism measured by oxygen consumption decreases during sleep. Another plausible reason is that humans and many other animals that sleep at night are highly dependent on visual information to find food and avoid predators. A recent idea about the advantage of sleep proposes that this is a period during which memories, in the form of changes in the strength of synaptic connections induced by experiences during waking hours, can be consolidated.

In mammals, sleep is evidently necessary for survival. Sleep-deprived rats lose weight despite increased food intake and progressively fail to regulate body temperature, as their core temperature increases several degrees. They also develop infections, suggesting some compromise of the immune system. Rats

Figure 28.2 Circadian rhythmicity of core body temperature, and of growth hormone and cortisol levels in the blood. In the early evening, core temperature begins to decrease whereas growth hormone begins to increase. The level of cortisol, which reflects stress, begins to increase toward morning and stays elevated for several hours.

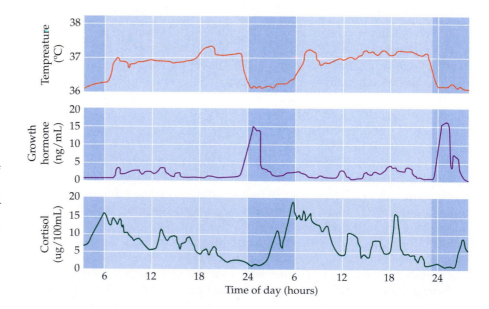

BOX 28A   Sleep Styles in Different Species

A wide variety of animals have a rest–activity cycle that often (but not always) occurs in a daily, or circadian, rhythm. Even among mammals, however, the organization of sleep depends very much on the species in question. As a general rule, predatory animals can indulge, as humans do, in long, uninterrupted periods of sleep that can be nocturnal or diurnal, depending on the time of day when the animal acquires food, mates, cares for its young, and deals with life's other necessities. The survival of animals that are preyed upon depends much more critically on continued vigilance. Such species—as diverse as rabbits and giraffes—sleep during short intervals that usually last no more than a few minutes. Shrews, the smallest mammals, hardly sleep at all.

An especially remarkable solution to the problem of maintaining vigilance during sleep is shown by dolphins and seals, in whom sleep alternates between the two cerebral hemispheres (see figure). Thus, one hemisphere can exhibit the electroencephalographic signs of wakefulness while the other shows the characteristics of sleep (see Box 28C and Figure 28.5). In short, although periods of rest are evidently essential to the proper functioning of the brain, and more generally to normal homeostasis, the manner in which rest is obtained depends on the particular needs of each species.

References

ALLISON, T. AND D.V. CICCHETTI (1976) Sleep in mammals: Ecological and constitutional correlates. *Science* 194: 732–734.

ALLISON, T. H. AND H. VAN TWYVER (1970) The evolution of sleep. *Natural History* 79: 56–65.

ALLISON, T., H. VAN TWYVER AND W. R. GOFF (1972) Electrophysiological studies of the echidna, *Tachyglossus aculeatus. Arch. Ital. Biol.* 110: 145–184.

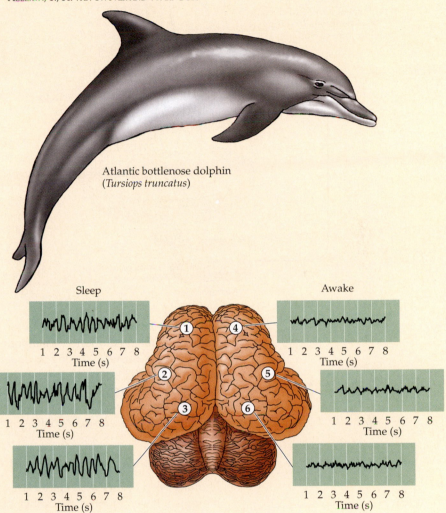

Atlantic bottlenose dolphin
(*Tursiops truncatus*)

Some animals can sleep one hemisphere at a time. These EEG tracings were taken simultaneously from left and right cerebral hemispheres of a dolphin. Slow-wave sleep is apparent in the left hemisphere (recording sites 1–3); the right hemisphere, however, shows low-voltage, high-frequency waking activity (sites 4–6). (After Mukhametov, Supin, and Polyakova, 1977.)

completely deprived of sleep die within a few weeks (Figure 28.3A,B). In humans, lack of sleep leads to impaired memory and reduced cognitive abilities and, if the deprivation persists, mood swings and often hallucinations. As the name implies, patients with the genetic condition *fatal familial insomnia* die within several years of onset. This rare disease, which appears in middle age, is characterized by hallucinations, seizures, loss of motor control, and the inability to enter a state of deep sleep (see the section "Stages of Sleep").

(A) Experiment setup

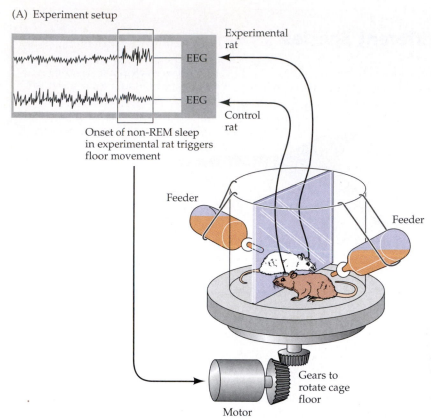

Onset of non-REM sleep
in experimental rat triggers
floor movement

Feeder

Feeder

Gears to
rotate cage
floor

Motor

(B) Experimental animals

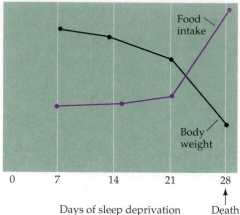

Days of sleep deprivation   Death

Figure 28.3   *Consequences of total sleep deprivation in rats. (A) In this apparatus, an experimental rat is kept awake because the onset of sleep (detected electroencephalographically) triggers movement of the cage floor. The control rat (brown) can thus sleep intermittently, whereas the experimental animal (white) cannot. (B) After 2–3 weeks of sleep deprivation, the experimental animals begin to lose weight, fail to control their body temperature, and eventually die. (After Bergmann et al., 1989.)*

The effects of sleep deprivation for shorter periods in humans seem mainly on behavioral performance, as anyone who has experienced a night or two of sleeplessness knows. The longest documented period of voluntary sleeplessness in humans is 453 hours, 40 minutes (approximately 19 days)—a record achieved without any pharmacological stimulation. The young man involved recovered within a few days, during which he slept more than normal but otherwise seemed none the worse for wear.

The Circadian Cycle of Sleep and Wakefulness

Human sleep occurs with **circadian** (*circa* = about; *dia* = day) periodicity, and biologists have explored a number of questions about this daily cycle. What happens, for example, when individuals are prevented from sensing the cues they normally use to distinguish night and day? This question has been addressed by placing volunteers in an environment such as a cave or bunker that lacks external time cues (Figure 28.4). In a typical experiment of this sort, subjects undergo a 5- to 8-day period of acclimatization that includes normal social interactions, meals at the usual times, and temporal cues (e.g., radio and television). During this period, the subjects typically arise and go to sleep at the usual times and maintain a 24-hour sleep–wake cycle. When the cues are removed, however, subjects awaken later each day, and under these conditions the clock is said to be "free running." It was originally thought that the cycle of sleep and wakefulness gradually lengthens to about 26 hours; however, subsequent work eliminating possible knowledge of time of day and exposure to room light found the length of the cycle is closer to 24 hours. Thus, humans (and many other animals; see Box 28B) have an internal "clock" that operates even in the absence of external information about the time of day.

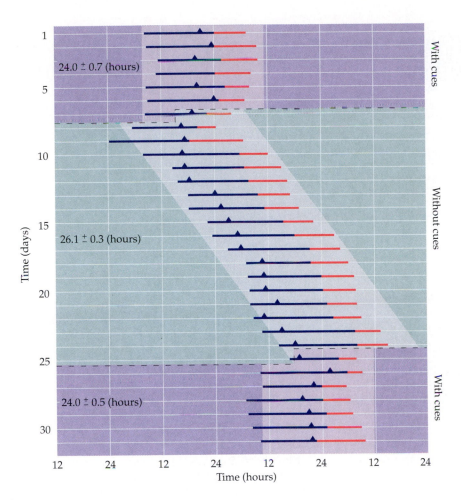

Figure 28.4 Rhythm of waking (blue lines) and sleeping (red lines) of a volunteer in an isolation chamber with and without cues about the day–night cycle. Numbers represent the mean ± standard deviation of a complete wake–sleep cycle in each condition. Triangles represent times when the rectal temperature was maximum. (After Aschoff, 1965, as reproduced in Schmidt et al., 1983.)

Presumably, circadian clocks evolved to maintain appropriate periods of sleep and wakefulness and to control other daily rhythms in spite of the variable amount of daylight and darkness in different seasons and at different locations on the planet. To synchronize, or **photoentrain**, physiological processes with the day-night cycle, the biological clock must be able to detect decreases in light levels as night approaches. The receptors that sense these light changes are, not surprisingly, in the outer nuclear layer of the retina, as demonstrated by the fact that removing or covering the eyes abolishes photoentrainment. The retinal detectors are not, however, rod or cone cells. Rather, the cells that convey this information lie within the ganglion cell layer of both primate and murine retinas. Unlike rods and cones that are hyperpolarized when activated by light (see Chapter 11), these special ganglion cells contain a novel photopigment called **melanopsin** and are depolarized rather than hyperpolarized by light (Figure 28.5A). The function of these unusual photoreceptors is evidently to encode environmental illumination and thus to set the biological clock. The axons of these neurons run in the retinohypothalamic tract (Figure 28.5B), which projects to the **suprachiasmatic nucleus (SCN)** of the anterior hypothalamus, the site of the circadian control of homeostatic functions.

Activation of the SCN in this way evokes responses in the paraventricular nucleus of the hypothalamus and ultimately the preganglionic sympathetic neurons in the intermediolateral zone of the lateral horns of the thoracic spinal cord. As described in Chapter 21, these preganglionic neurons modulate neurons in the superior cervical ganglia whose postganglionic axons project to the

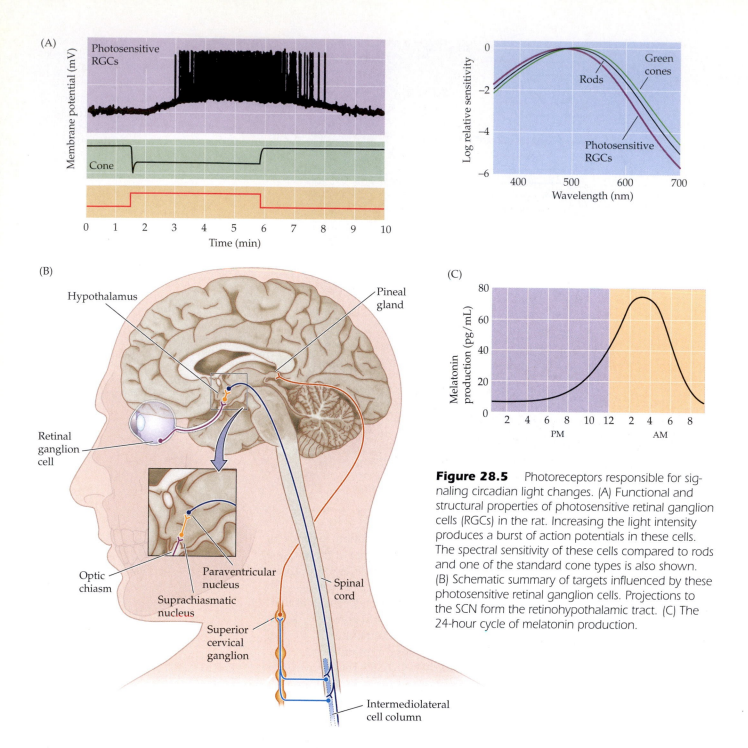

Figure 28.5 Photoreceptors responsible for signaling circadian light changes. (A) Functional and structural properties of photosensitive retinal ganglion cells (RGCs) in the rat. Increasing the light intensity produces a burst of action potentials in these cells. The spectral sensitivity of these cells compared to rods and one of the standard cone types is also shown. (B) Schematic summary of targets influenced by these photosensitive retinal ganglion cells. Projections to the SCN form the retinohypothalamic tract. (C) The 24-hour cycle of melatonin production.

pineal gland (*pineal* means "pinecone-like") in the midline near the dorsal thalamus (see Figure 28.5B). The pineal gland synthesizes the sleep-promoting neurohormone **melatonin** (*N*-acetyl-5-methoxytryptamine) from tryptophan. The pineal then secretes melatonin into the bloodstream, where it modulates the brainstem circuits that influence the sleep–wake cycle. Melatonin synthesis increases as light from the environment decreases, reaching a maximum between 2 A.M. and 4:00 A.M. (Figure 28.5C). In the elderly, the pineal gland produces less melatonin, perhaps explaining why older people sleep less at night and are more often afflicted with insomnia. Melatonin has been used to promote sleep in elderly insomniacs and to reduce disruption of the biological clocks that occurs with jet lag, but it remains unclear whether these therapies are really effective.

Most sleep researchers consider the suprachiasmatic nucleus to be the "master clock." Evidence for this conclusion is that removal of the SCN in experimental animals abolishes their circadian sleep–wake cycle. Furthermore, when SCN cells are placed in organ culture, they exhibit characteristic circadian rhythms (Box 28B). The SCN governs other functions that are synchronized with the sleep–wake cycle, including body temperature, hormone secretion (e.g., cortisol; see Figure 28.2), blood pressure, and urine production. In adults, urine production is reduced at night because of a circadian downregulation of antidieuretic hormone (ADH, also called vasopressin).

BOX 28B   Molecular Mechanisms of Biological Clocks

Virtually all animals (and many plants) adjust their physiology and behavior to the 24-hour day–night cycle under the governance of circadian clocks. Recent studies have revealed much about the genes and proteins that make up the molecular machinery of these clocks. The story began about 35 years ago, in the early 1970s, when Ron Konopka and Seymour Benzer at the California Institute of Technology discovered three mutant strains of fruit flies whose circadian rhythms were abnormal. Further analysis showed the mutations to be alleles of a single locus, which Konopka and Benzer called the *period* or *per* gene. In the absence of normal environmental cues (that is, in constant light or dark), wild-type flies have periods of activity geared to a 24-hour cycle; *per*[s] mutants have 19-hour rhythms, *per*[1] mutants have 29-hour rhythms, and *per*[0] mutants have no apparent rhythm.

Michael Young at Rockefeller University and Jeffrey Hall and Michael Rosbash at Brandeis University independently cloned the first of the three *per* genes in the early 1990s. Cloning a gene does not necessarily reveal its function, however, and so it was in this case.

(Continued on next page)

Diagram of the molecular feedback loop that governs circadian clocks. (After Okamura et al., 1999.)

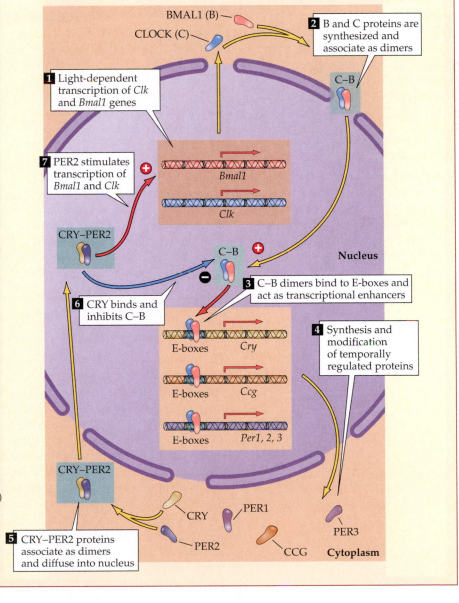

1. Light-dependent transcription of *Clk* and *Bmal1* genes

2. B and C proteins are synthesized and associate as dimers

3. C–B dimers bind to E-boxes and act as transcriptional enhancers

4. Synthesis and modification of temporally regulated proteins

5. CRY–PER2 proteins associate as dimers and diffuse into nucleus

6. CRY binds and inhibits C–B

7. PER2 stimulates transcription of *Bmal1* and *Clk*

BMAL1 (B)
CLOCK (C)
C–B
Nucleus
Bmal1
Clk
CRY–PER2
C–B
E-boxes   *Cry*
E-boxes   *Ccg*
E-boxes   *Per1, 2, 3*
CRY–PER2
CRY   PER1
PER2   PER3
CCG   Cytoplasm

BOX 28B *(Continued)*

Nonetheless, the gene product Per, a nuclear protein, is found in many *Drosophila* cells pertinent to the production of the fly's circadian rhythms. Moreover, normal flies show a circadian variation in the amount of *per* mRNA and Per protein, whereas *per⁰* flies, which lack a circadian rhythm, do not show circadian rhythmicity of gene expression.

Many of the genes and proteins responsible for circadian rhythms in fruit flies have now been discovered in mammals. In mice, the circadian clock arises from the temporally regulated activity of proteins (in capital letters) and genes (both abbreviations and full names in italics), including CRY (*cryptochrome*), CLOCK (*Clk, circadian locomotor output cycles kaput*), BMAL1 (*Bmal1, brain and muscle, ARNT-like*), PER1 (*Per1, Period1*), PER2 (*Per2, Period2*), PER3 (*Per3, Period3*), and vasopressin prepropressophysin (VP) (*clock-controlled genes; Ccg*). These genes and their proteins give rise to transcription/translation autoregulatory feedback loops with both excitatory and inhibitory components (see figure). The key points in this complex regulatory scheme are that: (1) the concentrations of BMAL1 and the three PER proteins cycle in counterphase; (2) PER2 is a

positive regulator of the BMAL1 loop; and (3) CRY is a negative regulator of the period and cryptochrome loops. The two positive components of this loop are influenced, albeit indirectly, by light and/or temperature.

At the start of the day, transcription of *Clk* and *Bmal1* commence and the proteins CLK and BMAL1 are synthesized in tandem. When the concentrations of CLK and BMAL1 increase sufficiently, they associate as dimers and bind to regulatory DNA sequences ("E-boxes") that act as a circadian transcriptional enhancers of the genes *Cry, Per1, Per2, Per3*, and *Ccg*. As a result, PER1, 2, and 3, CRY, and proteins such as VP are produced and diffuse from the nucleus into the cytoplasm, where they are modified.

The functions of PER1 and PER3 remain to be worked out. However, when cytoplasmic concentrations of PER2 and CRY increase, they associate as CRY–PER2 and diffuse back into the nucleus. In the nucleus, PER2 stimulates the synthesis of CLK and BMAL1, while CRY binds to C–B dimers, inhibiting their ability to stimulate the synthesis of the other genes. The complete time course of these feedback loops is approximately 24 hours.

References

CASHMORE, A. R. (2003) Cryptochromes: Enabling plants and animals to determine circadian time. *Cell* 114: 537–543.

DUNLAP, J. C. (1993) Genetic analysis of circadian clocks. *Annu. Rev. Physiol.* 55: 683–728.

KING, D. P. AND J. S. TAKAHASHI (2000) Molecular mechanism of circadian rhythms in mammals. *Annu. Rev. Neurosci.* 23: 713–742.

HARDIN, P. E., J. C. HALL AND M. ROSBASH (1990) Feedback of the *Drosophila period* gene product on circadian cycling of its messenger RNA levels. *Nature* 348: 536–540.

OKAMURA, H. AND 8 OTHERS (1999) Photic induction of *mPer1* and *mPer2* in *Cry*-deficient mice lacking a biological clock. *Science* 286: 2531–2534.

REN, D. AND J. D. MILLER (2003) Primary cell culture of suprachiasmiatic nucleus. *Brain Res. Bull.* 61: 547–553.

SHEARMAN, L. P. AND 10 OTHERS (2000) Interacting molecular loops in the mammalian circadian clock. *Science* 288: 1013–1019.

TAKAHASHI, J. S. (1992) Circadian clock genes are ticking. *Science* 258: 238–240.

VITATERNA, M. H. AND 9 OTHERS (1994) Mutagenesis and mapping of a mouse gene, clock, essential for circadian behavior. *Science* 264: 719–725.

Stages of Sleep

The normal cycle of human sleep and wakefulness implies that, at specific times, various neural systems are being activated while others are being turned off. For centuries—indeed, up until the 1950s—most people who thought about sleep considered it a unitary phenomenon whose physiology was essentially passive and whose purpose was simply restorative. In 1953, however, Nathaniel Kleitman and Eugene Aserinksy showed, by means of electroencephalographic (EEG) recordings from normal subjects, that sleep actually comprises different stages that occur in a characteristic sequence.

Over the first hour after retiring, humans descend into successive stages of sleep (Figure 28.6). These characteristic stages are defined primarily by electroencephalographic criteria (Box 28C). Initially, during "drowsiness," the frequency spectrum of the electroencephalogram is shifted toward lower values and the amplitude of the cortical waves increases slightly. This drowsy period, called **stage I sleep**, eventually gives way to light or **stage II sleep**, which is characterized by a further decrease in the frequency of the EEG waves and an increase in

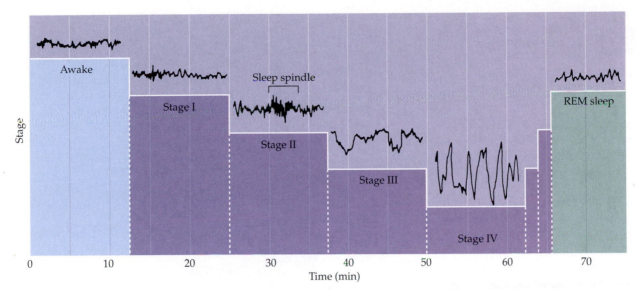

Figure 28.6 EEG recordings during the first hour of sleep. The waking state with the eyes open is characterized by high-frequency (15–60 Hz), low-amplitude activity (~30 μV) activity. This pattern is called beta activity. Descent into stage I non-REM sleep is characterized by decreasing EEG frequency (4–8 Hz) and increasing amplitude (50–100 μV), called theta waves. Descent into stage II non-REM sleep is characterized by 10–12 Hz oscillations (50–150 μV) called spindles, which occur period-ically and last for a few seconds. Stage III non-REM sleep is characterized by slower waves at 2–4 Hz (100–150 μV). Stage IV sleep is defined by slow waves (also called delta waves) at 0.5–2 Hz (100–200 μV). After reaching this level of deep sleep, the sequence reverses and a period of rapid eye movement sleep, or REM sleep, ensues. REM sleep is characterized by low-voltage, high-frequency activity similar to the EEG activity of individuals who are awake. (Adapted from Hobson, 1989.)

BOX 28C   Electroencephalography

Although electrical activity recorded from the exposed cerebral cortex of a monkey was reported in 1875, it was not until 1929 that Hans Berger, a psychiatrist at the University of Jena, first made scalp recordings of this activity in humans. Since then, the electroencephalogram, or EEG, has received a mixed press. Touted by some as a unique opportunity to understand human thinking, the EEG is denigrated by others as too complex and poorly resolved to allow anything more than a superficial glimpse of what the brain is actually doing. The truth probably lies somewhere in between, but no one disputes that electroencephalography has provided a valuable tool to

(*Continued on next page*)

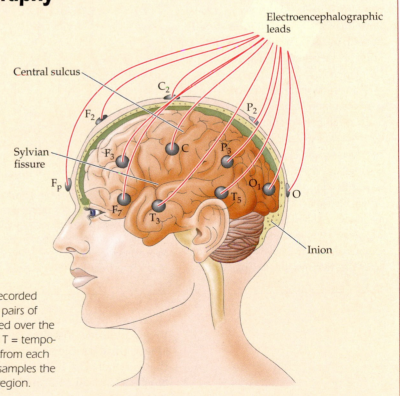

(A) The electroencephalogram represents the voltage recorded between two electrodes applied to the scalp. Typically, pairs of electrodes are placed in 19 standard positions distributed over the head. Letters indicate position (F = frontal, P = parietal, T = temporal, O = occipital, C = central). The recording obtained from each pair of electrodes is somewhat different because each samples the activity of a population of neurons in a different brain region.

BOX 28C   *(Continued)*

both researchers and clinicians, particularly in the fields of sleep physiology and epilepsy.

The major advantage of electroencephalography, which involves the application of a set of electrodes to standard positions on the scalp (Figure A), is its great simplicity. Its most serious limitation is poor spatial resolution, allowing localization of an active site only to within several centimeters. Four basic EEG phenomena have been defined in humans (albeit somewhat arbitrarily). The alpha rhythm is typically recorded in awake subjects with their eyes closed. By definition, the frequency of the alpha rhythm is 8–13 Hz, with an amplitude that is typically 10–50 mV. Lower-amplitude beta activity is defined by frequencies of 14–60 Hz and is indicative of mental activity and attention. The theta and delta waves, which are characterized by frequencies of 4–7 Hz and less than 4 Hz, respectively, imply drowsiness, sleep, or one of a variety of pathological conditions; these slow waves in normal individuals are the signature of stage IV non-REM sleep. The way these phenomena are generated is shown in Figures B and C.

Far and away the most obvious component of these various oscillations is the alpha rhythm. Its prominence in the occipital region—and its modulation by eye opening and closing—implies that it is somehow linked to visual processing, as was first proposed in 1935 by the British physiologist Edgar Adrian. In fact, evidence from large numbers of subjects suggests that at least several different regions of the brain have their own characteristic rhythms; for example, within

the alpha band (8–13 Hz), one rhythm, the classic alpha rhythm, is associated with visual cortex, one (the mu rhythm) with the sensory motor cortex around the central sulcus, and yet another (the kappa rhythm) with the auditory cortex.

In the 1940s, Edward Dempsey and Robert Morrison showed that these EEG rhythms depend in part on activity in the thalamus, since thalamic lesions can reduce or abolish the oscillatory cortical discharge (although some oscillatory

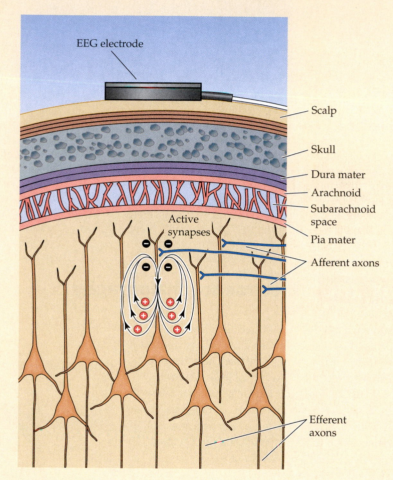

(B) An electrode on the scalp measures the activity of a very large number of neurons in the underlying regions of the brain, each of which generates a small electrical field that changes over time. This activity (which is thought to be mostly synaptic) makes the more superficial extracellular space negative with respect to deeper cortical regions. The EEG electrode measures a synchronous signal because many thousands of cells are responding in the same manner at more or less the same time. (Adapted from Bear et al., 2001.)

their amplitude, together with intermittent high-frequency spike clusters called **sleep spindles**. Sleep spindles are periodic bursts of activity at about 10–12 Hz that generally last 1–2 seconds and arise as a result of interactions between thalamic and cortical neurons (see below). In **stage III sleep**, which represents moderate to deep sleep, the number of spindles decreases, whereas the amplitude of

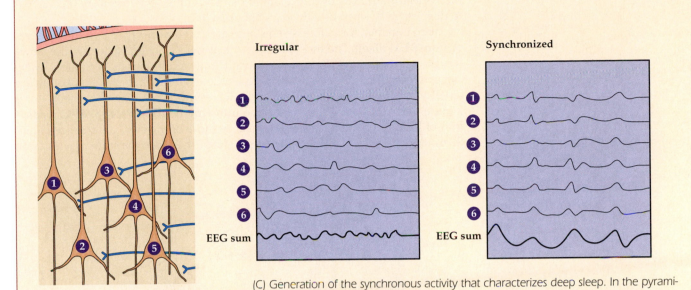

(C) Generation of the synchronous activity that characterizes deep sleep. In the pyramidal cell layer below the EEG electrode, each neuron receives thousands of synaptic inputs. If the inputs are irregular or out of phase, their algebraic sum will have a small amplitude, as occurs in the waking state. If, on the other hand, the neurons are activated at approximately the same time, then the EEG waves will tend to be in phase and the amplitude will be much greater, as occurs in the delta waves that characterize stage IV sleep. (Adapted from Bear et al., 2001.)

activity remains even after the thalamus has been inactivated). At about the same time, Horace Magoun and Giuseppe Moruzzi showed that the reticular activating system in the brainstem is also important in modulating EEG activity. For example, activation of the reticular formation changes the cortical alpha rhythm to beta activity, in association with greater behavioral alertness. In the 1960s, Per Andersen and his colleagues in Sweden further advanced these studies by showing that virtually all areas of the cortex participate in these oscillatory rhythms, which reflect a feedback loop between neurons in the thalamus and cortex (see text).

The cortical origin of EEG activity has been clarified by animal studies, which have shown that the source of the current that causes the fluctuating scalp potential is primarily the pyramidal neurons and their synaptic connections in the deeper layers of the cortex (see Figures B and C). (This conclusion was reached by noting the location of electrical field reversal upon passing an electrode vertically through the cortex from surface to white matter.) In general, oscillations come about either because membrane voltage of thalamocortical cells fluctuates spontaneously, or as a result of the reciprocal interaction of excitatory and inhibitory neurons in circuit loops. The oscillations of the EEG are thought to arise from the latter mechanism.

Despite these intriguing observations, the functional significance of these cortical rhythms is not known. The purpose of the brain's remarkable oscillatory activity is a puzzle that has defied electroencephalographers and neurobiologists for more than 70 years.

References

ADRIAN, E. D. AND K. YAMAGIWA (1935) The origin of the Berger rhythm. *Brain* 58: 323–351.

ANDERSEN, P. AND S. A. ANDERSSON (1968) *Physiological Basis of the Alpha Rhythm*. New York: Appleton-Century-Crofts.

CATON, R. (1875) The electrical currents of the brain. *Brit. Med. J.* 2: 278.

DA SILVA, F. H. AND W. S. VAN LEEUWEN (1977) The cortical source of the alpha rhythm. *Neurosci. Letters* 6: 237–241.

DEMPSEY, E. W. AND R. S. MORRISON (1943) The electrical activity of a thalamocortical relay system. *Amer. J. Physiol.* 138: 283–296.

NIEDERMEYER, E. AND F. L. DA SILVA (1993) *Electroencephalography: Basic Principles, Clinical Applications, and Related Fields*. Baltimore: Williams & Wilkins.

NUÑEZ, P. L. (1981) *Electric Fields of the Brain: The Neurophysics of EEG*. New York: Oxford University Press.

EEG activity increases further and the frequency continues to fall. In the deepest level of sleep, **stage IV sleep**, also known as **slow-wave sleep**, the predominant EEG activity consists of very low frequency (0.5–2 Hz), high-amplitude fluctuations called **delta waves**, the characteristic slow waves for which this phase of sleep is named. (Note that these can also be thought of as reflecting synchronized electrical activity of cortical neurons.) The entire sequence from drowsiness to deep stage IV sleep usually takes about an hour.

Taken together, these four sleep stages are called **non-rapid eye movement**, or **non-REM sleep**. The most prominent feature of non-REM sleep is the slow-wave stage (stage IV). It is more difficult to awaken people from slow-wave sleep, which is therefore considered to be the deepest stage of sleep. Following a period of slow-wave sleep, however, EEG recordings show that subjects enter a quite different state called **rapid eye movement (REM) sleep**. EEG recordings of REM sleep are remarkably similar to those of the awake state (see Figure 28.6). After about 10 minutes in REM sleep, the brain typically cycles back

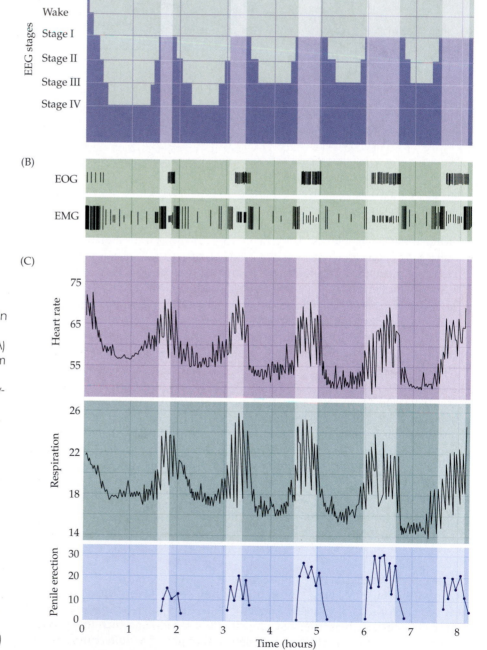

Figure 28.7 *Physiological changes in a volunteer during the various sleep states in a typical 8-hour sleep period. (A) The duration of REM sleep increases from 10 minutes in the first cycle to up to 50 minutes in the final cycle; note that slow-wave (stage IV) sleep is attained only in the first two cycles. (B) The electrooculo-gram (EOG, above) and movement of neck muscles, measured using an elec-tromyogram (EMG, below). Other than the few slow eye movements approach-ing stage I sleep, all other eye move-ments evident in the EOG occur in REM sleep. The greatest EMG activity occurs during the onset of sleep and just prior to awakening. (C) The heart rate (beats per minute) and respiration (breaths per minute) slow in non-REM sleep, but increase almost to the waking levels in REM sleep. Finally, penile erection (strain gauge units) occurs only during REM sleep. (After Foulkes and Schmidt, 1983.)*

through the non-REM sleep stages. Slow-wave sleep usually occurs again in the second round of this continued cycling, but there are generally only two slow-wave stages during any given night (Figure 28.7). On average, four additional periods of REM sleep occur, each having a longer duration.

In summary, the typical 8 hours of sleep experienced each night actually comprise several cycles that alternate between non-REM and REM sleep, and the brain is quite active during much of this supposedly dormant, restful time. The amount of daily REM sleep decreases from about 8 hours at birth to 2 hours at 20 years to only about 45 minutes at 70 years of age. The reasons for this change over the human lifespan are not known.

Physiological Changes in Sleep States

A variety of additional physiological changes take place during the different stages of sleep (Figure 28.7). The non-REM sleep stages are characterized by slow, rolling eye movements and by decreases in muscle tone, body movements, heart rate, breathing, blood pressure, metabolic rate, and temperature. All these parameters reach their lowest values during stage IV sleep. Periods of REM sleep, in contrast, are accompanied by increases in blood pressure, heart rate, and metabolism to levels almost as high as those found in the awake state. REM sleep, as the name implies, is also characterized by rapid, ballistic eye movements, pupillary constriction, paralysis of many large muscle groups (although obviously not the diaphragm and other muscles used for breathing), and the twitching of the smaller muscles in the fingers, toes, and the middle ear. Spontaneous penile erection also occurs during REM sleep, a fact that is clinically important in determining whether a complaint of impotence has a physiological or psychological basis. REM sleep has been observed in all mammals and in at least some birds; certain reptiles also have periods of increased brain activity during sleep that are arguably homologous to REM sleep in mammals.

Despite the similarity of EEG recordings obtained in REM sleep and in wakefulness, the two brain states are clearly not equivalent. For one thing, any awareness one has during REM sleep is **dreaming**, a peculiar state similar to hallucinations in the sense that the experience of dreams is not related to any corresponding sensory stimuli arising from the real world. Since most muscles are inactive during REM sleep, the motor responses to dreams are relatively minor. (Sleepwalking, which is most common in children from ages 4–12, and sleeptalking actually occur during non-REM sleep and are not usually accompanied or motivated by dreams.) The relative physical paralysis during REM sleep arises from increased activity in GABAergic neurons in the pontine reticular formation that project to inhibitory neurons that synapse in turn with lower motor neurons in the spinal cord (Figure 28.8). Increased activity of descending inhibitory projections from the pons to the dorsal column nuclei also causes a diminished response to somatic sensory stimuli.

Taken together, these observations have led to the aphorism that non-REM sleep is characterized by an inactive brain in an active body, whereas REM sleep is characterized by an active brain in an inactive body. Clearly, however, several sensory and motor systems are sequentially activated and inactivated during the different stages of sleep.

The Possible Functions of REM Sleep and Dreaming

Despite the wealth of descriptive information about the stages of sleep and an intense research effort over the last 50 years, the functional purposes of the various sleep states remain poorly understood. Whereas most sleep researchers

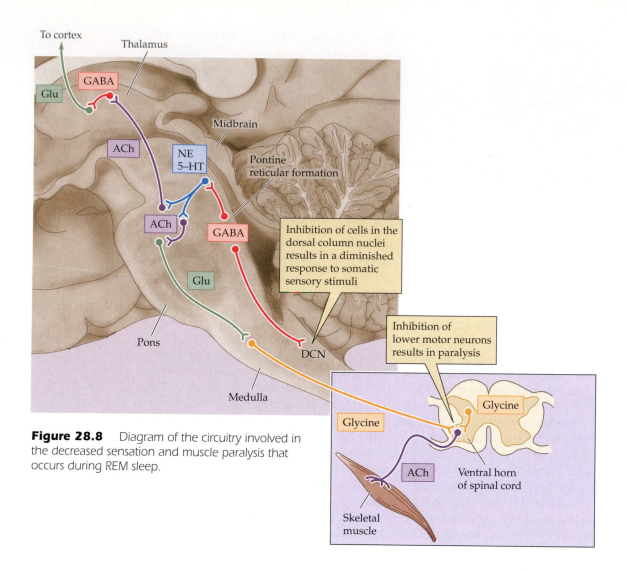

Figure 28.8 Diagram of the circuitry involved in the decreased sensation and muscle paralysis that occurs during REM sleep.

accept the idea that the purpose of non-REM sleep is at least in part restorative, the function of REM sleep remains a matter of considerable controversy.

A possible clue about the purposes of REM sleep is the prevalence of dreams during these epochs of the sleep cycle. The time of occurrence of dreams during sleep has been determined by waking volunteers during either non-REM or REM sleep and asking them if they were dreaming. Subjects awakened from REM sleep usually recall elaborate, vivid, and often emotional dreams; subjects awakened during non-REM sleep report fewer dreams, and when dreams do occur, they are more conceptual, less vivid, and less emotion-laden. Dreaming can also occur during light non-REM sleep, near the onset of sleep and before awakening, and is thus not entirely limited to REM sleep.

Dreams have been studied in a variety of ways, perhaps most notably within the psychoanalytic framework aimed at revealing unconscious thought processes considered to be at the root of neuroses. Sigmund Freud's *The Interpretation of Dreams*, published in 1900, speaks eloquently to the complex relationship between conscious and unconscious mentation. Specifically, Freud thought that during dreaming the conscious "ego" relaxes its hold on the "id," or subconscious. These ideas have been out of fashion in recent decades, but to give Freud his due, at the time he made these speculations little was known about neurobiology of the brain in general and sleep in particular. In fact, some recent evidence supports Freud's idea that dreams often reflect events and conflicts of the day (the "day residue" in his terminology) and may play a role in memory. A number of investigators have suggested that dreams help consolidate learned tasks, perhaps by further strengthening synaptic changes associated with recent experiences. The more general hypothesis that sleep is important in consolidating memories has been supported by studies of remembered spatial location in rodents, and by experiments in humans that show a sleep-dependent improvement in learning.

Another largely psychological rationalization of dreaming is that this process "releases" behaviors less commonly entertained in the waking state and rarely expressed in overt behavior (e.g., frank aggression). Whatever the merits of this idea, studies have found that about 60% of dream content is associated with sadness, apprehension, or anger; 20% with happiness or excitement; and (somewhat surprisingly) only 10% with sexual feelings or acts.

A quite different idea is that dreaming has evolved to dispose of unwanted memories that accumulate during the day. Francis Crick, for example, suggested that dreams might reflect a mechanism for expunging "parasitic" modes of thought that would otherwise become overly intrusive, as occurs in compulsive thought disorders. Finally, some experts such as Allan Hobson have taken the more skeptical view that dream content may be "as much dross as gold, as much cognitive trash as treasure, as much informational noise as a signal of something."

Adding to all this uncertainty about the purposes of REM sleep and dreaming is the fact that depriving human subjects of REM sleep for as much as two weeks has little or no obvious effect on their behavior. The apparent innocuousness of REM sleep deprivation contrasts markedly with the devastating effects of total sleep deprivation mentioned earlier. The implication of these findings is that we can get along without REM sleep, but we need non-REM sleep in order to survive. In summary, the questions of why we have REM sleep and why we dream basically remain unanswered. Nevertheless, most people, including most sleep researchers, give some credence to the significance of dreams as an important aspect of sleep physiology.

Neural Circuits Governing Sleep

From the descriptions of the various physiological states that occur during sleep, it is clear that periodic changes in the balance of excitation and inhibition must occur in many neural circuits. What follows is a brief overview of these still incompletely understood circuits and the interactions among them that govern sleeping and wakefulness.

In 1949, Horace Magoun and Giuseppe Moruzzi provided one of the first clues about the circuits involved in the sleep–wake cycle. They found that electrically stimulating a group of cholinergic neurons near the junction of the pons and midbrain causes a state of wakefulness and arousal. This region of the brainstem was given the name **reticular activating system** (Figure 28.9A; see also Box 17A). Their work implied that wakefulness requires special activating circuitry—that is, wakefulness is not just the result of adequate sensory experience. About the same time, the Swiss physiologist Walter Hess found that stimulating the thalamus in an awake cat with low-frequency pulses produced a slow-wave sleep (Figure 28.9B). These key experiments showed that sleep entails a patterned interaction between the thalamus and cortex.

Some further clues were provided by work on the circuitry underlying REM sleep. The saccade-like eye movements that define REM sleep are now known to arise because, in the absence of external visual stimuli, endogenously generated

Figure 28.9 *Activation of specific neural circuits triggers sleep and wakefulness. (A) Electrical stimulation of the cholinergic neurons near the junction of pons and midbrain (the reticular activating system) causes a sleeping cat to awaken. (B) Slow electrical stimulation of the thalamus causes an awake cat to fall asleep. Graphs show EEG recordings before and during stimulation.*

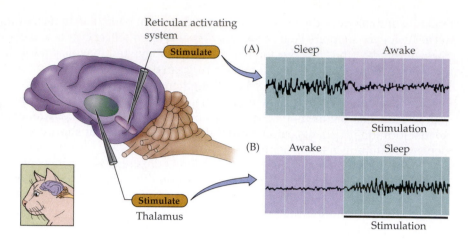

signals from the **pontine reticular formation** are transmitted to the motor region of the superior colliculus. As described in Chapter 20, collicular neurons project to the **paramedialpontine reticular formation (PPRF)** and the **rostral interstitial nucleus**, which coordinates timing and direction of eye movements. REM sleep is also characterized by EEG waves that originate in the pontine reticular formation and propagate through the lateral geniculate nucleus of the thalamus to the occipital cortex. These **pontine-geniculo-occipital**, or **PGO waves** provide a useful marker for the beginning of REM sleep; they also indicate yet another neural network by which brainstem nuclei can activate the cortex.

A further advance has come from fMRI and PET studies (see Box 1A) that have compared human brain activity in the awake state and in REM sleep, as

---

BOX 28D Consciousness

As the text explains, the mechanisms of sleep and wakefulness determine mental status at any moment on a continuum that normally ranges from stage IV sleep to high alert. There is, however, another way that "wakefulness" has been considered, namely from the perspective of *consciousness* in the sense of self awareness. Although the brainstem circuits and projections supporting consciousness are beginning to be understood, these neurological aspects of consciousness are—not surprisingly—insufficient to satisfy philosophers, theologians, and neuroscientists interested in the broader issues that the phenomenon of consciousness raises.

The common concern of these diverse groups is the more general basis of self-awareness, including whether other animals have this mental property, and whether machines could ever be self-aware in the way humans are. With respect to the first of these issues, despite a longstanding debate about consciousness in other animals, it would be unwise to assert that humans are alone in possessing this obviously useful biological attribute. However, from a purely logical vantage it is impossible, strictly speaking, to know whether any being other than one's own self is conscious; as philosophers have long pointed out, we must inevitably take the consciousness of others on faith (or on the basis of common sense).

Nonetheless, it is reasonable to assume that animals with brains structured much like ours (other primates and, to a considerable degree, mammals generally) have in some measure the same ability to be self-aware as we do. The ability to reflect on the past and plan for the future that is made possible by self-awareness is surely an advantage that evolution would have in some degree inculcated in at least the similar brains of higher primates. At what phylogenetic level this assumption about self-awareness falls below the definition of consciousness as we know it is unclear. But a good supposition would be that consciousness is present in animals in proportion to the complexity of their brains and behaviors—particularly those behaviors that are sophisticated enough to benefit from reflecting on past outcomes and future eventualities.

The question of whether machines can ever be conscious is a much more

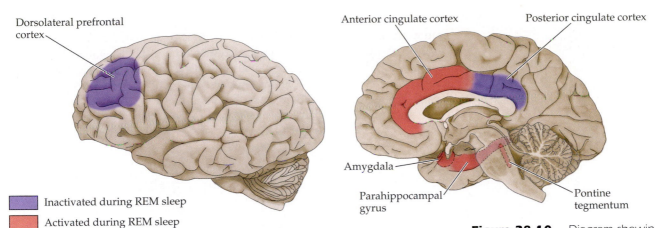

Dorsolateral prefrontal cortex

Anterior cingulate cortex

Posterior cingulate cortex

Amygdala

Parahippocampal gyrus

Pontine tegmentum

Inactivated during REM sleep

Activated during REM sleep

Figure 28.10 Diagram showing cortical regions whose activity is increased or decreased during REM sleep. (After Hobson et al., 1989.)

well as studies of the phenomenon of consciousness more generally (Box 28D). Activity in the amygdala, parahippocampus, pontine tegmentum, and anterior cingulate cortex all increase in REM sleep, whereas activity in the dorsolateral prefrontal and posterior cingulate cortices decreases REM sleep (Figure 28.10). The increase in limbic system activity, coupled with a marked decrease in the influence of the frontal cortex during REM sleep, presumably explains some characteristics of dreams (e.g., their emotionality and their often inappropriate social content; see Chapter 26 for the normal role of the frontal cortex in determining behavior that is appropriate to circumstances in the waking state).

On the basis of these discoveries and other studies using unit recording in experimental animals, it is now generally agreed that a key component of the

contentious issue, but is also subject to common sense informed by some knowledge of how brains work. If one rejects dualism (the Cartesian proposition that consciousness, or "mind," is an entity beyond the ken of physics, chemistry, and biology, and therefore not subject to the rules of these disciplines), it follows that a structure could be built that either mimicked our own consciousness (by being effectively isomorphic with brains), or achieved consciousness using physically different elements (e.g., computer elements). Such a structure could conceivably be sufficiently similar to biological brains to allow self-awareness.

A great deal of literature on the subject notwithstanding, these fascinating questions about consciousness are not readily subject to neurobiological investigation. Although a number of contemporary scientists have advocated the idea that neurobiology will soon reveal the "basis" of consciousness, such revelations are not likely. A more plausible scenario is that as information grows about the nature of other animals, about computers, and indeed about the brain, the question "What is consciousness?" may simply fade from center stage in much the same way that the question "What is life?" (which stirred up a similar debate early in the twentieth century) was asked less and less frequently as biologists and others recognized it as an ill-posed problem that admitted no definite answer. Many aspects of living organisms can already be created in the laboratory, and there seems no reason in principle why artificial neural networks capable of self-awareness should not also be possible someday.

References

CHURCHLAND, P. M. AND P. S. CHURCHLAND (1990) Could a machine think? *Sci. Am.* 262 (Jan.): 32–37.

CRICK, F. (1995) *The Astonishing Hypothesis: The Scientific Search for the Soul.* New York: Touchstone.

CRICK, F. AND C. KOCH (1998) Consciousness and neuroscience. *Cerebral Cortex* 8: 97–107.

PENROSE, R. (1996) *Shadows of the Mind: A Search for the Missing Science of Consciousness.* Oxford: Oxford University Press.

SEARLE, J. R. (2000) Consciousness. *Annu. Rev. Neurosci.* 23: 557–578.

TONONI, G. AND G. EDELMAN (1998) Consciousness and complexity. *Science* 282: 1846–1851.

reticular activating system is a group of **cholinergic nuclei** near the pons–midbrain junction that project to thalamocortical neurons (Figure 28.11). The relevant neurons in the nuclei are characterized by high discharge rates during both waking and REM sleep, and by quiescence during non-REM sleep. When stimulated, these nuclei cause "desynchronization" of the electroencephalogram (that is, a shift of EEG activity from high-amplitude, synchronized waves to lower amplitude, higher frequency, desynchronized ones; see Box 28C). These

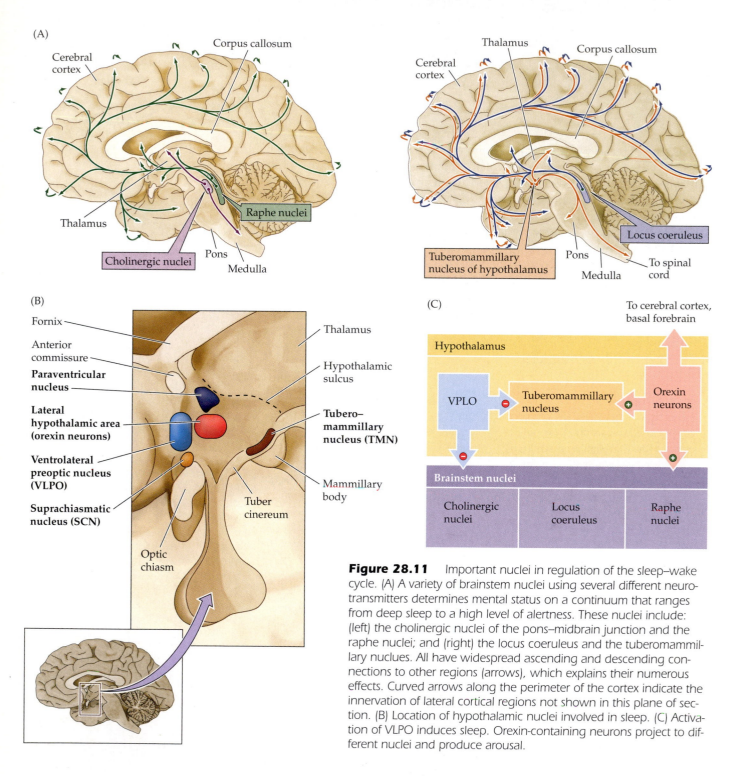

Figure 28.11 Important nuclei in regulation of the sleep–wake cycle. (A) A variety of brainstem nuclei using several different neurotransmitters determines mental status on a continuum that ranges from deep sleep to a high level of alertness. These nuclei include: (left) the cholinergic nuclei of the pons–midbrain junction and the raphe nuclei; and (right) the locus coeruleus and the tuberomammillary nuclues. All have widespread ascending and descending connections to other regions (arrows), which explains their numerous effects. Curved arrows along the perimeter of the cortex indicate the innervation of lateral cortical regions not shown in this plane of section. (B) Location of hypothalamic nuclei involved in sleep. (C) Activation of VLPO induces sleep. Orexin-containing neurons project to different nuclei and produce arousal.

features imply that activity of cholinergic neurons in the reticular activating system is a primary cause of wakefulness and REM sleep, and that their relative inactivity is important for producing non-REM sleep.

Activity of these cholinergic neurons is not, however, the only neuronal basis of wakefulness; also involved are the **noradrenergic neurons** of the locus coeruleus; the **serotonergic neurons** of the raphe nuclei; and the **histamine-containing neurons** in the tuberomammillary nucleus (TMN) of the hypothalamus (Figure 28.11). The activation of these cholinergic, monoaminergic, and histamine-containing networks together produces the awake state. The locus coeruleus and raphe nuclei are modulated by the TMN neurons located near the tuberal region that synthesize the peptide **orexin** (also called **hypocretin**). Orexin promotes waking and may have useful applications in jobs where operators need to stay alert. On the other hand, antihistamines inhibit the histamine-containing TMN network, thus tending to make people drowsy.

The three circuits responsible for the awake state are periodically inhibited by neurons in the **ventrolateral preoptic nucleus (VLPO)** of the hypothalamus (see Figure 28.11). Thus activation of VLPO neurons contributes to the onset of sleep, and lesions of VLPO neurons tend to produce insomnia. As if all this were not complicated enough, recent work suggests that adenosine neurotransmission in the basal forebrain is also involved in the regulation of sleep.

These complex interactions and effects are summarized in Table 28.1. In brief, both monoaminergic and cholinergic systems are active during the waking state and suppressed during REM sleep. Thus, decreased activity of the monoaminergic and cholinergic systems leads to the onset of non-REM sleep. In REM sleep, monoaminergic and serotonin neurotransmitter levels markedly decrease, while cholinergic levels increase to approximately the levels found in the awake state.

Given that so many systems and transmitters are involved in the different phases of sleep, it is not surprising that a wide variety of drugs can influence the sleep cycle (Box 28E).

TABLE 28.1 **Summary of the Cellular Mechanisms that Govern Sleep and Wakefulness**

Brainstem nuclei responsible	Neurotransmitter involved	Activity state of the relevant brainstem neurons
Wakefulness		
Cholinergic nuclei of pons-midbrain junction	Acetylcholine	Active
Locus coeruleus	Norepinephrine	Active
Raphe nuclei	Serotonin	Active
Tuberomammillary nuclei	Orexin	Active
Non-REM sleep		
Cholinergic nuclei of pons-midbrain junction	Acetylcholine	Decreased
Locus coeruleus	Norepinephrine	Decreased
Raphe nuclei	Serotonin	Decreased
REM sleep on		
Cholinergic nuclei of pons-midbrain junction	Acetylcholine	Active (PGO waves)
Raphe nuclei	Serotonin	Inactive
REM sleep off		
Locus coeruleus	Norepinephrine	Active

BOX 28E Drugs and Sleep

Many drugs can affect sleep patterns, largely because so many neurotransmitters (e.g., acetylcholine, serotonin, norepinepherine, and histamine) are involved in regulating the various states of sleep (see Table 28.1). A simple but useful way of looking at these effects is that in the waking state, the aminergic system is especially active (see Figure 28.14). During non-REM sleep, aminergic and cholinergic input both decrease, but aminergic activity decreases more, so that cholinergic inputs become dominant. Thus there are two major ways drugs alter the sleep pattern: by changing the relative activity of the inputs in any of the three states, or by changing when the different sleep states will commence. For example, insomnia will ensue if, during the waking state, the aminergic input is increased relative to the cholinergic input. In contrast, hypersomnia occurs when there is increased cholinergic activity relative to the aminergic input.

Because of the large number of people who suffer with sleep disorders, numerous drugs are available to treat these problems. One class of commonly used drugs is the benzodiazepines. As shown in the figure, these drugs decrease the time to onset of the deeper stages of sleep. Stimulant drugs that prevent sleep are also commonly used, especially caffeine, which is an adenosine receptor antagonist (adenosine induces sleep).

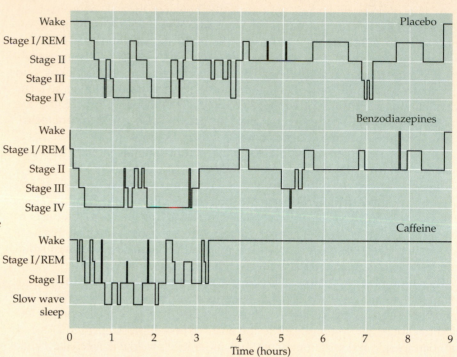

Compared to a placebo, benzodiazepines hasten the onset and depth of sleep, whereas caffeine has the opposite effect.

Thalamocortical Interactions

The effects of the neuronal activity (or its absence) in these brainstem nuclei are evidently achieved by modulating the rhythmicity of interactions between the thalamus and the cortex. Thus the activity of several ascending systems from the brainstem decreases both the rhythmic bursting of the thalamocortical neurons and the related synchronized activity of cortical neurons (hence the diminution and ultimate disappearance of high-voltage, low-frequency slow waves during waking and REM sleep; see Box 28C).

To appreciate how different sleep states reflect modulation of thalamocortical activity, it is useful to consider the electrophysiological responses of the relevant neurons. Thalamocortical neurons receive ascending projections from the locus coeruleus (noradregeneric), raphe nuclei (serotonergic), reticular activating system (cholinergic), TMN (histaminergic) and, as their name implies, project to cortical pyramidal cells. The primary characteristic of thalamocortical neurons is that they can be in one of two stable electrophysiological states: an intrinsic **oscillatory** or **bursting state**, and a **tonically active state** that is generated when the neurons are depolarized as occurs when the reticular activating sys-

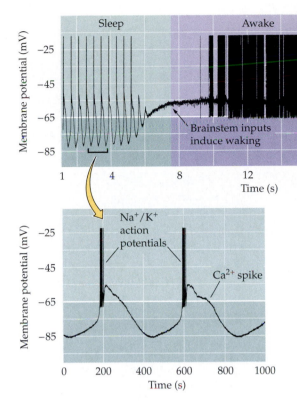

Figure 28.12 Recordings from a thalamocortical neuron, showing the oscillatory mode corresponding to a sleep state, and the tonically active mode corresponding to an awake state. An expanded view of oscillatory phase is shown below. Bursts of action potentials are evoked only when the thalamocortical neuron is hyperpolarized sufficiently to activate low-threshold calcium channels. These bursts account for the spindle activity seen in EEG recordings in stage II sleep (see Figure 28.6 and 28.13). Depolarizing the cell either by injecting current or by stimulating the reticular activating system transforms this oscillatory activity into a tonically active mode. (After McCormick and Pape, 1990.)

tem generates wakefulness; (Figure 28.12). In the tonically firing state, thalamocortical neurons transmit information to the cortex that is correlated with the spike trains encoding peripheral stimuli. In contrast, when thalamocortical neurons are in the oscillatory/bursting mode, the neurons in the thalamus become synchronized with those in the cortex, essentially "disconnecting" the cortex from the outside world. The disconnection is maximal during slow-wave sleep, when EEG recordings show the lowest frequency and the highest amplitude.

The oscillatory state of thalamocortical neurons can be transformed into the tonically active state by activity in the cholinergic or monoaminergic projections from the brainstem nuclei (Figure 28.13). Moreover, the oscillatory state is stabilized by hyperpolarizing the relevant thalamic cells. Such hyperpolarization can occur as a consequence of stimulation by GABAergic neurons in the thalamic reticular nucleus. These neurons receive ascending information from the brainstem and descending projections from cortical neurons, and they contact the thalamocortical neurons. When neurons in the reticular nucleus undergo a burst of activity, they cause thalamocortical neurons to generate short bursts of action potentials, which in turn generate spindle activity in cortical EEG recordings (indicating a lighter sleep state; see Figures 28.5 and 28.13).

In sum, the control of sleep and wakefulness depends on brainstem modulation of the thalamus and cortex. It is this modulation of the thalamocortical loop that generates the EEG signature of mental function along the continuum of deep sleep to high alert. The major components of the brainstem modulatory system are the cholinergic nuclei of the pons–midbrain junction; the noradrenergic cells of the locus coeruleus in the pons; the serotonergic raphe nuclei; and GABAergic neurons in the VLPO. All of these nuclei can exert both direct and indirect effects on the overall cortical activity that determines sleep and wakefulness. The relationship among the various sleep–wake states is summarized in the scheme shown in Figure 28.14.

(A)

(B)

(C)

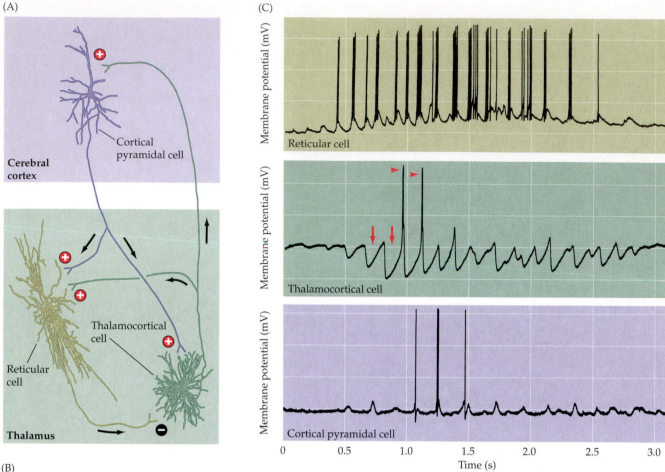

Figure 28.13 *Thalamocortical feedback loop and the generation of sleep spindles. (A) Diagram showing excitatory (+) and inhibitory (–) connections between thalamocortical cells, pyramidal cells in the cortex, and thalamic reticular cells, which provide the basis for sleep spindle generation. Inputs into thalamocortical and thalamic reticular cells are not shown. (B) EEG recordings illustrating sleep spindles (the bottom trace is filtered to accentuate the spindles). (C) The responses from individual thalamic reticular cells, thalamocortical cells, and cortical cells during the generation of the middle spindle (boxed in panel B). The bursting behavior of the thalamocortical neurons elicits spikes in cortical cells, which is then evident as spindles in EEG recordings. (After Steriade et al., 1993.)*

Sleep Disorders

As noted earlier, an estimated 40% of the U.S. population experiences some kind of sleep disorder at one time or another. Sleep problems, which range from simply annoying to life-threatening, occur more frequently with advancing age and are more prevalent in women than in men. The most significant problems are insomnia, sleep apnea, "restless legs" syndrome, and narcolepsy.

Insomnia is the inability to sleep for a sufficient length of time (or deeply enough) to produce a subjective sense of refreshment. This all-too-common problem has many causes. Short-term insomnia can arise from stress, jet lag, or simply drinking too much coffee. Another frequent cause is altered circadian rhythms associated with working night shifts. These problems can usually be prevented by improving sleep habits, avoiding stimulants like caffeine, and in

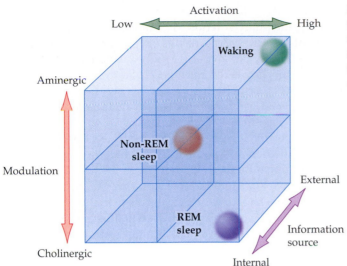

Figure 28.14   Summary scheme of sleep–wake states. In the waking state, activation is high, modulation is aminergic, and the information source is external. In REM sleep, activation is also high, the modulation is cholinergic, and the information source is internal. The other states can likewise be remembered in terms of this general diagram. (After Hobson, 1989.)

some cases taking sleep-promoting medications. More serious insomnia is associated with psychiatric disorders such as depression (see Chapter 29) that presumably affect the balance between the cholinergic, adrenergic, and serotinergic systems that control the onset and duration of the sleep cycles. Long-term insomnia is a particular problem in the elderly for reasons that are not well understood. Whatever the underlying physiology, the problem is aggravated by the fact that aged individuals are more prone to depression and often take medications that affect the relevant neurotransmitter systems.

Sleep apnea refers to a pattern of interrupted breathing during sleep that affects many people, most often obese middle-aged males. A person suffering from sleep apnea may wake up dozens or more times during the night, with the result that they experience little or no slow-wave sleep and spend less time in REM sleep (Figure 28.15). These individuals are continually tired and prone to depression as a result, which can compound the problem. In high-risk individuals, sleep apnea can lead to death from respiratory arrest. The underlying problem is that the airway in susceptible individuals collapses during breathing, blocking airflow. In normal sleep, breathing slows and muscle tone decreases throughout the body, including the tone of the pharynx. If the output of the brainstem circuitry regulating commands to the chest wall or to pharyngeal

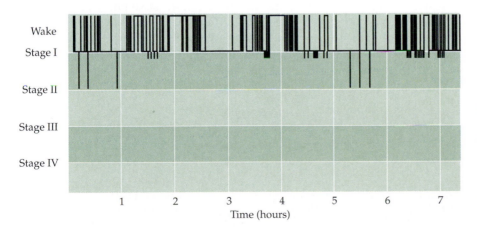

Figure 28.15   Sleep pattern of a patient with obstructive sleep apnea. In this condition, patients awake frequently and never descend into stage III or IV sleep. The brief descents below stage I in the record represent short periods of REM sleep. (After Carskadon and Dement, 1989, based on data from G. Niño-Murcia.)

muscles is decreased sufficiently, or if the airway is already compressed to some degree because of obesity, the pharynx tends to collapse as the muscles relax during the normal cycle of breathing. As a result, oxygen levels decrease and CO_2 levels rise. The rise in CO_2 causes an abrupt reflex inspiration, the force of which tends to wake up the individual. The most widely used remedy at present is a positive pressure mask worn at night that breaks this cycle.

A third sleep disorder is **restless legs syndrome**, a problem that affects many (again, mostly elderly) people. The characteristic of this syndrome is unpleasant crawling, prickling, or tingling sensations in one or both legs and feet, and an urge to move them about to obtain relief. These sensations occur when the person sits or lies down for prolonged periods. The result is constant leg movement during the day and fragmented sleep at night. The underlying cause of this problem is not understood, although it is more frequent in patients with chronic diseases. In mild cases, a hot bath, massaging the legs, or eliminating caffeine may alleviate the problem. In more severe cases, drugs such as benzodiazepines may help.

The best-understood sleep disorder is **narcolepsy**, a chronic disorder that affects about 250,000 people (mostly men) in the United States. It is the second leading cause of abnormal daytime drowsiness, ranking just behind sleep apnea. Individuals with narcolepsy have frequent "REM sleep attacks" during the day, in which they enter REM sleep from wakefulness without going through non-REM sleep. These attacks can last from 30 seconds to 30 minutes or more. The onset of sleep in such individuals can be abrupt, with potentially disastrous consequences; this latter phenomenon is called *cataplexy*, referring to a temporary loss of muscle control. Insights into the causes of narcolepsy have come from studies of dogs suffering from a genetic disorder similar to the human condition. In these animals, narcolepsy is caused by a mutation of the orexin-2 receptor gene (*Orx2*). (As already mentioned, orexins are neuropeptides homologous to secretin and are found exclusively in cells in the tuberal region of the hypothalamus, which project to target nuclei responsible for wakefulness; see Figure 28.11.) Evidence from both dogs and mice suggests that the *Orx2* mutation causes hyperexcitability of the neurons that generate REM sleep, and/or impairment of the circuits that inhibit REM sleep. Clinically, narcoleptics are treated with stimulants such as methylphenidate (Ritalin™) or amphetamines to increase their overall level of arousal.

Summary

All animals exhibit a restorative cycle of rest following activity, but only mammals divide the period of rest into distinct phases of non-REM and REM sleep. Why humans (and many other animals) need a restorative phase of suspended consciousness accompanied by decreased metabolism and lowered body temperature is not known. Even more mysterious is why the human brain is periodically active during sleep at levels not appreciably different from the waking state (that is, the neural activity during REM sleep). Despite the electroencephalographic similarities, the psychological states of wakefulness and REM sleep are obviously different. The highly organized sequence of human sleep states is actively generated by nuclei in the brainstem, most importantly the cholinergic nuclei of the pons–midbrain junction, the noradrenergic cells of the locus coeruleus, and the serotonergic neurons of the raphe nuclei. This control by the brainstem of the cortical states is mediated by modulation of activity in a thalamocortical loop. This complex physiological interplay involving brainstem, thalamus and cortex controls the degree of mental alertness on a continuum from deep sleep to waking attentiveness. These systems are in turn influenced by a circadian clock located in the suprachiasmatic nucleus and VLPO of the hypothalamus. The clock adjusts the overall periods of sleep and wakefulness to appropriate durations during the 24-hour cycle of light and darkness that is fundamental to life on Earth.

Additional Reading

Reviews

COLWELL, C. S. AND S. MICHEL (2003) Sleep and circadian rhythms: Do sleep centers talk back to the clock? *Nature Neurosci.* 10: 1005–1006.

DAVIDSON, A. J. AND M. MENAKER (2003) Birds of a feather clock together—sometimes: Social synchronization of circadian rhythms. *Curr. Opin. Neurobiol.* 13: 765–769.

HOBSON, J. A. (1990) Sleep and dreaming. *J. Neurosci.* 10: 371–382.

HOBSON, J. A., R. STRICKGOLD AND E. F. PACE-SCHOTT (1998) The neuropsychology of REM sleep and dreaming. *NeuroReport* 9: R1–R14.

LU J., M. A. GRECO, P. SHIROMANI AND C. B. SAPER (2000) Effect of lesions of the ventro-lateral preoptic nucleus on NREM and REM sleep. *J. Neurosci.* 20: 3830–3842.

MCCARLEY, R. W. (1995) Sleep, dreams and states of consciousness. In *Neuroscience in Medicine*, P. M. Conn (ed.). Philadelphia: J. B. Lippincott, pp. 535–554.

MCCORMICK, D. A. (1989) Cholinergic and noradrenergic modulation of thalamocortical processing. *Trends Neurosci.* 12: 215–220.

MCCORMICK, D. A. (1992) Neurotransmitter actions in the thalamus and cerebral cortex. *J. Clin. Neurophysiol.* 9: 212–223.

POSNER, M. I. AND S. DEHAENE (1994) Attentional networks. *Trends Neurosci.* 17: 75–79.

PROVENCIO, I. AND 5 OTHERS (2000) A novel human opsin in the inner retina. *J. Neurosci.* 20: 600–605.

SAPER, C. B. AND F. PLUM (1985) Disorders of consciousness. In *Handbook of Clinical Neurology*, Volume 1 (45): *Clinical Neuropsychology*, J. A. M. Frederiks (ed.). Amsterdam: Elsevier Science Publishers, pp. 107–128.

SIEGEL, J. M. (2000) Brainstem mechanisms generating REM sleep. In *Principles and Practice of Sleep Medicine*, 3rd Ed. M. H. Kryger, T. Roth and W. C. Dement (eds.). New York: W. B. Saunders.

STERIADE, M. (1992) Basic mechanisms of sleep generation. *Neurology* 42: 9–18.

STERIADE, M. (1999) Coherent oscillations and short-term plasticity in corticothalamic networks. *Trends Neurosci.* 22: 337–345,

STERIADE, M., D. A. MCCORMICK AND T. J. SEJNOWSKI (1993) Thalamocortical oscillations in the sleeping and aroused brain. *Science* 262: 679–685.

WILLIE, J. T. AND 13 OTHERS. (2003) Distinct narcolepsy syndromes in *orexin receptor-2* and *orexin* null mice: Molecular genetic dissection of non-REM and REM sleep regulatory processes. *Neuron* 38: 715–730

WILSON, M. A. (2002) Hippocampal memory formation, plasticity, and the role of sleep. *Neurobiol. Learn. Mem.* 3: 565–569.

Important Original Papers

ASCHOFF, J. (1965) Circadian rhythms in man. *Science* 148: 1427–1432.

ASERINSKY, E. AND N. KLEITMAN (1953) Regularly occurring periods of eye motility, and concomitant phenomena, during sleep. *Science* 118: 273–274.

CZEISLER, C. A. AND 11 OTHERS (1999) Stability, precision, and near-24-hour period of the human circadian pacemaker. *Science* 284: 2177–2181.

DEMENT, W. C. AND N. KLEITMAN (1957) Cyclic variation in EEG during sleep and their relation to eye movements, body motility, and dreaming. *Electroenceph. Clin. Neurophysiol.* 9: 673–690.

MORUZZI, G. AND H. W. MAGOUN (1949). Brain stem reticular formation and activation of the EEG. *Electroenceph. Clin. Neurophysiol.* 1: 455–473.

RIBEIRO, S. AND 7 OTHERS (2004) Long-lasting, novelty-induced neuronal reverberation during slow-wave sleep in multiple forebrain areas. *PLoS Biology* January 20: E24.

ROFFWARG, H. P., J. N. MUZIO AND W. C. DEMENT (1966) Ontogenetic development of the human sleep–dream cycle. *Science* 152: 604–619.

VON SCHANTZ, M. AND S. N. ARCHER (2003) Clocks, genes, and sleep. *J. Roy. Soc. Med.* 96: 486–489.

Books

FOULKES, D. (1999) *Children's Dreaming and the Development of Consciousness*. Cambridge, MA: Harvard University Press.

HOBSON, J. A. (2002) *Dreaming*. New York: Oxford University Press.

HOBSON, J. A. (1989) *Sleep*. New York: Scientific American Library.

LAVIE, P. (1996). *The Enchanted World of Sleep*. (Transl. by A. Barris.) New Haven: Yale University Press.

Chapter 29

Emotions

Overview

The subjective feelings and associated physiological states known as emotions are essential features of normal human experience. Moreover, some of the most devastating psychiatric problems involve emotional (affective) disorders. Although everyday emotions are as varied as happiness, surprise, anger, fear, and sadness, they share some common characteristics. All emotions are expressed through both visceral motor changes and stereotyped somatic motor responses, especially movements of the facial muscles. These responses accompany subjective experiences that are not easily described, but which are much the same in all human cultures. Because emotional expression is closely tied to the visceral motor system, it entails the activity of the central brain structures that govern preganglionic autonomic neurons in the brainstem and spinal cord. Historically, the higher order neural centers that coordinate emotional responses have been grouped under the rubric of the limbic system. More recently, however, several brain regions in addition to the classical limbic system have been shown to play a pivotal role in emotional processing, including the amygdala and several cortical areas in the orbital and medial aspects of the frontal lobe. This broader constellation of cortical and subcortical regions encompasses not only the central components of the visceral motor system but also regions in the forebrain and diencephalon that motivate lower motor neuronal pools concerned with the somatic expression of emotional behavior. Effectively, the concerted action of these diverse brain regions constitutes an emotional motor system. The same forebrain structures that process emotional signals participate in a variety of complex brain functions, including the regulation of goal-directed behavior, rational decision making, the interpretation and expression of social behavior, and even moral judgments. Unfortunately, these same brain systems are also subject to maladaptation when exposed to drugs of abuse or when life experiences interact with genetic factors to promote psychiatric illness.

Physiological Changes Associated with Emotion

The most obvious signs of emotional arousal involve changes in the activity of the visceral motor (autonomic) system, described in Chapter 21. Thus, increases or decreases in heart rate, cutaneous blood flow (blushing or turning pale), piloerection, sweating, and gastrointestinal motility can all accompany various emotions. These responses are brought about by changes in activity in the sympathetic, parasympathetic, and enteric components of the visceral motor system, which govern smooth muscle, cardiac muscle, and glands throughout the body. As discussed in Chapter 21, Walter Cannon argued that intense activity of the sympathetic division of the visceral motor system prepares the animal to fully utilize metabolic and other resources in challenging or threatening situations. Conversely, activity of the parasympathetic division (and the enteric division) promotes a building up of metabolic reserves. Cannon further suggested that the natural opposition of the expenditure and storage of resources is reflected in a parallel opposition of the emotions associated with these different physiological states. As Cannon pointed out, "The desire for food and drink, the relish of taking them, all the pleasures of the table are naught in the presence of anger or great anxiety."

Activation of the visceral motor system, particularly the sympathetic division, was long considered an all-or-nothing process. Once effective stimuli engaged the system, it was argued, a widespread discharge of all of its components ensued. More recent studies have shown that the responses of the autonomic nervous system are actually quite specific, with different patterns of activation characterizing different situations and their associated emotional states. Indeed, emotion-specific expressions produced voluntarily can elicit distinct patterns of autonomic activity. For example, if subjects are given muscle-by-muscle instructions that result in facial expressions recognizable as anger, disgust, fear, happiness, sadness, or surprise without being told which emotion they are simulating, each pattern of facial muscle activity is accompanied by specific and reproducible differences in visceral motor activity (as measured by indices such as heart rate, skin conductance, and skin temperature). Moreover, autonomic responses are strongest when the facial expressions are judged to most closely resemble actual emotional expression and are often accompanied by the subjective experience of that emotion. One interpretation of these findings is that when voluntary facial expressions are produced, signals in the brain engage not only the motor cortex but also some of the circuits that produce emotional states. Perhaps this relationship helps explain how good actors can be so convincing. Nevertheless, we are quite adept at recognizing the difference between a contrived facial expression and the spontaneous smile that accompanies a pleasant emotional state (Box 29A).

This evidence, along with many other observations, indicates that one source of emotion (but certainly not the only source) is sensory drive from muscles and internal organs. This input forms the sensory limb of reflex circuitry that allows rapid physiological changes in response to altered conditions. However, physiological responses can also be elicited by complex and idiosyncratic stimuli mediated by the forebrain. For example, an anticipated tryst with a lover, a suspenseful episode in a novel or film, stirring patriotic or religious music, or dishonest accusations can all lead to autonomic activation and strongly felt emotions. The neural activity evoked by such complex stimuli is relayed from the forebrain to visceral and somatic motor nuclei via the hypothalamus and brainstem reticular formation, the major structures that coordinate the expression of emotional behavior (see the next section).

In summary, emotion and sensorimotor behavior are inextricably linked. As William James put it more than a century ago:

> What kind of an emotion of fear would be left if the feeling neither of quickened heart-beats nor of shallow breathing, neither of trembling lips nor of weakened limbs, neither of goose-flesh nor of visceral stirrings, were present, it is quite impossible for me to think ... I say that for us emotion dissociated from all bodily feeling is inconceivable.
>
> William James, 1893 (*Psychology*: p. 379.)

The Integration of Emotional Behavior

In 1928, Phillip Bard reported the results of a series of experiments that pointed to the hypothalamus as a critical center for coordination of both the visceral and somatic motor components of emotional behavior (see Box 21A). Bard removed

BOX 29A   Facial Expressions: Pyramidal and Extrapyramidal Contributions

In 1862, the French neurologist and physiologist G.-B. Duchenne de Boulogne published a remarkable treatise on facial expressions. His work was the first to systematically examine the contributions of small groups of cranial muscles to the expressions that communicate the richness of human emotion. Duchenne reasoned that "one would be able, like nature herself, to paint the expressive lines of the emotions of the soul on the face of man." In so doing, he sought to understand how the coordinated contractions of groups of muscles express distinct, pan-cultural emotional states. To achieve this goal, he pioneered the use of transcutaneous electrical stimulation (then called "faradization," after the British chemist and physicist Michael Faraday) to activate single muscles and small groups of muscles in the face, dorsal surface of the head, and neck.

Duchenne also documented the faces of his subjects with another technological innovation: photography (Figure A). His seminal contribution was the identification of muscles and muscle groups, such as the obicularis oculi, that are not easily controlled by force of the will, but are mainly "put into play by the sweet emotions of the soul." Duchenne concluded that the emotion-driven contraction of these muscle groups surrounding the eyes, together with the zygomaticus major, convey the genuine experience of happiness, joy and laughter. In recognition of these insights, psychologists sometimes refer to this facial expression as the "Duchenne smile."

In normal individuals, such as the Parisian shoemaker shown here, the difference between a forced smile (produced by voluntary contraction or electrical stimulation of facial muscles) and a spontaneous (emotional) smile testifies to the convergence of descending motor signals from different forebrain centers onto premotor and motor neurons in the brainstem that control the facial musculature. In contrast to the Duchenne smile, the contrived smile of volition (sometimes called a "pyramidal smile") is driven by the motor cortex, which communicates with the brainstem and spinal cord via the pyramidal tracts. The Duchenne smile is motivated by motor areas in the anterior cingulate gyrus (see Box 17B) that access facial motor circuitry via multisynaptic, "extrapyramidal" pathways through the brainstem reticular formation.

Studies of patients with specific neurological injury to these separate descending systems of control have further differentiated the forebrain centers responsible for control of the muscles of facial expression (Figure B). Patients with unilateral facial paralysis due to damage of descending pathways from the motor cortex (upper motor neuron syndrome; see Chapter 17) have considerable difficulty moving their lower facial muscles on one side, either voluntarily or in response to commands, a condition called voluntary facial paresis (Figure B, left panels). Nonetheless, many such individuals produce symmetrical *involuntary* facial movements when they laugh,

(*Continued on next page*)

(A) Duchenne made use of early photography to study human facial expressions. (1) Duchenne and one of his subjects undergoing "faradization" of the facial muscles. (2) Bilateral electrical stimulation of the zygomaticus major mimicked a genuine expression of happiness, although closer examination shows insufficient contraction of the obicularis oculi (surrounding the eyes) compared to spontaneous laughter (3). In photograph (4), stimulation of the brow and neck produced an expression of "terror mixed with pain, torture … that of the damned"; however, the subject reported no discomfort or emotional experience consistent with the evoked contractions.

(A)

(1)

(2)

(3)

(4)

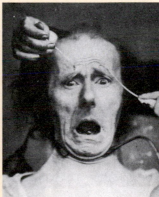

BOX 29A (Continued)

(B)

Facial motor paresis   Emotional motor paresis

Voluntary smile

Response to humor

(B) Left panels: Mouth of a patient with a lesion that destroyed descending fibers from the right motor cortex displaying voluntary facial paresis. When asked to show her teeth, the patient was unable to contract the muscles on the left side of her mouth (upper left), yet her spontaneous smile in response to a humorous remark is nearly symmetrical (lower left). Right panels: Face of a child with a lesion of the left forebrain that interrupted descending pathways from nonclassical motor cortical areas, producing emotional facial paresis. When asked to smile volitionally, the contractions of the facial muscles are nearly symmetrical (upper right). In spontaneous response to a humorous comment, however, the right side of the patient's face fails to express emotion (lower right).

References

DUCHENNE DE BOULOGNE, G.-B. (1862) *Mecanisme de la Physionomie Humaine.* Paris: Editions de la Maison des Sciences de l'Homme. Edited and translated by R. A. Cuthbertson (1990). Cambridge: Cambridge University Press.

HOPF, H. C., W. MÜLLER-FORELL AND N. J. HOPF (1992) Localization of emotional and volitional facial paresis. *Neurology* 42: 1918–1923.

TROSCH, R. M., G. SZE, L. M. BRASS AND S. G. WAXMAN (1990) Emotional facial paresis with striatocapsular infarction. *J. Neurol. Sci.* 98: 195–201.

WAXMAN, S. G. (1996) Clinical observations on the emotional motor system. In *Progress in Brain Research,* Vol. 107. G. Holstege, R. Bandler and C. B. Saper (eds.). Amsterdam: Elsevier, pp. 595–604.

frown, or cry in response to amusing or distressing stimuli. In such patients, pathways from regions of the forebrain other than the classical motor cortex in the posterior frontal lobe remain available to activate facial movements in response to stimuli with emotional significance.

A much less common form of neurological injury, called *emotional facial paresis,* demonstrates the opposite set of impairments—that is, the loss of the ability to express emotions by using the muscles of the face without loss of volitional control (Figure B, right panels). Such individuals are able to produce symmetrical pyramidal smiles, but fail to display spontaneous emotional expressions involving the facial musculature contralateral to the lesion. These two systems are diagrammed in Figure C.

(C)

VOLITIONAL MOVEMENT	NEURAL SYSTEMS FOR EMOTIONAL EXPRESSION
Descending "pyramidal" and "extrapyramidal" projections from motor cortex and brainstem	Descending "extrapyramidal" projections from medial forebrain and hypothalamus

Voluntary facial paresis   Emotional facial paresis

Motor neuron pools in facial nucleus

Pyramidal smile   Duchenne smile

Activation of facial muscles

(C) The complementary deficits demonstrated in Figure B are explained by selective lesions of one of two anatomically and functionally distinct sets of descending projections that motivate the muscles of facial expression.

both cerebral hemispheres (including the cortex, underlying white matter, and basal ganglia) in a series of cats. When the anesthesia had worn off, the animals behaved as if they were enraged. The angry behavior occurred spontaneously and included the usual autonomic correlates of this emotion: increased blood pressure and heart rate, retraction of the nictitating membranes (the thin connective tissue sheets associated with feline eyelids), dilation of the pupils, and erection of the hairs on the back and tail. The cats also exhibited somatic motor components of anger, such as arching the back, extending the claws, lashing the tail, and snarling. This behavior was called **sham rage** because it had no obvious target. Bard showed that a complete response occurred as long as the caudal hypothalamus was intact (Figure 29.1). Sham rage could not be elicited, however, when the brain was transected at the junction of the hypothalamus and midbrain (although some uncoordinated components of the response were still apparent). Bard suggested that whereas the subjective experience of emotion might depend on an intact cerebral cortex, the expression of coordinated emotional behaviors does not necessarily entail cortical processes. He also emphasized that emotional behaviors are often directed toward self-preservation (a point made by Charles Darwin in his classic book on the evolution of emotion), and that the functional importance of emotions in all mammals is consistent with the involvement of phylogenetically older parts of the nervous system.

Complementary results were reported by Walter Hess, who showed that electrical stimulation of discrete sites in the hypothalamus of awake, freely moving cats could also lead to a rage response, and even to subsequent attack behavior. Moreover, stimulation of other sites in the hypothalamus caused a defensive posture that resembled fear. In 1949, a share of the Nobel Prize in Physiology or Medicine was awarded to Hess "for his discovery of the functional organization of the interbrain [hypothalamus] as a coordinator of the activities of the internal organs." Experiments like those of Bard and Hess led to the important conclusion that the basic circuits for organized behaviors accompanied by emotion are in the diencephalon and the brainstem structures connected to it. Furthermore, their work emphasized that the control of the involuntary motor system is not entirely separable from the control of the voluntary pathways, an important consideration in understanding the motor aspects of emotion, as discussed below.

The routes by which the hypothalamus and other forebrain structures influence the visceral and somatic motor systems are complex. The major targets of the hypothalamus lie in the **reticular formation**, the tangled web of nerve cells and fibers in the core of the brainstem (see Box 17A). This structure contains over 100 identifiable cell groups, including some of the nuclei that control the brain states associated with sleep and wakefulness described in the previous chapter. Other important circuits in the reticular formation control cardiovascular function, respiration, urination, vomiting, and swallowing. The reticular neurons receive hypothalamic input from and feed into both somatic and autonomic effector systems in the brainstem and spinal cord. Their activity can therefore produce widespread visceral motor and somatic motor responses, often overriding reflex function and sometimes involving almost every organ in

(A) No "sham rage"

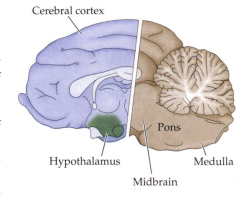

(B) "Sham rage" remains

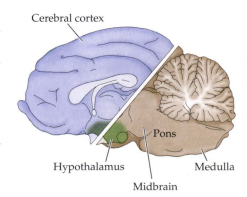

Figure 29.1   *Midsagittal view of a cat's brain, illustrating the regions sufficient for the expression of emotional behavior. (A) Transection through the midbrain, disconnecting the hypothalamus and brainstem, abolishes "sham rage." (B) The integrated emotional responses associated with "sham rage" survive removal of the cerebral hemispheres as long as the caudal hypothalamus remains intact. (After LeDoux, 1987.)*

(A)

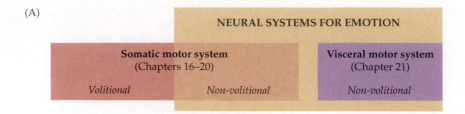

(B)

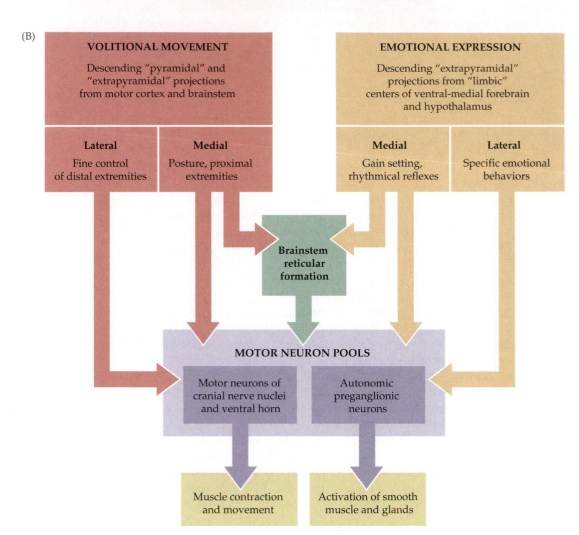

the body (as implied by Cannon's dictum about the sympathetic preparation of the animal for fight or flight).

In addition to the hypothalamus, other sources of descending projections from the forebrain to the brainstem reticular formation contribute to the expression of emotional behavior. Collectively, these additional centers in the forebrain are considered part of the **limbic system**, which is described in the following section. These descending influences on the expression of somatic and visceral motor behavior arise outside of the classic motor cortical areas in the posterior frontal lobe.

Thus, the descending control of emotional expression entails two parallel systems that are anatomically and functionally distinct (Figure 29.2). The voluntary motor component described in detail in Chapters 16 through 20 comprises

◀ **Figure 29.2** *Components of the nervous system that organize the expression of emotional experience. (A) The neural systems that help convey emotion include fore-brain centers that govern the nonvolitional expression of somatic motor behavior and the visceral motor system. (B) Diagram of the descending systems that control somatic and visceral motor effectors. Motor cortical areas in the posterior frontal lobe give rise to descending projections that, together with secondary projections arising in the brainstem, are organized into medial and lateral components. As described in Chapter 17, these descending projections account for volitional somatic movements. Function-ally and anatomically distinct centers in the forebrain govern the expression of nonvo-litional somatic motor and visceral motor functions, which are coordinated to meditate emotional behavior. "Limbic" centers in the ventral-medial forebrain and hypothalamus also give rise to medial and lateral descending projections. For both systems of descending projections, the lateral components elicit specific behaviors (e.g., volitional digit movements and emotional facial expressions), while the medial components sup-port and modulate the execution of such behaviors. The descending projections of both systems terminate in several integrative centers in the brainstem reticular forma-tion, as well as the motor neuronal pools of the brainstem and spinal cord. In addi-tion, the limbic forebrain centers innervate components of the visceral motor system that govern preganglionic autonomic neurons in the brainstem and spinal cord.*

the classical motor areas of the posterior frontal lobe and related circuitry in the basal ganglia and cerebellum. The descending pyramidal and extrapyramidal projections from the motor cortex and brainstem ultimately convey the impulses responsible for voluntary somatic movements. In addition to the descending systems that govern volitional movements, several cortical and subcortical structures in the medial frontal lobe and ventral parts of the forebrain, including related circuitry in the ventral part of the basal ganglia and hypothalamus, give rise to separate descending projections that run parallel to the pathways of the volitional motor system. These descending projections of the medial and ventral forebrain terminate on visceral motor centers in the brainstem reticular forma-tion, preganglionic autonomic neurons, and certain somatic premotor and motor neuron pools that also receive projections from volitional motor centers. The two types of facial paresis illustrated in Box 29A underscore this dual nature of descending motor control.

In short, the somatic and visceral activities associated with unified emotional behavior are mediated by the activity of both the somatic and visceral motor neurons, which integrate parallel, descending inputs from a constellation of forebrain sources. The remaining sections of the chapter are devoted to the organization and function of the forebrain centers that specifically govern the experience and expression of emotional behavior.

The Limbic System

Attempts to understand the effector systems that control emotional behavior have a long history. In 1937, James Papez (pronounced "Papes") first proposed that specific brain circuits are devoted to emotional experience and expression (much as the occipital cortex is devoted to vision, for instance). In seeking to understand what parts of the brain serve this function, he began to explore the medial aspects of the cerebral hemisphere. In the 1850s, Paul Broca used the term "limbic lobe" to refer to the part of the cerebral cortex that forms a rim (*limbus* is Latin for rim) around the corpus callosum and diencephalon on the medial face of the hemispheres (Figure 29.3). Two prominent components of this region are the **cingulate gyrus**, which lies above the corpus callosum, and the **parahippocampal gyrus**, which lies in the medial temporal lobe.

Figure 29.3   The so-called limbic lobe includes the cortex on the medial aspect of the cerebral hemisphere that forms a rim around the corpus callosum and diencephalon, including the cingulate gyrus (lying above the corpus callosum) and the parahippocampal gyrus. Historically, the olfactory bulb and olfactory cortex (not illustrated here) have also been considered to be important elements of the limbic lobe.

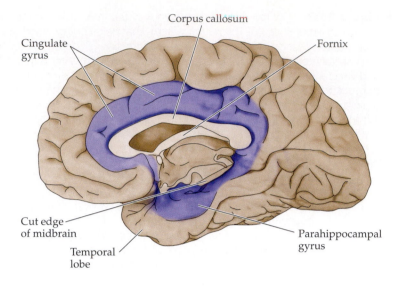

For many years, these structures, along with the olfactory bulbs, were thought to be concerned primarily with the sense of smell. Indeed, Broca considered the olfactory bulbs to be the principal source of input to the limbic lobe. Papez, however, speculated that the function of the limbic lobe might be more related to emotions. He knew from the work of Bard and Hess that the hypothalamus influences the expression of emotion; he also knew, as everyone does, that emotions reach consciousness and that higher cognitive functions affect emotional behavior. Ultimately, Papez showed that the cingulate cortex and hypothalamus are interconnected via projections from the **mammillary bodies** (part of the posterior hypothalamus) to the **anterior nucleus of the dorsal thalamus**, which projects in turn to the **cingulate gyrus**. The cingulate gyrus (and many other cortical regions as well) projects to the **hippocampus**. Finally, he showed that the hippocampus projects via the **fornix** (a large fiber bundle) back to the hypothalamus. Papez suggested that these pathways provided the connections necessary for cortical control of emotional expression, and they became known as the "Papez circuit."

Over time, the concept of a forebrain circuit for the control of emotional expression, first elaborated by Papez, has been revised to include parts of the **orbital and medial prefrontal cortex**, **ventral parts of the basal ganglia**, the **mediodorsal nucleus of the thalamus** (a different thalamic nucleus than the one emphasized by Papez), and a large nuclear mass in the temporal lobe anterior to the hippocampus, called the **amygdala**. This set of structures, together with the parahippocampal gyrus and cingulate cortex, is generally referred to as the **limbic system** (Figure 29.4). Thus, some of the structures that Papez originally described (the hippocampus, for example) now appear to have little to do with emotional behavior, whereas the amygdala, which Papez hardly mentioned, clearly plays a major role in the experience and expression of emotion (Box 29B).

About the same time that Papez proposed that these structures were important for the integration of emotional behavior, Heinrich Klüver and Paul Bucy were carrying out a series of experiments on rhesus monkeys in which they removed a large part of both medial temporal lobes, thus destroying much of the limbic system. They reported a set of abnormal behaviors in these animals that is now known as the Klüver-Bucy syndrome (Box 29C). Among the most prominent changes was visual agnosia: the animals appeared to be unable to recognize

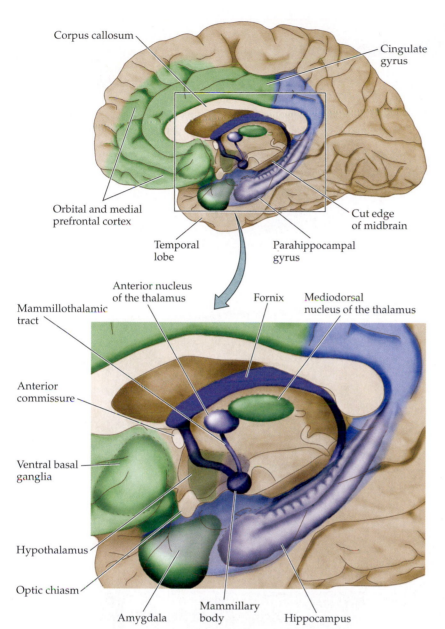

Figure 29.4 Modern conception of the limbic system. Two especially important components of the limbic system not emphasized in early anatomical accounts are the orbital and medial prefrontal cortex and the amygdala. These two telencephalic regions, together with related structures in the thalamus, hypothalamus and ventral striatum, are especially important in the experience and expression of emotion (green). Other parts of the limbic system, including the hippocampus and the mammillary bodies of the hypothalamus, are no longer considered important neural centers for processing emotion (blue).

objects, although they were not blind, a deficit similar to that sometimes seen in human patients following lesions of the temporal cortex (see Chapter 26). In addition, the monkeys displayed bizarre oral behaviors, putting objects into their mouths that normal monkeys would not. They exhibited hyperactivity and hypersexuality, approaching and making physical contact with virtually anything in their environment; most importantly, they showed marked changes in emotional behavior. Because they had been caught in the wild, the monkeys had typically reacted with hostility and fear to humans before their surgery. Postoperatively, however, they were virtually tame. Motor and vocal reactions generally associated with anger or fear were no longer elicited by the approach of humans, and the animals showed little or no excitement when the experimenters handled them. Nor did they show fear when presented with a snake—a strongly aversive stimulus for a normal rhesus monkey. Klüver and Bucy concluded that this remarkable change in behavior was at least partly due to the interruption of the

BOX 29B **The Anatomy of the Amygdala**

The amygdala is a complex mass of gray matter buried in the anterior-medial portion of the temporal lobe, just rostral to the hippocampus (Figure A). It comprises multiple, distinct subnuclei and cortical regions that are richly connected to nearby cortical areas on the ventral and medial aspect of the hemispheric surface.

The amygdala (or amygdaloid complex, as it is often called) can best be thought of in terms of three major functional and anatomical subdivisions, each of which has a unique set of connections with other parts of the brain (Figures B and C). The medial group of subnuclei has extensive connections with the olfactory bulb and the olfactory cortex. The basal-lateral group, which is especially large in humans, has major connections with the cerebral cortex, especially the orbital and medial prefrontal cortex of the frontal lobe and the associational cortex of the anterior temporal lobe. The central and anterior group of nuclei is characterized by connections with the hypothalamus and brainstem, including such visceral sensory structures as the nucleus of the solitary tract and the parabrachial nucleus (see Chapter 21).

The amygdala thus links cortical regions that process sensory information with hypothalamic and brainstem effector systems. Cortical inputs provide information about highly processed visual, somatic sensory, visceral sensory, and auditory stimuli. These pathways from sensory cortical areas distinguish the amygdala from the hypothalamus, which receives relatively unprocessed sensory inputs. The amygdala also receives sensory input directly from some thalamic nuclei, the olfactory bulb, and visceral sensory relays in the brainstem. Thus, many neurons in the amygdala respond to visual, auditory, somatic sensory, visceral sensory, gustatory, and olfactory stimuli.

Physiological studies have confirmed this convergence of sensory information.

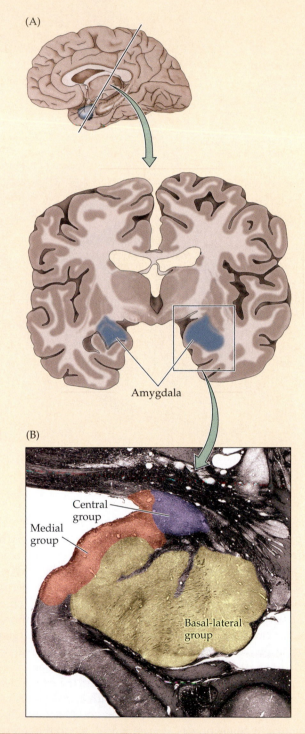

(A)

(B)

Amygdala

Central group

Medial group

Basal-lateral group

(A) Coronal section through the forebrain at the level of the amygdala. (B) Histological section through the human amygdala (boxed area in panel A), stained with silver salts to reveal the presence of myelinated fiber bundles. These bundles subdivide major nuclei and cortical regions within the amygdaloid complex. (B courtesy of Joel Price.)

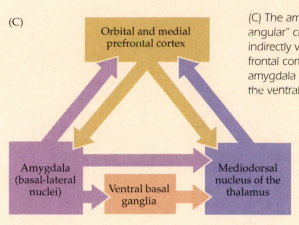

(C) The amygdala (specifically, the basal-lateral group of nuclei) participates in a "triangular" circuit linking the amygdala, the thalamic mediodorsal nucleus (directly and indirectly via the ventral parts of the basal ganglia), and the orbital and medial prefrontal cortex. These complex interconnections allow direct interactions between the amygdala and prefrontal cortex, as well as indirect modulation via the circuitry of the ventral basal ganglia.

26). In addition to sensory inputs, the prefrontal and temporal cortical connections of the amygdala give it access to more overtly cognitive neocortical circuits, which integrate the emotional significance of sensory stimuli and guide complex behavior.

Finally, projections from the amygdala to the hypothalamus and brainstem (and possibly as far as the spinal cord)

allow it to play an important role in the expression of emotional behavior by influencing activity in both the somatic and visceral motor efferent systems.

Reference

PRICE, J. L., F. T. RUSSCHEN AND D. G. AMARAL (1987) The limbic region II: The amygdaloid complex. In *Handbook of Chemical Neuroanatomy*, Vol. 5, *Integrated Systems of the CNS*, Part I, *Hypothalamus, Hippocampus, Amygdala, Retina*. A. Björklund and T. Hökfelt (eds.). Amsterdam: Elsevier, pp. 279–388.

Moreover, highly complex stimuli are often required to evoke a neuronal response. For example, there are neurons in the basal-lateral group of nuclei that respond selectively to the sight of faces, very much like the "face" neurons in the inferior temporal cortex (see Chapter

---

pathways described by Papez. A similar syndrome has been described in humans who have suffered bilateral damage of the temporal lobes.

When it was later demonstrated that the emotional disturbances of the Klüver–Bucy syndrome could be elicited by removal of the amygdala alone, attention turned more specifically to the role of this structure in the control of emotional behavior.

The Importance of the Amygdala

Experiments first performed in the late 1950s by John Downer at University College London vividly demonstrated the importance of the amygdala in aggressive behavior. Downer removed one amygdala in rhesus monkeys, at the same time transecting the optic chiasm and the commissures that link the two hemispheres (principally, the corpus callosum and anterior commissure; see Chapter 27). In so doing, he produced animals with a single amygdala that had access only to visual inputs from the eye on the same side of the head. Downer found that the animals' behavior depended on which eye was used to view the world. When the monkeys were allowed to see with the eye on the side of amygdala lesion, they behaved in some respects like those described by Klüver and Bucy (for example, they were relatively placid in the presence of humans). If, however, they were allowed to see only with the eye on the side of the intact amygdala, they reverted to their normal fearful and often aggressive behavior. Thus, in the absence of the amygdala, a monkey does not interpret the significance of the visual stimulus presented by an approaching human in the same way as a normal animal. Importantly, only visual stimuli presented to the eye on the side of the ablation produced this abnormal state; thus if the animal was touched on either side, a full aggressive reaction occurred, implying that somatic sensory information about both sides of the body had access to the remaining

BOX 29C   The Reasoning Behind an Important Discovery*

Paul Bucy explains why he and Heinrich Klüver removed the temporal lobes in monkeys:

When we started out, we were not trying to find out what removal of the temporal lobe would do, or what changes in behavior of the monkeys it would produce. What we found out was completely unexpected! Heinrich had been experimenting with mescaline. He had even taken it himself and had experienced hallucinations. He had written a book about mescaline and its effects. Later Heinrich gave mescaline to his monkeys. He gave everything to his monkeys, even his lunch! He noticed that the monkeys acted as though they experienced paraesthesias in their lips. They licked, bit and chewed their lips. So he came to me and said, "Maybe we can find out where mescaline has its actions in the brain." So I said, "OK."

We began by doing a sensory denervation of the face, but that didn't make any difference to the mescaline-induced behavior. So we tried motor denervation. That didn't make any difference, either. Then we had to sit back and think hard about where to look. I said to Heinrich, "This business of licking and chewing the lips is not unlike what you see in cases of temporal lobe epilepsy. Patients chew and smack their lips inordinately. So, let's take out the uncus." Well, we could just as well take out the whole temporal lobe, including the uncus. So we did.

We were especially fortunate with our first animal. This was an older female.... She had become vicious—absolutely nasty. She was the most vicious animal you ever saw; it was dangerous to go near her. If she didn't hurt you, she

would at least tear your clothing. She was the first animal on which we operated. I removed one temporal lobe.... The next morning my phone was ringing like mad. It was Heinrich, who asked, "Paul, what did you do to my monkey? She is tame!" Subsequently, in operating on non-vicious animals, the taming effect was never so obvious.

That stimulated our getting the other temporal lobe out as soon as we could evaluate her. When we removed the other temporal lobe, the whole syndrome blossomed.

*Excerpt from an interview of Bucy by K. E. Livingston in 1981. K. E. Livingston (1986) Epilogue: Reflections on James Wenceslas Papez, According to Four of his Colleagues. In *The Limbic System: Functional Organization and Clinical Disorders*. B. K. Doane and K. E. Livingston (eds.). New York: Raven Press.

amygdala. These anecdotal data, taken together with what is now a rich trove of empirical results and clinical observations in both experimental animals and humans show that the amygdala mediates neural processes that invest sensory experience with emotional significance.

To better understand the role of the amygdala in evaluating stimuli, and to define more precisely the specific circuits and mechanisms involved, several other animal models of emotional behavior have since been developed. One of the most useful is based on conditioned fear responses in rats. Conditioned fear develops when an initially neutral stimulus is repeatedly paired with an inherently aversive one. Over time, the animal begins to respond to the neutral stimulus with behaviors similar to those elicited by the threatening stimulus (i.e., it learns to attach a new meaning to the neutral stimulus). Studies of the parts of the brain involved in the development of conditioned fear in rats have begun to shed some light on this process. Joseph LeDoux and his colleagues at New York University trained rats to associate a tone with a mildly aversive foot shock delivered shortly after onset of the sound. To assess the animals' responses, they measured blood pressure and the length of time the animals crouched without moving (a fearful reaction called "freezing"). Before training, the rats did not react to the tone, nor did their blood pressure change when the tone was presented. After training, however, the onset of the tone caused a marked increase in blood pressure and prolonged periods of behavioral freezing. Using this paradigm, LeDoux and his colleagues worked out the neural circuitry that established the association between the tone and fear (Figure 29.5). First, they demonstrated that the medial geniculate nucleus is necessary for the develop-

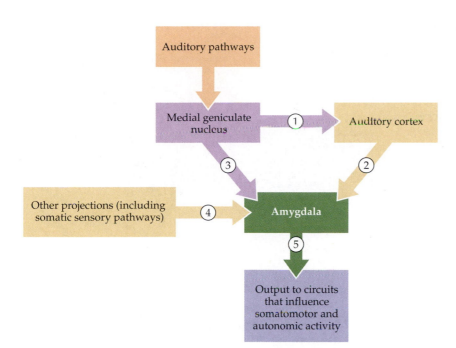

Figure 29.5   Pathways in the rat brain that mediate the association of auditory and aversive somatic sensory stimuli. Information processed by the auditory centers in the brainstem is relayed to the auditory cortex via the medial geniculate nucleus (1). The amygdala receives auditory information indirectly via the auditory cortex (2) and directly from one subdivision of the medial geniculate (3). The amygdala also receives sensory information about other sensory modalities, including pain (4). Thus, the amygdala is in a position to associate diverse sensory inputs, leading to new behavioral and autonomic responses to stimuli that were previously devoid of emotional content (5).

ment of the conditioned fear response. This result is not surprising, since all auditory information that reaches the forebrain travels through the medial geniculate nucleus of the dorsal thalamus (see Chapter 13). They went on to show, however, that the responses were still elicited if the connections between the medial geniculate and auditory cortex were severed, leaving only a direct projection between the medial geniculate and the basal-lateral group of nuclei in the amygdala. Furthermore, if the part of the medial geniculate that projects to the amygdala was also destroyed, the fear responses were abolished. Subsequent work in LeDoux's laboratory established that projections from the central group of nuclei in the amygdala to the midbrain reticular formation are critical in the expression of freezing behavior, while other projections from this group to the hypothalamus control the rise in blood pressure.

Since the amygdala is a site where neural activity produced by both tones and shocks can be processed, it is reasonable to suppose that the amygdala is also the site where learning about fearful stimuli occurs. These results, among others, have led to the broader hypothesis that the amygdala participates in establishing associations between neutral sensory stimuli, such as a mild auditory tone or the sight of inanimate object in the environment, and other stimuli that have some primary reinforcement value (Figure 29.6). The neutral sensory input can be stimuli in the external environment, stimuli communicated centrally via the special sensory afferent systems, or internal stimuli derived from activation of visceral sensory receptors. The stimuli with primary reinforcement value include sensory stimuli that are inherently rewarding, such as the sight, smell, and taste of food, or stimuli with negative valences such as an aversive taste, loud sounds, or painful mechanical stimulation. The associative learning process itself is probably a Hebbian-like mechanism (see Chapter 8) that strengthens the connections relaying the information about the neutral stimulus, provided that they activate the postsynaptic neurons in the amygdala at the same time as inputs pertaining to the primary reinforcer. The discovery that long-term potentiation (LTP) occurs in the amygdala provides further support for this hypothesis. Indeed, the acquisition of conditioned fear in rats is blocked by infusion into the amygdala of NMDA antagonists, which prevents the induction of LTP. Finally,

Figure 29.6   *Model of associative learning in the amyg-dala relevant to emotional function. Most neutral sensory inputs are relayed to principal neurons in the amygdala by projections from "higher order" sensory processing areas that represent objects (e.g., faces). If these sensory inputs depolarize amygdaloid neurons at the same time as inputs that represent other sensations with primary reinforcing value, then associative learning occurs by strengthening synaptic linkages between the previously neutral inputs and the neurons of the amygdala. The output of the amyg-dala then informs a variety of integrative centers responsi-ble for the somatic and visceral motor expression of emo-tion, and for modifying behavior relevant to seeking rewards and avoiding punishment. (After Rolls, 1999.)*

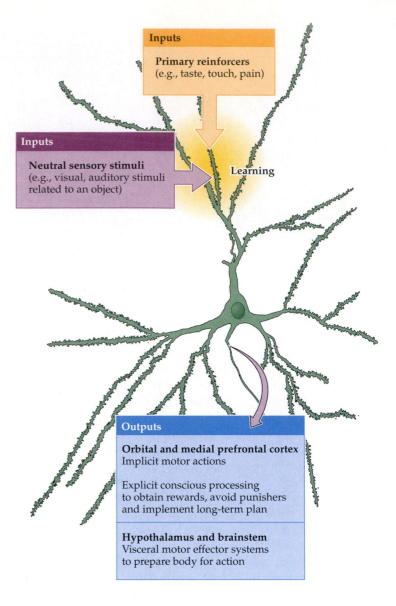

the behavior of patients with selective damage to the anterior-medial temporal lobe indicates that the amygdala plays a similar role in the human experience of fear (Box 29D).

The Relationship between Neocortex and Amygdala

As these observations on the limbic system (and the amygdala in particular) make plain, understanding the neural basis of emotions also requires under-standing the role of the cerebral cortex. In animals like the rat, most behavioral responses are highly stereotyped. In more complex brains, however, individual experience is increasingly influential in determining responses to special and even idiosyncratic stimuli. Thus in humans, a stimulus that evokes fear or sad-ness in one person may have little or no effect on the emotions of another. Although the pathways underlying such responses are not well understood, the amygdala and its interconnections with an array of neocortical areas in the pre-frontal cortex and anterior temporal lobe, as well as several subcortical structures, appear to be especially important in the higher order processing of emotion. In

BOX 29D   **Fear and the Human Amygdala: A Case Study**

Studies of fear conditioning in rodents show that the amygdala plays a critical role in the association of an innocuous auditory tone with an aversive mechanical sensation. Does this finding imply that the human amygdala is similarly involved in the experience of fear and the expression of fearful behavior? Reports of at least one extraordinary patient support the idea that the amygdala is indeed a key brain center for the experience of fear.

The patient (S.M.) suffers from a rare, autosomal recessive condition called Urbach-Wiethe disease, a disorder that causes bilateral calcification and atrophy of the anterior-medial temporal lobes. As a result, both of S.M.'s amygdalas are extensively damaged, with little or no detectable injury to the hippocampal formation or nearby temporal neocortex (Figure A). She has no motor or sensory impairment, and no notable deficits in intelligence, memory, or language function. However, when asked to rate the intensity of emotion in a series of photographs of facial expressions, she cannot recognize the emotion of fear (Figure B). Indeed, S.M.'s ratings of emotional content in fearful facial expressions were several standard deviations below the ratings of control patients who had suffered brain damage outside of the anterior-medial temporal lobe.

The investigators next asked S.M. (and brain-damaged control subjects) to draw facial expressions of the same set of emotions from memory. Although the subjects obviously differed in artistic abilities and the detail of their renderings, S.M. (who has some artistic experience) produced skillful pictures of each emotion, except for fear (Figure C). At first, she could not produce a sketch of a fearful expression and, when prodded to do so, explained that "she did not know what an afraid face would look like." After several failed

(Continued on next page)

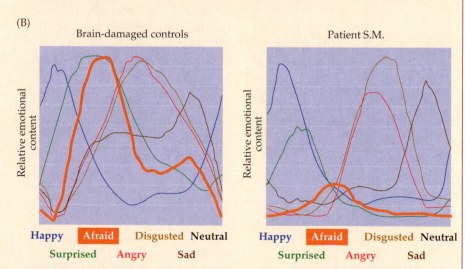

(A) MRI showing the extent of brain damage in patient S.M.; note the bilateral destruction of the amygdala and the preservation of the hippocampus. (B) Patients with brain damage outside of the anterior-medial temporal lobe and patient S.M. rated the emotional content of a series of facial expressions. Each colored line represents the intensity of the emotions judged in the face. S.M. recognized happiness, surprise, anger, disgust, sadness, and neutral qualities in facial expressions about as well as controls. However, she failed to recognize fear (orange lines). (A courtesy of R. Adolphs.)

BOX 29D (Continued)

attempts, she produced the sketch of a cowering figure with hair standing on end, evidently because she knew these clichés about the expression of fear. In short, S.M. has a severely limited concept of fear and, consequently, fails to recognize the emotion of fear in facial expressions, in part because she fails to seek out salient social information from the eye regions of human faces. Studies of other individuals with bilateral destruction of the amygdala are consistent with this account. As might be expected, S.M.'s deficiency also limits her ability to experience fear in situations where this emotion is appropriate.

Despite the admonition "have no fear," to truly live without fear is to be deprived of a crucial neural mechanism that facilitates appropriate social behavior, helps make advantageous decisions in critical circumstances, and, ultimately, promotes survival.

References

ADOLPHS, R., D. TRANEL, H. DAMASIO AND A. R. DAMASIO (1995) Fear and the human amygdala. *J. Neurosci.* 15: 5879–5891.

ADOLPHS, R., F. GOSSELIN, T. W. BUCHANAN, D. TRANEL, P. SCHYNS AND A. R. DAMASIO (2005) A mechanism for impaired fear recognition after amygdala damage. *Nature* 433: 68–72.

BECHARA, A., H. DAMASIO, A. R. DAMASIO AND G. P. LEE (1999) Differential contributions of the human amygdala and ventromedial prefrontal cortex to decision-making. *J. Neurosci.* 19: 5473–5481.

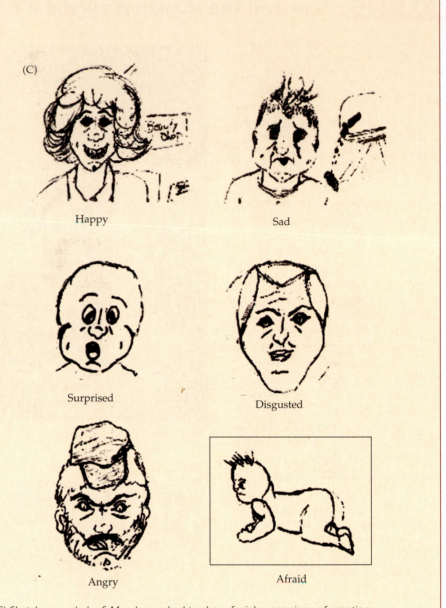

(C) Sketches made by S.M. when asked to draw facial expressions of emotion.

addition to its connections with the hypothalamus and brainstem centers that regulate visceral motor function, the amygdala has significant connections with several cortical areas in the orbital and medial aspects of the frontal lobe (see Box 29B). These prefrontal cortical fields associate information from every sensory modality (including information about visceral activities) and can thus integrate a variety of inputs pertinent to moment-to-moment experience. In addition, the amygdala projects to the thalamus (specifically, the mediodorsal nucleus), which projects in turn to these same cortical areas. Finally, the amygdala innervates neurons in the ventral portions of the basal ganglia that receive the major cortico-striatal projections from the regions of the prefrontal cortex thought to process emotions. Considering all these seemingly arcane anatomical connections, the amygdala emerges as a nodal point in a network that links together the cortical and subcortical brain regions involved in emotional processing.

Clinical evidence concerning the significance of this circuitry linked through the amygdala has come from functional imaging studies of patients suffering from depression (Box 29E), in which this set of interrelated forebrain structures displays abnormal patterns of cerebral blood flow, especially in the left hemisphere. More generally, the amygdala and its connections to the prefrontal cortex and basal ganglia are likely to influence the selection and initiation of behaviors aimed at obtaining rewards and avoiding punishments (recall that the process of motor program selection and initiation is an important function of basal ganglia circuitry; see Chapter 18). The parts of the prefrontal cortex interconnected with the amygdala are also involved in organizing and planning future behaviors; thus, the amygdala may provide emotional input to overt (and covert) deliberations of this sort (see the later section on "Emotion, Reason, and Social Behavior").

Finally, it is likely that interactions between the amygdala, the neocortex and related subcortical circuits account for what is perhaps the most enigmatic aspect of emotional experience: the highly subjective "feelings" that attend most emotional states. Although the neurobiology of such experience is not understood, it is reasonable to assume that emotional feelings arise as a consequence of a more general cognitive capacity for self-awareness. In this conception, feelings entail both the immediate conscious experience of implicit emotional processing (arising from amygdala–neocortical circuitry) and explicit processing of semantically based thought (arising from hippocampal–neocortical circuitry; see Chapter 31). Thus, feelings are plausibly conceived as the product of an emotional working memory that sustains neural activity related to the processing of these various elements of emotional experience. Given the evidence for working memory functions in the prefrontal cortex (see Chapter 26), this portion of the frontal lobe—especially the orbital and medial sector—is the likely neural substrate where such associations are maintained in conscious awareness (Figure 29.7).

Cortical Lateralization of Emotional Functions

Since functional asymmetries of complex cortical processes are commonplace (see Chapters 26 and 27), it should come as no surprise that the two hemispheres make different contributions to the governance of emotion.

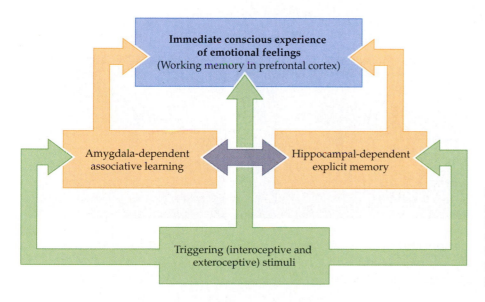

Figure 29.7 Neural model for the awareness of emotional feelings. The highly subjective feelings associated with emotional experience presumably arise from neural systems in the prefrontal cortex that produce awareness of emotional processing. (After LeDoux, 2000.)

BOX 29E Affective Disorders

Although some degree of disordered emotion is present in virtually all psychiatric problems, in affective (mood) disorders the essence of the disease is an abnormal regulation of the feelings of sadness and happiness. The most severe of these afflictions are major depression and manic depression. (Manic depression is also called "bipolar disorder," since such patients experience alternating episodes of depression and euphoria.) Depression, the most common of the major psychiatric disorders, has a lifetime incidence of 10–25% in women and 5–12% in men. For clinical purposes, depression (as distinct from bereavement or neurotic unhappiness) is defined by a set of standard criteria. In addition to an abnormal sense of sadness, despair, and bleak feelings about the future (depression itself), these criteria include disordered eating and weight control, disordered sleeping (insomnia or hypersomnia), poor concentration, inappropriate guilt, and diminished sexual interest. The personally overwhelming quality of major depression has been compellingly described by patient/authors such as William Styron, and by afflicted psychologists such as Kay Jamison. The depressed patient's profound sense of despair has been nowhere better expressed than by Abraham Lincoln, who during a period of depression wrote:

I am now the most miserable man living. If what I feel were equally distributed to the whole human family, there would not be one cheerful face on earth. Whether I shall ever be better, I cannot tell; I awfully forebode I shall not. To remain as I am is impossible. I must die or be better, it appears to me.

Indeed, about half of all suicides occur in individuals with clinical depression.

Not many decades ago, depression and mania were considered disorders that arose from external circumstances or a neurotic inability to cope. It is now universally accepted that these conditions are neurobiological disorders. Among the strongest lines of evidence for this consensus are studies of the inheritance of these diseases. For example, the concordance of affective disorders is high in monozygotic compared to dizygotic twins. It has also become possible to study the brain activity of patients suffering from affective disorders by non-invasive brain imaging (see Figure). In at least one condition (unipolar depression), abnormal patterns of blood flow are apparent in the "triangular" circuit interconnecting the amygdala, the mediodorsal nucleus of the thalamus, and the orbital and medial prefrontal cortex (see Box 29B). Of particular interest is the significant correlation of abnormal blood flow in the amygdala and the clinical severity of depression, as well as the observation that the abnormal blood flow pattern in the prefrontal cortex returns to normal when the depression has abated.

Despite evidence for a genetic predisposition and an increasing understanding of the brain areas involved, the cause of these conditions remains unknown. The efficacy of a large number of drugs that influence catecholaminergic and serotonergic neurotransmission strongly implies that the basis of the disease(s) is ultimately neurochemical (see Figures 6.11 and 6.12 for overviews of the projections of these neural systems). The majority of patients (about 70 percent) can be effectively treated with one of a variety of drugs (including tricyclic antidepressants, monoamine oxidase inhibitors, and selective serotonin reuptake inhibitors). Most successful are drugs that selectively block the uptake of serotonin without affecting the uptake of other neurotransmitters; these drugs are commonly known as selective serotonin reuptake inhibitors (SSRIs). Three such inhibitors—fluoxetine (Prozac®), sertraline (Zoloft®), and paroxetine (Paxil®)—are especially effective in treating depression and have few of the side

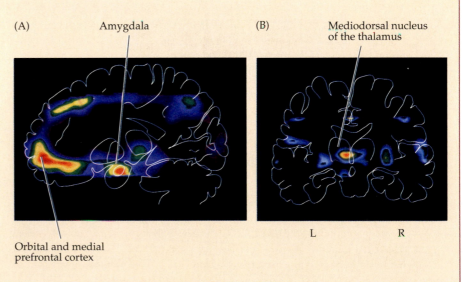

(A) Amygdala
Orbital and medial prefrontal cortex

(B) Mediodorsal nucleus of the thalamus
L   R

Areas of increased blood flow in the left amygdala, orbital, and medial prefrontal cortex (A) and in a location in the left medial thalamus consistent with the mediodorsal nucleus (B) from a sample of patients diagnosed with unipolar clinical depression. The "hot" colors indicate statistically significant increases in blood flow, compared to a sample of nondepressed subjects. (From Drevets and Raichle, 1994.)

effects of the older, less specific drugs. Perhaps the best indicator of the success of these drugs has been their wide acceptance: although the first SSRI's were approved for clinical use only in the late 1980s, they are now among the most prescribed pharmaceuticals.

Most depressed patients who use drugs such as the SSRI's report that they lead fuller lives and are more energetic and organized. Based on such information, these drugs are sometimes used not only to combat depression but also to "treat" individuals who have no definable psychiatric disorder. This abuse raises important social questions, similar to those posed by Aldous Huxley in his 1932 novel, where the mythical drug "Soma" was routinely administered to the inhabitants of his fictitious *Brave New World* to keep them content and docile. Presumably there is a middle ground between excessive suffering and excessive tranquility.

References

BREGGIN, P. R. (1994) *Talking Back to Prozac: What Doctors Won't Tell You about Today's Most Controversial Drug.* New York: St. Martin's Press.

DREVETS, W. C. AND M. E. RAICHLE (1994) PET imaging studies of human emotional disorders. In *The Cognitive Neurosciences*, M. S. Gazzaniga (ed.). Cambridge, MA: MIT Press, pp. 1153–1164.

FREEMAN, P. S., D. R. WILSON AND F. S. SIERLES (1993) Psychopathology. In *Behavior Science for Medical Students*, F. S. Sierles (ed.). Baltimore: Williams and Wilkins, pp. 239–277.

GREENBERG, P. E., L. E. STIGLIN, S. N. FINKELSTEIN AND E. R. BERNDT (1993) The economic burden of depression in 1990. *J. Clin. Psychiatry* 54: 405–424.

JAMISON, K. R. (1995) *An Unquiet Mind.* New York: Alfred A. Knopf.

JEFFERSON, J. W. AND J. H. GRIEST (1994) Mood disorders. In *Textbook of Psychiatry*, J. A. Talbott, R. E. Hales and S. C. Yudofsky (eds.). Washington: American Psychiatric Press, pp. 465–494.

ROBINS, E. (1981) *The Final Months: A Study of the Lives of 134 Persons Who Committed Suicide.* New York: Oxford University Press.

STYRON, W. (1990) *Darkness Visible: A Memoir of Madness.* New York: Random House.

WONG, D. T. AND F. P. BYMASTER (1995) Development of antidepressant drugs: Fluoxetin (Prozac®) and other selective serotonin uptake inhibitors. *Adv. Exp. Med. Biol.* 363: 77–95.

WONG, D. T., F. P. BYMASTER AND E. A. ENGLEMAN (1995) Prozac® (fluoxetine, Lilly 110140), the first selective serotonin uptake inhibitor and an antidepressant drug: Twenty years since its first publication. *Life Sciences* 57(5): 411–441.

WURTZEL, E. (1994). *Prozac Nation: Young and Depressed in America.* Boston: Houghton-Mifflin.

Emotionality is lateralized in the cerebral hemispheres in at least two ways. First, as discussed in Chapter 27, the right hemisphere is especially important for the expression and comprehension of the affective aspects of speech. Thus, patients with damage to the supra-Sylvian portions of the posterior frontal and anterior parietal lobes on the right side may lose the ability to express emotion by modulation of their speech patterns (recall that this loss of emotional expression is referred to as *aprosody* or *aprosodia*, and that similar lesions in the left hemisphere give rise to Broca's aphasia). Patients with aprosodia tend to speak in a monotone, no matter what the circumstances or meaning of what is said. For example, one such patient, a teacher, had trouble maintaining discipline in the classroom. Because her pupils (and even her own children) couldn't tell when she was angry or upset, she had to resort to adding phrases such as "I am angry and I really mean it" to indicate the emotional significance of her remarks. The wife of another patient felt her husband no longer loved her because he could not imbue his speech with cheerfulness or affection. Although such patients cannot express emotion in speech, they nonetheless do experience normal emotional feelings.

A second way in which the hemispheric processing of emotionality is asymmetrical concerns mood. Both clinical and experimental studies indicate that the left hemisphere is more importantly involved with what can be thought of as positive emotions, whereas the right hemisphere is more involved with negative ones. For example, the incidence and severity of depression (see Box 29E) is significantly higher in patients with lesions of the left anterior hemisphere compared to any other location. In contrast, patients with lesions of the right anterior hemisphere are often described as unduly cheerful. These observations suggest that lesions in the left hemisphere result in a relative loss of positive feelings, facilitating depression, whereas lesions of the right hemisphere result in a loss of negative feelings, leading to inappropriate optimism.

Figure 29.8 *Asymmetrical smiles on some famous faces. Studies of normal subjects show that facial expressions are often more quickly and fully expressed by the left facial musculature than the right, as suggested by examination of these examples (try covering one side of the faces and then the other). Since the left lower face is governed by the right hemisphere, some psychologists have suggested that the majority of humans are "left-faced," in the same general sense that most of us are right-handed. (After Moscovitch and Olds, 1982; images from Microsoft® Encarta Encyclopedia 98.)*

Hemispheric asymmetry related to emotion is also apparent in normal individuals. For instance, auditory experiments that introduce sound into one ear or the other indicate a right-hemisphere superiority in detecting the emotional nuances in speech. Moreover, when facial expressions are specifically presented to either the right or the left visual hemifield, the depicted emotions are more readily and accurately identified from the information in the left hemifield (that is, the hemifield perceived by the right hemisphere; see Chapters 12 and 27). Finally, kinematic studies of facial expressions show that most individuals more quickly and fully express emotions with the left facial musculature than with the right (recall that the left lower face is controlled by the right hemisphere, and vice versa) (Figure 29.8). Taken together, this evidence is consistent with the idea that the right hemisphere is more intimately concerned with both the perception and expression of emotions than is the left hemisphere. However, it is important to remember that, as in the case of other lateralized behaviors (language, for instance), both hemispheres participate in processing emotion.

Emotion, Reason, and Social Behavior

The experience of emotion—even on a subconscious level—has a powerful influence on other complex brain functions, including the neural faculties responsible for making rational decisions and the interpersonal judgments that guide social behavior. Evidence for this statement has come principally from studies of patients with damage to parts of the orbital and medial prefrontal cortex, as well as patients with injury or disease involving the amygdala (see Box 29D). Such patients often have impairments in emotional processing, especially of the emotions engendered by complex personal and social situations, and they have difficulty making advantageous decisions (see also Chapter 26). Adding to this body of evidence are results from brain imaging studies in normal subjects in which investigators have mapped the brain structures that participate in the necessary emotional and social appraisals.

Antonio Damasio and his colleagues at the University of Iowa have suggested that such decision-making entails the rapid evaluation of a set of possible outcomes with respect to the future consequences associated with each course of action. It seems plausible that the generation of conscious or subconscious mental images that represent the consequences of each contingency triggers emotional states that involve either actual alterations of somatic and visceral motor function, or the activation of neural representations of such activity. Whereas William James proposed that we are "afraid because we tremble," Damasio and his colleagues suggest a vicarious representation of motor action and sensory feedback in the neural circuits of the frontal and parietal lobes. It is these vicarious states, according to Damasio, that give mental representations of contingencies the emotional valence that helps an individual to identify favorable or unfavorable outcomes.

Experimental studies of fear conditioning have implied just such a role for the amygdala in associating sensory stimuli with aversive consequences. For example, the patient described in Box 29D showed an impaired ability to recognize and experience fear, together with impairment in rational decision-making. Similar evidence of the emotional influences on decision-making have also come from studies of patients with lesions in the orbital and medial prefrontal cortex. These clinical observations suggest that the amygdala and prefrontal cortex, as well as their striatal and thalamic connections, are not only involved in processing emotions, but also participate in the complex neural processing responsible for rational thinking. These same neural networks are engaged by sensory stimuli (e.g., facial expressions) that convey important cues pertinent to appraising social circumstances and conventions. Thus, when judging the trustworthiness of human faces—a task of considerable importance for successful interpersonal relations—neural activity in the amygdala is specifically increased, especially when the face in question is deemed untrustworthy (Figure 29.9). It is not surprising, then, that subjects with bilateral damage to the amygdala differ from control subjects in their appraisals of trustworthiness; indeed, individuals with such impairments often show inappropriately friendly behavior toward strangers in real-life social situations. Such evidence adds further weight to the idea that emotional processing is crucial for competent performance in a wide variety of complex brain functions.

Emotional Reinforcement and Addiction

Understanding the neurobiological basis of emotion and the contributions of affective neural processing to higher brain functions remains an important goal of neuroscience in the twenty-first century. The urgency of achieving this goal is highlighted by the pervasiveness and impact of drug abuse and addiction (see Box 6F), behaviors that reflect the vulnerability of emotional neural circuits to dysregulation when exposed to illicit drugs that misappropriate their normally adaptive operations in the guidance of goal-directed behavior. Under normal physiological conditions, emotional processing in the limbic system can signal the presence of or prospect for reward and punishment, as well as promote the activation of motor programs aimed at procuring beneficial rewards and avoiding punishment, as discussed above (see Figure 29.6). With this in mind, it is not surprising that most known drugs of abuse—including heroin, cocaine, alcohol, opiates, marijuana, nicotine, amphetamine and their synthetic analogs—act on elements of limbic circuitry. What may be surprising is that such diverse natural and synthetic compounds do so by altering the neuromodulatory influence of

Figure 29.9 *Activation of the amygdala during judgments of trustworthiness. (A) Functional MRI shows increased neural activation bilaterally in the amygdala when normal subjects appraise the trustworthiness of human faces; activity is also increased in the right insular cortex. (B,C) The degree of activation is greatest when subjects evaluate faces that are considered untrustworthy (Low, Med and High indicate ratings of trustworthiness; Low = untrustworthy). The same effect was observed when subjects were instructed to evaluate the trustworthiness of the faces (explicit condition) or whether the faces were those of high school or university students (implicit condition). (After Winston et al., 2002; A courtesy of J. Winston.)*

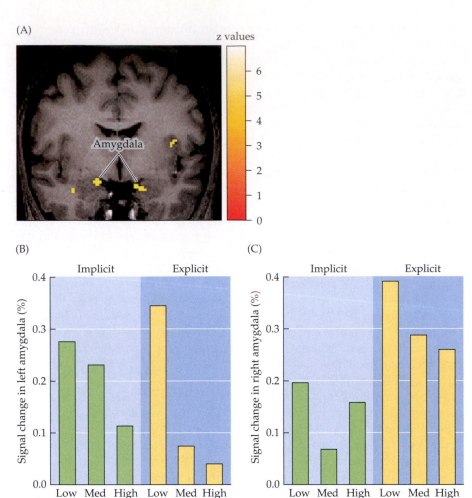

dopamine on the processing of reinforcement signals in the ventral divisions of the basal ganglia, which in turn leads to the consolidation of addictive behavior in limbic circuitry. To understand the neural mechanisms of addiction, therefore, it is necessary to return to the anatomy and physiology of the basal ganglia in the context of emotional reinforcement.

Recall from Chapter 18 that the dorsal divisions of the basal ganglia (dorsal caudate, putamen, and globus pallidus) are instrumental in gating the activation of thalamocortical circuits that initiate volitional movement. Also mentioned briefly in Chapter 18 is the existence of other, parallel processing streams that similarly gate the activation of non-motor programs, including those that pertain to cognition and affective processing in limbic circuits (see Box 18D). The organization of these non-motor processing streams is fundamentally comparable to the "direct pathway" for volitional movement: there are major excitatory inputs from cortex to striatum, neuromodulatory projections from midbrain dopaminergic neurons to striatum, internuclear connections from striatum to pallidum, and output projections from pallidum to thalamus. What distinguishes the "limbic loop" from the "motor loop" (discussed in detail in Chapter 18) is the source and nature of the cortical input, the relevant divisions of striatum and pallidum that processes this input, the source of dopaminergic projections from the midbrain, and the thalamic target of the pallidal output (Figure 29.10).

Central to the organization and function of the limbic loop are inputs from the amygdala, the subiculum (a ventral division of the hippocampal formation)

(A)   (B)

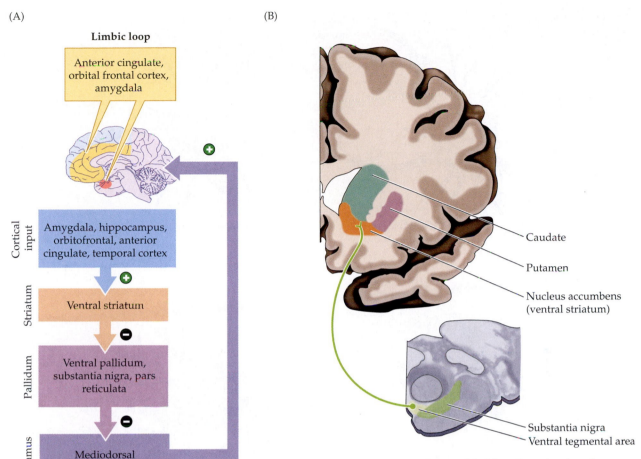

Figure 29.10 Functional and anatomical organization of the limbic loop through the basal ganglia. (A) This circuit comprises cortical inputs to striatum, internuclear projections from striatum to pallidum, pallidal output to thalamus, and thalamic projections back to cortex. (B) Coronal section through the rostral forebrain, showing the basal ganglia structures represented in (A) and the dopaminergic projection from the ventral tegmental area of the midbrain to the nucleus accumbens, the principal component of the ventral striatum.

and orbital-medial prefrontal cortex that convey signals relevant to emotional reinforcement to ventral divisions of the anterior striatum, the largest component of which is called the **nucleus accumbens**. Like the caudate and putamen, the nucleus accumbens contains medium spiny neurons that integrate excitatory telencephalic inputs under the modulatory influence of dopamine. However, unlike the larger dorsal division of the striatum, the nucleus accumbens receives its dopaminergic projections from a collection of neurons that lies just dorsal and medial to the substantia nigra, in a region of the midbrain called the **ventral tegmental area**. The nucleus accumbens and the ventral tegmental area are primary sites where drugs of abuse interact with the processing of neural signals related to emotional reinforcement; they do so by prolonging the action of dopamine in the nucleus accumbens or by potentiating the activation of neurons in the ventral tegmental area and nucleus accumbens (Figure 29.11).

Under normal conditions (i.e., when illicit drugs are absent), these dopaminergic neurons are only phasically active; when they do fire a barrage of action potentials, however, dopamine is released in the nucleus accumbens and medium spiny neurons are much more responsive to coincident excitatory input from telencephalic structures such as the amygdala and orbital-medial prefrontal cortex. These activated striatal neurons in turn project to and inhibit pallidal neurons in a region just below the globus pallidus called the **ventral pallidum**, as well as in the pallidal division of the substantia nigra (pars reticulata). The suppression of tonic activity in the pallidum then disinhibits the thalamic

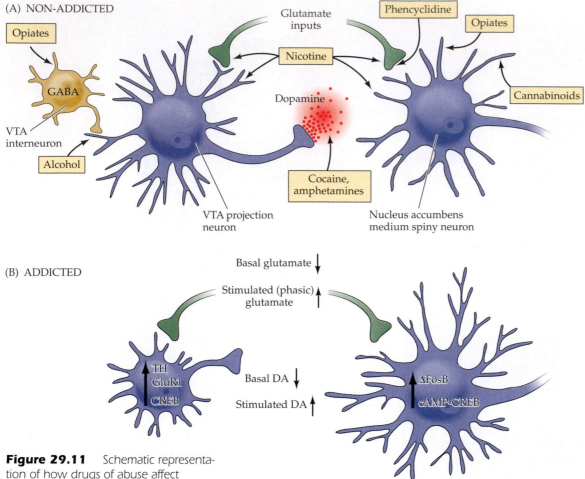

(A) NON-ADDICTED

Opiates

Glutamate inputs

Nicotine

Phencyclidine

Opiates

Dopamine

Cannabinoids

GABA

VTA interneuron

Alcohol

Cocaine, amphetamines

VTA projection neuron

Nucleus accumbens medium spiny neuron

(B) ADDICTED

Basal glutamate ↓

Stimulated (phasic) glutamate ↑

TH
GluR1
CREB

Basal DA ↓

Stimulated DA ↑

ΔFosB

cAMP-CREB

Figure 29.11 *Schematic representation of how drugs of abuse affect dopamine projections from the ventral tegmental area (VTA) to the nucleus accumbens. (A) Most drugs of abuse potentiate the activity of dopamine by interacting directly with dopamine synapses in the nucleus accumbens, or indirectly by modulating the activity of neurons in the ventral tegmental area. Other drugs may act directly on accumbens neurons to increase their responsiveness to telencephalic input. (B) Drug addiction is associated with cellular and molecular adaptations of this circuit (see text for details). The net effect of addiction is a chronic decrease in basal activity and an increase in the intensity of phasic activity in the presence of abusive drugs. (After Nestler, 2005.)*

target of the limbic loop, which is the mediodorsal nucleus, the thalamic nucleus that innervates cortical divisions of the limbic forebrain (see Figure 29.4). Activation of these cortical regions via the mediodorsal nucleus are reinforced by direct cortical projections of dopaminergic neurons in the ventral tegmental area and glutamatergic projections from the basolateral group of nuclei in the amygdala that target these same regions (see Box 29B).

Activation of these complex limbic circuits is believed to instantiate the rewarding effects of natural agents and experiences, such as food, water, micturition, sex, and more complex social rewards. However, the phasic release of dopamine is also subject to experience-dependent plasticity in a manner consistent with classical conditioning. During associative learning, for example, the activity of ventral tegmental neurons comes to signal the presence of the reward-predicting stimulus with diminished responsiveness to the presence of the primary reward itself (Figure 29.12). Interestingly, if a conditioned stimulus is not followed by reward delivery (e.g., because of an inappropriate behavioral response in an instrumental learning paradigm), ventral tegmental neurons are suppressed at precisely the same time that the neuronal response would have signaled the presence of a reward (Figure 29.12C). These observations indicate that the phasic release of dopamine signals the presence of reward relative to its prediction, rather than the unconditional presence of reward. The integration of such signals in the nucleus accumbens, orbital-medial prefrontal cortex, and amygdala leads to the activation of instrumental behaviors directed at obtaining and consolidating the benefits of the rewarding event.

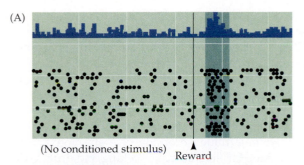

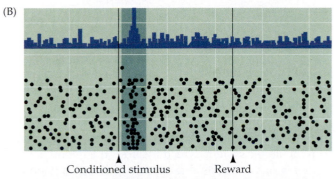

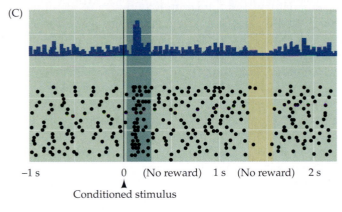

(A)

(No conditioned stimulus) ▲ Reward

(B)

▲ Conditioned stimulus   ▲ Reward

(C)

−1 s   0   (No reward)   1 s   (No reward)   2 s
                    ▲
           Conditioned stimulus

Figure 29.12   *Changes in the activity of ventral tegmental area dopamine neurons in an awake monkey during stimulus–reward learning. In each panel, poststimulus time histograms (blue, above) and data plots (below) report summed activity across trials and individual spikes within trials, respectively. (A) Before learning, the presentation of an unexpected juice reward evokes a burst of activity (darker-shaded zone). (B) After learning trials, the neurons respond to the presentation of visual or auditory cues (conditioned stimuli), but not to the reward itself. (C) In trials when a reward was predicted but never delivered, dopaminergic neurons become suppressed (light-shaded zone) at the time when the reward would have been available. These results show that the ventral tegmental area signals the occurrence of reward relative to its prediction. (After Schultz et al., 1997.)*

Unfortunately, the plastic potential of these limbic circuits can be co-opted by chronic exposure to drugs of abuse, leading to cellular and molecular changes that promote dysregulation (see Figure 29.11B). In the ventral tegmental area of addicted individuals, the activity of the dopamine-synthesizing enzyme tyrosine hydroxylase increases, as does the ability of these neurons to respond to excitatory inputs. The latter effect is secondary to increases in the activity of the transcription factor CREB (see Chapter 7) and the upregulation of GluR1, an important subunit of AMPA receptors for glutamate (see Chapter 6). In the nucleus accumbens, addiction is characterized by increases in another transcription factor, ΔFosB, in addition to induction of CREB with chronic exposure to at least some classes of addictive drugs. Activation of these molecular signaling pathways leads to a generalized reduction in the responsiveness of accumbens neurons to glutamate released by telencephalic inputs by regulation of different AMPA receptor subunits and/or changes in postsynaptic density proteins that alter the dynamics of receptor trafficking. However, during the phasic release of dopamine, the responsiveness of these striatal neurons is intensified, mediated in part by a shift in the expression of D1 and D2 classes of dopamine receptors and a coordinated upregulation of cAMP–PKA signaling pathways.

Clearly, the cellular and molecular maladaptations of ventral tegmental, striatal, and cortical neurons to chronic exposure of abusive drugs remains poorly understood. Nevertheless, the net effect of these and presumably other changes is that addiction dampens the response of this emotional reinforcement circuitry to less potent natural rewards, while intensifying the response to addictive drugs. At a systems level, these changes are likely reflected in the "hypofrontality"—reduced baseline activity in orbital-medial prefrontal cortex—commonly seen in human addicts. Taken together, these alterations in neural processing could account for the waning influence of adaptive emotional signals in the operation of decision making faculties as drug-seeking and drug-taking behaviors become habitual and compulsive.

Summary

The word "emotion" covers a wide range of states that have in common the association of visceral motor responses, somatic behavior, and powerful subjective feelings. The visceral motor responses are mediated by the visceral motor nervous system, which is itself regulated by inputs from many other parts of the brain. The organization of the somatic motor behavior associated with emotion is governed by circuits in the limbic system, which includes the hypothalamus, the amygdala, and several regions of the cerebral cortex. Although a good deal is known about the neuroanatomy and transmitter chemistry of the different parts of the limbic system and how they are affected by drugs of abuse, there is still a dearth of information about how this complex circuitry mediates specific emotional states. Similarly, neuropsychologists, neurologists, and psychiatrists are only now coming to appreciate the important role of emotional processing in other complex brain functions, such as decision-making and social behavior. A variety of other evidence indicates that the two hemispheres are differently specialized for the governance of emotion, the right hemisphere being the more important in this regard. The prevalence and social significance of human emotions and their disorders ensure that the neurobiology of emotion will be an increasingly important theme in modern neuroscience.

Additional Reading

Reviews

ADOLPHS, R. (2003) Cognitive neuroscience of human social behavior. *Nature Rev. Neurosci.* 4: 165–178.

APPLETON, J. P. (1993) The contribution of the amygdala to normal and abnormal emotional states. *Trends Neurosci.* 16: 328–333.

CAMPBELL, R. (1986) Asymmetries of facial action: Some facts and fancies of normal face movement. In *The Neuropsychology of Face Perception and Facial Expression*, R. Bruyer (ed.). Hillsdale, NJ: Erlbaum, pp. 247–267.

DAVIS, M. (1992) The role of the amygdala in fear and anxiety. *Annu. Rev. Neurosci.* 15: 353–375.

LEDOUX, J. E. (1987) Emotion. In *Handbook of Physiology*, Section 1, *The Nervous System*, Vol. 5. F. Blum, S. R. Geiger and V. B. Mountcastle (eds.). Bethesda, MD: American Physiological Society, pp. 419–459.

KAUER, J. A. (2004) Learning mechanisms in addiction: Synaptic plasticity in the ventral tegmental area as a result of exposure to drugs of abuse. *Annu. Rev. Physiol.* 66: 447–475.

NESTLER, E. J. (2005) Is there a common molecular pathway for addiction? *Nature Neurosci.* 8: 1445–1449.

SCHULTZ, W. (2002) Getting formal with dopamine and reward. *Neuron* 36: 241–263.

SMITH, O. A. AND J. L. DEVITO (1984) Central neural integration for the control of autonomic responses associated with emotion. *Annu. Rev. Neurosci.* 7: 43–65.

WISE, R. A. (2004) Dopamine, learning and motivation. *Nature Rev. Neurosci.* 5: 1–12.

Important Original Papers

BARD, P. (1928) A diencephalic mechanism for the expression of rage with special reference to the sympathetic nervous system. *Am. J. Physiol.* 84: 490–515.

DOWNER, J. L. DE C. (1961) Changes in visual agnostic functions and emotional behaviour following unilateral temporal pole damage in the "split-brain" monkey. *Nature* 191: 50–51.

EKMAN, P., R. W. LEVENSON AND W. V. FRIESEN (1983) Autonomic nervous system activity distinguishes among emotions. *Science* 221: 1208–1210.

KLÜVER, H. AND P. C. BUCY (1939) Preliminary analysis of functions of the temporal lobes in monkeys. *Arch. Neurol. Psychiat.* 42: 979–1000.

MacClean, P. D. (1964) Psychosomatic disease and the "visceral brain": Recent developments bearing on the Papez theory of emotion. In *Basic Readings in Neuropsychology*, R. L. Isaacson (ed.). New York: Harper & Row, pp. 181–211.

Mogenson, G. J., D. L. Jones and C.Y. Kim (1980) From motivation to action: Functional interface between the limbic system and the motor system. *Prog. Neurobiol.* 14: 69–97.

Papez, J. W. (1937) A proposed mechanism of emotion. *Arch. Neurol. Psychiat.* 38: 725–743.

Robinson, T. E. and K. C. Berridge (1993) The neural basis of drug craving: an incentive-sensitization theory of addiction. *Brain Res. Rev.* 18: 247–291.

Ross, E. D. and M.-M. Mesulam (1979) Dominant language functions of the right hemisphere? Prosody and emotional gesturing. *Arch. Neurol.* 36: 144–148.

Books

Appleton, J. P. (ed.) (1992) *The Amygdala: Neurobiological Aspects of Emotion, Memory and Mental Dysfunction*. New York: Wiley-Liss.

Corballis, M. C. (1991) *The Lopsided Ape: Evolution of the Generative Mind*. New York: Oxford University Press.

Damasio, A. R. (1994) *Descartes Error: Emotion, Reason, and the Human Brain*. New York: Avon Books.

Darwin, C. (1890) *The Expression of Emotion in Man and Animals*, 2nd Ed. In *The Works of Charles Darwin*, Vol. 23, 1989. London: William Pickering.

Hellige, J. P. (1993) *Hemispheric Asymmetry: What's Right and What's Left*. Cambridge, MA: Harvard University Press.

Holstege, G., R. Bandler and C. B. Saper (eds.) (1996) *Progress in Brain Research*, Vol. 107. Amsterdam: Elsevier.

James, W. (1890) *The Principles of Psychology*, Vols. 1 and 2. New York: Dover Publications (1950).

LeDoux, J. (1998) *The Emotional Brain: The Mysterious Underpinnings of Emotional Life*. New York: Simon and Schuster.

Rolls, E. T. (1999) *The Brain and Emotion*. Oxford: Oxford University Press.

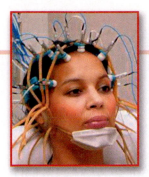

Chapter 30

Sex, Sexuality, and the Brain

Overview

The behaviors that facilitate reproduction and parenting are essential to any organism. Thus it is not surprising that the nervous systems of all animals, including humans, are highly specialized to control these behaviors, including the recognition and selection of mates, sexual intercourse, reproduction, and the feeding and rearing of offspring. For most animals, reproductive and parenting behaviors are unequally distributed between the two sexes. These differences include fundamental anatomical, behavioral, and neurobiological distinctions between the females and the males of any species, a phenomenon referred to as sexual dimorphism. Some of these differences reflect genetic mechanisms that establish male and female gonads and, subsequently, the external genitalia and other physical characteristics (referred to as secondary sex characteristics) that distinguish the two sexes. Corresponding differences in the central nervous system accommodate divergent reproductive and parenting functions. Beyond these clear sex differences lie behaviors, especially in humans, that are associated with sex but whose biological bases remain mostly unknown. These behaviors include the range of emotions and choices that define human sexuality as well as apparent sensory, motor, and cognitive differences that are not strictly associated with reproduction or parenting in males and females. The relationship between these sex- or gender-associated behavioral differences and the observable differences in the brains of males and females remains a highly controversial issue reflected by neurobiological uncertainty as well as literary, sociological, and cultural debate and speculation.

Sexual Dimorphisms and Sexually Dimorphic Behaviors

Sexual dimorphisms—clear and consistent physical differences between females and males of the same species—are seen throughout the animal kingdom, usually associated with functional differences that facilitate reproduction or rearing of offspring. These physical differences are often recognized in peripheral structures found in one sex but not the other. Quite often, divergent neural circuits in female and male brains parallel differences in female and male bodies. For example, in the tobacco hawk moth, *Manduca sexta*, the female and male antennae are strikingly different in size and structure: the female antennae are small and smooth, while the male antennae are larger and lined with rows of ciliated structures (Figure 30.1A). These anatomical specializations are essential for the distinct female and male reproductive behaviors: the female antennae sense specific odors from the tobacco plants that are optimal egg-laying sites, while male antennae are specialized for detecting extremely low concentrations of an airborne pheromone that identifies a nearby female as a potential mate. This physical dimorphism, essential for several aspects of successful reproduction in *M.*

Figure 30.1 Sexually dimorphic anatomy in the hawk moth, *Manduca sexta*. (A) The antennae of male and female moths are specialized for their different roles in courtship and mating behavior. (B) The physical dimorphism of the moth antennae is matched by dimorphism in the olfactory glomeruli of the brain's antennal lobe, which are specialized for odorant-mediated, sex-specific behaviors. The male-specific macroglomerular complex is essential for processing the female pheromone. (A, photos by C. Hitchcock, © Arizona Board of Regents.)

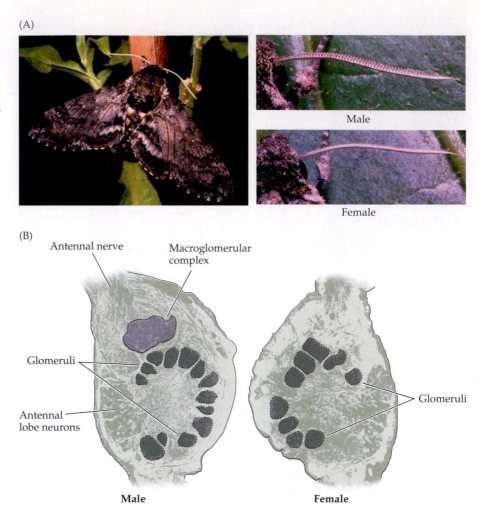

(A)

Male

Female

(B)

Antennal nerve

Macroglomerular complex

Glomeruli

Antennal lobe neurons

Glomeruli

Male

Female

sexta, is matched by dimorphic circuitry in the brain. In each sex, the antennae have olfactory receptor neurons whose axons project to glomerular structures in the antennal lobe (the equivalent of the olfactory bulb; see Chapter 15), whose distinct organization in males and females corresponds to the peripheral dimorphism (Figure 30.1B). Male moths have an additional large glomerulus, referred to as the macroglomerular complex, with neural circuitry specialized for responses to female pheromone. Moreover, the development of this glomerulus is dependent on the antennae that project to it: if a male antenna is transplanted onto the head of a female at a late larval stage, a macroglomerular complex develops in the female brain.

Equally dramatic sexual dimorphisms of body, behavior, and brain are seen throughout the animal kingdom. Among the best characterized are those of several songbird species in which males produce complex songs and females do not. The male song is critical for attracting mates and establishing territorial dominion. Such birdsong is also an important aspect of parenting: male hatchlings learn to sing from adult male "tutors" whose vocalizations they hear and mimic, thus determining their reproductive success. The peripheral structure that produces songs (the syrinx) is larger and more differentiated in male birds. Accordingly, in the songbird brain, the nuclei that control motor as well as sensory aspects of song production are larger in males than in females (see Chapter 24 and 25).

In all animals evaluated thus far—including humans—sexual dimorphism in the brain accords with peripheral differences in female versus male body structures. The dimorphisms in many of these structures in both body and brain are under the control of circulating male and female steroid hormones (especially androgens and estrogen; see below). Thus, female songbirds treated with male-specific levels of gonadal steroids during development acquire a highly differentiated, hypertrophic syrinx. The song-control nuclei in their brains also become masculinized, and these female birds—masculinized in the periphery as well as in the brain—sing like their male counterparts.

Brain and peripheral sexual dimorphism in songbirds and in *Manduca sexta* illustrate a fundamental concept in the neuroscience of sex differences: dimorphisms in body structure are essential for reproductive and parenting behaviors, and these somatic dimorphisms are paralleled by distinct structures and circuits within the brain.

Sex, Gonads, Bodies, and Brains

The word "sex" is perhaps one of the most evocative and complex words in any vocabulary. "Sex" has a broad range of meanings defined by biology, as well as the many definitions circumscribed by human tradition, history, and culture. **Chromosomal sex** is a biological term that refers specifically to an individual's sex chromosomes. Most species have two types of chromosomes: **autosomes**, which are identical in both sexes; and **sex chromosomes**, which are specific to male and female genomes. The number and/or identity of sex chromosomes determines chromosomal sex. In some species, males have three copies of the sex chromosomes, while females have only two. In others, including humans, there are distinct chromosomal identities; most commonly, a male-specific chromosome is present or absent.

Not surprisingly, the genes critical for the definition of primary sex characteristics—male and female gonadal tissues that support male and female gametes—are found on sex chromosomes. The physical state of the gonads and external genitalia is therefore a primary determinant of **phenotypic sex**. A range of secondary sex characteristics further defines an individual's phenotypic sex. These include mammary glands in females, sex-specific hair patterns in males and females, and musculoskeletal as well as organismal size differences. These phenotypic characteristics are more or less related to distinct reproductive and parenting functions in females or males.

In humans, the letters X and Y indicate the sex chromosomes (in contrast with the 22 pairs of autosomes, which are identified by numbers). With few exceptions (see the discussion of intersex conditions below), individuals with two X chromosomes are females, while those with one X and one Y chromosome are males (Figure 30.2A). The basic relationships between human genotype (i.e., chromosomal sex) and phenotype (i.e., phenotypic sex) for primary sex characteristics are best understood for the male genotype, XY. The defining gene for male genotypic and phenotypic sex is found on the Y chromosome: a single gene for a transcription factor known as testis-determining factor (TDF). This gene is commonly called *SRY* (an acronym for *sex reversal on the Y chromosome*, and a reference to how the gene was first identified; see the next paragraph). Remarkably, the *TDF/SRY* gene product is the sole determinant for establishing male gonadal tissues, and thus male phenotypic sex.

In most instances, the XX genotype leads to the development of ovaries, oviducts, uterus, cervix, clitoris, labia, and vagina; the XY genotype typically leads to a phenotype with testicles, epididymis, vas deferens, seminal vesicles, penis, and scrotum (Figure 30.2B). Rare apparently male individuals (and

Figure 30.2 *Chromosomal sex and primary sex determination in humans. (A) The SRY gene, located on the short arm of the Y chromosome, initiates a cascade of gene expression and hormonal signaling that results in masculinization of human genitalia. (B) The genitalia of the early human embryo (weeks 4–7 of gestation) are sexually indifferent. Under the influence of SRY gene product, testes develop and produce hormones that result in male genitalia (left diagrams) between the ninth and twelfth weeks of gestation. Without the influence of the Y chromosome and its SRY gene, the human gonads become ovaries, whose hormonal cascade results in female external genitalia (right).*

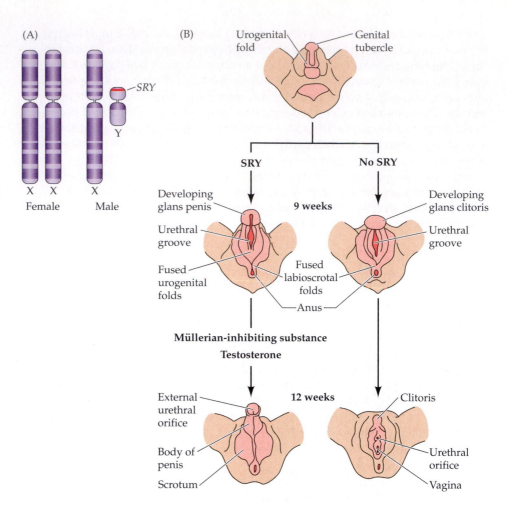

mutant mice, in which the *Sry* gene was first cloned and sequenced) carry two X chromosomes, but one of these X chromosomes has a translocated *SRY* gene (transferred to the paternal X chromosome from the paternal Y chromosome during meiotic recombination).

Despite the overwhelming majority of dual copies of X chromosome genes (and the corresponding lack of additional Y genes), *SRY* alone completely masculinizes these individuals: they become phenotypic males. In fact, whether *SRY* is translocated as an individual gene or transferred with the entire Y chromosome to an XX individual (resulting in an aneuploid XXY phenotype known as Kleinfelter's syndrome; see below), the result is the same: a phenotypic male. Other rare individuals carry a deletion of *SRY* on an otherwise intact Y chromosome and are phenotypic females. *SRY* is thought to be at the top of a genetic network that mediates the differentiation of male primary and secondary sex characteristics (see Figure 30.2B). Surprisingly, however, *SRY* is not expressed in the brain. Thus, as for the moth and the songbird, differences in male and female brains that reflect genotypic and phenotypic sex characteristics arise secondarily, in response to primary peripheral distinctions.

While compelling, this biological account of genotypic and phenotypic sex does not explain the full range of cognitive, emotional, and cultural experiences that humans recognize as part of sex. Accordingly, there is a vocabulary that describes additional facets of human sexuality. **Sexual identity** is the phrase used to define an individual's conscious perception of his or her phenotypic sex. **Sexual orientation** refers to the cognitive experience of emotions and attractions that are associated with human sexual relationships. Sexual orientation is not simply coordinated with many obvious genotypic, phenotypic, or gender-associated characteristics. Homosexual women are not phenotypically masculinized, nor are homosexual men phenotypically feminized. While all evidence indicates that sexual orientation is biologically established, it nevertheless entails self-appraisal and declaration in a social and cultural context.

Gender refers to an individual's subjective perception of their sex and their sexual orientation, and is harder to define—especially in strict biological terms—than chromosomal or phenotypic sex, or even sexual identity and orientation. Indeed, many people consider gender to be a social and political construct rather than a biologically constrained phenomenon. Clinical studies of transgendered individuals, however, challenge this view. These individuals report deeply held identities that oppose their genotypic and phenotypic sex. Their experiences, usually thoroughly assessed in rigorous diagnostic evaluation, often lead to hormonal or surgical reassignment of phenotypic sex. Current understanding of transgendered men and women thus reinforces a sense that an individual's perception of their sex and sexuality—their gender—has some biological components. Regardless of its biological basis, gender entails self-appraisal based on affinity for complex behavioral traits most often associated with one sex or the other, called gender traits. Such gender traits are not necessarily directly applicable to reproduction and parenting; they include the entire catalogue of behaviors and skills that females or males are arguably either more or less "interested in" or "better suited for." While such distinctions remain poorly understood (and in many cases the sources of great controversy), they are quite likely influenced by societal expectations and cultural norms as well as (or even instead of) by biology.

Hormonal Influences on Sexual Dimorphism

Chromosomal sex—particularly the differentiation of male versus female gonadal tissue—sets in motion a series of events that define major phenotypic dimorphisms both in the periphery and in the brain. The **gonadal steroid hormones** testosterone and estrogen (Figure 30.3A) are secreted by the testes and ovaries, respectively, and influence most aspects of sexual dimorphism. Accordingly, the initial differentiation of male and female gonads is the central event for the translation of chromosomal sex into phenotypic sex. Testes development depends on the presence of the *SRY* gene, and thus on the Y chromosome. The differentiation of ovaries is thought to be a "default" pathway regulated by the X chromosome as well as autosomal genes. The two gonadal tissue types secrete very different levels of testosterone and estrogen at distinct developmental times. At critical junctures, these differing hormone levels influence undifferentiated primordial gonadal structures, resulting in divergent developmental paths of the genitalia and, later, the secondary sex characteristics of males and females. This early influence of gonadal steroids on the divergent development of a variety of sexually dimorphic structures is sometimes referred to as the **organizational influence** of these hormones, reflecting their actions in guiding distinct male versus female differentiation in a variety of tissues, including the brain.

Males experience an early surge of testosterone, which, along with the peptide hormone Müllerian inhibiting substance (MIS), masculinizes the genitalia (Figure 30.3B). This early surge of testosterone may help to masculinize the nervous system as well, and ultimately may affect behavior; however, many of the specific effects of testosterone (and estrogen) on the developing male versus female brain happen at somewhat later stages of development, in response to

(A)

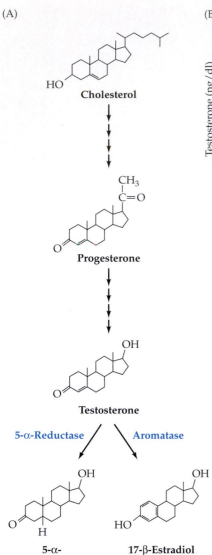

(B)

Testosterone (ng/dl)

350
300
250
200
150
100
50
0

0   5   10   15   20   25   30   35 38

Fetal age (weeks postconception)

Figure 30.3 Gonadal sex steroids and their organizational influence. (A) All sex steroids are synthesized from cholesterol. Cholesterol is first converted to progesterone, the common precursor, by four enzymatic reactions (represented by the four arrows). Progesterone can then be converted into testosterone via another series of enzymatic reactions; testosterone in turn is converted to 5-α-dihydrotestosterone via 5-α-reductase, or to 17-β-estradiol via an aromatase. 17-β-estradiol mediates most of the known hormonal effects in the brains of both female and male rodents. (B) Masculinization of the genitalia reflects increased secretion of testosterone by the immature male testes between the seventh and twentieth weeks of gestation. (B after Gustafson and Donahoe, 1994.)

differential levels of hormones, as well differing sensitivity of target tissues in the maturing fetal and neonatal brain and periphery (see below).

Paradoxically, many of the effects of testosterone in the male brain are really due to estrogens during midgestation (the second trimester of pregnancy in humans) due to the conversion of testosterone to estrogen in the brain. Neurons contain an enzyme called **aromatase** that converts testosterone to **estradiol**, a form of estrogen (see Figure 30.3A). Thus, the surge of testosterone in developing males also generates a surge of estradiol for the affected neurons. There are, however, instances where testosterone acts directly through its own receptors on developing as well as mature neurons. The conversion of testosterone to estrogen may not be as important in humans and other primates, where evidence suggests that sexual differentiation of the brain relies more on androgens acting on androgen receptors. Later in life, it is also androgens that bear most responsibility for stimulating the sex drive in females as well as males.

Estrogen, testosterone, and other steroids are highly lipophilic molecules and are usually transported via circulating carrier proteins in the blood. Surprisingly, many of the details of how the ligands are released from steroidogenic cells and diffuse through membranes of sensitive cells remain to be determined. It is clear, however, that fetuses of placental mammals are exposed to estrogens generated by the maternal ovary and placenta. Why doesn't this maternal estrogen interfere with sexual differentiation in the offspring? Developing mammals have a circulating protein called **α-fetoprotein** that binds circulating estrogens. The female brain is apparently kept from early exposure to large amounts of estrogens because estrogens are bound by α-fetoprotein; the male brain, however, *is* exposed to an early dose of masculinizing steroids via the early testosterone surge from the developing fetal testes (see Figure 30.3B). Testosterone is not affected by α-fetoprotein, and is aromatized to estradiol only once inside neurons.

Gonadal steroids act on cells by binding to specific receptors for either testosterone or estrogen (Figure 30.4A). These receptors are distributed in a fairly limited pattern in the mammalian brain, being especially concentrated at sites where reproductive and parenting functions are represented or sexual dimorphisms are seen (Figure 30.4B). These receptors belong to a larger family

(A)

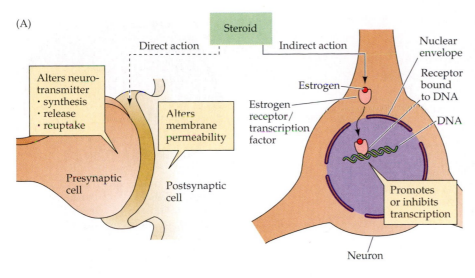

(B)

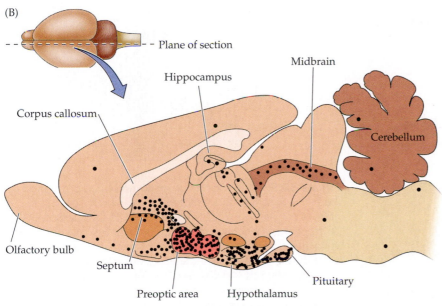

Figure 30.4 Sex steroid effects on neurons. (A) The left panel of this schematic the *direct* effects of steroid hormones on the pre- or postsynaptic membrane, which can alter neurotransmitter release and influence neurotransmitter receptors. The right side of the diagram indicates some *indirect* effects of these hormones, which bind to receptor/transcription factors that act in the nucleus to influence gene expression. (B) Distribution of estradiol-sensitive neurons in a sagittal section of the rat brain. Animals were given radioactively labeled estradiol; dots represent regions where the label accumulated. In the rat, most estradiol-sensitive neurons are located in the preoptic area, hypothalamus, and amygdala. (A after McEwen et al., 1978; B after McEwen, 1976.)

of proteins called the **steroid/thyroid nuclear receptors**, which includes the receptors for vitamin A (retinoic acid; see Chapter 22), vitamin D, and glucocorticoids (see Chapter 29). When testosterone (as 5-α-dihydrotestosterone) or estrogen (as 17β-estradiol) bind to their respective receptors, these receptors are then able to bind to DNA recognition sites and regulate gene expression.

Representation of Primary Sexual Dimorphisms in the Brain

As they exert their organizational influences, testosterone and estrogen each have specific targets, both in the periphery and in the brain. Not surprisingly, many of the central targets are neural structures that control the external genitalia or other specializations (such as mammary glands) that mediate functional dimorphisms for reproduction and parenting (see Figure 30.4B). Perhaps the best example of sexual dimorphism related to motor control of a reproductive behavior is the difference in size of a nucleus in the lumbar segment of the rodent spinal cord called the **spinal nucleus of the bulbocavernosus (SNB)**. The motor neurons of this

(A) Male rat pelvis

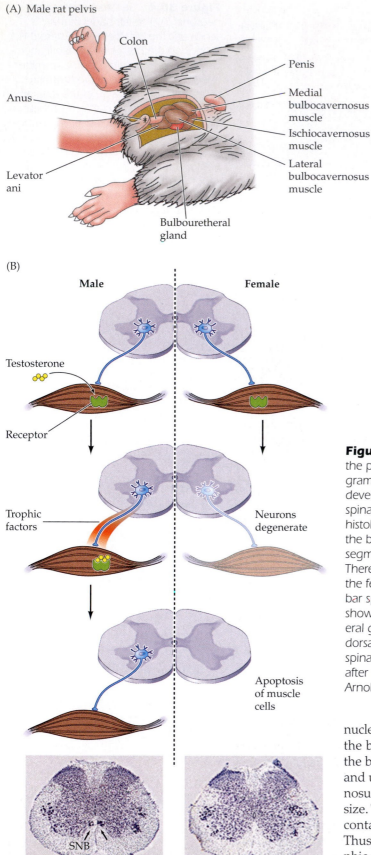

Colon

Penis

Medial bulbocavernosus muscle

Ischiocavernosus muscle

Lateral bulbocavernosus muscle

Anus

Levator ani

Bulbourethral gland

(B)

Male

Female

Testosterone

Receptor

Trophic factors

Neurons degenerate

Apoptosis of muscle cells

SNB

(C)

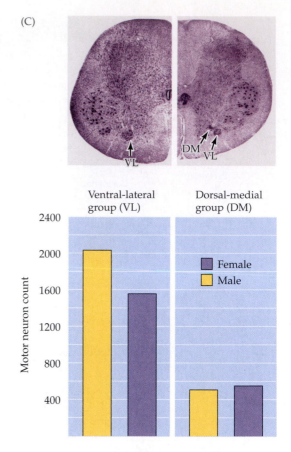

DM VL

VL

Ventral-lateral group (VL)

Dorsal-medial group (DM)

Motor neuron count

Female

Male

Figure 30.5 The number of spinal motor neurons related to the perineal muscles is different in females and males. (A) Diagram of the perineal region of a male rat. (B) Comparison of the developmental sequence for the bulbocavernosus muscle and spinal motor neurons that innervate it, in males and females. The histological cross sections show the dimorphic spinal nucleus of the bulbocavernosus (SNB), found in the fifth lumbar spinal cord segments. Arrows in the photo indicate the SNB in the male. There is no equivalent grouping of densely stained neurons in the female. (C) The micrograph shows Onuf's nucleus in the lumbar spinal cord of a male, and at right, in a female. Histograms show motor neuron counts in the dorsal-medial and ventral-lateral groups of Onuf's nucleus in human males and females. The dorsal-medial group is not visible in the section through the male spinal cord. (A after Breedlove and Arnold, 1984; B, diagram after Morris et al., 2004; photographs from Breedlove and Arnold, 1983; C from Forger and Breedlove, 1986.)

nucleus innervate two striated muscles of the male perineum: the bulbocavernosus and levator ani (muscles that attach at the base of the penis and are involved in both penile erection and urination; Figure 30.5A). In female rats, the bulbocavernosus is absent and the levator ani is dramatically reduced in size. The SNB is significantly larger in overall area in males, contains more neurons, and the neurons are larger in size. Thus, the male and female spinal cords are sexually dimorphic, in parallel with male and female genitalia.

This dimorphism is established via the influence of testosterone around birth in rodents; however, the primary target of gonadal steroid hormone action is not in the brain, but on the perineal muscles (the muscles that control the genitalia). At birth, undeveloped bulbocavernosus muscles are present in males and females, and their size is similar (Figure 30.5B). At this stage, there are also equivalent numbers of neurons in the SNB. The muscles in both the male and the female have androgen receptors (as do all muscles, much to the chagrin of the governing bodies of major sports organizations); however, only males have sufficient endogenous levels of testosterone to activate these receptors. The activation of testosterone receptors in the male perineal muscles spares these muscles from the apoptotic cell death in the bulbocavernosus muscle that occurs shortly after birth in females. Thus, the male peripheral target can support, via trophic interactions (see Chapter 24), the survival and differentiation of significantly more SNB neurons, while the loss of target in the female leads to trophic factor deprivation, cell atrophy, and cell death. If a female is artificially exposed to testosterone, the muscles are rescued, and the number of SNB neurons in this partially masculinized female approaches that of the male. This striking sexual dimorphism represents a special case of a more general developmental rule: structures in the central nervous system are matched to their peripheral targets during development based upon the level of trophic support provided by the target. In addition to this trophic mechanism, SNB neurons in the male spinal cord express testosterone receptors. Activation of these receptors is thought to secondarily regulate the growth of SNB neurons that are larger in males than females. The functional significance of this hormonally induced dimorphism in neuronal size in the male spinal cord is unknown.

As with most sexual dimorphisms, the analogous situation in humans is considerably less clear than in experimental animals. In humans, the spinal cord structure that corresponds to the SNB in rats is called **Onuf's nucleus**. Onuf's nucleus consists of two cell groups in the sacral spinal cord, the dorsal medial and the ventral lateral groups. The dorsal medial group is not sexually dimorphic; however, human females have fewer neurons in the ventral lateral group than males (Figure 30.5C). In contrast to rodents, human females retain a bulbocavernosus muscle throughout life (which serves to constrict the vagina), but the muscle is smaller than in the male. The difference in nuclear size and cell number in humans, as in rats, presumably reflects the difference in the number of bulbous cavernosus muscle fibers that Onuf's nucleus motor neurons innervate.

Thus, in a variety of mammals, including humans, the development, final size, and number of motor neurons that innervate muscles of the external genitalia are apparently responsive to gonadal steroid regulated dimorphic differentiation. This relationship provides a concrete example of the matching of sexual dimorphism in the periphery with that in the central nervous system.

Brain Dimorphisms and the Control of Reproductive Behaviors

In addition to the primary motor innervation of dimorphic genital muscles, there are brain dimorphisms that similarly reflect differences in reproductive function. A major site of these dimorphisms is the hypothalamus (see Chapter 21), presumably because of its central role in the control of visceral motor function—which includes the secretory, vascular, and smooth muscle control necessary for sexual functions in both males and females. The concentration of gonadal steroid receptors in the hypothalamus (see Figure 30.4B) reinforces this conclusion. Neurons in the medial preoptic area of the anterior hypothalamus, where steroid receptors are concentrated, mediate at least some of these behaviors, and in most mammals there are differences in size and cell number of subsets of nuclei in this region.

In rodents, one particular nuclear group, the **sexually dimorphic nucleus of the preoptic area** (**SDN-POA**) is consistently larger and has more neurons in males than in females (Figure 30.6A). The size and cell number of this nucleus are regulated by testosterone; a female exposed to elevated levels of androgens during early postnatal life will develop an enlarged SDN-POA with more neurons. This masculinization is due to diminished cell death in response to elevated gonadal steroids. In contrast to the spinal cord, however, the effects of testosterone on cell survival or death in the SDN-POA are direct, via the conversion to estrogen in the brain, and presumably the subsequent actions of estrogen via its receptors on SND-POA neurons. This dimorphism, like most others, can ultimately be linked to the differing ability of male and female gonadal tissues to provide distinct levels of testosterone and estrogen. Similar dimorphisms have been reported for a number of nuclei in the preoptic area of the human hypothalamus; however, their consistency and relationship to sexual behavior remains controversial (see below).

In a range of laboratory animals, the preoptic area in general, including the SDN-POA, has been implicated in dimorphic sexual behaviors. In male rats, lesions of the entire preoptic area abolish all copulatory behavior, while more discrete lesions of the SDN-POA diminish the frequency of mounting and copulation. In female rats, such lesions yield individuals that avoid male partners and do not display female-specific copulatory behaviors. Thus, the preoptic area

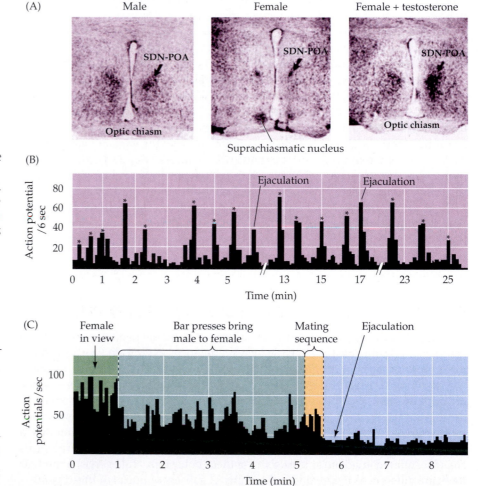

Figure 30.6 Hypothalamic nuclei are sexually dimorphic and their neuronal activity is associated with sexual behaviors. (A) The sexually dimorphic nucleus of the preoptic area (SDN-POA) is larger in male rats (left) than in female rats (center). This size difference can be approximated in genotypically female rats given testosterone perinatally (right). (B) Histograms chart the action potential activity of neurons in the female SDN-POA, correlated with time of sexual receptivity, intromission (insertion of the penis into the vagina; marked by asterisks), and ejaculation. These neurons fire in advance of and during sexual intercourse. (C) Neuronal activity recorded in the medial preoptic area in a male monkey exposed to a receptive female (see text). The firing rate of the neuron changes during different phases of sexual activity. (B after Kato and Sakuma, 2000; C after Oomura et al., 1983.)

is thought to mediate mate selection and the preparative behaviors for copulation, as well as some of the motor and visceral aspects of intromission, ejaculation, and female copulatory responses. Physiological recordings from neurons of the medial preoptic area of the anterior hypothalamus in both rats and monkeys support this view. These neurons fire during different phases of sexual arousal and partner seeking in females and males. Such recordings have been carried out in unrestrained rats with electrode arrays implanted in their brains, and in male monkeys in a flexible restraining apparatus that allows access to a receptive female by pressing a bar that brings the female close enough for the male to mount her. In the female, preoptic area neurons fire rapidly during specific phases of copulation; many of these neurons are maximally active during intromission (penile insertion into the vagina) as well as during the preparatory phases of mounting and the consummatory phase of ejaculation by the partner male (Figure 30.6B). Neurons in the medial preoptic area of the male hypothalamus fire rapidly before sexual behavior, but decrease their activity upon contact with the female, matings, and ejaculation (Figure 30.6B,C). Neurons in the dorsal anterior hypothalamus of both sexes, particularly the paraventricular nucleus, are maximally active during intercourse. Thus the neural structures and circuits that subserve male- and female-specific reproductive behaviors are both structurally and functionally dimorphic.

Structural and Functional Dimorphisms for Parenting Behaviors

Phenotypic characteristics based on the differing roles of males and female mammals in the care and rearing of offspring represent perhaps the next most obvious class of sexually dimorphic structures and behaviors. The most obvious dimorphism is the anatomic and cellular specialization of mammary glands in the periphery of female mammals. The control of lactation, its responsiveness to the hormonal state of a nursing mother, and its relationship to other parental behaviors including nesting, offspring recognition, and grooming provide examples of the **activational effects** (as opposed to organizational effects) of gonadal steroids on the brain and on sexually dimorphic behaviors. Cells sensitive to estrogen, androgens, and related hormones transiently change their functional properties in response to fluctuations in steroidal hormone levels, often activating a sexually dimorphic physiological response or behavior. Such changes are seen in brain regions that do not have dramatic structural dimorphisms.

In mammals, two hypothalamic nuclei, the **paraventricular nucleus** (**PVN**) and the **supraoptic nucleus** (**SON**) are specialized for the control of lactation (Figure 30.7A). These nuclei also influence general control of blood pressure as well as water balance via their control of vasopressin and oxytocin secretion (see Chapter 21). Although they are equivalent in size and shape in females and males, their sensitivity to gonadal steroids, and their ability to alter synaptic organization and function in response to changes in circulating hormones and reproductive state are dramatically different. In females, these nuclei are exquisitely sensitive to the hormone surge that occurs postpartum, and continue to retain functional differences that mediate maternal lactation from the time she gives birth through weaning of young.

For lactation, these changes in the brain are controlled not only by gonadal steroid levels, but by sensory feedback, including feedback from vaginal distension at birth, suckling of pups, and gastric distension that reflects additional food intake by nursing mothers. The information from such peripheral feedback is ultimately relayed by visceral sensory pathways to the PVN/SON in the hypothalamus, eliciting a distinct response in females who have recently given birth and thus need to provide nourishment for their offspring. Prior to pregnancy,

(A)

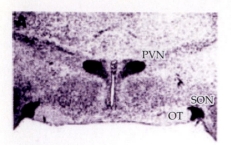

Figure 30.7 Hypothalamic regulation of lactation in nursing mothers. (A) Coronal section through the rat hypothalamus showing the locations of the paraventricular nucleus (PVN) and the supraoptic nucleus (SON) located above the optic tract (OT). These two nuclei are essential for regulating lactation. (B) Electron micrographs show changes in neurons of the rat supraoptic nucleus during lactation. Before birth (left), the relevant neurons and their dendrites are isolated from each other by astrocytic processes (blue). During nursing (right), the astrocytic processes withdraw, and neurons and their dendrites show close apposition (arrow pairs) that allows electrical synapses to form between adjacent neurons (see Chapter 5). (C) Correlation of the rate of action potential discharge (black histogram), presumably from electrically coupled neurons in the SON or PVN, pressure on the mammary region in the ventrum (blue histogram), and the ejection of milk (red trace). When pressure on the mammary gland is increased (as happens at the initiation of nursing by a pup), the SON cell fires a high frequency action potential burst, and shortly thereafter, milk is ejected. (B from Modney and Hatton, 1990; C after Poulain and Wakerley 1982.)

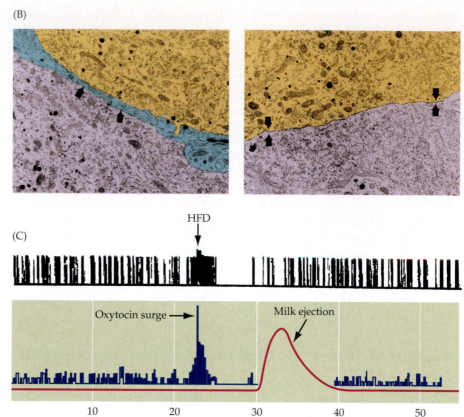

hypothalamic neurons in the PVN/SON are isolated from each other by thin astrocytic processes. Under the influence of the hormonal environment prevailing during birth and lactation, the glial processes retract and the oxytocin- and vasopressin-secreting neurons become electrically coupled by gap junctions (Figure 30.7B). Before females give birth, these neurons fire independently; during lactation, however, they fire synchronously, leading to a high frequency discharge that releases pulses of oxytocin into the maternal circulation. These surges of oxytocin cause the coordinated contraction of smooth muscles in the mammary glands, and hence milk ejection (Figure 30.7C).

Although some aspects of these functional changes are likely mediated by differing levels of gonadal steroids in females versus males (as well as versus virgin, non-pregnant, and pregnant females), there are other indirect influences. One of the most important neural influences arises from olfactory cues (see Chapter 15). The circuit changes that facilitate lactation can be induced in virgin females simply by placing them in the vicinity of pups, where the females recognize distinct pheromonal signals.

There are a number of other dimorphic structures and behaviors associated with lactation and related maternal behaviors. Beyond endocrine and hypothalamic control of these essential functions, other areas of the central nervous system are engaged, including the somatic sensory representation of the body surface in the cerebral cortex. In nursing females this representation dynamically accommodates the increased stimulation provided by nursing offspring. In rats, the cortical representation of the ventrum (the chest region that includes mammary tissue and nipples) enlarges transiently in the somatic sensory cortex of the lactating female. The representation of the ventrum, determined by electrophysiological mapping, is approximately twice as large in nursing females as in non-lactating controls. Moreover, the receptive field sizes of neurons representing the skin of the ventrum in lactating females are decreased to about a third of that of non-lactating females (Figure 30.8), suggesting increased sensitivity and resolution of the somatic sensory stimulation provided by nursing pups. Both the increase in cortical representation and the decrease in receptive field size highlight the fact that changes in behavior requiring increased sensitivity and more vigorous responses can be reflected in changes of cortical circuitry in adult animals (Box 30A). The extent to which these changes are under direct hor-

(A) Female rat

Location of nipples on ventrum

Primary somatic sensory cortex

(B) Nonlactating rat (18 days postpartum)

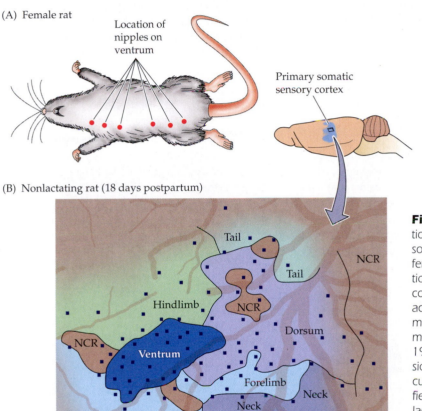

(C) Lactating rat (19 days postpartum)

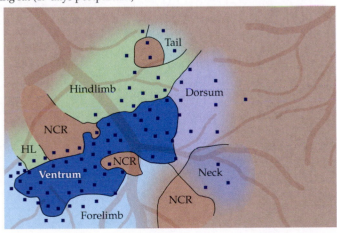

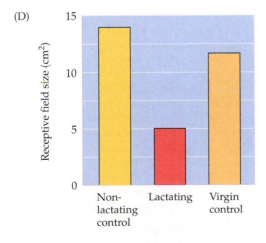

(D)

Figure 30.8 Changes in the cortical representation of the chest wall in the rat primary somatic sensory cortex during lactation. (A) Ventrum of the female rat; dots mark nipple positions. (B) Illustration of somatic sensory cortex in a nonlactating control rat, showing the amount of cortex normally activated by stimulation of the ventrum. Squares mark electrode penetrations; colors signify the estimated representation. (C) Similar diagram from a 19-day postpartum, lactating rat. Note the expansion of the representation of the ventrum. NCR, no cutaneous response. (D) Histogram of receptive field sizes of single neurons in nonlactating control, lactating, and virgin control rats. The receptive field sizes of neurons in lactating mothers are decreased. (B,C after Xerri et al., 1994.)

monal control versus more standard mechanisms of experience-dependent plasticity (see Chapter 24) remain to be determined.

Cellular and Molecular Basis of Sexually Dimorphic Structures and Behaviors

The establishment, maintenance, and plasticity of sexually dimorphic structures and behaviors clearly depends on differences in organizational as well as activational effects of circulating levels gonadal steroids in the brain. Thus, an important goal is to understand how gonadal steroids, their synthetic enzymes, receptors, and related genes influence neuronal structure and function.

One essential cellular target for estrogen and testosterone must be molecular pathways that regulate cell survival and cell death, since many sexually dimor-

---

BOX 30A The Good Mother

In Hollywood movies, fairy tales, and myths, mothers have been portrayed as both saintly (who can forget Stella Dallas?) and evil (remember Medea?). While these stories rarely address the source of maternal warmth and feeling (or lack thereof), recent observations suggest strongly that good mothers are made, not born.

The mothers in question are female rats, whose repertoire of motherly behaviors does not extend to sacrificing her life for her children (as did Stella Dallas) or simply sacrificing her children (Medea). Instead, the sign of a good rat mother is the amount of time she spends licking and grooming her pups when she enters the nest for nursing,

and her posture during nursing itself. The good rat mother arches her back distinctively (see figure), presumably allowing for better access for the pups without extreme spatial confinement. "Bad" rat mothers lick and groom much less frequently and do not arch their backs when nursing. Offspring reared by

high *licking-grooming/arched back nursing* ("LG/ABN") mothers grow up to have much greater adaptive response to stress and more modulated responses to fearful stimuli. When pups from low LG/ABN mice are transferred to high LG/ABN mothers, they acquire stress responses consistent with the maternal

The inset photo shows a "good" rat mother licking and grooming her pups; her back is arched to accommodate nursing. In the model diagrammed here, increased serotonin levels elicited by the increased tactile stimulation provided by a high LG/ABN mother may lead to a signaling cascade that ultimately alters expression of the glucocorticoid receptor (GR) gene. In this model, an alternative promoter for the GR gene is activated via DNA methylation, allowing the transcription factors NGF1-A and AP-2 to bind to GR, yielding increased levels of GR expression in the offspring of these mothers. (After Meaney and Szyf, 2005.)

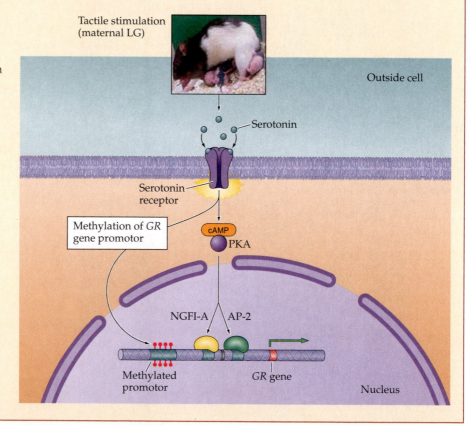

phic nuclei achieve their distinct cell numbers and cell sizes through apoptotic cell death and trophically mediated stabilization and growth of the surviving cells. It is not clear how estrogen and testosterone initially stimulate mechanisms that favor either apoptosis or cell survival and growth; however, recent evidence implies that the ultimate targets of the sex steroids are genes that regulate apoptotic cell death. Male or female mice overexpressing the anti-apoptotic gene *Bcl2* have more neurons in the SNB, while mice in which the pro-apoptotic gene *Bax* has been inactivated no longer display sexual dimorphisms normally seen in the basal forebrain. Thus, the regulation of structural sexual dimorphisms that match phenotypic sex is most likely dependent on gonadal steroid regulation of apoptosis, perhaps further modulated by trophic factors during brain development (see Figure 30.5).

---

qualities of their new mother. Thus, "good" mothering, perhaps as much or even more than genetic predisposition (at least in rats), makes for much better adjusted offspring and insures good mothering for the next generation.

Good maternal behavior apparently is truly essential for the health and well-being of offspring. Thus, the acquisition of motherly skills becomes key in understanding the transmission of adaptive stress responses from generation to generation. Michael Meaney and his colleagues at McGill University therefore asked whether good mother rats—high LG/ABN females—were determined genetically or acquired their maternal behaviors based upon their early experience. By cross-fostering offspring of established low LG/ABN mothers with high LG/ABN mothers and vice versa, they demonstrated that regardless of genetic identity, maternal skills depended on the skills of the mother that reared the pups. High LG/ABN mothers had foster daughters who were also high LG/ABN mothers—even if their birth mothers were low LG/ABN mothers. Similarly, female offspring of high LG/ABN mothers acquired low LG/ABN maternal skills when reared by a low LG/ABN foster mother. These observations suggest strongly that good mothers are made, not born, and that the foundation for maternal skills is established early in life—in part by the

maternal behaviors to which female pups are exposed.

Subsequent work from Michael Meaney's laboratory as well as others has shown that one of the key biological targets of the effects of early maternal behavior on offspring is the expression of glucocorticoid receptors in the hippocampus. These steroid-thyroid family receptors (see Figure 30.3) are key regulators of the stress response throughout the organism. Offspring of high LG/ABN mothers have higher levels of glucocorticoid receptor expression in the hippocampus, and thus are presumably are better equipped to deal with the deleterious effects of stress. Since the maternal behaviors that establish these differences are apparently not strictly encoded by the genome (they are not heritable, and can be acquired based upon experience), a central question that arises is how are differing levels of gene expression established in the offspring of high versus low LG/ABN mothers?

The tentative answer is that high LG/ABN behavior elicits altered levels of serotonergic signaling in the offspring. This signaling, via a specific serotonin receptor and subsequent signaling cascade, establishes differential expression of the glucocorticoid receptor via genomic imprinting: local modification of genomic DNA that leads to long-lasting changes in gene expression.

The details of this intriguing "epigenetic" mechanism for establishing essential differences in behavior remain to be determined. It appears that the glucocorticoid receptor gene has several alternative promoters, and that glucocorticoid receptor expression may be regulated from a particularly efficient promoter in the offspring of high LG/ABN mothers. Whatever the details of the molecular mechanism may be, this work illustrates that early experience can profoundly and—for better or worse—irreversibly alter an entire lifetime of essential behaviors, including parenting behaviors. The potential relevance of these observations to clinical psychology, psychiatry, and public policy are obvious.

Unfortunately, these studies also suggest that the happy endings enjoyed by certain fictional offspring who were fostered by wicked stepmothers—characters like Snow White, Cinderella, and Hansel and Gretl—may need to be rewritten to project a far more grim ever-after than previously imagined.

References

MEANEY, M. J. (2001) Maternal care, gene expression, and the transmission of individual differences in stress reactivity across generations. *Annu. Rev. Neurosci.* 24: 1161–1192.

MEANEY, M. J. AND M. SZYF (2005) Maternal care as a model for experience-dependent chromatin plasticity? *Trends Neurosci.* 28: 456–463.

Gonadal steroids not only influence cell death but in certain instances they can also act as trophic factors, directly regulating neuronal size as well as process growth. During development, and to some extent throughout life, estradiol stimulates brain dimorphisms by increasing cell size, nuclear volume, dendritic length, dendritic branching, dendritic spine density, and synaptic connectivity of the sensitive neurons, independent of cell survival or apoptosis (Figure 30.9A,B).

(A)

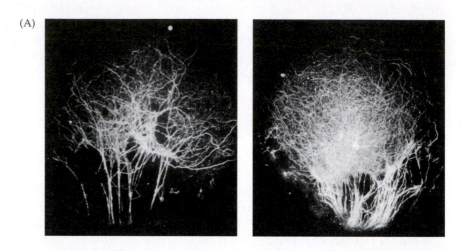

(B)

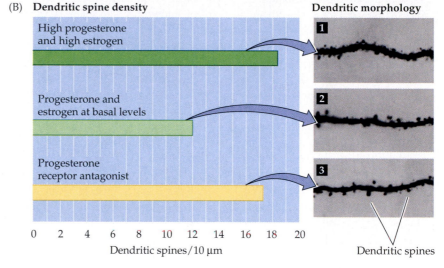

(C)

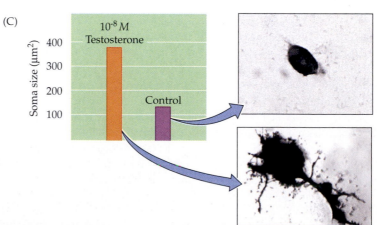

Figure 30.9 Estrogen and testosterone influence neuronal growth and differentiation. (A) A control explant (left) from the hypothalamus shows only a few silver-impregnated processes; an estradiol-treated explant (right) has many more neurites growing from its center. (B) Dendritic spine density in female rat hippocampal neurons in response to progesterone (a precursor of both estrogen and testosterone; see Figure 30.3A) and to estrogen. Recall that dendritic spines, which are small extensions from the dendritic shaft, are sites of synapses. Tracings at right are of representative apical dendrites from hippocampal pyramidal neurons. (1) After administration of progesterone and estrogen in high dosage. (2) After administration of progesterone and estrogen at basal levels. (3) After administration of a progesterone receptor antagonist. (C) The effects of testosterone on embryonic spinal cord neurons in cell culture. In response to testosterone, processes become thicker and more highly branched, and the cell soma grows in size. (A from Toran-Allerand, 1978; B after Woolley and McEwen, 1992; C from Meusberger and Keast, 2001.)

Testosterone can also influence neuronal size and differentiation, at least in vitro, in neurons that express testosterone receptors (Figure 30.9C,D); however, the extent to which these effects rely on the direct action of testosterone on its receptor versus aromatized 17β-estradiol via estrogen receptors is unclear.

Estradiol can also stimulate an increase in the number of synaptic contacts in adult animals. For example, during periods of high circulating estrogen in the estrous cycle of female rodents (or after administration of estrogens), there is an increase in the density of spines (and presumably synapses) on apical dendrites of pyramidal neurons in the hippocampus (see Figure 30.9B). These apparent changes in neuronal circuitry might underlie differences in learning and memory during the course of the estrous cycle. Such differences have been observed in rodents using tests of spatial navigation and memory; however, the relevance of these laboratory behaviors to significant functional differences in reproductive behavior engendered by estrous-dependent hippocampal changes is not understood.

Gonadal steroids can also modify electrical signaling between neurons in a variety of brain regions. Perhaps the most compelling example of this phenomenon is in the periventricular nucleus (PVN) of the hypothalamus, where fluctuating steroid levels facilitate the formation of gap junctions by regulating the transcription of relevant proteins that allow for neuronal synchrony correlated with lactation (see Figure 30.8). In addition, the influences of gonadal steroids, particularly estrogen, on neuronal activity have been evaluated in the hippocampus. The hippocampus was chosen since it is an established site of neuronal plasticity (see Chapters 24 and 31) and is sensitive to hormonal fluctuations, including those seen during estrous. Estrogen receptors are expressed by mature neurons, and often localized to synapses, as well as in the cytoplasm of the cell body (Figure 30.10A). Estrogen (17β-estradiol) can modify excitable properties of hippocampal neurons, including K^+ and Ca^{2+} conductances and the rate of action potential firing. Estrogen can also influence hippocampal synaptic signaling and plasticity. Estrogen at relatively high concentrations (arguably higher than those seen physiologically) can increase the amplitude of excitatory postsynaptic currents (EPSCs) over minutes to hours, and when coupled with high frequency stimulation that elicits long-term potentiation (LTP; see Chapter 24) results in a sustained change in excitatory postsynaptic potentials (EPSPs) (Figure 30.10B). It is tempting to speculate that such changes underlie some of the learned behaviors and memories associated with fluctuating gonadal steroid levels, and therefore associated with distinct reproductive behaviors; however, there is no solid evidence as yet that supports this speculation.

Steroid Receptors and Responses in the Adult Brain

Estrogens and androgens can influence neuronal and glial structure and function throughout life. Thus, the activation effects of gonadal steroids are not necessarily limited to specific reproductive or parenting events like menstruation, parturition, or lactation. The evidence of this influence in humans includes the consequences for behavior of therapeutic removal of the gonads (e.g., hysterectomy for medical reasons in women, or orchiotomy—removal of the testis—to treat testicular cancer in men), as well as responses to acute or chronic pharmacological manipulation of gonadal steroid levels in both humans and experimental animals. Perhaps best known among these apparent influences are the "neuroprotective" effects of estrogens for ischemic and other degenerative changes in neurons, particularly those associated with age or stress-related cognitive decline, including claims for effects of the course of Alzheimer's disease. The significance of these effects remains highly controversial due to difficulties in evaluating the

Figure 30.10 Estrogen influences synaptic transmission. (A) Electron micrograph showing localization of the estrogen receptor α (ERα; the dark, "electron-dense" label) in postsynaptic processes (presumably spines) in the hippocampus. (B) The effects of estrogen on excitatory postsynaptic potentials in individual neurons are augmented by high-frequency stimulation, suggesting that estrogen may modulate use-dependent plasticity in hippocampal synapses. (C) High-frequency stimulation in the presence of estrogen in hippocampal slices (see Chapter 8) results in enhanced long-term potentiation (measured as a change in the slope of the EPSP, which reflects excitatory threshold) following high-frequency stimulation. (After Woolley, 2007.)

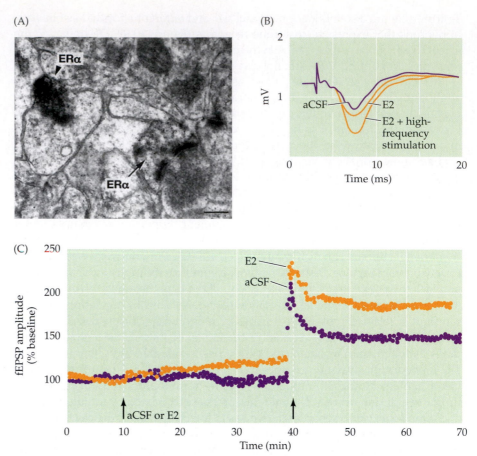

large, heterogenous population of women who have received estrogen replacement therapy.

The consequences of testosterone treatment following orchiotomy, or the effects of illicit steroid use on mood and behavior in athletes (especially aggressive behavior) indicates that the brain remains sensitive to this gonadal steroid throughout life. However, it is not certain whether these phenomena represent direct actions of testosterone via its receptors in the brain. The fairly widespread distribution of estrogen and androgen receptors in the adult brain most likely mediates ongoing activational effects on a wide range of behaviors beyond those directly related to reproduction and parenting (Figure 30.11). Aside from their high concentration in the hypothalamus (see Figure 30.4B), there are significant numbers of both estrogen and androgen receptors in the cerebral cortex, the amygdala, and the substantia nigra. These observations raise a number of important clinical issues, including the relationship between gonadal steroid levels and different responses in men and women to a broad range of medical and surgical therapies, in particular the potential side effects of therapies that manipulate gonadal steroid signaling, including estrogen and androgen antagonists for the treatment of breast and prostate cancer.

Human Genetic Disorders of Genotypic and Phenotypic Sex

Chromosomal sex, phenotypic sex, and gender are not always aligned, and genetic variations in humans challenge the usual definitions of female and male. The general term used to describe all of these inherited variations is **inter-**

Figure 30.11 *Distribution in the rat brain of the three major receptor/transcription factors that bind estrogen (ERα and ERβ) and androgen/testosterone (AR), eliciting corresponding changes in gene expression. ERα, ERβ, and AR tend to be expressed in the same subsets of brain structures. However, these structures are not restricted to the hypothalamic nuclei that control gonadal function, sexual behavior, and parenting behavior; they also include large regions of the cerebral cortex, amygdala, hippocampus, thalamus, substantia nigra, and cerebellum. The significance of gonadal steroid receptor expression and activity at sites beyond the hypothalamus is less well understood than their reproduction-specific functions. They may provide a substrate for the influence of these hormones on behaviors beyond reproduction and parenting, including cognition (cortex), learning and memory (cortex, hippocampus, amygdala), aggression and stress (hippocampus, amygdala), pain sensation (thalamus, brainstem), and motor control (substantia nigra, cerebellum).*

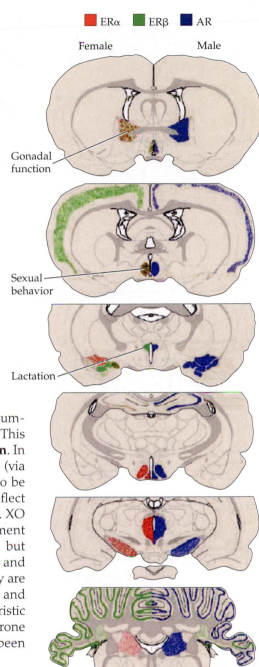

sexuality. Intersexuality is apparent in 1–2 percent of all live births. Affected individuals can experience sterility, sexual dysfunction, psychosocial conflicts over sexual identity gender, and other complications. The most obvious genomic variations that create a misalignment of chromosomal and phenotypic sex are **Turner's syndrome** (XO: fairly rare, between 1 in 2500 and 1 in 10,000 live births); **Klinefelter's syndrome** (XXY; more common, between 1 in 500 and 1 in 2500 live births); and **47-XYY syndrome** (between 1 in 325 and 1 in 1000 live births).

Recall that the *SRY* gene on the Y chromosome dictates gonadal differentiation. Thus an XXY individual will have male genitalia, but often has additional female secondary sex characteristics (e.g., mammary tissue), presumably due to an increased dosage of genes located on the X chromosome. This increased dosage may reflect failure of the normal process of **X-inactivation**. In XX females, one of each copy of most X chromosome genes is inactivated (via DNA modifications that silence a single allele while allowing the other to be expressed) to insure appropriate levels of expression (levels that must reflect those required for viability in males, who have only one X chromosome). XO individuals tend to be small in stature, have rudimentary gonad development and underdeveloped external genitalia (which usually appear female, but hypotrophic, perhaps due to dosage imbalance of X chromosome genes), and are sterile. XYY individuals are the least obviously abnormal, if indeed they are abnormal at all (aside from usually being sterile); their gonadal tissues and external genitalia are male, and their primary identifying physical characteristic is slightly increased height. Earlier studies suggesting that XYY males are prone to criminal behavior or diminished academic accomplishment have been refuted by well-controlled studies.

The above genetic anomalies are all examples of the broader class of genetic variations referred to as **aneuploidy**, reflecting the altered chromosomal number (i.e., ploidy) caused by nondisjunction of chromosomes at meiotic division during gametogenesis. Aneuploid disorders are not restricted to the X and Y chromosomes. They result in several other well-known conditions including Down syndrome, which is associated with duplications of some or all of chromosome 21. Other genetic disorders that result in intersexuality are caused by mutations in individual genes that encode metabolic enzymes that regulate steroid hormone production. One of the most prevalent examples is **congenital adrenal hyperplasia** (**CAH**), which occurs in approximately 1 in 5000 live births. The majority of CAH cases are the result of mutations in the gene encoding 21-hydroxylase, an enzyme responsible for synthesis of two additional steroids secreted by the adrenals: cortisol and aldosterone. In affected individuals, there is increased adrenal testosterone synthesis from metabolic intermediates that would normally yield cortisol and aldosterone, along with severe salt imbalance due to the lack of aldosterone. XY genotypes with CAH are dramatically masculinized individuals who are often very tall at an early age and undergo precocious puberty. In XX genotypes, CAH leads to overactive adrenal secretion of testosterone during development, resulting in abnormally high levels of circulating androgens and hence an ambiguous, masculinized sexual phenotype. In addition to having a large clitoris that resembles a penis and fused labia at birth, female CAH children often exhibit behavioral traits more often associated with boys than with girls. As adults these women may be more likely to form sexual relationships with female partners. By analogy with studies in rodents, the high levels of circulating androgens in CAH individuals may stimulate sexually dimorphic brain circuitry to have a male rather than female organization, leading to more aggressive, male associated behaviors and the eventual choice of a female sexual partner. This explanation, however, remains a hypothesis rather than proven fact.

Androgen insensitivity syndrome (AIS), also called **testicular feminization**, illustrates the results of genetic disruption of receptor-mediated responses to gonadal steroids. The best-studied cases of AIS are males who carry mutations in the gene encoding the receptors for testosterone or Müllerian-inhibiting hormone (see Figure 30.2B). In an XY individual with AIS, the testes form and secrete androgens, as do normal males. The deficiency of androgen receptors, however, leads to the development of female external genitalia with undescended, hypotrophic testes. Thus, chromosomal males with complete androgen insensitivity syndrome look like females and self-identify as female, even though they have a Y chromosome and testicular gonadal tissues. They are often unaware of their condition until puberty (when they fail to menstruate), and thus most AIS individuals see themselves as female and are experienced by the rest of society as female. The gender identity of androgen-insensitive individuals matches their external sexual phenotype, not their chromosomal sex. Androgen-insensitive individuals present one of the strongest arguments that brain circuits in primates are masculinized mainly by the action of androgens rather than by estrogens, which are the masculinizing agent in rodents. In the absence of androgen receptor-mediated responses, even though andogen is available, AIS individuals acquire female-associated behaviors, and presumably, a "feminized" brain.

Another variation in the alignment of chromosomal sex, phenotype, and gender occurs in chromosomal males who are apparently phenotypic females early in life, but whose sexual phenotype changes at puberty. As infants and young children, the genitalia of these individuals resemble those of females more than of males because they are deficient in one form of the enzyme **5-α-reductase**, which catalyzes the conversion of testosterone to the biologically active androgen dihydrotestosterone. This deficiency, which is restricted to the developing genitalia, results in a lack of androgens (which would normally promote the development of male genitalia) in genital tissues during early life. Since their genitalia are ambiguous and their testes have not descended at birth, these individuals are generally recognized and raised as females. At puberty, however, when testicular secretion of androgen increases (at this point synthesis of dihydrotestosterone is regulated by another variant of 5-α-reductase, encoded by a separate gene, that is more active in mature gonadal tissue as well as in the brain throughout life), the clitoris enlarges into a penis and the testes descend, changing these individuals into clearly phenotypic males.

The combined loss-of-function of one form of 5-α-reductase in the embryonic genitalia and maintenance of another in the brain has a provocative consequence: although the genitalia of these individuals are not masculinized in early life, their brains appear to be. Thus, such individuals tend to exhibit masculine behaviors as children and maintain these behaviors as adolescents and adults, despite having been raised as females for their first decade. Anecdotal reports indicate that many such young men retain their masculine gender identity as well as a heterosexual sexual orientation. In the Dominican Republic and Haiti, where this recessive congenital syndrome has been thoroughly studied in particularly consanguinous pedigrees, the condition is referred to colloquially as "testes-at-twelve." This condition is well recognized in areas where it is prevalent, and genetic testing can identify affected children, most of whom will then be raised from birth in a manner consistent with their genotypic sex. Even when the condition goes unrecognized at birth, most such individuals change their gender identification at puberty, assuming male-specific behaviors, sexual identity, and orientation.

The consequences for brain development in each instance of genetic intersexuality are generally clear; however, they lack detailed characterization. Sexual identity, gender identity, and gender-associated behaviors tend to follow the level of exposure of the brain to testosterone during fetal development, dictated by testicular and adrenal metabolism as well as androgen receptor-mediated responses. It is not yet clear, however, whether there are detectable changes in brain structure in intersex individuals consistent with well-established dimorphisms in male and females brains. Recent functional MRI studies have suggested some slight changes in general gray and white matter volume and in the size of discernible brain structures like the amygdala, especially in individuals with CAH. Unfortunately, the numbers of subjects studied thus far is small, and the results have not been consistent. Nevertheless, due to the long-range psychological and social consequences of conflict between genotypic sex, phenotypic sex, and sexual and gender identity, the rapid diagnosis of intersexuality at birth, and a variety of adjustments to assure appropriate matching of the sex of an individual's body and brain, are now standard clinical practice (Box 30B).

Sexual Orientation and the Brain: Molecular and Genetic Analysis

Much of the biology of sex and the brain focuses on the relationship between genotypic sex and phenotypic sex. Thus, the behavioral dimensions of the sex of a person is defined ultimately by the gonadal tissues found in each individual, the pattern of hormonal exposure these tissues provide for the brain, and the subsequent development of the brain to manage neural control of sexual function. This evidence, however, has not yet identified a compelling cellular or molecular basis for understanding a far more subtle, but central aspect of sexual behavior: individual sexual identity. Perhaps the most obvious example of sexual identity is sexual orientation. This most personal of topics has engendered many very public debates. Nevertheless, the biological basis upon which sexual identity, including sexual orientation, rests remains ill defined. Current efforts to define the biology of sexual identity focus on two issues: the genetics and molecular biology of sexual orientation (or more broadly defined, mate choice), and differences in brain structure and function between heterosexual and homosexual individuals (see below). Despite many uncertainties, this work has clearly placed human sexuality—both heterosexual and homosexual—on a much firmer biological foundation. This is a welcome advance over the not-too-distant past when sexual identity and behavior was commonly explained (or stigmatized) in social, psychoanalytical, or moralistic terms.

Ironically—but perhaps not surprisingly—the molecular and genetic control of sexual identity, behavior, and ultimately sexual orientation are perhaps best understood in the fruit fly *Drosophila,* where thorough genetic analysis is possible. These observations address the genetic control of male-specific courtship and mating behaviors, and the influence of the genome on the organization of

BOX 30B   The Case of Bruce/Brenda

One day in the early 1960s, identical XY twins were born to a Canadian couple. When the twins were 7 months old, the parents had them circumcised. The surgeon, performing the operation using an electrocautery knife, burned one of the twins' penises so severely that the organ was, in essence, destroyed. The medical consensus conveyed to the parents by the local physicians was that the disfigured twin would be unable to have a normal heterosexual life, would be shunned by his peers, and would suffer in a variety of other ways. Given this dire prognosis, the parents consulted an eminent sex researcher, John Money at Johns Hopkins University, to help them decide what should be done.

After meeting with the family, Money advocated that they surgically reassign the child's sex and raise the boy as a girl. The parents consented, and at age 17 months the child's testes were removed and his scrotum reshaped to resemble a vulva. The little boy, Bruce, became known as Brenda within the family and personal circle; Money's medical records and published papers used the pseudonyms "John" and "Joan."

The parents did everything they could to raise Brenda as a normal female. Although Money's published reports were optimistic, subsequent interviews with the family, including the child himself, indicated that the truth was far more complex, and indeed deeply problematic. In a detailed follow-up of the case, Milton Diamond of the University of Hawaii and Keith Sigmundson described the struggle that Brenda suffered from the earliest age. The child refused to wear dresses, urinated standing up, always felt that something was wrong, and refused to comply with the hormone treatments that were initiated at puberty. At the age of 14, Brenda demanded to know the truth, and her equally frustrated parents reluctantly gave an account of the early events that had resulted in the current situation.

Brenda was greatly relieved to learn the truth, which explained why "she" had always been subject to such deeply conflicting feelings—feelings that had made life so miserable "she" sometimes contemplated suicide. Brenda immediately reverted to male dress and behavior and started going by the name of David. David eventually underwent surgery to be reconfigured as a phenotypic male. He married, adopted his wife's children, and lived a relatively conventional life as a father and husband.

This case underscores the fact that, in the words of Diamond and Sigmundson, "the evidence [is] overwhelming that normal humans are not psychosocially neutral at birth but are, in keeping with their mammalian heritage, predisposed and biased to interact with environmental, familial, and social forces in either a male or female mode." Cases like David's raise serious moral and ethical questions about the assignment of gender when, for one reason or another, that option is open. Since there is no indication at birth how the brain has been shaped by early exposure to hormones, in many cases there is insufficient information to know with what sex the child, or the adult, will ultimately identify. In David's case, a grievous mistake was made by failing to understand the influence on the brain of circulating androgens during early sexual development. David, whose surname was Reimer, is the subject of a biography by J. Colapinto. Before his death in 2004, Reimer welcomed the opportunity to make his case known in the interest of preventing such mistakes in the future.

References

COLAPINTO, J. (2000) *As Nature Made Him: The Boy Who was Raised as a Girl*. New York: Harper Collins.

DIAMOND, M. AND H. K. SIGMUNDSON (1997) Sex reassignment at birth: Long-term review and clinical implications. *Arch. Ped. Adolesc. Med.* 151: 298–304.

DREGER, A. D. (1998) "Ambiguous sex" or ambiguous medicine? *Hastings Center Report* 28: 24–35

these behaviors. After extensive evaluation of the genetics of sex determination in flies, it became clear that the genes that control phenotypic sex (i.e., the fly equivalent of the human *SRY* gene) were insufficient to explain the differentiation of neural circuits that mediate male- or female-specific sexual behavior. Thus, when mutated, these sex-determining genes do not disrupt male and female courtship behaviors. In an effort to resolve this discrepancy between genetic control of gonadal differentiation and secondary sex characteristics versus the control of sex-specific reproductive behavior, a single male-specific transcription factor, **fruitless**, was identified.

This factor apparently exerts a major effect on organizing neural circuitry in the *Drosophila* brain that mediates male-specific behaviors. A male-specific form of the *fruitless* gene is expressed in subsets of neurons in the fly peripheral and central nervous system. It is not seen in gonadal tissue or other sexually dimorphic peripheral structures. Both males and females have a *fruitless* gene; however, due to alternative splicing, male- and female-specific transcripts are found only in the male or female brain, respectively. This molecular dimorphism is matched by behavioral dimorphisms in courtship and mating behavior, as well as by the ability of males to habituate to (and thus ignore) other males as potential mates. All of these behaviors rely upon chemical sensory, tactile, and auditory circuits. Nevertheless, there are no clear differences in brain size or neuron number between male and female flies. Instead, the male-specific splice variant of *fruitless* is preferentially expressed in a subset of neurons in the brains of male, but not female, flies (Figure 30.12A,B). This observation suggests that *fruitless* regulates the details of neural circuit structure and function. If the male-specific transcript is deleted, males fail to exhibit male-specific courtship. If *fruitless* is absent from all neurons, the flies fail to mate at all. In contrast, when the gene is inactivated only in olfactory neurons of the fly brain, males fail to habituate to other males based on olfactory cues and attempt to court and mate them (Figure 30.12C). Finally, when the male-specific form of *fruitless* is expressed in the female brain, it suppresses female-specific behaviors and

Figure 30.12 Distinct expression of sex-specific splice isoforms of the *Drosophila fruitless* gene correlate with male- and female-specific courtship and mating behaviors. (A) The fruitless protein (green label) is transcribed in subsets of neurons in the male *Drosophila* brain, but not in the female. (B) Alternative splicing of a female-specific exon in *fruitless* pre-mRNA leads to a transcript in which a premature stop codon prevents encoding of the protein. The male-specific transcript encodes a functional protein. (C) Deleting the male-specific transcript in male flies disrupts their ability to habituate to other males. The mutant flies continue to attempt courtship of other males rather than ignoring them after a short period of initial exposure. (D) Female flies do not normally compete for female mates (which defines the "mate preference index"). Similarly, male flies in which the *fruitless* transcript is deleted do not compete for female mates. Normal male flies compete for female mates, as do females in which the *fruitless* transcript has been expressed after experimental insertion of the appropriate male-specific promoter. (A from Demir and Dickson, 2005; C, D after Manoli et al., 2005.)

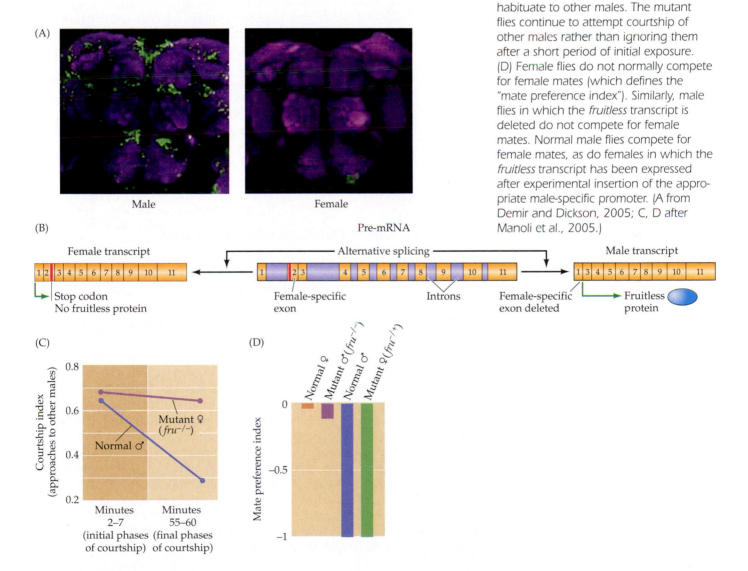

results in male-specific courtship in female flies (Figure 30.12D). These observations indicate that there is genetic control of circuits in the nervous system that mediate recognition and preference of opposite versus same sex, as well as behavioral responses that define—narrowly, in the case of the fly—heterosexual versus homosexual behavior.

Of course, understanding the genetics and molecular biology of human sexual orientation lacks the resolution afforded by extensive genetic analyses possible in *Drosophila*. In humans, the best evidence for a genetic component underlying complex behaviors such as sexual orientation is found in studies of monozygotic twins. Such studies indicate a slightly higher probability that both monozygotic twins will be homosexual than is the case for dizygotic twins or sibling pairs. While this outcome suggests a genetic dimension for sexual orientation, the concordance rate (30 percent; that is, in 30 out of every 100 pairs of monozygotic twins in which one twin is homosexual, the other twin will share that sexual orientation) is lower than for other complex traits or diseases (the concordance rate for schizophrenia, for example, is approximately 50 percent). Although additional genetic linkage studies have attempted to map chromosomal inheritance of male homosexuality, none have provided replicable, definitive linkages. There is, however, some indication that inheritance of male homosexuality may be maternally biased, leading to speculation of an X-linked trait.

Finally, a number of epidemiological observations suggest additional molecular explanations of the establishment of human sexual orientation. Foremost among these is the observation that male sibling birth order correlates strongly with homosexuality. Thus, for each subsequent male child born to a mother, the probability of homosexuality increases by approximately 30 percent. This observation suggested the intriguing hypothesis that women are exposed to male-specific antigens each time they give birth to a male child, possibly establishing an increased immunity that somehow disrupts the ability of subsequent sons to respond to masculinizing signals during intrauterine development. While provocative, this explanation, along with all others, remains speculative. At present, there are no genetic or molecular insights into the basis of sexual identity and behavior in humans that parallel those described in *Drosophila*.

Sexual Orientation and Human Brain Structure

In the early 1990s, several high-profile studies of postmortem brain samples reported anatomical dimorphisms between the brains of homosexual and heterosexual men. This issue was approached primarily in males, most likely because of the increased availability of postmortem brain samples from self-identified homosexual men who suffered AIDS-related deaths—a complicating factor for interpretation of these studies. Such analyses of anatomical dimorphisms are based on the notion that mechanisms resulting in a "homosexual brain" (if such a singular entity exists) would tend to make dimorphic structures more feminized in homosexual men and masculinized in homosexual women.

The structures chosen for analysis were a cluster of nuclei in the human anterior hypothalamus called the **interstitial nuclei of the hypothalamus (INAH)**. The INAH are a reasonable choice, since the animal studies described earlier in the chapter showed that this hypothalamic area (the SDN-POA is its equivalent in rodents) controls sexual arousal and intromissive behaviors. One of these human nuclei, INAH3, is consistently dimorphic in self-identified heterosexual women and men (approximately 40 percent larger in men than women), and was thus considered a likely candidate for "feminization" in homosexual men. Initial studies by Simon LeVay in a much-cited analysis of a sample of postmortem tissue from heterosexual and homosexual men indicated that such differences might exist. These observations, however, were only modestly significant and not absolutely predictive; the size of INAH3 alone was not a reliable indicator of sexual orientation in the sample reported by LeVay. Subsequent studies that have taken into account a significant tendency for degenerative changes in brain tissue from HIV-positive patients, regardless of sexual orientation, failed to replicate this suggested dimorphism. Indeed, several other anatomical dimorphisms of the hypothalamus in heterosexual and homosexual men (women remain largely unstudied) have failed to generate consistent results. The current evidence suggests, therefore, that the volume or number of neurons in sexually dimorphic hypothalamic nuclei alone does not reliably predict sexual orientation, if it has any correlation at all.

The application of functional imaging techniques to this question has added some clarity by mapping the relative activation of potentially dimorphic regions in the brains of heterosexual and homosexual men and women in response to behaviorally relevant stimuli. In these studies, the subjects have been carefully selected for comparable age, consistent sexual orientation (both heterosexual and homosexual), relationship status (a balanced percentage of both the heterosexual and homosexual subjects were in committed relationships), age, and HIV status (none of the subjects were HIV-positive). Heterosexual men and women show differential patterns of hypothalamic versus olfactory activation when presented with estrogens and androgens as odorants (see Chapter 15). Both types of compounds elicit olfactory percepts in men and women, and the conscious experience of each odorant (reported using a descriptive scale of pleasantness, familiarity, intensity, and irritability) is apparently the same. Nevertheless, androgens elicit activation in the anterior hypothalamus only in women; there is little or no focal activation in men. In contrast, estrogens elicit activation in the posterior hypothalamus in men, with little hypothalamic activation in women (there is, however, activation in the amygdala in response to estrogen as an odor in women). There is no noticeable difference in activation of other olfactory structures (e.g., the pyriform cortex), and sex-neutral odors do not elicit these distinct activation patterns.

These fMRI studies have revealed concordance (as high as 95 percent) between homosexual men and heterosexual women, as well as concordance between homosexual women and heterosexual men: androgens activate the anterior hypothalamus in homosexual men, as they do in heterosexual women; and estrogens activate the anterior hypothalamus in homosexual women, as in heterosexual men (Figure 30.13). The behavioral significance of this functional dimorphism is not clear. Nevertheless, these observations provide the best evidence to date for the existence of differences in the brain correlated with sexual identity or orientation. Moreover, these differences suggest that there may be some acquisition of female characteristics in the brains of homosexual men and male characteristics in the brains of homosexual women. As in fruit flies, these functional differences may reflect subtle distinctions in the activity or connections of subsets of neurons in brain regions the mediate sexual behaviors rather than gross dimorphisms related to more obvious differences in phenotypic sex.

Sex-Based Differences in Cognitive Functions

Differences in human brain and behavior that reflect differences in genotypic sex, phenotypic sex, and distinct reproductive and parenting behaviors are currently beyond debate: in this regard, it is easy to state that the brains of men and women are indeed different. When consideration of differences in brain and behavior moves beyond the realm of reproductive activities or lactation, however, the evidence becomes much less clear. Aside from conventional wis-

Figure 30.13 Distinct patterns of activation of estrogen and androgen in heterosexual and homosexual women and men. (A) Androgen elicits focal activation of the hypothalamus (red) in heterosexual women and homosexual men; there is no activation in the hypothalamus of heterosexual men. (B) Estrogen activates the cingulate cortex in heterosexual women, but not the hypothalamus. In lesbian women, estrogen elicits some activation in the hypothalamus, similar to that seen in heterosexual men.

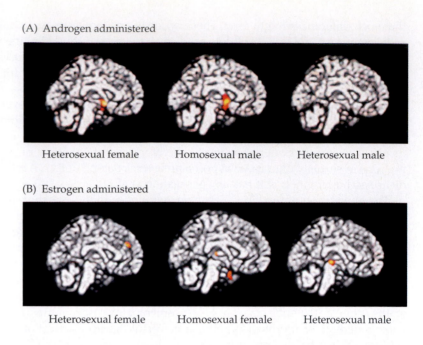

(A) Androgen administered

Heterosexual female   Homosexual male   Heterosexual male

(B) Estrogen administered

Heterosexual female   Homosexual female   Heterosexual male

dom (or prejudice) that boys and girls and/or men and women have greater aptitudes for, or interest in, certain tasks, there is little evidence that the cognitive abilities of men and women differ in ways that are strictly correlated with either genotypic or phenotypic sex. It is more likely that many of the apparent sex differences noted in cognitive tasks that entail language, learning, memory, or visuospatial ability reflect influences not directly related to genetically established sexual dimorphisms. Thus statistically significant differences in the performance of men and women on a variety of tasks are at least as likely to represent social or cultural influences that result in different patterns of learned behavior that may be indirectly related to genotypic or phenotypic sex. It is basically impossible to say if such influences were applied equally to both sexes, equality in performance of the tasks would not be seen.

The issue of brain structural differences beyond those in the spinal cord and hypothalamus has proven even more difficult to evaluate. Postmortem as well as structural MRI analyses suggest that several structures (including cerebral commissures like the corpus callosum and anterior commissure, nuclei like the amygdala, and a number of cortical regions) may differ in size or shape in men and women (Figure 30.14). Many of the relevant anatomical studies, however, are complicated by small samples and highly derived analyses that reveal only small differences in size and shape.

A great deal of interest has focused upon the amygdala as the most likely central site of sexual dimorphism beyond the spinal cord and hypothalamus, perhaps because of its established role in regulating the output of hypothalamic nuclei, as well as the hypersexuality observed in animals or patients with bilateral damage to the amygdala (see Chapter 29). Some MRI studies suggest that the male amygdala has a larger volume than the female; however, additional observations indicate that these differences are not significant when measurements are corrected for brain size or cranial volume. Brain size and cranial volume are robustly correlated with body size, which of course tends to differ in men and women. Studies of intersex subjects (Turner's syndrome, CAH, 47-XYY) do suggest that the amygdala may be sexually dimorphic, perhaps due to the influence of gonadal steroid levels altered by these conditions. These stud-

Cingulate gyrus

Corpus callosum

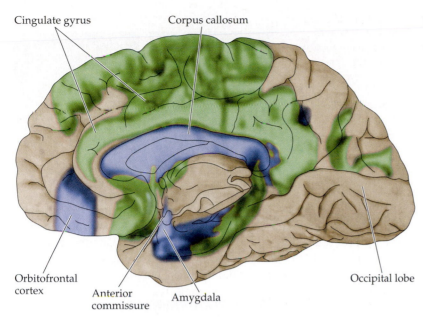

Orbitofrontal cortex

Anterior commissure

Amygdala

Occipital lobe

■ Structures that are larger in healthy female brain, relative to cerebrum size

■ Structures that are larger in healthy male brain, relative to cerebrum size

Figure 30.14 *Summary of brain regions beyond the hypothalamus whose size, on average, differs in the female versus the male brain. It is important to note that these representations are based on mean estimates of size differences. Individual variability makes it impossible to predict an individual's sex based solely on the sizes of the brain regions depicted here. (After Cahill, 2006.)*

ies, however, are complicated by the generally smaller total brain size in genetically intersex individuals. Nevertheless, the apparent trend in a small sample of patients implies that the extra X chromosome present in Turner's syndrome is indeed correlated with a smaller amygdala, and an extra Y chromosome in XYY individuals appears to correlate with a larger amygdala. In contrast, both men and women with CAH tend to have slightly smaller amygdala volumes than the norm, as measured by MRI.

In contrast to this somewhat equivocal anatomical data, there is a fairly robust functional distinction in the amygdala of men or women performing emotional memory tasks. In such tasks, subjects view aversive or frightening films or images that elicit an emotional response. Several weeks later, the subjects are evaluated for their memory of these images. In recalling emotionally charged content, the right amygdala is maximally activated in men, while the left is maximally activated in women (Figure 30.15). These functional differences suggest that *laterality of activation*, rather than size of the nucleus, is the most robust sexual dimorphism in the amygdala. The functional significance of this observation, however, is not known.

In addition to these observations, several other cognitive tasks have been suggested to be sexually dimorphic, based primarily on differential patterns of activation in men's and women's brains seen with fMRI. In each instance, however, the dimorphisms are statistical rather than absolute; therefore, performance on these tasks alone, or observation of the size or shape of the related structure, cannot reliably predict the sex of the individual. The tasks thought to elicit potentially dimorphic behavior include various visuospatial functions, working memory, and language. Although the language dimorphism remains controversial, recent studies have replicated the primary observation that men tend to process language information in a more lateralized manner, with greater activation of left hemisphere regions (see Chapter 27). Parallel anatomical stud-

Figure 30.15 *Activation of the amygdala in men and women differs in response to memory of images with defined emotional content. When men are presented with an image, seen previously, depicting a negative emotion, the right amygdala is focally activated. Women presented with the same image, again seen previously, show focal activation in the left amygdala. (From Cahill et al., 2004.)*

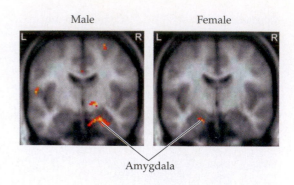

Male   Female

Amygdala

ies—many of which have not been consistently replicated—suggest differences in the hippocampus, prefrontal cortical areas (especially the orbitofrontal cortex, which may be proportionally larger in women), parietal cortex, cingulate cortex, and temporal cortices related to audition and language (see Figure 30.14). While suggestive, none of these observations demonstrate consistent and robust sex differences in cognitive functions or related brain regions. Certainly, none of these putative dimorphisms can be used reliably to distinguish or predict—or to limit—the performance of girls and boys (or women and men) on any behavioral task beyond those genotype/phenotype-dependent behaviors directly related to reproduction and parenting. This conclusion does not mean that such differences do not exist (indeed, many would argue that they do); however, it leave substantial doubt as to whether any observed behavioral differences other than those pertinent to reproduction and parenting are established as a direct consequence of genotypic and phenotypic sex.

Summary

Throughout the animal kingdom, the brains of male and female individuals are specialized for the division of behavioral tasks that deal with reproduction and the rearing of offspring. In mammals, the strongest determinant of these differences is the initial differentiation of gonadal tissues, under the control of the masculinizing transcription factor Sry. This factor determines an individual's genetic sex, and usually the phenotypic sex as well. The influence of Sry, however, is indirect; it is not expressed in the male or female brain. Instead, the development of male or female gonadal tissue, as a consequence of Sry-mediated masculinization, leads to differential production of circulating gonadal hormones (estrogen and testosterone in particular) that profoundly influence the development of the brain structures that subserve peripheral structures (genitalia, mammary glands) directly related to reproduction and parenting. These structures include groups of motor neurons in the spinal cord as well as distinct subsets of neurons in the anterior and medial hypothalamus. The activity of these neurons is directly related to mating behaviors (arousal, intromission, ejaculation) and/or primary parenting behaviors (nursing), and the structures in which these neurons are located are anatomically dimorphic in males and females. Some of these dimorphisms reflect trophic regulation of cell survival and death in relevant structures based on anatomical differences in the peripheral organs that these cells innervate or regulate (male and female genitalia, mammary glands in females, and related muscles). The existence and functional significance of dimorphisms related to more subtle distinctions in gender identity and sexual orientation remain far more controversial. These differences, while clearly biologically based, are not likely to be simply related to dimorphisms that coordinate the peripheral determination of genetic and phenotypic sex with the organization of the brain that subserves distinct reproductive and parenting functions in males and females.

Additional Reading

Reviews

BLACKLESS, M., A. CHARUVASTRA, A. DERRYCK, A. FAUSTO-STERLING, K. LAUZANNE AND E. LEE (2000) How sexually dimorphic are we? Review and synthesis. *Amer. J. Human Biol.* 12: 151–166.

MACLUSKY, N. J. AND F. NAFTOLIN (1981) Sexual differentiation of the central nervous system. *Science* 211: 1294–1302.

MCEWEN, B. S. (1999) Permanence of brain sex differences and structural plasticity of the adult brain. *Proc. Natl. Acad. Sci. USA* 96: 7128–7129.

MORRIS, J. A., C. L. JORDAN AND S. M. BREEDLOVE (2004) Sexual differentiation of the vertebrate nervous system. *Nature Neurosci.* 7: 1034–1039.

SMITH, C. L AND B. W. O'MALLEY (1999) Evolving concepts of selective estrogen receptor action: From basic science to clinical applications. *Trends Endocrinol. Metab.* 10: 299–300.

SWAAB, D. F. (1992) Gender and sexual orientation in relation to hypothalamic structures. *Horm. Res.* 38 (Suppl. 2): 51–61.

SWAAB, D. F. AND M. A. HOFMAN (1984) Sexual differentiation of the human brain: A historical perspective. In *Progress in Brain Research*, Vol. 61. G. J. De Vries (ed.). Amsterdam: Elsevier, pp. 361–374.

Important Original Papers

ALLEN, L. S., M. HINES, J. E. SHRYNE AND R. A. GORSKI (1989) Two sexually dimorphic cell groups in the human brain. *J. Neurosci.* 9: 497–506.

ALLEN, L. S., M. F. RICHEY, Y. M. CHAI AND R. A. GORSKI (1991) Sex differences in the corpus callosum of the living human being. *J. Neurosci.* 11: 933–942.

BEYER, C., B. EUSTERSCHULTE, C. PILGRIM, AND I. REISERT (1992) Sex steroids do not alter sex differences in tyrosine hydroxylase activity of dopaminergic neurons in vitro. *Cell Tissue Res.* 270: 547–552.

BREEDLOVE, S. M. AND A. P. ARNOLD (1981) Sexually dimorphic motor nucleus in the rat lumbar spinal cord: Response to adult hormone manipulation, absence in androgen-insensitive rats. *Brain Res.* 225: 297–307.

BYNE, W., M. S. LASCO, E. KERUETHER, A. SHINWARI, L. JONES AND S. TOBET (2000) The interstitial nuclei of the human anterior hypothalamus: Assessment for sexual variation in volume and neuronal size, density, and number. *Brain Res.* 856: 254–258.

BYNE, W. AND 8 OTHERS. (2002) The interstitial nuclei of the human anterior hypothalamus: An investigation of variation with sex, sexual orientation, and HIV status. *Horm. Behav.* 40: 86–92.

COOKE, B. M., G. TABIBNIA AND S. M. BREEDLOVE (1999) A brain sexual dimorphism controlled by adult circulating androgens. *Proc. Natl. Acad. Sci. USA* 96: 7538–7540.

DEVRIES, G. J. AND 9 OTHERS. (2002) A model system for study of sex chromosome effects on sexually dimorphic neural and behavioral traits. *J. Neurosci.* 22: 9005–9014.

FORGER, N. G. AND S. M. BREEDLOVE (1987) Motoneuronal death during human fetal development. *J. Comp. Neurol.* 264: 118–122.

FREDERIKSE, M. E., A. LU, E. AYLWARD, P. BARTA AND G. PEARLSON (1999) Sex differences in the inferior parietal lobule. *Cerebral Cortex* 9: 896–901.

GORSKI, R. A., J. H. GORDON, J. E. SHRYNE AND A. M. SOUTHAM (1978) Evidence for a morphological sex difference within the medial preoptic area of the rat brain. *Brain Res.* 143: 333–346.

GRON, G., A. P. WUNDERLICH, M. SPITZER, R. TOMCZAK AND M. W. RIEPE (2000) Brain activation during human navigation: Gender different neural networks as substrate of performance. *Nature Neurosci.* 3: 404–408.

LEVAY, S. (1991) A difference in hypothalamic structure between heterosexual and homosexual men. *Science* 253: 1034–1037.

LASCO, M. S., T. J. JORDAN, M. A. EDGAR, C. K. PETITO AND W. BYNE (2002) A lack of dimorphism of sex or sexual orientation in the human anterior commissure. *Brain Res.* 936: 95–98.

MEYER-BAHLBURG, H. F. L., A. A. EHRHARDT, L. R. ROSEN AND R. S. GRUEN (1995) Prenatal estrogens and the development of homosexual orientation. *Dev. Psych.* 31:12–21.

MODNEY, B. K. AND G. I. HATTON (1990) Motherhood modifies magnocellular neuronal interrelationships in functionally meaningful ways. In *Mammalian Parenting*, N. A. Krasnegor and R. S. Bridges (eds.). New York: Oxford University Press, pp. 306–323.

RAISMAN, G. AND P. M. FIELD (1973) Sexual dimorphism in the neuropil of the preoptic area of the rat and its dependence on neonatal androgen. *Brain Res.* 54: 1–29.

ROSSELL, S. L., E. T. BULLMORE, S. C. R. WILLIAMS AND A. S. DAVID (2002) Sex differences in functional brain activation during a lexical visual field task. *Brain Lang.* 80: 97–105.

SWAAB, D. F. AND E. FLIERS (1985) A sexually dimorphic nucleus in the human brain. *Science* 228: 1112–1115.

WALLEN, K. (1996) Nature needs nurture: The interaction of hormonal and social influences on the development of behavioral sex differences in Rhesus monkeys. *Horm. Behav.* 30: 364–378.

WOOLLEY, C. S. AND B. S. MCEWEN (1992) Estradiol mediates fluctuation in hippocampal synapse density during the estrous cycle in the adult rat. *J. Neurosci.* 12: 2549–2554.

XERRI, C., J. M. STERN AND M. M. MERZENICH (1994) Alterations of the cortical representation of the rat ventrum induced by nursing behavior. *J. Neurosci.* 14: 1710–1721.

ZHOU, J.-N., M. A. HOFMAN, L. J. G. GOOREN AND D. F. SWAAB (1995) A sex difference in the human brain and its relation to transsexuality. *Nature* 378: 68–70.

Books

FAUSTO-STERLING, A. (2000) *Sexing the Body*. New York: Basic Books.

GOY, R. W. AND B. S. MCEWEN (1980) *Sexual Differentiation of the Brain*. Cambridge, MA: MIT Press.

LEVAY, S. (1993) *The Sexual Brain*. Cambridge, MA: MIT Press.

LEVAY, S. AND S. M. VALENTE (2006). *Human Sexuality*, 2nd Ed. Sunderland, MA: Sinauer Associates.

Chapter 31

Memory

Overview

One of the most intriguing of the brain's complex functions is the ability to store information provided by experience and to retrieve much of it at will. Without this ability, many of the cognitive functions discussed in the preceding chapters could not occur. *Learning* is the name given to the process by which new information is acquired by the nervous system and is observable through changes in behavior. *Memory* refers to the encoding, storage, and retrieval of learned information. Equally fascinating (and important) is the normal ability to forget information. Pathological forgetfulness, or amnesia, has been especially instructive about the neurological underpinnings of memory; amnesia is defined as the inability to learn new information or to retrieve information that has already been acquired. The importance of memory in daily life has made understanding these several phenomena one of the major challenges of modern neuroscience, a challenge that has only begun to be met. The mechanisms of plasticity that provide plausible cellular and molecular bases for some aspects of information storage were considered in Chapter 8 and in Chapters 23 through 25. The present chapter summarizes the broader organization of human memory, surveys the major clinical manifestations of memory disorders, and considers the implications of these disorders for ultimately understanding human memory in more detailed terms.

Qualitative Categories of Human Memory

Humans have at least two qualitatively different systems of information storage, which are generally referred to as declarative memory and nondeclarative memory (Figure 31.1; see also Box 31A). **Declarative memory** is the storage (and retrieval) of material that is available to consciousness and can in principle be expressed by language (hence, "declarative"). Examples of declarative memory are the ability to remember a telephone number, a song, or the images of some past event. **Nondeclarative memory** (sometimes referred to as *procedural memory*), on the other hand, is not available to consciousness, at least not in any detail. Such memories involve skills and associations that are, by and large, acquired and retrieved at an unconscious level. Remembering how to dial the telephone, how to sing a song, how to efficiently inspect a scene, or making the myriad associations that occur continuously are all examples of memories that fall in this category. It is difficult or impossible to say how we do these things, and we are not conscious of any particular memory during their occurrence. In fact, thinking about such activities may actually inhibit the ability to perform them efficiently (thinking about exactly how to stroke a tennis ball or swing a golf club often makes matters worse).

Figure 31.1 The major qualitative categories of human memory. Declarative memory includes those memories that can be brought to consciousness and expressed as remembered events, images, sounds, and so on. Nondeclarative, or procedural, memory includes motor skills, cognitive skills, simple classical conditioning, priming effects, and other information that is acquired and retrieved unconsciously.

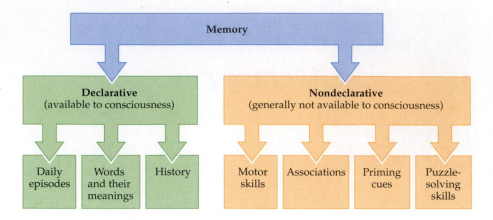

While it makes good sense to divide human learning and memory into categories based on the accessibility of stored information to conscious awareness, this distinction becomes problematic when considering learning and memory processes in animals. From an evolutionary point of view, it is unlikely that declarative memory arose *de novo* in humans with the development of language. Although some researchers continue to argue for different classifications in humans and other animals, recent studies suggest that similar memory processes operate in all mammals and that these memory functions are subserved by homologous neural circuitry. In other mammals, declarative memory typically refers to the storage of information which could, in principle, be declared through language (e.g., "the cheese is in the box in the corner"), and that it is dependent on the integrity of the medial temporal lobe and its associated structures (as discussed later in the chapter). Nondeclarative memory in other animals, as in humans, can be thought of as referring to the learning and storage of sensory associations and motor skills that are not dependent on the medial temporal portions of the brain.

Temporal Categories of Memory

In addition to the types of memory defined by the nature of what is remembered, memory can also be categorized according to the *time* over which it is effective. Although the details are still debated by both psychologists and neurobiologists, three temporal classes of memory are generally accepted (Figure 31.2). The first of these is **immediate memory**. By definition, immediate memory is the routine ability to hold ongoing experiences in mind for fractions of a second.

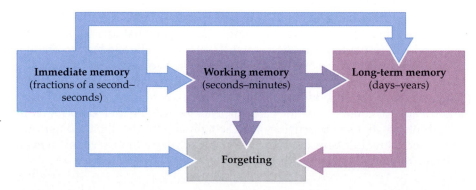

Figure 31.2 The major temporal categories of human memory. Information in both immediate and working memory can enter long-term memory, although most information is promptly forgotten.

BOX 31A Phylogenetic Memory

A category of information storage not usually considered in standard accounts is memories that arise from the experience of the species over the eons, established by natural selection acting on the cellular and molecular mechanisms of neural development. Such stored information does not depend on postnatal experience, but on what a given species has typically encountered in its environment. These "memories" are no less consequential than those acquired by individual experience and are likely to have much underlying biology in common with the memories established during an individual's lifetime. (After all, phylogenetic and ontogenetic memories are based on neuronal connectivity.)

Information about the experience of the species, as expressed by endogenous or "instinctive" behavior, can be quite sophisticated, as is apparent in examples collected by ethologists in a wide range of animals, including primates. The most thoroughly studied instances of such behaviors are those occurring in young birds. Hatchlings arrive in the world with an elaborate set of innate behaviors. First is the complex behavior that allows the young bird to emerge from the egg. Having hatched, a variety of additional behaviors indicate how much of its early life is dependent on inherited information. Hatchlings of precocial species "know" how to preen, peck, gape their beaks, and carry out a variety of other complex acts immediately. In some species, hatchlings automatically crouch down in the nest when a hawk

passes overhead but are oblivious to the overflight of an innocuous bird. Konrad Lorenz and Niko Tinbergen used hand-held silhouettes to explore this phenomenon in naïve herring gulls, as illustrated in the figure shown here. "It soon became obvious," wrote Tinbergen, "that … the reaction was mainly one to shape. When the model had a short neck so that the head protruded only a little in front of the line of the wings, it released alarm, independent of the exact shape of the dummy." Evidently, the memory of what the shadow of a predator looks like is built into the nervous system of this species. Examples in primates include the innate fear that newborn monkeys

have of snakes and looming objects.

Despite the relatively scant attention paid to this aspect of memory, it is probably the most important component of the stored information in the brain that determines whether or not an individual survives long enough to reproduce.

References

TINBERGEN, N. (1969) *Curious Naturalists*. Garden City, NY: Doubleday.

TINBERGEN, N. (1953) *The Herring Gull's World*. New York: Harper & Row.

LORENZ, K. (1970) *Studies in Animal and Human Behaviour*. (Translated by R. Martin.) Cambridge, MA: Harvard University Press.

DUKAS, R. (1998) *Cognitive Ecology*. Chicago: University of Chicago Press.

(A)

(B)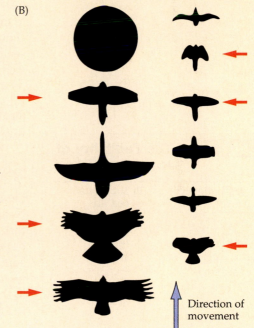

(A) Niko Tinbergen at work. (B) Silhouettes used to study alarm reactions in hatchlings. The shapes that were similar to the shadow of the bird's natural predators (red arrows) when moving in the appropriate direction elicited escape responses (crouching, crying, seeking cover); silhouettes of songbirds and other innocuous species (or geometrical forms) elicited no obvious response. (From Tinbergen, 1969.)

Direction of movement

The capacity of immediate memory is very large and each sensory modality (visual, verbal, tactile, and so on) appears to have its own memory register.

Working memory, the second temporal category, is the ability to hold and manipulate information in mind for seconds to minutes while it is used to achieve a behavioral goal. An everyday example of working memory is searching for a lost object; working memory allows the hunt to proceed efficiently, avoiding places already inspected. A conventional way of testing the integrity of working memory at the bedside is to present a string of randomly ordered digits, which the patient is then asked to repeat; surprisingly, the normal "digit span" is only 7–9 numbers.

The third temporal category is **long-term memory** and entails retaining information in a more permanent form of storage for days, weeks, or even a lifetime. Some of the information stored in both immediate and working memory is thought to enter into long-term memory, although most is evidently forgotten. There is general agreement that the so-called **engram**—the physical embodiment of the long-term memory in neuronal machinery—depends on long-term changes in the efficacy of transmission of the relevant synaptic connections, and/or the actual growth and reordering of such connections. As discussed in Chapter 8, there is good reason to think that both these varieties of synaptic change occur.

Evidence for a continual transfer of information from both immediate and working memory to long-term memory, or **consolidation**, is apparent in the phenomenon of **priming**. Priming is typically demonstrated by presenting subjects with a set of items to which they are exposed under false pretenses. For example, a list of words can be given with the instruction that the subjects are to identify some feature that is actually extraneous to the experiment (e.g., whether the words are verbs, adjectives, or nouns). Sometime thereafter (e.g., the next day) the same individuals are given a different test in which they are asked to fill in the missing letters of words with whatever letters come to mind. The test list actually includes fragments of words that were presented in the first test, mixed among fragments of words that were not. Subjects fill in the letters to make the words that were presented earlier at a higher rate than expected by chance, even though they have no conscious memory of the words that were seen initially; moreover, they are faster at filling in letters to make words that were seen earlier than new words. Priming shows that information previously presented is influential, even though we are entirely unaware of its effect on subsequent behavior. The significance of priming is well known—at least intuitively—to advertisers, teachers, spouses, and others who want to influence the way we think and act.

Despite the prevalence of such transfer, the information stored in this process is not particularly reliable. Consider, for instance, the list of words in Table 31.1A. If the list is read to a group of students who are immediately asked to identify which of several items were on the original list and which were not (Table 31.1B), the result is surprising. Typically, about half the students report that the word "sweet" was included in the list in Table 31.1A; moreover, they are quite certain about it. The mechanism of such erroneous "recognition" is presumably the strong associations that have previously been made between the words on the list in Table 31.1A and the word "sweet," which bias the students to think that "sweet" was a member of the original set. Clearly, memories, even those we feel quite confident about, are often false.

The Importance of Association in Information Storage

The normal human capacity for remembering relatively meaningless information is surprisingly limited (as noted, a string of about 7–9 numbers or other completely arbitrary items). This capacity, however, can be increased dramatically. For example, a college student who for some months spent an hour each day practicing the task of remembering randomly presented numbers was able to recall a string of up to about 80 digits (Figure 31.3). He did this primarily by making subsets of the string of numbers he was given signify dates or times at track meets (he was a competitive runner)—in essence, giving meaningless items a meaningful context. This same strategy of *association* is used by most

TABLE 31.1	**The Fallibility of Human Memory**[a]
(A) Initial list of words	**(B) Subsequent test list**
candy	taste
sour	point
sugar	sweet
bitter	chocolate
good	sugar
taste	nice
tooth	
nice	
honey	
soda	
chocolate	
heart	
cake	
eat	
pie	

[a]After hearing the words in list A read aloud, subjects were asked to identify which of the items in list B had also been on list A. See text for the results.

Figure 31.3 *Increasing the digit span by practice (and the development of associational strategies). During many months involving one hour of practice a day for 3–5 days a week, this subject increased his digit span from 7 to 79 numbers. Random digits were read to him at the rate of one per second. If a sequence was recalled correctly, one digit was added to the next sequence. (After Ericsson et al., 1980.)*

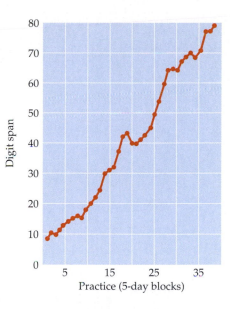

professional "mnemonists" who amaze audiences by apparently prodigious feats of memory. A challenge for many professional mnemonists is memorizing as many as possible of the infinite digits of the mathematical entity π; the current U.S. record is up to 13,000 decimal places and counting! The mnemonist who holds this record is musical, and accomplished the feat of memory by associating digits and musical notes.

Similarly, a good chess player can remember the position of many more pieces on a briefly examined board than an inexperienced player, presumably because the positions have much more significance for individuals who understand the intricacies of the game (Figure 31.4). Thus, the capacity of working memory very much depends on what the information in question means to the individual and how readily it can be associated with information that has already been stored.

(A)

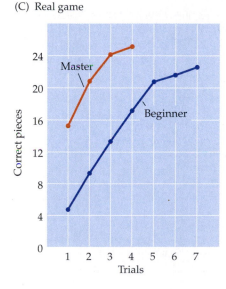

(B)

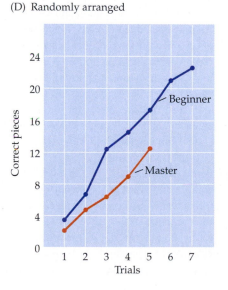

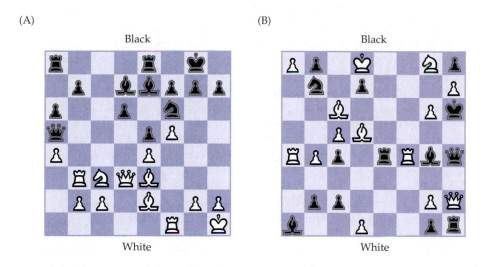

(C) Real game

(D) Randomly arranged

Figure 31.4 *The retention of briefly presented information depends on past experience, context, and its perceived importance. (A) Board position after white's twenty-first move in game 10 of the 1985 World Chess Championship between A. Karpov (white) and G. Kasparov (black). (B) A random arrangement of the same 28 pieces. (C, D) After briefly viewing the board from the real game, master players reconstruct the positions of the pieces with much greater efficiency than beginning players. With a randomly arranged board, however, beginners perform as well or better than accomplished players. (After Chase and Simon, 1973.)*

The ability of humans to remember significant information in the normal course of events is, in fact, enormous. Consider Arturo Toscanini, the late conductor of the NBC Philharmonic Orchestra, who allegedly kept in his head the complete scores of more than 250 orchestral works, as well as the music and librettos for some 100 operas. Once, just before a concert in St. Louis, the first bassoonist approached Toscanini in some consternation, having just discovered that one of the keys on his bassoon was broken. After a minute or two of deep concentration, the story goes, Toscanini turned to the alarmed bassoonist and informed him that there was no need for concern, since that note did not appear in any of the bassoon parts for the evening's program.

Feats of memory such as those performed by mathematical mnemonists, chess masters, and Arturo Toscanini are not achieved by rote learning, but are a result of the fascination that aficionados bring to their special interests (Box 31B). Such examples suggest that motivation plays an important role in mem-

BOX 31B   *Savant Syndrome*

A fascinating developmental anomaly of human memory is seen in rare individuals who until recently were referred to as *idiots savants*; the current literature tends to use the less pejorative phrase *savant syndrome*. Savants are people who, for a variety of poorly understood reasons (typically brain damage in the perinatal period), are severely restricted in most mental activities but extraordinarily competent and mnemonically capacious in one particular domain. The grossly disproportionate skill compared to the rest of their limited mental life can be striking. Indeed, these individuals—whose special talent may be in calculation, history, art, language, or music—are usually diagnosed as severely retarded.

Many examples could be cited, but a summary of one such case suffices to make the point. The individual whose history is summarized here was given the fictitious name "Christopher" in a detailed study carried out by psychologists Neil Smith and Ianthi-Maria Tsimpli. Christopher was discovered to be severely brain damaged at just a few weeks of age (perhaps as the result of rubella during his mother's pregnancy, or anoxia during birth; the record is uncertain in this respect). He had been institutionalized since childhood because he was unable to care for himself, could

not find his way around, had poor hand-eye coordination, and a variety of other deficiencies. Tests on standard IQ scales were low, consistent with his general inability to cope with daily life. Scores on the Wechsler Scale were, on different occasions, 42, 67, and 52.

Despite his severe mental incapacitation, Christopher took an intense interest in books from the age of about three, particularly those providing factual information and lists (e.g., telephone directories and dictionaries). At about six or seven he began to read technical papers that his sister sometimes brought home from work, and he showed a surprising proficiency in foreign languages. His special talent in the acquisition and use of language (an area in which savants are often especially limited) grew rapidly. As an early teenager, Christopher could translate from—and communicate in—a variety of languages in which his skills were described as ranging from rudimentary to fluent; these included Danish, Dutch, Finnish, French, German, modern Greek, Hindi, Italian, Norwegian, Polish, Portuguese, Russian, Spanish, Swedish, Turkish, and Welsh. This extraordinary level of linguistic accomplishment is all the more remarkable since he had no formal training in language even at the elementary school level, and could not

play tic-tac-toe or checkers because he was unable to grasp the rules needed to make moves in these games.

The neurobiological basis for such extraordinary individuals is not understood. It is fair to say, however, that savants are unlikely to have ability in their areas of expertise that exceeds the competency of normally intelligent individuals who focus passionately on a particular subject. Presumably, the savant's intense interest in a particular cognitive domain is due to one or more brain regions that continue to work reasonably well. Whether because of social feedback or self-satisfaction, savants clearly spend a great deal of their mental time and energy practicing the skill they can exercise more or less normally. The result is that the relevant associations they make become especially rich, as Christopher's case demonstrates.

References

MILLER, L. K. (1989) *Musical Savants: Exceptional Skill in the Mentally Retarded.* Hillsdale, New Jersey: Lawrence Erlbaum Associations.

SMITH, N. AND I.-M. TSIMPLI (1995) *The Mind of a Savant: Language Learning and Modularity.* Oxford, England: Basil Blackwell Ltd.

HOWE, M. J. A. (1989) *Fragments of Genius: The Strange Feats of Idiots Savants.* Routledge, New York: Chapman and Hall.

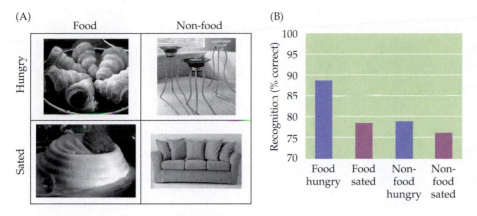

Figure 31.5 Motivated memory. (A) Subjects studied a set of pictures of food and nonfood (i.e., furniture) items, and were later tested for their ability to discriminate the pictures they had seen from a new set of pictures. In one condition, subjects were made hungry by withholding food for several hours. (B) Memory for food items was significantly enhanced when subjects were hungry, but there was no effect of hunger on memory for nonfood pictures. Results like these emphasize the importance of motivation and interest for memory performance. (After Morris and Dolan, 2001.)

ory. This does in fact seem to be true, even for mundane tasks like remembering a set of pictures in a laboratory experiment. In one such study, experimenters asked subjects to study a set of photographs that were either pieces of furniture or pieces of food (Figure 31.5). The subjects were later tested with a much larger set of photographs that included images from the previously studied set along with new ones; the subjects merely had to indicate whether a picture was "old" or "new." In one condition, the experimenters increased subjects' hunger by depriving them of food for several hours. Predictably, subjects were much more likely to remember pictures of food when they were hungry than they were to remember the same pictures when they were sated. There was no effect of motivation on memory for pictures of furniture.

Observations like these reveal the importance of motivation and interest for remembering and have important implications for the brain systems underlying memory. Although few can boast the mnemonic prowess of a Toscanini, the human ability to remember the things that deeply interest us—whether baseball statistics, soap opera plots, images of savory foods, or the details of brain structure—is amazing.

Forgetting

Some years ago, a poll showed that 84 percent of psychologists agreed with the statement "everything we learn is permanently stored in the mind, although sometimes particular details are not accessible." The 16 percent who thought otherwise should get the higher marks. Common sense indicates that, were it not for forgetting, our brains would be impossibly burdened with the welter of useless information that is briefly encoded in our immediate memory "buffer." In fact, the human brain is very good at forgetting. In addition to the unreliable performance on tests such as the example in Table 31.1, Figure 31.6 shows that the memory of the appearance of a penny (an icon seen thousands of times since childhood) is uncertain at best, and that people gradually forget what they have seen over the years (network television shows, in this case). Clearly, we forget things that have no particular importance, and unused memories deteriorate over time.

The ability to forget unimportant information may be as critical for normal life as retaining information that *is* significant. One sort of evidence for this presumption is rare individuals who have difficulty with the normal erasure of information. Perhaps the best-known case is a subject studied over several decades by the Russian psychologist A. R. Luria, who referred to the subject simply as "S." Luria's description of an early encounter gives some idea why S, then a newspaper reporter, was so interesting:

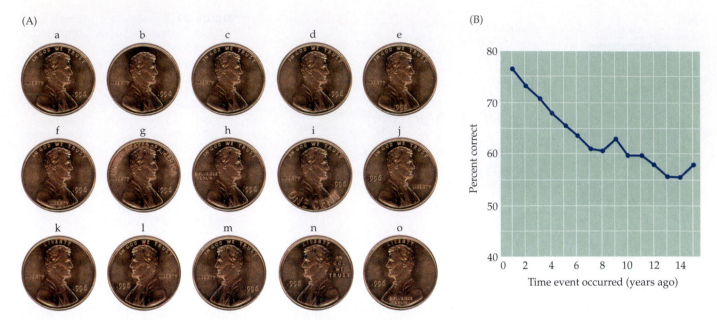

(A)

(B)

Figure 31.6 Forgetting. (A) Different versions of the "heads" side of a penny. Despite innumerable exposures to this familiar design, few people are able to pick out (a) as the authentic version. Clearly, repeated information is not necessarily retained. (B) The deterioration of long-term memories was evaluated in this example by a multiple-choice test in which the subjects were asked to recognize the names of television programs that had been broadcast for only one season during the past 15 years. Forgetting of stored information that is no longer used evidently occurs gradually and progressively over the years (chance performance = 25%). (A after Rubin and Kontis, 1983; B after Squire, 1989.)

I gave S a series of words, then numbers, then letters, reading them to him slowly or presenting them in written form. He read or listened attentively and then repeated the material exactly as it had been presented. I increased the number of elements in each series, giving him as many as thirty, fifty, or even seventy words or numbers, but this too, presented no problem for him. He did not need to commit any of the material to memory; if I gave him a series of words or numbers, which I read slowly and distinctly, he would listen attentively, sometimes ask me to stop and enunciate a word more clearly, or, if in doubt whether he had heard a word correctly, would ask me to repeat it. Usually during an experiment he would close his eyes or stare into space, fixing his gaze on one point; when the experiment was over, he would ask that we pause while he went over the material in his mind to see if he had retained it. Thereupon, without another moment's pause, he would reproduce the series that had been read to him.

A. R. Luria (1987), *The Mind of a Mnemonist*, pp. 9–10

S's phenomenal memory, however, did not always serve him well. He had difficulty ridding his mind of the trivial information that he tended to focus on, sometimes to the point of incapacitation. As Luria put it:

Thus, trying to understand a passage, to grasp the information it contains (which other people accomplish by singling out what is most important) became a tortuous procedure for S, a struggle against images that kept rising to the surface in his mind. Images, then, proved an obstacle as well as an aid to learning in that they prevented S from concentrating on what was essential. Moreover, since these images tended to jam together, producing still more images, he was carried so far adrift that he was forced to go back and rethink the entire passage. Consequently, a simple passage—a phrase, for that matter—would turn out to be a Sisyphean task.

Ibid., p. 113

Although forgetting is a normal and apparently essential mental process, it can also be pathological, a condition called **amnesia**. Some of the causes of memory loss are listed in Table 31.2. An inability to establish new memories following neurological insult is called **anterograde amnesia**, whereas difficulty

TABLE 31.2 Causes of Amnesia

Causes	Examples	Site of damage
Vascular occlusion of both posterior cerebral arteries	Patient R.B. (Box C)	Bilateral medial temporal lobe, the hippocampus in particular
Midline tumors	—	Medial thalamus bilaterally (hippocampus and other related structures if tumor is large enough)
Trauma	Patient N.A. (Box C)	Bilateral medial temporal lobe
Surgery	Patient H.M. (Box C)	Bilateral medial temporal lobe
Infections	Herpes simplex encephalitis	Bilateral medial temporal lobe
Vitamin B1 deficiency	Korsakoff's syndrome	Medial thalamus and mammillary bodies
Electroconvulsive therapy (ECT) for depression	—	Uncertain

retrieving memories established prior to the precipitating neuropathology is called **retrograde amnesia**. Anterograde and retrograde amnesia are often present together, but can be dissociated under various circumstances. Amnesias following bilateral lesions of the temporal lobe and diencephalon have given particular insight into where and how at least some categories of memory are formed and stored, as discussed in the next section.

Brain Systems Underlying Declarative Memory Formation

Three extraordinary clinical cases of amnesia have been especially revealing about the brain systems responsible for the short-term storage and consolidation of declarative information and are now familiar to neurologists and psychologists as patients H.M., N.A., and R.B. (Box 31C). Taken together, these cases provide dramatic evidence of the importance of midline diencephalic and medial temporal lobe structures—the hippocampus, in particular—in establishing new declarative memories (Figure 31.7). These patients also demonstrate that there is a different anatomical substrate for anterograde and retrograde amnesia, since in each of these individuals, memory for events *prior* to the precipitating injury was largely retained.

The devastating deficiency is the result of the inability to establish new memories. Retrograde amnesia—the loss of memory for events preceding an injury or illness—is more typical of the generalized lesions associated with head trauma and neurodegenerative disorders, such as Alzheimer's disease (see Box 31D). Although a degree of retrograde amnesia can occur with the more focal lesions that cause anterograde amnesia, the long-term storage of memories is presumably distributed throughout the brain (see the next section). Thus, the hippocampus and related diencephalic structures indicated in Figure 31.7 form and consolidate declarative memories that are ultimately stored elsewhere.

Other causes of amnesia have also provided some insight into the parts of the brain relevant to various aspects of memory (see Table 31.2). **Korsakoff's syndrome**, for example, occurs in chronic alcoholics as a result of thiamine (vitamin B_1) deficiency. In such cases, loss of brain tissue occurs bilaterally in the mammillary bodies and the medial thalamus, for reasons that are not well understood.

Figure 31.7 Brain areas that, when damaged, tend to give rise to declarative memory disorders. By inference, declarative memory is based on the physiological activity of these structures. (A) Studies of amnesic patients have shown that the formation of declarative memories depends on the integrity of the hippocampus and its subcortical connections to the mammillary bodies and dorsal thalamus. (B) Diagram showing the location of the hippocampus in a cutaway view in the horizontal plane. (C) The hippocampus as it would appear in a histological section in the coronal plane, at approximately the level indicated by the line in (B).

(A) **Brain areas associated with declarative memory disorders**

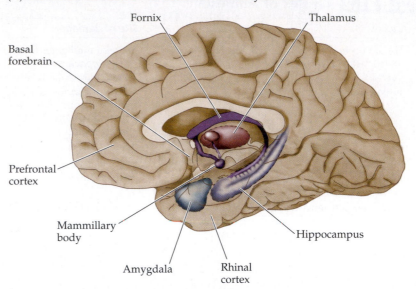

(B) **Ventral view of hippocampus and related structures with part of temporal lobes removed**

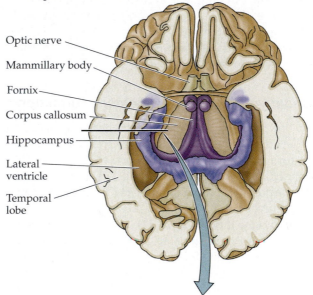

(C) **Hippocampus in coronal section**

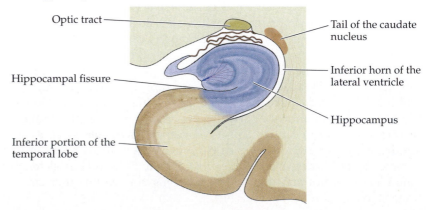

BOX 31C Clinical Cases Reveal the Anatomical Substrate for Declarative Memories

The Case of H.M.

H.M. had suffered minor seizures since age 10 and major seizures since age 16. In 1953, at the age of 27, he underwent surgery to correct his increasingly debilitating epilepsy. A high school graduate, H.M. had been working as a technician in a small electrical business until shortly before the time of his operation. His attacks involved generalized convulsions with tongue biting, incontinence, and loss of consciousness (all typical of grand mal seizures). Despite a variety of medications, the seizures remained uncontrolled and increased in severity. A few weeks before his surgery, H.M. became unable to work and had to quit his job.

On September 1, 1953, surgeons performed a bilateral medial temporal lobe resection in which the amygdala, uncus, hippocampal gyrus, and anterior two-thirds of the hippocampus were removed. At the time, it was unclear that bilateral surgery of this kind would cause a profound memory defect. Severe amnesia was evident, however, upon H.M.'s recovery from the operation, and his life was changed radically.

The first formal psychological exam of H.M. was conducted nearly 2 years after the operation, at which time a profound memory defect was still obvious. Just before the examination, for instance, H.M. had been talking to the psychologist; yet he had no recollection of this experience a few minutes later, denying that anyone had spoken to him. He gave the date as March 1953 and seemed oblivious to the fact that he had undergone an operation, or that he had become incapacitated as a result. Nonetheless, his score on the Wechsler-Bellevue Intelligence Scale was 112, a value not significantly different from his preoperative IQ. Various psychological tests failed to reveal any deficiencies in perception, abstract thinking, or reasoning; he seemed highly motivated and, in the context of casual conversation, normal. Importantly, he also performed well on tests of the ability to learn new skills, such as mirror writing or puzzle solving (that is, his ability to form nondeclarative memories was intact). Moreover, his early memories were easily recalled, showing that the structures removed during H.M.'s operation are not a permanent repository for such information. On the

(Continued on next page)

MRI images of the brain of patient H.M. (A) Sagittal view of the right hemisphere; the area of the anterior temporal lobectomy is indicated by the white dotted line. The intact posterior hippocampus is the banana-shaped object indicated by the white arrow. (B–D) Coronal sections at approximately the levels indicated by the red lines in (A). Image (B) is the most rostral and is at the level of the amygdala. The amygdala and the associated cortex are entirely missing. Image (C) is at the level of the rostral hippocampus; again, this structure and the associated cortex have been removed. Image (D) is at the caudal level of the hippocampus; the posterior hippocampus appears intact, although somewhat shrunken. Outlines below give a clearer indication of the parts of H.M.'s brain that have been ablated (black shading). (From Corkin et al., 1997.)

BOX 31C   *(Continued)*

Wechsler Memory Scale (a specific test of declarative memory), however, he performed very poorly, and he could not recall a preceding test-set once he had turned his attention to another part of the exam. These deficits, along with his obvious inability to recall events in his daily life, all indicate a profound loss of short-term declarative memory function.

During the subsequent decades, H.M. has been studied extensively, primarily by Brenda Milner and her colleagues at the Montreal Neurological Institute. His memory deficiency has continued unabated, and, according to Milner, he has little idea who she is in spite of their acquaintance for nearly 50 years. Sadly, he has gradually come to appreciate his predicament. "Every day is alone," H.M. reports, "whatever enjoyment I've had and whatever sorrow I've had."

The Case of N.A.

N.A. was born in 1938 and grew up with his mother and stepfather, attending public schools in California. After a year of junior college, he joined the Air Force. In October of 1959 he was assigned to the Azores as a radar technician and remained there until December 1960, when a bizarre accident made him a celebrated neurological case.

N.A. was assembling a model airplane in his barracks room while, unbeknownst to him, his roommate was practicing thrusts and parries with a miniature fencing foil behind N.A.'s chair. N.A. turned suddenly and was stabbed through the right nostril. The foil penetrated the cribriform plate (the structure through which the olfactory nerve enters the brain) and took an upward course into the left forebrain. N.A. lost consciousness within a few minutes (presumably because of bleeding in the region of brain injury) and was taken to a hospital. There he exhibited a right-sided weakness and paralysis of the right eye muscles innervated by the third cranial nerve. Exploratory surgery was undertaken and the dural tear

repaired. Gradually he recovered and was sent home to California. After some months, his only general neurological deficits were some weakness of upward gaze and mild double vision. He retained, however, a severe anterograde amnesia for declarative memories. MRI studies first carried out in 1986 showed extensive damage to the thalamus and nearby tracts, mainly on the left side; there was also damage to the right anterior temporal lobe. The exact extent of his lesion, however, is not known, as N.A. remains alive and well.

N.A.'s memory from the time of his injury over 40 years ago to the present has remained impaired and, like H.M., he fails badly on formal tests of new learning ability. His IQ is 124, and he shows no defects in language skills, perception, or other measures of intelligence. He also learns new nondeclarative skills quite normally. His amnesia is not as dense as that of H.M. and is more verbal than spatial. He can, for example, draw accurate diagrams of material presented to him earlier. Nonetheless, he loses track of his possessions, forgets what he has done, and tends to forget who has come to visit him. He has only vague impressions of political, social, and sporting events that have occurred since his injury. Watching television is difficult because he tends to forget the storyline during commercials. On the other hand, his memory for events prior to 1960 is extremely good; indeed, his lifestyle tends to reflect the 1950s.

The Case of R.B.

At the age of 52, R.B. suffered an ischemic episode during cardiac bypass surgery. Following recovery from anesthesia, a profound amnesic disorder was apparent. As in the cases of H.M. and N.A., his IQ was normal (111), and he showed no evidence of cognitive defects other than memory impairment. R.B. was tested extensively for the next 5 years, and, while his amnesia was not as

severe as that of H.M. or N.A., he consistently failed the standard tests of the ability to establish new declarative memories. When R.B. died in 1983 of congestive heart failure, a detailed examination of his brain was carried out. The only significant finding was bilateral lesions of the hippocampus—specifically, cell loss in the CA1 region that extended the full rostral–caudal length of the hippocampus on both sides. The amygdala, thalamus, and mammillary bodies, as well as the structures of the basal forebrain, were normal. R.B.'s case is particularly important because it suggests that hippocampal lesions alone can result in profound anterograde amnesia for declarative memory.

References

CORKIN, S. (1984) Lasting consequences of bilateral medial temporal lobectomy: Clinical course and experimental findings in H.M. *Semin. Neurol.* 4: 249–259.

CORKIN, S., D. G. AMARAL, R. G. GONZÁLEZ, K. A. JOHNSON AND B. T. HYMAN (1997) H. M.'s medial temporal lobe lesion: Findings from MRI. *J. Neurosci.* 17: 3964–3979.

HILTS, P. J. (1995) *Memory's Ghost: The Strange Tale of Mr. M. and the Nature of Memory.* New York: Simon and Schuster.

MILNER, B., S. CORKIN AND H.-L. TEUBER (1968) Further analysis of the hippocampal amnesic syndrome: A 14-year follow-up study of H.M. *Neuropsychologia* 6: 215–234.

SCOVILLE, W. B. AND B. MILNER (1957) Loss of recent memory after bilateral hippocampal lesions. *J. Neurol. Neurosurg. Psychiat.* 20: 11–21.

SQUIRE, L. R., D. G. AMARAL, S. M. ZOLA-MORGAN, M. KRITCHEVSKY AND G. PRESS (1989) Description of brain injury in the amnesic patient N.A. based on magnetic resonance imaging. *Exp. Neurol.* 105: 23–35.

TEUBER, H. L., B. MILNER AND H. G. VAUGHN (1968) Persistent anterograde amnesia after stab wound of the basal brain. *Neuropsychologia* 6: 267–282.

ZOLA-MORGAN, S., L. R. SQUIRE AND D. AMARAL (1986) Human amnesia and the medial temporal region: Enduring memory impairment following a bilateral lesion limited to the CA1 field of the hippocampus. *J. Neurosci.* 6: 2950–2967.

Studies of animals with lesions of the medial temporal lobe have largely corroborated these findings with human patients. For example, one test of the presumed equivalent of declarative memory formation in animals involves placing rats into a pool filled with opaque water, thus concealing a submerged platform; note that the pool is surrounded by prominent visual landmarks (Figure 31.8).

(A)

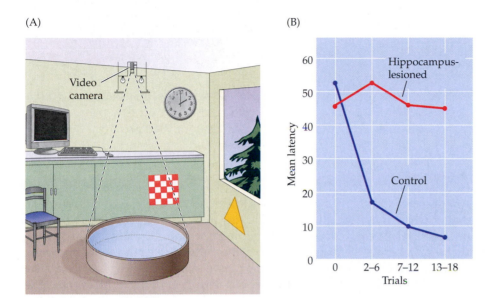

(B)

(C) Control rat

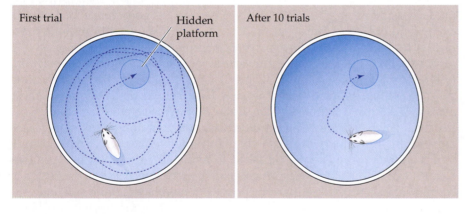

First trial   Hidden platform   After 10 trials

(D) Rat with hippocampus lesioned

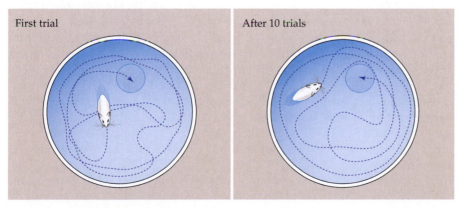

First trial   After 10 trials

Figure 31.8 Spatial learning and memory in rodents depends on the hippocampus. (A) Rats are placed in a circular tank about the size and shape of a child's wading pool filled with opaque (milky) water. The surrounding environment contains visual cues such as windows, doors, a clock, and so on. A small platform is located just below the surface. As rats search for this resting place, the pattern of their swimming (indicated by the traces in C) is monitored by a video camera. (B) After a few trials, normal rats rapidly reduce the time required to find the platform, whereas rats with hippocampal lesions do not. Sample swim paths of normal rats (C) and hippocampal lesioned rats (D) on the first and tenth trials. Rats with hippocampal lesions are unable to remember where the platform is located. (B after Eichenbaum, 2000; C,D after Schenk and Morris, 1985.)

Figure 31.9 Activation of the hippocampus and adjacent parahippocampal cortex predicts memory performance. Activation in these areas was much stronger for items that were later remembered. (After Wagner et al., 1998.)

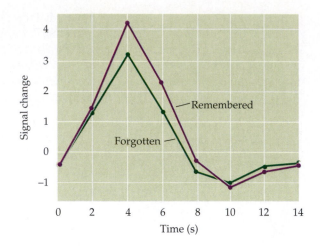

Normal rats at first search randomly until they find the submerged platform. After repeated testing, however, they learn to swim directly to the platform no matter where they are initially placed in the pool. Rats with lesions to the hippocampus and nearby structures cannot learn to find the platform, suggesting that remembering the location of the platform relative to the configuration of visual landmarks depends on the same neural structures critical to declarative memory formation in humans. Likewise, destruction of the hippocampus and parahippocampal gyrus in monkeys severely impairs their ability to perform delayed-response tasks (see Figure 26.14). These studies suggest that primates and other mammals depend on medial temporal structures such as the hippocampus and parahippocampal gyrus to encode and consolidate memories of events and objects in time and space, just as humans use these same brain regions for the initial encoding and consolidation of declarative memories.

Consistent with the evidence from studies of humans and other animals with lesions to the medial temporal lobe—in particular to the hippocampus and parahippocampal cortex—recent studies have shown that neurons in these areas are selectively recruited by the task of storing declarative memories. Early neuroimaging studies using positron emission tomography showed increased metabolism in the hippocampus in human subjects when they were studying information they would later be asked to recall. More recently, fMRI studies have shown that the hippocampus and parahippocampal cortex are activated in human subjects studying a list of items to be remembered. More strikingly, the amount of activity measured in these areas was higher for items that subjects subsequently remembered compared with items they later forgot (Figure 31.9).

These laboratory findings extend to real-life behavior. For example, anyone who has ridden in a taxi in a large city such as London or New York can appreciate the difficulty of negotiating the labyrinth of streets to finally arrive at the desired destination. And it seems equally obvious that some cabbies are much better than others at navigating city streets, a performance difference that seems likely to depend on experience. Amazingly, laboratory studies have found that the posterior hippocampus, which appears to be particularly useful in remembering spatial information, is larger in London cabbies than in age-matched controls (Figure 31.10A). Confirming the role of experience in memory performance, the size of the posterior hippocampus in cab drivers scales positively with the number of months spent driving a cab (Figure 31.10B). Together, such findings endorse the idea that neuronal activation within the hippocampus and closely allied cortical areas of the medial temporal lobe largely determines the

(A)

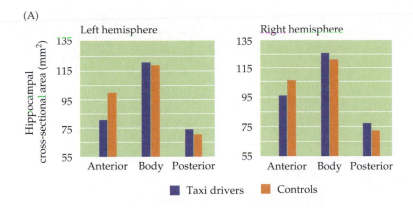

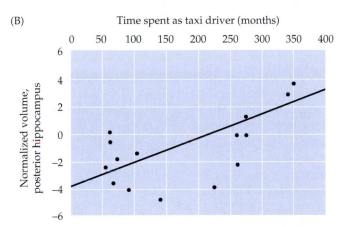

Figure 31.10   The case of the London taxi drivers. (A) Structural brain scans show that the posterior hippocampus, a region specialized for remembering spatial information, is larger in taxi drivers than in age-matched controls. (B) Hippocampus size scales positively with experience as a cabbie. (After Maguire et al., 2000.)

transfer of declarative information into long-term memory, and that the robustness with which such memories are encoded depends on structural and functional changes that occur with experience.

Brain Systems Underlying Long-Term Storage of Declarative Memory

Revealing though they have been, clinical studies of amnesic patients have provided relatively little insight into the long-term storage of declarative information in the brain (other than to indicate quite clearly that such information is *not* stored in the midline diencephalic and medial temporal lobe structures that are affected in anterograde amnesia). Nonetheless, a good deal of evidence implies that the cerebral cortex is the major long-term repository for many aspects of declarative memory.

One line of evidence comes from observations of patients undergoing electroconvulsive therapy (ECT). Individuals with severe depression are often treated by the passage of enough electrical current through the brain to cause the equivalent of a full-blown seizure. This remarkably useful treatment (performed under anesthesia, in well-controlled circumstances) was discovered because depression in epileptics was perceived to remit after a spontaneous seizure (see Box 8C). However, ECT often causes both anterograde and retrograde amnesia. Patients typically do not remember the treatment itself or the events of the preceding days, and even their recall of events of the previous 1–3 years can be affected. Animal studies (rats tested for maze learning, for example) have confirmed the amnesic consequences of ECT. The memory loss usu-

Figure 31.11 Connections between the hippocampus and possible declarative memory storage sites. The rhesus monkey brain is shown because these connections are much better documented in non-human primates than in humans. Projections from numerous cortical areas converge on the hippocampus and the related structures known to be involved in human memory; most of these sites also send projections to the same cortical areas. Medial and lateral views are shown, the latter rotated 180° for clarity. (After Van Hoesen, 1982.)

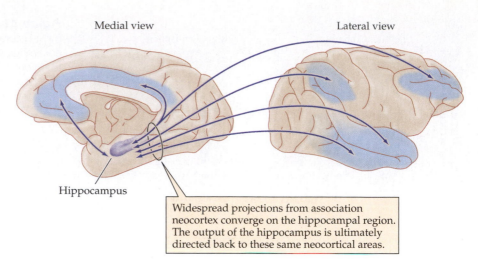

Medial view   Lateral view

Hippocampus

Widespread projections from association neocortex converge on the hippocampal region. The output of the hippocampus is ultimately directed back to these same neocortical areas.

ally clears over a period of weeks to months. However, to mitigate this side effect (which may be the result of excitotoxicity; see Box 6C), ECT is often delivered to only one hemisphere at a time. The nature of amnesia following ECT supports the conclusion that long-term declarative memories are widely stored in the cerebral cortex, since this is the part of the brain predominantly affected by this therapy.

A second line of evidence comes from patients with damage to association cortex outside the medial temporal lobe. Since different cortical regions have different cognitive functions (see Chapters 26 and 27), it is not surprising that these sites store information that reflects the cognitive function of that part of the brain. For example, the lexicon that links speech sounds and their symbolic significance is located in the association cortex of the superior temporal lobe, and damage to this area typically results in an inability to link words and meanings (Wernicke's aphasia; see Chapter 27). Presumably, the widespread connections of the hippocampus to the language areas serve to consolidate declarative information in these and other language-related cortical sites (Figure 31.11). By the same token, the inability of patients with temporal lobe lesions to recognize objects and/or faces suggests that such memories are stored there (see Chapter 26).

A third sort of evidence supporting the hypothesis that declarative memories are stored in cortical areas specialized for processing particular types of information comes from neuroimaging of human subjects recalling vivid memories. In one such study, subjects first examined words paired with either pictures or sounds. Their brains were then scanned while they were asked to recall whether each test word was associated with either a picture or a sound. Functional images based on these scans showed that the cortical areas activated when subjects viewed pictures or heard sounds were reactivated when these percepts were vividly recalled. In fact, this sort of reactivation can be quite specific. Thus, different classes of visual images—such as faces, houses, or chairs—tend to reactivate the same small regions of the visual association cortex that were activated when the objects were actually perceived (Figure 31.12).

These neuroimaging studies reinforce the conclusion that declarative memories are stored widely in specialized areas of the cerebral cortex. Retrieving such memories appears to involve the medial temporal lobe, as well as regions of the frontal cortex. Frontal cortical areas located on the dorsolateral and anterolateral aspect of the brain, in particular, are activated when normal subjects attempt to retrieve declarative information from long-term memory. Moreover, patients

(A)

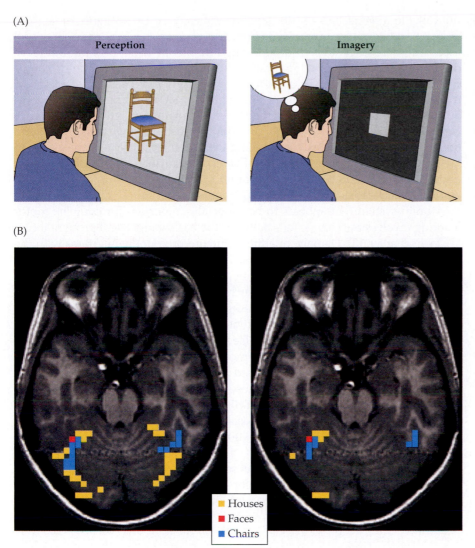

Figure 31.12 Reactivation of visual cortex during vivid remembering of visual images. (A) Subjects were instructed to view either images of objects (houses, faces, and chairs) (left) or imagine the objects in the absence of the stimulus (right). (B) At left, bilateral regions of ventral temporal cortex are specifically activated during perception of houses (yellow), faces (red), and chairs (blue). At right, when subjects recall these objects, the same regions preferentially activated during the perception of each object class are reactivated. (After Ishai et al., 2000.)

with damage to these areas often fail to accurately recall the details of a memory and sometimes resort to confabulation to fill in the missing information. Finally, whereas the ability of patients such as H.M. and R.B. to remember facts and events from the period of their lives preceding their lesions clearly demonstrates that the medial temporal lobe is not necessary for retrieving declarative information held in long-term memory, other studies have suggested that these structures may be important for recalling declarative memories during the early stages of consolidation and storage in the cerebral cortex.

Brain Systems Underlying Nondeclarative Learning and Memory

H.M., N.A., and R.B. had no difficulty establishing or recalling nondeclarative memories, indicating that this information is laid down by using an anatomical substrate different from that used in declarative memory formation. Nondeclarative memory apparently involves the basal ganglia, prefrontal cortex, amygdala, sensory association cortex, and cerebellum, but not the medial temporal lobe or midline diencephalon. In support of this interpretation, perceptual priming (the influence of previously studied information on subsequent performance, unavailable to conscious recall) depends critically on the integrity of sensory association cortex. For example, it should be obvious that lesions of the visual association cortex produce profound impairments in visual priming but leave declarative memory formation intact. Likewise, simple sensory-motor conditioning, such as learning to blink following a tone that predicts a puff of air directed at the eye, relies on the normal activation of neural circuits in the cerebellum. Ischemic damage to the cerebellum following infarcts of the superior cerebellar artery or the posterior inferior cerebellar artery cause profound deficits in classical eyeblink conditioning without interfering with the ability to lay down new declarative memories. Evidence from such double dissociations endorses the idea that independent brain systems govern the formation and storage of declarative and nondeclarative memories.

A brain system that appears to be especially important for complex motor learning involves the connections between the basal ganglia and prefrontal cortex (see Chapter 18). Damage to either structure profoundly interferes with the ability to learn new motor skills. Thus, patients with Huntington's disease, which causes atrophy of the caudate and putamen (see Figure 18.11B), perform poorly on motor skill learning tests such as manually tracking a spot of light, tracing curves using a mirror, or reproducing sequences of finger movements. Because the loss of dopaminergic neurons in the substantia nigra interferes with normal signaling in the basal ganglia (see Figure 18.11A), patients with Parkinson's disease show similar deficits in motor skill learning (Figure 31.13), as do patients with prefrontal lesions caused by tumors or strokes. Neuroimaging studies have largely corroborated these findings, revealing activation of the basal ganglia and prefrontal cortex in normal subjects performing these same skill-learning tests. Activation of the basal ganglia and prefrontal cortex has also been observed in animals as they are carrying out rudimentary motor learning and sequencing tasks.

The dissociation of memory systems supporting declarative and nondeclarative memory suggests the scheme for long-term information storage diagrammed in Figure 31.14. The generality of the diagram only emphasizes the

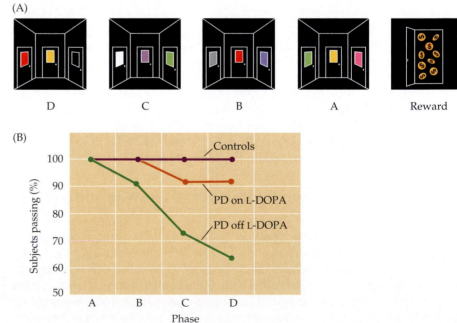

Figure 31.13 Parkinson's disease reveals a role for basal ganglia in nondeclarative memory. (A) Subjects performed a probabilistic learning task with four levels. They first learned that selecting a door of one color (say, pink) in condition A led to reward. Subjects then learned that selecting a differently colored door (say, red) in condition B would permit them to proceed to condition A, where they could select the rewarded door. This procedure was continued until subjects made choices from D → C → B → A → reward. (B) Parkinson's patients (PD) who were taking medication to replace depleted dopamine in the midbrain performed nearly as well as age-matched controls. However, Parkinson's patients who were not on dopamine replacement medication were dramatically impaired in their ability to learn the task. (After Shohamy et al., 2005.)

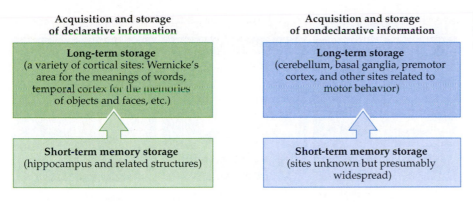

**Acquisition and storage
of declarative information**

| Long-term storage (a variety of cortical sites: Wernicke's area for the meanings of words, temporal cortex for the memories of objects and faces, etc.) |

| Short-term memory storage (hippocampus and related structures) |

**Acquisition and storage
of nondeclarative information**

| Long-term storage (cerebellum, basal ganglia, premotor cortex, and other sites related to motor behavior) |

| Short-term memory storage (sites unknown but presumably widespread) |

Figure 31.14   *Summary diagram of the acquisition and storage of declarative versus nondeclarative information.*

rudimentary state of present thinking about exactly how and where long-term memories are stored. A reasonable guess is that each complex memory is instantiated in an extensive network of neurons whose activity depends on synaptic weightings that have been molded and modified by experience.

Memory and Aging

Although it is all too obvious that our outward appearance changes with age, we tend to imagine that the brain is much more resistant to the ravages of time. Unfortunately, the evidence suggests that this optimistic view is not justified. From early adulthood onward, the average weight of the normal human brain, as determined at autopsy, steadily decreases (Figure 31.15). In elderly individuals, this effect can also be observed with noninvasive imaging as a slight but nonetheless significant shrinkage of the brain. Counts of synapses in the cerebral cortex generally decrease in old age (although the number of neurons probably does not change very much), suggesting that it is mainly the connec-

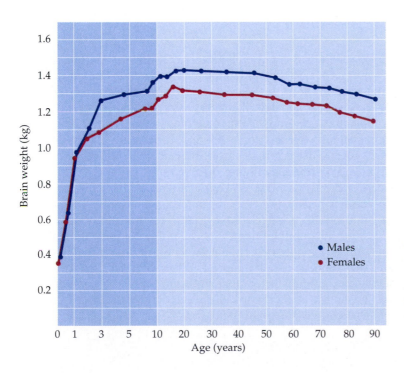

Figure 31.15   *Brain size as a function of age. The human brain reaches its maximum size (measured by weight in this case) in early adult life and decreases progressively thereafter. This decrease evidently represents the gradual loss of neural circuitry in the aging brain, which presumably underlies the progressively diminished memory function in older individuals. (After Dekaban and Sadowsky, 1978.)*

Figure 31.16 Compensatory activation of memory areas in high-functioning older adults. During remembering, activity in prefrontal cortex was restricted to the right prefrontal cortex (following radiological conventions, the brain images are left-right reversed) in both young participants and elderly subjects with poor recall. In contrast, elderly subjects with relatively good memory showed activation in both right and left prefrontal cortex. (After Cabeza et al., 2002).

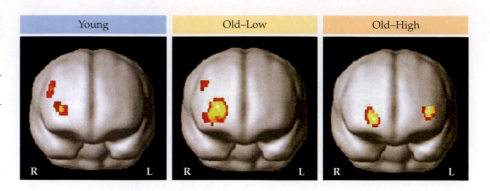

tions between neurons (i.e., neuropil) that are lost as humans grow old (consistent with the idea that the networks of connections that represent memories—i.e., the engrams—gradually deteriorate).

These several observations accord with the difficulty older people have in making associations (e.g., remembering names or the details of recent experiences) and with declining scores on tests of memory as a function of age. The normal loss of some memory function with age means that there is a large gray zone between individuals undergoing normal aging and patients suffering from age-related dementias such as Alzheimer's disease (Box 31D).

Just as regular exercise slows the deterioration of the neuromuscular system with age, age-related neurodegeneration and associated cognitive decline may be slowed in elderly individuals who make a special effort to continue using the full range of human memory abilities (i.e., both declarative and nondeclarative memory tasks). Although cognitive decline with age is ultimately inevitable, neuroimaging studies suggest that high-performing older adults may to some degree offset declines in processing efficacy through compensatory activation of cortical tissue that is less fully used during remembering in poorly performing older adults (Figure 31.16).

Summary

Human memory entails a number of biological strategies and anatomical substrates. Primary among these are a system for memories that can be expressed by means of language and can be made available to the conscious mind (declarative memory), and a separate system that concerns skills and associations that are essentially nonverbal, operating at a largely unconscious level (nondeclarative or procedural memory). Based on evidence from amnesic patients and knowledge about normal patterns of neural connections in the human brain, the hippocampus and associated midline diencephalic and medial temporal lobe structures are critically important in laying down new declarative memories, although not in storing them (a process that occurs primarily in the association cortices). In contrast, nondeclarative memories for motor and other unconscious skills depends on the integrity of the premotor cortex, basal ganglia, and cerebellum, and is not affected by lesions that impair the declarative memory system. The common denominator of these categories of stored information is generally thought to be alterations in the strength and number of the synaptic connections in the cerebral cortices that mediate associations between stimuli and the behavioral responses to them.

BOX 31D   Alzheimer's Disease

Dementia is a syndrome characterized by failure of recent memory and other intellectual functions. It is usually insidious in onset, but tends to progress steadily. Alzheimer's disease (AD) is the most common dementia, accounting for 60–80 percent of cases in the elderly. This unfortunate condition afflicts 5–10 percent of the U.S. population over the age of 65, and as much as 45 percent of the population over 85. The earliest signs are an impairment of recent memory function and attention, followed by failure of language skills, visual–spatial orientation, abstract thinking, and judgment. Inevitably, alterations of personality accompany these defects.

A tentative diagnosis of Alzheimer's disease is based on these characteristic clinical features and can only be confirmed by the distinctive cellular pathology evident on postmortem examination of the brain (Figure A). These histopathologcial changes consist of three principal features: (1) collections of intraneuronal cytoskeletal filaments called *neurofibrillary tangles*; (2) extracellular deposits of an abnormal protein (called amyloid) in so-called *senile plaques*; and (3) a diffuse loss of neurons. These changes are most apparent in neocortex, limbic structures (hippocampus, amygdala, and their associated cortices), and some brainstem nuclei (typically the basal forebrain nuclei).

The vast majority of AD cases occur only after age 60 ("late-onset" by definition) and arise sporadically. In contrast, early-onset forms occur relatively rarely and are caused by monogenic defects consistent with an autosomal dominant pattern of inheritance. Identification of the mutant genes in a few families with the early-onset form has provided considerable insight into the processes that go awry in Alzheimer's disease.

Investigators long suspected that a mutant gene responsible for familial AD might reside on chromosome 21, primarily because clinical and neuropathologic features similar to AD often occur in individuals with Down syndrome (caused by an extra copy of chromosome 21), but with a much earlier onset (about age 30 in most cases). A mutation of the gene encoding amyloid precursor protein (APP) emerged as an attractive candidate both because of the prominent amyloid deposits in AD together with isolation of a fragment of APP, Aβ peptide, from amyloid plaques. The gene that encodes APP was subsequently cloned by Dmitry Goldgaber and colleagues, and found to reside on chromosome 21. This discovery eventually led to the identification of mutations of the *APP* gene in almost 20 families with the early-onset, autosomal dominant form of AD. It should be noted, however, that only a few of the early-onset families

(and none of the late-onset families) exhibited these particular mutations.

The mutant genes underlying two additional autosomal dominant forms of AD have been subsequently identified (*presenilin 1* and *presenilin 2*). Interestingly, mutations of presenilin 1 and 2 modify processing of APP resulting in increased amounts of a particularly toxic form of Aβ peptide, Aβ42. Thus, mutation of any one of several genes appears to be sufficient to cause a heritable form of AD, and these converge on abnormal processing of APP.

In the far more common late-onset form of Alzheimer's, the disease is clearly not inherited in any simple sense (although the relatives of affected individuals are at a greater risk, for reasons that are not clear). The central role of

(Continued on next page)

(A)   Neurofibrillary tangle

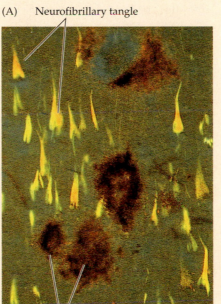

Amyloid plaque

(B)

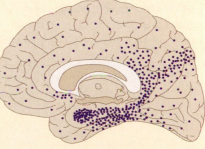

(A) Histological section of the cerebral cortex from a patient with Alzheimer's disease, showing characteristic amyloid plaques and neurofibrillary tangles. (B) Distribution of pathologic changes (including plaques, tangles, neuronal loss, and gray matter shrinkage) in Alzheimer's disease. Dot density indicates severity of pathology. (A from Roses, 1995, courtesy of Gary W. Van Hoesen; B after Blumenfeld, 2002, based on Brun and Englund, 1981.)

BOX 31D (Continued)

APP in the families with early-onset forms of the disease nonetheless suggested that APP might be linked to the chain of events culminating in the sporadic forms of Alzheimer's disease. Biochemists Warren Strittmatter and Guy Salvesen theorized that pathologic deposition of proteins complexed with Aα peptide might be responsible. To test this idea, they immobilized Aα peptide on nitrocellulose paper and searched for proteins in the cerebrospinal fluid of patients with AD that bound with high affinity. One of the proteins they detected was apolipoprotein E (ApoE), a molecule that normally chaperones cholesterol through the bloodstream.

This discovery was especially provocative in light of a discovery made by Margaret Pericak-Vance, Allen Roses, and their colleagues at Duke University, who found that affected members of some families with the late-onset form of AD exhibited an association with genetic markers on chromosome 19. This finding was of particular interest because a gene encoding an isoform of ApoE is located in the same region of chromosome 19 implicated by the association studies. As a result, they began to explore the relationship of the different alleles of ApoE with individuals with sporadic, late-onset form of AD. There are three major alleles of ApoE, ε2, ε3, and ε4. The frequency of allele ε3 in the general population is 0.78, and the frequency of allele ε4 is 0.14. The frequency of the ε4 allele in late-onset AD patients, however, is 0.52—almost 4 times higher than the general population. Thus, the inheritance of the ε4 allele is a risk factor for late-onset AD. In fact, people homozygous for ε4 are about 8 times more likely to develop AD compared to individuals homozygous for ε3. Among individuals with no copies of ε4, only 20 percent develop AD by age 75 compared to 90 percent of individuals with two copies of ε4.

In contrast to the mutations of *APP* or *presenilin 1* and *presenilin 2* that cause early-onset familial forms of AD, inheriting the ε4 form of ApoE is not sufficient to cause AD; rather, inheriting this gene simply increases the risk of developing AD. The cellular and molecular mechanisms by which the ε4 allele of ApoE increases susceptibility to late-onset AD are not understood, and elucidating these mechanisms is clearly an important goal.

Clearly, AD has a complex pathology and probably reflects a variety of related molecular and cellular abnormalities. The most apparent common denominator of this complex disease so far is abnormal APP processing. In particular, accumulation of the toxic Aβ42 peptide is thought to be a key factor. This conclusion has led to efforts to develop therapies aimed at inhibiting formation or facilitating clearance of this toxic peptide. It is unlikely that this important problem will be understood without a great deal more research, much hyperbole in the lay press notwithstanding.

References

ADAMS, R. D. AND M. VICTOR (2005) *Principles of Neurology*, 8th Ed. New York: McGraw-Hill, pp. 898–906.

CITRON, M. AND 8 OTHERS (1992) Mutation of the β-amyloid precursor protein in familial Alzheimer's disease increases β-protein production. *Nature* 360: 672–674.

CORDER, E. H. AND 8 OTHERS (1993) Gene dose of apolipoprotein E type 4 allele and the risk of Alzheimer's disease in late-onset families. *Science* 261: 921–923.

GOLDGABER, D., M. I. LERMAN, O. W. MCBRIDE, U. SAFFIOTTI AND D. C. GAJDUSEK (1987) Characterization and chromosomal localization of a cDNA encoding brain amyloid of Alzheimer's disease. *Science* 235: 877–880.

MURRELL, J., M. FARLOW, B. GHETTI AND M. D. BENSON (1991) A mutation in the amyloid precursor protein associated with hereditary Alzheimer's disease. *Science* 254: 97–99.

ROGAEV, E. I. AND 20 OTHERS (1995) Familial Alzheimer's disease in kindreds with missense mutations in a gene on chromosome 1 related to the Alzheimer's disease type 3 gene. *Nature* 376: 775–778.

SHERRINGTON, R. AND 33 OTHERS (1995) Cloning of a gene bearing missense mutations in early-onset familial Alzheimer's disease. *Nature* 375: 754–760.

SMALL, S. A. AND S. GANDY (2006). Sorting through the cell biology of Alzheimer's disease: Intracellular pathways to pathogenesis. *Neuron* 52: 15–31.

Additional Reading

Reviews

BUCKNER, R. L. (2000) Neuroimaging of memory. In *The New Cognitive Neurosciences*, M. Gazzaniga (ed.). Cambridge, MA: MIT Press, pp. 817–840.

BUCKNER, R. L. (2002) The cognitive neuroscience of remembering. *Nature Rev. Neurosci.* 2: 624–634.

CABEZA, R. (2001) Functional neuroimaging of cognitive aging. In *Handbook of Functional Neuroimaging of Cognition*, R. Cabeza and A. Kingstone (eds.). Cambridge, MA: MIT Press, pp. 331–377.

ERICKSON, C. A., B. JAGADEESH AND R. DESIMONE (2000) Learning and memory in the inferior temporal cortex of the macaque. In *The New Cognitive Neurosciences*, M. Gazzaniga (ed.). Cambridge, MA: MIT Press, pp. 743–752.

MISHKIN, M. AND T. APPENZELLER (1987) The anatomy of memory. *Sci. Am.* 256(6): 80–89.

PETRI, H. AND M. MISHKIN (1994) Behaviorism, cognitivism, and the neuropsychology of memory. *Am. Sci.* 82: 30–37.

SCHACTER, D. L. AND R. L. BUCKNER (1998) Priming and the brain. *Neuron* 20: 185–195.

SQUIRE, L. R. AND B. J. KNOWLTON (2000) The medial temporal lobe, the hippocampus, and the memory systems of the brain. In *The New Cognitive Neurosciences*, M. Gazzaniga (ed.). Cambridge, MA: MIT Press, pp. 765–779.

SQUIRE, L. R. (1992) Memory and hippocampus: A synthesis from findings with rats, monkeys, and humans. *Psych. Rev.* 99: 195–231.

THOMPSON, R. F. (1986) The neurobiology of learning and memory. *Science* 223: 941–947.

ZACKS, R. T., L. HASHER AND K. Z. H. LI (1999) Human memory. In *The Handbook of Aging and Cognition*. F. I. M. Craik and T. A. Salthouse (eds.). Mahwah, New Jersey: Lawrence Erlbaum Associates, pp. 293–357.

ZOLA-MORGAN, S. M. AND L. R. SQUIRE (1993) Neuroanatomy of memory. *Annu. Rev. Neurosci.* 16: 547–563.

Important Original Papers

CABEZA, R., N. D. ANDERSON, J. K. LOCANTORE AND A. R. MCINTOSH (2002) Aging gracefully: Compensatory brain activity in high-performing older adults. *NeuroImage* 17: 1394–1402.

GOBET, F. AND H. A. SIMON (1998) Expert chess memory: Revisiting the chunking hypothesis. *Memory* 6: 225–255.

ISHAI, A., L. G. UNGERLEIDER AND J. V. HAXBY (2000) Distributed neural systems for the generation of visual images. *Neuron* 28: 979–990.

SCOVILLE, W. B. AND B. MILNER (1957) Loss of recent memory after bilateral hippocampal lesions. *J. Neurol. Neurosurg. Psychiat.* 20: 11–21.

SQUIRE, L. R. (1989) On the course of forgetting in very long-term memory. *J. Exp. Psychol.* 15: 241–245.

ZOLA-MORGAN, S. M. AND L. R. SQUIRE (1990) The primate hippocampal formation: Evidence for a time-limited role in memory storage. *Science* 250: 288–290.

Books

BADDELEY, A. (1982) *Your Memory: A User's Guide*. New York: Macmillan.

CRAIK, F. I. M. AND T. A. SALTHOUSE (1999) *The Handbook of Aging and Cognition*. Mahwah, New Jersey: Lawrence Erlbaum Associates.

DUKAS, R. (1998) *Cognitive Ecology: The Evolutionary Ecology of Information Processing and Decision Making*. Chicago: University of Chicago Press.

GAZZANIGA, M. S. (2000) *The New Cognitive Neurosciences*, 2nd Ed. Cambridge MA: MIT Press.

GAZZANIGA, M. S., R. B. IVRY AND G. R. MANGUN (1998) *Cognitive Neuroscience: The Biology of the Mind*. New York: W. W. Norton & Company.

LURIA, A. R. (1987) *The Mind of a Mnemonist*. Translated by Lynn Solotaroff. Cambridge, MA: Harvard University Press.

NEISSER, U. (1982) *Memory Observed: Remembering in Natural Contexts*. San Francisco: W. H. Freeman.

PENFIELD, W. AND L. ROBERTS (1959) *Speech and Brain Mechanisms*. Princeton, NJ: Princeton University Press.

SAPER, C. B. AND F. PLUM (1985) *Handbook of Clinical Neurology*, Vol. 1(45): *Clinical Neuropsychology*, P. J. Vinken, G. S. Bruyn and H. L. Klawans (eds.). New York: Elsevier, pp. 107–128.

SCHACTER, D. L. (1997) *Searching for Memory: The Brain, the Mind, and the Past*. New York: Basic Books.

SCHACTER, D. L. (2001) *The Seven Sins of Memory: How the Mind Forgets and Remembers*. Boston: Houghton Mifflin Co.

SMITH, S. B. (1983) *The Great Mental Calculators: The Psychology, Methods, and Lives of Calculating Prodigies, Past and Present*. New York: Columbia University Press.

SQUIRE, L. R. (1987) *Memory and Brain*. New York: Oxford University Press, pp. 202–223.

ZECHMEISTER, E. B. AND S. E. NYBERG (1982) *Human Memory: An Introduction to Research and Theory*. Monterey, CA: Brooks/Cole Publishing.

Appendix

Survey of Human Neuroanatomy

Overview

Perhaps the major reason that neuroscience remains such an exciting field is the wealth of unanswered questions about the fundamental organization and function of the human brain. To understand this remarkable organ (and its interactions with the body that it governs), the myriad cell types that constitute the nervous system must be identified, their mechanisms of excitability and plasticity characterized, their interconnections traced, and the physiological role of the resulting neural circuits defined in behaviorally meaningful contexts. These challenges have been at the forefront of the five units of this textbook, where a broad range of questions about how nervous systems are organized and how they generate behavior have been addressed (albeit leaving many questions unanswered, especially those that pertain to distinctly human behaviors). This Appendix provides an anatomical framework for the integration of this knowledge and its application to the human nervous system. It provides a review of the basic terms and anatomical conventions used in discussing human neuroanatomy, and a general picture of the organization of the human forebrain, brainstem, and spinal cord. The Appendix is followed by an Atlas that presents a series of plates (modified from *Sylvius 4.0* atlases; see available media), where surface and cross-sectional views of the human central nervous system are presented and relevant neuroanatomical structures are identified.

Neuroanatomical Terminology

The terms used to specify *location* in the central nervous system are the same as those used for the gross anatomical description of the rest of the body (Figure A1). Thus, *anterior* and *posterior* indicate front and behind; *rostral* and *caudal*, toward the nose and tail (lower spinal region); *dorsal* and *ventral*, top and bottom (back and belly); and *medial* and *lateral*, at the midline or to the side. But the use of these coordinates in the body versus their use to describe position in the brain can be confusing, especially as these terms are applied to humans. For the entire body, these anatomical terms refer to the long axis, which is straight. The long axis of the human central nervous system, however, has a bend in it. In humans (and other bipeds), the rostral–caudal axis of the forebrain is tilted forward (because of the cephalic flexure that forms in embryogenesis; see Chapter 22), with respect to the long axis of the brainstem and spinal cord (Figure A1A). Once this forward tilt is appreciated, the other terms that describe position in the brain and the terms used to identify planes of section can be easily assigned.

   The proper assignment of the anatomical axes dictates the standard planes for histological sections or live images that are used to study the internal anatomy of the brain and to localize function (Figure A1B). **Horizontal** sections (also referred to as **axial** sections) are taken parallel to the rostral–caudal axis of

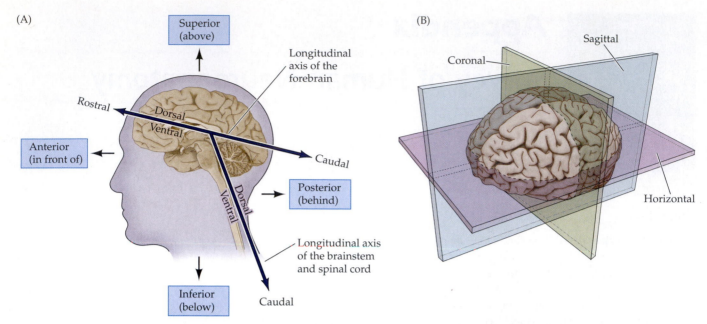

(A)

Superior
(above)

Longitudinal
axis of the
forebrain

Rostral

Dorsal

Ventral

Anterior
(in front of)

Caudal

Posterior
(behind)

Dorsal

Ventral

Longitudinal axis
of the brainstem
and spinal cord

Inferior
(below)

Caudal

(B)

Sagittal

Coronal

Horizontal

Figure A1   A flexure in the long axis of
the nervous system arose as humans
evolved upright posture, leading to an
approximately 120° angle between the
long axis of the brainstem and that of the
forebrain. The consequences of this flexure for anatomical terminology are indicated in (A). The terms *anterior, posterior,
superior,* and *inferior* refer to the long
axis of the body, which is straight. Therefore, these terms indicate the same direction for both the forebrain and the brainstem. In contrast, the terms *dorsal,
ventral, rostral,* and *caudal* refer to the
long axis of the central nervous system.
The dorsal direction is toward the back
for the brainstem and spinal cord, but
toward the top of the head for the forebrain. The opposite direction is ventral.
The rostral direction is toward the top of
the head for the brainstem and spinal
cord, but toward the face for the forebrain. The opposite direction is caudal.
(B) The major planes of section used in
cutting or imaging the brain.

the brain; thus, in an individual standing upright, such sections are parallel to
the ground. Sections taken in the plane dividing the two hemispheres are **sagittal**, and can be further categorized as **midsagittal** and **parasagittal**, according
to whether the section is near the midline (midsagittal) or more lateral (parasagittal). Sections in the plane of the face are called **coronal** or **frontal**.

Different terms are usually used to refer to sections of the brainstem and
spinal cord. The plane of section orthogonal to the long axis of the brainstem
and cord is called **transverse**, whereas sections parallel to this axis are called
longitudinal. In a transverse section through the human brainstem and spinal
cord, the dorsal–ventral axis and the posterior–anterior axis indicate the same
directions (see Figure A1A). This terminology is essential for understanding the
basic subdivisions of the nervous system and for discussing the locations of
brain structures in a common frame of reference.

Basic Subdivisions of the Central Nervous System

The central nervous system is usually considered to have seven basic parts
(Figure A2). From caudal to rostral ("tail to nose"), these are: the **spinal cord**,
the **medulla**, the **pons**, the **cerebellum**, the **midbrain**, the **diencephalon**,
and the **cerebrum** (**cerebral hemispheres**). Running through or between
these subdivisions are fluid-filled spaces called **ventricles**. These ventricles are
the remnants of the continuous lumen initially enclosed by the neural plate as
it rounded to become the neural tube during the process of neurulation in
early brain development (see Chapter 22). Variations in the shape and size of
the mature ventricular spaces are characteristic of each adult brain region. The
medulla, pons, and midbrain are collectively called the **brainstem** and they
underlie the **4th ventricle** (medulla and pons) and surround the **cerebral
aqueduct** (midbrain). The diencephalon and cerebral hemispheres are collectively called the **forebrain**, and they enclose the **3rd** and **lateral ventricles**,
respectively. A more detailed description of the ventricles is presented in a later
section.

The brainstem is one of the most complex regions of the central nervous system, and it too will be considered in greater depth below. At this juncture, suf-

(A)

(B)

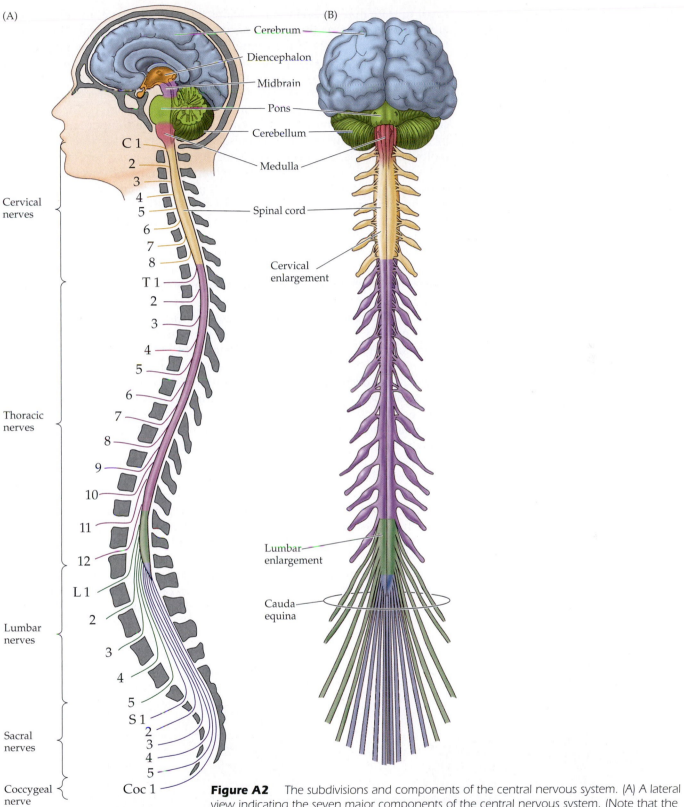

Cerebrum

Diencephalon

Midbrain

Pons

Cerebellum

Medulla

Spinal cord

C 1
2
3
4
5
6
7
8
T 1
2
3
4
5
6
7
8
9
10
11
12
L 1
2
3
4
5
S 1
2
3
4
5
Coc 1

Cervical nerves

Thoracic nerves

Lumbar nerves

Sacral nerves

Coccygeal nerve

Cervical enlargement

Lumbar enlargement

Cauda equina

Figure A2 The subdivisions and components of the central nervous system. (A) A lateral view indicating the seven major components of the central nervous system. (Note that the position of the brackets on the left side of the figure refers to the location of the spinal nerves as they exit the intervertebral foramina, not the position of the corresponding spinal cord segments.) (B) The central nervous system in ventral view, indicating the emergence of the segmental nerves, the cervical and lumbar enlargements and the cauda equina.

fice it to state that the brainstem contains the **cranial nerve nuclei** that send and receive signals via the **cranial nerves**, and that the brainstem is a conduit for several major tracts of the central nervous system that relay information between the forebrain and spinal cord, with the brainstem itself being a source and target of descending and ascending signals. Accordingly, detailed knowledge of the consequences of damage to the brainstem provides clinicians with an essential tool in the localization and diagnosis of brain injury. The brainstem contains numerous nuclei that are involved in myriad important functions, including the organization of rhythmical and stereotyped somatic and visceral motor behaviors, and transitions in states of conscious awareness and sleep.

Overlying the brainstem is the cerebellum, which extends over much of the dorsal aspect of the brainstem; together, the brainstem and cerebellum constitute the **hindbrain**. The cerebellum is essential for the coordination and planning of movements (see Chapter 19) as well as learning motor and cognitive tasks (see Unit V).

The forebrain has several anatomical subdivisions. The most obvious structures are the prominent **cerebral hemispheres** (Figure A3). In humans, the cerebral hemispheres (the outermost portions of which are continuous, highly folded sheets of cortex) are characterized by **gyri** (singular, *gyrus*), or crests of folded cortical tissue, and by **sulci** (singular, *sulcus*), which are the grooves that divide gyri from one another. Although gyral and sulcal patterns vary from individual to individual, several consistent landmarks divide the hemispheres into four **lobes**. The names of the lobes are derived from the cranial bones that overlie them: **occipital**, **temporal**, **parietal**, and **frontal**. A key feature of the surface anatomy of the cerebrum is the **central sulcus** located roughly halfway between the rostral and caudal poles of the hemispheres (see Figure A3A). This prominent sulcus divides the frontal lobe in the rostral half of the hemisphere from the more caudal parietal lobe. Other prominent landmarks that divide the cerebral lobes are the **lateral fissure** (also called the **Sylvian fissure**), which divides the temporal lobe inferiorly from the overlying frontal and parietal lobes, and the **parieto-occipital sulcus**, which separates the parietal lobe from the occipital lobe. The remaining major subdivisions of the forebrain are not visible from the surface; they comprise gray and white matter structures that lie deeper in the cerebral hemispheres and can only be appreciated in sectional views.

Next, the characteristic superficial features of these major subdivisions of the human central nervous system and their internal organization will be described in more detail from caudal to rostral, beginning with the spinal cord.

Figure A3 Surface anatomy of the cerebral hemisphere, showing the four lobes of the brain and the major fissures and sulci that define their boundaries. (A) Lateral view. (B) Midsagittal view.

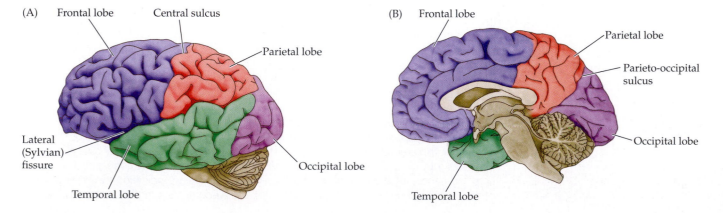

(A) Frontal lobe   Central sulcus   Parietal lobe   Lateral (Sylvian) fissure   Temporal lobe   Occipital lobe

(B) Frontal lobe   Parietal lobe   Parieto-occipital sulcus   Occipital lobe   Temporal lobe

External Anatomy of the Spinal Cord

The spinal cord extends caudally from the brainstem, running from the medullary-spinal junction at about the level of the first cervical vertebra to about the level of the twelfth thoracic vertebra (see Figure A2). The vertebral column (and the spinal cord within it) is divided into **cervical**, **thoracic**, **lumbar**, **sacral**, and **coccygeal** regions. The peripheral nerves (called the **spinal** or **segmental nerves**) that innervate much of the body arise from the spinal cord's 31 pairs of spinal nerves. On each side of the midline, the cervical region of the cord gives rise to 8 cervical nerves (C1–C8), the thoracic region to 12 thoracic nerves (T1–T12), the lumbar region to 5 lumbar nerves (L1–L5), the sacral region to 5 sacral nerves (S1–S5), and the coccygeal region to a single coccygeal nerve. The segmental spinal nerves leave the vertebral column through the intervertebral foramina that lie adjacent to the respectively numbered vertebral body. Sensory information carried by the afferent axons of the spinal nerves enters the cord via the **dorsal roots**, and motor commands carried by the efferent axons leave the cord via the **ventral roots** (Figure A4). Once the dorsal and ventral roots join, sensory and motor axons (with some exceptions) travel together in the segmental spinal nerves.

Two regions of the spinal cord are enlarged to accommodate the greater number of nerve cells and connections needed to process information related to the upper and lower limbs. The spinal cord expansion that corresponds to the arms is called the **cervical enlargement** and includes spinal segments C5–T1; the expansion that corresponds to the legs is called the **lumbar enlargement** and includes spinal segments L2–S3 (see Figure A2B). Because the spinal cord is considerably shorter than the vertebral column (see Figure A2A), lumbar and sacral nerves run for some distance in the vertebral canal before emerging, thus forming a collection of nerve roots known as the **cauda equina**. The cauda equina is the target for an important clinical procedure—the *lumbar puncture*—that allows for the collection of cerebrospinal fluid by placing a needle into the space surrounding these nerves to withdraw fluid for analysis. In addition, local anesthetics can be safely introduced into the cauda equina, producing spinal anesthesia; at this level, the risk of damage to the spinal cord from a poorly placed needle is minimal.

Internal Anatomy of the Spinal Cord

The arrangement of gray and white matter in the spinal cord is relatively simple: the interior of the cord is formed by gray matter, which is surrounded by white matter (Figure A5A). In transverse sections, the gray matter is conventionally divided into dorsal (posterior) and ventral (anterior) "horns." The neurons of the **dorsal horns** receive sensory information that enters the spinal cord via the dorsal roots of the spinal nerves (Figure A5B). The **lateral horns** are present primarily in the thoracic region, and contain the preganglionic visceral motor neurons that project to the sympathetic ganglia (visible in Figure A4). The **ventral horns** contain the cell bodies of motor neurons that send axons via the ventral roots of the spinal nerves to terminate on striated muscles. These major divisions of gray matter have been further subdivided according to the distribution of neurons in the dorsal–ventral axis.

Figure A4   Diagram of several spinal cord segments, showing the relationship of the spinal cord and spinal nerves to the vertebral column.

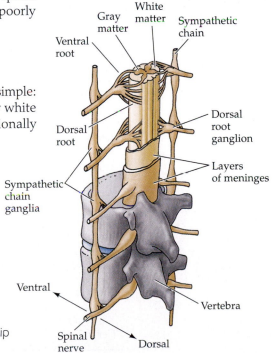

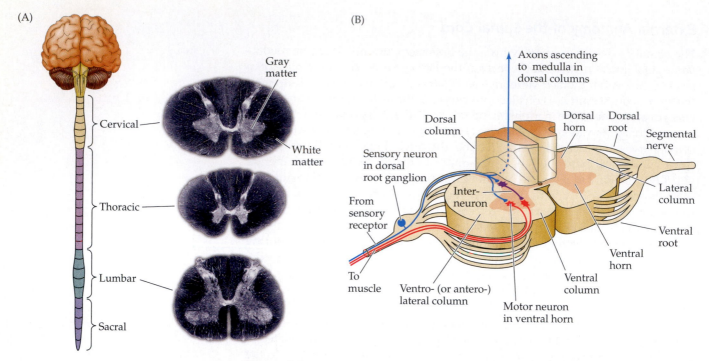

Figure A5 Internal structure of the spinal cord. (A) Transverse histological sections of the cord at three different levels, showing the characteristic arrangement of gray and white matter in the cervical, thoracic, and lumbar cord. The sections were acquired and processed to simulate myelin staining; thus, white matter appears darker and gray matter lighter. (B) Diagram of the internal structure of the spinal cord.

TABLE A1 Subdivisions of Spinal Cord Gray Matter

Division	Rexed lamina	Descriptive term	Significance
Dorsal horn	I	Marginal zone	Projection neurons that receive input from small diameter afferents; one source of anterolateral system projections
	II	Substantia gelatinosa	Interneurons that receive input mainly from small diameter afferents; integrates feedforward and feedback (descending) inputs that modulate pain transmission (see Chapter 10)
	III/IV	Nucleus proprius	Interneurons that integrate inputs from small and large diameter afferents
	V/VI	Base of dorsal horn	Projection neurons that receive input from both large and small diameter afferents and spinal interneurons; another source of anterolateral system projections
Intermediate zone (lateral horn)	VII	Intermediate gray	Mainly interneurons that communicate between dorsal and ventral horns; in the thoracic cord, also contains projection neurons of the dorsal nucleus of Clarke, a spinocerebellar relay (see Chapter 19), and the sympathetic preganglionic neurons of the intermediolateral cell column (underlying the lateral horn); in the sacral cord, also contains parasympathetic preganglionic nuclei (see Chapter 21)
Ventral horn	VIII	Motor interneurons	Interneurons in the medial aspect of ventral horn that coordinate the activities of lower motor neurons (see Chapter 16)
	IX	Motor neuron columns	Columns of lower motor neurons that govern limb musculature (see Chapter 16)
Central zone	X	Central gray	Interneurons surrounding the rudiment of the central canal

The Swedish neuroanatomist Bror Rexed recognized that neurons in the dorsal horn are organized into layers and neurons in the ventral horns, especially in the enlargements, are arranged into longitudinal columns (Figure A6). Rexed proposed a scheme for naming these subdivisions, since termed *Rexed's laminae*, which is still used by neuroanatomists and clinicians, although more descriptive terms are also applied (Table A1).

The white matter of the spinal cord is subdivided into dorsal (or posterior), lateral, and ventral (or anterior) columns, each of which contains axon tracts related to specific functions. The **dorsal columns** carry ascending sensory information from somatic mechanoreceptors (see Figure A5B). The **lateral columns** include axons that travel from the cerebral cortex to interneurons and motor neurons in the ventral horns; this important pathway is called the **lateral corticospinal tract** (see Chapter 17). The **ventral** (and **ventrolateral** or **anterolateral**) **columns** carry both ascending information about pain and temperature, and descending motor information from the brainstem and motor cortex.

(A)

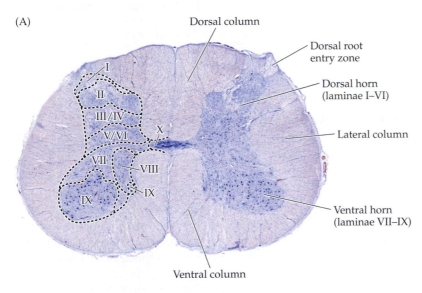

(B)

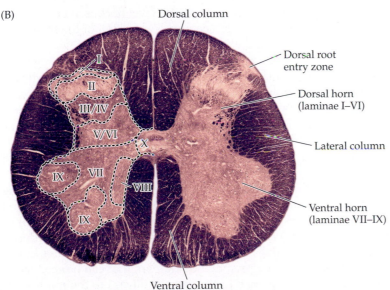

Figure A6 Demonstration of the internal histology of the human spinal cord in a lumbar segment. (A) Photomicrograph of a section stained for the demonstration of Nissl substance (showing cell bodies in a blue stain). (B) Photomicrograph of a section that was acquired and processed to simulate myelin staining. On the left side of both images, dotted lines indicate the boundaries between cytoarchitectonic divisions of spinal cord gray matter, known as *Rexed's laminae* (see Table A1 and Figure 10.3). Among the more conspicuous divisions are lamina II, which corresponds to the substantia gelatinosa (see Chapter 10), and lamina IX, which contains the columns of lower motor neurons that innervate skeletal muscle (see Chapter 16).

TABLE A2   **The Cranial Nerves and Their Primary Functions**

Cranial nerve	Name	Sensory and/or motor	Major function
I	Olfactory nerve	Sensory	Sense of smell
II	Optic nerve	Sensory	Vision
III	Oculomotor nerve	Motor	Eye movements; papillary constriction and accommodation; muscles of eyelid
IV	Trochlear nerve	Motor	Eye movements
V	Trigeminal nerve	Sensory and motor	Somatic sensation from face, mouth, cornea; muscles of mastication
VI	Abducens nerve	Motor	Eye movements
VII	Facial nerve	Sensory and motor	Controls the muscles of facial expression; taste from anterior tongue; lacrimal and salivary glands
VIII	Vestibulocochlear (auditory) nerve	Sensory	Hearing; sense of balance
IX	Glossopharyngeal nerve	Sensory and motor	Sensation from pharynx; taste from posterior tongue; carotid baroreceptors
X	Vagus nerve	Sensory and motor	Autonomic functions of gut; sensation from pharynx; muscles of vocal cords; swallowing
XI	Spinal accessory nerve	Motor	Shoulder and neck muscles
XII	Hypoglossal nerve	Motor	Movements of tongue

Brainstem and Cranial Nerves

The brainstem, comprising the midbrain, pons, and medulla, is continuous rostrally with the diencephalon (thalamus and hypothalamus); caudally it is continuous with the spinal cord. Although the medulla, pons, and midbrain participate in myriad specific functions, the integrated actions of these brainstem components give rise to three fundamental functions. First, the brainstem is the target or source for the cranial nerves that deal with sensory and motor function in the head and neck (Table A2). Second, the brainstem provides a "throughway" for all of the ascending sensory tracts from the spinal cord; the sensory tracts for the head and neck (the **trigeminal system**); the descending motor tracts from the forebrain; and local pathways that link eye movement centers. Finally, the brainstem is involved in regulating the level of consciousness, primarily though the extensive forebrain projections of a portion of the brainstem core, the **reticular formation** (see Box 17A).

Understanding the internal anatomy of the brainstem is generally regarded as essential for neurological diagnosis and the practice of clinical medicine. Brainstem structures are compressed into a relatively small volume that has a regionally restricted vascular supply. Thus, vascular accidents in the brainstem—which are common—result in distinctive, and often devastating, combinations of functional deficits (see below). These deficits can be used both for diagnosis and for better understanding of the intricate anatomy of the medulla, pons, and midbrain.

Unlike the spinal cord, which is relatively homogeneous in appearance along its length, the surface appearance of each brainstem subdivision is characterized by unique bumps and bulges formed by the underlying gray matter (nuclei) or white matter (tracts) (Figures A7 and A8). A series of swellings on the dorsal and ventral surfaces of the medulla reflects many of the major structures in this caudal part of the brainstem. One prominent landmark that can be seen laterally is the **inferior olive**. Just medial to the inferior olives are the **medullary**

Location of cells whose axons form the nerve	Clinical test of function
Nasal epithelium	Test sense of smell with standard odor
Retina	Measure acuity and integrity of visual field
Oculomotor nucleus in midbrain; Edinger-Westphal nucleus in midbrain	Test eye movements (patient can't look up, down, or medially if nerve involved); look for ptosis, pupillary dilation
Trochlear nucleus in midbrain	Can't look downward when eye abducted
Trigeminal motor nucleus in pons; trigeminal sensory ganglion (the gasserian ganglion)	Test sensation on face; palpate masseter muscles and temporal muscle
Abducens nucleus in midbrain	Can't look laterally
Facial motor nucleus; superior salivatory nuclei in pons; trigeminal (gasserian) ganglion	Test facial expression plus taste on anterior tongue
Spiral ganglion; vestibular (Scarpa's) ganglion	Test audition with tuning fork; vestibular function with caloric test
Nucleus ambiguus; inferior salivatory	Test swallowing; pharyngeal gag reflex
Dorsal motor nucleus of vagus; vagal nerve ganglion	Test above plus hoarseness
Spinal accessory nucleus; nucleus ambiguus; intermediolateral column of spinal cord	Test sternocleidomastoid and trapezius muscles
Hypoglossal nucleus of medulla	Test deviation of tongue during protrusion (points to side of lesion)

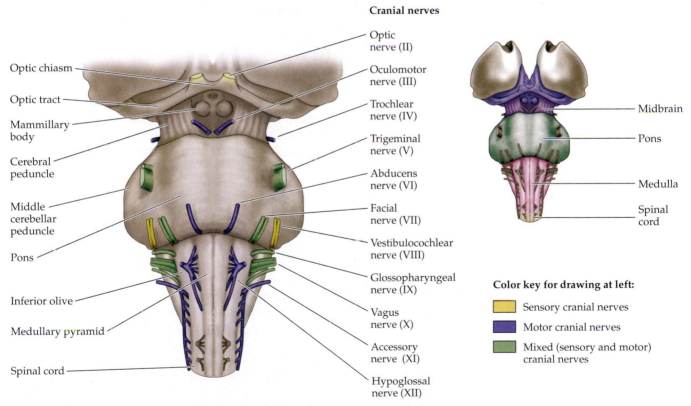

Cranial nerves

- Optic nerve (II)
- Oculomotor nerve (III)
- Trochlear nerve (IV)
- Trigeminal nerve (V)
- Abducens nerve (VI)
- Facial nerve (VII)
- Vestibulocochlear nerve (VIII)
- Glossopharyngeal nerve (IX)
- Vagus nerve (X)
- Accessory nerve (XI)
- Hypoglossal nerve (XII)

Optic chiasm
Optic tract
Mammillary body
Cerebral peduncle
Middle cerebellar peduncle
Pons
Inferior olive
Medullary pyramid
Spinal cord

Midbrain
Pons
Medulla
Spinal cord

Color key for drawing at left:

- Sensory cranial nerves
- Motor cranial nerves
- Mixed (sensory and motor) cranial nerves

Figure A7   At left is a ventral view of the brainstem showing the locations of the cranial nerves as they enter or exit the midbrain, pons, and medulla. Nerves that are exclusively sensory are indicated in yellow, motor nerves are in blue, and mixed sensory/motor nerves are in green. At right, the territories included in each of the brainstem subdivisions (midbrain, pons, and medulla) are indicated.

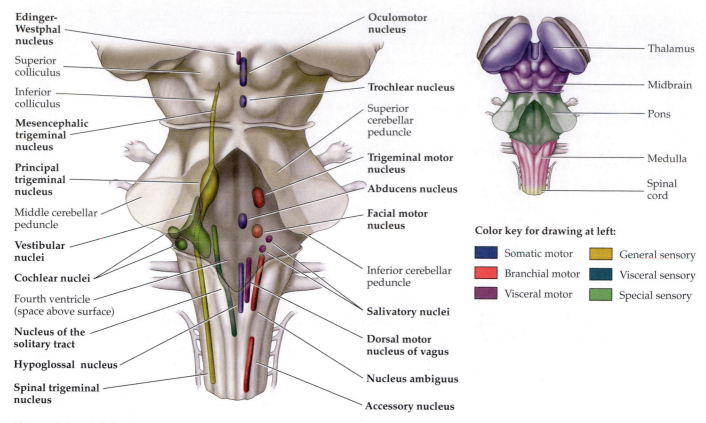

Figure A8 At left, a "phantom" view of the dorsal surface of the brainstem shows the locations of the brainstem cranial nerve nuclei that are either the target or the source of the cranial nerves. (See Table A2 for the relationship between each cranial nerve and cranial nerve nuclei and Table A3 for a functional scheme that localizes cranial nerve nuclei with respect to brainstem subdivision and sensory or motor function.) With the exception of the cranial nerve nuclei associated with the trigeminal nerve, there is fairly close correspondence between the location of the cranial nerve nuclei in the midbrain, pons, and medulla and the location of the associated cranial nerves. At right, the territories of the major brainstem subdivisions are indicated (viewed from the dorsal surface).

pyramids, prominent swellings on the ventral surface of the medulla that reflect the underlying descending corticospinal tract (see Chapter 17).

The **pons** is rostral to the medulla and is easily recognized by the mass of decussating fibers that cross (bridge) the midline on its ventral surface, giving rise to the name of this subdivision (*pons* literally means "bridge"). The **cerebellum** is attached to the dorsal aspect of the pons by three large white matter tracts: the **superior**, **middle**, and **inferior cerebellar peduncles**. Each of these tracts contains either efferent (superior and inferior) or afferent (inferior and middle) axons from or to the cerebellum (see Chapter 19).

The **midbrain** contains the **superior** and **inferior colliculi** defining its dorsal surface, or *tectum* (Latin for "roof"). Several midbrain nuclei, including the **substantia nigra**, lie in the ventral portion or tegmentum ("covering") of the midbrain. The other noteworthy anatomical feature of the midbrain is the presence of the prominent **cerebral peduncles** that are visible from the ventral surface.

The surface features of the midbrain, pons, and medulla can be used as landmarks for locating the source and termination of the majority of cranial nerves in the brainstem. Unlike the spinal nerves, the entry and exit points of the cranial nerves are not regularly arrayed along the length of the brainstem. Two cranial nerves, the **olfactory nerve** (I) and the **optic nerve** (II), enter the forebrain directly. The remaining cranial nerves enter and exit at distinct regions of the ventral (and in one case, the dorsal) surface of the midbrain, pons, and medulla (see Figure A7). The **oculomotor nerve** (III) exits into the space between the two cerebral peduncles on the ventral surface of the midbrain. The **trochlear nerve** (IV) associated with the caudal midbrain is the only cranial nerve to exit on the dorsal surface of the brainstem. The **trigeminal nerve** (V)—the largest

TABLE A3	Classification and Location of the Cranial Nerve Nuclei[a]					
Location	Somatic motor	Branchial motor	Visceral motor	General sensory	Special sensory	Visceral sensory
Midbrain	Oculomotor nucleus (III) Trochlear nucleus (IV)		Edinger-Westphal nucleus (III)	Trigeminal sensory: mesencephalic nucleus (V, VII, IX, X)		
Pons	Abducens nucleus (VI)	Trigeminal motor nucleus (V) Facial nucleus (VII)	Superior salivatory nucleus (VII) Inferior salivatory nucleus (IX)	Trigeminal sensory: principal nucleus (V, VII, IX, X)	Vestibular nuclei (VIII) Cochlear nuclei (VIII)	Nucleus of the solitary tract (VII, IX, X)
Medulla	Hypoglossal nucleus (XII)	Nucleus ambiguus (IX, X) Spinal accessory nucleus (XI)	Dorsal motor nucleus of vagus (X) Nucleus ambiguus (X)	Trigeminal sensory: spinal nucleus (V, VII, IX, X)		

[a] Associated cranial nerves are shown in parentheses.

cranial nerve—exits the ventrolateral pons through the middle cerebellar peduncle. The **abducens nerve (VI)**, **facial nerve (VII)**, and **vestibulocochlear nerve (VIII)** emerge in a medial to lateral manner, respectively, at the junction of the pons and medulla. The **glossopharyngeal nerve (IX)** and the **vagus nerve (X)** are associated with the lateral medulla, whereas the **hypoglossal nerve (XII)** exits the ventromedial medulla between the medullary pyramids and the inferior olive. The **spinal accessory nerve (XI)** does not originate in the brainstem but, as its name implies, exits the lateral portion of the upper cervical spinal cord. Table A2 describes the major functions of the cranial nerves.

Cranial nerve nuclei within the brainstem are the targets of cranial sensory nerves or the source of cranial motor nerves (Table A3; also see Figure A8). Cranial nerve nuclei that receive sensory input (analogous to the dorsal horns of the spinal cord) are located separately from those that give rise to motor output (which are analogous to the ventral horns). The primary sensory neurons that innervate these nuclei are found in ganglia associated with the cranial nerves— similar to the relationship between dorsal root ganglia and the spinal cord. In general, sensory nuclei are found laterally in the brainstem, whereas motor nuclei are located more medially (Figure A9).

There are three types of brainstem motor nuclei: **somatic motor nuclei** project to striated muscles; **branchial motor nuclei** project to muscles derived from embryonic structures called branchial arches (these arches give rise to the muscles—and bones—of the jaws and other craniofacial structures); and **visceral motor nuclei** project to peripheral ganglia that innervate smooth muscle or glandular targets, similar to preganglionic motor neurons in the spinal cord that innervate autonomic ganglia. Finally, the major ascending or descending tracts that carry sensory or motor information to or from the brain are found in the lateral and basal regions of the brainstem.

The rostral–caudal organization of the cranial nerve nuclei (all of which are bilaterally symmetric) reflects the rostral–caudal distribution of head and neck structures (see Figure A9). The more caudal the nucleus, the more caudally located the target structures in the periphery. For example, the spinal accessory

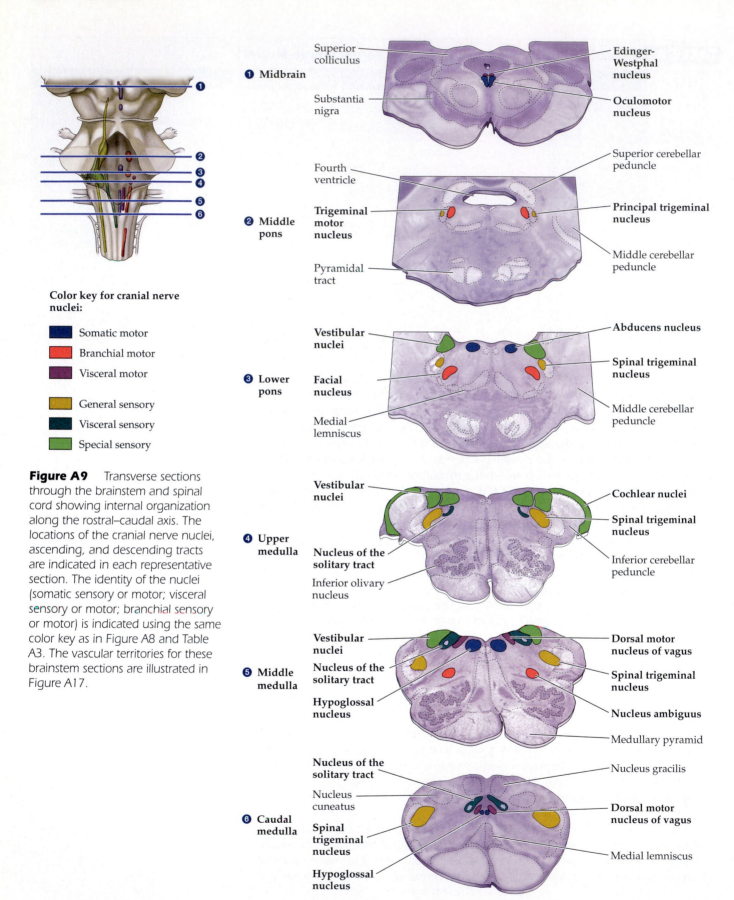

Color key for cranial nerve nuclei:

■ Somatic motor
■ Branchial motor
■ Visceral motor

■ General sensory
■ Visceral sensory
■ Special sensory

Figure A9 Transverse sections through the brainstem and spinal cord showing internal organization along the rostral–caudal axis. The locations of the cranial nerve nuclei, ascending, and descending tracts are indicated in each representative section. The identity of the nuclei (somatic sensory or motor; visceral sensory or motor; branchial sensory or motor) is indicated using the same color key as in Figure A8 and Table A3. The vascular territories for these brainstem sections are illustrated in Figure A17.

❶ Midbrain

Superior colliculus
Substantia nigra
Edinger-Westphal nucleus
Oculomotor nucleus

❷ Middle pons

Fourth ventricle
Trigeminal motor nucleus
Pyramidal tract
Superior cerebellar peduncle
Principal trigeminal nucleus
Middle cerebellar peduncle

❸ Lower pons

Vestibular nuclei
Facial nucleus
Medial lemniscus
Abducens nucleus
Spinal trigeminal nucleus
Middle cerebellar peduncle

❹ Upper medulla

Vestibular nuclei
Nucleus of the solitary tract
Inferior olivary nucleus
Cochlear nuclei
Spinal trigeminal nucleus
Inferior cerebellar peduncle

❺ Middle medulla

Vestibular nuclei
Nucleus of the solitary tract
Hypoglossal nucleus
Dorsal motor nucleus of vagus
Spinal trigeminal nucleus
Nucleus ambiguus
Medullary pyramid

❻ Caudal medulla

Nucleus of the solitary tract
Nucleus cuneatus
Spinal trigeminal nucleus
Hypoglossal nucleus
Nucleus gracilis
Dorsal motor nucleus of vagus
Medial lemniscus

nucleus in the cervical spinal cord and caudal medulla provides motor innervation for neck and shoulder muscles, and the motor nucleus of the vagus nerve provides preganglionic innervation for many enteric and visceral targets. In the pons, the sensory and motor nuclei are primarily concerned with somatic sensation from the face (the principal trigeminal nuclei); movement of the jaws and the muscles of facial expression (the trigeminal motor and facial nuclei). Further rostrally, in the mesencephalic portion of the brainstem, are nuclei concerned primarily with eye movements (the oculomotor and trochlear nuclei) and preganglionic parasympathetic innervation of the iris (the Edinger-Westphal nuclei). While this list is not complete, it indicates the basic order of the rostral–caudal organization of the brainstem.

Neurologists and other healthcare professionals assess combinations of cranial nerve deficits to infer the location of brainstem lesions, or to place the source of brain dysfunction either in the spinal cord or brain. The most common brainstem lesions reflect the vascular territories that supply subsets of cranial nerve nuclei as well as ascending and descending tracts (see below). For example, an occlusion of the posterior inferior cerebellar artery (PICA), a branch of the vertebral artery that supplies the lateral region of the mid- and rostral medulla, results in damage to three cranial nerve nuclei and several tracts (see the "Upper medulla" section in Figure A9). Accordingly, there are functional deficits that reflect the loss of the spinal trigeminal nucleus, the vestibular nucleus, and the nucleus ambiguus (which contains branchial motor neurons that project to the larynx and pharynx) on the same side as the lesion. In addition, ascending pathways from the spinal cord that relay pain and temperature from the contralateral body surface are disrupted, leading to a contralateral loss of these functions. Finally, the inferior cerebellar peduncle, which contains projections that relay information about body position to the cerebellum for postural control, is damaged. This loss results in ataxia (clumsiness) on the side of the lesion.

Anatomical relationships and shared vascularization, rather than any functional principle, unite these deficits and allow clinical localization of brainstem damage. For both clinicians and neurobiologists, understanding the brainstem requires integrating regional anatomical information with knowledge about functional organization and pathology.

Lateral Surface of the Brain

A lateral view of the human brain is the best perspective from which to appreciate all four lobes of the cerebral hemisphere (see Figure A3A). In this view, the two most salient landmarks are the deep lateral fissure that separates the temporal lobe from the overlying frontal and parietal lobes, and the central sulcus, which serves as the boundary between the frontal and parietal lobes (Figure A10). A particularly important feature of the frontal lobe is the **precentral gyrus**. (The prefix *pre-*, when used anatomically, refers to a structure that is in front of or anterior to another.) The cortex of the precentral gyrus is referred to as the **motor cortex** and contains neurons whose axons project to the lower motor neurons in the brainstem and spinal cord; these motor neurons innervate skeletal (striated) muscles (see Chapter 17).

The superior aspect of the temporal lobe contains cortex concerned with audition and language reception, and inferior portions of the lobe deal with highly processed visual information. Hidden beneath the frontal and temporal lobes, the **insular cortex** or **insula** can be seen only if these two lobes are pulled apart or removed (Figure A10B). The insular cortex is largely concerned with visceral and autonomic function, including taste. In the anterior parietal lobe just posterior to the central sulcus is the **postcentral gyrus**; this gyrus har-

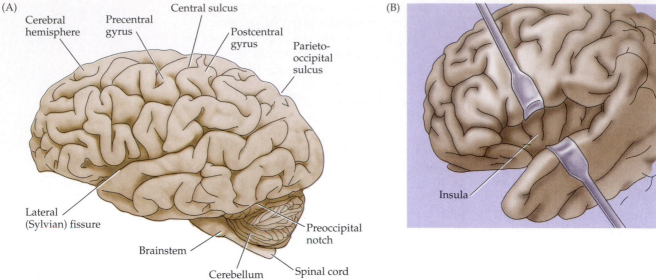

(A)

Cerebral hemisphere

Precentral gyrus

Central sulcus

Postcentral gyrus

Parieto-occipital sulcus

Lateral (Sylvian) fissure

Brainstem

Cerebellum

Preoccipital notch

Spinal cord

(B)

Insula

Figure A10   *Lateral view of the human brain. (A) Illustration of some of the major gyri and sulci from this perspective. (B) The banks of the lateral (Sylvian) fissure have been pulled apart to expose the insula.*

bors cortex that is concerned with somatic (bodily) sensation and is therefore referred to as the **somatic sensory cortex** (see Chapter 9).

The boundary between the parietal lobe and the occipital lobe, the most posterior of the hemispheric lobes, is a somewhat arbitrary line from the parieto-occipital sulcus to the preoccipital notch. The occipital lobe, only a small part of which is apparent from the lateral surface of the brain, is primarily concerned with vision. In addition to their role in primary and sensory processing, each cortical lobe participates in higher brain functions related to cognition (see Chapter 26). Thus, the frontal lobe is critical in planning responses to stimuli, the parietal lobe in attending to stimuli, the temporal lobe in recognizing stimuli, and the occipital lobe in all aspects of visual perception.

Dorsal and Ventral Surfaces of the Brain

Although the primary subdivisions of the cerebral hemispheres can be appreciated from a lateral view, other key landmarks are better seen from the dorsal and ventral surfaces. When viewed from the dorsal surface, the approximate bilateral symmetry of the cerebral hemispheres is apparent (Figure A11). Although there is some variation, major landmarks like the central sulci and parieto-occipital sulci are usually very similar in arrangement on the two sides. If the cortical hemispheres are spread slightly apart from the dorsal midline, another major structure, the **corpus callosum**, can be seen bridging the two hemispheres (see Figure A11C). This tract contains axons that originate from neurons in the cerebral cortex of both hemispheres that contact target neurons in the opposite cortical region.

The external features of the brain best seen on its ventral surface are shown in Figure A11B. Extending along the inferior surface of the frontal lobe near the midline are the **olfactory tracts**, which arise from enlargements at their anterior ends called the **olfactory bulbs**. The olfactory bulbs receive input from neurons in the epithelial lining of the nasal cavity whose axons make up the first cranial nerve (cranial nerve I is therefore called the **olfactory nerve**; see Table A2). On the ventral–medial surface of the temporal lobe, the **parahippocampal gyrus** conceals the **hippocampus**, a highly convoluted cortical structure that figures importantly in memory (see Chapter 31). Slightly more medial to the parahip-

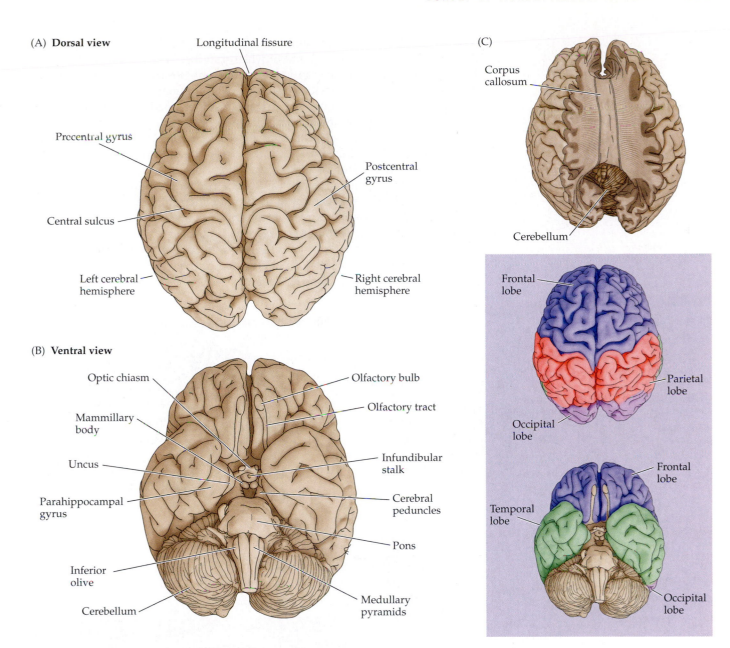

(A) **Dorsal view**

Longitudinal fissure

Precentral gyrus

Postcentral gyrus

Central sulcus

Left cerebral hemisphere

Right cerebral hemisphere

(B) **Ventral view**

Optic chiasm

Olfactory bulb

Mammillary body

Olfactory tract

Uncus

Infundibular stalk

Parahippocampal gyrus

Cerebral peduncles

Inferior olive

Pons

Cerebellum

Medullary pyramids

(C)

Corpus callosum

Cerebellum

Frontal lobe

Parietal lobe

Occipital lobe

Frontal lobe

Temporal lobe

Occipital lobe

Figure A11   Dorsal view (A) and ventral view (B) of the human brain, indicating some of the major features visible form these perspectives. (C) The cerebral cortex has been removed in this dorsal view (upper image) to reveal the underlying corpus callosum. Lower boxed images highlight the four lobes of the cerebral cortex. (C After Rohen et al., 1993.)

pocampal gyrus is the **uncus**, a slightly conical protrusion that includes the cortical divisions of the **amygdala**. At the most central aspect of the ventral surface of the forebrain is the **optic chiasm**, and immediately posterior, the ventral surface of the **hypothalamus**, including the **infundibulum** (also called the **pituitary stalk**, at the base of the pituitary gland) and the **mammillary bodies**. Posterior to the hypothalamus, the paired cerebral peduncles are located on either side of the ventral midline of the midbrain. Finally, the ventral surfaces of the pons, medulla, and cerebellar hemispheres can be seen on the ventral surface of the brain (see Figure A7).

Midsagittal Surface of the Brain

When the brain is hemisected in the midsagittal plane, all of its major subdivisions plus a number of additional structures are visible on the cut surface. In this view, the cerebral hemispheres, because of their great size, are still the most obvious structures. The frontal lobe of each hemisphere extends forward from the central sulcus, the medial end of which can just be seen (Figure A12A,B). The parieto-occipital sulcus, running from the superior to the inferior aspect of the hemisphere, is most obvious in this view of the hemisphere as it separates the parietal and occipital lobes. The **calcarine sulcus** divides the medial surface of the occipital lobe, running at very nearly a right angle from the parieto-occipital sulcus and marking the location of the **primary visual cortex** (see Chapter 12). A long, roughly horizontal sulcus, the **cingulate sulcus**, extends across the medial surface of the frontal and parietal lobes. The gyrus below it, the **cingulate gyrus**, is a prominent component of the **limbic system**, which comprise cortical and subcortical structures in the frontal and temporal lobes that form a medial rim of cerebrum roughly encircling the corpus callosum and diencephalon (*limbic* means border or rim). The limbic system is important in the experience and expression of emotion, as well as the regulation of visceral motor activity (see Chapter 29). Finally, ventral to the cingulate gyrus is the cut, midsagittal surface of the corpus callosum.

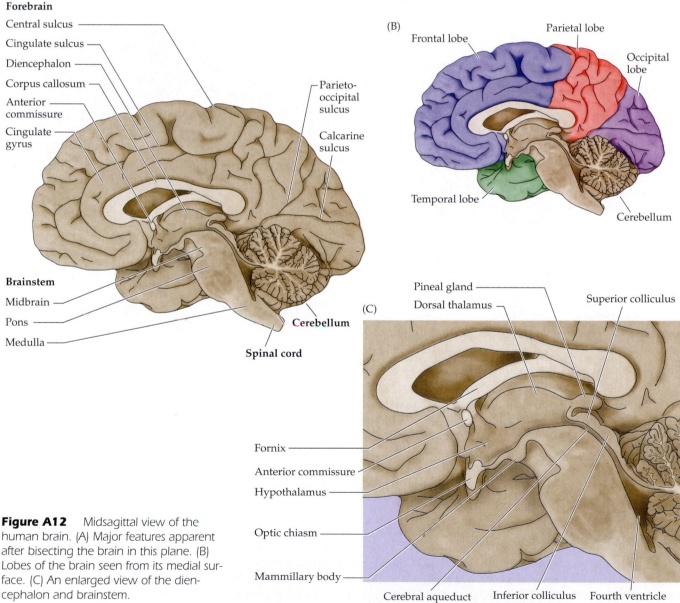

Figure A12   Midsagittal view of the human brain. (A) Major features apparent after bisecting the brain in this plane. (B) Lobes of the brain seen from its medial surface. (C) An enlarged view of the diencephalon and brainstem.

Although parts of the diencephalon, brainstem, and cerebellum are visible at the ventral surface of the brain, their overall structure is especially clear from the midsagittal surface (Figure A12C). From this perspective, the diencephalon can be seen to consist of two parts. The **dorsal thalamus**, the largest component of the diencephalon, comprises a number of subdivisions, all of which relay information to the cerebral cortex from other parts of the brain. The **hypothalamus**—a small but crucial part of the diencephalon—is devoted to the control of homeostatic and reproductive functions, among other diverse activities (see Box 21A). The hypothalamus is intimately related, both structurally and functionally, to the pituitary gland, a critical endocrine organ whose posterior part is connected to the hypothalamus by the infundibulum.

The midbrain tegmentum, which can be seen only in this view, lies caudal to the thalamus, and the pons is caudal to the midbrain. The cerebellum lies over the pons just beneath the occipital lobe of the cerebral hemispheres. From the midsagittal surface, the most visible feature of the cerebellum is the **cerebellar cortex**, a continuous layered sheet of cells folded into ridges and valleys called **folia**. The most caudal structure seen from the midsagittal surface of the brain is the medulla, which merges into the spinal cord.

Internal Anatomy of the Forebrain

A much more detailed neuroanatomical picture of the forebrain is apparent in gross or histological slices. In these slices (or sections), deep structures that are not visible from any brain surface can be identified. In addition, relationships between brain structures seen from the surface can be appreciated more fully. The major challenge to understanding the internal anatomy of the brain is to integrate the rostral–caudal, dorsal–ventral, and medial–lateral landmarks seen on the brain surface with the position of structures seen in brain sections taken in the horizontal, frontal, and sagittal planes. This challenge is not only important for understanding brain function; it is essential for interpreting noninvasive images of the brain, most of which are displayed as sections.

In any plane of section through the forebrain, the **cerebral cortex** is evident as a thin layer of neural tissue that covers the entire cerebrum. Most cerebral cortex is made up of six layers, and is referred to as **neocortex** (see Box 26A). Phylogenetically older cortex (**paleocortex**) with fewer cell layers occurs on the inferior and medial aspect of the temporal lobe within the parahippocampal gyrus. Cortex with even fewer layers (three), referred to as **archicortex**, occurs in the hippocampus and in the pyriform cortex (a major division of the olfactory cortex near the junction of the temporal and frontal lobes). The hippocampal cortex is folded into the medial aspect of the temporal lobe, and is therefore visible only in dissected brains or in sections (Figure A13).

The largest gray matter structures embedded within the cerebral hemispheres are the **caudate** and **putamen nuclei** (together referred to as **striatum**) as well as the **globus pallidus** (Figure A14). Collectively these several structures are referred to as the **basal ganglia** (the term *ganglia* does not usually refer to nuclei in the brain; the usage here is an exception). The basal ganglia are visible in horizontal sections through the mid-dorsal to mid-ventral portion of the forebrain, in frontal sections from just rostral to the uncus to the level of the posterior diencephalon, and in paramedian sagittal sections. The neurons of these large nuclei receive input from the cerebral cortex and participate in the organization and guidance of complex motor functions (see Chapter 18). In the base of the forebrain, ventral to the basal ganglia are several smaller clusters of nerve cells known as the **septal** or **basal forebrain nuclei**. These nuclei are of particular interest because they modulate neural activity in the cerebral cortex

Figure A13   *Major internal structures of the brain, shown after the upper half of the left hemisphere has been cut away.*

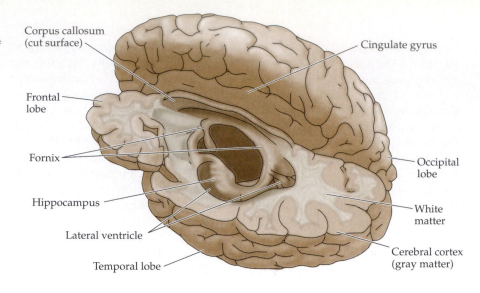

Corpus callosum (cut surface)

Cingulate gyrus

Frontal lobe

Fornix

Occipital lobe

Hippocampus

White matter

Lateral ventricle

Cerebral cortex (gray matter)

Temporal lobe

(A)

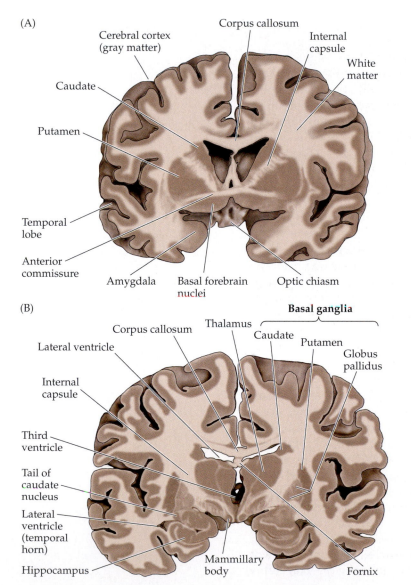

Cerebral cortex (gray matter)

Corpus callosum

Internal capsule

White matter

Caudate

Putamen

Temporal lobe

Anterior commissure

Amygdala

Basal forebrain nuclei

Optic chiasm

(B)

Corpus callosum

Thalamus

Basal ganglia

Caudate

Putamen

Globus pallidus

Lateral ventricle

Internal capsule

Third ventricle

Tail of caudate nucleus

Lateral ventricle (temporal horn)

Hippocampus

Mammillary body

Fornix

(C)

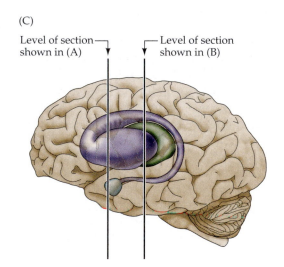

Level of section shown in (A)

Level of section shown in (B)

Figure A14   *Internal structures of the brain seen in coronal section. (A) This sections passes through the basal ganglia. (B) A more posterior section that also includes the thalamus. (C) A transparent view of the cerebral hemisphere showing the approximate locations of the sections in (A) and (B) relative to deep gray matter (the basal ganglia, thalamus and amygdala are represented). Notice that because the caudate nucleus has a "tail" that arcs into the temporal lobe, it appears twice in section (B); the same is true of other brain structures, including the lateral ventricle.*

and hippocampus and they are among the forebrain systems that degenerate in Alzheimer's disease. The other clearly discernible structure visible in sections through the cerebral hemispheres at the level of the uncus is the **amygdala**, a complex of nuclei and cortical divisions that lies in front of the hippocampus in the anterior pole of the temporal lobe.

In addition to these cortical and nuclear structures, the internal anatomy of the forebrain is characterized by a number of important axon tracts. As already mentioned, the two cerebral hemispheres and many of their component parts are interconnected by the corpus callosum; in some anterior sections, the smaller **anterior commissure** can also be seen (Figure A14A). Axons descending from (and ascending to) the cerebral cortex assemble into another large fiber bundle tract called the **internal capsule** (Figure A14A,B). The internal capsule lies just lateral to the diencephalon (forming a "capsule" around it), and many of its axons arise from or terminate in the dorsal thalamus. It is seen most clearly in frontal sections through the middle one-third of the rostral–caudal extent of forebrain, or in horizontal sections through the level of the thalamus. Other axons descending from the cortex in the internal capsule continue past the diencephalon to enter the cerebral peduncles of the midbrain. Axons in these corticobulbar and corticospinal tracts project to a number of targets in the brainstem and spinal cord, respectively (see Chapter 17). Thus, the internal capsule is the major pathway linking the cerebral cortex to the rest of the brain and spinal cord. Strokes or other injury to this structure interrupt the flow of ascending and descending nerve impulses, often with devastating consequences (see Box A). Finally, a smaller fiber bundle within each of the hemispheres, the **fornix**, interconnects the hippocampus and the hypothalamus and septal region of the basal forebrain.

Blood Supply of the Brain and Spinal Cord

Understanding the blood supply of the brain and spinal cord is crucial for neurological diagnoses and the practice of medicine, particularly for neurology and neurosurgery. Damage to major blood vessels by trauma or stroke results in combinations of functional defects that reflect both local cell death and the disruption of axons passing through the region compromised by the vascular damage. Thus, a firm knowledge of the major cerebral blood vessels and the neuroanatomical territories they perfuse facilitates the initial diagnoses of a broad range of brain damage and disease.

The entire blood supply of the brain and spinal cord depends on two sets of branches from the dorsal aorta. The **vertebral arteries** arise from the subclavian arteries; the **internal carotid arteries** are branches of the common carotid arteries. The vertebral arteries and the ten **medullary arteries** that arise from segmental branches of the aorta provide the primary vascularization of the spinal cord. These medullary arteries join to form the **anterior** and **posterior spinal arteries** (Figure A15). If any of the medullary arteries are obstructed or damaged (during abdominal surgery, for example), the blood supply to specific parts of the spinal cord may be compromised. The pattern of resulting neurological damage differs according to whether supply to the posterior or anterior artery is interrupted. As might be expected from the arrangement of ascending and descending neural pathways in the spinal cord described above, loss of the posterior supply generally leads to loss of sensory functions, whereas loss of the anterior supply more often causes motor deficits.

Anterior to the spinal cord and brainstem, the **internal carotid arteries** branch to form two major cerebral arteries, the **anterior** and **middle cerebral arteries**. The right and left **vertebral arteries** come together at the level of the

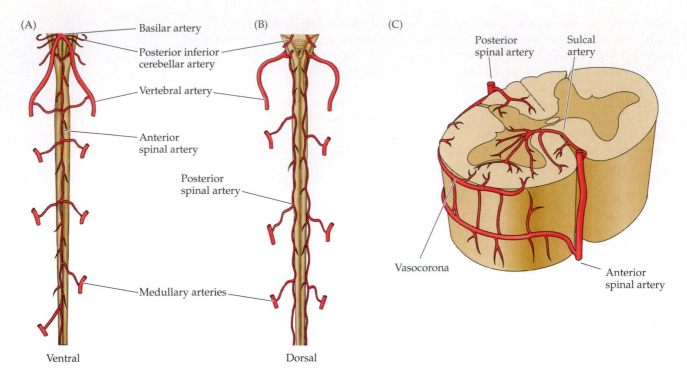

(A)

Basilar artery

Posterior inferior
cerebellar artery

Vertebral artery

Anterior
spinal artery

Medullary arteries

Ventral

(B)

Vertebral artery

Posterior
spinal artery

Medullary arteries

Dorsal

(C)

Posterior
spinal artery

Sulcal
artery

Vasocorona

Anterior
spinal artery

Figure A15 Blood supply of the spinal cord. (A) View of the ventral (anterior) surface of the spinal cord. At the level of the medulla, the vertebral arteries give off branches that merge to form the anterior spinal artery. Approximately 10 to 12 segmental arteries (which arise from various branches of the aorta) join the anterior spinal artery along its course. These segmental arteries are known as medullary arteries. (B) The vertebral arteries (or the posterior inferior cerebellar artery) give rise to paired posterior spinal arteries that run along the dorsal (posterior) surface of the spinal cord. (C) Cross section through the spinal cord, illustrating the distribution of the anterior and posterior spinal arteries. The anterior spinal arteries give rise to numerous sulcal branches that supply the anterior two-thirds of the spinal cord. The posterior spinal arteries supply much of the dorsal horn and the dorsal columns. A network of vessels known as the vasocorona connects these two sources of supply and sends branches into the white matter around the margin of the spinal cord.

pons on the ventral surface of the brainstem to form the midline **basilar artery**. The basilar artery joins the blood supply from the internal carotids in an arterial ring at the base of the brain (in the vicinity of the hypothalamus and cerebral peduncles) called the **circle of Willis**. The **posterior cerebral arteries** arise at this confluence, as do two small bridging arteries, the **anterior** and **posterior communicating arteries**. Conjoining the two major sources of cerebral vascular supply via the circle of Willis presumably improves the chances of any region of the brain continuing to receive blood if one of the major arteries becomes occluded.

The major branches that arise from the internal carotid artery—the anterior and middle cerebral arteries—form the **anterior circulation** that supplies the forebrain (Figure A16). These arteries branch from the internal carotids within the circle of Willis. Each gives rise to branches that supply the cortex and branches that penetrate the basal surface of the brain to supply deep structures. Particularly prominent are the **lenticulostriate** and **anterior choroidal arteries** that branch from the middle cerebral artery; these arteries supply the basal ganglia, internal capsule and hippocampus (see Figure A16D). The **posterior circulation** of the brain supplies the posterior cerebral cortex, thalamus, and the brainstem; it comprises arterial branches arising from the **posterior cerebral**, **basilar**, and **vertebral arteries**. The pattern of arterial distribution is similar for all the subdivisions of the brainstem: midline arteries supply medial structures, lateral arteries supply the lateral brainstem, and dorsolateral arteries supply dorsolateral brainstem structures and the cerebellum (Figures A16 and A17). Among the most important dorsolateral arteries (also called **long circumferential** arteries) are the **posterior inferior cerebellar artery** (**PICA**) and the **anterior inferior cerebellar artery** (**AICA**), which supply distinct regions of the medulla and pons. These arteries, as well as branches of the basilar artery that penetrate the brainstem from its ventral and lateral surfaces—the **paramedian** and **short circumferential** arteries—are especially common sites of occlusion and result in specific functional deficits of cranial nerve, somatic sensory, and motor function.

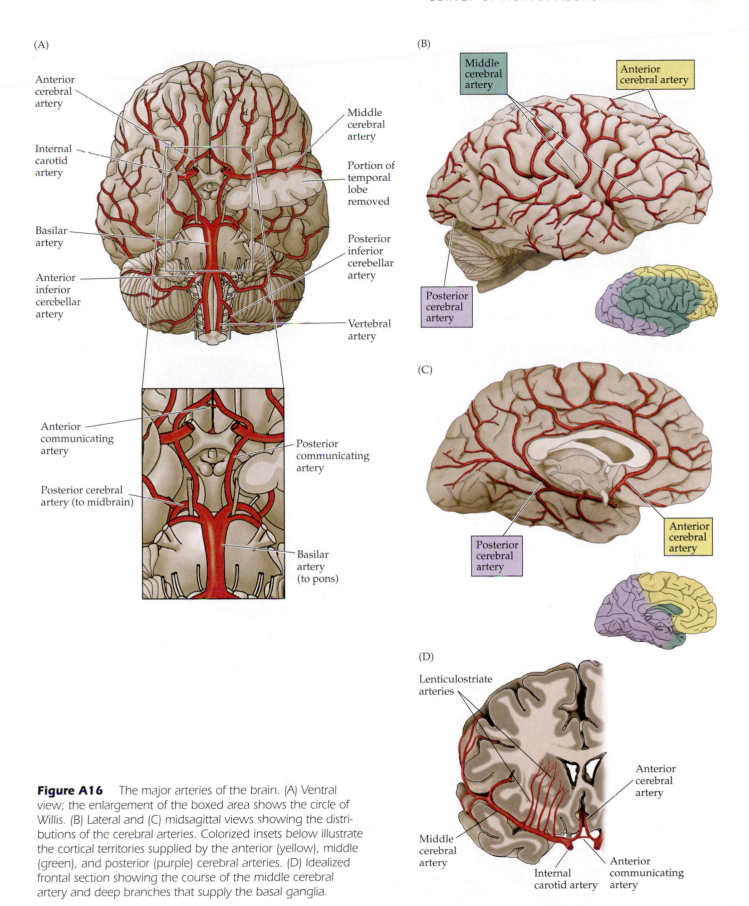

Figure A16   The major arteries of the brain. (A) Ventral view; the enlargement of the boxed area shows the circle of Willis. (B) Lateral and (C) midsagittal views showing the distributions of the cerebral arteries. Colorized insets below illustrate the cortical territories supplied by the anterior (yellow), middle (green), and posterior (purple) cerebral arteries. (D) Idealized frontal section showing the course of the middle cerebral artery and deep branches that supply the basal ganglia.

(A)

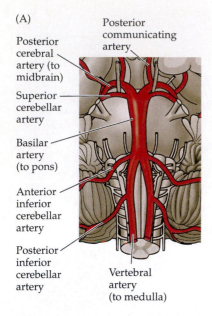

Posterior cerebral artery (to midbrain)

Posterior communicating artery

Superior cerebellar artery

Basilar artery (to pons)

Anterior inferior cerebellar artery

Posterior inferior cerebellar artery

Vertebral artery (to medulla)

Figure A17 Blood supply of the three subdivisions of the brainstem. (A) Illustration of the major blood supply derived from the basilar and vertebral arteries. (B) Sections through different levels of the brainstem indicating the territory supplied by each of the major brainstem arteries.

(B)

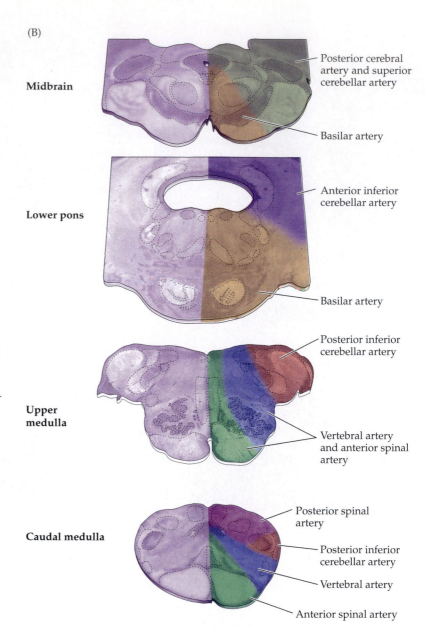

Midbrain

Posterior cerebral artery and superior cerebellar artery

Basilar artery

Lower pons

Anterior inferior cerebellar artery

Basilar artery

Upper medulla

Posterior inferior cerebellar artery

Vertebral artery and anterior spinal artery

Caudal medulla

Posterior spinal artery

Posterior inferior cerebellar artery

Vertebral artery

Anterior spinal artery

The physiological demands on the brain's blood supply are particularly significant because neurons are more sensitive to oxygen deprivation than cells with lower rates of metabolism. The high metabolic rate of neurons means that brain tissue deprived of oxygen and glucose as a result of compromised blood supply is likely to sustain transient or permanent damage. Even brief loss of blood supply (referred to as ischemia) can cause cellular changes that, if not quickly reversed, can lead to cell death. Sustained loss of blood supply leads much more directly to death and degeneration of the deprived cells. Strokes—an anachronistic term that refers to the death or dysfunction of brain tissue due to vascular disease—often follow the occlusion of (or hemorrhage from) the brain's arteries (Box A). Historically, studies of the functional consequences of strokes, and their relation to vascular territories in the brain and spinal cord, provided information about the location of various brain functions. The location of the major language functions in the left hemisphere, for instance, was discovered in this way in the latter part of the nineteenth century (see Chapter 27).

BOX A Stroke

Stroke is the most common neurological cause for admission to a hospital, and is the third leading cause of death in the United States (after heart disease and cancer). The term *stroke* refers to the sudden appearance of a limited neurological deficit, such as weakness or paralysis of a limb, or the sudden inability to speak. The onset of the deficit within seconds, minutes, or hours marks the problem as a vascular one. Brain function is exquisitely dependent on a continuous supply of oxygen, as evidenced by the onset of unconsciousness within about 10 seconds of blocking its blood supply (by cardiac arrest, for instance). The damage to neurons is at first reversible, but becomes permanent if the blood supply is not promptly restored.

Strokes can be subdivided into three main types: **thrombotic**, **embolic**, and **hemorrhagic**. The thrombotic variety is caused by a local reduction of blood flow arising from an atherosclerotic buildup in one of the cerebral blood vessels that eventually occludes it. A reduction of blood flow also can arise when an embolus (an object loose in the bloodstream) dislodges from the heart (or from an atherosclerotic plaque in the carotid or vertebral arteries) and travels to a cerebral artery (or arteriole), where it forms a plug and causes an embolic stroke. A hemorrhagic stroke occurs when a cerebral blood vessel ruptures, as can occur as a result of hypertension, a congenital aneurysm (bulging of a vessel), a congenital arteriovenous malformation, or a traumatic injury involving the meninges and/or the brain itself. The respective frequencies of thrombotic, embolic, and hemorrhagic strokes are approximately 50, 30, and 20 percent.

The diagnosis of stroke relies primarily on an accurate history and a competent neurological examination. Indeed, the neurologist C. Miller Fisher, a master of bedside diagnosis, remarked that medical students and residents should learn neurology "stroke by stroke." Understanding the portion of the brain supplied by each of the major arteries (see text) enables an astute clinician to identify the occluded blood vessel.

More recently, imaging techniques such as CT scans and MRI (see Box 1A) have greatly facilitated the physician's ability to identify and localize small hemorrhages and regions of permanently damaged tissue. Moreover, Doppler ultrasound, magnetic resonance angiography, and imaging of blood vessels by direct infusion of radio-opaque dye can now pinpoint atherosclerotic plaques, aneurysms, and other vascular abnormalities.

Several therapeutic approaches to strokes are feasible. Dissolving a thrombotic plug by tissue plasminogen activator (TPA) and other compounds is now standard clinical practice for selected stroke victims. Furthermore, recent understanding of some of the mechanisms by which ischemia injures brain tissue has made pharmacological strategies to minimize neuronal injury after stroke a potentially effective possibility (see Box 6C). Hemorrhagic strokes are treated neurosurgically by finding and stopping the bleeding from the defective vessel (when that is technically possible).

These approaches can minimize functional loss; however, strokes remain a serious health risk from which there is never full recovery. The inability of the mature brain to replace large populations of damaged or dead neurons, or to repair long axon tracts once they have been compromised, invariably prevents the complete restoration of lost functions. Despite these seemingly intractable limitations, novel strategies for neurorehabilitation continue to be investigated and introduced into clinical practice, offering some measure of hope to those afflicted with stroke and the disabilities that accompany cerebrospinal injury.

References

ADAMS, R. D., M. VICTOR AND A. H. ROPPER (2001) *Principles of Neurology*, 7th Ed. New York: McGraw-Hill, Ch. 34, pp. 821–924.

TAUB, E., G. USWATTE AND T. ELBERT (2002) New treatments in neurorehabilitation founded on basic research. *Nature Rev. Neurosci.* 3: 228–236.

Now, noninvasive functional imaging techniques based on blood flow have largely supplanted the correlation of clinical signs and symptoms with the location of tissue damage observed at autopsy (see Box 1A).

The Blood-Brain Barrier

In addition to their susceptibility to oxygen deprivation, brain cells are at risk from toxins circulating in the bloodstream. The brain is specifically protected in this respect, however, by the **blood-brain barrier**. The interface between the walls of capillaries and the surrounding tissue is important throughout the body, as it keeps vascular and extravascular concentrations of ions and molecules at appropriate levels in these two compartments. In the brain, this interface is especially significant—hence its unique and alliterative name. The special properties of the blood-brain barrier were first observed by the nineteenth-century bacteriologist Paul Ehrlich, who noted that intravenously injected dyes leaked out of capillaries in most regions of the body to stain the surrounding tissues; brain tissue, however, remained unstained. Ehrlich wrongly concluded that the brain had a low affinity for the dyes. It was his student, Edwin Goldmann, who showed that in fact such dyes do not traverse the specialized walls of brain capillaries.

The restriction of large molecules like Ehrlich's dyes (and many smaller molecules) to the vascular space is the result of tight junctions between neighboring capillary endothelial cells in the brain (Figure A18A). Such junctions are not found in capillaries elsewhere in the body, where the spaces between adjacent endothelial cells allow much more ionic and molecular traffic. The structure of tight junctions was first demonstrated in the 1960s by Tom Reese, Morris Karnovsky, and Milton Brightman. Using electron microscopy after the injection of electron-dense intravascular agents such as lanthanum salts, they showed that the close apposition of the endothelial cell membranes prevented such ions from passing (Figure A18B). Substances that traverse the walls of brain capillaries must move *through* the endothelial cell membranes. Accordingly, molecular entry into the brain should be determined by an agent's solubility in lipids, the major constituent of cell membranes. Nevertheless, many ions and molecules not readily soluble in lipids *do* move quite readily from the vascular space into brain tissue. A molecule like glucose, the primary source of metabolic energy for neurons and glial cells, is an obvious example. This paradox is explained by the presence of specific transporters for glucose and other critical molecules and ions.

In addition to tight junctions, astrocytic "end feet" (the terminal regions of astrocytic processes) surround the outside of capillary endothelial cells (see Figure A18A). The reason for this endothelial–glial allegiance is unclear, but may reflect an influence of astrocytes on the formation and maintenance of the blood-brain barrier.

The brain, more than any other organ, must be carefully shielded from abnormal variations in its ionic milieu, as well as from the potentially toxic molecules that find their way into the vascular space by ingestion, infection, or

(A)

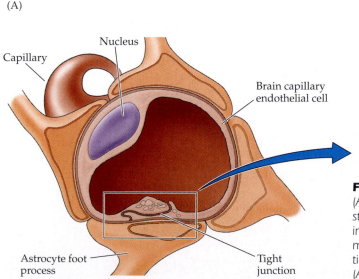

(B)

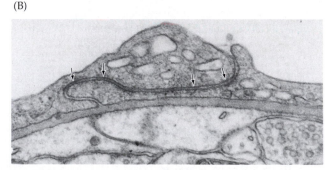

Figure A18   The cellular basis of the blood-brain barrier. (A) Diagram of a brain capillary in cross section and reconstructed views, showing endothelial tight junctions and the investment of the capillary by astrocytic end feet. (B) Electron micrograph of boxed area in (A), showing the appearance of tight junctions between neighboring endothelial cells (arrows). (A after Goldstein and Betz, 1986; B from Peters et al., 1991.)

other means. The blood-brain barrier is thus crucial for protection and homeostasis. It also presents a significant problem for the delivery of drugs to the brain. Large (or lipid-insoluble) molecules can be introduced to the brain only by transiently disrupting the blood-brain barrier with hyperosmotic agents such as the sugar mannitol.

The Meninges

The cranial cavity is conventionally divided into three regions called the anterior, middle, and posterior cranial fossae. Surrounding and supporting the brain within this cavity are three protective tissue layers, which also extend down the brainstem and the spinal cord. Together these layers are called the **meninges** (Figure A19). The outermost layer of the meninges is called the **dura mater** ("hard mother," referring to its thick and tough qualities). The middle layer is

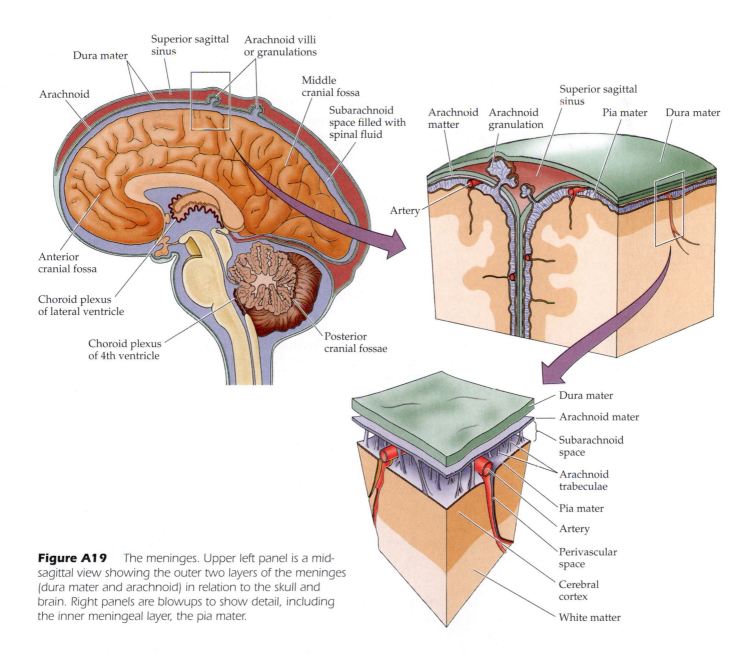

Figure A19 *The meninges. Upper left panel is a midsagittal view showing the outer two layers of the meninges (dura mater and arachnoid) in relation to the skull and brain. Right panels are blowups to show detail, including the inner meningeal layer, the pia mater.*

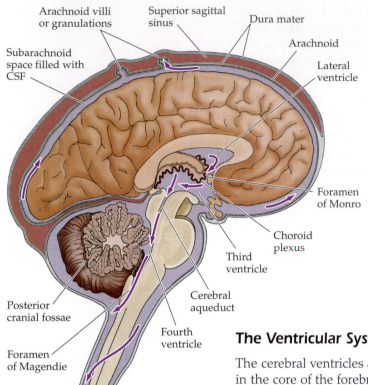

called the **arachnoid mater** because of spiderlike processes called arachnoid trabeculae, which extend from it toward the third layer, the **pia mater** ("tender mother"), a delicate layer of cells that closely invests the surface of the brain. Since the pia closely adheres to the brain as its surface curves and folds whereas the arachnoid does not, there are places, called **cisterns**, where the subarachnoid space enlarges to form significant collections of cerebrospinal fluid (the fluid that fills the ventricles; see the next section). The major arteries supplying the brain course through the subarachnoid space on the surface of the cerebrum where they give rise to branches that penetrate the substance of the hemispheres. The subarachnoid space is therefore a frequent site of bleeding following trauma. A collection of blood between the meningeal layers is referred to as a sub-dural or subarachnoid hemorrhage (or hematoma), as distinct from bleeding within the brain itself.

The Ventricular System

The cerebral ventricles are a series of interconnected, fluid-filled spaces that lie in the core of the forebrain and brainstem (Figures A20 and A21). These spaces are filled with **cerebrospinal fluid (CSF)** produced by a modified vascular structure called the **choroid plexus**, which is present in each of the ventricles. Cerebrospinal fluid percolates through the ventricular system and flows into the subarachnoid space through perforations in the thin covering of the fourth ventricle (see Figure A20); it is eventually passed through specialized structures called arachnoid villi or granulations along the dorsal midline of the forebrain (see Figure A19), and returned to the venous circulation.

The presence of ventricular spaces in the various subdivisions of the brain reflects the fact that the ventricles are the adult derivatives of the open space or lumen of the embryonic neural tube (see Chapter 22). Although they have no unique function, the ventricular spaces present in sections through the brain provide another useful guide to location (see Figure A21C). The largest of these spaces are the **lateral ventricles** (formerly called the first and second ventricles), one within each of the cerebral hemispheres. These particular ventricles are best seen in frontal sections, where their ventral and lateral surfaces are usually defined by the basal ganglia, their dorsal surface by the corpus callosum, and their medial surface by the **septum pellucidum**, a membranous tissue sheet that forms part of the midline sagittal surface of the cerebral hemispheres. The lateral ventricles, like several telencephalic structures, possess a "C" shape that is formed by the non-uniform growth of the cerebral hemispheres during embryonic development. CSF flows from the lateral ventricles through small openings (called the **interventricular foramina**, or the **foramina of Monro**) into a nar-

Figure A20   Circulation of cerebrospinal fluid (CSF). CSF is produced by the choroid plexus and flows from the lateral ventricles through the paired interventricular foramina (singular, foramen; foramina of Monro) into the third ventricle, through the cerebral aqueduct and into the fourth ventricle. CSF exits the ventricular system through several foramina associated with the fourth ventricle (e.g., foramen of Magendie) into the subarachnoid space surrounding the central nervous system. CSF is eventually passed through the arachnoid granulations and returned to the venous circulation.

(A)

Central part of left lateral ventricle

Frontal horn of lateral ventricle

Choroid plexus

Interventricular foramen of Monro

Third ventricle

Temporal horn of lateral ventricle

Right lateral ventricle

Left lateral ventricle

Occipital horn of lateral ventricle

Cerebral aqueduct

Fourth ventricle

Central canal

Choroid plexus

Figure A21 *The ventricular system of the human brain. (A) Location of the ventricles as seen in a transparent left lateral view. (B) Dorsal view of the ventricles. (C) Table showing the ventricular spaces associated with each of the major subdivisions of the brain. (See Chapter 22 for an account of brain development that more fully explains the origin of the ventricular spaces.)*

(B)

Interventricular foramen of Monro

Postcentral gyrus

Frontal horn of lateral ventricle

Third ventricle

Temporal horn of lateral ventricle

Central sulcus

Right cerebral hemisphere

Occipital horn of lateral ventricle

Fourth ventricle

Cerebral aqueduct

Left cerebral hemisphere

(C)

EMBRYONIC BRAIN		ADULT BRAIN DERIVATIVES	ASSOCIATED VENTRICULAR SPACE
Prosencephalon	Telencephalon (forebrain)	Cerebral cortex	Lateral ventricles
		Basal ganglia Hippocampus Olfactory bulb Basal forebrain	
	Diencephalon	Dorsal thalamus	Third ventricle
		Hypothalamus	
Mesencephalon		Midbrain (superior and inferior colliculi)	Cerebral aqueduct
Rhombencephalon	Metencephalon	Cerebellum	Fourth ventricle
		Pons	
	Myelencephalon	Medulla	Fourth ventricle
Spinal cord		Spinal cord	Central canal

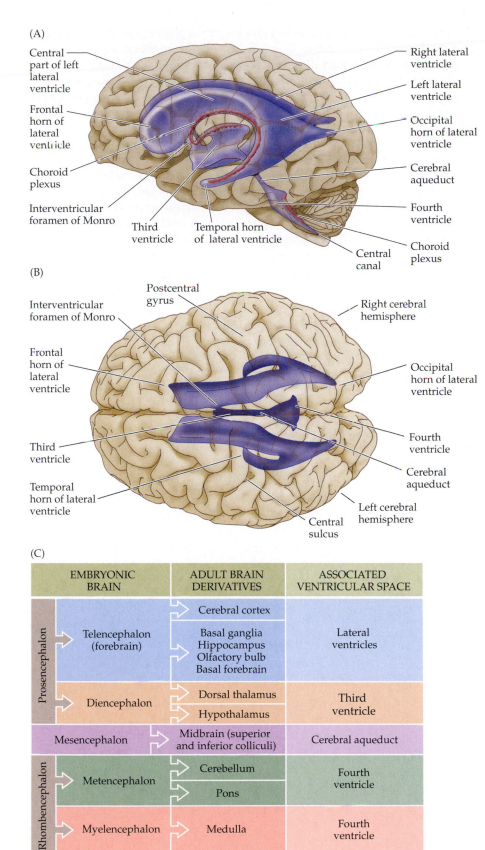

row midline space between the right and left diencephalon, the **third ventricle**. The third ventricle is continuous caudally with the **cerebral aqueduct** (also referred to as the **aqueduct of Sylvius**), which runs though the midbrain. At its caudal end, the aqueduct opens into the **fourth ventricle**, a larger space in the dorsal pons and medulla. The fourth ventricle, covered on its dorsal aspect by the cerebellum, narrows caudally to form the central canal of the spinal cord, which normally does not remain patent beyond the early postnatal period.

The normal total volume of CSF in the ventricular system and subarachnoid space is approximately 140 mL. The choroid plexus produces approximately 500 mL of CSF per day, so that the entire volume present in the system is turned over several times a day. Thus, obstruction of CSF flow results in an excess of cerebrospinal fluid in the intracranial cavity, a dangerous condition called **hydrocephalus** (literally, "water head") that can lead to enlargement of the ventricles and compression of the brain.

References

BLUMENFELD, H. (2002) *Neuroanatomy through Clinical Cases.* Sunderland, MA: Sinauer Associates.

BRIGHTMAN, M. W. AND T. S. REESE (1969) Junctions between intimately opposed cell membranes in the vertebrate brain. *J. Cell Biol.* 40: 648–677.

BRODAL, P. (2004) *The Central Nervous System: Structure and Function*, 3rd Ed. New York: Oxford University Press.

CARPENTER, M. B. AND J. SUTIN (1983) *Human Neuroanatomy*, 8th Ed. Baltimore, MD: Williams and Wilkins.

ENGLAND, M. A. AND J. WAKELY (1991) *Color Atlas of the Brain and Spinal Cord: An Introduction to Normal Neuroanatomy.* St. Louis: Mosby Yearbook.

GOLDSTEIN, G. W. AND A. L. BETZ (1986) The blood-brain barrier. *Sci. Amer.* 255(3):74–83

HAINES, D. E. (1995) *Neuroanatomy: An Atlas of Structures, Sections, and Systems*, 2nd Ed. Baltimore: Urban and Schwarzenberg.

MARTIN, J. H. (2003) *Neuroanatomy: Text and Atlas*, 3rd Ed. New York: McGraw-Hill.

NETTER, F. H. (1983) *The CIBA Collection of Medical Illustrations*, Vols. I and II.

PAXINOS, G AND J. K. MAI (2004) *The Human Nervous System*, 2nd Ed. Amsterdam: Elsevier Academic Press.

PETERS, A., S. L. PALAY AND H. DEF. WEBSTER (1991) *The Fine Structure of the Nervous System: Neurons and Their Supporting Cells*, 3rd Ed. Oxford University Press, New York.

REXED, B. (1952) The cytoarchitectonic organization of the spinal cord of the cat. *J. Comp. Neurol.* 96: 414–495.

SCHMIDLEY, J. W. AND E. F. MAAS (1990) Cerebrospinal fluid, blood-brain barrier and brain edema. In *Neurobiology of Disease*, A. L. Pearlman and R.C. Collins (eds.). New York: Oxford University Press, Chapter 19, pp. 380–398.

REESE, T. S. AND M. J. KARNOVSKY (1967) Fine structural localization of a blood-brain barrier to exogenous peroxidase. *J. Cell Biol.* 34: 207–217.

WAXMAN, S. G. AND J. DEGROOT (1995) *Correlative Neuroanatomy*, 22nd Ed. Norwalk, CT: Appleton and Lange.

Atlas

The Human Central Nervous System

This series of six plates presents labeled images of the human brain and spinal cord. The surface features of the brain are shown in photographs of a postmortem specimen after removal of the meninges and superficial blood vessels (Plate 1). Sectional views of the forebrain in each of three standard anatomical planes (see Figure A1) are derived from T1-weighted magnetic resonance imaging of a living subject (Plates 2–4). In these images, compartments filled with aqueous fluids, such as the ventricles, appear dark; tissues that are enriched with lipids, such as white matter, appear bright; and tissues that are relatively poor in lipid (myelin) and high in water content, such as gray matter, appear in intermediate shades of gray. Thus, the appearance of gray matter and white matter in the T1-weighted series is similar to what would be observed when dissecting a brain specimen obtained postmortem. The final images are transverse sections obtained from the major subdivisions of the brainstem (Plate 5) and spinal cord (Plate 6). Each of these histological images was acquired and processed to simulate myelin staining; thus, white matter appears dark, while gray matter and poorly myelinated fibers appear light.

The brain and spinal cord images in these plates are also present in the expanded atlases featured in the neuroanatomical media that accompanies this volume, *Sylvius 4.0: Interactive Atlas and Visual Glossary of the Human Central Nervous System*. *Sylvius 4.0* provides a unique computer-based learning environment for exploring and understanding the structure of the human central nervous system. The program features annotated surface views of the human brain, as well as interactive tools for dissecting the central nervous system and viewing annotated cross sections of these specimens. *Sylvius 4.0* also incorporates a comprehensive, searchable database of more than 500 neuroanatomical terms, all concisely defined and visualized in photographs; T1-weighted images; and illustrations from this text. For instructions on how to download and install *Sylvius 4.0*, see inside the front cover.

PLATE 1 PHOTOGRAPHIC ATLAS

(A)

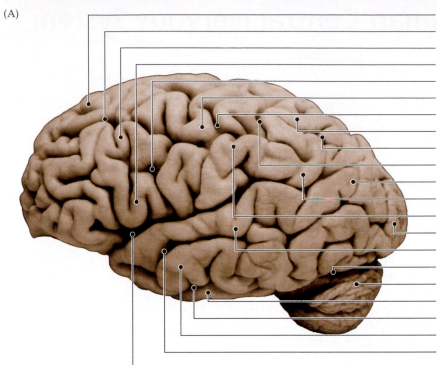

Superior frontal gyrus

Superior frontal sulcus

Middle frontal gyrus

Inferior frontal gyrus

Inferior frontal sulcus

Precentral gyrus

Central sulcus

Superior parietal lobule

Intraparietal sulcus

Postcentral sulcus

Angular gyrus

Supramarginal gyrus

Postcentral gyrus

Lateral occipital gyri

Superior temporal gyrus

Preoccipital notch

Cerebellar hemisphere

Inferior temporal gyrus

Inferior temporal sulcus

Middle temporal gyrus

Superior temporal sulcus

Lateral (Sylvian) fissure

(B)

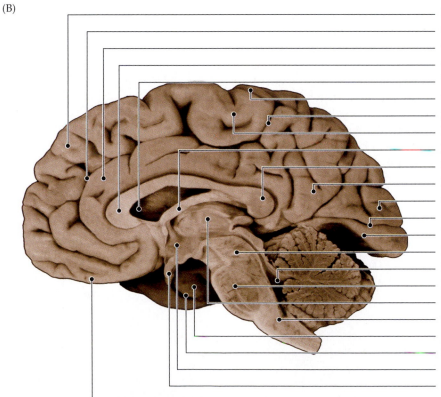

Superior frontal gyrus

Cingulate sulcus

Cingulate gyrus

Genu of corpus callosum

Lateral ventricle

Central sulcus

Marginal branch of cingulate sulcus

Paracentral lobule

Fornix

Splenium of corpus callosum

Precuneus gyrus

Cuneus gyrus

Calcarine sulcus

Lingual gyrus

Midbrain

Fourth ventricle

Pons

Thalamus

Medulla oblongata

Parahippocampal gyrus

Rhinal sulcus

Hypothalamus

Optic chiasm

Gyrus rectus

Surface features of a human brain specimen. (A) Lateral view of the left hemisphere. (B) Midsagittal view of right hemisphere. (C) Dorsal view. (D) Ventral view.

(C)

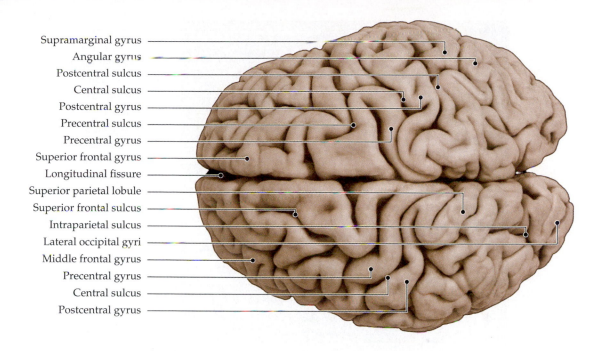

Supramarginal gyrus
Angular gyrus
Postcentral sulcus
Central sulcus
Postcentral gyrus
Precentral sulcus
Precentral gyrus
Superior frontal gyrus
Longitudinal fissure
Superior parietal lobule
Superior frontal sulcus
Intraparietal sulcus
Lateral occipital gyri
Middle frontal gyrus
Precentral gyrus
Central sulcus
Postcentral gyrus

(D)

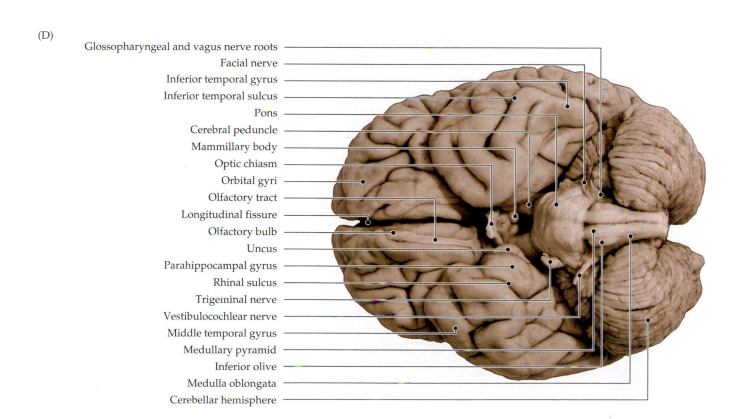

Glossopharyngeal and vagus nerve roots
Facial nerve
Inferior temporal gyrus
Inferior temporal sulcus
Pons
Cerebral peduncle
Mammillary body
Optic chiasm
Orbital gyri
Olfactory tract
Longitudinal fissure
Olfactory bulb
Uncus
Parahippocampal gyrus
Rhinal sulcus
Trigeminal nerve
Vestibulocochlear nerve
Middle temporal gyrus
Medullary pyramid
Inferior olive
Medulla oblongata
Cerebellar hemisphere

PLATE 2   CORONAL MR ATLAS

(A)

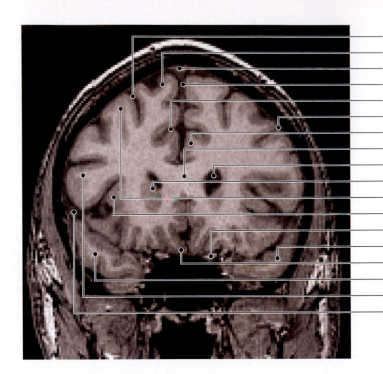

Superior frontal sulcus
Superior frontal gyrus
Superior sagittal sinus
Longitudinal fissure
Cingulate sulcus
Inferior frontal sulcus
Cingulate gyrus
Corpus callosum, genu
Lateral ventricle, anterior horn
Caudate
Middle frontal gyrus
Insular gyri
Optic nerve
Temporal pole
Gyrus rectus
Middle temporal gyrus
Inferior frontal gyrus
Lateral (sylvian) fissure

(B)

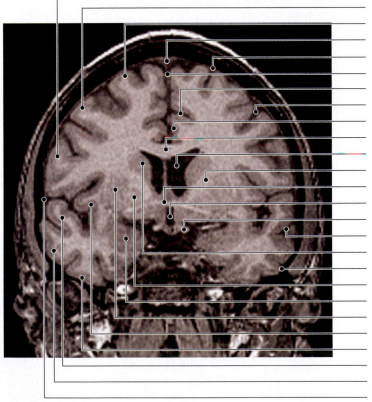

Inferior frontal gyrus
Middle frontal gyrus
Superior frontal gyrus
Superior sagittal sinus
Superior frontal sulcus
Longitudinal fissure
Cingulate sulcus
Inferior frontal sulcus
Cingulate gyrus
Corpus callosum, body
Lateral ventricle
Internal capsule
Anterior commissure
Third ventricle
Optic tract
Superior temporal sulcus
Caudate
Inferior temporal sulcus
Globus pallidus
Amygdala
Putamen
Insular gyri
Inferior temporal gyrus
Superior temporal gyrus
Middle temporal gyrus
Lateral (sylvian) fissure

Coronal sections of the human brain demonstrating internal forebrain structures in magnetic resonance images; images in (A)–(D) are arranged from rostral to caudal.

(C)

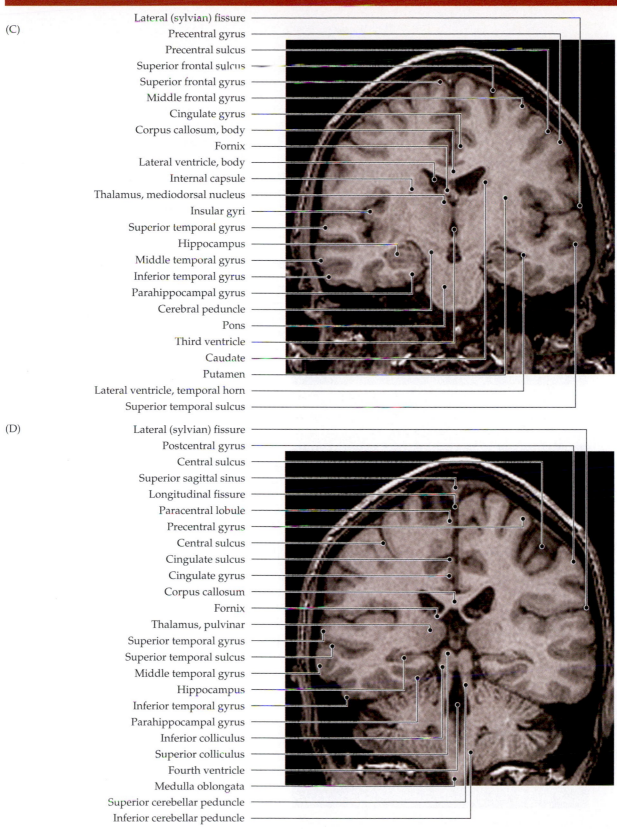

Lateral (sylvian) fissure
Precentral gyrus
Precentral sulcus
Superior frontal sulcus
Superior frontal gyrus
Middle frontal gyrus
Cingulate gyrus
Corpus callosum, body
Fornix
Lateral ventricle, body
Internal capsule
Thalamus, mediodorsal nucleus
Insular gyri
Superior temporal gyrus
Hippocampus
Middle temporal gyrus
Inferior temporal gyrus
Parahippocampal gyrus
Cerebral peduncle
Pons
Third ventricle
Caudate
Putamen
Lateral ventricle, temporal horn
Superior temporal sulcus

(D)

Lateral (sylvian) fissure
Postcentral gyrus
Central sulcus
Superior sagittal sinus
Longitudinal fissure
Paracentral lobule
Precentral gyrus
Central sulcus
Cingulate sulcus
Cingulate gyrus
Corpus callosum
Fornix
Thalamus, pulvinar
Superior temporal gyrus
Superior temporal sulcus
Middle temporal gyrus
Hippocampus
Inferior temporal gyrus
Parahippocampal gyrus
Inferior colliculus
Superior colliculus
Fourth ventricle
Medulla oblongata
Superior cerebellar peduncle
Inferior cerebellar peduncle

PLATE 3   AXIAL MR ATLAS

(A)

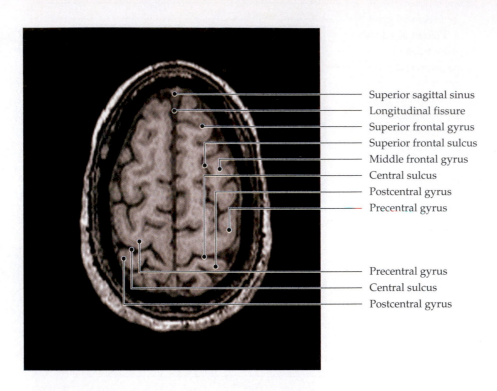

Superior sagittal sinus
Longitudinal fissure
Superior frontal gyrus
Superior frontal sulcus
Middle frontal gyrus
Central sulcus
Postcentral gyrus
Precentral gyrus

Precentral gyrus
Central sulcus
Postcentral gyrus

(B)

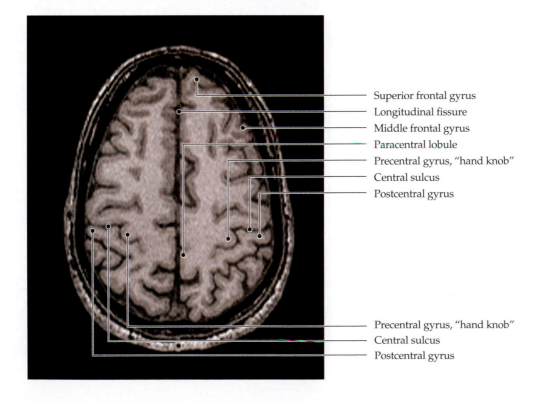

Superior frontal gyrus
Longitudinal fissure
Middle frontal gyrus
Paracentral lobule
Precentral gyrus, "hand knob"
Central sulcus
Postcentral gyrus

Precentral gyrus, "hand knob"
Central sulcus
Postcentral gyrus

Axial sections of the human brain demonstrating internal forebrain structures in T1-weighted magnetic resonance images; images in (A)–(H) are arranged from superior to inferior.

(C)

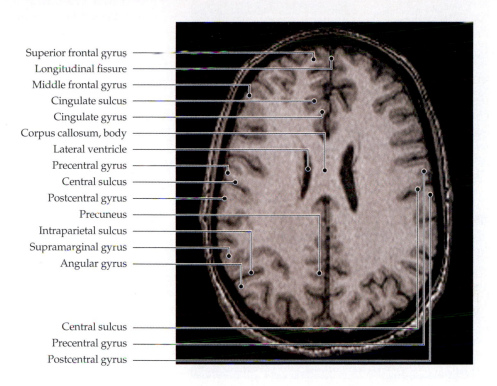

Superior frontal gyrus
Longitudinal fissure
Middle frontal gyrus
Cingulate sulcus
Cingulate gyrus
Corpus callosum, body
Lateral ventricle
Precentral gyrus
Central sulcus
Postcentral gyrus
Precuneus
Intraparietal sulcus
Supramarginal gyrus
Angular gyrus

Central sulcus
Precentral gyrus
Postcentral gyrus

(D)

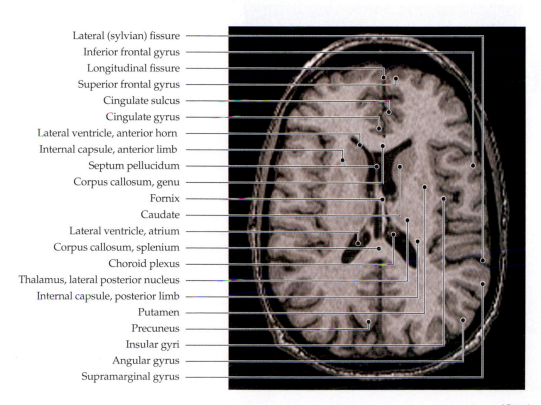

Lateral (sylvian) fissure
Inferior frontal gyrus
Longitudinal fissure
Superior frontal gyrus
Cingulate sulcus
Cingulate gyrus
Lateral ventricle, anterior horn
Internal capsule, anterior limb
Septum pellucidum
Corpus callosum, genu
Fornix
Caudate
Lateral ventricle, atrium
Corpus callosum, splenium
Choroid plexus
Thalamus, lateral posterior nucleus
Internal capsule, posterior limb
Putamen
Precuneus
Insular gyri
Angular gyrus
Supramarginal gyrus

(Continued on next page)

PLATE 3   AXIAL MR ATLAS (CONTINUED)

(E)

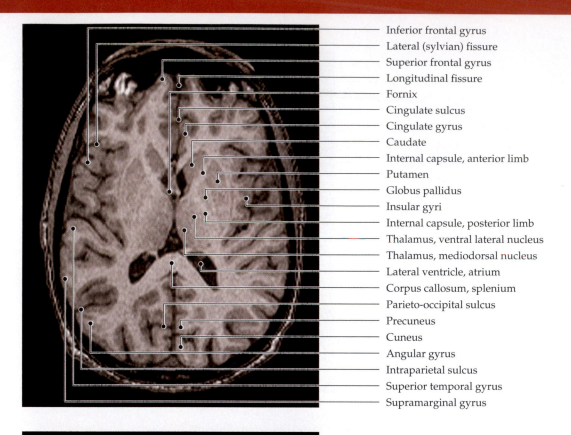

Inferior frontal gyrus
Lateral (sylvian) fissure
Superior frontal gyrus
Longitudinal fissure
Fornix
Cingulate sulcus
Cingulate gyrus
Caudate
Internal capsule, anterior limb
Putamen
Globus pallidus
Insular gyri
Internal capsule, posterior limb
Thalamus, ventral lateral nucleus
Thalamus, mediodorsal nucleus
Lateral ventricle, atrium
Corpus callosum, splenium
Parieto-occipital sulcus
Precuneus
Cuneus
Angular gyrus
Intraparietal sulcus
Superior temporal gyrus
Supramarginal gyrus

(F)

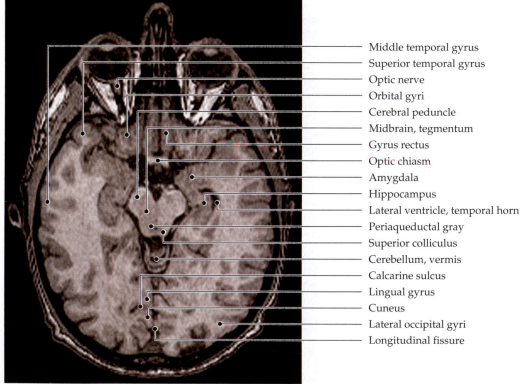

Middle temporal gyrus
Superior temporal gyrus
Optic nerve
Orbital gyri
Cerebral peduncle
Midbrain, tegmentum
Gyrus rectus
Optic chiasm
Amygdala
Hippocampus
Lateral ventricle, temporal horn
Periaqueductal gray
Superior colliculus
Cerebellum, vermis
Calcarine sulcus
Lingual gyrus
Cuneus
Lateral occipital gyri
Longitudinal fissure

Axial sections of the human brain demonstrating internal forebrain structures in T1-weighted magnetic resonance images; images in (A)–(H) are arranged from superior to inferior.

(G)

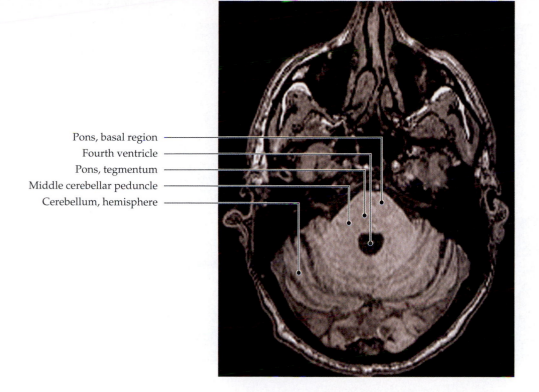

Pons, basal region

Fourth ventricle

Pons, tegmentum

Middle cerebellar peduncle

Cerebellum, hemisphere

(H)

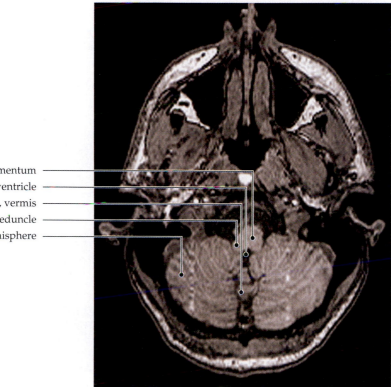

Medulla oblongata, tegmentum

Fourth ventricle

Cerebellum, vermis

Inferior cerebellar peduncle

Cerebellum, hemisphere

PLATE 4   SAGITTAL MR ATLAS

(A)

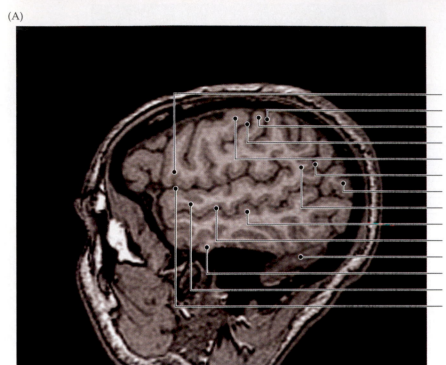

Inferior frontal gyrus
Postcentral sulcus
Postcentral gyrus
Central sulcus
Precentral gyrus
Intraparietal sulcus
Angular gyrus
Supramarginal gyrus
Middle temporal gyrus
Superior temporal sulcus
Cerebellum, hemisphere
Inferior temporal gyrus
Superior temporal gyrus
Lateral (sylvian) fissure

(B)

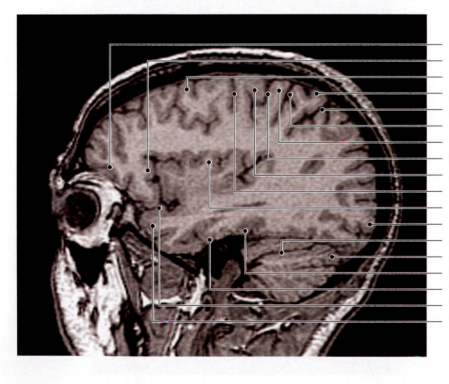

Orbital gyri
Inferior frontal gyrus
Middle frontal gyrus
Superior parietal lobule
Intraparietal sulcus
Postcentral sulcus
Postcentral gyrus
Central sulcus
Precentral gyrus
Precentral sulcus
Insular gyri
Lateral occipital gyri
Cerebellum, primary fissure
Cerebellum, hemisphere
Occipitotemporal gyrus
Inferior temporal gyrus
Lateral (sylvian) fissure
Superior temporal gyrus

Sagittal sections of the human brain demonstrating internal forebrain structures in T1-weighted magnetic resonance images; images in (A)–(D) are arranged from lateral to medial.

(C)

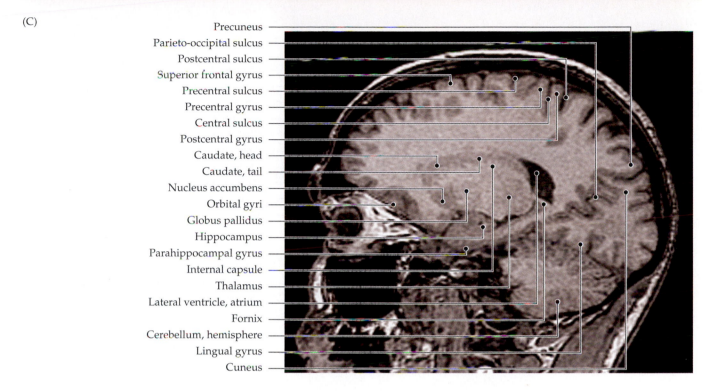

Precuneus
Parieto-occipital sulcus
Postcentral sulcus
Superior frontal gyrus
Precentral sulcus
Precentral gyrus
Central sulcus
Postcentral gyrus
Caudate, head
Caudate, tail
Nucleus accumbens
Orbital gyri
Globus pallidus
Hippocampus
Parahippocampal gyrus
Internal capsule
Thalamus
Lateral ventricle, atrium
Fornix
Cerebellum, hemisphere
Lingual gyrus
Cuneus

(D)

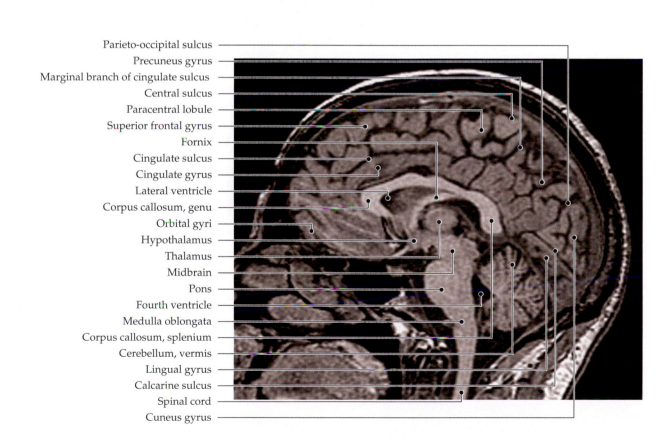

Parieto-occipital sulcus
Precuneus gyrus
Marginal branch of cingulate sulcus
Central sulcus
Paracentral lobule
Superior frontal gyrus
Fornix
Cingulate sulcus
Cingulate gyrus
Lateral ventricle
Corpus callosum, genu
Orbital gyri
Hypothalamus
Thalamus
Midbrain
Pons
Fourth ventricle
Medulla oblongata
Corpus callosum, splenium
Cerebellum, vermis
Lingual gyrus
Calcarine sulcus
Spinal cord
Cuneus gyrus

PLATE 5 BRAINSTEM ATLAS

(A)

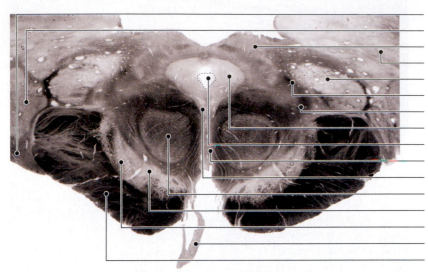
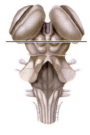

Optic tract
Lateral geniculate nucleus
Superior colliculus
Pulvinar
Medial geniculate nucleus
Anterolateral system
Medial lemniscus
Periaqueductal gray
Cerebral aqueduct
Raphe nuclei
Oculomotor complex
Red nucleus
Substantia nigra, pars compacta
Substantia nigra, pars reticulata
Oculomotor nerve
Cerebral peduncle

(B)

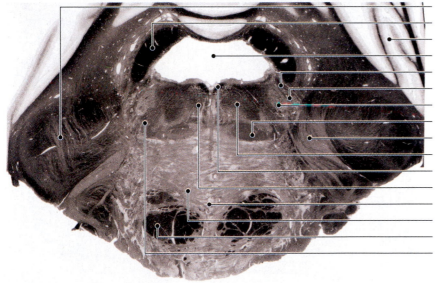
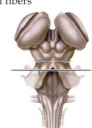

Middle cerebellar peduncle
Superior cerebellar peduncle
Cerebellum, cortex
Fourth ventricle
Mesencephalic trigeminal tract and nucleus
Chief sensory nucleus of the trigeminal complex
Trigeminal motor nucleus
Medial lemniscus
Trigeminal nerve roots
Central tegmental tract
Medial longitudinal fasciculus
Tectospinal fibers
Pontine nuclei
Pontocerebellar fibers
Corticobulbar and corticospinal fibers
Anterolateral system

(C)

Inferior cerebellar peduncle
External cuneate nucleus
Dorsal motor nucleus of vagus
Solitary tract
Nucleus of the solitary tract
Hypoglossal nucleus
Nucleus ambiguus
Medial longitudinal fasciculus
Tectospinaltract
Medial lemniscus
Medial vestibular nucleus
Spinal vestibular nucleus
Inferior olivary nucleus
Medullary pyramid
Spinal trigeminal nucleus
Anterolateral system
Spinal trigeminal tract

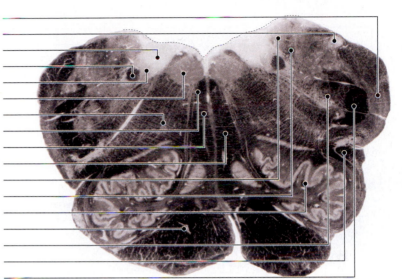

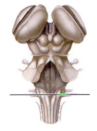

(D)

Gracile tract
Cuneate tract
Cuneate nucleus
Gracile nucleus
Pyramidal decussation
Spinal accessory nucleus
Anterolateral system
Spinal trigeminal nucleus, magnocellular layer
Spinal trigeminal tract
Spinal trigeminal nucleus, gelatinosa layer
Dorsal spinocerebellar tract

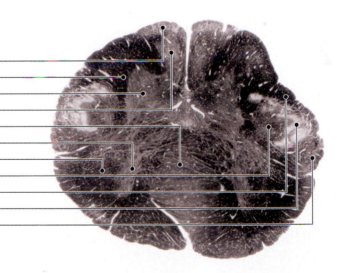

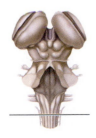

PLATE 6   SPINAL CORD ATLAS

(A)

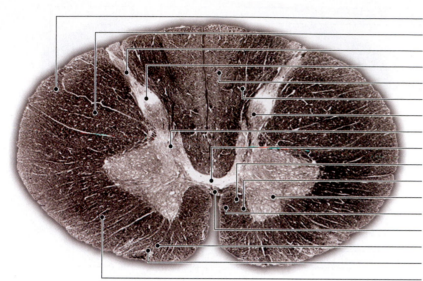

Dorsal spinocerebellar tract
Lateral corticospinal tract
Dorsolateral fasciculus
Substantia gelatinosa
Gracile tract
Cuneate tract
Dorsal horn
Intermediate gray
Central gray
Medial longitudinal fasciculus
Tectospinal tract
Ventral horn
Ventral corticospinal tract
Ventral white commissure
Lateral vestibulospinal tract
Reticulospinal tract
Anterolateral system

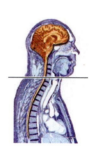

(B)

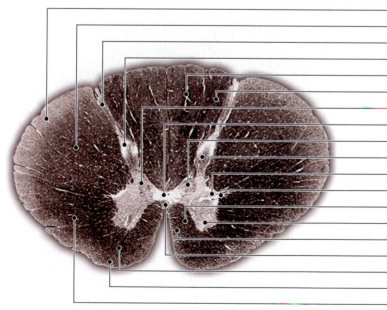

Dorsal spinocerebellar tract
Lateral corticospinal tract
Dorsolateral fasciculus
Substantia gelatinosa
Gracile tract
Cuneate tract
Intermediate gray
Central gray
Clarke's nucleus
Dorsal horn
Intermediolateral cell column
Lateral horn
Medial longitudinal fasciculus
Ventral horn
Ventral corticospinal tract
Ventral white commissure
Lateral vestibulospinal tract
Reticulospinal tract
Anterolateral system

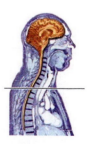

Transverse sections of the human spinal cord acquired and prepared to simulate myelin staining. (A) Cervical segment. (B) Thoracic segment. (C) Lumbar sement. (D) Sacral segment.

(C)

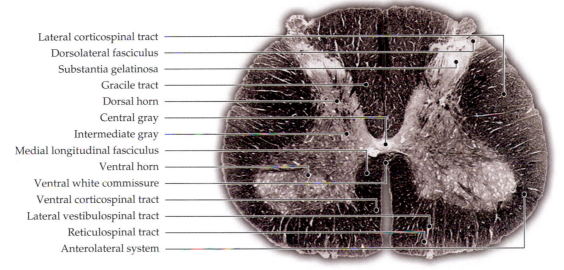
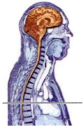

Lateral corticospinal tract
Dorsolateral fasciculus
Substantia gelatinosa
Gracile tract
Dorsal horn
Central gray
Intermediate gray
Medial longitudinal fasciculus
Ventral horn
Ventral white commissure
Ventral corticospinal tract
Lateral vestibulospinal tract
Reticulospinal tract
Anterolateral system

(D)

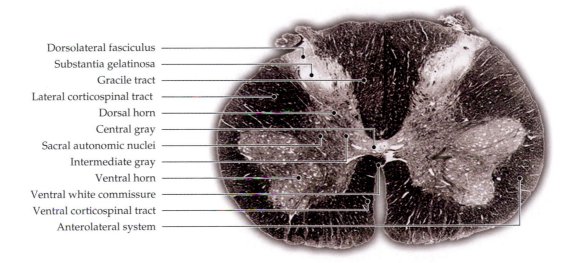
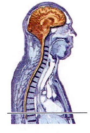

Dorsolateral fasciculus
Substantia gelatinosa
Gracile tract
Lateral corticospinal tract
Dorsal horn
Central gray
Sacral autonomic nuclei
Intermediate gray
Ventral horn
Ventral white commissure
Ventral corticospinal tract
Anterolateral system

Glossary

acetylcholine   Neurotransmitter at motor neuron synapses, in autonomic ganglia and a variety of central synapses; binds to two types of receptors: ligand-gated ion channels (nicotinic receptors) and G-protein-coupled receptors (muscarinic receptors).

action potential   The electrical signal conducted along axons (or muscle fibers) by which information is conveyed from one place to another in the nervous system.

activation   The time-dependent opening of ion channels in response to a stimulus, typically membrane depolarization.

adaptation   The phenomenon of sensory receptor adjustment to different levels of stimulation; critical for allowing sensory systems to operate over a wide dynamic range.

adenylyl cyclase   Membrane-bound enzyme that can be activated by G-proteins to catalyze the synthesis of cyclic AMP from ATP.

adhesion molecules   see cell adhesion molecules.

adrenaline   see epinephrine.

adrenal medulla   The central part of the adrenal gland that, under visceral motor stimulation, secretes epinephrine and norepinephrine into the bloodstream.

adrenergic   Refers to synaptic transmission mediated by the release of norepinephrine or epinephrine.

afferent   An axon that conducts action potentials from the periphery toward the central nervous system.9

agnosia   The inability to name objects.

alpha (α) motor neurons   Neurons in the ventral horn of the spinal cord that innervate skeletal muscle.

amacrine cells   Retinal neurons that mediate lateral interactions between bipolar cell terminals and the dendrites of ganglion cells.

amblyopia   Diminished visual acuity as a result of the failure to establish appropriate visual cortical connections in early life.

amnesia   The pathological inability to remember or establish memories; retrograde amnesia is the inability to recall existing memories, whereas anterograde amnesia is the inability to lay down new memories.

amphetamine   A synthetically produced central nervous system stimulant with cocaine-like effects; drug abuse may lead to dependence.

ampullae   The juglike swellings at the base of the semicircular canals that contain the hair cells and cupulae (see also cupulae).

amygdala   A nuclear complex in the temporal lobe that forms part of the limbic system; its major functions concern autonomic, emotional, and sexual behavior.

androgen insensitivity syndrome   A condition in which, due to a defect in the gene that codes for the androgen receptor, testosterone cannot act on its target tissues.

anencephaly   A congenital defect of neural tube closure, in which much of the brain fails to develop.

anosmia   Loss of the sense of smell.

anterior   Toward the front; sometimes used as a synonym for rostral, and sometimes as a synonym for ventral.

anterior commissure   A small midline fiber tract that lies at the anterior end of the corpus callosum; like the callosum, it serves to connect the two hemispheres.

anterior hypothalamus   Region of the hypothalamus containing nuclei that mediate sexual behaviors; not to be confused with region in rodent called the medial preoptic area, which lies anterior to hypothalamus and also contains nuclei that mediate sexual behavior (most notably the sexually dimorphic nucleus).

anterograde   A movement or influence acting from the neuronal cell body toward the axonal target.

anterolateral pathway   Ascending sensory pathway in the spinal cord and brainstem that carries information about pain and temperature to the thalamus.

antiserum   Serum harvested from an animal immunized to an agent of interest.

aphasia   The inability to comprehend and/or produce language as a result of damage to the language areas of the cerebral cortex (or their white matter interconnections).

apoptosis   Cell death resulting from a programmed pattern of gene expression; also known as "programmed cell death."

aprosodia   The inability to infuse language with its normal emotional content.

arachnoid mater   One of the three coverings of the brain that make up the meninges; lies between the dura mater and the pia mater.

areflexia   Loss of reflexes.

association cortex   Defined by exclusion as those neocortical regions that are not involved in primary sensory or motor processing.

associativity   In the hippocampus, the enhancement of a weakly activated group of synapses when a nearby group is strongly activated.

astrocytes   One of the three major classes of glial cells found in the central nervous system; important in regulating the ionic milieu of nerve cells and, in some cases, transmitter reuptake.

astrotactin   A cell surface molecule that causes neurons to adhere to radial glial fibers during neuronal migration.

athetosis   Slow, writhing movements seen primarily in patients with disorders of the basal ganglia.

ATPase pumps   Membrane pumps that use the hydrolysis of ATP to translocate ions against their electrochemical gradients.

atrophy   The physical wasting away of a tissue, typically muscle, in response to disuse or other causes.

attention   The selection of a particular sensory stimulus or mental process for further analysis.

auditory meatus   Opening of the external ear canal.

auditory space map   Topographic representation of sound source location, as occurs in the inferior colliculus.

autonomic nervous system   The components of the nervous system (peripheral and central) concerned with the regulation of smooth muscle, cardiac muscle, and glands (see also visceral motor system).

axon   The neuronal process that carries the action potential from the nerve cell body to a target.

axoplasmic transport   The process by which materials are carried from nerve cell bodies to their terminals (anterograde transport), or from nerve cell terminals to the neuronal cell body (retrograde transport).

baroreceptors   Sensory receptors in the visceral motor system that respond to changes in blood pressure.

basal ganglia   A group of nuclei lying deep in the subcortical white matter of the frontal lobes that organize motor behavior. The caudate and putamen and the globus pallidus are the major components of the basal ganglia; the subthalamic nucleus and substantia nigra are often included.

basal lamina   A thin layer of extracellular matrix material (primarily collagen, laminin, and fibronectin) that surrounds muscle cells and Schwann cells. Also underlies all epithelial sheets. Also called the basement membrane.

basilar membrane   The membrane that forms the floor of the cochlear duct, on which the cochlear hair cells are located.

basket cells   Inhibitory interneurons in the cerebellar cortex whose cells bodies are located within the Purkinje cell layer and whose axons make basketlike terminal arbors around Purkinje cell bodies.

bHLH proteins   Transcription factors (named for a shared *b*asic *h*elix-*l*oop-*h*elix amino acid motif that defines their DNA-binding domain) that have emerged as central to the differentiation of distinct neural and glial fates.

binocular   Referring to both eyes.

biogenic amines   The bioactive amine neurotransmitters; includes the catecholamines (epinephrine, norepinephrine, dopamine), serotonin, and histamine.

bipolar cells   Retinal neurons that provide a direct link between photoreceptor terminals and ganglion celldendrites.

bisexuality   Sexual attraction to members of both the opposite and the same phenotypic sex.

blastomere   A cell produced when the egg undergoes cleavage.

blastula   An early embryo during the stage when the cells are typically arranged to form a hollow sphere.

blind spot   The region of visual space that falls on the optic disk; due to the lack of photoreceptors in the optic disk, objects that lie completely within the blind spot are not perceived.

blood-brain barrier   A diffusion barrier between the brain vasculature and the substance of the brain formed by tight junctions between capillary endothelial cells.

bouton (synaptic bouton)   A swelling specialized for the release of neurotransmitter that occurs along or at the end of an axon.

bradykinesia   Pathologically slow movement.

brain-derived neutrophic factor (BDNF)   One member of a family of neutrophic factors, the best-known constituent of which is nerve growth factor.

brainstem   The portion of the brain that lies between the diencephalon and the spinal cord; comprises the midbrain, pons, and medulla.

Broca's aphasia   Difficulty producing speech as a result of damage to Broca's area in the left frontal lobe.

Broca's area   An area in the left frontal lobe specialized for the production of language.

cadherins   A family of calcium-dependent cell adhesion molecules found on the surfaces of growth cones and the cells over which they grow.

calcarine sulcus   The major sulcus on the medial aspect of the occipital lobe; the primary visual cortex lies largely within this sulcus.

cAMP response element binding protein (CREB)   A protein activated by cyclic AMP that binds to specific regions of DNA, thereby increasing the transcription rates of nearby genes.

cAMP response elements (CREs)   Specific DNA sequences that bind transcription factors activated by cAMP (see also cAMP response element binding protein).

carotid bodies   Specialized tissue masses found at the bifurcation of the carotid arteries in humans and other mammals that respond to the chemical composition of the blood (primarily the partial pressure of oxygen and carbon dioxide).

catecholamine   A term referring to molecules containing a catechol ring and an amino group; examples are the neurotransmitters epinephrine, norepinephrine, and dopamine.

cauda equine   The collection of segmental ventral and dorsal roots that extend from the caudal end of the spinal cord to their exit from the spinal canal.

caudal   Posterior, or "tailward."

caudate nucleus   One of the three major components of the basal ganglia (the other two are the globus pallidus and putamen).

cell adhesion molecules   A family of molecules on cell surfaces that cause them to stick to one another (see also fibronectin and laminin).

central nervous system (CNS)   The brain and spinal cord of vertebrates (by analogy, the central nerve cord and ganglia of invertebrates).

central pattern generator   Oscillatory spinal cord or brainstem circuits responsible for programmed, rhythmic movements such as locomotion.

central sulcus   A major sulcus on the lateral aspect of the hemispheres that forms the boundary between the frontal and parietal lobes. The anterior bank of the sulcus contains the primary motor cortex; the posterior bank contains the primary sensory cortex.

cerebellar ataxia   A pathological inability to make coordinated movements associated with lesions to the cerebellum.

cerebellar cortex   The superficial gray matter of the cerebellum.

cerebellar peduncles   The three bilateral pairs of axon tracts (inferior, middle, and superior cerebellar peduncles) that carry information to and from the cerebellum.

cerebellum   Prominent hindbrain structure concerned with motor coordination, posture, and balance. Composed of a three-layered cortex and deep nuclei; attached to the brainstem by the cerebellar peduncles.

cerebral achromatopsia Loss of color vision as a result of damage to extrastriate visual cortex.

cerebral aqueduct The portion of the ventricular system that connects the third and fourth ventricles.

cerebral cortex The superficial gray matter of the cerebral hemispheres.

cerebral peduncles The major fiber bundles that connect the brainstem to the cerebral hemispheres.

cerebrocerebellum The part of the cerebellar cortex that receives input from the cerebral cortex via axons from the pontine relay nuclei.

cerebrospinal fluid A normally clear and cell-free fluid that fills the ventricular system of the central nervous system; produced by the choroid plexus in the third ventricle.

cerebrum The largest and most rostral part of the brain in humans and other mammals, consisting of the two cerebral hemispheres.

c-fos A transcription factor, originally isolated from *cellular feline osteosarcoma* cells, that binds as a heterodimer, thus activating gene transcription.

chemical synapses Synapses that transmit information via the secretion of chemical signals (neurotransmitters).

chemoaffinity (chemoaffinity hypothesis) The idea that nerve cells bear chemical labels that determine their connectivity.

chemotaxis The movement of a cell up (or down) the gradient of a chemical signal.

chemotropism The growth of a part of a cell (axon, dendrite, filopodium) up (or down) a chemical gradient.

chimera An experimentally generated embryo (or organ) comprising cells derived from two or more species (or other genetically distinct sources).

cholinergic Referring to synaptic transmission mediated by acetylcholine.

chorea Jerky, involuntary movements of the face or extremities associated with damage to the basal ganglia.

choreoathetosis The combination of jerky, ballistic, and writhing movements that characterizes the late stages of Huntington's disease.

choroid plexus Specialized epithelium in the ventricular system that produces cerebrospinal fluid.

chromosome Nuclear organelle that bears the genes.

ciliary body Circular band of muscle surrounding the lens; contraction allows the lens to round up during accommodation.

cingulate cortex Cortex of the cingulate gyrus that surrounds the corpus callosum; important in emotional and visceral motor behavior.

cingulate gyrus Prominent gyrus on the medial aspect of the hemisphere, lying just superior to the corpus callosum; forms a part of the limbic system.

cingulate sulcus Prominent sulcus on the medial aspect of the hemisphere.

circadian rhythms Variations in physiological functions that occur on a daily basis.

circle of Willis Arterial anastomosis on the ventral aspect of the midbrain; connects the posterior and anterior cerebral circulation.

cisterns Large, cerebrospinal-fluid-filled spaces that lie within the subarachnoid space.

climbing fibers Axons that originate in the inferior olive, ascend through the inferior cerebellar peduncle, and make terminal arborizations that invest the dendritic tree of Purkinje cells.

clone The progeny of a single cell.

cochlea The coiled structure within the inner ear where vibrations caused by sound are transduced into neural impulses.

cognition A general term referring to higher order mental processes, the ability of the central nervous system to attend, identify, and act on complex stimuli.

collapsing A molecule that causes collapse of growth cones; a member of the semaphorin family of signaling molecules.

colliculi The two paired hillocks that characterize the dorsal surface of the midbrain; the superior colliculi concern vision, the inferior colliculi audition.

competition The struggle among nerve cells, or nerve cell processes, for limited resources essential to survival or growth.

concha A component of the external ear.

conduction aphasia Difficulty producing speech as a result of damage to the connection between Wernicke's and Broca's language areas.

conduction velocity The speed at which an action potential is propagated along an axon.

conductive hearing loss Diminished sense of hearing due reduced ability of sounds to be mechanically transmitted to the inner ear. Common causes include occlusion of the ear canal, perforation of the tympanic membrane, and arthritic degeneration of the middle ear ossicles. Contrast with sensorineural hearing loss.

cone opsins The three distinct photopigments found in cones; the basis for color vision.

cones Photoreceptor cells specialized for high visual acuity and the perception of color.

congenital adrenal hyperplasia Genetic deficiency that leads to overproduction of androgens and a resultant masculinization of external genitalia in genotypic females.

conjugate The paired movements of the two eyes in the same direction, as occurs in the vestibulo-ocular reflex (see also vergence movements and vestibulo-ocular reflex).

contralateral On the other side.

contralateral neglect syndrome Neurological condition in which the patient does not acknowledge or attend to the left visual hemifield or the left half of the body. The syndrome typically results from lesions of the right parietal cortex.

contrast The difference, usually expressed in terms of a percentage in luminance, between two territories in the visual field (can also apply to color when specified as spectral contrast).

convergence Innervation of a target cell by axons from more than one neuron.

cornea The transparent surface of the eyeball in front of the lens; the major refractive element in the optical pathway.

coronal Referring to a plane through the brain that runs parallel to the coronal suture (the mediolateral plane). Synonymous with frontal plane.

corpus callosum The large midline fiber bundle that connects the cortices of the two cerebral hemispheres.

corpus striatum General term applied to the caudate and putamen; name derives from the striated appearance of these basal ganglia nuclei in sections of fresh material.

cortex The superficial mantle of gray matter covering the cerebral hemispheres and cerebellum, where most of the neurons in the brain are located.

cortico-cortical connections Connections made between cortical areas in the same hemisphere or betweem the two hemispheres via the cerebral commissures (the corpus callosum and the anterior commissure).

corticospinal tract Pathway carrying motor information from the primary and secondary motor cortices to the brain stem and spinal cord.

co-transmitters Two or more types of neurotransmitters within a single synapse; may be packaged into separate populations of synaptic vesicles or co-localized within the same synaptic vesicles.

cranial nerve ganglia The sensory ganglia associated with the cranial nerves; these correspond to the dorsal root ganglia of the spinal segmental nerves.

cranial nerve nuclei Nuclei in the brainstem that contain the neurons related to cranial nerves III–XII.

cranial nerves The 12 pairs of nerves arising from the brainstem that carry sensory information toward (and sometimes motor information away from) the central nervous system.

CREB see cAMP response element binding protein.

crista The hair cell-containing sensory epithelium of the semicircular canals.

critical period A restricted developmental period during which the nervous system is particularly sensitive to the effects of experience.

cuneate nuclei Sensory relay nuclei that lie in the lower medulla; they contain the second-order sensory neurons that relay mechanosensory information from peripheral receptors in the upper body to the thalamus.

cupulae Gelatinous structures in the semicircular canals in which the hair cell bundles are embedded.

cytoarchitectonic areas Distinct regions of the neocortical mantle identified by differences in cell size, packing density, and laminar arrangement.

decerebrate rigidity Excessive tone in extensor muscles as a result of damage to descending motor pathways at the level of the brainstem.

declarative memory Memories available to consciousness that can be expressed by language.

decussation A crossing of fiber tracts in the midline.

deep cerebellar nuclei The nuclei at the base of the cerebellum that relay information from the cerebellar cortex to the thalamus.

delayed response task A behavioral paradigm used to test cognition and memory.

delta waves Slow (<4 Hz) electroencephalographic waves that characterize stage IV (slow-wave) sleep.

dendrite A neuronal process arising from the cell body that receives synaptic input.

denervation Removal of the innervation to a target.

dentate gyrus A region of the hippocampus; so named because it is shaped like a tooth.

depolarization Displacement of a cell's membrane potential toward a less negative value.

dermatome The area of skin supplied by the sensory axons of a single spinal nerve.

determination Commitment of a developing cell or cell group to a particular fate.

diencephalon Portion of the brain that lies just rostral to the midbrain; comprises the thalamus and hypothalamus.

differentiation The progressive specialization of developing cells.

dihydrotestosterone A more potent form of testosterone; masculinizes the external genitalia.

disinhibition Arrangement of inhibitory and excitatory cells in a circuit that generates excitation by the transient inhibition of a tonically active inhibitory neuron.

disjunctive eye movements Movements of the two eyes in opposite directions (see also vergence movements).

distal Farther away from a point of reference (the opposite of proximal).

divergence The branching of a single axon to innervate multiple target cells.

dopamine A catecholamine neurotransmitter.

dorsal Referring to the back.

dorsal column nuclei Second-order sensory neurons in the lower medulla that relay mechanosensory information from the spinal cord to the thalamus; comprises the cuneate and gracile nuclei.

dorsal columns Major ascending tracts in the spinal cord that carry mechanosensory information from the first-order sensory neurons in dorsal root ganglia to the dorsal column nuclei; also called the posterior funiculi.

dorsal horn The dorsal portion of the spinal cord gray matter; contains neurons that process sensory information.

dorsal root ganglia (DRG) The segmental sensory ganglia of the spinal cord; contain the first-order neurons of the dorsal column/medial lemniscus and spinothalamic pathways.

dorsal roots The bundle of axons that runs from the dorsal root ganglia to the dorsal horn of the spinal cord, carrying sensory information from the periphery.

dura mater The thick external covering of the brain and spinal cord; one of the three components of the meninges, the other two being the pia mater and arachnoid mater.

dynorphins A class of endogenous opioid peptides.

dysarthria Difficulty producing speech as a result of damage to the primary motor centers that govern the muscles of articulation; distinguished from aphasia, which results from cortical damage.

dysmetria Inaccurate movements due to faulty judgment of distance. Characteristic of cerebellar pathology.

dystonia Lack of muscle tone.

early inward current The initial electrical current, measured in voltage clamp experiments, that results from the voltage-dependent entry of a cation such as Na^+ or Ca^{2+}; produces the rising phase of the action potential.

ectoderm The most superficial of the three embryonic germ layers; gives rise to the nervous system and epidermis.

Edinger-Westphal nucleus Midbrain nucleus containing the autonomic neurons that constitute the efferent limb of the pupillary light reflex.

efferent An axon that conducts information away from the central nervous system.

electrical synapses Synapses that transmit information via the direct flow of electrical current at gap junctions.

electrochemical equilibrium The condition in which no net ionic flux occurs across a membrane because ion concentration gradients and opposing transmembrane potentials are in exact balance.

electrogenic   Capable of generating an electrical current; usually applied to membrane transporters that create electrical currents while translocating ions.

embryo   The developing organism before birth or hatching.

embryonic stem (ES) cells   Cells derived from pre-gastrula embryos. that have the potential for infinite self-renewal and can give rise to *all* tissue and cell types of the organism, including the germ cells that undergo meiosis and generate haploid gametes—whereas somatic stem cells mitotically generate only diploid, tissue-specific cell types.

endocrine   Referring to the release of signaling molecules whose effects are made widespread by distribution in the general circulation.

endocytosis   A budding off of vesicles from the plasma membrane, which allows uptake of materials in the extracellular medium.

endoderm   The innermost of the three embryonic germ layers. Gives rise to the digestive and respiratory tracts and the structures associated with them.

endogenous opioids   Peptides in the central nervous system that have the same pharmacological effects as morphine and other derivatives of opium.

endolymph   The potassium-rich fluid filling both the cochlear duct and the membranous labyrinth; bathes the apical end of the hair cells.

endorphins   One of a group of neuropeptides that are agonists at opioid receptors, virtually all of which contain the sequence Tyr-Gly-Gly-Phe.

end plate   The complex postsynaptic specialization at the site of nerve contact on skeletal muscle fibers.

end plate current (EPC)   Postsynaptic current produced by neurotransmitter release and binding at the motor end plate.

end plate potential (EPP)   Depolarization of the membrane potential of skeletal muscle fiber, caused by the action of the transmitter acetylcholine at the neuromuscular synapse.

engram   The term used to indicate the physical basis of a stored memory.

enkephalins   A general term for endogenous opioid peptides.

ependyma   The epithelial lining of the canal of the spinal cord and the ventricles.

ependymal cells   Epithelial cells that line the ventricular system.

epidermis   The outermost layer of the skin; derived from the embryonic ectoderm.

epigenetic   Referring to influences on development that arise from factors other than genetic instructions.

epinephrine (adrenaline)   Catecholamine hormone and neurotransmitter that binds to α- and β-adrenergic G-protein-coupled receptors.

epineurium   The connective tissue surrounding axon fascicles of a peripheral nerve.

epithelium   Any continuous layer of cells that covers a surface or lines a cavity.

equilibrium potential   The membrane potential at which a given ion is in electrochemical equilibrium.

estradiol   One of the biologically important C_{18} class of steroid hormones capable of inducing estrous in females.

excitatory postsynaptic potential (EPSP)   Neurotransmitter-induced postsynaptic potential change that depolarizes the cell, and hence increases the likelihood of initiating a postsynaptic action potential.

exocytosis   A form of cell secretion resulting from the fusion of the membrane of a storage organelle, such as a synaptic vesicle, with the plasma membrane.

explant   A piece of tissue maintained in culture medium.

external segment   A subdivision of the globus pallidus.

extracellular matrix   A matrix composed of collagen, laminin, and fibronectin that surrounds most cells (see also basal lamina).

extrafusal muscle fibers   Fibers of skeletal muscles; a term that distinguishes ordinary muscle fibers from the specialized intrafusal fibers associated with muscle spindles.

face cells   Neurons in the temporal cortex of rhesus monkeys that respond specifically to faces.

facilitation   The increased transmitter release produced by an action potential that follows closely upon a preceding action potential.

fasciculation   The aggregation of neuronal processes to form a nerve bundle; also refers to the spontaneous discharge of motor units after muscle denervation.

α-fetoprotein   A protein that actively sequestors circulating estrogens.

fetus   The developing mammalian embryo at relatively late stages when the parts of the body are recognizable.

fibrillation   Spontaneous contractile activity of denervated muscle fibers.

fibroblast growth factor (FGF)   A peptide growth factor, originally defined by its mitogenic effects on fibroblasts; also acts as an inducer during early brain development.

fibronectin   A large cell adhesion molecule that binds integrins.

filopodium   Slender protoplasmic projection, arising from the growth cone of an axon or a dendrite, that explores the local environment.

fissure   A deep cleft in the brain; distinguished from sulci, which are shallower cortical infoldings.

flexion reflex   Polysynaptic reflex mediating withdrawal from a painful stimulus.

floorplate   Region in the ventral portion of the developing spinal cord; important in the guidance and crossing of growing axons.

folia   The name given to the gyral formations of the cerebellum.

forebrain   The anterior portion of the brain that includes the cerebral hemispheres (includes the telencephalon and diencephalon).

fornix   An axon tract, best seen from the medial surface of the divided brain, that interconnects the hypothalamus and hippocampus.

fourth ventricle   The ventricular space that lies between the pons and the cerebellum.

fovea   Area of the retina specialized for high acuity in the center of the macula; contains a high density of cones and few rods.

foveola   Capillary and rod-free zone in the center of the fovea.

frontal lobe   One of the four lobes of the brain; includes all the cortex that lies anterior to the central sulcus and superior to the lateral fissure.

G-protein-coupled receptors   A large family of neurotransmitter or hormone receptors, characterized by seven transmembrane domains; the binding of these receptors by agonists leads to the activation of intracellular G-proteins. Also known as metabotropic receptors.

G-proteins Two large groups of proteins—the heterotrimeric G-proteins and the small-molecule G-proteins—that can be activated by exchanging bound GDP for GTP.

gamma (γ) motor neurons Class of spinal motor neurons specifically concerned with the regulation of muscle spindle length; these neurons innervate the intrafusal muscle fibers of the spindle.

Ganglion (ganglia) Collections of hundreds to thousands of neurons found outside the brain and spinal cord along the course of peripheral nerves.

ganglion cell A neuron located in a ganglion.

gap junction A specialized intercellular contact formed by channels that directly connect the cytoplasm of two cells.

gastrula The early embryo during the period when the three embryonic germ layers are formed; follows the blastula stage.

gastrulation The cell movements (invagination and spreading) that transform the embryonic blastula into the gastrula.

gender identification Self-perception of one's alignment with the traits associated with being a phenotypic female or male in a given culture.

gene Hereditary unit located on the chromosomes; genetic information is carried by linear sequences of nucleotides in DNA that code for corresponding sequences of amino acids.

genome The complete set of an animal's genes.

genotypic sex Sexual characterization according to the complement of sex chromosomes; XX is a genotypic female, and XY is a genotypic male.

germ cell The egg or sperm (or the precursors of these cells).

germ layers The three primary layers of the developing embryo from which all adult tissues arise: ectoderm, mesoderm, and endoderm.

glia The support cells associated with neurons (astrocytes, oligodendrocytes, and microglia in the central nervous system; Schwann cells in peripheral nerves; and satellite cells in ganglia). Also known as neuroglia.

globus pallidus One of the three major nuclei that make up the basal ganglia in the cerebral hemispheres; relays information from the caudate and putamen to the thalamus.

glomeruli Characteristic collections of neuropil in the olfactory bulbs; formed by dendrites of mitral cells and terminals of olfactory receptor cells, as well as processes from local interneurons.

glutamate-glutamine cycle A metabolic cycle of glutamate release and resynthesis involving both neuronal and glial cells.

G$_{olf}$ A G-protein found uniquely in olfactory receptor neurons.

Golgi tendon organs Receptors located in muscle tendons that provide mechanosensory information to the central nervous system about muscle tension.

Golgi stain see silver stain.

gracile nuclei Sensory nuclei in the lower medulla; these second-order sensory neurons relay mechanosensory information from the lower body to the thalamus.

gradient A systematic variation of the concentration of a molecule (or some other agent) that influences cell behavior.

granule cell layer The layer of the cerebellar cortex where granule cell bodies are found. Also used to refer to cell-rich layers in neocortex and hippocampus.

gray matter General term that describes regions of the central nervous system rich in neuronal cell bodies and neuropil; includes the cerebral and cerebellar cortices, the nuclei of the brain, and the central portion of the spinal cord.

green fluorescent protein (GFP) A reporter or marker protein that generates green fluorescent light. GFP is widely used as a genetic tag in fluorescence microscopy. Originally discovered in a light-emitting jellyfish, It allows observation of specific neurons and some of their components over time in living cells, or even in whole organisms.

growth cone The specialized end of a growing axon (or dendrite) that generates the motive force for elongation.

gyri The ridges of the infolded cerebral cortex (the valleys between these ridges are called sulci).

hair cells The sensory cells within the inner ear that transduce mechanical displacement into neural impulses.

helicotrema The opening at the apex of the cochlea that joins the scala vestibuli and scala tympani.

Hensen's node see primitive pit.

heterotrimeric G-proteins A large group of proteins consisting of three subunits (α, β, and γ) that can be activated by exchanging bound GDP with GTP resulting in the liberation of two signaling molecules—αGTP and the βγ dimer.

higher order neurons Neurons that are relatively remote from peripheral targets.

hindbrain see rhombencephalon.

hippocampus A cortical structure in the medial portion of the temporal lobe; in humans, concerned with short-term declarative memory, among many other functions.

histamine A biogenic amine neurotransmitter derived from the amino acid histidine.

holoprosencephaly Disrupted regional differentiation of the forebrain during development resulting in severe brain malformations.

homeotic genes Genes that determine the developmental fate of an entire segment of an animal. Mutations in these genes drastically alter the characteristics of the body segment (as when wings grow from a fly body segment that should have produced legs).

homologous Technically, referring to structures in different species that share the same evolutionary history; more generally, referring to structures or organs that have the same general anatomy and perform the same function.

homosexuality Sexual attraction to an individual of the same phenotypic sex.

horizontal cells Retinal neurons that mediate lateral interactions between photoreceptor terminals and the dendrites of bipolar cells.

horseradish peroxidase (HRP) A plant enzyme widely used to stain nerve cells (after injection into a neuron, it generates a visible precipitate by one of several histochemical reactions).

Hox genes A group of conserved homeotic genes characterized by a specific DNA sequence—"the homeobox"—and that specify body axis and segmentation patterns in the developing embryo. Hox genes are found among the vertebrates, in *Drosophila*, and in some other animal groups.

Huntington's disease An autosomal dominant genetic disorder in which a single gene mutation results in personality changes, progressive loss of the control of voluntary movement, and eventually death. Primary target is the basal ganglia.

hydrocephalus Enlarged cranium as a result of increased cerebrospinal fluid pressure (typically due to a mechanical outflow blockage).

hyperalgesia Increased perception of pain.

hyperkinesia   Excessive movement.

hyperpolarization   The displacement of a cell's membrane potential toward a more negative value.

hypokinesia   A paucity of movement.

hypothalamus   A collection of small but critical nuclei in the diencephalon that lies just inferior to the thalamus; governs reproductive, homeostatic, and circadian functions.

hypoxia   Lack of oxygen in the brain. Can be local, usually created by diminished blood flow (ischemia) due to local vascular occlusions; or global deprivation of oxygen after an event such as drowning or cardiac arrest.

imprinting   A rapid and permanent form of learning that occurs in response to early experience.

inactivation   The time-dependent closing of ion channels in response to a stimulus, such as membrane depolarization.

inducers   Chemical signals originating from one set of cells that influence the differentiation of other cells.

induction   The ability of a cell or tissue to influence the fate of nearby cells or tissues during development by chemical signals.

inferior colliculi (singular, colliculus)   Paired hillocks on the dorsal surface of the midbrain; concerned with auditory processing.

inferior olive   Prominent nucleus in the medulla; a major source of input to the cerebellum. Also called the inferior olivary nucleus.

infundibulum   The connection between the hypothalamus and the pituitary gland; also known as the pituitary stalk.

inhibitory postsynaptic potential (IPSP)   Neurotransmitter-induced postsynaptic potential change that tends to decrease the likelihood of a postsynaptic action potential.

innervate   Establish synaptic contact with a target.

innervation   Referring to all the synaptic contacts of a target.

input   The innervation of a target cell by a particular axon; more loosely, the innervation of a target.

input elimination   The developmental process by which the number of axons innervating some classes of target cells is diminished.

insula   The portion of the cerebral cortex that is buried within the depths of the lateral fissure.

integral membrane proteins   Proteins that possess hydrophobic domains that are inserted into membranes.

integration   The summation of excitatory and inhibitory synaptic conductance changes by postsynaptic cells.

integrins   A family of receptor molecules found on growth cones that bind to cell adhesion molecules such as laminin and fibronection.

intention tremor   Tremor that occurs while performing a voluntary motor act. Characteristic of cerebellar pathology.

internal arcuate tract   Mechanosensory pathway in the brainstem that runs from the dorsal column nuclei to form the medial lemniscus.

internal capsule   Large white matter tract that lies between the diencephalon and the basal ganglia; contains, among others, sensory axons that run from the thalamus to the cortex and motor axons that run from the cortex to the brainstem and spinal cord.

interneuron   Technically, a neuron in the pathway between primary sensory and primary effector neurons; more generally, a neuron whose relatively short axons branch locally to innervate other neurons. Also known as local circuit neuron.

interstitial nuclei of the anterior hypothalamus (INAH)   Four cell groups located slightly lateral to the third ventricle in the anterior hypothalamus of primates; thought to play a role in sexual behavior.

intrafusal muscle fibers   Specialized muscle fibers found in muscle spindles.

ion channels   Integral membrane proteins possessing pores that allow certain ions to diffuse across cell membranes, thereby conferring selective ionic permeability.

ion exchangers   Membrane transporters that translocate one or more ions against their concentration gradient by using the electrochemical gradient of other ions as an energy source.

ionotropic receptors   Receptors in which the ligand binding site is an integral part of the receptor molecule.

ion pumps   see transporters.

ipsilateral   On the same side of the body.

iris   Circular pigmented membrane behind the cornea; perforated by the pupil.

ischemia   Insufficient blood supply.

kinocilium   A true ciliary structure which, along with the stereocilia, comprises the hair bundle of vestibular and fetal cochlear hair cells in mammals (it is not present in the adult mammalian cochlear hair cell).

Korsakoff's syndrome   An amnesic syndrome seen in chronic alcoholics.

labyrinth   Referring to the internal ear; comprises the cochlea, vestibular apparatus, and the bony canals in which these structures are housed.

lamellipodia   The leading edge of a motile cell or growth cone, which is rich in actin filaments.

laminae (singular, lamina)   Cell layers that characterize the neocortex, hippocampus, and cerebellar cortex. The gray matter of the spinal cord is also arranged in laminae.

laminin   A large cell adhesion molecule that binds integrins.

late outward current   The delayed electrical current, measured in voltage clamp experiments, that results from the voltage-dependent efflux of a cation such as K^+. Produces the repolarizing phase of the action potential.

lateral columns   The lateral regions of spinal cord white matter that convey motor information from the brain to the spinal cord.

lateral (Sylvian) fissure   The cleft on the lateral surface of the brain that separates the temporal and frontal lobes.

lateral geniculate nucleus (LGN)   A nucleus in the thalamus that receives the axonal projections of retinal ganglion cells in the primary visual pathway.

lateral olfactory tract   The projection from the olfactory bulbs to higher olfactory centers.

lateral posterior nucleus   A thalamic nucleus that receives its major input from sensory and association cortices and projects in turn to association cortices, particularly in the parietal and temporal lobes.

lateral superior olive (LSO)   The auditory brainstem structure that processes interaural intensity differences and, in humans, mediates sound localization for stimuli greater than 3 kHz.

learning   The acquisition of novel behavior through experience.

lens   Transparent structure in the eye whose thickening or flattening in response to visceral motor control allows light rays to be focused on the retina.

lexical   The quality of associating a symbol (e.g., a word) with a particular object, emotion, or idea.

lexicon   Dictionary. Sometimes used to indicate region of brain that stores the meanings of words.

ligand-gated ion channels   Term for a large group of neurotransmitter receptors that combine receptor and ion channel functions into a single molecule.

limb bud   The limb rudiment of vertebrate embryos.

limbic lobe   Cortex that lies superior to the corpus callosum on the medial aspect of the cerebral hemispheres; forms the cortical component of the limbic system.

limbic system   Term that refers to those cortical and subcortical structures concerned with the emotions; the most prominent components are the cingulate gyrus, the hippocampus, and the amygdala.

lobes   The four major divisions of the cerebral cortex (frontal, parietal, occipital, and temporal).

local circuit neuron   General term referring to neurons whose activity mediates interactions between sensory systems and motor systems; interneuron is often used as a synonym.

locus coeruleus   A small brainstem nucleus with widespread adrenergic cortical and descending connections; important in the governance of sleep and waking.

long-term   Lasting days, weeks, months, or longer.

long-term depression   A persistent weakening of synapses based on recent patterns of activity.

long-term memory   Memories that last days, weeks, months, years, or a lifetime.

long-term potentiation (LTP)   A persistent strengthening of synapses based on recent patterns of activity.

lower motor neuron   Spinal motor neuron; directly innervates muscle (also referred to as α or primary motor neuron).

lower motor neuron syndrome   Signs and symptoms arising from damage to α motor neurons; these include paralysis or paresis, muscle atrophy, areflexia, and fibrillations.

macroscopic   Visible with the naked eye.

macroscopic currents   Ionic currents flowing through large numbers of ion channels distributed over a substantial area of membrane.

macula   The central region of the retina that contains the fovea (the term derives from the yellowish appearance of this region in ophthalmoscopic examination); also, the sensory epithelia of the otolith organs.

magnocellular   A component of the primary visual pathway specialized for the perception of motion; so named because of the relatively large cells involved.

mammillary bodies   Small prominences on the ventral surface of the diencephalon; functionally, part of the caudal hypothalamus.

map   The ordered projection of axons from one region of the nervous system to another, by which the organization of the body (or some function) is reflected in the organization of the nervous system.

mechanoreceptors   Receptors specialized to sense mechanical forces.

medial   Located nearer to the midsagittal plane of an animal (the opposite of lateral).

medial dorsal nucleus   A thalamic nucleus that receives its major input from sensory and association cortices and projects in turn to association cortices, particularly in the frontal lobe.

medial geniculate complex   The major thalamic relay for auditory information.

medial lemniscus   Axon tract in the brainstem that carries mechanosensory information from the dorsal column nuclei to the thalamus.

medial longitudinal fasciculus   Axon tract that carries excitatory projections from the abducens nucleus to the contralateral oculomotor nucleus; important in coordinating conjugate eye movements.

medial superior olive (MSO)   The auditory brainstem structure that processes interaural time differences and serves to compute the horizontal location of a sound source.

medium spiny neuron   The principal projection neuron of the caudate and putamen.

medulla   The caudal portion of the brainstem, extending from the pons to the spinal cord.

meduallary pyramids   Longitudinal bulges on the ventral aspect of the medulla that signify the corticospinal tracts at this level of the neuraxis.

medulloblastoma   Childhood brain tumor associated with mutations of *Sonic hedgehog* and other genes in the Shh signaling pathway.

Meissner's corpuscles   Encapsulated cutaneous mechanosensory receptors specialized for the detection of fine touch and pressure.

membrane conductance   The reciprocal of membrane resistance. Changes in membrane conductance result from, and are used to describe, the opening or closing of ion channels.

meninges   The external covering of the brain; includes the pia, arachnoid, and dura mater.

Merkel's disks   Encapsulated cutaneous mechanosensory receptors specialized for the detection of fine touch and pressure.

mesencephalon   see midbrain.

mesoderm   The middle of the three germ layers; gives rise to muscle, connective tissue, skeleton, and other structures.

mesopic   Light levels at which both the rod and cone systems are active.

metabotropic receptors   Refers to receptors that are indirectly activated by the action of neurotransmitters or other extracellular signals, typically through the aegis of G-protein activation.

Meyer's loop   That part of the optic radiation that runs in the caudal portion of the temporal lobe.

microglial cells   One of the three main types of central nervous system glia; concerned primarily with repairing damage following neural injury.

microscopic currents   Ionic currents flowing through single ion channels.

midbrain   The most rostral portion of the brainstem; identified by the superior and inferior colliculi on its dorsal surface, and the cerebral penduncles on its ventral aspect. Also called the mesencephalon.

middle cerebellar peduncle   Large white matter tract that carries axons from the pontine relay nuclei to the cerebellar cortex.

miniature end plate potential (MEPP)   Small, spontaneous depolarization of the membrane potential of skeletal muscle cells, caused by the release of a single quantum of acetylcholine.

mitral cells   The major output neurons of the olfactory bulb.

mnemonic   Having to do with memory.

modality   A category of function. For example, vision, hearing, and touch are different sensory modalities.

molecular layer   Layer of the cerebellar cortex containing the apical dendrites of Purkinje cells, parallel fibers from granule cells, a few local circuit neurons, and the synapses between these elements.

morphine   A plant alkaloid that gives opium its analgesic properties.

morphogen   A molecule that influences morphogenesis.

morphogenesis   The generation of animal form.

motor   Pertaining to movement.

motor cortex   The region of the cerebral cortex lying anterior to the central sulcus concerned with motor behavior; includes the primary motor cortex in the precentral gyrus and associated cortical areas in the frontal lobe.

motor neuron   By usage, a nerve cell that innervates skeletal muscle. Also called primary or α motor neuron.

motor neuron pool   The collection of motor neurons that innervates a single muscle.

motor system   A broad term used to describe all the central and peripheral structures that support motor behavior.

motor unit   A motor neuron and the skeletal muscle fibers it innervates; more loosely, the collection of skeletal muscle fibers innervated by a single motor neuron.

mucosa   Term referring the mucus membranes lining the nose, mouth, gut, and other epithelial surfaces.

muscarinic receptors   A group of G-protein-coupled acetylcholine receptors activated by the plant alkaloid muscarine.

muscle spindle   Highly specialized sensory organ found in most skeletal muscles; provides mechanosensory information about muscle length.

muscle tone   The normal, ongoing tension in a muscle; measured by resistance of a muscle to passive stretching.

myelin   The multilaminated wrapping around many axons formed by oligodendrocytes or Schwann cells.

myelination   Process by which glial cells wrap axons to form multiple layers of glial cell membrane that increase axonal conduction velocity.

myotatic reflex   A fundamental spinal reflex that is generated by the motor response to afferent sensory information arising from muscle spindles. The "knee jerk reaction" is a common example. Also called a stretch reflex.

myotome   The part of each somite that contributes to the development of skeletal muscles.

Na⁺/K⁺ transporter   A type of ATPase transporter in the plasma membrane of most cells that is responsible for accumulating intracellular K^+ and extruding intracellular Na^+. Also known as the Na^+ pump.

nasal   Referring to the region of the visual field of each eye in the direction of the nose.

near reflex   Reflexive response induced by changing binocular fixation to a closer target; includes convergence, accommodation, and pupillary constriction.

neocortex   The six-layered cortex that forms the surface of most of the cerebral hemispheres.

Nernst equation   A mathematical relationship that predicts the equilibrium potential across a membrane that is permeable to only one ion.

nerve   A collection of peripheral axons that are bundled together and travel a common route.

nerve growth factor (NGF)   A neurotrophic protein required for survival and differentiation of sympathetic ganglion cells and certain sensory neurons. Preeminent member of the neurotrophin family of growth factors.

netrins   A family of diffusible molecules that act as attractive or repulsive cues to guide growing axons.

neural cell adhesion molecule (N-CAM)   Molecule that helps bind axons together and is widely distributed in the developing nervous system. Structurally related to immunoglobin.

neural crest   A group of progenitor cells that forms along the dorsum of the neural tube and gives rise to peripheral neurons and glia (among other derivatives).

neural plate   The thickened region of the dorsal ectoderm of a neurula that gives rise to the neural tube.

neural stem cell   A precursor cell type that can give rise to the full complement of cell classes found in neural tissue—i.e., neurons, astrocytes, and oligodendroglia, as well as more neural stem cells.

neural tube   The primordium of the brain and spinal cord; derived from the neural ectoderm.

neurexin   An adhesion molecule of the presynaptic membrane in developing synapses. Binds to neuroligin in the postsynaptic membrane, promoting adhesion, and helps localize synaptic vesicles, docking proteins, and fusion molecules.

neurite   A neuronal branch (usually used when the process in question could be either an axon or a dendrite, such as the branches of isolated nerve cells in tissue culture).

neuroblast   A dividing cell, the progeny of which develop into neurons.

neurogenesis   The development of the nervous system.

neuroglial cells   see glia.

neuroleptics   A group of antipsychotic agents that cause indifference to stimuli by blocking brain dopamine receptors.

neuroligin   Postsynaptic binding partner of the presynaptic adhesion molecule neurexin. Promotes the clustering of receptors and channels of the postsynaptic density as the synapse matures.

neuromere   A segment of the rhombencephalon (synonym for rhombomere).

neuromuscular junction   The synapse made by a motor axon on a skeletal muscle fiber.

neuron   Cell specialized for the conduction and transmission of electrical signals in the nervous system.

neuronal geometry   The spatial arrangement of neuronal branches.

neuron-glia cell adhesion molecule (Ng-CAM)   A cell adhesion molecule, structurally related to immunoglobin molecules, that promotes adhesive interactions between neurons and glia.

neuropeptides   A general term describing a large number of peptides that function as neurotransmitters or neurohormones.

neuropil   The dense tangle of axonal and dendritic branches, and the synapses between them, that lies between neuronal cell bodies in the gray matter of the brain and spinal cord.

neurotransmitter   Substance released by synaptic terminals for the purpose of transmitting information from one nerve cell to another.

neurotrophic factors   A general term for molecules that promote the growth and survival of neurons.

neurotrophic hypothesis   The idea that developing neurons compete for a limited supply of trophic factors secreted by their targets.

neurotrophins   A family of trophic factor molecules that promote the growth and survival of several different classes of neurons.

neurula   The early vertebrate embryo during the stage when the neural tube forms from the neural plate; follows the gastrula stage.

neurulation   The process by which the neural plate folds to form the neural tube.

nociceptors   Cutaneous and subcutaneous receptors (usually free nerve endings) specialized for the detection of harmful (noxious) stimuli.

nodes of Ranvier   Periodic gaps in the myelination of axons where action potentials are generated.

non-rapid eye movement (non-REM) sleep   Collectively, those phases of sleep characterized by the absence of rapid eye movements.

norepinephrine   Catecholamine hormone and neurotransmitter that binds to α- and β-adrenergic receptors, both of which are G-protein-coupled receptors. Also known as noradrenaline.

notochord   A transient, cylindrical structure of mesodermal cells underlying the neural plate (and later the neural tube) in vertebrate embryos. Source of important inductive signals for spinal cord.

nucleus (nuclei)   Collection of nerve cells in the brain that are anatomically discrete, and which typically serve a particular function.

nucleus proprius   Region of the dorsal horn of the spinal cord that receives information from nociceptors.

nystagmus   Literally, a nodding movement. Refers to repetitive movements of the eyes normally elicited by large-scale movements of the visual field (optokinetic nystagmus). Nystagmus in the absence of appropriate stimuli usually indicates brainstem or cerebellar pathology.

occipital lobe   The posterior lobe of the cerebral hemisphere; primarily devoted to vision.

ocular dominance columns   The segregated termination patterns of thalamic inputs representing the two eyes in primary visual cortex of some mammalian species.

odorants   Molecules capable of eliciting responses from receptors in the olfactory mucosa.

olfactory bulb   Olfactory relay station that receives axons from cranial nerve I and transmits this information via the olfactory tract to higher centers.

olfactory epithelium   Pseudostratified epithelium that contains olfactory receptor cells, supporting cells, and mucus-secreting glands.

olfactory receptor neurons   Bipolar neurons in olfactory epithelium that contain receptors for odorants.

olfactory tracts   see lateral olfactory tract.

oligodendrocytes   One of three classes of central neuroglial cells; their major function is to elaborate myelin.

Onuf's nucleus   Sexually dimorphic nucleus in the human spinal cord that innervates striated perineal muscles mediating contraction of the bladder in males, and vaginal constriction in females.

opioid   Any natural or synthetic drug that has pharmacological actions similar to those of morphine.

opsins   Proteins in photoreceptors that absorb light (in humans, rhodopsin and the three specialized cone opsins).

optic chiasm   The junction of the two optic nerves on the ventral aspect of the diencephalon, where axons from the nasal parts of each retina cross the midline.

optic cup   see optic vesicle.

optic disk   The region of the retina where the axons of retinal ganglion cells exit to form the optic nerve.

optic nerve   The nerve (cranial nerve II) containing the axons of retinal ganglion cells; extends from the eye to the optic chiasm.

optic radiation   Portion of the internal capsule that comprises the axons of lateral geniculate neurons that carry visual information to the striate cortex.

optic tectum   The first central station in the visual pathway of many vertebrates (analogous to the superior colliculus in mammals).

optic tract   The axons of retinal ganglion cells after they have passed through the region of the optic chiasm en route to the lateral geniculate nucleus of the thalamus.

optic vesicle   The evagination of the forebrain vesicle that generates the retina and induces lens formation in the overlying ectoderm.

optokinetic eye movements   Movements of the eyes that compensate for head movements; the stimulus for optokinetic movements is large-scale motion of the visual field.

optokinetic nystagmus   Repeated reflexive responses of the eyes to ongoing large-scale movements of the visual scene.

orbital (medial) prefrontal cortex   Divisions of the prefrontal cortex that lie above the orbits in the most rostral and ventral extension of the sagittal fissure. Important in emotional processing and rational decision-making.

orientation selectivity   A property of many neurons in visual cortex in which they respond to edges presented over a narrow range of orientations.

oscillopsia   Inability to fixate visual targets while the head is moving as a result of vestibular damage.

ossicles   The bones of the middle ear

otoconia   The calcium carbonate crystals that rest on the otolithic membrane overlying the hair cells of the sacculus and utricle.

otolithic membrane   The gelatinous membrane on which the otoconia lie and in which the tips of the hair bundles are embedded.

otoliths   Literally, "ear stones." Dense calcific structures important in generating the vestibular signals pertinent to balance.

outer segment   Portion of photoreceptors made up of membranous disks that contain the photopigment responsible for initiating phototransduction.

oval window   Site where the middle ear ossicles transfer vibrational energy to the cochlea.

overshoot   The peak, positive-going phase of an action potential, caused by high membrane permeability to a cation such as Na^+ or Ca^{2+}.

oxytocin   A 9-amino acid neuropeptide that is both a putative neurotransmitter and a neurohormone.

Pacinian corpuscle   Encapsulated mechanosensory receptor specialized for the detection of high-frequency vibrations.

Papez's circuit   System of interconnected brain structures (mainly cingulate gyrus, hippocampus, and hypothalamus) in the medial aspect of the telencephalon and diencephalon described by James Papez. Participates in emotional processing, short-term declarative memory, and autonomic functions.

paracrine   Term referring to the secretion of hormone-like agents whose effects are mediated locally rather than by the general circulation.

parallel fibers   The bifurcated axons of cerebellar granule cells that synapse on dendritic spines of Purkinje cells.

paralysis   Complete loss of voluntary motor control.

paramedian pontine reticular formation (PPRF)   Neurons in the reticular formation of the pons that coordinate the actions of motor neurons in the abducens and oculomotor nuclei to generate horizontal movements of the eyes; also known as the "horizontal gaze center."

parasympathetic nervous system   A division of the visceral motor system in which the effectors are cholinergic ganglion cells located near target organs.

paresis   Partial loss of voluntary motor control; weakness.

parietal lobe   The lobe of the brain that lies between the frontal lobe anteriorly, and the occipital lobe posteriorly.

Parkinson's disease   Neurodegenerative disease of the substantia nigra that results in a characteristic tremor at rest and a general paucity of movement.

parvocellular   Referring to the component of the primary visual pathway specialized for the detection of detail and color; so named because of the relatively small cells involved.

passive current flow   Current flow across neuronal membranes that does not entail the action potential mechanism.

patch clamp   An extraordinarily sensitive voltage clamp method that permits the measurement of ionic currents flowing through individual ion channels.

periaqueductal gray matter   Region of brainstem gray matter that contains, among others, nuclei associated with the modulation of pain perception.

perilymph   The potassium-poor fluid that bathes the basal end of the cochlear hair cells.

perineurium   The connective tissue that surrounds a nerve fascicle in a peripheral nerve.

peripheral nervous system   All nerves and neurons that lie outside the brain and spinal cord.

phasic   Transient firing of action potentials in response to a prolonged stimulus; the opposite of tonic.

phenotype   The visible (or otherwise discernible) characteristics of an animal that arise during development.

phenotypic sex   The visible body characteristics associated with sexual behaviors.

photopic vision   Vision at high light levels that is mediated entirely by cones.

pia mater   The innermost of the three layers of the meninges, which is closely applied to the surface of the brain.

pigment epithelium   Pigmented coat underlying the retina important in the normal turnover of photopigment in rods and cones.

pineal gland   Midline neural structure lying on the dorsal surface of the midbrain; important in the control of circadian rhythms (and, incidentally, considered by Descartes to be the seat of the soul).

pinna   A component of the external ear.

pituitary gland   Endocrine structure comprising an anterior lobe made up of many different types of hormone-secreting cells, and a posterior lobe that secretes neuropeptides produced by neurons in the hypothalamus.

placebo   An inert substance that when administered may, because of the circumstances, have physiological effects.

planum temporale   Region on the superior surface of the temporal lobe posterior to Heschl's gyrus; notable because it is larger in the left hemisphere in about two-thirds of humans.

plasticity   Term that refers to structural or functional changes in the nervous system, or the ability to make such changes.

polarity   Referring to a continually graded organization along one of the major axes of an animal.

polymodal   Responding to more than one sensory modality.

polyneuronal innervation   A state in which neurons or muscle fibers receive synaptic inputs from multiple, rather than single, axons.

pons   One of the three components of the brainstem, lying between the midbrain rostrally and the medulla caudally.

pontine-geniculate-occipital (PGO) waves   Characteristic encephalographic waves that signal the onset of rapid eye movement sleep.

pontine relay nuclei   Collections of neurons in the pons that receive input from the cerebral cortex and send their axons across the midline to the cerebellar cortex via the middle cerebellar peduncle.

pore   A structural feature of membrane ion channels that allows ions to diffuse through the channel.

pore loop   An extracellular domain of amino acids, found in certain ion channels, that lines the channel pore and allows only certain ions to pass.

postcentral gyrus   The gyrus that lies just posterior to the central sulcus; contains the primary somatic sensory cortex.

posterior   Toward the back; sometimes used as a synonym for caudal or dorsal.

postganglionic   Referring to axons that link visceral motor neurons in autonomic ganglia to their targets.

postsynaptic   Referring to the component of a synapse specialized for transmitter reception; downstream at a synapse.

postsynaptic current (PSC)   The current produced in a postsynaptic neuron by the binding of neurotransmitter released from a presynaptic neuron.

postsynaptic density   A cytoskeletal junction in developing synapses that may serve to organize postsynaptic receptors and speed their response to neurotransmitter.

postsynaptic potential (PSP)   The potential change produced in a postsynaptic neuron by the binding of neurotransmitter released from a presynaptic neuron.

post-tetanic potentiation (PTP)   An enhancement of synaptic transmission resulting from high-frequency trains of action potentials.

precentral gyrus   The gyrus that lies just anterior to the central sulcus; contains the primary motor cortex.

prefrontal cortex   Cortical regions in the frontal lobe that are anterior to the primary and association motor cortices; thought to be involved in planning complex cognitive behaviors and in the expression of personality and appropriate social behavior.

preganglionic   Referring to neurons and axons that link visceral motor neurons in spinal cord and brainstem to autonomic ganglia.

premotor cortex   Motor association areas in the frontal lobe anterior to primary motor cortex; thought to be involved in planning or programming of voluntary movements.

pre-proproteins   The first protein translation products synthesized in a cell. These polypeptides are usually much larger than the final, mature peptide, and often contain signal sequences that target the peptide to the lumen of the endoplasmic reticulum.

presynaptic Referring to the component of a synapse specialized for transmitter release; upstream at a synapse.

pretectum A group of nuclei located at the junction of the thalamus and the midbrain; these nuclei are important in the pupillary light reflex, relaying information from the retina to the Edinger-Westphal nucleus.

prevertebral ganglia Sympathetic ganglia that lie anterior to the spinal column (distinct from the sympathetic chain ganglia).

primary auditory cortex The major cortical target of the neurons in the medial geniculate nucleus.

primary motor cortex A major source of descending projections to motor neurons in the the spinal cord and cranial nerve nuclei; located in the precentral gyrus (Brodmann's area 4) and essential for the voluntary control of movement.

primary neuron A neuron that directly links muscles, glands, and sense organs to the central nervous system.

primary sensory cortex Any one of several cortical areas receiving the thalamic input for a particular sensory modality.

primary visual cortex see striate cortex.

primary visual pathway Pathway from the retina via the lateral geniculate nucleus of the thalamus to the primary visual cortex; carries the information that allows conscious visual perception. Also known as the retinogeniculocortical pathway.

primate Any mammal in a group that includes lemurs, tarsiers, marmosets, monkeys, apes, and humans.

priming A phenomenon in which the memory of an initial exposure is expressed unconsciously by improved performance at a later time.

primitive pit The thickened anterior end of the primitive streak; an important source of inductive signals during early development.

primitive streak Axial thickening in the ectoderm of the gastrulas of reptiles, birds, and mammals; the mesoderm forms by the ingression of cells at this site.

procedural memory Unconscious memories such as motor skills and associations.

production aphasia Aphasia that derives from cortical damage to those centers concerned with the motor aspects of speech.

projection neuron A neuron with long axons that project to distant targets.

promoter DNA sequence (usually within 35 nucleotides upstream of the start site of transcription) to which the RNA polymerase and its associated factors bind to initiate transcription.

proproteins Partially processed forms of proteins containing peptide sequences that play a role in the correct folding of the final protein.

proprioceptors Sensory receptors (usually limited to mechanosensory receptors) that sense the internal forces acting on the body; muscle spindles and Golgi tendon organs are the preeminent examples.

prosencephalon The part of the brain that includes the diencephalon and telencephalon (derived from the embryonic forebrain vesicle).

prosody The emotional tone or quality of speech.

prosopagnosia The inability to recognize faces; usually associated with lesions to the right inferior temporal cortex.

proteoglycan Molecule consisting of a core protein to which one or more long, linear carbohydrate chains (glycosaminoglycans) are attached.

proximal Closer to a point of reference (the opposite of distal).

psychotropic Referring to drugs that alter behavior, mood, and perception.

pulvinar A thalamic nucleus that receives its major input from sensory and association cortices and projects in turn to association cortices, particularly in the parietal lobe.

pupil The perforation in the iris that allows light to enter the eye.

pupillary light reflex The decrease in the diameter of the pupil that follows stimulation of the retina.

Purkinje cell The large principal projection neuron of the cerebellar cortex. Its defining characteristic is an elaborate apical dendrite.

putamen One of the three major nuclei that make up the basal ganglia.

pyramidal tract White matter tract that lies on the ventral surface of the medulla and contains axons descending from motor cortex to the spinal cord.

pyriform cortex Component of cerebral cortex in the temporal lobe pertinent to olfaction; so named because of its pearlike shape.

radial glia Glial cells that contact both the luminal and pial surfaces of the neural tube, providing a substrate for neuronal migration.

ramus Branch; typically applied to the white and gray communicating rami that carry visceral motor axons to the segmental nerves.

Raphe nuclei A collection of serotonergic nuclei in the brainstem tegmentum important in the governance of sleep and waking.

rapid eye movement (REM) sleep Phase of sleep characterized by low-voltage, high-frequency electroencephalographic activity accompanied by rapid eye movements.

receptive field Region of a receptor surface (e.g., the body surface or the retina) that causes a sensory nerve cell (or axon) to respond.

receptor A molecule specialized to bind any one of a large number of chemical signals, preeminently neurotransmitters.

receptor neuron A neuron specialized for the transduction of energy in the environment into electrical signals.

receptor potential The membrane potential change elicited in receptor neurons during sensory transduction.

reflex A stereotyped (involuntary) motor response elicited by a defined stimulus.

refractory period The brief period after the generation of an action potential during which a second action potential is difficult or impossible to elicit.

reserpine An antihypertensive drug that is no longer used due to side effects such as behavioral depression.

resting potential The inside-negative electrical potential that is normally recorded across all cell membranes.

reticular activating system Region in the brainstem tegmentum that, when stimulated, causes arousal; involved in modulating sleep and wakefulness.

reticular formation A network of neurons and axons that occupies the core of the brainstem, giving it a reticulated appearance in myelin-stained material; major functions include control of respiration and heart rate, posture, and state of consciousness.

retina   Laminated neural component of the eye that contains the photoreceptors (rods and cones) and the initial processing machinery for the primary (and other) visual pathways.

retinoic acid   A derivative of vitamin A that acts as an inducer during early brain development.

retinotectal system   The pathway between ganglion cells in the retina and the optic tectum of vertebrates.

retrograde   A movement or influence acting from the axon terminal toward the cell body.

reversal potential   The membrane potential of a post-synaptic neuron (or other target cell) at which the action of a given neurotransmitter causes no net current flow.

rhodopsin   The photopigment found in rods.

rhombencephalon   The part of the brain that includes the pons, cerebellum, and medulla (derived from the embryonic hindbrain vesicle).

rhombomere   Segment of the developing rhombencephalon.

rising phase   The initial, depolarizing, phase of an action potential, caused by the regenerative, voltage-dependent influx of a cation such as Na^+ or Ca^{2+}.

rods   Photoreceptor cells specialized for operating at low light levels.

rostral   Anterior, or "headward."

rostral interstitial nucleus   Neurons in the midbrain reticular formation that coordinate the actions of neurons in the oculomotor nuclei to generate vertical movements of the eye; also known as the "vertical gaze center."

rostral migratory stream   A specific migratory route, defined by a distinct subset of glial cells, that facilitates migration of newly generated neurons from the stem cell niche of the anterior subventricular zone to the olfactory bulb.

saccades   Ballistic, conjugate eye movements that change the point of foveal fixation.

sacculus   The otolith organ that detects linear accelerations and head tilts in the vertical plane.

sagittal   Referring to the anterior-posterior plane of an animal.

saltatory conduction   Mechanism of action potential propagation in myelinated axons; so named because action potentials "jump" from one node of Ranvier to the next due to generation of action potentials only at these sites.

Scarpa's ganglion   The ganglion containing the bipolar cells that innervate the semicircular canals and otolith organs.

Schaffer collaterals   The axons of cells in the CA3 region of hippocampus that form synapses in the CA1 region.

Schwann cells   Neuroglial cells in the peripheral nervous system that elaborate myelin (named after the nineteenth-century anatomist and physiologist Theodor Schwann).

sclera   The external connective tissue coat of the eyeball.

scotoma   A defect in the visual field as a result of pathological changes in some component of the primary visual pathway.

scotopic   Referring to vision in dim light, where the rods are the operative receptors.

second-order neurons   Projection neurons in a sensory pathway that lie between the primary receptor neurons and the third-order neurons.

segment   One of a series of more or less similar anterior-posterior units that make up segmental animals.

segmentation   The anterior-posterior division of animals into roughly similar repeating units.

semaphorins   A family of diffusible, growth-inhibiting molecules (see also collapsin).

semicircular canals   The vestibular end organs within the inner ear that sense rotational accelerations of the head.

sensitization   Increased sensitivity to stimuli in an area surrounding an injury. Also, a generalized aversive response to an otherwise benign stimulus when it is paired with a noxious stimulus.

sensorineural hearing loss   Diminished sense of hearing due to damage of the inner ear or its related central auditory structures. Contrast with conductive hearing loss.

sensory aphasia   Difficulty in communicating with language that derives from cortical damage to those areas concerned with the comprehension of speech.

sensory ganglia   see dorsal root ganglia.

sensory system   Term sometimes used to describe all the components of the central and peripheral nervous system concerned with sensation.

sensory transduction   Process by which energy in the environment is converted into electrical signals by sensory receptors.

serotonin   A biogenic amine neurotransmitter derived from the amino acid tryptophan.

sexually dimorphic   Having two different forms depending on genotypic or phenotypic sex.

short-term memory   Memories that last from seconds to minutes.

silver stain   A classical method for visualizing neurons and their processes by impregnation with silver salts (the best-known technique is the Golgi stain, developed by the Italian anatomist Camillo Golgi in the late nineteenth century).

size principle   The orderly recruitment of motor neurons by size to generate increasing amounts of muscle tension.

sleep spindles   Bursts of electroencephalographic activity, at a frequency about 10–14 Hz and lasting a few seconds; spindles characterize the initial descent into non-REM sleep.

small-molecule neurotransmitters   Referring to the non-peptide neurotransmitters such as acetylcholine, the amino acids glutamate, aspartate, GABA, and glycine, as well as the biogenic amines.

smooth pursuit eye movements   Slow, tracking movements of the eyes designed to keep a moving object aligned with the fovea.

soma   The cell body.

somatic cells   Referring to the cells of an organism's body other than its germ cells.

somatic sensory cortex   The region of the cerebral cortex concerned with processing sensory information from the body surface, subcutaneous tissues, muscles, and joints. Located primarily in the posterior bank of the central sulcus and on the postcentral gyrus.

somatic sensory system   Components of the nervous system involved in processing sensory information about the mechanical forces active on both the body surface and on deeper structures such as muscles and joints.

somatic stem cell   A cell that can divide to give rise to more cells like itself, but also can divide to give rise to a new stem cell plus one or more differentiated cells of a distinct tissue type (e.g., a hematopoeitic stem cell that gives rise to all types of blood cells; see also neural stem cell).

somatotopic maps   Cortical or subcortical arrangements of sensory pathways that reflect the organization of the body.

somites   Segmentally arranged masses of mesoderm that lie alongside the neural tube and give rise to skeletal muscle, vertebrae, and dermis.

Sonic hedgehog (Shh)   An inductive signaling hormone essential for development of the mammalian nervous system; believed to be particularly important for establishing the identity of neurons in the ventral portion of the developing spinal cord and hindbrain.

specificity   Term applied to neural connections that entail specific choices between neurons and their targets.

spike timing-dependent plasticity (STDP)   Timing-dependent activity, probably the result of Ca^{2+} signaling in the postsynaptic cell, that is required for the establishment of some forms of synaptic plasticity.

spina bifida   A congenital defect in which the neural tube fails to close at its posterior end.

spinal cord   The portion of the central nervous system that extends from the lower end of the brainstem (the medulla) to the cauda equina.

spinal ganglia   see dorsal root ganglia.

spinal nucleus of the bulbocavernosus   Sexually dimorphic collection of neurons in the lumbar region of the rodent spinal cord that innervate striated perineal muscles.

spinal shock   The initial flaccid paralysis that accompanies damage to descending motor pathways.

spinal trigeminal tract   Brainstem tract carrying fibers from the trigeminal nerve to the spinal nucleus of the trigeminal complex (which serves as the relay for painful stimulation of the face).

spinocerebellum   Region of the cerebellar cortex that receives input from the spinal cord, particularly Clarke's column in the thoracic spinal cord.

spinothalamic pathway   see anterolateral pathway.

spinothalamic tract   Ascending white matter tract carrying information about pain and temperature from the spinal cord to the VP nuclear complex in the thalamus; also referred to as the anterolateral tract.

split-brain patients   Individuals who have had the cerebral commissures divided in the midline to control epileptic seizures.

sporadic   Cases of a disease that apparently occur at random in a population; contrasts with familial or inherited.

SRY   Gene on the Y chromosome whose expression triggers a hormone cascade that masculinizes the developing fetus

stem cell   see embryonic stem cell, neural stem cell, somatic stem cell

stem cell niche   A local environment in regenerating tissue that is conducive to the division and initial differentiation of the somatic stem cells that will reconstitute adult tissue.

stereocilia   The actin-rich processes that, along with the kinocilium, form the hair bundle extending from the apical surface of the hair cell; site of mechanotransduction.

stereopsis   The perception of depth that results from the fact that the two eyes view the world from slightly different angles.

strabismus   Developmental misalignment of the two eyes; may lead to binocular vision being compromised.

stria vascularis   Specialized epithelium lining the cochlear duct that maintains the high potassium concentration of the endolymph.

striate cortex   Primary visual cortex in the occipital lobe (also called Brodmann's area 17). So named because the prominence of layer 4 in myelin-stained sections gives this region a striped appearance.

striatum   see corpus striatum.

striola   A line found in both the sacculus and utricle that divides the hair cells into two populations with opposing hair bundle polarities.

subarachnoid space   The cerebrospinal fluid-filled space over the surface of the brain that lies between the arachnoid and the pia.

substance P   An 11-amino acid neuropeptide; the first neuropeptide to be characterized.

substantia nigra   Nucleus at the base of the midbrain that receives input from a number of cortical and subcortical structures. The dopaminergic cells of the substantia nigra send their output to the caudate/putamen, while the GABAergic cells send their output to the thalamus.

subthalamic nucleus   A nucleus in the ventral diencephalon that receives input from the caudate/putamen and participates in the modulation of motor behavior.

subventricular zones   Cell-dense regions adjacent to the ventricular spaces of cortical hemispheres.

sulcus (sulci)   Infoldings of the cerebral hemisphere that form the valleys between the gyral ridges.

summation   The addition in space and time of sequential synaptic potentials to generate a larger than normal postsynaptic response.

superior colliculus   Laminated structure that forms part of the roof of the midbrain; plays an important role in orienting movements of the head and eyes.

suprachiasmatic nucleus   Hypothalamic nucleus lying just above the optic chiasm that receives direct input from the retina; involved in light entrainment of circadian rhythms.

Sylvian fissure   see lateral fissure.

sympathetic nervous system   A division of the visceral motor system in vertebrates comprising, for the most part, adrenergic ganglion cells located relatively far from the related end organs.

synapse   Specialized apposition between a neuron and its target cell for transmission of information by release and reception of a chemical transmitter agent.

synapse elimination   see input elimination.

synaptic cleft   The space that separates pre- and postsynaptic neurons at chemical synapses.

synaptic depression   A short-term decrease in synaptic strength resulting from the depletion of synaptic vesicles at active synapses.

synaptic plasticity   Changes in the efficacy and local geometry of synaptic connections and transmission; one basis of learning, memory, and other forms of brain plasticity.

synaptic vesicle recycling   A sequence of budding and fusion reactions that occurs within presynaptic terminals to maintain the supply of synaptic vesicles.

synaptic vesicles   Spherical, membrane-bound organelles in presynaptic terminals that store neurotransmitters.

syncytium   A group of cells in protoplasmic continuity.

target   The object of innervation, which can be either non-neuronal targets, such as muscles, glands, and sense organs, or other neurons.

taste buds   Onion-shaped structures in the mouth and pharynx that contain taste cells.

tectorial membrane   The fibrous sheet overlying the apical surface of the cochlear hair cells; produces a shearing motion of the stereocilia when the basilar membrane is displaced.

tectum   A general term referring to the dorsal region of the brainstem (tectum means "roof").

tegmentum   A general term that refers to the central gray matter of the brainstem.

telencephalon   The part of the brain derived from the anterior part of the embryonic forebrain vesicle; includes the cerebral hemispheres.

temporal division   Referring to the region of the visual field of each eye in the direction of the temple.

temporal lobe   The hemispheric lobe that lies inferior to the lateral fissure.

terminal   A presynaptic (axonal) ending.

tetraethylammonium   A quaternary ammonium compound that selectively blocks voltage-sensitive K^+ channels; eliminates the delayed K^+ current measured in voltage clamp experiments.

tetrodotoxin (TTX)   An alkaloid neurotoxin, produced by certain puffer fish, tropical frogs, and salamanders, that selectively blocks voltage-sensitive Na^+ channels; eliminates the initial Na^+ current measured in voltage clamp experiments.

thalamus   A collection of nuclei that forms the major component of the diencephalon. Although its functions are many, a primary role of the thalamus is to relay sensory information from lower centers to the cerebral cortex.

thermoreceptors   Receptors specialized to transduce changes in temperature.

threshold   The level of membrane potential at which an action potential is generated.

tight junction   A specialized junction between epithelial cells that seals them together, preventing most molecules from passing across the cell sheet.

tip links   The filamentous structures that link the tips of adjacent stereocilia; thought to mediate the gating of the hair cell's transduction channels.

tonic   Sustained activity in response to an ongoing stimulus; the opposite of phasic.

tonotopy   The topographic mapping of frequency across the surface of a structure, which originates in the cochlea and is preserved in ascending auditory structures, including the auditory cortex.

transcription factors   A general term applied to proteins that regulate transcription, including basal transcription factors that interact with the RNA polymerase to initiate transcription, as well as those that bind elsewhere to stimulate or repress transcription.

transducin   G-protein involved in the phototransduction cascade.

transduction   see sensory transduction.

transforming growth factor (TGF)   A class of peptide growth factors that acts as an inducer during early development.

transmitter   see neurotransmitter.

transporters   Cell membrane molecules that consume energy to move ions up their concentration gradients, thus restoring and/or maintaining normal concentration gradients across cell membranes.

trichomatic   Referring to the presence of three different cone types in the human retina, which generate the initial steps in color vision by differentially absorbing long, medium, and short wavelength light.

tricyclic antidepressants   A class of antidepressant drugs named for their three-ringed molecular structure; thought to act by blocking the reuptake of biogenic amines.

trigeminal ganglion   The sensory ganglion associated with the trigeminal nerve (cranial nerve V).

Trk receptors   The receptors for the neurotrophin family of growth factors.

trophic   The ability of one tissue or cell to support another; usually applied to long-term interactions between pre- and post-synaptic cells.

trophic factor   A molecule that mediates trophic interactions.

trophic interactions   Referring to the long-term interdependence of nerve cells and their targets.

trophic molecules   see trophic factor.

tropic   An influence of one cell or tissue on the direction of movement (or outgrowth) of another.

tropic molecules   Molecules that influence the direction of growth or movement.

tropism   Orientation of growth in response to an external stimulus.

tuning curve   Referring to a common physiological test in which the receptive field properties of neurons are gauged against a varying stimulus such that maximum sensitivity or maximum responsiveness can be defined by the peak of the tuning curve.

tympanic membrane   The eardrum.

undershoot   The final, hyperpolarizing phase of an action potential, typically caused by the voltage-dependent efflux of a cation such as K^+.

upper motor neuron   A neuron that gives rise to a descending projection that controls the activity of lower motor neurons in the brainstem and spinal cord.

upper motor neuron syndrome   Signs and symptoms that result from damage to descending motor systems; these include paralysis, spasticity, and a positive Babinski sign.

utricle   The otolith organ that senses linear accelerations and head tilts in the horizontal plane.

vasopressin   A 9-amino-acid neuropeptide that acts as a neurotransmitter, as well as a neurohormone.

ventral   Referring to the belly; the opposite of dorsal.

ventral horn   The ventral portion of the spinal cord gray matter; contains the primary motor neurons.

ventral posterior complex   Group of thalamic nuclei that receives the somatic sensory projections from the dorsal column nuclei and the trigeminal nuclear complex.

ventral posterior lateral nucleus   Component of the ventral posterior complex of thalamic nuclei that receives brainstem projections carrying somatic sensory information from the body (excluding the face).

ventral posterior medial nucleus   Component of the ventral posterior complex of thalamic nuclei that receives brainstem projections related to somatic sensory information from the face.

ventral roots   The collection of nerve fibers containing motor axons that exit ventrally from the spinal cord and contribute the motor component of each segmental spinal nerve.

ventricles   The fluid-filled spaces in the vertebrate brain that represent the lumen of the embryonic neural tube.

ventricular zone   The sheet of cells closest to the ventricles in the developing neural tube.

vergence movements Disjunctive movements of the eyes (convergence or divergence) that align the fovea of each eye with targets located at different distances from the observer.

vesicle Literally, a small sac. Used to refer to the organelles that store and release transmitter at nerve endings. Also used to refer to any of the three dilations of the anterior end of the neural tube that give rise to the three major subdivisions of the brain.

vestibulocerebellum The part of the cerebellar cortex that receives direct input from the vestibular nuclei or vestibular nerve.

vestibulo-ocular reflex Involuntary movement of the eyes in response to displacement of the head. This reflex allows retinal images to remain stable while the head is moved.

visceral Referring to the internal organs of the body cavity.

visceral motor system The component of the motor system (also known as the autonomic nervous system) that motivates and governs visceral motor behavior.

visceral nervous system Synonymous with autonomic nervous system.

visual field The area in the external world normally seen by one or both eyes (referred to, respectively, as the monocular and visual binocular fields).

vital dye A reagent that stains cells when they are alive.

voltage clamp A method that uses electronic feedback to control the membrane potential of a cell, simultaneously measuring transmembrane currents that result from the opening and closing of ion channels.

voltage-gated Term used to describe ion channels whose opening and closing is sensitive to membrane potential.

Wallerian degeneration The process by which the distal portion of a damaged axon segment degenerates; named after Augustus Waller, a nineteenth-century physician and neuroanatomist.

Wernicke's aphasia Difficulty comprehending speech as a result of damage to Wernicke's language area.

Wernicke's area Region of cortex in the superior and posterior region of the left temporal lobe that helps mediate language comprehension. Named after the nineteenth-century neurologist, Carl Wernicke.

white matter A general term that refers to large axon tracts in the brain and spinal cord; the phrase derives from the fact that axonal tracts have a whitish cast when viewed in the freshly cut material.

working memory Memories held briefly in mind that enable a particular task to be accomplished (e.g., efficiently searching a room for a lost object).

Illustration Credits

Chapter 1 Studying the Nervous System
Opening Image Copyright © Max Delson/
istockphoto.com **Figure 1.3** PETERS, A., S. L.
PALAY AND H. DEF. WEBSTER (1991) *The Fine
Structure of the Nervous System: Neurons and
Their Supporting Cells*, 3rd Ed. Oxford University Press, New York. **Figure 1.4E** SALA, K., K.
FUTAI, K. YAMAMOTO, P. F. WORLEY, Y. HAYASHI
AND M. SHENG (2003) Inhibition of dendritic
spine morphogenesis and synaptic transmission by activity-inducible protein Homer1a. J.
Neurosci. 23: 6327–6337. **Figure 1.4F** MATUS,
A. (2000) Actin dynamics and synaptic plasticity. Science 290: 754–758. **Figure 1.5A–C**
JONES, E. G. AND M. W. COWAN (1983) The
nervous tissue. In *The Structural Basis of Neurobiology*, E. G. Jones (ed.). New York: Elsevier,
Chapter 8. **Figure 1.12** ZYLKA, M. J., F. L.
RICE AND D. J. ANDERSON (2005) Topographically distinct epidermal nociceptive circuits
revealed by axonal tracers targeted to Mrgprd.
Neuron 46: 17–25.

Chapter 2 Electrical Signals of Nerve Cells
Figures 2.7 & 2.8 HODGKIN, A. L. AND B. KATZ
(1949) The effect of sodium ions on the electrical activity of the giant axon of the squid. J.
Physiol. (Lond.) 108: 37–77.

**Chapter 3 Voltage-Dependent Membrane
Permeability**
Figures 3.1, 3.2, 3.3 & 3.4 HODGKIN, A. L. AND
A. F. HUXLEY (1952a) Currents carried by
sodium and potassium ions through the
membrane of the giant axon of *Loligo*. J. Physiol. 116: 449–472. **Figure 3.5** ARMSTRONG, C.
M. AND L. BINSTOCK (1965) Anomalous rectification in the squid giant axon injected with
tetraethylammonium chloride. J. Gen. Physiol.
48: 859–872. MOORE, J. W., M. P. BLAUSTEIN, N.
C. ANDERSON AND T. NARAHASHI (1967) Basis of
tetrodotoxin's selectivity in blockage of squid
axons. J. Gen. Physiol. 50: 1401–1410. **Figures
3.6 & 3.7** HODGKIN, A. L. AND A. F. HUXLEY
(1952b) The components of membrane conductance in the giant axon of *Loligo*. J. Physiol.
116: 473–496. **Figure 3.8** HODGKIN, A. L. AND
A. F. HUXLEY (1952d) A quantitative description of membrane current and its application
to conduction and excitation in nerve. J. Physiol. 116: 507–544. **Figure 3.10** HODGKIN, A. L.
AND W. A. RUSHTON (1938) The electrical constants of a crustacean nerve fibre. Proc. R. Soc.
Lond. B 133: 444–478.

Chapter 4 Channels and Transporters
Figure 4.1B,C BEZANILLA, F. AND A. M. CORREA (1995) Single-channel properties and gating of Na⁺ and K⁺ channels in the squid giant
axon. In *Cephalopod Neurobiology*, N. J. Abbott,
R. Williamson and L. Maddock (eds.). New
York: Oxford University Press, pp. 131–151.
Figure 4.1D VANDERBERG, C. A. AND F.
BEZANILLA (1991) A sodium channel model
based on single channel, macroscopic ionic,
and gating currents in the squid giant axon.
Biophys. J. 60: 1511–1533. **Figure 4.1E** CORREA, A. M. AND F. BEZANILLA (1994) Gating of
the squid sodium channel at positive potentials. II. Single channels reveal two open
states. Biophys. J. 66: 1864–1878. **Figure
4.2B–D** AUGUSTINE, C. K. AND F. BEZANILLA
(1990) Phosphorylation modulates potassium
conductance and gating current of perfused
giant axons of squid. J. Gen. Physiol. 95:
245–271. **Figure 4.2E** PEROZO, E., D. S. JONG
AND F. BEZANILLA (1991) Single-channel studies of the phosphorylation of K⁺ channels in
the squid giant axon. II. Nonstationary conditions. J. Gen. Physiol. 98: 19–34. **Figure 4.8**
DOYLE, D. A. AND 7 OTHERS (1998) The structure
of the potassium channel: Molecular basis of
K⁺ conduction and selectivity. Science 280:
69–77. **Figure 4.9A–C** LONG, S. B., E. B.
CAMPBELL AND R. MACKINNON (2005a) Crystal
structure of a mammalian voltage-dependent
Shaker family K⁺ channel. Science 309:
897–903. **Figure 4.9D** LONG, S. B., E. B.
CAMPBELL AND R. MACKINNON (2005b) Voltage
sensor of Kv1.2: Structural basis of electromechanical coupling. Science 309: 903–908.
Figure 4.9E LEE, A. G. (2006) Ion channels: A
paddle in oil. Nature 444: 697. **Figure 4.11**
HODGKIN, A. L. AND R. D. KEYNES (1955) Active
transport of cations in giant axons from *Sepia*
and *Loligo*. J. Physiol. 128: 28–60. LINGREL, J. B.,
J. VAN HUYSSE, W. O'BRIEN, E. JEWELL-MOTZ, R.
ASKEW AND P. SCHULTHEIS (1994) Structure-function studies of the Na, K-ATPase. Kidney
Internat. 45: S32–S38. **Figure 4.12** RANG, H.
P. AND J. M. RICHIE (1968) On the electrogenic
sodium pump in mammalian non-myelinated
nerve fibres and its activation by various external cations. J. Physiol. 196: 183–220. **Figure
4.13** LINGREL, J. B., J. VAN HUYSSE, W. O'BRIEN,
E. JEWELL-MOTZ, R. ASKEW AND P. SCHULTHEIS
(1994) Structure–function studies of the Na,
K-ATPase. Kidney Internat. 45: S32–S38. **Fig-

ure 4.14** TOYOSHIMA, C., H. NOMURA AND T.
TSUDA (2004) Luminal gating mechanism
revealed in calcium pump crystal structures
with phosphate analogues. Nature 432:
361–368.

Chapter 5 Synaptic Transmission
Figure 5.2B FURSHPAN, E. J. AND D. D. POTTER
(1959) Transmission at the giant motor
synapses of the crayfish. J. Physiol. (Lond.)
145: 289–324. **Figure 5.2C** BEIERLEIN, M., J. R.
GIBSON AND B. W. CONNORS (2000) A network
of electrically coupled interneurons drives
synchronized inhibition in neocortex. Nature
Neurosci. 3: 904–910. **Figure 5.4B,D** PETERS,
A., PALAY, S. L. AND H. WEBSTER (1991) *The Fine
Structure of the Nervous System: Neurons and
Their Supporting Cells*, 3rd Ed. Oxford University Press, New York. **Figure 5.6** FATT, P. AND
B. KATZ (1952) Spontaneous subthreshold
activity at motor nerve endings. J. Physiol.
(Lond.) 117: 109–127. **Figure 5.7** BOYD, I. A.
AND A. R. MARTIN (1955) Spontaneous subthreshold activity at mammalian neuromuscular junctions. J. Physiol. 132: 61–73. **Figure
5.8A,B** HEUSER, J. E., T. S. REESE, M. J. DENNIS,
Y. JAN, L. JAN AND L. EVANS (1979) Synaptic
vesicle exocytosis captured by quick freezing
and correlated with quantal transmitter
release. J. Cell Biol. 81: 275–300. **Figure 5.8C**
HARLOW, M. L., D. RESS, A. STOSCHEK, R. M.
MARSHALL AND U. J. MCMAHAN (2001) The
architecture of the active zone material at the
frog's neuromuscular junction. Nature 409:
479–484 **Figure 5.9** HEUSER, J. E. AND T. S.
REESE (1973) Evidence for recycling of synaptic
vesicle membrane during transmitter release
at the frog neuromuscular junction. J. Cell Biol.
57: 315–344. **Figure 5.10** AUGUSTINE, G. J.
AND R. ECKERT (1984) Divalent cations differentially support transmitter release at the
squid giant synapse. J. Physiol. 346: 257–271.
Figure 5.11A SMITH, S. J., J. BUCHANAN, L. R.
OSSES, M. P. CHARLTON AND G. J. AUGUSTINE
(1993) The spatial distribution of calcium signals in squid presynaptic terminals. J. Physiol.
(Lond.) 472: 573–593. **Figure 5.11B** MILEDI,
R. (1973) Transmitter release induced by injection of calcium ions into nerve terminals. Proc.
R. Soc. Lond. B 183: 421–424. **Figure 5.11C**
ADLER, E. M. ADLER, G. J. AUGUSTINE, M. P.
CHARLTON AND S. N. DUFFY (1991) Alien intracellular calcium chelators attenuate neuro-

transmitter release at the squid giant synapse. J. Neurosci. 11: 1496–1507. **Figure 5.13A** TAKAMORI, S. AND 21 OTHERS (2006) Molecular anatomy of a trafficking organelle. Cell 127: 831–846. **Figure 5.14A & Box C** SUTTON, R. B., D. FASSHAUER, R. JAHN AND A. T. BRÜNGER (1998) Crystal structure of a SNARE complex involved in synaptic exocytosis at 2.4 Å resolution. Nature 395: 347–353. **Figure 5.15A** MARSH, M. AND H. T. MCMAHON (1999) The structural era of endocytosis. Science 285: 215–219. **Figure 5.17** TAKEUCHI, A. AND N. TAKEUCHI (1960) On the permeability of endplate membrane during the action of transmitter. J. Physiol. 154: 52–67.

Chapter 6 Neurotransmitters and Their Receptors
Figure 6.3D TOYOSHIMA, C. AND N. UNWIN (1990) Three-dimensional structure of the acetylcholine receptor by cryoelectron microscopy and helical image reconstruction. J. Cell Biol. 111: 2623–2635. **Figure 6.9A** CHAVAS, J. AND A. MARTY (2003) Coexistence of excitatory and inhibitory GABA synapses in the cerebellar interneuron network. J. Neurosci. 23: 2019–2030. **Figure 6.16A,B** FREUND, T. F., I. KATONA AND D. PIOMELLI (2003) Role of endogenous cannabinoids in synaptic signaling. Physiol Rev. 83: 1017–1066. **Figure 6.16C** IVERSEN, L. (2003) Cannabis and the brain. Brain 126: 1252–1270. **Figure 6.17** OHNO-SHOSAKU, T., T. MAEJIMA AND M. KANO (2001) Endogenous cannabinoids mediate retrograde signals from depolarized postsynaptic neurons to presynaptic terminals. Neuron 29: 729–738.

Chapter 8 Synaptic Plasticity
Figure 8.1A,B CHARLTON, M. P. AND G. D. BITTNER (1978) Presynaptic potentials and facilitation of transmitter release in the squid giant synapse. J. Gen. Physiol. 72: 487–511. **Figure 8.1C** SWANDULLA, D., M. HANS, K. ZIPSER AND G. J. AUGUSTINE (1991) Role of residual calcium in synaptic depression and posttetanic potentiation: Fast and slow calcium signaling in nerve terminals. Neuron 7: 915–926. **Figure 8.1D** BETZ, W.J. (1970) Depression of transmitter release at the neuromuscular junction of the frog. J. Physiol. (Lond.) 206: 629–644. **Figure 8.1E** LEV-TOV, A., M. J. PINTER AND R. E. BURKE (1983) Posttetanic potentiation of group Ia EPSPs: Possible mechanisms for differential distribution among medial gastrocnemius motoneurons. J. Neurophysiol. 50: 379–398. **Figure 8.2A** KATZ, B. (1966) *Nerve, Muscle, and Synapse*. New York: McGraw-Hill. **Figure 8.2B** MALENKA, R. C. AND S. A. SIEGELBAUM (2001) Synaptic plasticity: Diverse targets and mechanisms for regulating synaptic efficacy. In *Synapses*. W. M. Cowan, T. C. Sudhof and C. F. Stevens (eds.). Baltimore: John Hopkins University Press, pp. 393–413. **Figures 8.3, 8.4 & 8.5** SQUIRE, L. R. AND E. R. KANDEL (1999) *Memory: From Mind to Molecules*. New York: Scientific American Library. **Figure 8.7A–C** MALINOW, R., H. SCHULMAN, AND R. W. TSIEN (1989) Inhibition of postsynaptic PKC or CaMKII blocks induction but

not expression of LTP. Science 245: 862–866. **Figure 8.7D** ABRAHAM, W. C., B. LOGAN, J. M. GREENWOOD AND M. DRAGUNOW (2002) Induction and experience-dependent consolidation of stable long-term potentiation lasting months in the hippocampus. J. Neurosci. 22: 9626–9634. **Figure 8.8** GUSTAFSSON, B., H. WIGSTROM, W.C. ABRAHAM, AND Y.Y. HUANG (1987) Long-term potentiation in the hippocampus using depolarizing current pulses as the conditioning stimulus to single volley synaptic potentials. J. Neurosci. 7: 774–780. **Figure 8.10** NICOLL, R. A., J. A. KAUER AND R. C. MALENKA (1988) The current excitement in long-term potentiation. Neuron. 1: 97–103. **Figure 8.12A,B** MATSUZAKI, M., N. HONKURA, G.C. ELLIS-DAVIES AND H. KASAI (2004) Structural basis of long-term potentiation in single dendritic spines. Nature 429: 761–766. **Figure 8.12C** LIAO, D., N. A. HESSLER AND R. MALINOW (1995) Activation of postsynaptically silent synapses during pairing-induced LTP in CA1 region of hippocampal slice. Nature 375: 400–404. **Figure 8.13** FREY, U. AND R. G. MORRIS (1997) Synaptic tagging and long-term potentiation. Nature 385: 533–536. **Figure 8.14A & 8.15** SQUIRE, L. R. AND E. R. KANDEL (1999) *Memory: From Mind to Molecules*. New York: Scientific American Library. **Figure 8.14B** ENGERT, F. AND T. BONHOEFFER (1999) Dendritic spine changes associated with hippocampal long-term synaptic plasticity. Nature 399: 66–70. **Figure 8.16B** SAKURAI, M. (1987) Synaptic modification of parallel fibre-Purkinje cell transmission in *in vitro* guinea-pig cerebellar slices. J. Physiol. (Lond) 394: 463–480. **Figure 8.17** BI, G. Q. AND M. M. POO (1998) Synaptic modifications in cultured hippocampal neurons: dependence on spike timing, synaptic strength, and postsynaptic cell type. J. Neurosci. 18: 10464–10472.

Chapter 9 The Somatic Sensory System
Box 9A and Table 9.1 ROSENZWEIG, M. R., S. M. BREEDLOVE AND A. L. LEIMAN (2002) *Biological Psychology*, 3rd ed. Sunderland, MA: Sinauer Associates. **Figure 9.3C** WEINSTEIN, S. (1968) Neuropsychological studies of the phantom. In *Contributions to Clinical Neuropsychology*, A. L. Benton (ed.). Chicago: Aldine Publishing Company, pp. 73–106. **Table 9.2** JOHNSON, K. O. (2002) Neural basis of haptic perception. In *Seven's Handbook of Experimental Psychology*, 3rd ed. Vol 1: *Sensation and Perception*. H. Pashler and S. Yantis (eds.). New York: Wiley, pp. 537–583. **Figure 9.5** JOHANSSON, R. S. AND A. B. VALLBO (1983) Tactile sensory coding in the glabrous skin of the human. Trends Neurosci. 6: 27–32. **Figure 9.6** PHILLIPS, J. R., R. S. JOHANSSON AND K. O. JOHNSON (1990) Representation of Braille characters in human nerve fibres. Exp. Brain Res. 81: 589–592. **Figure 9.7A** MATTHEWS, P. B. C. (1964) Muscle spindles and their motor control. Physiol. Rev. 44: 219–289. **Figure 9.10** BRODAL, P. (1992) *The Central Nervous System: Structure and Function*. New York: Oxford University Press, p. 151. JONES, E. G. AND D. P. FRIEDMAN (1982) Projection pattern

of functional components of thalamic ventrobasal complex on monkey somatosensory cortex. J. Neurophys. 48: 521–544. **Figure 9.11** PENFIELD, W. AND T. RASMUSSEN (1950) *The Cerebral Cortex of Man: A Clinical Study of Localization of Function*. New York: Macmillan. CORSI, P. (1991) *The Enchanted Loom: Chapters in the History of Neuroscience*, P. Corsi (ed.). New York: Oxford University Press. **Figure 9.13A** KAAS, J. H. (1989) The functional organization of somatosensory cortex in primates. Ann. Anat. 175: 509–517. **Figure 9.13C** SUR, M. (1980) Receptive fields of neurons in areas 3b and 1 of somatosensory cortex in monkeys. *Brain Res.* 198: 465–471. **Figure 9.14** MERZENICH, M. M., R. J. NELSON, M. P. STRYKER, M. S. CYNADER, A. SCHOPPMANN AND J. M. ZOOK (1984) Somatosensory cortical map changes following digit amputation in adult monkeys. J. Comp. Neurol. 224: 591–605.

Chapter 10 Pain
Figure 10.1 FIELDS, H. L. (1987) *Pain*. New York: McGraw-Hill. **Figure 10.2** FIELDS, H. L. (ed.) (1990) *Pain Syndromes in Neurology*. London: Butterworths. **Box 10C Figure B** WILLIS, W. D., E. D. AL-CHAER, M. J. QUAST AND K. N. WESTLUND (1999) A visceral pain pathway in the dorsal column of the spinal cord. Proc. Natl. Acad. Sci. USA 96: 7675–76710. **Box 10C Figure C** HIRSHBERG, R. M., E. D. AL-CHAER, N. B. LAWAND, K. N. WESTLUND AND W. D. WILLIS (1996). Is there a pathway in the dorsal funiculus that signals visceral pain? Pain 67: 291–305; NAUTA, H. J. W., E. HEWITT, K. N. WESTLUND AND W. D. WILLIS (1997) Surgical interruption of a midline dorsal column visceral pain pathway. J. Neurosurg. 86: 538–542. **Box 10D** SOLONEN, K. A. (1962) The phantom phenomenon in amputated Finnish war veterans. Acta. Orthop. Scand. Suppl. 54: 1–37.

Chapter 11 Vision: The Eye
Box 11A Figure D WESTHEIMER, G. (1974) In *Medical Physiology*, 13th Ed. V. B. Mountcastle (ed.) St. Louis: Mosby. **Figure 11.4A–C** HILFER, S. R. AND J. J. W. YANG (1980) Accumulation of CPC-precipitable material at apical cell surfaces during formation of the optic cup. Anat. Rec. 197: 423–433. **Figure 11.6A** OYSTER, C. W. (1999) *The Human Eye*. Sunderland, MA: Sinauer Assoc. **Figure 11.6B,C** YOUNG, R. W. (1971) Shedding of discs from rod outer segments in the rhesus monkey. J. Ultrastruc. Res. 34: 190–203. **Figure 11.7** SCHNAPF, J. L. AND D. A. BAYLOR (1987) How photoreceptors respond to light. Sci. Am. 256: 40–47. **Figure 11.12** BAYLOR, D. A. (1987) Photoreceptor signals and vision. Invest. Ophthalmol. Vis. Sci. 28: 34–49. **Figure 11.14B** HOFER, H., J. CARROLL, J. NEITZ, M. NEITZ AND D. R. WILLIAMS (2005) Organization of the human trichromatic cone mosaic. J. Neurosci. 25: 9669–9679. **Boxes 11E and 11F** PURVES, D. AND R. B. LOTTO (2003) *Why We See What We Do*. Sunderland, MA: Sinauer Associates. **Figure 11.16A** NATHANS, J. (1987) Molecular biology of visual pigments. *Annu. Rev. Neurosci.* 10:

163–194. **Figure 11.16B** DEEB, S. S. (2005) The molecular basis of variation in human color vision. Clin. Genet. 67: 369–377. **Figure 11.17** SAKMANN, B. AND O. D. CREUTZFELDT (1969) Scotopic and mesopic light adaptation in the cat's retina. Pflügers Arch. 313: 168–185.

Chapter 12 Central Visual Pathways
Figure 12.12B OHKI, K., S. GHUNG, P. KARA, M. HUBENER, T. BONHOEFFER AND R. C. REID (2006) Highly ordered arrangement of single neurons in orientation pinwheels. Nature 442: 925–928. **Figure 12.13D** HORTON AND E. T. HEDLEY-WHYTE (1984) Mapping of cytochrome oxidase patches and ocular dominance columns in human visual cortex. Philos. Trans. 304: 255–172. **Box 12A Figure A** WANDELL, B. A. (1995) Foundations of Vision. Sunderland, MA: Sinauer Associates. **Box C Figure B** BONHOEFFER, T. AND A. GRINVALD (1993) The layout of iso-orientation domains in area 18 of the cat visual cortex: Optical imaging reveals a pinwheel-like organization. J. Neurosci 13: 4157–4180. **Figure 12 15** WATANABE, M. AND R. W. RODIECK (1989) Parasol and midget ganglion cells of the primate retina. J. Comp. Neurol. 289: 434–454. **Figure 12.16A** MAUNSELL, J. H. R. AND W.T. NEWSOME (1987) Visual processing in monkey extrastriate cortex. Annu. Rev. Neurosci. 10: 363–401. **Figure 12.16B** FELLEMAN, D. J. AND D. C. VAN ESSEN (1991) Distributed hierarchical processing in primate cerebral cortex. Cereb. Cortex 1: 1–47. **Figure 12.17** SERENO, M. I. AND 7 OTHERS (1995) Borders of multiple visual areas in humans revealed by functional magnetic resonance imaging. Science 268: 889–893.

Chapter 13 The Auditory System
Figure 13.4 (inset) KESSEL, R. G. AND R. H. KARDON (1979) Tissue and Organs: A Text-Atlas of Scanning Electron Microscopy. San Francisco: W.H. Freeman **Figure 13.5** DALLOS, P. (1992) The active cochlea. J. Neurosci. 12: 4575–4585. VON BÉKÉSY, G. (1960) Experiments in Hearing. New York: McGraw-Hill. **Figure 13.7A** LINDEMAN, H. H. (1973) Anatomy of the otolith organs. Adv. Otorhinolaryngol. 20: 405–433. **Figure 13.7B** HUDSPETH, A. J. (1983) The hair cells of the inner ear. Sci. Amer. 248: 54–64. **Figure 13.7C** PICKLES, J. O. (1988) An Introduction to the Physiology of Hearing. London: Academic Press. **Figure 13.7D** FAIN, G. L. (2003) Sensory Transduction. Sunderland, MA: Sinauer Associates. **Figure 13.8** LEWIS, R. S. AND A. J. HUDSPETH (1983) Voltage- and ion-dependent conductances in solitary vertebrate hair cells. Nature 304: 538–541. **Figure 13.9A** SHOTWELL, S. L., R. JACOBS, AND A. J. HUDSPETH (1981) Directional sensitivity of individual vertebrate hair cells to controlled deflection of their hair bundles. Ann. NY Acad. Sci. 374: 1–10. **Figure 13.9B** HUDSPETH, A. J. AND D. P. COREY (1977) Sensitivity, polarity and conductance change in the response of vertebrate hair cells to controlled mechanical stimuli. Proc. Natl. Acad. Sci. USA 74: 2407–2411. **Figure 13.9C** PALMER, A. R. AND I. J. RUSSELL

(1986) Phase-locking in the cochlear nerve of the guinea-pig and its relation to the receptor potential of inner hair cells. Hear. Res. 24: 1–14. **Figure 13.11A** KIANG, N.Y. AND E. C. MOXON (1972) Physiological considerations in artificial stimulation of the inner ear. Ann. Otol. Rhinol. Laryngol. 81: 714–729. **Figure 13.11C** KIANG, N.Y. S. (1984) Peripheral neural processing of auditory information. In Handbook of Physiology: A Critical, Comprehensive Presentation of Physiological Knowledge and Concepts, Section 1: The Nervous System, Vol. III. Sensory Processes, Part 2, J. M. Brookhart, V. B. Mountcastle, I. Darian-Smith and S. R. Geiger (eds.). Bethesda, MD: American Physiological Society, pp. 639–674. **Figure 13.13** JEFFRESS, L. A. (1948) A place theory of sound localization. J. Comp. Physiol. Psychol. 41: 35–38.

Chapter 14 The Vestibular System
Figure 14.3 LINDEMAN, H. H. (1973) Anatomy of the otolith organs. Adv. Otorhinolaryngol. 20: 405–433. **Figure 14.6** GOLDBERG, J. M. AND C. FERNÁNDEZ (1976) Physiology of peripheral neurons innervating otolith organs of the squirrel monkey, Parts 1, 2, 3. J. Neurophys. 39: 970–1008. **Figure 14.9** GOLDBERG, J. M. AND C. FERNÁNDEZ (1971) Physiology of peripheral neurons innervating semicircular canals of the squirrel monkey, Parts 1, 2, 3. J. Neurophys. 34: 635–684.

Chapter 15 The Chemical Senses
Figure 15.1E & 15.2D ROLLS, E.T., M. L. KRINGELBACH, AND I. E.T. DE ARAUJO, (2003) Different representations of pleasant and unpleasant odours in the human brain. Eur. J. Neurosci. 18: 695–703. **Figure 15.2A** SHIER, D., J. BUTLER, AND R. LEWIS (2004) Hole's Human Anatomy and Physiology. Boston: McGraw-Hill. **Figure 15.2C** PELOSI, P. (1994) Odorant-binding proteins. Crit. Rev. Biochem. Mol. Biol. 29: 199–227. **Figure 15.3** CAIN, W. S. AND J. F. GENT (1986) Use of odor identification in clinical testing of olfaction. In Clinical Measurement of Taste and Smell, H. L. Meiselman and R. S. Rivlin (eds.). New York: Macmillan, pp. 170–186. **Figure 15.4A** MURPHY, C. (1986) Taste and smell in the elderly. In Clinical Measurement of Taste and Smell, H. L. Meiselman and R. S. Rivlin (eds.). New York: Macmillan, pp. 343–371. **Figure 15.4B** WANG, J., P. ESLINGER, M. B. SMITH, AND Q. X.YANG (2005) Functional magnetic resonance imaging study of human olfaction and normal aging. J. Gerontol. Med. Sci. 60A: 510–514. **15.5** SAVIC, I., H. BERGLUND, B. GULYAS AND P. ROLAND (2001) Smelling of odorous sex hormone-like compounds causes sex-differentiated hypothalamic activations in humans. Neuron 31: 661–668. **Figure 15.6A** ANHOLT, R. R. H. (1987) Primary events in olfactory reception. Trends Biochem. Sci. 12: 58–62. **Figure 15.6B** FIRESTEIN, S., F. ZUFALL AND G. M. SHEPHERD (1991) Single odor-sensitive channels in olfactory receptor neurons are also gated by cyclic nucleotides. J. Neurosci. 11: 3565–3572. **Figures 15.7A & 15.9A** MENINI, A. (1999) Calcium signalling and regulation in olfactory

neurons. Curr. Opin. Neurobiol. 9: 419–425. **Figure 15.7B** DRYER, L. (2000) Evolution of odorant receptors. BioEssays 22: 803–809. **Figure 15.8B–D** BOZZA, T., P. FEINSTEIN, C. ZHENG AND P. MOMBAERTS (2002) Odorant receptor expression defines functional units in the mouse olfactory system. J. Neurosci. 22: 3033–3043. **Figure 15.9B** BELLUSCIO, L., G. H. GOLD, A. NEMES, AND R. AXEL (1998) Mice deficient in Golf are anosmic. Neuron 20: 69–81; WONG, S. T. AND 8 OTHERS (2000) Disruption of the type III adenylyl cyclase gene leads to peripheral and behavioral anosomia in transgenic mice. Neuron 27: 487–497; Brunet, L., G. H. Gold and J. Ngai (1996) General anosmia caused by a targeted disruption of the mouse olfactory cyclic nucleotide–gated cation channel. Neuron 17: 681–693. **Figure 15.10** FIRESTEIN, P. (1992) Physiology of transduction in the single olfactory sensory neuron. In Sensory Transduction, D. P. Corey and S. D. Roper (eds.). New York: Rockefeller University Press, pp. 61–71. **Figure 15.11** BOZZA, T., P. FIRESTEIN, C. ZHENG AND P. ,MOMBAERTS (2002) Odorant receptor expression defines functional units in the mouse olfactory system. J. Neurosci. 22: 3033–3043. **Figure 15.12** GETCHILL, M. L. (1986) In Neurobiology of Taste and Smell, T. E. Finger and W. L. Silver (eds). New York: John Wiley and Sons, p. 112. **Figure 15.13A** LAMANTIA, A.-S., S. L. POMEROY AND D. PURVES (1992) Vital imaging of glomeruli in the mouse olfactory bulb. J. Neurosci. 12: 976–988. **Figure 15.13B,C** POMEROY, S. L., A.-S. LAMANTIA AND D. PURVES (1990) Postnatal construction of neural activity in the mouse olfactory bulb. J. Neurosci. 10: 1952–1966. **Figure 15.13E** MOMBAERTS, P. AND 7 OTHERS (1996) Visualizing an olfactory sensory map. Cell 87: 675–686. **Figure 15.14A** WANG, J. W., A. M. WONG, J. FLORES, L. B. VOSSHALL AND R. AXEL (2003) Two-photon calcium imaging reveals an odor-evoked map of activity in the fly brain. Cell 112: 271–282. **Figure 15.14B** BELLUSCIO, L. AND L. C. KATZ (2001) Symmetry, stereotypy, and topography of odorant representations in mouse olfactory bulbs. J. Neurosci. 21: 2113–2122. **Figure 15.15C** SCHOENFELD, M. A. AND 6 OTHERS (2004) Functional magnetic resonance tomography correlates of taste perception in human primary taste cortex. Neuroscience 127: 347–353. **Figure 15.17A** ROSS, M. H., L. J. ROMMELL AND G. I. KAYE (1995) Histology, A Text and Atlas. Baltimore: Williams and Wilkins. **Figure 15.19** ZHANG, Y. AND 7 OTHERS (2003) Coding of sweet, bitter, and umami tastes: Different receptor cells sharing similar signaling pathways. Cell 112: 293–301.

Chapter 16 Lower Motor Circuits and Motor Control
Figure 16.2 BURKE, R. E., P. L. STRICK, K. KANDA, C. C. KIM AND B. WALMSLEY (1977) Anatomy of medial gastrocnemius and soleus motor nuclei in cat spinal cord. J. Neurophys. 40: 667–680. **Figure 16.6** BURKE, R. E., D. N. LEVINE, M. SALCMAN AND P. TSAIRIS (1974) Motorunits in cat soleus muscle: Physiological,

histochemical and morphological characteristics. J. Physiol. (Lond.) 238: 503–514. **Figure 16.7** WALMSLEY, B., J. A. HODGSON AND R. E. BURKE (1978) Forces produced by medical gastrocnemius and soleus muscles during locomotion in freely moving cats. J. Neurophys. 41: 1203–1216. **Figure 16.9** MONSTER, A. W. AND H. CHAN (1977) Isometric force production by motor units of extensor digitorum communis muscle in man. J. Neurophys. 40: 1432–1443. **Figure 16.11** HUNT, C. C. AND S. W. KUFFLER (1951) Stretch receptor discharges during muscle contraction. J. Physiol. (Lond.) 113: 298–314. **Figure 16.12B** PATTON, H. D. (1965) Reflex regulation of movement and posture. In *Physiology and Biophysics*, 19th Ed., T. C. Ruch and H. D. Patton (eds.). Philadelphia: Saunders, pp. 181–206. **Figure 16.15** PEARSON, K. (1976) The control of walking. Sci. Amer. 235: 72–86.

Chapter 17 Upper Motor Neuron Control of the Brainstem and Spinal Cord
Figure 17.11 PORTER, R. AND R. LEMON (1993) *Corticospinal Function and Voluntary Movement.* Oxford: Oxford University Press. **Figure 17.12** GRAZIANO, M. S. A., T. N. S. AFLALO AND D. F. COOKE (2005) Arm movements evoked by electrical stimulation in the motor cortex of monkeys. J. Neurophysiol. 94: 4209–4223. **Figure 17.13** GEORGEOPOULOS, A. P., A. B. SWARTZ AND R. E. KETTER (1986) Neuronal population coding of movement direction. Science 233: 1416–1419. **Figure 17.14** GEYER, S., M. MATELLI AND G. LUPPINO (2000) Functional neuroanatomy of the primate isocortical motor system. *Anat. Embryol.* 202: 443–474. **Figure 17.15** RIZZOLATTI, G., L. FADIGA, V. GALLESE AND L. FOGASSI (1996) Premotor cortex and the recognition of motor actions. *Cogn. Brain Res.* 3: 131–141.

Chapter 18 Modulation of Movement by the Basal Ganglia
Figure 18.7 HIKOSAKA, O. AND R. H. WURTZ (1989) The basal ganglia. In *The Neurobiology of Eye Movements*, R. H. Wurtz and M. E. Goldberg (eds.). New York: Elsevier Science Publishers, pp. 257–281. **Figure 18.10** BRADLEY, W. G., R. B. DAROFF, G. M. FENICHEL AND C. D. MARSDEN (EDS.) (1991) *Neurology in Clinical Practice.* Boston: Butterworth-Heinemann. **Figure 18.11** DELONG, M. R. (1990) Primate models of movement disorders of basal ganglia origin. Trends Neurosci. 13: 281–285. **Box 18C, Figure A** Adapted from art by Lydia Kibiuk, in SFN Brain Briefings July 2004.

Chapter 19 Modulation of Movement by the Cerebellum
Figure 19.11A STEIN, J. F. (1986) Role of the cerebellum in the visual guidance of movement. Nature 323: 217–220. **Figure 19.12** THACH, W. T. (1968) Discharge of Purkinje and cerebellar nuclear neurons during rapidly alternating arm movements in the monkey. J. Neurophys. 31: 785–797. **Figure 19.13** OPTICAN, L. M. AND D. A. ROBINSON (1980) Cerebellar-dependent adaptive control of primate sac-

cadic system. J Neurophys. 44: 1058–1076. **Figure 19.15** VICTOR, M., R. D. ADAMS AND E. L. MANCALL (1959) A restricted form of cerebellar cortical degeneration occurring in alcoholic patients. Arch. Neurol. 1: 579–688. **Box 19B** RAKIC, P. (1977) Genesis of the dorsal lateral geniculate nucleus in the rhesus monkey: Site and time of origin, kinetics of proliferation, routes of migration and pattern of distribution of neurons. J. Comp. Neuro. 176: 23–52.

Chapter 20 Eye Movements and Sensory Motor Integration
Figure 20.1 YARBUS, A. L. (1967) *Eye Movements and Vision.* Basil Haigh, trans. New York: Plenum Press. **Box 20A** PRITCHARD, R. M. (1961) Stabilized images on the retina. Sci. Amer. 204 (June): 72–78. **Figures 20.4 & 20.5** FUCHS, A. F. (1967) Saccadic and smooth pursuit eye movements in the monkey. J. Physiol. (Lond.) 191: 609–630. **Figure 20.6** BAARSMA, E. AND H. COLLEWIJN (1974) Vestibulo-ocular and optokinetic reactions to rotation and their interaction in the rabbit. *J. Physiol.* 238: 603-625. **Figure 20.7** FUCHS, A. F. AND E. S. LUSCHEI (1970) Firing patterns of abducens neurons of alert monkeys in relationship to horizontal eye movements. J. Neurophys. 33: 382–392. **Figure 20.9** SCHILLER, P. H. AND M. STRYKER (1972) Single unit recording and stimulation in superior colliculus of the alert rhesus monkey. J. Neurophys. 35: 915–923. **Figure 20.10** SPARKS, D. L. AND L. E. MAYS (1983) Spatial localization of saccade targets. I. Compensation for stimulation-induced perturbations in eye position. *J. Neurophysiol.* 49: 45–63. **Figure 20.12** SCHALL, J. D. (1995) Neural basis of target selection. Reviews in the Neurosciences 6: 63–85. **Figure 20.13** KRAUZLIS, R. J. (2005) The control of voluntary eye movements: New perspectives. *Neuroscientist* 11: 124–137.

Chapter 21 The Visceral Motor System
Box 21C Figure A YASWEN, L., N. DIEHL, M. B. BRENNAN AND U. HOCHGESCHWENDER (1999) Obesity in the mouse model of pro-opiomelanocortin deficiency responds to peripheral melanocortin. Nature Medicine 5: 1066–1070. **Box 21C Figure B** O'RAHILLY, S., S. FAROOQI, G. S. H. YEO AND B. G. CHALLIS (2003) Human obesity: Lessons from monogenic disorders. Endocrinology 144: 3757–3764.

Chapter 22 Early Brain Development
Figure 22.2 SANES, J. R. (1989) Extracellular matrix molecules that influence neural development. Annu. Rev. Neurosci. 12: 491–516. **Box 22B Figure A** ANCHAN, R. M., D. P. DRAKE, C. F. HAINES, E. A. GERWE AND A.-S. LAMANTIA (1997) Disruption of local retinoid-mediated gene expression accompanies abnormal development in the mammalian olfactory pathway. J. Comp. Neurol. 379: 171–184. **Box 22B Figure B** LINNEY, E. AND A.-S. LAMANTIA (1994) Retinoid signaling in mouse embryos. Adv. Dev. Biol. 3: 73–114. **Figure 22.6A** GILBERT, S. F. (1994) *Develop-

mental Biology*, 4th Ed. Sunderland, MA: Sinauer Associates. **Figure 22.6B** INGHAM, P. (1988) The molecular genetics of embryonic pattern formation in *Drosophila*. Nature 335: 25–34. **Figure 22.6C** VERAKSA, A. AND W. MCGINNIS (2000). Developmental patterning genes and their conserved functions: From model organisms to humans. Molec. Genet. Metab. 69: 85-100. **Figure 22.8 & 22.12** RAKIC, P. (1974) Neurons in rhesus monkey visual cortex: Systematic relation between time of origin and eventual disposition. Science 183: 425–427. **Figure 22.9** KINTNER C. (2002) Neurogenesis in embryos and in adult neural stem cells. J. Neurosci. 22: 639–643. **Figure 22.10B,C** RUBIN, G. M. (1989) Development of the *Drosophila* retina: Inductive events studied at single-cell resolution. Cell 57: 519–520.

Chapter 23 Construction of Neural Circuits
Figure 23.1A TAKAHASHI, M., M. NARUSHIMA AND Y. ODA (2002) In vivo imaging of functional inhibitory networks on the Mauthner cell of larval zebrafish. J. Neurosci. 22: 3929–3938. **Figure 23.2B** DENT, E. W. AND K. KALIL (2001) Axon branching requires interactions between dynamic microtubules and actin filaments. J. Neurosci 21: 9757–9769. **Figure 23.2D & 23.4B** GOMEZ, T. M. AND J. O. ZHENG (2006) The molecular basis for calcium-dependent axon pathfinding. Nature Rev. Neurosci. 7: 115–117. **Figure 23.4A** HUBER, A. B., A. L. KOLODKIN, D. D. GINTY AND J. F. CLOUTIER (2003) Signaling at the growth cone: Ligand-receptor complexes and the control of axon growth and guidance. Annu. Rev. Neurosci. 26: 509–563. **Figure 23.4C** DONTCHEV, V. D. AND P. C. LETOURNEAU (2002) Nerve growth factor and semaphorin 3A signaling pathways interact in regulating sensory neuronal growth cone motility. *J. Neurosci.* 22: 6659–6669. **Figure 23.5A** SERAFINI, T., T. E. KENNEDY, M. J. GALKO, C. MIRZAYAN, T. M. JESSELL, M. TESSIER-LAVIGNE (1994) The netrins define a family of axon outgrowth-promoting proteins homologous to *C. elegans* UNC-6. Cell 78: 409–423. **Figure 23.5B** DICKINSON, B. J. (2001) Moving on. Science 291: 1910-1911. **Figure 23.5C** SERAFINI, T. AND 6 OTHERS (1996) Netrin-1 is required for commissural axon guidance in the developing vertebrate nervous system. Cell 87: 1001–1014. **Figure 23.6A,B** SPERRY, R. W. (1963) Chemoaffinity in the orderly growth of nerve fiber patterns and connections. Proc. Natl. Acad. Sci. USA 50: 703–710. **Figure 23.6C** WALTER, J., S. HENKE-FAHLE AND F. BONHOEFFER (1987) Avoidance of posterior tectal membranes by temporal retinal axons. Development 101: 909–913. **Figure 23.6D** WILKINSON, D. G. (2001) Multiple roles of EPH receptors and ephrins in neural development. Nat. Rev. Neurosci. 2: 155–164. **Figure 23.7A,B** WAITES, C. L., A. M. CRAIG AND C. C. GARNER (2005) Mechanisms of vertebrates synaptogenesis. Annu. Rev. Neurosci. 28: 251-274. **Figure 23.7C** DEAN, C. AND T. DRESBACH (2006)

Neuroligins and neurexins: Linking cell adhesion, synapse formation, and cognitive function. Trends Neurosci. 29: 21–29. **Figure 23.8A** SCHMUCKER, D. AND 7 OTHERS (2000) *Drosophila* Dscam is an axon guidance receptor exhibiting extraordinary molecular diversity. Cell 101: 671–684. **Figure 23.8C** PHILLIPS, G. R. AND 6 OTHERS (2003) Gamma-protocadherins are targeted to subsets of synapses and intracellular organelles in neurons. J. Neurosci. 23: 5096–5104. **Figure 23.9** HOLLYDAY, M. AND V. HAMBURGER (1976) Reduction of the naturally occurring motor neuron loss by enlargement of the periphery. J. Comp. Neurol. 170: 311–320; HOLLYDAY, M. AND V. HAMBURGER (1958) Regression versus peripheral controls of differentiation in motor hypoplasia. Amer. J. Anat. 102: 365–409; HAMBURGER, V. (1977) The developmental history of the motor neuron. The F. O. Schmitt Lecture in Neuroscience, 1970, Neurosci. Res. Prog. Bull. 15, Suppl. III: 1–37. **Figure 23.10** PURVES, D. AND J. W. LICHTMAN (1980) Elimination of synapses in the developing nervous system. Science 210: 153–157. **Figure 23.12A,B** PURVES, D. AND J. W. LICHTMAN (1985) *Principles of Neural Development.* Sunderland, MA: Sinauer Associates. **Figure 23.12C** CHUN, L. L. AND P. H. PATTERSON (1977) Role of nerve growth factor in the development of rat sympathetic neurons in vitro. III; Effect on acetylcholine production. J. Cell Biol. 75: 712–718. **Figure 23.12D** LEVI-MONTALCINI, R. (1972) The morphological effects of immunosympathectomy. In *Immonosympathectomy*, G. Steiner and E. Schönbaum (eds.). Amsterdam: Elsevier. **Figure 23.13A** MAISONPIERRE, P. C. AND 6 OTHERS (1990) Neurotrophin-3: A neurotrophic factor related to NGF and BDNF. Science 247: 1446–1451. **Figure 23.13B** BIBEL, M. AND Y.-A. BARDE (2000) Neurotrophins: Key regulators of cell fate and cell shape in the vertebrate nervous system. Genes Dev. 14: 2919–2937. **Figure 23.14A** CAMPENOT, R. B. (1981) Regeneration of neurites on long-term cultures of sympatic neurons deprived of nerve growth factor. Science 214: 579–581. **Figure 23.14B** LI, Y. AND 6 OTHERS (2005) Essential role of TRPC channels in the guidance of nerve grwoth cones by brain-derived neurotrophic factor. Nature 434: 894–898.

Chapter 24 Modification of Brain Circuits as a Result of Experience
Figure 24.1 PETTITO, L. A. AND P. F. MARENTETTE (1991) Babbling in the manual mode: Evidence for the ontogeny of language. Science 251: 1493–1496. **Table 24.1** HENSCH, T. K. (2004) Critical period regulation. Annu. Rev. Neurosci. 27: 549–579. **Figure 24.2A** SCHLAGGAR, B. L., T. T. BROWN, H. M. LUGAR, K. M. VISSCHER, F. M. MIEZIN AND S. E. PETERSEN (2002) Functional neuroanatomical differences between adults and school-age children in the processing of single words. Science 296: 1476–1479. **Figure 24.2B** JOHNSON, J. S. AND E. I. NEWPORT (1989) Critical period effects in

second language learning: the influences of maturational state on the acquisition of English as a second language. Cog. Psychol. 21. **Figure 24.3** LEVAY, S., T. N. WIESEL AND D. H. HUBEL (1980) The development of ocular dominance columns in Sillnormal and visually deprived monkeys. J. Comp. Neurol. 191: 1–51. **Figure 24.4A** HUBEL, D. H. AND T. N. WIESEL (1962) Receptive fields, binocular interaction and functional architecture in the cat's visual cortex. J. Physiol. 160: 106–154. **Figure 24.4B** HUBEL, D. H. AND T. N. WIESEL (1963) Receptive fields of cells in striate cortex of very young, visually inexperienced kittens. J. Neurophys. 26: 994–1003. **Figure 24.4C & 24.5** HUBEL, D. H. AND T. N. WIESEL (1970) The period of susceptibility to the physiological effects of unilateral eye closure in kittens. J. Physiol. 206: 419–436. **Figure 24.6A** HORTON, J. C. AND D. R. HOCKING (1999) An adult-like pattern of ocular dominance columns in striate cortex of newborn monkeys prior to visual experience. J. Neurosci. 16: 1791–1807. **Figure 24.6B** HUBEL, D. H., T. N. WIESEL AND S. LEVAY (1977) Plasticity of ocular dominance columns in monkey striate cortex. Phil. Trans. R. Soc. Lond. B. 278: 377–409. **Figure 24.7** ANTONINI, A. AND M. P. STRYKER (1993) Rapid remodeling of axonal arbors in the visual cortex. Science 260: 1819–1821. **Figure 24.9** HUBEL, D. H. AND T. N. WIESEL (1965) Binocular interaction in striate cortex of kittens reared with artificial squint. J. Neurophysiol. 28: 1041–1059. **Figure 24.10** WONG, R. O. AND A. GHOSH (2002) Activity-dependent regulation of dendritic growth and patterning. Nature Rev. Neurosci. 3: 803–812.

Chapter 25 Repair and Regeneration in the Nervous System
Figures 25.1 CASE, L. C. AND M. TESSIER-LAVIGNE (2005) Regeneration of the adult central nervous system. Curr. Biol. 15(18): R749–R753. **Figure 25.4B,C** PAN, Y. A., T. MISGELD, J. W. LICHTMAN AND J. R. SANES (2003) Effects of neurotoxic and neuroprotective agents on peripheral nerve regeneration assayed by time-lapse imaging in vivo J. Neurosci. 23: 11479–11488. **Figure 25.6A** SO, K. F. AND A. J. AGUAYO (1985) Lengthy regrowth of cut axons from ganglion cells after peripheral nerve transplantation into the retina of adult rats. Brain Res. 359: 402–406. **Figure 25.6B** BRAY, G. M., M. P. VILLEGAS-PEREZ, M. VIDAL-SANZ AND A. J. AGUAYO (1987) The use of peripheral nerve grafts to enhance neuronal survival, promote growth and permit terminal reconnections in the central nervous system of adult rats. J. Exp. Biol. 132: 5–19. **Figure 25.7B** PITTS, E. V., S. POTLURI, D. M. HESS AND R. J. BALICE-GORDON (2006) Neurotrophin and Trk-mediated signaling in the neuromuscular system. Int. Anesthes. Clin. 44: 21–76. **Figure 25.7C** NGUYEN, Q. T., J. R. SANES AND J. W. LICHTMAN (2002) Pre-existing pathways promote precise projection patterns. Nature Neurosci. 5: 861–867. **Figure 25.8A** BACK, S. A. AND 7 OTHERS (2002) Selective vulnerability of

late oligodendrocyte progenitors to hypoxia-ischemia. J. Neurosci. 22: 455–463. **Figure 25.9A** McGRAW, J., G. W. HIEBERT AND J. D. STEVENS (2001) Modulating astrogiosis after neurotrauma. J. Neurosci. Res. 63: 109–115. **Figure 25.9B** TAN, A. M., W. ZHANG AND J. M. LEVINE (2005) NG2: A component of the glial scar that inhibits axon growth. J. Anat. 207: 717–725. **Figure 25.9C** LADEBY, R. AND 6 OTHERS (2005) Microglial cell population dynamics in the injured adult CNS. Brain Res. Rev. 48: 196–206. **Figure 25.11A** OTTESON, D. C. AND P. F. HITCHCOCK (2003) Stem cells in the teleost retina: Persistent neurogenesis and injury-induced regeneration. Vis. Res. 43: 927–936. **Figure 25.11B** GOLDMAN, S. A. (1998) Adult neurogenesis: From canaries to the clinic. Trends Neurosci. 21: 107–114. **Figure 25.12** GAGE, F. H. (2000) Mammalian neural stem cells. Science 287: 1433–1438. **Figure 25.13A & 25.14C** ALVAREZ-BUYLLA, A. AND D. A. LIM (2004) For the long run: Maintaining germinal niches in the adult brain. Neuron 41: 683–686. **Figure 25.13B** COUNCILL, E. S. AND 7 OTHERS (2006) Limited influence of olanzapine on adult forebrain neural precursors in vitro. Neuroscience 140: 111–122. **Figure 25.14A,B** GHASHGHAEI, H. T., C. LAI AND E. S. ANTON (2007) Neuronal migration in the adult brain: are we there yet? Nature Rev. Neurosci. 8: 141–151.

Chapter 26 The Association Cortices
Figure 26.5A,B & 26.7 POSNER, M. I. AND M. E. RAICHLE (1994) *Images of Mind.* New York: Scientific American Library. **Figure 26.5C & 26.6B** BLUMENFELD, H. (2002) *Neuroanatomy through Clinical Cases.* Sunderland, MA: Sinauer Associates. **Figure 26.6A** HEILMAN, H. AND E. VALENSTEIN (1985) *Clinical Neuropsychology*, 2nd Ed. New York: Oxford University Press. **Figure 26.10A** LYNCH, J. C., V. B. MOUNTCASTLE, W. H. TALBOT AND T. C. YIN (1977) Parietal lobe mechanisms for directed visual attention. J. Neurophys. 40: 362–369. **Figure 26.10B** PLATT, M. L. AND P. W. GLIMCHER (1999) Neural correlates of decision variables in parietal cortex. Nature 400: 233–238. **Figure 26.11A** WOLDORFF, M. G. AND 10 OTHERS (1997) Retinotopic organization of early visual spatial attention effects as revealed by PET and ERPs. Hum. Brain Map, 5: 280–286. **Figure 26.11B** McADAMS, C. J. AND J. H. R. MAUNSELL (1999) Effects of attention on orientation-tuning functions of single neurons in macaque cortical area V4. J. Neuroscii. 19: 431–441. **Figure 26.12** DESIMONE, R., T. D. ALBRIGHT, C. G. GROSS AND C. BRUCE (1984) Stimulus-selective properties of inferior temporal neurons in the macaque. J. Neurosci. 4: 2051–2062. **Figure 26.13A** TANAKA, S. (2001) Computational approaches to the architecture and operations of the prefrontal cortical circuit for working memory. Prog. Neuro-Psychopharm. Biol. Psychiat. 25: 259–281. **Figure 26.13B** WANG, G., K. TANAKA AND M. TANIFUJI (1996) Optical imaging of functional organization in the monkey inferotemporal cortex. Sci-

ence 272: 1665–1668. **Figure 26.14** GOLD-MAN-RAKIC, P. S. (1987) Circuitry of the prefrontal cortex and the regulation of behavior by representational memory. In *Handbook of Physiology*. Section 1, *The Nervous System*. Vol. 5, Higher Functions of the Brain, Part I. F. Plum (ed.). Bethesda: American Physiological Society, pp. 373–417.

Chapter 27 Speech and Language
Figure 27.5A PENFIELD, W. AND L. ROBERTS (1959) *Speech and Brain Mechanisms.* Princeton, NJ: Princeton University Press, 1959) **Figure 27.5B** OJEMANN, G. A., I. FRIED AND E. LETTICH (1989) Electrocorticographic (EcoG) correlates of language. Electroencephalo. Clin. Neurophys. 73: 453–463. **Figure 27.6** POSNER, M. I. AND M. E. RAICHLE (1994) *Images of Mind.* New York: Scientific American Library. **Figure 27.7** DAMASIO, H., T. J. GRABOWSKI, D. TRANEL, R. D. HICHWA AND A. DAMASIO (1996) A neural basis for lexical retrieval. Nature 380: 499–505. **Figure 27.8** BELLUGI, U., H. POIZNER AND E. S. KLIMA (1989) Language, modality, and the brain. Trends Neurosci. 12: 380–388.

Chapter 28 Sleep and Wakefulness
Figures 28.1, 28.6, & 28.10 HOBSON, J. A. (1989) *Sleep.* New York: Scientific American Library. **Box 28A** MUKHAMETOV, L. M., A. Y. SUPIN AND I. G. POLYAKOVA (1977) Interhemispheric asymmetry of the electroencephalographic sleep patterns in dolphins. Brain Res. 134: 581–584. **Figure 28.3** BERGMANN, B. M., C. A. KUSHIDA, C. A. EVERSON, M. A. GILLILAND, W. OBERYMEYER AND A. RECHTSCHAFFEN (1989) Sleep deprivation in the rat: II. Methodology. Sleep 12: 5–12. **Figure 28.4** ASCHOFF, J. (1965) Circadian rhythms in man. Science 148: 1427–1432. **Box C Figures B & C** BEAR, M., M. A. PARADISO AND B. CONNORS (2001) *Neuroscience: Exploring the Brain*, 2nd Ed. Philadelphia: Williams & Wilkins/Lippincott. **Figure 28.7** FOULKES, D. AND M. SCHMIDT (1983) Temporal sequence and unit composition in dream reports from different stages of sleep. Sleep 6: 265–280. **Figure 28.12** MCCORMICK, D. A. AND H. C. PAPE (1990) Properties of a hyperpolarization-activated cation current and its role in rhythmic oscillation in thalamic relay neurons. J. Physiol. 431: 291–318. **Figure 28.13** STERIADE, M., D. A. MCCORMICK AND T. J. SEJNOWSKI (1993) Thalamocortical oscillations in the sleeping and aroused brain. Science 262: 679–685. **Figure 28.14** HOBSON, J. A. (1999) *Consciousness.* New York: Scientific American Library. **Figure 28.15** CARSKADON, M. A. AND W. C. DEMENT (1989) Normal human sleep: An overview. In *Principles and Practice of Sleep Medicine*, M. H. Kryger et al. (eds.). Philadelphia: Harcourt Brace Jovanovich, pp. 3–13.

Chapter 29 Emotions
Figure 29.1 LEDOUX, J. E. (1987) Emotion. In *Handbook of Physiology*, Section 1, *The Nervous System*, Vol. 5. F. Blum, S. R. Geiger, and V. B. Mountcastle (eds.). Bethesda, MD: American Physiological Society, pp. 419–459. **Figure 29.6** ROLLS, E. T. (1999) *The Brain and Emotion.* Oxford: Oxford University Press. **Figure 29.7** LEDOUX, J. E. (2000) Emotion circuits in the brain. Annu. Rev. Neurosci. 23: 155–184. **Figure 29.8** MOSCOVITCH, M. AND J. OLDS (1982) Asymmetries in spontaneous facial expressions and their possible relation to hemispheric specialization. Neuropsychologia 20: 71–81. **Figure 29.9** WINSTON, J. S., B. A. STRANGE, J. O'DOHERTY AND R. J. DOLAN (2002) Automatic and intentional brain responses during evaluation of trustworthiness of faces. Nature Neurosci. 5: 277–283. **Figure 29.11** NESTLER, E. J. (2005) Is there a common molecular pathway for addiction?, Nature Neurosci. 8: 1445–1449. **Figure 29.12** SCHULTZ, W., P. DAYAN AND R. P. MONTAGUE (1997) A neural substrate of prediction and reward. Science 275: 1593–1599.

Chapter 30 Sex, Sexuality, and the Brain
Figure 30.3B GUSTAFSON, M. L. AND P. K. DONAHOE (1994) Male sex determination: Current concepts of male sexual differentiation. Annu. Rev. Med. 45: 505–524. **Figure 30.4A** MCEWEN, B. S., P. G. DAVIS, B. PARSONS AND D. W. PFAFF (1978) The brain as a target for steroid hormone action. Ann. Rev. Neurosci. 2: 65–112. **Figure 30.4B** MCEWEN, B. S. (1976) Interactions between hormones and nerve tissue. Sci. Am. 235: 48–58. **Figure 30.4A** BREEDLOVE, S. M. AND A. P. ARNOLD (1984) Sexually dimorphic motor nucleus in the rat lumbar spinal cord: Response to adult hormone manipulation, absence in androgen-insensitive rats. Brain Res. 225: 297–307. **Figure 30.4B** BREEDLOVE, S. M. AND A. P. ARNOLD (1983) Hormonal control of a developing neuromuscular system. II. Sensitive periods for the androgen-induced masculinization of the rat spinal nucleus of the bulbocavernosus. J. Neurosci. 3: 424–432. **Figure 30.4D** FORGER, N. G. AND S. M. BREEDLOVE (1986) Sexual dimorphism in human and canine spinal cord: Role of early androgen. Proc. Natl. Acad. Sci. USA 83: 7527–7530. **Figure 30.6B** KATO, A. AND Y. SAKUMA (2000) Neuronal activity in female rat preoptic area associated with sexually motivated behavior, Brain Res. 862: 90–102. **Figure 30.6C** OOMURA, Y., H. YOSHIMATSU AND S. AOU (1983) Medial preoptic and hypothalamic neuronal activity during sexual behavior of the male monkey. Brain Res. 266: 340–343. **Figure 30.7B** MODNEY, B. K. AND G. I. HATTON (1990) Motherhood modifies magnocellular neuronal interrelationships in functionally meaningful ways. In *Mammalian Parenting*, N. A. Krasnegor and R. S. Bridges (eds.). New York: Oxford University Press, pp. 306–323. **Figure 30.7C** POULAIN, D. A. AND J. B. WAKERLEY (1982) Electrophysiology of hypothalamic magnocellular neurones secreting oxytocin and vasopressing. Neuroscience 7: 773–808. **Figure 30.8B–C** XERRI, C., J. M. STERN AND M. M. MERZENICH (1994) Alterations of the cortical representation of the rat ventrum induced by nursing behavior. J. Neurosci. 14: 1710–1721. **Figure 30.9A** TORAND-ALLERAND, C. D. (1978) Gonadal hormones and brain development. Cellular aspects of sexual differentiation. Amer. Zool. 18: 553–565. **Figure 30.9B** WOOLLEY, C. S. AND B. S. MCEWEN (1992) Estradiol mediates fluctuation in hippocampal synapse density during the estrous cycle in the adult rat. J. Neurosci. 12: 2549–2554. **Figure 30.9C** MEUSBURGER, S. M. AND J. R. KEAST (2001) Testosterone and nerve growth factor have distinct but interacting effects on neurotransmitter expression of adult pelvic ganglion cells in vitro. Neuroscience 108: 331–340. **Figure 30.10** WOOLEY, C. (2007) Acute effects of estrogen on neuronal physiology. Annu. Rev. Pharmacol. Toxicol. 47: 5.1–5.24. **Figure 30.12A** DEMIR, E. AND B. J. DICKSON (2005) *fruitless* splicing specifies male courtship behavior in *Drosophila*. Cell 121: 785–794. **Figure 30.12C** MANOLI, D .S., M. FOSS, A. VILLELLA, B. J. TAYLOR, J. C. HALL AND B. S. BAKER (2005) Male-specific *fruitless* specifies the neural substrates of *Drosophila* courtship behavior. Nature 436: 395–400. **Figure 30.14** CAHILL, L. (2006) Why sex matters for neuroscience. Nature Rev. Neurosci. 7: 477–484. **Figure 30.15** Cahill, L., M. Uncapher, L. Kilpatrick, M. T. Aikire and J. Turner (2004) Sex-related hemispheric lateralization of amygdala function in emotionally influenced memory: An fMRI investitation. Learn. Mem. 11: 261–266.

Chapter 31 Human Memory
Figure 31.3 ERICSSON, K. A., W. G. CHASE, AND S. FALOON (1980) Acquisition of a memory skill. Science. 208: 1181–1182. **Figure 31.4** CHASE W. G. AND H. A. SIMON (1973) *The Mind's Eye in Chess in Visual Information Processing*, W. G. Chase, ed. New York: Academic Press, pp. 215–281. **Figure 31.5** MORRIS, J. S. AND R. J. DOLAN (2001) Involvement of human amygdala and orbitofrontal cortex in hunger-enhanced memory for food stimuli. J. Neurosci. 21: 5304–5310. **Figure 31.6A** RUBIN, D. C. AND T. C. KONTIS (1983) A schema for common cents. Mem. Cog. 11: 335–341. **Figure 31.6B** SQUIRE, L. R. (1989) On the course of forgetting in very long-term memory. J. Exp. Psychol. 15: 241–245. **Figure 31.8B** EICHENBAUM, H. (2000). A cortical-hippocampal system for declarative memory. Nat. Rev. Neurosci. 1: 41–50. **Figure 31.8C,D** SCHENK, F. AND R. G. MORRIS (1985) Dissociation between components of spatial memory in rats after recovery from the effects of retrohippocampal lesions. Exp. Brain Res. 58: 11–28. **Figure 31.9** WAGNER, A. D. AND 7 OTHERS (1998) Building memories: Remembering and forgetting of verbal experiences. Science 281: 1188. **Figure 31.10** MAGUIRE, E. A. AND 6 OTHERS (2000) Navigation-related structural change in the hippocampi of taxi drivers. Proc. Natl. Acad. Sci. USA 97: 4398–4403. **Figure 31.11** VAN HOESEN, G. W. (1982) The parahippocampal gyrus. Trends Neurosci. 5: 345–350. **Figure 31.12** ISHAI, A., L. G. UNGERLEIDER, A. MARTIN AND J. V. HAXBY (2000) The representation of objects in the human occipital and temporal

cortex. J. Cog. Neurosci. 12 Suppl 2: 35–51. **Figure 31.13** SHOHAMY, D., C. E. MYERS, S. GROSSMAN, J. SAGE AND M. A. GLUCK (2005) The role of dopamine in cognitive sequence learning: Evidence from Parkinson's disease. Behav. Brain Res. 156: 191–199. **Figure 31.15** DEKABAN, A. S. AND D. SADOWSKY (1978) Changes in brain weights during the span of human life: Relation of brain weights to body heights and body weights. Ann. Neurol. 4: 345–356. **Figure 31.16** CABEZA, R., N. D. ANDERSON, J. K. LOCANTORE AND A. R. MCINTOSH (2002) Aging gracefully: compensatory brain activity in high-performing older adults. Neuroimage 17: 1394–1402. **Box 31D Figure A** ROSES, A. (1995) Apolipoprotein E and Alzheimer disease. Science & Medicine September/October 1995, 16–25. **Box 31D Figure B** BLUMENFELD, H. (2002) *Neuroanatomy through Clinical Cases*. Sunderland, MA: Sinauer Associates; BRUN, A. AND E. ENGLUND (1981) Regional pattern of degeneration in Alzheimer's disease: Neuronal loss and histopathalogical grading. Histopathology 5: 459–564.

Appendix and Atlas
Opening Image Copyright © Max Delson/istockphoto.com

Index

Citations in *italics* refer to information in a table or illustration.